A Dictionary of

Biomedicine

The Author

Dr John Lackie was formerly Senior Lecturer in Cell Biology at the University of Glasgow, then Director and Director of Research at the Yamanouchi Research Institute, Oxford (which sought new targets for drugs) and then Principal and CEO of the Westlakes Research Institute in Cumbria before setting up an independent consultancy company, Plumbland Consulting Ltd. He has edited four editions of the *Dictionary of Cell & Molecular Biology* (4th edition, 2007) and was Chief Editor of the *Chambers' Dictionary of Science and Technology* (2007).

The Advisory Editor

Dr Christopher O'Callaghan is a Reader and Honorary Consultant Physician in the Nuffield Department of Medicine in the University of Oxford. His current post combines clinical medicine with a research interest in immunology and molecular biology.

SEE WEB LINKS

Many entries in this dictionary have recommended web links. When you see the above symbol at the end of an entry go to the dictionary's web page at www.oup.com/uk/reference/resources/biomedicine, click on **Web links** in the Resources section and locate the entry in the alphabetical list, then click straight through to the relevant websites.

A Dictionary of

Biomedicine

JOHN LACKIE

Christopher O'Callaghan

OXFORD
UNIVERSITY PRESS

Great Clarendon Street, Oxford OX2 6DP

Oxford University Press is a department of the University of Oxford.
It furthers the University's objective of excellence in research, scholarship,
and education by publishing worldwide in

Oxford New York
Auckland Cape Town Dar es Salaam Hong Kong Karachi
Kuala Lumpur Madrid Melbourne Mexico City Nairobi
New Delhi Shanghai Taipei Toronto

With offices in
Argentina Austria Brazil Chile Czech Republic France Greece
Guatemala Hungary Italy Japan Poland Portugal Singapore
South Korea Switzerland Thailand Turkey Ukraine Vietnam

Published in the United States
by Oxford University Press Inc., New York

First published 2010

British Library Cataloguing in Publication Data
Data available

Library of Congress Cataloging in Publication Data
Data available

Typeset by SPI Publisher Services, Pondicherry, India
Printed in Great Britain
on acid-free paper by
Clays Ltd., St Ives plc

ISBN 978-0-19-954935-1

1 3 5 7 9 10 8 6 4 2

Contents

Preface

One of the most active research interfaces is that between medicine and the molecular biosciences, and rapid progress is being made in understanding the molecular basis of many diseases. Huge amounts of money have been, and are being, spent and there are probably more people working at this interface than in any other research area. Inevitably, the interface provides opportunities for confusion, since the vocabularies of clinicians and molecular bioscientists do not necessarily overlap. Biomedicine has no clear boundaries and yet this dictionary sets out to provide definitions of things that are relevant to those coming from either side. The approach has been inclusive rather than exclusive: some headwords may appear to belong to other disciplines—yet somewhere there will be a biomedical researcher using the term (I expect). Given my own background in molecular cell biology and drug discovery, there is probably a bias towards the biosciences but not so much that the bioscientist will not find many medical terms defined and the clinician should be able to use the dictionary to decode some of the jargon of the molecular biologists.

There is much more information available than can possibly be packed into a brief definition and there are astonishingly rich online resources, freely available, that can be consulted. What the dictionary aims to provide is a succinct definition, a digested version of the more extensive information that exists on the internet and in the literature. I have drawn heavily on the suite of databases managed by the US National Center for Biotechnology Information (NCBI), in particular PubMed (http://www.ncbi.nlm.nih.gov/sites/entrez) for accessing the primary literature and OMIM (Online Mendelian Inheritance in Man: http://www.ncbi.nlm.nih.gov/sites/entrez?db=omim) for information about those diseases for which a molecular basis is known (and some others). Another important source for information about proteins has been the Universal Protein Resource (UniProt: http://www.uniprot.org/uniprot). Generally I have trusted these databases, but if information has come from elsewhere then I have cross-checked. Some of the other resources are indicated in the online version.

Although the Human Genome Project has provided a rich source of data, linking nucleotide sequence to protein function and mutation to disorder requires considerable further effort and understanding the molecular basis of a disorder does not automatically lead to a therapy or a drug to treat the disease. For those diseases that involve an infectious agent, the molecular bioscientist must struggle with the development of resistance and the pharmaceutical 'arms race' with pathogens is far from over. The pace of research is such that by the time I completed the text, some definitions had already reached the stage of needing further revision in the light of new information: but there comes a stage at which a line needs to be drawn if the manuscript is to be 'finished'. This is, however, an admission that this will never be a finished product and the intention is to keep a record of terms that have been looked for and not found in the online version, and possibly to identify those terms that have never been looked at and could be culled to make space for new entries. The plan is that the dictionary will then evolve in response to the user community. If a definition is inadequate (for which I apologise and accept all responsibility) then I would be grateful for a correction (send to john@lackie.nildram.co.uk), so that it can be amended. If users make an input, then the whole of the biomedical community, or at least those who use the dictionary, will benefit.

John Lackie
March 2010

A note on conventions

Unless otherwise stated, the information in definitions relates to the human gene or protein. *Genes* are italicized, their products are not. The size of a protein is given in residues (the number of amino acids, e.g. 'ferroportin...571 aa'), but this is not always definitive since some proteins are post-translationally modified or proteolytically processed. In a few cases, particularly when a core protein is heavily glycosylated, I have given the size in kiloDaltons (kDa).

For the sake of brevity, I have often used what is strictly speaking an incorrect shorthand: mutations occur in genes and may lead to the expression of an altered protein—or the nonproduction of that protein, but quite frequently I have written of 'a mutation in protein X' rather than 'a mutation in the gene encoding protein X blocking its production or rendering it abnormal in structure or function'.

A note on drug names: Drugs that are used to treat patients have a scientific name, which is the name used in the peer-reviewed literature and the name under which they appear in this Dictionary, but they also have proprietary names (some of which are Registered, some Trademarked) and which may vary according the marketing territory in which the drug is sold. These proprietary names are often the ones used in the non-technical literature and have been included generically as 'tradenames', shown at the end of the entry as '*TN* ...' (e.g. '**famotidine** An [H2 blocker]. *TNs* Pepcid, Gaster.') They should be avoided in scientific writing. If the British Approved Name has a US spelling, then that is used (thus 'sulfasalazine' gets the main entry, not 'sulphasalazine').

Whether a protein is defined under its full name or its abbreviation depends on my prejudice as to which is the most likely to be first sought: I may not be right and it is probably worth trying the alternative even if there is not an explicit cross-reference.

Not everything is cross-referenced; usually only things that, if consulted, would add to the definition. I have assumed that the reader will guess that there are entries for 'actin' and 'gene' and so on. '*See*' and '*See also*' are stronger indications that there is additional and relevant information. '*But see*' is a signal that there is a potential source of confusion, '*compare*' often directs to an antonym or the converse phenomenon.

The boxed entries are signposting devices, they list the relevant headwords that relate to the disorders in question rather than being definitions. A few other entries are similar, e.g. the 'oncogene' definition lists all the oncogenes that have entries of their own.

The terms are in UK English, my native tongue, and although US spellings are given occasionally the expectation is that the reader will be able to cope with all 'heme-' entries being under 'haem-', having been told once. I have however, bowed to what I perceive to be standard practice and adopted a few neologisms. In a few cases the correct plural form is given, based on my recognition that they are very frequently misspelt. The singular form is usually the headword (so 'mitochondrion', not 'mitochondria') but quite a few singular names turn out to cover several forms of the same molecule (thus the headword is 'interferons' not 'interferon').

Boxed entries

Numerical entries

3M syndrome A disorder (gloomy face syndrome, Le Merrer's syndrome, dolichospondylic dysplasia, Yakut short stature syndrome) in which there is low birthweight, dwarfism, and facial abnormalities caused by mutation in the cullin-7 gene (*see* CULLINS).

5-HT *See* SEROTONIN.

5q– syndrome (5q minus syndrome) A *myelodysplasia characterized by a defect in erythroid differentiation in which there is a deletion of the long arm of chromosome 5 containing ~40 genes. The key deletion is that for the gene encoding ribosomal subunit protein RPS14, although several genes encoding cytokines or their receptors have been mapped to this region (αFGF, GM-CSF, IL-3, IL-4, IL-5, IL-9, both subunits of IL-12, the gene encoding IRF1 (interferon regulatory factor-1) and the genes encoding the receptor for M-CSF and PDGF, and the FGF receptor, FGFR4).

14-3-3 proteins A family of conserved regulatory adaptor molecules (~250 aa) that bind to phosphoserine-containing proteins, including a wide range of functionally diverse signalling proteins, including kinases, phosphatases, and transmembrane receptors. There are seven highly conserved human 14-3-3 proteins (β, γ, ε, η, σ, θ, ζ) and they are involved in cellular proliferation, checkpoint control, and apoptosis. For example, 14-3-3-σ (stratifin, 248 aa) is particularly abundant in stratified keratinizing epithelium and is strongly induced by γ-irradiation and other DNA-damaging agents.

α (alpha) The α prefix is either given as 'alpha' or ignored and the main portion of the name used as the headword.

A4 protein *See* AMYLOIDOGENIC GLYCOPROTEIN.

A20 A cytoplasmic zinc finger protein (790 aa) that inhibits *NFκB activity and TNF-mediated programmed cell death. The expression of the A20 mRNA is upregulated by TNFα. It is a dual function enzyme with deubiquitinating and ubiquitin ligase activity that triggers degradation of RIP (*receptor interacting protein kinase) an essential mediator of the TNF receptor-1 signalling complex.

A431 A human cell line established from an epidermoid carcinoma, often used in studies on *EGF.

AAA family A diverse superfamily (ATPases associated with various cellular activities) that have a characteristic AAA-domain. There are at least 300 known members in at least six families. Mutations in proteins of the AAA family cause a variety of human diseases including *cystic fibrosis, *Zellweger's syndrome, *adrenoleukodystrophy, and dystonia. Mutations in *paraplegin and *spastin cause two types of *hereditary spastic paraplegia. *See also* NSF and VALOSIN-CONTAINING PROTEIN.

SEE WEB LINKS

- The AAA family website.

AAMI Age-associated memory impairment.

Aarskog–Scott syndrome An X-linked developmental disorder (faciogenital dysplasia, FGDY) in which stature is disproportionately short and there are facial, skeletal, and urogenital abnormalities. The syndrome is caused by mutations in *FYVE, RhoGEF and PH domain-containing protein-1 (FGD1).

Ab Common abbreviation for antibody. *See* IGA, IGG, etc.

abamectin *See* AVERMECTINS.

ABC **1.** Antigen binding cell. **2.** Antigen binding capacity. **3.** The avidin–biotin peroxidase complex. Used in an immunocytochemical method in which biotinylated anti-immunoglobulin binds the primary antigen-specific antibody, then the ABC complex with excess biotin-binding capacity binds biotin on the second antibody and the peroxidase used to catalyse a colorimetric reaction generating brown staining. **4.** *See* ABC PROTEINS.

abciximab The Fab fragments of a monoclonal antibody directed against the glycoprotein IIb/IIIa receptor on platelets that blocks platelet aggregation. Used as a short-term treatment to prevent thrombus formation after angioplasty. *TN* ReoPro.

ABC proteins An extensive family of membrane proteins involved in active transport or regulation of ion channel function that have an ATP binding cassette. Many are prokaryotic, but important eukaryotic examples are the *multidrug transporter (P-glycoprotein), the *cystic fibrosis transmembrane conductance regulator (CFTR), *SUR, *TAPs. *See* ADRENOLEUCODYSTROPHY; CEREBRAL AMYLOID ANGIOPATHY; PSEUDOXANTHOMA ELASTICUM.

abdominal aortic aneurysm A multifactorial disorder in which there is permanent dilation of the abdominal aorta, usually due to degenerative changes in the aortic wall. There is chronic inflammation of the aortic wall with remodelling of the extracellular matrix, and depletion of vascular smooth muscle cells. Defects in collagen type III are one cause of susceptibility.

Abell–Kendall method A method for estimating total serum cholesterol levels that involves saponification of cholesterol ester by hydroxide, extraction with petroleum ether, and colour development with acetic anhydride–sulphuric acid. Other methods are assessed against this reference standard method.

Abelson leukaemia virus A replication-defective murine leukaemia virus (A-MuLV), a retrovirus, that requires coinfection of cells with a 'helper' virus to proliferate and to cause a rapidly progressive lymphosarcoma (the Moloney murine leukaemia virus is often used). The oncogenic effects are due to its acquisition of c-abl, a nonreceptor tyrosine kinase, that has a G-actin binding site and an independent F-actin binding site.

abenzyme *See* CATALYTIC ANTIBODY.

abetalipoproteinaemia An autosomal recessive defect (acanthocytosis, Bassen–Kornweig syndrome) in which there is total absence of *apolipoprotein B (a component of *LDL, *VLDL, and chylomicrons) caused by a defect in *microsomal triglyceride transfer protein. Common features are coeliac syndrome, pigmentary degeneration of the retina, progressive ataxic neuropathy, and the presence of *acanthocytes.

abl *See* ABELSON LEUKAEMIA VIRUS.

ABO blood group system The first of the blood group systems to be discovered. The H-locus, which encodes a fucosyl transferase, determines whether the H-antigen, a glycosyl chain attached to ceramide or protein, will be expressed on the erythrocyte surface (in the rare Bombay phenotype it is not expressed). If the H-antigen is expressed then the product of the A allele of the ABO locus, a glycosyltransferase, can add *N*-acetyl galactosamine to produce the A antigen, and the B-allele product can add galactose to produce the B-antigen. The O allele produces an inactive enzyme and the H-antigen remains unmodified. Antibodies to the ABH antigens occur naturally, making this an important set of antigens for blood transfusion. Transfusion of mismatched blood elicits a transfusion reaction in which there is complement-mediated haemolysis of the transfused red cells. The natural antibodies are usually IgM. Polymorphism in the ABO locus is thought to be maintained because of differential susceptibility to infectious diseases such as cholera and malaria and certainly the distribution of the ABO blood types varies considerably between populations. *See* DUFFY; KELL; MN BLOOD GROUP ANTIGENS; RHESUS BLOOD GROUP.

SEE WEB LINKS

- Blood Group Antigen Gene Mutation Database.

abortifacient Describing anything, but generally a drug, that induces abortion.

abortive infection A viral infection of a cell in which the virus fails to replicate. Sometimes replication-incompetent viruses will proliferate if a helper virus coinfects the cell. The term **abortive transformation** is used of cells transformed by a virus that does not integrate into the host genome.

ABP *See* ACTIN-BINDING PROTEINS.

abrin An *AB toxin from the seeds of the jequirity bean *Abrus precatorius*. It is lectin-like, having a binding site for galactose and related residues in carbohydrate, but is monovalent so does not act as an *agglutination. The A-chain (ABRaA) has *N*-glycosylase activity toward eukaryotic 28S rRNA and inhibits protein synthesis. There is no antidote.

abrineurin *See* BRAIN-DERIVED NEUROTROPHIC FACTOR.

abscess A pus-filled cavity within a tissue, usually caused by bacteria that resist killing by phagocytes.

absolute lethal concentration (LC_{100}) The concentration of a substance that kills 100% of test organisms or species under defined conditions.

absorption coefficient **1.** A coefficient that relates to the ability of a substance to absorb electromagnetic radiation or sound; may be linear or logarithmic and will depend upon the wavelength. In the case of light, relates to the opacity, an opaque substance having an infinite absorption coefficient. **2.** The ratio of uptake of a substance to the intake. *Compare* ADSORPTION COEFFICIENT.

absorption spectrum The spectrum of electromagnetic radiation (usually visible and ultraviolet light) that is absorbed (not transmitted) by a substance. Absorption involves the excitation of atoms (alone or within a molecule) from their ground state to an excited state by a photon at that particular wavelength.

AB toxins Multisubunit toxins that have an active (A) portion and a portion involved in binding (B) to target cells. The A portion is toxic if introduced directly into the cytoplasm. In several types (*cholera toxin, *diphtheria toxin, *pertussis toxin) the A subunit has *ADP-ribosylating activity. *See also* COLICINS; RICINUS COMMUNIS AGGLUTININS; SHIGA TOXIN; TETANUS TOXIN; VACA.

abzyme An antibody that has catalytic (enzymic) activity. Catalytic antibodies can be selected to catalyse a reaction for which there is no endogenous enzyme and act by stabilizing a transition state. For example, catalytic antibodies that cleave cocaine have been produced.

a

SEE WEB LINKS

- Article on nature, function, and design of catalytic antibodies.

ACADs A family of mitochondrial enzymes (acyl-coenzyme A dehydrogenases) that catalyse the first dehydrogenation step in the *beta-oxidation of fatty acyl-CoA derivatives. There are different ACADs for short-, medium-, long- and very long-chain fatty acids. The product of the ***ACAD8*** gene, also known as isobutyryl-coenzyme A (CoA) dehydrogenase (EC 1.3.99, 415 aa), converts isobutyryl-CoA into succinyl-CoA. At least four mutations in the *ACAD8* gene have been identified and lead to isobutyryl-CoA dehydrogenase deficiency in which there may be dilated *cardiomyopathy, anaemia, and carnitine deficiency.

acamprosate A drug used to treat alcohol dependence that may act by blocking glutamatergic *N*-methyl-D-aspartate (NMDA) receptors and activating γ-aminobutyric acid (*GABA) type A receptors. It is structurally similar to homotaurine. *TN* Campral.

Acanthamoeba A soil-dwelling amoeba 20–30 μm in diameter that has been extensively used in biochemical studies of cell motility. Can cause opportunistic infections, particularly keratitis if contact lenses are contaminated.

acanthocyte A cell that has spiky projections. In acanthocytosis (*see* ABETALIPOPROTEINAEMIA), erythrocytes show spiky deformation, a condition that can be caused experimentally by manipulating the lipid composition of the plasma membrane.

acanthoma A benign dermatological tumour composed of squamous or epidermal cells. Various forms are recognized including **pilar sheath acanthoma**, a follicular tumour usually of the upper lip and **clear cell acanthoma** (Degos acanthoma), found most frequently on the legs. *See* KERATOACANTHOMA.

acanthosis nigricans A rare disease in which there is thickening and hyperpigmentation of the skin, especially in flexural areas. It can be secondary to visceral carcinoma, drug-induced, associated with insulin resistance or with *Seip's syndrome, or may arise as a dominant hereditable condition.

acapnia A state (hypocapnia, hypocapnea, hypocarbia) in which blood concentrations of carbon dioxide are low, often a consequence of hyperventilation. *Compare* HYPERCAPNIA.

ACAT An evolutionarily conserved enzyme (acyl-coenzyme A:cholesterol acyltransferase, sterol *O*-acyltransferase) that catalyses cholesterol ester formation from cholesterol and fatty acyl-CoA substrates. The ***ACAT1*** gene encodes mitochondrial acetyl-CoA acetyltransferase (EC 2.3.1.9, 427 aa), deficiency in which causes 3-*ketothiolase deficiency. The ***ACAT2*** gene encodes cytosolic acetoacetyl-CoA thiolase (550 aa); mutation causes severe mental retardation and hypotonus.

accelerin Obsolete name for *coagulation factor V.

acceptable daily intake The amount of a specific substance that can be ingested over a lifetime without an appreciable health risk. Values tend to be adjusted downwards as risk aversion increases.

accessory cells A variety of cells, usually MHC class II positive (*see* HISTOCOMPATIBILITY ANTIGENS), that are involved in inducing an immune response through interactions with T cells. Include antigen presenting cells, antigen processing cells. Monocytes, macrophages, dendritic cells, *Langerhans cells, and B-cells may all act as accessory cells.

ACD *See* ACID–CITRATE–DEXTROSE.

ACE *See* ANGIOTENSIN CONVERTING ENZYME.

A cells (α cells) Cells of the *islets of Langerhans in the pancreas. They constitute around 20% of the population and secrete *glucagon. *See* BETA CELLS; D CELLS.

acellular Not composed of cells or not containing cells. The acellular slime moulds such as *Physarum* are multinucleate syncytia.

acentric Describing *chromosome fragments that lack a *centromere. *See* METACENTRIC.

acetaminophen *See* PARACETAMOL.

acetazolamide A carbonic anhydrase inhibitor that is used to treat glaucoma, epileptic seizures, benign intracranial hypertension, altitude sickness, cystinuria, and dural ectasia. *TN* Diamox.

acetonuria *See* KETONURIA.

acetylation The addition, either chemically or enzymically, of acetyl groups to a molecule.

acetylcholine (ACh) The acetyl ester of choline. An important neurotransmitter at the neuromuscular junction, in the autonomic nervous system and in the central nervous system as a neuromodulator. ACh can be either excitatory or inhibitory, and cholinergic receptors are classified as *nicotinic or *muscarinic, according to their pharmacology. *See* ACETYLCHOLINE ESTERASE.

acetylcholine esterase (AChE) The enzyme (EC 3.1.1.7, 614 aa) that hydrolyses acetylcholine released at the synapse so that the signal is short-lived. Agents such as nerve gases and insecticides that inhibit AChE prolong the time course of postsynaptic potentials. Various drugs (e.g. donezepil, rivastigmine) that inhibit the breakdown of acetylcholine in the central nervous system appear to slow the rate of decline in cognition in Alzheimer's disease.

acetylcholine receptor *See* MUSCARINIC ACETYLCHOLINE RECEPTORS; NICOTINIC ACETYLCHOLINE RECEPTOR.

acetyl CoA The acetylated form of *coenzyme A, an important carrier of acyl groups, particularly in the *tricarboxylic acid cycle.

acetylsalicylic acid *See* ASPIRIN.

ACh *See* ACETYLCHOLINE.

A chain The smaller of the polypeptides in a heteromeric protein. Often used for the 21 aa polypeptide chain of *insulin (the B chain has 30 aa). *But see* e.g. ABRIN; AB TOXINS; ACTIVIN; LAMININS; PLATELET-DERIVED GROWTH FACTOR; RELAXIN.

achalasia Failure to relax. 'Achalasia of the cardia' is a condition in which the sphincter between the oesophagus and the stomach fails to relax.

achlorhydria The absence of hydrochloric acid in gastric juice. In **achylia gastrica** both pepsin and hydrochloric acid are absent.

achondrogenesis *See* HYPOCHONDROGENESIS.

achondroplasia A form of dwarfism (chondrodystrophia fetalis) in which there is a failure of endochondral ossification caused by an autosomal dominant mutation in the gene for the *fibroblast growth factor receptor. *See* PSEUDOACHONDROPLASIA.

achromatopsia *See* COLOUR BLINDNESS.

achylia gastrica *See* ACHLORHYDRIA.

aciclovir (formerly **acyclovir)** A nucleoside analogue (hydroxyethoxymethylguanine) with antiviral properties against both type 1 and 2 herpes virus and varicella-zoster virus. The drug is converted to the monophosphate by viral *thymidine kinase and to the triphosphate by cellular enzymes. The triphosphate competitively inhibits viral DNA polymerase and can be incorporated into viral DNA where it blocks further replication. *TN* Zovirax.

acid–citrate–dextrose (ACD) A citrate-buffered glucose solution used as an anticoagulant for blood (the citrate complexes calcium ions).

acid hydrolases *Hydrolytic enzymes (EC 3) that operate best at an acid pH. The term is often used generically of the various lysosomal enzymes secreted during *phagocytosis.

acidic FGF *See* FIBROBLAST GROWTH FACTOR.

acidophilic **1.** Something easily stained with acid dyes. **2.** An organism that will flourish in an acidic environment.

acidophils Cells (alpha cells, A cells) of the pars distalis of the *adenohypophysis that secrete either *growth hormone or *prolactin.

acidosis A condition in which blood or tissues are more acidic than normal (blood pH <7.35). **Respiratory acidosis** is a consequence of the failure of the lungs to remove carbon dioxide; **metabolic acidosis** arises through overproduction of acidic compounds (e.g. lactic acid during anaerobic exercise, diabetic ketosis), the inability to form bicarbonate in the kidney, or failure of excretion.

acid phosphatase A lysosomal enzyme (EC 3.1.3.2, 423 aa) that catalyses the cleavage of inorganic phosphate from various substrates and that operates best at acid pH. Can be localized histochemically with the *Gomori procedure.

acid protease An imprecise name for a proteolytic enzyme that operates best at acid pH, usually a lysosomal enzyme. *See* PEPTIDASE.

acid secreting cells *See* OXYNTIC CELL.

aciduria A condition in which urine is acid (the normal pH is between 4.5 and 8.0). Aciduria can be caused by various metabolic disorders. *See* ETHYLMALONIC ENCEPHALOPATHY; HARTNUP'S DISEASE; KETOTHIOLASE DEFICIENCY; METHYLMALONYL-COENZYME A MUTASE; MEVALONIC ACIDURIA; UROCANIC ACID.

acinar cells Epithelial cells that are arranged as a ball around a small sac-like distension of the lumen of a gland, an acinus.

a

Acinetobacter A Gram-negative bacterium common in soil and water and on the skin. *Acinetobacter baumannii* (and other species) is an opportunistic pathogen, often acquired in hospitals and very resistant to antibiotics.

acipimox A *niacin derivative used to lower levels of cholesterol and other lipids in the blood by preventing the release of fatty acids from fatty tissue. *TN* Olbetam.

acne Inflammation of sebaceous glands which may be exacerbated by infection with the commensal bacterium *Propionibacterium acnes*. Acne vulgaris is the common type. *See* CHLORACNE; ROSACEA.

aconitase An enzyme (EC 4.2.1.3) of the *tricarboxylic acid cycle that catalyses isomerization of citrate/isocitrate. Isoforms are found both in cytoplasm (ACO1, 889 aa) and mitochondrial matrix (ACO2, 778 aa). ACO1 is bifunctional and not only an enzyme of the TCA cycle but also an *iron-responsive element (IRE)-binding protein involved in the control of iron metabolism.

ACPs Small acidic proteins (acyl carrier proteins) associated with fatty acid synthesis in prokaryotes and plants where fatty acid synthase (EC 2.3.1.85) is a multiprotein complex with acyl carrier protein and several monofunctional enzymes involved in the transfer of acyl groups from acetyl-CoA to long-chain fatty acids. In animals the acyl carrier and enzyme functions are all in a single protein (2504 aa).

acquired immune deficiency syndrome *See* AIDS.

acquired immunity (active immunity) Immunity arising as a result of mounting an immune response to a novel antigen and retaining a memory of the response, as opposed to immunity derived from a maternal source or by the transfer of immune serum or cells, or innate immunity which is not pathogen-specific.

acridines Heterocyclic compounds that are cell-permeable and will intercalate into double-stranded DNA and interact with RNA. They are fluorescent; acridine orange fluoresces green when intercalated in DNA (in cell nuclei) and orange when associated with cytoplasmic RNA. The emission spectrum is pH sensitive and can be used in studies on intracompartmental pH. Often used to detect dsDNA on gels. The acridines cause *frame-shift mutations and are cytostatic and antimicrobial.

acriflavine A compound derived from acridine that has antiseptic properties, used topically. Deep orange/brown in colour.

acrivastine An antihistamine drug that acts on the histamine H_1 receptor, used to treat allergies and hay fever.

acrocentric *See* METACENTRIC.

acrocyanosis A persistent blue (cyanotic) discoloration of the fingers and toes. Can be a symptom of various disorders or a benign neurohormonal condition more common in children and young adults.

acrodermatitis enteropathica A disorder in which there is diarrhoea and dermatitis with failure to thrive, caused by a mutation in the gene for the intestinal zinc-specific transporter SLC39A4.

acromegaly A disorder in adults in which the hands and feet enlarge and there is thickening of facial bone. Most commonly caused by growth hormone-producing tumours (somatotropinomas) of the pituitary which can arise as a result of mutation in the *aryl hydrocarbon receptor-interacting protein or sporadic mutations in the *GNAS locus.

acromesomelic dysplasia A group of disorders in which there is disproportionate shortening of skeletal elements, particularly the middle and distal parts of the limbs. **Maroteaux-type** acromesomelic dysplasia (AMDM) is caused by mutation in the receptor (NPR-B) for C-type *natriuretic peptide, which is important for bone growth. The **Hunter–Thompson type** of acromesomelic dysplasia (AMDH) and the **Grebe type** (AMDG) are caused by allelic mutations in the gene that encodes cartilage-derived morphogenetic protein-1, one of the *transforming growth factor β family of growth factors (growth/differentiation factor 5).

acroparaethesia Numbness and tingling of the fingers.

acrosin The major *serine peptidase (EC 3.4.21.10, 402 aa) in the *acrosome of the mature spermatozoan where it is stored as an inactive precursor, proacrosin, that is released and activated to assist penetration of the zona pellucida of the egg. The mRNA for proacrosin is synthesized only in the postmeiotic stages of spermatogenesis. **Acrosin-binding protein** (543 aa) is found in testis and expressed in a subset of tumours.

acrosome The Golgi-derived vesicle at the very front of the spermatozoon that releases hyaluronidase and *acrosin following initial contact with the zona pellucida of the egg.

acrosyringium The intraepidermal part of the duct of the sweat gland nearest to the surface of the skin.

ACT Artemisinin-based combination therapy. *See* ARTEMISININ; LUMEFANTRINE.

ActA The major surface protein (639 aa) of *Listeria monocytogenes* that spans both the bacterial membrane and the peptidoglycan cell wall. Once the bacterium is within the host cell ActA serves as a nucleating site for the Arp2/3-mediated assembly of a 'comet tail' of microfilaments from one pole, pushing the bacterium through the cytoplasm. A functionally similar protein, *IcsA, is found in *Shigella*.

ACTH *See* ADRENOCORTICOTROPIN.

actin A highly conserved and possibly the most abundant eukaryotic protein (377 aa), that constitutes 10–20% of the total cellular protein content. **Globular (G) actin** will assemble into **filamentous (F) actin** which is the basis of cytoplasmic *microfilaments and the *thin filaments of striated muscle. Not only are microfilament meshworks structural, the interaction of myosin with F-actin is the basis of the actomyosin motor system in the cytoplasm and in the *sarcomere. *Profilin, *thymosin $\beta4$, and *gelsolin prevent G-actin from polymerizing; other proteins stabilize F-actin. Extracellular actin is sequestered by *Gc protein and secreted gelsolin. **Actin α_1** is the skeletal muscle isoform, and is mutated in some *myopathies. **Actin αC1** is found in cardiac muscle and mutations are associated with dilated cardiomyopathy 1R and hypertrophic cardiomyopathy 11. **Actin α_2** is the smooth muscle form and mutations are associated with hereditary thoracic aortic aneurysm 6. **Actins β and γ** are found in nonmuscle cells. Mutations in actin β have been reported in twins with juvenile-onset dystonia; mutations in actin γ are associated with autosomal dominant deafness 20. *See also* ACTIN-BINDING PROTEINS; CENTRACTINS.

actin-binding proteins A diverse group of proteins that bind to *actin. Some interact with G-actin and render it assembly-incompetent, others bind to F-actin and stabilize microfilaments, others crosslink filaments to form cytoskeletal *actin meshworks, microfilament bundles (as in microvilli), or nucleate microfilaments, thereby affecting their distribution.

SEE WEB LINKS

- The web-based Encyclopedia of Actin-Binding Proteins.

actin depolymerizing factor An actin-binding protein of the ADF/*cofilin family. The human form (ADF, destrin,165 aa) has 95% homology to chick ADF. ADF will depolymerize microfilaments and bind G-actin, but does not cap filaments. Cofilin is similar in function but the product of a different gene. ADF has a nuclear localization domain and the interaction with actin is regulated by phosphoinositides. ADF and cofilin are both associated with *Hirano bodies. The **ADF domain** (ADF-homology domain) is an actin-binding motif found in three phylogenetically distinct classes of proteins, ADF/cofilins, *twinfilins and drebin/ABP-1s.

actinfilin A synaptic F-actin-binding protein (642 aa), one of the BTB–Kelch protein family (*see* MAYVEN; KELCH PROTEINS), expressed mainly in the brain. It binds GluR6, a kainate-type glutamate receptor subunit, and targets it for degradation through the E3 ubiquitin ligase pathway.

actinic keratosis *See* KERATOSES.

actin meshwork A mechanically resistive network of crosslinked microfilaments that are inserted proximally into the plasma membrane. Actin meshworks (cortical meshwork) support protrusions from cells, e.g. at the leading edge of a moving fibroblast.

actinodermatitis Inflammation of the skin, usually arising from overexposure to ultraviolet light (sunburn).

Actinomycetales An order of Gram-positive bacteria found in soil, compost, and aquatic habitats. Some are pathogenic (e.g. *Actinomyces, Corynebacterium, Mycobacterium*), others are important as the source of antibiotics (**Streptomyces*).

actinomycins A mixture of *antibiotics, actinomycins C1, C2, and D, produced by a species of **Streptomyces*. Actinomycin D binds to DNA and blocks the movement of *RNA polymerases in both prokaryotes and eukaryotes.

actinomycosis A chronic granulomatous infection, usually affecting the head and neck, caused by infection by *Actinomyces israelii* or other species of the genus.

actinotherapy A treatment involving exposure to actinic radiation, usually ultraviolet for dermatological conditions or infrared for muscle pain.

a

action potential A transient propagated depolarization of the membrane of an excitable cell, such as a *neuron, *muscle cell, fertilized egg, or certain plant cells. It is the result of opening of sodium channels, allowing the influx of sodium ions. The inside of the cell may reach +30 mV before reverting to the normal *resting potential of –50 to –90 mV. Action potentials in muscles last longer than those in nerves (up to 1 s as opposed to milliseconds). Action potentials are an example of an *all-or-nothing phenomenon and are ideally suited to digital encoding.

action spectrum The relationship between the wavelength of electromagnetic radiation and its ability to cause a specific chemical or biological effect.

activated leucocyte cell adhesion molecule The human homologue of a cell adhesion molecule first identified in the chick as DM-GRASP. A polypeptide (ALCAM, CD166, 500 aa) with five immunoglobulin (Ig) domains that is expressed on mitogen-activated leucocytes and a number of T-cell, B-cell, monocyte, and tumour-derived cell lines and is the ligand for CD6 on T cells. May also have a role in neural cell adhesion (*see* NEUROLIN).

activated macrophage A *macrophage that has been stimulated by *cytokines and exhibits enhanced cytotoxic and bactericidal capacity.

activation In fertilization, the first stage of development, triggered by contact between the spermatozoon and the egg membrane. Following fusion of the sperm with the egg there is a cortical reaction in which the fertilization membrane is elevated as a block to polyspermy. There are concomitant rapid changes in metabolic rate and an increase in protein synthesis from maternal mRNA.

activation energy The energy input that is needed for a system to reach the energy level at which a reaction will occur.

activation induced deaminase An RNA-editing deaminase (cytidine aminohydrolase, EC 3.5.4.5, 198 aa) involved in somatic hypermutation, gene conversion, and *class switching in the terminal differentiation of B cells. *See* ADARs.

activation loop A domain containing a conserved threonine residue that activates *AGC kinases when it is phosphorylated.

active immunity *See* ACQUIRED IMMUNITY.

active site The region of an enzyme that binds the target substrate and facilitates the reaction. The properties of the active site are determined by the amino acids juxtaposed in the folded protein.

active transport Directional (vectorial) transport of a substance up an *electrochemical potential gradient, brought about by coupling the transport to an energy-yielding process. In **primary active transport** systems the coupling is usually to *ATP hydrolysis within a multisubunit complex; in **secondary active transport** (cotransport, counter-transport) the coupling is to the movement of another molecule down a gradient generated by primary active transport.

active zone The presynaptic region where neurotransmitter release occurs, often located directly across the synaptic cleft from clusters of receptors. The active zone may be the sole location for presynaptic *calcium channels.

activin A dimeric growth factor composed of *inhibin β subunits, closely related to other *transforming growth factors (TGFβ). The two isoforms of inhibin β give rise to three forms of activin (A, βA/βA; AB, βA/βB; B, βB/βB). Activin stimulates the secretion of *follicle stimulating hormone (FSH) and has a range of other growth and differentiation-promoting effects. Activin receptors are serine/threonine kinases (activin receptor-like kinases, ALKs) of two types expressed in tissue-specific and developmental stage-specific ways. Activin binds to the type II receptor which then phosphorylates and activates type I receptors which autophosphorylate, then bind and activate *SMADs. ALK1 (503 aa) is highly expressed in endothelial cells and highly vascularized tissues in the same pattern as *endoglin, is a receptor for the TGFβ superfamily of ligands and is mutated in *hereditary haemorrhagic telangiectasia (Osler–Rendu–Weber syndrome 2). Mutations in **ALK2** (509 aa) are associated with *fibrodysplasia ossificans progressiva. **ALK3** (532 aa) is the type 1A receptor for *bone morphogenetic protein and is mutated in juvenile polyposis. **ALK4** (505 aa) is mutated in some pancreatic tumours. Mutations in activin receptor 2B (**ACVR2B**) are associated with a *left–right asymmetry disorder (visceral heterotaxy). **ALK6** (BMP receptor 2B) is mutated in type A2 *brachydactyly.

actomyosin **1.** Descriptor for any motor system that is based on *actin and *myosin in which myosin binds to an actin microfilament, undergoes a conformational change that moves the two proteins relative to one another, and then releases the contact, the release requiring ATP hydrolysis which also restores myosin to the configuration capable of binding. **2.** A viscous solution formed when actin and myosin solutions are mixed at high salt concentrations. Extruded threads will contract if ATP is added and the viscosity of the solution diminishes if ATP is present.

acumentin A protein originally thought to cap the pointed end of microfilaments, now known to be L-*plastin.

acute **1.** Sharp or pointed. **2.** Describing diseases that rapidly reach a crisis but do not persist (are not chronic). *See* INFLAMMATION; LEUKAEMIA (acute lymphoblastic or myeloblastic leukaemia).

acutely transforming virus A *retrovirus that causes the rapid *transformation of cultured cells. *Rous sarcoma virus was the first example discovered.

acute phase proteins A group of proteins that are found in plasma at higher levels during acute *inflammation. In particular *C-reactive protein, *orosomucoid, and *serum amyloid A protein. A range of other proteins also increase in level, including mannose-binding protein, α_1-antitrypsin, α_2-macroglobulin, blood clotting factors, complement components, ferritin, ceruloplasmin, and haptoglobin. Other proteins may decrease in concentration (negative acute phase proteins).

acute phase reaction The production of *acute phase proteins.

acute reference dose The maximum dose of a poison that does not (normally) cause harm. Higher doses can be expressed as multiples of the reference dose.

acute respiratory distress syndrome (ARDS) A major inflammatory reaction in the lung in which fluid may accumulate and prevent gaseous exchange. It may progress to cause multiple organ failure. Formerly known as 'adult respiratory distress syndrome' to differentiate it from infant respiratory distress syndrome in which there is deficiency of lung surfactant. *See* SEPTIC SHOCK.

ACV **1.** *See* ACICLOVIR. **2.** α-Amino-adipylcysteinylvaline, a precursor for isopenicillin synthesis.

acyclovir *See* ACICLOVIR.

acyl An organic radical or functional group derived from an organic acid. Common examples are formyl, acetyl, and benzoyl moieties, from formic, acetic, and benzoic acids respectively. **Acylation** is the addition of an acyl (RCO-) group into a molecule: e.g. the formation of an ester between glycerol and fatty acid to form mono-, di-, or tri-acylglycerol, or the formation of an aminoacyl-tRNA during protein synthesis. **Acyltransferases** (EC 2.3.1.-) catalyse the transfer of acyl groups from a carrier such as acetyl CoA to a reactant. Acyl CoA dehydrogenases, *see* ACADs; acyl CoA:cholesterol acyltransferase, *see* ACAT. Acyl carrier proteins, *see* ACPs.

ADA *See* ADENOSINE DEAMINASE.

adalimumab A human monoclonal antibody directed against *tumour necrosis factor α and approved for treatment of inflammatory disease. *Compare* INFLIXIMAB which is a humanized mouse monoclonal antibody and *etanercept which is a decoy receptor. *TN* Humira.

ADAM family A family (a disintegrin and metalloprotease family) of membrane-anchored peptidases that cleave proteins of the cell surface and extracellular matrix (ECM) and may release membrane-bound growth factors. Examples include α-*secretase, *fertilin, and *TACE. A similar and related family (with at least nineteen members) are the **ADAM metallopeptidases with thrombospondin type 1 motif** (ADAMTSs) that are thought to be anchored to ECM through interaction of the thrombospondin motif with *aggrecan or other matrix components. ADAMTS-2, -3, and -14 process the N-peptide of procollagens I, II and III. Mutations in **ADAMTS2** are associated with *Ehlers–Danlos syndrome VIIC, **ADAMTS10** mutations with *Weill–Marchesani syndrome. Mutation in **ADAMTS-like 2** is associated with *geleophysic dysplasia. *See* PUNCTIN (ADAMTS1 and 3).

adaptation **1.** A reduced sensitivity in sensory or excitable cells following repeated stimulation. Phasic cells show rapid adaptation, tonic cells respond slowly. **2.** More generally, describing changes in any system with time, e.g. the down-regulation of receptors (*tachyphylaxis).

adaptins *See* ADAPTOR PROTEINS.

adaptor proteins **1.** A general term for proteins that link multimolecular protein arrays. *See* TRANSMEMBRANE ADAPTOR PROTEINS. **2.** *Clathrin-associated adaptor protein complex 1 (**AP-1**) is a heterotetramer composed of two large adaptins (γ-type subunit AP1G1, 822 aa and β-type subunit AP1B1, 949 aa), a medium adaptin (μ-type AP1M1, or AP1M2, 423 aa) and a small adaptin (σ-type AP1S1, 2, or 3, 158 aa). The complex is involved in recruiting clathrin to the membrane and in recognition of sorting signals in the cytoplasmic domains of transmembrane cargo molecules. AP-1 is also involved in recycling of coated vesicles. **3.** Clathrin-associated adaptor protein complex 2 (**AP-2**) is functionally similar to AP-1 but with different adaptins: two large adaptins (α-type subunit AP2A1, 977 aa or AP2A2, 939 aa and β-type subunit AP2B1, 937 aa), a medium adaptin (μ-type subunit AP2M1, 435 aa) and a small adaptin (σ-type subunit AP2S1, 142 aa). **4.** Adaptor protein complex 3 (**AP-3**) has a similar heterotetrameric structure but is not clathrin associated and has different adaptin subunits (δ subunit AP3D1, 1153 aa, β subunit AP4B1, 1094 aa or AP3B2, 1082 aa; a medium adaptin, μ-type, AP3M1, 418 aa or AP3M2, 418 aa and a small adaptin σ-type AP3S1, 193 aa or AP3S2, 193 aa). AP-3 may be directly involved in trafficking to lysosomes. *See* HERMANSKY–PUDLAK SYNDROME; BLOCs. NB There is a transcription factor complex *AP-2.

ADARs A small family of adenosine deaminases (adenosine deaminases acting on RNA, EC 3.5.4.-, ADAR1–3, editases) that edit adenosine residues to inosine in double-stranded RNA (dsRNA). This editing will modify viral genomes and is also known to affect the mRNA for glutamate receptor subunits. **ADAR1** (1226 aa) is associated with the RNA surveillance protein HUPF1 (RENT1) in the supraspliceosome. **ADAR2** (ADARB1, 741 aa) is widely expressed in brain and other tissues, and knockout mice die young. **ADAR3** (739 aa) may have a regulatory role in RNA editing. Defects in ADAR1 are a cause of *dyschromatosis symmetrical hereditaria. *See* ADENOSINE DEAMINASE (ADAT1).

ADAT1 *See* ADENOSINE DEAMINASE.

ADCA *See* SPINOCEREBELLAR ATAXIA.

ADCC *See* ANTIBODY-DEPENDENT CELL-MEDIATED CYTOTOXICITY.

ADD *See* ATTENTION DEFICIT (HYPERACTIVITY) DISORDER.

addict An individual who is physiologically or psychologically dependent on a habit-forming substance (e.g. drug dependence) or process (e.g. gambling dependence).

Addison's disease A disorder (chronic adrenal insufficiency) in which there is underproduction of steroid hormones by the adrenal cortex, leading to weakness, wasting, low blood pressure, and pigmentation of the skin. Can arise in a variety of ways, e.g. through tuberculosis affecting the gland or specific autoimmune destruction of the *adrenocorticotropin-secreting cells (often due to antibodies against the enzyme steroid 21-hydroxylase). Defects in enzymes involved in steroid metabolism can also cause similar symptoms, as will adrenoleucodystrophy (*see* LEUKODYSTROPHY). **Addison's anaemia** is *megaloblastic anaemia.

additive effect A simple, non*synergistic, effect in which the effects of two or more agents add together. If there is synergy the combined effect is greater than the simple sum.

addressins *See* SELECTINS.

adducin A *calmodulin-binding protein in the erythrocyte membrane skeleton, that promotes the assembly of the spectrin–actin network. A heterodimer of αβ or αγ subunits (α, 737 aa; β, 726 aa; γ, 706 aa). A substrate for *protein kinase C.

adenine One of the bases (6-aminopurine) found in *nucleic acids and *nucleotides where it pairs with *thymine (in DNA) and *uracil (in RNA). Adenine is also found as part of adenosine triphosphate (*ATP) and the cofactors nicotinamide adenine dinucleotide (NAD) and flavin adenine dinucleotide (FAD).

adenine nucleotide translocator One of the solute carrier family of proteins (ANT, ADP/ATP translocator, SLC25A4), the most abundant mitochondrial protein, a homodimer of 298 aa subunits embedded asymmetrically in the inner mitochondrial membrane that forms a gated pore through which ADP is moved from the matrix into the cytoplasm. Mutations in this gene lead to *progressive external ophthalmoplegia type A2.

adeno- A prefix that indicates an association with, or similarity to, glandular tissue. An **adenoma** is a benign tumour with a gland-like structure or developed from glandular epithelium. An **adenocarcinoma** is a malignant tumour of glandular epithelium (a carcinoma) that retains a gland-like organization of cells. An **adenofibroma** is a benign tumour with gland-like elements embedded in connective tissue. **Adenoids** are lymphoid tissue in the nasopharynx that has become become enlarged as a result of repeated upper respiratory tract infection. **Adenitis** is inflammation of a gland, often a lymph gland. For **adenomyoma** and **adenomyosis**, *see* ENDOMETRIUM. **Adenopathy** is a general term for a disease or disorder of glandular tissue but usually refers to enlargement of lymphatic glands.

adenocarcinoma *See* CARCINOMA.

adenohypophysis The anterior lobe of the *pituitary gland that secretes various hormones, particularly *growth hormone and *prolactin.

adenomatous polyposis coli *See* POLYPOSIS COLI.

adenosine The *nucleoside (9-β-D-ribofuranosyladenine) formed by ribosylation of adenine. *See also* ADENOSINE DEAMINASE; ADENOSINE RECEPTORS; ADP; AMP; ATP; CYCLIC AMP.

adenosine deaminase An enzyme (ADA, EC 3.5.4.4, 363 aa) that deaminates adenosine and 2′-deoxyadenosine to inosine or 2′-deoxyinosine respectively. An autosomal recessive defect in ADA causes 20–30% of cases of *severe combined immunodeficiency (SCID) and was the first candidate disease for gene replacement therapy. A deficiency of ADA causes an increase of dATP, which inhibits *S*-adenosylhomocysteine hydrolase; *S*-adenosylhomocysteine accumulates and is lymphotoxic. *See also* ADAR. **Adenosine deaminase complexing protein 2** (ADCP2, dipeptidyl peptidase IV, EC 3.4.14.5, conversion factor, 766 aa) forms a complex with ADA and generates the tissue-specific forms. It is a serine exopeptidase that cleaves X-proline dipeptides from the N-terminus of polypeptides, an intrinsic membrane glycoprotein. ADCP1 is known, but there is some doubt whether it is actually involved in ADA complexes. **Adenosine deaminase, tRNA-specific 1** (ADAT1, 502 aa) is involved in editing tRNA and deaminates adenosine-37 to inosine in eukaryotic tRNA(ala).

adenosine diphosphate *See* ADP.

adenosine monophosphate *See* AMP; CYCLIC AMP.

adenosine receptors G-protein-coupled receptors for adenosine that regulate a variety of physiological functions including cardiac rate and contractility, smooth muscle tone, sedation, release of neurotransmitters, platelet function, lipolysis, renal function, and white blood cell function. **A_1** receptors are ubiquitously expressed in the central nervous system and in a range of peripheral tissues and agonists induce calcium release and a transient inositol 1,4,5-trisphosphate (IP3) increase. The **A_{2A}** receptor is abundant in basal ganglia, vasculature, and platelets, and stimulates adenylyl cyclase. It is a major target of caffeine. The **A_{2B}** receptor is actually a *netrin-1 receptor that interacts with *deleted in colorectal cancer and induces cAMP accumulation. The **A_3** receptor (319 aa) inhibits adenylyl cyclase.

adenosine triphosphate *See* ATP.

adenosquamous Describing a benign tumour of epithelial origin (an *adenoma) in which cells have flattened (squamous) morphology, rather than being cuboidal or columnar.

adenoviral vector A vector used for gene transfer based upon a replication-defective adenovirus with a deletion in the E1 region (early genes). Plasmids containing the defective adenovirus genome with the gene to be expressed integrated are introduced into cells that are constitutively expressing the E1A genes. Adenoviral vectors have been used for gene therapy but have caused problems.

Adenoviridae A large family of viruses first isolated from cultures of adenoids. They are nonenveloped icosahedral viruses with a double-stranded linear DNA genome. They are responsible for respiratory infections in humans although some are oncogenic in rodents and will transform cells in culture. There are many serotypes (~52) that infect humans. *See* ADENOVIRAL VECTOR.

adenylate cyclase (adenylyl cyclase) The membrane enzyme (EC 4.6.1.1) that produces the *second messenger *cyclic AMP (cAMP) from ATP and couples ligand binding to altered cytoplasmic concentrations of cAMP through *GTP-binding proteins which modulate the activity of the enzyme. There are multiple tissue-restricted isoforms, some calcium-sensitive (e.g. adenylate cyclase 1 is the brain form, adenylyl cyclase 9 is widely distributed and stimulated by β-adrenergic receptor activation).

ADF *See* ACTIN DEPOLYMERIZING FACTOR; ADULT T-CELL LEUKAEMIA-DERIVED FACTOR.

ADH *See* VASOPRESSIN.

adherens junction A specialized intercellular junction that resists mechanical disruption. Those that have microfilaments attached on the cytoplasmic face are also known as belt *desmosomes (*zonulae adherens), those with *intermediate filaments attached are spot desmosomes (maculae adherens).

adhesins **1.** A general term for molecules involved in adhesion. **2.** In microbiology, bacterial surface components that behave as lectins, binding to surface carbohydrates.

adhesion **1.** In medicine, the attachment of two normally separate parts by fibrous tissue laid down after an inflammatory response. **2.** An **adhesion plaque** is a *focal adhesion. **3.** An **adhesion site** is a rather imprecise term for any part of the cell surface specialized for adhesion. **4.** In Gram-negative bacteria, an adhesion site is a region where the outer membrane and the plasmalemma appear to fuse and may be important for the export of proteins or the entry of viruses. **5.** *See* CELL ADHESION.

a

adipocere A white or yellowish waxy substance (mortuary fat, grave wax) formed postmortem by hydrolysis of body fats, mostly composed of long-chain saturated fatty acids.

adipocyte A mesenchymal cell with large lipid-filled vesicles typical of fatty tissue. Those in *brown fat may be distinct. 3T3-L1 cells are often used as a model system because they will differentiate into adipocyte-like cells when treated with a combination of *dexamethasone, *insulin, and *IBMX.

adipofibroblasts Fibroblast-like cells formed by dedifferentiation of *adipocytes, or at least by the loss of conspicuous lipid vesicles.

adipokine A family of peptide hormones produced by *adipocytes and involved in metabolic regulation among other things. The family includes *adiponectin, *apelin, *chemerin, *leptin, *omentin, *resistin, *vaspin, and *visfatin.

adiponectin A homotrimeric *adipokine (AdipoQ, apM1, GBP28, adipocyte complement related protein 30, Acrp30, 244 aa) secreted by *adipocytes and involved in the control of fat metabolism and insulin sensitivity, with direct antidiabetic, antiatherogenic, and anti-inflammatory activities. It is secreted following activation of the nuclear receptor PPAR-γ. Normal serum levels are ~5–10 μg/mL, but levels are decreased in both obesity and type 2 diabetes. The serum level of adiponectin as a quantitative trait is associated with variation in the adiponectin gene and by various other loci. The C-terminal globular domain has significant homology to *tumour necrosis factor and adiponectin antagonizes TNFα by negatively regulating its expression in various tissues such as liver and macrophages, and also by counteracting its effects. It may also play a role in cell growth, angiogenesis, and tissue remodelling by binding and sequestering various growth factors.

adipophilin A membrane-associated protein (adipose differentiation related protein, 437 aa) that is rapidly up-regulated when adipocyte differentiation is triggered. Together with *perilipin, it is constitutively associated with lipid droplets and plays a role in sustained fat storage and regulation of lipolysis. One of the *PAT family (perilipin, adipophilin, and TIP47) of proteins.

adipose tissue Fibrous connective tissue with many fat-storing cells, *adipocytes.

adipsin A *serine peptidase (factor D, C3 convertase activator, properdin factor D, EC 3.4.21.46, 253 aa), synthesized by *adipocytes, that is part of the alternative pathway of *complement activation; factor D cleaves factor B, which is bound to C3b to produce the C3 convertase of the alternative pathway (C3bBb). Deficiency reduces the ability to mount a response to meningococci and altered levels are characteristic of some genetic and acquired obesity syndromes. *See* LIPODYSTROPHY.

adjuvant An additive that enhances the action of the main component, most commonly encountered in the context of additives that increase the primary immune response to an antigen. *See* FREUND'S ADJUVANT.

ADM *See* ADRENOMEDULLIN.

ADME Common shorthand term for information about the absorption, distribution, metabolism, and excretion of a drug.

A-DNA One of the double helical structures that can be adopted by DNA, a right-handed double helix fairly similar to *B-DNA, but with a shorter, more compact helical structure. Probably one of the three biologically active structures, along with B- and *Z-DNA.

adnexa Appendages; usually refers to ovaries and fallopian tubes but can be applied to accessory or adjoining organs.

adoptive immunity Immunity acquired as a result of the transfer of antigen-specific T cells from an immune donor. Can be used for transfer of other elements of the immune system (such as immunoglobulin, or primed antigen-presenting cells).

ADP Usually the nucleotide (adenosine diphosphate, 5′-ADP), *adenosine with a diphosphate (pyrophosphate) group ester linked at position 5′ of the ribose moiety. Adenosine 2′,5′- and 3′,5′-diphosphates also exist, the former as part of NADP (*nicotinamide adenine dinucleotide phosphate) and the latter in *coenzyme A.

ADP-ribosylation A *post-translational modification of proteins by the addition of the ADP-ribosyl moiety of *NAD. It is a normal regulatory mechanism but is also the way in which several *AB toxins have their effects. **ADP-ribosylation factor** (ARF) is an ubiquitous *GTP-binding protein (181 aa) that mediates the binding of non-clathrin coated vesicles and AP1 (adaptor-protein 1) of *clathrin-coated vesicles to Golgi membranes. At least six isoforms have been identified. It is also an allosteric activator of the *cholera toxin catalytic subunit. **ADP-ribosylation factor-binding proteins** (Golgi-associated γ-adaptin ear containing ARF-binding proteins, GGAs 1–3, 639 aa, 613 aa, and 723 aa respectively) mediate the ARF-dependent recruitment of clathrin to the trans-Golgi network.

adrenal gland An endocrine gland adjacent to the kidney. The cortex produces various *corticosteroids, the medulla releases *adrenaline and *noradrenaline.

adrenaline (epinephrine) A hormone secreted by the medulla of the *adrenal gland, and by adrenergic *neurons of the *sympathetic nervous system. Adrenaline induces the 'fight or flight' responses: increased heart function, an elevation in blood sugar, cutaneous vasoconstriction, and the raising of hairs on the neck.

adrenergic receptors (adrenoreceptors) *G-protein-coupled receptors for *noradrenaline and *adrenaline linked either to *adenylate cyclase or to the phosphoinositide second messenger pathway. There are three subgroups: α_1-adrenergic receptors linked to inhibitory Gi, α_2 linked to G_q and β-adrenergic receptors linked to stimulatory Gs. β_1-adrenergic receptors are located mainly in the heart and in the kidneys, β_2-receptors are found mainly in the lungs, gastrointestinal tract, liver, uterus, vascular smooth muscle, and skeletal muscle, β_3-receptors are located in fat cells.

adrenocorticotropin (ACTH) A pituitary hormone (39 aa) derived from a larger precursor, *pro-opiomelanocortin, by the action of an endopeptidase that also releases β-*lipotropin. *Corticotropin releasing factor (41 aa) is released by the hypothalamus in response to stress and causes an increase in ACTH production and thus, indirectly, the release of hormones from the adrenal cortex, mostly *glucocorticoids. The N-terminal 13 aa of ACTH can be cleaved to produce α-*melanocyte stimulating hormone (α-MSH).

adrenoleucodystrophy Addison's disease and cerebral sclerosis. *See* ADDISON'S DISEASE; LEUKODYSTROPHY.

adrenomedullin A hypotensive vasodilator peptide (ADM, 52 aa) found in blood in significant amounts. It has diuretic and natriuretic effects on the kidney where it inhibits aldosterone secretion and in the pituitary it inhibits basal ACTH secretion. Together with **adrenomedullin-binding protein-1** (AMBP-1) it has been shown to reduce tissue damage in inflammation and appears to protect cells against oxidative stress. **Adrenomedullin-2** (intermedin, 148 aa) may regulate gastrointestinal and cardiovascular activities through a cAMP-dependent pathway. The receptor is G-protein coupled. *See* PROADRENOMEDULLIN.

adrenomyeloneuropathy *See* LEUKODYSTROPHY.

adrenoreceptors *See* ADRENERGIC RECEPTORS.

adriamycin *See* DAUNORUBICIN.

ADRP *See* ADIPOPHILIN.

adseverin A calcium-dependent actin-severing protein (scinderin, 715 aa) isolated from adrenal medulla. It also has barbed end capping and nucleating activities, similar to those of *gelsolin, and its properties are regulated by phosphatidyl inositides and by calcium.

adsorption coefficient A constant that relates the binding of a molecule to a matrix (soil, column packing, etc.) as a function of the weight of matrix, under defined conditions. Adsorption implies binding to a surface rather than integration into a solid-phase material (absorption: in the case of radiation, changing the energy levels of electrons, in other cases dissolving in the material rather than binding to the surface). The term is also used for absorption of radiation by materials (more or less equivalent to the attenuation coefficient).

adult respiratory distress syndrome *See* ACUTE RESPIRATORY DISTRESS SYNDROME.

adult T-cell leukaemia-derived factor *See* THIOREDOXIN.

advanced glycation end products A diverse group of proteins (AGE) that have been modified by the formation of glycation adducts. They occur in greater levels in diabetes and contribute to the development of vascular disease. The receptor is *RAGE.

adventitia In general, the outer covering of an organ. Most commonly encountered in the context of the outermost vascularized loose connective tissue surrounding a vein or artery.

AEBP2 A zinc-finger DNA-binding transcriptional repressor (adipocyte enhancer-binding protein 2, 517 aa) that is a component of the *PRC2/EED–EZH1 complex.

Aedes An important genus of mosquitoes, several of which are vectors of human diseases. *Aedes aegypti* is the vector of *yellow fever virus.

aequorin A protein (196 aa), extracted from jellyfish *Aequorea victoria*, that emits light (emission peak 470 nm, blue) according to the calcium ion concentration. Calcium triggers the intramolecular oxidation of coelenterazine into coelenteramide with concomitant *bioluminescence. Related molecules are found in other coelenterates.

aerobes Organisms that depend on oxygen for aerobic respiration and on the controlled oxidation of carbohydrate to yield energy.

aerolysin A pore-forming cytolytic toxin (43 aa) produced by *Aeromonas hydrophila*, a Gram-negative bacterium associated with diarrhoeal diseases and deep wound infections. Similar to staphylococcal α-toxin.

Aeromonas A genus of Gram-negative bacteria containing some opportunistic pathogens.

aesculin A hydroxycoumarin extracted from the horse chestnut *Aesculus hippocastanum*, used as an anticoagulant in homeopathic medicine although it is toxic at higher doses. Aesculin is an additive in agar media for the isolation of glycopeptide-resistant enterococci.

aetiology The study of the causation of disease, although often used to mean the cause of a particular disease (or even 'the aetiology is unknown').

AF2 **1.** The activation function domain 2 (AF2) of steroid receptors. **2.** A major African genotype (Af2) of *JC virus. **3.** A nematode neuropeptide, AF2 (KHEYLRF-NH2), a FMRFamide-related peptide.

afadin A protein target (AF-6 protein, 1816 aa) of *ras that regulates cell–cell adhesions (*zonula adherens junctions) and is involved in the linkage of microfilaments to *nectins. It is fused with *MLL in leukaemias caused by t(6;11) translocations. Interacts with *Eph-related receptor tyrosine kinases.

afamin A protein (α-albumin, 599 aa) of the albumin family (the others being *albumin, *alpha-fetoprotein, and vitamin D-binding protein (*Gc protein)), present in small amounts in plasma (30 μg/mL).

AFAP-110 An actin-binding adaptor protein (actin filament-associated protein-110, 635 aa) that binds *src-family proteins through SH2 and SH3 domains. Phosphorylation of AFAP by src regulates self-association.

afebrile Not feverish (febrile).

affect A term used in psychology for the experience of feeling or emotion, mood, etc., as opposed to cognition and volition. Affective disorders are characterized by disturbance of mood, with feelings of elation or sadness being intense and unrealistic.

afferent Describing something that leads towards a target, e.g. afferent blood vessels lead to the heart. The opposite of *efferent.

affinity A measure of the strength of the interaction between two entities, e.g. between enzyme and substrate or receptor and ligand. The binding affinity is quantified by the equilibrium constant *association constant or *dissociation constant for the binding, the concentration at which 50% of the binding sites are occupied.

affinity chromatography A *chromatography method in which the immobile phase has a specific affinity for the substance to be separated or isolated. This can be achieved e.g. by immobilizing a specific antibody to purify an antigen-carrying protein, or a bait protein to isolate interacting proteins.

affinity labelling A method of labelling of the active site of an enzyme or the binding site of a receptor by using a reactive substance that binds and then forms a covalent linkage, sometimes triggered by a change in conditions, e.g. by illumination in photoaffinity labelling.

affinity maturation The process that occurs in the immune system after the first exposure to an antigen. There is selective differentiation of clones of cells that produce antibody of high affinity and memory lymphocytes that will produce high-affinity antibody upon re-exposure to the antigen.

afibrinogenaemia A rare autosomal recessive bleeding disorder in which there is a complete absence or extremely low levels of plasma and platelet fibrinogen, usually a result of truncating mutations in the *fibrinogen A or B genes.

aflatoxins A group of fungal toxins produced by *Aspergillus flavus* and other *Aspergillus* spp. Can be a problem in stored foodstuffs; the toxins will cause enlargement and death of liver cells if ingested, and may also be carcinogenic.

AFP *See* ALPHA-FETOPROTEIN.

agalactia (agalacia) Failure of milk production by the breast.

agammaglobulinaemia A sex-linked genetic defect that leads to the complete absence of immunoglobulins (IgG, IgM, and IgA) in the plasma and the absence of humoral immunity, although cell-mediated immunity is unimpaired.

Arises from a failure of B-cell differentiation and a failure of Ig heavy chain rearrangement because of mutations in the gene encoding *Bruton's tyrosine kinase.

agar (agar-agar) An unbranched *polysaccharide with galactose subunits, extracted from seaweed (Rhodophyceae) that forms a gel which is used as an inert support for the growth of cells, particularly bacteria. Many *transformed cells have the ability to grow in 'sloppy agar', indicating that the normal requirement for attachment to a mechanically resistive substratum has been lost.

agarose A galactan polymer purified from *agar that will produce a rigid gel. Used as a support matrix for the electrophoretic separation of biomolecules and as a macroporous chromatographic bed material. Sepharose™ is a crosslinked, beaded form of agarose.

agatoxins A group of toxins from the American funnel web spider *Agalenopsis aperta*. The ω-agatoxins (~100 aa) are antagonists of *voltage-sensitive calcium channels and block the release of neurotransmitters. For example, ω-agatoxin 1A selectively blocks *L-type calcium channels, ω-agatoxin 4B inhibits voltage-sensitive *P-type calcium channels. The μ-agatoxins act only on insect voltage-gated sodium channels.

AGC kinases A large subclass of protein kinases including protein kinases A, G, and C, also protein kinase B (PKB)/akt, p70 and p90 ribosomal S6 kinases, and phosphoinositide-dependent kinase-1 (PDK-1). Their characteristic feature is an activation by tyrosine phosphorylation.

SEE WEB LINKS

- An overview of protein kinase signalling.

AGE *See* ADVANCED GLYCATION END PRODUCTS.

agenesis A general term for the failure of development (or imperfect development) of an organ or tissue.

age-related macular degeneration A condition (ARMD) that is one of the commonest causes of visual impairment. Macular degeneration (damage in the central region of the retina) can arise through a variety of causes and different forms of the disorder are recognized. In '**dry ARMD**' debris (drusen) accumulates below the retina causing local damage and detachment; in '**wet ARMD**' there is neovascularization and frequently major retinal detachment following haemorrhage of the new vessels. **ARMD1** is associated with polymorphism in the fibulin-6 (hemicentin) gene, **ARMD2** with mutation in the *ABCR* gene (retina-specific ABC transporter), **ARMD3** with mutation in the fibulin-5 gene. **ARMD4** is mostly attributable to polymorphism in the complement factor H gene, **ARMD5** with mutation in the *ERCC6* gene that is part of the nucleotide excision repair (NER) pathway, **ARMD6** with mutations in the retinal homeobox *RAXL1* gene, **ARMD7** with mutation in the serine peptidase 11 gene, **ARMD8** with a SNP in the LOC387715 gene, and **ARMD9** with a SNP in the complement C3 gene. **ARMD10** is associated with a variant in the toll-like receptor-4 gene.

agglutination The formation of multicomponent masses (aggultinates, flocs) of particles or cells stimulated by the addition of a multivalent external agent (an agglutinin), e.g. an antibody or a lectin. *Compare* AGGREGATION.

aggrecan The major structural *proteoglycan of cartilage, the core protein (2316 aa) to which ~100 chondroitin sulphate chains and several keratan sulphate chains as well as *O*- and *N*-linked oligosaccharide chains are linked. It binds to a link protein (~40 kDa) and to hyaluronic acid to produce very large multimolecular aggregates. A mutation in aggrecan leads to Kimberley type *spondyloepiphyseal dysplasia. The linker domain between the N-terminal globular domains is highly sensitive to proteolysis by aggrecanases, peptidases of the *ADAM family.

SEE WEB LINKS

- Structure and function of aggrecan and versican.

aggregation The formation of multisubunit complexes in which direct adhesion between the particles is involved, rather than an extrinsic linker. Aggregation is usually slower than *agglutination.

aggresome *See* SEQUESTOSOME.

agitoxins A toxin, from the venom of the yellow scorpion *Leiurus quinquestriatus hebraeus*, that blocks *shaker-type voltage-regulated potassium channels. Closely related to *kaliotoxin.

aglycone (aglycon, aglucon) A *glycoside from which the sugar moiety has been removed.

agmatinase The bacterial enzyme (agmatine ureohydrolase, EC 3.5.3.11) that degrades *agmatine to *putrescine.

agmatine A metabolite of arginine produced by arginine decarboxylase, that can be metabolized to *putrescine by bacterial *agmatinase. Agmatine suppresses polyamine biosynthesis and polyamine uptake by cells by inducing *antizyme. It binds to *imidazoline

a

receptors, and may be the endogenous ligand for imidazoline (I_1) receptors, and to α_2-adrenoreceptors. It is thought to be a neurotransmitter involved in behavioural and visceral control.

agnoprotein A family of small, very basic proteins (71 aa) encoded by polyoma viruses (neurotropic *JC virus, *SV40, and BK virus). They apparently regulate viral gene expression, inhibit DNA repair after DNA damage, and interfere with DNA damage-induced cell cycle regulation. JC virus expresses agnoprotein during the lytic infection of glial cells in progressive multifocal leucoencephalopathy (PML), and in some neural tumours, particularly medulloblastoma.

agnosia The inability to recognize familiar things or people.

agonist **1.** In neurobiology, a *neuron or *muscle that assists the action of another. *Compare* ANTAGONISTS, which oppose one another's actions. **2.** In pharmacology, a compound that acts on the same receptor, and has similar effects, to the natural ligand. **3.** In ethology, 'agonistic behaviour' is aggressive behaviour directed towards a conspecific animal.

agouti **1.** A Central American rodent of the genus *Dasyprocta*, related to guinea pigs. **2.** A coat colour in mice in which there are alternate light and dark bands on individual hairs. The gene responsible encodes a secreted protein (131 aa) that regulates phaeomelanin synthesis in melanocytes, binds to the *melanocortin receptor 1 but does not antagonize α-*melanocyte stimulating hormone, and has antiproliferative effects on melanoma cells in culture. Mice with dominant mutation at the Ay locus develop diabetes and obesity. The dark agouti (DA) rat has a high susceptibility to developing arthritis. **Agouti-related peptide** (AGRP, 132 aa) is a neuropeptide that increases appetite and decreases metabolism, an endogenous antagonist of melanocortin type 3 and 4 receptors. Secretion of AGRP and *neuropeptide Y is stimulated when the amount of stored fat and *leptin decreases. A polymorphic variant has been proposed to play a role in late-onset obesity.

agranular vesicles Synaptic vesicles (40–50 nm diameter), characteristic of peripheral cholinergic *synapses, that do not have a granular appearance.

agranulocytosis A condition in which circulating granulocyte numbers are severely reduced.

agretope The region of a peptide that interacts with an *MHC molecule. Identifying and predicting agretopes is important in vaccine design.

agrin A secreted protein (1940 aa) that induces the aggregation of acetylcholine receptors and other postsynaptic proteins at the neuromuscular junction and stabilizes the junction. Many alternatively spliced isoforms exist. *See* MuSK.

AGS **1.** AGS cells are a line of human gastric epithelial cells derived from an *adenocarcinoma. **2.** *See* AICARDI–GOUTIERES SYNDROME.

AHNAKs A class of giant propeller-like proteins that interact with the β2 subunit of cardiac L-type calcium channels although predominantly present in the nucleus. **Ahnak1** (5890 aa), was first identified as being identical to a bovine desmosomal protein, desmoyokin; a similar protein (**Ahnak2**, 5795 aa) was identified later. *Dysferlin is involved in the recruitment and stabilization of Ahnak to the sarcolemma and Ahnak seems to play a role in the dysferlin membrane repair process. Ahnak and dysferlin are markers for *enlargeosomes, a type of cytoplasmic vesicle.

AHR *See* ARYL HYDROCARBON RECEPTOR.

Aicardi–Goutieres syndrome (AGS). A genetically heterogeneous autosomal recessive encephalopathy (Cree encephalitis, pseudo-TORCH syndrome, pseudotoxoplasmosis syndrome) phenotypically similar to *in utero* viral infection. **Type 1** is caused by mutation in the *TREX1* gene encoding a 3′-to-5′ exonuclease, **types 2, 3,** and **4** by mutations in the genes for subunits B, C, and A of ribonuclease H2 (RNAseH2). **Type 5** is an autosomal dominant form also caused by mutation in the *TREX1* gene. **Aicardi's syndrome** is a distinct disorder in which there is agenesis of the corpus callosum, spinal skeletal abnormalities, and chorioretinal abnormalities.

AIDS The immunodeficiency (acquired immune deficiency syndrome) caused by infection with *human immunodeficiency virus (HIV). Opportunistic infections are likely to occur, *tuberculosis being an increasing problem, and there is predisposition to certain types of tumour, particularly *Kaposi's sarcoma.

AIF *See* APOPTOSIS INDUCING FACTOR.

AIM **1.** Genes 'absent in melanoma': **AIM1** encodes a protein (1723 aa) of the βγ-crystallin superfamily that is up-regulated in association with tumour suppression in a model of human melanoma. **AIM2** encodes a deduced 344-aa protein that has a conserved sequence domain (~200 aa) shared with interferon-inducible genes.

Overexpression of AIM2 reversed the tumorigenic phenotype in a melanoma cell line. *See* PAAD DOMAIN. **2.** *See* AURORA KINASES.

AIP1 **1.** Actin interacting protein-1 (WD repeat-containing protein 1). The human homologue (606 aa) of an actin-binding protein originally identified in yeast. Induces disassembly of actin filaments in conjunction with *actin depolymerizing factor/*cofilin family proteins. **2.** ALG-2-interacting protein 1 (apoptosis-linked-gene-2 interacting protein 1, 868 aa), an adaptor protein that controls the production of endosomes and trafficking through the endosomal system.

AITR A receptor (activation-inducible TNFR family member, TNF receptor superfamily 18, glucocorticoid-induced TNFR family-related protein) expressed in lymph node and peripheral blood leucocytes, up-regulated in response to various activating stimuli. It appears to be involved in the regulation of T-cell receptor-mediated cell death, associates with TRAF1 (TNF receptor-associated factor 1), TRAF2, and TRAF3, and induces NFκB activation via TRAF2. The ligand is **AITRL**, a member of the TNF superfamily (TNFSF18, 177 aa) that is expressed in endothelial cells. AITRL regulates T-cell proliferation and survival, and is involved in the interaction between T cells and endothelial cells.

AKAP family An extensive family (A-kinase anchoring protein family) of scaffolding proteins that determine the cellular localization of different *protein kinase A isoforms which are tethered through a regulatory subunit. *Calcineurin and *protein kinase C also bind to some AKAPs; the PKC and calcineurin sites are different, so that both can be bound. *See* PERICENTRIN.

akasthisia A state of restlessness in which it is difficult to keep still. Can be a side effect of some neuroleptic drugs. **Restless leg syndrome** is similar but distinct. *See* EXTRAPYRAMIDAL SYSTEM.

A kinase *See* AKAP FAMILY; PROTEIN KINASE A.

akinesia The inability to initiate movement due to difficulty selecting or activating motor programmes in the central nervous system. Characteristic of diseases such as Parkinson's where there is diminished dopaminergic activity.

Akt (protein kinase B, PKB) A family of serine/threonine kinases activated by *PI-3-kinase downstream of insulin and other growth factor receptors. It is the product of the cellular homologue of v-*akt*, the transforming oncogene of AKT8 virus. Akt will phosphorylate *glycogen synthase kinase (GSK3) and is involved in stimulation of Ras and control of cell survival. Three members of the Akt/PKB family have been identified: **Akt1** (PKBα, EC 2.7.11.1, 480 aa), **Akt2** (PKBβ, 481 aa), and **Akt3** (PKBγ, 479 aa). Akt2 has been associated with ovarian carcinomas.

aladin *See* TRIPLE A SYNDROME.

Alagille's syndrome A disorder (arteriohepatic dysplasia) in which the bile ducts become reduced and there is cholestasis, cardiac disease, skeletal abnormalities, ocular abnormalities, and a characteristic facial phenotype. Alagille's syndrome 1 (**ALGS1**) is caused by mutation in *jagged-1, a ligand for the *Notch receptor; Alagille's syndrome 2 (**ALGS2**) is caused by mutation in the notch2 gene.

alalia *See* APHONIA.

alamethicin A peptide *ionophore, produced by the fungus *Trichoderma viride*, that forms relatively nonspecific anion or cation transporting pores that are sensitive to the transmembrane potential difference.

Aland Island eye disease An X-linked recessive retinal disease in which there is decreased visual acuity, progressive myopia, and defective dark adaptation caused by mutation in the retina-specific calcium channel α_1-subunit gene. Defects in this protein also cause a form of *stationary night blindness.

alanine (Ala, A) Usually, L-α-alanine (89 Da), the aliphatic amino acid found in proteins. The isomer β-alanine is a component of the vitamin pantothenic acid and *coenzyme A. *See* SITE-SPECIFIC MUTAGENESIS.

alanine aminotransferase A cytoplasmic enzyme in hepatocytes (EC 2.6.1.2, 498 aa) that catalyses the conversion of L-alanine and 2-oxoglutarate into pyruvate and L-glutamate. The enzyme was formerly known as serum glutamic–pyruvic transaminase (SGPT) and a raised *SGOT/SGPT ratio in blood is often taken as being indicative of liver damage, possibly alcoholic (non-A, non-B) hepatitis. In the brain it is a neuroprotectant against *glutamate excitotoxicity.

ALA synthase The first and rate-limiting enzyme in the haem biosynthetic pathway (5-aminolaevulinate synthase) that catalyses production of 5-aminolaevulinic acid. There are two tissue-specific mitochondrial isozymes: a housekeeping enzyme encoded by the ***ALAS1*** gene (EC 2.3.1.37, 640 aa) and an erythroid tissue-specific form (587 aa) encoded by ***ALAS2***. Defects in the *ALAS2* gene cause X-linked

a

sideroblastic anaemia, an anaemia associated with accumulation of iron and the presence of *sideroblasts in the bone marrow.

albamycin *See* NOVOBIOCIN.

albendazole An anthelmintic drug. *TN* Zentel.

Albers–Schoenberg disease *See* OSTEOPETROSIS.

albinism The condition in which there is decreased or absent pigmentation in the hair, skin, and eyes. **Oculocutaneous albinism type IA** (OCA1A) is caused by mutation in the tyrosinase gene essential for *melanin synthesis; the activity of the enzyme is reduced in OCA1B. OCA2, OCA3, and OCA4 are somewhat milder forms of the disorder. **OCA2** is caused by mutations in the *OCA2* gene, the product of which, P protein (838 aa), regulates the pH of melanosomes. **OCA3** is caused by mutation in tyrosinase-related protein (537 aa), a catalase that enhances melanin production (it is also defective in rufous oculocutaneous albinism). **OCA4** is caused by mutation in the solute carrier protein (SLC45A2). **Ocular albinism** (OA1) involves only the eyes.

alboaggregin A C-type lectin (alboaggregin A, a disulphide-linked heterotetramer of ~130 aa subunits), from the white-lipped tree viper *Trimeresurus albolabris,* that causes platelet aggregation by acting as a platelet GPIb agonist. Alboaggregin B is a heterodimer and also has platelet aggregating activity but is calcium-independent. Their action is similar to that of *botrocetin.

albolabrin *See* DISINTEGRINS.

Albright hereditary osteodystrophy A syndrome in which there is short stature, obesity, round facies, subcutaneous ossifications, brachydactyly, and other skeletal anomalies. There may also be mental retardation. AHO is often associated with pseudohypoparathyroidism, hypocalcaemia, and elevated PTH levels (PHP Ia). Patients with pseudopseudohypoparathyroidism have normal calcium metabolism and PTH levels. AHO, pseudohypoparathyroidism type Ia (PHP Ia), and pseudopseudohypoparathyroidism (PPHP) are all caused by mutations or imprinting defects in the gene coding for the α subunit of the Gs type of heterotrimeric G protein. *See* PROGRESSIVE OSSEOUS HETEROPLASIA.

albumen (ovalbumin) A major phosphoprotein (386 aa) of the white of avian eggs. *Compare* ALBUMIN.

albumin A major protein (serum albumin, 609 aa) constituting ~50% of the plasma protein content, that has multiple binding sites for various lipophilic metabolites, particularly fatty acids and bile pigments. There are numerous electrophoretically distinguishable variants. Analbuminaemic individuals show an apparently normal phenotype and are often only identified as a side consequence of blood analyses. The function of albumin (even although apparently dispensable) is carried out by *alpha-fetoproteins in the fetus. *See* AFAMIN; ALBUMEN; GC PROTEIN. **Albuminuria** is the condition in which there is albumin in the urine.

Alcaligenes A widespread genus of Gram-negative aerobic bacilli found in the digestive tract and on skin. Can be responsible for opportunistic infections.

alcaptonuria *See* ALKAPTONURIA.

Alcian blue A water-soluble histochemical stain (Alcian blue 8GX) for *glycosaminoglycans. Alcian green and Alcian yellow have comparable properties.

aldolase A glycolytic enzyme (fructose-1,6-bisphosphate aldolase, EC4.1.2.13, 363 aa) that catalyses the conversion of fructose 1,6-bisphosphate to glyceraldehyde 3-phosphate and dihydroxyacetone phosphate. There are three isozymes, aldolases A, B, and C, with tissue-specific distributions. Aldolase, like phosphofructokinase, enolase, and hexokinase, binds actin filaments, probably to cause local concentrations of enzymes and their substrates. The glucose transporter *GLUT4, is connected to actin via aldolase. **Aldolase A deficiency** is an autosomal recessive disorder associated with hereditary haemolytic anaemia, the enzyme being very thermolabile. Mutations in aldolase B are associated with fructose intolerance.

aldose reductase One of the monomeric, NADPH-dependent aldoketoreductase family of enzymes (aldehyde reductase 1, EC1.1.1.21, 316 aa) involved in glucose metabolism and osmoregulation. It is the first and rate-limiting enzyme of the polyol pathway of sugar metabolism, has a role in protecting against toxic aldehydes derived from lipid peroxidation and steroidogenesis, and has been implicated in diabetic complications.

aldosterone A mineralocorticoid produced by the *adrenal cortex, that increases the reabsorption of sodium and water and the secretion of potassium in the kidney and has an important role in control of blood pressure. Levels are diminished in *Addison's disease and increased in Conn's syndrome. *Atrial natriuretic factor (ANF) inhibits [aldosterone] secretion.

alemtuzumab A humanized rat monoclonal antibody directed against CD52, used in the treatment of B-cell chronic lymphocytic leukaemia and as an immunosuppressive prior to transplant surgery.

alendronic acid A bisphosphonate drug used to treat osteoporosis. The drug is adsorbed on to hydroxyapatite crystals in bone and inhibits osteoclast-mediated bone-resorption. *TN* Fosamax.

Alexander's disease Rare and usually fatal neurodegenerative disorder characterized by the development of megalencephaly in infancy accompanied by progressive spasticity and dementia. The features are similar to those of *Canavan's disease. It is caused by a mutation in the *glial fibrillary acidic protein (*GFAP*) gene. Astrocytes in Alexander's disease have protein aggregates (Rosenthal fibres) containing (alpha B-crystallin, heat shock protein 27, and GFAP.

***Alexandrium* spp.** A genus of dinoflagellates that produce the toxins responsible for shellfish poisoning.

alexia The inability to read due to a brain lesion (rather than a failure of learning).

algesis The sense of pain. **Algesia** is sensitivity to pain.

alginate Salts of alginic acids, linear polymers of mannuronic and glucuronic acids, found in the cell walls of some algae and used commercially in food processing, swabs, some filters, fire retardants, etc. Calcium alginates form gels.

algorithm A finite set of instructions for carrying out a calculation or process. Flow charts are visual representations of algorithms.

alimemazine A phenothiazine derivative (trimeprazine), an antihistamine, used as an antipruritic and as a sedative and antiemetic for prevention of motion sickness.

alimentary canal (alimentary tract) The more-or-less tubular cavity that opens at mouth and anus and through which food passes, usually rearwards.

aliphatic Describing carbon compounds that do not contain aromatic rings. They can be cyclic or linear, saturated or unsaturated. The naturally occurring amino acids with aliphatic side chains are glycine, alanine, valine, leucine, and isoleucine.

aliquot A small portion. Reagent solutions or samples are often subdivided into aliquots to reduce the handling of the whole (precious) sample.

aliskiren The first *renin inhibitor developed; used to treat essential hypertension. Crystal structure analyses of renin–inhibitor complexes combined with computational methods were used in the design process. *TNs* Rasilez, Tekturna.

alkaline phosphatase Membrane-bound glycoprotein enzymes (EC 3.1.3.1) that hydrolyse various monophosphate esters optimally at alkaline pH. There are three distinct forms: intestinal (**ALPI**, 528 aa), placental (**ALPP**, 532 aa), and liver/bone/kidney (tissue nonspecific isozyme, **ALPL**, 524 aa). Mutations in the gene for ALPL lead to *hypophosphatasia.

alkaloid A nitrogenous base. Many plant toxins (allelochemicals) are alkaloids. Examples include *atropine, *camptothecin, and *colchicine.

alkalosis A condition in which the pH of blood is more alkaline than normal (pH > 7.45). Can arise because of excessive loss of CO_2 (hyperventilation) or loss of stomach acid by vomiting. *Compare* ACIDOSIS.

alkaptonuria (alcaptonuria) A recessive disorder caused by mutation in the gene for homogentisate-1,2-dioxygenase leading to accumulation of *homogentisic acid and deposition of brown pigment in skin, the sclera of the eye, connective tissue, and joints (ochronosis). The urine blackens on standing.

ALKs Activin-like kinases. *See* ACTIVIN.

alkylating drug A cytotoxic drug that damages DNA: the addition of alkyl groups to guanine can lead to fragmentation during frustrated repair; nucleotides may be cross-bridged, blocking strand separation; mispairing can cause mutation during replication. There are six groups of alkylating agents: nitrogen mustards, ethylenimes, alkylsulphonates, triazenes, piperazines, and nitrosureas. Examples include *cyclophosphamide, *dacarbazine, *ifosfamide, *procarbazine, and *temozolomide.

ALL Acute lymphoblastic *leukaemia.

allantoin A derivative of uric acid found in the allantoic fluid and in the urine of some mammals, and an excretory product of some invertebrates. Used in various cosmetics, oral hygiene preparations, and eye drops.

allantois A ventral outgrowth from the hindgut of the embryo in reptiles, birds, and mammals

a

that serves as a store for nitrogenous waste. In the chick embryo it fuses with the chorion to form the *chorioallantoic membrane (CAM).

alleles (allelomorphs) Variants of a *gene at the same *locus on a *chromosome. They are assorted as a result of sexual reproduction and *genetic recombination alters their linkage to other alleles. There may be many possible variants and *polymorphism is not uncommon. **Allelic disorders** are clinically distinct conditions that arise because of different mutations in a gene: in some cases the mutation leads to complete absence of the gene product (e.g. through premature termination), in other cases the product may be less effective, though still present (e.g. a conservative amino acid substitution).

allelic exclusion The suppression of one allele in a diploid cell. In females one of the X chromosomes (either) is inactivated (*see* LYON HYPOTHESIS) so that individual cells express only one allelic form of the gene product and the body is effectively a mosaic. The process also occurs in *immunoglobulin genes so that a clone expresses only one of the two possible allelic forms of immunoglobulin.

allelic imbalance The condition in which one *allele of a heterozygous gene pair is lost (loss of heterozygosity) or amplified. This is a form of aneuploidy and is common in tumours, where it may cause dysregulation of oncogenes or tumour suppressor genes near the sites of imbalance.

allelochemical A substance produced by one species that has a deleterious effect on another (*allelopathy).

allelomorph *See* ALLELES.

allelopathy The harmful interaction between organisms or cells that are *allogeneic, although the term is often applied to interactions between *xenogeneic organisms.

allelotype A term used by analogy with *karyotype; the expression pattern of particular alleles, e.g. in a tumour cell population compared to normal tissue. The expression of particular microsatellite markers or isoenzymes, revealing the alleles being expressed, may show whether a tumour is polyclonal or monoclonal in origin and the extent to which further divergence (*aneuploidy) has occurred.

allergen A substance that will cause an allergic response (*see* ALLERGY). Usually applied to substances such as pollens or insect venoms that cause immediate-type *hypersensitivity reactions.

allergic encephalitis *See* EXPERIMENTAL ALLERGIC ENCEPHALOMYELITIS.

allergic rhinitis A response to inhaled allergens by the mucous membranes of the nose and upper respiratory tract, which become swollen. Hay fever is a common seasonal form.

allergy A *hypersensitivity response to an antigen (*allergen) to which a primary immune response has already been made. In extreme cases there may be *anaphylaxis.

allicin A strong-smelling antibacterial compound from garlic *Allium sativum*. The enzyme alliinase converts alliin in raw garlic to allicin. It is rapidly degraded, and the evidence for antibacterial efficacy is anecdotal.

alloantibody (alloserum) An antibody that binds to products of different alleles in other individuals of the same species (who are allogeneic). Alloantigens are common in multiparous women and individuals who have had many blood transfusions. The antigens are usually of the *MHC complex or *blood group antigens.

alloantigen *See* ALLOANTIBODY.

alloepitope *See* ALLOTOPE.

allogeneic Describing individuals (or strains) in which the genes at one or more loci differ in sequence. Allogenicity is usually specified with reference to a particular locus or loci.

allograft A graft made between individuals that are *allogeneic. Usually the important allogenicity is in the *MHC complex. *Compare* AUTOGRAFT; XENOGRAFT.

allometric growth A pattern of growth in which it is possible to express the mass or size of any organ or body part in relation to that of the entire organism according to the allometric equation $Y = ax^b$ where Y is the mass of the organ, x is the mass of the organism, b is the growth coefficient of the organ, and a is a constant. Allometric scaling is a method of data adjustment to allow either for changes in proportion of organs or tissues as the organism grows, or for differences between species of dissimilar size and shape.

allometry **1.** The study of the relationship between the growth rates of different parts of an organism. **2.** The changes in the proportions of body parts as the organism grows in size.

allopatric speciation The divergence of a population into two distinct species as a result of geographical isolation, which allows the accumulation of genetic differences that eventually make interbreeding impossible. Probably the most common cause of speciation.

allopolyploidy A condition in which the genomes that contribute to a polyploid organism are dissimilar. Such an organism is generally infertile, but a further doubling to produce an amphidiploid will restore fertility. Common in plants.

allopurinol An inhibitor of *xanthine dehydrogenase that reduces the production of uric acid from the degradation of purines. Used to treat *gout.

all-or-nothing Describing a process that has only two states, on or off, a digital rather than analogue response. Action potentials are of this form: once they are triggered they have a standard size and shape whatever the size of the stimulus.

alloserum *See* ALLOANTIBODY.

allosomes The chromosomes (accessory chromosomes, heterochromosomes, sex chromosomes) that carry genes for sex determination, the X and Y chromosomes in humans. *Compare* AUTOSOMES.

allospecific **1.** In taxonomy, having the status of a distinct species, genetically distinct and unable to interbreed with other similar species. **2.** In immunology, a shorthand term for 'allelle-specific' and thus antigenically distinct. Individuals, unless genetically identical, are allospecific and an immune response to the alloantigens (allotypic determinants) will lead to graft rejection.

allostasis The adaptive processes that actively maintain stability (homeostasis) through physiological or behavioural changes. In psychology this leads to the concept of '**allostatic load**', the long-term cost of handling stress.

allosteric Describing a shape change brought about by an interaction at another spatially distinct site on the same molecule. Many enzymes are regulated (activated or inhibited) by allosterically induced shape change: sometimes by binding of a regulator, sometimes by covalent modification (e.g. phosphorylation).

allotope (allotypic determinant, alloepitope) The region (epitope) of an antigen that differs from another allelic version (*allotype) of that antigen.

allotypes The products of different *alleles encoding a particular molecule and that can be distinguished as variants. Used mainly in the context of *immunoglobulins that are antigenically different; human light chains are of Km (Inv) allotypes, heavy chains of Gm allotypes. *See also* IDIOTYPE.

alloxan A toxic glucose analogue that will destroy pancreatic *beta cells in experimental animals, producing a model for studying type 1 *diabetes.

allozyme A variant of an enzyme coded by a different allele of a single gene. *Compare* ISOENZYMES (isozymes), which catalyse the same reaction but are the products of different genes.

aloisines Competitive inhibitors of ATP binding to the catalytic subunit of *cyclin-dependent kinases and *glycogen synthase kinase (GSK-3), that will arrest cells in both G_1 and G_2 phases of the *cell cycle.

alopecia Baldness. **Alopecia areata** is a genetically determined, immune-mediated disorder of the hair follicle leading to patchy hair loss from the head. The same disorder may become complete (**alopecia totalis**) or extend to the whole body (**alopecia universalis**). Alopecia universalis congenita is a distinct disorder caused by mutation in *hairless. **Androgenetic alopecia** (male pattern baldness) occurs in women as well as men and is caused by a shortening of the anagen (growth) phase and miniaturization of the hair follicle, which results in the formation of progressively thinner, shorter hair. The pattern of loss is defined. *See also* WOODHOUSE–SAKATI SYNDROME. Alopecia is associated with various other syndromes but the causes (and relevance of the alopecia) are generally unknown.

alpha (α) This prefix is generally, but not universally, ignored in alphabetization; look for the main portion of the word.

alpha-1-antiprotease *See* ALPHA-1-ANTITRYPSIN.

alpha-1-antitrypsin (α_1-antiprotease) A serine peptidase inhibitor (*serpinA1, 418 aa) that is a major protein in plasma (~3 mg/mL) and particularly important as an inhibitor of neutrophil elastase. Deficiency of this enzyme is associated with increased risk of developing chronic obstructive pulmonary disease and emphysema.

alpha-1-microglobulin *See* MICROGLOBULIN.

alpha-2-macroglobulin (α_2-macroglobulin, α_2M) *See* MACROGLOBULIN.

a

alpha-actinin (α-actinin) An actin-binding protein, a homodimer with the subunits (892 aa) arranged in antiparallel fashion so that it can link microfilaments with opposite polarity. A major protein of the *Z disc in striated muscle and the functional equivalents in smooth muscle. In nonmuscle cells it is distributed along microfilament bundles and associated with their anchorage at adherens-type junctions and focal adhesions. Various isoforms occur in muscle and nonmuscle cells. **α-Actinin-2** is found in all skeletal muscle fibres, whereas **α-actinin-3** expression is limited to a subset of type 2 (fast) fibres. A polymorphism in expression of α-actinin-3 is maintained and the absence of expression is common in endurance athletes and rare in elite sprinters. Mutation in the α-actinin-2 gene can lead to dilated cardiomyopathy (CMD1AA) and mutation in **α-actinin-4** to *focal segmental glomerulosclerosis. An actinin-like actin-binding domain is present in many actin-binding proteins, including spectrin, fodrin, dystrophin, filamin, and fimbrin. *See* CALPONINS (calponin homology domain).

alpha-amylase *See* AMYLASES.

alpha-B crystallinopathy An autosomal dominant form of desmin-related *myopathy (DRM), that results in weakness of the proximal and distal limb muscles, cardiomyopathy, and cataract. Patients with progressive myopathy characterized by myofibrillar degeneration that commences at the Z disc have been described. Mutations in α-*crystallin B truncate the C-terminal domain of the protein that is required for the chaperone function.

alpha-blockers (α-adrenoceptor blockers) Drugs that are antagonists of α-adrenoreceptors (α-*adrenergic receptors), blocking the effect of *noradrenaline and causing vasodilation of peripheral blood vessels. α_1-Blockers (e.g. *doxazosin, prazosin, tamsulosin) act at the α-1 receptors, α_2-blockers (e.g. atipamezole, efaroxan, *idazoxan, yohimbine, benzylpiperazine and, rauwolscine) act at the α_2-receptors (but are mostly used experimentally). *Phentolamine is nonspecific and acts at both α_1 and α_2 receptors.

alpha-cells (α cells) *See* A CELLS.

alpha-fetoprotein (α-fetoprotein) A major plasma protein (609 aa) in the fetus, probably the fetal equivalent of *albumin. Synthesis is switched off, and that of albumin switched on, at birth although it can persist into adulthood, at lower levels than in the fetus, without pathological consequences. In congenital nephrosis alpha-fetoprotein is increased in the maternal blood and amniotic fluid, and loss into the amniotic fluid in cases of spina bifida and anencephaly is the basis of a screening test.

alpha-glucosidase A lysosomal enzyme (acid maltase, EC 3.2.1.20, 952 aa) that catalyses the splitting of α-D-glucosyl residues from the nonreducing end of substrates to release α-glucose. Deficiencies cause infantile-onset *glycogen storage disease type II.

alpha-helix (α-helix) The right-handed helical protein folding pattern in which the carbonyl oxygens are all hydrogen-bonded to amide nitrogen atoms three residues further along the chain. This causes a regular repeat pattern with 3.6 residues per turn. α-Helical regions are found in both globular and fibrous proteins. *Compare* BETA-TYPE STRUCTURES.

alpha-neurotoxins (α-neurotoxins) Neurotoxins that act on postsynaptic sites, many of which are found in snake venoms, e.g. α-*bungarotoxin, α-*cobratoxin, *erabutotoxins.

alpha rhythm A regular pattern of waves (alpha waves, 8–12 Hz, 20–100 mV) observed in the electroencephalogram (EEG) of waking inactive adults.

alpha-sarcin (α-sarcin) A cytotoxic compound (177 aa) from *Aspergillus giganteus* that has ribonuclease activity and cleaves a single oligonucleotide from the 3′ end of 28S rRNA. One of a family that includes *mitogillin, *restrictocin, and AspFl.

alpha-secretase *See* SECRETASES.

alpha-synuclein *See* SYNUCLEIN.

alphaviruses *See* TOGAVIRIDAE.

Alport's syndrome An X-linked nephropathy caused by mutation in the gene for the α_5 chain of *basement membrane collagen (collagen 4A5). It is associated with nerve deafness and variable ocular disorders.

alprazolam A short-acting benzodiazepine drug used to treat moderate to severe anxiety disorders, panic attacks, and anxiety associated with moderate depression. *TNs* Niravam, Xanax, Xanor.

ALS *See* AMYOTROPHIC LATERAL SCLEROSIS.

Alsever's solution An isotonic citrate-buffered anticoagulant solution used to preserve red blood cells.

alsin A protein (1657 aa with an alternatively spliced 396-aa short form) that is defective in juvenile *amyotrophic lateral sclerosis-2 and juvenile *primary lateral sclerosis. It appears to be a guanine nucleotide exchange factor (*GEF) for rab5 and a member of the rho GTPase family.

ALT *See* ALANINE AMINOTRANSFERASE.

altenusin An antifungal penicillide isolated from *Alternaria* sp. It is a noncompetitive inhibitor of neutral sphingomyelinase, pp60src, and myosin light-chain kinase. Also has activity against HIV-1 integrase activity.

alteplase A recombinant form of tissue *plasminogen activator used to treat acute thrombotic events such as heart attack, stroke, and pulmonary thrombosis.

alternative pathway (alternate pathway) *See* COMPLEMENT.

alternative splicing *See* SPLICING.

Alu **1.** A type II *restriction endonuclease, isolated from *Arthrobacter luteus*. **2.** Highly repetitive short repeat sequences (**Alu repeats**, short interspersed elements, SINEs) found in large numbers (100 000–500 000) in the human genome, the most abundant mobile elements (retrotransposons) in the human genome. They were first identified because they were cleaved by the Alu endonuclease. The progenitor of the human Alu repeat may be 7SL RNA, a component of the signal recognition particle. They are useful markers in genomic studies because individuals will only share a particular Alu sequence insertion if they have a common sexual ancestor. *Compare* LONG INTERSPERSED NUCLEOTIDE ELEMENT.

SEE WEB LINKS

- Review on Alu repeats; *Nature* May 2002.

alveolar cells Mostly squamous cells that line the terminal dilations (alveoli) of the branched airways of the lung. A small proportion (great alveolar cells) secrete lung surfactant.

alveolar macrophages A subset of *macrophages found in the lung and that can be obtained by lung lavage. They are responsible for clearance of inhaled particles and lung surfactant.

alveolitis Inflammation of the alveoli of the lung.

Alzheimer's disease A presenile neurodegenerative disorder that leads to dementia, in which there are unusual intracellular helical protein filaments in nerve cells (neurofibrillary tangles) and the accumulation of extracellular amyloid in *senile plaques. Familial Alzheimer's disease-1 (**AD1**) is caused by mutation in the gene encoding the *amyloid precursor protein (*APP*). **AD2** is associated with the apolipoprotein E4 allele (ApoE4), **AD3** is caused by mutation in the *presenilin-1 gene (*PSEN1*) and **AD4** by mutation in the *PSEN2* gene. **Early-onset Alzheimer's disease** with cerebral amyloid angiopathy is associated with duplication of the *APP* locus. A range of other loci are associated as risk factors.

amacrine cell A class of *interneuron found in the middle layer of the *retina, with neurites parallel to the plane of the retina. They are probably involved in image processing by regulating bipolar cells.

amanitin A group of small peptide toxins (α-, β-, γ-amanitins) from the death cap toadstool *Amanita phalloides*. They block transcription by inhibiting *RNA polymerase II.

amantadine An antiviral drug approved for use against influenza A, that interferes with a viral ion channel required for uncoating of the virus once within the host cell. It produces some symptomatic relief in *parkinsonism although the mechanism is unclear. Rimantadine is a closely related derivative with similar biological properties. *TN* Symmetrel.

amastigote A stage in the life cycle of trypanosomatid protozoa, morphologically similar to the adult form of the genus *Leishmania*, in which the oval or round cell has a nucleus, kinetoplast, and basal body but no flagellum.

amaurosis Blindness without an obvious lesion to the eye. *See* LEBER'S CONGENITAL AMAUROSIS. **Amaurosis congenita** (UK retinal aplasia) is an autosomal dominant disorder with considerable genetic heterogeneity.

amber *See* TERMINATION CODONS.

amblyopia A disorder in which there is poor or indistinct vision in one eye, even though the eye is otherwise physically normal, usually caused by a sustained period in childhood in which the eye does not transmit information (or does so very badly). A disorder of brain processing rather than of the visual system.

AMD *See* AGE-RELATED MACULAR DEGENERATION.

amelia A developmental disorder in which one or more limbs is absent. Autosomal recessive tetra-amelia (affecting all four limbs) is caused by mutation in the **WNT3* gene; other developmental disorders are also present.

a

ameloblasts Columnar epithelial cells responsible for secreting the outer enamel layer of the tooth. The apical surface has a pyramidal process (Tomes process) that is embedded within the enamel matrix. *See* AMELOGENINS.

amelogenesis imperfecta A clinically and genetically heterogeneous defect of dental enamel formation. The **X-linked form** of hypoplastic amelogenesis imperfecta is caused by mutation in the gene encoding *amelogenin. **Types 1B and 1C** are caused by mutation in the *enamelin gene. **Type 3** (hypocalcified amelogenesis imperfecta) is caused by mutation in the *FAM83H* gene, the product of which (1179 aa) is expressed in ameloblasts and odontoblasts. **Type 4** is caused by mutation in the distal-less homoeobox 3 (*DLX3*) gene. The **pigmented hypomaturation type** of amelogenesis imperfecta can be caused by mutation in the matrix metalloproteinase-20 gene or in the *kallikrein-4 gene.

amelogenins Highly conserved extracellular matrix proteins (X, 191 aa; Y, 206 aa) of developing dental enamel that regulate the shape and size of hydroxyapatite crystallites during mineralization. They are hydrophobic and proline-rich and are secreted by *ameloblasts. There are amelogenin genes on both X and Y chromosomes and both are transcriptionally active, even though one X-amelogenin gene is suppressed in females.

amenorrhoea The absence or suppression of menstruation in a woman of reproductive age.

amentia Mental deficiency; failure of the mind to develop normally because of a deficiency in brain tissue.

Ames test A procedure used to test for potential carcinogenicity of compounds by testing for its capacity to cause mutagenesis in strains of *Salmonella typhimurium* engineered to be deficient in histidine synthesis, but capable of acquiring this ability by simple point or frame-shift mutations. Animal tissue (usually liver homogenate) may be used to generate active metabolites of the substance. Although false positives and negatives are known, the carcinogen–mutagen link is fairly strongly predictive and the test has the virtues of cheapness and simplicity.

amethocaine *See* TETRACAINE.

amethopterin *See* METHOTREXATE.

AMH *See* ANTIMÜLLERIAN HORMONE.

amiloride A potassium-sparing *diuretic that blocks the epithelial sodium channel (ENaC).

amiloride-sensitive sodium channels A family of voltage-insensitive sodium-permeable ion channels that are inhibited by *amiloride (*compare* VOLTAGE-GATED SODIUM CHANNELS). In epithelia they constitute the rate-limiting step for sodium ion reabsorption. They are heterotrimers of highly homologous α, β, and γ subunits which also have homology with *degenerins. Gain-of-function mutations in the β or γ subunits cause *Liddle's disease. A δ subunit that will assemble with the β and γ subunits to form a functional channel is also known and is expressed mainly in ovary, testis, pancreas, and brain. *See* APX/SHRM DOMAINS.

aminergic Describing receptors that respond to amines or neurons that release noradrenaline, dopamine, or serotonin. *Compare* ADRENERGIC RECEPTORS; CHOLINERGIC NEURONS.

amino acid An organic acid carrying an amino group. Proteins are linear polymers of the L-forms of ~20 common amino acids linked by *peptide bonds.

amino acid permeases A family of integral membrane proteins involved in the transport of amino acids into the cell. The term is usually applied to the family of transporters in yeast (*Saccharomyces*), which include general permeases that allow several different amino acids to pass and others that are more selective.

amino acid receptors An imprecise term usually referring to *ligand-gated ion channels that respond to the binding of *excitatory amino acids, although taste receptors for amino acids have been reported.

amino acid transmitters *See* EXCITATORY AMINO ACIDS.

aminoacylase A cytosolic, homodimeric, zinc-binding enzyme (EC 3.5.1.14, 408 aa) involved in the hydrolysis of most *N*-acylated or *N*-acetylated amino acids, possibly a salvage mechanism. Expression of the gene is reportedly reduced or undetectable in small-cell lung carcinoma lines and tumours. **Aminoacylase-1 deficiency** causes a metabolic disorder in which there is encephalopathy, psychomotor delay, muscular hypotonia and frequent epileptic seizures. Aspartoacylase (EC 3.5.1.15, aminoacylase-2, 313 aa) hydrolyses *N*-acetyl-L-aspartic acid and deficiency leads to *Canavan's disease.

aminoacylation The addition of an aminoacyl group to a substrate. The formation of *aminoacyl-tRNA is an ATP-dependent reaction involving amino acid-specific aminoacyl-tRNA synthetases.

aminoacyl transfer RNA The carrier complex of an *amino acid and its specific *transfer RNA, formed by the ATP-dependent action of aminoacyl tRNA synthetase.

aminocaproic acid (ε-aminocaproic acid, EACA) A lysine analogue and derivative that inhibits enzymes that bind at lysine residues. An inhibitor of the *plasminogen activators streptokinase and urokinase, it inhibits the activity of plasmin, and the production of plasmin from *plasminogen. Used clinically as an antifibrinolytic, antihaemorrhagic agent. *TN* Amicar.

aminoglycoside antibiotics A group of antibiotics, active against many aerobic Gram-negative and some Gram-positive bacteria, that inhibit bacterial protein synthesis by binding to a site on the 30S ribosomal subunit and altering codon–anticodon recognition. Common examples are amikacin, framycetin, *gentamicin, *kanamycin, neomycin, netilmicin, *streptomycin, tobramycin.

aminonaphthalimide An inhibitor of *poly (ADP-ribose) polymerase (PARP).

aminopeptidases Enzymes that catalyse the cleavage of amino acids from the N-terminus of a protein or peptide. Many, but not all, are zinc metalloenzymes. **Alanyl aminopeptidase** (aminopeptidase N, GP150, CD13, EC 3.4.11.2, 967 aa) is an integral membrane protein with a role in the final digestion of peptides generated from hydrolysis of proteins by gastric and pancreatic proteases and is also a regulator of angiogenesis. **Endoplasmic reticulum aminopeptidase 1** (ERAP1, adipocyte-derived leucine aminopeptidase, EC 3.4.11.-, 941 aa) is thought to be important in metabolism of peptides such as *angiotensin II involved in regulating blood pressure. **Glutamyl aminopeptidase** (E-AP, aminopeptidase A, EC 3.4.11.7, 957 aa) is a surface ectopeptidase present on various cells and probably important in regulating the response to bioactive peptides. **X-prolyl aminopeptidase** (EC 3.4.11.9, aminopeptidase P, 623 aa) specifically catalyses the removal of any unsubstituted N-terminal amino acid that is adjacent to a penultimate proline residue and may be important in the maturation and degradation of peptide hormones, neuropeptides, and tachykinins. Many other aminopeptidases are known with varying specificity and tissue distribution and undoubtedly different roles. *See* METHIONINE AMINOPEPTIDASE.

aminophylline An inhibitor of cAMP *phosphodiesterase used as a bronchodilator.

aminopterin A *folic acid analogue and inhibitor of *dihydrofolate reductase. Potently cytotoxic but now replaced by *methotrexate (amethopterin) in the treatment of acute *leukaemia. A component of *HAT medium.

amino-sugar A monosaccharide in which a hydroxyl group is replaced with an amino group; often acetylated. Common examples are D-*galactosamine, D-*glucosamine, **N*-acetylneuraminic acid.

aminotransferases A family of enzymes (transaminases, EC 2.6.1.x) that transfer an amino group from an amino acid to an α-keto acid. **Glutamate oxaloacetate transaminase** (EC 2.6.1.1, aspartate aminotransferase) catalyses the transfer from glutamate to oxaloacetic acid (producing aspartic acid and α-ketoglutarate) and is a pyridoxal phosphate-dependent enzyme which exists in both mitochondrial (430 aa) and cytosolic (413 aa) forms. **Branched chain aminotransferases** (EC 2.6.1.42, BCAT1, 393 aa; BCAT2) act on amino acids with nonlinear aliphatic side chains (leucine, isoleucine, and valine). Two clinical disorders are due to a defect of branched chain amino acid transamination, *hypervalinaemia and *hyperleucine–isoleucinaemia.

amiodarone An antiarrhythmic drug used to treat both ventricular and supraventricular (atrial) arrhythmias.

amitriptyline A tricyclic antidepressant drug used in the treatment of moderate to severe endogenous depression. *TNs* Elatrol, Elavil, Endep, Laroxyl, Redomex, Saroten, Trepiline, Triptyl, Tryptanol, Tryptizol.

AML Acute myeloblastic *leukaemia.

amlodipine A long-acting *calcium channel blocker. *TNs* Istin, Norvasc.

ammodytoxins (atx) Toxins (atx-A, -B, -C, ammodytin L) found in the venom of the southern European viper *Vipera ammodytes*. They are secreted *phospholipases A2 (122 aa) that act presynaptically at the neuromuscular junction.

amnesia Loss of memory. **Anterograde amnesia** can occur after injury to the brain, or mental trauma, and has little effect on information acquired previously. **Retrograde**

amnesia affects memories formed before the onset of the amnesia.

amnesic shellfish poisoning *See* DOMOIC ACID.

amniocentesis A procedure in which the amniotic fluid is sampled, usually between the 12th and 16th week of pregnancy. Cells in the fluid can be karyotyped and tested for various inherited disorders (prenatal genetic diagnosis). *See* CHORIONIC VILLUS SAMPLING; PREIMPLANTATION GENETIC DIAGNOSIS.

amniocyte Fetal cells found in the amniotic fluid.

amnion A fluid-filled sac formed by the outgrowth of the extraembryonic *ectoderm and *mesoderm as projecting folds which fuse to form two epithelial layers separated by mesoderm and *coelom. The inner layer is the amnion, which encloses the amniotic sac in which the embryo is suspended, the outer layer is the *chorion.

amniotes Tetrapod vertebrates with terrestrially adapted eggs which may be laid or carried internally in the female. They include mammals, birds, and reptiles. The embryo is protected by a system of membranes that include the amnion, chorion, and allantois.

amobarbital (formerly **amylobarbitone, amytal)** A barbiturate used as an anticonvulsant, pre-anaesthetic, and for short-term treatment of insomnia.

amoebiasis Dysentery caused by infection with *Entamoeba histolytica*.

amoeboid movement Crawling movement of a cell such as a leucocyte in which blunt *pseudopods are protruded from the front of the cell and form distal anchorages with the substratum.

SEE WEB LINKS

• A more detailed description.

amorolfine A topical antifungal drug used to treat infections of nails. *TNs* Curanail, Loceryl, Locetar.

amoxicillin (formerly **amoxycillin)** *See* BETA-LACTAM antibiotics. *TNs* Amoxil, Augmentin.

AMP (adenosine monophosphate) Usually the nucleotide (5′-AMP) with the phosphate at position 5 of the ribose moiety, but 2′ and 3′ derivatives also exist. *See also* CYCLIC AMP (adenosine 3′,5′-cyclic monophosphate).

AMPA *See* GLUTAMATE RECEPTORS.

amphetamine A powerful psychostimulant drug (α-methylphenethylamine) that increases wakefulness and decreases fatigue and appetite. Related to drugs such as methamphetamine, dextroamphetamine, and levoamphetamine, which act by increasing levels of noradrenaline, serotonin, and dopamine in the brain, inducing euphoria. Often used as a recreational drug. *TNs* Adderall, Dexedrine, Vyvanse.

amphibolic Describing a metabolic pathway that is involved in *catabolism, but that also provides precursors for *anabolic pathways.

amphipathic Describing a molecule that has both *hydrophobic and *hydrophilic regions. Detergents and phospholipids are standard examples, but many biological molecules have this property. *Antonymn* amphiphilic.

amphiphysin A synaptic vesicle-associated protein (695 aa) that forms a heterodimer with *bridging integrator 1 (BIN1, amphiphysin-2, Myc box-dependent-interacting protein 1, 593 aa) which binds to *synaptojanin-1, *dynamin 1, *clathrin, and *adapter-related protein complex 2 and may regulate the release of synaptic vesicles. Antibodies against amphiphysin are found in patients with *stiff person syndrome. *See* BAR PROTEINS.

amphiregulin A heparin-binding *growth factor (Schwannoma-derived growth factor, 84 aa) containing an *EGF-like domain. Binds to the EGF receptor, though with lower affinity than EGF, stimulates proliferation of human fibroblasts, and inhibits growth of some, but not all, human tumour cells. Expression is increased in psoriatic epidermis. *See* HEPARIN-BINDING EPIDERMAL GROWTH FACTOR.

ampholyte A substance with *amphoteric properties. Ampholytes are used in electrofocusing columns or gels.

amphoteric Describing a substance with both acidic and basic characteristics. Most proteins are amphoteric and the charges on their acidic and basic side groups balance at the *isoelectric point.

amphotericin B An antifungal antibiotic from *Streptomyces* spp. that interacts with cholesterol to form pores, leading to cytolysis. It only acts on membranes containing sterols and has a preference for ergosterol found in fungal membranes. *See* FILIPIN. *TN* Fungizone.

amphoterin A high mobility group protein (HMGB1, 215 aa) considered a structural protein in chromatin. It can be released from endotoxin or cytokine-stimulated macrophages and is toxic (levels are high in patients with sepsis). It is bound by *RAGE (receptor for advanced glycation end products) and inhibition of the RAGE–amphoterin interaction suppresses various systems linked to tumour proliferation, invasion, and expression of matrix metalloproteinases.

amphotropic virus *See* XENOTROPIC VIRUS.

ampicillin A semisynthetic *penicillin derivative, similar to amoxicillin, with a broad range of activity. Ampicillin resistance is often used as a marker for *plasmid transfer in genetic engineering; vectors such as *pBR322 confer resistance.

AMPK A heterotrimeric enzyme (AMP-activated protein kinase, EC 2.7.11.1) found in two major forms, AMPK1 and -2, with a catalytic α subunit (550 and 552 aa) and regulatory β (270 and 272 aa) and γ (331 and 569 aa) subunits. It is activated by phosphorylation by AMPK kinases (e.g. *LKB1), some of which are stimulated by *adiponectin. Phosphorylation by activated AMPK inhibits fatty acid and cholesterol synthesis and switches cells from ATP consumption towards ATP production.

amplicon The DNA product of a *polymerase chain reaction or any other amplification mechanism.

AMP-PNP A nonhydrolysable analogue of ATP.

ampulla A dilated region of a tubular structure.

amygdala (*pl.* amygdalae) One of a pair of almond-shaped nuclei within the medial temporal lobes of the brain. They are part of the limbic system and have a role in the processing and memory of emotional reactions.

amylases Enzymes that hydrolyse α(1–4) glycosidic linkages in glucose polymers such as starch. There are two isoenzymes of the endoglycosylase α-amylase (EC. 3.2.1.1, 511 aa) in serum, one produced by the salivary gland (ptyalin), one by the pancreas. β-Amylase (EC.3.2.1.2) also hydrolyses β-1,4 glycosidic bonds but only from the nonreducing end to yield maltose; it is not found in mammals.

amylin A polypeptide (islet amyloid polypeptide, IAPP, 37 aa) of the *calcitonin family, derived from an 89-aa precursor by proteolytic processing. It is produced by pancreatic *beta cells and cosecreted with insulin although it is found in fibrillar deposits (pancreatic amyloid) in type 2 diabetes. Amylin moderates the effects of insulin and controls nutrient influx to the blood by an inhibition of food intake, gastric emptying, and glucagon secretion.

amyloid A *glycoprotein deposited extracellularly in tissues in *amyloidosis. The glycoprotein may either derive from *light chain of immunoglobulin (**amyloid of immune origin,** AIO, 5–18 kDa glycoprotein; the product of a single clone of *plasma cells, the N-terminal part of λ or κ *L chain) or, in what used to be referred to as **amyloid of unknown origin** (AUO), from *serum amyloid A (SAA), one of the acute phase proteins that increases many-fold in inflammation. The polypeptides are organized as a beta-pleated sheet making the material rather inert and insoluble. Minor protein components are also found. Should be distinguished from β-amyloid deposited in the brain, which is derived from *amyloid precursor protein (*see* AMYLOIDOGENIC GLYCOPROTEIN).

amyloidosis A group of diseases in which there is deposition of *amyloid in tissues. **Primary amyloidosis** (AL, amyloidosis of immune origin, AIO) involves deposition of immunoglobulin light chain, usually as a consequence of multiple myeloma. **Secondary amyloidosis** (AA) arises as a result of another illness, such as multiple myeloma, chronic infection (e.g. tuberculosis), or chronic inflammatory disease (e.g. rheumatoid arthritis). The amyloid is derived from *serum amyloid protein. **Familial amyloidosis** (ATTR) is a rare inherited amyloidosis in which the deposits are of *transthyretin. Other forms of amyloidosis occur in which β_2-microglobulin is deposited or in which deposition is localized (as in Alzheimer's disease). *See* FAMILIAL PRIMARY LOCALIZED CUTANEOUS AMYLOIDOSIS; FIBRIN; GELSOLIN; ICELANDIC-TYPE CEREBROARTERIAL AMYLOIDOSIS.

amyloid precursor protein The membrane protein (APP, amyloid beta A4 protein, 770 aa) from which peptide B/A4, the principal component of amyloid fibrils in *senile plaques, is cleaved by *secretases.

amylose A linear polymer of glucose, mostly α-1,4 linked. Found in starch (together with amylopectin) and in glycogen (glycogen amylose).

amyotrophic lateral sclerosis A neurodegenerative disorder (ALS, Lou Gehrig's disease, motor neuron disease) characterized by the death of motor neurons in the brain, brainstem, and spinal cord, resulting in fatal paralysis. Only around 10% of cases are familial; of these, 15–20% (**ALS1**) are associated with mutations in the *superoxide dismutase-1 gene. One juvenile-onset form (**ALS2**) is associated with mutations in *alsin, another (**ALS4**) with mutations in the *senataxin gene and a locus on chromosome 15 with juvenile-onset **ALS5**. Autosomal dominant adult-onset forms are associated with mutations in genes for *VAMP-associated protein B (**ALS8**), *angiogenin (**ALS9**), TAR DNA-binding protein (**ALS10**), and SAC domain-containing inositol phosphatase 3 (**ALS11**). Other loci associated with the disorder are found on chromosomes 16, 18, and 20 and susceptibility has also been associated with mutations in the genes encoding the heavy neurofilament subunit, *peripherin, *dynactin, and *angiogenin.

amyotrophy Wasting (atrophy) of muscle.

amytal *See* AMOBARBITAL.

anabolic Describing a process, route, or reaction in which energy is expended in the synthesis of complex biomolecules. Metabolic pathways are conventionally subdivided into those of anabolism and *catabolism (degradation). **Anabolic steroids** are synthetic androgens that promote tissue growth, especially of muscle, and have been misused by athletes.

anaemia A condition in which the level of *haemoglobin in the blood is reduced for any of various reasons, including haemorrhage. *See* APLASTIC ANAEMIA; CONGENITAL DYSERYTHROPOIETIC ANAEMIA; DIAMOND–BLACKFAN ANAEMIA; ERYTHROBLASTOSIS FETALIS; FANCONI'S ANAEMIA; HAEMOLYTIC ANAEMIA; IRON-DEFICIENCY ANAEMIA; LEUCOERYTHROBLASTIC ANAEMIA; MEGALOBLASTIC ANAEMIA (pernicious anaemia); MICROANGIOPATHIC HAEMOLYTIC ANAEMIA; SICKLE CELL ANAEMIA; SPHEROCYTOSIS; WARM ANTIBODY HAEMOLYTIC ANAEMIA.

anaerobic Describing a habitat deficient in free oxygen or an organism that has a very low tolerance or requirement for oxygen. **Anaerobic respiration** is a set of energy-producing metabolic processes that operate in the absence of oxygen but require inorganic oxidizing agents or reduced coenzymes.

anaesthesia Strictly, the loss of feeling, but generally used for the techniques of pain relief using anaesthetics prior to surgery.

analeptic Describing something with restorative or strengthening properties. Doxapram hydrochloride is an analeptic drug used to stimulate breathing when artificial ventilation is impossible.

analgesia Describing a state of absence of pain without loss of consciousness. Analgesic drugs induce this state.

analogous Describing something (e.g. a gene or protein) that has a similar role in a different organism but is not *homologous.

analysis of variance (ANOVA) A powerful statistical technique that allows the estimation of variation due to specific causes and variation due to random factors.

analyte The substance or compound for which an analysis is being performed.

anamnestic response Obsolete term for a *secondary immune response or *immunological memory.

anandamide One of the endogenous agonists (arachidonyl ethanolamide) for *cannabinoid receptors.

anaphase The stage of mitosis or meiosis that begins with the separation of sister chromatids (or homologous chromosomes) and is then followed by their movement towards the spindle poles.

anaphase-promoting complex An E3 ubiquitin ligase (APC, cyclosome) that marks target proteins for degradation by the 26S proteasome, and is responsible for initiating sister chromatid separation and the inactivation of cyclin-dependent kinases at anaphase. It has 13 core subunits (total 1.5 MDa) and one of two loosely associated coactivators, Cdc20 and Cdh1. The binding of *cyclin A inactivates the APC during S-phase.

anaphylatoxin Historically, an antigen that reacted with an *IgE antibody and caused *anaphylaxis. Now restricted to *complement fragments C3a and C5a, which bind to mast cells and basophils and cause the release of inflammatory mediators.

anaphylaxis Strictly, a system or treatment that leads to damaging effects on the organism (as opposed to *prophylaxis), but used specifically for the inflammatory reactions that result from massive degranulation of mast cell and release of *histamine and histamine-like substances, causing localized or global immune responses. *See* HYPERSENSITIVITY.

anaplasia A lack of differentiation, a feature of some tumour cells.

Anaplasma phagocytophilum A Gram-negative bacterium (formerly *Ehrlichia phagocytophilum*) transmitted to humans by ticks, *Ixodes* spp. It is an intracellular parasite of neutrophils.

a

anaplerotic Describing reactions that replenish intermediates of the tricarboxylic acid cycle and allow respiration to continue. *Pyruvate carboxylase deficiency is an inherited metabolic disorder in which anaplerosis is greatly reduced.

anasarca Extreme generalized oedema of tissues.

anastomosis The fusion of two or more cell processes or multicellular tubules to form a branching system. The anastomosis of blood vessels provides alternative routes for blood flow.

anastrozole A nonsteroidal *aromatase inhibitor used in the treatment of advanced breast cancer. It significantly lowers serum oestradiol concentrations without affecting levels of adrenal corticosteroids or aldosterone. *TN* Arimidex.

anatoxins Small neurotoxic alkaloids from the cyanobacterium *Anabaena flos-aquae* and related species. Anatoxin-α and homoanatoxin-α irreversibly activate nicotinic acetylcholine receptors; anatoxin-α(s) is a natural organophosphate that inactivates acetylcholine esterase in the same way as organophosphate pesticides such as parathion and malathion. The effect is to cause overstimulation of muscle leading to death through respiratory failure.

ANCA Antibodies (antineutrophil cytoplasmic antibodies) found in various inflammatory disorders including *inflammatory bowel disease, *Wegener's granulomatosis, and hepatobiliary disorders. In **peripheral ANCA** (p-ANCA) the neutrophil antigen seems to be localized at the periphery of the nucleus, whereas in **cytoplasmic ANCA** (c-ANCA) the antigen is distributed throughout the cytoplasm.

anchorage dependence The phenomenon exhibited by many normal cells in culture which require attachment to a solid substratum in order to spread and grow. Spreading may be necessary in order to absorb rate-limiting growth factors from the culture medium. Loss of anchorage dependence is associated with independence from external growth control, and the ability to grow in soft agar (or suspension) is a good predictor of *tumorigenicity *in vivo.*

anchored PCR A variant of the *polymerase chain reaction in which a known 'anchor' sequence is added to the end of the DNA, either enzymically or by ligation. The reaction can then proceed using a gene-specific primer and the anchor primer.

anclyostoma *See* HOOKWORM.

ancovenin A *lantibiotic (16 aa) containing unusual amino acids such as threo-β-methyllanthionine, meso-lanthionine, and dehydroalanine, that inhibits *angiotensin I converting enzyme. Ancovenin is isolated from the culture broth of a *Streptomyces* species and a similar antibiotic, lanthiopeptin, is produced by *Streptoverticillium cinnamoneum.*

Andersen's syndrome An autosomal dominant multisystem channelopathy (long QT-7, Andersen cardiodysrhythmic periodic paralysis) characterized by periodic paralysis, ventricular arrhythmias, and distinctive dysmorphic facial or skeletal features. Caused by mutations in the *KCNJ2* gene that encodes the inwardly rectifying potassium channel, Kir2.1. **Andersen's disease** is *glycogen storage disease type IV. *Compare* ANDERSON'S DISEASE.

Anderson's disease A disorder of severe fat malabsorption causing failure to thrive in infancy. There is deficiency of fat-soluble vitamins, low blood cholesterol levels, and a selective absence of chylomicrons from blood. Like chylomicron retention disease, it is caused by mutation in the gene for *sar1B. *Compare* ANDERSEN'S SYNDROME.

androgen A general term for any male sex hormone. **Androgen insensitivity syndrome**, caused by mutation in the androgen receptor gene, is also referred to as testicular feminization, and genetic males are phenotypically female. *See* REIFENSTEIN'S SYNDROME.

androstenedione A steroid precursor of both testosterone and oestrone synthesized in the testis or ovary and also secreted into the circulation by the adrenal glands.

androsterone A steroid hormone with weak androgenic activity.

anemia *See* ANAEMIA.

anemone toxins A group of polypeptide toxins (~45–50 aa; more than 40 known examples) from sea anemones (anthozoans), most of which are neurotoxins acting on voltage-gated sodium or potassium channels.

anencephaly A developmental defect in which the skull and cerebral hemispheres are absent.

anergy **1.** A lack of energy. **2.** In immunology, a failure to produce a secondary immune response to an antigen that has been encountered before.

anesthesia *See* ANAESTHESIA.

aneuploid Describing a cell with a chromosome complement that is not an exact multiple of the haploid number. There may be multiple copies of one chromosome (as in trisomy) or one member of a homologous pair may be missing in an otherwise diploid cell. **Aneugenic agents** induce aneuploidy or polyploidy, rather than causing mutations.

aneurysm A balloon-like swelling of an artery.

ANF *See* ATRIAL NATRIURETIC PEPTIDE.

angel dust *See* PHENCYCLIDINE.

Angelman's syndrome A disorder in which there is severe mental retardation and ataxic movement, usually caused by the absence of a portion of maternal chromosome 15, in particular the absence of the β3 subunit of the *$GABA_A$ receptor. *Prader–Willi syndrome is a clinically distinct disorder resulting from deletion of the same 15q11–q13 region from the paternal genome. Some cases, however, result from defects in *genomic imprinting or by mutations in the gene encoding the ubiquitin-protein ligase E3A.

angina pectoris The abrupt onset of pain or crushing sensation in the chest, often provoked by exercise, and caused by narrowing of the coronary arteries and reduced blood supply to the heart. Often referred to simply as angina.

angioedema A condition (formerly angioneurotic oedema), usually associated with allergic reactions and bradykinin, in which large oedomatous welts develop in the dermis and subcutaneous tissue or in the mucosa and submucosal tissues. In contrast, urticaria is restricted to the dermis or mucosa and does not affect deeper tissues. **Hereditary angioedema** (HAE) types I and II are caused by mutation in the complement C1 inhibitor gene. **HAE type III** is caused by mutation in the gene encoding coagulation factor XII.

angiogenesis The process of *vascularization of a tissue involving invasion by proliferating endothelium.

angiogenin A polypeptide (EC 3.1.27.-, 147 aa) that potently induces neovascularization *in vivo*. It is one of the pancreatic ribonuclease A superfamily, and RNase activity is important for its angiogenic activity.

angiokeratoma *See* FABRY'S DISEASE.

angioma A benign tumour composed of disorderly arrays of blood or lymphatic vessels.

angiomatoid fibrous histiocytoma A variant of malignant *fibrous histiocytoma that typically occurs in children and adolescents. It is associated with somatic fusions of several genes, including *ATF1* (cAMP-dependent transcription factor-1), *EWS* (*Ewing's sarcoma gene), and **CREB1*.

angiomotin A protein (675 aa) that stimulates motility in endothelial cells and is found in tissues where angiogenesis is occurring. May be antagonized by *angiostatin.

angiomyolipoma A rare, slow-growing, benign lesion of the kidney composed of varying amounts of blood vessels, smooth muscle, and fat. A few cases are associated with *tuberous sclerosis.

angioneurotic oedema *See* ANGIOEDEMA.

angiopathy A generic term for a disorder affecting blood vessels. **Microangiopathy** affects small blood vessels and is a common complication of diabetes, particularly affecting the retinal vasculature. **Macroangiopathy** affects large blood vessels and can cause coronary artery disease, cerebrovascular disease, and peripheral vascular disease. **Hereditary angiopathy** with nephropathy, aneurysms, and muscle cramps (HANAC) is caused by missense mutations in the collagen 4A1 gene.

angioplasty A surgical procedure to widen an occluded blood vessel. In percutaneous transluminal coronary angioplasty (PTCA) a catheter is passed from the arm through to the coronary vessels and a balloon at the end of the catheter is then inflated. Often used as a treatment for *angina pectoris.

angiopoietins Members of the vascular endothelial growth factor family that have effects on blood vessels. **Angiopoietin-1** (498 aa) is the ligand for the endothelium-specific receptor tyrosine kinase *Tie2; **angiopoietin-2** (496 aa) is a natural antagonist. **Angiopoietin-4** (406 aa) is induced in endothelial cells by hypoxia and may have a protective function. It inhibits proliferation, migration, and tubule formation by endothelial cells, reduces vascular leakage and is involved in regulating glucose homeostasis, lipid metabolism, and insulin sensitivity. Various angiopoietin-like proteins are also known with miscellaneous tissue-specific functions. Not all are mitogenic for endothelial cells.

angiostatin A proteolytic fragment of plasminogen, containing the first three or four kringle domains, that inhibits angiogenesis. May antagonize the effects of *angiomotin.

angiotensin A peptide *hormone generated from the precursor **angiotensinogen** (485 aa) by the action of *renin which produces the inactive decapeptide **angiotensin I**, which is converted to the active form angiotensin II by *angiotensin converting enzyme. **Angiotensin II** is a potent vasoconstrictor and the production of angiotensin II raises blood pressure; it also stimulates *aldosterone release from the adrenal glands. Defects in angiotensinogen are associated with susceptibility to essential hypertension, and can cause *renal tubular dysgenesis, as can mutations in the angiotensin II receptor type 1. There are two distinct subtypes of **angiotensin receptors**, both G-protein coupled: **Type 1** (359 aa) mediates the major cardiovascular effects, **type 2** (363 aa) may antagonize type 1 and cause relaxation of blood vessels. Mutation in the type 2 receptor causes X-linked mental retardation 88. Angiotensin is broken down by angiotensinases such as prolylcarboxypeptidase (EC 3.4.16.2, 496 aa). **Angiotensin II receptor antagonists** (e.g. candesartan, irbesartan, losartan, valsartan) work by blocking binding of *angiotensin II to its receptor and thus have effects similar to those of *angiotensin converting enzyme inhibitors.

angiotensin converting enzyme (ACE) The dipeptidyl carboxypeptidase (**ACE1**, EC 3.4.15.1, 1306 aa) that converts *angiotensin I into the biologically active form, angiotensin II. Angiotensin converting enzyme 2 (**ACE2**, EC 3.4.17.-, 805 aa) is a carboxypeptidase which converts angiotensin I to angiotensin 1–9, a peptide of unknown function, and angiotensin II to angiotensin 1–7, a vasodilator. Polymorphisms in ACE1 are associated with susceptibility to ischaemic stroke and diabetic nephropathy and mutations can cause *renal tubular dysgenesis. Angiotensin converting enzyme inhibitors (**ACE inhibitors**) such as captopril and enalapril are used in the treatment of hypertension and heart failure.

Ångström unit A non-SI unit of measurement (10^{-10} m) named after the Swedish physicist and astronomer Anders Jonas Ångström (1814–74). Frequently used in early electron microscopy and often more convenient than the nanometre (10^{-9} m).

anhidrosis A condition in which there is an absence of sweating. **Autosomal dominant anhidrotic ectodermal dysplasia** with T-cell immunodeficiency is caused by mutation in the *NFKB1* gene, and an **X-linked** recessive form is caused by mutations in the *IκB kinase-γ gene. Congenital insensitivity to pain with anhidrosis (**CIPA**) is caused by mutation in the neurotrophic tyrosine kinase-1 **NTRK1* gene.

anidulafungin One of the *echinocandin antifungal drugs with broad spectrum activity against *Candida* and *Aspergillus*. *TNs* Ecalta, Eraxis.

anillin An actin-binding protein (1124 aa) that interacts with F-actin and is required for cytokinesis, where it seems to organize and stabilize the cleavage furrow. Ubiquitously expressed and overexpressed in many tumours.

animal pole The pole of the oocyte nearest the nucleus and from which the polar bodies are expelled at meiosis. There is a gradient of maternally derived material between animal and vegetal poles and this morphogenetic information is crucial for development.

anion exchangers A family of integral membrane proteins involved in the exchange of chloride and bicarbonate ions across the plasma membrane. *Band III of the erythrocyte membrane is the best-known example.

anionic detergent A detergent in which the hydrophilic moiety is an anionic group, e.g. sodium dodecyl sulphate (SDS, sodium lauryl sulphate).

aniridia A disorder that affects many tissues of the eye, although it is named for the obvious hypoplasia of the iris. **Aniridia II** is caused by an autosomal dominant mutation in the transcriptional regulator PAX6, as is Gillespie syndrome, in which there is aniridia, cerebellar ataxia, and mental deficiency. Other cases of aniridia can arise from a deletion of the short arm of chromosome 11, a region containing the genes for *Wilm's tumour, PAX6, and *brain-derived neurotrophic factor.

anisotropic Not *isotropic, not uniform in properties in all directions.

ANK repeat A conserved domain (~33 aa), originally identified in *ankyrin, and probably involved in protein–protein interactions. Also found in *notch, transcriptional regulators, cell cycle regulatory proteins, and a toxin produced by the black widow spider.

ankylosing spondylitis A rheumatic *arthritis (spondyloarthropathy, Marie-Strumpell spondylitis, Bechterew's syndrome) involving joints in the spine, which may become almost rigid. Susceptibility is associated with the histocompatibility antigen HLA-B27 and 90% of sufferers have HLA-B27. An additional susceptibility locus has been identified on chromosome 9.

ankylosis Stiffness of a joint, usually as a result of *chronic inflammation. In extreme cases the joint may become fused (**osseous ankylosis**). Ankylosis of the stapes bone in the ear (Teunissen–Cremers syndrome), causing deafness, can be caused by mutation in the gene encoding *noggin (there are other minor skeletal defects). *See* ANKYLOSING SPONDYLITIS.

ankylostomiasis (anclyostoma) *See* HOOKWORMS.

ankyrin A globular protein with multiple *ANK repeats that in the erythrocyte (**ankyrin 1**, 1881 aa) links *spectrin to *band III in the erythrocyte membrane. **Brain ankyrin** (Ank-2, ankyrin-B, 3924 aa) and **ankyrin G** (Ank-3, 4372 aa) are found elsewhere.

anlage The region of an embryo from which a specific organ will develop.

annealing **1.** In general, the toughening of a material (e.g. glass, steel) when it is cooled slowly. **2.** In the case of DNA, the formation of a duplex from strands that have been dissociated by heating, the rate of which is sensitive to complementarity. **3.** The end-to-end fusion of linear polymers such as microtubules or microfilaments.

annexins A widely distributed group of calcium-binding proteins that interact with acidic membrane phospholipids in membranes. The twelve mammalian annexin genes are classed as annexin-A 1–13 (*ANXA12* is missing). Most have four conserved repeats of a 61-aa domain. The multiple names for the annexins reflect the history of their discovery: e.g. annexin A1 has variously been named annexin-1, lipocortin-I, calpactin-2, chromobindin-9, p35, and phospholipase A2 inhibitory protein. Other names for annexins include *calelectrin, calphobindins, *endonexins, protein I, placental anticoagulant protein IV, synexin (annexin-7), vascular anticoagulant (annexins 5 and 8).

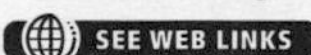

- The annexin homepage.

annulate lamellae Stacks of narrow membrane cisternae, often in parallel arrays and apparently extensions of *rough endoplasmic reticulum although they have many pores, indistinguishable from *nuclear pore complexes. They are frequently seen in virally infected cells where they may be sites of viral coat-protein glycosylation and storage.

annulus (*adj.* annular, annulate) A ring-like structure.

anoikis A type of *apoptosis seen in normal epithelial and endothelial cells that have become separated from the extracellular matrix to which they are normally attached.

anonychia The congenital absence of fingernails and toenails, without underlying skeletal abnormality, associated with mutations in the gene for *R-spondin 4.

anorectic Describing a substance that suppresses the sensation of hunger.

anorexia Lack of appetite. **Anorexia nervosa** is a severe disturbance in eating behaviour, usually beginning in late adolescence, characterized by obsessive fear of weight gain, severely restricted eating, and low body weight. A susceptibility locus for **anorexia nervosa-1** has been mapped to chromosome 1p and **anorexia nervosa-2** (like *bulimia nervosa-2) is associated with polymorphisms in the gene encoding *brain-derived neurotrophic factor.

anosmia The condition of being unable to smell. Electron microscopists sometimes develop anosmia as a result of inhaling fumes of osmic acid.

anosmin A glycosylated peripheral membrane protein (680 aa), proteolytically processed to yield a diffusible component that plays a key role in the migration and targeting of gonadotropin-releasing hormone neurons and olfactory nerves to the hypothalamus. *See* KALLMANN'S SYNDROME.

ANOVA *See* ANALYSIS OF VARIANCE.

anoxia A complete lack of oxygen. *Compare* HYPOXIA.

ANP *See* ATRIAL NATRIURETIC PEPTIDE.

ansamycins A family of antibiotics that have activity against Gram-positive and some Gram-negative bacteria and have some antiviral activity. Examples include the *streptovaricins, *geldanamycin, and *rifamycins.

ANT *See* ADENINE NUCLEOTIDE TRANSLOCATOR.

antagonist A substance that inhibits the effect of a hormone or drug. Can be competitive or noncompetitive, depending upon whether it binds to the same site as the agonist, reversible or irreversible depending upon the nature of the interaction with the target.

antennapedia One of the best-known of the *homeotic genes of *Drosophila* which, if mutated, causes a pair of second thoracic legs to develop on the head in place of antennae. The human homeobox gene *HOXA7* (Hox1A) is homologous but shows no lineage specificity, but promotes *HOXA9, HOXA10*, and *HOXA11* to induce differentiation along their respective pathways.

anterograde transport Describing transport of material away from the cell body of a neuron along axons or dendrites. The opposite movement is *retrograde axonal transport.

anthelmintic A drug used against parasitic worms.

anthopleurins Peptide toxins from *Anthopleura* sp., anthopleurins A, B, and C (49, 50, 47 aa) that affect voltage-gated sodium channels and increase the duration of the action potential.

anthracosis *Pneumoconiosis caused by inhalation of coal dust ('coal-miner's lung').

anthracyclines A class of antibiotics derived from *Streptomyces peucetius*, used in cancer chemotherapy. The first example was daunorubicin.

anthrax A very contagious disease of humans and domestic animals caused by *Bacillus anthracis*. The bacterium will persist for long periods in soil and the disease is rapid and often fatal. Anthrax toxins include **anthrax oedema factor**, a multisubunit AB toxin, the active subunit being a calmodulin-dependent adenylyl cyclase, and **anthrax lethal toxin** (LeTx) which is responsible for the shock-like effects of infection. There are two subunits, protective antigen (PA, 764 aa) that binds to the target cell and is processed to form an oligomeric pore through which lethal factor (LF, 809 aa, an endopeptidase) enters the cell where it induces an oxidative burst, the release of inflammatory cytokines and cytolysis.

antibiotic Any substance produced by one organism that will destroy or inhibit the growth of other organisms at low concentrations. Most antibiotics are products of bacteria or fungi, often chemically modified (semisynthetic antibiotics), and are used to treat infectious diseases of humans or domestic animals. The main types are the *aminoglycoside, *beta-lactam, *glutarimide, *macrolide, *polyene, *polypeptide, and *quinolone classes of antibiotics and the *cephalosporins, *lantibiotics, *sulfonamides and *tetracyclines. Increasing antibiotic resistance, probably a result of selection pressure for resistant mutants when antibiotic concentrations are low, makes the search for replacements a constant challenge. Antibiotic resistance genes are often engineered into cloning vectors to allow for selection of transfectants. In some cases the gene encodes an enzyme that degrades or excretes an antibiotic; e.g. *ampicillin resistance is conferred by expression of the *beta-lactamase gene. There are separate entries for: *actinomycins; *amphotericin B; *ansamycins; *anthracyclines; *antimycins; *aurovertins; *avilamycin; *azithromycin; *bafilomycin; *benzyl penicillin; *blasticidin; *bleomycin; *carbenicillin; *chloramphenicol; *ciprofloxacin; *clavulanic acid; *cloxacillin; *colistin; *coniosetin; *co-trimoxazole; *dalfopristin; *daptomycin; *demeclocycline; *doripenem; *dorrigocins; *doxorubicin; *doxycyclin; *echinosporin; *edeine; *erythromycin; *everninomicins; *fluoroquinolones; *framycetin; *fumagillin; *fusidic acid; *geldanamycin; *geneticin; *gentamicin; *gramicidin; *griseofulvin; *hadacidin; *hexacyclinic acid; *hygromycin B; *ionomycin; *isoniazid; *jinggangmycin; *kanamycin; *katanosins; *kosinostatin; *lacticin; *lantibiotic; *leptomycin B; *levofloxacin; *lincomycin; *linezolid; *magainins; *mannopeptimycins; *maytansine; *meticillin; *midecamycin; *mitomycin C; *monensin; *mupirocin; *nalidixic acid; *neomycins; *netropsin; *nigericin; *nitrofurantoin; *novobiocin; *nystatin; *pactamycin; *penicillin; *pivmecillinam; *plusbacins; *polymyxins; *psicofuranine; *puromycin; *pyocyanin; *quinupristin; *ramoplanin; *rapamycin; *rebeccamycin; *reticulol; *rifamycin; *ristocetin; *saccharomicins; *sideromycins; *sparsomycin; *streptogramins; *streptolydigin; *streptomycin; *streptovaricins; *streptozotocin; *tautomycin; *teicoplanins; *tetracyclines; *tetrocarcin; *ticarcillin; *trichosetin; *trichostatin A; *tunicamycin; *UK1; *vancomycin.

antibody *See* IgA; IgD; IgE; IgG; IgM.

antibody-dependent cell-mediated cytotoxicity (ADCC) The killing of target cells by cells that have antibody specific for the target cell bound to their surface via *Fc receptors. Usually lymphocytes or other leucocytes, although the cell may be just a passive carrier of the antibody.

antibody-directed drug therapy The targeting of a drug to particular cells by attaching the drug to an antibody directed against a surface marker.

antibody-induced lysis An imprecise term that should be avoided because it is ambiguous as to whether *complement lysis or *natural killer cells are involved.

antibody-producing cell A *B-cell (plasmacyte, *plasma cell) that synthesizes and secretes *immunoglobulin.

a

anticoagulant A substance that inhibits blood clotting. Calcium chelators (EDTA and citrate) are often used, so too is *heparin which interferes with the action of *thrombin, as does *hirudin. Various other anticoagulants or antithrombotics (e.g. *warfarin) act *in vivo* by interfering with clotting factors.

anticodon The nucleotide triplet on *transfer RNA that base-pairs with the complementary *codon of *messenger RNA.

anticonvulsants Drugs used to prevent or reduce the severity and frequency of seizures, which are most commonly due to epilepsy. Common examples are carbamazepine, phenytoin, valproate, and diazepam.

antidepressants Drugs that alleviate the symptoms of moderate to severe depression. The major classes are the *tricyclic antidepressants, *monoamine oxidase inhibitors, *selective serotonin reuptake inhibitors, and *lithium salts.

antidiuretic Describing something that inhibits the production of urine. Antidiuretic hormone is *vasopressin.

antidromic A term used in neurophysiology for action potentials travelling in the opposite direction to normal, from the presynaptic region towards the cell body.

antiemetics Drugs that are used to reduce nausea and vomiting induced by motion sickness, gastrointestinal disorders or chemotherapy. Examples include *hyoscine, *antihistamines, *phenothiazines, metoclopramide, and ondantseron.

antigen A substance that has elicited an immune response. Generally antigens need to be larger than ~1 kDa but antibodies can be produced that are specific for small molecules (haptens) that are attached to a larger carrier molecule. The antigenic determinant is termed an *epitope. Most antigens are foreign to the responder (except for autoantibodies: *see* AUTOIMMUNE DISEASES).

antigen–antibody complex (immune complex) The product of the interaction between *antigen and *immunoglobulin. A polyvalent antigen can bind to more than one antibody and the immune complex tends to be large and insoluble and may accumulate in the kidney (*see* GLOMERULONEPHRITIS). Immune complexes will activate the classical complement pathway. *See* ARTHUS REACTION; HYPERSENSITIVITY (type III).

antigenic determinant *See* EPITOPE.

antigenic variation A mechanism used by various parasites to evade host immune responses by changing the antigens expressed on the surface. This also makes it difficult to produce vaccines. Antigenic variation is seen in **Trypanosoma* and **Plasmodium* as well as in free-living protozoa and some bacteria. Trypanosomes may exhibit more than 100 different surface antigens, each encoded by a different gene, during an infection.

antigen presentation *See* ANTIGEN PRESENTING CELL.

antigen presenting cell (APC) Describing cells that bind antigen to MHC class I or class II *histocompatibility antigens on their surfaces and interact with T cells which 'see' the antigen in the restricted context of the MHC molecule (*see* MHC RESTRICTION). The major APCs are macrophages, endothelium, *dendritic cells, and *Langerhans cells of the skin.

antigen processing The cleavage or denaturation of an antigen before it is presented, bound to either MHC class I or II molecules, to T cells. Extrinsic substances are endocytosed by *accessory cells, processed and presented in association with MHC class II by *antigen presenting cells. Cytoplasmic proteins are processed by the *proteosome, and the peptides are presented in association with MHC class I.

antigen shift An abrupt form of *antigenic variation. **Antigen drift** is a more gradual change.

antihistamine A drug that blocks the action of *histamine, used in the treatment of allergic rhinitis, especially hay fever. A common side effect is drowsiness. *See* HISTAMINE RECEPTORS.

anti-idiotype antibody An antibody directed against the region of another antibody or T-cell receptor that is specific for an antigen. In principle, anti-idiotype antibodies should inhibit a specific immune response.

anti-inflammatory drugs Drugs that inhibit the inflammatory response. The two major classes are *NSAIDs and *glucocorticoids.

antilymphocyte serum A mixture of immunoglobulins raised *xenogeneically and directed against lymphocytes, particularly T cells. Used experimentally to cause *immuno-suppression.

antimetabolite A type of drug used in cancer chemotherapy; antimetabolites become incorporated into replicating DNA and block normal division. Common examples are cytoarabinose, *fluorouracil, and *methotrexate.

antimitotic drugs A general term for drugs that interfere with mitosis, often by blocking microtubule assembly so that chromosomes accumulate in metaphase. Examples include *colchicine and the *vinca alkaloids. Many antitumour drugs block proliferation by being antimitotic, rather than being cytotoxic.

antiMüllerian hormone A dimeric glycoprotein hormone (AMH, Müllerian inhibiting substance, MIS, 560 aa), of the TGFβ superfamily, produced by Sertoli cells during fetal and postnatal testicular development that causes regression of the Müllerian ducts that in females differentiate into the uterus and fallopian tubes. AMH works in conjunction with testosterone produced by Leydig cells to masculinize the fetus. Defects in AMH cause persistent Müllerian duct syndrome type 1 (PMDS1), a form of male pseudohermaphroditism in which the ducts do not regress. The **receptors** are serine/threonine kinases and mutations can cause PMDS2.

antimuscarinic drugs A subset of anticholinergic drugs that block the action of acetylcholine at the *muscarinic subclass of acetylcholine receptors. They will relax the smooth muscle of the gut (antispasmodics) or the airways (bronchodilators). An example is *tolterodine.

antimycins Antibiotics derived from *Streptomyces* spp. that inhibit mitochondrial electron transport.

antineoplaston A naturally occurring mixture of compounds that are claimed to have antitumour activity. There is no good evidence to support this.

antinuclear antibodies (ANA) Antibodies, usually IgG, directed against antigens in the nucleus, commonly dsDNA or histones although other nuclear components may be the antigen. ANA are characteristic of autoimmune diseases such as *systemic lupus erythematosus and *Sjögren's syndrome.

antioncogene *See* TUMOUR SUPPRESSOR.

antioxidant Any substance that inhibits oxidation, often by being more easily oxidized. Examples include *vitamin E and *vitamin C. They restrict the damage caused by reactive oxygen species produced in the *respiratory burst and may have some anti-ageing effects.

antipain A reversible inhibitor of *papain, *trypsin, and, to a lesser extent, *plasmin, isolated from actinomycetes.

antiparallel Describing structures with opposite *polarity, e.g. the two strands of a DNA duplex.

antiphospholipid syndrome An autoimmune disorder (familial lupus anticoagulant) in which there are antibodies directed against cellular phospholipid components. Clinically there is arterial and venous thrombosis, recurrent fetal loss, and immune thrombocytopenia. *See* APOLIPOPROTEIN H.

antiplasmin (α_2-antiplasmin) The major *plasmin inhibitor (serpin F2, 491 aa) in plasma. It will inhibit other serine peptidases such as factors XIa, XIIa, *plasma kallikrein, *thrombin, and *trypsin) and regulates the hydrolysis of fibrin clots (*fibrinolysis). Deficiency in α_2-plasmin inhibitor causes an autosomal recessive bleeding disorder.

antiplectic Describing a *metachronal pattern in a field of *cilia, in which the waves move in the direction opposite to that of the active stroke.

antiport A membrane transporter that moves two different ions or molecules in opposite directions. The transport may be active as with the *sodium-potassium ATPase, or it may be driven by a gradient of one of the molecules, as with sodium/proton antiport.

antiproteases (antiproteinases, antipeptidases) Substances that inhibit enzymes that hydrolyse polypeptides.

antipsychotic drugs (neuroleptic drugs) A class of drugs (major tranquillizers) used to calm or sedate disturbed patients, control acute symptoms of mania, and relieve severe positive symptoms of schizophrenia. Most reduce dopamine levels in the central nervous system (e.g. *chlorpromazine, flupentixol, *haloperidol), but the atypical antipsychotics (e.g. clozapine, *risperidone) act on *serotonin-based systems.

antipyretic Describing something that reduces fever.

antipyrine (phenazone) A compound with analgesic and antipyretic properties.

antisense A shorthand term for a nucleotide strand that is complementary to a coding sequence of DNA or mRNA. Antisense RNA binds to mRNA and leads to its inactivation.

antisepsis The prevention of infection by killing or inhibiting the growth of microorganisms with antiseptic agents (antiseptics). *Compare* ASEPSIS.

antiserum Serum that contains antibodies directed against specific *antigens.

antispasmodic drugs Drugs that cause the smooth muscle of the gut wall to relax, bringing symptomatic relief in indigestion, irritable bowel syndrome, and diverticular disease. Most are *antimuscarinic drugs.

antithrombin A plasma glycoprotein (antithrombin III, serpinC1, 464 aa) that inhibits serine proteases in the blood coagulation cascade (thrombin and factors IXa, Xa, and XIa). Its inhibitory activity is enhanced by *heparin. Antithrombin III deficiency is an important risk factor for hereditary thrombophilia, in which there is a tendency to recurrent thrombosis. Antithrombins I, II, and IV were terms used in early studies on coagulation and are obsolete.

antithymocyte globulin A polyclonal IgG that selectively destroys T cells and is used as an immunosuppressive agent.

antitoxin An *antibody directed against a toxin that leads to its inactivation and removal.

antitussive drug A cough suppressant, rather than a cure for a cough.

antiviral drugs Drugs that inhibit viruses. There are two main types: those which inhibit or interfere with viral replication (nucleoside analogues such as *aciclovir and *AZT) and those that interfere with virus-specific enzymes needed for processing of viral proteins to produce infective particles. The latter include antiproteases such as saquinavir, ritonavir, and indinavir and neuraminidase inhibitors such as zanamivir (*TN* Relenza) and oseltamivir (*TN* Tamiflu).

antizyme A polyamine-inducible protein (ornithine decarboxylase antizyme 1, 228 aa) involved in feedback regulation of cellular polyamine levels by repressing *ornithine decarboxylase (ODC). **Antizyme 2** (189 aa) and **antizyme 3** (187 aa) have similar properties and antizyme 3 is probably important in spermatogenesis. **Antizyme inhibitor 1** (448 aa) binds to and inhibits antizyme.

antolefinin *See* DREAM COMPLEX.

antrum **1.** A general term for a cavity or chamber. **2.** The lower third of the stomach (**pyloric antrum**). The mucosa of this region (antral mucosa, pyloric mucosa) has coiled and branching antral glands lined by mucus cells interspersed with endocrine cells (chiefly G and D types), and a few parietal (*oxyntic) cells. **3.** A fluid-filled space within the ovarian follicle. **4.** The antrum of Highmore is an air-filled cavity in the maxilla that is linked to the nasal cavity.

anucleate Without a nucleus.

anucleolate Literally, without nucleoli. A *Xenopus* anucleolate mutant, viable when heterozygous, was used in early nuclear transplantation experiments.

anuria (anuresis) Complete failure to secrete urine or, often, production of less that 50 mL of urine per day. Usually a result of renal failure.

anxiogenic Describing something that induces a state of anxiety.

anxiolytic Describing a drug that reduces anxiety. *Benzodiazepines and *barbiturates are commonly used.

aortic incompetence A condition, due to a defect in the aortic valve, that allows blood pumped into the aorta during ventricular systole to reflux back into the left ventricle causing dilation. Can be a result of rheumatic heart disease or infectious endocarditis and can be a complication in *Marfan's syndrome.

AP1 **1.** A *transcription factor, a heterodimer of the **fos* and **jun* proto-oncogene products. **2.** *See* ADAPTOR PROTEINS.

AP2 **1.** A family of *cis*-acting transcription activators often expressed in the same cell lineages but at different times or sites in the developing embryo. All bind to the same consensus sequence in DNA. Mutations in **AP2-**alpha (436 aa) cause *branchiooculofacial syndrome. AP2-beta and -gamma bind to the same DNA consensus sequence; AP2-beta (460 aa) is involved in control of differentiation of neuroectodermally derived cells and is mutated in *Char's syndrome; AP2-gamma (450 aa) and AP2-delta (452 aa) are also known. **2.** *See* ADAPTOR PROTEINS.

AP3, AP4, AP5 Selective antagonists for the *NMDA subclass of *glutamate receptors. *See also* ADAPTOR PROTEINS.

APA A low-affinity competitive antagonist (aminopimelic acid) of the *NMDA subclass of *glutamate receptors.

APAF-1 A protein (apoptosis protease activating factor 1, 1248 aa) that assembles into an oligomeric *apoptosome in the presence of cytochrome *c* and dATP. The apoptosome activates procaspase-9, initiating a cascade of events that end in apoptotic death of the cell. **APAF1-interacting protein** (242 aa) inhibits APAF1, has an antiapoptotic function and prevents ischaemic damage to muscle.

apamin A small basic peptide (18 aa), found in venom from the honey bee *Apis mellifera*, that selectively blocks calcium-activated *SK potassium channels and has an inhibitory action in the central nervous system.

APC **1.** *See* ANTIGEN PRESENTING CELL. **2.** Adenomatous *polyposis coli.

ape-1 An excision repair endonuclease (apurinic–apyrimidinic endonuclease 1, EC. 4.2.99.18, 318 aa) that initiates repair of apurinic/apyrimidinic (AP) sites in DNA by catalysing hydrolytic incision of the phosphodiester backbone immediately adjacent to the damage. *See* SET COMPLEX.

apelin The endogenous ligand for the G-protein-coupled receptor *APJ. The peptide (77 aa) is cleaved into various active fragments (apelins-13, -28, -31, and -36). It will inhibit HIV-1 entry in cells coexpressing CD4 and APJ. It is expressed in the brain and secreted by the mammary gland into colostrum and milk.

aperient A drug with a laxative or purgative effect.

Apert's syndrome A developmental disorder (Apert–Crouzon disease) characterized by craniosynostosis (premature closure of skull sutures), midface hypoplasia, and syndactyly of the hands and feet. It is caused by mutation in the gene encoding *fibroblast growth factor receptor-2.

Apgar score An assessment system devised by Virginia Apgar (1909–74) for the condition of a child at birth. A value of 0, 1, or 2 is given for five signs: colour, heart rate, response to stimulation, muscle tone, and breathing (appearance, pulse, grimace, activity, respiration). A score of 10 is desirable.

APH-1 An essential subunit (265 aa) of the γ-*secretase complex, where it probably stabilizes the presenilin homodimer. Widely expressed and mostly located as a multipass membrane protein in the endoplasmic reticulum and in the *cis*-Golgi.

aphagia Inability to swallow or to eat.

aphakia Absence of the lens of the eye. Congenital primary aphakia can be caused by a homozygous nonsense mutation in the *forkhead transcription factor FOXE3.

aphasia A loss or deficiency in language, either **receptive** (understanding speech or written words) or **expressive** (writing or speaking) according to the area of the brain that is affected.

aphidicolin A reversible inhibitor of eukaryotic *DNA polymerases isolated from *Cephalosporium*. It blocks the cell cycle at early S-phase and has some antiviral activity.

aphonia (alalia) Loss of voice. Can be caused by damage to the recurrent laryngeal nerve, which innervates the larynx, or may have a psychological basis.

aphthous ulcer A small oral ulcer.

apical ectodermal ridge A ridge of tissue at the apex of the developing limb bud in the embryo, a transient structure that is essential for limb outgrowth.

apical plasma membrane The cell membrane on the surface of transporting epithelial cells that is furthest from the basal lamina. The apical plasma membrane is isolated from the basolateral membrane by the *zonula occludens which prevents lateral diffusion of membrane proteins from one area to the other.

Apicomplexa A large phylum of parasitic protists that characteristically have an apical complex of microtubules within the cell in the sporozoite and *merozoite stages, but lack flagella. Includes **Babesia*, **Cryptosporidium*, **Plasmodium*, and **Toxoplasma*.

apigenin A yellow bioflavone (naringenin chalcone), naturally present in parsley and celery, claimed to be beneficial as an antioxidant, radical scavenger, and anti-inflammatory.

APJ An orphan G-protein-coupled receptor (angiotensin receptor-like 1, 380 aa) for which *apelin is the ligand. APJ can be a coreceptor with CD4 for HIV entry into cells.

aplasia Defective development of an organ or tissue so that there is total or partial absence from the body. *Compare* ATROPHY.

aplastic anaemia A type of *anaemia in which there is loss of most or all of the *haematopoietic tissue of bone marrow, usually affecting all cell types.

aplyronine A cytotoxic macrolide from the sea hare *Aplysia kurodai* that has F-actin depolymerizing activity and forms complexes with G-actin, inhibiting assembly. Has antitumour activity.

apnoea Cessation of breathing. Recognized to occur (temporarily) during sleep, particularly in obese individuals.

ApoA, B, C, D, etc. *See* APOLIPOPROTEINS.

apocrine A mode of secretion in which the apical portion of the cell is shed. The classic example is the secretion of lipids by the mammary gland which are released as droplets surrounded by plasma membrane.

apocynin A compound isolated from dogbane *Apocynum cannabinum* that blocks the production of reactive oxygen species by activated neutrophils by inhibiting the NADPH oxidase.

apoenzyme *See* APOPROTEIN.

apolipoprotein The protein component of plasma lipoproteins. *See* separate entries for the various classes.

apolipoprotein A Apolipoprotein A1 (ApoA1, 243 aa) is the major apoprotein of high-density lipoprotein (HDL) and a relatively abundant plasma protein (1.0–1.5 mg/mL). Defects in ApoA1 cause autosomal dominant high density lipoprotein deficiency type 2 (HDLD2, familial hypoalphalipoproteinaemia), and recessive HDLD1 (analphalipoproteinaemia, *Tangier disease) (TGD). Defects can also cause amyloid polyneuropathy–nephropathy Iowa type (van Allen type amyloidosis, familial amyloid polyneuropathy type III) and amyloidosis type 8 (systemic non-neuropathic amyloidosis, Ostertag-type amyloidosis). **Apolipoprotein A2** (100 aa) may stabilize HDL structure. **ApoA4** is a component of chylomicrons and high-density lipoproteins. Defects in **ApoA5** cause susceptibility to familial hypertriglyceridaemia and hyperlipoproteinaemia type 5. **ApoA1 binding protein** (288 aa) interacts with both ApoA1 and ApoA2. Apolipoprotein L1 interacts with ApoA1. **Apolipoprotein(a)** is the main constituent of *lipoprotein(a).

apolipoprotein B The main apolipoprotein of chylomicrons and low-density lipoproteins (LDL). There are two main forms in plasma, **ApoB48** synthesized by the gut, and **ApoB100** produced by the liver, although both are coded by the same gene differently spliced. Familial hypobetalipoproteinaemia (FHBL) is an autosomal dominant disorder of lipid metabolism in which there are very low plasma levels of apolipoprotein B although triglyceride levels are normal (in contrast to abetalipoproteinaemia due to microsomal triglyceride transfer protein deficiency, Bassen–Kornzweig syndrome or acanthocytosis).

apolipoprotein C A family of apolipoproteins (CI–CIV). **Apo-CI** (83 aa) constitutes ~10% of the protein of the VLDL (very low-density lipoprotein) and 2% of that of HDL (high-density lipoprotein) and appears to modulate the interaction of ApoE with VLDL. **ApoC-II** (101 aa) is a component of VLDL and is a necessary cofactor for the activation of lipoprotein lipase (EC 3.1.1.3). Mutations in the gene encoding ApoC2 cause hyperlipoproteinaemia type IB, an autosomal recessive disorder characterized by hypertriglyceridaemia, xanthomas, and susceptibility to pancreatitis and early atherosclerosis. **ApoC-III** (99 aa) forms 50% of the protein fraction of VLDL and inhibits lipoprotein lipase. Mutations cause deficiency of the apolipoprotein but without clinical consequences except perhaps a reduced risk of atherosclerosis. **ApoC-IV** (127 aa) is thought to be involved in lipoprotein metabolism.

apolipoprotein D A *lipocalin (189 aa) mainly a component of HDL, found in plasma as a homodimer or as a heterodimer with ApoA2. It is involved in the transport of various ligands, including bilin, and is also found in a macromolecular complex with lecithin–cholesterol acyltransferase.

apolipoprotein E An apolipoprotein found in all plasma liproteins that mediates their binding, internalization, and catabolism. ApoE is a ligand for the LDL (ApoB/E) receptor and for the specific ApoE receptor. Different alleles of the *APOE* gene are associated with variations in plasma cholesterol levels and the risk of coronary artery disease. Late onset *Alzheimer's disease is unusually frequent in individuals with the *APOE4* allele, although there seems to be no causative linkage. Defects in ApoE are a cause of *hyperlipoproteinaemia type III, *sea-blue histiocyte disease, and *lipoprotein glomerulopathy.

apolipoprotein F A minor apolipoprotein (lipid transfer inhibitor protein, 308 aa) mostly associated with low density lipoprotein and an inhibitor of cholesteryl ester transfer protein and cholesterol transport.

apolipoprotein H A multifunctional apolipoprotein (β_2-glycoprotein I, 345 aa) that binds to various kinds of negatively charged substances such as heparin, phospholipids, and dextran sulphate and may prevent activation of the intrinsic blood coagulation cascade by binding to phospholipids on the surface of damaged cells. The ApoH–phospholipid complex appears to the autoantigen in some cases of *systemic lupus erythematosus and in *antiphospholipid syndrome.

apolipoprotein L A family of proteins (ApoL1-6, 337–433 aa) that are important in cholesterol transport and are expressed in a tissue-specific manner. They are not free in plasma but are associated with ApoA lipoproteins. Some are up-regulated in response to TNF.

apollo One of several evolutionarily conserved proteins (DNA crosslink repair protein 1B, 532 aa) involved in repair of interstrand crosslinks in DNA. Binds to *telomeric repeat binding factor 2 (TRF2) and is one of the accessory proteins in the *shelterin complex.

apomorphic A term used in cladistics to describe phenotypic characteristics that arose relatively late in the evolution of a group of species and therefore are found only in some, not all, of the related taxa; the more recent the common ancestor, the more apomorphic traits are shared. Apomorphic characters for the primates, like humans, include a large brain size and binocular vision. **Plesiomorphic** characteristics are ancient homologies (e.g. pentadactyl limbs). **Symplesiomorphic** characteristics are shared between two or more taxa, but also shared with other taxa which have an earlier last common ancestor with the taxa under consideration. **Synaptomorphic** characters are shared between taxa but are derived from, and not present in, a common ancestor.

apomorphine An agonist at dopamine D_1 and D_2 receptors with expectorant and emetic properties. It is derived from morphine but does not bind opioid receptors and is not a narcotic.

apopain A peptidase (*caspase 3, Yama protein, SCA-1, EC 3.4.22.56), produced by proteolytic cleavage of a precursor (277 aa), that is involved in initiating apoptosis by cleaving *poly (ADP-ribose) polymerase (PARP). Will cleave and activate *sterol regulatory element binding proteins (SREBPs); caspases 6, 7, and 9; and *huntingtin.

apoplexy An obsolete medical term referring to a sudden loss of consciousness and paralysis, probably due to cerebral haemorrhage. Now generally used to mean a state of extreme agitation or rage.

apoprotein The protein portion of an active complex involving protein and nonprotein moieties. For example, *rhodopsin is composed of *retinal and the apoprotein *opsin. In enzymes that require a prosthetic group, the apoprotein is often referred to as an apoenzyme.

apoptin A nonstructural protein (121 aa) encoded by chicken anaemia virus, that will selectively induce apoptosis in cancer cells. Apoptin-induced apoptosis can be blocked by *bcl2. **Human apoptin-associating protein 1** (228 aa) is also called RING1 and YY1-binding protein and death effector domain-associated factor.

apoptosis An energy-requiring physiological process that leads to cell death without exciting an inflammatory response, unlike *necrosis. The process is characterized by cell shrinkage, cleavage of the DNA into fragments that give a 'laddering pattern' on gels, and condensation of chromatin. Apoptosis is distinct from *programmed cell death although the terms are often treated as interchangeable. *See also* APOPTOSIS INDUCING FACTOR; APOPTOSOME; BCL (BCL-2); CED GENES.

apoptosis inducing factor A group of proteins involved in initiation of apoptosis. **AIF1** (programmed cell death protein 8, 613 aa) is a mitochondrial oxidoreductase that acts as a caspase-independent mitochondrial effector of apoptotic cell death. It translocates to the nucleus where it binds to DNA in a sequence-independent manner and interacts with *XIAP. **AIF2** (373 aa) is similar in properties. **AIF-like protein** (AIF3, 605 aa) does not translocate from the mitochondrion to the nucleus and induces apoptosis through a caspase-dependent pathway.

apoptosome A multiprotein complex, the formation of which is triggered by cytochrome *c* released from mitochondria which binds *Apaf-1. The cytochrome-Apaf complex then recruits procaspase-9 which, when activated, starts the caspase cascade. *XIAP binds to the caspase 9 and inhibits further proteolytic activity.

applagin *See* DISINTEGRINS.

apple domain A consensus sequence (~90 aa) that includes six cysteines which form disulphide bonds; it can be drawn as apple-shaped. Found as a fourfold repeat in *plasma kallikrein and coagulation factor XI, both *serine peptidases. Mutations in apple 4 cause factor XI deficiency, an inherited bleeding disorder.

SEE WEB LINKS

- Image of the structure.

aprataxin A protein (forkhead-associated domain histidine triad-like protein, 356 aa) of the *histidine triad (HIT) superfamily that interacts with several nucleolar proteins, including

a

*nucleolin, *nucleophosmin, and upstream binding factor-1 (UBF-1), binds to DNA and is involved in DNA repair. The *APTX* gene is mutated in patients with the neurological disorder *ataxia with oculomotor apraxia type 1 (AOA1) and *coenzyme Q_{10} deficiency. Aprataxin and PNK-like factor (511 aa) are together involved in single-strand and double-strand DNA break repair.

apraxia Total or partial loss of the ability to perform coordinated movements but without any obvious motor or sensory impairment.

APRIL A member of the TNF ligand superfamily (a proliferation-inducing ligand, TNF-related death ligand 1, TNFSF13, CD256, 250 aa) highly expressed in various tumour cells, monocytes, and macrophages.

aprotinin An inhibitor (pancreatic trypsin inhibitor, 100 aa) of various peptidases (*trypsin, *chymotrypsin, *kallikrein, and *pepsin), originally isolated from bovine lung, now produced as a recombinant protein. *TN* Trasylol.

AP sites *See* APURINIC SITE.

aptamer (aptamere) A double-stranded DNA, single-stranded RNA, or peptide that binds to a specific molecular target. Sometimes generated by repeated rounds of *in vitro* selection.

APUD cells *Paracrine cells that show amine-precursor uptake and decarboxylation; *argentaffin cells are an example.

apurinic site A site of damage in DNA where one or more purines have been lost either by slow hydrolysis in physiological conditions or by rapid hydrolysis following alkylation. Specific endonucleases are involved in repair at these AP sites (AP sites can be apurininic or apyrimidinic).

APV **1.** Avian *polyoma virus. **2.** Amprenavir, an antiviral protease inhibitor. **3.** An *NMDA antagonist, (DL-2-amino-5-phosphonopentanoic acid).

Apx/Shrm domains Domains (ASD1, ASD2) found in *shroom family proteins and in an apical plasma membrane protein (Apx) that plays a role in the functional expression of the *amiloride-sensitive sodium channel in *Xenopus*. ASD2 is a common feature of the shroom proteins.

apyrase A multipass membrane protein that hydrolyses ATP or ADP to AMP (ectonucleoside triphosphate diphosphohydrolase, EC 3.6.1.5). Tissue-specific isoforms (510–530 aa) are found and may be responsible for regulating purinergic neurotransmission and regulating platelet aggregation (by hydrolysing ADP).

apyrexia The absence of fever.

aquaporins Proteins of the *major intrinsic protein family that form homotetrameric channels for the movement of water across membranes. **Aquaporin-1** (channel-like integral membrane protein, 28 kDa, CHIP28, 269 aa) was first found in erythrocytes and proximal tubules of the kidney; various other aquaporins have subsequently been discovered. **Aquaporin-2** (AQP2, 271 aa) is located in the renal collecting tubule and is *vasopressin-regulated. Defects in AQP2 are the cause of *diabetes insipidus nephrogenic type 2.

SEE WEB LINKS

- Review on 'The aquaporin family of molecular water channels'.

Arabidopsis thalliana The vascular plant (thale cress) that is the model organism for plant molecular biology. A remarkable number of *Arabidopsis* genes have human homologues.

arabinose A *pentose sugar found in plant polysaccharides. It is an epimer of ribose and is found in *cytosine arabinoside.

arachidonic acid A polyunsaturated fatty acid (5,8,11,14-eicosatetraenoic acid) that is the precursor for *prostaglandins, *thromboxanes, and hydroxyeicosatetraenoic acid derivatives including *leukotrienes. Arachidonic acid is the fatty acid moiety of many membrane phospholipids, especially *phosphatidylinositol, and its release from membranes by phospholipases may be the rate-limiting step in the formation of its active metabolites. It is an essential dietary component.

arachnodactyly Atypical elongation of fingers and toes, seen in *Marfan's syndrome.

arachnoid layer (arachnoid mater) *See* DURA MATER.

arboviruses A diverse group of enveloped single-stranded RNA viruses that are transmitted by arthropods. There are three major families: *Togaviridae, *Bunyaviridae, and *Arenaviridae.

Archaea (formerly **Archaebacteria**) One of two major subdivisions (domains) of the prokaryotes. There are three main orders, extreme *halophiles, methanobacteria, and sulphur-dependent extreme thermophiles. Archaea differ from *Bacteria (formerly Eubacteria) in ribosomal structure, the possession (in some cases) of *introns, and in a number of other features, including membrane composition. The peptidoglycan composition of the cell wall also differs from that in *Bacteria.

archenteron (archentron) The cavity formed by the process of *gastrulation that will eventually form the gastrointestinal tract. It opens to the exterior at the *blastopore.

archvillin *See* SUPERVILLIN.

arcuate Bent into a bow-like curve. The **arcuate nucleus** in the hypothalamus has two major types of neurons, one that secretes *neuropeptide Y and *agouti-related peptide, the other, POMC/CART neurons, that produce α-melanocyte-stimulating hormone (α-*MSH).

ARD-1 A GTP-binding protein (ADP-ribosylation factor domain-containing protein 1, tripartite motif-containing protein 23, RING finger protein 46, 574 aa) of unknown function.

ARE **1.** An *activin-response element to which *ARF binds. **2.** An *AU-rich element.

arecoline A pharmacologically active alkaloid, from betel palm *Areca catechu*, that is an agonist of *muscarinic acetylcholine receptors. Has also been used as an anthelmintic.

Arenaviridae A family of ssRNA viruses that includes *Lassa virus, lymphocytic choriomeningitis virus, and the *Tacaribe group of viruses. Although included in the arboviruses, not all are transmitted by arthropods.

areolar connective tissue The relatively loose *connective tissue that surrounds many organs. There is no particular orientation of fibres, nor unusually high content of any particular extracellular matrix components.

ARF *See* ADP-RIBOSYLATION.

arfaptins Putative cytosolic targets (ADP-ribosylating factor interacting proteins, arfaptin 1, 373 aa; arfaptin 2, 341 aa) of *ADP-ribosylation factor and *rac GTPases.

SEE WEB LINKS

- Website with details of BAR superfamily proteins including arfaptins.

arg An *oncogene (Abelson-related gene, ABL2, v-*abl* Abelson murine leukaemia viral oncogene homologue 2), related to **abl*, that encodes a *tyrosine kinase (1182 aa).

argentaffin cells Paracrine cells (*APUD cells) that secrete *serotonin and are found mostly in the epithelium of the gastrointestinal tract. The name derives from the observation that they contain cytoplasmic deposits of silver when stained with silver salts.

arginine (Arg, R) An essential amino acid (174 Da) found in proteins where it is important because it carries a permanent positive charge at physiological pH.

argonaute proteins Proteins (eukaryotic translation initiation factors 2C1–4, argonaute 1–4, AGO1–4, ~857 aa) that are concentrated in *GW bodies and are key components of RNA-induced silencing complexes (*RISCs) where they bind single-stranded small interfering RNA (siRNA) and microRNA. They are also important in peptide chain initiation. The argonaute proteins have *slicer activity and contain N-terminal PAZ (PIWI/Argonaute/Zwille) domains and C-terminal *piwi domains.

argyrophil cells Neuroendocrine cells (*APUD cells) that derive from neurectoderm and migrate to their final site. They will take up silver ions from a staining solution but require the addition of an external reducing agent to precipitate metallic silver (*compare* ARGENTAFFIN CELLS which occur in the same tissue locations).

ARIA A chick brain polypeptide (acetylcholine receptor inducing activity) that stimulates transcription of *acetylcholine receptor subunits. *Neuregulin 1 is the homologous human protein.

ARID domain A DNA-binding domain (AT-rich interaction domain, ~100 aa) conserved throughout the higher eukaryotes. Proteins with ARID domains are involved in processes including embryonic development, cell lineage gene regulation, and cell cycle control. The human *SWI/SNF complex protein p270 is an ARID family member.

arisostatins Antibiotics isolated from a strain of the actinomycete genus *Micromonospora*, that are active against Gram-positive bacteria and have antitumour activity.

Ark1 **1.** Aurora-related kinase 1: *see* AURORA KINASES. **2.** β-Adrenergic receptor kinase (BARK, βARK1, 689 aa), a cardiac *G-protein-coupled receptor kinase (GRK2) involved in desensitizing and down-regulating β-adrenoreceptors (βARs).

armadillo repeat A protein–protein interaction motif (42 aa), originally described in the *Drosophila* β-catenin family protein armadillo. The armadillo repeat-containing proteins can be divided into subfamilies based on the number of repeats, their overall sequence similarity, and the dispersion of the repeats throughout their sequences. Armadillo repeat proteins include *catenins and *plakophilins.

a

ARMD *See* AGE-RELATED MACULAR DEGENERATION.

ARNO *See* CYTOHESINS.

ARNT A basic helix–loop–helix (bHLH)-*PAS-domain protein (aryl hydrocarbon receptor nuclear translocator, ~790 aa) that forms heterodimeric transcriptional regulator complexes with other transcription factors. In association with *hypoxia-inducible transcription factors it is important in cellular adaptation to low-oxygen environments; it is also a partner for the *aryl hydrocarbon receptor.

aromatase A microsomal enzyme (EC 1.14.14.1, 503 aa) of the cytochrome P450 superfamily, encoded by the *CYP19* gene, that converts androgens to oestrogens and testosterone to oestradiol. **Aromatase inhibitors** are used to treat oestrogen-dependent breast cancer.

ARP-1 A member of the steroid/thyroid nuclear receptor family (apolipoprotein regulatory protein-1, transcription factor COUP2, 414 aa) that binds to regulatory elements of the *apolipoprotein A-I gene.

Arp2/3 complex A seven-membered complex that plays an important part in regulating actin assembly in cells. There are two actin-related proteins (Arp2 and Arp3) with five other subunits: ARC41 (ArpC1B), Arc34 (ARPC2), Arc21 (ARPC3), Arc20 (ARPC4), and Arc16 (ARPC5). Arp2/3 caps the pointed end of actin filaments and nucleates actin polymerization with low efficiency. *See also* ACTA; ICSA.

arrestins A family of proteins that bind to tyrosine-phosphorylated receptors and block their interaction with G proteins, thereby inhibiting signalling. **Arrestin-S** (S antigen; 405 aa, from retinal rods) competes with *transducin for light-activated *rhodopsin, thus inhibiting the response to light (adaptation); **arrestin-C** (cone arrestin (388 aa) is similar. Immune responses to arrestin-S lead to autoimmune uveitis. Similarly, β-arrestin binds to phosphorylated β-*adrenergic receptors, inhibiting their ability to activate the *G protein G_s (*see* ARK1).

Arrhenius plot A graphical plot of the logarithm of reaction rate against the reciprocal of absolute temperature. In the straightforward case of a single-stage reaction this gives a linear plot from which the activation energy and the frequency factor can be determined. Biological systems often show much more complex behaviour in response to temperature change and, although it is possible to interpret the complex Arrhenius plot as a series of linear elements, this is not necessarily justifiable.

arrhythmia The absence of a normal regular rhythm; in the heart, arrhythmia can be a prelude to cardiac arrest.

arrhythmogenic right ventricular dysplasia/cardiomyopathy (ARVD/ARVC) A range of autosomal dominant disorders in which there is arrhythmia of the right ventricle of the heart and replacement of muscle with fibrous tissue; they are an inherited cause of juvenile sudden death. The mutated gene product in **ARVD1** is TGFβ3; in **ARVD2**, *ryanodine receptor 2; in **ARVD5**, transmembrane protein 43; in **ARVD8**, *desmoplakin; in **ARVD9**, *plakophilin 2, in **ARVD10** *desmoglein 2; in **ARVD11** *desmocollin 2; in **ARVD12**, γ-*catenin.

arrowheads The description given to the ultrastructural appearance of *myosin molecules attached to an *F-actin filament, more easily visualized with the addition of tannic acid to the fixative. Conveniently, the arrowheads indicate the polarity of the filament: the barbed (attachment) end is where G-actin attaches; disassembly occurs at the pointed end.

ARSACS *See* AUTOSOMAL RECESSIVE SPASTIC ATAXIA OF CHARLEVOIX-SAGUENAY.

Artemether A rapid-acting antimalarial drug derived from *artemisinin and used in combination with *lumefantrine.

artemin A protein (neublastin, enovin, 113 aa) that is a ligand for *GDNF receptor α_3 and is important for survival of neurons from all peripheral ganglia and some neurons in the central nervous system.

artemis A protein (DNA crosslink repair protein 1C, 685 aa) with single-strand-specific 5′-to-3′ exonuclease activity regulated by DNA-dependent protein kinase and involved in V(D)J recombination in the immune system and in DNA repair. Mutation can lead to a form of *severe combined immunodeficiency syndrome that is complicated by radiation sensitivity.

artemisinin An antimalarial drug extracted from *Artemisia annua* (qinghaosu or sweet wormwood), used to treat uncomplicated falciparum malaria, often in combination therapy with *lumefantrine. Various derivatives have been developed.

artery A blood vessel that carries blood away from the heart. The walls of arteries have a layer of smooth muscle and are innervated by the *sympathetic nervous system. An **arteriole** is the smallest branch of an artery, just upstream of the capillary bed. **Arteriosclerosis** is a general and imprecise term for various arterial disorders, particularly hardening due to fibrosis or the deposition of calcium salts and not really a synonym for *atherosclerosis. **Arteritis** is inflammation of an artery.

arthr- A combining form indicating association with joints. **Arthralgia** is joint pain. **Arthritis** is a general term for inflammation of one or more joints. (*See* OSTEOARTHRITIS; RHEUMATOID ARTHRITIS.) **Arthrodesis** is immobilization of a joint by surgical fusion of the joint surfaces. **Arthropathy** is a general term for any disease affecting a joint and **arthrosclerosis** is stiffness of joints (not to be confused with *atherosclerosis). *See* ARTHROGRYPOSIS.

arthrogryposis A rare and heterogeneous disorder characterized by multiple joint contractures. **Distal arthrogryposis type 1**, affecting distal joints in hands and feet, can be caused by mutation in β-tropomyosin. Distal arthrogryposis type 2A (DA2A) can be caused by mutations in the *MYH3* gene which encodes embryonic myosin heavy chain. Mutations in *MYH3* can also cause DA type 2B (Sheldon–Hall syndrome). DA2B can also be caused by mutation in isoforms of troponin T and troponin I. Many other variants of distal arthogryposis, some with additional disorders, have been described, although the genetic defects have not been identified. Arthrogryposis, renal dysfunction, and cholestasis (ARC syndrome) is caused by mutations in the *VPS33B* gene that encodes a protein (617 aa) involved in vacuolar sorting.

Arthus reaction A localized haemorrhagic *inflammation that occurs if antigen is experimentally injected into an animal with a high level of circulating *antibody against that antigen. A similar reaction may occur in *farmer's lung, with inhalational exposure to antigen.

artificial selection The process of selective breeding, the conventional method of genetic engineering. The selection pressure is operated by the plant or animal breeder, rather than natural environmental pressures (*natural selection). The phenotypic diversity of dogs is a classic example of the variations that can be achieved.

ARVD *See* ARRHYTHMOGENIC RIGHT VENTRICULAR DYSPLASIA/CARDIOMYOPATHY.

aryl hydrocarbon receptor A cytosolic protein (~800 aa) with a basic helix-loop-helix motif that binds a range of aryl hydrocarbons and dioxin and is then translocated to the nucleus where it forms a dimer with *ARNT and binds to the *xenobiotic response element. The receptor plays an important role in the closure of the ductus venosus and when activated and translocated to the nucleus affects immune, endocrine, reproductive, developmental, cardiovascular, and central nervous system functions.

aryl sulphatases A group of enzymes that will hydrolyse sulphate esters of aromatic substrates. **Aryl sulphatase A** (EC 3.1.6.1, 507 aa) is a lysosomal enzyme that hydrolyses cerebroside 3-sulphate and is deficient in *metachromatic leucodystrophy. **Aryl sulphatase B** (*N*-acetylgalactosamine 4-sulphatase, EC 3.1.6.12, 533 aa) is involved in lysosomal hydrolysis of glycosaminoglycans and is deficient in *Maroteaux–Lamy syndrome. **Aryl sulphatase C** (steroid sulphatase, EC 3.1.6.2, 583 aa) hydrolyses several precursors for oestrogens, androgens, and cholesterol. Mutations in the gene causes X-linked *ichthyosis. X-linked recessive *chondrodysplasia punctata is caused by mutation in **aryl sulphatase E** (ARSE, 594 aa), the gene for which is located near those for arylsulphatases D and F. *See* MULTIPLE SULPHATASE DEFICIENCY.

AS-252424 A potent and selective inhibitor of phosphoinositide 3-kinase p110γ that inhibits C5a-mediated phosphorylation of *Akt in RAW 264.7 macrophages and neutrophil recruitment in a mouse model of peritonitis.

asbestosis Lung *fibrosis caused by the chronic inhalation of asbestos fibres, particularly those of crocidolite asbestos (blue asbestos). The needle-like asbestos fibres are phagocytosed by *alveolar macrophages but cause mechanical damage and kill the macrophage; the cycle is repeated. Mesothelioma, a tumour of the mesothelial lining of the pleura, is associated with asbestosis.

ASC An adaptor protein (apoptosis-associated speck-like protein containing a CARD, TMS1, 195 aa) containing a caspase activation and recruitment domain (*CARD domain) that is involved in the downstream signalling from *PYRIN-domain pathogen recognition receptors. May be a component of the *inflammasome.

Ascaris A genus of nematodes (Aschelminthes). *Ascaris lumbricoides* causes ascariasis in humans and is said to be the most common helminth infection of humans. Heavy infestations cause growth retardation in children.

Aschoff bodies Small *granulomas found in cardiac connective tissue following rheumatic

a

fever. They are composed of *macrophages, *lymphocytes, and multinucleate giant cells grouped around a necrotic centre.

ascites (hydroperitoneum) The accumulation of fluid in the peritoneal cavity. Can arise from various causes, including infection, portal hypertension and various tumours. **Ascites tumour cells** proliferate in suspension (are not *anchorage dependent) in the ascitic fluid and various strains of ascites tumour maintained by serial subculture in mice are used experimentally (e.g. *Ehrlich ascites). *Hybridomas are sometimes grown as ascites tumours, and the ascites fluid is then immunoglobulin-rich. *See* ASGP.

ascorbic acid *See* VITAMIN C.

asepsis The desirable condition in which harmful microorganisms are absent. *Compare* ANTISEPSIS.

asexual Describing reproduction that does not involve a sexual process, gametes are not produced, and there is no reassortment of genetic characters.

ASF/SF-2 One of several proteins (*see* SR PROTEINS) found in nuclear speckles and involved in pre-mRNA splicing. ASF/SF-2 (SRp30a, arginine/serine-rich splicing factor 1, ASF1, 248 aa) interacts with other spliceosomal proteins and plays a role in preventing exon skipping, ensuring the accuracy of splicing and regulating alternative splicing.The isoforms ASF-2 and ASF-3 act as splicing repressors.

ASGP A membrane-associated mucin (ascites sialoglycoprotein, mucin-4, 2169 aa) that is proteolytically processed into α and β chains and can be secreted. It has anti-adhesive properties and may play a role in tumour progression, acting as an intramembrane ligand for ErbB2 (*see* EPIDERMAL GROWTH FACTOR) which represses apoptosis and stimulates proliferation.

asialoglycoprotein A glycoprotein from which terminal *sialic acid residues have been removed by treatment with *neuraminidase. Asialo derivatives of some plasma proteins are specifically bound by the *scavenger receptor on liver cells.

ASIP 1. Agouti signalling protein (132 aa), the human homologue of mouse *agouti, which inhibits melanogenesis by binding to the melanocyte stimulating hormone (MSH) receptor and blocking α-MSH signalling. Variation in the gene encoding ASIP affects hair and eye pigmentation. ASIP may also affect adipocyte function. **2.** Atypical protein kinase C isotype-specific interacting protein interacts with PKC-(lambda and -zeta and when overexpressed, inhibits insulin-induced glucose uptake.

ASK1 A serine/threonine kinase (apoptosis signal-regulating kinase, mitogen-activated protein kinase kinase kinase 5, MAPKKK5, 1374 aa) that is up-regulated under conditions of cellular stress and activates MAP kinases MAP2K4 and MAP2K6, which in turn activate the JNK and p38 MAP kinases and cause apoptosis.

Askenazy cells *See* HURTHLE CELLS.

asparaginase An enzyme (EC 3.5.1.1, 305 aa) that hydrolyses L-asparagine to L-aspartate, which is an excitatory neurotransmitter and has been used as an antitumour agent (*TN* Elspar), to treat acute lymphoblastic leukaemia in which the cells cannot synthesize asparagine (unlike normal cells).

asparagine (β-asparagine, Asn, N) The β-amide of aspartic acid (132 Da); the L-form is one of the 20 amino acids directly coded in proteins, independently of aspartic acid. Responsible for the smell of urine passed after eating asparagus, in which it is very abundant.

aspartate (aspartic acid, Asp, D) An amino acid (L-aspartate, 133 Da) directly coded in proteins; in the free form it is an excitatory neurotransmitter. Asx is the three-letter abbreviation for either asparagine or aspartic acid.

aspartate beta-hydroxylase *See* JUNCTIN.

aspartate transaminase *See* SGOT.

aspartate transcarbamylase *See* CAD PROTEIN.

aspartic endopeptidases A clan of endopeptidases of the class EC 3.4.23 (aspartic proteases, aspartyl proteinases), that have an acid (<pH 5) optimum. Common examples are *cathepsins, *pepsin, and *renin.

aspartylglycosaminuria A lysosomal storage disease caused by deficiency of the *threonine peptidase, *N*-aspartyl-β-glucosaminidase (EC 3.5.1.26, 346 aa) that cleaves the asparagine from the residual *N*-acetylglucosamines produced by breakdown of glycoproteins. The Finnish form is characterized by severe mental retardation, sagging cheeks, and various skeletal abnormalities.

aspartyl protease *See* ASPARTIC ENDOPEPTIDASES.

aspergillin **1.** A name sometimes given to the black spore pigment produced by *Aspergillus niger*. **2.** A name occasionally used for toxins (ribonucleases) produced by *Aspergillus* spp.; examples include *α-sarcin, *mitogillin, and restrictocin.

aspergillosis A lung disease caused by the fungus *Aspergillus fumigatus*. There may be allergic responses to aspergillus antigens or actual infection, which may spread in immunocompromised individuals.

aspermia *See* AZOOSPERMIA.

asphyxia Severe deficiency of oxygen, usually a result of suffocation.

aspiration The process of taking up fluids or gases using suction. **Aspiration pneumonia** is caused by inhalation of food, drink, or gastric contents.

aspirin A *NSAID (acetylsalicylic acid) with analgesic and antipyretic properties that inhibits *cyclo-oxygenases and inhibits platelet clotting. Used prophylactically in low doses it reduces the chances of heart disease and stroke but can cause severe ulceration and bleeding of the gastric mucosa (exacerbated by the antithrombotic effects). It is often said that aspirin would not get approval were it submitted to regulatory authorities as a new entity.

asporin A class I *small leucine-rich repeat proteoglycan (periodontal ligament-associated protein 1, PLAP1, 380 aa) that has a unique stretch of aspartate residues at its N-terminus. The sequence has homology with decorin and biglycan and is associated with skeletal tissues. An allele with an increased number of Asp repeats seems to be associated with susceptibility to osteoarthritis.

association constant (K_a, K_{ass}) The reciprocal of the *dissociation constant (K_d) and thus a measure of the amount of reversible association between two molecular species at equilibrium.

astacin A zinc endopeptidase (EC 3.4.24.21, 251 aa), from crayfish *Astacus* sp., which is the prototype for the M12A family of metalloendopeptidases that includes *BMP-1, meprin A, *stromelysin 1, and *thermolysin.

aster The microtubules that radiate from the microtubule organizing centres at the spindle poles at the beginning of mitosis.

asthenia Lack of energy or strength, mostly encountered in compound forms such as *myasthenia.

asthma An inflammatory disease of the airways that involves eosinophil infiltration and airway remodelling. Asthmatic attacks can be triggered by allergic responses, physical exertion, inhaled chemicals, or stress. Multiple loci and candidate genes have been implicated in susceptibility to asthma and related traits, including variations in the prostaglandin D receptor, IL-1 receptor-associated kinase, various G-protein-coupled receptors, etc.; 50–80% of individuals with atopic dermatitis develop asthma.

astigmatism A condition in which the eye cannot bring a point object to a sharp focus on the retina. **Irregular astigmatism** is usually a result of corneal scarring; **regular astigmatism** arises from unequal curvatures and can be corrected with lenses.

astrin A microtubule-associated protein (sperm-associated antigen 5, mitotic spindle-associated protein p126, 1193 aa) that localizes with mitotic spindles in M phase and is necessary for normal spindle formation. Ubiquitous, but highly expressed in the testis.

astroblast An *astrocyte precursor.

astrocyte (astroglia) A subset of the *glial cells of the central nervous system that have a star-like shape. They have both mechanical and metabolic roles. *See* OLIGODENDROCYTE. **Astrocytomas** are tumours (gliomas) derived from astrocytes. **Astrogliosis** is hypertrophic growth of the astroglia, usually in response to injury.

astrotactins Neuronal adhesion molecules (astrotactin 1, 1302 aa; astrotactin-2, 1339 aa) that have a role in glial-guided migration of neuroblasts in cortical regions of the developing brain. There are multiple isoforms, presumably to specify particular developmental pathways.

Astroviridae Small (28–30 nm) spherical nonenveloped viruses with five- or six-pointed star-shaped surface appearance in the electron microscope. The genome is a positive-sense single-stranded RNA. They can cause mild diarrhoea in young children.

SEE WEB LINKS

- The astrovirus homepage.

asystole Complete cessation of heartbeat.

ATA *See* AURINTRICARBOXYLIC ACID.

a

ataxia Lack of coordination and unsteadiness of gait caused by loss of postural control, usually due to a lesion in the cerebellum. *See* ATAXIA TELANGIECTASIA.

ataxia–oculomotor apraxia-1 An early-onset autosomal recessive cerebellar ataxia with peripheral axonal neuropathy, oculomotor apraxia, and hypoalbuminaemia. It is caused by mutation in the gene encoding *aprataxin. A second locus associated with the disorder is known, although the gene involved has not yet been identified.

ataxia telangiectasia An autosomal recessive disorder (Louis Bar syndrome) characterized by cerebellar *ataxia, *telangiectases, immune defects, and a predisposition to malignancy. There is a high frequency of spontaneous chromosomal aberrations as a result of mutations in the **ATM* gene and the cells are abnormally sensitive to ionizing radiation. The **ataxia telangiectasia-like disorder** is caused by mutation in the *MRE11A* gene that encodes an endonuclease involved in recombination, repair, and genomic stability.

ataxins A group of proteins that are defective in various forms of *spinocerebellar ataxia (SCA) because of expanded numbers of trinucleotide (CAG) repeats (polyglutamine repeats). **Ataxin 1** (815 aa), defective in SCA1, binds RNA and may be involved in RNA metabolism. **Ataxin 2** (1313 aa) is defective in SCA2. **Ataxin 3** (Machado–Joseph disease protein 1, 376 aa) defective in SCA3, represses transcription by interacting with transcriptional regulators and binding to histones. **Ataxin 7** (892 aa), defective in SCA7, is a component of the *STAGA transcription coactivator-HAT complex. **Ataxin 10** (475 aa) is defective in SCA10. **Ataxin 1-like** (brother of ataxin-1, BOAT, 689 aa) will suppress the cytotoxicity of ataxin 1 in SCA1. **Ataxin-7-like protein 1** (833 aa) is known, but not its function. **Ataxin-7-like protein 3** (347 aa) is part of the *SAGA transcription regulatory complex.

ATCase Aspartate transcarbamylase. *See* *CAD protein.

ATCC The American Type Culture Collection, a nonprofit organization in the USA, linked to the UK Laboratory of the Government Chemist, that maintains an extensive collection of eukaryotic cell lines, bacteria, phages, cloning vectors, etc. which can be purchased. The website also provides definitive information about the various lines.

SEE WEB LINKS

- UK homepage for the Culture Collection.

atelectasis The collapse of all or part of one lung.

atelocollagen A preparation of collagen from calf skin that has very low immunogenicity because *telopeptides have been removed by pepsin treatment. Used in various tissue-engineering applications and as a complex with DNA or siRNA for transduction into cells.

atelosteogenesis A disorder of bone and cartilage development in which limbs are abnormally short and there are various other skeletal abnormalities (cleft palate, hitchhiker's thumb, etc.). **Types I and III** are caused by mutations in *filamin B. **Type II** (neonatal osseous dysplasia I) is caused by mutations in the *diastrophic dysplasia sulphate transporter gene.

atenolol A β-adrenoreceptor antagonist (*beta-blocker) with greater affinity for β_1-receptors. It is used as an antihypertensive drug. *TN* Tenormin.

ATF Usually: **1.** Activating transcription factor (ATF) one of the *CREB/ATF family of *basic leucine zipper (bZip) transcription factors involved in mediating transcription in response to intracellular signalling. Occasionally: **2.** Artificial transcription factor. **3.** Amino-terminal fragment.

ATG **1.** Cysteine proteases ATG4A–D (autophagins 2, 1, 3, and 4; ~393 aa) that are required for *autophagy. **2.** ATG genes and proteins are associated with autophagy. **3.** *See* ANTITHYMOCYTE GLOBULIN.

athanogene A name applied to a gene with an antiapoptotic function; only used in the context of the *BAG family.

atheroma *See* FIBROUS PLAQUE.

atherosclerosis A disorder of the walls of arteries in which there is a local accumulation of macrophages filled with lipid (*foam cells) to form an atheromatous plaque. There is also migration of smooth muscle cells from the media to intima, proliferation of these cells and extensive deposition of extracellular matrix, sometimes calcified. The lesion may occlude the lumen, a particular problem in the coronary arteries, or, if the endothelial lining is lost, thrombi may form and be released. Is considered by some to be a chronic inflammatory response, although the local cause is unclear.

athetosis A condition in which there are continuous slow writhing movements of fingers and toes, usually caused by a lesion in the corpus striatum or the basal ganglia of the brain.

AT hook A small DNA-binding motif centred around a Gly-Arg-Pro tripeptide that binds in the minor groove of AT-rich regions. AT hooks are found in various proteins, including the *high-mobility group proteins, and probably work in conjunction with other binding motifs.

SEE WEB LINKS

- Article on AT hooks in DNA-binding proteins.

atlastin A family of GTPases with homology to guanylate-binding protein-1 (GBP1), one of the *dynamin family of large GTPases. It is expressed at much higher level in brain, predominantly in vesicular tubular complexes and *cis*-Golgi cisternae, and **atlastin-1** (558 aa) is mutated in *hereditary spastic paraplegia type 3A (SPG3A). The N-terminal domain of *spastin (mutated in SPG4) binds directly to the C-terminal cytoplasmic domain of atlastin in yeast two-hybrid assays, suggesting that the two act in concert. **Atlastin-2** (ADP-ribosylation-like factor 6-interacting protein 2, 583 aa) and **atlastin-3** (541 aa) are closely related to atlastin-1, but are localized to the endoplasmic reticulum in nonbrain tissue.

ATM A serine/threonine protein kinase (3056 aa) that is encoded by the gene mutated in *ataxia telangiectasia. It is activated when recruited to double-strand breaks (DSB) in DNA by the *MRN complex. When bound at DSB, the homodimeric ATM dissociates and becomes enzymically active.

atony (atonia) Loss of strength or tone in a muscle.

atopy An allergic (*hypersensitive) response at a site remote from the antigen. Susceptibility is associated with variations in a number of different genes, including those for the PHD finger protein 11, the IL-21 receptor, the IgE-Fc receptor (all associated with IgE) and P-selectin. **Atopic hypersensitivity** is a term covering asthma, hay fever, and eczema, all of which tend to be linked and are under complex genetic control.

atorvastatin *See* STATINS.

ATP (adenosine 5′-triphosphate) A compound produced by energy-yielding processes such as glycolysis and oxidative phosphorylation that is hydrolysed to drive a wide range of metabolic processes. There is nothing peculiar about its phosphate bonds although the myth of 'high-energy bonds' in the 'energy currency' of the cell persists.

ATPase Any enzyme that will release the terminal (γ) phosphate from ATP, yielding *ADP and inorganic phosphate. In most cases the hydrolysis of ATP is linked to an energy-requiring process such as active transport or motility.

ATP binding site A consensus domain ('A' motif) found in many proteins that bind either ATP or GTP. Examples are found in *ATP synthase, *myosin heavy chain, *helicases, *thymidine kinase, *G protein α-subunits, GTP-binding *elongation factors, *Ras family members.

ATP-grasp proteins A superfamily of proteins that includes a diverse set of enzymes catalysing ATP-dependent carboxylate-amine ligation reactions. The ATP-grasp fold is an ATP-binding motif consisting of two sub-domains that jointly hold the nucleotide.

SEE WEB LINKS

- Pfam [Protein families database] page with details of AT-grasp proteins.

ATP synthase A proton-translocating *ATPase (F_1/F_0 ATPase) that is normally driven in reverse by the large *proton motive force generated by the *electron transport chain. It is located in the inner membrane of *mitochondria and chloroplasts and in the plasmalemma of bacteria and is responsible for synthesis of ATP. *See* CHEMIOSMOSIS; V-TYPE ATPASE; P-TYPE ATPASE.

atrial fibrillation *See* FIBRILLATION (2).

atrial natriuretic peptide A polypeptide hormone (ANP, atrial natriuretic factor, ANF, 153 aa) produced by cleavage of natriuretic peptide precursor and secreted from the *atrium of the heart in response to stretching caused by elevated blood pressure. ANP acts to reduce blood pressure by stimulating the excretion of sodium and water in the kidneys (reducing blood volume), by relaxing vascular smooth muscle (causing vasodilation), and through actions on the brain and adrenal glands. Mutations in the gene cause familial atrial fibrillation type 6. There are three related receptors: *see* NATRIURETIC PEPTIDES.

atrichia A disorder in which the hair follicles are defective and no hair is produced. In atrichia with papular lesions the defect is in *hairless.

atriopeptin Obsolete name used originally for the *natriuretic peptides ANP and BNP.

atrioventricular septal defect (AVSD) A heart defect that allows blood to pass from the right side of the heart directly to the left, bypassing the lungs. There is considerable heterogeneity with several loci implicated. **AVSD1** is associated with a locus on chromosome

1, **AVSD2** is caused by a mutation in CRELD1 (cysteine-rich with EGF-like domain protein 1, 420 aa). Other forms are associated with mutations in *connexin 43 (heart connexin) and with *Down's syndrome.

atrium (*pl.* atria) Can refer to any body cavity but is usually applied to the upper (inlet) chambers of the heart.

atrophin Atrophin 1 (ATN1, dentatorubral-pallidoluysian atrophy protein, 1185 aa) is expressed in all tissues, but is proteolytically cleaved in neuronal cells and may be a transcriptional corepressor. The poly-Gln region of ATN1 is highly polymorphic (7–23 repeats) in the normal population and is expanded to about 49–75 repeats in *dentatorubral-pallidoluysian atrophy. **Atrophin-1-related protein** (arginine-glutamic acid dipeptide repeats protein, 1566 aa) acts as a a transcriptional repressor during development and interacts with atrophin 1, the interaction being stronger if there are more repeats. There are a number of proteins that interact with atrophin including *MAGI (**ATN interacting protein 1**) and E3 ubiquitin-protein ligase Itchy homologue (**ATN-interacting protein 4**).

atrophy The wasting away of tissue with loss of mass and/or function. *Compare* APLASIA, in which the tissue does not develop.

atropine An alkaloid, isolated from deadly nightshade *Atropa belladonna,* that inhibits *muscarinic acetylcholine receptors. If topically applied to the eye, causes dilation of the pupil, enhancing beauty; hence 'bella donna' (*Ital.* beautiful woman).

attachment plaques Specialized structures that attach the ends (telomeres) of chromosomes to the nuclear envelope. They develop during early leptotene of meiosis and cluster at the leptotene/zygotene transition (causing the formation of a 'bouquet' of chromosomes).

attention deficit (hyperactivity) disorder (ADD/ADHD) A common childhood-onset behavioural disorder in which there is persistent inattention and/or hyperactive-impulsive behaviour. Boys are more likely to be affected than girls. At least six loci are associated with susceptibility.

attenuation The gradual loss of virulence that occurs if viruses are *passaged extensively. This can be used to generate vaccines.

ATX *See* AUTOTAXIN.

A-type channels Rapidly inactivating voltage-gated potassium channels (KCNA4), of the *shaker-related subfamily. They are activated by depolarization but only after a preceding hyperpolarization and are important for repetitive firing of cells at low frequencies.

auditory canal The duct (external auditory meatus) that connects the eardrum to the outside.

auditory disorders

Deafness can arise through defects in conduction through the outer or middle ear (conductive deafness), even though the sound-detection system is normal. A common cause is *otosclerosis. *See also* ANKYLOSIS. Sensorineural deafness is a result of insensitivity of the inner ear, the cochlea, or dysfunction in the auditory nervous system. Cochlear defects can arise through mutations in various genes, including those for *barttin, *catechol-*O*-methyltransferase, *claudin-14, *cochlin, *connexin, *cryopyrin (Muckle–Wells syndrome), *diaphanous-related formins, *espin, *harmonin, *M-channels, *otoferlin, *pejvakin, *radixin, *V-type ATPase, *whirlin, and *wolframin. Deafness can also be a feature of disorders that affect multiple systems, e.g. *Alport's syndrome, *CHARGE syndrome, *hypoparathyroidism, *Jervell and Lange-Nielsen syndrome, *LEOPARD syndrome, *Mohr–Tranebjaerg syndrome, *multiple epiphyseal dysplasia, *oculodentodigital dysplasia, *Pendred's syndrome, *Usher's syndrome, *Waardenburg's syndrome, and *Woodhouse–Sakati syndrome. Some of these conditions affect structural proteins, other are involved in ion channels or synaptic structure. The mouse mutant *diminuendo* is interesting because it affects a *miRNA. *See also* TUBBY.

AUF-1 A cytoplasmic protein (AU-rich element RNA-binding protein 1, 355 aa) involved in the post-transcriptional regulation of mRNA containing *AU-rich elements. There are four alternatively spliced AUF1 isoforms with different binding affinities.

AUG The *codon in *messenger RNA that specifies initiation, or within a chain, incorporation of a *methionine residue.

Augmentin Proprietary name for a mixture of *amoxicillin and *clavulanic acid (British Approved Name co-amoxiclav).

a

aura The sensation that precedes an epileptic seizure or migraine attack.

AU-rich element (ARE) Regions of RNA that are rich in adenine and uridine, usually located in the 3′ untranslated region (3′ UTR), that mark the mRNAs for rapid destruction. AREs are common in mRNA for cytokines and transcription factors. *See* AUF-1.

aurintricarboxylic acid A general inhibitor of nucleases, *restriction endonucleases, *topoisomerase, and protein–nucleic acid interactions. It inhibits platelet aggregation, apoptosis, viral neuraminidase, the replication of coronaviruses (e.g. *SARS) and various other biological activities. The ammonium salt is known as **aluminon**.

auristatins Inhibitors of tubulin polymerization derived synthetically from *dolastatin.

aurora kinases A family of centrosomal serine/threonine kinases (aurora A, B, and C) that are essential for spindle assembly, centrosome maturation, chromosomal segregation and cytokinesis during mitosis. Abnormalities in the mitotic process through overexpression/ amplification of aurora kinase have been linked to genomic instability. Overexpression of aurora kinases occurs in a wide range of tumours and has been implicated in oncogenic transformation. **Aurora B** (AIM-1, 344 aa) may be directly involved in regulating the cleavage of polar spindle microtubules and is a key regulator for the onset of cytokinesis. Inhibitors may have potential as antitumour drugs. *See* HEF1; TACCs.

aurosome A secondary lysosome containing gold, found only in patients treated with gold complexes.

aurovertins A family of related antibiotics from the fungus *Calcarisporium arbuscula* that inhibit the proton-pumping F-type *ATP synthase in mitochondria and in many bacterial species. Aurovertins B and D have identical biological properties and are more potent than aurovertin A.

auscultation Describing the act of listening to the sounds in the body.

Australia antigen An aggregate of the envelope proteins of *hepatitis B virus (HBsAg) found in serum of patients during phases of high infectivity.

autacoids Describing hormones that act locally rather than systemically. Examples are *angio-tensin, *eicosanoids, *histamine, *serotonin.

autapses (*adj.* autaptic) Synapses that a neuron forms with itself. They are found in interneurons of the central nervous system and are probably a feedback control mechanism.

autism A common shorthand for **infantile autism**, a pervasive developmental disorder, usually apparent by 3 years of age, characterized by limited or absent verbal communication; a lack of reciprocal social interaction or responsiveness; and restricted, stereotypical, and ritualized patterns of interests and behaviour. **Autism spectrum disorder** is a broader and less severe phenotype that includes Asperger's syndrome. A dozen or more loci have been identified as associated with susceptibility.

autoantibody *See* AUTOIMMUNE DISEASES.

autocatalytic Describing a reaction in which the product is also the catalyst or an enzymic reaction in which the product activates the enzyme.

autochthonous Indigenous. Describing something that is in the same place as that in which it was formed.

autoclave A sterilization apparatus, basically a large pressure cooker, that uses high-pressure steam to achieve a temperature of around 121 °C which is maintained for long enough (15–20 min) to kill even resistant bacterial spores. Also, as a verb, the use of such an apparatus.

autocoid *See* AUTACOIDS.

autocrine Describing a substance that stimulates the cell from which it is secreted, e.g. a growth factor which is produced by an activated proto-oncogene and allows the cell to escape normal growth control.

autofluorescence The phenomenon exhibited by some compounds and materials that will fluoresce without the addition of an exogenous fluorophore. Autofluorescence can cause problems in fluorescence microscopy and in assays involving detection of fluorescence as a read-out.

autogenous Describing something that happens without the need for external influence or input.

autograft Tissue transferred from one place to another in an individual and thus eliciting no immune response.

autoimmune diseases

Diseases in which autoreactive T cells or autoantibodies are found, although these can occur asymptomatically. Autoimmunity may

a

be because of a failure to develop self-tolerance by the deletion of self-reactive clones of cells during development or, in an alternative view (in which tolerance is the default condition), may be because normal antigens are presented as though they were dangerous (associated with 'danger' signals through the *toll-like receptors). A third possibility is that bacterial or viral antigens that elicit an immune response are sufficiently similar to self-antigens that there is a cross-reaction (*see* HUMAN ENDOGENOUS RETROVIRUSES). In some cases the autoantigen is clearly identified, in others it remains unknown or merely hypothetical. In some cases immunosuppressive drugs may be used (e.g. *ciclosporin, *methotrexate). Diseases with an autoimmune element include: *antiphospholipid syndrome, *Addison's disease, *Behçet's disease, primary sclerosing *cholangitis, *Churg–Strauss syndrome, *coeliac disease, *CREST syndrome, *cryoglobulinemia, *dermatomyositis, *epidermolysis bullosa acquisita, *Goodpasture's syndrome, *Graves' disease, *Guillain–Barré syndrome, *Hand–Schuller-Christian disease, *Hashimoto's thyroiditis, *Henoch–Schönlein purpura, *inflammatory bowel disease, *macrophage activation syndrome, *myasthenia gravis, *pemphigus, *polyarteritis nodosa, *polyendocrine syndrome, *rheumatoid arthritis, *scleroderma, *systematic sclerosis, *Sjögren's syndrome, *systemic lupus erythematosus, primary *thrombocytopenia purpura, *uveitis, *vitiligo, and *warm antibody haemolytic anaemia. There is an association between *complement deficiencies and *copy number variation in autoimmune disorders, and high levels of *interleukin-12 have also been implicated. *See also* TH17 CELLS and PYPAFs. *Experimental allergic encephalomyelitis, *Heymann nephritis and the *scurfy and *NOD mice are animal models for autoimmune disease.

autointoxication **1.** An outmoded view that ill-health could arise from toxins generated in the gut. **2.** A term occasionally used for suicide by self-administration of poison.

autologous Describing something derived from oneself. *Compare* HETEROLOGOUS; HOMOLOGOUS.

autolysis The release of lysosomal enzymes within a cell that leads to lysis of organelles, cells and eventually tissue. Occurs postmortem; limited autolysis occurs during the hanging of meat.

autonomic nervous system The part of the peripheral nervous system that is not under conscious control, subdivided into the *sympathetic and *parasympathetic nervous systems. The autonomic system maintains homoeostasis by regulating heart rate, breathing, sweating, etc.

autophagy The fusion of organelles with lysosomes, forming a *secondary lysosome in which they are digested. A common process during development and in senescence.

autophosphorylation The phosphorylation of a protein kinase by itself. This may occur following interaction with another protein and can activate the kinase.

autoradiography A technique used to locate radioactively labelled substances in a thin section of tissue or in a gel. The sample is overlaid by photographic emulsion and left for a time. When the emulsion is developed silver grains grow from point sources where radiation has interacted with the silver halide. The resolution depends on the path length of the radiation; tritium, which emits a low-energy β-radiation, is favoured.

autoregulation Feedback regulation, e.g. regulation of a gene encoding a *transcription factor by its own product.

autosomal dominant An allele on one of the *autosomes that has an effect when present as a single copy (heterozygous) and is dominant over the other allele. Can be the result of a *gain-of-function mutation.

autosomal dominant cerebellar ataxia *See* SPINOCEREBELLAR ATAXIA.

autosomal recessive An autosomal allele that is only deleterious if homozygous.

autosomal recessive spastic ataxia of Charlevoix-Saguenay (ARSACS) An early-onset spastic ataxia particularly prevalent in the Charlevoix-Saguenay–Lac-Saint-Jean region of Quebec. There is defective sensory nerve conduction, reduced motor nerve velocity, and hypermyelination of retinal neurons. The cause is mutation in the gene encoding *sacsin.

autosomes Chromosomes other than the sex chromosomes.

autotaxin A protein, originally named for its ability to stimulate motility in melanoma cells, that is in fact *lysophospholipase D (EC 3.1.4.39, 863 aa) which works through production of *lysophosphatidic acid (LPA). It is involved in angiogenesis, by stimulating migration of smooth muscle cells, and in neurite outgrowth.

autotroph An organism that can synthesize complex organic compounds from simple inorganic molecules using energy derived from oxidation of inorganic molecules (*chemotrophy) or from light (photosynthesis).

auxilin A protein tyrosine phosphatase (EC 3.1.3.48, 913 aa) that promotes uncoating of clathrin-coated vesicles.

auxotroph A mutant organism that requires a nutritional supplement for growth, unlike the wild type. A deficiency mutant.

auxotyping A method that uses nutritional requirements as markers to distinguish strains of bacteria. Originally developed for testing strains of *Neisseria*.

avascular Not connected to the blood vascular system.

avascular necrosis of femoral head A debilitating disease that usually leads to destruction of the hip joint in the third to fifth decade of life. It is caused by mutation in the collagen COL2A1 gene that is also mutated in *Perthes' disease.

avermectins (abamectin) A series of compounds that are potent acaricides, insecticides, and anthelmintics. Originally isolated from *Streptomyces avermitilis.*

avian erythroblastosis virus *See* AVIAN LEUKAEMIA VIRUS.

avian influenza (avian flu, bird flu) A very contagious strain of influenza that affects poultry and can switch species to infect humans. This was the source of the highly virulent 1918 pandemic strain, and the *H5N1 strain is currently (2009) causing considerable concern.

avian leukaemia virus A group of C-type RNA tumour viruses (*Oncovirinae) responsible for various avian leukaemias and tumours. Some are replication-defective and require helper viruses, although the avian lymphatic leukaemia viruses (ALV) are replication-competent. Avian erythroblastosis virus (AEV) has two transforming genes, v-*erbA* and v-*erbB* (c-*erbB* is the gene encoding the *epidermal growth factor receptor. Avian myeloblastosis virus (AMV) carries v-**myb* and causes a myeloid leukaemia; avian myelocytomatosis virus (AMV) carries v-**myc*.

avidin A protein (152 aa) from egg white that binds *biotin with very high affinity and can potentially cause biotin deficiency. The strength of the interaction is often used experimentally, e.g. to affinity-purify biotinylated proteins on an avidin affinity column.

avidity A measure of the strength of a binding interaction, usually one in which there are multiple weak interactions which cooperate in producing strong interaction. The commonest use is in describing the binding interaction of antibody with antigen. *Compare* AFFINITY which is used for simple receptor–ligand systems.

avilamycin An antibiotic used as an additive in animal feeds. *See* EVERNINOMICINS.

avirulent Describing an organism or virus that does not cause infection or disease.

avitaminosis A vitamin deficiency, usually specified (e.g. avitaminosis C).

AVSD *See* ATRIOVENTRICULAR SEPTAL DEFECT.

avulsion Tearing away.

axenic Describing an environment containing only one species. An **axenic culture** contains only one type of organism.

axial filaments The filaments of the periplasmic flagella of spirochaetes, polymers of proteins with homology to *flagellin, that rotate within the periplasmic space and cause the whole bacterium to flex like a corkscrew.

axial tomography *See* COMPUTED TOMOGRAPHY.

axilla (*adj.* axillary) The armpit.

axins Proteins (axin-1, 862 aa; axin-2, 843 aa) that control dorsoventral patterning by down-regulating β-*catenin to inhibit the *Wnt signalling pathway, ventralizing the embryo, and activating a Wnt-independent JNK signalling pathway that dorsalizes. Mutations in axin-1 are associated with hepatocellular carcinoma, and greater methylation of the promoter is reported in a case of a caudal duplication anomaly. Defects in axin-2 cause oligodontia–colorectal cancer syndrome in which there is severe tooth agenesis and colorectal cancer. Axin interaction partner and dorsalization antagonist (AIDA, 306 aa) inhibits the axin-mediated JNK activation by disrupting the axin homodimerization that is required.

***axl* oncogene** An oncogene encoding a receptor tyrosine kinase (EC 2.7.10.1, 887 aa) for which gas 6 (*see* GAS GENES) is a ligand. May mediate interactions between mesodermally-derived cells. It is overexpressed in chronic myelogenous leukaemia.

a

axolemma The *plasma membrane of an axon.

axon The long process extending from the cell body of a neuron that carries outgoing action potentials towards the target cell. *See* DENDRITE. The **axon hillock** is the tapering region between the cell body of a *neuron and the axon in which depolarizations from dendritic inputs are 'added up' and an action potential is initiated if the critical threshold is exceeded. **Axonogenesis** is the production of axonal processes by developing neurons.

axonal guidance Describing the various processes involved in directing axons to their correct target locations during development or regeneration. The guidance is achieved by regulating the motility of the nerve *growth cone.

axoneme The central *microtubule complex of eukaryotic *cilia and flagella that has a '9 + 2' arrangement of tubules in cross-section.

axonin *See* CONTACTINS.

axoplasm The *cytoplasm within an *axon.

azacytidine *See* AZACYTOSINE.

azacytosine An analogue of *cytosine, in which carbon 5 is replaced by a nitrogen. Azacytidine (5-azacytidine, β-ribofuranosyl 5-azacytidine) is the ribonucleoside. 5-Aza-2′-deoxycytidine was developed as a potential antineoplastic drug which can become incorporated into DNA and inhibits DNA methylation. Experimentally often used as an inhibitor of methylation.

azaserine A *glutamine analogue that irreversibly inhibits glutamine phosphoribosyl amidotransferase, an important enzyme in synthesis of purines. It competitively inhibits pathways in which glutamine is metabolized and has antitumour properties.

azathioprine An immunosuppressant drug that inhibits purine synthesis and is converted in the body to the active metabolites 6-mercaptopurine and 6-thioinosinic acid. *TNs* Azamun, Azasan, Imuran, Imurel.

azide Shorthand term, usually for the sodium salt (NaN_3), which inhibits electron transport and is often used to prevent microbial growth in biological materials that cannot be frozen.

azidothymidine *See* ZIDOVUDINE.

azithromycin A broad spectrum macrolide antibiotic. *TN* Zithromax.

azoospermia Complete absence of spermatozoa in semen. Nonobstructive spermatogenic failure is most commonly caused by interstitial deletions on the Y chromosome, particularly deletions of ubiquitin-specific protease 9 (EC 3.1.2.15, 2555 aa). A form of azoospermia involving perturbation of meiosis is caused by mutation in the gene encoding *synaptonemal complex protein 3 (236 aa).

azotaemia An excess of nitrogenous breakdown products in the blood as a result of kidney dysfunction.

azothioprine *See* AZATHIOPRINE.

AZT Azidothymidine. *See* ZIDOVUDINE.

azurin **1.** An intensely blue copper-containing protein from *Pseudomonas aeruginosa*.
2. A histochemical dye.

azurophil granules The primary lysosomes (primary granules), containing a range of hydrolytic enzymes, found in *neutrophil. *Compare* SPECIFIC GRANULES.

β (beta) The β prefix is either given as 'beta' or ignored and the main portion of the name used as the headword.

B7 A superfamily of costimulatory molecules that bind to CD28 or *CTLA-4 and regulate T-cell responses. They are part of the *immunoglobulin superfamily and have an extracellular domain related to variable and constant immunoglobulin domains. B7-1 is CD80, B7-2 is CD86. Other members are inducible costimulator ligand (ICOS-L), programmed death-1 ligands 1 and 2 (PD-L1, PD-L2), B7-H3, and B7-H4.

Babesia A genus of protozoa that are intracellular parasites found in erythrocytes. The disease (babesiosis, piroplasmosis), a haemolytic fever similar to malaria, is transmitted by ticks.

baby hamster kidney cells *See* BHK CELLS.

bacillaemia The presence of bacilli in the blood.

bacille Calmette–Guérin (BCG) An attenuated strain of bacteria derived from *Mycobacterium tuberculosis*, used in vaccination against tuberculosis. Extracts of the bacterium are potent *adjuvants.

bacilluria *See* BACTERURIA.

bacillus A cylindrical (rod-shaped) bacterium, generally 0.5–1.0 μm long and 0.3–1 μm wide.

Bacillus cereus A Gram-positive, facultatively aerobic spore-forming bacterium that is responsible for some cases of food poisoning. A diarrhoeal type of illness is caused by a high molecular weight protein, the vomiting (emetic) type by a low molecular weight, heat-stable peptide.

Bacillus megaterium A large Gram-positive soil bacterium found in the soil and extensively used in biotechnology because of its size and cloning abilities.

bacitracins Branched cyclic peptides produced by strains of *Bacillus licheniformis* that interfere with the synthesis of *peptidoglycan by Gram-positive bacteria.

BAC library *See* BACTERIAL ARTIFICIAL CHROMOSOME.

baclofen A *GABA derivative that is used as a skeletal muscle relaxant to relieve muscle spasm in trauma, multiple sclerosis and cerebral palsy. Selectively binds $GABA_B$ receptors. *TNs* Kemstro, Lioresal.

bacteraemia The presence of live bacteria in the blood, usually in fairly small numbers, in contrast to *septicaemia.

Bacteria (formerly **Eubacteria)** A major subdivision of the prokaryotes that includes most Gram-positive bacteria, cyanobacteria, mycoplasmas, enterobacteria, and pseudomonads. They are distinguished from the *Archaea (formerly Archaebacteria) by the presence of ester-linked lipids in the cytoplasmic membrane, *peptidoglycan in the cell wall, and the absence of *introns. Some taxonomists still favour a two-domain classification (eukaryotes and prokaryotes) but the three-domain scheme (Eukaryota, Bacteria, and Archaea) is now generally accepted. Bacteria are small (linear dimensions ~1 μm), noncompartmentalized, with circular DNA, and 70S ribosomes. Protein synthesis differs from that in eukaryotes, and many antibacterial antibiotics interfere with protein synthesis, but do not affect the infected host.

bacterial artificial chromosome (BAC) A type of vector used to construct a *genomic library, that has the sites required for it to be handled and replicated as a bacterial chromosome. The vector can contain long sequences (up to 200 kbp). *Compare* YEAST ARTIFICIAL CHROMOSOME.

bacterial cell wall The rigid wall that surrounds most bacteria and is essential for their survival. Cell walls that take up Gram stain (those of the Gram-positive bacteria) are ~50 nm thick and are composed of *teichoic acid and *peptidoglycan. The walls of Gram-negative bacteria are separated from the cell membrane by the periplasmic space, and are much thinner, contain a different peptidoglycan, and have an outer lipid bilayer (containing lipid A) resembling a membrane. Some bacteria, notably *mycoplasmas, do not produce a cell wall. *Archaea have cell walls but the peptidoglycan composition is different. Many antibiotics interfere with cell wall synthesis.

bacterial chemotaxis The response exhibited by motile bacteria in a gradient of attractant or repellent. Bacterial movement involves straight runs and brief tumbles during which the direction of flagellar rotation reverses and the bacterium changes direction. If the concentration of an attractant is increasing, the probability of a tumble is reduced and the bacterium will persist up-gradient. The converse is true if the concentration is decreasing, and the net effect is to bias movement towards the source of the attractant. Purists would therefore consider this a *klinokinesis with adaptation.

bacteriocide A substance that kills bacteria.

bacteriocins Bacterial *exotoxins, often encoded on *plasmids, that kill other bacteria but not eukaryotic cells. *See* COLICINS; LANTIBIOTICS.

bacteriophages (phages) Viruses that infect bacteria. The virulent phages take over all the machinery for nucleic acid and protein synthesis which becomes devoted entirely to production of 100–200 new viruses which self-assemble and are released when the host cell ruptures under the influence of virus-encoded lysozyme. The temperate phages, such as *lambda, may also show this lytic cycle when they infect a cell, but more frequently they induce *lysogeny in which the virus persists without killing the host. Lambda, a coliphage that infects *E. coli*, is extensively used as a *vector in recombinant DNA studies.

bacteriorhodopsin A photoprotein (248 aa), similar to *rhodopsin, found in 'purple patches' in the cytoplasmic membrane of the bacterium *Halobacterium halobium*. It has seven transmembrane helices, and acts as a light-driven proton pump which generates a proton gradient used (among other things) to drive the *chemiosmotic synthesis of ATP.

bacteriostatic Describing a substance that will inhibit bacterial proliferation without necessarily killing them.

bacteruria The presence of bacteria in the urine. Bacilluria if they are bacilli.

Baculovirus A virus that infects lepidopteran larvae, widely used as an eukaryotic *expression vector for proteins that need to be *post-translationally modified.

Bad A proapoptotic member of the *Bcl-2 family (Bcl2 antagonist of cell death, 168 aa) that is regulated by phosphorylation and seems to act as a link between growth factor receptor signalling and the apoptotic pathways.

BAF *See* BARRIER-TO-AUTOINTEGRATION FACTOR.

BAF complex A multisubunit complex in mammals that is functionally related to the *SWI/SNF complex found in *Drosophila* and involved in the chromatin remodelling that occurs during development. It consists of 9–12 subunits, (BRG1/brm-associated factors, BAFs), some of which are homologous to SWI/SNF subunits.

bafilomycin A macrolide antibiotic from *Streptomyces griseus* that inhibits *V-type ATPase and blocks lysosomal cholesterol transport in macrophages.

bagassosis A type III *hypersensitivity response, similar to *farmer's lung, caused by inhalation of dust from mouldy sugar cane.

BAG family A family of proteins (*Bcl-2 associated athanogene products) that act as cochaperones by linking molecules that recruit molecular chaperones to target proteins. **BAG-1** (345 aa, three major isoforms p50, p46, and p36 and a minor fourth, p29) facilitates ubiquitin-proteasome-mediated protein degradation. It is antiapoptotic and potentiates the antiapoptotic functions of *Bcl-2 to which it binds. It is deregulated in a variety of malignancies. At least four other BAG proteins are known. **BAG4** (silencer of death domains, SODD, 457 aa) associates with the TNF receptor 1 (TNFR1) but is displaced if TNF binds. **BAG5** (447 aa) inhibits *parkin and enhances dopaminergic neuron degeneration.

Bak A proapoptotic *bcl-2 family member (Bcl-2 homologous antagonist/killer, 211 aa), a mitochondrial membrane protein that can oligomerize and trigger the release of cytochrome *c*.

bakuchiol An inhibitor of tyrosine phosphatase 1B isolated from seeds of *Psoralea corylifolia* (a Chinese tree; used in traditional medicine). It is an antioxidant, inhibits mitochondrial lipid peroxidation and inducible nitric oxide synthase (iNOS/NOS II) expression, and has antimicrobial and cytotoxic activity.

BAL *See* DIMERCAPROL.

balano- Prefix relating to the glans penis. **Balanitis** is inflammation of the glans, **balanorrhagia** is a gonorrhoeal infection of the glans that discharges pus.

BALB/c An inbred strain of white (albino) mice, the source of one of the common 3T3 cell lines.

balloon cell A cell with abundant pale-staining cytoplasm. Can be a result of viral infection, as in viral hepatitis, or as a result of some storage diseases. The **balloon cell naevus** is characterized by large, vacuolated naevus cells with clear cytoplasm and central hyperchromatic nuclei, formed by degeneration of melanosomes. Balloon cells are found in various melanocytic proliferations, particularly intradermal melanocytic naevi and melanoma and also in dysplastic naevus.

Balo's concentric sclerosis A rare demyelinating disease (leucoencephalitis periaxialis concentrica) considered to be a variant of *multiple sclerosis in which the white matter of the brain is destroyed in concentric layers.

SEE WEB LINKS

- Article entitled 'Balo's concentric sclerosis: clinical and radiologic features of five cases' in *American Journal of Neuroradiology* (2001).

BALT *See* BRONCHUS ASSOCIATED LYMPHOID TISSUE.

Bamforth–Lazarus syndrome *See* HYPOTHYROIDISM.

BAN (British Approved Name) The formal name for a medicinal substance, usually identical to the recommended International Non-Proprietary Name (rINN). In this dictionary the BAN has been used as the main headword, with the former name as a synonym if necessary.

band 4.1 domain *See* FERM DOMAIN; PROTEIN 4.1.

band cells Immature *neutrophils released from the bone-marrow reserve if there is acute demand.

band III A dimeric transmembrane protein (911 aa), one of the solute carrier family 4 (member 1), the major anion transporter of the erythrocyte. Interacts with other *erythrocyte membrane proteins, including *ankyrin.

banding patterns The transverse bands of different staining intensity on chromosomes. The pattern is characteristic for each chromosome and the same in most tissues. **Q-bands** are produced by staining with quinacrine; **G-bands** by staining with *Giemsa. **R-bands** are the regions that are lightly-stained by Giemsa. Each band covers about 5–10% of the length of the chromosome, about 10^7 base pairs.

BAPTA A fluorescent compound that chelates calcium but has low affinity for magnesium. The absorption maximum changes when calcium is bound, so it can be used to measure intracellular calcium concentrations, although the chelation may perturb the system. *See* MAPTAM.

Barakat's syndrome *See* HYPOPARATHYROIDISM.

barbital (barbitone, veronal) A *barbiturate, usually used as the sodium salt.

barbiturates A class of drugs that depress the activity of the central nervous system and have sedative and hypnotic effects. Largely superseded by *benzodiazepines.

BARD1 A protein (BRCA1-associated RING domain protein 1, EC 6.3.2.-, 777 aa) that forms a heterodimeric complex with BRCA1 that mediates ubiquitin E3 ligase activity and coordinates a range of cellular pathways such as DNA damage repair, ubiquitination and transcriptional regulation. Also forms a heterodimer with CSTF1/CSTF-50, which modulates mRNA processing. Has been reported to have tumour suppressive functions independent of BRCA1.

Bardet–Biedl syndrome A genetically heterogeneous disorder in which there are renal abnormalities, retinal dystrophy, polydactyly, mental retardation, and mild obesity. At least twelve variants of the syndrome are known, with deletions on various chromosomes. The affected proteins are known in some, but not all, cases and are all associated with the basal body and cilia of the cell; the mutations affect ciliary function.

bariatrics The branch of medicine that deals with obesity.

baroreceptor (baroceptor) A sense organ that responds to pressure. High-pressure arterial baroreceptors respond to pressure in arteries whereas low-pressure baroreceptors (cardiopulmonary or volume receptors) are found in large veins and in the walls of the right atrium of the heart and are involved with the regulation of blood volume.

BAR proteins A superfamily of adaptor proteins (BIN1/amphiphysin/Rvs domain proteins) with a common conserved N-terminal BAR domain, that appear to integrate signal transduction pathways that regulate membrane dynamics, the F-actin cytoskeleton, and nuclear

b

processes. They all bind to membranes but differ in their preferences for membrane curvature. BAR/N-BAR modules bind to membranes of high positive curvature, **F-BAR** modules bind to a different range of positive membrane curvatures and *I-BAR* modules bind to negatively curved membranes. **BIN1** (*see* AMPHIPHYSIN) interacts with and inhibits the oncogenic properties of *myc. **BIN2** (546 aa) is mostly found in haematopoietic tissues, lacks tumour suppressor functions, and interacts with BIN1, but not amphiphysin. **BIN3** (253 aa) has only the BAR domain and is widely expressed. (Rvs is a yeast protein.)

SEE WEB LINKS

- Additional information on the BAR superfamily.

Barr body *See* SEX CHROMOSOMES.

barrier-to-autointegration factor A small conserved protein (BAF, breakpoint cluster region protein 1, 89 aa) that has an important role in nuclear assembly, chromatin organization, gene expression, and gonad development. It is exploited by retroviruses for inhibiting self-destructing autointegration of retroviral DNA.

Bartonella A genus of Gram-negative bacteria that are facultative intracellular parasites and opportunistic pathogens. They are transmitted by ticks, fleas, mosquitoes, etc., and various species cause a range of human illnesses ranging from cat scratch disease and benign lymph node enlargement to life-threatening systemic diseases. The commonest species causing infection are *B. henselae*, *B. quintana*, and *B. bacilliformis*.

Bartter's syndrome A group of disorders in which there is autosomal recessive transmission of impaired salt reabsorption in the thick ascending loop of Henle. Classic Bartter's syndome (**Bartter's syndrome type 3**) is caused by mutation in the kidney chloride channel B gene (*CLCNKB*). Antenatal **Bartter's syndrome type 1** is caused by mutation in the *SLC12A1* gene (solute carrier family 12 member 1; sodium-potassium-chloride transporter-2) and **type 2** by mutation in the *ROMK potassium channel (*KCNJ1*). **Infantile Bartter's syndrome** with sensorineural deafness (type 4) is caused by mutation in the *barttin gene or by simultaneous mutation in both the *CLCNKA* and *CLCNKB* genes that encode chloride channels.

barttin A protein (320 aa) found in the basolateral membranes of renal tubules and in epithelia of the inner ear that functions as an essential β subunit for chloride channels CLCKNA and CLCKNB. It enhances insertion into the plasma membrane and regulates the permeation and gating of the channel. Mutations can cause *Bartter's syndrome type 4 in which there is impaired renal ion absorption and sensorineural deafness.

basal body (kinetosome) The structure found at the base of eukaryotic *cilia and *flagella which is the extension of the nine outer sets of axonemal *microtubules with the addition of a C-tubule to form a triplet, as in the *centriole. In some protists the basal body is anchored by a system of fibres extending into the cytoplasm (the rootlet system).

basal cells The relatively undifferentiated stem cells of an epithelial sheet that give rise, by an unequal division, to another basal cell and a cell that progressively differentiates as it is displaced outward by cells produced in later divisions. In skin the basal cells of the epidermis (stratum basale) produce cells that progress through the spinous, granular, and horny layers, becoming progressively more keratinized, the outermost being shed as *squames. A **basal cell carcinoma** (BCC, rodent ulcer) is a relatively common *carcinoma, rarely metastatic, derived from the basal cells and often a result of exposure to sunlight, especially in people with fair skin. *See also* MYOEPITHELIAL CELL.

basal ganglia (basal nuclei) A group of distinct nuclei in the brain, including the putamen, the *caudate nucleus, and the globus pallidus, that are associated with motor control, cognition, emotions, and learning. Lesions of the basal ganglia occur in a variety of motor disorders including *Parkinson's disease and *Huntington's disease.

basal lamina The layer of extracellular matrix immediately below an epithelium and produced by the epithelial cells. Major components are *laminin and collagen type IV. *See* BASEMENT MEMBRANE; RETICULAR LAMINA.

basal metabolic rate The BMR is the amount of energy being expended when in a state of complete rest, but not asleep, and not involved in digestion. A simpler measurement is the resting metabolic rate.

BASC complex A multiprotein complex (BRCA1-associated genome surveillance complex) composed of *BRCA1, tumour suppressors and the DNA damage repair proteins MSH2, MSH6, MLH1, *ATM, BLM, and the *MRN complex (MRE11, RAD50, and *nibrin). It may coordinate the multiple activities required to maintain genomic integrity during DNA replication.

Basedow's disease Thyrotoxicosis. *See* HYPERTHYROIDISM.

base excision repair A DNA repair mechanism used to remove mutated bases. DNA glycosylase excises the mutated base but creates an apurinic/apyrimidinic (AP) site; this is recognized by AP endonuclease which cuts the strand upstream (5′) of the mutated base and polymerase I can then add a new (correct) base to the free 3′ end by pairing with the undamaged strand. Ligase then seals the join. In short-patch repair a single nucleotide is replaced; in long-patch repair 2–10 nucleotides are replaced.

basement membrane The layer of extracellular matrix material underlying an epithelium; not a membrane in the cell-biological sense. There are two distinct layers: the *basal lamina, immediately adjacent to the cells, and the *reticular lamina which is produced by fibroblasts of the underlying *connective tissue and contains fibrillar collagen.

base pairing The hydrogen bonding between *purines and *pyrimidines that leads to the formation of a duplex from two complementary single strands of nucleic acid. In DNA the pairs are the purines *adenine and *thymine, and the pyrimidines *guanine and *cytosine. In RNA thymine is replaced by *uracil.

basic leucine zipper proteins A family of transcription factors (bZIP proteins) with a basic DNA-binding domain and a *leucine zipper region involved in forming homo- or heterodimers. Examples include *AP-1, *ATF, and *CREB.

basilar membrane A thin layer of tissue that separates the cochlea from the scala tympani in the ear and that provides the base upon which the sensory hair cells are arranged. The complex physical properties of the basilar membrane are important in discriminating sounds.

basiphil *See* BASOPHIL.

basket cells **1.** Inhibitory cerebellar interneurons with many small dendritic branches that enclose the cell bodies of adjacent *Purkinje cells in a basket-like array. Similar cells are found in the hippocampus and cortex. **2.** *See* MYOEPITHELIAL CELL.

basolateral plasma membrane The plasma membrane of epithelial cells that is adjacent to the *basal lamina or to the surrounding cells. It is isolated from the *apical plasma membrane by *tight junctions which prevent diffusion of membrane proteins between the two domains. The phospholipid composition is also different.

basophil A *granulocyte with large heterochromatic basophilic granules containing *histamine bound to a protein and a heparin-like mucopolysaccharide matrix. They are nonphagocytic and resemble *mast cells, although it is not clear whether they have a common lineage.

basophilia **1.** An affinity for basic dyes. **2.** An excess of *basophils in the blood.

bassoon A large protein (3926 aa) involved in organizing (orchestrating) events at the presynaptic terminal along with others such as *piccolo. It is also involved in the cytoskeleton underlying the concentrations of vesicles at *ribbon synapses.

Bateman function An equation used in toxicology to express the build-up and decay in concentration of a substance, based on first-order uptake and elimination in a one-compartment model.

batrachotoxins A group of alkaloid neurotoxins, from the Columbian poison frog *Phyllobates* sp., that increase the permeability of sodium channels. More potent than curare.

batroxostatin *See* DISINTEGRIN.

battenin *See* BATTEN'S DISEASE.

Batten's disease A severe autosomal recessive neurodegenerative disorder of childhood (juvenile neuronal ceroid lipofuscinosis, JNCL, Spielmeyer–Vogt–Sjogren disease) characterized by progressive loss of vision, seizures, and psychomotor disturbances. There is lysosomal accumulation of hydrophobic material, mainly ATP synthase subunit C. It is caused by mutation in the *CLN3* gene which encodes **battenin** (438 aa), a multipass membrane protein. Battenin is evolutionarily conserved and localizes in lysosomes and/or mitochondria; the *S. pombe* homologue regulates vacuole homeostasis.

bavachin An inhibitor of acyl coenzyme A: cholesterol acyltransferase (ACAT), isolated from the herb *Psoralea corylifolia*. It is a weak antioxidant and reportedly stimulates bone formation.

bax A protein (192 aa) related to *bcl-2 that has a role in regulating apoptosis. Homodimers promote apotosis but heterodimers with Bcl-2 or Bcl-XL block cell death. Widely expressed and in multiple isoforms.

b

Bayesian statistics A statistical method, based on Bayes' decision rule, that provides an approach to decision-making based on relative payoffs of different outcomes. Bayes' theorem provides a mechanism for combining a prior probability distribution for a parameter with new sample information, the combined data giving a revised probability distribution, which can then be used as a prior probability with a future new sample, and so on. Some conventional statisticians consider the method unreliable, but it is often used (e.g.) in genetic counselling.

b–c1 complex Two cytochromes that accept electrons from *ubiquinone and pass them to *cytochrome *c* in the mitochondrial *electron transport chain.

BCA **1.** A CXC chemokine (CXCL13, B-lymphocyte chemoattractant, BLC) that regulates B-cell migration in lymphoid tissues and is the ligand for CXCR5. **2.** A protein estimation method, the benzethonium chloride assay (BCA).

BCAR3 A protein (breast cancer antioestrogen resistance-3, novel SH2-containing protein 2, NSP2, 825 aa) that is expressed in breast carcinoma cells that have developed *tamoxifen resistance. The mouse homologue (AND34, p130Cas-binding protein) is thought to be an adaptor protein that couples activated growth factor receptors to signalling molecules that regulate src-kinase activity and promote cell migration.

BCECF A fluorescent dye used to measure intracellular pH.

B-cell receptor The receptor on the surface of a B cell that binds an antigen. It is composed of two identical immunoglobulin heavy chains, two identical immunoglobulin light chains and a signalling subunit which is a heterodimer of the Ig-α (CD79a) and Ig-β (CD79b) proteins. Each B cell binds only one antigen; diversity exists in the population. Antigen binding stimulates a tyrosine kinase cascade that involves *lyn, *btk, *syk, etc., and triggers proliferation and the production of *plasma cells and memory B cells.

SEE WEB LINKS

- Signalling pathway shown diagramatically.

B cells **1.** **(β-cells)** *See* beta cells. **2.** Common name for B lymphocytes.

BCG *See* BACILLE CALMETTE–GUÉRIN.

bcl A miscellaneous group of oncogenes associated with B-cell lymphomas. **Bcl-1** (295 aa) interacts with cyclin-dependent kinases and when translocated may be a cause of B-lymphocytic malignancy, particularly mantle-cell lymphoma, multiple myeloma, and parathyroid adenomas. **Bcl-2** is an integral inner mitochondrial membrane protein (239 aa) that blocks the apoptotic death of pro-B-cells by controlling mitochondrial membrane permeability by acting on the VDAC porin channel. There are a family of **Bcl-2-related proteins**, some proapoptotic (Bax, Bad, Bak and Bok among others) or antiapoptotic (Bcl-2 itself, Bcl-xL, Bcl-w, and others). Bcl-2 family proteins are associated with various malignancies. **Bcl-3** (446 aa) interferes with the binding of NFκB to DNA by acting as an IκB homologue and if translocated into immunoglobulin gene regions can cause B-cell chronic lymphocytic leukaemia. **Bcl-6** (706 aa) is a transcriptional repressor. The **bcl-2 homology domain 3** (BH3) is characteristic of proapoptotic proteins such as Bim, Bmf, *Bik, *Bad, *Bid, Puma, Noxa, and Hrk.

bcr The region (breakpoint cluster region) of chromosome 22 that is involved in the *Philadelphia chromosome translocation. *BCR* is one of the two genes in the *BCR-ABL* complex and encodes a 160-kD phosphoprotein (1281 aa, EC 7.11.1) associated with serine/threonine kinase activity. It is also a GTPase-activator for rac1 and cdc42.

BDGF *See* BRAIN-DERIVED NEUROTROPIC FACTOR.

B-DNA The normal structural form of hydrated *DNA, a right-handed helix with about ten residues per turn and with major and minor grooves. The planes of the base pairs are perpendicular to the long axis. *Compare* A-DNA; Z-DNA.

BDNF *See* BRAIN-DERIVED NEUROTROPIC FACTOR.

beaded filaments (beaded-chain filaments) *See* FILENSIN; PHAKININ.

becatecarin A synthetic analogue of the glycoside antibiotic, *rebeccamycin, a potential antitumour drug. Becatecarin intercalates into DNA and stabilizes the DNA-topoisomerase I complex, thereby interfering with the topoisomerase I-catalysed DNA breakage–reunion reaction and initiating DNA cleavage and apoptosis.

Bechterew's syndrome *See* ANKYLOSING SPONDYLITIS.

Becker muscular dystrophy *See* MUSCULAR DYSTROPHY.

Becker's disease *See* MYOTONIA.

Beckwith–Wiedemann syndrome A rare developmental disorder (exomphalos-macroglossia-gigantism syndrome, EMG syndrome) with a complex pattern of inheritance caused by mutation or deletion of imprinted genes within the chromosome 11p15.5 region. Specific genes involved include p57 (*KIP2*), **H19*, and *LIT1* (long QT intronic transcript 1). There are various growth abnormalities, advanced ageing, and predisposition to childhood tumours. *Silver–Russell syndrome is caused by hypomethylation defects at 11p15.

becquerel (Bq) The Système Internationale (SI, MKS) unit of radioactivity, equal to one disintegration per second. Supersedes the curie (Ci). 1 Ci = 37 GBq.

Behçet's disease A disorder in which there are recurrent inflammatory lesions of the mouth, genitalia, and eyes. Both viral and autoimmune aetiologies have been proposed.

beige mouse A mouse strain that may be the murine equivalent of *Chediak–Higashi syndrome. The animals have beige hair, *lymphadenopathy, and giant lysosomal granules in their *leucocytes.

Bell's palsy A paralysis of facial muscles caused by a defect in cranial nerve VII (facial nerve). The cause is unknown, although there may be herpesvirus infection: in most cases it resolves spontaneously.

belt desmosome *See* ADHERENS JUNCTION.

benazepril A prodrug metabolized to benazeprilat, an *ACE inhibitor used to treat hypertension. *TN* Lotensin.

Bence Jones protein A dimer of *immunoglobulin light chains, excreted by the kidney, that are overproduced by neoplastic plasma cells (*myeloma). They are monoclonal and usually have κ rather than λ light chains.

benchmark concentration The concentration of a substance that produces a defined response (the benchmark response). Usually around 5–10% of the full response, but defined statistically.

bendroflumethiazide (formerly **bendrofluazide)** A thiazide *diuretic drug.

Benedict's test A test for glucose and other reducing disaccharides, in which the sugar is oxidized by an alkaline solution containing copper sulphate (Benedict's solution). A deep red copper(I) oxide precipitate is produced.

benign tumour A nonmalignant tumour composed of proliferating (*neoplastic) cells that does not invade locally or *metastasize. From a cell-biological perspective, *growth control has been lost but not positional control. Compression of adjacent tissue may produce a fibrous capsule of compressed tissue. Benign tumours can be life-threatening if the tumour is in a restricted space or causes obstruction.

benperidol A potent neuroleptic drug, used to treat antisocial sexual behaviour. *TN* Anquil.

benzamidine A potent *serine endopeptidase inhibitor.

benzatropine (benztropine) An antimuscarinic drug used in treatment of *Parkinson's disease. *TN* Cogentin.

benzhexol *See* TRIHEXYPHENIDYL.

benzocaine A local anaesthetic often used as a topical pain reliever.

benzodiazepines A class of anxiolytic or hypnotic drugs used widely but now recognized to be addictive. They act on inhibitory *$GABA_A$ receptors in the central nervous system. Common examples are diazepam (*TN* Valium), nitrazepam (*TN* Mogadon), and chlordiazepoxide (*TN* Librium).

benzonatate A drug that anaesthetizes stretch receptors in the airways and inhibits the coughing reflex. *TNs* Perles, Tessalon.

benzopyrene A polycyclic aromatic compound that will intercalate into DNA and is a potent mutagen and carcinogen.

benzotropine mesilate *See* BENZATROPINE.

benzyl penicillin The first of the penicillin family of antibiotics, still extensively used to treat streptococcal, gonococcal, and meningococcal infections.

Berardinelli–Seip congenital lipodystrophy *See* CONGENITAL GENERALIZED LIPODYSTROPHY.

beri-beri A *vitamin B_1 (thiamine) deficiency disorder that causes peripheral nerve lesions and/or heart failure.

Berk–Sharp technique (S1 mapping) An early method for genetic mapping in which *messenger RNA is hybridized with single-stranded DNA and *S1 nuclease used to digest nonhybridized regions. The DNA that hybridizes with the mRNA can then be characterized by *restriction mapping.

b

Bernard–Soulier syndrome A disorder (benign macrothrombocytopenia) in which platelets aggregate normally (*compare* GLANZMANN'S THROMBASTHENIA) but do not stick to the collagen of subendothelial basement membrane. Caused by deficiency in platelet membrane glycoprotein Ib (CD42), the receptor for *von Willebrand factor.

beryllicosis (berylliosis) A chronic allergic-type lung disease caused by occupational exposure to beryllium, which was used in early fluorescent lights.

Best's carmine A histochemical stain that gives a deep red colour with glycogen.

Best's disease An early-onset autosomal dominant form of vitelliform *macular dystrophy in which there are large deposits of lipofuscin-like material in the subretinal space, which creates characteristic macular lesions resembling the yolk of an egg (hence 'vitelliform'). It is caused by mutation in the *bestrophin gene.

bestatin A metal-chelating inhibitor of M17 aminopeptidases and a potent, irreversible inhibitor of leukotriene A_4 hydrolase, that has antimalarial activity.

bestrophins A family of anion channels, probably calcium-sensitive chloride channels. Bestrophin 1 (585 aa), the protein that is defective in *Best's disease, is localized to the basolateral plasma membrane of the pigmented retinal epithelium RPE. Three other related proteins (bestrophin-2, -3, and -4) share many structural features but have distinct tissue distributions.

beta-2-microglobulin (β_2-microglobulin) *See* MICROGLOBULIN.

beta-actinin *See* CAPZ.

beta-adrenoceptor blocking drugs (beta-blockers) A group of drugs that inhibit β-*adrenergic receptors. Some are relatively unselective (e.g. propranolol), others act primarily on β_1-receptors (e.g. atenolol). They are used in the treatment of angina, hypertension, migraine, thyrotoxicosis, and anxiety states.

beta-agonists *Sympathomimetic drugs that act on β_2-*adrenergic receptors, used by inhalation to produce a rapid bronchodilatory effect by acting on bronchial smooth muscle. Used to treat the symptoms of asthma; salmeterol is an example.

beta-amylase *See* AMYLASES.

beta-amyloid A fragment of *amyloid precursor protein (amyloid B, 39–43 aa) produced by the action of *secretases (*see* PRESENILINS) and the main component of plaques in the brain in *Alzheimer's disease.

beta-blockers *See* BETA-ADRENOCEPTOR BLOCKING DRUGS.

beta cells (β-cells) The insulin-secreting endocrine cells of the *islets of Langerhans in the pancreas.

betacellulin One of the EGF family of growth factors, synthesized primarily as a transmembrane precursor (178 aa), which is proteolytically cleaved to produce the mature molecule that binds to the EGF receptor. It is a potent mitogen for retinal pigment epithelial cells and vascular smooth muscle cells. Several alternatively spliced transcripts have been described.

beta-COP *See* COATED VESICLE.

beta-emitter (β-emitter) A radionuclide that decays with the emission of β particles, usually negatively charged electrons. The energy of the β emission varies, tritium (^{3}H) being a weak emitter, ^{32}P a strong one. Other commonly used β-emitters are ^{14}C and ^{35}S.

beta-galactosidase (β-galactosidase) A lysosomal hydrolase (EC 3.2.1.23) that cleaves the terminal β-galactose from ganglioside substrates and other glycoconjugates. Mutations cause a lysosomal storage disease, GM_1-gangliosidosis. The enzyme is also that encoded by the *LacZ* gene, widely used as a *reporter gene and incorporated in many plasmid *vectors to allow *blue-white colour selection. The action of the enzyme will produce coloured or fluorescent compounds from appropriate substrates (*Xgal produces a blue colour).

beta-glucosidase (β-glucosidase) **1.** A cytosolic enzyme (469 aa, EC 3.2.1.21) involved in the metabolism of dietary flavonoid glycosides. **2.** A lysosomal enzyme (glucosylceramidase, EC 3.2.1.45, 536 aa) that catalyses the breakdown of glucosylceramide to ceramide and glucose and is defective in *Gaucher's disease.

beta-interferon (interferon-β1b) A recombinant form (165 aa, nonglycosylated) of endogenous *interferon-β, produced in *E. coli* expressing a human fibroblast-derived gene. Used to treat relapsing *multiple sclerosis, although not all patients respond. *TN* Betaseron.

beta-lactam antibiotics A broad class of antibiotics that include penicillin derivatives, cephalosporins, monobactams and carbapenems. They inhibit bacterial cell wall synthesis but are destroyed by *β-lactamase or penicillinase, a common basis for bacterial resistance.

beta-lactamase A bacterial enzyme (penicillin amido-β-lactam hydrolase, EC3.5.2.6) that will hydrolyse *β-lactam antibiotics thereby conferring resistance; competitively inhibited by *clavulanic acid. Extended-spectrum β-lactamases (ESBLs) will degrade the extended spectrum cephalosporins (e.g. cefuroxime, cefotaxime, and ceftazidime) and are an increasing problem. ESBLs have been reported for *E. coli, Klebsiella, Enterobacter, Proteus, Pseudomonas, Salmonella,* and *Serratia* and more than 170 different ESBLs were known by 2007. Most are still inhibited by clavulanic acid but unfortunately many are carried on plasmids that also carry genes for resistance to aminoglycosides, tetracycline, chloramphenicol, and fluoroquinolones.

betamethasone A glucocorticoid steroid with moderate potency as an anti-inflammatory and immunosuppressive drug, often used topically.

beta-oxidation The process of fatty acid degradation in mitochondria or peroxisomes by sequential loss of two carbons as acetyl-CoA. *See* OMEGA-OXIDATION.

beta-secretase *See* SECRETASES.

beta-thymosin A small protein (β-thymosin/WH2 domain, ~44 aa) found in many organisms that consists of an isolated *WASP homology domain-2 (WH2 domain). β-Thymosin repeats are actin monomer-binding motifs found in many proteins involved in regulating the actin cytoskeleton. *See* THYMOSIN.

beta-type structures Various secondary structural regions of proteins formed from beta-strands (as opposed to alpha-helical regions). **Beta-strands** are stretches of amino acids (5–10 aa) in which the peptide backbone is almost fully extended. **Beta-arches** involve two adjacent antiparallel beta-strands from different sheets, joined by a coiled region usually forming a **beta-sandwich**. In **beta-barrels** a series of (typically *amphipathic) beta sheets are arranged around a central pore as in *voltage-gated ion channels. **Beta hairpins** are antiparallel beta strands lying adjacent in a sheet. A **beta-helix** is a large right-handed coil (or superhelix) containing two or three beta-sheets. **Beta pleated sheets** (beta-sheets) are formed from multiple strands—parallel, antiparallel, or a mixture—linked by interchain hydrogen bonds between peptide carboxyl and amino groups.

beta waves Waves of brain electrical activity at ~15–60 Hz.

Bethlem myopathy A rare autosomal dominant proximal myopathy characterized by early childhood onset and joint contractures that mainly affect the elbows and ankles. It is caused by mutations in the *collagen *COL6A3* gene.

bevacizumab A humanized monoclonal antibody directed against *VEGF, that blocks angiogenesis and is used in the treatment of colorectal carcinoma. *TN* Avastin.

BFA *See* BREFELDIN A.

bFGF *See* FIBROBLAST GROWTH FACTOR.

BFU-E *See* BURST FORMING UNIT.

BGH **1.** Bovine *growth hormone (bGH). **2.** *Brunner's gland hyperplasia.

BH3 **1.** Bcl-2 homology domain 3; *see* BCL. **2.** The borano (BH_3^-) group.

BHK cells A quasi-diploid line of cells (baby hamster kidney cells), descended from a clone (clone 13) isolated from a primary culture of newborn Syrian hamster kidney tissue. Although fibroblastic in appearance they express the smooth-muscle intermediate filament protein *desmin. They have been extensively used as a viral host, in studies of oncogenic transformation, and in cell physiology.

bHLH *See* HELIX–LOOP–HELIX.

Biacore Proprietary name for an instrument that uses surface plasmon resonance to detect the binding of a substance to the surface of a flow chamber. The machine can be used to measure on- and off-rates, e.g. for the binding of a ligand to an immobilized receptor.

bicuculline An inhibitor that acts on *GABA receptors, isolated from *Dicentra cucullaria* (Dutchman's breeches) and herbs of the genus *Corydalis*. *See* EXCITATORY AMINO ACIDS.

bid A proapoptotic member of the *bcl family (BH3-interacting domain death agonist, p22 BID, 195 aa) that is cleaved into smaller fragments which induce cytochrome *c* release, the production of ICE-like proteases, and apoptosis.

Bifidobacteria A genus of Gram-positive bacteria that are members of the normal microflora of the human lower intestine.

b

biglycan A small proteoglycan (bone/cartilage proteoglycan I, 368 aa) found in several connective tissues, especially in articular cartilages. The core protein, to which either chondroitin sulphate or dermatan sulphate may be attached, is very similar to the core protein of *decorin and *fibromodulin.

biguanides A class of drugs based on the biguanide molecule. Used as oral antihyperglycaemic drugs in the treatment of diabetes mellitus (e.g. *metformin) and as antimalarial drugs (e.g. proguanil).

bik A proapoptotic protein (Bcl-2-interacting killer, natural born killer, Nbk/Bik, 160 aa) of the *bcl-2 family, that antagonizes adenovirus E1B 19 kD-mediated inhibition of apoptotic cell death. Bik has a *BH3 domain and apparently regulates a *bax, *Bak-dependent endoplasmic reticulum pathway that contributes to mitochondrial apoptosis. Loss of Nbk/Bik is a common feature of clear-cell renal cell carcinoma.

bikunin A plasma protein (HI-30, urinary trypsin inhibitor, inhibitor subunit of inter-α-(trypsin) inhibitor, 147 aa), one of the superfamily of Kunitz-type protease inhibitors, derived from a precursor molecule (protein AMBP, 352 aa) that also gives rise to α_1-*microglobulin. Bikunin is a component of *inter-α-inhibitor.

bile A viscous alkaline liquid produced by the liver that contains bile salts, mucin, various yellow-green pigments, and a range of other organic molecules including cholesterol and phospholipids. It is important for the digestion of lipids and is also a route for excretion of haemoglobin breakdown products (e.g. bilirubin, which is reddish-yellow, and its oxidation product biliverdin, which is green) and excess cholesterol. Bile salts are sodium salts of the bile acids, particularly taurocholic and glycocholic acids which are powerful surfactants.

bilharzia *See* SCHISTOSOMIASIS.

bilirubin *See* BILE.

biliverdin *See* BILE.

bim A proapoptotic *Bcl-2 homology domain 3 (BH3)-only protein (Bcl-2 interacting mediator of death, 198 aa) that is important for eliminating most effector T cells at the end of an acute T-cell response.

BIN *See* BAR PROTEINS.

binary fission The division of a cell into two daughter cells, or an organism into two. Not a synonym for mitosis followed by cytokinesis but applicable to multinucleated cells or whole organisms and often applied to bacteria, mitochondria, and chloroplasts.

binovular twins Nonidentical twins (dizygotic twins) produced when two separate ova are released and fertilized.

binucleate Describing a cell with two nuclei.

bioaccumulation The accumulation of a substance in a living organism that occurs if the rate of intake exceeds the capacity to excrete or metabolize the substance. Bioaccumulation of toxins can be a particular problem for animals at the top of the food chain.

bioactivation The conversion of a *xenobiotic substance to a more toxic or active derivative within the body.

bioassay Any assay that involves using living material to test for the activity or potency of a substance.

bioavailability A measure of the relative amount of a drug (or other substance) that will reach the systemic circulation when taken by a route other than direct intravenous injection.

biochip **1.** An array of immobilized molecules on a substratum that can be used to identify interacting molecules in a test solution. An increasing range of interactions can be tested, not just between nucleic acids (*see* GENE CHIP), but also between proteins or between proteins and nucleic acids. The readout is often optical, with fluorescent reagents being added and microscopy allowing large numbers of samples (thousands) to be arrayed on each chip. **2.** A silicon chip implanted into and functioning as part of a human body.

biocide A substance that will kill living organisms.

biocompatible Describing a material that will not cause adverse effects when in contact with cells or tissues, sometimes for prolonged periods. The absence of toxicity is important, but more complex requirements need to be satisfied with materials that can be implanted into the body without exciting an inflammatory or thrombogenic response and which may be infiltrated by cells.

biodiversity The range of genetic, taxonomic, and ecosystem differences that exist in a given area or environment; this can, of course, extend to the whole planet.

bioengineering A broad range of diverse activities where the disciplines of engineering and biology interact. It can involve design of prosthetic devices, bioreactor design, production of biocompatible materials or *ex vivo* production of replacement tissue (a technique in its infancy). *Compare* GENETIC ENGINEERING.

biofilm A complex layer of bacteria that can develop on surfaces in contact with water or biological fluids. The film is a complex ecosystem in which there may be several different species of bacteria as well as fungi, algae, protozoa, debris, and corrosion products, all embedded in a mucilaginous slime. The behaviour and metabolism of bacteria in such films differs from that shown in normal suspension monoculture. The plaque formed on teeth is an example of a biofilm.

bioflavonoids A group of coloured phenolic pigments originally considered vitamins (vitamins P, C_2) and responsible for the red/purple colours of many higher plants. They have not been shown to have any nutritional role.

biogenic amines Various amines found in both animals and plants that are involved in signalling. Ethanolamine derivatives include *choline, *acetylcholine, and *muscarine; catecholamines include *adrenaline, *noradrenaline, and *dopamine; polyamines include *spermine; indolylalkylamines include tryptamine and *serotonin; betaines include *carnitine; polymethyline diamines include *cadaverine and *putrescine.

bioinformatics The increasingly important discipline of managing and analysing large amounts of biological data. A growing range of databases are available (for genes, proteins, metabolites, disease linkages, etc.) and these are not only reference sources but can be used for predictive purposes.

bioluminescence Light that is produced by a living organism. The best-known system is luciferase, an ATPase used naturally by fireflies to produce light signals and experimentally to assay ATP concentrations. *Chemiluminescence differs only in that the light-emitting molecule is a synthetic compound such as *luminol or lucigenin.

biomarker 1. Any biological feature that provides information about the status of a system, either an organism or a whole ecosystem. Biomarkers for the concentration of heavy metals in soil as pollution indicators are an example. **2.** A more restricted usage is for molecules (e.g. proteins, antigens) that may indicate disease if found at increased levels in blood.

biomaterials 1. Materials produced by living organisms, such as *chitin, fibroin, or bone. **2.** Any material used to replace tissues *in vivo*.

biometry (*adj.* **biometric**) The application of measurement, and usually statistical methodology, to biological problems.

biomimetic Describing processes, systems, devices, or substances that imitate, or are based upon, those found in biology.

bioprospecting A shorthand term for investigations carried out with the aim of discovering biological materials (e.g. pharmacologically active plant substances) that can be commercially exploited. The term **biopiracy** has been used when local inhabitants are not rewarded.

biopsy The process of removing a small sample of tissue for diagnostic examination.

biopterin The oxidized form of tetrahydrobiopterin, a coenzyme essential for the conversion of phenylalanine to tyrosine, tyrosine to L-dopa, and tryptophan to 5-hydroxytryptophan. It is also an essential cofactor for nitric oxide synthase (NOS). Defects in biopterin synthesis can lead to hyperphenylalaninemia, a disorder similar to phenylketonuria.

bioreactors Reaction vessels used to grow large numbers of cells that synthesize a product of interest. Originally developed for fermentation to produce antibiotics but more recently used to grow large numbers of selected or genetically engineered cells (e.g. for production of monoclonal antibodies or recombinant proteins).

biosecurity Describing methods or procedures designed to prevent harm from pathogenic organisms that are being handled for experimental purposes. Methods include containment (at different levels of stringency) and the use of strains that are engineered to be unlikely to survive in the natural environment.

biosynthesis Synthesis by a living system, as opposed to a chemical synthesis.

biotin *See* VITAMIN B (B_7, vitamin H).

BiP *See* GRP (2).

bipolar cells A class of retinal *interneurons that receive input from the photoreceptors and send their output to *ganglion cells. The response to light is graded (not an all-or-none response) and bipolar cells can be inhibited by adjacent cells (lateral inhibition), which increases contrast at a light/dark boundary.

b

bipolar disorder An affective disorder in which there are dramatic shifts from one emotional extreme (euphoria, intense activity) to another (depression). The term is used in preference to manic-depressive psychosis.

bipolar filaments Filaments with opposite polarity at the two ends, e.g. the *thick filaments of striated muscle.

Birbeck granules Inclusion bodies found in the *histiocytes of patients with *Langerhans cell histiocytosis.

birefringent Describing a material in which the refractive index differs according to the plane of polarization of the light. Birefringence can be detected by using crossed Nicol prisms (polarizers set at right angles in the incident and transmitted light beams respectively, so that no light will pass); the birefringent material rotates the plane of polarization and appears bright against a dark background. The birefringence can be due to structural anisotropy (**form birefringence**), through orientation of molecules as a result of mechanical stretching of the material (**stress birefringence**), or because molecules become aligned in flow (**flow birefringence**).

birthmark *See* NAEVUS.

bisoprolol fumarate A cardioselective *beta-adrenoceptor blocking drug (beta blocker) selective for β_1-adrenoreceptors and used to treat hypertension. *TNs* Zebeta, Ziac.

bisphosphoglycerate (BPG) A highly anionic compound present in erythrocytes at about the same molar ratio as haemoglobin. It binds to deoxyhemoglobin and allosterically alters the molecule, reducing its oxygen affinity and facilitating oxygen release in capillaries. The concentration of BPG is regulated by **bisphosphoglycerate mutase** (EC 5.4.2.4, 259 aa) which catalyses its production; it is hydrolysed by **bisphosphoglycerate phosphatase** (EC 3.1.3.13, 254 aa).

bisphosphonates A class of drugs used to prevent and treat *osteoporosis by inhibiting osteoclast activity. Examples include alendronic acid, etidronate, risedronate, and zoledronic acid.

bitistatin A *disintegrin from the venom of the puff adder *Bitis arietans*.

Bittner agent *See* MAMMARY TUMOUR VIRUS.

biuret reaction A colorimetric test for compounds with two or more peptide bonds (e.g. proteins) which form a purple colour with alkaline copper sulphate solution.

bivalent **1.** Describing a molecule with two binding sites (e.g. IgG). **2.** Shorthand term for a pair of homologous chromosomes in synapsis during *meiosis.

bivalirudin A synthetic anticoagulant peptide (20 aa), based on *hirudin, a specific and reversible thrombin inhibitor, that will bind to the catalytic site and to the anion-binding exosite of both circulating and clot-bound thrombin. *TN* Angiomax.

Bjornstad's syndrome An autosomal recessive disorder characterized by sensorineural hearing loss and twisted hair (pili torti) caused by mutation in the *BCS1L* gene that encodes a homologue of *S. cerevisiae* bcs1 protein which is involved in the assembly of complex III of the mitochondrial respiratory chain.

BK channels Calcium-activated *potassium channels (Maxi-K or slo1 channels) that allow calcium and potassium ions through cell membranes. They are activated (opened) by changes in membrane potential and/or intracellular Ca^{2+}. The pore-forming α subunit is the product of the *KCNMA1* gene, and there are four types of modulatory β subunits (encoded by *KCNMB1, KCNMB2, KCNMB3*, or *KCNMB4*); the pore has eight subunits with the stoichiometry $\alpha_4\beta_4$.

Black Death The colloquial name for the bubonic plague that swept Europe in the mid-14th century. Generally considered to have been caused by *Yersinia pestis*, although alternative pathogens (possibly a haemorrhagic virus) have been suggested.

black fever *See* ROCKY MOUNTAIN SPOTTED FEVER.

blackwater fever A complication of chronic infection with **Plasmodium falciparum*. Symptoms include fever, intravascular haemolysis, haemoglobinuria (red/black urine), and acute renal failure.

black widow spider venom *See* LATROTOXINS.

BLAST The Basic Local Alignment Search Tool (BLAST) is public domain software made available by the US government-funded National Center for Biotechnology Information (NCBI) that can be used to find regions of local similarity in DNA or protein sequences and calculates the statistical significance of matches with known sequences in the databases. It can be used to infer functional and evolutionary relationships between sequences as well as help identify members of gene families. The tool is constantly updated and extended and is an extremely valuable resource.

SEE WEB LINKS

- Homepage and starting point for searches.

blast cells General term for the proliferating cells in a particular cell lineage.

blastema 1. An organized mass of cells formed during regeneration, e.g. of an amphibian limb. **2.** A group of cells that will develop into a new individual by asexual reproduction (budding).

blasticidin (blasticidin S) An antibiotic produced by *Streptomyces griseochromogenes* that will inhibit prokaryotic and eukaryotic protein synthesis by interfering with peptide bond formation. Used experimentally to select transfected cells that have acquired blasticidin resistance genes (*bsr* or *BSD* genes) from the vector.

blastocoele (US blastocele) The cavity that forms within the mass of cells of the *blastula during the later stages of cleavage.

blastocyst The mammalian equivalent of the *blastula, a thin-walled hollow sphere formed before implantation occurs. The wall is the *trophoblast, and the embryo proper is a mass of cells at one side (inner cell mass).

blastoderm A layer of cells that forms on the surface of yolk-rich eggs through the process of meroblastic cleavage. In birds it is a flat disc of cells at one pole of the egg and in insects an outer layer of cells surrounding the yolk mass.

blastoma (*pl.* blastomas or blastomata) A neoplasm composed of immature *blast cells.

blastomere A cell produced by one of the early cell divisions of the fertilized egg, the cleavage divisions that produce the blastula or its equivalent.

blastopore The opening formed by the invagination of presumptive mesodermal and endodermal cells during gastrulation. The invagination produces the *archenteron which opens to the exterior through the blastopore which, in some species, becomes the anus. In some animals, e.g. chick, the cells move in along the *primitive streak, which may be termed a virtual blastopore.

blast transformation The changes in morphology and behaviour that occur in B and T cells that have been exposed to *antigen or a *mitogen. Receptor crosslinking triggers a cascade of signalling events that cause the cell to exit G_0 and enter the G_1 phase of the cell cycle and begin to proliferate.

blastula The stage of embryonic development of animals towards the end of cleavage but before the morphogenetic movements of *gastrulation begin. Where cleavage involves the whole egg, the blastula usually consists of a hollow ball of cells. *See* BLASTODERM; BLASTOCYST.

bleb A rounded protrusion from the surface of a cell which may be fluid-filled or be supported by a meshwork of microfilaments.

blebbistatin A small-molecule myosin inhibitor that blocks myosin II, but not myosin I, in the actin-detached state, but with ADP and phosphate still bound.

blenno- Prefix denoting mucus. **Blennorrhagia** (blennorrhoea) is the discharge of mucus, usually from the genital organs, caused by gonorrhoeal infection or urethitis.

bleomycin A family of structurally related glycopeptide antibiotics isolated from *Streptomyces verticillus*. Acts by induction of DNA strand breaks and blocks cell division in G_2. Used in treating Hodgkin's lymphoma, squamous cell carcinomas, and testicular cancer. *TN* Blenoxane.

blephar- Prefix indicating the eyelid. **Blepharism** is repeated uncontrolled blinking, **blepharitis** is chronic inflammation of the eyelids. **Blepharoplegia** is paralysis of the eyelid muscles resulting in drooping of the upper eyelid (**blepharoptosis**).

BLIMP1 *See* PRD1-BF1.

blindsight An unusual condition, caused by brain damage, in which responses are made to visual stimuli, even though the stimulus is not consciously perceived.

blind study A study or trial in which subjects do not know the expected result and may not even know the purpose of the study. The subjects are said to be 'blind' in expectations and there should be no placebo effect. *See* DOUBLE-BLIND TRIAL.

blister A thin-walled swelling on the skin filled with clear or bloodstained exudate. Can be caused by mechanical or chemical irritation.

blk A *src-family tyrosine kinase (B-lymphocyte kinase, EC 2.7.10.2, 505 aa) involved in B-cell maturation. *See* PREBCR.

b

blocking antibody An antibody used to prevent some other reaction taking place by physically masking a binding site, e.g. on a cell surface receptor.

BLOCs Protein complexes (biogenesis of lysosome-related organelles complexes) involved in the biogenesis of specialized lysosome-related organelles such as melanosomes and platelet dense granules. Subunits of **BLOC1** include *pallidin, *muted, dysbindin (*dystrobrevin-binding protein 1), *cappuccino, *snapin, and others. BLOC-1 interacts with *adaptor protein-3 (AP-3) to form complex that affects the targeting of SNARE and non-SNARE cargoes. **BLOC-2** and **BLOC-3** are related complexes. *See* HERMANSKY–PUDLAK SYNDROME.

blood The fluid that circulates throughout the vasculature, transporting oxygen, nutrients, hormones, waste products, and heat. In humans the relative density is about 1.054–1.060, and it has a normal pH of 7.4. The cells of the blood include *erythrocytes, *leucocytes, and *platelets and the acellular fluid (*plasma) contains a diverse range of proteins (e.g. *serum albumin), lipids, hormones, etc. Because blood is an easy tissue to sample, the properties in health and changes indicative of disease are relatively well understood. *See* BLOOD GROUP ANTIGENS; BLOOD DISORDERS, and entries prefixed haem- (e.g. *haemoglobin).

blood–brain barrier The physiological barrier to the entry of large molecules such as antibodies into the brain and retina. It is a consequence of relative impermeability of the endothelial layer of the vasculature in these tissues and, although the difference in endothelial permeability is not well understood, it has important implications for delivery of drugs to the brain and reduces immunological surveillance of the brain.

blood clotting factors The various proteins (coagulation factors) involved in blood clotting, numbered more or less in reverse order of their activation: the final step is the conversion of factor I, fibrinogen, into factor Ia (*fibrin) by *thrombin. Most of the factors are *zymogens of serine peptidases that are cleaved and activated by the activated factor upstream in the cascade, the whole system providing amplification and thus rapid haemostasis. Blood clotting can be triggered through the *contact activation pathway (formerly the intrinsic pathway), or the *tissue factor pathway (formerly known as the extrinsic pathway) triggered by the trauma-induced production of tissue factor. *See* HAGEMAN FACTOR; PROTEIN-C; PROTEIN-S; VON WILLEBRAND FACTOR. Deficiencies lead to various forms of *haemophilia.

blood count *See* HAEMATOCRIT.

blood disorders

*Blood serves several functions, apart from transporting oxygen and nutrients, including roles in wound healing, *inflammation, and immunity, by delivering components (cells and soluble factors such as antibodies and cytokines) to appropriate locations. The various cells (*erythrocytes, *leucocytes, and *platelets) are suspended in *plasma. The capacity of the blood to clot is important for *haemostasis and a complex cascade of *blood clotting factors (factors I–XII) are involved, together with *platelets (*see also* HAGEMAN FACTOR; PROTEIN C; PROTEIN S; THROMBOMODULIN; VON WILLEBRAND FACTOR). The fluid left after clotting has occurred is *serum. Clot formation can be initiated by the *contact activation pathway or the *tissue factor pathway. Blood clots (*thrombi) can, however, form inappropriately and cause *embolisms or in extreme cases *disseminated intravascular coagulation. Tests for clotting time are standardized using the *international normalized ratio. Various bleeding disorders (*haemophilias), are caused by deficiencies in components of the clotting system (*see* CHRISTMAS DISEASE). Defects in platelet function can also lead to bleeding disorders or excessive tendency to clotting (*see* BERNARD–SOULIER SYNDROME; GLANZMANN'S THROMBASTHENIA; HEREDITARY HAEMORRHAGIC TELANGIECTASIA; THROMBOPHILIA; THROMBOTIC THROMBOCYTOPENIC PURPURA). Red blood cell numbers (*see* HAEMATOCRIT) can be reduced in *anaemia, reducing the oxygen-transport capacity, or increased in *polycythemia. Mutations in haemoglobin (*see* THALASSAEMIA) can also affect oxygen-carrying capacity. Leucocyte and platelet numbers can be abnormally low (leucopenia) or raised (leucocytosis) in various forms of *leukaemia, *myelodysplasia, and *thrombocythaemia. Various other disorders or abnormalities are recognized: *see* ACIDOSIS; ALKALOSIS; AZOTAEMIA; BACILLAEMIA; BACTERAEMIA; GALACTOSAEMIA; HYPERAEMIA; HYPERCALCAEMIA; HYPERCAPNIA; HYPERCHOLANEMIA; HYPERCHOLESTEROLAEMIA; HYPERCKMIA; HYPERFERRITINAEMIA–CATARACT SYNDROME;

HYPERGAMMAGLOBULINAEMIA; HYPERGLYCAEMIA; HYPERKALAEMIA; HYPERLEUCINE–ISOLEUCINAEMIA; HYPERLIPIDAEMIA; HYPERPHOSPHATAEMIA; HYPERTENSION; HYPERTRIGLYCERIDAEMIA; HYPERURICAEMIA; HYPERVALINAEMIA; HYPOCALCAEMIA; HYPOGAMMAGLOBULINAEMIA; HYPOGLYCAEMIA; HYPONATRAEMIA; HYPOTENSION; HYPOVOLAEMIA; VIRAEMIA. *See also* BLOOD GROUP ANTIGENS and entries with the prefix haem-.

blood group antigens The cell surface antigens found chiefly, but not exclusively, on cells in the blood; many being carbohydrate moieties of membrane glycoproteins or glycolipids. More than fifteen different blood group systems are recognized in humans. Naturally occurring antibodies are found to some, even without immunization, and matching blood groups is important for safe transfusion. *See* ABO BLOOD GROUP SYSTEM; DUFFY; KELL; LEWIS BLOOD GROUP; MN BLOOD GROUP ANTIGENS; RHESUS BLOOD GROUP.

SEE WEB LINKS

- Blood group antigen mutation database.

blood pressure The pressure of blood in the arteries which rises to a maximum as blood is pumped out by the left ventricle (systole) and drops to a minimum in diastole. The systolic/diastolic pressure is normally ~120/80 mmHg in a young adult.

blood smear A thin air-dried film of blood on a microscope slide. The process of smearing spreads the cells out as a monolayer and by staining with a Romanovsky-type (e.g. Giemsa) the proportions of different leucocytes can be counted and blood-borne parasites seen.

blood substitutes Fluids used to increase blood volume after blood loss and when whole blood is not available. They range from simple isotonic solutions (plasma expanders) to complex solutions with oxygen-carrying capacity, although none of the latter have yet been particularly successful.

blood sugar Shorthand term for the concentration of glucose in the blood, usually ~3.2–5.2 mmol/L in the fasting state.

blood vessels The system of tubes, lined with *endothelium, through which blood circulates. *See* ARTERY; CARDIOVASCULAR DISORDERS; VASCULARIZATION; VEIN.

Bloom's syndrome An autosomal recessive disorder in which there is proportionate growth deficiency, sun-sensitivity, immunodeficiency, a predisposition to malignancy and chromosomal instability. Caused by mutation in the gene encoding DNA *helicase (RecQ protein-like-3, Bloom's syndrome protein, EC 3.6.1.-, 1417 aa) that is involved in DNA replication and repair.

blotting A general term for transferring molecules from an acrylamide or agarose gel in which they have been separated electrophoretically, to a paper-like membrane, usually nylon or nitrocellulose, maintaining the spatial arrangment. The binding interaction with the membrane can then be stabilized and the bound molelecules can be detected at high sensitivity by *hybridization (in the case of DNA and RNA), or antibody labelling (in the case of protein). RNA blots are called **northern blots**, DNA blots are **Southern blots**, (named after E. Southern, who developed the technique), protein blots are **western blots**. In **northwestern blotting** protein is transferred but is probed with specific RNA. A simpler procedure, without a separation step, is the **dot** or **slot blot** in which a droplet or line of solution is put directly on to nitrocellulose paper and is then probed.

blue-green algae *See* CYANOBACTERIA.

blue naevus *See* NAEVUS.

Bluescript A widely-used proprietary plasmid which can be used for *blue-white colour selection of successfully transfected clones.

bluetongue virus One of the *Reoviridae, responsible for a serious disease (bluetongue) of sheep, cattle, and pigs but not humans. Transmitted by biting flies.

blue–white colour selection A common method for identifying bacterial clones that have been successfully transfected with a plasmid vector carrying the sequence of interest. The plasmid has a LacZ gene encoding *beta-galactosidase disabled by insertion of the sequence; empty (religated) vectors produce blue colonies when grown on plates containing *IPTG and *Xgal, but colonies with the insert do not produce functional (beta-galactosidase, and are white.

blunt end The end of a DNA duplex that has been cut by a *restriction enzyme at the same site on both strands so that there is no single-stranded overhang. *Compare* STICKY ENDS.

B lymphocyte (B cell) *See* LYMPHOCYTE.

BM-40 *See* OSTEONECTIN.

BMAL1 A protein (brain and muscle ARNT-like protein 1, 583 aa) that has a Per–ARNT–Sim (PAS)-domain that is found in several other transcription factors and that dimerizes with the *clock protein; the heterodimer apparently drives the oscillating transcription of **period*.

***B*$_{max}$** A parameter (analogous to *V_{max} in enzyme kinetics) that is a measure of the binding of a drug and the number of receptors present in the sample. Usually derived from a *Scatchard plot.

bmf A proapoptotic protein (Bcl-2-modifying factor, 184 aa) of the *Bcl-2 family.

BMPs *See* BONE MORPHOGENETIC PROTEINS.

BMR *See* BASAL METABOLIC RATE.

BNLF-1 *See* LATENT MEMBRANE PROTEINS.

BNP *See* NATRIURETIC PEPTIDES.

BN-PAGE A modification (blue-native polyacrylamide gel electrophoresis) of *polyacrylamide gel electrophoresis (PAGE) in which the binding of *Coomassie Blue to nondenatured protein confers negative charge which can be used to separate the proteins. A further electrophoretic separation can be carried out using denaturing SDS (*see* DODECANOIC ACID).

BOAA *See* LATHYRISM.

BODIPY A group of fluorescent dyes used as probes in fluorescence microscopy. They can be conjugated to a range of molecules and some show a large pH-dependent enhancement of emission.

body louse *Pediculosis humanus humanus*, a parasite that infests humans, particularly in overcrowded insanitary conditions, and is a vector for typhus.

body scanner Any scanning device that can be used for the whole body. May be used, now usually with computer enhancement, for X-ray or ultrasound imaging or radiation detection.

Bohr effect The decrease in the binding affinity of haemoglobin for oxygen when the pH decreases or the concentration of carbon dioxide increases.

boil (furuncle) An inflamed and swollen hair follicle infected with *Staphylococcus aureus*.

Bolivian haemorrhagic fever *See* MACHUPO VIRUS.

bombesin A tetradecapeptide neurohormone that has paracrine and autocrine effects, first isolated from the skin of the fire-bellied toad *Bombina bombina*. The subfamily of bombesin-like peptides includes *litorin, *phyllolitorin, and *ranatensin. The mammalian equivalent is *gastrin-releasing peptide (GRP) and bombesin cross-reacts with GRP receptors. Both are mitogenic for Swiss 3T3 fibroblasts at nanomolar levels. Neuropeptides of this type are found in many tissues and at high levels in pulmonary (small cell carcinoma) and thyroid tumours. *See* NEUROMEDINS.

bone disorders

Bone is mineralized connective tissue that during development may start as cartilage and become mineralized (endochondral ossification) or arise in mineralization centres within mesodermal tissue as in the case of the skull bones. *Dysostosis, *hypophosphatasia, and *rickets are failures of mineralization (*see* LIM MINERALIZATION PROTEIN; VITAMIN D). Excessive mineralization is a feature of *osteopetrosis. The organic matrix is of type I collagen (defective in *Caffey's disease), glycosaminoglycans, *matrix Gla protein, *osteocalcin, *osteonectin, *osteopontin, and *tenascin W. *Osteoblasts are involved in synthesis and remodelling of bone; *osteoclasts break down bone and are regulated by *osteoclast differentiation factor, *osteoprotegerin, and *parathyroid hormone. Many skeletal abnormalities are a result of developmental defects in bone morphogenesis, which is under the control of various hormones and growth factors (especially *bone morphogenetic proteins, BMPs). Antagonists of BMPs affect bone development and include *chordin, *gremlin, *noggin, and *sclerostin. Bone disorders include *acromegaly, *acromesomelic dysplasia, *ankylosing spondylitis, *atelosteogenesis, *brachydactyly, *campomelic dysplasia, *Camurati–Engelmann disease, various *chondrodysplasias, *ectromelia, *hypochondrogenesis, *Mona, *osteodystrophia fibrosa, *otosclerosis, *Paget's disease of bone, *Rothmund–Thomson syndrome, *Saethre–Chotzen syndrome, *sclerosteosis, and *van Buchem's disease. Age-related disorders include *osteoporosis, the latter often being treated with *bisphosphonates. Neoplastic disorders affecting bone include *chordoma, *enchondroma, *Ewing's sarcoma, *exostosis, and *sarcoma (osteosarcoma). *See also* PROGRESSIVE OSSEOUS HETEROPLASIA.

bone marrow The tissue found in the centre of bones. Red marrow is the site of *haematopoiesis and is the most radiation-sensitive tissue of the body, whereas yellow marrow is composed mostly of fat cells. In **bone marrow grafting**, cells derived from donor marrow and purged of lymphocytes that could cause a *graft-versus-host response are transfused into the circulation and home to the marrow of the recipient which has been therapeutically destroyed; this technique can be used to treat aplastic anaemia and leukaemia.

bone morphogenetic proteins A family of multifunctional cytokines, the BMPs belong to the *transforming growth factor β superfamily. Many are *growth and differentiation factors (GDFs) and may be regulated by BMP-binding proteins such as *noggin, *chordin, and *follistatin. The **receptors** are serine/threonine kinase receptors (types I and II) that link with *MAD-related proteins. **BMP2** and **BMP3** (osteogenin) are involved in regulating bone formation; **BMP4** acts during development as a regulator of mesodermal induction and is overexpressed in *fibrodysplasia ossificans.

bongkrekic acid An inhibitor (531 Da) of the mitochondrial adenine nucleotide translocator that will also inhibit apoptosis by preventing *PARP cleavage. It is produced in fermented coconut contaminated by the bacterium *Burkholderia gladioli* pathovar *cocovenenans* and is highly toxic.

bootstrap analysis A statistical method for testing the variability in a dataset by resampling to produce pseudoreplicates; the method makes no assumptions about the distribution of the data and thus has some advantages. It is extensively used in phylogenetic analyses.

Bordetella pertussis A small, aerobic, Gram-negative bacillus that causes whooping cough. The bacterium produces various toxins including a dermonecrotizing toxin, an adenyl cyclase, an *endotoxin, and *pertussis toxin.

borealin A protein (cell division cycle-associated protein 8, pluripotent embryonic stem cell-related gene 3 protein, Dasra-B, 280 aa) that is one of the components of the *chromosomal passenger complex.

Borna disease A neurological disorder caused by Borna disease virus, a negative-strand RNA virus that will infect a range of domestic animals and humans. It causes behavioural disorders in animals and may be associated with some psychiatric conditions including bipolar disorder and depression.

SEE WEB LINKS

- Article on Borna disease in *Emerging Infectious Diseases* (1997).

Bornholm disease *See* PLEURODYNIA.

Borrelia burgdorferi A Gram-negative spirochaete that causes *Lyme disease.

bortezomib A low molecular weight inhibitor of the 26S *proteosome, used in treatment of *multiple myeloma. *TN* Velcade.

Botox Proprietary name for *botulinum toxin type A, used in cosmetic surgery as an inhibitor of dermal muscles that cause wrinkles.

botrocetin venom coagglutinin A heterodimeric C-type lectin (α, 133 aa; β, 125 aa) in the venom of a pit viper *Bothrops jararaca*, that forms an activated complex with *von Willebrand factor, which then binds to platelet GPIb causing agglutination.

botulinolysin A *cholesterol binding toxin from *Clostridium botulinum*.

botulinum toxin A neurotoxin (~1296 aa, 7 distinct serotypes, A–G) produced by certain strains of *Clostridium botulinum* (*see* BOTULISM). The toxin is proteolytically cleaved to an enzymically active A component and a larger B subunit that binds to gangliosides and stimulates internalization. The A subunit (EC 3.4.24.69) acts proteolytically on *SNAP-25, an essential component of the mechanism for release of acetylcholine-containing vesicles at the neuromuscular synapse. *See* BOTOX. **Botulinum C2 toxin** is an AB-type toxin (431 and 721 aa) that ADP-ribosylates G-actin and blocks the formation of microfilaments. **Botulinum toxin C3** is an ADP-ribosyl transferase that inactivates ras-related proteins, *racs (ras-related C3 botulinum toxin substrates) and is used experimentally, although it needs to be injected into cells.

botulism A severe form of food poisoning caused by the anaerobic bacterium *Clostridium botulinum* that has contaminated tinned or processed meat. The effects, often fatal, are due to *botulinum toxin.

Bouin's solution A fixative containing picric acid, acetic acid, and formaldehyde, bright yellow in colour, in which specimens can be stored indefinitely.

bouton A small swelling at the end of an axon, or along the length of an individual axon (bouton en passant) where it makes a synaptic contact.

b

box An informal term for a DNA sequence involved in binding a regulatory protein. Examples include *homeobox, *TATA box, and *CAAT box.

Boyden chamber A simple design of two-compartment chamber used to test for *chemotaxis, especially of leucocytes. The upper compartment, in which cells are placed, is separated from the lower chamber containing the chemotactic factor by a Millipore-type filter (3–8 μm pore size, ~150 μm thick) across which a chemotactic gradient develops. The distance cells have moved into the filter is measured after an incubation period sufficiently short to avoid the gradient decaying. Interpretation is complicated by the fact that most chemotactic factors are also *chemokinetic. *See also* CHECKERBOARD ASSAY.

Bq *See* BECQUEREL.

brachiocephalic Pertaining to the arm and head, e.g. describing the artery (the brachiocephalic or innominate artery) that supplies blood to the right arm and the head and neck. *Compare* BRACHYCEPHALIC.

brachycephalic Describing a skull with a breadth that is at least four-fifths of the length.

brachydactyly A feature of several developmental disorders in which there is premature closure of the epiphyses, particularly of terminal limb bones, so that fingers and toes are shorter than normal. **Type A1** brachydactyly is caused by mutation in the *Indian hedgehog gene, or mutation at a locus on chromosome 5. **Type A2** is caused by mutation in bone morphogenetic protein (BMP) receptor B1 or in *growth/differentiation factor-5. **Type B1** is caused by mutation in the *ROR2* gene which encodes a receptor tyrosine kinase (NTRKR2, 943 aa) that is also mutated in autosomal recessive *Robinow's syndrome. **Type C** brachydactyly is caused by mutation in the growth/differentiation factor-5 gene, **Types D** and **E** by mutation in the homeobox D13 (*HOXD13*) gene. Other forms of brachydactyly are known, although their molecular basis is unclear.

brachyolmia A heterogeneous group of skeletal dysplasias, primarily affecting the spine. Type 3 is caused by a mutation in the *transient receptor potential channel, TRPV4.

brachyury A mouse gene that encodes a T-box transcription factor important in tissue specification, morphogenesis, and organogenesis.

Bradford assay A standard protein assay based on the change in absorbance of Coomassie Brilliant Blue G-250 when it binds to arginine and aromatic residues in the protein. It is important to standardize the assay with a protein of comparable arginine content, and for proteins with few aromatic residues (particularly collagen) the results are very inaccurate.

brady- A prefix meaning slow. In **bradycardia** the heart beats unusually slowly (usually <50 beats per minute). *Compare* TACHYCARDIA. In **bradykinesia**, body movements are slow.

bradykinin A vasoactive nonapeptide derived from kininogens and similar to *kallidin, which has an additional N-terminal lysine. It is a potent vasodilator, increases the permeability of postcapillary venules, and activates phospholipase A2 in endothelial cells.

brain The anterior portion of the neural tube that is massively increased in size and complexity of function in humans. The lumen of the embryonic neural tube persists in the adult brain as the cerebral ventricles and the central canal of the spinal cord, which are filled with *cerebrospinal fluid. The forebrain consists of the cerebral hemispheres, the basal ganglia, and the diencephalon which forms the thalamus and hypothalamus. The two hemispheres of the cerebrum, which are massively expanded, are connected by the corpus callosum and their surfaces are folded into complex ridges (gyri) and valleys (sulci). The outer layer (cerebral cortex) is responsible for higher order functions such as memory, consciousness, and abstract thought; the deeper layers (basal ganglia) include the caudate nucleus and putamen (collectively the striatum), amygdaloid nucleus, and hippocampus. The hypothalamus controls endocrine function and regulates hunger, thirst, emotion, and sleep; the thalamus coordinates sensory input and pain perception. The midbrain, by contrast, is relatively small and is subdivided into the corpora quadrigemina and the cerebral peduncle. The hindbrain consists of the anterior metencephalon and the posterior myelencephalon, nearest the spinal cord. The metencephalon contains the cerebellum, responsible for sensory input and coordination of voluntary muscles, and the pons. The myelencephalon contains the medulla oblongata, which regulates blood pressure, heart rate, and other basic involuntary functions, and dorsally the choroid plexus.

brain-derived growth factor (BDGF) *See* BRAIN-DERIVED NEUROTROPIC FACTOR.

brain-derived neurotropic factor (BDNF, BDGF, abrineurin) One of the *neurotrophin family, a small basic protein (247 aa) that promotes the survival of neurons in or connected to the central nervous system. BDNF is a regulator of synaptic transmission and plasticity in adults and is important in *long-term potentiation, long-term depression and some forms of short-term synaptic plasticity. *Huntingtin upregulates transcription of BDNF and transport of BDNF-containing vesicles along microtubules. Polymorphisms in the BDNF gene are associated with susceptibility to *bulimia nervosa 2 and *anorexia nervosa 2.

brain natriuretic peptide (BNP) *See* NATRIURETIC PEPTIDES.

brain small vessel disease with haemorrhage An autosomal dominant disorder that is responsible for 20–30% of ischaemic strokes. It is caused by a defect in *collagen type IV (COL4A1). *Porencephaly is an allelic disorder with more severe manifestations.

branchiooculofacial syndrome A developmental defect that affects the bronchial system, face, and eyes, caused by mutation in the *TFAP2A* gene that encodes transcription factor *AP-2A.

BRAP1 *See* BRIDGING INTEGRATOR 2.

BRCA Genes (breast cancer genes, *BRCA1*, *BRCA2*) associated with an inherited predisposition to breast–ovarian carcinoma syndrome, although other genes are also involved. The nuclear protein **BRCA1** (1863 aa) is part of a large multiprotein tumour-suppressor complex (BRCA1-associated genome surveillance complex, *BASC) associated with DNA damage repair, ubiquitination, and transcriptional regulation to maintain genomic stability. **BRCA2** (Fanconi anaemia group D1 protein, 3418 aa) is involved in DNA double-strand break repair and homologous recombination. Variations in BRCA2 may also be responsible for susceptibility to uveal *melanoma. *See* BARD; BRCT DOMAIN; BRIDGING INTEGRATORS.

BRCT domain A tandem repeat domain (BRCA C-terminal domain, ~100 aa) involved in binding phosphoserine proteins and found predominantly in proteins that are involved in cell cycle checkpoint functions responsive to DNA damage.

SEE WEB LINKS

- BRCT domain structure.

Brdu *See* BUDR.

breakpoint cluster region *See* BCR.

breast cancer genes *See* BRCA.

brefeldin A A macrocyclic lactone synthesized by several fungi including *Penicillium brefeldianum*. It inhibits the process of protein secretion at a pre-Golgi stage by binding to the *ARF1/GDP/*Sec7 complex and blocking guanine nucleotide exchange (*GEF). A useful experimental tool in investigations of membrane traffic.

bretylium tosylate An antiarrhythmic drug that blocks the release of noradrenaline from the peripheral sympathetic nervous system.

brevetoxins Neurotoxins produced by the dinoflagellate *Ptychodiscus* (formerly *Gymnodinium*) *brevis* that are responsible for shellfish poisoning. They reduce the threshold for depolarization of excitable membranes by binding to the sodium channel.

bridging integrators A family of adaptor proteins containing BAR domains (*see* BAR PROTEINS). **Bridging integrator 1** (BIN1, myc box-dependent-interacting protein 1, 593 aa) is an *amphiphysin-like protein that may be involved in regulation of synaptic vesicle endocytosis and may act as a tumour suppressor. Mutations are the cause of autosomal recessive *centronuclear myopathy (autosomal recessive myotubular myopathy). **Bridging integrator 2** (BRAP1, breast cancer-associated protein 1, 565 aa) is preferentially expressed in haematopoietic tissues. **Bridging integrator 3** (253 aa) has a role in the organization of F-actin during cytokinesis.

Bright A transcription factor (B-cell regulator of immunoglobulin heavy chain transcription, 593 aa) that binds as a homodimer to AT-rich sequences in the enhancer regions of the immunoglobulin heavy chain locus.

bright-field microscopy *See* MICROSCOPY.

Bright's disease An obsolete name for a type of nephritis.

Brilliant Cresyl Blue A histological dye (Brilliant Blue C) used to stain bone marrow smears.

British Approved Name *See* BAN.

brittle cornea syndrome A disorder (Ehlers–Danlos syndrome type VIB) in which the sclerae of the eye are blue, the cornea is prone to rupture after minor trauma, there is keratoconus or keratoglobus (bulging of the cornea), hyperelasticity of the skin, and hypermobility of the joints. It is caused by mutation in the zinc finger protein encoded by the *ZNF469* gene.

b

Broca's area A region located in the left inferior convolution of the frontal lobe of the brain that is involved in the production, but not comprehension, of speech. Damage to this region causes Broca's aphasia.

bromelain Members of the clan of cysteine (thiol) endopeptidases from the pineapple *Ananas comosus*. There are slightly different enzymes in the stem (EC 3.4.22.32) and the fruit (EC 3.4.22.33).

bromocriptine A dopamine agonist that acts on D_2 receptors and that is used in the treatment of pituitary tumours and Parkinson's disease. *TN* Parlodel.

bromodomain A conserved domain (~70 aa) that binds to acetylated lysine residues such as those on the N-terminal tails of histones. Bromodomains are found in DNA-associated proteins from a wide range of organisms and are thought to be involved in protein–protein interactions in transcriptional activation complexes.

bromophenol dyes Dyes used as pH indicators: bromophenol blue changes from yellow to blue in the pH range 3.0–4.6, bromophenol red changes from yellow to red in the range 5.2–6.8, bromthymol blue changes from yellow to blue in the range 6.0–7.6.

broncho- Prefix indicating association with the bronchial system. **Bronchiectasis** is an obstructive lung disease in which there is localized, irreversible dilation of part of the bronchial tree. It can be congenital (e.g. Kartagener's syndrome and cystic fibrosis) or acquired, probably as a result of recurrent inflammation associated with bacterial infections. **Bronchitis** is inflammation of the lung airways (bronchi). Acute bronchitis is usually associated with infection, e.g. by cold viruses; chronic bronchitis is a result of persistent irritation of the bronchi. **Bronchodilators** are drugs that dilate airways by relaxing the smooth muscle of the bronchial wall, thus relieving breathlessness caused acutely by asthma or chronically by obstructive pulmonary disease. Often administered by inhalation. Examples include *sympathomimetic drugs (e.g. salbuterol), *antimuscarinic drugs, and xanthines (e.g. theophylline). **Bronchoscopy** is endoscopic examination of the tracheobronchial tree.

bronchus associated lymphoid tissue (BALT) A subset of *mucosal associated lymphoid tissue (MALT) consisting of lymphoid nodules in the lamina propria of the bronchus. In humans it develops in response to infection.

bronzed diabetes *See* HAEMOCHROMATOSIS.

brown adipose tissue Highly vascularized adipose tissue found in restricted locations in the body, particularly in mammals that hibernate but also in human infants. Brown adipose tissue is important for heat production and regulation of body temperature. *See* THERMOGENIN.

brownian motion The random motion of microscopic objects caused by intermolecular collisions, first described by the Scottish microscopist Robert Brown (1773–1858). A simple form of stochastic process.

Brucella A genus of Gram-negative aerobic bacteria that are intracellular parasites or pathogens in humans and other animals. *Brucella abortus* (*Brucella melitensis* biovar *abortus*) causes spontaneous abortion in cattle and undulant fever (brucellosis), a persistent recurrent acute fever, in humans.

Bruch's membrane The innermost layer of the *choroid composed of five layers, the basal lamina of the retinal pigment epithelium, an inner collagenous zone, a central layer rich in elastin fibres, an outer collagenous zone, and the basal lamina of the *choriocapillaris.

Brugada syndrome An autosomal dominant disorder (sudden unexplained nocturnal death syndrome). **Type 1** is caused by mutation in the pore-forming α subunit of *voltage-gated sodium channel $NaV_{1.5}$, leading to cardiovascular problems including idiopathic ventricular fibrillation and the risk of sudden cardiac arrest. **Type 2** is caused by a defect in the *GPD1L* gene (glycerol-3-phosphatase dehydrogenase 1-like) which affects the properties of the ion channel mutated in type 1.

bruise An area of skin discolouration caused by the rupture of blood vessels and extravasation of blood as a result of a blow that does not break the skin.

Brunner's glands (duodenal glands) Small, branched, coiled tubular glands in the submucosa of the first part of the duodenum that secrete an alkaline mucus that partly neutralizes gastric acid.

brush border The densely packed *microvilli that are found on the apical surface of some epithelia, e.g. in the intestine, where they increase the surface area for absorption.

Bruton's agammaglobulinaemia A sex-linked *agammaglobulinaemia caused by defective *B cell function because of a mutation in the gene encoding Bruton's tyrosine kinase (btk, 659 aa), one of the *Tec family. Btk interacts with other tyrosine kinases such as *fyn, *lyn, and *hck, which are activated upon stimulation of B- and T-cell receptors. *Itk is the T-cell homologue of btk. The exact mechanism by which btk regulates B-cell differentiation is unclear, but btk interacts with the phosphatidyl inositol signalling pathway. *Sab selectively inhibits btk.

Bruton's tyrosine kinase *See* BRUTON'S AGAMMAGLOBULINAEMIA.

Brx **1.** A-kinase anchor protein 13 (breast cancer nuclear receptor-binding auxiliary protein, lymphoid blast crisis oncogene, 2813 aa), one of the *Dbl family that increases oestrogen receptor activity. It anchors cAMP-dependent protein kinase (PKA, A-kinase) and acts as an adaptor protein to couple Gα-13 and rho for which it is a guanine nucleotide exchange factor (*GEF). **2.** Cysteine-rich hydrophobic domain 1 protein (CHIC1, brain X-linked protein, BRX, 224 aa), a palmitoylated membrane protein preferentially expressed in brain and possibly mutated in an X-linked mental retardation syndrome.

bryostatin A group of compounds, isolated from bryozoans but probably produced by symbiotic bacteria, that activate *protein kinase C (PKC). Have been investigated as potential antitumour agents and memory activators.

BSE Bovine *spongiform encephalopathy. A transmissible encephalopathy that affected large numbers of cattle in the UK during the 1990s. Similar to scrapie, but distinct; a link with new variant *Creutzfeldt–Jakob disease is suspected.

BTB/POZ domain A structurally well-conserved domain (bric-a-brac tramtrack broad/poxvirus and zinc finger complex, POZ/BTB domain, BTB domain, ~120 aa) involved in protein–protein interactions and found in several *zinc finger and poxvirus proteins. Mediates homo- or heterodimer formation of complexes involved in transcriptional regulation, cytoskeleton dynamics, ion channel assembly and gating, and targeting for ubiquitination.

SEE WEB LINKS

- Structural details of the BTB/POZ domain.

BTC *See* BETACELLULIN.

btk *See* BRUTON'S AGAMMAGLOBULINAEMIA.

BTX *See* BATRACHOTOXINS.

bubo An inflamed and swollen lymphatic gland.

bubonic plague An infection with *Yersinia pestis*, characterized by swelling of lymphatic glands, particularly in the groin.

budesonide A synthetic corticosteroid with strong *glucocorticoid and weak *mineralocorticoid activity, used for treatment of asthma, non-infectious rhinitis and inflammatory bowel disease. *TNs* Entocort, Pulmicort, Rhinocort.

BUdR (BRdU) A thymidine analogue, the deoxynucleoside of 5-bromouracil (bromodeoxyuridine), that induces point mutations because of its tendency to tautomerization leading to mispairing with guamine (G) instead of adenine (A). It is used as a marker for DNA synthesis because the incorporation of BUdR can be recognized by staining or with a monoclonal antibody.

buffer A general term for a system that minimizes the change in concentration of a specific chemical species in solution when the species is added or removed. pH buffers are weak acids or weak bases in aqueous solution, metal ion buffers are ion chelators, e.g. *EDTA, partially saturated by the metal ion.

buffy coat The thin yellow-white layer of leucocytes that overlies the mass of red cells when whole blood is centrifuged.

bufotenine (mappine) An indole alkaloid similar to serotonin that has hallucinogenic effects, originally isolated from the skin glands of toads (*Bufo* sp.). Found in many plant species.

Buggy Creek virus *See* CHIKUNGUNYA VIRUS.

bulimia nervosa A psychiatric disorder characterized by episodes of binge eating and compensatory behaviour (self-induced vomiting or laxative abuse) combined with an excessive concern over weight and body shape. A locus for susceptibility to bulimia nervosa-1 has been mapped to chromosome 10p. Both bulimia nervosa-2 and *anorexia nervosa-2 have been associated with polymorphisms in the gene encoding *brain-derived neurotropic factor.

bulla (*pl.* bullas or **bullae)** A blister or bleb containing clear fluid, larger than a vesicle.

bullous impetigo *See* EXFOLIATIN.

bullous pemphigoid A form of *pemphigus in which blisters (bulli) form on the skin and there are circulating antibodies (usually IgG) to specific hemidesmosomal antigens of stratified epithelium. The antibody titre does not correlate with the severity of the disease.

b

bumetanide A potent loop *diuretic that inhibits sodium reabsorption in the ascending limb of the loop of Henle.

bundle of His A group of cardiac muscle cells specialized for electrical conduction that transmit electrical impulses from the atrioventricular node to the point of the apex of the fascicular branches and from there to the Purkinje fibres that regulate heartbeat.

bungarotoxins Neurotoxins from the venom of the multibanded krait *Bungarus multicinctus*. **α-Bungarotoxin** (74 aa) binds to the α subunits of the postsynaptic nicotinic acetycholine receptors (nAChR) causing irreversible blockage of the neuromuscular junction and is used experimentally to identify, quantify, and localize these receptors. **β-Bungarotoxin** is a heterodimeric *phospholipase A2 neurotoxin that acts at the presynaptic site of motor nerve terminals and blocks transmitter release. The A subunit (EC 3.1.1.4, 147 aa) is structurally homologous to other PLA2s, the B subunit (85 aa) has homology with serine peptidase inhibitors (but no inhibitory activity) and *dendrotoxins. **κ-Bungarotoxin** is a homodimeric toxin (bungarotoxin 3.1, Toxin F, neuronal bungarotoxin, 87 aa) similar to α-bungarotoxin but binds more strongly to neuronal receptors than to muscle receptors.

bunion An enlarged and deformed joint of the big toe.

Bunyaviridae A family of enveloped viruses (90–100 nm diameter) that infect vertebrates and arthropods. The genome is a single strand of negative sense RNA molecules. Some genera (*Hanta viruses, *Nairovirus) cause serious disease.

bupivacaine A local anaesthetic.

buproprion hydrochloride An antidepressant, unrelated to *tricyclic antidepressants and *selective serotonin reuptake inhibitors, that is used to assist in smoking cessation. *TN* Wellbutrin.

Burkholderia pseudomallei *See* MELIOIDOSIS.

Burkitt's lymphoma A rare malignant tumour of *lymphoblasts derived from B cells that affects children in tropical Africa. *Epstein–Barr virus and immunosuppression due to malarial infection are involved but the cause is mutation in the **myc* gene or translocations involving *myc* and immunoglobulin genes.

burr cells *See* ECHINOCYTES.

bursa A small sac, lined by synovial membrane and filled by synovial fluid, that acts as a cushion between bones and tendons or muscles around a joint. **Bursitis** is an inflammation of a bursa.

bursa of Fabricius A haematopoietic tissue in birds that is also necessary for development of B cells (and the origin of their designation as B cells). In mammals the bone marrow serves the same function.

burst forming unit (BFU-E) A *stem cell lineage in the bone marrow that is detected in culture by a proliferative response to *erythropoietin and production of erythrocytes after about twelve mitotic cycles.

Buruli ulcer A tropical disease caused by infection with *Mycobacterium ulcerans* that leads to the formation of large ulcers, usually on the legs or arms. It is the third most common mycobacteriosis of humans worldwide, after tuberculosis and leprosy. Most of the tissue damage is due to the bacterial toxin *mycolactone, which also inhibits the immune response.

buspirone HCl An anxiolytic non-sedating drug that acts as a serotonin receptor agonist. It is unrelated to benzodiazepines and thought to be nonaddictive. *TNs* Anxiron, BuSpar, Narol, Spamilan, Spitomin, Sorbon.

butobarbital (butobarbitone) A barbiturate hypnotic and sedative drug used to treat seizure disorders and severe insomnia, also as a sedative before surgery. *TNs* Butomet, Soneryl.

butyrophilins A subfamily of integral membrane proteins (~525 aa) of the immunoglobulin superfamily, found in the secretory epithelial cells of the breast where they may act as receptors for the association of cytoplasmic droplets with the apical plasma membrane. They become incorporated into the membrane of milk fat globules.

butyrylcholinesterase A family of enzymes (BChE, choline esterase II, pseudocholinesterase, EC 3.1.1.8, 602 aa), produced mainly in the liver, that hydrolyse esters such as *procaine and *suxamethonium. There are several variants, dependent on four alleles, with different enzymatic activity; the variants that are expressed determine sensitivity to suxamethonium. Defects in the enzyme cause **butyrylcholinesterase deficiency**, an autosomal recessive metabolic disorder in which there is prolonged apnoea after the use of certain anesthetic drugs, including the muscle relaxant succinylcholine.

byssinosis An asthma-like occupational respiratory disease caused by exposure to vegetable fibres (cotton, jute, flax).

bystander effect The phenomenon in which mammalian cells irradiated in culture produce damage-response signals which are communicated to their unirradiated neighbours. The effect has important implications although the mechanism is unclear.

bystander help Nonspecific T-helper effect in which T cells stimulated by one antigen enhance the response of B cells stimulated by different antigens, presumably through the release of cytokines.

bystin A protein (306 aa) that mediates the interaction between *trophinin and tastin in the formation af an adhesion complex involved in implantation of the blastocyst to the uterine wall. Binding to trophinin and tastin is enhanced when cytokeratins 8 and 18 are present.

bZip *See* BASIC LEUCINE ZIPPER PROTEINS.

c- Prefix used for the normal cellular form of a viral oncogene, e.g. c-*src*, the normal homologue of the viral oncogene v-*src*.

C_0t curve The graphical representation of the kinetics of renaturation of DNA that has been heated to make it single-stranded and then cooled to allow reformation of duplex strands. The process can be followed spectroscopically and the rate is affected by the length of the DNA fragments and by the extent of repetition. Highly repetitive DNA anneals much faster than unique sequence DNA (single copies of sequences).

C1–C9 *See* COMPLEMENT.

C2-kinin A vasoactive peptide thought to be generated by cleavage of complement C2b fragment by plasmin in hereditary angioedema; the major mediator is now considered likely to be bradykinin.

C3G A rap1 guanine nucleotide exchange factor (1077 aa) ubiquitously expressed in adult and fetus that transduces signals from *crk to activate *ras.

C3 nephritic factor An autoantibody that binds to and stabilizes the C3 convertase enzyme of the alternate *complement pathway (C3bBb). Deficiencies in C3 nephritic factor are associated with type 2, dense-deposit membranoproliferative glomerulonephritis and partial *lipodystrophy.

C57BL An inbred strain of black mice.

CAAT box A nucleotide sequence found in the *promoter region of many genes, usually ~90 bp upstream. It is the motif to which *NF-1 binds.

cabergoline An ergot derivative that is a potent dopamine D_2 receptor agonist and used in treating Parkinson's disease. *TNs* Cabaser, Dostinex.

cabin A nuclear *calcineurin-binding protein (calcineurin-binding protein 1, 2200 aa) with a regulatory role in the T-cell-receptor signalling pathway. It binds *MEF2B, and probably all the MEF family of transcription factors. If intracellular calcium levels rise, activated calmodulin will displace the MEFs, allowing them to act.

cachectin An early name given to the factor that causes wasting (cachexia) in cancer, now known to be *tumour necrosis factor (TNF).

cachexia The wasting and emaciation associated with cancer and other serious diseases such as AIDS.

Caco cells A cell line with epithelial morphology that was derived from a primary colonic carcinoma of a 72-year-old white man.

CAD **1.** A multifunctional protein (2225 aa) with the enzymatic activities required for the first three steps of the *de novo* pyrimidine synthesis pathway: glutamine-dependent carbamoyl-phosphate synthase (EC 6.3.5.5), aspartate carbamoyltransferase (EC 2.1.3.2), and dihydro-orotase (EC 3.5.2.3). It is allosterically regulated and loses feedback inhibition when phosphorylated. **2.** *See* DNA FRAGMENTATION FACTOR.

CADASIL A disorder (cerebral autosomal dominant arteriopathy with subcortical infarcts and leukoencephalopathy) characterized by relapsing strokes with neuropsychiatric symptoms that affects relatively young adults of both sexes. Caused by a mutation in *notch-3 which has various effects but particularly on vascular smooth muscle, hence perhaps the arteriopathy.

cadaverine A strong-smelling substance formed in decaying meat and fish by microbial decarboxylation of lysine. Like other diamines (e.g. *putrescine) it has effects on cell proliferation and differentiation.

cadherins Integral transmembrane proteins that form homodimers that are calcium-sensitive links between cells at *adherens junctions although they may occasionally form heterodimers. **Cadherin-1** (E-cadherin, uvomorulin, CD324, 882 aa) is found in epithelia, **cadherin-2** (N-cadherin, 906 aa) in neural cells, and **cadherin-3** (CDH3, P-cadherin,

829 aa) in placenta. **Cadherin-4** (916 aa) is retinal cadherin and **cadherin-5** is vascular endotherial cadherin (784 aa). Many other tissue-specific cadherins are known and the cadherin superfamily includes cadherins, protocadherins, desmogleins, and desmocollins. Cell adhesion defects caused by mutation in cadherin-1 are associated with many carcinomas and cadherin-1 interacts with the *catenins. Defects in CDH3 are the cause of *hypotrichosis with juvenile macular dystrophy and *ectodermal dysplasia with ectrodactyly and macular dystrophy.

Caenorhabditis elegans A nematode that has been extensively used as an experimental animal. The number of nuclei is determined (there are 959 cells), an advantage for lineage studies, and the nervous system is relatively simple. The genome was sequenced at an early stage and many mutations are known. A remarkable number of *C. elegans* genes have human homologues.

caeruloplasmin *See* CERULOPLASMIN.

caesium chloride A simple salt used to produce aqueous solutions of high density, used in ultracentrifugation studies on macromolecules.

CAF-1 *See* CHROMATIN ASSEMBLY FACTOR.

caffeine A methyl *xanthine that inhibits cAMP phosphodiesterase. The stimulatory effects of coffee are presumably due to an increase in cellular cAMP levels somewhere.

Caffey's disease A disorder of *collagen type 1 (infantile cortical hyperostosis) in which there is an episode of massive subperiosteal new bone formation that typically involves the diaphyses of the long bones, mandible, and clavicles. The bone changes usually begin before 5 months of age and resolve before 2 years of age.

caged ATP A biologically inactive form of ATP that can be 'released' by cleavage of a photosensitive bond.

caisson disease Decompression sickness ('the bends').

Cajal bodies (coiled bodies) Spherical suborganelles (0.1–2.0 μm) found in the nucleus of transcriptionally active and rapidly proliferating cells. There may be between one and ten per nucleus. They contain proteins involved in RNA splicing, ribosome biogenesis, and transcription, such as *coilin.

CAK *See* CYCLIN-DEPENDENT KINASE ACTIVATING KINASE.

calbindins Members of the family of calcium-binding proteins containing an *EF-hand motif that includes calmodulin, parvalbumin, troponin C, and S100 protein. **Calbindin 1** (261 aa) is found in primate striate cortex and other neuronal tissues. Calbindin-positive neurons in the thalamus are thought to play a role in arousal by stimulating the cortex. **Calbindin 2** (calretinin) is similar to calbindin 1 but slightly larger (271 aa) and found in cerebellar Purkinje cells. **Calbindin 3** (intestinal calcium-binding protein, 79 aa) is found in the intestine and is induced by vitamin D.

calcein (fluorexon) A calcium chelator that fluoresces brightly when calcium is bound. Can be transported into cells as the acetomethoxy derivative (calcein AM) and becomes trapped when de-esterified. Used as a viability test and for short-term marking of cells.

calcicludine A toxin (60 aa), from the eastern green mamba *Dendroaspis angusticeps*, that blocks L-, N-, or P-type calcium channels. Has structural homology with Kunitz-type serine peptidase inhibitors and *dendrotoxins.

calciferol *See* VITAMIN D.

calcineurin A calmodulin-stimulated protein phosphatase (EC 3.1.3.16, 521 aa) found in brain. The phosphatase activity is inhibited by binding of an *immunophilin–ligand complex.

calcinosis The deposition of calcium salts in tissues, usually as hydroxyapatite crystals or amorphous calcium phosphate.

calciosome A membrane compartment proposed to be the equivalent of the sarcoplasmic reticulum in nonmuscle cells but distinct from the endoplasmic reticulum (probably obsolete).

calcipressins A family of endogenous *calcineurin inhibitors. **Calcipressin 1** (Down's syndrome critical region protein 1, DSCR1, myocyte-enriched calcineurin-interacting protein 1, MCIP1, Adapt 78, 252 aa) inhibits calcineurin-dependent transcriptional responses by binding to the catalytic domain of calcineurin A and is highly expressed in heart, brain, and skeletal muscle. When phosphorylated it is more active but has a shorter half-life. **Calcipressin 2** (197 aa) and **calcipressin 3** (241 aa) are similar but with different tissue distribution.

calciseptine A toxin (60 aa), from the black mamba *Dendroaspis polylepis*, that is a specific blocker of L-type calcium channels, causes relaxation of smooth muscle, and inhibits contractility in cardiac muscle but has no effect on skeletal muscle.

calcitermin *See* CALGRANULINS.

calcitonin A polypeptide hormone (32 aa) produced by parafollicular (C cells) of the thyroid that causes a reduction of blood calcium ion levels, *compare* PARATHYROID HORMONE. Alternative splicing gives rise to **calcitonin gene-related peptide** (CGRP), a neuropeptide (37 aa) that colocalizes with *substance P in neurons, regulates neuromuscular junctions, antigen presentation, vascular tone, and sensory neurotransmission. The **calcitonin family peptides** are a group of small (32–51 aa) highly homologous peptides that are ligands for G-protein-coupled receptors. The family includes *adrenomedullin, *amylin, calcitonin, and CGRP. The receptors are themselves regulated by *RAMPs.

calcitriol *See* VITAMIN D (D_3).

calcium ATPase Any ATPase involved in active transport of calcium ions, but usually refers to the *sarcoplasmic–endoplasmic reticulum calcium ATPase.

calcium-binding proteins Proteins that bind calcium: some are similar to *calmodulin (*EF-hand proteins), others bind calcium and phospholipid (e.g. *endonexin) and are generically called *annexins. Other proteins can also bind calcium, usually through a domain with homology to the EF-hand.

calcium blocker A compound (calcium antagonist) that blocks or inhibits movement of calcium ions across membranes. Most are potent vasodilators and some are antiarrhythmic. Examples include nifedipine, verapamil, and diltiazem.

calcium/calmodulin-dependent kinases Widely distributed calcium-regulated kinases (CaMKs). **CaMK1** (EC 2.7.11.17, 370 aa) phosphorylates *synapsins, *CREB, and *CFTR and is involved in cellular processes such as transcriptional regulation, hormone production, translational regulation, regulation of actin filament organization, and neurite outgrowth. It is involved in calcium-dependent activation of the ERK pathway. **CaMK II** (EC 2.7.1.37) is composed of four different chains, α, β, γ, and δ (all ~500 aa), which assemble into homo- or heteromultimeric holoenzymes composed of eight to twelve subunits. It has been implicated in various neuronal functions, including *synaptic plasticity and is highly concentrated in the postsynaptic region. Other CaM kinases with restricted tissue distribution are also known. **Calcium/calmodulin-dependent protein kinase kinase 1** (CaMKK1, EC 2.7.11.17, 505 aa) phosphorylates CaM kinases and is part of a proposed calcium-triggered signalling cascade.

calcium channel A transmembrane channel that is specific for calcium ions. *See* VOLTAGE-SENSITIVE CALCIUM CHANNELS; RYANODINE (ryanodine receptor). The **calcium current** is the result of influx of calcium through calcium channels which is important for the discharge of neurotransmitter from vesicles in the presynaptic terminal. **Calcium channel blockers** such as diltiazem and the *dihydropyridines are used for treatment of hypertension and angina.

calcium phosphate precipitation A method for introducing DNA or chromosomes into cells (transfection) by coprecipitation with calcium phosphate.

calcium pump A *transport protein responsible for moving calcium out of the cytoplasm. *See* SARCOPLASMIC–ENDOPLASMIC RETICULUM CALCIUM ATPASE.

calcium sensing receptor A plasma membrane G-protein-coupled receptor that is expressed in the *parathyroid hormone-producing cells of the parathyroid gland and the cells lining the kidney tubule. By sensing blood calcium levels it regulates parathyroid hormone secretion and thus calcium homeostasis. Mutations can lead to *hypoparathyroidism.

calcofluor A fluorochrome that exhibits antifungal activity and has a high affinity for yeast cell wall chitin. Also used as a whitening agent in paper, textiles, etc.

calculus (*pl.* calculi) A stone formed in an organ or duct; a concretion, usually of mineral salts, that forms in the body (the process of lithiasis). Kidney stones are usually of calcium oxalate.

calcyclin A protein (S100-A6, 90 aa) associated with the *prolactin receptor, and one of the *S100 calcium-binding proteins that is regulated through the cell cycle. Binds to annexin II (p36) and to glyceraldehyde-3-phosphate dehydrogenase.

calcyphosine (calcyphosin) A cytoplasmic *EF-hand calcium-binding protein (thyroid protein p24, 189 aa) that is involved in both Ca^{2+}-phosphatidylinositol and cAMP signal cascades. **Calcyphosine-2** (557 aa) is similar and expressed in many tissues. **Calcyphosine-like protein** (208 aa) is also known.

caldesmon A protein (793 aa) with several tissue-specific isoforms that is implicated in the regulation of actomyosin interactions in smooth muscle and nonmuscle cells by stimulating the binding of tropomyosin to actin filaments. In muscle, inhibits the actomyosin ATPase in a calcium-sensitive fashion by binding to F-actin. Has an important role in mitosis and receptor capping. Caldesmon can block the effect of *gelsolin on F-actin and will dissociate actin–profilin complexes.

calelectrin *Annexin A6 (673 aa), which may regulate the release of calcium from intracellular stores.

calexcitin A calcium and GTP-binding protein (191 aa) found in invertebrates that interacts with and activates the ryanodine receptor; it also activates the calcium ATPase. It is implicated in neuronal excitation and plasticity in the squid (*Loligo* sp.).

calgizzarin A calcium-binding protein (S100 calcium-binding protein A11, 105 aa) of the *S100 family, originally isolated from chick gizzard, implicated in the differentiation and cornification of keratinocytes.

calgranulins Members of the *S100 calcium-binding protein family. **Calgranulin-A** (S100-A8, 93 aa) is expressed by macrophages in chronic inflammation and is also expressed in epithelial cells constitutively or induced during dermatoses. May interact with intermediate filaments in monocytes and epithelial cells. **Calgranulin-B** (S100-A9, migration inhibitory factor-related protein 14, 114 aa) is expressed in inflammation and may inhibit protein kinases. **Calgranulin-C** (S100-A12, 92 aa) is proteolytically cleaved to calcitermin, which possesses antifungal and some antibacterial activity.

calicin An actin-binding protein (588 aa) found in the postacrosomal calyx region of vertebrate sperm. Has multiple *Kelch repeats.

caliciviruses A family of positive-sense, single-stranded RNA viruses that are a major cause of gastrointestinal upsets in humans. Rabbit haemorrhagic fever is caused by a calicivirus and was released (prematurely) to control rabbit numbers.

calitoxins Toxins (~79 aa), from the sea anemone *Calliactis parasitica*, that inhibit sodium channels.

callosity (callus) A skin thickening caused by persistent irritation or friction.

callus 1. A mass of new bony trabeculae and cartilaginous tissue formed by *osteoblasts at an early stage in the healing of bone. **2.** *See* CALLOSITY.

calmegin A calcium-binding membrane protein (610 aa) of the *calreticulin family, similar to *calnexin, that plays a role in spermatogenesis.

calmidazolium An inhibitor of calmodulin-regulated enzymes that also blocks sodium channels and voltage-sensitive calcium channels.

calmodulin A ubiquitous and highly conserved calcium-binding protein (149 aa) with four *EF-hand motifs that regulates a wide range of enzymes. The probable ancestor of proteins such as *troponin C and *parvalbumin.

calnexin A calcium-binding lectin-like protein (MHC class I antigen-binding protein p88, 592 aa) of the *calreticulin family, that couples glycosylation of newly synthesized proteins in the endoplasmic reticulum with their folding, acting as a chaperone and preventing degradation.

calpactins *See* ANNEXINS.

calpains Calcium-activated cytoplasmic thiol endopeptidases (formerly EC 3.4.22.17) with broad specificity, involved in cytoskeletal remodelling and signal transduction. The larger catalytic subunit has EF-hand calcium-binding motifs and is associated with a smaller regulatory subunit that has a calmodulin-like domain. **Calpain-1** (EC 3.4.22.52, CAPN1, 714 aa) is activated by micromolar calcium, and is inhibited by *calpastatin. **Calpain-2** (EC 3.4.22.53) is activated by millimolar calcium. Defects in **calpain-3** (EC 3.4.22.54, CAPN3, 821 aa) are the cause of limb-girdle *muscular dystrophy type 2A. Genetic variations in **calpain-10** are associated with susceptibility to a type of diabetes. Various tissue-specific calpains are known.

calpastatin A cytoplasmic inhibitor (708 aa) of *calpain, more effective against calpains-1 and -2 than calpain-3.

calpeptin A cell-permeable *calpain inhibitor.

calphobindins *Annexins V and VI (320 aa, 673 aa) found in the placenta.

calphostin C An inhibitor of *protein kinase C isolated from *Cladosporium cladosporioides*. It will inhibit other kinases at higher concentrations.

calponins Thin-filament-associated proteins implicated in the regulation and modulation of smooth muscle contraction. They will bind actin, calmodulin, troponin C, and tropomyosin and will inhibit myosin ATPase activity when not phosphorylated. Various isoforms are known:

calponin-1 (calponin H1, basic calponin, 297 aa), **calponin-2** (calponin H2, neutral calponin, 309 aa), **calponin-3** (acidic calponin, 329 aa). **Calponin homology domains** (CH domains) are actin-binding domains found in cytoskeletal and signal transduction proteins; actinin-type actin-binding domain are a subclass of the CH domains.

calregulin *See* CALRETICULINS.

calreticulins *Calcium-binding proteins with lectin-like domains that are found in the *endoplasmic reticulum where they act as chaperones for newly synthesized proteins, in conjunction with *calnexin. There are two forms, **calreticulin-1** (calregulin, high-affinity calcium-binding protein, HACBP, 417 aa) and **calreticulin-2** (calreticulin-3, 384 aa).

calretinin *See* CALBINDINS.

calsarcin *See* MYOZENIN.

calsequestrin A family of proteins with tissue-specific isoforms that are high-capacity, moderate affinity calcium-binding proteins that sequester calcium in the sarcoplasmic reticulum. **Calsequestrin-1** (390 aa) is the skeletal muscle isoform, **calsequestrin-2** that of cardiac muscle.

calstabin An *immunophilin (FK506 binding protein 1B, FKBP1B) that acts as a channel-stabilizing protein (80 aa) for the cardiac *ryanodine receptor channel complex. Depletion can cause an intracellular calcium leak and trigger a fatal cardiac arrhythmia.

calumenin (crocalbin) A widely expressed protein (315 aa) with *EF-hand motifs that binds seven calcium ions with a low affinity and is involved in the regulation of vitamin K-dependent carboxylation of multiple N-terminal glutamate residues. A member of the *CREC family of proteins.

calvarium The upper part of the skull, made up of the frontal, two parietals, occipital, and two temporal bones. Calvaria from experimental animals are used in organ culture to investigate bone metabolism.

calvasculin *See* S100 CALCIUM-BINDING PROTEINS.

Calvé's disease **1.** A rare aseptic necrosis (juvenile osteochondrosis of spine; Scheuermann's disease) of a vertebral body (vertebra plana), usually in children and probably the result of an eosinophilic granuloma. **2.** *See* PERTHES' DISEASE.

calycin superfamily. A protein superfamily sharing structural features. It includes the *lipocalins, *fatty acid-binding proteins (FABPs), *avidins, triabin (an anticoagulant produced by the triatomine bug *Triatoma pallidipennis*) and a group of bacterial metalloprotease inhibitors (MPIs).

calyculin A A serine/threonine protein phosphatase inhibitor isolated from the marine sponge *Discodermia calyx*. Calyculin A is a potent tumour promoter.

CAM *See* CELL ADHESION MOLECULE.

cambium The inner region of the periosteum from which *osteoblasts differentiate. NB. In plants the cambium is a meristematic layer of cells that produce daughter cells that differentiate into various tissues.

camera lucida An aid to drawing objects viewed under the microscope, now rarely used. It allows a simultaneous view of the object and the viewer's hand and pen.

CaM kinase *See* CALCIUM/CALMODULIN-DEPENDENT KINASES.

cAMP *See* CYCLIC AMP.

campomelic dysplasia A developmental disorder characterized by congenital bowing and angulation of long bones, together with other skeletal and extraskeletal defects. Many XY individuals have genital defects or may develop as phenotypic females. Caused by mutation in *sox 9.

camptothecin A cytotoxic alkaloid that inhibits DNA *topoisomerase I, originally isolated from *Camptotheca acuminata* (cancer tree, tree of life). Various derivatives such as CPT-11 (*Irinotecan) are used as antitumour drugs.

Campylobacter A genus of Gram-negative microaerophilic motile bacteria with a single flagellum at one or both poles that are a common cause of food poisoning and can cause opportunistic infections. **Helicobacter pylori* was previously known as *Campylobacter pyloridis*.

Camurati–Engelmann disease A disorder in which there is osteitis (excessive bone growth), particularly of the lower limbs, caused by mutation in *TGFB1*, the gene for *transforming growth factor β. There is a variant of the disease in which the TGFβ gene is unaffected.

canal cell A cell type, probably endothelial in origin, that lines Schlemm's canal, a circular channel in the eye that collects aqueous humour from the anterior chamber and delivers it into the bloodstream.

canaliculi Small channels (canals). In bone the canaliculi link lacunae in the calcified matrix that contain *osteocytes. In liver, small channels between *hepatocytes through which bile flows to the bile duct.

canavanine A non-protein amino acid found in plants, especially legumes and their seeds. L-Canavanine is a selective inhibitor of inducible *nitric oxide synthase (iNOS).

Canavan's disease A severe neurological disorder involving cerebral spongy degeneration, demyelination, and leukodystrophy, fatal in early life, due to mutation in the gene encoding *aminoacylase-2.

cancellous bone (trabecular bone) The type of bone (spongy bone) found at the end of long bones consisting of regularly ordered mineralized collagen fibres in a looser array than in the lamellar bone of the shaft.

cancer Originally descriptive of carcinoma, now a colloquial term for malignant neoplastic diseases.

cancer susceptibility gene *See* TUMOUR SUPPRESSOR.

candesartan A selective AT_1-subtype *angiotensin II receptor antagonist used to treat hypertension; marketed as the prodrug candesartan cilexetil which is metabolized by esterases during absorption from the intestine. *TNs* Amias, Atacand, Blopress, Ratacand.

candidiasis Infection by the fungus *Candida albicans* (formerly *Monilia*), commonly on mucous membranes ('thrush'), but can cause widespread opportunistic infection in immunocompromised individuals.

canicola fever A disease caused by infection by *Leptospira canicola*, normally a pathogen of dogs.

cannabinoids A group of related compounds found in *cannabis that mimic the actions of endogenous agonists *anandamide and palmitoyl ethanolamine. The major cannabinoids are cannabidiol, cannabidol, and various tetrahydrocannabinols (THCs). The receptors are G-protein coupled; the CB_1 type is mostly found in brain where it mediates psychotropic activities, the CB_2 receptors are more peripheral and found extensively in the immune system.

cannabis General name for various parts of the plant *Cannabis sativa* or *Cannabis indica* (Indian hemp) from which hashish, marihuana, or bhang are derived, and a source of *cannabinoids. Most recreational use is for the psychotropic effects but it is also used (often illegally) to reduce muscular spasm and pain in multiple sclerosis.

canonical Describing something classical, archetypal, or prototypic, often applied to a sequence of nucleotides or amino acids that reflects the most common choice of base or amino acid at each position. A canonical pathway is the orthodox or accepted pathway, a term often used e.g. for the *wnt signalling pathway.

cantharidin (cantharadin) A vesicant (blistering agent) prepared from the dried wing-cases (elytra) of the Spanish fly (blister beetle) *Lytta vesicatoria*. Formerly used topically in the treatment of warts and *molluscum contagiosum but is dangerously toxic. Inhibits *protein phosphatase 2A.

CAP **1.** *See* CYCLASE ASSOCIATED PROTEIN. **2.** *See* CAP PROTEINS **3.** *See* CAP-BINDING PROTEIN.

CAP-18 *See* CATHELICIDIN.

capacitance flicker A phenomenon seen in *patch clamping, caused by brief transitions of ion channels between the open and closed state.

capacitation The changes that occur in sperm after they are exposed to secretions in the female genital tract and that are necessary before the sperm can fertilize an egg.

cap-binding protein A protein (eukaryotic initiation factor 4E, eIF4E, 217 aa) that binds to the 5′-cap of mRNA and, together with other initiation factors, regulates binding of mRNA to the 40S ribosomal subunit.

capecitabine An oral prodrug used in tumour chemotherapy that is enzymatically converted to *fluorouracil. *TN* Xeloda.

capG One of the gelsolin family of proteins (macrophage capping protein, MCP, gCap39, 348 aa) that binds to the barbed ends of microfilaments. Has ~50% sequence homology with *gelsolin, and like gelsolin responds to calcium and to phosphoinositides; unlike gelsolin it does not sever filaments. Originally isolated from macrophages but widely distributed and is the only gelsolin-related actin-binding protein found in both nucleus and cytoplasm. It is also secreted into plasma, though does not have a signal sequence. Not the same as *capZ, though functionally similar.

capillary Small blood vessel that links an arteriole with a venule. The lumen, which may be enclosed by a single endothelial cell, is sufficiently small that erythrocytes must deform in order to pass through. There may be a precapillary sphincter that regulates transient opening.

capillary electrophoresis A highly sensitive electrophoretic method for separating compounds by electro-osmotic flow through a long (~100 cm), narrow (<100 μm) silica column.

capnine A sulphonolipid derived from ceramide, found in the envelope of the *Cytophaga/Flexibacter* group of gliding Gram-negative bacteria.

capnophilic Describing an organism that thrives in concentrations of carbon dioxide higher than normally found in air.

capping **1.** The accumulation of crosslinked cell-surface material at the rear of a moving cell, or a region above the nucleus. **2.** The intracellular accumulation of intermediate filament protein in the pericentriolar region when microtubules are disrupted by colchicine or vinca alkaloids. **3.** The binding of a protein to the free (assembly) end of a linear polymer such as actin, blocking further subunit addition. *See also* CAP-BINDING PROTEIN.

CAP proteins Obsolete name for *Fas-associated via death domain (FADD) proteins (cytotoxicity-dependent APO-1-associated proteins).

cappuccino **1.** The product of a gene mutated in a murine model of *Hermansky–Pudlak syndrome that is involved in the development of lysosome-related organelles, such as melanosomes and platelet-dense granules. The human homologue (217 aa) interacts with *pallidin and *muted proteins in the *BLOC-1 complex. **2.** An actin-nucleation factor that regulates the onset of ooplasmic streaming in *Drosophila*, a maternal-effect locus important for pattern formation in the early embryo. There are multiple isoforms ranging from 819 aa to 1361 aa.

capsaicin The active principle in chillies that is the ligand for the *vanilloid receptor-1 (VR_1) and causes the sensation of hotness. It will stimulate the release of neurogenic peptides (substance P, neurokinins) from sensory neurons and can be used to desensitize and eventually kill nociceptors to which it binds.

capsazepine A competitive antagonist of *capsaicin.

capsid The protein coat that surrounds the nucleoprotein core or nucleic acid of a virion. The coat is composed of subunits that self-assemble into symmetrical structures which may sometimes be enclosed within an envelope (as in the *Togaviridae). **Capsomeres** are polymers of the subunit protein which then assemble into the complete coat.

capsule **1.** A layer of hydrophilic polysaccharide that covers the surface of some Gram-positive and Gram-negative bacteria, giving colonies a 'smooth' appearance. The hydrophilicity inhibits phagocytosis by leucocytes and thus contributes to pathogenicity. **2.** A multicellular layer of cells surrounding a foreign object too large to be phagocytosed. In some cases the cells may fuse to form multinucleate giant cells. **3.** A dense connective tissue sheath around an organ.

captopril An *angiotensin converting enzyme (ACE) inhibitor, used to treat hypertension.

capZ A heterodimeric microfilament capping protein (capping protein, cap32/34, α, 286 aa; β, 277 aa) that binds to the barbed (assembly) end, blocking further G-actin addition. It is calcium-insensitive and does not sever filaments. There are multiple isoforms, some tissue-specific, and in muscle it is found in the *Z-disc (sometimes called β-actinin). **CapZ-interacting protein** (protein kinase substrate CapZIP, 416 aa) is phosphorylated in stress conditions and then binds to, and inhibits, capZ.

carbachol (carbamoyl choline) A substituted form of acetylcholine (with a carbamoyl group, $-CONH_2$, in place of the acetyl moiety) that is not hydrolysed by *acetylcholine esterase and acts on both *muscarinic and *nicotinic acetylcholine receptors. Mostly used for treating glaucoma. *TNs* Carbastat, Carboptic, Isopto Carbachol, Miostat.

carbamazepine An anticonvulsant and mood-stabilizing drug used primarily in the treatment of epilepsy, bipolar disorder that is unresponsive to lithium, trigeminal neuralgia, and phantom limb pain.

carbamoylcholine *See* CARBACHOL.

carbamoyl phosphate synthetase An enzyme activity (CPS, glutamine-dependent carbamoyl-phosphate synthase, EC 6.3.5.5) present in one of the domains of *CAD protein, responsible for a key step in pyrimidine biosynthesis.

carbamyl Obsolete variant of carbamoyl.

carbenicillin A semisynthetic penicillin-class antibiotic sometimes used as a selective agent in molecular biology.

carbenoxolone A synthetic derivative of *glycyrrhizin used to treat oesophageal ulceration and inflammation.

carbidopa An inhibitor of aromatic L-amino acid decarboxylase (DOPA decarboxylase) often used in combination with *levodopa in the treatment of Parkinson's disease. *TN* Lodosyn.

carbimazole A prodrug that is converted into the active agent methimazole, which inhibits *thyroxine synthesis and is used to treat hyperthyroidism.

carbohydrate-deficient-glycoprotein syndrome type 1 *See* JAEKEN'S SYNDROME.

carbohydrates An abundant and diverse group of compounds that have the general formula $C_n(H_2O)_n$. The smallest are monosaccharides like glucose, the largest are complex polymers such as starch, cellulose, and glycogen that are indeterminate in length.

carbonic anhydrase The enzyme (EC 4.2.1.1, carbonate dehydratase, 260 aa) that catalyses reversible hydration of carbon dioxide to form carbonic acid. It is an intracellular enzyme of the erythrocyte and is important for carbon dioxide transport. Defects in the enzyme cause autosomal recessive *osteopetrosis type 3 (Guibaud–Vainsel syndrome, marble brain disease), associated with renal tubular acidosis. There are multiple isoforms, and defects in carbonic anhydrase 4 cause *retinitis pigmentosa type 17.

carbon monoxide *See* CARBOXYHAEMOGLOBINAEMIA.

carbon replica A surface replica of a specimen produced by vacuum-depositing carbon on to a surface that is to be examined in the electron microscope; the film can then be shadowed with heavy metal. The specimen itself is removed using an appropriate solvent.

carboprost A synthetic analogue of prostaglandin $PGF_{2\alpha}$ that has *oxytocin-like properties.

carboxyglutamate (γ-carboxyglutamate) An amino acid, formed by post-translational carboxylation of glutamate, found in some proteins, particularly those that bind calcium.

carboxyhaemoglobinaemia The potentially life-threatening condition in which carbon monoxide is bound to haemoglobin and blocks its oxygen-carrying capacity; haemoglobin has greater affinity for CO than for O_2. There can be asphyxia without cyanosis because haemoglobin complexed with CO is pink.

carboxypeptidases A group of enzymes that remove the C-terminal amino acid from a protein or peptide. **Carboxypeptidase A** (EC 3.4.17.1) will remove any amino acid; **carboxypeptidase B** (EC 3.4.17.2) is specific for terminal lysine or arginine. Many of the carboxypeptidases are involved in processing of proteins. The nomenclature is complex and may be based on mechanism (metallocarboxypeptidases, EC 3.4.17.-; serine carboxypeptidases, EC 3.4.16.- and cysteine carboxypeptidases, EC 3.4.18.-) or on substrate specificity (e.g. glutamate carboxypeptidase).

carboxy terminus *See* C-TERMINUS.

carbuncle A staphylococcal infection of subcutaneous tissue that produces an abscess larger than a *boil.

carcinoembryonic antigen (CEA) A group of immunoglobulin superfamily cell surface proteins involved in adhesion and found in the blood of patients suffering from some carcinomas and sometimes used as a tumour marker. Normally restricted to fetal tissue and absent in adults. **CEA1** (carcinoembryonic antigen-related cell adhesion molecule 1, biliary glycoprotein 1, CD66a, 526 aa) was the first identified: many others are now recognized.

carcinogen An agent that can stimulate the development of a tumour. Most carcinogens are chemicals but electromagnetic radiation can be carcinogenic. **Carcinogenesis** is the process by which normal epithelial cells become neoplastic and malignant. Although strictly carcinogenesis should refer only to the production of carcinomas, the term is often used as a synonym for *tumorigenesis.

carcinoid An intestinal tumour (argentaffinoma) that derives from paracrine neuroendocrine secretory cells (*APUD cells). The primary tumour is most often in the appendix, where it is clinically benign; hepatic secondaries may release large amounts of vasoactive amines such as serotonin to the systemic circulation causing **carcinoid syndrome**. Some intestinal carcinoid tumours can result from mutation in the gene that encodes subunit D of *succinate dehydrogenase. Some confusion exists over terminology because in 2000 the World Health Organization redefined carcinoids as neuroendocrine tumours that are benign or have uncertain metastatic potential, in distinction to neuroendocrine cancers (which are malignant), but this redefinition has not been universally accepted.

carcinoma (*pl.* carcinomata) A malignant tumour (*see* NEOPLASIA) derived from an epithelial cell. They are the commonest type of tumour, possibly because epithelia retain proliferative capacity in the adult, unlike mesodermally derived tissues such as bone and muscle. Tumours derived from glandular tissue are often called adenocarcinomas. **Carcinoma en cuirasse** is a metastatic infiltration of cutaneous tissue, in which there is hardening and thickening (to produce a 'breastplate' effect, a cuirasse), often in the chest as a result of a recurrence of breast cancer. **Carcinoma *in situ*** is a tumour that has not invaded or metastasized and remains in the site of origin. A **carcinosarcoma** is a rare and aggressive tumour with distinct histological features of both carcinoma and sarcoma; the commonest location is the uterus. *See* BASAL CELLS; GORLIN'S SYNDROME; HEPATOCARCINOMA; MERKEL CELL CARCINOMA; NASOPHARYNGEAL CARCINOMA; OAT CELL CARCINOMA; SCIRRHOUS CARCINOMA.

carcinomatosis (carcinosis) A tumour that has metastasized extensively, giving rise to secondary tumours in many sites.

carcinosis *See* CARCINOMATOSIS.

CARD domain A protein motif (caspase activation and recruitment domain) found in a wide array of proteins, particularly those involved in inflammation and apoptosis, where it is involved in the formation of multiprotein complexes through interaction with other CARD domains. Many of the CARD domain-containing proteins are members of the *MAGUK family, e.g. **CARD4** (nucleotide-binding oligomerization domain protein 1, Nod-1, 953 aa) is expressed abundantly in adult heart, spleen, and lung. It is also found in numerous cancer cell lines and fetal tissues. The protein interacts with the CARD domain of *RICK and is a potent activator of NFκB. **CARD12** (CARD, LRR, and NACHT domain-containing protein, CLAN, ICE protease-activating factor, IPAF, 1024 aa with shorter isoforms) is ubiquitously expressed and has a role in promoting apoptosis. At least seven other CARDs are known. **CARD 8** is *cardinal.

cardiac Describing something pertaining to the heart or to the upper part of the stomach. A **cardiac aneurysm** is a dilatation of a heart ventricle due to destruction of cardiac muscle. **Cardiac asthma** (paroxysmal dyspnea, paroxysmal nocturnal dyspnea) is a sudden, severe shortness of breath at night when asleep, closely associated with congestive heart failure. **Cardiac cells** are, strictly speaking, cells of the cardium of the heart, but loosely applies to all heart cells. **Cardiac jelly** is the gelatinous extracellular matrix material that lies between the endocardium and the myocardium in the embryo. **Cardiac muscle** is the striated but involuntary muscle found only in the walls of the heart that is responsible for a lifetime of pumping blood. Unlike skeletal muscle it is not a syncytium and individual muscle cells are joined by *intercalated discs. **Cardiac tamponade** is a condition in which fluid or blood fills the pericardial sac and interferes with the pumping action of the heart. By contrast, the **cardiac sphincter** (cardia) is the sphincter that regulates the opening of the oesophagus into the stomach. *See also* CARDIO-.

cardiac ankyrin repeat protein A protein (CARP, ankrd1) found in cardiac muscle that is considered to play an important role in cardiac remodelling by transcriptional regulation and direct involvement in myofilament assembly. Expression is up-regulated in wounds and CARP may be important in revascularization. There are two close homologues, **ankrd2** (Arpp) and **DARP**, forming a conserved family of muscle ankyrin repeat proteins. *See* MYOPALLIDIN.

cardiac glycosides Plant-derived inhibitors of the *sodium-potassium ATPase especially that in heart muscle, e.g. *strophanthin.

cardif *See* MAVS.

cardinal A component of some *inflammasomes (caspase recruitment domain-containing protein 8, apoptotic protein NDPP1, 431 aa) that inhibits NFκB activation. Has a single *CARD domain.

cardio- A prefix indication association with the heart. A **cardioblast** is an embryonic mesodermal cell which will differentiate into heart tissue. **Cardiocentesis** is the puncture of the heart with a needle. A **cardiogram** (electrocardiogram, ECG) is the recording, on a **cardiograph**, of the electrical activity generated during a heartbeat. **Echocardiograms** are produced using ultrasound. **Cardiology** is the branch of medicine concerned with the function and diseases of the heart. **Cardiomalacia** is a pathological weakening of the heart muscle. **Cardiomegaly** is an abnormal enlargement of the heart. **Cardiomyopathy** is any disease affecting the heart muscle. A rare infantile form (histiocytoid cardiomyopathy) is caused by mutation in the gene encoding mitochondrial cytochrome *b*. *See* DANON'S DISEASE; DILATED CARDIOMYOPATHY; HYPERTROPHIC

CARDIOMYOPATHY; ISCHAEMIC CARDIOMYOPATHY; NAXOS DISEASE. A **cardiopulmonary bypass** is a procedure in which blood is diverted through a pump and an oxygenator to maintain the circulation during operations on the heart. (The term can also describe the equipment required.) **Cardiospasm** is a spasm of the cardia (or cardiac sphincter) of the stomach.

cardiofaciocutaneous syndrome A disorder (CFC syndrome) in which there is a distinctive facial appearance, heart defects, and mental retardation. Can be caused by mutation in K-*ras* (*See* RAS), BRAF (*See* RAF), *MEK1, or MEK2, all of which feed into the RAS–ERK pathway.

cardiolipin A diphosphatidyl glycerol, found in the membrane of *Treponema pallidum*. It can be purified from beef heart and is the antigen detected by the Wasserman test for syphilis.

cardiomyopathy *See* DANON'S DISEASE; DILATED CARDIOMYOPATHY; HYPERTROPHIC CARDIOMYOPATHY; ISCHAEMIC CARDIOMYOPATHY.

cardiotoxins Strictly speaking, toxins that act directly on the heart, but the term is actually applied to toxins (cobramine A, cobramine B, cobra cytotoxin, gamma-toxin, membrane-active polypeptide) from the venom of the cobra *Naja* spp., although they do not seriously affect the heart in humans. They are basic polypeptides of 57–62 aa with cytolytic activity.

SEE WEB LINKS

- University of Adelaide website with information about toxins.

cardiotropin-1 A cytokine (CT-1, 201 aa) of the *IL6 cytokine family. Binds to and activates the IL-6 signal transducer/gp130 receptor and will induce the synthesis of acute phase proteins in hepatocytes. *In vitro* will induce hypertrophy of cardiac myocytes.

cardiovascular Describing something associated with the heart and the vascular system.

cardiovascular disorders

The circulatory system is subject to various disorders, not only in various functions of the blood (*see* BLOOD DISORDERS) but also in the vessel wall and in the heart (*see* CARDIOMYOPATHY). A common problem is *hypertension which increases the risk of *myocardial infarction and stroke. Disorders can occur through developmental defects (*see* ENDOGLIN; JUMONJI; MYOCYTE ENHANCER FACTOR 2; SITUS INVERSUS) or defects in the connective tissue of the vessel wall (e.g. *Marfan's syndrome; *Williams–Beuren syndrome). Other cardiovascular disorders include *Brugada syndrome, *Danon's disease, and *carditis. Neovascularization in the adult can cause problems in the retina, because the new vessels are prone to haemorrhage (diabetic retinopathy, *see also* FIBROPLASIA) and can enhance tumour growth (*see* TUMOUR ANGIOGENESIS FACTOR). Accumulation of lipid in cells of the vessel wall causes *fatty streaks and *fibrous plaques and narrowing of the lumen (*atherosclerosis).

cardiovirus A genus of viruses in the family *Picornaviridae, nonenveloped viruses with a positive sense, single-stranded RNA genome, isolated mostly from rodents and responsible for encephalitis and myocarditis.

carditis Inflammation of the heart. Depending on whether the enveloping outer membrane, the muscle or the inner lining is affected, may be called pericarditis, myocarditis, or endocarditis.

caries (*adj.* carious) Progressive destruction of bone or teeth (dental caries).

carminative Describing a medicine that relieves gastric flatulence.

Carney complex An autosomal dominant multiple neoplasia syndrome in which there are cardiac, endocrine, cutaneous, and neural myxomatous tumours, as well as a variety of pigmented lesions of the skin and mucosae. Multiple endocrine glands may be involved, as in *multiple endocrine neoplasia syndrome and there are some similarities to McCune-Albright syndrome (*See* GRANINS). **Carney complex type 1** (CNC1) is caused by mutation in the gene encoding protein kinase A regulatory subunit-1α. **Carney complex type 2** is genetically heterogenous. The Carney complex variant associated with distal arthrogryposis is caused by mutation in the myosin heavy chain-8 (*MYH8*) gene.

carnitine A compound, biosynthesized in the liver and kidneys from lysine or methionine, that transports long-chain fatty acids into the mitochondrion, where they will be oxidized, Sometimes referred to as vitamin B_t or Vitamin B_7. The **carnitine palmitoyltransferase system** (CPT system) regulates fatty acid oxidation/ketogenesis in the liver and is itself switched off by malonyl CoA.

carnosine A dipeptide formed from β-alanine and histidine found at high concentration in muscle and brain tissues. Has been claimed to have beneficial effects in various disorders but there is no good evidence.

caronte A secreted antagonist (273 aa) of *bone morphogenic protein (BMP), one of the cysteine-knot *cer/dan gene family. Mediates the sonic hedgehog-dependent induction of left-specific genes in the lateral plate mesoderm of chick, but it is not clear whether the human homologue has the same role.

carotenes A range of brightly coloured compounds (terpenes) found in plants and responsible for yellow/orange/red colouration. They have an auxiliary function in photosynthesis. β-Carotene is composed of two retinyl groups and is the precursor of *vitamin A.

carotid body A well-vascularized mass of tissue close to the carotid sinus that contains chemoreceptors sensitive to the oxygen, carbon dioxide, and pH levels of the blood and involved in their regulation. **Carotid body cells**, the chemosensory cells, are derived from the neural crest. The **carotid sinus** is the dilated portion of the common carotid artery at its division into internal and external branches. The wall of the sinus has *baroreceptors (the sensory endings of the glossopharyngeal nerve) which have an important homeostatic function.

carrageenan (carrageenin) A sulphated cell-wall polysaccharide found in certain red algae and used commercially as an emulsifier and thickener in foods. Sometimes used to induce a sterile inflammatory lesion when injected into experimental animals where it probably activates *complement.

carrier **1.** In genetics, an individual heterozygous for a recessive disorder, phenotypically unaffected but capable of passing the allele to the next generation. **2.** A nonradioactive form of a radioactively labelled compound, used to dilute the radiotracer. **3.** A molecule or molecular system that transports a solute across a cell membrane either by active transport or by facilitated diffusion. **4.** An organism harbouring a parasite and, though showing no symptoms of disease, possibly infectious to others. **5.** A more or less inert material used either as a diluent or vehicle for the active ingredient, e.g. of a fungicide, or as a support, e.g. for cells in a bioreactor. **6.** *See* CARRIER PROTEIN.

carrier protein **1.** A protein to which a specific ligand or hapten has been conjugated, used as an immunogen to produce an antibody directed against the hapten. **2.** Unlabelled protein added at relatively high concentrations into an assay system that behaves in the same way as the labelled analyte protein that is present at low concentration. **3.** Protein that is added to block the nonspecific binding of reagents to surfaces, molecules in the sample, etc. Albumin (bovine serum albumin, BSA) is often used for this purpose. **4.** A membrane protein involved in the transport of small molecules from one side to the other.

CART **1.** Cocaine- and amphetamine-regulated transcript protein. A hypothalamic neuropeptide (116 aa) associated with the actions of *leptin and *neuropeptide Y that has anorectic properties, inhibiting normal and starvation-induced feeding and blocking the feeding response induced by neuropeptide Y and regulated by leptin. It promotes neuronal development and survival *in vitro*. A defect in CART is associated with reduced resting energy expenditure and cosegregates with obesity phenotype. **2.** ALX homeobox protein 1 (cartilage homeoprotein 1, CART-1, 326 aa), a paired homoebox transcription factor that may play a role in chondrocyte differentiation and may also influence cervix development. Functions as a repressor with the rat prolactin promoter *in vivo*.

cartilage A flexible and compressible *connective tissue that has extensive extracellular matrix with *collagen type II and large amounts of *proteoglycan, particularly *chondroitin sulphate. Cartilage is often an early skeletal framework that is progressively mineralized during childhood and adolescence. Cartilage is produced by *chondrocytes that come to lie in small lacunae surrounded by the matrix rich in *cartilage oligomeric matrix protein.

cartilage oligomeric matrix protein One of the *thrombospondin family of proteins (757 aa), present at high levels in the matrix surrounding *chondrocytes. There are five identical glycoprotein subunits, each with EGF-like and calcium-binding domains. Mutations cause *pseudoachondroplasia or a milder allelic disorder, epiphyseal dysplasia.

caruncle Any small fleshy excrescense.

carvedilol A nonselective *beta-blocker/alpha-1 blocker used to treat mild to moderate congestive heart failure. *TNs* Carloc, Coreg, Dilatrend, Eucardic.

caryotype *See* KARYOTYPE.

Cas An adaptor protein (p130Cas, Crk associated protein, breast cancer anti-oestrogen resistance protein 1, 870 aa) that has central coordinating role in tyrosine-kinase-based signalling related to cell adhesion and implicated in induction of cell migration and tumour invasiveness. When tyrosine-phosphorylated, Cas acts as a docking protein for binding SH2 domains of *src-family kinases. One of a family of adaptor proteins that includes human enhancer of filamentation 1 (*HEF1) and src-interacting protein (Sin)/Efs.

casamino acid The product of acid hydrolysis of *casein that is added as a nitrogen source in some culture media.

cascara sagrada The dried bark of cascara buckthorn *Rhamnus purshiana*, formerly used as a laxative.

caseation The process of becoming cheese-like (caseous), e.g. in tissue infected with tubercle bacillus where the cells break down into an amorphous cheese-like mass.

case–control study A type of epidemiological study in which subjects with the condition being investigated ('cases') are compared with matched 'controls'. A major problem is that such studies do not give any indication of the absolute risk of the factor in question.

casein A group of proteins that are the main protein component of milk. The α- and β-caseins (185 and 226 aa) form micellar polymers that precipitate in presence of calcium. κ-Casein (182 aa) is a glycoprotein that stabilizes the micelles of α- and β-casein. A proteolytic heptapeptide fragment of α-S-casein, casoxin D, acts as an opioid antagonist and has vasorelaxing activity mediated by bradykinin B_1 receptors.

casein kinases A group of kinases that preferentially phosphorylate acidic proteins such as caseins but can phosphorylate a large number of proteins. **Casein kinase 1** (EC 2.7.11.1, multiple isoforms, ~415 aa) participates in *Wnt signalling and is a central component of the circadian clock. **Casein kinase II** (phosvitin, 215 aa) is part of a tetrameric complex that is also involved in Wnt signalling. It is thought to regulate a broad range of transcription factors and in the brain has been associated with *long-term potentiation by phosphorylating proteins important for neuronal plasticity.

caseous necrosis *See* CASEATION.

CASK A member of the membrane-associated guanylate kinase (MAGUK) family of proteins (calcium/calmodulin-dependent serine protein kinase; vertebrate LIN2 homologue, 926 aa), present in the nervous system where it interacts with *syndecan-2 and the actin-binding band 4.1 protein and presynaptic *neurexin.

caspases A family of cysteine-aspartic acid peptidases involved in apoptosis, necrosis, and inflammation, eleven of which are identified in humans. **Caspase 1** (EC 3.4.22.36, interleukin-1β-converting enzyme, ICE, 404 aa) is important in the initiation of inflammation by producing the active form of *interleukin-1. **Caspase 2** (EC 3.4.22.55, ICH-1 protease, 452 aa) is one of the **initiator caspases** (CASP2, CASP4, CASP8, CASP9, and CASP10) that cleave inactive pro-forms of effector caspases, thereby activating them. **Effector caspases** (CASP3, CASP6, CASP7), once activated, cleave other protein substrates within the cell to trigger the apoptotic process. *See* CARD DOMAIN.

caspofungin *See* ECHINOCANDINS.

CASPRs Members of the *neurexin family of transmembrane adhesion/signalling proteins, that may have a role in the forming the functionally distinct domains critical for saltatory conduction in myelinated nerve fibres. Together with *contactin are involved in axon–glia communication. **CASPR1** (contactin-associated protein 1, neurexin 4, 1384 aa) is one of four known CASPRs.

cassette mechanism A mechanism for interchanging genes, e.g. the *a* and α genes that determine mating type in yeast. A double-stranded *nuclease makes a cut at a specific point in the MAT (mating type) locus and the old gene is replaced with a copy of a silent gene (a different cassette) from one or other flanking regions, and the new copy becomes active. Similar mechanisms are used in some cloning vectors.

CAST A protein (CAZ-associated structural protein 1) of the active zone of synapses (a *CAZ protein) that binds directly to others such as *RIMs1, *bassoon, and *piccolo.

castanospermine An alkaloid, isolated from seeds of blackbean *Castanospermum australe*, that inhibits the processing of *N*-linked oligosaccharides and leaves them in the *high-mannose precursor form. Acts by inhibiting α- and β-glucosidases. Reported to have antiviral properties and to inhibit angiogenesis.

Castleman's disease A rare disorder in which there is nonmalignant lymph node swelling, hypergammaglobulinaemia, increased levels of *acute phase proteins, and increased numbers of platelets. There are various forms of the disease, all probably caused by excess production of *interleukin-6.

castration The removal or surgical destruction of the testicles rendering the individual infertile and deprived of male sex hormones.

CAT *See* CHLORAMPHENICOL; COMPUTED TOMOGRAPHY.

catabolin A protein that stimulates the breakdown of extracellular matrix in connective tissue, subsequently shown to be *interleukin-1.

catabolism The metabolic processes involved in degradation rather than synthesis (anabolism).

catabolite repression The process of inhibiting the production of some enzymes when a more favourable carbon source becomes available. An important regulatory mechanism that has been extensively studied in various microorganisms, the classic example being the *lactose operon in *E. coli.*

catalase A tetrameric haem enzyme (EC 1.11.1.6, 527 aa) that breaks down hydrogen peroxide and protects cells from reactive oxygen species.

catalepsy (*adj.* cataleptic) The condition in which there is muscular rigidity, fixity of posture regardless of external stimuli, and decreased sensitivity to pain. Can occur due to Parkinson's disease, epilepsy, and some forms of schizophrenia. A similar state can be induced by deep hypnosis.

catalytic antibody (abenzyme, abzyme) An antibody that will catalyse a chemical reaction, usually a monoclonal antibody directed against a transition-state analogue, which is thereby stabilized.

catalytic RNA An RNA molecule that can catalyse the cleavage or transesterification of the phosphodiester linkage. Catalytic RNAs are important in the self-splicing of group I and group II introns and in the maturation of various tRNA species.

cataplerosis The net loss of intermediates in a biochemical cycle as a result of consumption or degradation. The opposite of *anaplerosis.

cataplexy A sudden and transient episode of loss of muscle tone, often triggered by emotions. It is very rare and often associated with narcolepsy.

cataract Loss of vision as a result of opacity of the lens of the eye. This can arise through exposure to radiation, to other environmental factors or to various mutations. **Congenital cataracts** cause 10–30% of all blindness in children and **age-related cataracts** are one of the leading causes of visual impairment in older people. In some cases the genes involved are known (many being genes for the major proteins of the lens, *crystallins); in others there are susceptibility loci but no identified genetic lesions. **Lamellar cataract** (zonular cataract, perinuclear cataract, Marner-type cataract) is the most common type of infantile cataract and is caused by mutation in the gene for heat-shock transcription factor-4 (463 aa). Lamellar cataract type 2 can be caused by mutation in the crystallin βA4 gene. **Autosomal dominant juvenile-onset cataracts** arise from mutations in *phakinin. **Juvenile cataract with microcornea and renal glucosuria** is caused by heterozygous mutation in the *SLC16A12* solute-carrier gene. **Zonular pulverulent cataract** is caused by mutation in the α8 subunit of the gap junction protein. **Coppock-like cataract** can be caused by mutation in the crystallin γC or the crystallin βB2 genes; mutations in the latter are also associated with **congenital cerulean cataract type 2** (CCA2, blue-dot cataract). **Autosomal recessive congenital nuclear cataract** is caused by mutation in crystallin βB3. **Cerulean type congenital cataract-1** (CCA1) has been mapped to chromosome 17q24, CCA3 is caused by mutation in the γD-crystallin gene. CCA4 is caused by mutation in the proto-oncogene *maf.* **Posterior polar cataract type 2** (CTPP2) is caused by mutation in the gene for crystallin αB, CTPP3 by mutation in the gene for chromatin-modifying protein 4B, and CTPP4 by mutation in the *PITX3* homeobox gene. Loci for CTPP1 and CTPP5 are identified, but not the affected genes. **Autosomal dominant nonnuclear congenital polymorphic cataract** is caused by mutation in the crystallin γD (*CRYGD*) gene. **Aculeiform cataract** is caused by a different mutation in the *CRYGD* gene. **Cataract with microcornea** is caused by mutation in the *GJA8* (*connexin 50) gene. Multiple types of autosomal dominant cataract can be caused by mutation in the gene encoding *beaded filament structural protein-2 although there are phenotypically and genotypically distinct forms. A mutation in the same protein can also cause juvenile-onset cortical cataract. **Autosomal recessive congenital cataract** can be caused by mutation in the βB1 crystallin gene.

catechins The principal polyphenolic compounds in green tea, the four major ones being epicatechin (EC), epigallocatechin (EGC), epicatechin gallate (ECg), and epigallocatechin gallate (EGCg). They occur in a range of plants and are present in wine, chocolate, and many traditional herbal remedies. Remarkable health benefits are claimed for these compounds.

catecholamines A class of *biogenic amines derived from tyramine that includes *adrenaline, *noradrenaline, and *dopamine.

catechol-*O*-methyltransferase An enzyme (COMT, EC 2.1.1.6, 271 aa) that catalyses the *O*-methylation and inactivation of catecholamine neurotransmitters and catechol hormones. There are two polymorphic variants (COMT-val and COMT-met) with high and low enzyme activity, but the functional consequence is unclear. There is also a transmembrane form of the enzyme (catechol-*O*-methyltransferase 2, LRTOMT2, 291 aa) that is defective in nonsyndromic deafness.

catenins Proteins that are associated with the cytoplasmic domain of adhesion molecules such as *cadherins and link them to the cytoskeleton. A second role is in regulating transcription factors. **α-Catenin** (906 aa) is a *vinculin-related protein involved in adherens junction-mediated intercellular adhesion; abnormalities are associated with cancer invasion and metastasis. **β-Catenin** (781 aa) binds E-cadherin and N-cadherin. A second pool of β-catenin is cytoplasmic and coimmunoprecipitates with the adenomatous *polyposis coli (APC) tumour suppressor protein, interacts with Tcf and Lef transcription factors, and is an essential member of the Wingless *wnt signal transduction pathway. Ubiquitination of β-catenin is greatly reduced in Wnt-expressing cells. *See* DAPPER. Mutations in catenin B1 are associated with colorectal cancer, ovarian and prostate carcinomas, hepatoblastoma, hepatocellular carcinoma, *pilomatrixoma, and *medulloblastoma. **γ-Catenin** (desmoplakin-3, plakoglobin, 745 aa), is a major component of desmosomes. Mutations are associated with *Naxos disease and familial arrhythmogenic right ventricular dysplasia type 12. **δ-Catenin-1** (p120, p120ctn, cadherin-associated src substrate, 938 aa) is involved in cell–cell adhesion complexes together with E-*cadherin, α-, β-, and γ-catenins. It is tyrosine-phosphorylated by ligand-activated EGF-, PDGF-, and CSF1-receptors and loss of expression is associated with some invasive tumours. **δ-Catenin-2** (neural plakophilin-related armadillo repeat protein, δ-catenin, neurojungin, CTNND2, 1225 aa) is found in neural tissues and, together with *kaiso and myogenic transcription factors, regulates synapse-specific transcription of *rapsyn. Hemizygosity of the gene is associated with *cri-du-chat syndrome.

cathartic Describing something that has laxative or purgative properties.

cathelicidins A group of cationic antimicrobial peptides mostly found in the secretory (azurophil) granules of neutrophils, although may be found elsewhere. The antimicrobial activity is in a structurally varied C-terminal cationic region that is joined to a conserved *cathelin-like propiece from which it is proteolytically released. Examples are LL-37/hCAP18, CAP-18, SMAP-29, BMAP-27, and BMAP-28, the *protegrin PG-1, and indolicidin (13 aa).

cathelin A protein (96 aa) isolated from porcine neutrophils that was incorrectly thought to be a cysteine peptidase inhibitor but is probably a microbicidal peptide. Cathelin-like sequences are a common feature of various antimicrobial peptides (protegrins). *See* CATHELICIDINS.

cathepsins A large family of peptidases mostly found in lysosomes. Most are processed before becoming active. **Cathepsin A** (lysosomal protective protein, carboxypeptidase C, EC 3.4.16.5, 480 aa) has a protective role for β-galactosidase and neuraminidase and is involved in their activity. Defects cause galactosialidosis (*Goldberg's syndrome). **Cathepsin B** (EC 3.4.22.1, 339 aa) is a cysteine endopeptidase thought to be involved in intracellular degradation and turnover of proteins. **Cathepsin C** (EC 3.4.14.1, 463 aa) is a dipeptidyl peptidase. Defects are a cause of *Papillon–Lefevre syndrome, *Haim–Munk syndrome, and juvenile periodontitis. **Cathepsin D** (EC 3.4.23.5, 412 aa) has pepsin-like specificity. Defects cause neuronal ceroid lipofuscinosis Type 10. **Cathepsin E** (EC 3.4.23.34, 401 aa) is an endosomal peptidase thought to be involved in the processing of antigenic peptides during MHC class II-mediated antigen presentation. **Cathepsin F** (EC 3.4.22.41, 484 aa) is involved in intracellular protein degradation and has also been implicated in tumour invasion and metastasis. **Cathepsin G** (EC 3.4.21.20, 255 aa) is a serine protease with trypsin- and chymotrypsin-like specificity. **Cathepsin H** (EC 3.4.22.16, 335 aa) has aminopeptidase activity. **Cathepsin K** (cathepsin O, EC 3.4.22.38, 329 aa) is involved in osteoclastic bone resorption and defects cause *pycnodysostosis. Several other cathepsins (L, N, S, W, Z) with different substrate specificities are also known.

catheter A rigid or flexible tube that is inserted into a blood vessel or an organ (such as the bladder).

cathomycin *See* NOVOBIOCIN.

cationic proteins A rather general term, used specifically of arginine-rich proteins found in the azurophil granules of neutrophils. **Eosinophil cationic protein** (160 aa), found in the granules, has weak ribonuclease activity but more importantly is helminthotoxic and will damage *schistosomula. It is also responsible for neurotoxicity (the *Gordon phenomenon).

cationized ferritin *See* FERRITIN.

cat scratch disease A zoonotic infection caused by the bacterium **Bartonella henselae*. Can cause swelling of lymph nodes but is usually mild and self-limiting.

CATSPER A putative sperm cation channel with a sequence resembling a single six-transmembrane-spanning repeat of the sort found in voltage-dependent calcium channels (which have four repeats; *two-pore channels have two). It is important for cAMP-mediated calcium influx in sperm, sperm motility, and fertilization.

caudate nucleus The frontal portion of the *basal ganglia in the brain. Damage to neurons in this region is characteristic of *Huntington's disease and other motor disorders.

causalgia A complex regional pain syndrome in which there is burning pain and touch hypersensitivity in the area innervated by an injured peripheral nerve.

caustic Describing something that is destructive or corrosive to tissue.

caveola (*pl.* caveolae) A small invagination of the plasma membrane. The membrane is enriched in cholesterol and sphingolipids and may be the efflux route for newly synthesized lipids; it contains integral membrane proteins, *caveolins, but not *clathrin.

caveolins A group of integral membrane proteins that are structural components of *caveolae and that interact with G-protein α subunits and functionally regulate their activity. There are three closely related human isoforms, caveolin-1, -2, and -3 (178, 162, 151 aa), that are differentially expressed in various cell types. Defects in **caveolin-1** cause *congenital generalized lipodystrophy type 3. Defects in **caveolin-3** cause limb-girdle *muscular dystrophy type 1C, *hyperCKmia, rippling muscle disease, familial *hypertrophic cardiomyopathy, *long QT syndrome type 9, and some cases of sudden infant death syndrome. **Caveolin scaffolding peptide-1** (CSP-1) is a peptide corresponding to the caveolin-1 scaffolding domain through which it interacts with signalling molecules. The peptide will block the ability of noradrenaline and histamine to induce changes in the internal calcium ion concentrations of vascular smooth muscle cells by inhibiting the activation of phospholipase Cβ3 and MAPK.

CAZ proteins A set of multidomain proteins associated with the cytoskeletal matrix at the active zone (CAZ) of synapses in the central nervous system where synaptic vesicles dock, fuse, release their contents, and are recycled. Includes *bassoon, *CAST, *munc-13, *piccolo, and *RIM.

C banding (centromeric banding) A method used to identify chromosomes by staining with *Giemsa and looking at the transverse bands in the heterochromatin of the *centromere. Can be combined with Giemsa banding (G banding) or quinacrine banding (Q banding). *See* BANDING PATTERNS.

CBC A common abbreviation for complete blood count: the numbers of erythrocytes, leucocytes, and platelets per unit volume of blood.

CBHA A synthetic inhibitor of *histone deacetylases.

cbl The product (E3 ubiquitin-protein ligase CBL, EC 6.3.2.-, 906 aa) of an oncogene (Casitas B-lineage lymphoma proto-oncogene), originally identified in a murine retrovirus that induces lymphomas and leukaemias, an adaptor protein that functions as a negative regulator of many signalling pathways. It is a RING-finger type E3 *ubiquitin-conjugating enzyme and binds to *grb-2, *Nck, *crk, and p85 of *PI-3-kinase. The transforming oncogene v-*Cbl* lacks the ubiquitin ligase domain. *See* CIN85.

CBP *See* CREB.

CC16 *See* UTEROGLOBIN.

CCAAT box (CAAT box, CAT box) The consensus sequence, found about 80 bases upstream of the transcription start site to which various transcription factors bind. Less well conserved than the *TATA box.

CCCP An *uncoupling agent that causes the proton gradient across the mitochondrial membrane to dissipate.

CCK *See* CHOLECYSTOKININ.

CCN family A family of proteins involved in cell differentiation and function, that interact with integrins and modulate their activity. The family includes cysteine-rich 61 (CYR61/CCN1, cysteine-rich angiogenic inducer 61, 381 aa), connective tissue growth factor (*CTGF/CCN2), nephroblastoma overexpressed (*nov/CCN3), and Wnt-induced secreted proteins (*WISP) 1, 2, and 3. (CCN4, CCN5, and CCN6).

CCR (chemokine receptors) *See* CHEMOKINES.

CCVs (clathrin-coated vesicles) *See* CLATHRIN.

CD2-associated protein A multifunctional adaptor-type molecule (CAS ligand with multiple SH3 domains, 639 aa), which is localized in the cytoplasm, membrane ruffles, and leading edges of cells. It colocalizes with F-actin and *p130cas and is involved in regulating the distribution of CD2, the T-cell surface marker involved in adhesion to antigen-presenting cells (the CD2–LFA3 interaction). Mutations are associated with *focal segmental glomerulosclerosis 3.

CD3 complex A group of cell surface molecules of the immunoglobulin superfamily that associate with the *T-cell receptor (TCR). The TCR/CD3 complex consists of either a TCR alpha/beta or TCR gamma/delta heterodimer coexpressed with the invariant CD3 gamma (CD3γ, 182 aa), delta (CD3δ, 171 aa), epsilon (CD3ε, 207 aa), zeta (CD247, 164 aa) and (in the mouse) eta (CD247, 206 aa) subunits. Defects in CD3delta cause severe combined immunodeficiency (autosomal recessive, T-cell-negative, B-cell-positive, NK-cell-positive type).

CD antigens The 'cluster of differentiation' antigens that were originally marker antigens found on different classes of lymphocytes using monoclonal antibodies. The CD nomenclature was introduced as a means of bringing consistency because the same CD antigen can be recognized by different monoclonal antibodies; international workshops are run to cross-compare antibodies against standard antigens. The CD antigens are no longer restricted to the immune system and the list covers a wide range of cell surface antigens, some of which are transiently expressed at particular stages of differentiation or following stimulation of cells by particular agents. In mid-2009 the list extended to CD350.

SEE WEB LINKS

- Details of all CD antigens in Prof. Dr H Ibelgaufts's COPE encyclopedia.
- A free updated table of these is produced by Abcam.

cdc* genes** Cell cycle control genes, the products of which are essential for cell division. Many are known and they are highly conserved, with human genes able to restore function to yeast cells in which the homologue is mutated. The product of the ***cdc42 gene is a *rho GTPase family member (191 aa) involved in control of cell morphology, migration, endocytosis, and cell cycle progression. It induces the formation of filopodia by activating the *Arp2/3 complex thereby affecting the actin cytoskeleton. The product of oncogene **dbl* stimulates the dissociation of GDP from cdc42, allowing activation to occur when GTP binds. *See* CYCLINS; DREAM COMPLEX.

CDEP A protein (chondrocyte-derived ezrin-like protein, FERM, RhoGEF, and pleckstrin domain-containing protein 1, FARP1, 1045 aa), thought to act as a rho-guanine nucleotide exchange factor, expressed in differentiated chondrocytes and various fetal and adult tissues.

CDGSH iron–sulphur domain proteins A family of proteins (CISD proteins) containing iron–sulphur (Fe–S) clusters and a unique 39-aa CDGSH domain. **CDGSH1** (MitoNEET, 105 aa) is located in the outer mitochondrial membrane and may be involved in the transport of iron into the mitochondrion. **CDGSH2** (MitoNEET-related 1, miner1, 135 aa) is involved in calcium homeostasis in the endoplasmic reticulum and is mutated in one form of *Wolfram's syndrome. **CDGSH3** (mitoneet-related 2, miner2, 127 aa) is mitochondrial.

cdk *See* CYCLIN-DEPENDENT KINASES.

cDNA *See* COMPLEMENTARY DNA.

CDP (cytidine 5′-diphosphate) *See* CYTIDINE.

CDR *See* COMPLEMENTARITY DETERMINING REGION.

CDRP (calcium-dependent regulator protein) Early name, now obsolete, for *calmodulin.

cdt1 A so-called 'licensing protein' (Cdc10-dependent transcript 1, 546 aa) that binds to the origin of DNA replication and facilitates the binding of cell division cycle 6 (Cdc6) and mini-chromosome maintenance proteins (Mcms). The *Drosophila* homologue is Double-parked (Dup). Once replication has been initiated at the end of G1, Cdt1 is exported out of the nucleus and degraded. *See* GEMININ.

CEA *See* CARCINOEMBRYONIC ANTIGEN.

CEBP (C/EBP) A family of transcription factors (CCAAT-box enhancer binding proteins) that promote the expression of proteins involved in a range of cellular activities including the control of cellular proliferation, growth, and differentiation and have a role in the development and maintenance of metabolically important processes including adipocyte differentiation. They form homo- or heterodimers that bind to the CCAAT box motif in DNA and then recruit coactivators.

cecropins Antimicrobial peptides (30–40 aa) isolated from haemolymph of silk moth *Hyalophora cecropia* pupae. Synthetic analogues may have therapeutic potential.

***ced* genes** Genes associated with cell death in **Caenorhabditis elegans* and implicated in the control of *apoptosis. *See* INTERLEUKIN-1 CONVERTING ENZYME.

cef- A prefix for *cephalosporin-type antibiotics such as cefaclor, cefadroxil, cefalexin, ceftazidime, and cefuroxime.

celecoxib A *NSAID that selectively inhibits *cyclo-oxygenase-2 (COX-2) thereby inhibiting prostaglandin, but not thromboxane synthesis, and thus having no effect on platelet aggregation or blood clotting. There is, however, an increased risk of heart attack and stroke and some drugs in this class have been withdrawn. *TN* Celebrex.

C. elegans *See* CAENORHABDITIS ELEGANS.

celiac *See* COELIAC.

cell In principle, an autonomous self-replicating unit that may be an organism (a unicellular organism) or a component of a multicellular organism, in which case autonomy is sacrificed though self-replication can occur (and autonomy would involve escape from controls, as in tumours). It is generally held that all living organisms are composed of one or more cells and, on this basis, that viruses are nonliving since they cannot replicate independently.

cell adhesion *See* ADHESINS; CADHERINS; CELL ADHESION MOLECULE: INTEGRINS; SORTING OUT; *see also* various specialized junctions: ADHERENS JUNCTION; DESMOSOME; FOCAL ADHESIONS; GAP JUNCTION; ZONULA OCCLUDENS.

cell adhesion molecule (CAM) Although this could mean any molecule involved in cellular adhesive phenomena, it has acquired a more restricted sense; CAMs are cell surface molecules of the immunoglobulin superfamily that were first identified by the use of blocking antibodies. Examples are **LCAM** (liver cell adhesion molecule) and **NCAM** (neural cell adhesion molecule), both named from the tissues in which they were first detected, although they actually have a wider tissue distribution. Others include *LECAM, *SynCAM, *VCAM, and *activated leucocyte cell adhesion molecule (ALCAM).

cell aggregation The formation of clumps of cells mediated by cell surface adhesion molecules, not external agglutinating agents.

cell-attached patch *See* PATCH CLAMPING.

cell bank A store of frozen cells that can be used to reinitiate cell cultures or, in the case of stem cells, to restore normal function to the organism from which they were removed. Freezing is usually done in liquid nitrogen, with cryoprotectant added to the cell suspension to prevent ice crystal formation, although storage may be in an ultra-deep freeze at around –80 °C.

cell behaviour A term used by analogy with animal behaviour; cellular activities such as movement, adhesion, and proliferation.

cell body Describing the main portion of a neuron around the nucleus and excluding long processes such as axons.

cell centre The pericentriolar *microtubule organizing centre (MTOC) of the cell.

cell culture The maintenance of cell strains or lines in the laboratory.

cell cycle The sequence of events that happen between mitotic divisions. The cell cycle is conventionally divided into G_0, G_1, (G for gap), S (synthesis phase during which the DNA is replicated), G_2 and M (mitosis), followed by cytokinesis. The G_0 stage is the state in which terminally differentiated cells that will not divide again are considered to exist; transition from G_0 to G_1 commits the cell to completion of the cycle.

cell death Non-accidental death seems to occur either when cells have completed a fixed number of division cycles (~60; the Hayflick limit) or at some earlier (programmed) stage, as occurs in limb development to separate the digits. It is still unclear whether death at this limit is a consequence of accumulated errors or a result of telomere shortening or some other 'timing' programme. Transformed cells appear to have escaped the Hayflick limit and can divide indefinitely. *See* APOPTOSIS.

cell division The separation of one cell into two daughter cells; nuclear division (*mitosis) is followed by cytoplasmic division (*cytokinesis).

cell electrophoresis A method for estimating the surface charge of a cell by looking at its rate of movement in an electrical field. Most cells have a net negative surface charge, mostly due to carboxyl groups on terminal carbohydrate chains of glycoproteins and glycolipids. The cell is surrounded by a fluid layer in which the ionic composition is affected by the fixed charges on the membrane and the electrical potential measured (the electrokinetic or zeta potential) is actually some distance away from the plasma membrane. By varying the pH of the suspension fluid it is possible to identify the pK of the charged groups.

cell fate Describing the differentiated tissues that will be formed by the progeny of an early embryonic cell. As differentiation progresses the range diminishes as the cells become progressively determined (*see* DETERMINATION). A 'fate map' can be produced for the early embryo.

cell fractionation **1.** The separation of homogeneous subset of cells from a mixed population, e.g. the separation of lymphocytes from whole blood or particular lymphocyte subsets by *flow cytometry. **2.** The separation of subcellular components from disrupted cells by differential centrifugation, to give nuclear, mitochondrial, microsomal, and soluble fractions.

cell-free system A system in which a cellular activity is reconstituted in the absence of whole cells, e.g. *in vitro* translation systems in which protein synthesis from mRNA occurs using a lysate of cells that are active in protein synthesis (rabbit reticulocytes or wheatgerm). The systems have been progressively refined so that it is now possible to use more defined systems with relatively pure proteins rather than a crude homogenate.

cell fusion The fusion of two or more cells to form a mutinucleated cell (as in skeletal muscle) or the fusion of egg and sperm to form a zygote (in which case there is also nuclear fusion). Cell fusion can be induced artificially using *Sendai virus or fusogens such as polyethylene glycol. The fusion of dissimilar cells produces a *heterokaryon.

cell growth Strictly speaking, a term that should mean an increase in the size of an individual cell, but usually used to mean an increase in the size of a population of cells, the sizes of the individual cells remaining relatively constant.

cell junctions *See* ADHERENS JUNCTION; DESMOSOME; GAP JUNCTION; TIGHT JUNCTION.

cell line An established population of cells that will proliferate indefinitely in culture provided the medium is replenished and there is space available (to prevent *density dependent inhibition). A cell line has escaped the Hayflick limit and become immortalized, whereas a **cell strain** is restricted to a finite number of doublings. Cell lines are more easily established from some species than others, and human cell lines are relatively difficult to produce. The escape from the normal limits on proliferation is considered by many cell biologists to indicate that the cells are already abnormal and are on the route to becoming tumour-like (neoplastic). *See* CELL DEATH; TRANSFORMATION.

cell lineage The differentiated products of a *stem cell.

cell locomotion The movement of a cell from one place to another. *Compare* CELL MOVEMENT.

cell-mediated immunity An immune response involving effector T cells rather than the production of humoral antibody. Cell-mediated responses are involved in *allograft rejection, delayed *hypersensitivity, and defence against viral infection and intracellular protozoan parasites.

cell membrane A term usually describing the outer limiting *plasma membrane rather than intracellular membranes.

cell migration The movement of a population of cells from one place to another, e.g. the migration of embryonic germ cells to the gonads.

cell movement A generic term covering shape-change, cytoplasmic streaming, and movement of appendages, as well as *cell locomotion.

cell polarity **1.** In epithelia, the differentiation of apical and basal specializations. The apical and basolateral regions of the plasma membrane may differ in lipid and protein composition, and tight junctions prevent exchange of membrane molecules from one domain to the other. The apical membrane may have active transport functions or may be the only site for exocytosis. **2.** In motile cells the internal polarity that defines 'front' and 'back' and allows locomotion. The polarity probably derives from the microtubular cytoskeleton associated with the *microtubule organizing centre, and is disrupted by drugs such as colchicine.

cell proliferation An increase in the size of a population of cells.

cell signalling The interchange of information between cells, usually mediated by the release of soluble factors but sometines by contact or by exchange of small molecules or ions through *gap junctions.

cell sorting *See* FLOW CYTOMETRY; SORTING OUT.

cell strain *See* CELL LINE.

C

cell surface marker A molecule characteristic of the plasma membrane of a cell type that can be detected by external reagents. Enzymatic markers have largely been superseded by cell-type specific antigens (e.g. *CD antigens) that can be detected with labelled monoclonal antibodies.

cell synchronization The process of obtaining a population of cells that are simultaneously in one phase of the cell cycle. This can be achieved by selection (e.g. by shaking off and collecting the relatively nonadherent rounded cells that are undergoing mitosis) or by imposing, and then releasing, a block on cell cycle progression.

cell trafficking The movement of leucocytes through tissues, in particular the recirculation of lymphocytes from blood into tissues and back via the lymph as a means of immune surveillance.

cellubrevin A *synaptobrevin-like protein (VAMP-3, 100 aa), involved in regulating vesicle fusion, that is the target for *tetanus toxin.

cellular immunity A nonspecific immune response involving enhanced phagocytic activity, rather than *cell-mediated immunity, but because of the possibility of confusion the term should probably be avoided.

cellular retinoic acid binding protein One of the *calycin superfamily of *fatty acid binding proteins that regulates the transport of retinoic acid to the nuclear *retinoic acid receptors. There are several isoforms: CRABP-1 (137 aa), CRABP-2 (138 aa), and CRABP-4 (retinoid-binding protein 7, 134 aa).

cellulitis *See* ERYSIPELAS.

cell wall Extracellular material that surrounds a cell and serves a structural role. Most animal cells, with the exception of protozoa, do not have a wall. Bacterial cell walls are complex and differ between Gram-positive and Gram-negative bacteria, the basis of *Gram staining. Removal of the bacterial wall leaves a vulnerable *protoplast or spheroplast.

cenexins Alternatively spliced products of the gene that encodes Odf2 (*see* OUTER DENSE FIBRES) that are recruited to the mature *centriole and have a scaffolding function.

CENPs A large and diverse group of *centromere-associated proteins. **CENP-A** (140 aa) is a histone H3 variant that recruits a nucleosome associated multiprotein complex. **CENP-E** (2701 aa) is a kinesin-related microtubule motor protein required for chromosome movement during mitosis. **CENP-F** (3210 aa) is a microtubule-binding protein that, among other things, is required for kinetochore localization of dynein, *Lis1, *NDE1, and NDE-L1, and regulates recycling of the plasma membrane by acting as a link between recycling vesicles and the microtubule network. Many react strongly with antibodies from *CREST sera (CENP antigens).

centaurins A large family of proteins involved in regulating cytoskeletal and membrane trafficking events by acting as GTPase-activating proteins for the *ADP-ribosylation factor family and binding signalling molecules from the phosphatidylinositol system. Examples include **centaurin-α2** (Arf-GAP with dual PH domain-containing protein 2, 381 aa), **centaurin-β1** (ACAP-1, ARFGAP with coiled-coil, ANK repeat, and PH domain-containing protein 1, 740 aa), **centaurin-δ1** (ARAP-2, Arf-GAP, Rho-GAP domain, ANK repeat, and PH domain-containing protein 2, PARX protein, 1704 aa), and **centaurin-γ3** (AGAP-3, Arf-GAP, GTPase, ANK repeat, and PH domain-containing protein 3, CRAM-associated GTPase, 875 aa).

SEE WEB LINKS

- Further details about this and other BAR superfamily proteins.

centractins Actin-related proteins, centractin-α (Arp1A, 376 aa) and centractin-β (Arp1B, 376 aa) with ~70% sequence homology to cytoplasmic actin and involved in microtubule-linked cytoplasmic movement as components of the *dynactin complex.

central core disease A myopathy, characterized by the presence of central core lesions extending the length of type I muscle fibres, that causes 'floppy infant' syndrome, in which there is mutation in the *ryanodine receptor-1 (*RYR1*) gene. The cores are areas in which the sarcomeres are disorganized and few mitochondria are present.

central lymphoid tissue Primary *lymphoid tissue.

central nervous system (CNS) The *brain and spinal cord, enclosed in the *meninges.

centrifugation A standard method for separating substances or objects on the basis of differences in density or size by applying a centrifugal force. In the simplest method particulate material is sedimented at the bottom of a tube, leaving a soluble supernatant. More complex methods involve the use of a *density gradient and separating on the basis of size (which affects the rate of movement through a viscous solution), and density (which affects the equilibriuim position in the gradient). At very high *g* values (ultracentrifugation) it is possible to separate macromolecules (often on a caesium chloride density gradient), although this is now rarely done. In continuous centrifugation the supernatant is removed continuously as it is formed. *See* SVEDBERG UNITS.

centrins A family of *EF-hand calcium-binding phosphoproteins involved in the structure and function of the *microtubule organizing centre. There are three isoforms: **centrin-1** (caltractin isoform 2, 172 aa), **centrin-2** (caltractin isoform-1, 172 aa), which is essential for centriole duplication and correct spindle formation and interacts with calmodulin and *centriolin, and **centrin-3** (167 aa).

centriolar region *See* CENTROSOME; PERICENTRIOLAR REGION.

centriole A subcellular structure composed of two orthogonally arranged cylinders of nine microtubule triplets, very similar to to the *basal body of a cilium. The pericentriolar material, but not the centriole, is the major *microtubule organizing centre of the cell. The centrioles and its associated material divides prior to mitosis and the daughter centrioles come to lie at the poles of the spindle. Plants have a similar microtubule organizing centre but no centriole. A range of proteins are associated with the centriole: *see* CENTRINS; CENPs; CENTRIOLIN.

centriolin A centriolar protein (110-kDa centrosomal protein, CEP110, 2325 aa) that is involved in cell cycle progression and cytokinesis. During the late steps of cytokinesis, it anchors *exocyst and SNARE complexes at the midbody. A chromosomal translocation that allows the formation of a fusion protein of CEP110 and the fibroblast growth factor receptor-1 may be a cause of *stem cell myeloproliferative disorder.

centroblast A B cell that has been activated by exposure to antigen and stimulated to proliferate. There is little surface immunoglobulin but somatic mutation and class-switching of immunoglobulin genes occurs in centroblasts. *See* CENTROCYTE.

centrocyte A differentiated cell derived from a *centroblast that does not proliferate but expresses surface immunoglobulin. Centrocytes are probably positively or negatively selected on the basis of the affinity of the surface immunoglobulin for antigen.

centromere The region of the chromosome where daughter chromatids are joined together, adjacent to the *kinetochore. The region has large arrays of repetitive DNA. *See* C BANDING (centromeric banding); CENPs.

centronuclear myopathy A group of muscle disorders (myotubular myopathy) in which there is slowly progressive muscular weakness and wasting. The **autosomal dominant form** can be caused by mutation in the gene encoding *dynamin-2, the **autosomal recessive form** by mutations in the *bridging integrator 1 gene. The **X-linked form** involves mutation in the *myotubularin gene.

centrophilin Obsolete name for *nuclear mitotic apparatus protein.

centrosome The *microtubule organizing centre of the pericentriolar region that divides and moves to the poles of the mitotic spindle; in the interphase cell it organizes the microtubular cytoskeleton and influences *cell polarity in motile cells.

cephalic index The ratio of the maximum breadth of skull divided by the maximum length, used by anthropologists to categorize human populations as dolichocephalic (long headed), mesocephalic (moderate headed), or brachycephalic (broad headed), although the value of the classification is doubtful.

cephalocele The protrusion of membranes of the brain through a hole in the skull.

cephalometry The measurement of fetal head dimensions *in utero* as an aid to gauging developmental stage. Usually done using ultrasound.

cephalosporins A group of broad-spectrum β-lactam antibiotics isolated from the fungus *Cephalosporium* sp. although semisynthetic derivatives are now produced (e.g. cephaloridine), most named with either ceph- or cef- as a prefix. Like *penicillins, they block cell wall synthesis in Gram-positive bacteria.

ceramides A family of lipids composed of *N*-acyl *sphingosine and a fatty acid; a component of sphingomyelin and the lipid moiety of *glycosphingolipids.

cerberus One of the *cer/dan family of cysteine-knot proteins (DAN domain family member 4, 267 aa), that are antagonists of *bone morphogenetic proteins. May play a role in anterior neural induction and somite formation during embryogenesis and can regulate *Nodal signalling during gastrulation.

CERC *See* CREC FAMILY.

cer/dan family Secreted proteins that are antagonists of *bone morphogenetic proteins (BMPs) and play a role in complex morphogenetic processes by fine-tuning or mediating signals. The proteins have a characteristic cysteine-knot motif. Examples are *caronte, *cerberus, *dan, *gremlin, and *sclerostin.

cerebellopontine angle The region at the base of the brain from where the facial nerve arises, the space between the cerebellum and the pons, a common site for the growth of acoustic neuromas. Tumours in this region can cause a range of problems secondary to compression of nearby cranial nerves (cerebellopontine angle syndrome).

cerebellum The region of the hindbrain that controls motor function, muscle tone, and the maintenance of balance. Cerebellar *granule cells account for nearly half of the neurons in the central nervous system and their migratory pathway during brain development has been extensively studied. Granule cell migration fails in the mouse *weaver mutant.

cerebral amyloid angiopathy A disease of small blood vessels in the brain in which there are deposits of amyloid protein in the vessel walls leading to stroke, brain haemorrhage, or dementia. The commonest form is associated with ageing but there are genetic forms. In the **Dutch** and **Flemish types** there is mutation in *amyloid precursor protein, in the **Icelandic type** mutation in *cystatin C, in the **British type** the mutation is in integral membrane protein 2B (266 aa). It can also be associated with high-density *lipoprotein deficiency caused by mutation in the *ABCA1* gene.

cerebral cavernous malformations Autosomal dominant disorders characterized by venous sinusoids that predispose to intracranial haemorrhage. **Cerebral cavernous malformation type 1** (CCM1) can be caused by mutation in the **KRIT1* gene, **CCM2** by mutation in the *CCM2* gene encoding *malcaverin, and **CCM3** by mutation in the *PDCD10* gene (*programmed cell death 10).

cerebral dominance The dominance of one hemisphere of the brain in the control of speech, language, and movement control (handedness); usually the left hemisphere, although this does not necessarily correlate with being right-handed.

cerebral haemorrhage *See* CEREBROVASCULAR ACCIDENT.

cerebral palsy A nonprogressive disorder of posture or movement, caused by damage to the motor area of the brain that becomes obvious at the stage of rapid brain development. Occurrence is around 2 per 1000 live births and is usually a consequence of perinatal damage, but rare cases (~2%) have a hereditable cause. **Autosomal recessive spastic cerebral palsy** is caused by mutation in the gene encoding *glutamate decarboxylase-1. Approximately 50% of **ataxic cerebral palsy**, which accounts for 5–10% of all cases, may be inherited as an autosomal recessive trait.

cerebral thrombosis *See* THROMBOSIS.

cerebroside A glycosphingolipid that is a major lipid of the brain (>10% of dry matter), a glycosylated form of *ceramide. A defect in the degradation of glucocerebrosides causes *Gaucher's disease, and the corresponding defect for galactocerebrosides is *Krabbe's disease.

cerebrospinal fluid The clear colourless fluid which fills the subarachnoid space, the ventricles of the brain, and the central lumen of the spinal cord and in which the brain effectively floats. It is produced by modified ependymal cells in the choroid plexus and is isolated from other tissue fluids by the *blood–brain barrier.

cerebrotendinous xanthomatosis A lipid storage disease characterized clinically by progressive neurologic dysfunction, premature atherosclerosis, and cataracts. Large deposits of cholesterol and cholestanol are found in virtually every tissue, particularly the Achilles tendons, brain, and lungs. It is caused by a deficiency of the mitochondrial sterol 27-hydroxylase CYP27A1.

cerebrovascular accident A euphemistic phrase for a sudden interruption to the blood supply to the brain, either by haemorrhage or thrombosis, causing a stroke.

cereolysin A *thiol-activated cytolysin produced by *Bacillus cereus*.

cernunnos A DNA repair protein (nonhomologous end-joining factor 1, 299 aa) that interacts with the XRCC4/DNA-ligase IV complex in repairing double-strand breaks in DNA and V(D)J recombination that generates diversity in the immune system. Defects in cernunnos lead to a severe immunodeficiency condition associated with microcephaly and other developmental defects. *See* NONHOMOLOGOUS END-JOINING.

ceroid lipofuscinosis *See* NEURONAL CEROID LIPOFUSCINOSES.

ceruloplasmin (caeruloplasmin) A blue copper-binding ferroxidase (*dehydrogenase) (EC 1.16.3.1, 1065 aa) with six or seven cupric ions per molecule, found in plasma (200–500 mg/mL). Apparently involved in copper detoxification and storage, and in iron transport across the cell membrane. Mutation in the ceruloplasmin gene causes aceruloplasminemia and is associated with iron accumulation in tissues. *See* WILSON'S DISEASE.

cervical smear A small sample of cells from the uterine cervix that are examined microscopically to detect cancer or preneoplastic changes in cells.

cesium chloride *See* CAESIUM CHLORIDE.

cetuximab A humanized monoclonal antibody that binds to the extracellular domain of the human *epidermal growth factor receptor (EGFR) and blocks signalling. Used to treat colorectal carcinoma. *TN* Erbitux.

CFS *See* CHRONIC FATIGUE SYNDROME.

CFTR *See* CYSTIC FIBROSIS.

CFU *See* COLONY-FORMING UNIT.

CGD *See* CHRONIC GRANULOMATOUS DISEASE.

CG island *See* CpG ISLAND.

cGMP *See* CYCLIC GMP.

CGRP (calcitonin gene-related peptide) *See* CALCITONIN.

chaconines Bitter-tasting toxic glycoalkaloids (alpha- and beta-chaconine), related to *solanine and solanidine, found in potatoes (*Solanum tuberosum*), where they are produced in increased amounts as a stress response. The cytotoxic effects have been investigated for chemotherapy and alpha-chaconine is reported to induce apoptosis of HT-29 cells through inhibition of *ERK and activation of *caspase-3. It also has anti-cholinesterase activity.

chaetoglobosins A large group of fungal metabolites produced by *Chaetomium globosum* and related to *cytochalasins. Chaetoglobosin J will inhibit addition of G-actin to the barbed end of an actin microfilament.

Chagas' disease South American trypanosomiasis, infection with *Trypanosoma cruzi*, that is transmitted by blood-sucking reduviid bugs such as *Rhodnius prolixus*.

chalaza One of two spirally twisted cords of dense albumen that anchor the yolk of an avian egg to the shell membrane.

chalazion A cyst (meibomian gland lipogranuloma) in the eyelid caused by inflammation of a blocked meibomian gland.

chalcone An intermediate in the biosynthesis of flavanones, *flavones, and anthocyanidins by plants. Synthetic derivatives have some anti-inflammatory properties.

chalone A tissue-specific inhibitor of cell proliferation produced by cells and thought to regulate the size of the population and thus the size of an organ or tissue. Their existence has been controversial, but *myostatin could be described as a chalone.

Chanarin–Dorfman syndrome *See* ICHTHYOSIFORM ERYTHRODERMA.

chancre The hard swelling which is the primary lesion in syphilis. **Chancroid** is a sexually transmitted nonsyphilitic ulceration of the genital organs caused by infection with Gram-negative *Haemophilus ducreyi*.

Chang liver cells A human cell line with epithelial morphology that will grow to high density, extensively used in virology and biochemistry. Subsequently shown to be a subline of *HeLa cells.

channel-forming ionophore A small molecule (*ionophore) that will produce a hydrophilic pore, usually cation-selective, in a membrane.

channel gating *See* GATING CURRENTS.

channelopathies

Disorders caused by the defective function of an ion channel, either because of mutation in the channel-forming gene or in genes encoding regulatory subunits. Channelopathies involving *potassium, *sodium, *chloride, and *calcium channels are known; others involve the *acetylcholine receptor, the *glycine receptor, and other receptors for neurotransmitters. Channelopathies can cause diverse diseases; e.g. the calcium channelopathies include familial hemiplegic *migraine, *malignant hyperthermia, *episodic ataxia type 2, *spinocerebellar ataxia type 6, hypokalaemic *periodic paralysis type I, *central core disease, congenital night blindness, and *stationary night blindness. Comparable diversity is found in disorders associated with other types of channels. *See* ANDERSEN'S SYNDROME; BRACHYOLMIA; BARTTER'S SYNDROME; BRUGADA SYNDROME; CONGENITAL INSENSITIVITY TO PAIN; ERYTHROMELALGIA; HYPERINSULINISM; JERVELL AND LANGE-NIELSEN SYNDROME; LIDDLE'S DISEASE; LONG QT SYNDROME; MYASTHENIC SYNDROME; MYOTONIA; OSTEOPETROSIS; PSEUDOHYPOALDOSTERONISM; SHORT QT SYNDROME; STARGARDT DISEASE; TIMOTHY'S SYNDROME. Many extremely potent toxins from spiders, scorpions, snakes, etc. act on ion channels.

channel protein A protein that forms a multimeric pore that facilitates the passive diffusion of hydrophilic molecules or ions across a lipid membrane.

channelrhodopsin Light-gated ion (proton) channels (ChR1 and ChR2), originally isolated from *Chlamydomonas reinhardtii*, that can be used experimentally to activate cells by pulses of light. The channels open rapidly after absorption of a photon to generate a large permeability for monovalent and divalent cations.

SEE WEB LINKS

- Article on channelrhodopsin in the ion channel newsletter.

chaotropic (chaeotropic) Describing an agent that causes chaos, usually in the context of disrupting or denaturing macromolecules. Urea is commonly used to denature proteins and guanidium isothiocyanate is used in RNA extractions because it prevents the action of *RNAases.

chaperones A diverse set of proteins, found in both prokaryotes and eukaryotes, that bind to nascent or unfolded polypeptides and ensure their correct folding or transport. They do not bind covalently to their targets and do not form part of the finished product. The **chaperonins** are a subset of chaperones found in prokaryotes, mitochondria, and plastids and include prokaryotic GroEL and eukaryotic heat shock protein 60 (hsp60). Two other major families are the hsp70 family and the hsp90 family. Other proteins with similar functions including *nucleoplasmin, secB, and T-cell-receptor-associated protein. *See* HEAT SHOCK PROTEINS.

CHAPS A zwitterionic detergent used to solubilize membranes.

Charcot–Leyden crystals Naturally occurring hexagonal bipyramidal crystals of a protein with lysophospholipase activity, subsequently shown to be *galectin-10. They are found in human tissues and secretions in association with increased numbers of peripheral blood or tissue eosinophils in parasitic and allergic processes.

Charcot–Marie–Tooth disease (CMT) A group of motor and sensory neuropathies affecting peripheral nerves, mostly inherited as autosomal dominants. They are a heterogeneous set of diseases, both clinically and genetically, and taken together are the most common inherited disorders of the peripheral nervous system. The disease is subdivided: type 1 (**CMT1**) is the demyelinating form with slow motor nerve conduction velocity (NCV); type 2 (**CMT2**) is the axonal form, with a normal or slightly reduced NCV. A third type (distal hereditary motor neuropathy, also known as **spinal CMT**) is characterized by normal motor and sensory NCV and degeneration of spinal cord anterior horn cells. There are also recessive (**CMT4**) and X-linked forms (**CMTX**). The commonest, type 1A, (**CMT1A**), is caused by duplication or mutation of the gene for peripheral myelin protein-22. **CMT1B** is caused by mutation in the gene encoding myelin protein zero, **CMT1C** is caused by mutation in the *LITAF* gene, **CMT1D** by mutation in the *early growth response gene-2 (*EGR2*), **CMT1F** by mutation in the neurofilament light polypeptide (*NEFL*) gene. **CMT2** and dominant intermediate CMT can be caused by mutation in the myelin *protein-zero gene (*MPZ*), **CMT2A1** results from mutation in the *kinesin-like protein, KIF1B, **CMT2A2** from mutation in the *mitofusin-2 gene. Many other variants of type 2 are ascribed to different genes. **CMT4A** is caused by mutation in the gene encoding ganglioside-induced differentiation-associated protein-1, **CMT4B1** by mutation in the gene encoding the *myotubularin-related protein-2, **CMT4B2** by mutation in the *SBF2* gene encoding myotubularin-related protein 13, **CMT4C** by mutation in the *SH3TC2* gene that encodes SH3 domain and tetratricopeptide repeats-containing protein 2, **CMT4D** by mutation in the *N-*myc* downstream-regulated gene-1 (*NDRG1*), **CMT4E**, congenital hypomyelinating neuropathy, by mutation in the early growth response-2 (*EGR2*) gene or the MPZ gene, **CMT4F** (*Dejerine–Sottas neuropathy), by mutations in genes encoding myelin protein zero, peripheral myelin protein 22, *periaxin, or mutation in the early growth response gene (*EGR2*). **CMT4G** (Russe type) is linked to locus 10q23.2, **CMT4H** to mutation in the gene for *frabin. The X-linked form, **CMTX1**, is caused by mutation in the *connexin-32 *GJB1* gene. Loci, but not genes, are identified for **CMTX 2–5**.

CHARGE syndrome A complex set of abnormalities, the CHARGE association (*coloboma, heart anomaly, *choanal atresia, retardation of mental and somatic development, genital hypoplasia, ear abnormalities and/or deafness) that are due to mutation in the gene encoding the *chromodomain *helicase DNA-binding protein-7 or the *semaphorin-3E gene. The **PAX2* gene is involved in all the developmental primordia that are affected, but does not seem to be mutated.

Char's syndrome A developmental disorder in which there is a patent ductus arteriosus and unusual facial appearance caused by missense mutations in the gene for transcription factor *AP-2β, acting in a dominant-negative manner. The effect seems to be largely on *neural crest cells.

chartins Obsolete name for a class of three families of *microtubule-associated proteins (MAPs) of 64, 67, and 80 kDa, distinct from *tau protein, that were isolated from neuroblastoma cells.

charybdotoxin A toxin (37 aa), isolated from the venom of the scorpion *Leiurus quinquestriatus hebraeus*, that selectively blocks high-conductance calcium-activated potassium channels.

CHD proteins A family of ATP-dependent chromatin remodelling factors (chromo-ATPase/helicase-DNA-binding proteins) belonging to the SNF2 *DNA helicase superfamily, that have conserved double *chromodomains. They can act as transcriptional repressors or activators, probably depending on other proteins with which they are associated. **CHD7** (2997 aa) is mutated in *CHARGE syndrome, and variations in CHD7 are associated with susceptibility to idiopathic *scoliosis type 3, *Kallmann's syndrome type 5, and idiopathic hypogonadotropic *hypogonadism.

checkerboard assay An improved methodology for investigating the potential chemotactic or chemokinetic properties of substances using a *Boyden chamber assay but with positive, negative, and zero-gradient conditions. In nongradient conditions a chemokinetic effect will enhance leucocyte penetration into filters; only if the movement in gradient conditions is greater is there evidence for gradient perception and directed movement (chemotaxis).

checkpoint A stage in the *cell cycle at which progress can be halted if the appropriate conditions are not met. Major checkpoints, at which *cdc proteins act, are at the G_1/S and G_2/M boundaries. **Checkpoint kinases** (CHK1, CHK2, EC 2.7.11.1) are serine/threonine kinases that act downstream of *ATM in response to detection of DNA damage. **Chk 1** (476 aa) is essential for normal cell division and Chk 2 is mutated in many tumours. **Chk2** (543 aa), along with ATM regulates the transcription factor *p53 by preventing its ubiquitination by the RING E3 ligase *Mdm2 and is defective in *Li–Fraumeni syndrome-2.

Chediak–Higashi syndrome An autosomal recessive disorder, caused by mutation in the *lysosomal trafficking regulator gene, resulting in giant lysosomal vesicles in phagocytes and poor bactericidal function as a result. Some perturbation of microtubule dynamics seems to be involved. Reported from humans, albino Hereford cattle, mink, beige mice, and killer whales.

cheil- A prefix indicating association with the lip. **Cheilitis** is inflammation of the lip.

cheiloschisis *See* CLEFT PALATE.

cheiropompholyx A dermatological disorder (dyshidrotic eczema) characterized by the sudden eruption of vesicles filled with clear fluid on the hands and feet.

chelating agent A compound that will reversibly bind metal ions with high affinity. The best-known example is *EDTA, which chelates calcium and other heavy metals and can be used to treat lead poisoning.

chelerythrine A red alkaloid, from the greater celandine *Chelidonium majus*, that inhibits *protein kinase C.

chemerin An *adipokine (137 aa, generated from a 164-aa precursor) whose expression is up-regulated *in vitro* by the synthetic retinoid tazarotene. It is a potent chemoattractant specific for antigen-presenting cells which express chemokine-like receptor 1 (CMKLR1) and also potentiates insulin-stimulated glucose uptake.

chemical potential A term with different connotations in thermodynamics, theoretical chemistry, and particle physics, but for biological systems the important usage is the chemical potential difference, the work (in J mol^{-1}) required to bring one mole of a substance from a solution at one concentration to another solution at a different concentration: if the second concentration is higher, then work must be done (going 'up-gradient' in transport terms); if the second concentration is lower then energy is produced (and could potentially be harnessed to drive some energy-requiring process, as in some cotransport systems where down-gradient movement of sodium ions is used to drive glucose uptake). For charged molecules, the electrical potential difference must be considered as well as the concentration (*see* ELECTROCHEMICAL POTENTIAL).

chemical synapse A neuron–neuron or nerve–muscle junction where the signal passes from one cell to the other by means of a neurotransmitter substance released from the presynaptic cell that diffuses across the intervening extracellular space and binds to a receptor on the postsynaptic cell, causing that cell to respond (e.g. by opening ion channels and producing an action potential). An important feature is that signals pass in only one direction.

chemiluminescence Light produced as a chemical reaction proceeds, e.g. the production of light by firefly *luciferase when it hydrolyses ATP (used as a sensitive assay system). Another assay involving chemiluminescence is for the production of reactive oxygen species by activated leucocytes in which bystander substrates (*luminol or *lucigenin) emit light when oxidized. *See* BIOLUMINESCENCE.

chemiosmosis The mechanism by which ATP synthesis in the mitochondrion is driven by the flow of protons down an electrochemical potential gradient across the inner mitochondrial membrane. The theory, first proposed by P.D. Mitchell (1920–92), is now generally accepted as being correct. In more general terms, chemiosmosis is the coupling of one enzyme-catalysed reaction to another using the transmembrane flow of an intermediate species.

chemoattractant A substance that causes cells to accumulate, sometimes causing a directional motile response (chemotaxis), although other mechanisms can achieve the same end result. *See* CHEMOKINESIS.

chemoautotroph An organism (chemotroph, chemotrophic autotroph) for which the energy source is inorganic molecules, not light, and carbon is obtained from carbon dioxide, not organic molecules.

chemokines A diverse family of small secreted proteins (~100 aa) that are chemotactic for leucocytes, a subset of the *cytokines. They are subdivided on the basis of conserved cysteine residues. The **α-chemokines** (*interleukin-8, NAP-2, Gro-α, Gro-γ, ENA-78, and GCP-2) have a conserved CXC motif and are mainly chemotactic for neutrophils; the **β-chemokines** (MCP-1–5, MIP-1α, MIP-1β, eotaxin, RANTES) have adjacent cysteines (CC) and attract monocytes, eosinophils or basophils; the **γ-chemokines** have only one cysteine pair and are chemotactic for lymphocytes (lymphotactin); the **δ-chemokines** are structurally rather different, being membrane-anchored and having a CXXXC motif, and are restricted (so far) to brain (*neurotactin). The **chemokine receptors** are G-protein coupled and are named according to the class of chemokine they bind, thus CCRs bind CC-chemokines (β-chemokines), CXCR bind CXC-chemokines (α-chemokines), etc. Some are also coreceptors for the binding of immunodeficiency viruses (*HIV, SIV, *FIV) to leucocytes. CXCR4 is a coreceptor for T-tropic viruses, CCR5 for macrophage-tropic (M-tropic) viruses. Individuals deficient in particular CCRs seem to be resistant to HIV-1 infection.

chemokinesis A response to a soluble chemical that involves a change in the motile behaviour of a cell. A change in speed is a positive or negative *orthokinesis, a change in the frequency or magnitude of turning behaviour is a *klinokinesis. In the extreme case of a strong negative orthokinesis, cells can become trapped because their speed drops to zero, and they accumulate at the top of a gradient of chemokinetic factor, although there is no directional response by individual cells.

chemoreceptor **1.** A receptor on a cell that binds a specific chemical substance and elicits an intracellular response; the term is usually restricted to receptors for small inorganic molecules rather than those for hormones, neurotransmitters, cytokines, etc. **2.** A group of cells in a multicellular organism that are specialized for detecting chemical substances in the environment.

chemorepellant A substance that repels cells. *Compare* CHEMOATTRACTANT.

chemosis Swelling of the conjunctiva, in response to allergy, infection, or mechanical trauma.

chemostat An apparatus in which a bacterial population can be maintained in the exponential phase of growth by regulating the input of a rate-limiting nutrient and the removal of exhausted medium and cells.

chemosynthesis The synthesis of organic compounds using energy derived from oxidation of inorganic molecules rather than photosynthesis. *See* AUTOTROPH; CHEMOTROPHY.

chemotaxis A response of motile cells or organisms in which the direction of movement is affected, positively or negatively, by a gradient of a diffusible substance (chemotactic factor). The response is vectorial, unlike the scalar responses of *chemokinesis, because the gradient alters the probability of motion in one direction, rather than the rate or frequency of random motion.

chemotherapy The treatment of a disease with drugs that kill the causative organism or the abnormal cells in a tumour.

chemotrophy A metabolic system that derives energy from endogenous chemical reactions rather than from food or light. Examples of chemotrophic organisms include *Archaea that oxidize sulphur in deep-sea hot springs ('black smokers'). *See* CHEMOAUTOTROPH.

chemotropism Describing the growth or possibly bending of an organism, usually a plant, in response to an external chemical gradient. Should not be confused with *chemotaxis or *chemokinesis in which the cell or organism moves from one place to another.

chenodeoxycholic acid (chenodesoxycholic acid) A major species of bile acid synthesized in the liver from cholesterol. Used pharmaceutically to decrease the cholesterol saturation of bile and bring about gradual dissolution of cholesterol gallstones.

CHF **1.** Congestive *heart failure or chronic heart failure. **2.** *See* CHICK HEART FIBROBLASTS.

chiasma (*pl.* **chiasmata)** A junction point or crossing-point. **1.** In anatomy, any *crossing over of nerves or ligaments, in particular the crossing of optic nerves at the optic chiasm. **2.** In genetics, used specifically for the junction of nonsister *chromatids at the first *diplotene of *meiosis, the consequence of a crossing-over event between maternal and paternally derived *chromatids. A chiasma also has a mechanical function and is required for normal equatorial alignment at meiotic *metaphase.

chickenpox A highly infectious so-called 'childhood disease' caused by infection with varicella zoster virus (human herpesvirus type 3, *see* HERPESVIRIDAE) that causes a vesicular skin rash, mostly on the trunk, face, upper arms, and thighs. Following a childhood infection the virus can remain latent in nervous tissue and become reactivated as *shingles if the immune system is compromised.

chick heart fibroblasts (CHF) The cells that emigrate from an explant of embryonic chick heart maintained in culture. Because of the relative ease of obtaining them and the fact that they have probably not divided in culture they have been considered as archetypal normal cells and were used as the 'normal' reference cells by Michael Abercrombie (1912–79) in his studies of contact inhibition.

Chikungunya virus (Buggy Creek virus) A positive-sense, single-stranded RNA virus, an alphavirus of the Togaviridae family. Carried by *Aedes aegypti* and *A. albopictus* (tiger mosquito), and present in parts of Africa, South-East Asia, and the Indian subcontinent. The virus causes flu-like symptoms including a severe headache, chills, fever, joint pain, nausea, and vomiting. The name comes from the Makonde word describing the positions that victims take to relieve the joint pain. Cases occurred in Italy in 2007.

SEE WEB LINKS

- European Centre for Disease Control fact-sheet.

chilblains *See* ERYTHEMA.

chimera An organism composed of two genetically distinct types of cells. Chimeras can be formed by the fusion of two early blastula-stage embryos, in which case the ratios of the two lineages is likely to be similar; if an organ is of only one genetic type, this suggests that there was only one stem cell (or a small number). Reconstitution of the bone marrow in an irradiated recipient produces a chimera, as does any allogeneic transplantation of tissue. Females are in a sense chimeric since one X chromosome or the other is inactivated more or less randomly.

chimerins A family of GTPase-activating proteins for *rac that have high affinity for phorbol esters and *diacylglycerol. **N-chimerin** (alpha-chimerin, 459 aa) may be important in neuronal signal-transduction mechanisms. Mutations in (alpha-chimerin are responsible for *Duane retraction syndrome 2. Beta-chimerin (468 aa) is found mostly in the brain and pancreas, and expression is much reduced in malignant gliomas.

Chinese hamster ovary cells A cell line (CHO cells) derived in 1957 from a Chinese hamster *Cricetulus griseus*. CHO cells are extensively used for experimental studies and for the production of recombinant therapeutic proteins.

CHIP **1.** The C-terminus of Hsp70-interacting protein (C-terminus of Hsp70-interacting protein, STIP1 homology, and U box-containing protein 1, EC 6.3.2.-, E3 ubiquitin-protein ligase CHIP, 303 aa) A cochaperone and ubiquitin ligase that modulates the activity of several chaperone complexes, including Hsp70, Hsc70, and Hsp90. **2.** Aquaporin-1 (aquaporin-CHIP, CHIP-28, 269 aa), a transmembrane protein that forms a water-specific channel in erythrocytes and epithelial cells of the kidney proximal tubule. *See* AQUAPORINS.

chi-squared (χ-squared) A common statistical test used to determine whether the observed values of a variable differ significantly from those expected on the basis of a null hypothesis.

chitin A polymer of *N*-acetyl-D-glucosamine that is extensively crosslinked and is the major structural component of arthropod exoskeletons and fungal cell walls. Chitinase (EC 3.2.1.14) hydrolyses the 1,4-β linkages.

chitosan A polymer of 1,4-β-D-glucosamine and *N*-acetyl-D-glucosamine found in some fungal cell walls. It is sold as a health supplement, manufactured by deacetylation of chitin from crustacean shells, although the benefits are unproven.

Chk (checkpoint kinases, choline kinases) *See* CHECKPOINT; CHOLINE; C-TERMINAL SRC-KINASE HOMOLOGOUS KINASE.

Chlamydia A genus of small prokaryotes that are obligate intracellular parasites that replicate in cytoplasmic vacuoles within eukaryotic cells. *C. trachomatis* causes the eye disease *trachoma and is also a common sexually transmitted infection that can cause pelvic inflammatory disease in women and infertility. The genome is small, about one-third that of *E. coli,* and the phylum Chlamydiae is evolutionarily ancient.

chloracne A form of acne caused by exposure to halogenated hydrocarbons such as chlorinated dioxins and dibenzofurans. The condition progresses to cause severe inflammation and scarring.

chloramphenicol An antibiotic isolated from *Streptomycetes venezuelae* that inhibits protein synthesis in prokaryotes, and in mitochondria and chloroplasts, by acting on the 50S ribosomal subunit. It has a broad spectrum of activity against Gram-positive and Gram-negative bacteria, *Rickettsia, Mycoplasma,* and *Chlamydia* but because of its toxic side effects is rarely used. **Chloramphenicol acetyltransferase** (CAT, EC 2.3.1.28, 219 aa in *E.coli*) inactivates chloramphenicol by acetylation and is widely used as a *reporter gene.

chlordiazepoxide A mild tranquillizer, a benzodiazepine derivative. *TN* Librium.

chlorhexidine An antiseptic and disinfectant used for dressing minor skin wounds or burns, for skin cleansing (*TN* Hibiscrub), and in mouthwash (*TN* Oral B). Acts by disrupting bacterial membranes.

chloride channels Ion channels that are selective for chloride ions. There are a range of different types including *ligand-gated Cl-channels at synapses (the *GABA- and *glycine-activated channels), as well as *voltage-gated Cl-channels. The voltage-gated channels have ten or twelve transmembrane helices and each protein forms a single pore. They are unrelated to known cation channels or other types of anion channels and there are three CLC subfamilies: **CLC-1** is involved in setting and restoring the resting membrane potential of skeletal muscle, whereas other channels are important in solute concentration (mechanisms in the kidney. *See also* CYSTIC FIBROSIS (CFTR), MDR. Mutations in CLCN1 are associated with various forms of *myotonia congenita. *See* CHLORIDE INTRACELLULAR CHANNEL PROTEINS. Chloride channels can be blocked by substitution of other anions (I^-, Br^-, or SO_4^{2-}) or compounds such as 5-nitro-2-(3-phenylpropylamino) benzoic acid (NPPB), indanyloxyacetic acid 94, niflumic acid, and 4,4-dinitrostilbene-2,2-disulphonic acid (DNDS) as well as some toxins (*chlorotoxin, *picrotoxin).

chloride current A current due to the flow of chloride ions through *chloride channels. Chloride conductance has an important role in regulating excitability and the equilibrium between excitation and inhibition.

chloride intracellular channel proteins A family of proteins (CLIC proteins, ~240–250 aa) expressed in a wide variety of tissues that may have diverse functions in addition to their role as chloride channels. They are expressed in the nucleus and in cytoplasmic locations. **CLIC1** is primarily expressed in the nucleus, **CLIC5** interacts with the cortical actin cytoskeleton in placental microvilli. Members of the family include *parchorin and bovine *p64.

chloroquine An antimalarial drug that accumulates in lysosomes and interferes with parasite metabolism, although the exact mechanism is not understood. Chloroquine resistance, which is widespread, seems to be due to an enhanced ABC transporter activity (*Plasmodium falciparum* chloroquine resistance transporter and *Pf* – multidrug resistance genes) that pump chloroquine from the cell. Chloroquine is also used to treat some autoimmune diseases such as rheumatoid arthritis and systemic lupus erythematosus and may have potential activity against *Chikungunya virus.

chlorothiazide A thiazide *diuretic used to treat oedema and hypertension.

chlorotoxin An intracellular blocker (36 aa) of small-conductance chloride channels of epithelial cells, found in the venom of the scorpion *Leiurus quinquestriatus*.

chlorpromazine A phenothiazine antipsychotic drug, used to treat schizophrenia, bipolar disorder, and uncontrollable hiccups. It is thought to act primarily as a dopamine antagonist, but also antagonizes α-adrenergic, H_1 histamine, muscarinic, and serotonin receptors. *TNs* Thorazine, Largactil.

chlorpropamide A *sulphonylurea.

chlortalidone *See* CHLORTHALIDONE.

chlortetracycline A bright yellow *tetracycline derivative, isolated from *Streptomyces aureofaciens*; hence the proprietary name Aureomycin.

chlorthalidone (chlortalidone) A thiazide-like *diuretic with a long duration of action.

choanal atresia A developmental defect in which the back of the nasal passage (choana) is blocked by abnormal tissue overgrowth. *See* CHARGE SYNDROME.

CHO cells *See* CHINESE HAMSTER OVARY CELLS.

chol- A prefix denoting something associated with bile. **Cholaemia** (cholemia) is the presence of bile pigments in the blood. A **cholagogue** increases the release of bile. **Cholangiocytes** are cuboidal epithelial cells that form the lining of the bile ducts and are involved in the absorption and secretion of water, ions, and solutes.
A **cholangiocarcinoma** is a malignant tumour, usually an adenocarcinoma, of the bile duct, derived from cholangiocytes. **Cholangitis** (cholangeitis) is inflammation of the bile duct. **Ascending cholangitis** is caused by bacterial infection, **primary sclerosing cholangitis** by an autoimmune inflammation and is often associated with *inflammatory bowel disease. **Cholecystectomy** is the surgical removal of the gallbladder. **Cholecystitis** is an inflammation of the wall of the gallbladder caused by *Salmonella typhi*. **Cholelithiasis** is the condition of having gallstones. **Cholestasis** is a condition in which excretion of bile is blocked. In **choluria** there are bile pigments in the urine. *See* CHOLIC ACID; CHOLESTEROL.

cholecystokinin Peptide hormones (CCK, pancreozymin, 58 aa and 33 aa), synthesized and secreted by I-cells of the duodenum, that stimulate the secretion of digestive enzymes by the pancreas and release of bile from the gallbladder. There are several forms, all derived from preprocholecystokinin by proteolysis. The C-terminal octapeptide (CCK8) is found in some dorsal root ganglion neurons where it presumably acts as a peptide neurotransmitter. The **receptors** are G-protein coupled (CCK-AR, 428 aa; CCK-BR, 447 aa) and also bind *gastrin, but with much lower affinity. CCK-BR is distributed throughout the central nervous system and thought to modulate anxiety, analgesia, arousal, and neuroleptic activity.

cholera An acute bacterial infection by *Vibrio cholerae* that is characterized by severe vomiting and diarrhoea leading to dehydration, often fatal. Many of the effects are the result of **cholera toxin** acting on intestinal epithelial cells and causing hypersecretion of chloride and bicarbonate followed by water. The toxin is an *AB toxin in which the toxic A subunit (EC 2.4.2.36, 258 aa) activates adenylate cyclase irreversibly by *ADP-ribosylation of a G_s protein. The B subunit (124 aa) has five identical monomers, binds to GM1 *ganglioside, and facilitates passage of the A subunit across the cell membrane. The bacteria adhere to intestinal epithelium by an unknown mechanism, although it may be relevant that those with type O blood are the most susceptible, while those with type AB are the most resistant, and A and B types are intermediate. The bacteria also produce enzymes (neuraminidase, proteases) that facilitate access of the bacterium to the epithelial surface.

cholesteatoma An abnormal accumulation of keratinizing squamous epithelium in the middle ear and/or mastoid process, a keratoma rather than a tumour. A **cholesteatoma cyst** consists of desquamating layers of epithelium and may contain cholesterol crystals. Most are a consequence of otitis media, but there are rare primary or congenital cholesteatomas.

cholesterol The major *sterol lipid of higher animals, an important component of cell membranes, especially of the plasma membrane, where it has buffering effects on membrane fluidity. It is synthesized in the liver (*see* HMG-COA REDUCTASE) and transported in the esterified form via plasma lipoproteins. **Cholesteraemia** (cholesterolemia) is the condition in which there is an excess of cholesterol in the blood, an important risk factor for atherosclerosis. (*See* HYPERCHOLESTEROLAEMIA). Cholesterol is oxidized by the liver into a variety of bile acids.

cholesterol binding toxins A family of pore-forming peptide toxins produced by various genera of bacteria. They bind to cholesterol and oligomerize to form a pore which leads to cytolysis. *See* STREPTOLYSIN. Other examples include *botulinolysin, *cereolysin O, *listeriolysin O, *perfringolysin O, *pneumolysin, *tetanolysin, and *thuringolysin O.

cholesteryl ester transfer protein A *phospholipid transport protein (lipid transfer protein I, 493 aa) that transfers phospholipids between lipoprotein particles. Deficiencies in the protein do not seem to have particularly deleterious effects, although there is hyperalphalipoproteinaemia (an excess of high density *lipoprotein) which may actually be antiatherogenic.

cholestyramine (colestyramine) An anion exchange resin that binds bile and forms insoluble complexes that are excreted. As a result bile levels drop and stimulate the conversion of cholesterol to bile acids, thereby reducing plasma cholesterol levels. Used as a treatment for *hypercholesterolaemia. *TNs* Cholybar, Questran.

cholic acid A major bile acid with strong detergent properties. It can be used to replace membrane lipids and generate soluble complexes of membrane proteins.

choline A saturated amine that is found as an ester in the head group of phospholipids (*phosphatidyl choline and *sphingomyelin) and acetylated as *acetylcholine. It is also a major source for methyl groups via its metabolite, trimethylglycine (betaine) that participates in the *S-adenosyl methionine synthesis pathways. **Choline kinases** (EC 2.7.1.32, CHKa, 457 aa; CHKb, 395 aa) catalyse the phosphorylation of choline in the biosynthesis of *phosphatidylcholine.

cholinergic neurons Neurons for which acetylcholine is the neurotransmitter.

chondro- A prefix denoting something associated with *cartilage. A **chondroblast** is an embryonic cartilage-producing cell, a **chondrocyte** is a differentiated cell that secretes the extracellular matrix of cartilage. A **chondroma** is a tumour composed of cartilage cells. **Chondromalacia** is the condition in which there is damage to the cartilage at the back of the patella (kneecap). A **chondrosarcoma** is a malignant tumour composed of sarcoma cells and cartilage.

chondrocalcin A calcium-binding protein (246 aa), the C-propeptide of type II procollagen cleaved off by *ADAMTS3, found in developing fetal cartilage matrix and in growth plate cartilage. It appears to play a role in enchondral ossification.

chondrodysplasia

A clinically and genetically diverse group of rare diseases in which bone growth is affected, usually being much reduced. **Grebe-type chondrodysplasia**, in which all four limbs are markedly shortened and end in tiny digits, is caused by mutation in the *CDMP1* gene which encodes cartilage-derived morphogenetic factor-1. **Metaphyseal chondrodysplasia**, Jansen type, is caused by constitutively active mutations in the *parathyroid hormone receptor whereas inactivating mutations cause Blomstrand-type chondrodysplasia. **Schmid-type metaphyseal chondrodysplasia** is caused by mutation in the *COL10A1* gene that encodes a short-chain minor collagen of cartilage. **McKusick-type metaphyseal chondrodysplasia** (cartilage–hair hypoplasia) is caused by mutations in the *RMRP* gene for mitochondrial RNA-processing endoribonuclease. **Chondrodysplasia punctata** (Conradi–Hunermann syndrome) is a clinically and genetically heterogeneous disorder. The X-linked dominant form is caused by mutation in the gene encoding δ(8)-δ(7) sterol isomerase emopamil-binding protein. An X-linked recessive form is caused by mutation in the *ARSE* gene (*arylsulphatase E). There is a possible autosomal form, brachytelephalangic chondrodysplasis punctata, and an autosomal dominant tibia–metacarpal type. It can also be caused by maternal vitamin K deficiency or warfarin teratogenicity. **Rhizomelic chondrodysplasia punctata** is a rare multisystem developmental disorder in which there are stippled foci of calcification in hyaline cartilage along with a variety of other symptoms depending upon the subtype. Type 1 is caused by mutations in the *PEX7* gene, which encodes the peroxisomal type 2 targeting signal (PTS2) receptor, and leads to a deficiency in various peroxisomal enzymes and reduced synthesis of plasmalogens. Other foms of chondrodysplasia are described but the mutated gene(s) unknown. *See* ACROMESOMELIC DYSPLASIA; KNIEST'S SYNDROME; PLATYSPONDYLIC LETHAL SKELETAL DYSPLASIAS.

chondroitin sulphates Major components of the extracellular matrix, especially cartilage. They are polymers of variable length consisting of glucuronic acid and sulphated *N*-acetyl glucosamine (with one or more sulphate covalently linked in the 2, 4, or 6 position), highly hydrophilic and anionic, and associated with matrix proteins. 'Chondroitin' is sold as a dietary supplement, supposedly to relieve osteoarthritis, but it is unclear whether it would ever reach the joints.

chondronectin A protein distinct from fibronectin and laminin that was reported to be involved in the attachment of *chondrocytes to type II *collagen but that has disappeared from the recent literature and does not appear in the UniProt database.

chordae tendineae Cord-like collagen-rich tendons that connect the papillary muscles to the tricuspid valve and the mitral valve in the heart, restricting their movement and preventing backflow of blood into the atrium.

chordamesoderm The region of the embryonic mesoderm that will give rise to the *notochord.

CHORD domain A highly conserved zinc-binding domain (cysteine and histidine rich domain, ~60 aa) with uniquely spaced cysteine and histidine residues. **CHORD domain-containing protein 1** (chp1, 332 aa) may play a role in the regulation of NOD1 (*see* NOD PROTEINS) through an interaction with HSP90AA1.

chordin An important protein (dorsalizing factor, 955 aa) that binds to the ventralizing TGFβ family of *bone morphogenetic proteins (BMPs) and antagonizes their effects during development. Chordin mRNA and protein levels are down-regulated in osteoarthritic chondrocytes. Other BMP antagonists are *follistatin, *gremlin, *noggin, and *sclerostin.

chordoma A rare type of slow-growing primary bone cancer in the skull and spinal column that develops from the remnants of embryonic *notochord.

chordotomy The surgical severing of nerve fibres in the spinal cord to relieve severe pain.

chorea Involuntary repetitive jerky movements that are a feature of several neurological diseases, including *Sydenham's chorea and *Huntington's disease. **Choreoathetosis** describes involuntary movements in a combination of chorea and *athetosis.

chorioallantoic membrane The extraembryonic membrane that surrounds the embryo of amniotes. It comprises a highly vascularized epithelial layer derived from the trophoblast by the fusion of the *allantois and the inner face of the *chorion. The chorioallantoic membrane of hen's eggs is used experimentally to test the invasive capacity of tumour cells.

choriocapillaris A layer of capillaries immediately adjacent to *Bruch's membrane in the choroid.

choriocarcinoma A malignant tumour derived from the trophoblast, often preceded by *hydatidiform mole.

chorion **1.** One of the extraembryonic membranes that surrounds the early embryo of *amniotes. It consists of two layers, the outer formed by the primitive ectoderm or trophoblast, the inner formed by the somatic mesoderm. At a later stage in development, it fuses with the *allantois to form the *chorioallantoic membrane. **2.** A protective membrane that surrounds the eggs of insects and fishes. **3.** *Compare* CHOROID.

chorionic gonadotropin A glycoprotein hormone, produced by the early embryo and later the placenta, that promotes the maintenance of the *corpus luteum. It is heterodimeric with an α subunit (116 aa) common to *luteinizing hormone (LH), *follicle-stimulating hormone (FSH), and *thyroid stimulating hormone (TSH), and a unique β subunit (165 aa).

chorionic villus sampling A method used as an alternative to *amniocentesis, in which small biopsy samples of fetally derived chorionic villi are removed for chromosomal analysis or DNA-based testing for genetic disorders. Usually done in the sixth to tenth week of gestation.

choroid The highly vascularized layer of tissue that lies between the *retina and the sclera of the eye. It is pigmented and may reflect light back on to the retina, especially in those animals that have a tapetum, an irridescent layer within the choroid. Not to be confused with the *choroid plexus. **Choroiditis** is an inflammation of the choroid; **chorioretinitis** is inflammation involving both the choroid and retina.

choroideraemia An X-linked disease (tapetochoroidal dystrophy) that leads to the degeneration of the choriocapillaris, the retinal pigment epithelium, and the photoreceptors of the eye. It is caused by mutation in the *CHM* gene which encodes *rab escort protein-1 (653 aa), which is involved in membrane trafficking.

choroid plexus A highly vascularized region found in each of the ventricles of the brain and separated from the ventricles by choroid epithelial cells. It is responsible for the secretion of *cerebrospinal fluid and for the absorption of molecules destined for excretion.

chp **1.** Calcium-binding protein p22 (calcineurin homologous protein, CHP, 195 aa), a ubiquitously expressed *EF-hand protein required for constitutive membrane traffic. It inhibits GTPase-stimulated Na^+/H^+ exchange and inhibits calcineurin phosphatase activity. **2.** *CHORD domain-containing protein 1, chp1.

Christmas disease (haemophilia B) A congenital deficiency of *blood clotting factor IX (first described in a patient called Stephen Christmas in the Christmas issue of the *British Medical Journal*, 1952). An X-linked disorder that can only be distinguished from *haemophilia A by assay of the coagulation factors VIII and IX.

Christ–Siemens–Touraine syndrome *See* HYPOHIDROSIS.

chromaffin cells Two populations of cells found in the chromaffin tissue of the adrenal medulla. One type produces adrenaline, the other noradrenaline. The *catecholamines are associated with carrier proteins (chromogranins) in membrane vesicles (chromaffin granules). *See* GRANINS. A **chromaffinoma** (*phaeochromocytoma) is a tumour derived from chromaffin cells. *See* PARAGANGLIA (chromaffin bodies).

chromagen Any substance that gives rise to a coloured product when acted upon by an enzyme or the products of enzymic activity. Various chromagens are used in spectrophotometric assay systems.

chromatid A single chromosome containing only a single DNA duplex. Chromatids are replicated in the S phase of the cell cycle and become visible at mitotic metaphase, though they are present throughout the cell cycle.

chromatin The nucleic acid and associated proteins, organized as *nucleosomes, that can be stained with appropriate dyes in the interphase nucleus. **Euchromatin** is loosely packed and is potentially being transcribed by RNA polymerases, whereas **heterochromatin** is highly condensed and probably inaccessible to the transcriptional machinery. The **chromatin body** is the condensed X chromosome in female cells. *See* SEX CHROMOSOMES.

chromatin assembly factor (CAF-1) A heterotrimeric protein of the CAF-1 complex that is thought to mediate chromatin assembly in DNA replication and DNA repair. The complex comprises **CAF1A** (CHAF1A, 956 aa), **CAF1B** (CHAF1B, 559 aa), and **CAF1C** (histone-binding protein RBBP4, *retinoblastoma-binding protein 4, RBBP4, 425 aa) and is responsible for assembling histone octamers. CAF1A binds to histones H3 and H4, CAF1B binds to CAF1A. During mitosis CAF1B is hyperphosphorylated and displaced into the cytosol; it is then progressively dephosphorylated from G_1 to S and G_2 phase. CAF1A and CAF1B are massively down-regulated during quiescence in several cell lines and their presence has been used as a marker of the proliferative state.

chromatography A separation method based upon the differential retention of dissolved molecules as the solution passes through a matrix (now usually packed in a column, although the original methods used paper) to which they bind to a greater or lesser extent. A range of different approaches can be used, e.g. using an eluant which is increasingly alkaline or a matrix that traps molecules according to their size (size exclusion chromatography). *See* AFFINITY CHROMATOGRAPHY; HPLC; ION-EXCHANGE CHROMATOGRAPHY.

chromatophores Pigment-containing cells of the dermis, particularly in teleosts and amphibians. The intracellular distribution of granules can be varied and in the dispersed state they contribute strongly to the overall colour, whereas when concentrated in a small spot their impact is minimal. In melanophores the pigment is melanin; mammalian *melanocytes contain melanin granules but these cannot be redistributed for camouflage purposes.

chromobindin *See* ANNEXINS.

chromoblast An embryonic cell that will give rise to a *chromatophore.

chromobox proteins A class of proteins containing *chromodomains that are involved in binding to histones. **Chromobox protein homologue 1** (CBX1, heterochromatin protein 1 homologue β, 185 aa) binds histone H3 tails that are methylated at Lys-9; chromobox protein homologue 2 (**CBX2**, 532 aa) is a component of the *polycomb repressive complex-1 (PRC1). Chromobox protein homologue 3 (**CBX3**, 183 aa) also binds methylated histone H3 and is involved in the formation of a functional kinetochore through interaction with *MIS12 complex proteins. Chromobox protein homologue 4 (**CBX4**, E3 *SUMO-protein ligase, 558 aa) facilitates SUMO1 conjugation by UBE2I and is a component of PRC1. At least four other chromobox protein homologues are known.

chromocentre A region of centromeric heterochromatin that stains strongly and contains highly repetitive DNA.

chromodomains Protein domains (chromatin organization modifier domains) implicated in the recognition of lysine-methylated histones. **CHD proteins** (for chromo-ATPase/helicase/DNA-binding) regulate ATP-dependent nucleosome assembly and mobilization through their conserved double chromodomains. *See* CHROMOBOX PROTEINS; CHROMOSHADOW DOMAIN.

chromogranins *See* CHROMAFFIN CELLS; GRANINS.

chromomere **1.** A granular region of condensed *chromatin seen in chromosomes at the leptotene and zygotene stages of meiosis. **2.** Condensed regions at the base of loops on *lampbrush chromosomes. **3.** Condensed bands in the *polytene chromosomes of Diptera.

chromophore The portion of a coloured molecule that absorbs light at specific wavelength and gives rise to the colour. Although it normally refers to molecules that absorb in the visible spectrum, it can be extended to include absorption at longer or shorter wavelengths (infrared or ultraviolet). *Compare* CHROMATOPHORES.

chromoshadow domain A domain (40–70 aa) found in some, but not all, proteins that have a *chromodomain; it is involved in mediating dimerization, transcriptional repression, and interaction with multiple nuclear proteins. Related to the *chromodomain.

chromosomal passenger complex A multiprotein complex consisting of *aurora B kinase, inner centromere protein (INCENP), *borealin, and *survivin that acts as a key regulator of mitosis. It has essential functions at the centromere in ensuring correct chromosome alignment and segregation and is required for chromatin-induced microtubule stabilization and spindle assembly.

SEE WEB LINKS

- An article from the journal *Cell Division* (2008) gives more details.

chromosome A strand of DNA duplex associated with histones and other proteins that forms part of the DNA complement of the cell. In humans there are 22 pairs of autosomes and two sex chromosomes ranging from 51 million to 245 million base pairs in length, each with a *centromere. During the S phase of the cell cycle the DNA duplex is replicated and the two daughter chromatids condense during the prophase of mitosis. Chromosomes can be identified by characteristic *banding patterns when stained.

SEE WEB LINKS

- Much additional information on chromosomes.

chromosome map The sequence of genes along a chromosome. The first such maps were based upon the probability of recombination between loci (*see* GENETIC LINKAGE) but using fluorescence *in situ* hybridization techniques (*see* FISH) more direct mapping is possible.

chromosome painting *See* FLUORESCENCE (fluorescence *in situ* hybridization (FISH)).

chromosome segregation The process by which daughter chromatids are separated during mitosis so that each daughter cell has an identical set of chromosomes.

chromosome synapsis *See* SYNAPSIS.

chromosome translocation The aberrant condition in which part of one chromosome has become fused to another, often in a reciprocal exchange process. The result can be activity of genes that would normally be quiescent. Translocations seem to be sporadic and random, although there are some that occur often enough to be clinically significant. *See* PHILADELPHIA CHROMOSOME; SPECTRAL KARYOTYPING.

chromosome walking Describing the procedure that was used to find a gene on the basis of its known position on a chromosome as determined by *genetic linkage. The starting point was a nearby known (sequenced) gene which was used to produce a probe to select clones from a *genomic library; these in turn were used to produce probes to select further clones, so that the known sequence was gradually extended, in a stepwise fashion, towards the gene of interest.

chronaxie The shortest time that an electric current of twice the threshold strength must flow to elicit a response in a nerve or muscle.

chronic Describing something that is persistent or long-lasting, as opposed to *acute. Chronic *inflammation is generally considered to be a response to a persistant antigenic stimulus, even if the antigen is not always identifiable.

chronic fatigue syndrome A poorly understood disorder in which there is inexplicable fatigue, often accompanied by transient joint and muscle pains and various neuropsychological complaints. *See* MYALGIC ENCEPHALOMYELITIS.

C

chronic granulomatous disease (CGD) A disease, usually fatal in childhood, in which phagocytes are unable to produce the reactive oxygen species used to kill bacteria. The absence of the oxygen-dependent killing mechanism seriously compromises the primary defence system and there is recurrent infection. At least three mutations can cause the disease, the commonest being an X-linked defect in plasma membrane cytochrome (*see* PHOX).

chronic infantile neurologic cutaneous and articular syndrome A rare congenital inflammatory disorder (neonatal onset multisystem inflammatory disease, NOMID) in which there is neonatal onset of cutaneous symptoms, chronic meningitis, and joint manifestations with recurrent fever and inflammation. It is caused by mutations in *cryopyrin.

chronic lymphocytic leukaemia *See* LEUKAEMIA.

chronic myelogenous leukaemia *See* LEUKAEMIA.

chronic obstructive pulmonary disease (COPD) A disorder of the lungs in which there is narrowing of the airways and increased difficulty in breathing. *See* EMPHYSEMA.

chronic wasting disease A transmissible *spongiform encephalopathy of cervids (deer).

chrysotherapy The treatment, usually of rheumatoid arthritis, by the injection of gold salts (sodium aurothiomalate).

CHUK The kinase (conserved helix-loop-helix ubiquitous kinase, IκB kinase 1, IκBKA, EC 2.7.11.10, 745 aa) that is a component of the *IκB-kinase (IKK) core complex which consists of CHUK, IκBKB, and IκBKG. The IKK core complex seems to associate with regulatory or adapter proteins to form a IKK-signalosome which phosphorylates *IκB, causing it to dissociate from NFκB. This frees NFκB to move to the nucleus and the phosphorylated IκB is eventually degraded.

Churg–Strauss syndrome A nonhereditable autoimmune vasculitis (allergic granulomatosis) that affects mid-sized to small blood vessels. There is no complement consumption and no deposition of immune complexes, but antineutrophil cytoplasmic antibodies (*ANCA) may be present. Has some similarities to *polyarteritis nodosa but is distinct and distinguished by eosinophilia.

CHX *See* CHLORHEXIDINE; CYCLOHEXIMIDE.

chyle A milky fluid consisting of a mixture of lymphatic fluid with chylomicrons. The appearance of chyle in the peritoneal cavity (**chyloperitoneum**) can be caused by obstruction of the abdominal lymphatics. **Chylothorax** is the presence of chyle in the pleural cavity; **chyluria** is the presence of chyle in the urine.

chylomicron A large *lipoprotein particle involved in the transport of dietary lipids through in blood or lymph. Chylomicrons are of very low density and have a high triacylglyceride content. Other lipid transporters are *HDL, IDL, *LDL, *VLDL. **Chylomicron retention disease** is caused by mutation in *sar1B (*see* ANDERSON'S DISEASE).

chymase A serine peptidase (EC 3.4.21.39, mast cell protease, 247 aa) that is important in generating angiotensin II in response to injury of vascular tissues, and converts big endothelin 1 to the 31 aa peptide *endothelin 1.

chymosin An aspartyl peptidase (EC 3.4.23.4, 381 aa) from the abomasum (fourth stomach) of unweaned calves that will cleave casein to paracasein. Used in cheesemaking.

chymostatin A potent inhibitor of many proteases, including chymotrypsin, papain, chymotrypsin-like serine peptidases, chymases, and lysosomal cysteine peptidases such as *cathepsins A, B, C, D, H, and L. It weakly inhibits human leucocyte elastase. It is of microbial origin and several variants are found, all small molecules around 600 Da.

chymotrypsin A secreted serine peptidase (EC 3.4.21.1, 228 aa) of the S1 endopeptidase family, a digestive enzyme produced in the pancreas.

ciboulot A G-actin binding protein (129 aa) from *Drosophila* that has a role in axonal growth during brain metamorphosis.

cicatrization The process of forming a scar, the closure of a wound by the contraction of fibroblasts that have invaded the granulation tissue and that will secrete collagen to restore the mechanical properties of the tissue. Excessive contraction can cause severe distortion and be seriously disfiguring.

ciclosporin (formerly **cyclosporine)** A cyclic undecapeptide isolated from the fungus *Tolypocladium inflatum* (formerly *Beauveria nivea*), that is used as an immunosuppressive drug to prevent transplant rejection and to treat some autoimmune diseases. It binds to *cyclophilin in T cells and the complex inhibits *calcineurin and blocks the production of various cytokines, particularly IL-2, thus inhibiting cell-mediated immune responses. *TNs* Neoral, Sandimmune.

ciguatera A form of food poisoning caused by eating seafood containing natural ciguatoxins that are produced by certain strains of the dinoflagellate *Gambierdiscus toxicus* and accumulate in the higher levels of the food chain. The toxin is heat stable and activates voltage-gated sodium channels in the peripheral nervous system.

ciliary body The circumferential tissue inside the eye composed of the **ciliary muscles** that act on the lens to produce accommodation and the **ciliary processes** that produce aqueous humour. The ciliary body is covered by ciliary epithelium that is continuous with the *pigmented retinal epithelium, and neural retina. The **ciliary ganglion** acts as a relay between parasympathetic neurons of the oculomotor nucleus in the midbrain and the muscles regulating the diameter of the pupil of the eye.

ciliary neurotropic factor A *neurotropin (CNTF, 200 aa) that is a survival factor for various neuronal cell types. It is considered to be one of the *interleukin-6 cytokine family since it acts through a receptor containing gp130.

cilium (*pl.* cilia) A hair-like projection from the surface of a cell, supported internally by a characteristic '9 + 2' arrangement of microtubules, an *axoneme, that is linked to a *basal body. Bending of the cilium is caused by the activity of a dynein ATPase that causes differential sliding of doublets relative to one another. The asymmetric beat pattern of cilia may move the cell or, if the cell is fixed, cause the flow of fluid (or mucus) over the surface. A eukaryotic *flagellum has a similar structure but is generally longer and with a different beat pattern.

cimetidine An *H_2 blocker. *TN* Tagamet.

CIN85 A multiadaptor protein (Cbl-interacting protein of 85 kDa, SH3 domain-containing kinase-binding protein 1, 665 aa) involved in different cellular functions including the down-regulation of activated receptor tyrosine kinases such as the *EGF receptor and the MET/hepatocyte growth factor receptor. It is also involved in the regulation of endocytosis. *Cbl (ubiquitin ligase) binds to the activated receptor; further binding of the CIN85 /*endophilin complex leads to clathrin-mediated internalization.

CINCA *See* CHRONIC INFANTILE NEUROLOGIC CUTANEOUS AND ARTICULAR SYNDROME.

cinchocaine (dibucaine) A long-lasting local anaesthetic.

cingulin A rod-shaped homodimeric protein (1197 aa) found in the cytoplasmic domain of vertebrate tight junctions that interacts with *ZO-1. **Cingulin-like protein 1** (CGNL1, paracingulin, 1302 aa) may be involved in anchoring the *tight junctions of the apical junctional complex to the actin cytoskeleton. A chromosomal inversion in which the promoter of the *CGNL1* gene is brought upstream of the aromatase coding region causes aromatase excess syndrome in which there is oestrogen excess due to an increased aromatase activity.

cinnamycin *See* DURAMYCIN.

cinnarizine An antihistaminic drug mainly used for the control of motion sickness. *TNs* Stugeron, Stunarone.

CIP/KIP *See* CYCLIN-DEPENDENT KINASE INHIBITORS.

ciprofibrate *See* FIBRATES.

ciprofloxacin A synthetic *quinolone antibiotic.

circadian rhythm A regular cycle of behaviour with a period of approximately 24 h. The suprachiasmatic nuclei of the hypothalamus are thought to contain the master clock that regulates most circadian rhythms in mammals. The endogenous clock is usually entrained to 24 h by environmental cues (zeitgebers), often light (*see* PHOTOLYASES). The biochemical basis is complex but involves oscillations in transcription regulated by positive and negative inputs and that has downstream effects on hormonal systems, body temperature, etc. *See* CLOCK; PERIOD; TIMELESS.

circular DNA A DNA molecule that forms a closed circle rather than being linear. The resulting topological problems involved in replication are solved with *DNA topoisomerase. Circular DNA is found in prokaryotes, mitochondria, chloroplasts, and some viral genomes.

cirrhosis An irreversible condition affecting the liver in which there is cell loss, inflammation, and loss of normal tissue architecture that leads, eventually, to hepatic failure.

cirsoid aneurysm A mass of tortuous and dilated arteries (racemose aneurysm) due to a congenital malformation in which there is arteriovenous shunting. Can arise in the scalp from abnormal connections between the external carotid artery and scalp veins.

CIS Cytokine-inducible immediate early gene. *See* SOCS.

C

***cis*-activation** Activation brought about by something on the same cell or on the same chromosome. There are examples of *cis*-activation of adhesion molecules through protein–protein contacts that occur in the plane of the membrane but do not occur if the two molecules are in different cells. *Cis*-activation of a gene is brought about by activators binding to upstream regions of the chromosome, whereas *trans*-activation would involve a diffusible product. *Compare* TRANSACTIVATION.

cisapride A serotonin $5HT_4$ receptor agonist that was used to treat gastroesophageal reflux disease and to increase gastric emptying in people with diabetic gastroparesis. Now withdrawn because of side effects. *TNs* Prepulsid, Propulsid.

***cis*-dominance** The effect of a promoter or gene on the activity of genes that are on the same chromosome, in contrast to *trans* effects when a gene or promoter on one DNA molecule can affect genes on another.

***cis*-Golgi** *See* GOLGI APPARATUS.

cisplatin A platinum-based cytotoxic drug (*cis*-diamminedichloroplatinum(II)) used in tumour chemotherapy. It induces crosslinking of DNA which leads to apoptosis. Related drugs are carboplatin and oxaliplatin.

***cis*-regulatory modules** Short stretches of DNA that may regulate gene expression and may be up to 1 million bases away from the genes they regulate, but are on the same chromosome.

cisternae The membrane-enclosed regions of the rough and smooth *endoplasmic reticulum and the *Golgi. The contents of the intracisternal space are destined for secretion after a complex transition through the *GERL during which various post-translational modifications occur.

***cis–trans* test** A *complementation test in which the interacting genes are put in *cis* and in *trans* relationships to each other. The test is used to determine whether two mutations which affect the same phenotype are on the same functional unit (a *cis* configuration) or on different functional units (a *trans* arrangement). If the offspring of a mating between an individual with one mutation and an individual with the other mutation are normal, then the mutations are *trans*.

cistron A segment of DNA that contains all the information for production of single polypeptide, the coding region and the various control elements. Originally defined using the *cis–trans* *complementation test. In eukaryotic organisms it is effectively the same as a gene.

citric acid cycle *See* TRICARBOXYLIC ACID CYCLE.

Citrobacter freundii A Gram-negative, facultatively anaerobic, coliform bacterium of the family Enterobacteriaceae that can cause opportunistic infections and has been associated with neonatal meningitis.

citrulline An α-amino acid not found in proteins but an intermediate in the *urea cycle.

CJD *See* CREUTZFELDT–JAKOB DISEASE.

CKIs *See* CYCLIN-DEPENDENT KINASE INHIBITORS.

CLA **1.** Conjugated linoleic acid. **2.** (*occasionally*) Cutaneous lymphocyte-associated antigen.

cladistics A taxonomic method based on shared ancestry and thus based on evolutionary relationships. Members of a **clade** (a single organism and all of its descendants) possess shared derived characteristics. A **cladogram** is a branching diagram (dendrogram) that illustrates the relationships between groups of organisms determined by the methods of cladistics, which tend to involve objective, quantitative analysis, and often computational methods.

clans of peptidases A classification used in the *MEROPS peptidase database; peptidases that can be grouped on the basis of evolutionary relationship. Each clan is identified by two letters, the first representing the catalytic type of the families included in the clan: Aspartic (A), Cysteine (C), Glutamic (G), Metallo (M), Mixed (P), Serine (S), Threonine (T), Unknown (U). The mixed clan contains families of more than one of the catalytic types. Some clans are divided into subclans where there is evidence of ancient divergence.

Clara cells The epithelial cells that line small airways and secrete *surfactant proteins and *uteroglobin. They protect against environmental agents and regulate inflammatory and immune responses in the lung. **Clara cell secretory protein** is idential to *uteroglobin.

clarins There are three clarin paralogues, clarins 1-3, that appear to belong to the hyperfamily of small integral membrane glycoproteins with four transmembrane domains (tetraspanins) that includes *connexins and *claudins. There is some sequence similarity to *stargazin. Mutations in the gene encoding clarin-1 (Usher's syndrome type-3 protein, 232 aa) are associated with *Usher's syndrome type 3A.

class switching The change that occurs in the type of immunoglobulin being produced as an immune response matures. B cells that start by making IgM switch to producing IgG with the same antigenic specificity. Switching also occurs between other immunoglobulin classes.

clastogen A substance that causes chromosome breakage. Examples include acridine yellow, benzene, ethylene oxide, arsenic, phosphine, and mimosine.

clathrin A multimeric protein with three heavy chains (clathrin heavy chain 1, 1675 aa; clathrin heavy chain 2, 1640) and three light chains (248 aa or 229 aa), that forms the polyhedral meshwork of *triskelions around a *coated vesicle. The two light-chain genes are alternatively spliced in a tissue-specific manner. Mutations in clathrin heavy chain 2 may be involved in the hypotonia seen in *velocardiofacial syndrome. **Clathrin-coated vesicles** are important in receptor-mediated endocytosis and mediate the transport of cargo from the *trans*-Golgi network to the endosomal/lysosomal compartment. *See* COATAMER.

claudication An impairment in walking (limping) because of fatigue and pain in the legs, usually due to poor blood flow in the calf muscles. *See* INTERMITTENT CLAUDICATION.

claudins A family of integral membrane proteins that have four transmembrane domains and are important structural components of tight junctions where they occlude the intercellular space. There are 24 human claudins that are expressed in a tissue-specific pattern. **Claudin-1** (senescence-associated epithelial membrane protein, 211 aa) forms homopolymers and heteropolymers with some other claudins. It is also a coreceptor for *hepatitis C entry into liver cells. Mutations in the claudin-1 gene are the cause of *ichthyosis-sclerosing cholangitis neonatal syndrome. Mutations in **claudin-14** are associated with autosomal recessive deafness-29. Mutations in **claudin-19** cause a form of renal magnesium wasting with hypercalcinosis and progressive renal failure and severe ocular involvement. *See* PARACELLIN (paracellin-1).

clavulanic acid A β-lactamase inhibitor isolated from *Streptomyces clavuligerusa*, often used in combination with antibiotics such as *amoxicillin that are susceptible to degradation by lactamase.

cleavage In embryology, the early divisions of the fertilized egg to form blastomeres.

cleft palate A relatively common (1 in 600–800 live births) failure in craniofacial development (palatoschisis) in which elements of the palate fail to fuse in the midline. May be accompanied by cleft lip (cheiloschisis, hare lip). Can be corrected by surgery soon after birth. **Cleft lip/palate–ectodermal dysplasia syndrome** (Zlotogora–Ogur syndrome, Margarita Island ectodermal dysplasia) is a disorder caused by mutations in *nectin 1 in which there is cleft lip/palate, ectodermal dysplasia, developmental defects of the hands and, in some cases, mental retardation.

Cleland's reagent *See* DITHIOTHREITOL.

CLI *See* CLUSTERIN.

CLIC proteins *See* CHLORIDE INTRACELLULAR CHANNEL PROTEINS.

clindamycin *See* LINCOMYCIN.

clinical trials The testing of a new drug or treatment, though the latter is done less often than perhaps it should be. Trials can be subdivided into two main types: retrospective and prospective. **Retrospective trials** use existing records and information about environmental exposure, etc. and attempt to correlate these with the clinical outcome. A correlation is not the same as a cause, however, and interpretation is often difficult. In **prospective trials**, two groups of patients, matched for age, sex, clinical stage, etc. are used: one group receiving the test drug or treatment, the other receiving the conventional treatment or sometimes a *placebo. Ideally neither patients nor clinicians should know which patient is in which group (a double-blind trial, the only really reliable approach) and the clinical outcome is statistically analysed.

clioquinol An antifungal and antiprotozoal drug (iodochlorhydroxyquinoline), used to treat infection with *Entamoeba histolytica* and topically to treat skin infections.

CLIPs Microtubule-associated proteins that link endocytic vesicles to microtubules. **CLIP-1** (cytoplasmic linker protein-170, CAP-Gly domain-containing linker protein 1, Reed–Sternberg intermediate filament-associated protein, restin, 1427 aa) is highly expressed in *Reed–Sternberg cells. CLIP-1 binds to growing ends of microtubules that have bound plus end-binding proteins such as *EB1 and is part of the microtubule plus end tracking system that regulates microtubule dynamics. **CLIP3** (CLIP-170 related 59-kDa protein, 547 aa) is also a cytoplasmic linker protein and is involved in *trans*-Golgi/endosome dynamics.

CLL *See* LEUKAEMIA.

CLN3 *See* BATTEN'S DISEASE.

clobazam A benzodiazepine derivative with anxiolytic and anticonvulsant effects. *TNs* Frisium, Urbanol.

clock A gene encoding a basic helix–loop–helix transcription factor (846 aa) that, as a heterodimer with *BMal1, acts on the transcription of **period* in a manner that can be regulated by input from cryptochromes. Period and *timeless proteins block clock's ability to activate period and timeless promoters in a negative-feedback loop. *See* CIRCADIAN RHYTHM.

clofibrate A lipid-lowering *fibrate. *TN* Atromid-S.

clomifene (clomiphene) A selective oestrogen receptor modulator that inhibits the action of oestrogen on the hypothalamus, causing increased release of pituitary gonadotropins (FSH, LH) which stimulates ovulation. *TNs* Clomid, Milophene, Serophene.

clomipramine A tricyclic antidepressant used to treat depression and obsessive-compulsive disorder. *TN* Anafranil.

clonal deletion A fairly generally accepted hypothesis for immune tolerance and the nonrecognition of self-antigens, the programmed death of autoreactive T cells.

clonal selection The selection of cells that express a particular gene from a mixed population. The term is commonly used in reference to *B cells that produce antibody to a particular antigen and are stimulated to proliferate.

clonazepam A benzodiazepine derivative with highly potent anticonvulsant, muscle relaxant, and anxiolytic properties. *TNs* Klonopin, Rivotril.

clone A population of cells or organisms derived from a single progenitor and therefore genetically identical.

clonic phase The third phase in a grand mal seizure following the prefiguring aura and the tonic phase in which muscles tense up and the body is rigid and unmoving. In the clonic phase muscles contract and relax rhythmically while the body jerks in violent spasms. *See* EPILEPSY.

clonidine An antihypertensive drug that acts on the neurons in the brainstem that regulate arterial pressure. Although generally described as an α_2-adrenoreceptor agonist, it also has agonist activity for *imidazoline receptors (I_1-R). Moxonidine and rilmenidine are similar but more specific for I_1 receptors. Clonidine is also used at lower doses in the treatment of attention-deficit hyperactivity disorder (ADHD). *TNs* Catapres, Dixarit.

SEE WEB LINKS

- Article (*American Journal of Physiology* (1997)) about the use of clonidine.

cloning The process of establishing clones of cells or organisms. Often used to mean *gene cloning, which involves the isolation of clones of bacteria expressing the sequence of interest. **Cloning vectors** are plasmid *vectors used to transfer DNA from one cell type to another, usually designed to have convenient restriction sites for splicing in the DNA sequence of interest and selectable markers expressed when the cloned sequence is being expressed.

Clonorchis sinensis A trematode parasite, the Chinese liver fluke, that can infect humans and can cause biliary obstruction. Raw or undercooked fish is often the source of the infection.

clopidogrel An inhibitor of platelet aggregation that irreversibly inhibits the P2Y12 ADP receptor and blocks activation of the glycoprotein GPIIb/IIIa complex. Used as an antithrombotic. *TN* Plavix.

Clostridium A genus of Gram-positive anaerobic spore-forming bacilli. Many species produce potent exotoxins such as *botulinum toxin (from *C. botulinum*) and *tetanus toxin (from *C. tetani*). *C. perfringens* produces *perfringolysin (theta toxin), an α-toxin (phospholipase C), β-, ε-, and ι-toxins (which act on vascular endothelium to cause increased vascular permeability), δ-toxin (a haemolysin), and κ-toxin (a collagenase). *C. difficile* secretes enterotoxin A (2710 aa) that enters eukaryotic cell by receptor-mediated endocytosis and glucosylates small G proteins of the *rho family, inactivating them and causing the loss of microfilament bundles. Cytotoxin B is similar.

clotrimazole An antifungal drug used to treat oral candidiasis and fungal skin infections.

cloxacillin A semisynthetic penicillin-type antibiotic that is resistant to staphylococcal penicillinase.

clozapine An atypical neuroleptic drug used to treat schizophrenia but only in patients unresponsive to conventional antipsychotic drugs, because of the risk of agranulocytosis as a side effect. *TNs* Clopine, Clozaril, Denzapine, Fazaclo, Froidir, Leponex, Klozapol, Zaponex.

clusterin A glycoprotein (complement-associated protein SP-40, complement cytolysis inhibitor, CLI, apolipoprotein J, sulphated glycoprotein 2, SGP-2, dimeric acid glycoprotein, DAG, glycoprotein III, GpIII, 449 aa) of uncertain function. It is expressed in a variety of tissues; binds to cells, membranes, and hydrophobic proteins; and has been associated with apoptosis.

cluster of differentiation antigens *See* CD ANTIGENS.

cluster of orthologous group (COG) A classification system for genes of orthologous sequence, a derivative of the bidirectional best BLAST hits (BDBH) system. A COG consists of genes from different species that are identified by sequence similarity searches using *BLAST.

c-maf The cellular proto-oncogene homologous to v-*maf*, the musculoaponeurotic fibrosarcoma oncogene. It encodes a *basic leucine zipper (bZip) transcription factor (373 aa) that can act either as an activator or repressor and that is expressed during the development of various organs and tissues.

CMC **1.** Cell-mediated cytotoxicity, the killing of cells by effector T cells or *NK cells. The target cells may express a particular antigen (cytotoxic T-cell killing) or be marked with antibody–antigen complexes (antibody-dependent cell-mediated cytotoxicity).
2. Carboxymethylcellulose, a negatively charged form of cellulose used as a cation-exchange resin in *ion-exchange chromatography and as a thickening agent in a range of products (food, paint, drilling mud, etc.).

CML *See* LEUKAEMIA.

CMT *See* CHARCOT–MARIE–TOOTH DISEASE.

CMV *See* CYTOMEGALOVIRUS.

c-myc tag A commonly used *epitope tag (EQKLISEEDL) derived from the c-myc protein. *See* MYC.

CNFs *See* CYTOTOXIC NECROTIZING FACTORS.

CNP **1.** C-type natriuretic peptide, *see* NATRIURETIC PEPTIDES. **2.** *See* CNPASE.

CNPase (CNP) A marker enzyme (2′, 3′-cyclic nucleotide 3′-phosphodiesterase, EC 3.1.4.37, 421 aa) for oligodendrocytes. It is reported to interact with *nogo.

CNQX A specific antagonist of the AMPA subtype of *glutamate receptors.

CNS *See* CENTRAL NERVOUS SYSTEM.

CNTF *See* CILIARY NEUROTROPHIC FACTOR.

CNTs (concentrative nucleoside transporters) A family (SLC28) of sodium dependent transporters (concentrative nucleoside transporters) that operate in conjunction with the equilibrative (sodium-independent) transporters (*ENTs), although they belong to structurally unrelated protein families. There are three subtypes, CNT1, CNT2, and CNT3 (SLC28A1, SLC28A2, and SLC28A3, respectively), that transport both naturally occurring nucleosides and synthetic nucleoside analogues used in chemotherapy. **CNT1** (649 aa), found primarily in epithelia, is pyrimidine-nucleoside preferring; **CNT2** (658 aa) is purine-nucleoside preferring; and **CNT3** (691 aa) transports both pyrimidine and purine nucleosides. The latter two are more widely distributed in tissues. They are important in nucleotide salvage pathways.

coactosin A protein of the ADF/*cofilin family (coactosin-like protein, CLP, 142 aa), similar to that found in *Dictyostelium*, that binds to F-actin in a calcium-independent manner but does not appear to affect microfilament depolymerization. Binds 5-lipoxygenase and F-actin through different sites.

coagulase An enzyme produced by *Staphylococcus aureus* that reacts with prothrombin to form staphylothrombin, which will convert *fibrinogen to fibrin. It is also produced by **Yersinia pestis*.

coagulation factors *See* BLOOD CLOTTING FACTORS.

coated vesicle A membrane-bounded vesicle surrounded by a protein meshwork, in some cases formed as an invagination of the plasma membrane (a **coated pit**). The proteins (**coatomers**) that enclose the vesicle are of three types, **COPI** (required for retrograde transport from the *trans*-Golgi apparatus to *cis*-Golgi and endoplasmic reticulum), **COPII** (involved in anterograde transport from ER to the *cis*-Golgi) and *clathrin (with associated *adaptins). Coat protein complex I (COPI) is a cytosolic complex with seven equimolar subunits (alpha (alpha-COP, 1224 aa; *see* XENIN), beta (953 aa), beta′ (906 aa), gamma (874 aa), delta (archain, 411 aa), epsilon (308 aa), and zeta (177 aa or 210 aa). Coatomer can only be recruited by membranes in association with *ADP-ribosylation factors (ARFs). Coat protein complex II (COPII) is composed of at least five proteins: the Sec23/24 complex, the Sec13/31 complex, and Sar1; Sec24 and its two nonessential homologues Sfb2p and Sfb3p may serve in cargo selection. *Sar1p is involved in generating membrane curvature and vesicle formation. Mutations in Sec23A are associated with *craniolenticulosutural dysplasia.

C

coatomer *See* COATED VESICLE.

coat protein complex II (COPII) *See* COATED VESICLE.

cobalamin *See* VITAMIN B (vitamin B_{12}).

cobratoxin (α-cobratoxin) A peptide toxin (long neurotoxin 1, neurotoxin 3, 71 aa) from the cobra *Naja kaouthia.* It causes paralysis by binding, with high affinity, to nicotinic *acetylcholine receptors.

cobra venom factor The complement-activating protein (CVF, 1642 aa) found in the venom of the monocled cobra *Naja kaouthia.* It is a structural and functional analogue of complement component C3 which will bind factor B and is then cleaved by factor D to form a C3/C5 convertase, CVF,Bb. Its potency is probably due to insensitivity to human C3b inactivator. It is used experimentally to deplete the complement system.

cocaine An alkaloid, derived from the coca plant *Erythroxylum coca,* that increases neurotransmitter levels in the midbrain by inhibiting the uptake transporters for dopamine, noradrenaline, and serotonin. The psychostimulatory effects are the reason for its recreational use and its addictive properties.

cocarcinogens Substances that potentiate the activity of a carcinogen, although not particularly carcinogenic themselves. Unlike *tumour promotors, they need to be present at the same time as the primary carcinogen.

cocci Bacteria that are spherical.

coccidiosis An enteritis caused by infection by protozoa of the genera *Eimeria* and *Isospora,* an important veterinary problem in poultry and other domestic animals. Can cause infection in humans, but most coccidia are fairly species-specific and are not a major medical problem.

cocculin *See* PICROTOXIN.

cochlear hair cells The sound-sensing cells of the organ of Corti in the inner ear which have stereocilia (hairs: *see* STEREOVILLUS) on their apical surfaces that convert mechanical distortion (shear forces in the endolymph of the cochlea caused by sound waves) into an electrical signal, by opening ion channels in the membrane. The frequency of the sound is detected partly by the position of the responding cells in the cochlea, although individual hair cells apparently resonate at particular frequencies, enhancing the sensitivity of the auditory system. The inner hair cells are responsible for sound perception, the outer hair cells oscillate in response to electrical potential changes induced by the inner hair cells (*see* PRESTIN) and improve the frequency discrimination.

cochlin A secreted protein (550 aa) expressed in the cochlea and vestibule of the inner ear. Mutations in cochlin cause autosomal dominant sensorineural deafness type 9 (DFNA9) and may contribute to *Ménière's disease.

Cockayne's syndrome A disorder characterized by slow growth and abnormal development from an early age, leading to 'cachectic dwarfism' but also to a range of defects in many systems. The syndrome is caused by defects in the DNA *excision repair system, **type A** by mutation in the group 8 excision-repair cross-complementing protein (ERCC8), **type B** by mutation in the *ERCC6* gene.

co-codamol An analgesic mixture of *codeine and *paracetamol.

coculture The culture of two distinct cell types in a single culture, often used as a strategy to enable cells to grow at low density, the second cell type serving as a *feeder layer producing growth factors. The feeder cells can be macrophages or irradiated cells that will not proliferate. Coculture is, of course, a much more realistic model of the real environment in which cells proliferate *in vivo.*

codanin A protein identified on the basis of a gene mutated in type 1 *congenital dyserythropoietic anaemia that is a putative *O*-glycosylated protein (1227 aa). There are no obvious transmembrane domains and it might be involved in nuclear envelope integrity.

codeine An opioid drug (methylmorphine), widely used as an analgesic. Approximately 10% of codeine is converted to morphine in the body.

codocyte *See* TARGET CELL.

codominant Describing genes in which both alleles are expressed in the heterozygote, e.g. genes for the AB blood group in which both A and B antigens are present. *See* INCOMPLETE DOMINANCE.

codon The triplet of bases in DNA that specifies a triplet in mRNA which is in turn recognized by anticodons in transfer RNA. Most codons specify an amino acid, with the exception of the *termination codons, and there is degeneracy, particularly in the third (3′) position, with some amino acids being specified by several different codons (e.g. alanine is specified by GCU, GCC, GCA, and GCG).

coelenterazine An imidazolopyrazine derivative that emits light when oxidized by an enzyme such as luciferase. Luciferin is coelenterazine disulphate.

coeliac disease An autoimmune disorder in which there is hypersensitivity to *gluten from wheat, barley, and rye leading to inflammation and tissue remodelling of the intestinal mucosa. It is genetically determined by possession of specific HLA-DQ alleles on chromosome 6p21.3, although there are other susceptibility loci and the inheritance is complex and multifactorial.

coelom The body cavity found in most multicellullular animals (the coelomates) that develops within the embryonic mesoderm (*see* SOMATIC MESODERM; splanhnic mesoderm at SPLANCHNIC) and is lined by the peritoneum, an epithelial sheet derived, unlike most, from mesoderm.

coenzyme A confusing term that can refer either to a low molecular weight intermediate that is involved in a reaction catalysed by an enzyme (e.g. NAD), which might perhaps be better termed a *cofactor, or a tightly coupled nonprotein component such as *haem (a prosthetic group) that in the case of an enzyme forms the holoenzyme (apoenzyme plus coenzyme). **Coenzyme A** (CoA) is a derivative of adenosine triphosphate and pantothenic acid (vitamin B_5) that is involved in the transfer of acyl groups in a range of metabolic pathways, e.g. the *tricarboxylic acid cycle and in fatty acid oxidation. **Coenzyme Q** is *ubiquinone.

cofactor A nonprotein compound or substance that is required for the biological activity of a protein. There are two main types: organic cofactors, such as *flavin or *haem, and inorganic cofactors such as the metal ions Mg^{2+}, Cu^{2+}, Mn^{2+} or iron–sulphur clusters (*see* CDGSH IRON–SULPHUR DOMAIN PROTEINS). If the activity is of an enzyme, loosely bound cofactors are sometimes referred to as coenzymes and tightly bound cofactors as prosthetic groups.

Coffin–Lowry syndrome A rare form of X-linked mental retardation characterized by skeletal malformations, growth retardation, hearing deficit, paroxysmal movement disorders, and cognitive impairment caused by mutations in the *RSK2* gene for ribosomal S6 kinase, a growth factor-regulated serine/threonine kinase.

cofilin An actin-binding protein that controls actin polymerization and depolymerization in a pH-sensitive manner. There are two isoforms: **cofilin-1** (166 aa) found in nonmuscle cells and **cofilin-2** (166 aa) in striated muscle. Very similar to (*actin depolymerizing factor (ADF).

COG *See* CLUSTER OF ORTHOLOGOUS GROUP.

cohesins A family of proteins involved in the **cohesin complex**, a large proteinaceous ring that holds sister chromatids together. The complex is cleaved at anaphase which allows sister chromatids to segregate. The cohesin complex may also play a role in spindle pole assembly during mitosis. The components of the complex are a heterodimer between a *SMC1 protein (structural maintenance of chromosomes protein, SMC1A or SMC1B) and SMC3, which are attached via their hinge domain, and *RAD21 which links them at their heads, and one of the three STAG proteins. *See* CORNELIA DE LANGE SYNDROME.

cohort A group of people or other organisms that share a common characteristic, e.g. year of birth or similar work environments. The comparison of a cohort with a matched group is important in statistical and epidemiological studies.

coiled body *See* CAJAL BODIES.

coilin A protein (p80-coilin, 576 aa) found in *Cajal bodies (coiled bodies). A **coilin-interacting nuclear ATPase** (172 aa) has been reported.

coisogenic A strain of animal that differs from an inbred partner strain at a single locus. Various strains of mice that are coisogenic for mutations in disease-related alleles are important experimental models.

colchicine An alkaloid (400 Da), isolated from the autumn crocus *Colchicum autumnale*, that binds to the *tubulin heterodimer (but not to tubulin) and blocks assembly of microtubules. By preventing the assembly of new spindle microtubules it will block cells in metaphase. **Colcemid** is a methylated derivative of colchicine.

cold agglutinins Antibodies (IgM) that react with blood groups I and i (precursors of the ABH and Lewis blood group substances), and *agglutinate red blood cells on cooling, causing *Raynaud's disease *in vivo*. Cold agglutinin disease is an extreme form in which there are high levels of IgM that cause haemolytic anaemia.

cold insoluble globulin An obsolete name originally given to *fibronectin that was prepared from *cryoprecipitate.

colestipol A bile acid sequestering compound used to increase bile acid excretion (rather than recycling) and lower blood cholesterol levels because more cholesterol is diverted to the synthesis of bile acids. *TN* Colestid.

C

colestyramine *See* CHOLESTYRAMINE.

colforsin *See* FORSKOLIN.

colic A sharp abdominal pain with an acute onset that can be due to contraction of the smooth muscle of the intestinal, renal or biliary tracts or blockage with kidney stones or gallstones. Some babies, particularly during their first 3 months, seem to suffer a form of intestinal colic.

colicins Bacterial exotoxins (*bacteriocins) that are toxic to other bacteria. Most are plasmid-coded AB toxins or channel-forming transmembrane peptides.

coliform Describing: **1.** Any rod-shaped bacterium. **2.** Any Gram-negative enteric bacillus. **3.** Bacteria of the genera *Klebsiella* or *Escherichia*.

colipase A protein cofactor (90 aa) for pancreatic triglyceride lipase (PNLIP) required for efficient hydrolysis of dietary lipid. *See* ENTEROSTATIN; PROCOLIPASE.

colistin (polymyxin E) A polypeptide antibiotic produced by *Bacillus polymyxa* var. *colistinus*; effective against most Gram-negative bacilli.

colitis A chronic digestive disorder in which there is inflammation of the colon. **Coloenteritis** (enterocolitis) involves both colon and small intestine.

collagen A family of proteins that are major structural components of the extracellular matrix and constitute probably around 30% of the whole-body protein content. They are unusual proteins in their structure and in their amino acid composition (very rich in glycine (30%), proline, *hydroxyproline, lysine, and *hydroxylysine but with no tyrosine or tryptophan (which has implications for standard protein assays)). Three collagen molecules, not all identical, form a triple helical structure ~300 nm long (*tropocollagen) and these can associate in fibril-forming types with a characteristic quarter-stagger overlap between them to produce tension-resisting fibrils. **Type I collagen**, one of the group I fibril-forming collagens, is trimeric (one α2(I) (COL1A2) and two α1(I) (COL1A1) chains; 1366 aa and 1464 aa respectively) and forms the fibrils of tendon, ligaments, and bones. In bones the fibrils are mineralized with calcium hydroxyapatite. Defects in collagen α1(I) are the cause of *Caffey's disease, some forms of *Ehlers–Danlos syndrome, and *osteogenesis imperfecta. Variations are associated with involutional *osteoporosis, and a chromosomal translocation of the gene causes *dermatofibrosarcoma protuberans. **Collagen type II** (COL2A1, 1487 aa) is a homotrimer of α1 (II) collagen (chondrocalcin) specific for cartilaginous tissues; defects are the cause of a variety of chondrodysplasias, including *hypochondrogenesis and *osteoarthritis. Defects are also responsible for congenital *spondyloepiphyseal dysplasia, *Perthes' disease, *Kniest's syndrome, primary *avascular necrosis of femoral head, osteoarthritis with mild chondrodysplasia, *platyspondylic lethal skeletal dysplasia Torrance type, *multiple epiphyseal dysplasia with myopia and conductive deafness, *spondyloperipheral dysplasia, *Wagner's syndrome type II, some forms of *Stickler's syndrome, and *rhegmatogenous retinal detachment. **Collagen type III** occurs in most soft connective tissues along with type I collagen and is composed of three identical α1(III) (COL3A1, 1466 aa) chains; defects are associated with various forms of Ehlers–Danlos syndrome and osteogenesis imperfecta, also susceptibility to *abdominal aortic aneurysm. **Collagen type IV** does not form fibrils and is characteristic of *basal lamina. Defects in COL4A1 are a cause of brain small-vessel disease with haemorrhage; hereditary *angiopathy with nephropathy, aneurysms, and muscle cramps; and porencephaly type 1. Autoantibodies against the NC1 domain of the alpha 3 isoform are found in Goodpasture's syndrome. Defects in COL4A4 cause autosomal recessive *Alport's syndrome and benign *familial haematuria. **Collagen type V** is fibril forming. **Collagen type VI** acts as a cell-binding protein and defects in COL6A2 and COL6A3 are a cause of *Bethlem myopathy and Ullrich congenital *muscular dystrophy,; defects in COL6A2 also cause autosomal recessive *myosclerosis. Defects in **collagen type X** (COL10A1) cause Schmid type metaphyseal *chondrodysplasia. Defects in **collagen type XI** (COL11A1) are the cause of *Marshall's syndrome and variants of Stickler's syndrome; defects in COL11A2 cause of Stickler's syndrome type 3, autosomal recessive *otospondylomegaepiphyseal dysplasia, Weissenbacher–Zweymueller syndrome, nonsyndromic sensorineural deafness autosomal dominant types 13 and 53. **Collagen type XVII** (COL17A1) is the *bullous pemphigoid antigen 2. Defects in **collagen type XVIII** (COL18A1) are a cause of *Knobloch's syndrome. Many other forms are known, often with restricted tissue distribution, some glycosylated (glucose-galactose dimer on the hydroxylysine), and nearly all types can be crosslinked through lysine side chains. Defects in post-translational processing of collagen can arise if there is vitamin C deficiency, causing *scurvy. *See* FACIT COLLAGENS.

collagenase A peptidase that will hydrolyse native collagen, a protein which is resistant to many peptidases. Once the initial cleavage is made, other less specific peptidases can continue the degradation. Collagenases from mammalian cells (EC 3.4.24.7) are metalloenzymes and are collagen-type specific. Bacterial collagenases (EC 3.4.24.3) are used to help extract cells from tissues.

collapsin A *semaphorin isolated from chick brain that is a repellent guiding molecule for developing axonal processes, causing the collapse of the nerve growth cone, and that inhibits the regeneration of mature neurons. In humans there are a family of cytosolic **collapsin response mediator proteins** (CRMPs), e.g. **CRMP-1** (dihydropyrimidinase-related protein 1, 572 aa) that are necessary for signal transduction by class 3 semaphorins that causes remodelling of the cytoskeleton. *Rho kinase phosphorylates **CRMP-2** during growth cone collapse and blocks its ability to bind tubulin dimers. **CRMP-4** colocalizes with F-actin. Although CRMPs resemble liver dihydropyrimidinase in sequence and quaternary structure, they do not hydrolyse several DHPase substrates.

collectins A family of C-type *lectins that are involved in innate immune responses by binding to viruses and by opsonizing yeasts and bacteria. In humans there are three collectins: **collectin-10** (collectin liver protein 1, 277 aa) that binds galactose > mannose = fucose > *N*-acetylglucosamine > *N*-acetylgalactosamine; **collectin-11** (collectin kidney protein 1, 271 aa) that binds to LPS and binds fucose > mannose but does not bind to glucose, *N*-acetylglucosamine and *N*-acetylgalactosamine; **collectin-12** (collectin placenta protein 1, scavenger receptor class A member 4, 742 aa) that promotes binding and phagocytosis of Gram-positive and Gram-negative bacteria and yeast. it is also involved in the recognition, internalization, and degradation of oxidatively modified low density lipoprotein (oxLDL) by vascular endothelial cells. The collectins have a collagen-like region and a C-type lectin domain and are part of a family that also includes pulmonary surfactant proteins A and D (SP-A, SP-D), CL-43, serum *mannan-binding lectin (MBP), and conglutinin. Complement C1q is also structurally related.

collectrin A homologue of *angiotensin converting enzyme 2 that is expressed in pancreatic beta cells and renal proximal tubular and collecting duct cells.

colliculus A small elevation. There are four colliculi in the tectum of the midbrain, two caudal (inferior) and two rostral (superior) containing the visual and auditory reflex centers (corpora quadrigemina).

colligative properties The properties of a solution that are determined by the numbers of molecules present rather than on their chemistry, e.g. the osmotic pressure.

collodion A solution of cellulose tetranitrate in a mixture of ethanol and ethoxyethane, used for coating materials and for sealing wounds and dressings.

colloid goitre An abnormal enlargement of the thyroid gland caused by the accumulation of viscous fluid which may cause pressure-related atrophy of the thyroid epithelium. Commonly caused by iodine deficiency in the diet.

collybistin A neurospecific rho-GTPase guanine nucleotide exchange factor (516 aa) of the *Dbl family that is specific for Cdc42. It is negatively regulated by *gephyrin. Defects cause *startle disease with epilepsy.

coloboma A developmental defect of the eye, particularly one affecting the lens, iris, or retina.

colonization factors The pili on the surface of *E. coli* or other gut bacteria that facilitate adhesion (adherence factors) to receptors (probably GM_1 gangliosides) on gut epithelial cells and contribute to pathogenicity.

colony-forming unit (CFU) **1.** An indirect measure of the numbers of stem cells in bone marrow used to reconstitute the immune system of mice that have been irradiated. The grafted cells form colonies in the spleen (**CFU-S**), each colony a clonal derivative of a single stem cell. **CFU-E** are the derivatives of a partially committed stem cell of the erythoid lineage. **2.** A measure of the number of live bacteria or fungi in a (diluted) sample which is put into appropriate culture conditions, e.g. on a nutrient-agar plate; each colony represents a single organism in the sample. A similar approach can be used with viruses by plating them on to a culture of host cells.

colony-stimulating factor A *cytokine involved in promoting the growth and differentiation of leucocytes in culture. Examples include *MCSF and *interleukin-3.

colostrum The first milk produced by a lactating animal, which is important for the transfer of immunity from mother to neonate. In some species colostrum is rich in maternal antibodies that are absorbed from the neonatal gut; in other species the transfer is transplacental.

colour blindness The inability to detect the full range of colours as a result of the lack of one or more of the cone pigments (opsins). Approximately 8% of western European males are colour blind, about 75% with a defect in the deutan (green) series (daltonism, deuteranopia) and about 25% with a defect in the protan (red) series. If the defect is in the absorption spectrum of the pigment then the disorder is referred to as protanomaly and deuteranomaly. Most red–green colour vision defects arise as a result of unequal crossing-over between the red and green pigment genes. *See* DEUTERANOPIA; DICHROMATISM; PROTANOPIA; TRITANOPIA. Complete colour blindness (achromatopsia) and some cases of incomplete achromatopsia are caused by a defect in the α subunit of the cone photoreceptor cGMP-gated cation channel; cones are present but dysfunctional.

colposcopy The endoscopic examination of the vagina using a colposcope.

columnar cells Epithelial cells in which the area in contact with the basal lamina is less than the area in contact with adjacent cells; they are 'taller than they are wide'. *Compare* CUBOIDAL EPITHELIUM; SQUAMOUS EPITHELIUM.

coma A state of complete unconsciousness, unresponsive to external stimulation.

combination therapy The treatment of a disease with two or more drugs, as in the treatment of malaria with artemisinin and lumefantrine, or tuberculosis with a cocktail of different antibiotics.

combinatorial chemistry A general term for synthetic methods that generate large numbers of related compounds, often for the purposes of screening. The simplest methods generate a mixture of compounds; more sophisticated methods are more correctly termed high-speed parallel synthesis, in which each reaction chamber contains only one compound.

combined immunodeficiency *See* SEVERE COMBINED IMMUNODEFICIENCY DISEASE.

combining site The region of a molecule that binds a particular compound. Used particularly for the antigen-binding site on an immunoglobulin.

combretastatins A family of compounds, originally derived from the African bush willow tree *Combretum caffrum*, used in cancer chemotherapy where they inhibit tumour vascularization by binding to tubulin in endothelial cells. Synthetic variants are emerging with slightly different efficacy. **Combretastatin A-4** is the most potent naturally occurring combretastatin and is currently (2009) in clinical trials as a combination therapy with carboplatin and paclitaxel for thyroid carcinoma.

comedo Colloquially, a blackhead. A collection of cells, sebum, and bacteria blocking the opening of a sebaceous gland near a hair follicle.

comet assay A sensitive method (single cell gel (SCG) electrophoresis) to examine DNA damage and repair at individual cell level. Single cells are embedded in agarose on a microscope slide, are lysed and then electrophoresed *in situ*. The slide is then stained with fluorescent dye (e.g. acridine orange) and examined under the microscope. Intact DNA remains in the cell, as the comet's 'head', fragments of different sizes form the tail.

COMMD A multifunctional protein (copper metabolism MURR1 domain-containing protein 1, MURR1, 186 aa) that inhibits NFκB by affecting the association of NFκB with chromatin and has a copper metabolism (MURR1) gene domain. It is a candidate for diseases of copper deficiency (*Menkes' disease) or copper accumulation (*Wilson's disease). It is the archetype of a family of conserved proteins (**COMM domain-containing 1–10**) that form multimeric complexes and share the MURR1 domain which is an interface for protein–protein interactions.

committed cells Cells that have become committed to a particular pathway of differentiation (*see* DETERMINATION). Embryonic stem cells are completely uncommitted.

common cold An acute viral infection of the mucous membranes of the upper respiratory tract, usually involving rhinoviruses (picornaviruses) or coronaviruses.

communicating junction *See* GAP JUNCTION.

comorbid Describing diseases or disorders that exist simultaneously.

compactin (mevastatin) An intermediate for the production of pravastatin (one of the *statins), a metabolite of *Penicillium citrinum*.

compaction The morphological rearrangement that occurs during the morula stage of embryogenesis in which the *blastomeres become less spherical, increase their cell–cell contact area, and develop gap junctions.

compartment **1.** A part of the body (organs, tissues, cells, or fluids) treated as an independent entity when modelling the distribution and clearance of a substance, a process referred to as compartmental analysis. **2.** In developmental biology, a clonal territory in which the cells are expressing a particular set of genes. Whether such compartments exist in the vertebrate embryo is uncertain.

competence stimulating peptide A small signalling peptide (CSP-1, 41 aa) involved in the quorum-sensing mechanism that regulates competence for genetic transformation in *Streptococcus pneumoniae*. Similar peptides are found in other streptococcal species.

competent cells **1.** Bacterial cells with enhanced ability to take up exogenous DNA and thus to be transformed, an important property for molecular biologists. Some bacteria (*Pneumococcus, Bacillus* and *Haemophilus* spp.) acquire competence naturally; in others, such as *E. coli*, it can be induced by treatment with calcium chloride. **2.** Embryonic cells that will differentiate in response to an inducer.

competitive inhibition The inhibition of an enzyme or the binding capacity of a receptor that occurs when an inactive compound competes with the normal substrate or ligand for the active site. If the natural ligand is at sufficiently high concentration, the inhibition disappears. A competitive inhibitor will alter the K_m or K_a but the V_{max} is unchanged.

complement A heat-labile enzyme cascade system in plasma that is important in the response to injury. The cascade can be activated either through the classical pathway, involving antibody–antigen complexes, or through the alternate pathway in response to foreign surfaces. The first step of the **classical pathway** is the activation of the complement C1 complex (C1q, two molecules of C1r, and two molecules of C1s) when C1q binds to IgM or IgG complexed with antigen. C1r and C1s are serine peptidases and when activated, they cleave C4 and then C2 to produce C4a, C4b, C2a, and C2b. The activity of C1r and C1s is regulated by C1 inhibitor. C4b and C2a bind (C4b2a complex) and act as C3 convertase, which cleaves C3 into C3a (anaphylatoxin) and C3b, an opsonin. C3b forms a C4bC2aC3b complex which acts on C5 to release C5a (an anaphylatoxin and chemotactic factor) and C5b, which combines with C6789 to form a cytolytic *membrane attack complex (MAC). The **alternate pathway** converges at C3 cleavage, which occurs without the involvement of C142, and can be activated by IgA, endotoxin, or polysaccharide-rich surfaces (e.g. yeast cell wall, zymosan). Factor B combines with C3b to form a different C3 convertase that is stabilized by factor P, generating a positive feedback loop, and acts on C5. C3 convertase can be inhibited by *decay accelerating factor. A third pathway, the lectin pathway, is activated by *mannan-binding lectin (MBL) bound to surface of a pathogen. This binding activates the MBL-associated serine proteases, *MASP-1, and MASP-2, which cleave C4 and C2 in the same way as C1r and C1s in the classical pathway. Although the alternate pathway sounds subsidiary it is probably the basic pathway, to which the classical pathway adds the sophistication of antibody recognition. The net result of activating the cascade is to produce a massive amplification, and various by-products (C3a, C5a) recruit and activate leucocytes and act as opsonins; the MAC is involved directly in killing. The whole system is regulated, and deficiencies in various components have serious consequences. *See* COMPLEMENT DEFICIENCIES.

complementarity determining region The hypervariable region of genomic sequence that encodes the equivalently hypervariable protein sequence within the antigen binding site of immunoglobulin molecules and T-cell receptors and determines the epitope that is bound.

complementary base pairs The nucleotides that form hydrogen bonds between the two strands of the double helix of DNA and some double-stranded regions of RNA; guanine (G) and cytosine (C) are complementary, as are adenine (A) and thymine (T) (uracil, U, in RNA).

complementary DNA DNA that is synthesized from an RNA template, usually mRNA, by *reverse transcriptase. A cDNA library is a *genomic library based upon the genes that are being expressed as mRNA, the 'transcriptome'.

complementation In genetics, the ability of two homozygous strains that have mutations in the same metabolic or developmental pathway to produce heterozygous progeny that are phenotypically normal. This happens only if the two strains have mutations in different steps in the pathway so that the heterozygous progeny have one normal allele at each locus. If the two mutated strains have defects in the same gene then they will not complement. *See* CIS–TRANS TEST.

complement cytolysis inhibitor *See* CLUSTERIN.

complement deficiencies

A range of relatively rare disorders of the *complement system, in some cases because of inadequate secretion, in others because there is mutation in the gene encoding one of the components. Deficiencies of the classic pathway cause a predisposition to immune complex diseases because of the role of complement in clearance of immune complexes. A general consequence is also impaired opsonization with a greater chance of infection. C1s deficiency is associated with early-onset multiple autoimmune diseases; homozygous deficiency of C4a is associated with systemic lupus erythematosus and with type I diabetes; homozygous deficiency of C4b is associated with susceptibility to bacterial meningitis; complement C9 deficiency is associated with increased susceptibility to meningococcal infections; *complement factor H deficiency can cause various phenotypes, including asymptomatic, recurrent bacterial infections and renal failure. *See also* C3 NEPHRITIC FACTOR.

complement factor H A serum glycoprotein (CFH, β-1H globulin, 1231 aa) that regulates the alternate complement pathway by binding to C3b, accelerating the decay of the alternate pathway convertase C3bBb, and acting as a cofactor for complement factor I. Genetic variations in CFH are associated with basal laminar *drusen and with *age-related macular degeneration type 4. Deficiency of CFH allows uncontrolled activation of the alternate pathway and is associated with a number of renal diseases, including membranoproliferative glomerulonephritis and atypical *haemolytic uraemic syndrome. There is also increased susceptibility to meningococcal infections.

complement factor I A serine proteinase in plasma (CFI, EC 3.4.21.45, 583 aa, post-translationally cleaved into disulphide-linked light and heavy chains) that cleaves and inactivates *complement C4b and C3b. CFI deficiency is an autosomal recessive condition associated with frequent pyogenic infections and may also be associated with or predispose to *haemolytic uraemic syndrome.

complement fixation The binding and activation of *complement as a result of its interaction with immune complexes or particular surfaces.

complement receptors (CR1, etc.) A diverse set of cell surface molecules that bind different complement components or fragments. **CR1** (CD35, 2039 aa) binds complement C3b and C4b and is involved in the phagocytosis of bacteria and the uptake of immune complexes. It is found on neutrophils, mononuclear phagocytes, B cells, Langerhans cells, follicular dendritic cells, and glomerular podocytes. **CR2** (CD21, 1033 aa) binds C3d and is found on B cells and follicular dendritic cells. It is the site to which the *Epstein–Barr virus binds. **CR3** (CD11b/CD18, Mac-1, integrin α-M,1152 and integrin β-2, 769 aa) binds C3bi (iC3b) and is found on neutrophils and mononuclear phagocytes. **CR4** is another β-2 integrin (CD11c/CD18) and binds C3dg, the fragment that remains when C3b is cleaved to C3bi. It is thought to be present on monocytes, macrophages, and neutrophils.

complexins A family of proteins involved in regulating exocytosis, particularly the release of neurotransmitter at synapses. **Complexin-1** (synaphin-2, 134 aa) binds to the *SNARE core complex (SNAP25, *synaptobrevin, and *syntaxin) and positively regulates a late step in synaptic vesicle exocytosis by inhibitory neurons. It is also involved in glucose-induced secretion of insulin by pancreatic beta cells. It is overexpressed in the substantia nigra in Parkinson's disease. **Complexin-2** (synaphin-1, 134 aa) is also involved in regulating synaptic vesicle exocytosis, but mainly in excitatory neurons, and is involved in mast cell exocytosis. It is down-regulated in brain cortex in *Huntington's disease, *bipolar disorder, and major *depression and is down-regulated in the cerebellum in schizophrenia. **Complexin-3** (158 aa) and **complexin-4** (160 aa) also regulate synaptic vesicle release.

Compton–North congenital myopathy An autosomal recessive myopathy, lethal within the first year of life. It is caused by mutation in *contactin-1 and characterized by a secondary loss of β_2-syntrophin and α-dystrobrevin from the muscle sarcolemma, central nervous system involvement, and fetal akinesia.

computed tomography (computed axial tomography, computer assisted tomography, CAT scanning, CT) A technique in which X-ray images taken from different angles are computationally integrated to produce a three-dimensional image. Whole-body CT scanning is becoming increasingly useful as the resolution improves with greater computing power available.

COMT *See* CATECHOL-O-METHYLTRANSFERASE.

Con A *See* CONCANAVALIN A.

conalbumin (ovotransferrin) A nonhaem iron-binding protein (705 aa) found in chick plasma and at high concentration in egg white. It can be allergenic in humans.

conantokins An extensive group of peptide toxins, some around 17–21 aa, other larger (~100 aa) from cone shells *Conus* spp. They inhibit the NMDA class of *glutamate receptors. Several are in clinical trials for various neurological disorders.

concanamycin A (folimycin) A macrolide isolated from *Streptomyces* sp. that is a specific inhibitor of vacuolar H^+-ATPase.

concanavalin A (Con A) A *lectin, isolated from the jack bean *Canavalia ensiformis*, that binds to the mannose residues of many different glycoproteins and glycolipids. These are sometimes inaccurately referred to as Con A receptors though they should really be considered ligands for the lectin.

concatamer Two or more identical linear molecular units such as nucleic acids covalently linked in series. Also used of artificial polymers.

concretion In medicine, a hard mass, often mineralized, in a tissue.

condensation **1.** The process in which water vapour forms droplets on a cold surface, or the product of the process. **2.** In chemistry, a condensation reaction is one in which two molecules combine with the loss of a small molecule, sometimes water. **3.** In cell biology, the condensation of chromosomes is the thickening and shortening that occurs in prophase of mitosis as a consequenceof increased supercoiling of the chromatin.

condensing vacuole A membrane-bounded secretory vesicle formed by the coalescence of smaller granules (vesicles) derived from the *Golgi apparatus, often a feature of discontinously secreting cells. The contents of the condensing vacuole may be concentrated and become semicrystalline and may be inactive precursors, as in *zymogen granules.

conditional mutation A mutation that is only expressed under certain conditions, e.g. temperature-sensitive mutants.

conditioned medium Cell culture medium in which cells have already been cultured and that is partially depleted of some components but enriched with cell-derived material, such as growth factors. Conditioned medium is often used to grow cells at low density, as in cloning.

condyloma (*pl.* **condylomata)** Inflammatory papules on the external genitalia. **Condylomata acuminata** are genital warts (venereal warts) caused by some subtypes of human *papillomavirus. **Condylomata lata** are flat, whitish lesions characteristic of secondary syphilis, not restricted to the anogenital area.

cone cell *See* RETINAL CONE.

cone–rod dystrophy A form of retinal dystrophy (CORD) in which there is an initial loss of colour vision and visual acuity (loss of retinal cones), followed by night blindness and loss of peripheral vision as rods are lost. The molecular basis for many subtypes are known: in **CORD2** the defect is in the cone-rod homeobox-containing gene (*CRX* gene) that encodes a 299-aa protein similar to the OTX1 and OTX2 homeodomain proteins. In **CORD3**, a retina-specific ABC transporter (ABCA4). In **CORD5**, a membrane-associated phosphatidylinositol transfer protein (PITPNM3, 974 aa). In **CORD6**, a guanylate cyclase (GUCY2D, EC 4.6.1.2, 1123 aa) of the rod outer segment membrane. In **CORD7**, a defect in rab3A-interacting molecule-1 (RIM1, 1692 aa), a scaffold protein that coordinates different stages of the secretory process. In **CORD10**, a defect in *semaphorin 4a. In **CORD11**, the retina and anterior neural fold homeobox-like protein (RAXL1, 184 aa) is defective. In **CORD12** the defect is in *prominin-1 and in **CORD13** in the retinitis pigmentosa GTPase regulator-interacting protein (RPGRIP1, 1286 aa). Loci for other forms of CORD are mapped but the defective proteins not yet identified.

confluent culture A cell culture in which the cells are all in contact and the entire surface of the culture vessel is covered. This is not necessarily the maximum population density. *See* DENSITY-DEPENDENT INHIBITION.

confocal microscopy *See* MICROSCOPY.

conformational change A change in the tertiary structure of a protein that may be a result of alteration in the environment (pH, temperature, ionic strength), the binding of a ligand to a receptor or a substrate to an enzyme.

congenic Describing organisms that differ at only one locus, produced by repeated back-crossing. Strictly, they are conisogenics.

congenital cataracts, facial dysmorphism, and neuropathy (CCFDN) syndrome An autosomal recessive developmental disorder that is characterized by a complex clinical phenotype involving multiple organs and systems. Abnormalities include congenital cataracts and microcorneae, hypomyelination of the peripheral nervous system, impaired physical growth, delayed early motor and intellectual development, facial dysmorphism, and hypogonadism. The cause is mutation in the *CTD phosphatase.

congenital deformity An abnormality present at birth, a result of either a genetic defect or environmental factors *in utero*.

congenital dyserythropoietic anaemia A rare group of inherited disorders associated with dysplastic changes in late erythroid precursors leading to macrocytic anaemia, ineffective erythropoiesis, and secondary haemochromatosis. Type 1 is caused by mutation in the gene for *codanin 1.

congenital generalized lipodystrophy Autosomal recessive disorders in which there is almost no adipose tissue, extreme insulin resistance, hypertriglyceridaemia, hepatic steatosis, and early onset of diabetes. **Type 1** congenital generalized lipodystrophy is caused by mutation in the *AGPAT2* gene that encodes lysophosphatidic acid acyltransferase (LPAAT, EC 2.3.1.51), **type 2** by mutation in the gene encoding *seipin, **type 3** (Berardinelli–Seip congenital lipodystrophy) by mutation in the gene for *caveolin-1.

congenital insensitivity to pain Autosomal recessive congenital indifference to pain can be caused by *loss-of-function mutations in the *SCN9A* gene encoding a *voltage-gated sodium channel. *Gain-of-function mutations in the same gene cause autosomal dominant primary *erythermalgia. Although indifference and insensitivity are considered distinct, with pathological changes in peripheral nerves a feature of insensitivity, the absence of the ion channel is not morphologically obvious and the indifference is likely to be because of the absence of the sensation of pain. An autosomal dominant form has also been reported. **Congenital insensitivity to pain with anhidrosis** (CIPA) is caused by defects in the neurotrophic tyrosine kinase-1 receptor (a *nerve growth factor receptor).

congenital nephrosis *See* ALPHA-FETOPROTEIN.

congenital severe combined immunodeficiency *See* SEVERE COMBINED IMMUNODEFICIENCY DISEASE.

congestion **1.** The accumulation of blood in a part of the body. **2.** Colloquially, blockage of nasal sinuses by mucus.

congestive heart failure *See* HEART FAILURE; CARDIOVASCULAR DISORDERS.

Congo red A pH sensitive dye, blue-violet at pH 3, red at pH 5, used as a vital stain, and to stain amyloid.

coniosetin A tetramic acid antibiotic with antibacterial and antifungal properties, from the ascomycete *Coniochaeta ellipsoidea*, DSM 13856.

conjugate A molecule produced in a biological system by the covalent linkage of two chemical moieties from different sources or the linkage e.g. of an antibody with a fluorochrome.

conjugation The joining of two gametes or cells leading to the transfer of genetic material. Conjugation occurs in many Gram-negative bacteria (*Escherichia, Shigella, Salmonella, Pseudomonas, Streptomyces*) and involves adhesion of a F^+ bacterium (with F-pili) and an F^-, followed by the transfer of the F-plasmid through the sex pilus. In Hfr mutants the F-plasmid is integrated into the chromosome and so chromosomal material is transferred as well.

connectin *See* TITIN.

connective tissue A nonspecific term for mesodermally derived tissue that is rich in extracellular matrix (*collagen, *proteoglycan, etc.) and that surrounds more highly ordered tissues and organs. Cartilage, bone, and blood can be considered as specialized connective tissues. *See* AREOLAR CONNECTIVE TISSUE. The so-called connective tissue diseases (rheumatic diseases) are a group of diseases including rheumatoid arthritis, systemic lupus erythematosus, rheumatic fever, scleroderma, and others, that affect many other tissues and may be autoimmune in origin.

connective tissue-activating peptide III A *chemokine (CTAP III, CXCL7, low affinity platelet factor-4, 85 aa), produced from *platelet basic protein, that is chemotactic for various leucocytes and mitogenic for fibroblasts. Further cleavage generates other active molecules such as neutrophil activating peptide (NAP).

connective tissue growth factor One of the *CCN family of cysteine rich regulatory proteins (CTGF, 349 aa) that is secreted by vascular endothelial cells and stimulates the proliferation and differentiation of chondrocytes; induces angiogenesis; promotes cell adhesion of fibroblasts, endothelial, and epithelial cells; and binds to *insulin-like growth factor, *transforming growth factor (TGFβ1), and *bone morphogenetic protein-4. It enhances *fibroblast growth factor-induced DNA synthesis. **Connective tissue**

growth factor-like (CTGFL/WISP-2, Wnt-1-induced signalling pathway protein-2, CCN5, 250 aa) is expressed in primary osteoblasts, fibroblasts, ovary, testes, and heart. It promotes adhesion of osteoblasts, inhibits the binding of fibrinogen to integrin receptors, and inhibits osteocalcin production.

connexins A family of four-membrane-pass proteins that form the subunits of *gap junctions. A homohexamer of connexins forms a **connexon**, apposed connexons in the adjacent cells forming an aqueous channel. Currently 21 human genes for connexins have been identified, each with tissue- or cell-type-specific expression. There are two different naming conventions, one based on molecular weight (e.g. connexin-43 with a size of 43 kDa) the other based upon evolutionary relationships with two major subclasses, α and β (connexin-43 is α-1, GJA1, 382 aa). Most organs and many cells express more than one connexin. Mutations in connexin genes are associated with peripheral neuropathies, cardiovascular diseases, dermatological diseases, hereditary deafness, and cataract. For example, defects in **GJA1** cause autosomal dominant *oculodentodigital dysplasia, syndactyly type III, and hypoplastic left heart syndrome; defects in **GJA3** cause zonular pulverulent *cataract type 3; defects in **GJB3** cause *erythrokeratodermia variabilis and autosomal dominant nonsyndromic sensorineural deafness type 2B.

connexon *See* CONNEXINS.

Conn's syndrome Hyperaldosteronism due to an aldosterone-secreting adenoma or in some cases adrenal hyperplasia and adrenal carcinoma.

conotoxins A group of small peptide toxins from cone shells *Conus* spp. The α-conotoxins are competitive inhibitors of nicotinic acetylcholine receptors, the μ-conotoxins bind *voltage-gated sodium channels in muscle, causing paralysis and the ω-conotoxins inhibit *voltage-sensitive calcium channels and block synaptic transmission. *See* CONANTOKINS.

consensus sequence A sequence that reflects the most common choice of base or amino acid at each position in a nucleic acid or protein. Highly conserved sequences are often functional domains.

conservative substitution The substitution (as a result of a point mutation in the coding DNA) of one amino acid in a protein with another that has similar properties, so that the function is unlikely to be seriously affected.

consolidation The physiological changes in the brain that are thought to be associated with memory storage.

constant region (C-region) The C-terminal half of the light or the heavy chain of an immunoglobulin molecule that is identical in all antibodies of the same isotype. The sequence of the variable region determines the antigen specificity. Heavy chains α, γ, and δ (in IgA, IgG, and IgD) have a constant region of ~450 aa with three tandem Ig-type domains, and a hinge region; heavy chains μ and ε (in IgM and IgE) have four immunoglobulin domains in the constant region. The constant regions of light chains are smaller (~215 aa).

constitutive Describing things that are constantly present, whether or not there is demand. Some enzymes are constitutive, other can be induced when needed.

constitutive transport element A conserved motif in the RNA of simian type D retroviruses that allows nuclear export of unspliced viral RNAs by recruiting host *TAP.

constriction ring The equatorial ring of *microfilaments that contracts and disassembles as *cytokinesis proceeds, causing separation of the daughter cells.

constrictive pericarditis A chronic inflammatory response in the pericardium that mechanically interferes with heart function.

consumption *See* TUBERCULOSIS.

contact activation pathway (intrinsic pathway) The process in which blood clotting is activated by exposure of platelets to collagen to which they bind via the Gp Ia/IIa receptor. Platelet adhesion is strengthened further by recruitment of *von Willebrand factor (vWF) and triggers release of platelet granule contents which also play a part in initiating the complex cascade of events that culminate in cleavage of *factor I to form fibrin. *See* TISSUE FACTOR (tissue factor pathway) (formerly the 'extrinsic pathway').

contact guidance A directed locomotory response of cells caused by anisotropy of the environment, e.g. the tendency of fibroblasts to align parallel to the alignment of collagen fibres in a stretched gel because the fibres resist deformation axially.

contact-induced spreading The consequence of contact between two *epithelial cells; a stable adhesion is formed between the two and they spread so that their combined area is greater after contact than the area they occupied when on their own.

contact inhibition of growth/ division *See* DENSITY-DEPENDENT INHIBITION OF GROWTH.

C

contact inhibition of locomotion/ movement A response in which the direction of motion of a cell is altered when it comes in contact with another cell. The outcome of contact may be symmetrical, with both cells responding or may be non-reciprocal in which case the non-responsive cell type may be able to invade territory occupied by the other. **Type I contact inhibition** involves paralysis of the locomotory machinery, **type II** is a consequence of preferential adhesion to the substratum rather than to the upper surface of the other cell.

contact inhibition of phagocytosis A phenomenon observed in kidney epithelial cells that lose their weak phagocytic activity when confluence occurs. It may be a result of reduced adhesiveness of the cell once it is no longer making locomotory protrusions.

contactins A family of GPI-anchored membrane-associated proteins that mediate cell surface interactions during development of the nervous system. **Contactin-1** (CNTN1, glycoprotein gp135, 1018 aa) is a ligand for *notch-1 and defects in CNTN1 cause *Compton–North congenital myopathy. **Contactin-2** (axonin-1, transient axonal glycoprotein 1, tag-1, tax-1, 1040 aa) may be important for initial growth and guidance of axons. **Contactin-3** (brain-derived immunoglobulin superfamily protein-1, BIG-1, Plasmacytoma-associated neuronal glycoprotein, 1028 aa), **contactin-4** (BIG-2), **contactin-5** (neural recognition molecule NB-2, 1100 aa) and **contactin-6** neural recognition molecule NB-3, 1028 aa) are thought to have similar roles. Contactin-associated proteins: *See* CASPRS.

contact sensitivity An allergic response to contact with an irritant, usually a *hypersensitivity response.

contig A colloquialism for a DNA sequence assembled from overlapping shorter sequences forming a contigous sequence.

contractile ring *See* CONSTRICTION RING.

contractile vacuole A specialized vacuole of eukaryotic cells that is important for osmoregulation and perhaps excretion in single-celled organisms, particularly protozoa. The vacuole fills with fluid from the cytoplasm and intermittently discharges to the exterior.

contrast medium A substance used to increase the differences between the different voxels (three-dimensional pixels) in the body in terms of their ability to absorb or reflect energy from electromagnetic radiation or ultrasound. Positive contrast media for X-rays often contain elements of high atomic number (barium, iodine); negative contrast can be obtained by introducing air or other gases into a cavity. For MRI, gadolinium compounds have been used. Contrast media for ultrasonography are being developed and usually consist of suspensions or emulsions that have different acoustic properties from normal tissues.

control element (control region) *See* REGULATORY SEQUENCE.

controlled drugs Drugs, usually those with the potential to cause addiction and dependence, that can only be prescribed under guidelines laid down in law.

convulsion A condition (colloquially a 'fit') in which muscles contract and relax rapidly and repeatedly, causing the body to shake. Epileptic seizures are a form of convulsion, but not all convulsions are due to epilepsy.

convulxin A toxin from the South American rattlesnake *Crotalus durissus terrificus*, an octomer of four α (158 aa) and four β (148 aa) disulphide-linked subunits. It binds and activates platelets through glycoprotein VI, the collagen receptor, which signals through *syk and phospholipase C (PLCγ2).

Coomassie Brilliant Blue A blue dye (Coomassie Brilliant Blue G-250, Brilliant Blue R, Acid Blue 90, Kenacid Blue) that binds nonspecifically to proteins, a property exploited by the *Bradford assay for proteins and in staining gels. The absorbance shifts from 465 nm to 595 nm when the dye is bound to protein.

Coombs' test A diagnostic test to determine whether erythrocytes are coated with autoantibodies or immune complexes which can cause *haemolytic anaemia. The direct Coombs' test involves the addition of anti-immunoglobulin antibody to washed erythrocytes, a positive result being agglutination. The indirect Coombs' test is for unbound antibodies against erythrocytes present in the serum of pregnant women by adding erythrocytes of known antigenicity to determine whether there is a risk of *erythroblastosis fetalis (haemolytic disease of the newborn).

cooperativity A phenomenon that can occur with enzymes or receptors that have multiple binding sites when binding of one ligand alters the affinity of the other site(s). Both positive and negative cooperativity can occur and cooperativity is one possible cause of nonlinearity in binding data.

COP9 signalosome A conserved regulatory complex (constitutive photomorphogenic-9 complex) present in diverse organisms. The complex has eight subunits (CSN1–8; total 450 kDa) with some homologies to the proteosome. Subunits include *Gps1, *Jab1 (a coactivator of AP-1 transcription factor), CSN5 (a isopeptidase and a candidate oncogene in human breast cancer), TRIP15 (thyroid hormone receptor interactor-15), SGN3 (which has homology to the 26S proteosome S3 regulatory subunit), and others. Remarkably, human *Gps1 can substitute for FUS6 (the plant homologue) in the COP9 complex that represses photomorphogenesis in *Arabidopsis thaliana*. Inactivation of the COP9 signalosome will impair T-cell development.

COPD *See* CHRONIC OBSTRUCTIVE PULMONARY DISEASE.

cophenotrope A mixture of an opioid (diphenoxylate HCl) that reduces gut motility, and atropine sulphate, which is antispasmodic, used to treat diarrhoea. *TN* Lomotil.

COPI, COPII *See* COATED VESICLE.

copy number The number of identical molecules on or in a cell or part of a cell. Usually refers to specific genes, or to plasmids within a bacterium. **Copy number polymorphisms** (CNPs) are variations between individuals in the number of copies they possess of a particular gene, either because of duplication or deletion. Using microarrays for *SNPs it is becoming clear that CNPs are relatively common, and individuals differ by eleven CNPs on average. Copy number variation has been found in 70 different genes and can have both positive and negative consequences. Increased numbers of EGF receptor genes are found in non-small-cell lung cancer, but a higher copy number of the cytokine CCL3L1 gene is associated with reduced susceptibility to HIV infection. Copy number variation has also been associated with *autism, *autoimmune diseases, schizophrenia, and idiopathic learning disability.

CORD *See* CONE–ROD DYSTROPHY.

cord blood Blood taken from the umbilical cord after birth, a potential source of stem cells.

cord factor An important virulence factor in *Mycobacteria*, a glycolipid (trehalose 6,6′-dimycolate) found in the cell walls that causes them to grow in serpentine cords. Cord factor on its own will induce granulomatous reactions.

Cori's disease *See* GLYCOGEN STORAGE DISEASES.

C

corn A localized overgrowth of the keratinized layer of the skin as a result of mechanical irritation. A corn tends to have a focal centre, unlike the more generalized thickening of a *callosity.

cornea The layer of transparent tissue at the front of the eye. The cornea is largely composed of collagen laid down in orthogonal arrays with very few cells and no vascularization. There is a thin outer squamous epithelium and an inner endothelial layer next to the aqueous humour. The optical properties depend on the regularity of the spacing of the collagen fibrils. **Cornea plana** is a disorder in which the corneal radius of curvature is larger than normal, causing severe loss of visual acuity and extreme *hypermetropia. The autosomal recessive form (CNA2) is associated with mutations in *keratocan, but this gene is normal in the milder autosomal dominant form (CNA1). The late-onset form of Fuchs' **endothelial corneal dystrophy** has been mapped to chromosome 13 and the rare early-onset form is associated with mutation in the gene for the α-2 chain of collagen type VIII (COL8A2), as is posterior polymorphous corneal dystrophy. **Corneal endothelial dystrophy-2** is an autosomal recessive disorder caused by mutation in the *SLC4A11* gene, which encodes a sodium borate cotransporter. Autosomal recessive corneal dystrophy and perceptive deafness are caused by mutation in the same gene.

Cornelia de Lange syndrome A disorder with by characteristic facial features, prenatal and postnatal growth retardation, mental retardation, and often upper limb anomalies. **Type 1** is caused by mutation in the *NIPBL* gene, which encodes a component of the *cohesin complex, **types 2** and **3** are due to mutations in other components of the complex.

corneocyte The heavily keratinized dead cell type found in the outer layer of the skin, the *stratum corneum (cornified epithelium).

cornification The progressive keratinization of skin cells that occurs as they progress from the basal layer towards the outside where they are found as *corneocytes and are eventually shed as squames.

cornulin One of the *'fused' gene family of proteins, 490 aa, expressed in the granular and lower cornified cell layers of scalp epidermis and foreskin, as well as in calcium-induced differentiated cultured keratinocytes. Considered to be a marker of late epidermal differentiation.

coronal section A section of the brain taken 'where the edge of a crown would touch'.

coronal suture The serrated junction of the frontal and parietal bones in the skull.

corona radiata **1.** A layer, two or three cells deep, of follicular cells that surrounds the developing mammalian ovum. **2.** In neuroanatomy, a sheet of white matter that contains both descending and ascending axons carrying neural traffic to and from the cerebral cortex.

coronary Describing things related to the heart, although colloquially implying a coronary thrombosis in which a coronary artery is blocked by a thrombus. **Coronary heart disease** is a general term for heart disorders caused by disease of the coronary arteries and includes *angina pectoris and *myocardial infarction.

Coronaviridae A family of enveloped, single-stranded, positive-sense RNA viruses with club-shaped projections radiating outwards that give a charateristic corona appearance to negatively stained virions. They are responsible for respiratory diseases including *SARS.

cor pulmonale An alteration of the structure and function of the right ventricle of the heart (pulmonary heart disease) as a result of a respiratory disorder such as chronic bronchitis or emphysema. In chronic pulmonary heart disease there is usually right ventricular hypertrophy, in acute cases there is more generalized dilation as a result of increased pressure in the right ventricle.

corpus callosum A wide bundle of axons beneath the cortex that is responsible for communication between the right and left cerebral hemispheres.

corpus luteum (*pl.* corpora lutea) A temporary endocrine structure in the ovary, formed from the Graafian follicle after the ovum is released, responsible for the secretion of *progesterone.

corpus striatum A compound structure with a striated appearance consisting of the striatum and the globus pallidus in the brain.

corralling Describing the way in which membrane proteins are restricted in their long-range diffusion in the plane of the membrane so that some areas have different properties, e.g. in the expression of ion channels.

cortactin A protein (amplaxin, 550 aa) that is a substrate for *src-kinase and has proline-rich and SH3 domains. It binds F-actin and redistributes to membrane ruffles when *rac1 is activated by growth factors. It interacts with the adapter proteins *SHANK2 and SHANK3 via its SH2 domain and with FGD1 (*FYVE, RhoGEF, and PH domain-containing protein 1) and PLXDC2 (*plexin domain-containing protein 2, 529 aa). Overexpression of cortactin increases cell motility and invasiveness.

cortex A general histological term for the outer region of an organ. *Compare* MEDULLA.

cortical granules Specialized secretory granules that lie immediately below the plasma membrane of the egg and fuse with the membrane following fertilization, forming the fertilization membrane and blocking further sperm fusion. They have been extensively studied in sea urchin eggs.

cortical inhibition The blocking of neural impulses by 'higher level' centres in the cortex.

cortical meshwork (cortical layer) The meshwork of microfilaments, anchored to the plasma membrane by their barbed ends, that lies immediately below the plasma membrane The meshwork contributes to the mechanical properties of the cell and probably restricts the access of cytoplasmic vesicles to the plasma membrane.

corticostatin A defensive peptide (97 aa) of the α-defensin family that has antibiotic, antifungal, and antiviral activity. It also inhibits corticotropin (ACTH) stimulated corticosterone production.

corticosteroids The *steroid hormones secreted by the *adrenal cortex that regulate carbohydrate metabolism (glucocorticoids such as cortisol and cortisone) and salt/water balance (minerocorticoids, e.g. aldosterone). The production of corticosteroids is regulated by *adrenocorticotropin (ACTH). Glucocorticoids are important natural anti-inflammatory agents.

corticotrophic *See* CORTICOTROPIC.

corticotropic (corticotrophic, adrenocorticotrophic) Describing something that stimulates the adrenal cortex. *See* ADRENOCORTICOTROPIN; CORTICOTROPIN RELEASING HORMONE.

corticotropin releasing factor *See* CORTICOTROPIN RELEASING HORMONE; ADRENOCORTICOTROPIN.

corticotropin releasing hormone A peptide hormone (CRH, corticotropin releasing factor, CRF, corticoliberin, 41 aa derived from a 191-aa preprohormone) produced by the hypothalamus that stimulates corticotropic cells in the anterior lobe of the pituitary to produce ACTH (*adrenocorticotropin) and other biologically active substances (e.g. β-endorphin). It is an important regulator of the *hypothalamic–pituitary–adrenal (HPA) axis, produced by both the placenta and fetal membranes at term and released in response to stress. The **receptors** (CRHR1, 444 aa; CRHR2, 411 aa) are G-protein coupled. **Corticotropin releasing hormone-binding protein** (CRH-BP, 322 aa) is a secreted glycoprotein that binds CRH and *urocortin with high affinity and is structurally unrelated to the CRH receptors. It is an important modulator of CRH activity.

cortisol The major adrenal *glucocorticoid that stimulates conversion of proteins to carbohydrates, raises blood sugar levels, and promotes glycogen storage in the liver. **Hydrocortisone** is a natural metabolic derivative and has immunosuppressive and anti-inflammatory activity. **Cortisone** (11-dehydroxycortisol) is a natural glucocorticoid that is inactive until converted into hydrocortisone in the liver.

cortistatin (CST) A neuropeptide (17 aa) structurally homologous to *somatostatin and able to bind to all human somatostatin receptor subtypes. It is restricted to *GABA-containing cells in the cerebral cortex and hippocampus, where it may modulate neural activity.

Corynebacteria A genus of *Gram-positive nonmotile rod-like bacteria, mostly facultative anaerobes. *C. diphtheriae* (Klebs–Löffler bacillus) is the causative agent of *diphtheria.

COS cells A line of simian fibroblasts (CV-1 cells) transformed by *SV40 that is deficient in the origin of replication. Will express vectors with a normal SV40 origin at high level and is often used to produce recombinant proteins.

cosegregation The tendency of two or more linked genes to be inherited together. *See* GENETIC LINKAGE.

cosmid A vector derived from bacteriophage lambda that can carry a longer DNA insert than *plasmid vectors; used for construction of *genomic libraries.

cos sites The sites, twelve nucleotides long, that are required for integration of a *cosmid vector into host DNA.

costamere *See* DYSTROPHIN-ASSOCIATED PROTEIN COMPLEX.

cot death *See* SUDDEN INFANT DEATH SYNDROME.

cotinine A metabolite of nicotine that persists in the body for 2–4 days after tobacco use. Serum and urine cotinine is a more reliable indicator of smoking than any interview technique.

cotranslational transport The standard mechanism by which most proteins destined for secretion are translated in eukaryotic cells. The first part of the message (*See* SIGNAL SEQUENCE) is translated and causes the ribosome to become attached to the rough endoplasmic reticulum; subsequent translation causes the nascent polypeptide to move wholly or partially across the membrane into the intracisternal space. Proteins that are released into the lumen then progress through the GERL and are post-translationally modified in various ways.

cotransport The coupling of the transport of one species (generally sodium ions) to another (e.g. a sugar or amino acid). One of the partners is moved down a gradient of *electrochemical potential, providing the energy to drive the other molecule up-gradient. *See* ACTIVE TRANSPORT (secondary active transport).

co-trimoxazole A mixture of *trimethoprin, which is antibacterial, and a sulfonamide antibiotic, sulphamethoxazole, that is used to treat *Pneumocystis jirovecii* pneumonia, toxoplasmosis, and nocardiasis. There appears to be a synergistic effect when the two drugs are used in conjunction.

cotton-wool patches Areas of white opacity in the retina, often associated with hypertensive retinopathy.

Coulter counter A proprietary name for an electronic particle counter that detects changes in electrical conductance as fluid containing suspended particles is drawn through a small aperture. The particles are nonconductive and alter the effective cross-section of the aperture. By varying the aperture size, bacteria or eukaryotic cells can be counted.

coumarin A compound released from plants as they wilt, probably responsible for the smell of hay. It has anticoagulant activity by competing with vitamin K. Various derivatives such as *esculin, *fraxin, *scopoletin, and umbelliferone have anti-inflammatory and antimetastatic properties.

counter ion A mobile ion having the opposite charge to a fixed ionized residue such as the carboxyl group of a terminal sialic acid residue on a membrane glycoprotein. Because fixed charges on the cell surface are predominantly negative the immediate local environment is enriched in cations.

counterstain A stain used in conjunction with a histochemical reagent of greater specificity to provide contrast; e.g. Light Green is a counterstain in the Mallory procedure.

countertransport A transport system in which the movement of a molecule across a membrane is matched by the movement of a different molecule in the opposite direction. If both have the same charge then no potential gradient will develop, if equal numbers move in each direction. The opposite of *cotransport.

coupled transporter **1.** A membrane transport system in which movement of one molecule or ion down an electrochemical gradient is linked to that of another molecule in the same or the opposite direction. *See* ANTIPORT; COTRANSPORT; COUNTERTRANSPORT; SYMPORT. **2.** Used rather inaccurately of a controlled transport system, e.g. one that is regulated by G proteins.

coupling The linking of two independent processes by a common intermediate. Coupling factors are proteins that link transmembrane potentials to ATP synthesis in chloroplasts and *mitochondria.

COUP-TFs Transcription factors (chick ovalbumin upstream promoter-transcription factors) of the steroid/thyroid hormone receptor family; but orphan receptors, since the ligands are unidentified. COUP-TFs play a role in glial cell development and central nervous system myelination. **COUP-TF2** (414 aa) regulates the transcription of the apolipoprotein A-I gene. *See* ARP-1. **COUP-TF-interacting protein 1** (B-cell lymphoma/leukaemia 11A, 835 aa) functions as a myeloid and B-cell proto-oncogene; chromosomal aberrations involving the gene may be a cause of lymphoid malignancies. **COUP-TF-interacting protein-2** (894 aa) is a tumour-suppressor protein involved in T-cell lymphomas.

Cowden's disease A rare autosomal dominant multiple-*hamartoma syndrome caused by mutations in *PTEN.

cowpox *See* VACCINIA.

COX **1.** *See* CYCLO-OXYGENASE. **2.** COX1 is sometimes used as an alternative name for cytochrome *c* oxidase subunit 1 (EC 1.9.3.1) and many cytochrome oxidase subunits are designated as COX, e.g. COX-8C, cytochrome *c* oxidase subunit 8C.

coxib General suffix for cyclo-oxygenase (COX) inhibitors, NSAIDs.

Coxsackieviruses Enteroviruses of the family *Picornaviridae, first isolated in Coxsackie, New York. Coxsackie A viruses are responsible for mild childhood infections (hand, foot, and mouth disease) but can, rarely, cause more serious illnesses in adults including acute haemorrhagic conjunctivitis (A24 specifically), herpangina, and aseptic meningitis. Coxsackie B viruses can cause infectious myocarditis, infectious pericarditis, *pleurodynia, and immune complex-mediated glomerulonephritis.

CPA An ambiguous abbreviation: **1.** cardiopulmonary arrest. **2.** carboxypeptidase A. **3.** *See* CEREBELLOPONTINE ANGLE; CYCLOPENTYLADENOSINE; CYCLOPHOSPHAMIDE; CYCLOPIAZONIC ACID; CYPROTERONE ACETATE.

CPE Usually an abbreviation for cytopathic effect, but occasionally: **1.** *Cytoplasmic polyadenylation element. **2.** *Clostridium perfringens* *enterotoxin.

CpG island (CG island) A region of genomic DNA at least 200 bp in length and with a CG content greater than 50%. Found in ~70% of the promoters of human genes. The region is unmethylated in genes that are expressed (*housekeeping genes) and methylation persists through cell divisions, affecting the level of transcription (*see* DNA METHYLATION).

CPK *See* CREATINE KINASE.

C-polysaccharide (C substance) A highly immunogenic species-specific polysaccharide that is a component of the cell wall of capsular pneumococci. The structure includes phosphocholine. *See* C-REACTIVE PROTEIN.

C-proteins Myosin-binding proteins (~1141 aa) of the immunoglobulin superfamily located in the cross-bridge region (C-zone) of vertebrate striated muscle A bands that may modulate muscle contraction or may play a more structural role. They are structurally related to various other myosin-binding proteins such as *titin, *myosin light chain kinase, and *myomesin.

CPS *See* CARBAMOYL PHOSPHATE SYNTHETASE.

CPSF A multisubunit complex (cleavage and polyadenylation specificity factor) that binds the AAUAAA conserved sequence (*cytoplasmic polyadenylation element) in pre-mRNA and is essential for splicing. Subunits range in size from CPSF-1 (1443 aa) to CPSF-5 (227 aa), with at least seven being involved, some recognizing RNA, others involved in cleavage and religation.

CPT *See* CAMPTOTHECIN; CARNITINE (palmitoyltransferase).

CR1, CR2, CR3, CR4 *See* COMPLEMENT RECEPTORS.

CR16 *See* VERPROLINS.

CRABP *See* CELLULAR RETINOIC ACID-BINDING PROTEIN.

cranial nerves The twelve paired nerves that have their origin in the brain the: olfactory, optic, oculomotor, trochlear, trigeminal, abducens, facial, auditory/vestibulocochlear, glossopharyngeal, vagus, accessory, and hypoglossal nerves. A polite mnemonic is 'Old officers often trust the army for a glory vague and hypothetical'.

craniolenticulosutural dysplasia An autosomal recessive disorder of craniofacial development in which there is delayed fusion of bones in the skull. It is caused by a mutation in the *SEC23A* gene (*see* COATED VESICLE) leading to abnormal endoplasmic reticulum-to-Golgi trafficking.

CRD domain *See* C-TYPE LECTINS.

cre A recombinase from an *E. coli* *bacteriophage that mediates *site-specific recombination at loxP sites. *See* LOX-CRE SYSTEM.

C-reactive protein A *pentraxin (pentaxin 1, PTX1, 224 aa) found in plasma at increased levels during the acute phase of an immune response as a result of cytokine-triggered synthesis by hepatocytes. It is a homopentamer and binds to polysaccharides present in cell walls of bacteria or fungi, to cell surfaces, to *lecithin, and to phosphoryl- or choline-containing molecules such as the *C-polysaccharide of pneumococci (the latter property being the source of its name). It is related in structure to serum *amyloid. *See also* ACUTE PHASE PROTEINS.

creatine kinase A dimeric enzyme (CPK, creatine phosphokinase, EC 2.7.3.2, 381 aa) that reversibly catalyses the transfer of phosphate between ATP and various phosphogens such as creatine phosphate. It is important in tissues with high energy demands such as muscle. There are two isoforms, MM being the major form in skeletal muscle and myocardium, MB existing in myocardium, and BB in many tissues, especially brain. The mitochondrial form is found on the inner mitochondrial membrane as an octamer composed of four CKMT homodimers with a specific isoform in sarcomeric mitochondria (CKMT2, 419 aa) and a ubiquitous isoform (CKMT1, 417 aa).

creatine phosphate (phosphocreatine) A compound found abundantly in muscle where it serves as an energy reserve. *See* CREATINE KINASE.

creatinuria The abnormal presence of creatinine in the urine.

CREB A *basic leucine zipper (bZip) transcription factor (cAMP response element binding factor, CREB-1, 341 aa) that is activated when it is phosphorylated by cAMP-dependent protein kinase (PKA). Phosphorylation allows DNA binding and interaction with *MECT1 and with **CREB binding protein** (EC 2.3.1.48, 2442 aa), a transcriptional coactivator. Several other forms of CREB are known (CREB3, CREB5) and various CREB-regulated transcriptional coactivators. A chromosomal aberration involving CREB1 is associated with *angiomatoid fibrous histiocytoma.

CREC family A family of low affinity, calcium-binding, multiple *EF-hand proteins that includes *reticulocalbins, *calumenin, and other endoplasmic proteins. They appear to have a range of functions in the secretory pathway, have chaperone activity, and are involved with signal transduction. They are thought to be potential disease biomarkers or targets for therapeutic intervention. NB Sometimes 'CERC' for Cab-45, Erc-55, reticulocalbin, and calumenin.

C-region *See* CONSTANT REGION.

CREM One of a large family of transcriptional activators (cAMP response element modulators, ATF/CREM/CREB family).

crenation An abnormality of erythrocyte shape that can be a result of ATP depletion or excess free fatty acids in plasma. The erythrocytes have regular spiky protrusions and are referred to as *echinocytes (burr cells).

crepitation (crepitus) A crackling sensation or noise that is heard when rheumatic joints are moved or broken bones grate together. A similar noise in the chest can occur in some diseases of the pleura or lungs.

CREST syndrome A complex syndrome with calcinosis, Reynaud's phenomenon, oesophageal dysmotility, sclerodactyly, and telangielactasia; a variant of *scleroderma. There are probably both genetic and environmental factors involved. Most patients have autoantibodies to kinetochore proteins.

cretinism A congenital deficiency of thyroid hormones, sometimes because of iodine insufficiency, that causes failure of physical and mental development. Has become a term of general abuse and is better avoided.

Creutzfeldt–Jakob disease (CJD) A rare fatal presenile dementia of humans, similar to *kuru and other transmissible *spongiform encephalopathies. Three categories are distinguished: **sporadic CJD**, the most common type (85% of cases) where there is no known cause; **hereditary CJD**, in which there is a mutation in the *prion protein; **acquired CJD**, in which the disease is transmitted by exposure to brain or nervous system tissue, usually through certain medical procedures. A **new variant CJD** (nvCJD) was recognized in the late 1990s and linked, speculatively, with bovine spongiform encephalopathy.

CRF 1. Chronic renal failure. **2.** *See* CORTICOTROPIN RELEASING HORMONE.

CRIB 1. A consensus sequence (cdc42/rac interactive binding sequence) found in various signalling proteins (e.g. PAK kinases) and involved in the binding of cdc42. **2.** Clinical Risk Index for Babies, a tool for assessing initial neonatal risk.

SEE WEB LINKS

- Clinical Risk Index for Babies homepage.

cribriform Describing something perforated or sievelike.

cribrostatin A family of compounds, isolated from the blue marine sponge *Cribrochalina* sp., that inhibit the growth of cancer cells and have antimicrobial activity.

cri-du-chat syndrome A severe developmental disorder caused by deletion of part of the short arm of chromosome 5 which carries, among other genes, those for telomerase reverse transcriptase (*see* HTERT) and δ-*catenin-2. Affected infants have a characteristic cry, said to resemble that of a cat.

Crigler–Najjar syndrome *See* GILBERT'S SYNDROME.

Crimean–Congo haemorrhagic fever A serious tick-borne disease caused by one of the *Nairovirus group.

crinophagy The process of digestion of the contents of a secretory vesicle after it has fused with a lysosome.

CRISP *See* CYSTEINE-RICH SECRETORY PROTEINS.

critical concentration The concentration of a substance that causes functional changes in a cell or an organism. Often used in the context of toxicity (critical dose).

critical point drying A method used to prepare specimens for scanning electron microscopy. Water in the specimen is replaced e.g. by liquid CO_2, which is then allowed to vaporize at a temperature above the critical point, the temperature at which the liquid state no longer occurs, thereby avoiding shrinkage.

crk The homologue of an oncogene originally identified in an avian *sarcoma. The alternatively spliced products, **crk1** (204 aa) and **crk2** (304 aa) are adaptor proteins that link cytoplasmic proteins to receptor tyrosine kinases through SH2 and SH3 domains. Crk-2 has less transforming ability than crk-1. **Crk-like protein** (CRKL, 303 aa) has similar adaptor properties.

CRM 1. Chromosomal region maintenance 1 (*see* EXPORTINS (exportin-1)). **2.** Certified reference materials. **3.** Colorectal metastasis. **4.** *See* CIS-REGULATORY MODULES.

CRMP 1. Chromium-mesoporphyrin (CrMP), an inhibitor of haem oxygenase. **2.** *See* COLLAPSIN (collapsin response-mediated protein).

crocalbin *See* CALUMENIN.

Crohn's disease *See* INFLAMMATORY BOWEL DISEASE.

crossing-over The exchange of DNA between homologous chromatids in meiosis, giving rise to *chiasmata. *See* RECOMBINATION.

cross-sectional study An observational study carried out at a single time-point, e.g. looking at individuals of different ages in order to sample the population, as opposed to a longitudinal study [or follow-up study] where the same subjects are studied at different times. A **cross-sequential study** involves a combination of cross-sectional and longitudinal studies.

croton oil Oil extracted from the seeds of the tropical plant *Croton tiglium* (Euphorbiaceae) that can cause severe skin irritation and contains *phorbol esters which are tumour promoters.

crotoxin A peptide neurotoxin from the venom of the South American rattlesnake *Crotalus durissus terrificus*. The precursor (crotoxin acid chain, crotapotin, 138 aa) is cleaved into three fragments, two of which (the acidic and basic chains) act together as a phospholipase A2 which blocks acetylcholine release at the neuromuscular junction.

croup A hoarse, croaky cough that is a feature of childhood inflammation of the larynx and trachea.

CRP *See* CALRETICULIN; C-REACTIVE PROTEIN.

crush syndrome A serious medical condition (traumatic rhabdomyolysis or Bywaters' syndrome) in which there is shock and renal failure following crushing injuries to skeletal muscle that cause the release of myoglobin and potassium.

cryofixation A method used to prepare tissue for microscopy, particularly scanning electron microscopy, carried out at very low temperatures (liquid nitrogen or helium) with rapid cooling (>10 000 deg °C/s) so that there is vitrification rather than ice crystal formation.

cryoglobulins Abnormal immunoglobulins (IgG or IgM) that precipitate when serum is cooled. In **type 1 cryoglobulinaemia** there is monoclonal IgG or IgM produced by a lymphoproliferative disorder. **Type 2** is the most common cryoglobulinaemia and is usually associated with hepatitis C infection. It may also be associated with infections, lymphoproliferative disorders, plasma cell dyscrasias and autoimmune diseases. **Type 3** is often associated with liver disease (especially hepatitis C), autoimmune diseases, and chronic infections.

cryomicroscopy Any form of microscopy (light or electron) in which samples have been prepared by *cryofixation. Frozen sections can be prepared and examined rapidly and are important for fast pathological examination of biopsy material.

cryoprecipitate A precipitate formed when plasma is frozen and then thawed. It is rich in *fibronectin and *blood clotting factor VIII.

cryoprotectant A substance that will prevent or reduce the formation of ice crystals during freezing. For freezing cells, *DMSO and glycerol are the commonest cryoprotectants.

cryopyrin A protein (NACHT, LRR, and PYD domains-containing protein 3, NALP3, cold autoinflammatory syndrome 1 protein, PYRIN-containing APAF1-like protein-1, angiotensin/vasopressin receptor AII/AVP-like, 1034 aa) that is strongly expressed in polymorphonuclear cells and osteoblasts. It interacts selectively with *ASC, may be an upstream activator of NFκB signalling, and activates caspase-1 in response to a number of triggers including bacterial or viral infection which leads to processing and release of IL-1B and IL-18. Mutations cause familial cold urticaria (*familial cold autoinflammatory syndrome type 1), *Muckle–Wells syndrome, and *chronic infantile neurological cutaneous and articular syndrome.

cryotherapy The local or general use of low temperatures in therapy. **Cryosurgery** uses extreme cold to kill tissues (e.g. warts).

crypt A deep infolding of the epithelium of the small intestine, at the base of which stem cells proliferate. The differentiated progeny move progressively towards the top of the villus and are shed 3–5 days later.

cryptochrome Human **cryptochrome 1** (Cry1, photolyase 1, 586 aa) is 25% identical to plant blue-light receptors and may function as blue-light photoreceptor in humans, linked to the circadian rhythm generating system (*See* CIRCADIAN RHYTHM. **Cry2** (589 aa) has substantial homology with Cry1 but is located in the nucleus rather than mitochondria. Cry1 and Cry2 act as light-independent inhibitors of a heterodimeric transcriptional activator consisting of *clock and *Bmal1, that drives transcription of *period. Neither Cry 1 nor Cry2 has photolyase repair capacity.

cryptococcosis Infection by the yeast *Cryptococcus neoformans* (formerly *Torula histolytica*), usually affecting the lungs although it can opportunistically cause meningitis, especially in immunocompromised individuals.

cryptophycins Macrolides isolated from the cyanobacterium *Nostoc* that have potent antimitotic activity. They bind strongly to β-tubulin (sharing a binding site with *dolastatin 10, *hemiasterlin, and *phomopsin A), interfering with microtubule dynamics and deactivate the *Bcl-2 protein causing rapid apoptosis. May have potential in chemotherapy.

Cryptosporidium A genus of protozoal parasites that cause intestinal disorders, usually *Cryptosporidium parvum* or *C. hominis*. Ingested sporulated oocysts release sporozoites in the ileum that penetrate intestinal epithelial cells where they divide to form merozoites which can infect other cells and can proliferate asexually or differentiate into either macro- or microgametocytes. Microgametocytes penetrate the macrogametocytes to form a zygote that undergoes meiotic division, producing sporozoites within a resistant oocyst that is then released in faeces, completing the cycle. The disease is usually self-limiting but in immunocompromised individuals infections can be much more serious, leading to dehydration and malnutrition.

crystallins The major proteins of the vertebrate lens, α-, β-, and γ-crystallins, are water soluble and present in high concentrations in the cytoplasm of lens fibre cells. They do not turn over during the life of an individual, unlike most proteins. The **α-crystallins** (α-crystallin-A, CRYAA, 173 aa; α-crystallin-B, CRYAB, 175 aa) are members of the small heat shock protein (HSP20) family and contribute to the optical properties of the lens, but have additional functions in other tissues. Defects in CRYAA cause zonular central nuclear *cataract; defects in CRYAB are the cause of *α-B crystallinopathy. The **β-** and **γ-crystallins** form a separate group. Mutations in various of the crystallins lead to cataracts.

crystal violet The darkest of the methyl violet dyes (Gentian violet; methyl violet 10B), used in *Gram staining, for the metachromatic staining of amyloid, and as an enhancer for bloody fingerprints. Also a pH indicator, turning yellow below pH 1.8. Its use to treat skin infections is inadvisable since it is probably carcinogenic.

CSD *See* CHROMOSHADOW DOMAIN.

CSF-1 Colony-stimulating factor-1. *See* MCSF.

CSIF *See* INTERLEUKIN-10.

csk Protein tyrosine kinase (c-src kinase, EC 2.7.10.2, 450 aa) that specifically phosphorylates Tyr-504 on *lck, which has an inhibitory effect. Can act similarly on *lyn and *fyn kinases.

CSP An ambigous abbreviation for: **1.** *Caveolin scaffolding protein. **2.** Circumsporozoite protein of *Plasmodium vivax*. **3.** Common salivary protein, a protein secreted by the parotid gland. **4.** carotid sinus pressure. **5.** *See also* CASPASES; CICLOSPORIN; COMPETENCE STIMULATING PEPTIDE; CYSTEINE (cysteine string proteins).

CST *See* CORTISTATIN.

C-subfibre The third partial microtubule that is associated with the A- and B-tubules of the outer axonemal doublets to form a triplet in the *basal body and in the *centriole.

C substance *See* C-POLYSACCHARIDE.

CT *See* COMPUTED TOMOGRAPHY

CT-1 *See* CARDIOTROPIN-1.

CTAB A cationic detergent (cetyltrimethylammonium bromide) used to solubilize membranes.

CTACK A CC (β) *chemokine (CCL27, cutaneous T-cell attracting chemokine, ESkine, 112 aa) that attracts skin-associated memory T cells and mediates the homing of lymphocytes to cutaneous sites. The receptor is CCR10. Levels are up-regulated during wound healing and in psoriasis and atopic dermatitis.

CTAP III *See* CONNECTIVE TISSUE-ACTIVATING PEPTIDE III.

CTCF A transcriptional repressor (CCCTC-binding factor, 727 aa) that binds to promoters of c-*myc*, polo-like kinase, and *PIM1* genes but is a transcriptional activator of the gene for amyloid precursor protein. May act as a tumour suppressor and mutations in the gene have been associated with invasive breast cancers, prostate cancers, and *Wilms' tumours.

CTD A domain (C-terminal domain) unique to RNA polymerase II that is an important binding site for the various protein factors involved in RNA-processing. Phosphorylated **CTD-interacting factor 1** (704 aa) may play a role in transcription elongation or in coupling transcription to pre-mRNA processing. The *CTDP1* gene encodes RNA polymerase II subunit A. **CTD phosphatase** (C-terminal domain phosphatase, EC 3.1.3.16, 961 aa) dephosphorylates serines in the heptad repeats YSPTSPS found in the CTD which promotes the activity of RNA polymerase II. Mutations in CTDP1 are a cause of *congenital cataracts, facial dysmorphism, and neuropathy syndrome.

C-terminal binding proteins *See* RIBEYE.

C-terminal src kinase homologous kinase A *src-family kinase (Csk homologous kinase, CHK) that negatively regulates the chemokine receptor CXCR4. Also thought to inactivate src-family tyrosine kinases (SFKs) by phosphorylating their consensus C-terminal regulatory tyrosine and noncatalytically by binding to hck, lyn, and src to form stable protein complexes. Not to be confused with *checkpoint kinases (Chks).

C-terminus The carboxy-terminal end of a polypeptide or protein.

CTF A transcription factor (CCAAT box-binding transcription factor, CCAAT/enhancer-binding protein ζ, 1054 aa) that stimulates transcription from the HSP70 promoter.

CTGF (CTGFL/WISP-2) *See* CONNECTIVE TISSUE GROWTH FACTOR.

CTL *See* CYTOTOXIC T CELLS.

CTLA-4 A transmembrane protein (CD152, 223 aa) of the immunoglobulin superfamily, found on activated T cells and similar to the T-cell costimulatory CD28. Both CTLA-4 and CD28 bind to B7-1 (CD80) and B7-2 (CD86) on antigen-presenting cells but CTLA4 transmits an inhibitory signal to T cells, whereas CD28 is stimulatory. The CTLA-4 cytoplasmic domain interacts with SH2 domain of *Shp2 (protein tyrosine phosphatase) and possibly with PI3-kinase. Genetic variation in CTLA4 influences susceptibility to systemic *lupus erythematosus, *diabetes type 1, *coeliac disease type 3, hepatitis B virus infection, and possibly susceptibility to *Graves' disease.

CTX **1.** Crosslinked C-telopeptide of *collagen type I, a marker for bone degradation. **2.** Cefotaxime, a beta-lactam antibiotic. **3.** Bacterial genes (*CTX-M* genes) that encode extended-spectrum beta-lactamases that confer antibiotic resistance. **4.** *See also* CARDIOTOXINS; CEREBROTENDINOUS XANTHOMATOSIS; CHOLERA TOXIN; CIGUATERA; CONOTOXINS; CROTOXIN (Ctx); CYCLOPHOSPHAMIDE.

C-type lectins One of the two classes of *lectin produced by animal cells. C-type lectins require disulphide-linked cysteines and calcium ions for binding to carbohydrate. The carbohydrate recognition domain (CRD) that confers specificity is ~130 aa with 18 invariant residues including cysteines in a highly conserved pattern. All identified C-type lectins are extracellular proteins although some are integral membrane-bound proteins, such as the *asialoglycoprotein receptor. *Compare* S-TYPE LECTINS.

C-type virus An oncogenic RNA virus (one of the *Oncovirinae) that buds from the plasma membrane of the host cell starting as a characteristic electron-dense C-shaped crescent, hence the name. Examples include feline leukaemia virus and murine leukaemia and sarcoma viruses.

CUB domain A motif (complement subcomponent C1r /C1s, embryonic sea urchin protein Uegf, bone morphogenetic protein 1 domain, 110 aa) found in many functionally diverse, but often developmentally regulated proteins. CUB domains have been found in bone morphogenetic protein 1, a family of spermadhesins, *complement subcomponents Cls/Clr, and the neuronal recognition molecule A5, peptidases of the M12A (astacin) and S1A (chymotrypsin) families.

cuboidal epithelium An epithelium in which the area of each cell in contact with the basal lamina is comparable to the area of lateral cell–cell contact. *Compare* COLUMNAR CELLS; EPITHELIUM; SQUAMOUS EPITHELIUM.

cucurbitacins A group of bitter-tasting compounds, found in a range of plants, that have potential antitumour activity by targeting the JAK–STAT signalling pathway. Some are selective for JAKs, others for STATs.

cuffing The accumulation of white cells immediately around a blood vessel, seen in some infections of the nervous system.

cullins A family of proteins (at least eight) involved in cell cycle control that when mutated may contribute to tumour progression. **Cullin-1** (cdc53, 776 aa) is part of the *SCF* ubiquitin protein ligase complex (*See* SKP). Conjugation of *nedd8 to cullin-1 renders it unstable. Mutations in **cullin-4B** (895 aa) are associated with Cabezas' X-linked mental retardation syndrome. **Cullin-5** (780 aa) is the arginine vasopressin (AVP)-activated calcium-mobilizing receptor-1 (VACM1), a cell surface protein. The *3M syndrome is caused by mutation in **cullin-7** (1698 aa) which is a component of an E3 ubiquitin ligase-complex.

cumulative median lethal dose An estimate of the total administered amount of a substance that causes the death of 50% of animals when given repeatedly at doses that are fractions of the median lethal dose.

cumulus A tightly packed multilayer of cells that surround the developing ovum in the Graafian follicle and become more dispersed in the preovulatory period. Fancifully thought to resemble the eponymous clouds.

curare A general name for a variety of plant alkaloids, isolated from *Strychnos toxifera*, used as arrow poisons. They include D-tubocurarine, alloferine, toxiferine, and protocurarine. They act as competitive inhibitors for acetylcholine at the neuromuscular junction and cause muscle relaxation.

curcumin The yellow pigment in turmeric; a polyphenol, that seems to have some anti-inflammatory and antioxidant actions and is extensively used in traditional medicine.

CURL The compartment for uncoupling of receptors and ligands, a collection of vesicles derived from the plasma membrane and enriched in receptor–ligand complexes: the ligands are released and the receptors recycled.

Currarino's syndrome A developmental disorder (sacral agenesis) in which there is partial sacral agenesis with an intact first sacral vertebra, a presacral mass, and anorectal malformation. Some cases are caused by mutation in the *HLXB9* homeobox gene, others may be due to mutations in **RHEB* which is mapped to the same chromosomal region.

Cushing's syndrome An endocrine disorder (hyperadrenocorticism, hypercorticism) in which there is overproduction of *cortisol in response to excessive secretion of *adrenocorticotropic hormone (ACTH), usually from a pituitary adenoma. Symptoms include weight gain, hypertension, and a range of other changes.

cutis laxa A connective tissue disorder characterized by loose, sagging, and inelastic skin. Autosomal dominant cutis laxa can be caused by mutations in the *elastin gene. Mutations in the gene encoding *fibulin-5 can cause either autosomal dominant or autosomal recessive cutis laxa and the latter may also arise through mutation in fibulin-4 or in a subunit of the *V-type ATPase.

CV-1 cells A line of fibroblast-like pseudodiploid cells from the African green monkey *Cercopithecus aethiops* that are used in transfection experiments and studies on viral infection.

C value paradox The paradoxical observation that the amount of DNA in the haploid genome (the C value) of two closely related species can vary enormously and there does not seem to be any correlation between genomic size and organismal complexity. Genome sequencing has not really resolved the paradox, although it is now clear that much DNA can be noncoding.

CVF *See* COBRA VENOM FACTOR.

CVS *See* CHORIONIC VILLUS SAMPLING.

CWCV domain A rare amino acid sequence (Cys-Trp-Cys-Val) found in extracellular and cell surface molecules, thought to have binding properties.

CWD *See* CHRONIC WASTING DISEASE.

Cyanobacteria (formerly **blue-green algae**) A phylum of photosynthetic bacteria that include unicellular and colonial species. Some filamentous colonies are differentiated into vegetative photosynthetic cells and thick-walled heterocysts, which are responsible for nitrogen fixation. Chloroplasts in eukaryotic cells are thought to be derived from cyanobacteria by endosymbiosis.

cyanocobalamin *See* VITAMIN B (B_{12}).

cyanogen bromide A compound (CNBr) that will cleave peptide bonds at methionine residues; the fragments can then be analysed (*see* PEPTIDE MAP).

cyanosis A blue tinge to the skin caused by insufficient oxygenation of blood in capillaries. May be a natural response to cold or a pathological response, e.g. to cyanide.

cyclase associated protein Actin-binding proteins (adenylyl cyclase-associated protein, CAP-1, 475 aa; CAP-2, 477 aa) that regulate microfilament dynamics and may be involved in various developmental and morphological processes, the localization of mRNA, and the establishment of cell polarity.

cyclic ADP-ribose A compound (cADPR, adenosine 5′-cyclic diphosphoribose), synthesized by the multifunctional transmembrane ectoenzyme CD38, that acts as a calcium-mobilizing second messenger. It is the endogenous modulator of *ryanodine receptors.

cyclic AMP (cAMP) An important *second messenger, the 3′,5′-cyclic ester of AMP produced from ATP by the action of *adenylate cyclase that is coupled to hormone receptors by *GTP-binding proteins and is hydrolysed by a phosphodiesterase to produce 5′-AMP. It activates a cAMP-dependent protein kinase (PKA). It is also a morphogen (acrasin) for some slime moulds.

cyclic GMP (cGMP) A *second messenger (the 3′,5′-cyclic ester of GMP) generated by guanylyl cyclase. *See* ATRIAL NATRIVRETIC PEPTIDE; NITRIC OXIDE.

cyclic inositol phosphates Cyclic intermediates in the hydrolysis of phosphatidylinositol 4,5 bisphosphate by PI-specific phospholipase C. They have been suggested as possible *second messengers in hormone-activated pathways.

cyclic nucleotide phosphodiesterases A group of enzymes (EC 3.1.4.-) often called simply phosphodiesterases (PDEs), that exhibit tissue-type and developmental-stage specificity and are controlled in various ways. **PDE-1** (535 aa) is calcium/calmodulin regulated and important in the central nervous system and vasorelaxation, **PDE-2** (941 aa) is cGMP-dependent and hydrolyses cAMP. **PDE-3** (1112 aa) regulates smooth muscle in airways and blood vessels and has effects on platelet aggregation, cytokine production, and lipolysis; **PDE-4** (700–800 aa) is inhibited by rolipram and is involved in the control of airway smooth muscle and the release of inflammatory mediators as well as regulating gastric acid secretion and having a role in the central nervous system; variations in $PDE4_D$ (809 aa) may be associated with susceptibility to stroke. **PDE-5** (875 aa) and **PDE-6** (860 aa) are cGMP-specific, PDE-5 is involved in platelet aggregation, PDE-6 is regulated by interaction with *transducin in photoreceptors; defects in PDE6 are a cause of *retinitis pigmentosa and one form of *stationary night blindness. **PDE-7** (482 aa) is abundant in skeletal muscle and present in heart and kidney. **PDE-8** (829 aa) is found in thyroid, **PDE-9** (593 aa) is a high affinity cGMP-specific phosphodiesterase, **PDE10** (779 aa) is found mainly in the central nervous system. Defects in **$PDE11_A$** (dual 3′,5′-cAMP and cGMP phosphodiesterase 11A, 934 aa) are the cause of *primary pigmented nodular adrenocortical disease type 2. There are multiple subtypes of many of the PDEs. *CNPase hydrolyses 2′,3′-cyclic-nucleotide 3′-phosphodiesterase.

cyclic phosphorylation A process in which a phosphatide ester is linked to an adjacent hydroxyl group to produce a cyclic diester.

cyclin-dependent kinase activating kinase A kinase complex with cdk7 as the catalytic subunit and cyclin-H as the regulator, that activates *cdks (cdc2/cdk1, cdk2, cdk4, and cdk6) by threonine phosphorylation.

cyclin-dependent kinase inhibitors (CKIs) Two classes of CKIs are known in mammals, the cip1/kip1 family (including p27KIP1, *p57KIP2, and *Waf-1) that inhibit G1/S *cyclin-dependent kinases (cdks), and the *INK4 class that bind and inhibit only cdk4 and cdk6. The p21CIP1 inhibitor is transcriptionally regulated by *p53 tumour suppressor and is important for the G1 DNA-damage checkpoint.

cyclin-dependent kinases (cdks) A family of *serine/threonine kinases that are only active in a complex with a *cyclin and are involved in regulation of the cell cycle. The cyclin–cdk complexes are themselves regulated by phosphorylation by cdk-activating kinase. Cdk9 is atypical (*see* CYCLINS (cyclin-T1)).

cyclins Proteins that change in concentration as the cell cycle progresses, rising to a threshold level and then declining rapidly. They interact with *cyclin-dependent kinases (CDKs) and products of the *cdc genes. **G1/S cyclins** control the G_1/S transition, **G2/M cyclins** control the transition into mitosis. **Cyclin A** (A1, 465 aa; A2, 432 aa) interacts with CDK2 and *cdc2 and is essential for the control of the cell cycle at the G_1/S and the G_2/M transitions; the transition from G_1 to S phase also involves **cyclin D** (295 aa) interacting with CDK4 and CDK6 and **cyclin E** (410 aa) interacting with CDK2. **Cyclin B1** (433 aa) is essential for the G_2/M transition and interacts with cdc2 to form a serine/threonine kinase holoenzyme complex (maturation promoting factor, MPF); it is ubiquitinated and destroyed abruptly at the end of mitosis. Cyclin B interacting with CDK1 regulates progression from the S to G_2 phase. Various other cyclins are known: **cyclin C** (283 aa) is a component of the *mediator complex. **Cyclin F** (786 aa) is probably involved in control of the cell cycle during S phase and G_2. **Cyclin G1** (295 aa) is associated with G_2/M phase arrest in response to DNA damage. **Cyclin H** (323 aa) regulates CDK7, the catalytic subunit of the CDK-activating kinase (CAK) enzymatic complex. **Cyclin I** (377 aa) does not change in concentration during the cell cycle, **cyclin J** (372 aa) has an unknown role but is phylogenetically ancient. **Cyclin K** (580 aa) may play a role in transcriptional regulation, **cyclin L1** is a transcriptional regulator involved in the pre-mRNA splicing process, **cyclin M** (cyclin-related protein FAM58A, 248 aa) may have a role in regulating the cell cycle; defects cause *STAR syndrome. **Cyclin O** (EC 3.2.2.-, uracil-DNA glycosylase 2, 350 aa) will remove uracil residues accidentally incorporated into DNA or arising by deamination of cytosine. **Cyclin T1** (726 aa) is the regulatory subunit of the CDK9/cyclin–T1 complex (positive transcription elongation factor B, P-TEFβ), which is proposed to facilitate the transition from abortive to productive elongation by phosphorylating the *CTD (C-terminal domain) of the large subunit of RNA polymerase II. **Cyclin Y** (cyclin fold protein 1, 341 aa) is membrane associated.

cyclitis Inflammation of the *ciliary body.

cyclizine A histamine H_1-receptor antagonist used to treat motion sickness.

cyclodextrins *See* DEXTRINS.

cycloheximide An inhibitor of eukaryotic protein synthesis (281 Da) isolated from *Streptomyces griseus*. Acts by preventing initiation and elongation on 80S ribosomes.

cyclolysin A protein toxin (1706 aa) from **Bordetella pertussis* that is both an *adenylate cyclase (EC 4.6.1.1) and a *haemolysin.

cyclo-oxygenase (COX) An enzyme complex present in most tissues that produces various prostaglandins and thromboxanes from arachidonic acid and that is inhibited by nonsteroidal anti-inflammatory drugs (*NSAIDs) such as aspirin. Three isoforms are known, COX-1, COX-2, and COX-3. **COX-1** (prostaglandin G/H synthase 1, EC 1.14.99.1, 599 aa) and **COX-2** (604 aa) are both inhibited by aspirin. Many selective COX-2 inhibitors such as celecoxib have the analgesic and anti-inflammatory activity but do not affect the gastric mucosa in the same way as the nonselective NSAIDs; they have other side effects, however, and some have been withdrawn. In humans, a putative **COX-3** mRNA is expressed in cerebral cortex and heart; paracetamol (acetaminophen) may inhibit this COX variant.

cyclopamine An alkaloid, produced by the skunk cabbage *Veratrum californicum*, that is a potent teratogen and causes major developmental defects (e.g. cyclopia) by blocking the *hedgehog signalling pathway. Cyclopamine will block the oncogenic effects of mutations of *patched (*see* GORLIN'S SYNDROME).

cyclopenthiazide A thiazide *diuretic.

cyclopentyladenosine A specific agonist of the adenosine A_1 receptor.

cyclophilin An *immunophilin (754 aa) that binds *ciclosporin A.

cyclophosphamide A nitrogen mustard alkylating agent used as an immunosuppressant and in tumour chemotherapy. *TNs* Cytoxan, Endoxan, Neosar, Procytox, Revimmune.

cyclopiazonic acid A mycotoxin from *Penicillium cyclopium* that is a selective inhibitor of the sarcoplasmic–endoplasmic reticulum calcium ATPase (SERCA).

cycloplegia Paralysis of the muscles in the *ciliary body.

cyclosome *See* ANAPHASE PROMOTING COMPLEX.

cyclosporine (cyclosporin A) *See* CICLOSPORIN.

cyesis Pregnancy.

cylindrospermopsin A toxin produced by the cyanobacterium *Cylindrospermopsis raciborskii* and other freshwater cyanobacteria. It is toxic to liver and kidney tissue and is thought to inhibit protein synthesis and to covalently modify nucleic acid.

CYP1A *See* CYTOCHROME P450.

cypermethrin A synthetic pyrethroid which acts on sodium channels in neurons. Used as an insecticide but highly toxic to aquatic organisms, including fish.

cypin An alternatively spliced form of *guanine deaminase (cytosolic PSD-95 interactor) that in neurons promotes an increase in dendrite numbers, an activity dependent on enzyme activity. It interacts with PSD-95 (*See* POSTSYNAPTIC PROTEIN), and binds tubulin via its *CRMP homology domain to promote microtubule assembly, an interaction blocked by *snapin.

cyproterone acetate An antiandrogen (CPA) that blocks androgen receptors and suppresses luteinizing hormone, thereby reducing testosterone levels. It is used to treat prostate cancer, benign prostatic hyperplasia, priapism and hypersexuality. *TN* Cyprostat.

cyst 1. A fluid-filled sac bounded by a multicellular wall that may result from a wide range of insults (e.g. *hydatid disease, *sebaceous cyst) or be of embryological origin. **2.** A resting stage of many prokaryotes and eukaryotes in which a cell or several cells are surrounded by a tough protective wall.

cyst- A prefix indicating bladder-related. **Cystitis** is inflammation of the bladder. **Cystocele** is herniation of the bladder. A **cystoscope** is an endoscope for inspecting the interior of the bladder.

cystatins A superfamily of structurally homologous cysteine peptidase inhibitors divided into family I (the *stefins); family II (the cystatins), and family III (the *kininogens). **Cystatin C** (cystatin 3, CSTC, 120 aa) is expressed in most tissues whereas **cystatins D** (cystatin 5), **S** (CST4), **SN** (CST1), and **SA** (CST2) are secreted in saliva. Cystatin C is excreted in urine and the serum concentration correlates with the glomerular filtration rate and has been suggested as a marker in testing for renal dysfunction. Mutations in CSTC are associated with *Icelandic-type cerebroarterial amyloidosis.

cysteine (Cys, C) The only amino acid to contain a thiol (SH) group, important because of the potential to form disulphide bridges. Often found at the active site of enzymes, as in the **cysteine endopeptidases** (thiol peptidases, EC 3.4.22), a clan of endopeptidases that includes papain and several cathepsins. Natural inhibitors are *alpha-2-macroglobulin and *cystatins. Cysteine string proteins (Csps) are membrane-associated proteins containing a sequence of palmitoylated cysteine residues and a *J-domain that have been functionally implicated in regulated exocytosis.

cysteine-rich secretory proteins A family of proteins characterized by 16 conserved cysteine residues. **Cysteine-rich secretory protein-1** (CRISP-1, 249 aa) is found in the male reproductive tract and may have a role in sperm–egg fusion; **CRISP-2** (243 aa) is found in testis and epididymis. **CRISP-3** (245 aa) is found in the salivary gland, pancreas, and prostate, also in the specific granules of neutrophils. A number of other members of the family are known, some with *LCCL domains (e.g. CRISP-10, trypsin inhibitor Hl, 500 aa). *See* HELOTHERMINE; RESISTIN.

cystic fibrosis An autosomal recessive disorder of exocrine gland secretion that affects multiple systems, most notably the lung, where the abnormally thick mucus is not cleared well and there is a predisposition to respiratory infections. There are also effects on the pancreas, where blockage of the ducts can cause cysts and there may be a shortage of digestive enzymes. There are also effects in the bowel, biliary tree, and sweat glands. The cystic fibrosis transmembrane conductance regulator (CFTR) is an *ABC protein, an epithelial chloride channel that is defective in the disorder. A range of different mutations occur, making preimplantation genetic diagnosis difficult.

cystine The amino acid dimer formed when the sulphydryl (SH) groups of two *cysteines interact to form a disulphide bond. Within a protein each component is referred to as a half-cystine.

cystinosin A lysosomal protein (367 aa) responsible for cystine export. Defects lead to nephropathic *cystinosis and mutations in the cystinosin (*CTNS*) gene are the most common cause of inherited renal *Fanconi's syndrome. Highly conserved in mammals.

cystinosis An autosomal recessive lysosomal storage disorder caused by mutation in *cystinosin and defective lysosomal transport of cystine. The classical nephropathic form is characterized by renal failure by 10 years of age; an intermediate form has later onset of renal disease, and benign (non-nephropathic) cystinosis has only corneal crystals and photophobia.

cytarabine *See* CYTOSINE ARABINOSIDE.

cytidine The nucleoside consisting of D-ribose and the pyrimidine, cytosine. Cytidine 5′-diphosphate (CDP) derived from CTP is important in phosphatide biosynthesis; activated choline is CDP-choline.

cytisine A natural alkaloid that is the toxin in laburnum and several other related plants. It binds with high affinity to the $\alpha_4\beta_2$-nicotinic ACh receptor.

cytoband *See* CYTOGENETIC BAND.

cytochalasins A group of fungal metabolites that inhibit the addition of G-actin to the preferred assembly site on the microfilament and perturb labile microfilament arrays. Cytochalasin B has effects on glucose transport at higher concentrations; cytochalasin D does not have this disadvantage.

cytochemistry The branch of histochemistry associated with the localization of cellular components by specific staining methods, now done mostly with fluorescently tagged monoclonal antibodies (immunocytochemistry).

cytochrome oxidase The terminal enzyme (EC 1.9.3.1.) in the electron transport chain, an integral membrane protein of the inner mitochondrial membrane composed of several metal prosthetic sites and 13 protein subunits (COX1, COX2, etc.). It oxidizes cytochrome *c* by accepting electrons which are then transferred to molecular oxygen.

cytochrome P450 (cytochrome *m*, CYP) A family of oxidases (EC 1.14.14.1) of the cytochrome *b* type. The nomenclature is systematic and based roughly on sequence. The genes and proteins are named 'Cyp' suffixed by family, subfamily, and specific identifiers, e.g. *CYP3A5* and *CYP3A7* are two genes in the 3A subfamily that encode closely related proteins (CYP3A5, CYP3A7). Members of the Cyp2, Cyp3, and Cyp5 families are primarily involved in xenobiotic detoxification in the liver, others are involved in core metabolic functions such as steroid biosynthesis. Variation in the P450 enzymes can have major effects on the rate of drug metabolism.

cytochromes Highly conserved enzymes of the electron transport chain that have *haem prosthetic groups.

cytogenetic band (cytoband) A subregion (band) on a chromosome that is visible after staining. *See* BANDING PATTERNS.

cytogenetics The study of genetics at the level of the cell, in particular the study of chromosomes and their abnormalities.

cytohesins A family of guanine nucleotide exchange factors (GEFs) for ADP-ribosylation factor (*ARF) GTPases. **Cytohesin-1** (PH, SEC7 and coiled-coil domain-containing protein-1, SEC7 homologue B2-1, 398 aa) is ubiquitous, and acts on ARF1 and ARF5. **Cytohesin-2** (ARF nucleotide-binding site opener, ARNO, 400 aa) acts on ARF1, ARF3, and ARF6. **Cytohesin-3** (general receptor of phosphoinositides 1, Grp-1, 400 aa) acts on ARF1 and binds with high affinity to the inositol head group of phosphotidylinositol 3,4,5-trisphosphate via its PH domain. **Cytohesin-4** (394 aa) is found predominantly in peripheral blood leucocytes. **Cytohesin-interacting protein** (359 aa) is a scaffolding and modulator protein that binds to cytohesin-1 and modifies the activation of ARF.

cytokeratins The proteins that form the intermediate filaments in epithelial cells.

cytokines A rather loose category of small proteins (~5–20 kDa) that are released by cells and that affect the behaviour of other cells. Normally taken to include *interleukins, lymphokines, *chemokines and several related signalling molecules such as *tumour necrosis factor and *interferons but not hormones or growth factors except perhaps *transforming growth factor β (TGFβ).

SEE WEB LINKS

- A very valuable encylopaedia of cytokines and cells (22 600 entries), created, developed, and maintained by Prof. Dr H Ibelgaufts.

cytokinesis The division of the cytoplasm of a cell after nuclear division (mitosis) is complete.

cytology The study of cells, originally mostly morphological studies using light or electron microscopy. Most cell biologists would not consider themselves cytologists.

cytolysis Cell *lysis.

cytomegalovirus One of the Herpetoviridae group of viruses (human herpesvirus 5) that can cause abortion or stillbirth or lead to various congenital defects, although can be opportunistic in the immunocompromised host. A very common virus that remains latent in most cases. Infected cells have a characteristic nuclear inclusion body composed of virus particles. *See* VIPERIN.

cyton The region of a neuron that contains the nucleus and most cellular organelles.

cytonectin An ion-insensitive adhesion protein (35 kDa) that is reported to be present in most tissues and overexpressed in the entorhinal cortex in Alzheimer's disease There is only one literature reference (2002) but a patent application is filed and antibodies are commercially available. Not in UniProtKB.

cytoneme A long, thin projection from an epithelial cells, more commonly a filopodium.

cytoplasm The contents of a cell, with the exception of the nucleus in eukaryotes.

cytoplasmic bridge A thin, direct cytoplasmic connection between two cells through which large molecules can pass (unlike *gap junctions). Common in plants (plasmodesmata) but rare in animals except between *nurse cells and developing gametes.

cytoplasmic determinants Any feature or property of the cytoplasm that determines how a process or activity proceeds, but more often used to describe asymmetrically distributed components of the maternally derived oocyte cytoplasm that determine the fate of blastomeres.

cytoplasmic inheritance The inheritance of parental characters through a nonchromosomal mechanism. The major example is the inheritance of mitochondrial DNA, which is not segregated at mitosis and derives completely from the maternal side. In a more general sense the continuity of structures from one generation to the next; membranes derive from other membranes in most cases, and centrioles divide. *See* MATERNAL INHERITANCE.

cytoplasmic polyadenylation element (CPE) A uridine-rich sequence in the 3′ untranslated region of mRNA to which the cytoplasmic polyadenylation element binding protein (CPEB) binds. CPEB binding recruits a protein complex, called *CPSF, which interacts with CPEB and the *nuclear polyadenylation hexanucleotide sequence in the mRNA. CPSF recruits poly-A polymerase to the mRNA which adds a poly-A tail. This is important in the reactivation of mRNA in the oocyte, during early development and at postsynaptic site in neurons (mRNAs may be transported to dendrites in a translationally dormant form). *Maskin is thought to repress the translation of CPE-containing mRNA. Four different CPEBs (ranging from 526 aa to 729 aa) are found with differing tissue distribution.

cytoplast Fragment of cell from which the nucleus has been removed as a *karyoplast, usually by treating the cell with cytochalasin B and centrifuging on a step gradient.

cytosine The pyrimidine base found in DNA and RNA that is complementary to guanine. The glycosylated base is *cytidine.

cytosine arabinoside (cytarabine, araC) A cytotoxic antimetabolite that blocks DNA synthesis and is used in chemotherapy, particularly of acute myeloid *leukaemia, and to treat viral infections.

cytoskeleton A general term for the cytoplasmic elements that confer mechanical properties and motility. The main components are long polymers of actin (*microfilaments) that form a meshwork, *microtubules, and the *intermediate filaments (polymers of various proteins), together with a range of anchoring, crosslinking, and motor proteins.

cytosol The fluid portion of cytoplasm from which organelles and membranes have been removed.

cytosome **1.** The body of a cell without the nucleus. **2.** A multilamellar body found in some cells.

cytotactin *See* TENASCINS.

cytotoxic drug A general term for drugs that kill cells, but more specifically for drugs used in the treatment of cancer that are more or less selective for rapidly proliferating cells. They include alkylating drugs, cytotoxic antibiotics (e.g. doxorubicin, bleomycin), antimetabolites, and antimitotic drugs (eleutherobin, maytansine, taxanes, vinca alkaloids). Other examples are bakuchiol, camptothecin, cisplatin, equisetin, gemcitabine, hadacidin, hexacyclinic acid, kosinostatin, pentostatin, psoralidin, reticulol, and rottlerin.

cytotoxic necrotizing factors Protein toxins produced by various pathogenic *E. coli* and *Yersinia* strains that constitutively activate small GTPases of the Rho family.

cytotoxic T cells (CTL) The subset of T-cells (mostly $CD8^+$) that kill cells (virus infected cells, tumour cells, or allogeneic cells) expressing antigens recognized by the T-cell receptor in the context of class I *histocompatibility antigens.

cytotrophic Describing a substance that promotes the growth or survival of cells. Rare except for the tissue-specific example of *neurotrophic factors. *Compare* CYTOTROPIC.

cytotropic Describing something that has an affinity for cells. *Compare* CYTOTROPHIC. Cytotropism is the movement of cells towards or away from other cells.

cytovillin *See* EZRIN.

C

D_2O (deuterium oxide) *See* HEAVY WATER.

DAB *See* DIAMINOBENZIDINE.

DAB1 A receptor (mammalian disabled) for *reelin that is required for layering of neurons in the cerebral cortex and cerebellum. Mutations in murine homologue are responsible for various behavioural changes ('scrambler' and 'yotari' mutants). DAB1 is a phosphoprotein (588 aa) that binds nonreceptor tyrosine kinases and interacts with lipoprotein receptor-related protein-8 (*see* LDL-RECEPTOR).

DABA *See* DIAMINOBENZOIC ACID.

dacarbazine A cytotoxic alkylating agent (DTIC) used in chemotherapy, particularly of malignant *melanoma. One of the ABVD cocktail of drugs (adriamycin, bleomycin, vinblastine, and dacarbazine) used to treat *Hodgkin's lymphoma.

dacryo- A prefix denoting tears (from *Gk* dakry, a tear). **Dacryoadenitis** is inflammation of the lacrimal (tear) gland. A **dacryagogue** (*adj.* dacryagogic) is a substance that promotes the formation of tears. **Dacryocystitis** is inflammation of the lacrimal sac. *See* LACRIMATION.

dactyl (*adj.* **dactylar)** A digit. **Dactylitis** is inflammation of a finger or toe.

DAF 1. *See* DECAY ACCELERATING FACTOR. **2.** A compound (diaminofluorescein) that reacts with nitric oxide (NO) in the presence of oxygen to produce a highly fluorescent compound, triazolofluorescein; used for the real-time detection of NO *in vivo*.

DAG *See* DIACYLGLYCEROL.

daidzein An isoflavone, the aglycone of daidzin, classified as a phytoestrogen, a plant-derived nonsteroidal compound with oestrogen-like biological activity, also an antioxidant. Mainly found in legumes, such as soybeans and chickpeas. Often used to treat conditions affected by oestrogen levels in the body (e.g. hormone-dependent breast carcinoma, osteoporosis). The average daily intake of isoflavones in Japan is ~200 mg and there is no evidence of adverse effects. *See also* GENISTEIN.

DAL-1 *See* DIFFERENTIALLY EXPRESSED IN ADENOCARCINOMA OF THE LUNG.

dalfopristin A *streptogramin antibiotic (type A streptogramin) often used in combination with the similar antibiotic *quinupristin to treat infections with vancomycin-resistant *Enterococcus faecium*.

Dalton (Da) The unit of atomic mass.

daltonism *See* COLOUR BLINDNESS.

damaged DNA-binding proteins Proteins (DDBs) that are involved in the initial recognition of ultraviolet-damaged DNA and mediate recruitment of nucleotide *excision repair factor. *In vivo*, UV-DDB plays an important role in the p53-dependent response of mammalian cells to DNA damage. Mutations in the *DDB2* gene inactivate UV-DDB in individuals from complementation group E of *xeroderma pigmentosum. The damaged DNA-binding protein complex consists of a heterodimer of p127 (DDB1) and p48 (DDB2) subunits.

DAN The gene (differential screening-selected gene aberrant in neuroblastoma) that encodes a *bone morphogenetic protein (BMP)-antagonist that regulates BMP activity spatially and temporally during patterning and partitioning of the medial otic tissue in ear development. It has been suggested that the *cer/dan family members control diverse processes in growth, development, and the cell cycle and that DAN is a neurotransmitter in the primary sensory nerve fibres for pain sensation.

danazol A synthetic steroid derivative that was used to treat endometriosis, now largely superseded by GnRH agonists. It suppresses the pituitary–ovarian axis and depresses the output of both follicle stimulating hormone (FSH) and luteinizing hormone (LH). *TN* Danocrine.

dander Small flakes of skin, fur, or feathers shed by animals and often responsible for exciting an allergic response in sensitive individuals.

Dane particle A 42-nm particle identified by electron microscopy; the complete infective virion of *hepatitis B virus.

Danon's disease An X-linked dominant disorder (vacuolar cardiomyopathy and myopathy, X-linked lysosomal glycogen storage disease without acid maltase deficiency, glycogen storage cardiomyopathy, pseudoglycogenosis II, Antopol disease; glycogen storage disease IIb) predominantly affecting cardiac muscle with variable myopathy and mental retardation. It is a lysosomal *storage disease with characteristic intracytoplasmic vacuoles containing autophagic material and glycogen in skeletal and cardiac muscle. The defect is in *LAMP-2, a lysosomal membrane structural protein, not an enzyme.

dansyl chloride A compound that reacts with amines to produce a fluorescent product, the emission spectrum of which is sensitive to the local environment. Often used as a label in immunofluorescence methods or to label the N-terminal amino acid of a peptide.

dantrolene A muscle relaxant that acts on the *ryanodine receptor to block excitation–contraction coupling in muscle. *TNs* Dantrium, Dantrolen.

DAP **1.** Diabetes related peptide. *See* ISLET AMYLOID PEPTIDE. **2.** *See* DEATH-ASSOCIATED PROTEIN.

DAP-12 A signalling adaptor (DNAX-activating protein-12, 113 aa), expressed by NK cells, B cells, and T cells, that resembles the γ-chain of FcεRI and the ζ chain of the T-cell receptor. It interacts through an immunoreceptor tyrosine-based activation motif (ITAM) with *zap-70 and *syk and thus activates NK cells, although it has been implicated in inhibitory signalling in mouse macrophages and dendritic cells. Defects cause *Nasu–Hakola disease. *See* TREM-2.

DAPI stain A fluorescent dye that binds to DNA and is used to detect DNA and to stain the nucleus in fluorescence microscopy and FACs analysis.

DAP kinase *See* DEATH-ASSOCIATED PROTEIN KINASE.

dapper An antagonist of β-*catenin (frodo, 836 aa) that inhibits activation of *JNK by *dishevelled. Dapper-2 (774 aa) has a different tissue distribution.

dapsone A drug related to the sulphonamides that is used to control the dermatological symptoms of *dermatitis herpetiformis and other skin diseases characterized by neutrophil infiltration. Dapsone is also used alone or in combination therapy for leprosy and is bacteriocidal and bacteriostatic for *Mycobacterium leprae.* May act by inhibiting folate synthesis; its anti-inflammatory effects may be due to inhibition of calcium-dependent functions of neutrophils including release of tissue-damaging oxidants and proteases in affected skin.

daptomycin A cyclic lipopeptide antibiotic that is active against Gram-positive bacteria including methicillin-resistant *Staphylococcus aureus,* glycopeptide-intermediate and -resistant *S. aureus,* and vancomycin-resistant *Enterococcus faecalis* and *E. faecium* (VRE). Acts by inhibiting lipoteichoic acid synthesis as a consequence of membrane binding in the presence of calcium.

dardarin A mixed-lineage kinase (*MLK) (leucine-rich repeat kinase 2, LRRK2, 2482 aa) that may play a role in the biogenesis or regulation of membranous intracellular structures in the brain. Mutation in the gene is associated with *Parkinson's disease type 8 (PARK8).

Darier's disease An autosomal dominant skin disorder (Darier–White disease, keratosis follicularis) characterized by warty papules and plaques in seborrheic areas. Caused by mutations in the gene encoding the sarcoplasmic reticulum calcium ATPase (*SERCA2*) which affects *desmosome integrity.

dark current A current in the retina caused by constant influx of sodium ions through cGMP-gated sodium channels into the rod outer segment of *retinal rods, and that is blocked by light (leading to hyperpolarization). The plasma membrane sodium channel is controlled through a cascade of amplification reactions initiated by photon capture by *rhodopsin in the disc membrane. The so-called 'dark current' depolarizes the cell to around −40 mV.

Darling's disease *See* HISTOPLASMOSIS.

darwinian fitness General term used as a measure of the fitness of an individual under selection pressure. Basically, the number of offspring of that individual that will survive to reproduce in the next generation thereby passing on alleles of genes that contributed to the selective advantage. More rigorously, calculated relative to the phenotype with the highest absolute reproductive success (thus normalizing the values) so that the maximum darwinian fitness is 1.0.

database A collection of records or of data stored in a computer system. Such collections allow (e.g.) all examples of records with a particular attribute to be extracted. This dictionary was prepared using a relational database that has information (in addition to that in the printed version) about the origin of the headword, date of writing, notes, and links to references and that also keeps a record of the total number of entries and the word count.

dATP The reduced form of *ATP, produced by the action of ribonucleotide reductase, the deoxyribonucleotide that is incorporated into DNA.

Datura stramonium A poisonous plant (loco weed, Jamestown weed, thorn apple, angel's trumpet, zombie's cucumber) of the nightshade family (Solanaceae), native to North America, that is the source of *scopolamine and other alkaloids (atropine, hyoscyamine).

Daudi A well-characterized and established B lymphoblast cell line, derived in 1967 from a 16-year-old black male with Burkitt's lymphoma. The cells have surface complement receptors and IgG and are *EBV marker positive, but negative for β_2-microglobulin. They have been employed extensively in studies of mechanisms of leukaemogenesis.

daunomycin *See* DAUNORUBICIN.

daunorubicin, A cytotoxic drug (cerubidine, daunomycin) from *Streptomyces peucetius*, used in cancer chemotherapy, particularly for acute myelogenous *leukaemia (AML) and acute lymphoblastic leukaemia (ALL). Intercalates into DNA causing uncoiling of the helix, ultimately inhibiting DNA synthesis and DNA-dependent RNA synthesis. Inhibits topoisomerase II activity by stabilizing the DNA-topoisomerase II complex, preventing the religation phase of the topoisomerase-catalysed ligation-religation reaction. May also act by inhibiting polymerase activity, affecting regulation of gene expression and generating free radicals. Cytotoxic activity is cell cycle phase nonspecific, although effects are maximal in S phase. *TN* Adriamycin.

DAXX *See* DEATH-ASSOCIATED PROTEIN (6).

dbl An *oncogene (cell line-derived transforming sequence, MCF2) originally identified by transfection of NIH-3T3 cells with DNA from human diffuse B-cell lymphoma. The product, Dbl, is a guanine nucleotide exchange factor (*GEF) for *rho family members and has a Dbl-homology (DH) domain. The DH domain is a cytoskeletal modulation-type domain (~150 aa) arranged as an extended bundle of alpha-helices that is found in Dbl, Vav, and a family of other Dbl-family proteins (Lfc, Lsc, Ect2, Dbs, Brx). DH domains are invariably located immediately N-terminal to a *PH domain that is required for normal function. The Dbl family of related proteins with similar functions and containing a DH domain includes Dbl itself, Dbs, Brx, Lfc, Lsc, Ect2, DRhoGEF2, and Vav.

- For details of the dbl domain.

DBP *See* DIBUTYL PHTHALATE.

DCC protein *See* DELETED IN COLORECTAL CANCER.

D cells Cells (δ-cells, delta cells) that constitute about 5% of the pancreas and that contain small argentaffin-positive granules. The granules, as in gastric D cells of the oxyntic and pyloric mucosa, contain *somatostatin, probably as a precursor form.

DCIP A commonly used electron acceptor dye (Tillman's reagent), which can accept electrons. DCIP is blue in neutral solution and pink in acidic solution; the reduced form is colourless. Used in a test kit for the detection of haemoglobin E and for laboratory testing of vitamin C levels.

DCX *See* DOUBLECORTIN.

DDAVP A synthetic analogue (deamino-8-D-arginine vasopressin) of the natural pituitary hormone 8-arginine *vasopressin (antidiuretic hormone, ADH), used in management of *diabetes insipidus and also for bedwetting. *See* DESMOPRESSIN.

DDB complex *See* DAMAGED DNA-BINDING PROTEINS.

ddNTP Generic name for dideoxy nucleotide triphosphates used in *dideoxy sequencing. They are essentially the same as nucleotides but have a hydrogen group on C-3 instead of a hydroxyl group.

DD-PCR *See* DIFFERENTIAL DISPLAY PCR.

DDS *See* DAPSONE.

DEA *See* DEHYDROEPIANDROSTERONE.

deacetylase Any enzyme that removes an acetyl group. *See* HISTONE DEACETYLASES.

DEAD-box A four amino acid consensus, -D-E-A-D- (Asp-Glu-Ala-Asp), that resembles an ATP binding site of the Walker B type. DEAD-box helicases are a family of ATP-dependent DNA or RNA *helicases belonging to the second superfamily of helicases (SFII), the DExD/H-box RNA helicases (*see* DEAH-BOX PROTEINS). They are important in RNA metabolism in both eukaryotes and prokaryotes and may have important functions in mediating microbial pathogenesis. Humans have 36 putative DEAD-box helicases including *p68, a nuclear protein involved in cell growth. DEAD proteins have multiple DEAD motifs, and several hundreds of these proteins can be identified in databases. Not all have helicase activity and it may be that their role is to alter the conformation of RNA, thus regulating its binding interaction with proteins.

DEAE A positively charged side group (diethylaminoethyl) often linked to cellulose or Sephadex to produce an anion exchange matrix for *chromatography.

deafness *See* AUDITORY DISORDERS.

DEAH-box proteins A family of RNA helicases, within helicase superfamily II, that have the conserved DEAH-motif (Asp-Glu-Ala-His). They appear to be particularly involved in mRNA splicing. Human homologues of the *S. cerevisiae* proteins have been identified. *See* DEAD-BOX.

deamination The removal of an amine group from a molecule, an important first step in the breakdown of amino acids, which occurs mostly in the liver. Deamination of nucleic acids is the spontaneous loss of the amino groups of cytosine to produce uracil, methyl cytosine to produce thymine, or adenine to produce hypoxanthine.

DEAS *See* DEHYDROEPIANDROSTERONE.

death Simplistically, the complete and permanent cessation of life, although defining the exact point at which it occurs is difficult in complex multicellular organisms. The **death domain** is a conserved domain of approximately 80 aa, found in the cytoplasmic tail of death receptors and essential for generating signals that lead to apoptosis. **Death receptors** are a superfamily of receptors including tumour necrosis factor receptor-1, CD95, and TRAMP, that trigger apoptotic cell death through interaction with *caspases such as FLICE, via adapter proteins such as *FADD and *TRADD. Death receptor 3 (DR3) is more comonly duplicated in individuals with rheumatoid arthritis. The **death-effector domain** (DED) is a region at the C-terminus of FADD and the N-terminus of FLICE. Interaction of these regions leads to the assembly of the **death-inducing signalling complex** (DISC) which activates other caspases.

death-associated protein A basic, proline-rich protein (DAP-1, 102 aa) thought to be a positive mediator of programmed cell death induced by interferon-γ. **Death-associated protein-3** (DAP3, 398 aa) is a component of the 28S mitochondrial ribosome subunit. **Death-associated protein 6** (Fas-binding protein, Daxx, 760 aa) is an interferon-induced protein that binds to the Fas death domain and enhances Fas-mediated apoptosis and is a component of nuclear promyelocytic leukaemia protein (PML) oncogenic domains (PODS). Acts downstream of apoptosis signal-regulating kinase (*ASK1). **Death-associated protein kinase-1** (DAP-kinase, EC 2.7.11.1, 1430 aa) is one of a proapoptotic family of multidomain calcium/calmodulin (CaM)-dependent Ser/Thr *protein kinases that are phosphorylated upon activation of the ras-extracellular signal-regulated kinase (ERK) pathway. Loss of DAP kinase expression in sporadic pituitary tumours is associated with invasive behaviour. **DAP-Kinase-3** is identical to zip kinase. **DAP-related apoptotic kinase-2** (DRAK2, 372 aa) is thought to be a positive regulator of apoptosis in thymocyte selection and lymphoid maturation.

debridement Cleaning a wound by removal of necrotic, infected, or foreign material.

Debye–Hückel limiting law A method for arriving at the activity of a molecule in solution, taking into account the influence of the ions in the solution and making some assumptions about behaviour of the solute. In general terms the activity is proportional to the concentration multiplied by a factor (the activity coefficient). The law holds for dilute solutions of known ionic strength and is therefore appropriate for many biological fluids.

dec 1. Regulatory proteins of the basic helix–loop–helix transcription-factor superfamily; **DEC1** is BHLH-B2 (412 aa); **DEC2** (482 aa) is BHLH-B3. DEC1 and DEC2 are regulators of the mammalian molecular clock and show striking circadian oscillation in the suprachiasmatic nucleus. DEC2 inhibits transcription from the Per1 (*See* PERIOD) promoter induced by Clock/Bmal1. **2.** Deleted in esophageal cancer 1 (70 aa), a candidate tumour suppressor.

***decapentaplegic* (*dpp*)** A *Drosophila* gene involved in pattern formation in imaginal discs (of which there are fifteen, hence decapenta-) the product of which is a growth factor homologous to mammalian *bone morphogenetic protein (BMP). *See* MAD-RELATED PROTEINS.

SEE WEB LINKS

- The Flybase entry for this gene.

decapping Removal of the mRNA 5′ cap structure, a key step in mRNA turnover, is catalysed by the Dcp1–Dcp2 complex (the decapping complex). Decay is initiated by poly-A shortening, and oligoadenylated mRNAs (but not polyadenylated mRNAs) are selectively decapped allowing their further degradation by the 5′ to 3′ exonuclease XRN1. The decapping activator complex is made up of the highly conserved heptameric Lsm1p–7p complex (made up of the seven *Like Sm proteins, Lsm1p–Lsm7p) and their interacting partner, Pat1p. It activates decapping by an unknown mechanism and localizes with other decapping factors to the P-bodies in the cytoplasm. *See* GW BODIES.

decay accelerating factor A plasma protein (DAF, 381 aa) that regulates the *complement cascade by blocking the formation of the C3bBb complex (the C3 convertase of the alternate pathway). A glycolipid-anchored form of DAF is expressed on the surface of cells that are in contact with plasma complement proteins and it is a deficiency in the addition of the GPI anchor in somatically mutated stem cells that leads to *paroxysmal nocturnal haemoglobinuria.

decerebrate Describing an animal lacking a cerebrum or in which the cerebrum has been destroyed. **Decerebrate posture** is an abnormal body posture with rigid extension of the arms and legs, downward pointing of the toes, and backward arching of the head, usually as a result of severe injury to the brain.

decibel A dimensionless logarithmic unit of measurement for the magnitude of a physical quantity (usually power or intensity) relative to a reference level. It is most commonly encountered in the context of acoustics where it is used to quantify sound levels relative to the 0-dB reference level, typically the threshold of auditory perception. The association with acoustics is historically appropriate since the bel was the unit used by American Bell Telephone Company laboratory engineers to quantify the reduction in audio level over 1 mile of standard telephone cable.

decidua *See* ENDOMETRIUM.

decorin A small *proteoglycan (proteoglycan II) similar to *biglycan, that 'decorates' collagen fibres in many connective tissues. The core protein is ~359 aa and there is one *glycosaminoglycan chain, either chondroitin sulphate or dermatan sulphate and *N*-linked oligosaccharides. A mutation in decorin has been reported in a family with congenital stromal corneal dystrophy.

decorticated Describing an organ from which the outer layer (cortex), or sometimes the capsule, has been removed.

decoy receptor In general, a receptor that competes for binding of a ligand but does not induce a response, thereby diluting the signal's effect. More specifically, **decoy receptors DcR1** (259 aa) and **DcR2** (386 aa) compete with death receptor 4 (DR4) or DR5 for binding to the ligand (*TRAIL); **decoy receptor 3** (DcR3, 300 aa) and *osteoprotegerin are soluble receptors for *Fas ligand and apparently act as decoy receptors. Other soluble receptors, which bind ligand but cannot transduce a signal, are being identified and this may be a widespread mechanism for damping signalling systems.

dectin-1 A small type II membrane receptor (dendritic cell-associated molecule, 247 aa) with an extracellular C-type lectin-like domain fold and a cytoplasmic domain with an immunoreceptor tyrosine-based activation motif (*ITAM). Originally reported as a dendritic cell-specific β-glucan receptor, it was subsequently demostrated on all macrophage classes.

decubitus ulcer An ulcer or bedsore that develops because of restricted blood flow in areas of the body compressed for long periods.

decussation A point at which two structures cross, used particularly of the crossing-over of nerve tracts in the central nervous system.

DEDD superfamily A superfamily of RNases and DNases with four conserved acidic residues, DEDD, responsible for binding two metal ions involved in catalysis. The DEDD superfamily can be divided into two subgroups, DEDDY and DEDDH, according to whether a fifth conserved residue is tyrosine (Y) or histidine (H).

dedifferentiation The loss of differentiated characteristics and the restoration of an undifferentiated and implicitly multipotential state. Dedifferentiation is common in plants but remains controversial in animals, even during regeneration, e.g. in the tail of amphibians. Loss of differentiated characteristics may not be accompanied by loss of *determination and many tissues contain a reserve of uncommitted cells for repair.

deep vein thrombosis *See* THROMBOSIS.

deer-fly fever *See* TULARAEMIA.

defective virus A virus that is unable to replicate unless it coinfects a host cell together with a wild-type 'helper' virus. Most acute transforming *retroviruses are defective, possibly because the oncogene has displaced essential viral genetic information.

defensins Small (30–35 aa) cysteine-rich cationic proteins, often membranolytic pore-formers, active against bacteria, fungi and enveloped viruses. They are very variable but can be divided into two main subfamilies, α- and β-defensins. α-Defensins are found primarily in neutrophils, macrophages, and Paneth cells, β-defensins are found in a wider range of tissues including some epithelial cells. Defensins may constitute up to 5% of the total protein. Insect defensins have some sequence homology with the vertebrate forms. *See* HEPCIDIN ANTIMICROBIAL PEPTIDE.

SEE WEB LINKS

- Homepage for the defensins database.

defined medium A medium for cell culture in which the components are fully characterized. This requires the substitution of known protein components such as insulin, transferrin, and specific growth factors, in place of fetal calf serum.

definitive erythroblast Erythrocyte-producing cells found in fetal liver and bone marrow that take over the role of haematopoiesis from primitive erythroblasts of the yolk-sac at ~6 weeks of development. They are smaller than primitive erythroblasts, have a lower cytoplasmic/nuclear ratio and are $CD36^+$. As they mature they extrude their nuclei.

degeneracy A casual term for the mismatch between the number of different amino acids that could potentially be specified by a three-letter code (64) and the limited number that are actually coded (20). Three nucleotide triplets are used as stop signals, but several amino acids are specified by several different codons, e.g. six in the case of leucine.

degenerate PCR Describing a *polymerase chain reaction (PCR) in which the primers are not specific but are deliberately mixed (degenerate primers—*see* DEGENERACY) to amplify a sequence that has been deduced from the peptide sequence (so the exact codon usage is unknown) or to isolate homologous genes. A similar approach, using deliberately degenerate sequences, is also used in screening *genomic libraries.

degenerins Products of the *deg-1*, *mec-4* and *mec-10* genes in *Caenorhabditis elegans* which have homology with *amiloride-sensitive sodium channels. Mutations cause neuronal degeneration, probably by disrupting ion fluxes. A related protein, the product of *unc-105*, interacts with collagen and may be a stretch-activated channel. Mammalian homologues (BNAC1 and -2) are amiloride-sensitive brain-specific sodium channels similar to, but distinct from, the epithelial amiloride-sensitive channels.

degradosome A multienzyme complex (RNA degradosome) in *E. coli* that degrades mRNA, thus regulating protein production. The complex contains exoribonuclease, polynucleotide phosphorylase (PNPase), endoribonuclease E (RNAase E), enolase, and Rh1B (a member of the DEAD-box family of ATP-dependent RNA helicases).

degranulation Secretion by fusion of membrane-bounded granules with the plasma membrane.

degrees of freedom An important parameter in statistics, the number of values in the final calculation of a statistic that are free to vary. Often one less that the number of measurements, but advice from a professional statistician is usually a wise precaution.

dehydration **1.** Excessive loss of water from the tissues of the body. **2.** Removal of water as in preparing a fixed tissue specimen for embedding or a histological section for clearing and mounting. Usually done using a series of ethanols of increasing purity.

dehydroepiandrosterone (DHEA, DEA) A precursor of steroid hormones produced from cholesterol by the adrenal glands, gonads, adipose tissue, brain, and in the skin. It is in turn a precursor of androstenedione from which testosterone and oestrogens are produced. DHEA is reversibly converted to the sulphated form (dehydroepiandrosterone sulphate, DHEAS) by sulphotransferase (EC 2.8.2.2) which is predominant in blood. Considered to play an important immunomodulatory role; the decline in DHEA levels with age correlates with reduced immune competence. Has been shown to have tumour suppressive and antiproliferative effects in rodent tumours and is a potent *sigma-1 receptor agonist.

dehydrogenases Broad class of enzymes (oxidoreductases, EC 1.-) that oxidize a substrate by transferring hydrogen to an acceptor that is usually either $NAD^+/NADP^+$ or a flavin enzyme, although other acceptors may be used. Most biological oxidations are of this type. The term oxidase is restricted to enzymes in which the oxidation reaction has molecular oxygen (O_2) as the electron acceptor.

d

Deiter's cells *See* PHALANGEAL CELLS. Deiter's neurons are large multipolar nerve cells in the lateral vestibular nucleus.

Dejerine–Sottas neuropathy A severe degenerating early-onset neuropathy (Charcot–Marie–Tooth disease type 4F) with both autosomal recessive and dominant forms. It can be caused by mutations in the *MPZ* gene (myelin protein zero, peripheral myelin protein), the *PMP22* gene (peripheral myelin protein 22), the *periaxin gene, and the early growth response gene, *EGR2*. Mutations in the *connexin 32 gene may contribute to the phenotype.
See CHARCOT–MARIE–TOOTH DISEASE.

delayed rectifier channels Neuronal ion channels that change their conductance for potassium ions with a brief delay following a voltage step. They control action potential repolarization, interspike membrane potential, and action potential frequency in excitable cells and are also important in stimulus–secretion coupling. A-type potassium channels can be converted into delayed rectifier-type channels by alterations in membrane lipid, and arachidonic acid and *anandamide will change delayed rectifier channels into rapid voltage-dependent inactivation-type channels.

SEE WEB LINKS

- Additional detail in 'Functional conversion between A-type and delayed rectifier K^+ channels by membrane lipids' in *Science* (2004).

delayed-type hypersensitivity *See* HYPERSENSITIVITY.

deleted in colorectal cancer The *DCC* gene, deleted in 70% of colorectal cancers, encodes a protein of the immunoglobulin superfamily with considerable homology to neural cell adhesion molecule-14. May be involved in regulation of cell–cell or cell–substratum interactions with a potential functional role in the control of cell growth, cell differentiation, and development of metastases. It is one of a family of receptors that includes *Drosophila* Frazzled which binds *netrin in association with one of the *UNC-5 family of receptors.

deleted in malignant brain tumours-1 A glycoprotein (DMBT1, Gp-340, surfactant pulmonary-associated D-binding protein, hensin, salivary agglutinin, 2413 aa) that is a candidate tumour suppressor gene and may be important in mucosal defence with a broad calcium-dependent binding spectrum against both Gram-positive and Gram-negative bacteria. It is found in both secreted and membrane-associated forms. CRP-ductin is the mouse homologue.

deletion mutation A mutation caused by the loss of one or more nucleotides. If the number lost is not a multiple of three and is in a coding region, then the effect may be a *frame shift.

Delhi boil Cutaneous *leishmaniasis (oriental sore, tropical sore, Baghdad boil).

delirium An acute disturbance of mental processes that may involve delusions, hallucinations, incoherence, etc. Can be the result of fever or intoxication. **Delirium tremens** (the 'DTs') is characteristic of chronic alcoholism with insomnia, restlessness, hallucinations, and spatial and temporal disorientation.

delta chains (δ-chains) *See* IMMUNOGLOBULIN.

delta-like The homologues (DLL1, 3, 4) of the *Drosophila* protein delta which is a ligand for *notch. DLLs, like *jagged, are processed by a *presenilin-dependent intramembranous γ-secretase to generate soluble intracellular derivatives. Mutations in **DLL3** are associated with autosomal recessive spondylocostal dysostosis. **DLL4**-mediated notch signalling is largely restricted to the vascular compartment.

delta sleep-inducing peptide A nonapeptide (DSIP) found in neurons, peripheral organs, and plasma that will induce mainly delta-type sleep in mammals. DSIP has effects in pain, adaptation to stress, and epilepsy, and also anti-ischaemic effects. **Delta sleep-inducing peptide immunoreactive peptide** (DIP; TSC22 domain family, member 3, 77 aa) has close homology to the product of the *Drosophila* gene *shortsighted* and was originally isolated from porcine brain (pDIP) using polyclonal antibodies against the delta sleep-inducing peptide (DSIP). There is, however, no sequence homology. DIP has a *leucine zipper domain but no DNA-binding domain and has a role in the anti-inflammatory and immunosuppressive activity of steroids and IL-10.

delta virus Hepatitis D virus. A defective virus requiring coinfection with hepatitis B virus for replication. The agent consists of a particle 35 nm in diameter consisting of the delta antigen surrounded by an outer coat of HBsAg. The genome is very small and consists of a single-stranded RNA. Found endemically throughout the world. *See* HEPATITIS VIRUSES.

delta waves Waves of high amplitude and low frequency (1–3 Hz) that can be detected in electroencephalographic recordings and that are characteristic of deep sleep. Delta-type sleep can be induced by *delta sleep-inducing peptide.

deltoid Describing any structure that has the approximate shape of an equilateral triangle, e.g. the deltoid muscle of the shoulder.

dematin An actin-bundling protein (band 4.9) originally found in the human erythroid membrane skeleton. There are two palmitoylated polypeptide chains of 405 aa and 518 aa with *SH3 domains. In solution, dematin exists as a trimer and the actin-bundling activity is abolished if it is phosphorylated by PKA.

demeclocycline A tetracycline antibiotic from *Streptomyces aureofaciens. TNs* Declomycin, Declostatin, Ledermycin.

dementia Degeneration of various central nervous system functions, including motor reactions, memory, learning capacity, problem solving, etc. Although age-related decline is normal, some dementia syndromes result from pathological organic deterioration of the brain. **Dementia praecox** is an obsolete term for schizophrenia.

demulcent Describing a substance that is soothing or that will allay irritation.

demyelinating diseases Diseases in which the myelin sheath of nerves is destroyed or deficient. In some diseases there is an autoimmune component. Examples are *Charcot–Marie–Tooth disease type 1, *multiple sclerosis, acute disseminated encephalomyelitis (a complication of acute viral infection), *experimental allergic encephalomyelitis, *Guillain–Barré syndrome, *encephalitis periaxialis, *subacute sclerosing panencephalitis.

denaturation Loss of higher order (secondary, tertiary, or quaternary) structure of large macromolecules as a result of exposure to nonphysiological conditions (heat, pH, salt concentration, solvents, etc.). In some cases the denaturation is reversible and function can be restored.

dendr-, dendro- Prefixes meaning tree-like or branched; from the *Gk.* dendron, a tree.

dendrite (dendron) A branched outgrowth from a *neuron with which *axons make synapses. The simplistic concept is that neurons have multiple short dendrites (collecting input) and a single long axon (delivering output) but the existence of synapses between axons (axon–axon) and dendrites (dendrite–dendrite) makes the definition rather imprecise. **Dendritic spines** are microfilament-rich protrusions on the dendrites of interneurons, especially in the cortical regions, where excitatory afferent neurons form synapses. Usually described as being wineglass- or mushroom-shaped, they contain postsynaptic densities (PSD). The so-called **dendritic tree** is the tree-like pattern of outgrowths of dendrites. A **dendritic ulcer** is a branching corneal ulcer caused by herpes virus.

dendritic cells **1.** In general, cells with a branched morphology. Many are involved in antigen presentation although some are named simply on the basis of morphology. **2.** Dendritic cells found in the red and white pulp of the spleen and lymph node cortex are positive for class II MHC antigens and are involved with stimulating T-cell proliferation. **3.** Follicular dendritic cells are found in germinal centres of spleen and lymph nodes and retain antigen for long periods. **4.** Dendritic epidermal cells (DECs) are T lymphocytes found in epidermis and express predominantly $\gamma\delta$-T-cell receptors. **5.** Immature dendritic cells (Langerhans cells) derived from bone marrow are also found in the basal layers of the epidermis and are distinct from dendritic epidermal cells. They are strongly MHC class II positive and are responsible for *antigen processing and *antigen presentation. Having been exposed to antigen they migrate to the lymph nodes. **6.** Myeloid dendritic cells (mDC), of which there are two subsets, are most similar to monocytes. mDC-1 stimulate T cells, mDC2 are rare and may have a role in combatting wound infection. Langerhans cells are probably a subset of mDCs. **7.** Plasmacytoid dendritic cells (pDC) resemble plasma cells but are able to produce high amounts of interferon-α and were originally known as interferon-producing cells (IPC). It is claimed that they can be converted into myeloid dendritic cells following viral infection. **8.** There are DOPA-positive dendritic cells (melanocytes) in the basal part of epidermis that are derived from neural crest and are distinct from dendritic epidermal cells (4, above).

dendro- *See* DENDR-.

***Dendroaspis* natriuretic peptide** A natriuretic peptide (DNP, 38 aa) that has structural similarities with atrial, brain and C-type *natriuretic peptides and that was isolated from the venom of the green mamba *Dendroaspis angusticeps*. Selective for atrial natriuretic peptide receptor A (natriuretic peptide receptor 1).

dendrogram A branching diagram, like a family tree, used to illustrate relationships between species.

dendron *See* DENDRITE.

dendrotoxins A family of small proteins (57–60 aa) isolated from various *Dendroaspis* (mamba) venoms that are selective blockers of *voltage-gated potassium channels in a variety of tissues and cell types. They have sequence similarity with Kunitz-type serine peptidase inhibitors. The kalicludine toxins from the sea anemone *Anemonia sulcata* are structurally homologous to the dendrotoxins.

d

denervation The removal of nerve supply to a tissue, usually by damaging axons.

dengue An acute fever caused by a flavivirus (one of the *arboviruses), transmitted by the mosquito *Aedes aegypti*. **Dengue shock syndrome** is a haemorrhagic fever, probably an immune complex hypersensitivity as a result of re-exposure to the virus.

denileukin diftitox An engineered protein that combines IL-2 with the enzymatic and translocation domains of *diphtheria toxin (DT). In effect the B subunit of DT has been replaced with IL-2 so that cells overexpressing the IL-2 receptor are targeted, the construct is internalized, processed, translocated, and the target cell killed. It has some promise as an antineoplastic agent against lymphomas. *TN* Ontak.

dense bodies Electron-dense areas found within smooth muscle cells that are anchorage points for the barbed ends of actin microfilaments (thin filaments). Some are associated with the plasma membrane, others are cytoplasmic and together with thin filaments and thick filaments constitute a contractile unit.

density-dependent inhibition of growth The cessation of cell division that occurs once cells in culture reach a critical density (topoinhibition). The number of cells per unit area at which this occurs is higher than the density at which a monolayer is formed (*confluent culture) indicating that this is not a contact-mediated phenomenon, unlike *contact inhibition of locomotion. Cells that are not *anchorage dependent may not exhibit density dependence.

density enhanced phosphatase-1 A transmembrane protein (Dep-1, CD148, protein-tyrosine phosphatase -η, HPTPε, EC 3.1.3.48, 1337 aa) with a single cytoplasmic tyrosine phosphatase domain. Dep-1 negatively regulates growth factor receptor signalling, may act as a tumour suppressor and may have a role in density dependent inhibition of growth.

density gradient A column of liquid in which the density increases with depth. Step gradients may be formed by layering or a continous gradient may be established by progressive mixing of solutions of different density. With *caesium chloride solutions a gradient can be established by centrifugation. Separation on a gradient may depend upon density differences between particles, in which case the positions differ at equilibrium, or differences in size, in which case the rate of sedimentation is the critical factor.

dental caries Decay and disintegration of tooth enamel and dentine due to acids produced by bacterial plaque. Plaque is an example of a biofilm, and bacterial behaviour may be quite different in such an environment.

dental formula A convention used to describe the dentition of a mammal. For humans the adult formula is I 2/2, C 1/1, P 2/2 M 3/3 = 16 × 2 = 32 where I is incisor, C is canine, P is premolar, and M is molar; the first number is for the upper jaw, the second for the lower (the same, in the case of humans) and the value (16) is for half of the mouth. For the deciduous dentition of a child the formula is I 2/2, C 1/1, P 0/0, M 2/2 = 10 × 2 = 20.

dentate nucleus The largest of the four deep cerebellar nuclei, a region where the cell bodies of neurons involved in the planning, initiation, and control of volitional movements are clustered. The other nuclei are the emboliform, globose, and fastigial nuclei. These deep cerebellar nuclei are in the centre of the cerebellum, embedded in the white matter, and are the origin of most of the output fibres of the cerebellum. They receive inhibitory inputs from Purkinje cells in the cerebellar cortex and excitatory inputs from mossy fibre and climbing fibre pathways.

dentatorubral–pallidoluysian atrophy A syndrome (Haw river syndrome, Naito–Oyanagi disease) of myoclonic epilepsy, dementia, ataxia, and *choreoathetosis as a result of degeneration of the dentatorubral and pallidoluysian systems caused by an expanded trinucleotide repeat (CAG repeat) in the gene encoding *atrophin-1. As with other trinucleotide repeat disorders, the longer the expansion the earlier the onset and the more severe the effects.

dentin (dentine) The major component of teeth, a structural biocomposite of needle-like crystals of hydroxyapatite embedded in a fibrous collagen matrix, very similar to compact bone but secreted by *odontoblasts. **Dentin matrix protein** (DMP-1, 513 aa) is a serine-rich acidic matrix protein that has a key role in mineralization. Oligomers of DMP-1 temporarily stabilize calcium phosphate nanoparticle precursors by sequestering them and preventing their further aggregation and precipitation. It binds to RGD sequence-dependent *integrins. Dentin sialophosphoprotein (DSPP, 1301 aa) is cleaved to produce two major acidic matrix proteins of dentin, dentin sialoprotein (DSP, 447 aa) and dentin phosphoprotein (DPP, phosphophoryn, 839 aa). Mutations in *DSPP* are a causative factor in *dentinogenesis imperfecta.

dentinogenesis imperfecta A disorder that is distinct from *osteogenesis imperfecta, being restricted to teeth. Type 1 is caused by mutation in the *DSPP* gene that encodes *dentin phosphoprotein and dentin sialoprotein.

Denys–Drash syndrome A syndrome in which there is pseudohermaphroditism, *Wilms' tumour, hypertension, and degenerative renal disease. Caused by mutation in the *WT1* gene that encodes a zinc finger DNA-binding protein that can be a transcriptional activator or repressor depending on context, and is required for normal formation of the genitourinary system. Meacham's syndrome and Frasier's syndrome are allelic disorders with similar clinical features.

deoxycholate A secondary bile acid produced from the salts of glycocholic and taurocholic acid by the action of bacterial enzymes in the intestine. The sodium salt is used as a biological detergent to lyse cells and solubilize cellular and membrane components.

deoxyglucose A glucose analogue that is taken up by cells but cannot be metabolized. The hydroxyl on C-2 of glucose is replaced by a hydrogen.

deoxyhaemoglobin Haemoglobin which does not have bound oxygen.

deoxynojirimycin A glucose analogue, an imino sugar in which the ring oxygen is replaced with a nitrogen atom. Deoxynojirimycin and nojirimycin act as a saccharide decoys and inhibit the ceramide glucosyltransferase (EC 2.4.1.80) involved in the first step of glucosphingolipid synthesis and thus the production of cell surface glycoproteins. Occurs naturally in mulberry plants, *Streptomyces*, and *Bacillus* sp.

deoxyribonuclease (DNAase, DNase) An *endonuclease that hydrolyses phosphodiester bonds of DNA. The **DNase type I enzymes** (EC 3.1.21.-) cleave adjacent to a pyrimidine nucleotide, yielding 5′-phosphate terminated polynucleotides with a free hydroxyl group on position 3′, whereas **type II DNase** (EC 3.1.22.1, formerly EC 3.1.4.6) is a lysosomal enzyme that yields products with 3′-phosphates, operates better at acid pH, and will degrade both native and denatured DNA. Type I enzymes act on single-stranded DNA, double-stranded DNA, and chromatin, although the sensitivity of chromatin to digestion by DNase I depends on its state of organization, transcriptionally active genes being much more sensitive than inactive genes. **DNase γ** is involved in apoptotic cleavage of DNA (*See* DNA LADDERING). DNase-1 (EC 3.1.21.1) binds to G actin and to the pointed end of actin filaments with high affinity; this has been experimentally useful but the physiological relevance of the actin binding is unclear.

deoxyribonucleic acid *See* DNA.

deoxyribonucleoside A purine or pyrimidine base linked to 2-deoxy-D-ribofuranose by a *N*-glycosidic bond. Phosphate esters of nucleosides are nucleotides.

deoxyribonucleotide A phosphate ester of a *deoxyribonucleoside.

deoxyribose A deoxy sugar derived from the pentose sugar ribose by the replacement of the hydroxyl group at the 2 position with hydrogen. Linked by 3′-5′ phosphodiester bonds, it forms the backbone of DNA.

Dep-1 *See* DENSITY-ENHANCED PHOSPHATASE-1.

DEP domain A globular domain (disheveled, EGL-10, pleckstrin homology domain, ~80 aa) of unknown function. Found in both eukaryotic and prokaryotic proteins. Mammalian regulators of G-protein signalling contain these domains and the domain may be important for binding interactions with the C-terminal tail of a G-protein-coupled receptor.

dephosphorylation The enzymatic removal of a phosphate group by a phosphatase.

depolarization An alteration in the resting potential of a cell towards electroneutrality. Most excitable cells have a resting potential of around −70 mV and an influx of sodium ions will depolarize the cell to around −50 mV before stimulating an *action potential. Hyperpolarization is the reverse.

depolarizing muscle relaxant A drug that binds to the acetylcholine receptors at the neuromuscular junction and activates them, thereby causing depolarization. Acetylcholine would have the same effect if present at very high (nonphysiological) concentrations. Suxamethonium, the only commonly used drug of this type, will produce a brief period of paralysis and relaxation.

depressant A drug that reduces functional activity. Anaesthetics are depressants of the nervous system.

depressor A muscle that lowers a part or organ.

d

depsipeptides Natural or synthetic compounds composed of amino- and hydroxy-carboxylic acid residues (usually α-amino and α-hydroxy acids). In many cases, although not all, the two types alternate and in cyclodepsipeptides, the residues are connected in a ring. Cyclic depsipeptides are common metabolic products of microorganisms, often with antibiotic activity (e.g. *actinomycin; enniatins; *valinomycin).

depurination The spontaneous loss of purine residues from DNA, which occurs at an appreciable rate *in vivo* (estimated to be ~10 000 per cell per day). Although depurination does not disrupt the structural integrity of the phosphodiester backbone of DNA, constant repair by the base excision repair pathways is required to avoid irreversible loss of information.

derepression Activation by removal or inactivation of a repressor, a common feature in gene regulation.

derivatization The alteration of a molecule in order to change its properties (e.g. solubility) to facilitate analysis, or to enable it to be followed through a process, e.g. by addition of a fluorescent moiety. The molecule is said to have been derivatized.

dermal Relating to or associated with the *dermis.

dermamyotome The embryonic region that gives rise to the *dermis and to the axial musculature.

dermaseptins A large and growing family of antibacterial peptides (27–34 aa) isolated from the skin of the South American frogs of the genus *Phyllomedusa*. They have broad-spectrum antibacterial activity, and some have antifungal and antiprotozooal activities. Structurally they are polycationic (Lys-rich), alpha-helical, and amphipathic which may allow them to interact with membrane bilayers, causing cytolysis, although they do not lyse mammalian erythrocytes. **Dermaseptin S9** acts on both Gram-positive and Gram-negative bacteria. **Dermaseptin O1** from *Phyllomedusa oreades* is also active against *Trypanosoma cruzi*.

SEE WEB LINKS

- More information in Cope with Cytokines online encyclopaedia.

dermatan sulphate The *glycosaminoglycan composed of repeating units of D-glucuronic acid-*N*-acetyl-D-galactosamine or L-iduronic acid-*N*-acetyl-D-galactosamine that is found in the *extracellular matrix of skin, blood vessels, and heart. There are one or two sulphates per unit and the overall size varies from 15 to 40 kDa. Accumulates in *Hurler's syndrome and *Hunter's syndrome where there is a defect in L-iduronidase.

dermatitis Inflammation of the *dermis which can be a consequence of *contact sensitivity. **Dermatitis herpetiformis** (Duhring's disease) is a chronic, very itchy rash, with vesicles and papules, associated with coeliac disease. There is also IgA deposition in the dermis. The **familial form**, an autosomal dominant, is associated with HLA DQA1 and B1.

dermatofibrosarcoma protuberans A rare infiltrative skin tumour of intermediate malignancy in which there is unregulated PDGF production as a result of a chromosomal translocation that generates a collagen-1/PDGF-B fusion protein that is processed to generate mitogenically-active PDGF-B.

dermatographia (dermographia) A form of *urticaria in which light pressure on the skin will produce a reddish weal and it is possible to write on the skin with a blunt point.

dermatological disorders

The skin is a complex tissue that serves as a barrier to fluid loss and to the entry of pathogenic organisms; it must be flexible but nevertheless withstand mechanical stress and abrasion. The outermost layer (epidermis) is a stratified *squamous epithelium; daughter cells from the stem cells (*basal cells) become progressively filled with *keratin and are eventually shed as *squames. Dysfunction in the keratinization process (*cornification) can cause *palmoplantar keratoderma, *Naxos disease, and *Vohwinkel's syndrome whereas simple mechanical irritation can cause the formation of *calluses or *corns. Neoplasia of basal cells, the major cell type, can produce *basal cell carcinoma or *Gorlin's syndrome. Other cells of the epidermis are *melanocytes and cells involved in immune surveillance (*dendritic cells, *Langerhans cells, *skin associated lymphoid tissue). Underlying the epidermis is the vascularized connective tissue of the *dermis containing fibroblasts which are important in wound healing. Because they are visible, skin diseases are well recognized. Some have an inflammatory basis (e.g. atopic dermatitis; *cheiropompholyx; *eczema; *psoriasis); in others the defects are in cellular components such as *desmosomes that are important for mechanical resistance (*see* ECTODERMAL DYSPLASIA; EPIDERMOLYSIS BULLOSA; HAILEY–HAILEY DISEASE; PEMPHIGUS). *See also* ACANTHOMA; DERMATOFIBROSARCOMA PROTUBERANS; DERMATOGRAPHIA; DERMATOMYOSITIS; DERMATOPHYTES; FAMILIAL PRIMARY LOCALIZED CUTANEOUS AMYLOIDOSIS; HYPERKERATOSIS; ICHTHYOSIFORM ERYTHRODERMA; ICHTHYOSIS; KERATOACANTHOMA; LICHEN PLANUS; MERKEL CELL CARCINOMA; MOLLUSCUM CONTAGIOSUM VIRUS; MYCOSIS; PIGMENTATION; RINGWORM; ROTHMUND–THOMSON SYNDROME; TUBEROUS SCLEROSIS. The synthesis of *vitamin D in the skin is important.

dermatomyositis An inflammatory myopathy in which the muscle weakness is preceded by a skin rash. The cause is probably an autoimmune reaction and may be associated with diseases such as lupus.

Dermatophagoides pteronyssinus The house dust mite. The major allergen to which many asthmatics react is a cysteine peptidase, Der p 1. Related molecules Der p 2 and Der p 5 are also found in mite faeces and may contribute to the allergenicity.

dermatophytes Fungi that can infect the skin, hair, and nails e.g. ringworm and athlete's foot.

dermatopontin One of the tyrosine-rich acidic extracellular matrix (ECM) proteins (TRAMPs) found in many metazoa. Dermatopontin (201 aa) was first identified in skin extracts and because it binds both to cells and to dermatan sulphate proteoglycans, was named dermatopontin by analogy with *osteopontin. It is found in many tissues, interacts with other ECM components, especially *decorin, and is thought to regulate ECM formation and collagen fibrillogenesis. In *leiomyoma development there is frequently reduced expression of dermatopontin.

dermatosclerosis *See* SCLERODERMA.

dermicidin A naturally occurring fungicidal protein (110 aa) produced by eccrine sweat glands of the skin. Levels are apparently reduced in individuals with atopic dermatitis.

dermis The *connective tissue that lies below the epithelium of the skin. **Dermatitis** is inflammation of the dermis.

dermographia *See* DERMATOGRAPHIA.

dermoid cyst A cystic teratoma, usually benign, that contains developmentally mature skin with hair follicles and sweat glands. Many ovarian tumours are dermoid cysts. Often loosely referred to simply as dermoid.

DES *See* DIETHYLSTILBOESTROL.

DeSanctis–Cacchione syndrome An autosomal recessive disorder in which there is *xeroderma pigmentosum, often of complementation group A, together with mental deficiency, dwarfism, and gonadal hypoplasia. Has also been associated with mutation in the *ERCC6* gene, part of the nucleotide *excision repair (NER) pathway that is defective in Cockayne syndrome type B.

Descemet's membrane The basal lamina of the corneal endothelium that separates the cornea from the aqueous humour.

descending Describing a structure that runs from anterior to posterior, even in quadrupeds.

desensitization Generally, a decrease in responsiveness following repeated exposure to a stimulus. In immunology, a method of reducing or abolishing the effects of a known allergen by giving gradually increasing doses until tolerance develops.

desert hedgehog One of the *hedgehog family of conserved signalling molecules that coordinate various patterning processes during early embryonic development. Desert hedgehog (Dhh, 396 aa) is restricted, in the developing mouse, to Sertoli cells of developing testes and Schwann cells of peripheral nerves. The precursor molecule is autocatalytically cleaved and the C-terminal product covalently attaches a cholesterol moiety to the soluble N-terminal product that contains the signalling activity, restricting it to the cell surface. Defects in Dhh have been associated with partial gonadal dysgenesis (PGD) accompanied by minifascicular polyneuropathy.

desferrioxamine A naturally occurring trihydroxamic acid (desferoxamine, desferal, DFO, DFOA, 560 Da) that will chelate trivalent ions such as iron and aluminium; it is used to treat iron poisoning and occasionally aluminium toxicity. *Streptomyces coelicolor* and *Streptomyces pilosus* produce desferrioxamine siderophores and the latter is used for the production of desferrioxamine B which is used clinically.

desmin A muscle-specific type III intermediate filament protein (470 aa) that appears to mediate attachments between the terminal Z disc and the junctional membrane, as well as linking Z discs in series in each sarcomere. Colocalizes with *synemin, paranemin, and *plectin in the appropriate cell types. Defects lead to various forms of *myopathy.

desmocollins Desmosomal glycoproteins (DSCs, ~900 aa) of the cadherin superfamily of calcium-dependent cell adhesion molecules. In the human three types of desmocollin have been identified, some with alternatively-spliced variants. **DSC2** mutations are a cause of arrhythmogenic right ventricular dysplasia-11 (ARVD-11) and normal levels of DSC2 are crucial for normal cardiac desmosome formation, early cardiac morphogenesis, and cardiac function. *See* DESMOGLEINS.

desmogleins Transmembrane glycoproteins of the *cadherin family that, together with *desmocollins, form the desmosome. The major proteins of the urea-insoluble core are desmoglein-1 (DG 1, DSG-1, 1049 aa) and the related proteins desmogleins-2 and -3. In *pemphigus foliaceus DSG-1 is the autoantigen; DSG3 is the antigen target in pemphigus vulgaris.

desmoid tumour A histologically benign fibrous neoplasm tumour of connective tissue (aggressive fibromatosis, grade I fibrosarcoma) caused, in some familial cases, by mutation in the adenomatous *polyposis coli (*APC*) gene and in sporadic cases somatic mutation in the β-*catenin gene has been reported. Although nonmalignant they are locally aggressive and have a tendency to recur.

desmoplakins Proteins isolated from the urea-soluble plaque region of desmosomes. **Desmoplakin 1** (DSP1, 2871 aa) is found in all desmosomes; **DSP-2** is a smaller splice variant found in stratified epithelia. They are long, flexible, rod-like molecules ~100 nm long with two polypeptide chains in parallel. **Desmoplakin 3** (*plakoglobin) is smaller (745 aa). Autoantibodies against the desmoplakins are important in paraneoplastic pemphigus, a disorder associated with lymphoid malignancies, thymomas, and poorly differentiated sarcomas.

desmoplasia Excessive growth of fibrous or connective tissue, e.g. around some tumours or in the formation of adhesions (scar tissue) within the abdomen after abdominal surgery.

desmopressin A synthetic analogue (deamino-D-arginine-vasopressin, DDAVP, *TN* Minirin) of antidiuretic hormone *vasopressin that is often used in treating bedwetting and to supplement vasopressin which is produced in insufficient quantities in *diabetes insipidus. It has also been used in the treatment of type I *von Willebrand's disease, mild factor VIII deficiency, and intrinsic platelet function defects.

desmosine An intramolecular crosslinking amino acid found in elastin (very similar to isodesmosine) formed from four side chains of lysine. Urinary desmosine and isodesmosine may be a marker of lung damage in *chronic obstructive pulmonary disease.

desmosome (macula adherens) The characteristic specialized adhesion structure of epithelia (spot desmosomes), particularly those that are subject to mechanical stress, that links the intermediate filaments (tonofilaments) of adjacent cells. The gap between plasma membranes is of the order of 25–30 nm and is bridged by interacting adhesion molecules of the cadherin family, *desmogleins, and *desmocollins. On the cytoplasmic face *desmoplakins link the transmembrane cadherins to the cytokeratin tonofilaments. Various disorders of desmosomes are known: autoimmune diseases that involve desmosome components (e.g. *pemphigus vulgaris and pemphigus foliaceus), congenital diseases that affect intracellular calcium channels (e.g. *Hailey–Hailey disease and *Darier's disease) and congenital defects in desmosomal structural components. Defects in desmoplakin or desmoglein are responsible for autosomal dominant striate palmoplantar keratoderma, plakoglobin is defective in *Naxos disease, and *plakophilin defects are associated with autosomal recessive skin fragility–*ectodermal dysplasia syndrome.

desmoyokin A desmosomal plaque protein (5890 aa) from bovine muzzle *keratinocytes that is thought to be dumbell-shaped and ~170 nm in length. The desmoyokin gene is identical to the human **AHNAK* gene, which is expressed ubiquitously but usually considered a nuclear protein, although also associated with membranes of *enlargeosomes. Desmoyokin/Ahnak null mutant mice do not show any marked phenotype. The disparities continue to be puzzling.

desogestrel A synthetic progestogen used as an oral contraceptive, sometimes in combination with oestradiol.

desorption The release of a substance from or through a surface: the greater the desorption, the less the substance will be retained, e.g. on a *chromatography column. Thermal desorption is a process often used to remove, although not destroy, organic contaminants from soil.

desoxy- *See* DEOXY-.

desquamation The release of the outer layer of dead cells (squames) from the surface of skin or any stratified squamous epithelium.

destrin *See* ACTIN DEPOLYMERIZING FACTOR.

desynapsis Describing the separation of paired homologous chromosomes that takes place at the *diplotene stage of meiotic prophase I.

detergents Surface-active molecules that will interact with water through a hydrophilic polar region and with hydrophobic molecules through a nonpolar domain. Their amphipathic nature is important in solubilizing nonpolar molecules such as lipids. Examples include sodium dodecyl sulphate, fatty acid salts, the Triton family, octyl glycoside.

determinate cleavage The pattern of the earliest cleavage that is typical of molluscs and other mosaic embryos in which each blastomere will give rise to specific tissues, in contrast to the situation in so-called regulating embryos such as those of mammals where the early blastomeres remain competent to produce a complete embryo (although this capacity is lost after the first few divisions).

determination A phenotypically invisible but generally irreversible commitment of a cell to differentiate in a particular fashion.

detoxification reactions Reactions that inactivate toxins. These generally occur in the liver, where diverse cytochrome P450 enzymes play an important part in oxidizing toxins (or other xenobiotics) as a first step in their inactivation and removal. Oxidized compounds are then conjugated to polar compounds by transferases (e.g. glutathione *S*-transferase) and then pumped out, sometimes after further modification. Not all toxins can be handled in this way and a range of other methods for inactivating toxins are known.

detrusor Although in principle anything that presses down, common usage implies the smooth muscle that surrounds the bladder (detrusor urinae), contraction of which is important for releasing urine.

deuteranopia Defect of colour vision (daltonism, green *colour blindness) in which red, yellow, and green are confused. The problem arises because of mutations in the medium-wavelength (530 nm) sensitive *opsin, the green cone pigment, for which there are several X-linked genes in tandem array, the number varying between individuals with normal colour vision although probably only one is expressed.

deuterium oxide (D_2O) *See* HEAVY WATER.

developmental disorders

The process of morphological development, involving *differentiation of specialized cells and tissues, complex morphogenetic movements (*see* PLANAR CELL POLARITY PATHWAY) and interactions between tissues at specific stages (*see* EPITHELIAL MESENCHYMAL TRANSITION) is fraught with hazards and many developmental disorders are known, even if the molecular basis is not always clear. Major defects in basic systems are probably lethal at an early stage and the defects that are consistent with survival are relatively restricted in their effects, even if they are seriously disabling. Many involve *homeotic genes (*see* PAX GENES; SOX; T-BOX GENES; TWIST), others affect signalling pathways or growth factors. Homeotic genes are affected in *brachydactyly, *campomelic dysplasia, *Currarino's syndrome, *holoprosencephaly, *Mowat–Wilson syndrome, *Rieger's syndrome, and probably many others. *Amelia is caused by defects in the *wnt signalling pathway (*See* DAN). *Growth factor or growth factor-receptor deficiencies are responsible for various growth abnormalities, e.g. *achondroplasia, *Apert's syndrome, (*see* ACTIVIN; BONE MORPHOGENETIC PROTEINS). Faulty *genomic imprinting is associated with the developmental defects in *Beckwith–Wiedemann syndrome. Many developmental disorders are referred to as dysplasias (*see* CHONDRODYSPLASIA) and many of the *bone disorders have a developmental basis. Other developmental disorders include *Aarskog–Scott syndrome, *atelosteogenesis, *branchiooculofacial syndrome, *Char's syndrome, *CHARGE syndrome, *choanal atresia, *cleft palate, *congenital cataracts facial dysmorphism, and neuropathy syndrome, *cri-du-chat syndrome, *craniolenticulosutural dysplasia, *ectodermal dysplasia with ectrodactyly and macular dystrophy, *exomphalos, *Fraser's syndrome, *Goldberg–Shpritzen megacolon syndrome, *hypotonia-cystinuria syndrome, *Meckel's syndrome, *ocular coloboma, *oculodentodigital dysplasia, *Patau's syndrome, *Rieger's syndrome, *Robinow's syndrome, *Rothmund–Thomson syndrome, *situs inversus, *Smith–Magenis syndrome, *spina bifida, *STAR syndrome, *Treacher–Collins' syndrome, *Wolfram's syndrome. Defects may be induced by *teratogens such as *cyclopamine and *thalidomide.

d

devitrification Loss of transparency, particularly important in the lens of the eye where devitrification (opacity) occurs in cataract.

Devoret test An old name for a carcinogenicity test that continues to be used under a variety of different names (Microscreen Assay, Inductest). The readout is the induction of prophage lambda in bacteria, detected as plaques which occurs if carcinogens cause DNA damage and activate the *SOS repair system which turn induces the phage. The correlation between carcinogenicity and prophage induction is good and it is claimed to be more sensitive than the *Ames test.

dexamethasone A synthetic corticosteroid used as an anti-inflammatory drug. *TNs* Decadron, Dexone, Maxidex.

dexamfetamine (formerly **dexamphetamine)** A sympathomimetic amine of the amphetamine group. Although it is a stimulant in the central nervous system it is useful in the treatment of *attention deficit hyperactivity disorder (ADHD) in children. *TN* Dexedrine.

dexiocardia *See* DEXTROCARDIA.

dexosome An exosome released by a dendritic cell. A small heat-stable membrane-bounded vesicle ~60–90 nm in diameter. A characteristic set of proteins are present, including tetraspanins such as CD9 and CD81, all the known antigen presenting molecules (MHC class I and II, CD1 a, b, c, and d) and the costimulatory molecule CD86. Dexosomes transfer antigen-loaded MHC class I and II molecules to naive dendritic cells.

dextrans Polysaccharides of variable molecular weight composed of D-glucose linked mostly by α-1,6 bonds (and a few α-1,3 and α-1,4 bonds). Dextran 75 (average 75 kDa) has a colloid osmotic pressure similar to that of blood plasma and solutions are used as plasma expanders. In laboratory practice dextrans are useful because they will shield surface charge and facilitate flocculation of erythrocytes which sediment, leaving a leucocyte-rich plasma from which white-cell fractions can more easily be prepared. *Sephadex is crosslinked dextran.

dextrins Low-molecular-weight mixtures of linear α-1,4-linked D-glucose polymers starting with an α-1,6 bond, produced by the hydrolysis of starch by α-amylase. Cyclodextrins are cyclic structures formed by six to eight α-1,4-linked D-glucose residues and produced by enzymatic degradation of starch by bacteria, e.g. *Bacillus macerans*. The toroidal structure allows them to act as hydrophilic carriers of hydrophobic molecules.

dextrocardia (dexiocardia) A congenital anomaly in which the heart is a mirror image of normal and the heartbeat is on the right-hand side of the chest. The abdominal viscera may be similarly transposed, in which case the condition is termed *situs inversus. Perhaps surprisingly, the incidence of left-handedness is no more frequent than in the general population.

dextromethorphan An oral cough suppressant chemically related to codeine but without the analgesic properties. An ingredient in many proprietary cough medicines that are sometimes abused as recreational drugs.

dextrose Synonym for *glucose.

dextrose/nitrogen ratio (D/N ratio) A measure used in early experimental studies of diabetes. Normally little dextrose is excreted in urine.

DFF *See* DNA FRAGMENTATION FACTOR.

DFF45 *See* ICAD.

DFNA5 A gene that is mutated in autosomal dominant nonsyndromic sensorineural deafness; the product is related to pejvakin.

D gene segment (diversity gene segment) One of the genes for immunoglobulin *heavy chain, that codes for part of the *hypervariable region of the variable heavy (V_H) domain (*D region) and is located between the V_H and J_H segments. There are twelve different D gene segments in humans.

DH5 A disabled strain of *E. coli* K-12 used experimentally when its escape would be dangerous.

DHA *See* DOCOSAHEXAENOIC ACID.

DHAP **1.** 3,4-Dihydroxyacetophenone or 3,4-*dihydroxyacetophynone, one of the constituents of a traditional Chinese herbal medicine. **2.** Chemotherapeutic drugs used in combination to treat lymphoma: *dexamethasone, cytarabine (Ara C), and *cisplatin. **3.** *See* DIHYDROXYACETONE PHOSPHATE.

DHAP-AT An enzyme (dihydroxyacetone phosphate acyltransferase, glycerone-phosphate *O*-acyltransferase, EC 2.3.1.42, 680 aa), found in peroxisomes, that catalyses the first step in ether-phospholipid biosynthesis. A partial deficiency of DHAPAT occurs in most patients with *rhizomelic chondrodysplasia punctata.

DH domain *See* DBL.

DHEA *See* DEHYDROEPIANDROSTERONE.

DHFR *See* DIHYDROFOLATE REDUCTASE.

Dhh *See* DESERT HEDGEHOG.

DHT *See* DIHYDROTESTOSTERONE.

diabetes A term covering two main types of disorder: diabetes insipidus and diabetes mellitus. In **diabetes insipidus** there is poor water reabsorption in the kidney, leading to the excessive production of dilute urine. This may be caused by reduced *antidiuretic hormone (ADH, vasopressin) production in the pituitary, defects in the receptors for ADH or defects in the *aquaporins of the collecting duct. In **diabetes mellitus** carbohydrate metabolism is uncontrolled as a result of reduced production of *insulin. In juvenile onset (**type 1**) diabetes mellitus (formerly called insulin-dependent diabetes mellitus, IDDM), caused by autoimmune destruction of pancreatic beta cells, the insulin deficiency tends to be almost total, whereas in adult onset (**type 2**) diabetes (formerly non-IDDM, NIDDM) there seems to be no immunological component but an association with obesity. **Maturity onset diabetes of the young** (**MODY**) is an autosomal dominant form occurring before 25 years of age and caused by defects in insulin secretion. There are several forms: **MODY1** is caused by mutation in the hepatocyte nuclear factor-4α gene, **MODY2** by mutation in the glucokinase gene, **MODY3** by mutation in the hepatocyte nuclear factor-1-(alpha gene. **MODY4** is caused by mutation in the insulin promoter factor-1 gene, **MODY5** by mutation in the gene encoding hepatic transcription factor-2, **MODY6** by mutation in the NeuroD transcription factor. , **MODY7** is caused by mutation in the Krüppel-like transcription factor-11 gene, **MODY8** (diabetes-pancreatic exocrine dysfunction syndrome) by mutation in the carboxyl-ester lipase (cholesterol esterase) gene, and **MODY9** by mutation in the *PAX4 transcription factor. *See also* HAEMOCHROMATOSIS (bronzed diabetes).

diabetes related peptide *See* ISLET AMYLOID PEPTIDE.

diabetic coma A condition that can arise in diabetics for a variety of reasons, including severe hypoglycaemia, or if blood sugar is not controlled and there is an accumulation of ketones in the blood causing metabolic acidosis (ketoacidosis).

diablo/smac (smac/diablo) A dimeric protein (second mitochondria-derived activator of caspases/direct IAP-binding protein with low isoelectric point) released from mitochondria in response to apoptotic stimuli that can remove the constraint exercised by IAPs (inhibitor of apoptosis proteins). It seems that binding of diablo to IAP sterically hinders binding and inhibition of effector caspases. Smac/diablo export from mitochondria into the cytosol is provoked by cytotoxic drugs and DNA damage, as well as by ligation of the CD95 death receptor, and is a caspase-dependent response. Release can be reduced by overexpression of Bcl-2.

diacetoxyscirpenol A *trichothecene mycotoxin (DAS, anguidine) produced by *Fusarium* spp. that is an important contaminant of animal feedstuffs. Cytotoxicity of T cells arises from apoptosis initiated by caspase-8 activation and interruption of cell cycle progression caused by down-regulation of cdk4 and cyclin B1.

diacylglycerol (DAG) Glycerol with long chain fatty acyl residues substituted on the 1 and 2 hydroxyl groups. DAG is an intermediate in the biosynthesis of phosphatidyl phospholipids. DAG released by phospholipase C activity from phosphatidylinositol polyphosphates, acts as a *second messenger and activates *protein kinase C by stabilizing its catalytically active complex with membrane-bound phosphatidylserine and calcium.

diakinesis The final stage of the first *prophase of *meiosis in which the chromosomes become maximally condensed, the nucleolus disappears, and remaining fragments of the nuclear envelope disperse.

diallyl trisulphide A pungent constituent of garlic extracts. Reportedly has antibiotic, antiatherosclerotic, immunomodulatory, renoprotective, antimutagenic, and anticarcinogenic properties, although the mechanism of action is unclear.

dialysis A method for separating large molecules from small ones. Dialysis membranes have a pore size that prevents the passage of macromolecules but allows water, ions, and small molecules to pass into the dialysate on the far side of the membrane.

diaminobenzidine (DAB) An artificial substrate (chromagen) from which *peroxidase will generate a coloured reaction product. A potent carcinogen.

diaminobenzoic acid (DABA) A compound used in fluorimetric determination of DNA content. Not the same as *diaminobenzidine.

Diamond–Blackfan anaemia A rare congenital disorder of erythrocyte production (congenital erythroid hypoplastic anaemia, erythrogenesis imperfecta, Aase–Smith syndrome II) which presents in early childhood and in many cases (30–40%) is associated with other disorders, particularly of the upper limb and craniofacial regions. It results from defective erythropoiesis and lack of nucleated erythrocytes in the bone marrow. Approximately 25% of cases of Diamond–Blackfan anaemia appear to be caused by mutation in the gene on chromosome 19 that encodes ribosomal protein S19, although this might have been expected to have a much more dramatic systemic impact.

diamorphine hydrochloride A semisynthetic *opioid, the 3,6-diacetyl ester of morphine (diacetylmorphine, heroin). A very effective, but highly addictive, analgesic. Taken orally it is a prodrug for morphine; when injected it crosses the blood–brain barrier more readily than morphine itself. Binds to μ-opioid receptors.

diapedesis The emigration of leucocytes across vascular endothelium.

diaphanous The product of the *Drosophila* gene *diaphanous*, the first member of a family of proteins, the **diaphanous-related formins** (Drfs; DRF1 and DRF2 in humans) that act as *rho GTPase effectors during cytoskeletal remodelling. Mutations in **DRF1** (DIA1) are associated with deafness, probably affecting the actin core of hair cells and mutations in **DRF2** (DIA2) affect spermatogenesis or oogenesis and lead to sterility. *See* FORMINS. The activity of DRFs is inhibited by an intramolecular interaction between their N-terminal regulatory region (the **diaphanous inhibitor domain**, DID) and a conserved C-terminal segment of 20–30 aa termed the **diaphanous autoinhibitory domain** (DAD).

diaphoresis Sweating. A **diaphoretic** (sudorific) is a substance that stimulates perspiration.

diaphragm A transverse partition that subdivides a cavity. More specifically, the muscular partition that divides the thoracic cavity from the abdominal cavity in mammals.

diaphysis The central shaft of a limb bone that characteristically has a wall of heavily mineralized bone and a marrow-filled cavity. *See* EPIPHYSIS. **Diaphysitis** is inflammation of the diaphysis.

diarrhoea (US diarrhea) The passage, often frequently and urgently, of very soft or liquid faeces. A potential cause of dehydration.

diarthrosis A freely moving joint between two bones. The bearing surfaces are often encased in a capsule that is filled with synovial fluid.

diastase A mixture of enzymes, amylases, that hydrolyse starch. Amylases produced during the germination of barley are commercially important in the production of malt for brewing and are occasionally used as supplements to aid digestion.

diastasis In pathology, the separation of parts of the body that are normally joined together; in physiology, the mid phase of diastole.

diastema A natural gap in a row of teeth.

diastole The period between two contractions of the heart during which the ventricle fills with blood. A **diastolic murmur** is an abnormal noise heard during diastole and usually an indication of a valvular disorder.

diastrophic dysplasia A autosomal recessive form of short-limb dwarfism caused by mutation in a sulphate transporter (SLC26A2) in which there is scoliosis, bilateral defomity of feet, hitchhiker's thumb, and various abnormal cartilage calcifications. *See* ATELOSTEOGENESIS (TYPE II).

diathermy The use of high-frequency electric currents to heat parts of the body. In diathermic surgery (electrosurgery) an electric current, either between two parts of an instrument or between the instrument and a large external electrode via the body, is used to destroy tissue and cauterize small blood vessels.

diazepam A *benzodiazepine used for treatment of anxiety and insomnia and, because it is long-lasting, as a preoperative relaxant. *TN* Valium.

diazoxide A potassium channel activator, which causes local relaxation in smooth muscle. It is used as a vasodilator in the treatment of acute hypertension and in treating persistent hypoglycaemia due to oversecretion of insulin.

dibucaine *See* CINCHOCAINE.

dibutyl phthalate A compound (DBP, 1,2-benzenedicarboxylic acid, dibutyl ester; Palatinol C; phthalic acid di-n-butyl ester) used extensively as a plasticizer but also in applications ranging from use as an antifoaming agent to an ingredient of skin cream. There is little evidence to support concern about health risks.

dibutyryl cyclic AMP (db-cAMP) An analogue of cAMP often used experimentally because it probably enters cells more readily and is less rapidly hydrolysed.

dicentric Describing a chromosome with two centromeres.

dicer A highly conserved ribonuclease that cleaves double-stranded RNA (dsRNA) or the pre-miRNAs produced by the *microprocessor complex into 21- to 23-nt small interfering RNAs (siRNAs) which guide the RNA-induced silencing complex (*RISC), to destroy specific mRNAs. The human *DICER* gene can partially rescue dicer-null *S. pombe*, indicating the degree of conservation.

dichromatism A form of *colour blindness in which only two of the three primary colours are recognized, although this is probably the norm for most mammals other than primates.

dickkopf A family of secreted proteins (220–350 aa) that are inhibitors of *Wnt signalling. Originally identified in *Xenopus* but human homologues (DKK1–4) have subsequently been identified. There is a **DKK3-related protein** designated 'soggy', (Dickkopf-like-1, DKKL1, 242 aa).

Dick test An obsolete skin test for immunity to *Streptococcus pyogenes*, the causative agent of scarlet fever.

diclofenac A NSAID. Induces fatal kidney failure in vultures and has caused a major decline in their numbers. *TNs* Diclomax, Motifene, Volraman, Voltarol.

Diconal® A combination of an opioid, dipipanone hydrochloride and an antiemetic, cyclizine hydrochloride, used to relieve moderate to severe pain.

dicrotic Describing a process with a double beat, e.g. the pulse felt as a double beat for every heart contraction in severe fever.

dictyostatin A 22-member macrolactone natural product, from the marine sponge *Corallistidae* sp., that stabilizes microtubules and causes cell cycle arrest in G_2/M at nanomolar concentrations. It is effective in paclitaxel-resistant cell lines and is structurally similar to *discodermolide.

dictyotene The prolonged *diplotene of meiosis during which the paired chromosomes appear to elongate. Oocyte nuclei remain at this stage while yolk is produced.

dicyclomine hydrochloride A drug used to relieve smooth muscle spasm of the gastrointestinal tract. It has a specific anticholinergic effect on muscarinic acetylcholine receptors and a direct (musculotropic) effect on smooth muscle by antagonizing bradykinin- and histamine-induced spasms. *TNs* Bentyl, Carbellon, Kolanticon, Merbentyl.

didanosine An antiretroviral drug (2′,3′-dideoxyinosine) that is converted intracellularly to dideoxyadenosine triphosphate (ddATP), and acts as a chain terminator thus inhibiting viral reverse transcriptase. *TN* Videx.

dideoxynucleotide A nucleoside triphosphate in which the hydroxyl groups on C-2 and C-3 of the pentose have been substituted by hydrogen. Dideoxynucleotides can be added during DNA synthesis but block further chain extension. This is important in the dideoxy sequencing method (Sanger–Coulson method) that uses four differently labelled dideoxy nucleotides.

DIDS An anion transport inhibitor (4,4-diisothiocyanato-2,2-disulphonic acid stilbene).

dieldrin A cyclodiene insecticide that acts on the $GABA_A$ receptor–chloride channel complex. Dieldrin was extensively used in agriculture and for the control e.g. of tsetse flies, but its persistence in soil leads to accumulation in the food chain and as a result it is now generally banned.

diencephalon A region of the forebrain consisting of the thalamus, hypothalamus, subthalamus, and epithalamus. It is a key relay zone for transmitting information about sensation and movement and is important for homeostasis.

diestrus *See* DIOESTRUS.

diethylcarbamazine An anthelmintic drug used to treat filarial nematode infections. It inhibits filarial arachidonic acid metabolism.

diethylstilboestrol (diethylstilbestrol, DES) An orally active synthetic nonsteroidal oestrogen that was used therapeutically for a variety of conditions until it became clear that exposure to DES *in utero* could cause clear-cell adenocarcinoma of the vagina and cervix in women, increased the subsequent risk of breast cancer, and increased the risk of testicular cancer. Now recognized to be a teratogen and no longer used.

difference threshold The increase or decrease in a stimulus that is just detectable (the just noticeable difference, JND).

d

differential adhesion An hypothesis advanced to explain the sorting-out of mixed cell aggregates into territories occupied by only one cell type. It was proposed that quantitative differences between homotypic and heterotypic adhesion would suffice although tissue specific *cell adhesion molecules are now known to exist.

differential display PCR A method, based on the *polymerase chain reaction, used to identify genes differentially expressed in various tissues or in normal vs pathological states. The *messenger RNA from the two samples is reverse transcribed, amplified with rather nonspecific primers, and run on a high-resolution gel. Bands unique to single samples are considered to be differentially expressed and can be used to clone the full-length cDNA. Superseded by gene chip technologies. *See also* DIFFERENTIAL SCREENING.

differentially expressed in adenocarcinoma of the lung A gene (*DAL1*) located on chromosome 18p11.3, which is lost in ~60% of nonsmall cell lung carcinomas, and exhibits growth-suppressing properties in lung cancer cell lines. The product is a member of the *protein 4.1 family.

differential screening A range of techniques used to identify genes that are expressed in different tissues or in a tissue under different environmental conditions. Methods included *subtractive hybridization, differential hybridization, and *differential display PCR but these have been largely superseded by arrays (*gene chips).

differential screening-selected gene aberrative in neuroblastoma *See* DAN.

differential stain In histology, either a stain that is selective for some component (e.g. nucleic acid) or a stain that binds reversibly and will give greater discrimination if the specimen is over-stained and then progressively washed to differentiate the regions of greatest binding affinity.

SEE WEB LINKS

- Useful resource on histological stains.

differentiation The developmental process in which cells become specialized for particular functions within a multicellular organism. In molecular terms, the selective expression of a subset of genes, either temporarily until differentiation is completed or permanently throughout the lifetime of the cells concerned. Differentiation may be preceded by *determination, a commitment to differentiate that is not phenotypically obvious. In oncology the degree of differentiation of a tumour is often important in prognosis; tumours that have lost most differentiated characteristics of the tissue of origin are likely to be proliferating rapidly and are more likely to be unresponsive to positional control and thus malignant. A **differentiation antigen** is a macromolecule that can be detected immunologically and is a marker for a particular cell type or types. Many differentiated cells have characteristic sets of marker antigens although their expression is a result of differentiation, not a cause. *See* CD ANTIGENS.

diffusion limitation The proposition that cell proliferation in culture is rate-limited by the diffusion of some essential factor, probably a growth factor, from the bulk medium and through a diffusion boundary layer (boundary layer hypothesis) immediately adjacent to the cell surface. By spreading the cell increases its capacity to obtain sufficient of this critical factor and once cells become crowded they will not progress from G_0 to G_1 in the cell cycle—a *density dependent inhibition of proliferation.

diffusion potential An electrical potential difference (a junction potential) that arises at the boundary of dissimilar fluids because of differential mobility of ions.

diflunisal A NSAID used in treatment of arthritis and osteoarthritis, a difluorophenyl derivative of salicylic acid that is a prostaglandin synthetase inhibitor. *TN* Dolobid.

digenetic Describing a parasite having two hosts, notably the Digenean trematodes (flukes). Can be used more generally of a pattern of reproduction involving alternation of sexual and asexual cycles in successive hosts.

di George syndrome *See* THYMIC HYPOPLASIA.

digestion The process by which food is broken down.

digestive vacuole *See* SECONDARY LYSOSOME.

digitalis A mixture of pharmacologically active compounds, positive inotropes, derived from the foxglove *Digitalis purpurea* and used to treat congestive heart failure and supraventricular arrhythmias. The active substances, now more often used alone, are the cardiac glycosides, digoxin, and digitoxin. **Digoxin** is usually extracted from *Digitalis lanata*, and is an inhibitor of the sodium-potassium ATPase. **Digitoxin** is similar to digoxin but has longer-lasting effects and is used less commonly. **Digoxigenin** is the aglycone of digoxin and is used as a molecular probe to label DNA or RNA with subsequent detection by enzymes linked to antidigoxygenin antibodies.

digitonin *See* SAPONIN.

diglyceride An ester of two fatty acids and glycerol usually with a saturated fatty acid at the C-1 position and an unsaturated fatty acid at the C-2 position of the glycerol.

digoxin *See* DIGITALIS.

digoxygenin *See* DIGITALIS.

dihydrocodeine A powerful opioid analgesic often combined with paracetamol in co-dydramol.

dihydrofolate reductase The enzyme (DHFR, EC 1.5.1.3, 187 aa) that converts dihydrofolate into tetrahydrofolate, the active form of *folic acid in humans, by transferring hydrogen from NADP to dihydrofolate. Inhibitors (e.g. *methotrexate) are used as anticancer drugs.

dihydropyridines The most smooth muscle selective class of *calcium channel blockers, used to treat hypertension. Examples include nifedipine and nitrenidine. More lipophilic derivatives (e.g. lercanidipine, lacidipine) have been developed.

dihydrotestosterone A biologically active metabolite of *testosterone (DHT, androstanolone) synthesized in the prostate gland, testes, hair follicles, and adrenal glands by 5-alpha reductase. Binds to and activates specific nuclear androgen receptors.

dihydroxyacetone *See* GLYCERONE.

dihydroxyacetone phosphate (DHAP) An intermediate in many biochemical pathways, notably glycolysis. Formerly glycerone phosphate.

dihydroxyacetophynone A component (3,4-dihydroxyacetophenone) of a traditional Chinese herbal medicine (Qingxintong), isolated from leaves of tumaodongqing (*Ilex pubescens*), with anti-inflammatory properties.

dihydroxyphenylalanine (L-DOPA, levodopa) The precursor of *dopamine, made from L-tyrosine by tyrosine 3-mono-oxygenase. Used as a treatment for *Parkinson's disease.

dihydroxypurine *See* XANTHINE.

Dii ('di-i') A highly lipophilic derivative of indocarbocyanine iodide that is weakly fluorescent in water but highly fluorescent and quite photostable when incorporated into membranes. Used as a probe for membrane fluidity measurements and also as a general stain for membranes.

dilated cardiomyopathy A set of disorders in which there is cardiac dilation and reduced systolic function. It is the result of a heterogeneous group of inherited and acquired disorders. These include myocarditis, coronary artery disease, systemic diseases, and myocardial toxins in approximately half the cases; familial forms, in which there is autosomal dominant inheritance, account for ~25% of the remaining idiopathic cases. *See* HYPERTROPHIC CARDIOMYOPATHY; PHOSPHOLAMBAN.

diltiazem HCl A *calcium channel blocker. *TN* Tiazac.

dilution cloning A simple method for cloning cells by diluting the suspension to the point where the chosen inoculum is likely to contain only a single cell.

dimenhydrinate An antihistamine used to prevent motion sickness; closely related to *diphenhydramine HCl. *TNs* Dramamine, Gravol, Vertirosan.

dimercaprol (British anti-Lewisite, BAL) A drug originally synthesized as an antidote for the chemical warfare gas Lewisite, now used to treat heavy metal poisoning and *Wilson's disease.

SEE WEB LINKS

- Further information in the IPCS Inchem safety database.

dimethicone An viscoelastic compound (polydimethylsiloxane), a component of Silly Putty. Used in a very wide range of preparations as a defoaming agent, a mould release agent, and as a constituent of lubricants, polishes, cosmetics, hair conditioners, etc. As a food additive it is E900.

dimethyl formamide (DMF) A polar solvent, miscible with water and most organic solvents, sometimes used as an alternative to *dimethyl sulphoxide (DMSO) for making stock solutions of rather hydrophobic compounds; the concentrated solution can then be diluted. Also an important industrial solvent.

dimethyl sulphoxide (DMSO) An unpleasant-smelling colourless liquid often used as a solvent for poorly soluble compounds, and as a cryoprotectant when freezing cells. It is used clinically for the treatment of arthritis, although its efficacy is disputed.

diminuendo A mouse mutant that exhibits progressive loss of hearing and hair cell anomalies. Unlike other mutations leading to deafness, the defect is in the gene encoding a microRNA and 96 mRNA transcripts are affected in the homozygous mutant.

dimorphic (dimorphous) Describing organelles, organs, or individuals that exist in two morphologically distinct forms.

dinitrophenol A small molecule (2,4-dinitrophenol) often used as a hapten, for which purpose it needs to be coupled to a carrier protein. Will also act as an uncoupler of oxidative phosphorylation.

dinoprostone The natural eicosanoid *prostaglandin E_2, used to stimulate labour and for various other clinical interventions. *TNs* Cervidil, Latanoprost, Prostin E2, Propess.

dioestrus (diestrus) Describing the period between oestrus cycles; in humans, the postovulatory, premenstrual period.

dioptre The unit for the power of a lens, the reciprocal of focal length in metres. A first set of reading glasses might typically have a power of 1 dioptre.

dioxin Strictly, the name for heterocyclic compounds with two oxygen and four carbon atoms in the ring; can be either 1,2-dioxin (*o*-dioxin) or 1,4-dioxin (*p*-dioxin) but used casually as a shorthand term for chlorinated derivatives of dibenzo-*p*-dioxin, the most toxic of which is 2,3,7,8-tetrachlorodibenzo-*p*-dioxin. The polychlorinated dibenzodioxins are lipophilic, will bioaccumulate in humans, and have been shown to have teratogenic, mutagenic, and carcinogenic properties.

dioxygenase An oxidoreductase that catalyses the addition of two oxygen atoms to the compound being oxidized. They are in the EC 1.13.11 class.

diphenhydramine hydrochloride An antihistamine drug that has antiemetic, sedative, and hypnotic effects. It also has anticholinergic activity and is used to treat acute dystonic reactions caused by neuroleptic drugs. An ingredient of many nonprescription treatments to relieve the symptoms of the common cold and often used as a mild sleeping draught. *TNs* Benadryl, Dimedrol, Nytol.

diphenoxylate hydrochloride An opiate rapidly and extensively metabolized to produce biologically active diphenoxylic acid (difenoxine). Its effect is due to an increase in muscle tone in the small intestine. Lomotil™, a common antidiarrhoeal drug, is a mixture of diphenoxylate hydrochloride and atropine sulphate, the latter added to prevent recreational abuse of the opiate.

diphtheria The disease caused by the Gram-positive bacillus *Corynebacterium diphtheriae*. Bacteria are generally confined to the throat but diphtheria toxin can cause myocarditis, polyneuritis, and other systemic effects. **Diptheria toxin** is an AB exotoxin (567 aa) coded by β-corynephage of virulent *Corynebacterium diphtheriae* strains but regulated by bacterial genes. The B subunit binds to binds to a specific receptor (heparin-binding EGF receptor) on susceptible cells and enters by receptor-mediated endocytosis. Endosomal processing releases the enzymically active A subunit (193 aa) which is translocated into the cytoplasmic compartment where it ADP-ribosylates *elongation factor 2, thereby halting mRNA translation. **Diphtheria toxoid** is formaldehyde-treated toxin which is enzymatically inactive but retains antigenicity and can be used for immunization.

diphyodont Describing animals that, like humans, have a deciduous or milk dentition that is later replaced by permanent dentition.

dipicolinic acid (DPA) An important constituent (as much as 10%) of bacterial spores, where it plays a role in survival of wet heat and ultraviolet radiation. It is also a chelating agent.

dipipanone hydrochloride A strong synthetic opioid analgesic, chemically similar to methadone.

diplococcus A strain of bacteria in which spherical cocci occur in pairs. **Streptococcus pneumoniae* (formerly *Diplococcus pneumoniae*) is a well-known example.

diploid Describing a cell that has two copies of genomic information. Diploid cells have pairs of homologous chromosomes and are usually described as being 2n where n is the haploid chromosome number. In mammals, only gametes are haploid.

diplonema *See* DIPLOTENE.

diplont Describing those organisms in which vegetative cells are haploid and only the zygote is diploid.

diplophase The phase of the life cycle in which nuclei are diploid. For mammals, this is the majority of the lifespan.

diplopia Double vision.

diplornavirus Obsolete name proposed for double-stranded RNA viruses. Taxonomically unsound.

diplotene (diplonema) The stage in the first prophase of meiosis when all four *chromatids can be distinguished and homologous chromosomes begin to move away from one another except at *chiasmata.

dipyridamole A drug that indirectly inhibits platelet adhesion, probably through a combination of inhibition of the erythrocyte adenosine transporter and inhibition of cGMP-dependent phosphodiesterase. Used to treat angina pectoris and prevent blood clotting. *TN* Persantine.

directed evolution **1.** Scientifically, a powerful approach to the production of molecules with desirable properties through sequential rounds of synthesis, each round being informed by data on the properties of the previous set. The method has been applied to producing novel peptide ligands and various enzymes with improved characteristics.
2. Theologically, the creationist view that biological systems are so complex that there must have been an external director (i.e. a creator deity). There is no evidence to support this belief.

disaccharide Any sugar in which there are two monosaccharide units linked by a glycosidic bond. Trehalose-type disaccharides are formed from two nonreducing sugars, the maltose type from two reducing sugars.

DISC *See* DEATH (death-inducing signalling complex); DISRUPTED-IN-SCHIZOPHRENIA (1).

disc gel A gel, of any shape, in which there is a discontinuity of some sort, e.g. in gel concentration.

discodermolide A microtubule-stabilizing drug (a polyhydroxylated alkatetraene lactone) that promotes the formation of microtubule bundles. Probably binds to the same site as *paclitaxel and used similarly in chemotherapy. Isolated from the marine sponge *Discodermia dissoluta*.

disease-modifying antirheumatic drug *See* DMARD.

dishevelled-1 The homologue of the product of the *Drosophila* gene *dishevelled*, a scaffold phosphoprotein that regulates cell proliferation, and that is involved in the *Wnt signalling system. Interacts with *MuSK in the formation of the neuromuscular junction. *See* DAPPER.

disintegrins Inhibitors of integrins (45–84 aa) that were originally isolated from various snake (viper) venoms. Almost all have a conserved *RGD motif and work only on integrins that recognize this (the β_1 and β_3 classes). Their biological role is probably to inhibit blood clotting by competitive inhibition of integrin binding. Examples include albolabrin, applagin, batroxostatin, bitistatin, echistatin, elegantin, flavoridin, halysin, kistrin, triflavin, and trigramin.

disjunction The separation of the two members of each pair of homologous chromosomes that occurs during meiotic anaphase. Disjunction mutants arise when chromosomes are partitioned unequally between daughter cells as a result of nondisjunction.

disopyramide A type 1 antiarrhythmic drug that acts directly on cardiac muscle depolarization and has no effect on α- or β-adrenergic receptors. *TN* Norpace.

dispermy Condition that arises if an ovum fuses with two spermatozoa causing diandric triploidy, which probably accounts for about half of the triploidy that is found in 6% of spontaneous abortions.

disrupted-in-schizophrenia 1 A susceptibility gene in the 1q42 chromosomal region for mood disorders and schizophrenia, originally identified in a large Scottish family. The **DISC1** gene is predicted to encode a multifunctional 854-aa protein involved in intracellular transport, the determination of neurite architecture, and/or neuronal migration. Highly expressed in olfactory bulb and dentate gyrus of hippocampus, both areas in which adult neurogenesis occurs. Binds NDEL1. **DISC2** apparently specifies a single exon, thought to be a noncoding RNA molecule that is antisense to DISC1.

dissecting aneurysm A tear in the inner wall of the aorta allowing blood to split the layers of the aortic wall longitudinally, usually with fatal consequences.

disseminated intravascular coagulation Systemic clotting of the blood probably as a result of endotoxin from

Gram-negative bacteria acting on on neutrophils. A complication of septic shock.

disseminated sclerosis *See* MULTIPLE SCLEROSIS.

dissociation constant (K_d) A constant that describes the relationship of the activities (approximately, concentrations) of reactants and products in a chemical equilibrium of form $A + B = AB$. The dissociation constant, K_d, is the product of the activities of A and B divided by the activity of the product, AB, and has dimensions of concentration. The acid dissociation constant, K_a, is the special case where one of the reactants is a proton (H^+) and is usually expressed as pK_a (which is $-\log_{10}K_a$).

distal Describing something that is furthest from the point of attachment or of reference. The opposite of proximal.

distemper virus A morbillivirus (paramyxovirus) that causes disease in dogs. A variant affect seals (phocavirus), and the human equivalent is measles virus.

distigmine An anticholinesterase.

disulphide bond A linkage between the sulphydryl (-SH) groups of two *cysteine moieties. Each cysteine then becomes a half-*cystine residue. Disuphide bonds may form between residues in a single protein or between those in different peptide chains and stabilize, but do not determine, the secondary structure of proteins. Disulphide bonds are disrupted by small molecules with -SH groups and are absent in cytosolic proteins because the cytosol has a high concentration of *glutathione which has a free -SH residue.

disulphiram A drug that blocks the oxidation of alcohol at the acetaldehyde stage and causes extreme nausea if alcohol is consumed; used in aversion therapy for alcoholism. *TN* Antabuse.

diterpene A large class of natural products (~5000 are known) composed of four isoprene units and derived from geranylgeranyl pyrophosphate. The basic structure is of twenty carbon atoms. Diterpenes form the basis for biologically important compounds such as *retinol.

dithioerythritol *See* DITHIOTHREITOL.

dithiothreitol A reagent (Cleland's reagent, dithioerythritol) used to protect sulphydryl groups from oxidation or to reduce disulphides to sulphydryl groups.

diuretics Drugs that stimulate diuresis. **Thiazide diuretics** inhibit sodium/chloride reabsorption from the distal convoluted tubules of the kidney and cause a loss of potassium. They are often used to treat hypertension; examples include bendroflumethiazide (bendrofluazide), chlortalidone (chlorthalidone), benzthiazide, clopamide, cyclopenthiazide, hydrochlorothiazide, xipamide, indapamide, hydroflumethiazide, and metolazone. **Loop diuretics** potently inhibit resorption from the loop of Henle in the renal tubule and thus increase the rate of fluid excretion. Examples are furosemide, bumetanide, and ethacrynic acid. **Potassium-sparing diuretics** are a class of mild diuretics that act on the kidney to promote excretion of water without loss of potassium ions. Examples are amiloride, triamterene, and spironolactone.

diurnal Describing something that occurs in daylight hours, the opposite of nocturnal, but commonly misused (e.g. diurnal rhythm) to mean having a 24-hour cycle, which is more correctly a *circadian rhythm.

divergence In evolution, the process by which organs of different form and function arise from the same original structure. Often encountered in the term divergent evolution in contrast to convergent evolution. *See* HOMOLOGY.

diversity gene segment *See* D GENE SEGMENT.

diverticulum (*pl.* diverticula) A side branch of a cavity or channel, e.g. a blind-ending outgrowth from the gut. **Diverticulitis** is inflammation of diverticula of the colon.

dizygotic Describing twins that arise from the simultaneous fertilization of two ova by two spermatozoa. Dizygotic twins, unlike monozygotic twins, are genetically nonidentical.

DKK *See* DICKKOPF.

DLGs A family of proteins that are homologues of the *Drosophila* 'discs large' tumour suppressor protein, Dlg, the prototype of the *MAGUKs. **DLG1** (Discs large homologue-1, synapse-associated protein 97, SAP-97, 904 aa) is a multidomain scaffolding protein that recruits channels, receptors, and signalling molecules to discrete plasma membrane domains, regulates the excitability of cardiac myocytes by modulating the functional expression of Kv4 potassium channels, and functionally regulates the Kv1.5 channel. Several other human DLG proteins are known.

DLL *See* DELTA-LIKE.

DMARD Common abbreviation for a disease-modifying antirheumatic drug, any drug used for treating rheumatoid arthritis that does more than relieve symptoms. Examples include gold, penicillamine, sulphasalazine, and chloroquine.

DMBT1 *See* DELETED IN MALIGNANT BRAIN TUMOURS-1.

DMEM A tissue culture medium (Dulbecco modified Eagle's medium) that will support the growth of a wide range of mammalian cell types.

dmf **1.** An antagonist of the PCB126 nuclear *aryl hydrocarbon receptor, 3′,4′-dimethoxyflavone. **2.** *See* DIMETHYL FORMAMIDE.

DM-GRASP *See* ACTIVATED LEUCOCYTE CELL ADHESION MOLECULE.

dmp **1.** *See* DENTIN (dentin matrix protein (DMP-1)). **2.** *See* DEXTROMETHORPHAN. **3.** 2,6-dimethylphenol. **4.** 2′,6′-dimethylphenylalanine. **5.** dimethylphosphate. **6.** 4,6-diamino-2-mercaptopyrimidine. **7.** Disease management programme. **8.** Diabetes management programme.

DMSO *See* DIMETHYL SULPHOXIDE.

DNA (deoxyribonucleic acid) The nucleotide polymer that is the genetic material of all cells and many viruses. There are four monomer components, each consisting of phosphorylated 2-deoxyribose linked to one of four bases *adenine, *cytosine, *guanine, and *thymine (A, C, G, and T) which are linked together by 3′,5′-phosphodiester bridges to form a linear array. The four units define, by their sequence, the sequence of complementary RNA and in the case of mRNA the amino acid sequence of proteins. In the Watson–Crick double-helix model two complementary strands are wound in a right-handed helix and held together by hydrogen bonds between *complementary base pairs, A–T and G–C. Three major conformations exist: *A-DNA, *B-DNA (the form described in the original Watson–Crick model), and *Z-DNA.

DNA-activated protein kinase A nuclear DNA-dependent serine/threonine protein kinase (DNA-PK) consisting of a catalytic subunit (4096 aa) and the *Ku autoantigen which directs it to DNA and activates the kinase. Phosphorylates various transcription factors, probably modulating their activity.

DNA adduct A general term for DNA that has been modified by the covalent addition of another moiety. Adducts are usually considered to be markers for carcinogenic modification of the cell.

DNA annealing The reassociation of single strands of DNA, that have been separated by heating, to form double-stranded DNA in which there is complementary pairing of bases. The rate at which double strands re-form depends upon the degree of repetition, and is slowest for unique sequences (*see* C_0T CURVE).

DNAase *See* DEOXYRIBONUCLEASE.

DNA-binding domain Protein motifs found in *DNA-binding proteins that are involved in the DNA binding interaction, usually conferring specificity for a particular nucleotide sequence. Examples are found in transcription factors and include *zinc finger proteins, *helix–turn–helix proteins, and the *leucine zipper proteins.

DNA-binding proteins A general term for proteins that interact with DNA, although with the implication that the binding is not a substrate–enzyme type interaction, i.e. that nucleases and polymerases are not in this category. Many bind to specific sequences and as a result of binding affect the probability of transcription (*transcription factors); these often contain conserved motifs (*DNA-binding domains). A different set of DNA-binding proteins are those that have a structural role in chromatin, particularly *histones. *See also* DAMAGED DNA-BINDING PROTEINS.

DNA chips A regularly arranged set (DNA array, DNA microarray, gene array, biochip) of different DNA fragments or polynucleotides bound to a solid support (often microscope-slide sized) in such a way as to allow hybridization with added labelled single-stranded RNA or DNA. Such arrays can be used to look for expression of particular genes or to identify novel sequences. Each array may have thousands of different fragments, and with careful washing the array may be used for multiple experiments.

DNA fingerprinting *See* RESTRICTION FRAGMENT LENGTH POLYMORPHISM.

DNA footprinting *See* FOOTPRINTING.

DNA fragmentation factor A magnesium-dependent endonuclease (DFF, caspase-activated DNase, CAD) specific for double-stranded DNA. In nonapoptotic cells it is found in the nucleus as a heterodimer, composed of an inhibitory chaperone (DFF45; inhibitor of CAD, ICAD-L, 331 aa) and a latent nuclease (DFF40/CAD, 338 aa). The nuclease is activated by caspase-3 and possibly other caspases.

DNA glycosylase (DNA glycosidase) A class of enzymes that recognize altered bases in DNA and catalyse their removal by cleaving the glycosidic bond between the base and the deoxyribose sugar. At least twenty such enzymes occur in cells. Thymine-DNA glycosylase, for example, corrects G/T mismatches to G/C base pairs; single-strand-selective monofunctional uracil-DNA glycosylase 1 (SMUG) removes uracil from single- and double-stranded DNA in nuclear chromatin as part of the process of *base excision repair.

DNA gyrase An enzyme, a type II topoisomerase (EC 5.99.1.3) that can induce or relax supercoiling in circular DNA, a mechanochemical activity that requires energy from ATP hydrolysis. In bacteria it is essential for DNA replication and gyrases are the target of antibiotics such as the *quinolones.

DNA helicase A family of motor proteins (unwindases) found in both prokaryotes and eukaryotes that move along a DNA double helix separating the strands at the *replication fork to allow copying to occur. Two molecules of ATP are required for each nucleotide pair of the duplex. *See* CHD PROTEINS; *RECQ-like helicases; *RNA helicases.

DNA hybridization *See* HYBRIDIZATION.

DNA ladder A casual term for the molecular weight (base pair) standards run in parallel with DNA samples to calibrate an electrophoretic gel. Each rung of the ladder represents a different size of polynucleotide. **DNA laddering** is the term applied to the pattern of DNA fragmentation seen in apoptotic cells and refers to the pattern seen on a gel. The activity of the endonuclease DNase γ (endo G) generates fragments that are multiples of the 180-bp nucleosomal unit (*see* DNA FRAGMENTATION FACTOR).

DNA library *See* GENOMIC LIBRARY.

DNA ligases Enzymes that join DNA molecules together and are important in DNA replication and repair. In DNA replication in *E. coli*, ligase is important in linking precursor fragments (*Okazaki fragments) generated on the lagging strand. The ligase from phage T4 will join two DNA molecules that having completely base-paired (blunt) ends. *See* NICK TRANSLATION. DNA ligase D (LigD) performs end-remodelling and end-sealing reactions during nonhomologous end joining (*NHEJ) in bacteria.

DNA markers **1.** Any unique sequence that can be used to identify a gene or chromosomal region including, among other things, *single nucleotide polymorphisms (SNPs) and *short tandem repeats (STRs). **2.** Oligonucleotides of known length, often colour-labelled, used as size markers in gel electrophoresis.

DNA methylation The enzymatic addition of methyl groups to certain nucleotides in genomic DNA, thereby reducing the ease with which it can be transcribed and thus restricting the range of genes that are active in a particular differentiated tissue. The pattern of methylation is hereditable and is copied on to the daughter strands at mitosis by maintenance DNA methylases. Methylation may also allow targeting of specific regions by methyl-CpG-binding domain proteins which then induce binding of other regulatory proteins such as histone deacetylases and silencing of the region. In bacteria, methylation is important in the restriction systems because many *restriction enzymes will not cut methylated sequences. *See* GENOMIC IMPRINTING.

DNA-PK *See* DNA-ACTIVATED PROTEIN KINASE.

DNA polymerases Key enzymes (EC 2.7.7.7) involved in template-dependent synthesis of DNA, usually from DNA although retroviruses possess a unique and important DNA polymerase (*reverse transcriptase) that uses an RNA template. Five DNA polymerase classes are known in *E. coli*; I will edit out unpaired bases at the end of growing strands, III appears to be the main replicative enzyme and the others are mostly involved in repair. Eukaryotic cells have α, β, and γ polymerases all of which use DNA as the template. **DNA polymerase-α** (pol α; p180-p68) forms a four-subunit complex with *DNA primase (prim-2; p58-p48) and is the only enzyme able to start DNA synthesis *de novo*. The major role of the DNA polymerase α-primase complex (pol-prim) is in the initiation of DNA replication at chromosomal origins and in the discontinuous synthesis of *Okazaki fragments on the lagging strand of the replication fork. **Polymerase-β** is implicated in repairing DNA and **polymerase-γ** replicates mitochondrial DNA. The DNA **polymerase-δ** complex is involved in DNA replication and repair and consists of *proliferating cell nuclear antigen (PCNA), the multisubunit replication factor C, and the four-subunit polymerase complex: POLD1-4. Many others are known.

DNA primases Despite the name, these are RNA polymerases (RNA primase, EC 2.7.7.6) that catalyse the synthesis of short (~10 bases) RNA primers on single-stranded DNA templates. These primers are used by DNA polymerase to initiate the synthesis of *Okazaki fragments on the lagging strand. Eukaryotic DNA primase is a p60/p50 heterodimer with little homology to prokaryotic primases and forms a complex with DNA polymerase-α.

DNA probe A short labelled DNA sequence used to detect a complementary nucleotide sequence to which it hybridizes. A range of labelling method are available and they are important reagents for the identification of pathogens and for genetic screening.

DNA profiling The comparison of DNA samples for forensic purposes—a preferable term to DNA fingerprinting.

DNA renaturation *See* DNA ANNEALING.

DNA repair The key mechanism for error correction in the genetic code. *See* CERNUNNOS; DAMAGED DNA-BINDING PROTEINS; DNA LIGASE; DNA POLYMERASE; EXCISION REPAIR; KARP-1; MISMATCH REPAIR; NONHOMOLOGOUS END-JOINING; PHOTOLYASE; POLY(ADP-RIBOSE) POLYMERASE; REC PROTEINS; SOS SYSTEM; *see also* diseases associated with deficiencies in error repair: ATAXIA TELANGIECTASIA; XERODERMA PIGMENTOSUM.

DNA replication The process of copying a DNA molecule, and thus duplicating the genetic information it contains. The parental double-stranded DNA molecule is separated (*See* DNA HELICASE) and used as a template for the synthesis by *DNA-polymerase of daughter strands. Replication is said to be semiconservative since at the end of the process there are two identical DNA duplex strands, each composed of one parental strand and one newly synthesized strand. Accurate replication is essential for transmission of unchanged genetic information and a range of error-correcting mechanisms ensure remarkable fidelity.

DNase *See* DEOXYRIBONUCLEASE.

DNA sequencing The process of analysing the nucleotide sequence of a length of DNA. Usually requires cloning of the sequence to produce sufficient material and then reacting the sample in either the Sanger *dideoxy sequencing method or, less commonly, the *Maxam–Gilbert method. The reaction products are separated on a large sequencing gel that allows resolution of polynucleotides differing in length by a single nucleotide. This classical approach is being rapidly superseded by rapid automated methods using column chromatography to separate reaction products, which will work with amounts of starting sample that can be generated by PCR rather than requiring cloning. As the cost per nucleotide drops, the possibility of routinely genotyping individuals gets ever closer.

DNA synthesis In principle, the polymerization of deoxyribonucleotide triphosphates to form DNA which *in vivo* occurs mostly as template-directed *DNA replication although some repair synthesis does occur. In retroviruses, DNA synthesis is directed by an RNA template and the *reverse transcriptase enzyme has been used *in vitro* to produce DNA (complementary DNA; cDNA) from mRNA. *In vitro* synthesis can also be done by sequential operator-controlled reactions that generate an oligonucleotide of the desired sequence.

DNA topoisomerases Enzymes (EC.5.99.1.2) that control and alter the topologic states of DNA in both prokaryotes and eukaryotes. Type I topoisomerases break a single DNA strand, whereas the type II enzymes break both strands. **Topoisomerase I** in eukaryotic cells normally functions during transcription. **Topoisomerase II** catalyses the relaxation of supercoiled DNA molecules, catenation, decatenation, knotting, and unknotting of circular DNA. The **topoisomerase III** gene is commonly deleted in patients with the *Smith–Magenis syndrome; the product of this gene apparently specifically reduces the number of supercoils in highly negatively supercoiled DNA and is found in many somatic tissues. There is a marked reduction of topoisomerase II in some cell lines from patients with ataxia-telangiectasia. Topoisomerase II of *E. coli* is also known as *DNA gyrase.

DNA transfection The experimental introduction of genes or gene fragments directly into cells in the form of isolated DNA. Various methods such as *electroporation and the use of reagents such as *lipofectamine, are used to facilitate the passage of the DNA across the plasma membrane.

DNA tumour virus A DNA virus that will cause tumours in animals. One of the best known examples is *SV40, one of the Polyomaviridae, but a range of viruses in the *Papillomaviridae, *Herpesviridae (a family that includes *Epstein–Barr virus and Kaposi's sarcoma-associated herpesvirus (HHPV-8)), *Hepadnaviridae, and *Adenoviridae will cause tumours, usually because the virus carries a gene that overrides normal cell cycle control mechanisms in one way or another.

DNA vaccine Vaccine that does not itself consist of antigenic material but of a DNA sequence that will produce antigen when transiently expressed in the host. At present they are only in the experimental stage of development, although some clinical trials have been done.

DNA virus A virus in which the genomic material is in the form of double- or single-stranded DNA. **Group I viruses** possess double-stranded DNA and include, among other families, the *Adenoviridae, *Herpesviridae,

*Papillomaviridae, *Polyomaviridae, *Poxviridae, *Mimivirus, and many tailed bacteriophages. **Group II** viruses possess single-stranded DNA and include families such as the *Parvoviridae and bacteriophage M13. *See* DNA TUMOUR VIRUS.

DNP *See* DENDROASPIS NATRIURETIC PEPTIDE.

d

D/N ratio *See* DEXTROSE/NITROGEN RATIO.

DNS A common abbreviation for dansyl. *See* DANSYL CHLORIDE.

dobutamine hydrochloride A synthetic catecholamine that acts through β-adrenergic receptors in the heart. Used for short-term treatment of organic heart disease or after heart surgery.

docetaxel A semisynthetic *taxane, an esterified product of 10-deacetyl baccatin III, which is extracted from the European yew *Taxus baccata*. It is slightly more potent than paclitaxel (taxol). *TN* Taxotere.

docking protein *See* SIGNAL RECOGNITION PARTICLE-RECEPTOR.

docosahexaenoic acid (DHA) Generically, any straight-chain fatty acid with 22 carbon atoms and 6 double bonds. It is a major omega-3 fatty acid in human neural tissue and a range of health benefits from dietary supplementation with the all-Z isomer found in fish oil are claimed. *See* PROTECTIN (D1).

dodecanoic acid A saturated fatty acid (lauric acid) with 12 (dodeca) carbon atoms. The sodium salt (sodium dodecyl sulphate, SDS, or sodium lauryl sulphate) is extensively used as a detergent.

dok proteins A family of adaptor proteins downstream of (tyrosine) kinases consisting of five members, including Dok, Dok-R, DokL, Dok4, and Dok5, structurally rather similar to the insulin receptor substrate family of proteins. When tyrosine-phosphorylated they link clustered receptors to other signalling molecules including Nck, CrkL, Cas, and rasGTPase-activating protein (rasGAP). Dok, Dok-R, and DokL have all been shown to to play a negative role in receptor tyrosine kinase and cytokine signalling. DokR, also known as p56dok/FRIP, is involved in cytokine and immunoreceptor signalling in myeloid and T cells. Dok4 and Dok5 may be a separate subgroup and have been shown to potentiate signals emanating from the c-ret receptor.

dolastatins A family of small peptides with diverse activities originally isolated from the marine nudibranch mollusc *Dolabella auricularia* (sea hare), although they may actually be of cyanobacterial origin. **Dolastatin 10** is a potent antimitotic pentapeptide that inhibits microtubule assembly and is being tested in cancer therapy. **Dolastatin 11**, a depsipeptide, binds to actin and stabilizes F-actin *in vitro*, like *phalloidin and *jasplakinolide although acting at a different site. **Tasidotin**, a synthetic analogue of dolastatin 15, inhibits microtubule assembly and induces a G_2/M block in treated tumour-derived cells, ultimately resulting in apoptosis. **Lyngbyastatin 4** is a depsipeptide isolated from the marine cyanobacterium *Lyngbya confervoides* and is an analogue of dolastatin 13; it selectively inhibits elastase and chymotrypsin *in vitro*. *See* DOLICULIDE; PHOMOPSIN A.

dolichol Any of a group of long-chain mostly unsaturated organic compounds with a varying numbers (13–24) of isoprene units and a terminal phosphorylated hydroxyl group, biosynthetized by the general isoprenoid pathway from acetate via mevalonate and farnesyl pyrophosphate. Dolichol can be post-translationally added to proteins and also has an important role as a transmembrane carrier for glycosyl units in the biosynthesis of glycoproteins and glycolipids. Dolichol accumulation in tissues has been suggested as a biomarker for ageing.

doliculide A macrocyclic depsipeptide that has antitumour activity, originally extracted from the sea hare *Dolabella auricularia*. Stimulates actin assembly in the same way *jasplakinolide and competes with *phalloidin for binding. *See* DOLASTATINS.

dolioform Barrel-shaped.

domain A part or region of a molecule or structure that has particular physicochemical features or properties, e.g. hydrophobic, polar, globular, α-helical domains, DNA-binding domain, ATP-binding domain. A range of domains are recognized on the basis of their sequence and their presence can be a helpful indicators of the likely function or location of a molecule.

SEE WEB LINKS

- Useful website with links to details of many domains.
- The NCBI's Conserved Domains database.

dominant An allele that has an effect when present as a single copy in a diploid organism, when heterozygous. Can also be used to describe a character due to a dominant gene. *Compare* RECESSIVE. In change-of-function mutations the product has novel function; in [dominant negative] mutations the product suppresses the activity of the normal allelic product by forming an inactive complex or by competing with the normal protein so that the overall activity is below a critical threshold level and function is abnormal (haploinsufficiency).

domoic acid A tricarboxylic acid toxin originally isolated from the macroscopic red alga *Chondria armata*, known in Japan as domoi, that is used as an anthelmintic in a traditional herbal medicine. An agonist for the kainate subclass of ionotropic *glutamate receptors of the central nervous system. Domoic acid is the cause of amnesic shellfish poisoning, the source of the toxin being the diatom *Pseudo-nitzschia* (previously *Nitzschia*) *pungens* forma *multiseries*.

domperidone An antiemetic drug that works primarily by blocking dopamine receptors, particularly D_2 and D_3. *TN* Motilium.

donepezil hydrochloride A reversible acetylcholine esterase inhibitor used to treat symptoms of mild to moderate Alzheimer's dementia, where it is thought to act by enhancing cholinergic neurotransmission without acting on the underlying disorder. *TN* Aricept.

Donnai–Barrow syndrome An autosomal recessive disorder (facio-oculo-acoustico-renal (FOAR) syndrome), caused by mutation in the *LDL-receptor-related protein 2, in which there are facial anomalies, ocular anomalies, sensorineural hearing loss, diaphragmatic hernia, absent corpus callosum, and proteinuria.

donor splice junction The junction between an *exon and an *intron at the 5′ end of the intron. When the intron is removed during processing of *hnRNA the donor junction is spliced to the acceptor junction at the 3′ end of the exon. Mutations in the invariant GU start or AG termination of an intron have serious consequences for RNA processing; e.g. a patient with a severe form of haemophilia B was shown to have a point mutation in a donor splice junction of the factor IX gene.

DOPA *See* DIHYDROXYPHENYLALANINE; DOPAMINE.

dopamine A *catecholamine (153 Da), formed by decarboxylation of dihydroxyphenylalanine (DOPA). An important *neurotransmitter and *hormone in its own right and a precursor of *adrenaline and *noradrenaline. Dopamine receptors are *G-protein coupled and fall into two classes, D_1-like (D_1, D_3, D_4; activators of adenylyl cyclase) and D_2-like (D_2 and D_5; inhibitors of adenylyl cyclase). Most antipsychotic drugs are dopamine receptor antagonists and most neuroleptics were developed as D_2 receptor antagonists. **Dopaminergic neurons** are ones for which dopamine is the neurotransmitter. The **dopamine uptake transporter** (sodium-dependent dopamine transporter, solute carrier family 6 member 3, 620 aa), which is inhibited by *cocaine, terminates the action of dopamine by reuptake into presynaptic terminals.

doppel A prion-like protein (Dpl, 179 aa) related to the *prion protein (PrP) and encoded by a locus, *PRND*, 16 kb downstream from the PrP_C-coding gene on chromosome 20. It is expressed during development but in the adult is scarcely detectable in the brain, unlike PrP.

doripenem A synthetic broad-spectrum carbapenem antibiotic, structurally related to beta-lactam antibiotics, effective against both Gram-positive and Gram-negative bacteria. Used as a single agent for the treatment of complicated intra-abdominal or urinary tract infections. *TN* Doribax.

dorrigocins Glutarimide antibiotics that have moderate antifungal activity and modify the morphology of ras-transformed NIH/3T3 cells from a transformed phenotype to a normal one.

dorsal horn Part of the grey matter of the spinal cord, consisting of five zones (lamina I–V) where pain and other sensory information from the skin, striated muscles, or joints, or from blood vessels and internal organs, begins to be processed.

dorsalin-1 A protein of the *TGFβ family, originally isolated from chick, that regulates cell differentiation within the neural tube in conjunction with distinct ventralizing signals from the notochord and floor plate. The human homologue is probably growth differentiation factor 2 (*BMP9) which is mostly expressed in liver, though during development it is thought to have a role in regulating the cholinergic phenotype of some spinal cord neurons.

dorsal root ganglion (DRG) The thickened region of the dorsal root of the spinal cord that contains the cell bodies of afferent spinal nerve neurons leading into the *dorsal horn.

dorzolamide An inhibitor of carbonic anhydrase used topically in treatment of glaucoma because inhibition of carbonic anhydrase in the ciliary processes of the eye decreases aqueous humour secretion and reduces intraocular pressure. *TN* Trusopt.

dosage compensation A regulatory mechanism that allows genes to be expressed at a similar level irrespective of the number of copies at which they are present. This is important for genes on the X chromosome which need to be expressed to the same extent in XX females and XY males and seems to be achieved by randomly inactivating one of the X chromosomes in females. *See* LYON HYPOTHESIS. If a more general mechanism operated, trisomy (as in Down's syndrome) would not be a problem.

dose–response curve The relationship between dose (concentration of substance introduced into the system) and the effect (enzyme activity, membrane potential, mortality, etc.) that is being measured, expressed in graphical format. Dose-response curves are generally sigmoidal, as are receptor-binding curves, though exceptions can occur if there is cooperativity or if a second response begins to dominate at higer levels. A dose–response curve with a standard slope has a *Hill coefficient of 1.0.

dosimeter A device used to record radiation levels to which the individual has been exposed. A common form is a film badge in which a piece of film is variously shielded so that different types of radiation exposure (β, γ, etc.) can be determined.

dot blot *See* BLOTTING.

DOTS A shorthand for 'directly observed therapy, short-course', a World Health Organization standard strategy for drug treatment of tuberculosis in which patients are closely supervised to ensure the treatment regime is completed.

double-blind trial The only reliable testing method for the effectiveness of a drug or treatment, the gold standard. In the case of a double-blind drug trial neither the patient nor the experimenter scoring the effect knows whether the active agent or an inactive control (*placebo) is being taken and the possibility of bias is much reduced (there can be a problem in some cases when the effect is so obvious that the patient inevitably realizes that it is the active compound, not a placebo).

doublecortin A protein (441 aa) that seems to be expressed only in fetal brain, where it is important in neuronal migration. The protein is associated with microtubules and is often used as a marker for newly generated neurons. The gene is on the X chromosome and mutations lead, in males, to *lissencephaly with severe mental retardation and epilepsy and in heterozygous females to subcortical laminar heterotopia (SCLH), also known as 'double cortex' syndrome, associated with milder mental retardation and epilepsy. A family of **doublecortin like proteins** (including the retinitis pigmentosum 1 (RP1) gene product and doublecortin-like kinase 1 (Dclk)) has been identified, all with **doublecortin-like (DCX) domains** (usually in tandem). The domain is a microtubule-binding module and is involved in protein–protein interactions. Phylogenetically ancient, with homologues in invertebrates and unicellular organisms.

double diffusion A method used for detecting an antibody response in which both antigen and antibody are allowed to diffuse through an agar or agarose gel towards one another. Where the antibody/antigen ratio is balanced, a precipitin line will form and can be seen as an arc of opacity in the gel. Multiple samples can be arranged around a single antibody well and identity of antigens is revealed if precipitation lines coincide.

double helix A molecular conformation resembling a ladder twisted into a helix, famously the conformation of the DNA molecule in which the rungs are the bonds between complementary bases.

double minute *See* MDM.

double minute chromosome Small, paired extrachromosomal bodies comprising circular DNA, associated with many tumours where there may be multiple copies. A range of different genes can be amplified on double minute chromosomes and multiple copies of growth factor genes are associated with some tumours. Can also be formed by amplification of *DHFR* genes in response to methotrexate treatment.

double mutant Any organism in which the occurrence of two mutations allows an effect to be seen, because dual pathways mask a mutation in either one, or in some cases a mutation that reverses the effect of the first, restoring the normal phenotype.

double recessive A condition of homozygosity for a recessive allele so that the phenotype is expressed.

double strand break Damage to DNA involving breakage of both strands. *See* MRN COMPLEX; NIBRIN.

double-stranded RNA (dsRNA) Ribonucleic acid in which complementary base-pairing generates duplex regions similar to those seen in DNA. Some viruses have double-stranded RNA as their genetic material. Long double-stranded RNA can trigger RNA interference and in some cases the production of *interferon.

doublet microtubules A name often given to the outer nine sets of microtubules in the *axoneme, although strictly-speaking only one (the A tubule) has a complete set of thirteen protofilaments. The B tubule has only ten or eleven protofilaments, and shares the remainder with the A tubule. *Dynein attached to the A tubule makes transient connections to the B tubule of the adjacent doublet and undergoes a conformational change before detaching: this generates sliding movement between doublets which is transformed into bending of the cilium or flagellum.

doubling time The time taken for a population of cells to double in number and thus the time taken to complete the cell cycle if all the cells are actively dividing.

Douglas bag A bag in which expired air is collected for experimental determination of oxygen consumption.

down-regulation A term often used rather imprecisely. The process by which a cell reduces the response to the second and subsequent stimuli. Often due to a reduction in the number of receptors expressed on the surface.

Down's syndrome A congenital disorder caused by trisomy of chromosome 21. It is common (1 in 700 live births) and incidence increases with maternal age. The cause is usually nondisjunction at meiosis but occasionally a translocation of fused chromosomes 21 and 14. A whole range of complications arise from the excess in the products of the chromosome. The transient myeloproliferative disorder and megakaryoblastic leukaemia of Down's syndrome are associated with mutations in the gene for the *GATA1 transcription factor. The syndrome was formerly referred to as mongolism because of the supposed similarity of the facial features with those of the eponymous ethnic group. The **Down's syndrome critical region** (DSCR1) on chromosome 21 is the location of the genes that, triplicated, are responsible for the features of the syndrome. Among the genes in this region is that for *calcipressin-1.

downstream **1.** A casual term for things that happen late in a sequence, e.g. in a signalling cascade. **2.** Describing DNA or RNA sequences that are at a distance from the initiation sites and that will be translated or transcribed later.

doxapram hydrochloride A drug that stimulates respiration through peripheral carotid chemoreceptors and at higher doses by acting on central respiratory centers in the medulla. Occasionally used in the treatment of severe respiratory failure.

doxazosin A quinazoline compound that is a selective inhibitor (an *alpha-blocker) of the α_1-adrenergic receptors, used to treat hypertension and benign prostatic hyperplasia. *TN* Cardura.

doxepin A tricyclic antidepressant. *TNs* Prudoxin, Sinequan, Xepin, Zonalon.

doxorubicin (adriamycin) A cytotoxic anthracycline antibiotic isolated from *Streptomyces peucetius*, structurally very similar to daunomycin and used in clinical oncology. It blocks *topoisomerase and *reverse transcriptase by intercalating into the DNA. *TNs* Adria, Doxil.

doxycyclin (doxycycline) A broad-spectrum *tetracycline antibiotic, synthetically derived from oxytetracycline, used for treatment of various Gram-positive and Gram-negative infections and for chlamydial, rickettsial, and mycoplasmal infections. Used prophylactically for drug-resistant malaria. Antimalarial activity is a result of blocking the expression of apicoplast genes. *TN* Vibramycin.

Doyne honeycomb retinal dystrophy An autosomal dominant disease (malattia leventinese) caused by mutation in *fibulin-3 and characterized by yellow-white deposits known as drusen that accumulate beneath the retinal pigment epithelium.

DP_1, DP_2 **1.** Receptors for prostaglandin D_2. The two G-protein-coupled receptors seem to have opposite effects: e.g. neuroprotection by PGD is mediated by DP_1 whereas activation of DP_2 promotes neuronal loss. DP_2 is preferentially expressed on type 2 lymphocytes, eosinophils, and basophils and is thought to be important in the promotion of Th2-related inflammation. **2.** Transcription factors (E2F dimerization partners) that form heterodimers with *E2F and regulate the cell cycle. DP-1 is highly expressed in the ectoderm and all epidermal layers during embryogenesis; there are additional isoforms, DP-1α, 278 aa; DP-1β, 357 aa. DP2 exists in several alternatively-spliced forms. **3.** A pneumococcal bacteriophage (Dp-1) that encodes a lytic enzyme, *N*-acetylmuramoyl-L-alanine amidase.

DPA *See* DIPICOLINIC ACID.

Dpl *See* DOPPEL.

dpm Disintegrations per minute, a measure of radioactivity.

DPN (diphosphopyridine nucleotide) An obsolete name for *nicotinamide adenine dinucleotide (NAD).

dpp *See* DECAPENTAPLEGIC.

DPPC Dipalmitoyl-phosphatidylcholine.

d

DPT *See* TRIPLE VACCINE.

DR3 *See* DEATH (death receptors).

DRAK *See* DAP KINASE.

DRB One of the most polymorphic loci in the DR region of the class II MHC gene cluster, with over 100 reported alleles. Other loci are DQ and DP.

DREAM complex A multiprotein complex that represses cell cycle-dependent (*cdc) genes in quiescent cells. Subunits of the complex include the transcription factors E2F4, E2F5, MYBL1, MYBL2, Dp1, and Dp2, together with *retinoblastoma-like proteins, RBL1 and RBL2, the histone-binding protein, RBBP4 (retinoblastoma-binding protein 4, *chromatin assembly factor 1 subunit C, 425 aa), and LIN9, LIN37 (antolefinin), LIN52, and LIN54. The complex dissociates in S phase when the LIN proteins form a subcomplex that binds to MYBL2. *Compare* LINC COMPLEX.

D region A short sequence of amino acids in the variable region of immunoglobulin heavy chains which is coded by one or another of several separate genes. The combination of variable (V) and diversity (D) genes, joined by one or the other of the joining (J) region genes produces a VDJ recombinant and thus contributes to antibody diversity. Germ-line DNA has many (up to 200) different V region genes, in addition to 12 D region genes and 4 J region genes.

DRG *See* DORSAL ROOT GANGLION.

dropsy *See* OEDEMA.

drosha The nuclear RNase III (EC 3.1.26.3, 1374 aa) that initiates *microRNA (miRNA) processing. Originally described in *Drosophila*, but the human homologue is ubiquitiously expressed in tissues. Drosha interacts with the double-stranded-RNA-binding protein DGCR8 (deleted in *di George syndrome) and they are both present within the 650-kDa *microprocessor complex. Drosha is also a component of a larger complex containing multiple classes of RNA-associated proteins including RNA helicases, proteins that bind double-stranded RNA, novel heterogeneous nuclear ribonucleoproteins, and the Ewing's sarcoma family of proteins.

drug A substance with a physiological action used for the treatment of disease or the alleviation of pain. So-called 'recreational' drugs are diverse, and according to custom their use may be considered as drug abuse (opiates) or acceptable behaviour (alcohol in many cultures). Prolonged use often leads to addiction. Drug resistance can arise in infectious organisms (as is increasingly the case for bacteria involved in hospital infections, in TB, and in malaria) or in tissues, where tumour cells become resistant to cytotoxic drugs. *See* ANTIBIOTIC RESISTANCE.

drug delivery The means whereby a drug reaches the appropriate compartment of the body, which depends on the properties of the drug and the site where it is required. Delivery may be oral if it resists degradation at high pH and can be absorbed by the intestine, or may be by inhalation, injection, or transdermal diffusion. Topical application is appropriate for some superficial sites, in which case systemic absorption may well be undesirable.

drusen Extracellular deposits that accumulate below the retinal pigment epithelium on *Bruch's membrane. **Basal laminar drusen** (drusen of Bruch's membrane or cuticular drusen) are a type of early adult-onset drusen, uniformly small, slightly raised yellow subretinal nodules randomly scattered in the macula. They may progressively enlarge and eventually cause retinal detachment and loss of vision. Variations in *complement factor H are associated with basal laminar drusen.

DSB **1.** 2,6-di-*sec*-butyl phenol, an anaesthetic. **2.** 3-*O*-(3′,3′-dimethylsuccinyl)-betulinic acid, a small-molecule inhibitor of the proteolytic cleavage of HIV Gag protein. **3.** The disulphide bond formation protein (dsb) involved in processing of bacterial periplasmic proteins. **4.** *See* DOUBLE STRAND BREAK.

DSE *See* SERUM RESPONSE ELEMENT.

DSH *See* DYSCHROMATOSIS SYMMETRICA HEREDITARIA.

DSIP *See* DELTA SLEEP-INDUCING PEPTIDE.

DTE Dithioerythritol. *See* DITHIOTHREITOL.

DTH Delayed-type hypersensitivity. *See* HYPERSENSITIVITY.

DTIC *See* DACARBAZINE.

DTLET A synthetic selective agonist (D-Thr$_2$-Leu-enkephalin-Thr, deltakephalin) of the G-protein-coupled δ *opioid receptor.

DTNB A dithiol-oxidizing agent (Ellman's reagent, 5,5′- dithiobis-(2-nitrobenzoic acid)) that reacts with sulphydryl groups on proteins and releases 5-sulphido-2-nitrobenzoic acid which is spectrophotometrically detectable by absorption at 412 nm.

dual recognition hypothesis The hypothesis, now known to be incorrect, that antigens were recognized by two distinct receptors on T cells, one for the antigen, one for the associated MHC molecule; association with class I producing cytotoxic T cells, with class II producing T helper cells.

Duane retraction syndrome A congenital disorder of eye movement caused by developmental abnormality in cranial nerve VI. Duane retraction syndrome 1 (**DURS1**) maps to chromosome 8 and **DURS2** is caused by mutation in *chimerin-α. A third form (**DURS3**) involves deficiency in both adductor and abductor function.

Duchenne muscular dystrophy *See* MUSCULAR DYSTROPHY.

duct A general term for a tube formed of cells or lined with cells. A **ductule** is a small duct.

ductin Highly conserved multifunctional proteins (16 kDa) involved in the formation of membrane channels. When oriented with N- and C-termini on the cytoplasmic face they are components of *gap junctions; in the opposite orientation they are subunits of vacuolar-type proton pumps (C subunit of *V-type ATPases). This has, however, been controversial.

ductless gland An *endocrine gland.

ductus arteriosus The blood vessel that links the pulmonary artery to the aorta in the fetus and reduces blood flow through the fluid-filled lungs. Normally closes at birth or very shortly after; persistence (**patent ductus arteriosus**) causes shunting of blood from the aorta to the pulmonary artery with eventual problems due to the inefficiency of the circulation.

Duffy One of the *blood group antigens determined by a single gene on chromosome 1. A glycoprotein (gp-Fy; CD234; 35 to 45 kDa), named GPD, on the surface of erythrocytes and endothelium of postcapillary venules has the antigenic determinants. There are five phenotypes (Fy-a, Fy-b, Fy-o, Fy-x, and Fy-y) and five antigens. Nearly all white people are Duffy-positive, whereas most people of African descent are Duffy-negative, probably because the Fy(a–, b–) phenotype confers resistance to infection by *Plasmodium vivax*. The **Duffy blood group antigen** (Duffy antigen receptor for chemokines, DARC) is also the receptor for the chemokines interleukin-8 and melanoma growth stimulatory activity (CXCL1) and these chemokines compete for binding with the malarial parasite ligand.

SEE WEB LINKS

- Blood Group Antigen Gene Mutation Database.

duodenum The region of the small intestine immediately downstream from the stomach. **Duodenal glands** : *See* BRUNNER'S GLANDS.

duplex Casual term for double-stranded nucleic acid.

duplicon Chromosome-specific low-copy repeats (2–10+) found in various regions of the human genome, varying in size from a few kilobases in length to hundreds of kilobases. Recombination between duplicated segments can lead to deletion, further duplication, inversion, etc. and dosage imbalance caused by duplication can cause abnormalities. There appear to be two general classes of duplicons, sometimes separated by internal $(TTAGGG)_n$-like islands; large and highly similar centromerically positioned subtelomere duplications and more abundant, dissimilar distal duplicons. More stringent definitions are used in detailed genomic analyses (e.g. >90% nucleotide sequence identity and >1 kb in size).

Dupuytren's contracture A disease of the soft tissues of the palm and fingers characterized by a progressive fibroma-like thickening of connective tissue underlying the palm of the hand that causes a flexion contracture. The condition is commoner in men than in women and is hereditable. It has been associated with multiple fibroproliferative conditions, suggesting an underlying defect in wound repair.

dural ectasia A widening or ballooning of the dural sac that surrounds the spinal cord, usually at the lumbosacral level. Can occur in *Marfan's syndrome.

dura mater The outer meningeal membrane that covers the brain and spinal cord. The *pia mater adjacent to the brain is separated from the dura mater by the arachnoid membrane (arachnoid mater) and the subarachnoid space.

duramycin A group of type B *lantibiotics. Duramycin A was isolated from *Streptoverticillium hachijoense*, duramycins B and C from *Streptoverticillium* strain R2075 and *Streptomyces griseoluteus* respectively. A structurally similar compound, cinnamycin, has been isolated from *Streptomyces longisporoflavus*. All are potent inhibitors of phospholipase A2

dutasteride A selective inhibitor of type 1 and type 2 isoforms of steroid 5α-reductase, an enzyme that converts testosterone to 5α-dihydrotestosterone (DHT) and used to treat benign prostatic hyperplasia. *TNs* Avodart, Duagen.

d

DVT Deep vein thrombosis. *See* THROMBOSIS.

DXM *See* DEXTROMETHORPHAN.

dyad Generally, any two individuals or units that form a pair, e.g. the homologous chromosomes before they separate during mitosis.

dyad symmetry element (DSE) *See* SERUM RESPONSE ELEMENT.

dydrogesterone A synthetic analogue of *progesterone used in the treatment of menstrual disorders and endometriosis and in hormone replacement therapy. Withdrawn from use in the UK in March 2008. *TN* Duphaston.

dye coupling A description of the intercellular communication, usually through *gap junctions, detected by the passage of microinjected fluorescent dye (e.g. *lucifer yellow) from one cell to another. *See also* ELECTRICAL COUPLING.

Dyggve–Melchior–Clausen syndrome A rare autosomal recessive disorder characterized by short trunk dwarfism, microcephaly, and psychomotor retardation. The rough endoplasmic reticulum is dilated and there are enlarged and aberrant vacuoles and numerous vesicles. Caused by a defect in *dymeclin.

dymeclin A protein (Dyggve-Melchior-Clausen syndrome protein, 669 aa) associated with the Golgi. Defects are associated with *Dyggve–Melchior–Clausen syndrome and Smith–McCort dysplasia (SMC). SMC is a rare autosomal recessive osteochondrodysplasia characterized by short limbs and trunk with barrel-shaped chest. The radiographic phenotype includes platyspondyly, generalized abnormalities of the epiphyses and metaphyses, and a distinctive lacy appearance of the iliac crest, features identical to those of Dyggve–Melchior–Clausen syndrome

dynactin A stable 20S multiprotein assembly (10 MDa) that is required for dynein-driven movement of cytoplasmic organelles along microtubules. There are ten distinct polypeptides including dynactin 1 (1278 aa), *dynamitin, and *centractins. Null mutations are lethal but heterozygotes have been reported with progressive neuropathies and mutations in the dynactin 1 gene may be a susceptibility factor for *amyotrophic lateral sclerosis.

dynamic equilibrium The situation in which forward and reverse reaction rates are exactly balanced. This may be a rather transient state, although most methods for measuring rate constants of reactions assume that such an equilibrium has been reached. Alterations in reaction rates alter the concentrations of reactants and products.

dynamin A subfamily of GTP-binding proteins of which at least three are found in mammalian systems (dynamins 1–3). **Dynamin 1** is a large GTPase (EC 3.6.5.5, 864 aa) that is thought to act mechanochemically in constricting the neck of forming endocytotic vesicles, particularly in synapses. **Dynamin 2** has a similar role but in T-cell interactions and in renal cells. It associates with *amphiphysin in cells in culture. **Dynamin 3** is highly expressed in brain and spinal cord. **Dynamin-binding protein** (scaffold protein tuba, 1577 aa) belongs to the DBL family of guanine nucleotide exchange factors and an association with late-onset Alzheimer's disease has been reported.

dynamitin An essential subunit (p50, Jnm1p, 401 aa) of the *dynactin complex. If overexpressed it will interfere with the *dynein–dynactin interaction involved in vesicle transport. *Immunophilins reportedly link to dynein indirectly via dynamitin. The dynactin subunit, p24, binds to dynamitin and affects oligomerization.

dyneins A family of motor ATPases that interact with microtubules. There are two main classes, cytoplasmic and axonal, the former responsible for transport of vesicles and various other intracellular movements towards the minus ends of microtubules, the latter involved in generating sliding of axonemal doublets relative to one another and thus for the active bending of cilia and flagella. Conventional **cytoplasmic dyneins** have two identical heavy chain polypeptides (4307 aa) that undergo conformational change on ATP hydrolysis, and a number of intermediate (~520 aa) and light (~350 aa) chains. **Axonemal dynein** is attached to the A tubule of the nine peripheral microtubule doublets and interacts with the adjacent B tubule. At least fourteen human axonemal dynein heavy chains are known, with different tissue distributions, and there are several intermediate and light chain variants. Mutations in axonemal dynein are responsible for *primary ciliary dyskinesia (*see also* KARTAGENER'S SYNDROME).

dynorphin An family of endogenous opiate peptides with a leucine-enkephalin structure that are agonists for the κ opioid receptor. They are all derived from the prodynorphin (preproenkephalin B) gene which is within a cluster of genes for peptides such as arginine *vasopressin and *oxytocin. Produced in various parts of the brain, particularly the hypothalamus. The dynorphins include dynorphin A, dynorphin B, α- and β-neoendorphin, and big dynorphin.

dyrks A highly conserved family of protein kinases (dual-specificity tyrosine phosphorylation-regulated kinases) with at least seven mammalian isoforms that autophosphorylate a tyrosine residue in their activation loop but phosphorylate exogenous substrates on serine/threonine residues. The closest relative of Dyrk1 in *Drosophila* is the minibrain (*mnb*) gene that has a role in postembryonic neurogenesis. Human **Dyrk1A** maps to the Down's syndrome critical region on chromosome 21 and may be involved in pathogenesis of certain phenotypes of *Down's syndrome. **Dyrk2** shares 46% identity with dyrk1 but is on chromosome 12 and regulates p53 to induce apoptosis in response to DNA damage. It may also phosphorylate NFAT, thereby keeping it in the cytoplasmic compartment. **Dyrk3**, the product of a gene on chromosome 1, is very similar to Dyrk2 but expressed mostly in testis.

dys- Prefix generally denoting abnormal, difficult, or impaired, even dysfunctional.

dysautonomia A general term for any disease or malfunction of the autonomic nervous system which can manifest in a wide variety of ways. **Familial dysautonomia** (*hereditary sensory and autonomic neuropathy type 3), is caused by mutations in the gene that encodes IκB-kinase associated protein. There is demyelination in nerves in various parts of the brain and loss of cells in autonomic ganglia.

dyschromatosis symmetrica hereditaria An autosomal dominant disorder (DSH, reticulate acropigmentation of Dohi) in which there are regions of hyper- and hypopigmentation on the face and dorsal regions of hands and feet that appear in childhood and persist into adulthood, but with no adverse effect. Can be caused by mutation in the **ADAR1* gene. **Dyschromatosis universalis hereditaria** (DUH) occurs most commonly in Japanese people but differs in that the pigmented macules are found mostly on the trunk. There is evidence of linkage to a locus on chromosome 6.

dyscrasia A term used for an illness in which there are abnormal materials, usually abnormal cells, in the blood.

dysentery A severe diarrhoea caused by infection of the gut with *Shigella dysenteriae*, *Shigella flexneri*, *Shigella boydii*, or *Shigella sonnei* (shigellosis or **bacillary dysentery**) or with *Entamoeba histolytica* (**amoebic dysentery**).

dysferlin One of a family of proteins similar to *Caenorhabditis elegans* *ferlin. They are all transmembrane proteins with multiple C2 domains, which are implicated in calcium-dependent membrane fusion events. Dysferlin is required for vesicle fusion during calcium-induced muscle membrane repair and mutations in the dysferlin gene (*DYSF*) underlie two main muscle diseases: limb girdle muscular dystrophy (LGMD)-2B and Miyoshi myopathy (MM). Other ferlin homologues are found in myoblasts (myoferlin, Fer1L3) and cochlea (*otoferlin). *See* MUSCULAR DYSTROPHY.

dyskeratosis congenita A bone marrow failure syndrome that can be a result of mutations in the genes for the RNA (TERC) or reverse transcriptase (TERT) components of *telomerase or in *TIN2.

dyskinesia A condition in which the ability to control voluntary movements is impaired; a common side effect of some *antipsychotic drugs.

dysostosis A defect in the process of ossification of cartilage (sometimes dyschondroplasia: *see* ENDOCHONDROMA) or the occurrence of bone in inappropriate locations.

dysplasia Literally 'wrong form'. Usually used to denote early precancerous changes in epithelia, particularly cervical intraepithelial neoplasia, the precursor lesion of cervical carcinoma. *Compare* METAPLASIA; NEOPLASIA.

dyspnoea Difficulty with breathing. There can be various underlying causes, some of which are reversible. Four major disease groups may cause dyspnoea: *chronic obstructive pulmonary disease (COPD), chronic heart diseases, neurological diseases (e.g. amyotropic lateral sclerosis, multiple sclerosis) that weaken the muscles involved and cancer (often through blockage of airways by primary or secondary tumours). Other causes include asthma and pneumonia.

dyssynergia A collection of rare, degenerative, neurological disorders (dyssynergia cerebellaris progressiva, dentate cerebellar ataxia, dentatorubral atrophy, primary dentatum atrophy (formerly Ramsay Hunt syndrome I)) characterized by epilepsy, cognitive impairment, myoclonus (muscle spasm), and progressive ataxia. Symptoms include seizures, tremor, and reduced muscle coordination. Onset of the disorder generally occurs in early adulthood. Tremor may begin in one extremity and later spread to involve the entire voluntary muscular system. Arms are usually more affected than legs. Some cases are due to mitochondrial abnormalities.

d

SEE WEB LINKS

- Further details in the National Institute of Neurological Disorders and Stroke database.

dystrobrevins A family of *dystrophin-associated proteins with α and β isoforms that have significant sequence homology with several protein-binding domains of the dystrophin C-terminal region. Five distinct mRNA transcripts of **α-dystrobrevin** are expressed in tissue-specific ways; **β-dystrobrevin** is the nonmuscle form and is abundantly expressed in brain and other tissues. Dystrobrevin interacts with *kinesin and may play a role in the transport and targeting of components of the dystrophin-associated protein complex to specific sites in the cell. **Dystrobrevin-binding protein 1** (dysbindin, 351 aa) is ubiquitously expressed and through dystrobrevin is linked to the dystrophin-associated protein complex (DPC), which appears to be involved in signal transduction pathways, and to *BLOC-1. The gene has been suggested as a potential susceptibility gene for schizophrenia. α- and β-Dystrobrevin colocalize at the sarcolemma of most muscle fibres and in large blood vessels and in axon bundles in the brain. Normal function is important for platelet-dense granule and melanosome biogenesis. *See* HERMANSKY–PUDLAK SYNDROME.

dystroglycan A component of the *dystrophin-associated protein complex, dystrophin-associated glycoprotein, that is proteolytically processed to produce **α-dystroglycan** (624 aa) which is an extracellular laminin-binding glycoprotein that remains associated with transmembrane **β-dystroglycan** (242 aa) that associates with *dystrophin in the cytoplasm. The actin cytoskeleton is linked, via dystrophin and dystroglycans, to the extracellular matrix. α-Dystroglycan also binds *agrin and *neurexin. Post-translational modification of α-dystroglycan by the *LARGE acetylglucosaminyltransferase is essential for normal function and to prevent muscle degeneration. Hypoglycosylation of dystroglycan is a feature of *Fukuyama muscular dystrophy and *muscle–eye–brain disease.

dystrophia adiposogenitalis A range of endocrine abnormalities (adiposogenital dystrophy, Fröhlich's syndrome) believed to result from damage to the hypothalamus that resembles *Prader–Willi syndrome but without an identified genetic lesion. Characterized primarily by obesity and hypogonadotrophic hypogonadism in adolescent boys.

dystrophia myotonica A multisystem autosomal dominant disorder (myotonic dystrophy), the most common form of *muscular dystrophy in adults. **Dystrophia myotonica 1** (DM1) is caused by an amplified trinucleotide repeat in the 3′-untranslated region of the dystrophia myotonica protein kinase gene (the kinase, 624 aa, has homology with cAMP-dependent serine/threonine kinases). It is thought that the mutant RNA transcript interferes with expression or processing of other RNA species. **DM2** (Ricker's syndrome) is caused by expansion of a CCTG repeat in intron 1 of the zinc finger protein-9 gene.

dystrophin A protein (3685 aa) that is missing in Duchenne *muscular dystrophy (DMD) and present in abnormal truncated form in the less severe Becker form. In X-linked *dilated cardiomyopathy dystrophin is mutated but skeletal muscle is normal. Dystrophin represents only about 0.002% of total striated muscle protein and is part of the linkage between cytoplasmic actin and extracellular matrix that involves the *dystrophin-associated protein complex and *dystroglycan. In the brain, dystrophin is localized subcellularly to the postsynaptic density and DMD is sometimes associated with mental retardation. There are sequence homologies between dystrophin and both nonmuscle α-actinin and spectrin.

dystrophin-associated protein complex A multiprotein complex (DPC, costamere) that links the actin cytoskeleton with extracellular matrix. According to the tissue, *syntrophin, *dystrobrevin, *sarcoglycan, and *dystroglycan isoforms are involved and defects in any of the components of the chain can cause disease. In muscle, defects lead to *muscular dystrophies (*see also* dystrophin). The role of the complex may be to cluster and anchor signalling proteins and ion and water channels.

dystrophy A condition in which there is impaired or imperfect nutrition of an organ or tissue. *See* DYSTROPHIA ADIPOSOGENITALIS; FUCHS' ENDOTHELIAL DYSTROPHY; MUSCULAR DYSTROPHY.

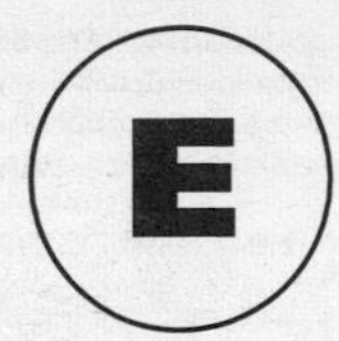

E Single-letter code for *glutamic acid.

***E1* (*E1A, E1B*)** Adenoviral oncogenes that interact with the Rb *tumour suppressor gene product. E1A (~290 aa) inhibits transcriptional activation by both p73α and p73β. E1B (19 kDa, 175 aa) is antiapoptotic and is essential for cell transformation by adenovirus and for the regulation of viral early gene transcription.

E1–E2-type ATPase A superfamily of ion-pumping ATPases (P-type ATPases), present in prokaryotes and eukaryotes, that can be divided into four main groups:
1. calcium-transporting ATPases (e.g. of sarcoplasmic reticulum), 2. sodium-potassium ATPases of plasma membrane and hydrogen-potassium ATPase of gastric mucosa, 3. plasma membrane proton pumps of plants, fungi, and lower eukaryotes and 4. all bacterial P-type ATPases, except the magnesium ATPase of *Salmonella typhimurium*. The catalytic (α) subunit or region generally has multiple membrane-spanning domains and the second (β) subunit is involved in stabilizing and transporting the α subunit to the membrane. The pumps are generally inhibited by ATP analogues such as vanadate, but selectively by other inhibitors (e.g. ouabain for the sodium-potassium ATPase, thapsigargin for the calcium ATPase). A general feature is that an aspartyl-phosphoryl enzyme intermediate is formed during the catalytic cycle. *Compare* V-TYPE ATPASE

E1 enzymes *See* UBIQUITIN CONJUGATING ENZYMES.

E2 enzymes *See* UBIQUITIN CONJUGATING ENZYMES.

E2F A family of transcription factors that bind to promoters for various genes involved in the G_1 and S phases of the cell cycle and are regulated by interaction with *retinoblastoma (Rb) proteins. **EMA** (E2F-binding site modulating activity) is a transcriptional repressor similar to E2F but without the C-terminal activation domain.

E3 ligase *See* UBIQUITIN CONJUGATING ENZYMES.

E5, E6, E7 Papillomavirus oncogenes. The *E5* product binds and blocks the 16-kDa (155 aa) *proteolipid of the *V-type ATPase so that growth factor receptors are not processed properly and growth factor signal transduction pathways are stimulated. E6 and E7 proteins drive cell proliferation through their association with *PDZ domain proteins and the *retinoblastoma gene product.

E 64 An inhibitor of most cysteine peptidases in clan CA but, despite early claims, not a general inhibitor of cysteine peptidases: some other peptidases are reversibly inhibited and some cysteine peptidases are not.

EAA *See* EXCITATORY AMINO ACIDS.

EaA cells An insect cell line used for baculovirus-based expression. Derived from the salt marsh caterpillar *Estigmene acrea*.

EACA *See* AMINOCAPROIC ACID (ε-aminocaproic acid).

EAC rosettes *Scc* E ROSETTING.

Eadie–Hofstee plot A graphical method for analysing enzyme kinetics. The velocity of reaction (*V*) is plotted on the ordinate, *V*/*S* on the abscissa, *S* being the initial substrate concentration. The intercept on the ordinate is V_{max}, the slope is $-K_m$. A better method than using the *Lineweaver–Burke plot.

EAE *See* EXPERIMENTAL ALLERGIC ENCEPHALOMYELITIS.

early antigens Virus-coded cell surface antigens that appear soon after the infection of a cell by virus, but before virus replication has begun. They are among the products of early genes found in the early region of the viral genome, the first genes to be expressed.

early growth response genes A family of genes (*Egr* genes) that encode Cys(2)-His(2)-type zinc finger proteins; transcription factors that are up-regulated after a variety of stresses. The **EGR1** gene product (KROX-24) acts upstream of TGFβ1, **Egr-2** (KROX-20) has similar but distinguishable characteristics, **Egr-3** regulates the early stages of thymocyte proliferation and mediates retina-derived stimulation of the suprachiasmatic nucleus. *See also* IMMEDIATE EARLY GENES.

EA rosettes *See* *E ROSETTING.

EAST Epidermal growth factor receptor-associated protein with SH3 and TAM domains. *See* EPIDERMAL GROWTH FACTOR RECEPTOR.

e

EB1 A member of the RP/EB family of proteins (adenomatous polyposis coli-binding protein EB1, microtubule-associated protein, RP/EB family member 1, MAPRE1, 268 aa) that was isolated in a two-hybrid screen of the APC protein (*see* ADENOMATOUS POLYPOSIS COLI), to which it binds. During interphase it is bound to plus-ends of microtubules and during mitosis is associated with the centrosomes and spindle microtubules. Associates with components of the *dynactin complex and the intermediate chain of cytoplasmic *dynein.

EBBP *See* OESTROGEN-RESPONSIVE B BOX PROTEIN.

Ebola virus A *Filovirus, morphologically similar to Marburg virus, that causes severe haemorrhagic fever, frequently fatal. The genome is a single-stranded negative-sense RNA. Originally identified in the Democratic Republic of the Congo (Ebola valley); outbreaks have so far been mostly confined to Africa. Various subtypes are now recognized.

E box The nucleotide motif (CANNTG, where N is any nucleotide) that is the binding site for basic helix–loop–helix transcription factors, found in numerous promoters and enhancers.

EBP50 A widely distributed protein (ERM-binding phosphoprotein 50 kDa, Na^+/H^+ exchanger regulatory factor, solute carrier family 9, isoform A3, 358 aa) present at higher levels in tissues with polarized epithelia and microvilli. Associates with *ezrin, and with the *cystic fibrosis transmembrane conductance regulator (CFTR) linking them to the cortical actin cytoskeleton.

Ebstein's anomaly A rare abnormality in the tricuspid valve such that it leaks blood backwards into the right atrium. It may be associated with rearrangements of chromosomal region 11q where there are several genes thought to be involved in cardiac development.

eburnation Ivory-like hardening of bone at joint surfaces, a common anatomical marker of chronic arthropathy. The eburnated surface is less extensive in osteoarthritis than in advanced rheumatoid arthritis.

SEE WEB LINKS

- Abstract of review on eburnation in *Clinical Rheumatology* (2005/6).

EBV *See* EPSTEIN–BARR VIRUS.

EC_{50} The concentration of a substance required to produce 50% of the maximal effect or, sometimes, an effect in 50% of the organisms treated.

EC cells **1.** Endocrine cells. **2.** *See* EMBRYONAL CARCINOMA CELLS.

ecchondroma (osteochondroma) A protrusion, usually benign, growing from the surface of bone, capped with cartilage. In some cases ecchondrosis may be associated with arthritis.

ecchymosis A discoloured patch of skin due to extravasation of blood, usually as a result of mechanical trauma (bruising).

eccrine Describing a gland in which the secretory product is excreted from the cells. Sweat glands are the classic example.

ecdemic A disease that originates elsewhere, not *endemic.

ecdysis Shedding of skin: a term more commonly encountered in the context of arthropods.

ECE *See* ENDOTHELIN CONVERTING ENZYMES.

ECF *See* EOSINOPHIL CHEMOTACTIC PEPTIDE.

ECG *See* ELECTROCARDIOGRAM.

Echinacea A genus of North American perennial plants (e.g. purple coneflower) from which herbal remedies are prepared. Such extracts are claimed to boost the immune system.

echinocandins A class of antifungal agents that inhibit, noncompetitively, the synthesis of cell-wall 1,3-β-glucans. They are used in clinical treatment of candidiasis and aspergillosis. Examples include caspofungin, micafungin, mulundocandin.

SEE WEB LINKS

- Additional information on caspofungin in the Fungal Research Trust database.
- Further information, chiefly on micafungin.

Echinococcus A genus of cestodes that normally parasitize canids where as adults they are intestinal worms. Eggs are ingested by the secondary host (usually herbivores such as sheep, but occasionally humans), hatch to produce oncospheres which migrate into lung, liver, or brain where they form *hydatid cysts. The larvae proliferate within the cyst (or within secondary cysts that can form if the primary cyst bursts) and are eaten by the primary host to complete the cycle. *Echinococcus granulosus* and *E. multilocularis* are the two species that can cause hydatid disease.

echinocytes Erythrocytes (burr cells) that have adopted a shrunken configuration with a spiky appearance as a result of being placed in hypertonic medium.

echinosporin An antibiotic isolated from *Streptomyces echinosporus* and other *Streptomyces* spp. Has weak activity against Gram-positive and -negative microorganisms and blocks tumour cell proliferation.

echistatin A *disintegrin present in the venom of the saw-scaled viper *Echis carinatus*.

echocardiogram *See* CARDIO-.

echocardiography A method that uses reflected pulsed ultrasound to assess heart function.

ECHO viruses A group of picornaviruses (enteric cytopathic human orphan viruses) that are a common cause of enteric infections, particularly in children, and can occasionally cause aseptic meningitis.

ECL **1.** Electrochemiluminescence. **2.** Enhanced chemiluminescence.

eclampsia Convulsions, probably due to reduced cerebral blood flow, that occur during or immediately after pregnancy. It is rare for pre-eclampsia to progress to eclampsia.

E classification A generally accepted scheme for enzyme classification, as recommended by the International Union of Biochemistry. The first number shows the broad type of enzyme (1, oxidoreductase; 2, transferase; 3, hydrolase; 4, lyase; 5, isomerase; 6, ligase (synthetase)). The second and third numbers indicate subsets within these broad categories, and the last number, unique to each enzyme, is assigned arbitrarily.

SEE WEB LINKS

- E classification database.

ecm *See* EXTRACELLULAR MATRIX.

EC number *See* E CLASSIFICATION.

E. coli *See* ESCHERICHIA COLI.

econazole An imidazole with broad-spectrum antifungal properties that acts by inhibiting ergosterol synthesis. *TN* Spectazole.

Economo's disease (von Economo's disease) *See* ENCEPHALITIS.

Eco RI, RII Type II *restriction endonucleases isolated from **Escherichia coli*. **EcoRI** cuts the sequence GAATTC between G and A, **EcoRII** cuts the sequence CC(T/A)GG in front of the first C: both produce 5′ *sticky ends.

ecotropic viral integration site 1 A gene (*Evi-1*) expressed at high level during the development of the mouse urinary system, Müllerian ducts, lung, and heart, but at low level in most of the adult tissues; the human homologue is expressed abundantly during development of kidney, lung, pancreas, and ovaries, and to a lesser extent elsewhere. Inappropriate expression of Evi-1 in the adult is associated with acute and chronic myelogenous leukaemia and *myelodysplastic syndrome. Evi-1 (1051 aa) is a DNA-binding protein with multiple sets of *zinc finger motifs.

ecotropic virus Generally, a virus that will only replicate in its original host species. Most commonly refers to murine retroviruses, others of which infect only heterologous species (xenotropic viruses); a third category (amphotropic viruses) infect cells of both original and heterologous species. The distinction is important when considering the risks of xenografting; pig endogenous retroviruses (PERVs) have been shown to infect human cells.

ECP *See* EOSINOPHIL CATIONIC PROTEIN.

ecstasy Colloquial name for the recreational drug MDMA (3,4-methylenedioxymethamphetamine), which has effects similar to methamphetamine (a stimulant) and mescaline (a hallucinogen).

ECT *See* ELECTROCONVULSIVE THERAPY.

ecthyma A dermal ulcer caused by group Ab haemolytic streptococci.

ectoderm The outer of the three germ layers of the embryo that gives rise to epidermis and neural tissue.

ectodermal dysplasia Hypohidrotic ectodermal dysplasia, a congenital disorder of teeth, hair, and eccrine sweat glands, is usually inherited as an X-linked recessive trait, although rarer autosomal dominant and autosomal recessive forms are known. *See* IκB. McGrath's syndrome (ectodermal dysplasia syndrome) is caused by mutations in the *plakophilin-1 gene. *See* DESMOSOME; ECTODYSPLASIN.

ectodermal dysplasia with ectrodactyly and macular dystrophy A heterogeneous group of autosomal recessive disorders (EEM syndrome, Albrectsen–Svendsen syndrome, Ohdo–Hirayama–Terawaki syndrome) in which there is abnormal development of two or more ectodermal structures; caused by defects in *cadherin-3.

e

ectoderm-neural cortex-1 A putative oncogene (*ENC-1*, p53-induced gene 10) highly overexpressed in group I vs group II parathyroid adenomas. The product is an actin-binding protein of the *Kelch family that is an early and highly specific marker of neural induction in vertebrates. Expression of *ENC1* has been shown to induce the formation of neuronal processes. Also expressed in adipose tissue, where it appears to play a regulatory role early in adipocyte differentiation.

ectodysplasin A protein expressed in keratinocytes, hair follicles, sweat glands, and in other adult and fetal tissues, which is defective in X-linked anhidrotic ectodermal dysplasia. Isoform 1 (135 aa) is predicted to contain a single transmembrane domain. The larger isoform 2 (391 aa) may be the active protein. Probably a member of the TNF-related ligand family and involved in early epithelial-mesenchymal interactions. Binds to a specific receptor (EDAR) that can activate NFκB, JNK, and cell death pathways. *See* HYPOHIDROSIS.

ectoenzyme An enzyme that acts on extracellullar substrates either because it is secreted or because it is located on the outer surface of the cell.

ectoglycosidase *See* EXOGLYCOSIDASE.

ectopia (ectopy, *adj.* **ectopic)** Displacement from normal position. **Ectopia vesicae** is a congenital abnormality in which the anterior wall of the bladder is absent. An **ectopic pregnancy** is one in which the ovum is fertilized and develops outside the uterus.

ectopy *See* ECTOPIA.

ectromelia **1.** The congenital absence or gross shortening of the long bones of one or more limbs. **2.** Ectromelia virus (EV, mousepox) is a dsDNA virus of the poxvirus family that has been used as a model for generalized infection and evasion of the immune response.

eczema An inflammatory skin condition usually due to immune hypersensitivity to food or environmental allergens.

eczematous conjunctivitis *See* PHLYCTENULAR CONJUNCTIVITIS.

ED1 **1.** An antibody directed against a lysosomal membrane protein (similar to CD68) that is often used as a marker for monocytes/macrophages in the rat. **2.** A gene associated with X-linked hypohidrotic ectodermal dysplasia. *See* ECTODYSPLASIN.

ED_{50} The median effective dose required to produce 50% of the maximal response or a response in 50% of individuals treated with the substance.

edeines Basic polypeptide antibiotics produced by *Bacillus brevis* Vm4. They inhibit bacterial DNA synthesis, but not RNA synthesis. In cell-free systems they inhibit bacterial protein synthesis.

SEE WEB LINKS

- Link to edeine on a site with useful and interesting microbiology informational resources.

edema *See* OEDEMA.

editosome The multiprotein complex involved in *RNA processing.

Edman degradation *See* EDMAN REAGENT.

Edman reagent A reagent (phenyl isothiocyanate) used in peptide sequencing. The N-terminal residue of the peptide reacts with the reagent and is removed as a phenylthiohydantoin derivative; the process is then repeated.

EDRF *See* NITRIC OXIDE.

edrophonium An anticholinesterase with a short half-life used in diagnosis of *myasthenia gravis: the drug will cause a transient increase in muscle contractility by overcoming the deficit in muscarinic acetylcholine receptors. *TNs* Enlon, Reversol, Tensilon.

EDTA A chelating agent (ethylenediaminetetraacetic acid) for divalent cations, often used as the disodium salt. Binding affinity is pH sensitive and at neutral pH the affinity for calcium is greater than for magnesium. *See* EGTA.

Edward's syndrome One of the commonest congenital abnormalities (1 in 6000 live births), trisomy 18, that produces a wide range of problems and has a very poor prognosis.

EEA **1.** Early endosome antigen 1 (EEA1). Autoantibodies to EEA1 were first reported in the serum of a patient with subacute cutaneous lupus erythematosus. **2.** *Euonymus europaeus* agglutinin, a lectin from the European spindle tree that binds blood group B oligosaccharides and high mannose *N*-glycans. **3.** End-to-end anastomosis, a surgical procedure.

EED A *polycomb group (PcG) protein (WD protein associating with integrin cytoplasmic tails 1,WAIT-1, 441 aa) that is a component of the *PRC2/EED-EZH complexes. Expression of EED peaks at the G_1/S phase boundary and it is overexpressed in breast and colon cancer.

EEG *See* ELECTROENCEPHALOGRAM.

EEM syndrome *See* ECTODERMAL DYSPLASIA WITH ECTRODACTYLY AND MACULAR DYSTROPHY.

EF-1 *See* ELONGATION FACTOR.

E face A term referring to the inner face of the outer lipid monolayer of a membrane as visualized by the technique of freeze fracture. The **P face** is the complementary surface. In the case of the plasma membrane, the E face is the surface that would be seen from within the cell if the inner leaflet of phospholipids were stripped away, the P face the surface that would be seen from outside the cell if the outer monolayer of phospholipid were removed.

efalizumab A recombinant humanized monoclonal antibody that binds to CD11a (LFA-1) and acts as an immunosuppressant. Used to treat psoriasis. *TN* Raptiva.

efferent Describing (e.g.) a nerve or a blood vessel that leads away from a centre or source. The opposite of *afferent.

effusion An abnormal release of fluid into the tissues or body cavities.

EF-hand A calcium-binding motif that involves a loop of twelve amino acids with α-helices at each end. Members of the family include *aequorin, *α-actinin, *calbindin, *calcineurin, *calcyclin, *calcyphosin, *calmodulin, *calpain, *diacylglycerol kinase, *fimbrin, *myosin regulatory light chains, *oncomodulin, *osteonectin, *spectrin, *troponin C.

eflornithine An inhibitor of ornithine decarboxylase, used in the treatment of trypanosomiasis and of hirsutism.

eformoterol *See* FORMOTEROL.

EF-Tu *See* ELONGATION FACTOR.

egasyn A glycoprotein (567 aa) in the lumen of the endoplasmic reticulum that binds β-*glucuronidase and retains it in the GERL compartment. It has a second function as an enzyme, being identical to esterase 22, one of a family of carboxylesterases (EC 3.1.1.1) with broad substrate specificities.

EGF *See* EPIDERMAL GROWTH FACTOR.

EGFP *See* GREEN FLUORESCENT PROTEIN.

EGF receptor *See* EPIDERMAL GROWTH FACTOR.

eglin C A serine-peptidase inhibitor (70 aa) originally isolated from the medicinal leech *Hirudo medicinalis* but now produced as a recombinant protein. Particularly effective as an inhibitor of neutrophil elastase and cathepsin G.

Egr-1, Egr-2, Egr-3 *See* EARLY GROWTH RESPONSE GENES.

EGS *See* EXTERNAL GUIDE SEQUENCE.

EGTA A chelator of various divalent cations (ethylene glycol tetraacetic acid) that has greater selectivity for calcium than magnesium at pH 7 than *EDTA. Because protons compete for binding, the apparent association constant is very pH sensitive.

EH domain *See* EPS15.

Ehlers–Danlos syndrome A group of congenital disorders in which genes for collagen are defective. Ehlers–Danlos syndrome **types I and II** affect alpha-1(V), alpha-2(V) or alpha-1(I) collagen and are characterized by loose-jointedness and fragile, bruisable skin. **EDS III** is a benign form of classic EDS. In **type IV** there is a defect in the gene for type III collagen (an autosomal dominant disorder). **Type VIIC** is caused by a defect in the peptidase that removes the N-terminal extension peptides from collagen I. Other forms are also recognized with defects in various aspects of collagenous connective-tissue production.

Ehrlich ascites A mouse carcinoma-type cell line, originally derived from mammary carcinoma, that has been extensively used as a model system. The cells will grow either as a solid tumour or in ascites form.

Ehrlichia A genus of rickettsia that cause serious tick-borne human zoonoses, infecting mononuclear cells and granulocytes. A wide range of other mammals are also susceptible to ehrlichiosis.

EHS cells *See* ENGLEBRETH–HOLM–SWARM SARCOMA CELLS.

eicosanoids Generic term for compounds derived from arachidonic acid; includes *leukotrienes, prostacyclin, prostaglandins, and *thromboxanes.

eIF-1, eIF-2, etc. *See* EUKARYOTIC INITIATION FACTOR.

Eisenmenger's syndrome An anatomical defect of the heart in which a left-to-right shunt causes increased flow through the pulmonary vasculature, causing pulmonary hypertension. This in turn can cause an increase in pressure in the right side of the heart so that deoxygenated blood is forced to the left and into the peripheral circulation where it causes cyanosis. *See* ATRIOVENTRICULAR SEPTAL DEFECT.

EJC *See* EXON JUNCTION COMPLEX.

Ekbom's syndrome A disorder (restless legs syndrome) characterized by uncomfortable and unpleasant sensations in the legs that induce an irresistible desire to move the legs. A number of loci on various chromosomes have been associated with an increased probability of the syndrome. Most patients respond favourably to treatment with dopaminergic drugs.

EKLF *See* ERYTHROID KRÜPPEL-LIKE FACTOR.

ektacytometry A method used to measure deformability, usually of erythrocytes. The cell suspension is exposed to increasing shear-stress and the laser diffraction pattern through the suspension shifts from circular to elliptical as shear increases and the cells deform.

elafin/SKALP A low molecular weight leucocyte elastase inhibitor (skin-derived antileukoproteinase, proteinase inhibitor 3) derived originally from psoriatic skin and found in other inflammatory skin lesions where it can be demonstrated immunochemically in suprabasal differentiated keratinocytes. It is also secreted in the lung and has antimicrobial activity.

ELAM-1 *See* SELECTINS.

elastase Serine endopeptidases that will digest various components of extracellular matrix such as *elastin and *collagen type IV. A range of elastases are known; the **pancreatic elastase** (EC 3.4.21.36, formerly EC 3.4.4.7) is perhaps the commonest; the elastase of neutrophil granules differs, as does that from macrophages. It is inhibited by α_1-protease inhibitor of plasma.

elastin The protein (786 aa) that is randomly coiled and crosslinked to form the elastic fibres of connective tissue. The amino acid composition is unusual, with 30% of residues being glycine and with a high proline content. Crosslinking involves the formation of *desmosine from four lysine side groups. **Elastin microfibril interface-located protein 1** (EMILIN1, 1016 aa) may anchor smooth muscle cells to elastic fibres.

elastonectin An *elastin-binding glycoprotein (120 kDa) of the extracellular matrix; not in current databases.

ELAV proteins *See* EMBRYONIC LETHAL ABNORMAL VISUAL PROTEINS.

electrical coupling (ionic coupling) The coupling of cells by *gap junctions or *electrical synapses so that electrical current (and small molecules) will pass readily from one to the other. Can be detected by intracellular microelectrode recording or, indirectly, by observing the passage of dye from one cell to others. Not confined to excitable cells. *See also* METABOLIC COOPERATION.

electrical synapse The *electrical coupling of two excitable cells allowing very rapid signal transmission. Unlike a chemical synapse the signal can pass in either direction, although there are some rectifying electrical synapses.

electrocardiogram (ECG) A record (electrocardiograph) of the electrical activity of the heart.

electrochemical potential The work, in joules/mole, done in bringing 1 mole of an ion from a standard state (infinitely separated) to a specified concentration and electrical potential. Generally the measurement of interest is, however, the electrochemical potential difference between two points (e.g. inside and outside the cell). If there is no electrical potential difference, or if the molecule is uncharged, the electrochemical potential reduces to the *chemical potential difference. If the system is at equilibrium the electrochemical potential difference will be zero and the *Nernst equation can be used.

electrochemiluminescence (ECL) Light emission stimulated by an electrochemical reaction (electrogenerated chemiluminescence). In screening assays it is convenient to arrange that the electrochemical reaction will only occur if the components are physically adjacent, e.g. when labelled ligand binds to an immobilized receptor: Following electrical stimulation the light emission is then a measure of binding. A common label is tris(bipyridine)ruthenium(II), NHS ester.

electroconvulsive therapy Treatment (electroshock) involving induction of seizures by the application of electrical shocks to the brain. Used for treating severe clinical depression that does not respond to drugs, although the basis for the proven theraputic benefit is unclear.

electrodermal effect A change in the electrical resistance of the skin associated with an emotional reaction.

electroencephalogram (EEG) A record of the electrical activity of the brain made using an electroencephalograph.

electrofocusing *See* ISOELECTRIC FOCUSING.

electrogenic pump An ion pump that generates net flow of charge. An important example is the sodium-potassium exchange pump which transports two potassium ions into the cell for every three sodium ions transported out, a net outward current that makes the inside of the cell negative.

electrokinetic potential *See* CELL ELECTROPHORESIS.

electrolyte A compound that dissociates into ions when dissolved.

electromyogram A record obtained by electromyography, of electrical activity in muscle.

electron microscopy (EM) Microscopy in which an electron beam is used instead of light, thereby increasing the resolution and allowing the ultrastructure of specimens to be studied. Electrons are diffracted or absorbed in transmission electron microscopy (TEM) or scattered and reflected from the surface of the specimen in scanning electron microscopy (SEM) and the image is formed either on a phosphorescent screen or photographically. In high-voltage electron microscopy (HVEM) the use of a high voltage source shortens the wavelength of the electrons, and thus increases resolving power and also increases the penetrating power of the beam, so that thicker specimens can be used. By tilting the specimen and getting stereoscopic views, a three-dimensional image can be reconstructed.

electron transport chain The cascade of compounds through which electrons are transferred to an eventual donor in respiration (oxidative phosphorylation; the respiratory chain) or photosynthesis. In the mitochondrion, electrons from NADH or FADH are transferred via *ubiquinone, cytochromes, and various other compounds, eventually to oxygen.

electrophoresis A separation method based on the mobility of molecules in an electric field. Most techniques use a gel such as polyacrylamide, starch, or agar to support the fluid phase, so frictional resistance affects the rate of movement as well as the charge of the molecule. Many variations and elaborations of the technique have been developed. *See also* ISOELECTRIC FOCUSING; POLYACRYLAMIDE GEL ELECTROPHORESIS; PULSE-FIELD ELECTROPHORESIS.

electrophoretogram The result or record of an electrophoretic separation.

electroplax The specialized noncontractile muscle fibres that generate extremely high voltages in the electric organs of so-called 'electric eels'. The electroplax is a rich source of *sodium-potassium ATPases, *acetylcholine receptors, *sodium gates and other ion channels.

electroporation A method for making cells temporarily permeable to large molecules, as in *transfection. A millisecond pulse with a potential gradient of ~700 V/cm is often required.

electroretinogram A record of electrical activity in the retina made using external electrodes.

electrospray mass spectroscopy A variant of mass spectroscopy in which the sample is introduced as a fine spray from a highly charged needle. Increasingly being used in protein sequencing and for the analysis of biological samples.

electrotonus The state of a nerve while an electrical current is passing.

electuary A drug formulation in which the active ingredient is mixed with honey or syrup.

elegantin A *disintegrin from the venom of the pit viper *Trimeresurus elegans*.

eleidin A keratin precursor or derivative found in the stratum lucidum of skin.

Elejalde's syndrome A disorder (acrocephalopolydactylous dysplasia) in which there is excessive birthweight, a swollen globular body with thick skin, apparently short limbs, polydactyly, acrocephaly, omphalocele, and abnormal facial features. It may be due to a defect in a fibroblast growth factor receptor gene. **Elejalde's disease** (neuroectodermal melanolysosomal disease) is distinct from Elejalde's syndrome and, like Griscelli syndrome type 1, is caused by mutation in the *MYO5A* gene that encodes myosin heavy chain 12 (myoxin).

elementary bodies **1.** The infectious extracellular form of *Chlamydia trachomatis*, small electron-dense structures ~0.3 µm in diameter, spherical or pear-shaped. Once inside a cell the elementary body germinates and converts into the proliferative reticulate form. **2.** In the older literature, cell inclusions, often of virus particles.

elephantiasis A condition (lymphatic filariasis) in which blockage of lymph nodes by filarial nematode parasites, particularly *Brugia malayia* and *Wuchereria bancrofti*, leads to gross enlargement of the limbs, or of the scrotum.

e

eleutherobin A microtubule-stabilizing agent (a diterpene glycoside), originally isolated from a marine soft coral *Eleutherobia aurea*, but subsequently synthesized. Cytotoxic for cancer cells with an IC_{50} similar to that of *paclitaxel.

eleutherosides Triterpenoid saponins with anti-inflammatory and immunostimulatory activity, isolated from the roots of Siberian ginseng *Eleutherococcus senticosus* and other medicinal herbs. Eleutherosides B and E have been most extensively studied.

ELF Eph ligand family. *See* EPHRINS.

elimination Removal of a substance by metabolism, secretion, or excretion. The rate of elimination of a pharmacologically active substance is an important aspect of its pharmacological or toxicological profile. *See* ADME.

ELISA A sensitive method (enzyme-linked immunosorbent assay) used to detect small amounts of protein or other antigenic substance. Various configurations of the assay are used, one being to immobilize the protein (antigen), add specific antibody, and then use a peroxidase-coupled anti-IgG to quantify the amount of immune complex that has been formed.

elixir A sweetened alcohol/water extract or tincture.

elixophyllin Proprietary name for an elixir in which the active ingredient is *theophylline.

Elk-1 An *Ets-related member (p62 ternary complex factor) of the winged helix-turn-helix superfamily of transcription factors that binds to the *serum response element (SRE) and induces transcription of *fos once phosphorylated by MAP kinase.

Elk-L *See* EPHRINS.

Elk proteins A family of cell surface receptor tyrosine kinases (Eph-like kinases) similar to the *Eph and Eck kinases, but restricted to brain and testis. *Elk-1 is unrelated.

ELL *See* ELONGATION FACTOR ELL.

elliptocytosis An autosomal dominant disorder in which erythrocytes are elliptical and there is some haemolytic anaemia. Caused by various mutations in genes coding for erythrocyte membrane skeleton. The Rhesus-linked form is a defect in the gene for band 4.1, the rhesus-unlinked forms are defects in genes for α-spectrin, β-spectrin, or band III.

Ellman's reagent A reagent (dithionitrobenzoic acid, DNTB) used in estimating free sulphydryl groups.

elongation factor ELL An elongation factor (elongation factor 'eleven-nineteen lysine-rich leukaemia', 621 aa) that increases the rate at which RNA polymerase II transcribes DNA by suppressing transient pausing. The human ELL gene on chromosome 19 undergoes frequent translocations with the trithorax-like *MLL* gene on chromosome 11 in acute myeloid leukaemias. Functionally similar to *elongin.

elongation factors (EFs) Ribosomal components that catalyse formation of the acyl bond between the incoming amino acid residue and the peptide chain (peptidyl transferases). **EF1a** (462 aa, EF-Tu in prokaryotes) binds GTP and aminoacyl-tRNA, delivering it to the A site of ribosomes. **EF-1b** (225 aa, EF-Ts) is an exchange factor that regenerates GTP-EF1a. **EF-2** (858 aa, EF-G, translocase) binds GTP and peptidyl-tRNA and translocates it from the A site to the P site. Diphtheria toxin ADP-ribosylates EF-2 and blocks eukaryotic protein synthesis.

elongin A transcription elongation factor, which increases the overall rate at which RNA polymerase II transcribes duplex DNA, first identified in mammalian systems, and composed of three subunits, elongin A, B, and C. **Elongin A** (transcription elongation factor B; 772 aa) is the active subunit of the elongin complex, an F-box protein. **Elongins B** (118 aa) and **C** (112 aa) are regulatory subunits which form a stable binary complex that interacts with elongin A and strongly induces its transcriptional activity. **Elongin C** has homology to Skp1. The von Hippel–Lindau (VHL) tumour suppressor protein will bind the elongin BC complex preventing it from activating elongin A.

elotuzumab A humanized monoclonal antibody that binds to the CS1 cell-surface glycoprotein (CD319); may be effective against multiple myelomas which express this at much higher levels than other cells.

elutriation A method for separating particles by differential sedimentation.

EMA **1.** Epithelial membrane antigen; *see* EPISIALIN. **2.** E2F-binding site modulating activity: *see* E2F.

EMAP-II *See* ENDOTHELIAL MONOCYTE-ACTIVATING POLYPEPTIDE II.

Embden–Meyerhof pathway (Embden–Meyerhof–Parnas pathway) The main pathway (glycolysis) for carbohydrate degradation. Starch or glycogen is hydrolysed firstly to glucose-1-phosphate and then a series of intermediates to produce *pyruvate or lactate with a yield of two ATP molecules per glucose.

embedding The process for preparing specimens for the cutting of thin sections and subsequent microscopic examination. Wax, for standard histology, or plastic, for electron microscopy, is infiltrated into the specimen and provides mechanical support.

embolism The occlusion of a blood vessel, usually by a blood clot or thrombus (embolus) from a remote part of the circulation.

embryo The developmental stage of an animal while it is isolated from the external environment by egg membranes, fetal membranes, etc.

embryogenesis The process of development from the egg to the final embryonic stage at which parturition occurs.

embryoid In animals, an aggregate of cells (embryoid body), derived from embryonic stem cells, that will exhibit some differentiation *in vitro*.

embryoma A mass of rapidly growing cells (an embryonal tumour) that originates in embryonic tissue. The term is also applied to tumours developing later in life but thought to derive from residual embryonic tissue. Embryomas may be benign or malignant; examples include neuroblastomas and Wilms' tumour.

embryonal carcinoma cells Cell lines (EC cells) from germ-line tumours (*teratocarcinomas) that arise more commonly in the testis than the ovary. They are pluripotent and may differentiate to produce a variety of tissue types and are an important experimental model system for the study of cancer and of differentiation.

embryonic induction The induction of differentiation in one tissue as a result of proximity to another tissue, often arising from the morphogenetic movements of gastrulation. Epithelial-mesenchmyal interactions are particularly important, e.g. the induction of the neural tube by the underlying chordamesoderm. *See* EPITHELIAL MESENCHYMAL TRANSITION.

embryonic lethal abnormal visual proteins A family of RNA-binding proteins (ELAV proteins) that regulate mRNA stability, highly conserved in vertebrates. There are four human members: HuR (326 aa) is expressed in all proliferating cells, Hel-N1 (359 aa), HuC (367 aa), and HuD (380 aa) are expressed in terminally differentiated neurons. *See* AREs.

embryonic stem cell A totipotent cell (ES cell) cultured from an early embryo. They have excited much interest because of the possibility of modifying them by genetic engineering *in vitro* and then producing a chimeric embryo, a *transgenic animal. Their therapeutic potential in degenerative disease has been much discussed, though more rarely demonstrated.

emergent Describes the novel and unpredictable properties that can appear in a complex system. Consciousness may be an emergent property of brain neurophysiology.

emerin A type II integral nuclear membrane protein (254 aa) that forms part of a nuclear protein complex consisting of the *barrier-to-autointegration factor (BAF), the nuclear lamina, nuclear actin, and other associated proteins; apparently links A-type *lamins to the inner nuclear envelope. In the heart emerin is also associated with intercalated discs. Emerin is missing or defective in Emery–Dreifuss muscular dystrophy.

Emery–Dreifuss muscular dystrophy *See* MUSCULAR DYSTROPHY.

emesis Vomiting. Nausea is a common side effect of cancer chemotherapy and antiemetic drugs are often used.

emetine An alkaloid derived from ipecac root (variously named *Cephaelis ipecacuanha, Uragoga ipecacuanha, Psychotria ipecacuanha*); used in the treatment of amoebiasis and as an emetic. Irreversibly blocks protein synthesis by inhibiting translocation of peptidyl-tRNA from the A-site to the P-site on the ribosome. Inhibits DNA replication in the early S phase and induces apoptosis in leukaemic cells. Emetine-resistant *Chinese hamster ovary cell lines have been extensively studied.

EMG *See* ELECTROMYOGRAM.

emmetropia (*adj.* **emmetropic)** The normal condition of the refractive system of the eye.

Objects at infinity are in sharp focus on the retina in the relaxed eye.

emperipolesis A phenomenon in which lymphocytes appear to be phagocytosed by macrophages (histiocytes) in the lymph node. This may be an artefact, since leucocytes moving underneath spread cells often appear, erroneously, to be within the larger flattened cells. In the early literature lymphocytes and leucocytes were thought to enter the cytoplasm of endothelial cells in order to cross the wall of the postcapillary venule and this was termed emperipolesis; subsequent ultrastructural studies have shown that this does not happen and leucocytes move between the endothelial cells.

emphysema A *chronic obstructive pulmonary disease in which there is a loss of elasticity in the lung tissue. It may be a result of excessive leucocyte elastase activity because α_1-antiprotease, which would normally inhibit the enzyme, is inactivated by reactive oxygen species produced in inflammation.

empyema The accumulation of pus.

EMT *See* EPITHELIAL MESENCHYMAL TRANSITION.

ENA-78 An α-chemokine (epithelial derived neutrophil activating peptide-78, CXCL5, 114 aa) that is chemotactic for activated T cells, binds to CXCR2, and affects the growth, movement, or activation of cells involved in immune and inflammatory responses. TNFα enhances its production by dermal fibroblasts and vein endothelial cells and it is induced by inflammatory cytokines in human colonic enterocyte cell lines. Levels are raised in the colon of patients with *inflammatory bowel disease (IBD). The mouse homologue is LIX.

enalapril An *ACE inhibitor used to treat hypertension, usually in conjunction with thiazide-type diuretics. Following oral administration it is hydrolysed to the active form, enalaprilat.

enamelin A protein that comprises ~5% of the total enamel matrix protein in developing teeth. The precursor (1142 aa) is processed into several polypeptides. Mutations lead to *amelogenesis imperfecta types 1B and 1C.

enamelysin A metallopeptidase (matrix metallopeptidase-20, MMP-20, 483 aa) that has specificity for *amelogenin.

enanthema A rash on a mucous membrane, e.g. in the mouth.

enantiomer Either of a pair of the stereoisomers of a chiral compound.

enaptin One of the family of giant proteins that associate with the F-actin cytoskeleton and with the nuclear envelope. The human enaptin gene spreads over 515 kb and gives rise to several splicing isoforms (*nesprin-1, myne-1, synaptic nuclear envelope protein 1 (syne-1), CPG2), the largest of which is predicted to be >1 MDa. It is highly homologous to nuance (nucleus and actin connecting element, SYNE2, nesprin-2) and has multiple spectrin repeats, a nuclear localization signal, a transmembrane domain within the C-terminal Klarsicht homology region, and several *N*-glycosylation and phosphorylation sites.

Enc-1 *See* ECTODERM-NEURAL CORTEX-1.

encephalitis Inflammation of the brain, caused by viral or bacterial infection. **Encephalitis lethargica** (sleepy sickness; von Economo's disease) is an acute viral encephalitis that emerged as a new infectious disease near the end of the First World War but by 1940 had apparently disappeared and is now probably only of historical interest. **Encephalitis periaxialis**, *see* SCHILDER'S DISEASE. **Japanese encephalitis** is the major cause of viral encephalitis in Asia, caused by a flavivirus and transmitted by mosquitoes. *See also* AICARDI–GOUTIERES SYNDROME; EXPERIMENTAL ALLERGIC ENCEPHALOMYELITIS.

encephalo- Prefix indicating something associated with the brain. **Encephalomalacia** is softening (malacia) of the brain. **Encephalomyelitis** is a diffuse inflammation of the brain and the spinal cord. **Encephalopathy** is a generalized disease of the brain often a result of poisoning (lead encephalopathy is caused by lead poisoning, hepatic encephalopathy is associated with severe liver disease and the build-up of toxic substances in the body). *See also* ENCEPHALITIS.

encephalopsin A photoreceptor molecule (panopsin, opsin3), homologous to vertebrate retinal and pineal opsins, found primarily in the brain, not the retina, that may play a role in nonvisual but light-dependent processes such as the entrainment of circadian rhythms or the regulation of pineal melatonin production. *See* OPSIN (opsin subfamilies).

enchondroma (endochondroma) A common benign cartilagenous tumour of bone, usually close to, or in continuity with, growth plate cartilage. When associated with haemangiomata, the condition is known as Maffucci's syndrome. **Multiple**

enchondromatosis is also known as Ollier's disease. The suggestion that the cause is mutation in the receptor for parathyroid hormone has not been supported by examination of recent cases.

encysted Describing something enclosed in a cyst or a sac.

endarteritis Chronic inflammation of the intima of arteries. **Endarteritis obliterans** is the complete occlusion of an artery as a result of such inflammation.

endemic Ocurring commonly or frequently in a particular geographical area or population.

endobrevin *See* SYNAPTOBREVINS.

endocannabinoids The natural (endogenous) ligands for the *cannabinoid receptors of which there are two types, CB_1 and CB_2, the former mostly in the brain and responsible for the psychotropic effects, the latter in the immune system. Anandamide (arachidonylethanolamide) and 2-arachidonyl glycerol (2-AG) are the main endocannabinoids and bind to both receptor types.

endocarditis Inflammation of the lining of the heart (endocardium), particularly that over the valves, which seems to be particularly susceptible. Often a result of rheumatic fever but may be caused by viral or bacterial infection. Bacterial endocarditis affecting the valves is associated with *Staphylococcus epidermidis* which apprears to escape immune clearance.

endochondral Describing something situated within, or occurring within, cartilage. Endochondral bone formation occurs through mineralization of a cartilagenous scaffold.

endochondroma *See* ENCHONDROMA.

endocrine gland A gland (ductless gland) from which the secretion passes directly into the blood rather than into a duct. Examples are the pituitary, thyroid, and adrenal glands, the ovary and testis, placenta, and beta cells of pancreas. An **endocrinopathy** is a disease caused by abnormal function of endocrine glands.

endocytosis Uptake of material into a cell in a membrane-bound endocytotic vesicle.

endocytotic vesicle *See* ENDOCYTOSIS.

endoderm The germ layer that gives rise to internal tissues such as gut and in the embryo is surrounded by *mesoderm and *ectoderm.

endogenous A product or an activity that arises in the body or cell, in contrast to exogenous agents or stimuli that come from outside.

endogenous pyrogen The substance, now known to be *interleukin 1, that is responsible for inducing fever by acting on the hypothalamic thermoregulatory centre.

endoglin A homodimeric glycoprotein (CD105, 658 aa) that is part of the TGFβ receptor complex, abundantly expressed on endothelial cell surfaces but also present on activated macrophages, fibroblasts, and smooth muscle cells. Endoglin is important in cardiovascular development and vascular remodelling and its cytoplasmic domain interacts with the actin cytoskeleton via zyxin-related protein 1 (ZRP-1). Defects in the endoglin gene cause *hereditary haemorrhagic telangiectasia type 1.

endoglycosidase An enzyme of the subclass EC 3.2 that hydrolyses nonterminal glycosidic bonds in oligosaccharides or polysaccharides. They are often used experimentally to determine the role of carbohydrate portions of glycoproteins, particularly endoglycosidases F and H. **Endo-F** cleaves high mannose or complex type glycans at the link to asparagine in the protein; **Endo H** is an endo-β-*N*-acetyl-glucosaminidase.

endolymph Fluid of the membranous labyrinth in the inner ear.

endolyn A sialomucin (CD164, 197 aa), present in membranes of endosomes and lysosomes, that regulates the adhesion of haematopoietic stem cells ($CD34^+$) to bone marrow stroma, stimulates proliferation of these cells and associates with the CXCR4 chemokine receptor.

endometrium The mucous membrane that lines the uterus and thickens during the menstrual cycle ready for implantation of the embryo. If implantation does not occur the endometrium returns to its previous state and the excess tissue is shed at menstruation. If implantation does occur the endometrium becomes the decidua and is not shed until after parturition. An **endometrioma** is an adenomyoma derived from the endometrium and consisting of glandular elements and cellular connective tissue. **Endometriosis** is a condition in which fragments of tissue resembling endometrium occur in other tissue or organs. It is subject to the same menstrual changes as normal endometrium and causes severe pain and discomfort during menstruation. **Endometritis** is an inflammation of the endometrium.

e

endomitosis Replication of chromosomes without subsequent mitotic division so that the cell becomes polyploid. If the daughter chromosomes remain associated (synapsed) then giant polytene chromosomes are produced.

endomorphins Endogenous peptides (endomorphin-1, YPWF-NH2; endomorphin-2, YPFF-NH2) that have high and selective affinity for the μ-opiate receptor. The precursor for these peptides remains unidentified. *See* MORPHICEPTIN.

SEE WEB LINKS

- Article 'Search of the human proteome for endomorphin-1 and endomorphin-2 precursor proteins' in *Life Sciences* (2007).

endomyocarditis Acute or chronic inflammation of cardiac muscle and the inner connective-tissue lining of the heart. *See* ENDOCARDITIS. **Endomyocardial fibrosis** is a cardiomyopathy, generally restricted to tropical regions, in which there is fibrosis of endocardium and myocardium leading to severely restricted function. The causes are unclear.

endomysium The connective tissue sheath that surrounds individual muscle fibres.

endoneurium The connective tissue sheath that surrounds individual nerve fibres in a nerve bundle.

endonexins Calcium-dependent membrane-binding proteins (320 aa) that act as indirect inhibitors of the thromboplastin-specific complex and have anticoagulant activity. **Endonexin I** (annexin A4, lipocortin IV, chromobindin-4, placental anticoagulant protein I, 35-β calcimedin, 319 aa) and **endonexin II** (annexin A5, lipocortin V, calphobindin I, placental anticoagulant protein I, thromboplastin inhibitor, vascular anticoagulant-alpha, anchorin CII, 320 aa) can modify membrane properties and endonexin II forms voltage-gated divalent-cation selective channels. Analogous calcium-dependent membrane-binding proteins are *synexin and p36 (annexin II), a component of brush-border membrane that is a target for the *src-gene tyrosine kinase. Beta3-Endonexin (centromere protein R, CENP-R, nuclear receptor-interacting factor-3, 111 aa) is a selective binding partner of the cytoplasmic tail of beta3- integrin that modifies its affinity.

endonuclease Any enzyme that cleaves nucleic acids within the chain. A wide range have been identified, some of which cut both RNA and DNA (e.g. S1 nuclease, EC.3.1.30.1, which is specific for single-stranded molecules), some specific for RNA (*ribonucleases), others for DNA (*deoxyribonucleases). Bacterial *restriction endonucleases cleave double-stranded DNA at highly specific sites and are important tools for molecular biology.

endopeptidase Any enzyme that cleaves protein at positions within the amino acid chain, a subclass of the *peptidases. Strictly speaking they are peptidyl-peptide hydrolases, often casually called *proteinases or *proteolytic enzymes.

endophilin A family of proteins with a *BAR domain involved in regulating clathrin-mediated endocytosis during synaptic vesicle recycling. **Endophilin I** (352 aa) is involved in the recruitment of *synaptojanin and dynamin-binding proteins to the membrane through their interaction with endophilin's SH3 domain. **Endophilin B1** (365 aa) is required for the maintenance of normal mitochondrial morphology. The enzymatic (LPAAT) activity reported for endophilin was due to contamination.

endoplasmic reticulum (ER) A cytoplasmic system of parallel membranes separated by 50–200 nm. The compartment enclosed by these membranes, the *cisternal space, is distinct from the rest of the cytoplasm and it is into this compartment that nascent peptides destined for export are cotranslationally synthesized on the rough ER. The Golgi region, together with ER and lysosomes, constitute the GERL system. *See also* ROUGH ENDOPLASMIC RETICULUM; SMOOTH ENDOPLASMIC RETICULUM.

endoplasmin The primary chaperone of the endoplasmic reticulum, the most abundant protein in microsomal preparations from mammalian cells, and one of the *hsp90 family of *heat shock proteins. A glycoprotein (tumour rejection antigen gp96, TRA1, glucose-regulated protein, grp94, 803 aa) with calcium-binding properties that acts as a natural adjuvant by stimulating antigen-presenting cells to express cytokines. In the adaptive immune response endoplasmin has peptide-binding activity and a role in antigen presentation. It is also important in embryonic development, probably through an involvement with the production of insulin-like growth factors.

endoradiosonde An analytical device incorporating a radio transmitter that is small enough to swallow and that transmits data about the gut contents.

endorphins A family of peptide hormones derived from *pro-opiomelanocortin, that bind to *opioid receptors and induce analgesia and sedation. Their release is regulated and they are rapidly inactived by peptidases.

endoscope An instrument, now usually with flexible fibre-optics, for illuminating, inspecting, and sometimes obtaining biopsy material from internal cavities of the body. **Endoscopy** is the use of such an instrument.

endosome An endocytotic vesicle (receptosome) formed from invaginated plasma membrane, but more specifically an acidic nonlysosomal compartment in which receptor–ligand complexes dissociate.

Endostatin The proprietary name for a natural antiangiogenic protein (20-kDa C-terminal fragment derived from type XVIII collagen) that will inhibit endothelial cell proliferation and neovascularization of tumours. Has shown promise in clinical trials. An endostatin analogue derived from collagen XV has similar but not identical properties.

endothelial monocyte-activating polypeptide II A multifunctional polypeptide (small inducible cytokine subfamily E, member 1, SCYE1, ARS-interacting multifunctional protein 1, AIMP1, 312 aa) with both cytokine and tRNA-binding activities. The precursor is cleaved to an active fragment that is proinflammatory, activates endothelial cells and is chemotactic for neutrophils and mononuclear phagocytes. Also has some antiangiogenic activity. Reportedly identical to the C-terminal domain of the p43 subunit of the aminoacyl-tRNA synthetase complex.

endothelin converting enzymes Integral membrane proteins (ECEs) that are zinc-binding metallopeptidases of the same family as *neprilysin with a role in processing various neuropeptides. **ECE-1** (EC 3.4.24.71, 770 aa) converts big endothelin-1 (ET-1) to active ET-1 and is important in regulating blood pressure. Isoforms of endothelin-converting enzyme-1 (ECE-1a–d) are present in early endosomes, where they degrade various neuropeptides and regulate postendocytic sorting of receptors. Defects in the ECE-1 gene are associated with *Hirschsprung's disease. **ECE-2** (883 aa) has been implicated in Alzheimer's disease and knockout mice show deficiencies in learning and memory.

endothelins Potent vasoconstrictor hormones (21 aa) that are released by endothelial cells. They are structurally related to snake venom *sarafotoxins. The precursor, pre-pro-endothelin-1 (203 aa) is cleaved to inactive **big endothelin-1** (92 aa) and then to active **endothelin-1** by *endothelin converting enzyme. ET-1 is produced by endothelial cells and is the predominant form, ET-2 and ET-3 are produced by various other tissues. The endothelins have a range of activities besides being vasoconstrictive: they have *inotropic and *mitogenic properties, influence salt and water balance, alter central and peripheral sympathetic activity and stimulate the *renin–*angiotensin–aldosterone system. Though ET-1 acts as a vasoconstrictor through binding to type A receptors it is vasodilatory when bound to ET(B) because it stimulates the release of endothelial derived relaxing factor (*nitric oxide). The G-protein-coupled endothelin receptors are present on vascular smooth muscle cells, where they mediate vasoconstriction, and on endothelium, where they regulate *nitric oxide release. The **type A receptor** (ET(A)) shows preferential binding of ET-1 whereas **type B** (ET (B)) binds all the endothelins with equal affinity.

endothelioma *See* ENDOTHELIUM.

endothelium The layer of cells that lines blood vessels, lymphatics, and other fluid-filled cavities. Generally a simple *squamous epithelium although modified to cuboidal form in areas where there is lymphocyte traffic (*see* HIGH ENDOTHELIAL VENULE). Unlike most epithelia it is mesodermally derived. An **endothelioma** is a tumour, usually benign, derived from endothelium.

endothelium-derived relaxation factor (EDRF) *See* NITRIC OXIDE.

endothermic Describing a process or reaction that requires a source of external energy in order to proceed.

endotoxic shock *See* ENDOTOXIN.

endotoxin A general term for bacterial toxins that are bound rather than released, but usually refers to the lipopolysaccharide (LPS) of the outer membrane of Gram-negative bacteria. The LPS molecule comprises *lipid A (six fatty acid chains linked to two glucosamine residues) that binds it to the membrane, a core oligosaccharide chain of ten sugars, and a polysaccharide side chain (variable, but up to 40 sugar units in smooth forms) that can be removed without affecting the toxicity (rough LPS). Endotoxin is responsible for much of the virulence of Gram-negative bacteria. **Endotoxic shock** is the serious condition that arises when there is systemic infection with endotoxin-producing bacteria. Fever is accompanied by hypotension and leucopenia.

endovanilloids A group of compounds that are the endogenous ligands of transient receptor potential vanilloid type 1 (TRPV1) channels.

Examples are *N*-arachidonoyl-dopamine, *N*-oleoyl-dopamine and *anandamide. *See* TRANSIENT RECEPTOR POTENTIAL CHANNELS; VANILLOID RECEPTOR-1.

end plate (motor end plate) *See* NEUROMUSCULAR JUNCTION.

end plate potential (epp) The sum of *miniature end plate potentials (mepp) in the sarcolemma at a neuromuscular junction. Mepps are caused by the quantal release of acetylcholine from the motor nerve and if the sarcolemma becomes sufficiently depolarized, muscle contraction will be initiated. *Curare, by blocking the production of the end plate potential, effectively paralyses the muscle.

enduracidin *See* RAMOPLANIN.

enflurane A halogenated ether formerly used as a general inhalational anaesthetic.

Englebreth–Holm–Swarm sarcoma cells A murine cell line (EHS cells) that produces large amounts of basement lamina-type extracellular matrix. Often used as a source of *laminin, collagen type IV, *nidogen, and heparan sulphate.

enhanced chemiluminescence (ECL) A sensitive method for detecting specific proteins on gels. Peroxidase-coupled antibody to the protein oxidizes luminol and the light can be detected on film.

enhancer A DNA *control element that enhances the levels of expression of the gene when a specific transcription factor is bound. Unlike a *promoter, does not stimulate expression of the gene on its own. Frequently found in the upstream (5′) region of the gene. **Enhancer trapping** is a technique that has been used to investigate the pattern of gene expression in which a *transposon carrying a *reporter gene linked to a very weak *promoter is induced to 'jump' randomly within the genome.

enkephalins Natural *opiate peptides that modulate pain; found in the thalamus of the brain and in some parts of the spinal cord, although the richest source is the adrenal gland. Leu-enkephalin (YGGFL) and Met-enkephalin (YGGFM) bind strongly to δ opiate receptors.

enlargeosomes Cytoplasmic organelles that are identified on the basis of having *desmoyokin/*Ahnak associated with their membranes. Their contents are released by exocytosis and enlargeosomes are particularly prominent in a neurosecretion-incompetent clone of *PC12 cells, although their function remains rather unclear.

enovin *See* ARTEMIN.

enoximone An imidazole phosphodiesterase inhibitor with positive inotropic effects used in treatment of congestive heart failure. *TN* Perfan.

ENT Common abbreviation for 'ear, nose, and throat'.

entactin *See* NIDOGEN.

Entamoeba histolytica The protozoon that is the causative agent of amoebiasis (amoebic dysentery), a disease responsible for considerable morbidity and mortality in humans (estimated to be responsible for 50 000–100 000 deaths every year worldwide). The life cycle involves an infective cyst stage and a multiplying trophozoite stage and is transmitted by faecal contamination of water. 'Nonpathogenic' entamoebiasis is due to infection with another species, *E. dispar*.

enteral (enteric) Describing something within the intestine, or a substance given by this route. **Enteritis** is inflammation of the small intestine.

enteric-coated Describing a drug tablet or capsule that is coated so that it will pass through the stomach and not release its contents until within the intestine.

Enterobacter A genus of enteropathic bacilli of the *Klebsiella* group found in the environment and in the human intestinal tract. The most common pathogenic species are *E. cloacae* and *E. aerogenes*, which can cause opportunistic infections in immunocompromised patients. *E. sakazakii* has been associated with infection in neonates due to contamination of infant formula. Not to be confused with the family *Enterobacteriaceae of which they are members.

Enterobacteriaceae A large family of Gram-negative bacilli found in the large intestine. The commonest is *Escherichia coli* which is generally harmless (with the exception of strains such as *E. coli* O157) but others, e.g. *Salmonella* and *Shigella*, can cause intestinal disease.

enterobactin *See* ENTEROCHELIN.

enterobiasis An infection with the nematode *Enterobius vermicularis*, the threadworm or pinworm. Relatively common in children, and the cause of anal itching.

enterochelin (enterobactin) An iron-chelating compound (a *siderophore, a cyclic trimer of 2,3-dihydroxybenzoylserine) found in *E. coli* and *Salmonella* spp. that will remove iron bound by *lactoferrin and *transferrin, thus giving the bacteria access to an essential factor for growth.

enterochromaffin cells Neuroendocrine cells (Kulchitsky cells, EC cells) that produce and store the majority of the body's serotonin. Most are found in the epithelial lining of the gastrointestinal tract. **Enterochromaffin-like cells** synthesize and secrete histamine in response to stimulation by the hormones gastrin and pituitary adenylyl cyclase-activating peptide but do not produce serotonin. Most are located in the gastric mucosa, particularly in the acid-secreting regions of the stomach.

Enterococcus A genus of Gram-positive bacteria (cocci) that are facultative anaerobes commonly found in the bowel of normal healthy individuals. Most enterococcal infections of humans are due to *E. faecalis* or *E. faecium* and include a range of illnesses including urinary tract infections, bacteraemia, and wound infections. Glycopeptide-resistant enterococci (GRE), mostly *E. faecium*, are resistant to glycopeptide antibiotics such as vancomycin and teicoplanin.

enterocyte The major cell type of the intestinal epithelium: columnar cells specialized for absoption with a brush border, numerous mitochondria, abundant smooth and rough endoplasmic reticulum, and Golgi.

enterogastrone An obsolete term for a hormone that affected gastric secretion and motility. None of the known hormones affecting these activities has exactly the activities ascribed in the older literature.

enterogenous cyanosis An apparent cyanosis caused by the absorption of nitrites or other toxic materials (including acetanilide, phenacetin, acetaminophen, and dapsone) from the intestine with the formation of methaemoglobin or sulphaemoglobin. High levels of methaemoglobin or sulphaemoglobin respectively can arise through a rare congenital deficiency in NADH-cytochrome b_5 reductase or an unstable haemoglobin variant.

enteropathy *See* GASTROENTEROLOGICAL DISORDERS.

enterostatin An appetite-regulating peptide that is the N-terminal portion (VPDPR) of *procolipase, generated when it is cleaved to *colipase. Enterostatin alters 5-HT release in the brain, and 5-HT_{1B} receptor antagonists block the anorectic response to enterostatin. Experimentally it has been shown to suppress fat intake after peripheral and central administration. Release of enterostatin varies in a circadian fashion in some animals.

enterotoxins Bacterial *exotoxins produced by enterobacteria that perturb ion and water transport in the intestinal mucosa and induce diarrhoea. *Cholera toxin is a well-known example.

enterovirus RNA viruses of the *Picornaviridae family that preferentially replicate in the mammalian intestinal tract. It includes the *polioviruses and in the nonpoliovirus category *coxsackieviruses, echoviruses, and other enteroviruses.

entopic Describing something that develops or is located in the normal place, the opposite of *ectopic.

entrainment Synchonization of an endogenous clock-driven rhythm with an external environmental rhythm. *See* ZEITGEBER.

ENTs *See* EQUILIBRATIVE NUCLEOSIDE TRANSPORTERS.

enucleation **1.** Complete removal of a tumour or globular swelling. **2.** Removal of the nucleus of a cell.

env The retroviral gene that encodes the glycoproteins of the viral envelope. Since these viral proteins interact with host cell-surface molecules they determine host range, and rapid changes (as is the case with the HIV-1 *env* gene) make vaccine development very difficult.

envelope **1.** The outer layer of some viruses that derives from the plasma membrane of the host cell when the virus buds off. **2.** The plasma membrane and cell wall complex of a bacterium.

envoplakin A membrane-associated protein (2033 aa) expressed in human stratified squamous epithelia. It is transglutaminase crosslinked and deposited under the plasma membrane of keratinocytes during the formation of the stratum corneum. Has sequence homology with *desmoplakin, bullous pemphigoid antigen 1, and *plectin.

enzyme induction The induced synthesis of an enzyme in response to change in environmental conditions. The best-known example is the activation of the *lactose operon in *E. coli* when galactose is added to the medium.

enzyme nomenclature *See* E CLASSIFICATION.

eosin A red dye (tetrabromofluorescein) used extensively as a histological counterstain.

eosinophil A polymorphonuclear leucocyte (granulocyte) in which the granules stain red with eosin. Phagocytic, particularly associated with helminth infections and with hypersensitivity reactions. Normal eosinophil counts are in the range of 4–40 × 10^3 mL^{-1} but **eosinophilia** (increased numbers in the blood) may arise as a result of infection or allergy.

eosinophil cationic protein A member of the RNase A superfamily, an arginine-rich protein (ECP, EC 3.1.27.-, ribonuclease 3, 160 aa) found in the granules of eosinophils and released when the cells are activated. The protein plays a role in the defence against helminths, e.g. it will damage schistosomula *in vitro*. Not the same as the MBP (major basic protein) of the granules. Serum levels of ECP have been used as a marker for allergic inflammation.

eosinophil chemotactic peptides Tetrapeptides (eosinophil chemotactic factor, ECF of anaphylaxis, ECF-C, VGSE, and AGSE) released by mast cells that were said to both attract and activate eosinophils. Have not featured in recent literature.

eosinophilic **1.** Having affinity for the red dye *eosin. **2.** Describing an inflammatory lesion in which there are many *eosinophils.

eosinophilopoietin Outmoded name for a substance that specifically induced proliferation of eosinophils (an eosinophil colony-stimulating factor). The major eosinophilopoetin is now know to be IL-5 released by $CD4^+CD8^-$ T cells and overproduced in hypereosinophilic syndrome.

eotaxin A chemokine (CCL11, 97 aa), originally described as being eosinophil-specific but subsequently shown to attract IL-2- and IL-4-stimulated T cells which, unlike naive T cells, express the CCR3 chemokine receptor.

epalons Endogenous neuroactive steroids that act as positive allosteric modulators of the $GABA_A$ receptor/chloride ion channel complex. The name is a contraction of 'epiallopregnanolone', a progesterone metabolite. Epalons have anxiolytic, anticonvulsant, and sedative-hypnotic properties and a synthetic variant, ganaxolone, has been tested for the treatment of epilepsy.

ependymal cells The subclass of glial cells that form the cuboidal/columnar epithelial lining (ependyma) of brain ventricles and produce *cerebrospinal fluid. **Ependymitis** is inflammation of the ependyma; an **ependymoma** is a tumour considered to arise from these cells.

ephedra (ma huang) A natural plant extract usually, but not exclusively, prepared from *Ephedra sinica*, used in traditional Chinese medicine for the treatment of asthma, hay fever, and the common cold. The principal active ingredient is *ephedrine. Ephedra 'dietary supplements' are considered by the US FDA to present an unreasonable risk of illness or injury and are banned.

ephedrine An alkaloid extracted originally from plants of the genus *Ephedra*. It is a structural analogue of epinephrine (*adrenaline) and has comparable pharmacological effects.

ephelides Freckles. Nearly everybody with freckles has at least one *melanocortin-1 receptor variant.

ephexin An exchange factor (Eph-interacting exchange factor that links the Eph receptor with rho GTPases and thus regulates actin remodelling in 'forward signalling'. Ephexin1 binds to the kinase domain of the FGF-receptor mainly through its DH and PH domains. FGFR-mediated phosphorylation enhances the exchange activity toward rhoA. *See* EPH-RELATED RECEPTOR TYROSINE KINASES.

Eph-related receptor tyrosine kinases A family of receptor tyrosine kinases (e.g. EphB2, formerly Nuk/Cek5/Sek3, 482 aa) that are involved in control of axonal navigation, neuronal fasciculation and vascular assembly. **EPH receptors** are activated by the binding of their cell-bound ligands (*ephrins) on an adjacent cell. The ephrins in turn respond to binding of the Eph receptor, so signalling is bidirectional.

ephrins Membrane-achored ligands for the *Eph-related receptor tyrosine kinases. **Class A ephrins** are GPI-anchored and bind to EphA receptors, **ephrinBs** are transmembrane and bind to EphB receptors and to EphA4. The consequence of an ephrin–Eph interaction depends on the cell type, but is often a repulsion cue, e.g. of the nerve growth cone. Binding leads to the formation of clusters of Ephs (on one cell) and ephrins (on the other) and the transmembrane ephrins becomes tyrosine phosphorylated on their cytoplasmic domains when they are bound by the receptor: both receptor and ligand respond to the binding interaction. This also seems to be true for the GPI-linked ephrins, although the reverse signalling mechanism presumably involves association with other signalling molecules in a lipid domain.

SEE WEB LINKS

- The Panther database gives additional information on the ephrin family.

epi- Prefix indicating something on, above or near. **Epithelia** cover (are on top of) other tissues.

epiblast The outer germinal layer of an embryo at the blastula stage that will give rise to all adult tissues. The other layer is the *hypoblast.

epicanthus The fold of skin (epicanthic fold) that partially covers the inner angle of the eye in the Mongolian races and often seen in individuals with Down's syndrome.

epicatechin One of the *catechins, a flavonoid found in chocolate and as a component of green tea, thought to be beneficial for blood vessel function, as an antioxidant, and reportedly with insulinomimetic properties.

epichromosomal Describing genetic material that does not become integrated into the host chromosomes but proliferates in tandem. An example would be an adenovirus used as a vector.

epidemic An outbreak of infectious disease that spreads rapidly and affects a high proportion of susceptible people in a region. **Epidemiology** is the study of such outbreaks, and the theoretical requirements for epidemic spread, in terms of infectivity, transmission rates, susceptibility, etc., are well understood. Also used as an adjective.

epidemiology *See* EPIDEMIC.

epidermal cell A cell of the epidermis, the outermost layer of the *dermis in animals.

epidermal growth factor (EGF) A mitogenic polypeptide (53 aa) cleaved from the 1207-aa pro-EGF (urogastrone) initially isolated from male mouse submaxillary gland and also found in urine. The name derives from the early bioassay, but EGF is active on a variety of cell types, especially but not exclusively epithelial. A family of similar growth factors are now recognized. The **EGF receptor** is a tyrosine kinase encoded by c-*erbB1* and is a member of the type I family of growth factor receptors that also includes the TGFα and heregulin receptors (HER-1). **EGF receptor-associated protein with SH3 and TAM domains** (EAST) is an actin-binding protein that is phosphorylated by the EGF receptor and is enriched at focal adhesions and some adherens-type junctions. The **EGF-like domain** is a region of 30–40 aa with six cysteines found originally in EGF but also present in a range of proteins involved in cell signalling including TGFα, *amphiregulin, *urokinase, *tissue plasminogen activator, *complement C6–C9, *fibronectin, *laminin, *nidogen, and *selectins as well as various *Drosophila* gene products.

epidermin *See* LANTIBIOTICS.

epidermis The outer epithelial layer of an animal (or plant). In humans, a complex stratified squamous epithelium.

epidermoid cyst *See* SEBACEOUS CYST.

epidermolysis bullosa A very rare genetic condition in which the skin and internal body linings blister at the slightest knock or rub. EB is subdivided into: 1. **Epidermolysis bullosa simplex** (EBS): a collection of keratin disorders characterized by intraepidermal blistering with relatively mild internal involvement. The more severe EBS subtypes include Koebner, Dowling–Meara, and Weber–Cockayne forms. Most cases of EBS are associated with mutations of the genes coding for keratins 5 and 14 (components of the intermediate filaments in basal keratinocytes). 2. **Junctional epidermolysis bullosa** (JEB) a variant in which there is separation in the lamina lucida or central basal membrane zone (BMZ). Primary subtypes include a lethal subtype termed Herlitz or JEB lethalis (a severe defect in laminin 5), a nonlethal subtype termed JEB mitis, and a generalized benign type termed generalized atrophic benign EB (GABEB). Mutations in genes coding for laminin 5 subunits (α_3 chain, laminin β_3 chain, laminin γ_2 chain), collagen XVII (BP180), α_6 integrin, and β_4 integrin have been demonstrated. 3. **Dystrophic epidermolysis bullosa** (DEB) in which the separation is in the sublamina densa BMZ. This is associated with mutations of the gene coding for type VII collagen. 4. **Hemidesmosomal epidermolysis bullosa** (HEB) which involves blistering at the hemidesmosomal level in the most superior aspect of the BMZ. There are two subtypes of this rare disease. The first arises from a disorder of the protein *plectin (HD1) and is associated with muscular dystrophy. The second arises from a defect of the $\alpha_6\beta_4$ integrin receptor and is associated with pyloric atresia. Each disease shows intraepidermal blistering at the most basal aspect of the lower cell layer. 5. **Epidermolysis bullosa acquisita** is a chronic autoimmune condition with subepidermal blistering disease of the skin and mucus membranes. Immunologically, EBA is characterized by the presence of IgG autoantibodies against the noncollagenous (NC1) domain of type VII collagen, the major component of anchoring fibrils that connect the basement membrane to dermal structures.

epididymis The convoluted tubule that connects the vas efferens to the vas deferens and in which sperm matures and is stored. **Epididymitis** is inflammation of the epididymis.

e

epidural anaesthesia Loss of sensation (anaesthesia) brought about by injection of anaesthetic into the epidural space that surrounds the spinal cord: the effect is only posterior to the site of injection.

epigastrium The central upper region of the abdomen between the umbilicus and sternum.

epigen One of the *epidermal growth factor (EGF) superfamily (152 aa), a ligand for the EGF receptors, ErbB1 and ErbB2, and a potent epithelial mitogen implicated in wound healing. The transmembrane precursor is processed by the metallopeptidase ADAM17 which releases the soluble bioactive ectodomain. Production can be stimulated by phorbol esters, phosphatase inhibitors, and calcium influx.

epigenetics The study of mechanisms by which phenotypic complexity arises during morphogenesis. In the epigenetic view of differentiation (**epigenesis**), cells make a series of choices, not always phenotypically obvious, that commit them to their eventual differentiated state (*see* DETERMINATION). The processes involve selective gene repression or derepression at an early stage, thereby restricting the range of final possible fates. Epigenetic mechanisms may involve DNA methylation, chromatin restructuring, or changes in transcription factor levels that have wider-ranging effects. *See* EPIGENOMICS.

epigenomics The analysis of genome-wide consequences of *epigenetic modifications. The Human Epigenome Project is an international collaboration 'to identify, catalogue and interpret genome-wide DNA methylation patterns of all human genes in all major tissues' and results are made freely available on the internet as they accumulate.

SEE WEB LINKS

- The Human Epigenome Project homepage.

epiglycanin An extensively glycosylated transmembrane glycoprotein originally found in mouse mammary carcinoma cells. The human orthologue, MUC21, is expressed in lung, large intestine, thymus, and testis. Functionally analogous to *episialin but without sequence homology in the protein.

epilepsy A disorder of cerebral function accompanied by recurrent seizures and sometimes loss of consciousness. The severity varies (grand mal, Jacksonian epilepsy, *MERRF syndrome, *myoclonic epilepsy of Unverricht and Lundborg, petit mal, etc.) but can usually be controlled with *anticonvulsant drugs. *See* DOUBLECORTIN; EPALONS. A condition described as **epileptiform** resembles epilepsy. **Epileptogenic agents** induce an epileptic attack.

epiligrin *See* KALININ.

epilipoxins The 5R-epimers of *lipoxins with a comparable anti-inflammatory role.

epilysin *See* MATRIX METALLOPEPTIDASES.

epinephrine *See* ADRENALINE.

epiphysis **1.** The region at the end(s) of long (limb) bones that is separated from the shaft (diaphysis) by cartilage until skeletal growth is complete and full ossification occurs. The outer region of compact bone is relatively thin and the remainder is composed of trabecular bone with red marrow. **Epiphysitis** is inflammation of this region. **2.** The pineal body (epiphysis cerebri).

epiplakin A member (>700 kDa) of the *plakin family of cytolinker proteins but somewhat atypical, having only plakin repeat domains and none of the other domains found in other plakins. It was originally identified as an autoantigen in serum from a patient with subepidermal blistering disease and it probably interacts with intermediate filaments. Its function is, however, unclear since its absence in mice does not lead to severe skin dysfunction.

epiregulin A member of the *epidermal growth factor (EGF) family that activates ErbB1 and ErbB4 homodimers and all possible heterodimeric ErbB complexes. Will stimulate the proliferation of fibroblasts, hepatocytes, smooth muscle cells, and keratinocytes but inhibits the growth of several tumour-derived epithelial cell lines. Epiregulin and *amphiregulin are not expressed in normal colonic mucosa, but are clearly detectable in adenomas and carcinomas. Epiregulin-deficient mice develop chronic dermatitis although wound healing is normal, and epiregulin may have a role in immune/inflammatory-related responses of keratinocytes and macrophages.

episialin A heavily glycosylated membrane glycoprotein (mucin-1, polymorphic epithelial mucin, PEM, epithelial membrane antigen, EMA) encoded by the *MUC1* gene; has a molecular weight of ~300 kDa, more than half of which is *O*-linked glycan. There is a 69 aa cytoplasmic domain and the extracellular N-terminal portion has variable numbers of 20-aa tandem repeats that are extensively modified by *O*-linked glycans and may extend hundreds of nanometres beyond the plasma membrane. The normal role may be to inhibit bacterial adhesion to mucous membranes but increased expression in carcinoma cells may reduce their adhesion and mask the antigenic properties of the cells. Similar functions are ascribed to *ASGP, *epiglycanin, and *leukosialin.

episodic ataxia A range of disorders characterized by intermittent episodes of ataxia, often accompanied by other neurological disturbances. **Episodic ataxia type 1** (EA1) is caused by mutation in the potassium channel gene *KCNA1*; **type 2** (EA2) by mutation in the calcium ion channel gene *CACNA1A*; **type 3** has been linked to a locus on chromosome 1; **type 4** (periodic vestibulocerebellar ataxia) is autosomally dominant but as yet unlinked; **type 5** (EA5) is caused by mutation in the calcium ion channel gene *CACNB4*; **type 6** (EA6) by mutation in the *SLC1A3* glutamate-transporter gene and **type 7** has been mapped to chromosome 19q13.

episome A fragment of hereditary material that can exist either as free, autonomously replicating DNA or be integrated into the chromosome of the cell. When integrated it is replicated as part of the chromosome. *Bacteriophages such as lambda can be considered episomes, so too are transposons and insertion sequences (known as mobile genetic elements).

epistasis The nonreciprocal interaction of nonallelic genes which can be synergistic (positive) or antagonistic (negative); e.g. a gene that blocks hair development will 'mask' genes for hair colour, an extreme form of antagonistic epistasy. The interaction can be complex and the expression of many genes is undoubtedly affected by the overall genomic environment, hence the variability in the effects of some mutations.

epistasy *See* EPISTASIS.

epitectin An epithelial mucin (MUC-1 glycoprotein) found on the surface of bladder epithelium and in the urine. Also found on the surface of some tumour cells (CA antigen). Like other mucins, it is a transmembrane protein with the extracellular N-terminal subunit having variable numbers of 20 aa tandem repeats that are extensively *O*-glycosylated. Probably normally exists as aggregates of three or four monomers, each of 350–400 kDa. *See* EPISIALIN.

epithelial membrane antigen (EMA) *See* EPISIALIN.

epithelial mesenchymal transition A developmental process in which epithelial cells are apparently reprogrammed and exhibit completely different behaviour. There is reduced adhesion, increased cell mobility, and loss of E-cadherin expression. The transition in behaviour is important in mesoderm formation and neural tube formation. The process is stimulated by various growth factors such as IGF-1 and TGFβ-1 and is regulated through the Ras-MAPK and Wnt signalling pathways. Ras-MAPK activates two related transcription factors, *Snail and Slug, that are transcriptional repressors of E-cadherin and their expression induces EMT. *Twist, another transcription factor, has also been shown to induce EMT, and is implicated in the regulation of metastasis.

epithelioid cells **1.** Cells that have an epithelial morphology. **2.** The very flattened macrophages found in granulomas (e.g. in tubercular lesions).

epithelioma A malignant tumour derived from epithelium, a synonym for a carcinoma.

epithelium (*pl.* epithelia) A simple tissue consisting of a sheet of cells that are adherent to one another. The cells may be flattened, as in *squamous epithelia, the cell–cell contact area may be approximately the same as the cell–basal lamina contact area, as in **cuboidal epithelia**, or the cells may be vertically elongated, as in **columnar epithelia**. More complex examples involve multilayers of cells, not all of which are in contact with the basal lamina (**stratified epithelia**, of which skin is the classic example), although **pseudostratified epithelia** do occur. A characteristic feature is that the cells have polarity: the apical surface of the cells may have microvilli or cilia, and is separated from the basolateral domain by a band of tight junctions that prevent lateral diffusion of integral membrane proteins so that the two domains can have a different complement of transporters or receptors. Epithelia may be specialized for transport (e.g. in kidney), for absorption (e.g. in intestine), or for mechanical protection (e.g. skin). Diagnostic features are *cytokeratins and specialized junctional complexes (*see* DESMOSOME; GAP JUNCTION; TIGHT JUNCTION).

epitope The region of a large molecule that is recognized and bound by a specific antibody or to which the *T-cell receptor responds. In proteins the determinant will usually involve four or five residues, in polysaccharides three to five sugars. Most biomolecules have multiple epitopes and epitope mapping can be used to determine which portion of a protein is involved in a particular activity. Methods include the use of *epitope libraries, synthetic peptides, protein *footprinting (using monoclonal antibody to protect the epitope), or the isolation and characterization of the peptide bound to MHC. Since some epitopes may depend upon the tertiary structure of the protein or arise through post-translational modification, the process is not straightforward. *See also* AGRETOPE; HAPTEN.

epitope library A large set of peptides, potential *epitopes, that can be used to identify epitopes on a protein. The **phage display library** method involves inserting randomly mutated DNA into a phage (a *genomic library) in such a way that the products are expressed in a chimeric phage surface protein. By sequencing the phages that bind to the antibody the peptide sequence of the epitope can be deduced.

epitope tag A short peptide sequence that is engineered into a transgenic protein to assist with tracking its expression etc. The sequence is an *epitope for an existing antibody. A commonly used example is the 'myc' tag, the peptide sequence of which is EQKLISEED. *See also* FLAG TAGGING.

epitrichium (periderm) The outer layer of the epidermis in an embryo or fetus, usually only a single cell-layer thick.

eplin A LIM domain protein (epithelial protein lost in neoplasm, Lima-1) that mediates linkage of the cadherin–catenin complex to F-actin and is important in assembly and stability of the adherens junction. Eplin increases the number and size of actin stress fibers and inhibits membrane ruffling induced by Rac. There are two isoforms, EPLIN-α (600 aa) and EPLIN-β (759 aa) with different tissue distribution. Eplin is down-regulated in neoplastic cells.

SEE WEB LINKS

- Article 'EPLIN mediates linkage of the cadherin–catenin complex to F-actin and stabilizes the circumferential actin belt' in PNAS (2007/8).

EPM *See* MYOCLONIC EPILEPSY OF UNVERRICHT AND LUNDBORG.

EPO *See* ERYTHROPOIETIN.

epoprostenol *See* PROSTACYCLIN.

epothilones Microtubule-stabilizing compounds (16-member macrolides: epothilone A and B) originally isolated from myxobacterium *Sorangium cellulosum* Str 90. Their mode of action is similar to that of taxanes but the binding site is different and they can be used in treatment of tumours that are resistant to paclitaxel. Epothilone B has a threefold higher affinity for the taxoid binding site on β-tubulin than paclitaxel. A range of synthetic and semisynthetic analogues have been produced.

epp *See* END PLATE POTENTIAL.

Eps15 An adaptor protein involved in endocytosis and trafficking of the epidermal growth factor receptor. The **EH domain** (Eps15-homology domain) is a highly conserved motif of ~100 aa found in many proteins involved in regulation of the endocytic pathway. The ligand of EH domains is the tripeptide Asp-Pro-Phe (NPF).

epsilon-aminocaproic acid *See* AMINOCAPROIC ACID.

epsins A conserved family of proteins (550–650 aa) that are essential components of the endocytotic machinery, bind phosphoinositides through an epsin N-terminal homology (ENTH) domain and interact with *ubiquitin, *clathrin, and endocytic scaffolding proteins through the C-terminal region. Epsin is thought to directly modify the curvature of membranes when it binds to phosphatidylinositol 4,5-bisphosphate (PIP2) A range of epsin-like proteins have been identified.

EPSP *See* POSTSYNAPTIC POTENTIAL.

Epstein–Barr virus (EBV) One of the herpesvirus group (Herpetoviridae), that is responsible for glandular fever (infective mononucleosis) and, in the presence of other factors, tumours such as *Burkitt's lymphoma and nasopharyngeal carcinoma. The virus binds to the C3d,g receptor (CR2/CD21) on B cells through the viral envelope glycoprotein gp350 although it is known to infect epithelial cells as part of its normal cycle of persistence in a human host, and will sometimes infect T cells, NK cells, smooth muscle cells, and possibly monocytes.

SEE WEB LINKS

- Article 'Epstein–Barr Virus Infection' in *New England Journal of Medicine* (2000).

eptifibatide A small molecule (832 Da) that reversibly inhibits platelet aggregation by preventing the binding of fibrinogen, *von Willebrand factor, and other ligands to GP IIb/IIIa. It is a cyclic heptapeptide containing six amino acids and one mercaptopropionyl (desaminocysteinyl) residue. It is used clinically to treat patients with acute coronary syndrome and those undergoing percutaneous coronary intervention (PCI). *TN* Integrilin.

equatorial plate The region of the mitotic spindle that lies midway between the poles and where chromosomes become aligned at metaphase.

equilibrative nucleoside transporters Integral membrane proteins (ENTs) with eleven transmembrane domains that transport nucleotides in a sodium-independent fashion. The plasma membrane monoamine transporter (PMAT (ENT4)) is one of the family and functions as a polyspecific organic cation transporter and transports monoamine neurotransmitters and the neurotoxin 1-methyl-4-phenylpyridinium (MPP^+). Inhibitors of mammalian ENTs include dilazep and dipyridamole. *See* CNTs (concentrative nucleoside transporters).

equilibrium constant (equilibrium dissociation constant, dissociation constant) The ratio of the reverse and forward rate constants for a reaction of the type A + B = AB that describes the position at equilibrium. The equilibrium constant (K) has dimensions of concentration and is equals (concentration A × concentration B)/(concentration AB) ([A][B]/[AB]). The affinity (association) constant (K_a) is the reciprocal of the equilibrium constant.

equilibrium dialysis A method used to measure the binding affinity of small-molecule ligand to a macromolecule. The macromolecule is confined in a dialysis chamber and allow to equilibrate with diffusible ligand.

equinatoxins Peptide toxins (actinoporins, ~200 aa), from the sea anemone *Actinia equina*, that form cation-selective pores and are cytolytic.

equisetin Fungal metabolite isolated from *Fusarium equiseti* that has antibiotic and cytotoxic activity. It is a potent inhibitor of HIV-1 integrase and of mitochondrial ATPases.

equivalence A description, sometimes applied to high-avidity interactions such as those between antibody and antigen, of the situation where the two interacting molecular species are at concentrations such that all binding sites are occupied.

ER *See* ENDOPLASMIC RETICULUM.

ERAB A mitochondrial enzyme (endoplasmic reticulum associated binding protein, hydroxyacyl-CoA dehydrogenase II (HADH2), 17-β-hydroxysteroid dehydrogenase X, short-chain 3-hydroxyacyl-CoA dehydrogenase (SCHAD), amyloid β-binding alcohol dehydrogenase (ABAD), EC 1.1.1.178, 262 aa) that acts on a wide spectrum of substrates, including steroids, alcohols, and fatty acids. Interacts with β-amyloid in the brain tissue of patients with Alzheimer's disease, and the complex may be responsible for neurotoxicity. Reduced expression of the HADH2 protein probably causes an X-linked mental retardation syndrome (MRXS10) and deficiency causes problems with isoleucine degradation.

erabutotoxins Polypeptide toxins (62 aa), from the venom of the sea snake *Laticauda semifasciata*, that bind to nicotinic acetylcholine receptors and have a similar effect to *curare.

erb An oncogene originally identified in avian erythroblastosis virus. In fact there are two oncogenes, *erbA* and *erbB*, and the human cellular homologues of *erbB* are structural genes for receptor tyrosine kinases (*see* EPIDERMAL GROWTH FACTOR (EGF receptor); c-ErbB2 is also known as HER2/neu) and the *erbA* products are members of the steroid/thyroid hormone receptor family. Mutations in the T_3-binding domain of c-ErbAβ cause generalized resistance to thyroid hormones (GRTH) syndrome.

e

ERBIN A protein (ErbB2 receptor-interacting protein, 1412 aa) of the LAP (leucine-rich repeat (LRR) and PDZ domain) family that binds to *p0071 and ErbB2 and colocalizes with *PAPIN on the lateral membrane of epithelial cells. Appears to play a regulatory role in the activation by *epidermal growth factor of the Ras–Raf–MEK pathway.

erbstatin A compound isolated from *Streptomyces* MH435-hF3 that inhibits the autophosphorylation of the EGFR receptor. *Tyrphostins are derived from erbstatin.

ergocalciferol *See* CALCIFEROL.

ergosterol The biological precursor of *vitamin D_2 (provitamin D_2), a sterol found in membranes of ergot, yeasts, and other fungi where it functions in the same way as cholesterol in membranes of animal cells. Conversion to vitamin D_2 occurs in the skin as a result of the absorption of ultraviolet light.

ergot Product of the infection of rye *Secale cornutum* with the fungus *Claviceps purpurea*. The 'ergot' that replaces the grain of the rye is a dark purplish sclerotium, in which the fungus overwinters. A mycotoxin (*ergotamine) produced by the fungus contaminates rye flour and causes ergotism.

ergotamine tartrate The mycotoxin produced by *Claviceps purpurea*, a natural alkaloid that acts as an α-adrenergic blocking agent with a stimulatory effect on vascular smooth muscle of peripheral and cranial blood vessels and also causes depression of vasomotor centres. Ergotamine has serotonin antagonist properties. Used to relieve migraine by constricting cranial arteries.

ergotism *See* ERGOT.

ERKs *See* MAP KINASES.

erlotinib An small-molecule (423 Da) inhibitor of the EGF-receptor type-1 tyrosine kinase (HER1) that has been used to treat non-small-cell lung cancer. *TN* Tarceva.

ERM **1.** Ezrin/radixin/moesin (ERM) proteins: *See* FERM DOMAIN. **2.** A transcription factor of the *Ets family (Erm). **3.** Bacterial genes (*ermA, ermB,* etc.) that code for dimethyl- or methyl-transferases and that confer drug resistance.

e

E rosetting A technique used to show T cells in leucocyte mixture. Sheep erythrocytes ('E') will bind to form a 'rosette' around T cells whereas B cells, which have Fc receptors, require the erythrocytes to be antibody (A) treated and will then form **EA rosettes**. Antibody-coated erythrocytes with bound complement (C) will form **EAC rosettes** around cells with C3b or C3bi receptors (*CR1 or *CR3). EAC rosettes form more easily than E or EA rosettes.

error-prone repair *See* SOS SYSTEM.

ERV **1.** Endogenous retroviruses: *see* RETROVIRIDAE. **2.** Expiratory reserve volume.

Erwinase An L-asparaginase (crisantaspase) derived from *Erwinia chrysanthemi*, the bacterium responsible for soft rot in plants. Approved for treatment of acute lymphoblastic leukaemia, especially when an allergic response to *E. coli*-derived asparaginase has developed.

Eryf1 *See* ERYTHROID TRANSCRIPTION FACTOR.

erysipelas (cellulitis) A streptococcal infection of skin and subcutaneous tissue, particularly of the face, neck, forearm, and hands. There is inflammation and possibly an allergic reaction to products of the causative organism, *Streptococcus pyogenes*.

erysipeloid A rare skin infection by the Gram-positive bacterium *Erysipelothrix rhusiopathiae*, usually acquired from contaminated meat or fish.

Erysipelothrix rhusiopathiae *See* ERYSIPELOID.

erythema A superficial flushing or redness of the skin caused by capillary dilation. **Erythema infectiosum** (fifth disease; slapped cheek disease) is a benign childhood disease caused by parvovirus B19. **Erythema multiforme** is characterized by raised red patches, particularly on the upper part of the body, and is a type of hypersensitivity reaction in response to medications or infections (usually with herpes simplex or mycoplasma). The systemic form of the disease (Stevens–Johnson syndrome) can be serious. In **erythema nodosum** there are pink or red nodules, usually on the lower limbs and often a result of streptococcal infection. **Erythema pernio** (chilblain) is caused by constriction of small blood vessels in cold weather.

erythermalgia *See* ERYTHROMELALGIA.

erythraemia *See* POLYCYTHAEMIA (erythrocytosis).

erythrasma Chronic infection of the stratum corneum of skin with *Corynebacterium minutissimum*.

erythremia *See* POLYCYTHAEMIA (erythrocytosis).

erythritol A four-carbon sugar alcohol (polyol) formed from erythrose and used as a low-calorie sweetener.

erythroblast (normoblast) A nucleated cell of the myeloid series in the bone marrow that gives rise to erythrocytes. *See also* BFU-E; CFU-E at COLONY-FORMING UNIT; DEFINITIVE ERYTHROBLASTS; PRIMITIVE ERYTHROBLASTS. **Erythroblastosis** is the pathological condition in which there are erythroblasts in the blood.

erythroblastosis fetalis A rare but severe haemolytic anaemia of the newborn caused by transplacental passage of maternal IgG directed against fetal erythrocytes. *See* RHESUS BLOOD GROUP system.

erythrocyte (red blood cell, rbc) In humans, and most mammals, a non-nucleated blood cell with few cytoplasmic organelles and large amounts of haemoglobin in the cytoplasm. Approximately 7 μm in diameter, biconcave, and very deformable. Erythrocyte counts vary but the normal range in blood is ~4.2–5.4 × 10^6/mL (women), 4.7–6.1 × 10^6/mL (men), 4.6–4.8 × 10^6/mL (children). An **erythrocyte ghost** is the erythrocyte devoid of cytoplasmic contents, but with the membrane and cytoskeletal elements preserving the original morphology. The membrane proteins have been studied in great detail. The **erythrocyte sedimentation rate** (ESR) is a standard haematological test that measures the rate of settling of erythrocytes in anticoagulant treated blood, usually the number that settle in an hour. Sedimentation rates are more rapid in inflammation and in pregnancy.

erythrocyte membrane proteins A well-studied group of proteins because of the ease with which erythrocyte ghosts, essentially plasma membranes with associated cytoskeletal elements, can be prepared. The initial nomenclature was based upon the position of bands on an SDS polyacrylamide gel with band 1 (protein 1) being at the top, the largest protein.

Subsequently the proteins in these bands became identified. Bands 1 and 2 are *spectrin α and β respectively, band 2.1 is *ankyrin, *band III is the major anion transporter (solute carrier family 4, member 1), bands 4.1 and 4.2 are cytoskeletal linkers (*see* PROTEIN 4.1; PROTEIN 4.2). Band 4.5 is a nucleoside transporter, band 4.9 is *dematin, band 5 is actin, band 6 is glyceraldehyde 3-phosphate dehydrogenase, and band 7 is *stomatin.

erythrocytosis (erythraemia) *See* POLYCYTHAEMIA.

erythrogenic toxin *See* STREPTOCOCCAL PYROGENIC EXOTOXINS.

erythroid cell A cell that gives rise to erythrocytes; the stages of erythropoesis and the control of differentiation have been extensively studied and a range of erythroid or erythroleukaemic (transformed) lines that will differentiate in culture (e.g. *Friend murine erythroleukaemia cells) have been developed.

erythroid Krüppel-like factor An erythroid cell-specific zinc finger transcription factor (EKLF, 362 aa) that is important in differentiation of the megakaryocyte-erythroid progenitor cells: it inhibits the formation of megakaryocytes while at the same time stimulating erythroid differentiation. EKLF is essential for establishing high levels of adult β-globin expression.

erythroid transcription factor General term for a transcription factor involved in erythroid cell differentiation. Important examples are *GATA-1 and *erythroid Krüppel-like factor, but as more and more have been discovered it is essential to be more specific.

erythrokeratodermia variabilis A genodermatosis in which there are transient erythematous patches of skin and hyperkeratosis, usually localized. It is caused by defects in the *connexin GJB3 (270 aa).

erythromelalgia **1. Primary erythromelalgia** (erythermalgia, acromelalgia, Mitchell's disease, red neuralgia) is an autosomal dominant disorder in which there is episodic symmetrical red congestion, vasodilatation, and burning pain of the feet and lower legs. It is caused by mutation in the *SCN9A* gene that codes for the α subunit of the *voltage-gated sodium channel-9 (*see* CONGENITAL INSENSITIVITY TO PAIN). **2. Secondary erythromelalgia** is an acquired disorder, associated with thrombocythaemia or myeloproliferative disorders, caused by platelet aggregation in end-arteriolar circulation, leading to ischaemia.

erythromycin A macrolide antibiotic, produced by a strain of *Saccharopolyspora erythraea* (formerly *Streptomyces erythraeus*), that has a wide spectrum of action, similar to that of penicillin but often used in patients who are allergic to the penicillins. Inhibits protein synthesis by binding to the 23S rRNA molecule in the 50S bacterial ribosome subunit, blocking the exit of the growing peptide chain. Resistance can arise through production of degradative enzymes, enhanced export, or modification of the rRNA to reduce binding.

erythropoiesis The process of erythrocyte production that in adults occurs within the bone marrow and, in the fetus, in the yolk sac mesoderm and later in the liver. A pluripotent stem cell (*colony forming unit, CFU) produces, by a series of divisions, committed stem cells (*burst forming units-erythrocytic, BFU-Es) that give rise to CFUs-erythrocytic (CFU-Es), cells that will divide only a few more times to produce mature erythrocytes. Each stem cell can give rise to 211 mature red cells. The process is regulated by *erythropoeitin and probably also by *hepcidin.

erythropoietin (EPO) A glycoprotein hormone (193 aa) that regulates *eythropoiesis and has cardioprotective and neuroprotective effects. EPO is mostly produced in the kidney, gene expression being induced by hypoxia-inducible transcription factors (HIF). Recombinant EPO is used therapeutically in patients, particularly those with kidney failure. The **erythropoietin-receptor** is a type I transmembrane protein of the cytokine-receptor family and has sequence homology with IL-2 receptor β. Like IL-2-Rβ, a second subunit is probably required to produce an active signalling unit. Defects in the receptor cause familial erythrocytosis-1 (*see* POLYCYTHAEMIA).

ES cells *See* EMBRYONIC STEM CELL.

***Escherichia coli* (*E. coli*)** A rod-shaped Gram-negative bacillus (0.5 × 3–5 μm) abundant in the large intestine (colon) of mammals. Normally nonpathogenic, but the *E. coli* O157 strain, common in the intestines of cattle, produces a potent *verotoxin and has caused a number of deaths. *E. coli* is the biochemists' standard experimental organism, although what is true for *E. coli* is not necessarily true for all species (*see* OPERON).

***Escherichia coli* haemolysins** A family of membrane-permeabilizing exotoxins (~303 aa) of the *RTX family produced by many *E. coli* strains that are responsible for nonintestinal infections. Lytic activity depends on binding of calcium ions to the toxin in solution, causing a conformational change and allowing it to produce a functional pore when it binds to membranes.

esculin A *coumarin derivative used in pharmaceutical manufacture (it is said to have venotonic, capillary-strengthening, and antiphlogistic activity). In diagnostic microbiology it is a component of **bile esculin agar** : Group D streptococci hydrolyse esculin to esculetin and dextrose and the esculetin can be detected by the formation of a blackish-brown coloured complex when it reacts with ferric citrate. Extracted from the bark of flowering ash *Fraxinus ornus* or the leaves and bark of the horsechestnut tree *Aesculus hippocastanum*.

- Additional information from chem-online.org.

E selectin *See* SELECTINS.

eserine (physostigmine) An alkaloid with anticholinesterase activity, used in treatment of glaucoma. Isolated from the Calabar bean *Physostigma venenosum*, although it can be synthesized.

E site The *ribosomal site that binds deacylated tRNA after it leaves the *P-site and before it leaves the ribosome.

ESL **1.** Endothelial surface layer. **2.** E-*selectin ligand-1 (ESL-1).

espins Small actin bundling proteins that are highly enriched in the microvilli of certain chemosensory and mechanosensory cells, often with other actin bundling proteins. Various isoforms generated by alternative splicing occur in a tissue-specific manner and range in size from 854 aa to 322 aa. They cause elongation of actin bundles and thus help determine the steady-state length of microvilli and stereocilia. Espins bind actin monomer via their WH2 domain and some isoforms can also bind phosphatidylinositol 4,5-bisphosphate, profilins, or SH3 proteins. Mutations in espin are associated with DFNB36 (autosomal recessive neurosensory deafness 36).

espundia An ulcerative infection (mucocutaneous leishmaniasis), by *Leishmania brasiliensis*, of the skin and of the mucous membranes, particularly those of the nose and mouth.

ESR *See* SEDIMENTATION TEST.

essential amino acids Amino acids that cannot be synthesized and must therefore be present in the diet. Anthropocentrically, those amino acids required by humans (Ileu, Leu, Lys, Met, Phe, Thr, Try, and Val), although rats need Arg and His as well. Vegan diets can be deficient in some of these essential components unless care is taken.

essential fatty acids The three fatty acids that are required for growth in mammals but cannot be synthesized: arachidonic, linolenic, and linoleic acids. Provided the diet contains linoleic acid the others can be synthesized.

EST *See* EXPRESSED SEQUENCE TAG.

established cell line *See* CELL LINE.

estradiol (follicular hormone) US name and *BAN for *oestradiol.

estrogen *See* OESTROGEN.

estrone *See* OESTROGEN.

ET *See* ENDOTHELINS.

etanercept An anti-TNF drug used in treatment of rheumatoid arthritis as an alternative to the anti-TNF monoclonal antibodies infliximab and adalimumab. It consists of the extracellular ligand-binding portion of the human 75 kDa (p75) *tumour necrosis factor receptor linked to the Fc portion of human IgG1 and is effectively a decoy receptor. *TN* Enbrel.

ethacrynic acid A glutathione *S*-transferase inhibitor and potent loop diuretic (acts on the ascending limb of the loop of Henle and on the proximal and distal tubules). Ethacrynic acid acts rapidly and inhibits reabsorption of a much greater proportion of filtered sodium than most other diuretic agents. Used in the treatment of oedema and oliguria due to renal failure. Has been shown to inhibit signalling by NFκB. Experimentally often used as a chloride pump blocker. *TN* Edecrin.

ethambutol A drug used, usually in combination with others, to treat tuberculosis. Mode of action seems to be through inhibition of arabinogalactan and arabinomannan biosynthesis, resulting in the accumulation of decaprenyl-*P*-arabinose. *TN* Myambutol.

ethidium bromide A dye that intercalates into DNA and to some extent RNA and when intercalated is intensely fluorescent. Often used to detect nucleic acid on gels, but since it is very probably mutagenic is increasingly being replaced by other dyes such as SYBR Green. Ethidium bromide added to DNA prior to ultracentrifugation on a caesium chloride gradient allows separation of linear (nuclear) and circular (mitochondrial or bacterial) DNA on the basis of density; less intercalates into the circular DNA which is therefore more dense.

ethmoid cells Small sinuses between the eyes, which number from around five to more than fifteen, but not cells in the cell-biological sense. **Ethmoiditis** is inflammation of the ethmoid cells.

ethosuximide An anticonvulsant drug used for control of absence seizures (petit mal-type epilepsy) as an alternative to valproic acid. *TN* Zarontin.

ethylmalonic encephalopathy A syndrome characterized by ethylmalonic and methylsuccinic aciduria and lactic acidaemia associated with developmental delay, acrocyanosis, petechiae, and chronic diarrhoea. It is caused by mutation in the *ETHE1* gene, which encodes a mitochondrial matrix protein, although the nature of the metabolic disorder is unclear.

etidronate A first-generation bisphosphonate drug used in treatment of osteoporosis and in the treatment of Paget's disease, heterotopic ossification, and malignant hypercalcaemia. *TN* Didronel.

etiology *See* AETIOLOGY.

etoposide (VP16) A semisynthetic derivative of *podophyllotoxin that is an inhibitor of *topoisomerase II and rapidly induces apoptosis. Used in chemotherapy of various tumours such as Ewing's sarcoma, lung cancer, testicular cancer, and nonlymphocytic leukaemia. *TNs* Etopophos, Vepesid.

ets **1.** An *oncogene originally found in E26 transforming *retrovirus of chickens (e26 transformation-specific sequence) that had the unusual property of inducing both myeloid and erythroid leukaemias. Subsequently found in a range of species including humans and was the 'founder member' of the ETS family of oncogenes and transcription factors defined by the presence of a conserved DNA binding domain recognizing a canonical...GGAA/T...sequence. The **ETS domain** is a DNA-binding domain, formed of three *alpha-helices found in the DNA-binding domain of the human ETS-1 transcription factor. **2.** External transcribed spacers. *See* INTERNAL TRANSCRIBED SPACERS.

SEE WEB LINKS

- Article 'Ets and retroviruses' in *Oncogene* (2000).

etynodiol A progestogen (ethynodiol diacetate) which is used in progestogen-only oral contraceptive pills.

Eubacteria *See* BACTERIA.

eucaryote *See* EUKARYOTA.

euchromatin The chromosomal regions that are diffuse during interphase and condensed at the time of nuclear division in contrast to *heterochromatin.

Eukaryota One of the major subdivisions of living organisms. The cells of eukaryotes have linear DNA organized into chromosomes with nucleosomal structure involving histones, with the nucleus separated from the cytoplasm by a two-membrane envelope, and compartmentalization of various functions in distinct cytoplasmic organelles. A range of other characteristics distinguish them from the *prokaryotes, which are now generally considered to be separated into two domains, *Bacteria and *Archaea (formerly Eubacteria and Archaebacteria).

eukaryotic initiation factor (eIF) A multiprotein initiator factor complex (43S). **eIF-1** is a low molecular weight factor (113 aa) critical for stringent AUG selection, **eIF-2** has three subunits: α (315 aa), β (333 aa, a guanine nucleotide exchange factor), and γ (472 aa), and forms a ternary complex with Met-tRNAi and GTP. **eIF-3** contains at least eight distinct polypeptides and plays a role in recycling of ribosomal subunits. **eIF-4** is trimeric (eIF-4A, eIF-4E, and eIF-4G) and associates with the 5′ cap of mRNA. **eIF-5** interacts with the 40S ribosomal initiation complex and promotes the hydrolysis of bound GTP and subsequent release of eIF-3 from the 40S subunit. The 40S subunit is then able to interact with the 60S ribosomal subunit to form the functional 80S initiation complex.

eumelanin *See* MELANIN.

euploidy A form of polyploidy in which the whole chromosome set is replicated as in diploid, tetraploid, etc. *Compare* ANEUPLOID.

eutely Describing the situation in which all individuals of a species have the same number of cells (or nuclei in a coenobium). An important example is in Nematoda where the fixed number of cells has made lineage studies a little easier.

euthyroid Describing a normal level of thyroid activity.

Evans blue A diazo dye that binds to albumin and that can be used to estimate blood volume by injecting a known amount of dye and sampling the concentration after it has bound and been distributed throughout the blood. If there is leakage of plasma protein from blood vessels, e.g.

in a site of inflammation, this can easily be visualized.

evening primrose oil An oil extracted from seeds of *Oenothera biennis*, rich in γ-linolenic acid, one of the *essential fatty acids and the precursor for *prostaglandin synthesis. Remarkable claims are made for the therapeutic value of this oil as a dietary supplement and there is no evidence that it is hazardous.

everninomicins (evernimicins) Class of oligosaccharide antibiotics with high activity against Gram-positive organisms. Although known for some time they are now being used to treat antibiotic-resistant *S. aureus*. Several variants have been isolated from the fermentation broth of *Micromonospora carbonacea* var *africana*. Everninomicins and *avilamycin bind to the same site on the bacterial 50S ribosomal and resistance can arise either through sequence variation in ribosomal protein L16 or changes in the peptidyltransferase domain of 23S ribosomal RNA.

Evi-1 *See* ECOTROPIC VIRAL INTEGRATION SITE 1.

Ewing's sarcoma A tumour that develops in bone marrow, considered by James Ewing (1866–1943) to be a diffuse endothelioma of bone. There is evidence for a determinant on the long arm of chromosome 22 and the protein encoded by this region, termed EWS (656 aa) has RNA-binding activity. This region, fused with transcriptional activators or repressors (e.g. Fli-1 or ATF 1) by chromosomal translocation, is found in a range of different tumours including Ewing's sarcoma, related primitive neuroectodermal tumours, malignant melanoma of soft parts, and desmoplastic small round cell tumours.

excipient A pharmacologically inactive substance used as a carrier for a drug. May stabilize the drug, mask bitterness, enable formulation as a syrup or as a pill, etc.

excision repair A DNA repair mechanism (nucleotide excision repair) in which the site of damage is recognized, a short sequence surrounding the damage is excised by an *endonuclease, the correct sequence is copied from the complementary strand by a *polymerase, and the ends of this correct sequence are joined to the rest of the strand by a *ligase. Deficiencies in the repair mechanism cause serious diseases such as *xeroderma pigmentosum and *Cockayne's syndrome.

excitable cell A cell in which membrane depolarization leads to an *action potential thereby amplifying and propagating the depolarization. The main examples are neurons and muscle cells but electrical excitability is also found in fertilized eggs, some plants, and glandular tissue. The response involves *voltage-gated ion channels.

excitation–contraction coupling The sequence of events that couples electrical excitation of a muscle by the arrival of a nervous impulse at the *motor end plate to the contractile process, the sliding of the filaments of the *sarcomere. The coupling is brought about by the release of calcium from the sarcoplasmic reticulum, and the analogy is often drawn between this and *stimulus-secretion coupling, in which calcium release into the cytoplasm is also involved.

excitatory amino acids (EAA) Excitatory neurotransmitters in the central nervous system, the amino acids L-glutamate and L-aspartate, that may be involved in *long-term potentiation, and that can act as *excitotoxins. At least three classes of **EAA receptor** have been identified; the agonists of the **N-type** receptor are L-aspartate, *NMDA, and ibotenate; the agonists of the **Q-type** receptor are L-glutamate and quisqualate; agonists of the **K-type** are L-glutamate and kainate. All three receptor types are widely distributed through the central nervous system, particularly in the telencephalon; N- and Q-type receptors tend to occur together, and their distribution is complementary to that of the K-type receptors. The ion fluxes through the Q and K receptors are relatively brief, whereas the flux through the N-type is longer, and carries a significant amount of calcium. The N-type receptor shows *voltage-gated ion channel properties, leading to a regenerative response, and this may be important for long-term potentiation. Invertebrate glutamate receptors differ significantly from those of vertebrates. There are also inhibitory amino acids: γ-aminobutyric acid (*GABA) and *glycine are fast inhibitory transmitters in the central nervous system, their effects being blocked by *bicuculline and *strychnine respectively.

excitatory synapse A synapse between neurons that, when activated, increases the probability of a response in the postsynaptic cell, making it more easily excited. Excitatory synapses can be either chemical or electrical and have the opposite effect to *inhibitory synapses.

excitotoxin A substance that causes damage to neurons by inducing overactivity. The *excitatory amino acids can act as excitotoxins and will produce brain lesions similar to those of *Huntington's disease or *Alzheimer's disease. Excitotoxicity may well contribute to neuronal cell death associated with stroke.

exein *See* INTEINS.

exenatide An *incretin mimetic approved for treatment of type 2 *diabetes. It is a synthetic derivative of exendin-4, an agonist of receptors of glucagon-like peptide-1 (GLP-1) and is resistant to inactivation by dipeptidylpeptidase-4. *TN* Byetta.

exendin Group of peptide hormones (39 aa), related to the *glucagon family, found in the saliva of Gila monsters *Heloderma suspectum, H. horridum.* Helospectin is exendin-1; helodermin is exendin-2. Exendin-3 activates VIP receptors and stimulates amylase release from pancreatic acinar cells, whereas exendin-4 activates GLP-1 (glucagon-like peptide-1) receptors but has no effect on VIP receptors.

exfoliatin An epidermolytic toxin, of which there are two serotypes, ET-A and ET-B, produced by some strains of *Staphylococcus aureus.* Both are serine peptidases of the S1 (chymotrypsin) family and are *superantigens. They cause intraepidermal splitting by specific cleavage of *desmoglein 1 and disruption of the desmosomes of the stratum granulosum. **Staphylococcal scalded skin syndrome** (SSSS) is a spectrum of blistering skin diseases caused by these toxins; bullous impetigo, the localized form of SSSS, is the most common.

exocrine Describing glands that release their products into ducts rather than into the bloodstream. The pancreas and sweat glands are examples. *See* ENDOCRINE GLAND.

exocyst A multiprotein complex involved in the docking of exocytic vesicles with fusion sites on the plasma membrane. There are eight subunits (EXOC1–8, ranging from 725 aa to 974 aa) that interact with *centriolin. The EXOC subunits are members of the SEC3 family, homologous to the yeast SEC3 that is involved in exocytosis.

exocytosis Process by which material is released from a cell by the fusion of a vesicle (an exocytotic vesicle) with the plasma membrane; effectively the reverse of phagocytosis.

exodus-2 A β-chemokine (CCL21, EC, SLC, CKb9, TCA4, 6Ckine, SCYA21, MGC34555, 134 aa) that is a ligand for CCR7 and is chemotactic *in vitro* for thymocytes and activated T cells, but not for B cells, macrophages, or neutrophils. Shows preferential activity towards naive T cells. May play a role in mediating homing of lymphocytes to secondary lymphoid organs. Unusually, also reported to be a functional ligand for endogenously expressed CXCR3 in human adult microglia. A potent inhibitor of the proliferation of normal haematopoietic progenitors.

exoenzyme 1. An enzyme that has its active site on the outer surface of a cell (an ectoenzyme) or is released from the cell into the extracellular space. **2.** An enzyme that cleaves terminal residues from a polymer but will not cut internally, e.g. an *ectoglycosidase (in contrast to an endoenzyme, e.g. an endonuclease).

exoglycosidase An enzyme that cleaves the glycosidic bonds of terminal sugar moieties. Sequential cleavage by exoglycosidases can be used in the sequencing of carbohydrates. Can also be used to refer to glycosidases that act on the exterior of a cell (ectoglycosidases) which does not remove ambiguity.

exomphalos A congenital defect (umbilical hernia) in which the abdominal wall fails to close completely during fetal development and abdominal contents protrude into the umbilicus. **Exomphalos–macroglossia–gigantism syndrome** *see* BECKWITH–WIEDEMANN SYNDROME.

exon In eukaryotes, the DNA sequence that codes for amino acid sequence in a protein. Most proteins are the product of several exons spliced together (*see* ALTERNATIVE SPLICING). The primary RNA transcript contains exons separated by *introns and the introns are removed before the mature *messenger RNA leaves the nucleus. **Exon shuffling** is a proposed mechansim by which protein evolution could be facilitated. If exons encode functional domains (as many do) then a novel multifunctional protein can be generated by a novel assemblage of exons ('modules'). **Exon skipping** is a common cause of alternative splicing in which one or more exons are omitted during processing of the primary RNA transcript. In some cases this may be the omission of an exon that would lead to destruction of the message, an indirect mechanism to down-regulate the gene (*see* NONSENSE-MEDIATED MRNA DECAY). **Exon trapping** is a method for identifying regions of genomic DNA that are part of an expressed gene. The genomic sequence is cloned into an intron, flanked by two exons, and expressed through a strong promoter. If the genomic fragment does encode an exon the processed mRNA will be detectably larger.

exon junction complex (EJC) A complex of proteins that assembles on newly spliced RNA 20 nt upstream of the exon–exon junction and is a central effector of mRNA functions. It contains factors involved in mRNA export, cytoplasmic localization, and nonsense-mediated mRNA decay and it is claimed that the presence of an EJC enhances the translatability of mRNAs, although the mechanism is not yet understood.

exonuclease An enzyme that cleaves the terminal nucleotide from nucleic acid (*compare* ENDONUCLEASE). The cleavage is usually end-specific and there are 5′- and 3′-exonucleases.

exopeptidase A peptidase of the class EC 3.4.11–19.- that cleaves a bond not more than three residues from the free N- or C-terminus of a peptide. Exopeptidases can be further subdivided into aminopeptidases (EC 3.4.11), carboxypeptidases (EC 3.4.16–18), dipeptidyl- and tripeptidyl-peptidases (EC 3.4.14), peptidyl-dipeptidases (EC 3.4.15), dipeptidases (EC3.4.13), and acylaminoacyl-peptidases (EC3.4.19).

SEE WEB LINKS

- Enzyme nomenclature homepage.

exophthalmic goitre *See* GRAVES' DISEASE.

exosome 1. A multimolecular complex composed of 3′–5′-exoribonucleases involved in various RNA surveillance, processing, and degradation pathways. A significant proportion of nuclear exosomes are present in larger complexes (60–80S) whereas cytoplasmic exosomes are all ~10S. Exosomes are found in eukaryotes and archaea. **2.** A type of antigen-presenting vesicle secreted by some 'professional' *antigen-presenting cells. Exosomes are membrane-bounded and enriched in MHC class I and II proteins. Those released by dendritic cells are referred to as *dexosomes. **3.** In the older literature an exosome was a DNA fragment taken up by a cell that did not become integrated with host DNA but transformed the cell and continued to replicate. This usage is probably redundant.

exostosins Products (Ext1 and Ext2) of tumour suppressor genes involved in the expression of proteoglycans on the cell surface and in extracellular matrix. The two proteins (746 aa, 718 aa) form a hetero-oligomeric complex with glycosyltransferase activity that accumulates in the Golgi. Mutations in Ext1 or Ext2 are associated with multiple exostes type 1 or 2 respectively (autosomal dominant disorders in which there are multiple projections of bone capped by cartilage (*exostoses), particularly, but not exclusively, in metaphyses of long bones). May also regulate the *hedgehog signalling pathway.

exostosis (*pl.* exostes) A bony tumour growing outwards from a bone. *See* EXOSTOSINS.

exotoxins Toxins released into the environment by Gram-positive and Gram-negative bacteria. Important examples involved in human disease are *cholera, *diphtheria, and *pertussis toxins which are specific and highly toxic. *Compare* ENDOTOXIN, which are integral to the bacterial cell wall.

experimental allergic encephalomyelitis (EAE) An animal model for demyelinating diseases in which an autoimmune response is induced by the injection of homogenized brain or spinal cord in *Freund's adjuvant. The main antigen appears to be myelin basic protein, and the response is characterized by focal areas of lymphocyte and macrophage infiltration into the brain, with demyelination and destruction of the blood-brain barrier. The validity of the model is uncertain.

exportins A family of proteins (in the karyopherin superfamily) that are involved in regulating the export of proteins from the nucleus to the cytoplasm, the reverse of the task carried out by *importins. Exportins bind to proteins with nuclear export signal sequences in association with ranGTP and pass into the cytoplasm through the *nuclear pore complex. The protein cargo is released when the GTP is hydrolysed and the exportins diffuse back into the nucleus. *Leptomycin B inhibits export by blocking the binding interaction of exportin with the nuclear export signal. Exportin-1 is CRM-1 (chromosome region maintenance-1, 1071 aa), exportin-6 (exp6, 1125 aa) seems to be specific for profilin-actin complexes and for maintaining the nucleus actin-free, exportin-T (962 aa) mediates the nuclear export of aminoacylated tRNAs.

expressed sequence tag (EST) A short DNA sequence (500–800 bp) derived from mRNA that has been reverse transcribed and incorporated in a cDNA library. In principle the set of ESTs from a particular cDNA library represent a snapshot of the genes that are being expressed by the cells of interest, although some of the ESTs may be from mRNA destined for degradation (*see* NONSENSE-MEDIATED MRNA DECAY) and others may be derived from different portions of the same mRNA (so that the numbers may represent an overestimate of the numbers of genes being transcribed. Approximately 43 million ESTs were available in public databases in 2007.

SEE WEB LINKS

- Database of ESTs.

expression cloning A method of *gene cloning in which cells are *transfected with *complementary DNA in an *expression vector, and the transfected cells are then screened for the phenotype of interest (e.g. expression of a particular surface receptor or aberrant behaviour if an oncogene is being expressed).

expression profiling A shorthand term for 'gene expression profiling', a technique in which DNA microarrays are used to give an indication of the genes being expressed under different conditions, e.g. following an environmental insult, following hormonal stimulation, or at a particular developmental stage.

expression vector A *vector that will allow the gene that it is carrying to be expressed in the host cell into which it has been transfected. The vector usually contains a strong promoter, often inducible (e.g. under the control of the lac repressor and inducible with IPTG) and optimized for the host cell.

extended-spectrum beta-lactamase *See* BETA-LACTAMASE.

external guide sequence A RNA oligonucleotide that binds to mRNA in a sequence specific manner (by virtue of complementarity) and targets it for degradation by cytoplasmic *RNAse P.

external transcribed spacers *See* INTERNAL TRANSCRIBED SPACERS.

extinction coefficient Alternative name for the *absorption coefficient.

extracellular matrix (ecm) In common usage, the noncellular portion of animal tissue, usually produced by the resident cells of the tissue. In connective tissue the extracellular matrix is extensive and its composition largely determines the mechanical properties. Fibrous elements (particularly *collagen, *elastin, or *reticulin) provide tensile strength and the number, fibre properties, and orientation of fibrils determine the direction(s) in which tensile strength is greatest. The organization of fibrous elements is partly determined by link proteins (e.g. *fibronectin; *laminin); these also anchor cells within the matrix and hold space-filling molecules (usually *glycosaminoglycans) in place. The matrix may be mineralized to resist compression (as in bone) or be flexible (as in ligament or tendon). Epithelial cells tend to have more restricted extracellular matrix, the *basal lamina. Not only is matrix produced by cells but it can influence the behaviour of cells, an important consideration when growing cells in culture and for tissue engineering.

extrachromosomal element Any heritable element not integrated into a chromosome, usually a *plasmid, although the term can be applied to the DNA of organelles such as mitochondria and chloroplasts.

extrapyramidal system A neural network that is distinguished from the tracts of the motor cortex that reach their targets by travelling through the pyramids of the medulla (the *pyramidal system). Mainly important for stereotyped reflex movements. Reduction in dopamine levels in this system are responsible for motor defects of *parkinsonism, *akasthisia, and *dyskinesia.

extrasystole A premature ventricular contraction of the heart, the most common form of arrhythmia in which the contraction is triggered from the ventricle rather than from the normal pacemaker region.

extravascular Describing something that happens, or is located, outside a blood vessel.

extrinsic pathway *See* TISSUE FACTOR PATHWAY.

exudate The fluid which accumulates in tissues or in body cavities as a result of altered vascular permeability in the inflammatory response. Cells, particularly leucocytes, also migrate into the inflamed tissue and may be present in large number in the exudate.
A **peritoneal exudate** induced by saline with complement-activating glycogen is an experimental source of large numbers of neutrophil leucocytes, if drained in the early stages of inflammation, and of mononuclear cells at later stages.

exudative vitreoretinopathy *See* LDL RECEPTOR (LDL receptor-related protein-5).

eyepiece graticule (micrometer eyepiece, US ocular micrometer) A grid or scale incorporated in the eyepiece of an optical instrument to assist in counting or measurement. A wide variety of different patterns can be used and are usually engraved on a glass disc that can be temporarily inserted into the eyepiece. They need to be calibrated against a scale (micrometre slide) for each objective of a microscope.

EZH Polycomb group (PcG) proteins (enhancer of zeste homologues) that are the catalytic component of the *polycomb repressive complexes (PRC2/EED-EZH complexes), histone-lysine *N*-methyltransferases, EC 2.1.1.43). **EZH2** (746 aa) is a marker for aggressiveness in prostate and breast carcinoma.

ezrin An actin-binding protein (villin-2, cytovillin, 586 aa) that links the plasma membrane and microfilament bundles in microvilli of intestinal epithelium and elsewhere. Cytovillin (which is identical) is expressed strongly in placental syncytiotrophoblasts and in certain human tumours. *See* FERM DOMAIN.

F1 *See* NEUROMODULIN.

F1 hybrid An organism that is the product (the first filial generation) of crossing two dissimilar parents. F1 hybrids are often sterile although may have desirable hybrid vigour.

Fab An antibody fragment produced by papain treatment. Fab fragments (45 kDa) contain one antigen binding site and consist of one light chain (111 aa) linked through a disulphide bond to a portion (127 aa) of the heavy chain; they are effectively univalent antibodies. $Fab_{(2)}$ fragments (90 kDa) are produced by pepsin treatment and consist of two linked Fab fragments. They are divalent but lacks the complement-fixing (Fc) domain.

FABP *See* FATTY ACID BINDING PROTEINS.

Fabry's disease (angiokeratoma) An X-linked *storage disease due to mutation in α-galactosidase (ceramide trihexosidase, EC 3.2.1.22, 429 aa). The pathology is due to systemic accumulation of glycosphingolipids which is particularly damaging in the eyes, kidneys, autonomic nervous system, and cardiovascular system. Even in heterozygotes the deficiency can cause serious disease.

facilitated diffusion A process (passive transport) involving shuttles or pores that allows hydrophilic molecules to diffuse more rapidly across the hydrophobic barrier of a membrane than they would unassisted. *See* ANTIPORT; SYMPORT; UNIPORT.

facilitates chromatin transcription *See* FACT.

facilitation (synaptic facilitation) A phenomenon in neuronal systems in which successive presynaptic impulses produce progressively greater effects, usually because there is increased transmitter release probably as a result of residual calcium from the earlier impulses.

facilitator neuron A neuron that enhances the effect of a second neuron on a third, thereby modulating the system. In the well-studied examples in invertebrates the facilitator neuron synapses with the second (sensory) neuron near the axon terminal and modifies the ion channels (in some cases by phosphorylation) so that larger depolarizations occur and more neurotransmitter is released from the sensory neuron.

faciogenital dysplasia *See* AARSKOG–SCOTT SYNDROME.

FACIT collagens A subgroup of nonfibril-forming collagens (fibril associated collagens with interrupted triple helices). Includes *collagen types IX, XII, XIV, XIX, and XXI.

FACS *See* FLOW CYTOMETRY.

FACT An essential accessory factor for transcription when DNA is packaged in chromatin. Originally identified as an agent (facilitates chromatin transcription) required for RNA polymerase II to transcribe from nucleosomal DNA *in vitro*, but now recognized to be important in DNA replication and repair as well as transcription. It can also be considered as a histone chaperone. The highly conserved complex (~230 kDa) has two subunits, SSRP1 (structure-specific recognition protein, 709 aa), one of the high mobility group (HMG) protein family, and Cdc68 (SUPT16H, homologue of yeast Spt16, 1047 aa).

F-actin Filamentous *actin.

factors I–XII *See* BLOOD CLOTTING FACTORS.

factor C An endotoxin-sensitive, intracellular serine peptidase zymogen (994 aa) which initiates the coagulation cascade in the haemolymph of horseshoe crabs **Limulus polyphemus* and *Carcinoscorpius rotundicauda*. It binds *endotoxin and is heterodimeric with an 80-kDa heavy chain with several *sushi domains and a 52-kDa light chain.

factor D *See* ADIPSIN.

facultative heterochromatin The *heterochromatin which is condensed in some cells and not in others. Since gene expression from heterochromatin is probably neglegible, this represents a stable difference in activity. In females one of the two X chromosomes is normally inactivated by condensation to form heterochromatin (*Lyonization).

FAD Flavin adenine dinucleotide. *See* FLAVIN NUCLEOTIDES.

FADD Fas-associated via death domain. *See* FAS.

FAF1 Fas-associated factor 1. *See* FAS.

FAK Focal adhesion kinase. *See* FOCAL ADHESIONS.

falciparum malaria *See* PLASMODIUM.

Fallot's tetralogy *See* TETRALOGY OF FALLOT.

false positive A positive result in an assay or test that arises for some spurious reason. This may be simply noise or may be because of a confounding factor. Inhibition of proliferation might arise because the test compound inhibits a positive switching system (a true positive) or be general toxicity (a false positive). For this reason it is often preferable to have assays with a positive readout (e.g. survival) rather than a negative (death).

familial adenomatous polyposis *See* POLYPOSIS COLI.

familial cold autoinflammatory syndrome A group of rare autosomal dominant systemic inflammatory diseases (familial cold urticaria) characterized by episodes of rash, arthralgia, fever and conjunctivitis after generalized exposure to cold. **Type 1** is caused by mutation in *cryopyrin, **type 2** is caused by mutation in NLRP12 (NACHT, LRR, and PYD domains-containing protein 12).

familial haematuria Benign familial haematuria (BFH, thin basement membrane disease) is an autosomal dominant disorder in which there is persistent haematuria. The glomerular basement membrane is unusually thin, as shown by ultrastructural studies, as a result of a defect in type IV collagen.

familial hemiplegic migraine *See* MIGRAINE.

familial hypercholesterolaemia *See* HYPERCHOLESTEROLAEMIA.

familial hyperinsulinaemic hypoglycaemia *See* NESIDIOBLAST.

familial Mediterranean fever An autosomal recessive disorder (familial paroxysmal polyserositis) in which there are recurrent attacks of fever and inflammation in the peritoneum, synovium, or pleura. The defect is in *pyrin. Most of the affected individuals originate from Mediterranean or Middle Eastern countries.

familial primary localized cutaneous amyloidosis An autosomal dominant disorder associated with chronic skin itching and the deposition of epidermal keratin filament-associated amyloid material in the dermis. It is caused by a mutation in the *oncostatin M receptor.

familial startle disease A disorder (hyperexplexia) in which affected individuals show greatly exaggerated startle responses. Autosomal dominant and recessive forms of the disease are caused by mutation in the gene for the α_1 subunit of the *glycine receptor. It can also arise through mutations in the β subunit gene and in the gene for the presynaptic glycine transporter-2.

famotidine An *H_2 blocker. *TNs* Pepcid, Gaster.

Fanconi anaemia An anaemia caused by a defect in thymine-dimer excision from DNA predisposing to development of leukaemia and arising from mutation in any one of the thirteen Fanconi anaemia complementation group genes, *FANCA–FANCN*. Not only are all bone-marrow derived cells affected but there are cardiac, renal, and limb malformations as well as dermal pigmentary changes. Depending on the affected gene, symptoms vary.

Fanconi's syndrome (Fanconi–Bickel syndrome) A rare autosomal recessive disease in which there is hepatorenal glycogen accumulation, proximal renal tubular dysfunction, and impaired utilization of glucose and galactose. The defect is in transport rather than metabolism and appears to be due to mutation in the GLUT2 glucose transporter (solute carrier family 2, member 2 (SLC2A2)).

Farber's disease A *storage disease (Farber's lipogranulomatosis) caused by deficiency of acid *ceramide degrading enzyme (*N*-acylsphingosine amidohydrolase, EC 3.5.1.23, 395 aa). The effects are extensive and death usually occurs within the first 2 years.

farmer's lung A type III *hypersensitivity response to the bacterium *Micropolyspora faeni* found in mouldy hay.

farnesylation The post-translational modification of a protein by the enzymatic addition (by farnesyl transferase, EC 2.5.1.58, 379 aa) of a linear grouping of three isoprene units, a farnesyl group, that probably acts as a membrane-attachment device. Only proteins with a C-terminal CAAX motif become farnesylated.

far Western blot A blotting method to assess protein–protein interactions. Proteins are transferred from a gel to a membrane (often nitrocellulose), allowed to renature and are then probed with the putative interacting protein. Binding is usually detected with antibody to the binding protein.

f

SEE WEB LINKS

- Methodological details on blotting techniques.

Fas A transmembrane protein (Fas antigen, Fas receptor, TNFRSF6, 319 aa) of the *tumour necrosis factor receptor superfamily (*see* TUMOUR NECROSIS FACTOR receptors) with homology to the nerve growth factor receptor that is important in negative selection of autoreactive T-cells in the thymus. **Fas-associated factor 1** (FAF-1, 650 aa) is involved in negative regulation of NFκB activation probably both by interacting with the *pyrin domains of several *PYPAFs and by suppressing IκB kinase (IKK) activation by interacting with IKKβ and blocking assembly of the IKK complex, in response to proinflammatory stimuli (e.g. TNFα, IL-1β, LPS). **Fas-associated via death domain** (FADD, Mort1, 208 aa) is an adaptor protein that mediates signalling from all known death domain-containing members of the TNF receptor superfamily *death receptors to *caspases involved in apoptotic cell death. The 11q13.3 region where FADD is located is amplified in several human malignancies. **Fas ligand** (FasL) is a type II transmembrane protein (281 aa), similar to tumour necrosis factor, expressed on activated splenocytes and thymocytes. Fas ligand has three binding sites for Fas and thus stimulates trimer formation; subsequent intracellular signalling leads to apoptotic death of the FasR-bearing cell. The metalloprotease ADAM10 will cleave FasL in T cells to produce an N-terminal fragment without the receptor-binding extracellular domain. This FasL fragment is further processed by a signal peptide peptidase-like enzyme (SPPL2a) that liberates a small unstable fragment mainly containing the intracellular FasL domain which translocates to the nucleus and will inhibit transcription.

fascia (*pl.* fasciae, fascias) Connective tissue arranged as layers or sheets and found in most parts of the body. Deep fascia envelops organs, muscles and muscle groups. Superficial fascia is found below the dermis. **Fasciitis** (fascitis) is inflammation of the fascia. *See* NECROTIZING FASCIITIS.

fascicle A bundle, e.g. of neurons. *See* FASCICULATION (2).

fasciculation **1.** A local involuntary muscle contraction (twitch) due to spontaneous contraction of a small bundle (fascicle) of muscle fibres. Usually benign, but can be a symptom of disease of motor neurons. Also called fibrillation. **2.** The tendency of developing *neurites to form bundles or *fascicles, in part because nerve growth cones respond to the guidance cue of the existing bundle. The process also involves *axon-associated cell adhesion molecules such as *L1, *contactin, *neurofascin, and transient axonal glycoprotein TAG-1. In insect nervous systems a related group of molecules, the fasciclins, are involved.

fascin An actin filament-bundling protein, first identified in sea urchin but a human homologue (493 aa) was subsequently identified. An important structural component of microspikes, membrane ruffles, and stress fibres. The actin binding ability is regulated by phosphorylation. There is a photoreceptor-specific paralogue (retinal fascin, FSCN2, 492 aa) and a testis-specific isoform (FSCN3, 498 aa).

Fasciola hepatica A trematode, the common liver fluke, an economically important parasite of sheep and cattle that sometimes infects humans. The intermediate host is a freshwater snail from which are released cercariae which encyst as resting metacercariae on vegetation. When eaten, the larvae excyst and migrate through the intestinal wall to the liver where they mature and come to lie in the bile duct where they release eggs into the gut. Contamination of water with faeces leads to infection of snails, completing the life cycle. **Fascioliasis** (fasciolosis) is the disease caused by the fluke. Treatment is usually with triclabendazole, a benzimidazole anthelmintic drug.

Fasciolopsis A genus of trematodes (flukes) that will infect the small intestine and can cause quite serious disease (fasciolopsiasis) with complications of malnutrition. The disease is common in Asia. The intermediate host is an aquatic snail and infection is usually from eating contaminated plants such as water chestnut or cress.

FAST Forkhead activin signal transducer. *See* FORKHEAD.

FASTA A suite of bioinformatics programs that can be used for a fast protein comparison or a fast nucleotide comparison against databases of protein and nucleotide sequences. FASTA stands for FAST-ALL. The programs are made available through the European Bioinformatics Institute.

SEE WEB LINKS
- The FASTA homepage.

fast twitch muscle *See* SLOW MUSCLE.

fasudil A selective inhibitor of *rho kinase (ROCK1).

fatal familial insomnia A spongiform encephalopathy associated with mutation (D178N with methionine at position 129) in the *prion protein gene (D178N with valine at 129 results in familial *Creutzfeldt–Jakob disease). It is an autosomal dominant disorder characterized by neuronal degeneration limited to selected thalamic nuclei.

fat cell *See* ADIPOCYTE.

fat droplets Small aggregates, mostly of triglycerides, visible within the cytoplasm. In the mammary gland fat droplets are released into milk surrounded by apical plasma membrane.

fate map A diagram indicating which tissues the cells of each embryonic region will produce. Fate maps have generally been produced by vital staining of small groups of cells in the blastula and following the location of the dye as development proceeds.

fats The storage lipids of animal tissues, mostly triglyceride esters of long-chain fatty acids.

fatty acid binding proteins Cytosolic proteins (FABPs, ~130 aa) involved in transport of fatty acids within cells and in regulation of lipid metabolism. They bind long-chain fatty acids, fatty-acid acyl CoA and acyl-L-carnitine. At least seven different types of human FABP occur, each with a specific tissue distribution and possibly with a distinct function. Fatty acid binding protein 3 (FABP3, 133 aa) is also known as mammary-derived growth inhibitor.

fatty acids Organic acids with long, straight hydrocarbon chains, commonly with an even number of carbon atoms. Low concentrations are present in tissues but the majority are constituents of *triglycerides and *phospholipids.

fatty degeneration The pathological accumulation of fat, usually as droplets within cells, as a result of poor diet, excess alcohol consumption, or hypoxia.

fatty streak A region of the artery wall in which cholesterol accumulates in *foam cells. Often considered to be precursors of atherosclerotic plaques or *fibrous plaques.

favism A severe haemolytic anaemia that can arise in individuals who are glucose-6-phosphate dehydrogenase-deficient (an X-linked genetic trait) if they eat fava beans (from the broad bean *Vicia fava*). The active principal is probably DOPA-quinone. G6PD deficiency is the commonest metabolic deficiency, probably because the defect confers resistance to malaria.

favus A contagious ringworm of the scalp caused by infection with the fungus *Trichophyton schoenleinii* (formerly *Achorion schoenleinii*).

F-box A protein motif (~50 aa) generally found in the N-terminal half of proteins and involved in protein–protein interactions, particularly linking proteins with the ubiquitin-driven proteolytic system. The motif is present in several eukaryotic regulatory proteins, e.g. SCF ubiquitin–ligase complexes, and at least 38 human F-box proteins are known. F-box proteins often have other motifs potentially involved in protein–protein interaction, particularly WD and leucine-rich repeats.

SEE WEB LINKS
- Review of the F-box protein family.

Fc Region of the immunoglobulin molecule that binds to cells when the antigen-binding sites are occupied. Treatment of antibody with pepsin will release the Fc portion (fragment crystallizable) leaving the *Fab_2 fragment. Fc regions from the various antibody classes differ in their properties. **Fc receptors** are found on the surfaces of various cells involved in the immune response. There are five receptors for IgG: Fcγ-RI, -RII-A, -RII-B2, -RII-B1, and -RIII (CD16). FcγRII-B1 and FcγRII-B2 are inhibitory and their cytoplasmic domains have *ITIM motifs. FcεRI binds IgE, FcαRI binds IgA. The consequences of receptor occupancy differ according to the receptor subtype.

FCS *See* FETAL CALF SERUM.

FDA *See* FOOD AND DRUG ADMINISTRATION.

febrifuge A drug or remedy that reduces fever.

febrile Describing the condition of being feverish or something that causes, or is caused by, fever.

Fechner's law (Weber–Fechner law) That sensory systems require a logarithmic change in the intensity of a stimulus to give a linear change in the perceived intensity: if a stimulus varies as a geometric progression (doubles, triples, etc.), the corresponding perception increases in an arithmetic progression. The law is approximately true for a significant proportion of the sensitivity range of acoustic and photoreceptor systems and for weight estimation.

Fechtner's syndrome *See* MAY–HEGGLIN ANOMALY.

feedback regulation A control mechanism in which the rate of a process is regulated by the consequences of that process. Thus if the rate of a reaction reduces as the products begin to accumulate (product inhibition) then there is negative feedback; if there is positive feedback and products cause the reaction to accelerate then the reaction increases exponentially, even explosively. A good biological example of the latter is the generation of an action potential, where the voltage-gated sodium channels show positive feedback. A general feature of feedback-regulated systems is their tendency to resonate (oscillate) unless there is some damping.

feeder layer A layer of cells used to condition culture medium to allow other fussier cells to grow at low or clonal density, probably by producing growth factors or matrix components.

Fehling's reagent A solution used to differentiate water-soluble aldehyde and ketone functional groups. A red precipitate of cuprous oxide indicates the presence of an aldehyde, whereas most ketones do not react. Used to detect glucose in urine as a test for diabetes but now largely superseded by *Benedict's solution.

feline immunodeficiency virus (FIV) An endemic virus that causes immunodeficiency and neurological defects in domestic cats *Felix catus.* FIV is a lentivirus related to two other feline retroviruses, feline leukaemia virus (FeLV), and feline foamy virus (FFV). FIV infects $CD4^+$ and $CD8^+$ T cells, B cells, and macrophages, using feline CD134 and CXCR4 as primary receptor and coreceptor respectively. The immunodeficiency may be due to failure to generate an IL-12-dependent type I response and excessive IL-10 production.

feline sarcoma oncogene homologue (fgr) The homologue of the cytoplasmic tyrosine kinase found in Gardner–Rasheed feline sarcoma virus. The gene is activated when B cells are infected by the *Epstein–Barr virus. Fgr is involved in integrin signalling in myeloid leucocytes and will increase phosphorylation of the p85 subunit of PI_3 kinase.

Felty's syndrome A syndrome characterized by rheumatoid arthritis, splenomegaly, and neutropenia. Autoantibodies to elongation factor EF1A-1 occur in 66% of patients.

Femara *See* LETROZOLE.

fenbufen A NSAID acting on cyclo-oxygenases. *TN* Lederfen.

fenfluramine A serotonin reuptake inhibitor that also stimulates release of serotonin from the synaptosomes. Was used, in conjunction with phentermine, to treat obesity but has been withdrawn because of adverse side effects on heart valves. *TN* Pondimin.

fenoprofen A NSAID used in treatment of arthritis. Has analgesic and antipyretic activities and probably works by inhibiting prostaglandin synthetase. *TNs* Fenopron, Nalfon.

fentanyl An opioid analgesic that is an agonist for the μ opiate receptors. More potent than morphine and used clinically for anaesthesia and analgesia. Several analogues have been produced, some of them substantially more potent.

FEP *See* FREE ERYTHROCYTE PROTOPORPHYRIN.

ferlin A protein required for fusion of specialized vesicles with the sperm plasma membrane during spermiogenesis in *C. elegans.* Subsequently homologous proteins with a transmembrane domain and multiple C2 domains that commonly mediate calcium-dependent lipid-processing events were found in mammals. *See* DYSFERLIN; MYOFERLIN; OTOFERLIN.

FERM domain A conserved domain (~150 aa) found in various proteins that link plasma membrane proteins to the cytoskeleton. The name derives from the four proteins in which it was first found: protein 4.1 (F), *ezrin (E), *radixin (R), and *moesin (M). The ERM (ezrin–radixin–moesin) proteins are structurally and functionally highly homologous, with *merlin/schwannomin, the NF2 tumour suppressor protein.

ferredoxins Small proteins (~180 aa) that contain one or more iron–sulphur clusters and are generally involved in electron transfer. They are classified on the basis of the iron–sulphur clusters and sequence similarity.

ferrichromes (siderochromes) Any of a group of growth-promoting ferric iron chelates, cyclic hexapeptides, formed by various genera of microfungi. Synthetic ferrichromes have been shown to inhibit *in vitro* growth of *Plasmodium falciparum* by scavenging intracellular iron.

ferritin An iron storage protein that has dual functions of iron detoxification and iron reserve, found in liver, spleen, and bone marrow. Ferritin molecules from vertebrates are composed of two types of subunit, heavy (H,183 aa) and light

(L, 175 aa). The three-dimensional structure is highly conserved with 24 protein (apoferritin) subunits arranged to give a hollow shell with an 8-nm diameter cavity that is capable of storing up to 4500 Fe(III) atoms as an inorganic complex. Because of its characteristic shape and electron density it has been used as a label in electron microscopy, and cationized ferritin that has been treated with dimethyl propanediamine to give it a positive charge is used to show the distribution of negative charge on cell surfaces. A mutation in the iron-responsive element (IRE) in the ferritin light-chain gene is associated with the *hyperferritinaemia–cataract syndrome.

SEE WEB LINKS

- Review 'The ferritins: molecular properties, iron storage function and cellular regulation' in *Biochimica et Biophysica Acta* (1996).

ferroportin A transmembrane iron transporter (SLC40A1, solute carrier family 40, member 1, 571 aa) with highest level of expression in human placenta (probably transporting iron from mother to fetus), liver, spleen, and kidney and also also at the basolateral surface of duodenal enterocytes (where it is important for iron uptake from the intestine). Mutations in the ferroportin gene cause *haemochromatosis type 4. *See* HEPCIDIN ANTIMICROBIAL PEPTIDE.

fertilin Type I membrane glycoproteins (PH-30, ADAMs) in sperm membrane that are members of the *ADAM family and bind to an integrin on the egg membrane, thus mediating sperm–egg adhesion. They are heterodimers; the α subunit (probably ADAM20, 726 aa, in humans, *ADAM1* being a pseudogene) has homology to viral fusion peptides, the β subunit (ADAM2, 735 aa) has a tripeptide sequence involved in integrin binding.

fertilization The fusion of two mature *haploid gametes, sperm (spermatozoon) and egg (ovum), to produce a diploid zygote. In some animals, notably the sea urchin, the membrane of the fertilized egg changes very rapidly after entry of the sperm to produce a **fertilization membrane**. There is a wave of electrical depolarization, an influx of calcium, the fusion of cortical granules, and an expansion of the plasma membrane. The net result is a block to polyspermy. No morphological equivalent of the fertilization membrane forms in humans, but similar events trigger changes in the *zona pellucida and plasma membrane that prevent polyspermy. *See* IN VITRO FERTILIZATION.

fes/fps An oncogene, originally identified in avian and feline *sarcomas, for which there is a human homologue located on chromosome 15 that is probably involved in acute promyelocytic leukaemia where there is a translocation between chromosomes 15 and 17. The gene product is a cytoplasmic protein tyrosine kinase involved in downstream signalling from various cytokines involved in haematopoiesis (e.g. IL-3, GM-CSF, and EPO) and has a role in regulating the innate immune response. Mutations in the Fes/Fps kinase have been identified in human colorectal carcinomas.

festuclavine One of the ergot alkaloids produced by *Aspergillus fumigatus*.

fetal calf serum (FCS) Serum from fetal calves used in culture media for many animal cells, often at concentrations of ~10%.

fetuin Human fetuin is α2-Heremans-Schmid glycoprotein (α2-HS-glycoprotein, AHSG, 367 aa) which is cleaved to produce A (282 aa) and B (27 aa) polypeptides. The plasma concentration falls progressively through childhood. AHSG promotes endocytosis, has opsonic properties, and has a high affinity for calcium, probably playing some role in bone metabolism and calcium homeostasis. Fetuin is a major component in *fetal calf serum.

fetus (foetus) In human reproduction, an unborn child from the eighth week of development onwards.

Feulgen reaction A cytochemical staining procedure in which Schiff or Schiff-like reagents prepared from basic fuchsin (a pararosaniline, rosaniline, and magenta III mixture) react with aldehyde groups produced by mild hydrolysis of the deoxyribose of DNA. The reaction product is purple (magenta) and the amount of DNA can be quantified by spectrophotometry.

fever An increase in body temperature as a result of infection. *See* ENDOGENOUS PYROGEN.

fexofenadine A histamine H_1-receptor antagonist, orally available as the hydrochloride, used to treat hay fever. *TN* Allegra.

FFA 1. Free fatty acids. 2. (*less commonly*) fundus fluorescein angiography, a technique used to examine the retinal vasculature.

F-factor A *plasmid, first described in *E. coli.*, that confers the ability to conjugate and carries the *tra* genes (transfer genes).

FGD *See* FYVE, RHOGEF, AND PH DOMAIN-CONTAINING PROTEINS.

FGDY *See* AARSKOG–SCOTT SYNDROME.

FGF *See* FIBROBLAST GROWTH FACTOR.

fgr *See* FELINE SARCOMA ONCOGENE HOMOLOGUE.

FHIT *See* FRAGILE HISTIDINE TRIAD PROTEIN.

FIAU A nucleoside analogue (fluoroiodoarabinofuranosyluracil) that can be labelled with ^{3}H, ^{14}C, ^{125}I, or ^{18}F and is used to image the site of expression of viral thymidine kinase.

fibrates A group of cholesterol-lowering drugs effective in lowering triglyceride and, to a lesser extent, in increasing HDL-cholesterol levels. Examples are gemfibrozil and fenofibrate. Fibrates alter the transcription of several genes involved in *lipoprotein metabolism and are synthetic ligands for *peroxisome proliferator-activated receptor alpha (PPARα).

fibrillar centres Clear fibrillar zones found in nucleoli of many cells and considered the nucleolar counterparts of the chromosomal nucleolar-organizer regions. Transcribed rRNA genes are located in the fibrillar centres of the nucleolus.

fibrillarin The component in a small nucleolar ribonucleoprotein (snoRNP) particle that catalyses the site-specific 2′-hydroxyl methylation of ribose moieties in pre-ribosomal RNA, site-specificity being provided by a guide RNA. It is a highly conserved protein (321 aa), associates with U3, U8, and U13 small nuclear RNAs and is found in the *coiled body of the nucleolus. Antibodies to fibrillarin are found in various autoimmune diseases including *scleroderma, *systematic sclerosis, *CREST syndrome, and other connective tissue diseases.

fibrillation **1.** Twitching of individual muscle fibres, or bundles of fibres, symptomatic of some neurological diseases. Also called fasciculation. **2.** Rapid and disorganized contraction of the heart, or regions of the heart (atrial fibrillation, ventricular fibrillation). May be idiopathic or caused by any of a number of conditions. There are rare familial forms of atrial fibrillation, some caused by defects in voltage-gated potassium channels (e.g. *short QT syndrome).

fibrillin The major component (FBN1, 2871 aa) of the extracellular 10-nm microfibrils that are found in both elastic and nonelastic connective tissue throughout the body. Mutations in **fibrillin-1** are responsible for *Marfan's syndrome. Mutations in **fibrillin-2** are probably responsible for the somewhat similar condition congenital contractural arachnodactyly (CCA, Beal's syndrome, distal arthrogryposis type 9). *See* TROPOELASTIN.

fibrin The protein that forms the basis of a blood clot at a wound. Fibrin is the end product of the blood clotting cascade and is produced from fibrinogen by the action of *thrombin which cleaves the aspartic acid and glutamate-rich **fibrinopeptides** from *fibrinogen in the presence of calcium. Two peptides (16 aa and 14 aa) are produced from each fibrinogen molecule, leaving fibrin (fibrin A, FGA, 831 aa and FGB, 447 aa). Monomeric fibrin polymerizes to produce insoluble fibres that have a characteristic 23-nm repeat; the fibrous gel is further stabilized by covalent crosslinking as a result of the action of factor XIII, plasma *transglutaminase. **Fibrinolysis** is the digestion of a clot by the peptidase *plasmin.

fibrinogen A soluble protein, present at ~2–3 mg/mL in blood plasma, that is the source of *fibrin. Fibrinogen has six peptide chains, two each of α (887 aa), β (491 aa), and γ (453 aa). *See* FIBRIN. Defects in any of the fibrinogen genes can cause congenital *afibrinogenaemia and *amyloidosis type 8 (AMYL8, systemic non-neuropathic amyloidosis, Ostertag-type amyloidosis).

fibrinolysis *See* FIBRIN.

fibrinopeptides *See* FIBRIN.

fibroadenoma A benign and usually encapsulated tumour in which there is proliferation of both glandular and stromal (fibrous) elements. Quite frequently found in breast tissue.

fibroblast The mesodermally derived cells of connective tissue that are responsible for the synthesis and secretion of procollagen, fibronectin, collagenase, and extracellular matrix components. Many cells in culture adopt a fibroblastic morphology, even if they are not truly fibroblasts.

fibroblast growth factor (FGF) An important growth factor for mesodermal and neuroectodermal cells and important in developmental signalling. There are two structurally related forms, acidic FGF (heparin-binding growth factor-1, HBGF-1, 155 aa) and basic FGF (HBGF-2, 288 aa). The secretory pathway is ill-defined and neither has the classical signal sequence characteristic of secreted proteins. The FGF receptor is a receptor tyrosine kinase.

fibrocystin (polyductin) A large (4074 aa), receptor-like protein, localized predominantly at the apical domain of polarized epithelial cells, that is mutated in autosomal recessive *polycystic kidney disease.

fibrodysplasia ossificans progressiva A rare disorder in which there is intermittent progressive ectopic ossification. It can be caused by mutation in the *ACVR1* gene that encodes an *activin receptor.

fibroid **1.** A benign tumour (leiomyoma, fibromyoma), quite commonly asymptomatic, found in the uterus and derived from smooth muscle. The cells often retain oestrogen receptors and may enlarge during pregnancy and regress after the menopause. **2.** Describing tissue that has fibrous elements.

fibroma Any benign tumour composed of fibrous tissue.

fibromodulin A member of a family of small proteoglycans that also includes *decorin, *biglycan, and *lumican. The core protein (376 aa) has a central region with multiple repeats of domains of ~25 aa with leucine-rich motifs. There are four keratan sulphate chains attached to *N*-linked oligosaccharides. Fibromodulin is particularly abundant in articular cartilage, tendon, and ligament where it interacts with type I and type II collagen fibrils. Animals in which the fibromodulin and lumican genes are knocked out exhibit defects similar to *Ehlers–Danlos syndrome.

fibromuscular dysplasia An occlusive disease of arteries in children or young adults that can cause stroke, hypertension, claudication, or myocardial infarction, although the effects are variable. Probably an autosomal dominant disorder.

fibromyoma *See* FIBROID.

fibromyositis Inflammation of fibromuscular tissue.

fibronectin A major glycoprotein (two chains, disulphide-bonded, each of 2386 aa) of connective tissue and plasma. In connective tissue it is present in fibrillar form, interacting with collagen and with resident fibroblasts. In plasma it is present in soluble form (originally called cold-insoluble globulin), probably an alternatively spliced isoform. Fibronectins are multidomain proteins with an RGD domain recognized by *integrins, and two repeats of an *EGF-like domain. There are 15–17 fibronectin type III domains (FnIII), each ~90 aa long, and this domain is common in many cell surface proteins.

fibroplasia The production of fibrous tissue, e.g. in wound healing. **Retrolental fibroplasia** (retinopathy of prematurity) is retinal detachment as a result of overproduction of fibrous tissue associated with excessive vascularization of the retina and subsequent scarring. Excessive oxygen is a risk factor but not the main cause.

fibrosarcoma A malignant tumour derived from fibroblasts.

fibrosis The deposition or formation of excess collagenous fibrous tissue at a wound-repair site or in an area of chronic inflammation. The wound is often well vascularized at first (*granulation tissue) but later becomes avascular and is dominated by extracellular matrix, forming a scar. Excessive contraction, brought about by fibroblasts, and hyperplasia may lead to the formation of a *keloid. *But see* CYSTIC FIBROSIS.

fibrositis Inflammation of fibrous connective tissue.

fibrous histiocytoma A benign neoplasm characterized by an aggregation of spindle-shaped, fibroblast-like cells, usually found in the skin of the extremities and probably derived from *histiocytes. Malignant forms are the commonest soft-tissue sarcomas in adults.

fibrous lamina *See* NUCLEAR LAMINA.

fibrous plaque A thickening of the wall of an artery as a result of the accumulation of smooth muscle cells and fibrous tissue in the intima. The cells often have many fat globules (*foam cells) and there may be extracellular lipid; *fatty streaks may be precursors. If there is necrosis the thickening may be considered an atheroma. In extreme cases the lumen of the vessel may become occluded.

fibrous tissue An imprecise term that can be applied to any connective tissue with well-defined fibrillar elements, although often used in referring to the consequences of *fibrosis.

fibulins Calcium-binding, cysteine-rich glycoproteins found in extracellular matrix and in plasma. **Fibulin-1** (FBLN-1, 703 aa) occurs in three alternatively spliced forms and has inhibitory effects on cell adhesion and motility mediated by fibronectin but will mediate attachment of platelets to exposed blood vessel wall basal lamina via a bridge of fibrinogen. Synpolydactyly-2 (SPD2) is caused by disruption of the fibulin-1 gene. **Fibulin-2** is a considerably larger protein (1184 aa) that has no immunological cross-reactivity with fibulin-1 (although it has 43% sequence identity with fibulin-1) and is present in extracellular matrix and only at very low levels in plasma. Mutations in **fibulin-3** (EGF-containing fibulin-like extracellular matrix protein-1, EFEMP1, 493 aa) cause Doyne honeycomb retinal dystrophy and fibulin-3 has substantial sequence homology with *fibrillin. **Fibulin-4** (EFEMP2, 443 aa) may be associated with other retinopathies. **Fibulin-5** (FBLN5, 448 aa) has also been called DANCE (developmental arteries and neural crest EGF-like). Mutations in FBLN5 are associated with *age-related macular degeneration-3 (ARMD3) and a severe autosomal recessive form of *cutis laxa. **Fibulin-6** (5635 aa) was originally called hemicentin and may be associated with ARMD1. **Fibulin 7** (439 aa) is an adhesion molecule with a role in tooth development.

Fick's law Actually two laws relating to diffusion. The first law is used in steady state conditions and is that the flux (J) is proportional to the product of the diffusion constant (D) and the concentration gradient ($\mathrm{d}C/\mathrm{d}x$), i.e. $J = -D.\mathrm{d}C/\mathrm{d}x$. The second law applies in non-steady-state conditions (where the concentrations are changing with respect to time) and requires solution of a partial differential equation. Diffusion coefficients of most ions are in the range of 0.6×10^{-9} to 2×10^{-9} m^2/s. For biological molecules the diffusion coefficients normally range from 10^{-11} to 10^{-10} m^2/s.

Fick's principle A method originally used to measure cardiac output but of greater applicability. Three variables need to be measured: 1. the amount of substance taken up per unit time (oxygen by the whole body in the original application), 2. input concentration (concentration in arterial blood), 3. output concentration (concentration in venous blood). This approach can be used to estimate rate of uptake of a substance if the blood flow to the organ is known, or to estimate the blood flow to an organ, e.g. to the kidney, by measuring urinary excretion of a marker.

ficolins A family of molecules (L-ficolin/P35, H-ficolin/Hakata antigen, M-ficolin, ~300 aa) that, together with mannose-binding lectin (MBL), act as immune sensors and activate the lectin complement pathway. When they bind to markers on microbial surfaces they trigger the cleavage of complement C4 by the serine peptidase MASP2 (MBL-associated serine protease-2). They are structurally similar to *collectins, conglutinin, and *surfactant proteins A and D and have a central collagen-like triple-helical region with a C-terminal fibrinogen-like domain. L- and H-ficolin are found in serum, M-ficolin in leucocytes. Polymorphisms in the ficolin 1 gene (*FCN1*) are associated with susceptibility to the development of rheumatoid arthritis.

Ficoll™ A high molecular weight branched polysaccharide used to produce solutions of high viscosity and low osmotic pressures for the density-gradient separation of leucocytes from whole blood. **Ficoll-Paque** is premixed Ficoll and Hypaque (diatrizoate) with a density greater than that of all blood cells other than lymphocytes (1.077 g/cm^3). This allows a one-step centrifugation process for isolating lymphocytes which remain at the plasma–Ficoll-Paque boundary.

FIH-1 *See* HYPOXIA-INDUCIBLE FACTOR-1.

filaggrins A major histidine-rich protein component of the keratohyalin granules (filament-aggregating protein) that is present in the suprabasal cells of mammalian epidermis. It is produced as a large polyprotein precursor (4061 aa) which is proteolytically cleaved into ten to twelve individual functional filaggrin peptides of ~50 aa as a final step in differentiation. Filaggrin peptides cause aggregation of the keratin cytoskeleton and the granular cells collapse into flattened squames. The aggregated keratin is further stabilized by crosslinking by a transglutaminase. Mutations in filaggrin are responsible for *ichthyosis vulgaris and are a major predisposing factor for atopic dermatitis, but not for asthma.

filamentous phage A rod-shaped bacteriophage (inovirus) with a circular single-stranded DNA genome. The best-known example is bacteriophage M13 which is used as a cloning and sequencing vector.

filaments *See* AXIAL FILAMENTS; BEADED FILAMENTS; INTERMEDIATE FILAMENTS; MICROFILAMENT; THICK FILAMENTS; THIN FILAMENTS.

filamins Actin-binding proteins that crosslink F-actin calcium-independently to form the meshwork that is the actin cytoskeleton and also link this meshwork to the plasma membrane. **Filamin A** (filamin 1; ABP280, 2647 aa) is widely expressed and interacts with integrins, transmembrane receptor complexes, and second messengers. Mutation in the X-linked *FLNA* gene can cause the neurologic disorder *periventricular heterotopia although there is some compensation by filamin B. **Filamin B** (filamin β; ABP 276/278, 2602 aa) has considerable similarity with filamin A (70% similarity at sequence level) but the gene is on chromosome 3 and the expression pattern is different. Four distinct skeletal disorders result from mutation in the *FLNB* gene, spondylocarpotarsal syndrome, autosomal dominant Larsen's syndrome, type I atelosteogenesis, and type III atelosteogenesis. **Filamin C** (filamin 2; ABP280A, 2725 aa) is very similar to both filamin 1 and filamin B and almost identical in the actin-binding region. Interacts with sarcoglycans and with the muscular dystrophy Ky protein.

filariasis Infestation with filarial nematodes (Filariidae) particularly *Wuchereria bancrofti*, *Brugia malayi*, and *B. timori*, that are transmitted by mosquitoes. Blockage of lymphatic vessels can cause elephantiasis. Two approaches to treatment are possible, either the use of anthelmintics such as albendazole and ivermectin, or antibiotics that kill the symbiotic bacteria, *Wolbachia* spp., and thus indirectly kill the nematode.

filensin The protein (beaded filament structural protein 1, 665 aa) that coassembles with *phakinin to form the lens-specific intermediate filament system (beaded filaments). Mutations are responsible for some forms of juvenile-onset cortical cataract.

filiform papillae Cone-shaped projections on the upper surface of the tongue consisting of a core of connective tissue covered by keratinized epithelium. Not involved in taste and may serve to increase the roughness for mechanical reasons. The interspersed fungiform papillae do have taste buds.

filipin The collective name given to four isomeric polyene macrolides isolated from cultures of *Streptomyces filipinensis*. Filipin binds to cholesterol in membranes and the complexes can be visualized by freeze-fracture and electron microscopy. The filipin–cholesterol complexes produce pores and cause cytolysis.

filopodium (*pl.* filopodia) A long thin spike-like protrusion from the front of a cell. Filopodia are supported by microfilaments and seem to be an alternative to the leading lamella as a means of achieving the forward protrusion that is a prerequisite for cell locomotion.

Filoviridae Single-stranded RNA viruses, Marburg and Ebola viruses being the only two known genera in the family. Morphologically they are long filaments (up to 14 mm, 70 nm thick), sometimes branched, with the RNA surrounded by a cell-derived envelope. Infection causes major haemorrhagic complications with activation of the blood clotting cascade, and is frequently fatal.

fimbria (*pl.* fimbriae) *See* PILUS.

fimbrillin The protein that forms bacterial *pili (fimbriae). Fimbriae have been most extensively studied in *Porphyromonas* (*Bacteroides*) *gingivalis*, the bacterium responsible for periodontal disease. The main fimbrillin (~195 aa) is the product of the *fimA* gene of which there are five variants; the virulence varies according to the type being expressed. A second type of fimbrillin (~350 aa, product of the *mfa1* gene) is also known. Fimbriae are required for attachment (through integrins) and invasion of the gingival mucosa. Fimbriae are not restricted to *P. gingivalis*, are important virulence factors for some strains of *E. coli*, and are found in *Salmonella*, *Shigella*, and various other pathogenic bacteria.

fimbrins *See* PLASTINS.

finasteride A drug used to treat benign prostatic hyperplasia and carcinoma of the prostate. Inhibits 5-α-reductase (EC 1.3.1.22) the enzyme that converts testosterone to dihydrotestosterone in the prostate.

fingerprinting A general approach to identifying large molecules by cutting them into fragments using appropriate enzymes, and then running the fragments on a gel. The pattern is a characteristic 'fingerprint' provided the cleavage sites are specific. Can be applied to proteins (peptide mapping), viruses, or nucleic acids (*see* RESTRICTION FRAGMENT LENGTH POLYMORPHISM). Not to be confused with *footprinting.

Firmicutes A division of the Bacteria (Eubacteria) that has three classes, the anaerobic Clostridia, the aerobic Bacilli, and the Mollicutes (mycoplasmas). They differ from Actinobacteria in having low G+C levels in their DNA. Most, but not all, are Gram-positive.

FISH analysis (fluorescence *in situ* hybridization) *See* FLUORESCENCE.

fish-eye disease A disorder of lipoprotein metabolism due to mutation in the *LCAT* gene in which there are diffuse corneal opacities, target cell haemolytic anaemia, and proteinuria with renal failure.

fission yeast *See* SCHIZOSACCHAROMYCES POMBE.

fistula (*pl.* fistulas, fistulae; *adj.* fistulous) An abnormal connection between two compartments or territories that are lined by epithelium, e.g. between anus and skin (anal fistula) or between rectum and vagina (obstetric fistula).

FITC *See* FLUORESCEIN.

FIV *See* FELINE IMMUNODEFICIENCY VIRUS.

fixation In histology, the process of stabilizing tissue for processing, usually by crosslinking and denaturing proteins making them insoluble.

FK506 (tacrolimus) An immunosuppressive drug that inhibits activation of the transcription factor *NFAT in T-cells. Like *ciclosporin, binds to an *immunophilin. *See* GLOMULIN.

FKBP FK506 binding protein. *See* IMMUNOPHILINS.

FKHR (forkhead homologue in rhabdomyosarcoma) *See* FORKHEAD.

flagella and basal body proteome An open resource of proteomic data relating to cilia and basal bodies.

SEE WEB LINKS

- The ciliary proteome homepage (only via Firefox).

flagellin The protein subunit (498 aa in *E. coli*) of the bacterial flagellum.

flagellum (*pl.* flagella) A long, thin projection from a cell that is used in propulsion. **Eukaryotic flagella** (like *cilia) have a characteristic core of microtubules arranged in a 9+2 array (*axoneme) and bend actively by dynein-mediated sliding of the outer doublets. **Prokaryotic flagella** are made of polymerized *flagellin and are rotated by the basal motor.

flag tagging An octapeptide tag (N-DYKDDDDK-C) that is incorporated into a recombinant expressed protein to facilitate purification or localization using labelled antibody. Because it is hydrophilic it is likely to be accessible on the surface of the protein. Triple FLAG-tags are sometimes used.

FLAP A protein (5-lipoxygenase activating protein, 161 aa) that is necessary for the activity of 5-lipoxygenase (EC 1.13.11.34) the enzyme responsible for the production of 5-HPETE from arachidonic acid, and leukotriene A_4 from 5-HPETE. An Icelandic study suggested that a 4-SNP haplotype in the gene conferred an approximate doubling of the risk of myocardial infarction and stroke, although this has not held up in larger studies.

flavin adenine dinucleotide *See* FLAVIN NUCLEOTIDES.

flavin nucleotides A general term for both flavin adenine dinucleotide (FAD) or flavin mononucleotide (FMN), cofactors for flavin enzymes.

Flaviviridae A family of enveloped RNA arboviruses containing only the genus *Flavivirus*. Responsible for yellow fever (hence the name), dengue, Japanese encephalitis, tick-borne encephalitis, and West Nile fever.

flavonoids A general term for a large group of polyphenolic compounds that includes flavonols, flavones, flavanones, isoflavones, catechins, anthocyanidins, and chalcones. They may be valuable diet-derived antioxidants and a remarkable range of beneficial properties have been ascribed to them.

flavoproteins Enzymes or proteins that have a *flavin nucleotide as a coenzyme or prosthetic group. They catalyse oxidation–reduction reactions and are involved in a wide range of processes, including bioluminescence, photosynthesis, DNA repair, and apoptosis.

flavoridin *See* DISINTEGRINS.

flecainide A class 1 antiarrhythmic drug that has local anesthetic activity and membrane stabilizing effects and that is used to prevent and treat tachyarrhythmias. It produces a dose-related decrease in intracardiac conduction by blocking the $NaV_{1.5}$ voltage-gated sodium channel, causing prolongation of the cardiac action potential and with the greatest effect on the His–Purkinje system. This is the sodium channel affected in *long QT syndrome-3, *Brugada syndrome, and idiopathic ventricular fibrillation. *TN* Tambocor.

FLICE FADD-like interleukin converting enzyme, caspase 8 (479 aa). *See* CASPASES.

FLIP A protein (FLICE-inhibitory protein, caspase-8 related protein, CASPER, 480 aa) that interacts with caspase 3, caspase 8 (FLICE), FADD, TRAF1, and TRAF2 through different domains and by inhibiting caspase 8 may block the cascade that leads to apoptotic death. A range of alternatively spliced variants have been found, not all of which inhibit caspases. **Viral FLIPs** (v-FLIPs) are produced by various herpesviruses.

flk-1 *See* VASCULAR ENDOTHELIAL GROWTH FACTOR.

flotillins Proteins isolated from caveolin-rich membrane domains. **Flotillin 1** (reggie-2, 427 aa) and **flotillin-2** (reggie-1, epidermal surface antigen-1, 428 aa) copurify with caveolin. *See* CAVEOLA. They have a *prohibitin homology (PHB) domain, and are part of the SPFH (stomatin–prohibitin–flotillin–HflC/K) superfamily.

flow cytometry Casual term for the analysis of cell populations by means of a fluorescence-activated cell sorter (FACS). The cell population that is to be analysed is labelled with a fluorescent dye that binds selectively to some cell marker (e.g. with a fluorescently labelled antibody that binds to a specific CD antigen) and the cell suspension is passed through a dropping orifice so arranged that there is one cell per droplet. A laser-based detector excites fluorescence and droplets containing a fluorescent cell are counted. If the cells are to be separated (sorted) then the droplets with labelled cells are given a small positive charge which allows them to be electrostatically separated from negative droplets, but this is rarely done. A second laser detection system can give an estimate of cell size (by orthogonal light scattering) or detect a second label with a different fluorophore. Because very large numbers of cells are analysed individually, the results are statistically very robust.

flow-mediated dilation The dilation of blood vessels in response to increased flow. Commonly measured by temporary occlusion of flow in the brachial artery using a cuff and then quantifying the flow rate following release using noninvasive Doppler imaging. Dilation is mediated by *endothelium-derived relaxing factor and is a response to short-term increases in local shear stress; reduced responsiveness is considered an early indicator of atherogenesis.

flt Receptor tyrosine kinases (fms-related tyrosine kinases) related to *src. **Flt-1** (1338 aa) is VEGF receptor-1; **flt-3** (Flk2, stem cell tyrosine kinase, STK-1, CD135, 993 aa) has a ligand of 235 aa similar to c-kit ligand and M-CSF; **flt-4** (1298 aa) is VEGFR-3. flt-2 does not appear in PubMed. *See* VASCULAR ENDOTHELIAL GROWTH FACTOR.

flu *See* INFLUENZA VIRUS.

flucloxacillin A rather narrow-spectrum semisynthetic penicillin used to treat infections caused by susceptible Gram-positive bacteria, particularly those which produce *beta-lactamase. *TN* Floxapen.

fluconazole A synthetic triazole antifungal drug that is a selective inhibitor of fungal cytochrome P-450 ergosterol synthesis. Used to treat candidiasis and cryptococcal menigitis. *TN* Diflucan.

fluctuation analysis A method originally introduced in 1943 by Salvador Luria (1912–1991) and Max Delbrück (1906–1981) to estimate mutation rates in bacterial populations but subsequently applied to a variety of systems. A modified form of fluctuation analysis has been used to demonstrate that pre-existent and highly metastatic variant cells develop spontaneously within a tumour cell population. More sophisticated variations have been developed, e.g. detrended fluctuation analysis, which can be used to reveal long-range correlations in time series; nonstationary fluctuation analysis, which can be used to determine single-channel contributions to a macroscopic current even though the single channels cannot be studied directly.

SEE WEB LINKS

- More information on non-stationary fluctuation analysis.

flucytosine An antimetabolite that interferes with fungal DNA synthesis and is used against some strains of *Candida* and *Cryptococcus*. It is taken up via the cytosine permease and rapidly converted into *fluorouracil by the enzyme cytosine deaminase once in the cytoplasm. *TNs* Ancobon, Ancotil.

fludarabine A derivative of *vidarabine, an antiviral adenosine analogue, that is less susceptible to inactivation by adenosine deaminase. The active metabolite is 2-fluoro-ara-ATP which inhibits DNA polymerase α, ribonucleotide reductase and DNA primase. Used for treatment of adults with B-cell chronic lymphocytic leukaemia that has not responded to other drugs.

fludrocortisone acetate A synthetic steroid with *mineralocorticoid and *glucocorticoid properties similar to those of hydrocortisone but more potent and longer-lasting. Used in replacement therapy for adrenocortical insufficiency (Addison's disease). *TN* Florinef acetate.

fluid bilayer model *See* PHOSPHOLIPID BILAYER.

flumazenil (flumazepil) A competitive inhibitor of benzodiazepine binding to the $GABA_A$ receptor used in treatment of benzodiazepine overdose. *TNs* Anexate, Lanexat, Mazicon, Romazicon.

flunisolide An anti-inflammatory corticosteroid often administered as a nasal spray for the treatment of allergic rhinitis. *TNs* AeroBid, Nasalide, Nasarel.

fluocinolone acetonide A synthetic hydrocortisone analogue used topically for the treatment of dermal inflammation and itching.

fluorescein A commonly used fluorophore. Fluorescein diacetate can be used alone as a vital stain, or fluorescein can be linked covalently to proteins using fluorescein isothiocyanate (FITC). Fluorescein will emit green-yellow light (450–490 nm) if excited by light at 365 nm; the emission spectrum is pH sensitive and it can be used as an intracellular pH indicator.

fluorescence The emission of light from a fluorochrome as a result of absorbing light at a different (shorter) wavelength. Usually one photon is emitted for every photon absorbed and emission occurs within 10^{-8} s, whereas in phosphorescence there is a relatively long delay in re-radiation. Some materials show intrinsic fluorescence and this autofluorescence can be a nuisance in microscopy. **Fluorescence activated cell sorter** (FACS), *see* FLOW CYTOMETRY. **Fluorescence energy transfer** (fluorescence resonance energy transfer) is the transfer of energy between closely apposed fluorochromes. The phenomenon relies on the emission wavelength from the fluorochrome activated by incident light being of an appropriate wavelength to activate the second fluorochrome and since it will only work over distances of a few nanometres successful transfer indicates close proximity, and probably interaction, of the two labelled species. **Fluorescence *in situ* hybridization** (FISH) is a microscopical technique in which a fluorescently labelled DNA probe is used as a sequence-specific stain to bind, by hybridizing, and to reveal the location of a particular gene on a *metaphase spread of chromosomes. Using multiple fluorochromes different genes can be visualized and the colourful results lead to the casual term 'chromosome painting'. **Fluorescence microscopy** is any type of microscopy in which the illuminating light causes fluorescence but is generally applied to situations in which fluorescently labelled reagents (e.g. fluorescein-conjugated antibody) are used to locate molecules (antigens) of interest in the specimen. Illumination is usually with ultraviolet light, often shone onto rather than through the specimen (epi-illumination, **epifluorescence**), and narrow band-pass filters are used to allow only fluorescently emitted light into the eyepieces so that the specimen appears dark except where label has bound. **Fluorescence recovery after photobleaching** (FRAP) is a technique that makes use of the fact that most fluorochromes are bleached by continued excitation: if a region of a specimen is bleached it will recover fluorescence if unbleached molecules are free to diffuse into the area, and the kinetics of the recovery indicate the mobility of the fluorescently labelled molecules, the viscosity of the environment in which they are moving, etc.

fluorexon *See* CALCEIN.

fluoridation The addition of inorganic fluorides (usually sodium fluoride at ~1 ppm) to drinking-water supplies. It has been claimed that fluoridation has been the most successful public health measure ever carried out and the effect in reducing tooth decay is dramatic, probably because bacterial metabolism is very sensitive to low fluoride levels.

fluorochromes *See* FLUORESCENCE.

fluorography A method for visualizing radioactively labelled molecules by adding scintillant or fluor to the separation medium or on to the blot.

fluoroquinolones A group of broad-spectrum antibiotics related to *nalidixic acid that act by inhibiting DNA gyrase and topoisomerase. They can be divided into two groups, based on antimicrobial spectrum and pharmacology: the older group includes ciprofloxacin, norfloxacin, and ofloxacin, and the newer group gatifloxacin, gemifloxacin, levofloxacin, moxifloxacin, and trovafloxacin.

fluorouracil A fluorinated pyrimidine (5-fluorouracil, 5-FU) that acts as a toxic antimetabolite by interfering with DNA synthesis. Used for the topical treatment of actinic or solar keratoses and superficial basal cell carcinomas and systemically for various solid tumours, especially colorectal and breast carcinoma. *TN* Efudex.

fluoxetine A selective serotonin reuptake inhibitor (SSRI) used as an antidepressant. *TN* Prozac.

flupentixol (flupenthixol) A typical antipsychotic neuroleptic drug that acts by antagonism of dopamine receptors D_1 and D_2 (as well as serotonin) and is given intramuscularly as a long-lasting depot injection in schizophrenic patients unlikely to be compliant about taking medication. *TNs* Depixol, Fluanxol.

fluphenazine A phenothiazine derivative intended for the treatment of schizophrenia that blocks postsynaptic dopaminergic D_1 and D_2 receptors in the brain and depresses the release of hypothalamic and hypophyseal hormones. More recently has been shown to be capable of blocking neuronal sodium channels. *TNs* Modecate, Permitil, Prolixin.

flurazepam A benzodiazepine derivative used for short-term treatment of insomnia, although it has a long half-life and may remain in the body for several days. *TNs* Dalmadorm, Dalmane.

flurbiprofen A NSAID also used to prevent contraction of the pupil during eye surgery. *TNs* Froben, Ocufen.

flutamide An antiandrogen used in treatment of prostatic carcinoma, a competitive inhibitor of testosterone binding to androgen receptors. *TNs* Drogenil, Eulexin, Flutamin.

fluticasone proprionate A synthetic trifluorinated corticosteroid of moderate potency that is used by inhalation for the treatment of allergic rhinitis and orally for asthma. *TNs* Flonase, Flovent.

flutter A disturbance of heartbeat similar to *fibrillation but less marked.

fluvoxamine A selective serotonin reuptake inhibitor (SSRI) often used to treat obsessive–compulsive disorder. Has the highest affinity among SSRIs for sigma receptor subtype 1. *TNs* Dumyrox, Faverin, Fevarin, Luvox.

FlyBase The definitive web-based resource for *Drosophila* genetics.

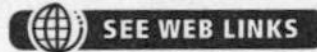

- The FlyBase homepage.

FMD **1.** *See* FIBROMUSCULAR DYSPLASIA. **2.** *See* FLOW-MEDIATED DILATION. **3.** Foot-and-mouth disease.

f-Met-Leu-Phe (fMLP) *See* FORMYL PEPTIDES.

fMLP *See* FORMYL PEPTIDES.

FMN *See* FLAVIN NUCLEOTIDES.

Fmoc An abbreviation for the fluorenylmethyloxycarbonyl group that is used as a blocking group to protect the amino group of the amino acid that is being added to the growing peptide chain during chemical synthesis. Once the peptide bond has been made the Fmoc group is removed by treatment with a mild base and the next Fmoc-protected amino acid can be added.

fMRI A noninvasive imaging technique (functional *MRI) that detects neural activity through monitoring changes in blood flow and thus energy demand, in different regions of the brain while specific tasks are carried out by the experimental subject. The method depends on the fact that haemoglobin is diamagnetic when oxygenated but paramagnetic when deoxygenated.

fms An *oncogene, originally identified in a feline *sarcoma but with a human homologue that is the receptor for *colony stimulating factor-1 (CSF-1), a receptor tyrosine kinase. Mutations that disrupt regulation of the kinase are associated with myelodysplasia and myeloid leukaemia.

foam cells Cells found in *fatty streaks on the arterial wall that have a foamy appearance in histological sections because of abundant lipid inclusions. More probably macrophages than smooth muscle cells, although controversy still exists.

focal adhesions (focal contacts) Punctate adhesions (~1 × 0.2 μm) of cells with matrix-coated substratum, associated with intracellular microfilament bundles (stress fibres) and a complex of proteins such as *vinculin and *talin on the cytoplasmic face. They are oriented with the long axis parallel to the direction of movement and tend to be characteristic of slow-moving cells. **Focal adhesion kinase** (FAK; cytoplasmic protein tyrosine kinase, PTK2, EC 2.7.10.2, 1052 aa) is a kinase that is concentrated in focal adhesions. It is activated by phosphorylation, primarily by *src-kinase, and activity is important for spreading and motility of fibroblasts, neurites and myocytes. A related kinase is *PYK2.

focal segmental glomerulosclerosis A feature of several renal disorders, characterized by proteinuria and progressive decline in renal function. The histological picture is of localized scarring of glomeruli. Hereditable forms can be caused by mutations in α-actinin-4 (**FSGS1**), transient receptor potential channel C6 (**FSGS2**) and haploinsufficiency in CD2-associated protein (**FSGS3**).

focus Describing a morphologically distinct group of cells in a cell monolayer in culture. **Focus forming units** were a measure of the number of active viruses present on the basis of an assay of their ability to generate foci of transformation or infection when added to cell cultures.

fodrin The nonerythroid isoform of *spectrin, a tetrameric (2α2β) protein (fodrin-α, 2472 aa; fodrin-β, 2364 aa).

foetus, foetal UK spelling for *fetus, *fetal, now obsolete in the medical and scientific literature.

FOG1 A zinc finger protein (friend of GATA1, 1006 aa) that interacts with the amino finger of *GATA1 and GATA2 and acts synergistically with GATA1 to promote differentiation. Both GATA1 and Fog1 are necessary for erythroid and megakaryocyte development.

folic acid (vitamin B_9) *See* VITAMINS.

folimycin *See* CONCANAMYCIN A.

Folin–Ciocalteau reagent A solution of copper tartarate with added detergent (sodium dodecyl sulphate) to which is added Folin's phenol reagent, a phosphotungstic phosphomolybdic acid. The reagent is used in the *Lowry assay.

follicle A small sac or vesicle. The **hair follicle** is an infolding of epidermis surrounding the hair root. An **ovarian follicle** is a small aggregate of cells consisting of an oocyte surrounded by *ovarian granulosa cells that matures into the fluid-filled vesicular Graafian follicle that contains the germinal vesicle and bursts at ovulation to release the oocyte. A **folliculoma** is a tumour derived from cells of the Graafian follicle. **Follicle stimulating hormone** (FSH; follitropin) is a pituitary hormone that induces development of ovarian follicles and stimulates the release of oestrogens. The alpha subunit (92 aa) is identical to that of *luteinizing hormone, *thyroid stimulating hormone, and human *chorionic gonadotrophin; the beta subunit (129 aa) is specific and is the portion of the dimer that interacts with the receptor. The FSH-receptor is a G-protein protein-coupled receptor (695 aa) found in ovary, testis, and uterus.

follicular dendritic cells The cells found in germinal centres that present intact antigen to potential memory cells. B cells with high affinity B-cell receptors (BCR) bind and survive, the others die apoptotically.

follistatin An activin antagonist (344 aa) that inhibits the secretion of follicle stimulating hormone. Not a member of the inhibin family. Has three follistatin domains, which are presumed to be growth factor binding motifs. **Follistatin-like 1** (follistatin-related protein, 308 aa) appears to act on macrophages to increase TNF and IL-1β production and IL-6 activity. **Follistatin-like-3** (263 aa) is an extracellular matrix-associated glycoprotein thought to bind morphogens or growth/differentiation factors and regulate their activity during development.

Fong's disease *See* NAIL–PATELLA SYNDROME.

Fontana stain (Fontana–Masson stain) A stain used for melanin that relies on the melanin granules reducing ammoniacal silver nitrate to produce a black deposit. Argentaffin, chromaffin, and some other lipochrome pigments will also stain black.

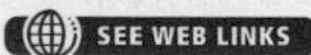

- Useful resource on histological stains.

Food and Drug Administration (FDA) The US regulatory authority that scrutinizes prescription drugs. FDA approval is particularly important for drug development, because of the size of the US market.

footprinting A general approach for identifying the binding site of one molecule on another in which the binding interaction protects the binding site (the footprint region) from hydrolysis and hydrolysates with and without the binding partner present will be different. Has most often been used to determine where proteins such as transcription factors interact with upsteam nucleotides (DNA footprinting).

Forbes' disease *See* GLYCOGEN STORAGE DISEASES.

foreign body giant cell A multicellular syncytium of macrophages that have fused in response to a large indigestible particle (e.g. talc, silica, or asbestos fibres). Up to 100 cells may contribute and the nuclei are randomly distributed. Langhans cells, found at the centre of tuberculous lesions, are similar but the nuclei tend to be peripherally located.

forensic medicine Strictly speaking, any aspect of medicine which has civil or statutory legal implications, but increasingly used to mean the application of medical and biomedical techniques in the investigation of crime or suspected crime.

forkhead Originally the product of the *Drosophila* homeobox gene *fork head*, a transcription factor. Subsequently many related transcription factors have been identified, and are now reclassified as **forkhead box proteins** (Fox proteins) with subclasses ranging from FoxA to FoxS. All are winged-helix transcription factors containing the forkhead box motif that binds DNA, and are involved in cell growth, proliferation, differentiation, longevity and embryonic development. The **forkhead activin signal transducer** (FAST, FoxH1, 365 aa) is expressed in most human tissues including breast, colon, thymus, and muscle, and in several cancer cell lines. It binds to *SMAD2 and activates an *activin response element. **Forkhead homologue in rhabdomyosarcoma** (FKHR, forkhead receptor, FoxO1, 655 aa) is translocated and fused with PAX3 or PAX7 in alveolar rhabdomyosarcomas. Activation in many cells induces cell cycle arrest followed by apoptosis but this is blocked when FKHR is phosphorylated by protein kinase B (Akt) which causes it to be transported to the cytoplasm. FKHR can interact with both steroid and nonsteroid nuclear receptors and, depending on the receptor, can act as either a coactivator or corepressor. Forkhead transcription factor in rhabdomyosarcoma like-1 (**FKHRL1**, 673 aa) is related to FKHR and has a role in the regulation of the proliferation of vascular smooth muscle. *See* Fox.

formaldehyde A compound with fixative and antibacterial properties. As a histological fixative it needs to be well buffered, otherwise it will cause excessive crosslinking and denaturation of proteins. Formaldehyde and hydrochloric acid vapour together are potently carcinogenic.

formiminotransferase-cyclodeaminase A dual function enzyme (EC 2.1.2.5–EC 4.3.1.4, 541 aa) involved in the histidine degradation pathway and the autoantigen (liver cytosol antigen-1, LC-1) in nearly 30% of patients with autoimmune hepatitis type II.

formins A conserved protein family originally identified as alternatively spliced products of the *ld* (limb deformity) locus of the mouse, mutations in which disrupt pattern formation, cause limb deformities and renal aplasia. Formins are involved in regulating actin and microtubule networks, signalling to the nucleus and in embryonic development. A characteristic feature are the **formin homology domains** (FH1 and FH2) that cooperate in rapidly assembling profilin–actin into long filaments although binding to microtubules involves other regions. Mammalian formins are mDia1 and mDia2. *See* DIAPHANOUS-related formins.

formoterol (eformoterol) A long-lasting sympathomimetic drug (β_2-agonist) used by inhalation as a bronchodilator in chronic asthma and chronic obstructive pulmonary disease. *TNs* Foradil, Foradile, Oxis.

formyl peptides Casual name for small peptides that are potent chemokinetic and chemotactic factors for human leucocytes, particularly neutrophils. The first commonly used example was formyl-methionyl-leucyl-phenyalanine (fMet-Leu-Phe; fMLP), thought to be a synthetic analogue of the bacterial protein start sequence, although the requirement is simply for a formylated N-terminus and usually a hydrophobic amino acid at the C-terminal end. Specific receptors of high affinity are found on human cells but not on those of many domestic animals (e.g. pig, chick). At concentrations around the K_d they elicit locomotory activity, at higher concentrations they stimulate secretion and the metabolic burst.

fornix Anatomically, any arch-like structure. There are two common examples: **1.** A nerve-fibre tract connecting the hippocampus to the septal nuclei and mammillary bodies in the brain. **2.** Dilations of the vagina on either side of the cervix (vaginal fornices).

foroxymithine An inhibitor of *angiotensin-converting enzyme, produced by actinomycetes. It is also a potent chelator of the ferric ion.

forskolin (colforsin) A diterpenoid, originally extracted from *Coleus forskohlii*, that will stimulate adenylate cyclase and raise intracellular levels of cAMP, especially if used in conjunction with inhibitors of phosphodiesterases.

Forssman antigen A glycolipid *heterophile antigen (globopentosylceramide) expressed on cell surfaces during embryonic and adult life in rodents and other mammals but probably not in humans. Anti-Forssman antibodies directed against the terminal sugar moiety are commonly found in plasma and may be involved in Guillain–Barré syndrome. Anti-Forssman antibodies also reportedly disrupt tight junction formation, apical–basal polarization, and cell adhesion.

fortilin *See* TRANSLATIONALLY CONTROLLED TUMOUR PROTEIN.

fortimicin A An aminoglycoside antibiotic isolated from *Micromonospora* sp. strain 40027.

fos The leucine zipper protein (380 aa) that dimerizes with *jun to form the AP-1 transcription factor. Originally identified as the product of the v-*fos* oncogene of the Finkel–Biskis–Jenkins and Finkel–Biskis–Reilly murine osteogenic sarcoma retroviruses. The *fos* gene family has four members: FOS, FOSB, fos-related antigen-1 (FRA1; fos-like antigen-1, Fosl-1), and fos-related antigen-2 (FRA2), all of which can dimerize with members of the jun family. Fra-1 lacks the transactivation domains so will suppress AP-1 activation.

foscarnet An analogue of inorganic pyrophosphate that inhibits replication of herpesviruses such as *cytomegalovirus and herpes simplex virus types 1 and 2 by binding to virus-specific DNA polymerases. Used in treating some viral infections in AIDS patients. *TN* Foscavir.

Fouchet's stain A histological stain for bile pigments that produces a blue/green colour. Van Gieson's stain is often used as a counterstain.

founder cell A cell that is the precursor of a clone that constitutes a tissue. Generally there are multiple founder cells, which can be demonstrated by analysing chimeric embryos (e.g. females heterozygous for an enzyme isoform specified on the X chromosome), but single founder cells for the intestine and germ line are found in *Caenorhabditis elegans.* Muscle founder cells are uniquely specified myoblasts that fuse with neighbouring myoblasts to generate the body wall muscles in the *Drosophila* embryo; each founder cell specifies a single muscle.

fovea (*pl.* **foveae)** Anatomically, a small pit on the surface of a structure or organ. The fovea centralis of the retina is particularly rich in cones and is the region of greatest visual acuity and colour sensitivity.

Fox Prefix for proteins of the *forkhead family of which many are now recognized, ranging from FoxA to FoxS. The **FoxO** (foxo) subfamily are key components of the insulin signalling cascade. Insulin and insulin-like growth factor, through the PI3kinase signalling system, stimulate phosphorylation of FoxO by protein kinase B (PKB/Akt) so that FoxO is retained in the cytoplasm and cannot act as a transcription factor. **FoxP2** is involved in brain development and mutations result in disruption of neural pathways essential for human speech, although there are other effects. Defects in FoxP2 in a family with speech dysfunction led journalists to claim it as the gene for speech. Mutation in the mouse *FoxP3* gene gives rise to the *'scurfy' mouse.

FPLC Fast protein liquid *chromatography.

fps *See* FES/FPS.

fra-1 *See* FOS.

frabin *See* FYVE, RHOGEF, AND PH DOMAIN-CONTAINING PROTEINS.

fractalkine An unusual membrane-anchored *chemokine (FK, CX3CL1, SCYD1, neurotactin, 397 aa) that differs from others in having a CXXXC motif. It is mostly expressed in the brain where it is up-regulated on capillary endothelium and microglia in inflammation and *EAE. It binds to cells expressing the chemokine receptor CX3CR1. Soluble active fragments (76 aa) generated by TNFα converting enzyme (TACE) are chemotactic for T cells, NK cells, and monocytes.

fractionation Any method for separating and purifying molecules. *See also* CELL FRACTIONATION.

fragile histidine triad protein (FHIT) An AP3A (diadenosine triphosphate) hydrolase thought to be involved in regulation of DNA replication and signalling stress responses. The conserved histidine triad residues are required for enzymatic activity. FHIT is inactivated in various tumours and FHIT is a target for *src-kinase.

fragile X syndrome A condition that accounts for ~50% of cases of X-linked mental retardation in which the X chromosome tends to show breaks in appropriately prepared lymphocytes. The cause is an expanded trinucleotide repeat in the 5′-untranslated region of the *FMR1* gene, often with more than 200 repeats.

fragilysin Zinc metallopeptidase (EC 3.4.24.74, 405 aa) of the M10 family that may be a virulence factor of some strains of *Bacteroides fragilis.* The enzyme damages intestinal epithelium by cleaving E-cadherin and causes diarrhoea, but the correlation between fragilysin production and pathogenicity is weak.

fragmentins Fragmentin-1 was originally identified as a 32-kDa granule protein from rat NK cells and T cells that induced rapid DNA fragmentation and cell death by apoptosis and is identical to *granzyme A (262 aa in humans). Fragmentin-2 (247 aa) is granzyme B and fragmentin-3 (264 aa) is granzyme K.

fragmin **1.** The proprietary name for dalteparin sodium, a low molecular weight heparin used prophylactically to prevent deep vein *thrombosis. **2.** An F-actin severing and capping protein (536 aa) from the slime mould *Physarum polycephalum.*

framboesia *See* YAWS.

frameshift mutation A mutation that arises as a result of the insertion or deletion of a number of bases not divisible by three in an open reading frame. Generally results in the appearance of a nonsense codon downstream and thus premature termination.

framycetin A broad spectrum aminoglycoside antibiotic that inhibits bacterial protein synthesis by binding to the 30S subunit of the bacterial ribosome and inhibiting peptide chain elongation. Mostly used for skin and subcutaneous tissue infections.

FRAS1 A gene encoding a putative extracellular matrix protein (4007 aa) expressed in many adult tissues, with highest levels in kidney, pancreas, and thalamus. *FRAS1* is mutated in *Fraser's syndrome

Fraser's syndrome An autosomal recessive developmental disorder in which there are multiple congenital malformations including cryptophthalamos (complete fusion of eyelids), syndactyly, and renal defects, that can be caused by mutation in the **FRAS1* gene or in the **FREM2* gene.

Frasier's syndrome *See* DENYS–DRASH SYNDROME.

Frat proteins Products of the human homologues of the murine proto-oncogene (frequently rearranged in advanced T-cell lymphomas), originally identified in transplanted tumours of Moloney murine leukaemia virus (M-MuLV)-infected Emu-Pim1 transgenic mice. Together with GBP (*glycogen synthase kinase (GSK3)-binding protein), Frat-1 and Frat-2 constitute a family of GSK3-binding proteins that inhibit the phosphorylation of β-*catenin, preventing its degradation by the ubiquitin-proteasome pathway, thus potentially playing a role in *Wnt signal transduction. A Frat-null mouse is, however, viable.

frataxin A protein (210 aa) that is involved in mitochondrial iron metabolism and is the product of the *X25* gene that is mutated in *Friedreich's ataxia. Frataxin will form a homopolymer that binds up to ten iron atoms per protein molecule and is important in resisting oxidative stress.

fraxin A compound with anti-inflammatory and antimetastatic properties, extracted from the ash *Fraxinus excelsior*, that inhibits production of the prostanoid, 5-HETE.

free erythrocyte protoporphyrin An analytical measurement, now generally superseded as a test for lead poisoning but still used to assess the long-term effects of high-level lead exposure. Free protoporphyrin is noncomplexed, nonhaem, protoporphyrin that accumulates because lead inhibits incorporation of iron into haem.

free-running Describing a rhythmic process that is not subject to entrainment or resetting by external cues. A free-running circadian rhythm in activity is often shown by animals kept in unvarying conditions, e.g. of constant light.

freeze cleavage *See* FREEZE FRACTURE.

freeze drying A method for preserving biological material by rapid freezing and sublimation of the water. If this is done appropriately denaturation can be avoided and the dried material can be stored for an extended period before being reconstituted simply by the addition of water.

freeze etching A method for preparing *freeze fractured specimens by allowing sublimation of water from areas of the specimen that are not protected by the presence of a lipid layer before shadowing. Some etching occurs in standard preparation, but by extending the delay in shadowing deeper structures such as cytolasmic filaments can be revealed.

freeze fracture A method of specimen preparation in which rapidly frozen tissue is cracked to produce a fracture plane. The newly exposed surface is then coated with heavy metal, often from one side so as to give a shadowing effect, then further coated with carbon and the biological material digested away, leaving a replica that can be examined in the transmission electron microscope. The fracture plane tends to pass along the centre of lipid bilayers, and so the pattern of integral membrane proteins can be seen. *Freeze etching reveals deeper structures.

FREM Genes encoding extracellular matrix protein related to that encoded by *FRAS1 (FRAS1-related extracellular matrix protein, 2179 aa). Mutations in FREM2 (3169 aa) can lead to *Fraser's syndrome. The protein has a predicted transmembrane domain and multiple chondroitin sulphate proteoglycan repeats.

frequenin One of a large family of myristoyl-switch calcium-binding proteins (neuronal calcium sensor, NCS-1, 190 aa) that modulates synaptic activity and secretion in response to calcium ions. Highly conserved and homologous to *recoverin and visinin.

frequently rearranged in advanced T-cell lymphomas *See* FRAT PROTEINS.

Freund's adjuvant A water-in-oil emulsion to which antigen is added when raising antibody in experimental animals. Complete Freund's adjuvant also contains heat-killed tubercle bacilli and will elicit a severe granulomatous reaction in humans.

Friedreich's ataxia One of the commonest forms of autosomal recessive ataxia caused by an excessive number of trinucleotide (GAA) repeats within an intron of the gene for *frataxin, and deficiency in that protein. There is progressive damage to the nervous system with symptoms ranging from gait disturbance and speech problems to heart disease.

Friend murine leukaemia virus A replication-competent murine leukaemia virus that causes an erythroblastosis-like leukaemia, in which there are large numbers of nucleated red cells in blood. Appears to induce tumours by insertion into the proviral integration site, Fli-1, on mouse chromosome 9. **Friend helper virus** is a murine lymphoid leukaemia virus originally thought to assist replication of Friend virus. **Friend murine erythroleukaemia cells** (Friend cells) are erythroblast lines transformed by the virus that can be induced to differentiate *in vitro*, as shown by the production of haemoglobin. **Friend spleen focus-forming virus** is a defective virus found in some strains of Friend helper virus. Will form foci in infected mouse spleen and the leukaemia is associated with polycythaemia rather than anaemia.

f

frizzled A family of ten G-protein-coupled receptors involved in the *wnt signalling pathway and other signalling pathways. Ligand binding induces activation of *dishevelled proteins in the cytoplasm and eventually to activation of transcription factors. Variation in the cytoplasmic domain links the receptors to various pathways. Mutations in **frizzled 4** (FRZ4, 537 aa) lead to exudative vitreoretinopathy; the **FZD9** gene is located within the *Williams–Beuren syndrome common deletion region. **Secreted frizzled-related proteins** (SFRPs) appear to modulate Wnt signalling by competing with membrane-bound frizzled receptors for the binding of secreted Wnt ligands.

Fröhlich's syndrome *See* DYSTROPHIA ADIPOSOGENITALIS.

Froin's syndrome Condition in which *cerebrospinal fluid, obtained by lumbar puncture, is yellow and the protein content is raised. Caused by obstruction of the spinal canal.

frontotemporal dementia A group of adult-onset behavioural disturbances that progress to frontal lobe dementia, parkinsonism, and amyotrophy in variable proportions. They can be caused by mutation in the gene for *tau or *presenilin-1 (*see* PICK'S DISEASE). Ubiquitin-positive frontotemporal dementia is caused by mutation in the gene for *progranulin. Other loci have also been identified.

frozen stock Cells frozen in the presence of cryoprotectant and stored at −70°C or in liquid nitrogen. The stock can be used to produce cells of comparable passage number, avoiding the inevitable drift in properties that occurs with subculturing, and also as an insurance against loss or contamination. Frozen sperm is stored in a similar way.

frusemide *See* FUROSEMIDE.

FSH (follicle-stimulating hormone) *See* FOLLICLE.

F-spondin One member of a small subfamily of thrombospondin type 1 (TSR) class molecules, defined by two domains of homology, the FS1/FS2 and TSR domains. Spondin-1 (F-spondin, vascular smooth muscle cell growth-promoting factor, VSGP, 807 aa) is expressed in the midline of the developing embryo, primarily in the floor plate of the neural tube where it attaches to the extracellular matrix and provides neuronal guidance cues. It also binds to amyloid precursor protein (APP) and inhibits cleavage by β-secretase. The other subfamily member, spondin-2 (331 aa), is also known as *mindin. *See also* R-SPONDIN.

F-type ATPase The multisubunit proton-transporting ATPase (F_1/F_o ATPase, ATP synthase, F-ATPase), that is normally driven in reverse by a proton gradient to produce ATP (*chemiosmosis). It is found in the inner membrane of *mitochondria and chloroplasts, and in bacterial plasma membranes. The transmembrane F_o sector of the enzyme contains a rotary motor that is fuelled by the proton gradient; rotation is transmitted via a central stalk to the catalytic subunits in the soluble F_1 domain, where conformational changes induced by the rotation enable the synthesis and subsequent release of ATP. Vacuolar (V-type) ATPases are related but operate in reverse and use ATP to pump protons.

SEE WEB LINKS

- Additional detail in a website produced as an assignment for an undergraduate course at Davidson College, North Carolina.

Fuchs' endothelial dystrophy (corneal endothelial dystrophy) *See* CORNEA.

fuchsin A synthetic rosaniline dye used in *Schiff's reagent and in carbol-fuchsin as a topical antifungal agent.

fucose A monosaccharide (6-deoxy-L-galactose) that is a component of *N*-glycan chains of glycoproteins and the only common L-form of sugar in mammalian systems.

fucosyl transferase An enzyme responsible for transferring fucosyl residues from GDP-fucose to glycan chains. Glycoprotein-fucosylgalactoside (α-*N*-acetylgalactosaminyl transferase, EC 2.4.1.40, 354 aa) is the key enzyme of the *ABO blood group antigen system.

Fugu rubripes A fish (Japanese puffer fish) that is famous (or infamous) for the *tetrodotoxin found in lethal amounts in the poison gland and at low levels elsewhere. More importantly, the genome has very low levels of repetitive DNA, suggesting that much of the repetitive DNA found in other organisms is indeed junk.

Fujian flu **1.** Influenza caused by a strain of the human H3N2 subtype of the influenza A virus (A/Fujian, H3N2), responsible for outbreaks in 2003–4. **2.** An avian influenza strain of the H5N1 subtypeA from the province of Fujian, China.

fukutin A secreted protein (461 aa) of the extracellular matrix that is important for the normal glycosylation of *dystroglycan and its localization and function. **Fukutin-related protein** (495 aa) is also involved in the glycosylation. Disruption in the dystroglycan glycosylation pathway leads to Fukuyama *muscular dystrophy, where the mutation is in fukutin, *muscle-eye-brain disease, limb-girdle muscular dystrophy type 2I (where the mutation is in fukutin-related protein) and *Walker–Warburg syndrome.

Fukuyama muscular dystrophy A form of progressive *muscular dystrophy with mental retardation and various brain malformations caused by mutation in the *fukutin gene. The changes in muscle are similar to those in Duchenne dystrophy.

fumagillin An antibiotic isolated from *Aspergillus fumigatus* that inhibits endothelial cell proliferation (angiogenesis) by binding to *methionine aminopeptidase 2.

fumigaclavines A family of *ergot alkaloids that can cause colitis and liver damage and have structural similarities to lysergic acid. Fumigaclavine A has been isolated from *Aspergillus* sp. Fumigaclavines B and C, and other similar alkaloids, have been isolated from conidia of *Aspergillus fumigatus*.

fumonisins A group of mycotoxins produced by various *Fusarium* species. Fumonisin B1, probably the most toxic, inhibits sphingolipid biosynthesis and causes fatal leucoencephalomalacia in horses.

functional cloning A cloning method (expression cloning) that relies upon some property of the gene product to select the expression vector (derived from a cDNA library) that carries the gene and thus to allow sequencing and identification. For example, if the gene produces a cell-surface receptor then cells expressing the gene product can be identified by binding of labelled ligand.

fundus **1.** The base of a hollow organ, the region most distant from the opening. **2.** That part of the retina, opposite the pupil, which can be seen with an ophthalmoscope. *See* STARGARDT DISEASE (fundus flavimaculatus, fundus albipunctatus).

fura-2 A dye used for studying intracellular calcium levels. The acetoxymethyl ester (fura-2 AM) diffuses readily into cells but becomes hydrolysed and trapped: the fluorescence spectrum shifts in a concentration-dependent manner when calcium is bound. *See* QUIN2.

furin A serine endopeptidase (EC 3.4.21.75, 794 aa, a *kexin) that processes various secretory proproteins (including some growth factors) by cleavage at paired basic amino acids. Will cleave the HIV envelope glycoprotein (gp160) into two portions, gp120 and gp41, making the virus fusion-competent. It is inhibited by serpin B8. *See* NOTCH.

furosemide (frusemide) Potent loop diuretic that is used to treat oedema associated with congestive heart failure, cirrhosis of the liver, and renal disease. Furosemide increases the excretion of sodium, potassium, and chloride ions, and inhibits their resorption in the proximal and distal renal tubules and the loop of Henle. *TN* Lasix.

furuncle *See* BOIL.

***Fusarium* mycotoxins** Important contaminants of various food products, particularly maize and wheat that have been infected with ear-rot or scab. Include zearalenone, *diacetoxyscirpenol, T-2 toxin, neosolaniol monoacetate, deoxynivalenol, nivalenol, *fumonisin B1, fumonisin B2, moniliformin, fusarenon-X, HT-2 tioxin, and β-zearalenol.

fused gene family A family of proteins that are associated with keratin intermediate filaments and are partially crosslinked to the cell envelope in keratinocytes. Members of the family include *profilaggrin, *trichohyalin, *repetin, *hornerin, and *cornulin. The calcium binding *EF-hand domain shows significant homology with that of *S100 proteins, and the family is thought to have arisen through fusion between S100-type calcium-binding proteins and proteins involved in cornification.

fusidic acid A bacteriostatic antibiotic derived from the fungus *Fusidium coccineum* that blocks translocation of the elongation factor G (EF-G) from the ribosome during bacterial protein synthesis. Only effective against Gram-positive bacteria such as *Staphylococcus* and *Corynebacterium*. Used topically for skin and eye infections.

fusin The chemokine receptor CXCR4 (352 aa) that binds stromal cell-derived factor-1

(SCDF-1). It is the coreceptor, along with CD4, for T-cell tropic strains of HIV-1.

fusion protein An engineered chimeric protein containing sequences from two different genes that have been combined in an expression vector. A common strategy is to produce a fusion protein in which the added protein acts as a label, e.g. the addition of *green fluorescent protein.

futile cycles Enzyme-catalysed reactions in which the forward and reverse processes, involving different enzymes, are constantly active so that there is continual turnover. The classic example is the phosphorylation and dephosphorylation of phosphatidyl inositol derivatives in cell membranes.

FVB An inbred strain of mice, derived from Swiss mice and susceptible to *Friend leukaemia virus B, often used in transgenic research.

Fx A homodimeric NADP(H)-binding protein (tissue-specific transplantation-antigen 3, TSTA3, EC 1.1.1.271, a dimer of 321 aa subunits), first isolated from erythrocytes and the enzyme responsible for the last steps of GDP-L-fucose synthesis from GDP-D-mannose. Up-regulation of Fx expression, leading to an increase in GDP-L-fucose levels, may be a marker for hepatocellular carcinoma.

fyn An oncogene related to *src and thought to be important in brain development (fyn mutation inhibits positive turning responses to *netrin) and function (fyn mutants seem to have defects in long-term potentiation). Also has a role in T-cell signalling in association with *lck. It encodes a nonreceptor *tyrosine kinase. *See* RAK (fyn-related kinase).

FYVE domain A domain (~70–80 aa) named after the four proteins in which it was first found (Fab1, YOTB/ZK632.12, Vac1, and EEA1) that coordinates two zinc ions and binds PI3P but not more highly phosphorylated phosphoinositides. More than 60 proteins have been shown to have such domains and all are involved in membrane trafficking and phosphoinositide metabolism.

FYVE, RhoGEF, and PH domain-containing proteins A family of proteins containing zinc fingers, GDP-exchange proteins involved in regulating actin-based motility. For example, FYVE, RhoGEF, and PH domain-containing protein-1 (**FGD1**, faciogenital dysplasia 1 protein, 961 aa) catalyses the replacement of GDP by GTP on the rho family GTPase, cdc42 and has a role in regulating the actin cytoskeleton and cell shape. Mutations in FGD1 are involved in faciogenital dysplasia (*Aarskog–Scott syndrome). **FGD4** (FGD1-related F-actin-binding protein, frabin, 766 aa) has an additional actin-binding domain and when overexpressed *in vitro*, induces microspike formation FGD4 is thought to be is important in myelination of the peripheral nervous system; defects are the cause of *Charcot–Marie–Tooth disease type 4H. There are at least four other members of the family.

G_0 The 'resting' phase of the cell cycle and the state in which nonproliferating cells exist. If the cell is stimulated into division then the cell enters *G_1 and becomes committed to completing the *cell cycle. Not to be confused with *Go.

G_1 The first 'gap' phase of the eukaryotic *cell cycle, between the end of a mitiotic division and the start of another round of DNA synthesis. Cell that are not going to divide are generally thought to be in *G_0 and only enter G_1 if stimulated into a further round of division.

G_2 The second 'gap' phase of the cell cycle that occurs after DNA replication (*S phase) is complete and before mitosis (*M phase) commences.

gab-1 Grb2-associated binder-1. *See* GRB-2.

GABA An amino acid (γ-aminobutyric acid) that is a very important inhibitory neurotransmitter in the central nervous system and retina (and also in many invertebrates). The **GABA receptor** is one of a family of neurotransmitter receptors that includes the *glycine receptor and the *nicotinic acetylcholine receptor and is the target for a variety of drugs. There are two main classes, *ionotropic ($GABA_A$ and $GABA_C$ receptors) and *metabotropic ($GABA_B$ receptor). The **$GABA_A$ receptor** is a ligand-gated chloride channel specifically blocked by *bicuculline and *picrotoxin, a hetero-oligomer with (probably) five subunits, generally two pairs of α and β subunits and a γ subunit, probably arranged as a tight group with the chloride channel in the centre. The α and γ chains are needed for binding of *benzodiazepine and the β chains bind GABA. Phosphorylation will alter the receptor's properties. The metabotropic **$GABA_B$ receptor** is a G-protein-coupled receptor found in the brain and is negatively coupled to adenylate cyclase through a Go protein and thus acts indirectly on N-type calcium channels reducing catecholamine release. The **$GABA_C$ receptor** resembles $GABA_A$ but is restricted to the retina.

gabapentin An anticonvulsant drug used in the treatment of epilepsy and neuropathic pain. Apparently acts on voltage-gated *N-type calcium channels, not *GABA receptors. *TN* Neurontin.

G-actin Globular *actin, the soluble protomer that polymerizes to form fibrillar actin (*F-actin) or *microfilaments.

***gadd* genes** A set of genes (growth arrest- and DNA damage-inducible genes) that are activated when DNA is damaged. **Gadd-34** (protein phosphatase 1 regulatory subunit 15A, PPP1R15A, 674 aa) mediates growth arrest and apoptosis following DNA damage or negative growth signals. **Gadd-45** (165 aa) is strongly induced by X-rays. There are actually three related proteins, gadd-45 α, β, and γ, all induced by stress. Gadd-45α associates with the cyclin dependent kinase inhibitor p21Cip1 and *proliferating cell nuclear antigen, stimulates DNA excision repair *in vitro* and inhibits entry of cells into S phase. **Gadd-153** (C/EBP-homologous protein, CHOP, 169 aa) is not induced by X-rays but by alkylating agents and ultraviolet light and acts as an inhibitor of the transcription factors C/EBP and LAP. It also seems to activate Gadd-34. The Gadd-153 gene is consistently rearranged in myxoid liposarcomas.

gadolinium (Gd) On of the lanthanide elements used experimentally because the trivalent ion blocks T-type voltage-sensitive calcium channels and stretch-activated ion channels in a concentration-dependent manner.

GAG *See* GLYCOSAMINOGLYCANS.

gag protein (group specific antigen) A retroviral protein that forms the shell of the nucleocapsid that encloses the genomic RNA. The gag protein binds a specific RNA motif in the 5′ UTR of the genome.

gain-of-function mutation A mutation that alters the properties of the protein product so that it has novel properties or has greater activity because a regulatory site has been lost. Gain-of-function mutations are less common than loss-of-function mutations, but if the mutation confers selective advantage the mutant allele will become more common in later generations.

galact- A prefix indicating association with milk. A **galactocele** is a cyst-like swelling in the breast as the result of a blockage of a milk duct. **Galactophoritis** is inflammation of the milk ducts. **Galactopoiesis** is an increase in milk secretion. **Galactorrhoea** is the excessive secretion of milk causing it to overflow through the nipple. **Galactotrophic** (galactotropic) substances stimulate the secretion of milk and are *galactogogues.

g

galactagogue A substance or drug that promotes the secretion of breast milk. *Domperidone and *metoclopramide are used clinically and many herbal remedies are said to be effective, including fenugreek, brewer's yeast, blessed thistle, and alfalfa.

galactans Galactose polymers, mostly found in plants, which may be branched or unbranched. A sulphated D-galactan from the red alga *Botryocladia occidentalis* has anticoagulant properties comparable to *heparin.

galactocerebroside (galactosylceramide) A ceramide (GalC) that has a galactose residue at the 1-hydroxyl moiety. It is the major glycolipid in myelin and is a cell-surface marker for *oligodendrocytes. Galactocerebrosides can accumulate if there is a defect in β-galactosidase, the cause of galactosylceramide lipidosis or globoid cell leukodystrophy.

galactosaemia An autosomal recessive disorder in which the enzyme galactose-1-phosphate uridyl transferase (GALT, EC 2.7.7.12, 379 aa), that converts galactose-1-phosphate into glucose-1-phosphate, is absent. Excess galactose-1-phosphate accumulates in the blood and a variety of problems result.

galactosamine An amino sugar (2-amino-2-deoxygalactopyranose) that is a common component of some glycolipids, chondroitin sulphate, and dermatan sulphate. *N*-Acetyl galactosamine is the terminal carbohydrate that constitutes blood group A antigen and is typically the first monosaccharide that connects serine or threonine in *O*-glycosylation.

galactose A hexose sugar identical to glucose except that orientation of -H and -OH on carbon 4 are exchanged. A component of *cerebrosides and *gangliosides, and many glycoproteins. *Lactose consists of galactose joined to glucose by a β-glycosidic link.

galactosylceramide *See* GALACTOCEREBROSIDE.

galactosyl transferase A glycosyl-transferase (EC 2.4.-.-) that catalyses the transfer of galactose units from UDP-galactose to an acceptor, commonly *N*-acetylglucosamine, in a glycan chain. The key step in biosynthesis of galactocerebrosides is the enzymatic transfer of galactose to ceramide catalysed by UDP-galactose ceramide galactosyltransferase (EC 2.4.1.45, 541 aa, cerebroside synthase).

galanin A neuropeptide (29 aa) widely distributed in the peripheral and central nervous systems. In the brain, the highest concentrations are in the hypothalamus and the median eminence. It stimulates luteinizing hormone secretion and also regulates gut motility and the activity of endocrine pancreas. The **galanin receptors** (GALR) receptors are *G-protein coupled; GALR1 inhibits adenylyl cyclase via a G protein of the Gi/Go family, GALR2 stimulates inositol phospholipid turnover and intracellular calcium mobilization through a Gq/G11-type and GALR3 couples to a G protein of the Gi/Go class but has a different profile from GALR1. **Galanin-like peptide** (GALP, 60 aa) preferentially binds and activates galanin receptor-2.

galantamine (galanthamine) A plant alkaloid that is a specific, competitive, and reversible *acetylcholine esterase inhibitor shown to have mild cognitive and global benefits for patients with Alzheimer's disease. Originally isolated from Voronov's snowdrop *Galanthus woronowii*, but now synthesized. *TNs* Nivalin, Razadyne, Reminyl.

galectins Family of conserved S-type β-galactoside-binding lectins that bind to cell surface glycoconjugates. They are soluble and do not have transmembrane domains or signal sequences, although found associated with the outer cell surface. Some are homodimers (galectins 1–3), others are monomeric with two carbohydrate-binding sites. They have been identified in a large variety of metazoan phyla and are involved in many biological processes such as morphogenesis, control of cell death, immunological response, and cancer. **Galectin-1** (galaptin, 135 aa) is a homodimer and is abundant in smooth and skeletal muscle. It mediates cell–cell and cell–substratum adhesion and plays a role in immune regulation. Recombinant galectin-1 will induce apoptosis in T cells. **Galectin-2** (132 aa) binds to lymphotoxin-α. A SNP in galectin-2 is associated with susceptibility to myocardial infarction. **Galectin-3** (IgE binding protein, Mac-2, e-BP, 250 aa) is usually considered proinflammatory and can act as an immunomodulator by inducing apoptosis in T cells. It is part of the AGE–receptor (*RAGE) complex. Aberrant expression of galectin-3 involved in various aspects of tumour progression and p53-induced apoptosis is associated with transcriptional repression of Gal-3. **Galectin-4** (323 aa) is monomeric with two carbohydrate binding sites, restricted to small intestine, colon, and rectum. Expression is altered in colorectal carcinoma. **Galectins 5 and 6** have not been identified in humans although known in other mammals. **Galectin-7** (136 aa) is confined to keratinocytes. **Galectin 8** (prostate carcinoma tumour antigen-1, PCTA1, 316 aa) was isolated from a prostate carcinoma library and is related to galectin-4. **Galectin-9** (355 aa) plays a role in thymocyte–epithelial interactions relevant to the biology of the thymus and an allelic variant, ecalectin, is a potent eosinophil chemoattractant produced by antigen-activated T cells. **Galectin-10** (142 aa) is the *Charcot–Leyden crystal protein. Galectin-12 and galectin-13 have also been identified.

gallamine A nondepolarizing muscle relaxant that competes with acetylcholine at the neuromuscular junction. Similar drugs are *tubocurarine, alcuronium, pancuronium, atracurium, and vecuronium. *See* SUXAMETHONIUM. *TN* Flaxedil.

gallidermin A Type A *lantibiotic produced by *Staphylococcus gallinarum*, very similar to *epidermin, and effective against Propionobacteria.

gallstones Concretions (biliary calculi) that occur pathologically in the gallbladder and bile duct. The components vary but can include cholesterol, calcium carbonate, and calcium bilirubinate or mixtures of these.

GAL promoter The promoter region of the operon that encodes various proteins involved in galactose metabolism, including *β-galactosidase ('beta-gal'). The operon is inducible and once β-galactosidase is expressed it will act on the chromogenic substrate *Xgal to produce a blue colour. This colour change is a convenient marker of colonies that have been transformed by a vector carrying the beta-gal gene. Both yeast and bacterial forms of the promoter are used experimentally and GAL4, the heterodimeric yeast transcription factor that binds to the UASG (upstream activation site G) promoter, is commonly used in yeast *two-hybrid screening.

GALT *See* GUT-ASSOCIATED LYMPHOID TISSUE.

galvanotaxis Directed cell movement in response to an applied voltage. Movement is generally directed toward the cathode, and requires fields of ~100 mV/mm. Galvanotaxis is thought to be important in wound repair and morphogenetic movements. Neuronal responses to electrical fields are strictly galvanotropisms because the cell body remains stationary and neuronal processes grow towards the cathode.

galvanotropism *See* GALVANOTAXIS.

gamete A haploid cell produced by meiosis and involved in sexual reproduction. Male gametes (spermatozoa) are small, motile, and produced in large numbers, whereas female gametes (oocytes) are larger and nonmotile.

gametocyte **1.** A cell that produces gametes. Primary gametocytes are diploid and undergo meiosis; secondary gametocytes, the products of the meiotic division of primary gametocytes, are haploid and divide mitotically to produce gametes. **2.** The sexual reproductive stage of the malaria parasite that develops within erythrocytes.

gametogenesis The sequence of events that culminates in the production of *gametes.

gamma-aminobutyric acid *See* GABA.

gamma-delta cells (γδ-T-cells) A small subset (~5%) of T cells in which the T-cell receptor has γ and δ subunits (unlike the more common α and β). They are found in the gut mucosa, within the intraepithelial lymphocyte population, and also in epithelial sites. Their exact role is unclear although they may be an important first-line defence. There is considerable diversity in the receptor subunits and most have neither CD4 nor CD8.

gamma-glutamyltransferase A highly glycosylated heterodimeric enzyme (γ-glutamyl transpeptidase, EC 2.3.2.2, 662 aa) attached to the outer surface of the plasma membrane that transfers a γ-L-glutamyl residue (usually from glutathione) to the amino group of an amino acid, e.g. in the conversion of leukotriene LTC_4 to LTD_4. The enzyme is widely distributed in tissues involved in absorption and secretion and there appear to be several γ-glutamyltransferase genes located on chromosome 22. Elevated plasma levels are used as a diagnostic marker of hepatic disorder and pancreatitis.

gamma-haemolysin *See* GAMMA-TOXIN.

gamma-secretase *See* SECRETASES.

gamma-toxin An exotoxin (γ-haemolysin) produced by most isolates of *Staphylococcus aureus*, forming a protein family with *leucocidins and *staphylococcal α-toxin. Two active cytolytic toxins (AB and CB) can be formed, combining one of the class-S components, HlgA or HlgC, with the class-F component HlgB. γ-Haemolysins form cation-selective pores similar to those produced by α-toxin.

gammopathy The presence in serum or urine of monoclonal immunoglobulin light chain (*paraprotein, M-protein) in apparently healthy persons (monoclonal gammopathy, monoclonal gammopathy of unknown significance, MGUS). Most cases seem to be relatively benign, although a proportion (up to 25%) may progress to multiple myeloma, Waldenström's macroglobulinemia, primary amyloidosis, B-cell lymphoma, or chronic lymphocytic leukaemia.

g

ganciclovir A synthetic analogue of 2′-deoxyguanosine that is used in treatment of cytomegalovirus (CMV) infections. *In vivo* it is phosphorylated first to the monophosphate form by a CMV-encoded kinase and then further phosphorylated by cellular kinases. It accumulates in CMV-infected cells and probably inhibits viral DNA synthesis by competitive inhibition of viral DNA polymerases and by termination of viral DNA elongation. *TNs* Cymevene, Cytovene.

ganglion **1.** A physical cluster of *neurons. In vertebrates, the ganglia are appendages to the central nervous system; in invertebrates, the majority of neurons are organized as separate ganglia. **2.** A **ganglion cyst** is a swelling on or around joints and tendons in the hand or foot. **3.** A **ganglion cell** is an *interneuron that conveys information from the *bipolar, horizontal, and *amacrine cells of the retina to the brain. **4.** A **ganglioneuroma** is a tumour composed of nerve cells and fine fibrous tissue, usually arising in connection with sympathetic nerves. **5.** Ganglionitis is iflammation of a nerve ganglion.

gangliosides Complex *glycosphingolipids that have oligosaccharide chains containing *N*-acetylneuraminic acid attached to a ceramide lipid moiety. GM forms have a single sialic acid, GD forms have two. They constitute ~5% of the plasma membrane lipid in neuronal cells. A **gangliosidosis** is a disease in which there is a defect in metabolism of breakdown of gangliosides. *See* TAY–SACHS DISEASE.

gangrene Decay of a part of the body following death of the tissue, usually as a result of loss of blood supply. Gas gangrene is usually caused by infection of the dead tissues with *Clostridium perfringens*.

gankyrin An antiapoptotic oncoprotein (PSMD10, p28) containing seven *ankyrin repeats that is overexpressed early in hepatocarcinogenesis and in hepatocellular carcinomas. Gankyrin interacts with S6 ATPase of the 19S regulatory particle of the 26S proteosome, regulates the phosphorylation of the *retinoblastoma protein (pRb) by CDK4, enhances the ubiquitylation of *p53 by the RING ubiquitin ligase MDM2, and inhibits NFκB activity.

GAP *See* GTPASE ACTIVATING PROTEIN; GROWTH ASSOCIATED PROTEINS.

GAPDH *See* GLYCERALDEHYDE-3-PHOSPHATE DEHYDROGENASE.

gap junction (communicating junction) A specialized region of contact between two cells that has pores which permit the free passage of molecules up to ~900 Da in size. The pores (connexons) are hexagonal arrays of six transmembrane proteins (*connexins) in each plasma membrane that align to form a pore through which hydrophilic molecules can pass. *Electrical synapses are gap junctions, and *metabolic cooperation depends on the formation of gap junctions.

GAPT Growth factor receptor-bound protein 2 (grb2-binding adaptor protein, transmembrane) *See* LAT ADAPTOR.

Gardner's syndrome A variant of familial adenomatous *polyposis coli.

Gardnerella vaginalis A facultatively anaerobic Gram-variable rod (can appear Gram-negative in some situations) of the Bifidobacteriaceae family (formerly *Haemophilus vaginalis, Corynebacterium vaginalis*). The organism that is commonly responsible for bacterial vaginosis, although it does not always cause disease.

gargoyle cells Fibroblasts cultured from individuals with *Hurler's disease (gargoylism) that have large intracellular metachromatic deposits of mucopolysaccharide.

GARPs Multivalent proteins (glutamic-acid-rich proteins), that interact with the cGMP phosphodiesterase, guanylate cyclase, and the retina-specific ATP-binding cassette cation transporter (the cGMP-gated channel) that together comprise the amplification system of the rod photoreceptors, sometimes called the *transducisome. They exist as two soluble forms,

GARP1 and GARP2, and as a large N-terminal cytoplasmic domain (GARP′ part). **GARP2**, the predominant splice variant of the β subunit, associates with light-activated phosphodiesterase and inhibits the breakdown of cGMP. Mutations in the gene can lead to autosomal recessive *retinitis pigmentosa.

GASC1 A putative oncogene (gene amplified in squamous cell carcinoma 1, JMJD2C) that encodes a histone trimethyl demethylase. One of the jumonji family, containing two *jumonji domains.

gasdermins A family of genes expressed specifically in cells at advanced stages of differentiation in the upper epidermis. The function of the products is unknown. Expression of gasdermin (445 aa) is suppressed in gastric cancer.

***gas* genes** A large set of disparate genes (growth arrest specific genes) associated with cellular quiescence. **Gas1** protein (345 aa) appears to be a component of the *sonic hedgehog signalling system and is a potential locus for human craniofacial malformations. **Gas2** (313 aa) colocalizes with microfilaments at the cell border and along the stress fibres in growth-arrested mouse fibroblasts. The **Gas2-related gene** encodes two alternatively spliced proteins of 36 kDa (GAR22α) and 73 kDa (GAR22β), both of which have actin and microtubule-binding domains. **Gas3** (peripheral myelin protein 22, PMP22, 160 aa) makes up 2–5% of peripheral nervous system myelin: mutations and duplications in the gene are responsible for various peripheral neuropathies. The **gas5** gene is a non-protein-coding gene, the products of which are multiple small nucleolar RNAs, and apparently plays an essential role in normal growth arrest in both T-cell lines and nontransformed lymphocytes. **Gas6** (721 aa) is a secreted ligand for the *AXL oncogene product, a receptor tyrosine kinase, and is a vitamin K-dependent protein structurally related to anticoagulant protein S, but without anticoagulant activity. **Gas7** (476 aa) is found primarily in terminally differentiated brain cells, particularly in cerebellar Purkinje neurons. **Gas11** (478 aa) is probably a tumour suppressor, by homology with the mouse protein, and may associate with the Golgi. Homologues have a role in regulating axonemal dynein. Other gas genes are known in nonhuman species but human homologues have not (yet) been described.

GAS motif A DNA motif (interferon-γ activation site/sequence) found in various interferon-γ regulated genes. Most *STATs bind to tandem or multiple GAS motifs.

gastr- A prefix indicating association with the stomach. **Gastric juice**, the fluid found in the stomach, is mostly water but has 0.02% free HCl. **Gastritis** is inflammation of the mucous membrane of the stomach and in **gastroenteritis** the intestinal membranes are also affected. A **gastrocele** is a hernia of the stomach, not to be confused with the gastrocoel, the *archenteron. **Gastroparesis** is a disorder, often associated with type 1 diabetes, in which paralysis of the stomach muscles delays the passage of food. Compound forms (e.g. gastroduodenal, gastrojejunal) indicate association with stomach and some other organ.

gastric cancer Carcinoma of the stomach, probably the second most common carcinoma. Several genes may be associated with an increased risk of gastric cancer and there are various inherited cancer predisposition syndromes, including hereditary nonpolyposis colon cancer (*See* GTBP); familial adenomatous polyposis (*polyposis coli); *Peutz–Jeghers syndrome; *Cowden's disease; *Li–Fraumeni syndrome). One form of familial gastric cancer is caused by germline mutations in the E-cadherin gene, others by somatic mutations in a range of genes. Only ~10% of cases have a clear hereditable component and other important factors include diet, infection with **Helicobacter pylori*, and gastritis.

gastric inhibitory polypeptide A peptide hormone (GIP, glucose-dependent insulinotropic polypeptide, 42 aa) that stimulates insulin release and inhibits the release of gastric acid and pepsin. It is a member of a family of structurally related hormones (incretins) that includes *secretin, glucagon, vasoactive intestinal peptide, and growth hormone releasing factor. It is produced by intestinal K cells and the receptor is G-protein coupled.

gastrin A peptide hormone, secreted by the mucosal cells in the gastric antrum (*G cells and by the D cells of the pancreatic islets, that stimulates secretion of HCl by the gastric mucosa. Human gastrin I (17 aa) is identical to gastrin 2 except that the latter has a sulphate ester of tyrosine in position 12. Both are overproduced by pancreatic tumours in the Zollinger–Ellison syndrome. Gastrin is competitively inhibited by *cholecystokinin and there is feedback regulation by HCl. **Gastrin-releasing peptide** (GRP, 27 aa) is a neuropeptide, the human homologue of *bombesin. It stimulates the release of gastrin, pancreatic polypeptide, glucagon, gastric inhibitory peptide, and insulin. It has other effects, causing bronchoconstriction and vasodilation in the respiratory tract, and regulating cell proliferation through a G-protein-coupled receptor.

gastrocoel *See* ARCHENTERON.

gastroenterological disorders

Manydisorders of the gastrointestinal tract are caused by toxins (enterotoxins) in food or produced by infectious agents such as **Enterobacter*, **Vibrio cholerae*, *enteroviruses (e.g. *norovirus, *Reoviridae) or parasites (e.g. **Cryptosporidium*, **Hymenolepis*, **Toxocara*, **Trichuris* spp.). More serious disorders may be due to a defects in intestinal transport systems (*acrodermatitis enteropathica, *Hartnup's disease, *sitosterolaemia) or hypersensitivity (*coeliac disease or may have an autoimmune aetiology (*inflammatory bowel disease). The gut is protected by the *gut-associated lymphoid tissue, *intestinal L cells, and secreted *IgA which can inhibit bacterial adhesion to the gut mucosa (*See* COLONIZATION FACTORS). Other relevant entries are for *cholera, *colitis, *dysentery, *gastric cancer, *guanylin, **Helicobacter pylori*, hereditary nonpolyposis colon cancer (*See* GTBP; HIRSCHSPRUNG'S DISEASE; LACTOSE; POLYPOSIS COLI.

Gastrografin A contrast medium used in diagnostic radiography of the gastrointestinal tract. It is a water-based solution of sodium amidotrizoate and meglumine amidotrizoate.

gastrointestinal stromal tumours Tumours found in the gastrointestinal tract that originate from the interstitial cells of Cajal, the pacemaker cells that regulate peristalsis. They are caused either by mutation in **kit* or in the gene for the platelet derived growth factor receptor A (*PDGFRA*).

gastrula The embryonic stage, following the *blastula stage, at which major morphogenetic movements occur (gastrulation) and the triploblastic structure of the embryo is established with the formation of the three primary germ layers, *ectoderm, *mesoderm, and *endoderm.

GATA transcription factors A family of zinc finger transcription factors that bind to the 'GATA' nucleotide motif. **GATA1** protein (NFE1, Eryfl, GF-1, or EFya, 413 aa) regulates the switch from fetal to adult haemoglobin in conjunction with **friend of GATA1** (FOG1, 1004 aa). Mutations in this gene have been associated with X-linked dyserythropoietic anaemia and thrombocytopenia. **GATA2** is similar to GATA1 and is expressed in haematopoietic progenitors and also in nonhaematopoietic embryonic stem cells. **GATA-3** is abundantly expressed in the T-cell lineage and is involved in T-cell receptor gene activation and necessary for the expression of many T-cell-derived cytokines. Haploinsufficiency in GATA-3 in humans leads to HDR syndrome (hypoparathyroidism, sensorineural deafness, and renal anomaly syndrome). **GATA4** is expressed in adult vertebrate heart, gut epithelium, and gonads; mutations in this gene have been associated with cardiac septal defects. **GATA5** expression is associated with differentiation in a human colon cancer cell line. **GATA6** has an expression pattern similar to GATA4 but is also found in lung and liver. It may regulate the differentiation of vascular smooth muscle cells. **GATA-like protein-1** (GLP1, 271 aa) acts as a transcriptional repressor of GATA6 and is important in germ cell development in the gonads.

gated ion channel Ion channels that are regulated either by transmembrane potential (*voltage-gated ion channels) or the binding of specific ligands (*ligand-gated ion channels).

gating currents Movement of a charge within a membrane (a current) that precedes and signals the opening of voltage-gated ion channels that permit a much larger current flow.

Gaucher's disease An autosomal recessive defect of glucocerebrosidase (*β-glucosidase, EC 3.2.1.45, 927 aa) that leads to the intracellular accumulation of glucosylceramide (GlcCer) particularly within cells of the mononuclear phagocyte system. These cells are the so-called '**Gaucher cells**' found in most tissues. Three types of disease are recognized although the affected enzyme is the same in all: **type I** in which there is no neuropathological abnormality, **type II** in which there is acute neuronopathy and the enzyme is probably present but unstable, and **type III** which is a subacute neuronopathic form. There is also a perinatal lethal variant, probably a severe form of type II. Enzyme replacement therapy with macrophage-targeted recombinant human glucocerebrosidase is effective in neuronopathic and non-neuronopathic forms although it will not reverse the damage to neurons.

gavage Forced feeding through a tube that passes into the stomach. Also a means of administering drugs to uncooperative experimental animals.

***gax* gene** A homeobox gene (growth arrest-specific homeobox gene, mesenchyme homeobox-2, Mox2) associated with growth arrest (*See* GAS GENES) that is restricted to mesodermally derived tissues. It is downregulated by mitogens.

G banding (Giemsa banding) A method for identifying individual chromosomes on metaphase spreads by staining with *Giemsa after a brief trypsin treatment. The pattern of Giemsa banding is different from the banding produced when quinacrine is used as a stain (Q bands).

gCAP39 *See* CapG.

GC box A nucleotide binding motif (GGGCG) recognized by the mammalian *transcription factor Sp1.

G cell A cell type found in the stomach mucosa that secretes *gastrin. They are innervated by the vagus nerve which releases *gastrin releasing peptide.

GCIP A helix-loop-helix protein (grap2 cyclin-D-interacting protein, CCNDBP1, 360 aa) that is a putative tumour suppressor. Interacts with *p29 and with *GRB2-related adaptor protein 2 (Grap2).

GCNF *See* GERM CELL NUCLEAR FACTOR.

GCP **1.** Good clinical practice. **2.** *Granulocyte chemotactic protein 2 (GCP2). **3.** Glutamate carboxypeptidase II (GCP2).

Gc protein A plasma protein (group-specific component, vitamin D-binding protein, 474 aa) of the *albumin family, also found in cerebrospinal fluid and urine and on the surface of many cell types It is involved in sterol transport and scavenging extracellular *actin. It is highly polymorphic and more than 80 variants have been described.

GDF (GDI displacement factor; growth (and) differentiation factor *See* GDI; GROWTH/DIFFERENTIATION FACTORS.

GDGF Glioma-derived growth factor. *See* GDNF.

GDI A protein (GTP dissociation inhibitor) that inhibits GDP/GTP exchange on *ras-like GTPases so that they remain in the active GTP-bound form. Different families of GTPases have specific GDIs (e.g. rho-GDI). Mutations in the rab-GDI (GDI-1, oligophrenin, *See* P21 ACTIVATED KINASES) are associated with some types of X-linked mental retardation. GDI-D4 is the rho-GDIb. The **GDI displacement factor** (GDF) is a membrane-associated protein (~25 kDa) that is involved in the dissociation of GDI from prenylated *rab following membrane binding.

gDNA Genomic DNA.

GDNF A neurotrophic growth factor (glial-cell line derived neurotrophic factor, astrocyte-derived trophic factor 1, formerly glioma-derived growth factor, GDGF, 211 aa) originally derived from glioma cells. It is a disulphide-linked homodimer, structurally related to *artemin, *neurturin, and *persephin, belonging to the cysteine-knot superfamily of growth factors. It is specific for midbrain dopamine neurons and has been tested for effects in Parkinson's disease. The **GDNF receptor** (GFRα-1, 465 aa) is one of a family of GPI-linked receptors: GFRα-1 is the receptor for GDNF, GFRα-2 for neurturin, GFRα-3 for artemin and GFRα-4 for persephin. The receptors forms a heterodimeric (α/β) receptor complex with the tyrosine kinase *ret, the latter being the β subunit.

GDP Guanosine diphosphate.

GEFs A family of proteins (guanine nucleotide exchange factors) involved in the cycle of activation and inactivation that occurs with the small G-proteins such as ras, rap, and rho. The exchange factors stimulate the release of bound GDP, allowing GTP to bind and activate the G protein. Inactivation occurs when the GTP is hydrolysed to GDP, stimulated by *GTPase activating proteins (GAPs). Members of the family include *cytohesin, *ARNO, Gea-1 and 2, kalirin, and yeast *Sec7.

gel A semisolid material formed by the coagulation of a colloidal liquid. Gels formed from collagen have a fibrous matrix with fluid-filled interstices and behave as viscoelastic materials that will resist some mechanical stress without deformation. Other examples are the gels formed by large molecules such as denatured collagen (gelatin), agarose, acrylamide, and starch.

gelatin Heat-denatured collagen.

gelatinous lesion A rather imprecise term but often applied to lesions of the intima of the blood vessel wall that are considered to be precursors of the characteristic *fibrous plaques of atherosclerosis.

geldanamycin A benzoquinone ansamycin antibiotic produced by *Streptomyces hygroscopicus* that binds to the ATP binding pocket of Hsp 90 and inhibits its chaperone function. Has antitumour effects but is too hepatotoxic for clinical use, although derivatives have better toxicological profiles. Structurally similar to *herbimycin A.

gel electrophoresis *See* ELECTROPHORESIS.

geleophysic dysplasia A disorder caused by mutation in the **ADAMTS2* gene in which the affected children appear to have 'happy faces' (*Gk.* gelios = happy). The disorder is apparently a mucopolysaccharidosis and progressive changes to extracellular matrix prove fatal in adolescence or earlier.

gel filtration A method for separating molecules according to their size. The solution containing the molecules of interest is passed down a column packed with polymer beads that have pores similar in size to those of the molecules of interest. Molecules that enter the pores have a longer path-length to traverse than those that are excluded and pass between the beads. The polymers are gel-like and the extent of gel crosslinking determines the pore size and the separation range. For protein separation, gels made from polyacrylamide, or from flexible (Sephadex) or rigid (agarose, Sepharose) sugar polymers are used.

gel mobility shift *See* GEL RETARDATION ASSAY.

gelonin An RNA-*N*-glycosidase (316 aa) isolated from a Himalayan plant, *Gelonium multiflorum* (Euphorbiaceae), that depurinates RNA in ribosomes, inhibiting protein synthesis. It has been used in immunotoxins for the treatment of tumours.

gel retardation assay A method (gel mobility shift assay, bandshift assay) often used in studying DNA–protein interactions, although in principle applicable to any binding interaction. DNA that has bound a protein such as a transcription factor has a larger size and thus runs more slowly on a gel (i.e. is retarded). The method can be used to estimate the amount of a DNA-binding protein present by quantifying the proportion of radiolabelled DNA that is retarded.

gelsolin Actin-binding protein (brevin, actin depolymerizing factor, 782 aa) that nucleates actin polymerization and severs microfilaments at sub-micromolar calcium concentrations. A variant with an additional N-terminal 23 aa is found in plasma and probably clears actin from the circulation. The amyloid protein in the Finnish type of hereditary *amyloidosis is a fragment of gelsolin with a single mutated amino acid.

gemcitabine An antitumour prodrug agent used alone, or in combination with other cytotoxic drugs, to treat non-small-cell lung cancer, pancreatic, bladder, and breast cancer. It is a deoxycytidine analogue, and is metabolized intracellularly to the active diphosphate (dFdCDP) and triphosphate (dFdCTP) nucleosides and blocks progression through the G_1/S-phase boundary. *TN* Gemzar.

gemeprost A synthetic analogue of prostaglandin E_1.

gemfibrozil *See* FIBRATES.

geminin A protein (209 aa) that inhibits DNA replication and is degraded during the mitotic phase of the cell cycle. It inhibits the formation of the prereplication complex by binding to the replication initiation factor *Cdt1 so that the minichromosome maintenance proteins are not recruited to chromatin. Geminin is ubiquitinated by the anaphase-promoting complex *in vitro* and thereby targeted for destruction.

gemins A family of proteins involved in the multiprotein *SMN complex where they apparently modulate the activity of the complex. Six of the seven gemins are involved inthe SMN complex: gemin2 (formerly SIP1), gemin3 (a DEAD-box RNA helicase), gemin4, gemin5 (a WD-repeat protein), gemin6 and gemin7, but several of the gemins are also found outside the SMN complex.

gene A surprisingly difficult term to describe because the meaning has changed with increasing knowledge and the original 'one gene, one polypeptide' definition is no longer accurate. One view is that there are now two concepts: 1. an abstract 'gene', 'something that brings about a certain phenotypic trait e.g. a 'gene for cystic fibrosis', even though it is the mutated form of the gene that is responsible for the condition. 2. A physical gene, a specific sequence of DNA (RNA in some viruses) that has a specific locus on a chromosome, can exist in different allelic forms, and is transcribed. However, since only certain parts of the nucleic acid sequence actually contribute to the product it is debatable whether regulatory sequences and *introns are part of the gene since they do not code for a product (whether an RNA or a polypeptide).

gene amplification A selective increase in the numbers of particular genes so that multiple copies of the gene exist in the cell. An example is the amplification of ribosomal RNA genes in the maturing oocyte of *Xenopus*, but amplification also occurs in some tumour cells, e.g. *MDM2* gene amplification has been demonstrated in 42% of colonic carcinomas. Regions of chromosomes that are prone to amplification may have palindromic sequences of DNA that allow the formation of hairpin loops, although there are various mechanisms for achieving amplification. An increase in the number of copies of a gene can have serious consequences, as seen in *trisomy 21.

gene chip An immobilized oligonucleotide array used to screen an RNA sample (after reverse transcription) to determine which genes are being expressed. The oligonucleotides may be specific probes for known genes (e.g. G-protein-coupled receptors) or a random set of defined sequences where the genes will be unknown but the pattern of hybridization is indicative of the expression profile.

gene cloning An experimental method for producing large numbers of copies of a gene by incorporating the sequence into a cloning vector that is then propagated.

gene conversion The conversion of one allele to another that can occur if the template for repairing a gene by homologous recombination differs slightly so that the repaired gene acquires the donor sequence. The effect becomes obvious when alleles segregate in a 3:1 not a 2:2 ratio in meiosis.

gene duplication In simple terms, the amplification of a gene so that two copies exist in the haploid genome. Provided the increase in dosage is not immediately deleterious the 'extra' copy is free to acquire new functions in future generations and through time a family of genes with related sequences but different functions can evolve.

gene expression The *transcription and usually *translation of a gene to produce either RNA or protein and thus express the phenotype encoded by the gene. Control of gene expression can occur by regulation of transcription, through control of the stability of the mRNA, regulation of translation or modification of the stability of the protein. Different tissues express different genes, and in the process of development the timing, pattern, and extent of expression are all important.

gene family (multigene family) A set of genes that have strong sequence similarity (homology), and may well have arisen through successive rounds of *gene duplication, and that code for proteins with distinct funtions. The divergence of function is probably a mark of the time that has elapsed since divergent evolution began to operate on the duplicated genes. The divergence of function is greater in a superfamily.

general adaptation syndrome The biological response to severe stress, usually subdivided into three stages: the first **alarm stage** with shock and counter-shock phases, the second **resistance stage** during which recuperation begins and, in extreme cases, the third stage, **exhaustion**, may ensue and may be fatal.

general import pore (GIP) A multisubunit pore complex (400 kDa) in the outer mitochondrial membrane through which cytoplasmically synthesized proteins pass and are then transported by the TIM23 complex (translocase of the inner membrane) into either the matrix, inner membrane, or intermembrane space. The channel protein of the outer mitochondrial membrane is Tom40 (translocase of the outer mitochondrial membrane-40) which is tightly associated with the receptor protein Tom22 along with Tom7, Tom6, and Tom5. The receptor proteins Tom70 and Tom20 are more weakly associated and are involved in the recognition of preproteins. Precursors of signal-anchored proteins, however, appear not to be inserted via the general import pore.

general paralysis of the insane An historically important disorder (general paresis of the insane) caused, it was eventually realized, by *syphilis. Diagnosis was based upon abnormalities in eye pupil reflexes (Argyll Robertson pupil) with later the development of muscular reflex abnormalities, seizures, memory impairment (dementia), and psychosis.

general receptor for phosphoinositides-1 A nucleotide exchange factor (Grp1, cytohesin-3, 400 aa) containing a *pleckstrin homology domain that has been reported to have a role as a corepressor for the formation of thyroid hormone receptor complexes on thyroid response elements of genes regulated by thyroid hormone.

generation time The population doubling time which is equivalent to the average length of the cell cycle.

gene regulatory protein Any protein that regulates the expression of a gene, although generally restricted to proteins that regulate transcription.

generic drug A drug that is no longer covered by patent and can be manufactured by any pharmaceutical company. Generic drugs are usually much cheaper and must contain bioequivalent levels of the active compound. The generic name of a drug is usually the one with a main entry in this dictionary.

gene therapy Treatment of a genetic disorder by stable *transfection of the cells of the organism with the normal gene. An extremely attractive idea, but the technical problems are considerable and only a few successes have been recorded so far (*See* ADENOSINE DEAMINASE).

genetically modified organism (GMO) In general usage, an organism that has been altered by genetic engineering rather than by normal selective breeding methods, although both are forms of genetic modification.

genetic burden *See* GENETIC LOAD.

genetic code The relationship between the sequence of bases in *nucleic acid and the sequence of amino acids in the polypeptide encoded by that DNA. Each amino acid is specified by at least one triplet of bases (a *codon), although there is degeneracy in the code and some amino acids are specified by more than one codon.

g

genetic drift A change in the frequency of particular alleles in a population which may occur randomly in a small population or one that is isolated.

genetic engineering General term for the techniques of modern molecular biology that include the methods for synthesizing novel DNA, for moving genes from one organism to another, for cloning genes, for knocking out particular genes, and so on.

Geneticin A highly toxic aminoglycoside antibiotic (G418) related to gentamicin but for which there are resistance genes. The resistance genes, though bacterial in origin, can be expressed and are effective in eukaryotic cells. Vectors are often engineered to have the resistance genes so that transformed cells are selected by their survival.

genetic linkage A characteristic of genes that are physically close so that the alleles at these loci tend to remain together (are coinherited); parental combinations of linked alleles are more likely to be found in offspring than would be expected on the basis of random assortment of characteristics. The closer two genes are, the less likely there is to be a recombination (separation) event and this is the basis for quantifying linkage. Unlinked genes show 50% recombination. *See* LINKAGE DISEQUILIBRIUM; SYNTENIC.

genetic load (genetic burden) Loosely, the cost, in terms of loss of fitness, of accumulated deleterious mutations in the population. More specifically, the average number of *recessive lethal mutations estimated to be present in heterozygous form in the genome of an individual in a population.

genetic locus The chromosomal location of a gene responsible for a particular phenotypic character or its location on a linkage map, which relates to its physical location on a chromosome. In some cases the locus for a particular disease or phenotypic trait may be known although the gene (or genes) that are responsible may not (yet) have been identified.

genetic recombination New combinations of alleles that arise as a result of *crossing-over between homologous chromosomes in meiosis or experimentally, as a result of *genetic engineering techniques. The probability of recombination is reciprocally related to *linkage.

genetic transformation *See* TRANSFORMATION.

gene transfer Insertion of allogeneic or xenogeneic genes into a cell or organism by *transfection.

genistein A soy-derived isoflavone that is an antioxidant but has a range of other activities, including interaction with oestrogen receptors (a phytoestrogen). Competes for the ATP binding site of many protein tyrosine kinases and also inhibits the mammalian hexose transporter GLUT1 and topoisomerase-II. A potent vasorelaxant, probably by binding and opening chloride channels (*CFTR and *NKCC1).

genitourinary Pertaining to the genital and the urinary systems.

genome The complete set of genes carried by an individual or cell. The Human Genome Project is a coordinated programme to identify the full set of human genes, sequence the full DNA complement, and produce a database together with the tools to manage, interrogate, and annotate the information. The first draft was produced by 2001 and a 'complete' version by 2003, but it continues to be refined and, increasingly, function assigned to the genes, of which there seem to be approximately 20 000–25 000, rather fewer than anticipated. Comparable genome projects for other organisms are complete or in progress and comparative studies are increasingly possible.

SEE WEB LINKS

- The Human Genome Project homepage.

genomic clone A clone derived from genomic DNA that includes the nontranscribed sequence with promoter sites, introns, etc.in contrast to clones generated by reverse transcription from mRNA. The latter are, however, known to be transcribed.

genomic imprinting The pattern of gene expression or repression that is transmitted from parent to offspring so that function may depend upon whether it is a maternal or paternal copy.

Silencing of imprinted genes, often in clusters (*See* DNA METHYLATION), is brought about by *imprinting control regions (ICRs) that are set differently during gametogenesis in male and female germ lines. Increasing numbers of imprinting effects are being identified, including several of importance in the pathology of diseases, e.g. altered paternal expression of genes within the 15q11–13 region giving rise to *Prader–Willi syndrome and altered maternal expression of the *UBE3A* (ubiquitin protein ligase E3A) gene causing *Angelman's syndrome. Since imprinting is an epigenetic phenomenon the effects are not deducible from the genomic sequence, but imprinting maps for the mouse have been produced.

SEE WEB LINKS

- Imprinting map for the mouse genome:

genomic instability The phenomenon in which there is an abnormal rate of genetic change in a cell population which becomes evident as proliferation continues. It is a feature of some rapidly proliferating tumours and may be a consequence of incompetent repair mechanisms; mutation in these systems gives rise to a range of instability syndromes including *ataxia telangiectasia, *Nijmegen breakage syndrome, *Werner's syndrome, *Bloom's syndrome, *Fanconi's anaemia, *xeroderma pigmentosa, and *Cockayne's syndrome. Associated with increased predisposition to tumours is a tendency to immunodeficiency because DNA breakage and repair is part of the mechanism for generating immunological diversity. A delayed response to ionizing radiation may occur if there is mutation in a repair system that maintains normal genomic stability.

SEE WEB LINKS

- Review on genomic instability syndromes.

genomic islands Mobile genetic elements 10–200 kb in size that have become integrated and fixed. They are found in many bacteria where they may code for a range of functions, e.g. the multidrug-resistance phenotype in *Salmonella* that depends upon an antibiotic resistance gene cluster in a 43-kb genomic island named salmonella genomic island 1 (SGI1). They allow the rapid spread of a trait between phylogenetically distinct organisms. Islands often have an atypical G+C content, are frequently associated with tRNA-encoding genes and tend to be flanked by repeat structures.

genomic library A collection of DNA sequences incorporated in vectors that allow it to be propagated (a DNA library) and that contains genomic DNA fragments rather than DNA generated from mRNA by reverse transcription (cDNA). Because of the size of the fragments they generally require different vectors such as *bacterial artificial chromosomes, *yeast artificial chromosomes, or *cosmids.

genotoxin A toxin that acts on DNA, often by forming an adduct, and disrupts gene function. Environmentally important examples are aromatic hydrocarbons, aflatoxin, and vinyl chloride.

genotype The genetic constitution of an organism, as opposed to the expressed features, the *phenotype.

gentamicins *Aminoglycoside antibiotics produced by Gram-positive *Micromonospora* spp., and effective against Gram-negative bacteria although there is a risk of nephrotoxicity and neurotoxicity, especially in the auditory system.

gephyrin A cytoplasmic peripheral membrane scaffold protein (736 aa with smaller splice variants) that anchors the inhibitory glycine receptor and a subset of $GABA_A$ receptors to synapses by interacting with the cytoskeleton. Gephyrin has microtubule-binding activity but also links indirectly to the actin cytoskeleton and may be active in the clustering process as well as involved in anchoring receptors once they are localized. Mutations are associated with *molybdenum cofactor deficiency syndrome. Gephyrin negatively regulates *collybistin.

geranyl The geranyl group is geraniol (a monoterpenoid) lacking the terminal -OH, a prenyl group that as geranyl pyrophosphate is an intermediate in the HMG-CoA reductase pathway which leads to the synthesis of farnesyl pyrophosphate, geranylgeranyl pyrophosphate and cholesterol. The geranyl-geranyl group is post-translationally added by geranyl transferase (farnesyl pyrophosphate synthetase; EC 2.5.1.10, 419 aa) to some cytoplasmic proteins, generally at a CAAX motif (where X is usually leucine) and serves as a membrane attachment device. *See* FARNESYLATION.

GERL Common abbreviation for the *Golgi–endoplasmic reticulum–lysosome system.

germ cell nuclear factor A nuclear orphan receptor (GCNF, retinoid receptor-related testis-associated receptor, RTR, 480 aa), encoded by the *NR6A1* gene, that functions as a transcriptional repressor, mainly in the germ cells of testis in adults. It is reported to be a negative regulator of the peroxisome proliferator-activated receptor δ (PPARδ) gene, represses several pluripotency genes and is transiently expressed in mammalian carcinoma cells during retinoic acid-induced neuronal differentiation.

germ cells The cells that give rise to *haploid gametes. The germ cell line is formed very early in embryonic development and is 'protected' from the hazards of multiple cell division required to generate the soma. *See* WEISMANN'S GERM PLASM THEORY.

germinal centre Morphologically distinct regions that develop within lymphoid organs following activation of B cells. They contain large numbers of antigen presenting cells (follicular dendritic cells) as well as proliferating B cells (centroblasts), some nondividing B cells (centrocytes) and a few T cells. Plasma cells and memory B cells are generated in the germinal centres. **Germinal centre kinase** (mitogen-activated protein kinase kinase kinase kinase, MAP4K2, EC 2.7.11.1, 820 aa) is a STE20-like serine/threonine kinase that is activated by TNFα, activates the SAPK pathway, and will inactivate CLH-3 chloride channel. The germinal centre kinases (GCKs) are a subfamily of STE20-like kinases that include *MINK1, Traf2-, and Nck-interacting kinase (TNIK), *SOK1. **Germinal centre kinase-like kinase** (HGK, hepatocyte progenitor kinase-like kinase, MAP4K4, 1239 aa) activates *JNK and is upstream of *mTOR.

germ layers In mammals, and most multicellular animals with the exception of coelenterates, the three main tissue types that are recognizable from the early stages of gastrulation; ectoderm, mesoderm, and endoderm. The ectoderm gives rise to external epithelia and to the nervous system, the mesoderm to muscle, connective tissue, and the cardiovascular, blood, and immune systems. Endoderm produces internal epithelia that line the intestinal tract and to organs such as liver and lung, although some tissues contain elements derived from two layers.

germ-line therapy Gene therapy in which new genes are engineered into *germ cells and the changes are therefore hereditable. Generally abjured on the grounds of ethics and risk. *See* GERM-LINE TRANSFORMATION.

germ-line transformation The procedure by which *transgenic organisms are produced. The transforming DNA is microinjected into an early embryo, becomes incorporated into the *germ cell line and is stably inherited in subsequent generations. Successful transformation is often marked by the expression of a reporter gene associated with the tranforming vector.

German measles *See* RUBELLA.

gerodermia osteodysplastica An autosomal recessive disorder characterized by wrinkly skin and osteoporosis caused by loss-of-function mutations in the *golgin SCYL1BP1, which is highly expressed in skin and osteoblasts. SCYL1BP1 interacts with *rab6.

Gerstmann–Straussler–Scheinker syndrome A familial *spongiform encephalopathy caused by mutation in the *prion protein gene. Transgenic mice with the mutant form of the prion develop a degenerative brain disease similar, but not identical, to *scrapie.

GFAP *See* GLIAL FIBRILLARY ACIDIC PROTEIN.

GFP *See* GREEN FLUORESCENT PROTEIN.

GFRα Glial cell line-derived neurotrophic factor (GDNF)-family receptor α. *See* GDNF.

GGT **1.** Galactosyl-hydroxylysine glucosyltransferase. **2.** *See* GAMMA-GLUTAMYLTRANSFERASE.

ghrelin A peptide hormone (28 aa) that is the endogenous ligand for the G-protein-coupled growth hormone secretagogue (GHS) receptor. It is released from the secretory granules of X/A-like cells found in the submucosal layer of the stomach; levels rise in fasting animals and apparently signal the necessity for increased metabolic efficiency as well as stimulating appetite. Ghrelin is derived from preproghrelin, which also generates a second peptide called *obestatin. Variants of ghrelin are associated with susceptibility to obesity.

GHRH *See* GROWTH HORMONE.

GI **1.** Common abbreviation for gastrointestinal (GI). **2.** Gi: *see* GTP-BINDING PROTEIN. **3.** (glycaemic index) *See* GLYCAEMIC.

giant axonal neuropathy A severe autosomal recessive sensorineural neuropathy in which both peripheral nerves and the central nervous system are affected. There are characteristic accumulations of neurofilaments apparently due to mutation in *gigaxonin.

giant axons Very large unmyelinated axons found in invertebrates. The squid giant axon, which is almost 1 mm in diameter, has high conduction speed and has been a popular experimental system.

giantism (gigantism) Excessive growth of the body, usually due to overproduction of growth hormone by the pituitary. Can arise because of a benign pituitary tumour or, occasionally, multiple endocrine neoplasia type 1 (MEN-1), McCune-Albright syndrome (MAS), neurofibromatosis, or Carney complex.

Giardia lamblia A flagellate protozoan that is the cause of acute or chronic diarrhoea in humans (giardiasis), often associated with sulphurous breath. Infection is usually from contaminated water.

Giemsa An improved Romanovsky-type stain that contains both methylene blue (basic) and eosin (acidic) and will therefore differentiate acid and basic granules in granulocytes. Often used to stain blood films that are suspected to contain protozoan parasites. Also used for staining chromosomes (*G banding).

gigaxonin A protein (597 aa) that is thought to be important in the crosstalk between the intermediate filament and microtubule networks. One of the cytoskeletal BTB/kelch (broad-complex, tramtrack, and bric-a-brac) repeat family. It controls protein degradation and is essential for neuronal function and survival. Gigaxonin binds to the ubiquitin-activating enzyme E1 and interacts with microtubule-associated protein-1B. Mutations lead to *giant axonal neuropathy.

Gilbert's syndrome A benign disorder (hyperbilirubinemia 1) causing mild jaundice resulting from mutation in the UDP-glucuronosyltransferase gene (*UGT1A1*) that encodes a key enzyme for glucuronidation, an important mechanism to enhance the elimination of many lipophilic xenobiotics and endogenous substances such as bilirubin, leading to reduced clearance of bilirubin. In Gilbert's syndrome there are reduced levels of the enzyme; more severe deficiencies lead to Crigler–Najjar syndrome types I and II in which higher serum levels of bilirubin are found. *See* IRINOTECAN.

gingivitis Inflammation of the gums, usually induced by bacterial biofilms (plaque) on the surfaces of teeth.

GINS complex A multiprotein complex (go-ichi-ni-san complex, 100 kDa) essential for the initiation and elongation phases of genome duplication. The four subunits, Sld5, Psf1 (partner of Sld5-1), Psf2, and Psf3, are ubiquitous and evolutionarily conserved in eukaryotic organisms. The complex mediates the loading of the enzymatic replication machinery at a step after the action of the S-phase kinases and may be a part of the replication machinery itself.

GIP **1.** *Gastric inhibitory polypeptide which is the same as glucose-dependent insulinotropic polypeptide. **2.** *See* GENERAL IMPORT PORE. **3.** GTPase inhibitory protein, a protein that interferes with the binding of *GAP to ras, or enhances nucleotide exchange (a *GEF). **4.** *See* GIP2.

gip2 A putative oncogene that encodes GTPase-deficient α subunits of small inhibitory G proteins (Gi2) and appears to be tissue-selective. Has been identified in tumours of the ovary and adrenal cortex, although more recent reports suggest it is not found in ovarian granulosa tumours and its status is uncertain.

girdin A protein (girders of actin filament, coiled-coil domain-containing protein 88A, Akt phosphorylation enhancer, 1871 aa) that enhances phosphoinositide 3-kinase (PI3K)-dependent phosphorylation and kinase activity of AKT1/PKB, though not itself a kinase. It is essential for the integrity of the actin cytoskeleton and for cell migration, is required for the formation of actin stress fibers and lamellipodia, and may be involved in membrane sorting in the early endosome.

GIRKs A family of homo- and hetero-oligomeric potassium channels (G-protein-gated inward rectifying potassium channels) composed of different subunits (GIRK1–4 in mammals). GIRK4 (KIR3.4) and GIRK1 (KIR3.1) are found mainly in the atrium, whereas neuronal cells predominantly express the GIRK1, GIRK2, and GIRK3 isoforms. They are activated through direct interactions of their cytoplasmic N- and C-terminal domains with G protein β-γ subunits. The anaesthetic halothane is thought to interact with these channels. A point mutation in the *GIRK2* gene in *weaver (wv) mutant mice causes neurological and reproductive defects.

GITR One of the tumor necrosis factor receptor superfamily (glucocorticoid-induced tumor necrosis factor receptor, TNFRSF18, 241 aa) involved in the regulation of T-cell receptor-mediated cell death.

GLA **1.** γ-linolenic acid. **2.** Galactosidase-α (EC 3.2.1.22, GLA, 429 aa), the gene for which is defective in *Fabry's disease. **3.** Gla proteins are found in bone (bone Gla protein or *osteocalcin) and matrix (matrix Gla protein). Gla proteins are post-translationally modified by glutamate carboxylase, which mediates the conversion of glutamate to γ-carboxyglutamate (Gla) with vitamin K as a cofactor. The γ-carboxylation of the Gla proteins is essential for the proteins to attract calcium ions and to incorporate these into hydroxyapatite crystals. The **gla-domain**, none to twelve residues of Gla distributed over a ~45-aa peptide sequence, is found in vitamin

K-dependent coagulation plasma proteins where it is important in binding calcium.

gland General term for an organ that is specialized for secretion. Although the geometry is usually complex, glands are basically epithelial sheets arranged either as an acinus with a duct or as a tubule. If the secreted product is released to a free surface the gland is exocrine, if it is released directly into the circulation the gland is endocrine.

glanders *See* MELIOIDOSIS.

glandular fever (infectious mononucleosis) A disease, often contracted in late adolescence, caused by infection with *Epstein–Barr virus and characterized by the presence of many large lymphoblasts in the circulation.

Glanzmann's thrombasthenia An autosomal recessive bleeding disorder in which platelet aggregation is deficient because the β_3 integrin (glycoprotein IIb/IIIa) is absent or present at reduced levels. The disorder can be caused by mutation in the gene of either one of the two glycoproteins that constitute the fibrinogen receptor. A possible autosomal dominant form has been reported.

glatiramer An immunomodulatory drug, a random polymer of four amino acids that are found in *myelin basic protein, used to reduce the frequency of relapses in relapsing-remitting *multiple sclerosis. The mechanism of action is unclear, although it may be acting as a decoy. *TN* Copaxone.

glaucoma A generic term for a range of disorders of the eye in which the intraocular pressure rises, causing damage to optic nerve fibres. It is the major cause of blindness after the age of 45. The commonest form is **primary open angle glaucoma** (POAG), characterized by open anterior chamber angles (channels that link the aqueous and vitreous chambers), visual field abnormalities, and raised pressure. The condition is multifactorial and a range of factors probably combine to cause the problem. Mutations in various genes can give rise to the condition, including mutations in *myocilin, optineurin, WD repeat-containing protein-36, and at a number of loci where the gene is not yet identified. Susceptibility to **normal tension glaucoma** in which the intraocular pressure is within normal range is associated with a particular intronic polymorphism of the *optic atrophy-1 gene and mutation in the optineurin gene. **Acute narrow angle glaucoma** occurs in individuals with a reduced space between cornea and iris: as the lens increases in size with age the angle becomes blocked and pressure can rise. *Nail–patella syndrome has open angle glaucoma as one of many clinical features.

GlcNAc-phosphotransferase A multisubunit enzyme (*N*-acetylglucosamine-1-phosphotransferase, EC 2.7.8.15, 2 α, 2 β, and 2 γ subunits) which catalyses the first step in the synthesis of the mannose 6-phosphate determinant required for efficient intracellular targeting of newly synthesized lysosomal hydrolases. The α and β subunits are coded on a single cDNA derived from a gene on chromosome 12, which is proteolytically processed to generate two peptides. The γ subunits are coded on a different gene on chromosome 16. Deficiency in the enzyme leads to *mucolipidosis.

Gleevec (Glivec) *See* IMATINIB MESYLATE.

glial cells (neuroglia) Non-neuronal cells found in nervous tissue where they perform a variety of support functions. **Microglia** constitute about 15% of cells in the nevous system and are a macrophage subset; the **macroglia** are subdivided into *astrocytes, *oligodendrocytes, *ependymal cells, and *radial glia in the central nervous system and *Schwann cells in the peripheral system.

glial fibrillary acidic protein An intermediate filament protein (GFAP) found in *astrocytes where it polymerizes to form glial filaments. Mutations in the gene cause *Alexander's disease.

glibenclamide (glyburide) *See* SULPHONYLUREAS.

glicentin An N-terminal peptide (69 aa) derived from proglucagon by post-translational processing in the intestine. The *glucagon sequence is within glicentin but the C-terminal region of glicentin is *oxyntomodulin and there is an N-terminal 'glicentin-related pancreatic peptide' extension. Although glicentin has physiological effects these may well be glucagon-mediated and no glicentin receptor has been identified.

gliclazide A *sulphonylurea, *TN* Diamicron MR.

glioblastoma An aggressive primary brain tumour (glioblastoma multiforme) derived from glial cells. *See* GLIOMAS.

gliomas A general term for tumours that arise from neuroglia and includes astrocytomas, oligodendroglioma, and ependymoma derived from *astrocytes, *oligodendrocytes, and *ependymal cells respectively. Although there is local invasion they do not metastasize. **Glioblastoma** is an aggressive form of glioma. Gliomas are known to occur in association with several well-defined hereditary tumour syndromes such as *neurofibromatosis, *tuberous sclerosis, *Li–Fraumeni syndrome, and *Turcot's syndrome but many other loci are implicated. **Gliomatosis cerebri** (infiltrative diffuse astrocytosis) is a rare primary brain tumour in which there is diffuse infiltration of the brain with neoplastic glia.

***gli* oncogenes** Oncogenes (glioma-associated oncogenes) that encode zinc finger transcription factors of the Krüppel family. **GLI1** (1106 aa) is colocalized in cytoplasm with *SUFU. In mice three related proteins, Gli1, Gli2, and Gli3, have been implicated in the transduction of sonic hedgehog signalling. A variety of disorders are associated with mutations in the Gli transcription factors, including foregut malformations. Mutations in **gli2** are associated with basal cell carcinomata and *holoprosencephaly, mutations in **gli3** are found in *Pallister–Hall syndrome and *Greig cephalopolysyndactyly syndrome.

gliostatin *See* THYMIDINE PHOSPHORYLASE.

glipizide A *sulphonylurea, *TN* Glucotrol.

Glivec *See* IMATINIB MESYLATE.

globin The protein part of haemoglobin. In the adult human two α (141 aa) and two β (146 aa) globin chains, each associated with a haem moiety, associate to form functional haemoglobin. Mutant β-globin causes *sickle cell anaemia and other defects in globins cause *thalassemia.

globoside A glycosphingolipid (cytolipin K) with *N*-acetylgalactosamine as the side chain, found in kidney and erythrocytes. It can be broken down by β-hexosaminidase and mutations in this enzyme lead to *Sandhoff's disease.

globular protein Any protein that adopts a compact, roughly spherical morphology in solution or within a membrane.

globus pallidus *See* BASAL GANGLIA.

glomerulonephritis (glomerulitis) An inflammatory response in the kidney glomerulus that can arise from a variety of primary or secondary immune responses. A general problem is that immune complexes cannot traverse the basement membrane of the fenestrated epithelium and are trapped, leading to the accumulation of neutrophils on the immune complexes, which may also activate complement, further exacerbating the inflammatory response. The immune complexes are irregularly distributed, in contrast to the picture in *Goodpasture's syndrome. In Berger's nephropathy (IgA disease), the immune complexes involve IgA; *Henoch–Schönlein purpura is a variant of IgA disease.

glomulin A 48-kDa protein (FKBP-associated protein) that interacts with *FK506 binding proteins and is associated with *glomus tumours. Binding to the *immunophilin is inhibited by FK506 and rapamycin in a dose-dependent manner and glomulin may be the natural ligand.

glomus tumour **1.** Benign cutaneous neoplasms (glomuvenous malformations, venous malformations with glomus cells, glomangioma) with rounded glomus cells (derived from modified perivascular smooth muscle cells) around distended vein-like channels. Caused by mutations in the *glomulin gene and inherited as an autosomal dominant disorder. **2.** A name sometimes given to *multiple paragangliomata.

gloomy face syndrome *See* 3M SYNDROME.

gloss- A prefix indicating association with the tongue. Thus **glossitis** is inflammation of the tongue, **glossodynia** is pain in the tongue, and **glossoplegia** is paralysis of the tongue.

GLP-1 **1.** A transcriptional activator that is a *GARP homologue found in the protozoan *Giardia lamblia*. **2.** One of the *glutaredoxin-like proteins. **3.** (glucagon-like peptide 1) *See* GLUCAGON.

glucagon A peptide hormone (29 aa) produced by the alpha cells (*A cells) of pancreatic islets in response to a fall in blood sugar levels. It induces *hyperglycaemia and opposes the action of insulin in peripheral tissues. In the liver the insulin/glucagon ratio determines the rates of gluconeogenesis and glycogenolysis. A single glucagon gene encodes proglucagon, which is processed in a tissue-specific manner to produce glucagon in the pancreas, *glicentin, *oxyntomodulin, and glucagon-like peptides in the intestine, with an intermediate profile of peptides liberated in the central nervous system. The glucagon receptor (477 aa) signals through adenylate cyclase and also mediates an increase in intracellular calcium. Polymorphisms in the receptor are observed with higher frequency in some but not all diabetic populations. **Glucagon-like peptides 1 and 2** (GLP-1, GLP-2) are the products of the same gene and are 36-aa hormones produced by *intestinal L cells that act to augment insulin secretion.

SEE WEB LINKS

- Web resource for the study of glucagon-like peptides.

glucans Polysaccharides with glucose subunits that include cellulose, *callose, *laminarin, starch, and *glycogen.

glucocorticoid receptor interacting protein 1 A member of the steroid receptor coactivator (SRC) family of transcriptional regulators (GRIP1, nuclear receptor coactivator-2, 1128 aa) that interacts with steroid receptors once they have bound ligand and moved from the cytoplasm to the nucleus; the complex activates transcription. GRIP1 interacts with all five steroid receptors and class II nuclear receptors, including thyroid receptor α, vitamin D receptor, retinoic acid receptor α, and retinoid X receptor α. Degradation of GRIP1 by the 26S proteosome may regulate its availability.

glucocorticoids A group of steroid hormones that promote *gluconeogenesis, in contrast to *mineralocorticoids and sex steroids which are similar but have different receptors, targets, and effects. They also have important anti-inflammatory and immunosuppressive activity and a range of potent synthetic derivatives have been produced. The reference compound is hydrocortisone (cortisol); dexamethasone is 25–80 times more potent, other common examples are cortisone, prednisone, prednisolone, betamethasone, and triamcinolone. The **glucocorticoid receptor** (nuclear receptor subfamily-3, group-C, member-1; NR3C1, 777 aa) is a ligand-activated transcription factor that is ubiquitously distributed. The nonactivated form is cytoplasmic and complexed with Hsp90, Hsp70, FK binding protein, and other proteins: once cortisol binds it is released and is transported into the nucleus where it binds to specific response elements. Mutations in the gene give rise to familial glucocorticoid resistance.

gluconeogenesis The production of glucose from noncarbohydrate precursors, such as pyruvate, amino acids, and glycerol, rather than by hydrolysis of glycogen. Mostly occurs in the liver and maintains blood glucose under conditions of starvation or intense exercise.

glucosamine An amino-sugar that occurs most frequently in the acetylated form (β-D-*N*-acetylglucosamine). It is a component of *chitin, *heparan sulphate, *chondroitin sulphate, and other complex polysaccharides. Its role in connective tissues is the basis for its use as a dietary supplement and it is claimed to be beneficial in treating osteoarthritis.

glucosaminoglycan *See* GLYCOSAMINOGLYCAN.

glucose (dextrose) A six-carbon sugar (aldohexose), the main product of photosynthesis, that is a major energy source for metabolic processes. It is broken down by glycolysis, the three-carbon products of which feed into the *TCA cycle. It is stored in plants as starch, in animals as glycogen, and in microorganisms as *dextran. *See* GLUCONEOGENESIS. Glucose 1-phosphate is produced when glycogen is hydrolysed by phosphorylase and is converted by phosphoglucomutase to glucose 6-phosphate which is an intermediate in the glycolytic pathway (where it is next converted to fructose 6-phosphate), and of the NADPH-generating *pentose phosphate pathway.

glucose-6-phosphate dehydrogenase An ubiquitous enzyme (EC 1.1.1.49, 515 aa), present in prokaryotic and eukaryotic cells, that catalyses the first step in the in the hexose monophosphate pathway, using NADP as a cofactor and generating NADPH. *See* HEXOSE MONOPHOSPHATE SHUNT. The hexose monophosphate pathway is the only NADPH-generating process in mature erythrocytes and deficiency (X-linked) is the basis of *favism, primaquine sensitivity, and some other drug-sensitive haemolytic anaemias, anaemia and jaundice in the newborn, and chronic nonspherocytic haemolytic anaemia. Deficiency may, however, confer some resistance to malaria and may account for this being the most common enzyme deficiency in the world. Many polymorphisms in the enzyme have been described and, being X-linked, can be used as a marker of which X chromosome is active in single cells derived from a heterozygous female (used to confirm the *Lyon hypothesis).

glucose-dependent insulinotropic polypeptide (GIP) *See* GASTRIC INHIBITORY POLYPEPTIDE.

glucose-regulated protein (GRP94) *See* ENDOPLASMIN.

glucose transporter Any protein that transports glucose. In mammals they are a family of twelve-transmembrane integral transporters. **GLUT1** is ubiquitously expressed with particularly high levels in erythrocytes and in the endothelial cells in the brain. **GLUT2** is a low-affinity transporter present in liver, intestine, kidney, and pancreatic beta cells. **GLUT3** is expressed primarily in neurons and GLUT1 in conjunction with GLUT3 allows glucose to cross the blood–brain barrier and enter neurons. The

GLUT4 isoform is the major insulin-responsive transporter, predominantly restricted to striated muscle and adipose tissue. Defects in GLUT-4 may be associated with some forms of *diabetes. **GLUT5** is a fructose transporter, expressed in insulin-sensitive tissues (skeletal muscle and adipocytes) of humans and rodents. High-fructose diets lead to up-regulation of the transporter. **GLUT6** is a pseudogene on chromosome 5. **GLUTX1** appears to be important in early blastocyst development.

glucosuria *See* GLYCOSURIA.

glucosylation The transfer of glucose residues, usually from the nucleotide-sugar derivative uridine UDP-glucose to a target molecule by a glycosyl transferase. Enzymic glucosylation of collagen is a standard post-translational modification that generates the glucosyl-galactosyl moiety on hydroxylysine residues. Nonenzymic glucosylation of proteins has been speculatively linked with ageing (*See* AGE) and the level of glucosylation of serum proteins can provide a sensitive, short-term integrative measure of blood glucose homeostasis in diabetes. Glucosylation is also a step in the synthesis of a variety of small molecules to produce glucosides (e.g. the glucosylation of scopoletin to scopolin).

glucuronic acid (GA, GlcA) An acid formed by oxidation of the OH group of glucose in position 6. D-Glucuronic acid is a subunit in many oligosaccharides.

glucuronidase ((beta-glucuronidase) A tetrameric glycoprotein (EC 3.2.1.31) composed of identical 651 aa subunits that is involved in the degradation of glucuronic acid-containing glycosaminoglycans. Often used as a *reporter gene. Mutations in the gene lead to mucopolysaccharidosis type VII (Sly's syndrome), an autosomal recessive lysosomal storage disease. *See* EGASYN.

GLUT1, GLUT2, etc. *See* GLUCOSE TRANSPORTER.

glutamate An *excitatory amino acid that is involved in fast excitation in the *central nervous system. Excessive dietary intake can cause Chinese restaurant syndrome. The **glutamate receptors** belong to a superfamily of amino acid receptors and are implicated in many important brain functions including *long-term potentiation (LTP). At least four major glutamate-gated *ion channel subtypes are distinguished on pharmacological grounds, named after their most selective agonists: *N*-methyl-D-aspartate (NMDA), implicated in memory and learning, neuronal cell death, ischemia, and epilepsy; *kainic acid (KA); quisqualate/AMPA, and L-2-amino-4-phosphobutyrate (APB). A fifth glutamate receptor subtype with the selective agonist *trans*-1-amino-cyclopentane 1,3 dicarboxylate, (APCD), differs in being G-protein coupled. **Glutamate receptor interacting proteins** (GRIPs, ~1128 aa) contain multiple *PDZ domains and through them interact with AMPA-type glutamate receptors, and may be responsible for targeting them to the postsynaptic terminal of excitatory synapses.

glutamate carboxypeptidase II A membrane-associated zinc metalloenzyme (GCP-2, NAAG peptidase, EC 3.4.17.21, 750 aa) expressed in many tissues, not just brain, that catalyses the hydrolysis of the neurotransmitter *N*-acetyl-L-aspartyl-L-glutamate (*NAAG) to *N*-acetyl-L-aspartate and L-glutamate; it also hydrolyses polyglutamylated folates in the intestine (and is thus also folylpolyglutamate carboxypeptidase, FGCP) and is identical to prostate-specific membrane antigen (PSM). GCP-2 inhibitors have been shown to provide some neuroprotection in preclinical models of stroke, amyotrophic lateral sclerosis, and neuropathic pain.

glutamate decarboxylases The enzymes that catalyse the production of the neurotransmitter *GABA (glutamate decarboxylase 1, GAD67, EC 4.1.1.15, 594 aa; glutamate decarboxylase 2, GAD65, 585 aa). Defects in the glutamate decarboxylase-1 gene are responsible for autosomal recessive symmetric spastic *cerebral palsy.

glutamate pyruvate transaminase *See* ALANINE AMINOTRANSFERASE.

glutamic acid (Glu, E) An α-amino acid (147 Da) that is commonly found in proteins and is important in amino acid metabolism, being a precursor of *glutamine, *proline, and *arginine. *Glutamate is a excitatory neurotransmitter and the product of its decarboxylation is the inhibitory neurotransmitter *GABA.

glutamic peptidases A subset of peptidases formerly termed 'pepstatin-insensitive carboxyl proteinases', fungal endopeptidases.

glutamine (Gln, Q) An amino acid (146 Da) commonly found in proteins, the amide at the γ-carboxyl of *glutamate. Crosslinks between glutamine and lysine side groups, formed by *transglutaminase, are important in stabilizing the structure of protein aggregates such as fibrin. Tissue culture media for animal cells are commonly enriched tenfold for glutamine, the excess presumably acting as a general carbon source.

glutaraldehyde A fixative used especially for electron microscopy. It forms crosslinks between protein by the interaction of its two aldehyde groups with amino groups of the proteins. Glutaraldehyde-fixed proteins are less likely to retain their native configuration than formaldehyde-fixed proteins.

glutaredoxins A ubiquitous family of proteins (thiol transferases) which catalyse the reduction of disulphide bonds in their substrate proteins by use of reduced glutathione. Glutaredoxin-like proteins are diverse and their functions are unclear, but various forms have been found in prokaryotes, *Plasmodium*, and higher plants.

glutarimide antibiotics A class of antibiotics that block peptide chain elongation by interacting with the 60S ribosomal subunit. They include *cycloheximide, acetoxycycloheximide, streptimidone, streptovitacins, inactone, epiderstatin, actiketal, and dorrigocin.

glutathione A tripeptide, γ-glutamylcysteinylglycine, that is the main sulphydryl buffer in animal cells. Glutathione participates in *leukotriene synthesis and is a cofactor for **glutathione peroxidases** (EC 1.11.1.9, ~200 aa) detoxifying enzymes that eliminates hydrogen peroxide and organic peroxides. The active site has a selenocysteine residue. The reaction causes glutathione (GSH) to become oxidized to glutathione disulphide (GSSG) which is reduced to GSH by **glutathione reductase** (EC 1.8.1.9, 754 aa) in an NADH-dependent reaction. The GSH/GSSG ratio in the cytoplasm is maintained at 500:1 which tends to destabilize disulphide bonds. **Glutathione synthase** (GSS; EC 6.3.2.3, 474 aa) is responsible for synthesis of glutathione from ATP, γ-L-glutamyl-L-cysteine, and glycine. A deficiency in GSS causes 5-oxoprolinuria, an autosomal recessive disorder characterized by urinary excretion of 5-oxoproline, metabolic acidosis, haemolytic anaemia, and central nervous system damage. Severe forms of the deficiency are fatal in the neonatal period.

glutathione *S*-transferases A family of enzymes (GSTs, EC 2.5.1.18) that catalyse the coupling of glutathione to many xenobiotics as the first step in their detoxification. There are four main classes of soluble human GSTs: α, μ, π, and θ with minor classes κ and σ. The most active GSTs of liver are the products of two autosomal loci, GST1 (GSTμ) and GST2 (GSTα) both of which are polymorphic. Glutathione *S*-transferase-π (GSTπ; GST3) is overexpressed in some multidrug-resistant cells, enabling rapid detoxification of drugs. GST is often used experimentally to enable one-step purification of expressed proteins. The gene of interest is engineered as a GST fusion protein, expressed, and then the bacterial lysate is passed down a glutathione affinity column and then eluted with free glutathione. The GST is then proteolytically cleaved.

gluten A mixture of the proteins gliadin and glutenin associated with starch in the grain of various grasses but particularly wheat and barley. The mechanical properties of dough depend on the gluten content and gluten is a major protein source for many people. *Coeliac disease is an allergic response to gluten.

glyburide *See* SULPHONYLUREAS.

glycaemic Describing something that causes blood sugar concentrations to rise. The **glycaemic index** (GI) is a ranking of foods based on their overall effect on blood glucose levels. Food that is absorbed slowly has a low glycaemic index and extensive listings are available. The **glycaemic load** (GL) of a specific food portion is calculated by taking the percentage of carbohydrate content per portion and multiplying it by its glycaemic index value.

SEE WEB LINKS

- Links to information on glycaemic index values for foods.

glyceraldehyde-3-phosphate dehydrogenase The enzyme (GAPDH, GAPD, G3PD, EC 1.2.1.12, 335 aa) that catalyses the reversible oxidative phosphorylation of glyceraldehyde 3-phosphate, the three-carbon intermediate of the glycolytic pathway formed by the cleavage of fructose 1,6-bisphosphate by *aldolase. GAPDH interacts with various elements of the *cytoskeleton, and with the trinucleotide repeat in the *huntingtin gene. A testis-specific isoform (GAPD2) has been identified. Genetic variation in members of the *GAPD* gene family may be risk factors in Alzheimer's disease.

glycerination An experimental technique used to permeabilize the plasma membrane by incubation in aqueous glycerol at low temperature. Glycerinated muscle will contract if exogenous ATP and calcium are added.

glycerol (glycerine) A sugar alcohol that is used in formulation of medicines (tinctures), as an additive in cosmetics and foods, as a cryoprotectant, and in various synthetic application. It is the core structural component of the major classes of biological lipids, *triglycerides, and phosphatidyl phospholipids.

glycerone The simplest ketose (1,3-dihydroxy-2-propanone, dihydroxyacetone, DHA) that in its phosphate form, dihydroxyacetone phosphate (DHAP), takes part in glycolysis. Used as a tanning product because it reacts chemically with keratin in a Maillard-like reaction to produce brownish pigments, melanoidins, in the outer layers of the skin.

glyceryl trinitrate (nitroglycerine) A short-acting vasodilator that is used in the treatment of angina pectoris. It is a prodrug and generates *nitric oxide (endothelial-derived relaxing factor).

glycine (Gly, G) The simplest amino acid (75 Da), a common residue in proteins, especially collagen and elastin, and also a major inhibitory *neurotransmitter. The **glycine receptor** is a chloride-channel forming receptor composed of five subunits (four α and one β) surrounding a central pore, each subunit having four alpha-helical transmembrane segments. It can be activated by glycine, β-alanine, and taurine, and is selectively blocked by *strychnine. Congenital *stiff person syndrome (hereditary hyperexplexia) is caused by mutations in the α_1 subunits, and mutations in the gene can also cause *familial startle disease.

glyco- Prefix indicating association with sugars or sugar residues.

glycocalyx An obsolete term for the carbohydrate-rich region adjacent to the plasma membrane as visualized by electron microscopy. The carbohydrate is covalently attached to lipids (*See* GLYCOSPHINGOLIPIDS) or integral membrane glycoproteins.

glycocholate The anion of glycholic acid (*N*-cholylglycine) found in bile. Usually present as the sodium salt and important for emulsifying dietary fats, acting as a detergent.

glycoconjugate Any biological macromolecule with a carbohydrate moiety. A generic term for *glycolipids, *glycoproteins, and *proteoglycans.

glycodelin A protein (progestagen-dependent endometrial protein, pregnancy-associated endometrial α_2-globulin, placental protein-14, 162 aa) secreted by the human endometrium under the influence of progesterone but also found, in a differently glycosylated form, in seminal plasma. It is highly homologous to β-*lactoglobulin and may be angiogenic.

glycogen The major glucose-storage polysaccharide of animals, very abundant in liver and muscle. It is a branched D-glucose polymer, mostly α(1–4)-linked, but α(1–6) linked at branch points. The size varies but there may be up to 105 glucose units. The linear portions are metabolized by glycogen *phosphorylase but *glycogen debranching enzyme and an α(1–6) glucosidase are needed to break down branch regions. Glycogen has a characteristic star-like appearance in electron micrographs.

glycogen debranching enzyme A monomeric protein (1515 aa) with two catalytic activities, amylo-1,6-glucosidase (EC 3.2.1.33) and 4-α-glucanotransferase (EC 2.4.1.25), with different catalytic sites. Both activities are necessary to break down branched regions of *glycogen. Deficiencies in the enzyme lead to glycogen storage disease III.

glycogenin A self-glucosylating protein (333 aa), a glycosyl transferase, expressed prominently in human skeletal muscle and heart (**glycogenin 1**), where it primes glycogen granule synthesis. Multiple isoforms of **glycogenin 2** are found in the liver. Glycogenin binds glucose from UDP-glucose, further glucose units are added until the chain is around eight units long, then glycogen synthetase continues the process.

glycogenoses *See* GLYCOGEN STORAGE DISEASES.

glycogenosis *See* GLYCOGEN STORAGE DISEASES.

glycogen phosphorylase The enzyme (EC 2.4.1.1) that catalyses the sequential removal of glycosyl residues from unbranched glycogen, producing glucose 1 phosphate. Activity is regulated allosterically and by *phosphorylase kinase.There are tissue-specific isozymes in muscle (GPMM, 842 aa), liver (GPLL, 847 aa) and brain (GPBB, 862 aa). GPMM is released from cardiac muscle following ischaemia and is a marker for acute myocardial infarction. Mutations in the liver enzyme lead to glycogen storage disease VI (*Hers' disease), in the muscle enzyme to *McArdle's disease.

glycogen storage diseases (glycogenoses) A range of *storage diseases caused by mutations in genes required for glycogen metabolism or its regulation. **Type 0** is deficiency in either liver or muscle glycogen synthase-2. **Type I** (Von Gierke's disease) is a defect in glucose-6-phosphatase (EC 3.1.3.9). **Types Ib and Ic** can result from mutation in the glucose 6-phosphate transporter gene. **Type II** (*Pompe's disease) is a defect in α-1,4-glucosidase; (*Danon's disease was originally considered a variant of glycogen storage disease II). **Type III** (Forbes' disease, Cori's disease) is due to defects in the *glycogen debranching enzyme. **Type IV** (Andersen's disease) is caused by mutation in the gene encoding the glycogen branching enzyme. **Type V** (*McArdle's disease) is caused by mutation in the gene encoding muscle *glycogen phosphorylase. **Type VI** (*Hers' disease) is due to a deficiency in liver glycogen phosphorylase. **Type VII** (Tarui's disease) is caused by mutation in the gene encoding muscle phosphofructokinase. A very mild form, **type IXa** (formerly type VIII), is caused by deficiency of liver phosphorylase kinase (EC 2.7.1.38). Glycogen storage disease of the heart can be caused by mutation in the gene for the noncatalytic γ_2 subunit of AMP-activated protein kinase.

glycogen synthase kinase A serine/threonine protein kinase (GSK, EC 2.7.11.26) present in two isoforms, GSK3-α (483 aa) and -β (420 aa), that is constitutively active in resting cells and is a negative regulator in various intracellular signalling pathways. Levels of GSK are rapidly reduced by insulin and various growth factors, releasing a constraint, and GSK is involved in the control of glycogen metabolism and protein synthesis by insulin, the modulation of *AP1 and *CREB, the specification of cell fate in *Drosophila* and the regulation of dorsoventral patterning in *Xenopus*. In *Drosophila* **GSK3** (zeste-white 3/shaggy) functions as a negative regulator of the Wnt/β-catenin pathway. GSK3 itself is regulated by *PKB. GSK3 is identical to tau protein kinase I and may therefore be important in the phosphorylation of *tau that is known to occur in the neurofibrillary tangles of Alzheimer's disease.

glycolipid A lipid to which oligosaccharides are covalently attached through a glycosidic bond. *See* GLYCOSPHINGOLIPIDS.

glycolipid transfer protein *See* PHOSPHOLIPID TRANSFER PROTEINS.

glycolysis *See* EMBDEN–MEYERHOF PATHWAY.

glycomics By analogy with genomics and proteomics, the study and analysis of the carbohydrate moieties of complex macromolecules (glycolipids and glycoproteins) in an organism. Although the structure of complex oligosaccharides attached to proteins is determined by the activity of the enzymes of the synthetic pathways involved, it is less tightly 'coded' than the sequence of a protein and technically considerably more difficult to sequence.

glycophorins A family of sialylated transmembrane glycoproteins of the human erythrocyte. **Glycophorin A** is an integral membrane protein (131 aa) that is highly *O*-glycosylated and is rich in terminal sialic acid. The peptide chain carries the *MN blood group antigens at its N-terminus. **Glycophorin B** carries Ss and U antigens. **Glycophorin C** (CD236/CD236R, glycoprotein β, glycoconnectin, 128 aa) constitutes ~4% of the erythrocyte glycoprotein. It is the receptor for the *Plasmodium falciparum* protein PfEBP-2 (erythrocyte binding protein 2, baebl, EBA-140). Glycophorin C has a role in the maintenance of red cell shape, probably through an interaction with *band 3. **Glycophorin D** is a shorter form of glycophorin C and together the two encode the Gerbich (Ge) antigens. **Glycophorin E**, like glycophorin B, is derived from ancestral glycophorin A by gene duplication.

glycoprotein Proteins with covalently attached oligosaccharide. Two major forms exist, those in which the sugar moiety is attached to the -OH group of serine or threonine (*O*-glycosylated), and those in which attachment is to the amide -NH_2 of asparagine (*N*-glycosylated). Most secreted proteins, with the exception of serum albumin, are glycoproteins and many integral membrane proteins are also glycosylated. Common subunits are mannose, *N*-acetylglucosamine, *N*-acetylgalactosamine, galactose, fucose, and sialic acid.

glycosaminoglycans (formerly mucopolysaccharides) The polysaccharide side chains of *proteoglycans composed of many repeating disaccharide units (>100) of amino sugars, at least one of which has a negatively charged side group (carboxylate or sulphate). The commonest are *hyaluronate, *chondroitin sulphate, *dermatan sulphate, *keratan sulphate, and *heparan sulphate. The side chains, except in hyaluronic acid, are covalently attached to the core protein at about every twelfth residue, producing a **proteoglycan**; proteoglycans are noncovalently attached by link proteins to hyaluronate to generate a huge, highly hydrated space-filling polymer found in *extracellular matrix. The extent of sulphation is variable and the structure allows tremendous diversity. The glycosaminoglycan attachment site in *proteoglycans is a serine residue at the consensus pattern S-G-X-G.

glycosidase (glycosylase) General and imprecise term for any enzyme that degrades polysaccharides and can include any of the EC 3.2 class of hydrolases that cleave glycosidic bonds. Substrate specificity is not very high. *See* ENDOGLYCOSIDASE.

glycoside Any molecule in which a sugar group is bonded through its anomeric carbon to another group via an *O*-glycosidic bond or an

S-glycosidic bond (thioglycosides). Conventionally the link is between a sugar and a nonsugar, thereby excluding polysaccharides; the sugar group is the **glycone** and the nonsugar group is the **aglycone**. Glycosides are often storage forms of molecules that are active only when the glycosidic bond is hydrolysed. Glycosides can be classified either on the basis of the sugar and its linkage (e.g. glucosides where the sugar is glucose, α-glycosides where the aglycone is α-linked) or on the nature of the aglycone (e.g. anthroquinone glycosides, steroidal glycosides, as in digitalis).

glycosome A membrane-bounded organelle found in protozoa of the Kinetoplastida (e.g. trypanosomes) in which the enzymes involved in glucose and glycerol metabolism are compartmentalized.

glycosphingolipids Glycolipids of the plasma membrane with the lipid moiety embedded in the bilayer and the carbohydrate protruding into the aqueous phase. In glycosphingolipids the lipid is *sphingosine with the amino group acylated by a fatty chain, forming a *ceramide. Most of the oligosaccharide chains belong to one of four series, the *gangliosides, *globosides, lacto-type 1, and lacto-type 2 series. Blood group antigens are GSLs.

glycosuria (glucosuria) A condition in which glucose is excreted in the urine, usually because of untreated diabetes. Glucose is excreted if the plasma level exceeds the threshold of ~8.9–10 mmol/L (*compare* normal fasting levels of 3.9–5.8 mmol/L).

glycosylation The enzyme-mediated transfer of glycosyl units to a lipid or protein. There are two types of glycosylation, **N-linked glycosylation** where the linkage is to the amide nitrogen of asparagine side chains and **O-linked glycosylation** where the sugar is attached to the hydroxyl side group of serine or threonine. **Glycation** is nonenzymatic transfer. **Glycosyl transferases** are the class of ectoenzymes (EC 2.4) that catalyse the transfer.

glycosyl phosphatidylinositol *See* GPI-ANCHOR.

glycylcyclines Synthetic derivatives of *tetracycline that are effective against bacteria with efflux and ribosomal protection mechanisms of resistance, and act in the same way, binding to the bacterial 30S ribosomal subunit and inhibiting protein synthesis. The most developed glycylcycline is the 9-tert-butyl-glycylamido derivative of minocycline, otherwise known as tigecycline.

glycyrrhizin The main sweet tasting compound (a glycoside) in liquorice root, used as an artificial sweetener.

glyoxalines *See* IMIDAZOLES.

glypiation *See* GPI-ANCHOR.

glypicans A family of cell surface heparan sulphate proteoglycans (the glypican-related integral membrane proteoglycans, GRIPS) in which the core protein (500–600 aa) is GPI-anchored. **Glypican-1** and **glypican-4** are low-affinity receptors for *endostatin. **Glypican-3** appears to form a complex with insulin-like growth factor-2 and is implicated in growth control of mesodermal tissues during embryogenesis. Deletions and translocations involving the glypican-3 gene have been identified in patients with *Simpson–Golabi–Behmel syndrome) and the glypican-3 gene is frequently upregulated in hepatocellular carcinoma. **Glypican-5** is reported to be amplified in rhabdomyosarcomas and **glypican-6** has been identified.

GM130 One of a subfamily of Golgi-associated proteins (*golgins), a large coiled-coil protein (golgin 95, 990 aa) of the *cis*-Golgi that is involved in the docking and fusion of *coatamer (COP I) coated vesicles to the Golgi membranes. GM130 regulates the fragmentation and subsequent reassembly of the Golgi complex during mitosis and regulates the stability and localization of *GRASP65 a Golgi-associated protein required for bipolar spindle formation and cell cycle progression.

Gm allotypes Antigenic variants (altered amino acid epitopes; 25 are known) of the constant γ heavy chains of the IgG subclasses IgG_1, IgG_2 and IgG_3. Gm allotypes have an influence on IgG subclass levels but their significance in predisposition to disease is unclear and the proportion of various allotypes in different populations is very variable. Comparable allotypes (Km) are found in the κ light chain.

GM-CSF A cytokine (granulocyte-macrophage colony stimulating factor, CSF2, 127 aa, glycosylated) that stimulates the formation of granulocyte or macrophage colonies from myeloid stem cells and is secreted by activated macrophages and T-cells and secretion can be induced from endothelial cells and fibroblasts. It can also be expressed as an integral membrane protein. Recombinant protein has been expressed in yeast (and called sargramostim or molgramostim) and used clinically to stimulate bone marrow regeneration after transplantation or chemotherapy (*TNs* Leukine, Leukomax,

Prokine). There are two **receptors**, a low-affinity one, CSF2Rα (400 aa) and a higher-affinity one (CSF2Rβ, 889 aa) that is also the high-affinity subunit for the IL-5 receptor (IL-5Rβ). A soluble form of the low-affinity receptor is also found.

GMO *See* GENETICALLY MODIFIED ORGANISM.

GNAS locus A complex *genomically imprinted locus that produces multiple transcripts through the use of alternative promoters and alternative splicing. Among the transcripts is that for the α subunit of trimeric *G proteins.

gnotobiotic Describing an organism or environment that is completely or almost completely depleted of all organisms or all organisms but one. Specific pathogen-free (SPF) animals are gnotobiotic.

GnRH *See* GONADOTROPIN-RELEASING HORMONE.

GNRP (guanine nucleotide releasing protein) A persistent synonym for guanine nucleotide exchange factors (*GEFs), proteins that facilitate the cycling of signalling proteins between active (GTP-bound) and inactive (GDP-bound) states. Examples include *sos, rasGRP, and *C3G.

Go One of the three types of heterotrimeric *GTP-binding proteins, those that are neither stimulatory (Gs) nor inhibitory (Gi) but 'other' (Go). They are the most abundant brain G proteins and their effects are mostly mediated by the G βγ dimer, with the specific functions of the α subunit remaining rather unclear, although it may mediate interaction with the PKA signalling pathway. Not to be confused with *G_0, the resting phase of the cell cycle.

goblet cells Columnar mucus-secreting epithelial cells found in the mucosa of the small intestine and in other locations where there is mucus production, such as the lining of the trachea. The name arises from their shape, which is a result of the apical region being distended with secretory granules. Not all mucus-secreting cells have this morphology.

goitre (*adj.* goitrous) A swelling in the neck caused by enlargement of the thyroid gland. This can arise through iodine deficiency or a variety of disorders such as *Graves' disease, *Hashimoto's thyroiditis, other thyroid disorders, or more generalized inflammation.

Golber–Hogness box *See* TATA BOX.

Goldberg–Shpritzen megacolon syndrome An autosomal recessive trait with complex developmental abnormalities, often associated with *Hirschsprung's disease and caused by mutation in the *KIAA1279* gene that encodes a 621-aa protein expressed in different parts of the adult central nervous system.

Goldberg's syndrome A condition caused by deficiency in neuraminidase and β-galactosidase which has features of a mucopolysaccharidosis and a sphingolipidosis and could therefore be considered a *mucolipidosis. The defect appears to be in lysosomal cathepsin A, which is required for normal processing of the two deficient enzymes, rather than a defect in the neuraminidase and galactosidase genes. The protective enzyme functions as cathepsin A at acidic pH and as a deamidase/esterase at neutral pH and the severity of the syndrome depends upon the amount of residual activity.

Goldmann equation An equation (Goldmann constant field equation, Hodgkin–Huxley equation), derived from the *Nernst equation, with the simplifying assumption the electrical field across the membrane is constant, and also that *active transport is not involved. From the equation a good approximation to the membrane potential can be calculated on the basis of the concentrations on either side of the membrane and the relative permeabilities of the main ions present in biological fluids (typically sodium, potassium, and chloride).

Golgi apparatus (Golgi body, Golgi vesicles) An intracellular stack of membrane-bounded vesicles that forms part of the *GERL complex. Proteins synthesized on the endoplasmic reticulum and destined for secretion are glycosylated as they pass sequentially through the *cis*- and *trans*-Golgi and then are transported to the plasma membrane or to lysosomes. The Golgi apparatus in plants is referred to as the dictyosome and in flagellate protozoa as the parabasal body.

Golgi cells Inhibitory interneurons found in the granular layer of the cerebellum. They receive excitatory input from mossy fibres and synapse onto the soma of granule cells and unipolar brush cells. The neurotransmitter is GABA.

golgins A family of coiled-coil proteins associated with the Golgi apparatus, that are involved in the tethering and targeting of vesicles, the reconstitution of the Golgi after mitosis and probably other interactions. *See* GM130.

golli proteins A group of proteins that are the product of a region of three exons upstream of the classic myelin basic protein (MBP) exons. They are found in both nervous and immune systems. Oligodendrocytes express golli mRNA primarily during intermediate stages of differentiation and golli proteins may promote proliferation of oligodendrocyte progenitor cells. In a murine experimental system a **golli-interacting protein** (GIP) was identified and appears to be the murine homologue of human nuclear *LIM interactor-interacting factor and thus a potential regulator of differentiation. One of the golli proteins (golli BG21) can serve as a negative regulator of signalling pathways in T cells.

Gomori procedure A staining method used to demonstrate the localization of *acid phosphatases. These enzymes will produce phosphate ions from organic phosphoesters such as β-glycerophosphate which will then form a precipitate with lead salts. The lead phosphate is then converted to brown sulphide of lead by the action of yellow ammonium sulphide. The method has frequently been modified, and because heavy metal salts are involved it can be used in electron microscopy. There is also a **Gomori trichrome stain** that is used for connective tissues.

gonadorelin A synthetic form of *gonadotropin-releasing hormone. A variety of gonadorelin analogues (buserelin, *goserelin, leoprorelin, and triptorelin) have greater activity and are used in treatment of endometriosis, fibroids and prostate carcinoma. *TN* Factrel.

gonadotropin-releasing hormone A decapeptide hormone (GnRH, luteinizing hormone releasing hormone), derived from a 92 aa precursor, produced by hypothalamic neurons and that acts on the gonadotrope cells within the pituitary to stimulate production of *luteinizing hormone and *follicle stimulating hormone which regulate gametogenic and hormonal functions of the gonads. The **receptor** is G protein-coupled and calcium-dependent, and mutations in the receptor lead to hypogonadotropic hypogonadism. *See also* GRAS (GnRH-receptor activating sequence).

gonadotropins (gonadotrophins) A general term for pituitary hormones such as *follicle-stimulating hormone, *luteinizing hormone, and *chorionic gonadotropin that stimulate growth of the gonads and the secretion of sex hormones. The traditional UK spelling, gonadotrophin, is obsolescent.

gonioscope An instrument used to view the anterior chamber of the eye, the region between the cornea and the iris, in particular to measure the irideocorneal angle, an important diagnostic test for *glaucoma.

gonococcus Casual name for the Gram-negative diplococcus *Neisseria gonorrhoeae*, the causative agent of gonorrhoea.

gonorrhoea A sexually transmitted disease in which there is infection of the mucous membrane of the genital tract and sometimes (particularly in neonates born to infected mothers) the conjunctiva with the gonococcus *Neisseria gonorrhoeae.*

Goodpasture's syndrome An autoimmune disease of lung and kidney in which immune complexes are deposited, especially in the glomerulus, inducing an inflammatory response. The antigen is the (alpha-3 chain of type IV collagen, although approximately one- quarter of cases are ANCA-positive and there are also antibodies to myeloperoxidase. Patients with Goodpasture's disease have an increased incidence of HLA-DR2 compared to control populations.

goosecoid A homeodomain transcription factor named for the *Drosophila* homeodomain proteins gooseberry and bicoid with similar DNA-binding characteristics. In *Xenopus* it is expressed in the dorsal lip of the blastopore and may be the molecular marker of Spemann's organizer. Homologous genes have been isolated from human, mouse, zebrafish, and chick, and it is likely to play an important part in early embryonic development. **Goosecoid-like** is a related gene product, also expressed early in development, and a candidate for the abnormalities seen in DiGeorge's syndrome (*thymic hypoplasia) and *velocardiofacial syndrome.

Gordon phenomenon Cerebellar Purkinje cell degeneration that occurs after intracerebral injection of eosinophil granulocytes or eosinophil cationic protein and may be the cause of the syndrome in humans which is characterized by ataxia, muscular rigidity, paralysis, and tremor that may lead to death.

Gorlin's syndrome An autosomal dominant disorder (naevoid basal cell carcinoma) due to mutation in the **patched* gene (*PTCH1*), the product of which is part of the patched/hedgehog/smoothened signalling pathway. Mutations in **smoothened* (*SMO*) have been isolated from human basal cell carcinomas and somatic mutations in the *PTCH2* gene have been identified in basal cell carcinoma and in *medulloblastoma.

goserelin A synthetic decapeptide analogue of luteinizing hormone releasing hormone (LHRH) that acts as a potent inhibitor of pituitary gonadotropin secretion. In males it reduces testosterone levels (in effect, pharmacological castration) and is used in treatment of prostate carcinoma; in females it is used to treat endometriosis and hormone-sensitive carcinomata. *TN* Zoladex.

***GOS* genes** A set of genes (G_0/G_1 switch genes) that were identified as being overexpressed in lymphocytes entering G_1. In the current literature only GOS19-1 is encountered, its product being *macrophage inflammatory protein-1α (MIP-1α).

g

gout An acute metabolic arthritis, most frequently in the joints of the large toe, caused by the accumulation of monosodium urate crytals that induce inflammation. May arise from overproduction or underexcretion of uric acid. Usually treated with *allopurinol, although classically *colchicine was used, probably acting by inhibiting lysosome-phagosome fusion.

Gp-Ib An integral membrane protein (glycoprotein Ib) of platelets that is a component of the receptor for *von Willebrand factor, a disulphide-linked heterodimer of Gp-Ibα (626 aa) and Gp-Ibβ (206 aa) and noncovalently associated platelet glycoproteins GpIX and GP5. Defects in any of the components lead to deficiency in thrombus formation (*Bernard–Soulier syndrome).

gp41 An *HIV envelope glycoprotein (41 kDa, 861 aa), encoded by the *env* gene, that promotes fusion of the virus with the plasma membrane of the host cells and is essential for infectivity. It is noncovalently bound to gp120, another envelope glycoprotein. The ectodomain has an N-terminal fusion domain that is highly hydrophobic, a membrane-spanning domain, and an endomain with a Tyr-based motif that interacts with the AP-2 clathrin adaptor protein.

gp100 A melanosomal matrix protein (661 aa) involved in melanin synthesis that was originally identified as a *melanoma-associated antigen recognized by tumour-infiltrating lymphocytes associated with cancer regression. Vaccines directed against gp100 show some therapeutic promise.

gp130 *See* INTERLEUKIN-6.

GPBB *See* GLYCOGEN PHOSPHORYLASE.

GPCR *See* G-PROTEIN-COUPLED RECEPTORS.

GPI *See* GENERAL PARALYSIS OF THE INSANE.

GPI-anchor A post-translational modification (glypiation) of proteins by the addition of glycosyl phosphatidyl inositol (GPI) to the C-terminal amino acid of the protein. The lipid portion ('pigtail') is embedded in the membrane, thereby anchoring the protein.

G protein **1.** The spike glycoprotein of *vesicular stomatitis virus. **2.** *See* GTP-BINDING PROTEINS.

G-protein-coupled receptor kinases A family of kinases involved in down-regulating receptors. **GRK1** (rhodopsin kinase) is involved in deactivation of *rhodopsin, **GRK2** (β-adrenergic receptor kinase, BARK, *Ark1) desensitizes β-adrenergic receptors.

G-protein-coupled receptors A large family of eukaryotic cell surface receptors (GPCRs, G-protein-linked receptors, GPLR) that are coupled to intracellular heterotrimeric G proteins (*GTP-binding proteins). All have seven membrane-spanning domains (and are sometimes referred to as serpentine receptors, seven transmembrane-domain receptors, or heptahelical receptors) and high-resolution crystal structure has been obtained for the human β_2-adrenergic receptor. Ligand binding induces a conformational change which then activates a G-protein signalling cascade; the effect depends upon which G protein is coupled. Multiple classes are recognized, on the basis of the ligand, and as the identification of GPCRs progresses this classification will undoubtedly evolve. Ligands include neurotransmitters, chemokines, cytokines, light-sensitive molecules (in the case of the rhodopsin class), odorants, and hormones. They have proved excellent targets for drugs because the receptor binding site is accessible.

SEE WEB LINKS

- Signal transduction animation.

GPS **1.** G protein pathway suppressor-1 (Gps1, CSN1) and Gps2 are similar proteins of ~500 aa that suppress *ras and *MAPK-mediated signalling and interfere with *JNK activity. Gps1 is one of the subunits of the *COP9 signalosome. **2.** *See* GREY PLATELET SYNDROME.

GPT *See* GLUTAMATE PYRUVATE TRANSAMINASE.

Graafian follicle *See* FOLLICLE.

graft-versus-host reaction A complication of allogeneic bone-marrow grafting in which the functional immune cells of the graft mount an attack on host tissues. **Acute GVH** disease is characterized by selective damage to the liver, skin, mucosa and the gastrointestinal tract although other tissues are also affected. **Chronic**

GVH disease involves a different set of effector cells and affects a broader range of tissues.

gramicidin A mixture of antibiotics isolated from *Bacillus brevis*. Gramicidins A, B, and C (collectively called gramicidin D) are linear pentadecapeptides that will form a dimeric pore across membranes that is selective for monovalent cations. Gramicidin S is a cyclic decapeptide and has been used as a topical antibiotic.

grammotoxin A peptide toxin (-grammotoxin, 36 aa), from the Chilean rose tarantula *Grammostola spatulata*, that inhibits P/Q and N-type *voltage-gated calcium channels by binding at a site associated with channel gating, not the binding region of pore blockers such as -*conotoxins and -*agatoxin-IIIA. It also binds to potassium channels with lower affinity.

Gram staining A staining method that differentiates bacteria into two main classes. **Gram-positive** bacteria have thick cell walls containing *teichoic and *lipoteichoic acid complexed to the *peptidoglycan. **Gram-negative** bacteria have thinner *peptidoglycan walls that are bounded by an outer membrane containing *endotoxin (lipopolysaccharide). A few exceptions do exist and some bacteria with walls having a high lipid content, notably *mycobacteria, show acid-fast staining—strong acid does not remove the stain. The standard procedure involves staining a heat-fixed smear with crystal violet (methyl violet) which is then treated with 3% iodine/potassium iodide solution, washed with alcohol and counterstained.

SEE WEB LINKS

- Useful resource on histological stains:

grand mal A convulsive epileptic seizure in which there is loss of consciousness. *See* EPILEPSY.

granins A family of related acidic proteins (chromogranins, secretogranins, 400–600 aa) found in many endocrine cell secretory vesicles where they may have a role in granulogenesis and packaging, and are the source of several physiologically active peptide fragments. **Chromogranin A** (parathyroid secretory protein) is stored and coreleased with catecholamines and parathyroid hormone from storage granules in the adrenal medulla and the parathyroid gland respectively. **Chromogranin B** (secretogranin 1) was initially characterized in a rat pheochromocytoma cell line. **Chromogranin C** (secretogranin II) was originally described in the anterior pituitary. Four other members of the family, secretogranins III, IV, V and VI are known although the role of **secretogranins III and IV** are unclear. **Secretogranin V** (pituitary polypeptide 7B2; secretory granule neuroendocrine protein-1; 212 aa) is a specific chaperone for prohormone convertase-2. **Secretogranin VI** is the α subunit of the stimulatory trimeric G protein, Gs, and mutations are associated with a range of disorders such as Albright hereditary osteodystrophy, McCune-Albright syndrome, and pseudohypoparathyroidism Ib, as well being mutated in various pituitary tumours. Peptides derived from chromogranins (e.g. *pancreastatin; *vasostatins) have autocrine, paracrine, and endocrine activities.

granisetron An antiemetic drug that is an antagonist of serotonin 5-HT$_3$ receptors and is used to treat nausea and vomiting following chemotherapy. *TN* Kytril.

granulation tissue Highly vascularized fibrous connective tissue that replaces the initial fibrin clot in a wound. Endothelial cells from adjacent capillaries invade to produce the vasculature and fibroblasts within the granulation tissue will gradually synthesize extracellular matrix. There are usually many phagocytes present, particularly in the early stages of repair.

granule cells Small neurons found within the granular layer of the cerebellum, layer 4 of cerebral cortex, the dentate gyrus of the hippocampus, and in the olfactory bulb. **Cerebellar granule cells**, which are very numerous, receive excitatory input from mossy fibres originating from pontine nuclei and form excitatory synapses, with glutamate as a neurotransmitter, to the intermediate and distal dendrites of Purkinje cells. **Olfactory bulb granule cells** are GABAergic and axonless; granule cells in the dentate gyrus have glutamatergic axons. Granule cells of layer 4 in the cerebral cortex receive inputs from the thalamus.

granulins *See* PROGRANULIN.

granulocyte A leucocyte with conspicuous cytoplasmic granules. In humans the granulocytes are also classified as polymorphonuclear leucocytes and are subdivided according to the properties of the granules when stained with a *Romanovsky-type stain into *eosinophils, *basophils, and *neutrophils. **Granulocytopenia** is a condition in which there are low granulocyte numbers in circulating blood. If the deficiency is in neutrophils then the condition may be described as neutropenia. **Granulopoiesis** is the production of granulocytes in the bone marrow.

g

granulocyte chemotactic protein 2 An α-chemokine (GCP-2, CXCL6, chemokine α3, CKA-3, SCYB6, small inducible cytokine B6 precursor, 114 aa) originally isolated from cytokine-stimulated osteosarcoma cells. It complements the activity of IL-8 as neutrophil chemoattractant and activator in humans. Signals through the IL-8 receptor and CXCR1 and CXCR2.

granulocyte/macrophage colony stimulating factor *See* GM-CSF.

granuloma A nodule that is a chronic inflammatory lesion with large numbers of cells of various types (macrophages, lymphocytes, fibroblasts, giant cells) together with fibrous connective tissue and sometimes surrounded by a cuff of lymphocytes. May arise through infection or form around persistent nondegradable foreign material. Granulomas are characteristic of various diseases such as *Crohn's disease, tuberculosis, leprosy, sarcoidosis, berylliosis, and syphilis. It is also a feature of autoimmune diseases such as *Wegener's granulomatosis and *Churg–Strauss syndrome. *See* CHRONIC GRANULOMATOUS DISEASE.

granulomatous disease **1.** Any disease characterized by the formation of *granulomas. **2.** *See* CHRONIC GRANULOMATOUS DISEASE.

granulysin A *saposin-like protein (129 aa) present in cytotoxic granules of cytolytic T cells and NK cells and released upon antigen stimulation.

granzymes A family of serine endopeptidases found in cytotoxic T cells and NK cells and involved in *perforin-dependent cell killing. **Granzyme A** (cytotoxic T-lymphocyte-associated serine esterase-3, CTLA3; EC 3.4.21.78, 262 aa) induces caspase-independent apoptosis but with single-stranded DNA nicking. A target of granzyme A is the *SET complex. **Granzyme B** (fragmentin-2; CTLA1; EC 3.4.21.79, 247 aa) is crucial for the rapid induction of target cell apoptosis by cytotoxic T cells. It enters target cells in a perforin-independent manner by binding to cell surface cation-independent *mannose-6-phosphate receptor (IGF2R). **Granzyme K** (granzyme 3, tryptase II, 264 aa) is similar to granzyme A and also found in lymphocytes and NK cells. Granzymes H and M are similar serine peptidases.

GRAS An upstream regulatory element (GnRH-receptor activating sequence) that is involved in the control of expression of the receptor for gonadotropin releasing hormone which is enhanced by *activin and inhibited by *follistatin (a downstream activin regulatory element or DARE is also involved). Control is complex with feedback from gonadotropin to the expression of its receptor and the GRAS regulatory element interacts with multiple classes of transcription factors including *Smads, *AP-1, and a *forkhead DNA binding protein. Most of the studies have been done on gonadotrope-derived αT3-1 murine cells but the principles are probably conserved.

GRASP **1.** Golgi reassembly stacking protein (Golgi membrane-associated protein). A family of proteins involved in the stacking of Golgi cisternae and the reassembly of the Golgi post-mitosis. **GRASP65** (440 aa) is a dimer that can directly link adjacent surfaces through *trans*-oligomerization in a mitotically regulated manner. **GRASP55** (452 aa), is structurally related to GRASP65, and forms a rab2 effector complex with *golgin-45 on medial-Golgi that is essential for normal protein transport and Golgi structure. **2.** Grp1 (*general receptor for phosphoinositides-1)-associated scaffold protein. The product of a retinoic acid-induced gene that enhances Grp1 (*cytohesin-3) association with the plasma membrane. (This usage is less common.) **3.** *See* ATP-GRASP PROTEINS.

Graves' disease *See* HYPERTHYROIDISM.

Grawitz's tumour Obsolete name for *renal cell carcinoma.

grb-2 An adaptor protein (growth factor receptor bound protein 2, 217 aa), the mammalian homologue of *Caenorhabditis elegans* Sem-5 and *Drosophila* Drk that links the cytoplasmic domain of growth factor receptors to downstream elements of the signalling pathway through its SH2 and SH3 domains. The SH2 domain binds autophosphorylated residues on the growth factor receptor's cytoplasmic tail and the SH3 domain binds nucleotide exchange factors such as *sos1. **Grb2-associated binder-1** (gab-1, 694 aa) binds to Grb-2 and has homology with IRS-1 (insulin receptor substrate-1). It is a substrate for the EGF receptor and may integrate signals from different receptors into the control of cellular responses. **Grb2-related adaptor protein-1** (GRAP1; 217 aa) interacts *in vitro* with c-kit (stem cell factor receptor) and the erythropoietin receptor, linking them to the ras signalling pathway. **Grb2-related adaptor protein-2** (GRAP2, GADS, 330 aa) is another of the Grb2/Sem5/Drk family and reportedly involved in leucocyte-specific protein tyrosine kinase signalling.

Greek key A protein folding pattern named after the pattern that was common on Greek pottery. Three up-and-down beta-strands connected by hairpins are followed by a longer connection to the fourth strand, which lies adjacent to the first. An example is in γ-crystallin which has two Greek key motifs.

green fluorescent protein A protein (GFP) isolated from the luminous jellyfish *Aequorea victoria*. When excited by blue light it emits green light. The gene has been cloned and mutagenized (e.g. enhanced GFP (EGFP)) to give brighter fluorescence, and different colour variants have also been developed. GFP and its derivatives are often used as reporter genes.

Greig cephalopolysyndactyly syndrome A disease characterized by unusual skull shape and syndactyly (fused digits) that is caused by mutations in **gli-3*, the glioma-related oncogene that encodes a zinc finger DNA-binding transcription factor from the GLI-Krüppel gene family.

gremlin A secreted glycoprotein (184 aa) that binds and antagonizes the actions of *bone morphogenetic proteins (BMP)-2, -4, and -7. It opposes BMP effects on osteoblastic differentiation and function *in vitro*. A related protein, gremlin 2 (protein related to *Dan and *cerberus, PRDC, 168 aa) has been identified. Other BMP antagonists are *noggin, *sclerostin, and *chordin.

grey platelet syndrome A rare inherited disorder in which platelets have few or no α-granules and platelet-specific α-granule proteins. The defect is apparently megakaryocyte-specific since Weibel–Palade bodies in endothelial cells, which have the same contents (von Willebrand factor and P-selectin) as α-granules, are normal. Mutations in the X-linked transcription factor GATA-1 cause an abnormal platelet morphology which is similar but distinct.

GRF (growth hormone releasing factor) *See* GROWTH HORMONE.

GRIP *See* GLUTAMATE (GLUTAMATE RECEPTOR INTERACTING PROTEIN); GLUCOCORTICOID RECEPTOR INTERACTING PROTEIN 1; GLYPICANS.

Griscelli's syndrome A rare autosomal recessive disorder in which skin and hair pigmentation is reduced, probably because of altered melanosome transport. There is also neurological impairment. In **Griscelli's syndrome type 1** (GS1) there is a mutation in the myosin-5A gene, but there is no immunological impairment. **GS2**, caused by mutation in the *RAB27A* gene, the product of which is ras-related protein from megakaryocytes, is complicated by immune dysfunction. In **GS3** there is hypomelanosis without immunologic or neurologic manifestations, which may be caused either by mutation in the gene for *melanophilin or that for myosin-5A.

griseofulvin An antifungal polyketide antibiotic isolated from *Penicillium griseofulvum* that blocks microtubule assembly and thus mitosis. Because it binds to keratin it concentrates in skin and is taken up by dermatophytic fungi.

GRKs The family of serine/threonine protein kinases (G-protein-receptor kinases, EC 2.7.11.16) that phosphorylate agonist-occupied G-protein-coupled receptors. **GRK-1** is rhodopsin kinase; **GRK-2** is β-adrenergic receptor kinase-1, BARK-1, *See* ARK1) and many others are known. Mutations in the gene for GRK1 cause *Oguchi's disease. *See* ARRESTINS.

gRNA *See* GUIDE RNA.

gro *See* MELANOMA GROWTH STIMULATORY ACTIVITY protein.

groEL *See haperonins at* CHAPERONES.

Gronblad–Strandberg syndrome *See* PSEUDOXANTHOMA ELASTICUM.

ground-glass cells A histopathological term to describe the appearance of liver cells infected with hepatitis B in which the hepatitis B surface antigen appears as fine granules either diffusely spread throughout the cytoplasm or concentrated in the cytoplasm peripherally to the sinusoid space. It is also used to describe the appearance of multinucleated giant cells in multicentric reticulohistiocytosis and cells of glassy cell carcinoma of the endometrium.

group specific antigen *See* GAG PROTEIN.

growth arrest specific genes *See* GAS GENES.

growth-associated proteins (GAPs) A rather general term for developmentally regulated polypeptides from neural tissues. More specifically GAP-43 (*neuromodulin) is an acidic membrane phosphoprotein that regulates neural cell adhesion molecule (NCAM)-mediated neurite outgrowth. It is a substrate for protein kinase C (originally identified as pp46) and is now known to be identical to B-50 that regulates phosphotidylinositol turnover in mature presynaptic membranes and F1 found in the *hippocampus during *long-term potentiation. It is not restricted to neurons but is also found in astrocytes. Another member of the family, also involved in regulating neurite outgrowth, is brain acid-soluble protein (BASP1), also known as CAP-23; the rat homologue is the 22-kD neuronal tissue-enriched acidic protein, NAP-22. BASP1 is also present in significant amounts in kidney, testis, and lymphoid tissues.

growth cone The motile tip of a growing *neurite that moves towards the target cell with the neuron being extended behind it. The growth cone responds to environmental cues including gradients of *semaphorins, adhesiveness, growth factors, neurotransmitters, and electric fields (*galvanotropism). The front of the growth cone extends long *filopodia that differentially adhere to surfaces in the embryo. Not only are there positive cues that encourage extension but there are also specific molecules (*collapsins) that inhibit the motility of particular growth cones, and that are important in establishing correct pathways in developing nervous systems.

g

growth control The control of the number of cells in a population (or organ), the control of cell division not the control of cell size.

growth/differentiation factors A subfamily of the TGFβ superfamily, involved in growth and differentiation (Gdfs). **Gdf1** (372 aa) acts early in the pathway of gene activation that leads to the establishment of left-right asymmetry and mutations are associated with congenital heart defects such as transposition of the great arteries and the *tetralogy of Fallot. **Gdf2** (429 aa) is *bone morphogenetic protein (BMP)-9 which may be a differentiation factor for cholinergic central nervous system neurons. **Gdf3** (364 aa) has been implicated in testis carcinoma and deposition of adipose tissue. **Gdf4** is unreported. **Gdf5** (BMP-14; cartilage-derived morphogenetic protein-1, 501 aa) is important in bone repair; **Gdf6** (455 aa) is BMP13; **Gdf7** (450 aa) is BMP12;. In mice Gdf5, Gdf6, and Gdf7 are all required for normal formation of bones and joints in the limbs, skull, and axial skeleton. **Gdf8** (375 aa) is *myostatin. **Gdf9** (454 aa) plays an important role in somatic cell differentiation and the formation of primordial follicles in hamster ovary. **Gdf10** (478 aa) is BMP3B, closely related to BMP3 but with different tissue distribution. **Gdf11** (407 aa) is BMP11, important in anterior/posterior patterning of the axial skeleton. The next reported Gdf of the series is **Gdf15** (308 aa), macrophage inhibitory cytokine-1 (MIC-1).

growth factor receptor bound protein *See* GRB-2.

growth factors A diverse group of proteins that are important in the regulation of cell proliferation (growth) and differentiation. The distinction between growth factors and *cytokines is blurred since some cytokines act as growth factors and some cytokines, originally described as important in the haematopoietic lineages, act on a broader range of cell types. Autonomous growth factor production or altered responsiveness to growth factors is a common characteristic of many neoplastic cells which thereby lose growth control. The receptors are diverse and are detailed in the entries for individual growth factors. Growth factors (GFs) include *activin, *amphiregulin, *betacellulin, *bone morphogenetic proteins, *brain-derived neurotrophic factor, the *CCN family, *CSF-1, *CTGF, *epidermal GF, *epiregulin, *fibroblast GF, *GDNF, *growth/differentiation factors, *HB-EGF, *hepatocyte GF, *heregulins, *insulin, *insulin-like GFs, *keratinocyte GF, *midkine, *neuregulin, *nerve GF, *neurotrophins, *nodal, *oncomodulin, *p185, *platelet-derived GF, *schwannoma-derived GF, *somatomedin, *transforming growth factor, and *vascular endothelial GF.

SEE WEB LINKS

- Downloadable poster of human growth factors.

growth hormone A polypeptide hormone (somatotropin, somatotropin, 191 aa) synthesized by acidophilic or somatotropic cells of the anterior pituitary that stimulates liver to produce *somatomedins 1 and 2. Deficiencies in growth hormone production lead to proportionate short stature, with decreased growth velocity and delayed bone maturation, although this can now be treated with recombinant hormone. Although important during childhood and adolescence for the control of growth, the hormone continues to be produced throughout life and is important in preventing bone loss. Growth hormone in the circulation is bound to **growth hormone binding protein** (GHBP, 246 aa), the soluble extracellular portion of the **receptor** (638 aa), which buffers the level of GH. Binding of hormone to two G-protein-coupled growth hormone receptors causes dimerization which activates a signalling pathway that leads to the synthesis and secretion of *insulin-like growth factor I. Mutations in the growth hormone receptor lead to *Laron's syndrome. Synthesis and release are controlled by family of genes that include the transcription factors PROP1 and PIT1 which regulate differentiation of somatotrophs and regulated by **growth hormone releasing hormone** (GHRH; growth hormone-releasing factor, GRF; somatocrinin; somatoliberin, 108 aa) produced in the hypothalamus. Release of GHRH is inhibited by *somatostatin.

GRP **1.** GRP-1 is *cytohesin-3. **2.** Grp78 is glucose-regulated protein-78 (immunoglobulin heavy chain-binding protein, BiP), an endoplasmic reticulum-specific chaperone of the heat-shock protein-70 (HSP-70) family. **3.** *See* ENDOPLASMIN (glucose-regulated protein-94). **4.** *See* GASTRIN-RELEASING PEPTIDE; GENERAL RECEPTOR FOR PHOSPHOINOSITIDES-1 (Grp1).

G_s *See* GTP-BINDING PROTEIN.

GSG domain A domain (GRP33/Sam68/GLD-1 domain), also found in *STAR proteins (signal transduction and activator of RNA proteins) of around 200 aa that mediate dimerization and RNA binding of several regulatory proteins such as *sam68 and *quaking.

GSK *See* GLYCOGEN SYNTHASE KINASE.

GSS **1.** Global Symptom Score. **2.** Glaucoma Staging System, for classifying visual field defects in glaucoma. **3.** *See* GLUTATHIONE SYNTHASE. **4.** *See* GERSTMANN–STRAUSSLER–SCHEINKER SYNDROME. **5.** The Genome Survey Sequences (GSS) database, a database with DNA sequences based on genomic rather than expressed (cDNA) sequences.

SEE WEB LINKS

- Genome Survey Sequences database homepage.

GST *See* GLUTATHIONE S-TRANSFERASES.

GTBP A protein (G/T binding protein, MutS homologue-6, MSH6, 1068 aa) that forms a heterodimer with MSH2 (*See* MSH GENES) which will bind mismatches where G is paired with T. This is the first step in *mismatch repair. Absence of either protein predisposes to tumours. The *GTBP* gene is located near that for MSH2, and GTBP can be considered one of the MSH family. Mutations in GTBP are responsible for some cases of hereditary nonpolyposis colon cancer, although more are due to mutation in other MSH genes, and some cases of mismatch repair cancer syndrome.

GTP (guanosine 5′-triphosphate) The triphosphate of guanosine that is involved in a distinct set of energy-requiring processes. GTP is required in protein synthesis, the assembly of *microtubules, and for the activation of regulatory G proteins *GTP-binding proteins where the binding of GTP activates and slow hydrolysis to GDP inactivates.

GTPase-activating protein (GAP) A protein that stimulates the GTPase activity of a small *GTP-binding protein and switches it to the inactive (GDP-bound) state. The first GAP was first purified as a 125-kDa protein from bovine brain and activated GTPase activity of *ras-p21; it would now be termed ras-GAP. GAPs may be regulated by phospholipids and by phosphorylation on a tyrosine residue by growth factor receptors (PDGF-R, EGF-R). The *neurofibromatosis type 1 gene (*NF1*) codes for a protein homologous to ras-GAP. The action of a GAP on a GTP-binding protein is reversed by *GEFs, **GTP-exchange factors** that stimulate release of GDP and allow binding of GTP; this can be blocked by *GDIs (guanosine nucleotide dissociation inhibitors) that prevent release of GDP.

GTP-binding proteins (G proteins) Although many proteins bind GTP (e. g. actin, tubulin) the GTP-binding proteins are considered special cases because of their role in signalling. There are two classes of these G proteins: heterotrimeric G proteins that associate with *G-protein-coupled receptors, and the small cytoplasmic G proteins. The **heterotrimeric G proteins** have a Gα subunit (39–52 kDa) with slow GTP-ase activity: when GTP binds the α subunit dissociates from the βγ subunits (β, 35–36 kDa: γ, 6–10 kDa) and is able to interact with various *second messenger systems, either inhibiting (Gi), stimulating (Gs), or otherwise (*Go). Once the GTP is hydrolysed the heterotrimer re-forms and activation ceases. The βγ subunits, most of which are membrane associated through post-translational myristoylation or isoprenylation, may have direct activating effects in their own right, though they are less diverse than the α subunits. Stimulatory G proteins (Gαs-βγ) are permanently activated by *cholera toxin, inhibitory ones (Gαi-βγ) by *pertussis toxin. *Transducin (Gαt-βγ) was one of the first of the heterotrimeric G proteins to be identified. The **small G proteins** are a diverse group of monomeric GTPases that include *ras, *rab, *rac, and *rho and play an important part in regulating many intracellular processes including cytoskeletal organization and secretion. Their GTPase activity is regulated by activators (*GTPase-activating protein, GAPs) and inhibitors (GIPs) that determine the duration of the active (GTP-bound) state. *See also* GEFs; RAS-LIKE GTPASES.

guanethidine An antihypertensive drug that reduces the release of catecholamines. Does not cross the blood–brain barrier so does not affect the central nervous system and is used topically for the treatment of glaucoma and to counteract eyelid retraction in Graves' disease. *TN* Ismelin.

guanidinium chloride Chloride salt of guanidinium (guanidine hydrochloride, $(C(NH_2)_3)^+$), a strong chaotropic agent used for the denaturation and solubilization of proteins by disruption of hydrogen bonds. It is also used in

g

RNA isolation procedures to dissociate the nucleic acid and protein moieties.

guanine One of the constituent bases (2-amino-6-hydroxypurine) of nucleic acids, nucleosides, and nucleotides. It binds to *cytosine through three hydrogen bonds.

guanine deaminase A ubiquitous enzyme (EC 3.5.4.3, 454 aa) that catalyses the first step of purine metabolism by hydrolytic deamination of guanine, producing xanthine. The alternatively spliced form in neurons is *cypin.

guanosine A nucleoside (9-β-D-ribofuranosyl guanine) formed by linking ribose to guanine.

guanosine 5′-triphosphate *See* GTP.

guanylate cyclase (guanylyl cyclase) A heterodimeric protein (with α and β subunits) that catalyses the synthesis of the second messenger guanosine 3′,5′- cyclic monophosphate (cGMP) from GTP. The enzyme is the main receptor for *nitric oxide (NO) and nitrovasodilator drugs which stimulate the production of cGMP. The soluble form is found in brain and has α_1 (690 aa) and β_1 (619 aa) subunits. The plasma membrane form of guanylate cyclase is an integral membrane protein with an extracellular receptor for peptide hormones, a transmembrane domain, a *protein kinase-like domain, and a guanylate cyclase domain. Various isoforms are known, some being receptors for *natriuretic peptides. Mutations in human photoreceptor guanylate cyclase 2D (1103 aa) cause some forms of *Leber congenital amaurosis and a dominant form of *cone-rod dystrophy. **Guanylate cyclase-activating protein** (GCAP1, 201 aa) is a calcium-binding protein of the calmodulin superfamily, expressed exclusively in photoreceptor inner segments, that stimulates synthesis of cGMP. Mutations in this cause *cone dystrophy and may be responsible for other types of Leber amaurosis. *Guanylin is **GCAP2A**, uroguanylin is GCAP2B.

guanylin A small peptide (guanylate cyclase-activating protein-2A, GCAP2A, 15 aa) found in vertebrate gut that binds to the intestine-specific receptor *guanylate cyclase known as *STaR (the receptor that binds bacterial heat-stable enterotoxins such as *E.coli* ST toxin; STa) and activates cGMP production. As a result of the elevated cGMP there is inhibition of salt absorption and stimulation of chloride secretion, accompanied by a massive accumulation of water in the gut, producing diarrhea and dehydration. Guanylin is generated from inactive 94-aa proguanylin. Uroguanylin (GCAP2B) is a similar peptide found in urine that may be involved in the regulation of electrolyte homeostasis by the kidney.

guanylyl cyclase *See* GUANYLATE CYCLASE.

Guarnieri body An eosinophilic inclusion body found in cells infected with smallpox or *vaccinia virus; composed of viral particles and proteins.

Guibaud–Vainsel syndrome *See* CARBONIC ANHYDRASE.

guidance *See* CONTACT GUIDANCE.

guide RNA (gRNA) **1.** A name originally applied to small RNA molecules (60–80 nucleotides) involved in RNA editing within the mitochondria of kinetoplastid organisms and found in the *editosome. **2.** A more generic use of the term is for any RNA sequence involved in sequence-specific recognition of target RNA (*See* RNA INTERFERENCE).

Guillain–Barré syndrome An acute inflammatory demyelinating polyneuropathy (Landry–Guillain–Barré syndrome) with symmetric limb weakness and loss of tendon reflexes. It is probably an autoimmune disorder and is associated with preceding *Campylobacter jejuni* infection. Antibodies formed against ganglioside-like epitopes in the lipopolysaccharide (LPS) layer of some infectious agents will cross-react with the ganglioside of peripheral nerves There are rare familial cases, caused by mutation in the peripheral myelin protein (PMP22, *gas-3, growth arrest-specific-3) gene, but most cases are thought to be multifactorial with both genetic and environmental factors involved.

gumma (*adj.* gummatous) A granuloma most commonly found in the liver (**gumma hepatis**), but also in brain, heart, skin, bone, testis, and other tissues, that is caused by syphilitic infection in the late or tertiary stage.

Gunther's disease *See* PORPHYRIA.

GUS *See* GLUCURONIDASE.

gustducin A *GTP-binding protein expressed in taste buds of papillae. The Gα subunit resembles *transducin and appears to play a key role in both bitter and sweet taste transduction.

Gut associated lymphoid tissue (GALT) The subset of mucosal lymphoid tissue (MALT) that is associated with the intestinal tract. It comprises Peyer's patches, tonsils (Waldeyer's ring), adenoids (pharyngeal tonsils), mesenteric lymph nodes, the appendix, and diffusely distributed lymphoid cells and plasma cells in the lamina propria of the gut. GALT may play a role in the persistence of HIV even in individuals who have been taking antiviral drugs for a long period.

GVH *See* GRAFT-VERSUS-HOST DISEASE.

GW bodies (P bodies) Small, generally spherical, cytoplasmic domains that contain proteins such as GW182 and *argonaute 1 and 2 that are involved in mRNA degradation, translational repression, mRNA surveillance, and RNA-mediated gene silencing, together with their mRNA targets. The number and size of the bodies varies in various mammalian cell types. **GW182** is an RNA-binding protein (1962 aa) that was originally identified as the target of an autoantibody in serum from a patient with a *sensory ataxic polyneuropathy.

gynaecology That branch of medicine which deals with the health of the female reproductive system (uterus, vagina, and ovaries). It is usually combined with obstetrics as a medical speciality.

gyrate atrophy A progressive condition, starting in late childhood, in which there is night blindness with sharply demarcated circular areas of retinal atrophy. It is an autosomal recessive condition caused by deficiency in ornithine aminotransferase (EC 2.6.1.13) and can be partly treated by dietary restriction of arginine.

gyrin *See* SYNAPTOGYRINS.

gyrus A ridge-like fold of the cerebral cortex.

H2 antigens The murine equivalent of the HLA antigens, products of the H2 complex, the *MHC of mice. Divided into class I and class II antigens.

H_2 blocker *See* HISTAMINE RECEPTORS.

H5N1 A strain of avian influenza in which the haemagglutinin (H or HA, of which there are 16 subtypes) is H5 and the neuraminidase (N or NA, of which there are 9 subtypes) is N1. Haemagglutinin and neuraminidase are transmembrane glycoproteins of the viral envelope. A similar avian flu strain gave rise to the 1918 flu pandemic and if the avian virus becomes highly infectious (transmissible) in humans there is the potential for a major *pandemic.

***H19* gene** An imprinted, maternally expressed gene (adult skeletal muscle gene) that is tightly linked and coregulated with the imprinted, paternally expressed gene of insulin-like growth factor 2 (*IGF2*). Control of which allele is expressed is by the *imprinting centre 1 (IC1). The product of H19 is an oncofetal RNA that is not translated and may be a tumour suppressor. Hypermethylation of *H19* is associated with *Beckwith–Wiedemann syndrome and hypomethylation of the *H19* promoter with Silver–Russell syndrome.

H89 A specific inhibitor of *protein kinase A (PKA) although it has been reported to inhibit *rho kinases (ROCKs).

HACBP High affinity calcium-binding protein. *See* CALRETICULIN.

hadacidin An antibiotic originally isolated from *Penicillium frequentans* and produced by *Penicillium* species. It is an aspartate analogue and a competitive inhibitor of adenylosuccinate synthetase with some cytotoxic effects; named for its action against human adenocarcinoma.

haem The prosthetic group of cytochromes and oxygen carriers such as haemoglobin. Consists of iron complexed in a porphyrin (tetrapyrrole) ring. Various different haems are found with different porphyrins. Cytochrome *a* contains haem A, cytochrome *c* contains haem C.

haem-, haemato- Prefixes indicating something related to blood. NB The UK English form is used in this dictionary rather than the US **heme-, hemato-**.

haemagglutination The agglutination of red blood cells (*erythrocytes), often used as a test for the presence of haemagglutinins which may be antibodies directed against erythrocyte surface antigens, lectins that bind sugar residues on glycoproteins or glycolipids or viral haemagglutinins. Agglutination requires crosslinking so at least two binding sites must be present.

haemangioblasts The earliest haematopoietic precursors that can give rise to both haematopoietic and endothelial cells. A marker, *flk-1 (vascular endothelial growth factor receptor), is first observed in the human embryo at the beginning of week 4. Early haematopoiesis from primitive haemangioblasts in the extraembryonic mesoderm of the yolk sac generates mostly erythroid and some myeloid lineages, whereas the later definitive haemangioblasts from intraembryonic mesoderm give rise to a wider range of cells including cells of the lymphoid series. It is not clear for mammals whether definitive haemangioblasts develop by maturation of primitive haemangioblasts, or arise from a separate mesodermal cell population, or from a combination of both. *Lindau's tumour is a haemangioblastoma.

haemangioma A benign tumour (angioma) produced by proliferation of capillary endothelial cells. Usually sporadic but there are some familial cases probably caused by mutations in **flt4*. **Cutaneous haemangiomas** include capillary naevi, port wine stains, vin rosé patches, cavernous haemangiomata, Campbell de Morgan spots, and telangiectasias. **Liver haemangiomas** are relatively common but asymptomatic.

haematocoele A swelling caused by blood accumulating in a cavity within the body.

haematocrit (blood count) A measure of the concentration of erythrocytes in the blood. An average figure for humans is 45%, i.e. a packed red cell volume of 45 mL in 100 mL of blood. Usually measured by centrifuging heparinized blood in a capillary tube or by using a cell analyser which will give cell numbers and volumes from which the haematocrit can be calculated.

haematoma A swelling in which blood has diffused into connective tissue. Bruises are a form of haematoma, as are petechiae (haematomas <3 mm in diameter) and ecchymoses. They can occur internally, and in the tissues surrounding the brain can cause compression and be lethal.

haematopoiesis (haemopoiesis) The production of blood cells by proliferation and differentiation of haematopoietic *stem cells. In adults the major site is bone marrow. Pluripotent stem cells can give rise to all lineages, committed stem cells (derived from the pluripotent stem cell) only to some.

haematopoietic cell kinase A protein tyrosine kinase (hck, haemopoietic cell kinase, 526 aa) of the *src family found in lymphoid and myeloid cells and that is bound to B-cell receptors in unstimulated B cells. The *hck* gene is located in a region affected by interstitial deletions in some acute myeloid leukaemias and myeloproliferative disorders. The catalytic activity of hck is regulated, both positively and negatively, by tyrosine phosphorylation: phosphorylation of the Tyr499 residue in the COOH terminus of hck by the protein kinase *Csk inactivates hck, whereas autophosphorylation of Tyr388 within the kinase domain upregulates kinase activity.

haematoxylin A basophilic stain (Natural Black 1, C.I. 75290) extracted from the wood of the logwood tree that, when oxidized, produces **haematein**, a compound with rich blue-purple colour. Extensively used in histopathology, usually in combination with an acidophilic dye such as eosin ('H & E' is haematoxylin and eosin) which stains the cytoplasm pink whereas the haematoxylin stains nuclei blue. Various mordants (usually aluminium or iron salts) are used to form lakes (coloured complexes) which are blue with aluminium salts, blue-black with iron. Various staining procedures and mixtures have been developed. **Heidenhain's iron haematoxylin** stains nuclei black and this can be helpful for automated image processing.

haematuria *See* PAROXYSMAL COLD HAEMOGLOBINURIA; PAROXYSMAL NOCTURNAL HAEMOGLOBINURIA.

haemochromatosis A relatively common iron-overload disorder in which *haemosiderin is deposited in excess in various organs. Estimates of incidence of iron overload due to homozygous haemochromatosis may be as high as 1 in 300 among Australians. Clinical features include cirrhosis of the liver, enlargement of the spleen, diabetes (bronzed diabetes), skin pigmentation, and heart failure. Primary hepatocellular carcinoma, complicating cirrhosis, causes approximately 30% of deaths in affected homozygotes. Most forms are autosomal recessive: **classic haemochromatosis** (hereditary haemochromatosis; HFE) is most often caused by mutation in the *HFE* gene; haemochromatosis **type 2A** (HFE2A) by mutation in the *HJV* (hemojuvelin) gene; **HFE2B** by mutation in the gene for *hepcidin antimicrobial peptide; haemochromatosis **type 3** by mutation in the gene encoding transferrin receptor-2. Haemochromatosis **type 4** is an autosomal dominant disorder caused by mutation in the *ferroportin gene.

haemocyanin A blue copper-containing protein found in the blood of molluscs and crustacea, where it acts as an oxygen carrier. It is a very large protein (20–40 subunits, 2–8 MDa) with a characteristic cuboidal appearance under the electron microscope that made it useful when conjugated with antibody for localizing antigen (now largely superseded by immunogold techniques). Keyhole limpet *Megathura crenulata* haemocyanin (KLH) is widely used as a carrier in the production of antibodies.

haemocytoblast A pluripotential haematopoietic stem cell that gives rise to various committed stem cells. *See* HAEMATOPOIESIS.

haemocytometer A special glass slide used for counting blood cells, etc. under the microscope. The slide is arranged so that a defined thickness of fluid is trapped under a thick coverslip, and a grid engraved on the slide then demarcates a known volume. Knowing the volume being viewed, the number of particles (cells) can be counted and the number per unit volume calculated. Various grid patterns are available (Neubauer and Fuchs–Rosenthal being common) and a smaller-scale device (Helber cell) is used for bacterial counting.

SEE WEB LINKS

- Description of a haemocytometer with diagrams.

haemodialysis A method used when kidney function is inadequate. Essentially, the dialysis of blood against a physiological saline so that unwanted metabolites are removed but larger molecules in plasma are retained.

haemoglobin A four-subunit globular oxygen-carrying protein of vertebrates and some invertebrates. In adult humans there are two α and two β chains (very similar to myoglobin); a haem moiety (an iron-containing substituted porphyrin) is firmly held in a nonpolar crevice in each peptide chain. **Fetal haemoglobin** differs in having γ chains instead of β chains and has higher oxygen affinity. Mutations in the genes for the haemoglobins cause haemoglobinopathies, the best known of which is sickle-cell disease (in which valine is substituted for glutamate in position 6 of the β chain), and *thalassemias where there is underproduction of normal and sometimes abnormal haemoglobins. **Haemoglobinaemia** is the presence of haemoglobin in plasma as a result of destruction of red blood cells and is often the cause of **haemoglobinuria**, a condition in which there is haemoglobin in the urine.

haemolysins Bacterial *exotoxins that cause **haemolysis**, the lysis of erythrocytes.

haemolytic anaemia An *anaemia arising from reduced erythrocyte survival time, either because of intrinsic erythrocyte defects (hereditary *spherocytosis or ellipsocytosis), enzyme defects, *haemoglobinopathies), or an extrinsic damaging agent, e.g. autoantibody (autoimmune haemolytic anaemia), isoantibody, parasitic invasion of the cells (*malaria), bacterial or chemical haemolysins, or mechanical damage to erythrocytes.

haemolytic disease of the newborn *See* ERYTHROBLASTOSIS FETALIS.

haemolytic plaque assay A method used to detect cells secreting antibody *in vitro*. Sheep red cells, treated so that they will bind the antibody and be lysed, are mixed with the cell suspension, immobilized in agarose, and incubated. Antibody-producing cells will lyse adjacent erythrocytes, forming a plaque of haemolysis. The assay method can be adapted to identify other types of cells, e.g. those secreting a protein that will generate an immune complex with added antibody, activate complement and cause haemolysis.

haemolytic uraemic syndrome A disorder of the microvasculature associated with mutations in *complement factor H that leads to microangiopathic haemolytic anaemia associated with distorted erythrocytes ('burr cells'), thrombocytopenia, and acute renal failure. Both dominant and recessive modes of inheritance have been reported.

haemopexin A globulin (β-glycoprotein, 462 aa), structurally related to *vitronectin and some *collagenases, that forms ~1.4% (~1 mg/mL) of total serum protein. Hemopexin has very high affinity for haem and the haem-haemopexin complex is bound by receptors on hepatocytes and the iron salvaged.

haemophagocytic lymphohistiocytosis A rare autosomal recessive disorder in which there is massive infiltration of several organs by activated lymphocytes and macrophages and anaemia caused by phagocytosis of erythrocytes. **Type 1** is linked to chromosome 9 but the defective gene is not yet identified. In familial haemophagocytic lymphohistiocytosis **type 2** (FHL2) there is a mutation in the gene encoding *perforin. **Type 3** is caused by mutation in the *UNC13D* gene, the product of which (1090 aa) is essential for the priming step of cytolytic granule secretion from cytotoxic T cells. **Type 4** is caused by mutation in the syntaxin-11 gene.

haemophilia A set of diseases in which there is an inherited deficiency in the blood clotting system leading to a variable phenotype of haemorrhage into joints and muscles, easy bruising, and prolonged bleeding from wounds. Classical **haemophilia A** is an X-linked recessive disorder caused by a deficiency in coagulation factor VIII. In **haemophilia B** (*Christmas disease), which is also X-linked, there is factor IX deficiency, and in *von Willebrand's disease the deficiency is in the eponymous factor.

***Haemophilus influenzae* (Pfeiffer's bacillus)** A nonmotile Gram-negative coccobacillus associated with influenza virus infections that causes pneumonia and meningitis; most strains are opportunistic pathogens. ***H. influenzae* type b** (Hib) causes bacteraemia and acute bacterial meningitis and is a major cause of lower respiratory tract infections in unvaccinated infants and children. It was the first free-living organism to have its entire genome sequenced.

haemopoiesis *See* HAEMATOPOIESIS.

haemorrhage The escape of blood from a ruptured blood vessel, usually with the implication that the bleeding is substantial. Anything that causes haemorrhage is referred to as **haemorrhagic**.

haemorrhoids Varicose dilations of the haemorrhoidal veins at the lower end of the rectum and the anus. Colloquially, piles.

haemosiderin A yellow-brown iron compound (iron(III) oxide/iron hydroxide) deposited in tissues in various diseases. Not in itself harmful, but a sign of bleeding and subsequent haemolysis, or iron overload. *See* HAEMOCHROMATOSIS.

haemostasis The cessation of bleeding as a result of blood clotting and local vasoconstriction. Anything that assists this process is **haemostatic**.

haem oxygenases Enzymes (HSP32, EC 1.14.99.3), of which various isoforms are known, that metabolize haem to form carbon monoxide, iron, and *biliverdin. Increased haem-derived CO inhibits nitric oxide synthase and is a putative neurotransmitter. Haem oxygenase-1 (HO-1, 288 aa) is an inducible heat-shock/stress protein and regulates inflammation by attenuating adhesive interactions and cellular infiltration. HO-2 (316 aa) is constitutive and present in endothelial cells, adventitial nerves of blood vessels and neurons in autonomic ganglia. The role of HO-3 is not clearly defined.

Hageman factor (*not* Hagemann factor) Blood-clotting factor XII, a β-globulin (615 aa), that is activated by contact with surfaces, forming factor XIIa, which then activates factor XI. Factor XIIa also generates *plasmin from plasminogen and *kallikrein from prekallikrein. Both plasmin and kallikrein activate the complement cascade. Hageman factor is important both in clotting and activation of the inflammatory process. Named after John Hageman, who was the first person to be identified with a deficiency in the factor. Mutations in the factor XII gene are the cause of hereditary angioedema (HAE) type III.

Hailey–Hailey disease A skin-blistering disease (familial chronic benign pemphigus) due to a defect in epidermal cell junctions, even though apparently normal desmosomes and adherens junctions can be assembled. Transmitted as an autosomal dominant and caused by mutation in a calcium-transporting ATPase (ATP2C1) that is highly expressed in human epidermal keratinocytes. It is not clear how this leads to the adhesion problem.

Haim–Munk syndrome An autosomal recessive disorder (Cochin Jewish disorder) characterized by palmoplantar keratosis, onychogryphosis, and periodontitis. The cause is a defect in *cathepsin C.

hair cells Epithelial cells found in the lining of the labyrinth of the inner ear. The cells have *stereovilli up to 25 μm long on their apical surfaces that restrict the plane in which deformation of the apical membrane of the cell by fluid movement induced by sound can be brought about. Movement of the single *stereocilium transduces mechanical movements into electrical receptor potentials.

hairless A transcriptional corepressor (1189 aa) for the thyroid hormone receptor, produced by the human homologue of the mouse hairless gene. It may regulate any one of the three phases of hair growth: anagen (growth phase), catagen (shortening phase), and telogen (resting phase) but the exact mode of action is unclear. Mutations lead to *alopecia universalis congenita and congenital atrichia.

hairpin A protein *motif in which two adjacent regions of a polypeptide chain lie alongside each other in an antiparallel manner. If the peptide is in α-helical form then it is an **α-hairpin**, if in β-sheet configuration, then it is a **β-hairpin**.

hairy cell leukaemia *See* LEUKAEMIA.

Hakata antigen *See* FICOLINS.

half-life ($t_{½}$) The time taken for the activity or concentration of a specified chemical or element to fall to half its original activity or concentration. Most commonly encountered in the context of the decay of radioactive isotopes but applicable to other situations.

Hallervorden–Spatz disease *See* INFANTILE NEUROAXONAL DYSTROPHY.

hallucinogen A drug or chemical which induces hallucinations.

haloperidol An antipsychotic drug of the butyrophenone class with a strong central antidopaminergic action, used in the treatment of schizophrenia. A decanoate ester is used as a long-acting injectable.

halothane An inhalational anaesthetic that was widely used but has now been superseded by halogenated ethers (e.g. isoflurane).

halysin *See* DISINTEGRINS.

hamartin *See* TUBERIN; TUBEROUS SCLEROSIS.

hamartoma A non-neoplastic malformation of tissue that is disordered in structure but grows at about the same rate as surrounding normal tissue. Examples are *haemangiomas and the pigmented naevus (mole) but they also occur in deep tissues where they may remain unnoticed.

hammerhead ribozyme A *ribozyme in which there are three helical regions radiating from a central core, fancifully resembling a hammerhead.

HAMP *See* HEPCIDIN ANTIMICROBIAL PEPTIDE.

hanatoxin Peptide toxins (35 aa), from the venom of the Chilean tarantula *Grammostola spatulata*, that inhibits the $KV_{2.1}$ voltage-gated potassium channel by altering the energetics of gating. Hanatoxin1 (HaTx1) and hanatoxin2 (HaTx2) are unrelated in primary sequence to other potassium channel inhibitors.

Hand–Schuller–Christian disease A rare disease, probably an autoimmune disorder, a type of *Langerhans cell histiocytosis, with onset between 3 and 5 years of age. Granulomatous histiocytic lesions are seen in bone and visceral tissues.

Hanks' balanced salt solution (HBSS) A physiological saline made up according to the recipe given originally by John H. Hanks, who devised it to be suitable for sterilization by autoclaving. It is phosphate-buffered to pH 7.0–7.2 and also contains bicarbonate. Phenol red is usually added as an indicator. Suitable for mammalian and avian cells but is not a growth medium and is only appropriate for temporary storage.

Hansen's disease *See* LEPROSY.

Hanta virus Viruses of the Bunyaviridae family that are responsible for haemorrhagic fevers and exist in various serotypes with different pathogenicity, varying from asymptomatic infection to fatal disease. The viruses persist in rodents and infection occurs through inhalation of aerosolized excreta. First identified near the Hantaan river in South Korea.

haploid The condition of possessing a single set of unpaired *chromosomes, a single set of genes. *Gametes are haploid.

haploinsufficiency The condition that arises when both copies of a gene (in a diploid organism) are required for normal function and one allele is defective.

haplotype **1.** One of the two sets of alleles possessed by a diploid organism. **2.** The set of alleles on one *chromosome or a part of a chromosome, i.e. one set of alleles of linked genes. This is commonly used in reference to the linked *MHC genes.

hapten An antigenic determinant, a moiety against which an antibody is directed, effectively an isolated *epitope. An isolated hapten may not induce an immune response and may need to be conjugated to a carrier protein. A classic example is dinitrophenol (DNP), against which anti-DNP antibodies are produced if it is conjugated to bovine serum albumin. Because haptens are monovalent they will block immune complex formation by competing with the hapten–carrier conjugate for antibody binding sites. (An immune complex will only form if there is more than one hapten per carrier.) Competitive inhibition by soluble small molecules is sometimes referred to as **haptenic inhibition**, and the term has been applied to lectin-mediated haemagglutination where monosaccharides that bind to the lectin will block haemagglutination.

haptic Pertaining to the sense of touch.

haptoglobin An *acute phase protein, a plasma glycoprotein (1–1.5 mg/mL) that will scavenge oxyhaemoglobin, the complex being removed in the liver by the monocyte/macrophage *scavenger receptor (*compare* HAEMOPEXIN which scavenges haem). Haptoglobin is a tetrameric α_2-globulin (two α subunits; α_1 of 83 aa, α_2 of 142 aa: two β subunits of 245 aa) with a common polymorphism in the α chain so that heterozygotes may have three isoforms of haptoglobin (Hp1-1, Hp2-1, and Hp2-2). Each β chain can bind an α–β heterodimer of haemoglobin so that each haptoglobin tetramer binds one haemoglobin tetramer. Neither deficiency (**hypohaptoglobinaemia**) nor complete absence of haptoglobin (anhaptoglobinaemia; **ahaptoglobulinaemia**) has any clear pathological consequences.

haptotaxis A directed (tactic) response of cells in a gradient of adhesion, but often misused to explain situations in which cells accumulate in more adhesive regions, where they are trapped because their speed drops to zero.

Ha-ras Harvey-ras. *See* RAS.

Hardy–Weinberg law One of the key principles of population genetics, that genotype frequencies in a population remain constant or are in equilibrium from generation to generation unless specific disturbing influences are introduced. The law can be formally expressed for alleles *A*, *a* with frequencies of p and q respectively: then $p + q = 1$ and the ratio $AA:Aa:aa$ is $p^2:2pq:q^2$. Deviation from this equilibrium distribution suggests that survival benefits are associated with one allele and that a new equilibrium will eventually be reached.

harlequin ichthyosis *See* ICHTHYOSIS.

harmonin A scaffolding protein (552 aa) that will bundle actin filaments and has multiple *PDZ domains. Harmonin forms complexes with cadherin-23 (CDH23, *otocadherin) through two of the PDZ domains and the two are coexpressed in the stereocilia of *hair cells and in photoreceptors in the retina. Harmonin anchors

CDH23 to the microfilaments of the stereocilia. Mutations in harmonin are the cause of *Usher's syndrome type 1C (USH1C), a rare, autosomal recessive syndrome of congenital deafness and progressive blindness; mutations in CDH23 are responsible for Usher's syndrome 1D. Mutations in harmonin are also responsible for autosomal recessive neurosensory deafness-18. **Harmonin-interacting, ankyrin repeat-containing protein** (HARP; 460 aa) is the human homologue of mouse *SANS protein.

Harris–Benedict equation An equation that uses basal metabolic rate and an activity factor to determine (approximately) total daily energy expenditure in calories. Activity factors range from 1.2 for sedentary individuals to 1.9 for extra-active individuals with a hard physical job and who also take hard exercise.

Hartnup's disease A disorder characterized by a light-sensitive rash, cerebellar ataxia, emotional instability, and aminoaciduria. The defect is in the amino-acid transporter SLC6A19 (solute carrier family-6) gene leading to an excessive loss of neutral amino acids (particularly tryptophan) in the urine, and poor absorption in the gut. If diet is rich and these amino acids are abundant then the disorder is not evident. *See* IMINOGLYCINURIA.

Harvey sarcoma virus *See* RAS.

Hashimoto's thyroiditis An autosomal dominant disease in which there are autoantibodies directed against thyroid peroxidase (EC 1.11.1.7), the enzyme involved in thyroid hormone biosynthesis, and against *thyroglobulin and *thyroid-stimulating hormone (TSH) receptors leading to destruction of the thyroid. **Hashimoto's encephalopathy** is a rare autoimmune disease associated with Hashimoto's thyroiditis in which α-enolase (EC 4.2.1.11) is the autoantigen.

Hassell's corpuscles Small spherical or ovoid bodies (thymic corpuscles, 20–50 μm diameter) found in the medulla of the thymus, composed of concentrically arranged whorls of flattened keratinized or hyaline epithelia cells with a central core of dead cells. Their function is unclear.

HA tag An *epitope tag from influenza virus haemagglutinin (HA) protein, residues 97–115.

HAT medium A growth medium for animal tissue cells that contains hypoxanthine, *aminopterin (a folate antagonist that blocks *de novo* DNA synthesis), and thymine. In HAT medium, cells depend on the exogenous bases, via the salvage pathways, for synthesis of purines and pyrimidines. Cells that lack enzymes such as *HGPRT or *thymidine kinase (TK) cannot survive. This provides a basis for selecting hybrid cells, e.g. in the production of monoclonal antibodies, because HGPRT-negative myeloma cells will be eliminated, the B cells have a short lifespan, and only hybrids will grow.

Haversian canals Small channels running along the length of bones, that surround blood vessels and nerves and link to osteocytes through canaliculi. Each Haversian canal is surrounded by concentric layers of compact bone, the whole forming an osteon.

HAX A bcl2 family-related protein (HS1-associated protein X1, 279 aa) that interacts with the haematopoietic and lymphoid-restricted intracellular protein (HS1) and with cortactin. HAX1 is critical to maintaining the inner mitochondrial membrane potential and protecting against apoptosis in myeloid cells; mutation leads to recessive severe congenital neutropenia (*Kostmann's disease).

hay fever A common form of seasonal allergic rhinitis caused by inhaled allergens, usually pollens. There is acute nasal catarrh and conjunctivitis caused by a type I immediate *hypersensitivity reaction. The predisposition tends to be inherited.

Hayflick limit *See* CELL DEATH; CELL LINE.

HB-EGF *See* HEPARIN-BINDING EPIDERMAL GROWTH FACTOR.

HBGF Heparin-binding growth factor. *See* FIBROBLAST GROWTH FACTOR-1; MIDKINE.

HBSS *See* HANKS' BALANCED SALT SOLUTION.

HBV Hepatitis B virus. *See* HEPATITIS VIRUSES.

HCC1 An RNA-binding protein (hepatocellular carcinoma protein 1, transcription coactivator CAPER, RNA binding motif protein 39, 530 aa) and possible splicing factor that colocalizes with core spliceosomal proteins. A mouse protein with high sequence similarity may act as a transcriptional coactivator for JUN/AP1 and oestrogen receptors. Multiple transcript variants have been found.

HCG (hCG) Human chorionic gonadotropin. *See* CHORIONIC GONADOTROPIN.

H chain An immunoglobulin heavy chain; *see* IGG; IGM; etc.

hck *See* HAEMATOPOIETIC CELL KINASE.

HCP **1.** Health care provider. **2.** *See* HISTIDINE-RICH CALCIUM-BINDING PROTEIN.

HCT **1.** A human colon cancer cell line, of which there are many variants, HCT-15, HCT-116, etc. **2.** Haematopoietic cell transplantation. **3.** *See* HAEMATOCRIT (Hct).

HCV Hepatitis C virus. *See* HEPATITIS VIRUSES.

HCV core protein One of 10 virally encoded proteins, a viral structural protein (hepatitis C core protein, 3010 aa) that affects multiple cellular processes, blocks the activity of caspase-activated DNase and thus inhibits apoptotic cell death. Maturation of core protein requires coordinate cleavage by signal peptidase and an intramembrane protease, signal peptide peptidase. An alternate reading frame (ARF) overlaps with the core protein-encoding sequence and encodes the ARF protein (ARFP).

h

HDAC *See* HISTONE DEACETYLASES.

HDF **1.** Haemodiafiltration, an alternative to conventional haemodialysis combining convection and diffusion. **2.** Human dermal fibroblasts. **3.** Human diploid fibroblasts.

HDL High density *lipoprotein.

hdm2 Human double minute 2. The human homologue (MDM2) of a negative regulator for the tumour suppressor *p53 found originally in mouse. The nuclear phosphoprotein coded by the *MDM2* gene is an ubiquitin *E3 ligase that ubiquitinylates p53, leading to its destruction. Overexpression of hdm2 leads to increased susceptibility to tumours.

HDR syndrome Hypoparathyroidism, sensorineural deafness, and renal anomaly syndrome. *See* GATA TRANSCRIPTION FACTORS (Gata-3).

Heaf test A skin test used to determine whether there had been prior exposure to tuberculin. The test involved intradermal injection of tuberculin with a multiple-puncture apparatus. A positive reaction indicated T-cell reactivity to mycobacterial products. Has now been superseded by the *Mantoux test.

heart block A disorder of the intrinsic pacemaker system of the heart that can arise through defects in various electrical conduction pathways.

heartburn A burning sensation in the midline of the chest, usually a result of indigestion.

heart failure (congestive heart failure) Inability of the heart to pump sufficient blood to other tissues. It can arise through a variety of causes, usually chronic, the commonest being hypertension and coronary artery disease.

heart muscle *See* CARDIAC (cardiac muscle).

heart valve implant An artificial valve used to replace diseased valves; it may be mechanical or bioprosthetic. The former may cause clotting and activation of complement, although modern blood-compatible materials have reduced this problem, whereas the latter, often made from glutaraldehyde-fixed porcine valves, are less durable.

heat exhaustion A heat-related illness that can develop after several days of exposure to high temperatures and poor salt and water replenishment. It is less severe than **heat stroke** (heat hyperpyrexia) in which there may be coma, convulsions, and elevated core tempertaure.

heat shock factors A family of DNA-binding proteins (HSP1, 529 aa; HSP2, 536 aa, and others) that bind heat shock promoter elements (HSE) and activate transcription. In nonstress conditions are bound by *heat shock proteins and retained in the cytoplasm.

heat shock proteins (Hsps) A group of proteins, conserved through pro- and eukaryotes, that might more accurately be termed stress-induced proteins, since a variety of environmental stressors will induce their expression (although a few are constitutively expressed). Four main families of hsps are recognized, particularly Hsp60, Hsp70, Hsp90, and small hsps, all named for their approximate molecular weights. Most act as molecular *chaperones. Hsp90 complexes with inactive steroid hormone receptor and is displaced upon ligand binding. Hsps have been suggested to act as major immunogens in many infections.

heavy chain In general, the larger subunit of a multimeric protein. The immunoglobulin heavy chain is of 50 kDa, the light chain of 22 kDa, whereas in myosin the heavy chain is ~220 kDa and the light chain a mere 20 kDa. **Heavy chain diseases** are rare B-cell proliferative disorders characterized by the synthesis and secretion of aberrant immunoglobulin heavy chains. They can be subdivided according to the heavy chain type (μ, γ, etc.) and the extent to which the peptide is truncated or altered.

heavy water (deuterium oxide, D_2O) Water with deuterium (D) as the hydrogen isotope. Because substitution of heavy water for normal water alters the thermodynamics of polymerization it will affect some equilibrium systems, e.g. it will stabilize *microtubules.

Heberden's nodes Small bony knobs that form most commonly in distal interphalangeal joints of elderly people, usually a sign of *osteoarthritis. The is evidence of hereditable predisposition and the condition is more common in women than men.

hebetude A condition of mental dullness and lethargy.

HECT domain A catalytic domain (homologous to E6-AP C-terminus, ~350 aa) found at the C-terminus of Hect-class E3 ubiquitin protein ligases. The domain binds specific E2 enzymes, accepts ubiquitin, and transfers it either directly to the substrate or to the growing end of multiubiquitin chains. *See* UBIQUITIN CONJUGATING ENZYMES.

hedgehog A secreted morphogen originally discovered in *Drosophila*; various mammalian homologues have subsequently been identified (*sonic hedgehog (462 aa), *Indian hedgehog (411 aa), and *desert hedgehog (396 aa)), all of which are involved in patterning of the embryo. The protein precursors are autoprocessed to produce ~20-kDa N-terminal portions that are active in the hedgehog signalling pathway which, if perturbed, can disrupt organ development and cause various degenerative and neoplastic human diseases. Further processing of the active portion involves palmitoylation and the addition of cholesterol. The primary receptor is the twelve-transmembrane protein *patched (Ptc) but a second receptor, a seven-transmembrane protein called smoothened (Smo), actually transduces the signal. Smo is held in a dormant state by Ptc in normal cells and Ptc is internalized when it binds hedgehog, thereby releasing the inhibition. The signal is up-regulated by the *gli* oncogene family of transcription factors which activate *forkhead box (FOX) transcription factors in a complex cascade. *See* HEDGEHOG SIGNALLING COMPLEX.

hedgehog signalling complex The multiprotein complex downstream of *smoothened that has been most extensively characterized in *Drosophila*, but which probably has a homologue in humans. It is a tetrameric microtubule-associated complex involving the kinesin-related protein Costal2 (Cos2), the protein kinase fused (Fu), and the transcription activator cubitus interruptus (Ci), the mammalian homologues of which are the Gli family. *See* HEDGEHOG; SMOOTHENED; SONIC HEDGEHOG.

SEE WEB LINKS

- A mini-review of the hedgehog signalling system.

HEF1 A multifunctional docking protein (human enhancer of filamentation 1, neural precursor cell expressed, developmentally downregulated-9, NEDD-9, Cas-L, 835 aa) of the *Cas family that is involved in both integrin and growth factor signalling pathways. There are two isoforms, p105 and p115, the larger resulting from serine/threonine phosphorylation of p105HEF1. It can stimulate the formation of neurite-like processes and has been shown to interact with *aurora A kinase at the basal body of cilia causing phosphorylation and activation of a tubulin deacetylase (HDAC6), thereby promoting ciliary disassembly.

hefutoxin *See* KAPPA TOXIN.

Heidenhain's Azan A trichrome histochemical staining method that results in chromatin, erythrocytes, and neuroglia being stained red, mucus blue, collagen sharp blue, and cytoplasmic granules red, yellow, or blue. Aesthetically very pleasing, but rarely used.

Heidenhain's iron haematoxylin *See* HAEMATOXYLIN.

Heidenhain's Susa A general-purpose histological fixative that contains mercuric chloride.

HEK-293 cells A line of human embryonic kidney cells transformed by adenovirus type 5 DNA. They stain strongly and specifically with antibodies to several neurofilament proteins and may be derived from a neuronal cell. Although probably not a useful model for kidney, they are easily transfected and are used extensively in biotechnological applications.

HeLa cells An established line of human epithelial cells derived from a cervical carcinoma in the patient Henrietta Lacks. Because they grow readily they have cross-contaminated many other cell lines.

Helber cell *See* HAEMOCYTOMETER.

helicase *See* DNA HELICASE.

Helicobacter pylori **(formerly *Campylobacter pyloridis*)** An S-shaped Gram-negative bacterium (0.5–0.9 × 3.0 μm) that infects the stomach and is a major factor predisposing to gastric and probably duodenal ulcers, although infection in most people is asymptomatic. Can be treated with antibiotics, but resistant strains are developing.

helix–loop–helix A conserved DNA-binding domain (40–50 aa) found in many *transcription factors consisting of two amphipathic helices joined by a variable-length linker region that forms a loop. The DNA sequence-specific binding requires dimerization. Most of the HLH proteins have an extra basic region of about 15 amino acid residues adjacent to the HLH domain and are referred to as **basic HLH** (bHLH) proteins and bind variations on the core sequence 'CANNTG' (E-box motif). Examples include myoblast MyoD1, c-myc, *sterol regulatory element binding proteins (SREBPs). Not the same as *helix-turn-helix.

helix–turn–helix A DNA-recognition motif found in some *transcription factors. It consists of two *amphipathic α-helices separated by a short nonhelical sequence. The C-terminal recognition helix enters the major groove and interacts with specific bases in the DNA duplex; the other helix interacts with other proteins. Examples include lambda repressor, tryptophan repressor, catabolite activator protein (CAP), octamer transcription factor 1 (Oct-1), and heat shock factor (HSF). Not the same as *helix-loop-helix.

HELLP syndrome A variant of pre-eclampsia characterized by haemolytic anaemia, elevated liver enzymes and low platelet count. *See* HUMAN ENDOGENOUS RETROVIRUSES.

helminthiasis Infestation with parasitic worms (helminths).

helospectin *See* EXENDIN.

helothermine A protein toxin (242 aa) stabilized by eight disulphide bridges, from the venom of the Mexican beaded lizard *Heloderma horridum horridum*, that inhibits the *ryanodine receptor in skeletal and cardiac muscle. Has no homology with other channel-blocking toxins but appears to be a member of a family of *cysteine-rich secretory proteins (CRISP) described in human and mouse testis.

helper factor An obsolete term for a group of factors that were invoked to explain some of the cell–cell interactions involved in immune responses. Many cytokines and interleukins act as helper factors, but the term itself is probably better avoided.

helper T cell *See* T-HELPER CELLS.

helper virus A virus that permits a defective virus to replicate if there is coinfection. The intact helper virus produces a protein that is deficient in the defective virus.

hema-, hemo-, hemato- US form of UK English *haema-, *haemo-, *haemato-; the relevant UK form is used in this dictionary.

heme *See* HAEM.

hemi- Prefix indicating half, although often used to mean a condition restricted to one side of the body. Thus hemithyroidectomy is the surgical removal of one half of the thyroid gland but **hemichorea** is *chorea, repetitive jerky movement, restricted to one side of the body, **hemiparesis** is weakness confined to one side of the body, and **hemiplegia** is unilateral paralysis.

hemiasterlin A tripeptide derived from marine sponges (among them *Hemiasterella minor*) that will block polymerization of microtubules by binding to the vinca-alkaloid binding site on β-tubulin. The amino acids are unusual, and synthetic derivatives of hemiasterlin have been produced for chemotherapeutic use. In standard tubulin polymerization assays it is more potent than vincristine, marginally less active than *dolastatin-10, and approximately equal in activity to *cryptophycin-1. *See* PHOMOPSIN A.

hemicentin *See* FIBULINS (fibulin-6).

hemicrania *See* MIGRAINE.

hemidesmosome A specialized adhesion structure formed by an epithelial cell with its basal lamina that provides mechanical linkage between the intermediate filament cytoskeleton of cytokeratin and the extracellular matrix. Hemidesmosomes are morphologically similar to half a desmosome but the proteins involved are different with *integrins replacing *cadherins.

hemizygote A cell or organism that has only one set of a normally *diploid set of genes. Males are hemizygous for the *X chromosome.

hemojuvelin A glycosylphosphatidylinositol-linked protein (426 aa) that undergoes a partial autocatalytic cleavage during its intracellular processing. It is a coreceptor for the *bone morphogenetic proteins 2 and 4 and enhances BMP-induced *hepcidin expression. Hemojuvelin also binds *neogenin, a membrane protein widely expressed in many tissues, and neogenin apparently regulates the shedding of hemojuvelin. The majority of characterized cases of juvenile *haemachromatosis involve mutations in HJV/hemojuvelin.

Henderson–Hasselbalch equation An equation used to calculate the pH of solutions where the concentrations of the protonated (HA) and non-protonated (A^-) forms of a weak acid are known. The equation is of the form: $pH = pK_a + \log([A^-]/[HA])$, where pK_a is the acid dissociation constant.

Henoch–Schönlein purpura An IgA-mediated autoimmune condition in which there is a generalized vasculitis involving the small vessels of the skin, accompanied by abdominal pain, IgA deposition, arthritis or arthralgia, or renal involvement. It is the most common vasculitis in children. In adults the predominant feature may be glomerulonephritis as a result of immune complex deposition.

HEp2 cells A cell line of epithelioid morphology originally established from a laryngeal carcinoma but probably contaminated by and now derived from *HeLa cells. Extensively used in viral studies and susceptible to arboviruses and measles virus. They are also routinely used in tests for antinuclear antibodies (ANA).

Hepadnaviridae A family of small DNA viruses (Baltimore class VII) that cause liver disease. The genome consists of partially double-stranded, partially single-stranded circular DNA that replicates via an RNA intermediate which is reverse-transcribed into DNA. The best known example is *hepatitis B virus.

heparanase A lysosomal endoglycosidase (HSE1, HPA, EC 3.2.1.-, 543 aa) that will cleave *heparan sulphate into characteristic large molecular weight fragments. Heparanase apparently plays an important role in degrading extracellular matrix heparan sulphate to facilitate extravasation of tumour cells and leucocytes and, by degrading heparan sulphate, will alter the binding of various signalling molecules.

heparan sulphate A linear glycosaminoglycan (usually ~50 kDa) closely related in structure to heparin. The commonest disaccharide unit is glucuronic acid (GlcA) linked to *N*-acetylglucosamine (GlcNAc) but there are small numbers of *N*- and *O*-linked sulphated sugars. Heparan sulphate is a constituent of membrane-associated *proteoglycans, particularly *syndecans and GPI-linked *glypicans, and is present in extracellular matrix bound to perlecan, agrin, and collagen XVIII core proteins. Heparan sulphate is involved in the binding of various proteins; e.g. in fibroblast growth factor (FGF) signalling, HS forms part of a ternary complex that involves both FGF and its receptor. *Endostatin–HS complexes apparently block the proangiogenic activities of FGFs and vascular endothelial growth factor (VEGF).

heparin A highly sulphated glycosaminoglycan (3–40 kDa), similar to *heparan sulphate but with the commonest disaccharide being 2-*O*-sulphated iduronic acid linked to *N*-sulphated glucosamine. It is found in the granules of mast cells and inhibits the action of *thrombin on *fibrinogen by potentiating antithrombins, thereby interfering with the blood clotting cascade. Platelet factor IV will neutralize heparin. Heparin is widely used as an anticoagulant and can be used to coat surfaces that will come into contact with blood.

heparin-binding epidermal growth factor (HB-EGF) An *EGF-like growth factor (208 aa) that stimulates smooth muscle cell proliferation. The factor is synthesized as a type I transmembrane protein (proHB-EGF) and expressed on the cell surface; the ectodomain is shed following cleavage by a metallopeptidase and will bind to and activate the EGF receptors HER-1 and HER-4. Soluble HB-EGF will bind heparin and other connective tissue macromolecules such as heparan sulphate and this immobilization may produce a local concentration greater than that of soluble factors. The cytoplasmic domain of proHB-EGF (HB-EGF-cyto) interacts with transcriptional repressors to reverse their repressive activities. The precursor is the binding site for diptheria toxin.

heparin-binding growth factor (HBGF) *See* FIBROBLAST GROWTH FACTOR (fibroblast growth factor-1).

hepatic lipase An important enzyme in triglyceride metabolism (hepatic triglyceride lipase, EC 3.1.1.3, 499 aa) that is bound to and acts at the endothelial surfaces of hepatic tissues. Mutations in hepatic lipase result in abnormally triglyceride-rich low- and high-density lipoproteins. *See* LIPOPROTEIN LIPASE.

hepatic stellate cells *See* ITO CELLS.

hepatitis Inflammation of the liver. Often, but not always, caused by *hepatitis viruses.

hepatitis viruses A disparate group of viruses that cause inflammation of the liver. **Hepatitis A** virus is a small (27 nm diameter) single-stranded RNA virus with some resemblance to *enteroviruses such as polio; it causes 'infectious hepatitis'. **Hepatitis B** virus (HBV) causes 'serum hepatitis' and is a problem because the virus can persist for long periods in asymptomatic carriers. The virion (Dane particle) is 42 nm in diameter, with an outer sheath enclosing an inner 27-nm core particle with the circular viral DNA. Aggregates of the envelope proteins are found in plasma and are referred to as **hepatitis B surface antigen** (HBsAg; previously called Australia antigen). Integrated virus is associated with hepatocellular carcinoma. **Hepatitis non-A, non-B** is caused by a virus that is neither hepatitis A or B and has no antigenic cross-reaction with either; often, but not always, hepatitis C virus. **Hepatitis C** virus (HCV) is an enveloped RNA virus of the *Flaviviridae family with a particle size ~50 nm in diameter and a positive-sense RNA genome of 9600 nucleotides. Chronic HCV infection causes hepatocellular carcinoma although the mechanism is unknown. **Hepatitis D** virus (delta virus) is a small circular RNA (single-stranded, negative sense) virus that is replication defective and cannot propagate without a *helper virus, usually HBV. **Hepatitis E** virus (HEV) is a spherical, nonenveloped, single-stranded RNA virus and is responsible for acute enterically transmitted non-A, non-B hepatitis worldwide. **Hepatitis F** virus may be the cause of a viral infection, common in East Asia, which is neither HBV nor HCV. **Hepatitis G** virus (HGV, GB virus C, GBV-C) is a blood-borne flavivirus, an RNA virus similar to hepatitis C although there is some doubt as to whether it causes disease in its own right. **Hepatitis H** virus may be emerging—there is already a candidate non-A, non-E virus.

hepato- A prefix denoting something associated with the liver. A **hepatolith** is a biliary stone (gallstone) in the liver, rather than in the gallbladder. A **hepatoma** is a carcinoma that is derived from liver cells, although hepatocarcinoma is a preferable term. **Hepatomegaly** is enlargement of the liver. **Hepatosplenomegaly** is concurrent enlargement of liver and spleen (splenomegaly).

hepatocarcinoma A malignant tumour (hepatocellular carcinoma) derived from hepatocytes and associated with *hepatitis B or C virus in 80–90% of cases.

hepatocyte The main cell type of the liver, constituting 70–80% of the overall mass. Hepatocytes are epithelial cells, often considered the archetypal animal cell, and are responsible for a range of metabolic functions including detoxification of xenobiotics. Hepatocytes are in direct contact with blood because the blood vessels of the liver have fenestrated endothelium.

hepatocyte growth factor A mitogen for hepatocytes (HGF, scatter factor, hepatopoietin A, lung fibroblast-derived mitogen) derived, in the liver, from nonparenchymal cells. It is mitogenic for several cell types and is found in many cells outside the liver, including platelets. HGF is synthesized as a single chain precursor (728 aa) that is cleaved to a heavy chain (463 aa) and a light chain (234 aa) that remain disulphide-linked. Both the uncleaved precursor and the two-chain form of HGF are biologically active and HGF is generally isolated as a mixture of the two forms. HGF also alters cell motility and is identical to *scatter factor. The **HGF receptor** gene, the *met oncogene product, is on the same region of chromosome 7 (7q) as HGF itself so that polysomy of chromosome 7 can confer invasive and proliferative capacity through an autocrine effect, and polysomy of chromosome 7 is the most common abnormality in human malignant *gliomas. Differentiation of pre-pro-B cells to pro-B cells requires signalling through the IL-7 receptor by IL-7 in association with the β chain of HGF and heparin sulphate. HGF-stimulation of hepatocytes through the MET receptor makes cells susceptible to infection by sporozoites of *Plasmodium* and damage caused by sporozoite migration causes release of HGF.

hepatocyte growth factor-like protein *See* MACROPHAGE STIMULATING PROTEIN.

hepatocyte nuclear factor *See* HNF.

hepatolenticular degeneration *See* WILSON'S DISEASE.

hepatoma transmembrane kinase A receptor tyrosine kinase (HTK, EphB4, 987 aa) of the *EPH class that is expressed abundantly in placenta and in a range of primary tissues and malignant cell lines. May be involved in differentiation of monocytes. The ligand is *Htk-L.

hepcidin antimicrobial peptide A hormone peptide (HAMP, liver-expressed antimicrobial peptide, LEAP1) produced in the liver that is essential to the negative regulation of iron. It regulates duodenal iron absorption and iron trafficking in the reticuloendothelial system by inhibiting the cellular efflux of iron by binding to and inducing the degradation of *ferroportin, the sole iron exporter in iron-transporting cells. The active C-terminal 25 aa peptide is cleaved from a 84 aa precursor and contains a unique 17-aa stretch with eight cysteines forming four disulphide bridges. HAMP has antimicrobial properties against Gram-positive bacteria, and inhibits the growth of certain yeast and Gram-negative species with activity similar to β-*defensin.

HEPES One of the zwitterionic buffers (hydroxyethyl-piperazine-ethane-sulphonic acid) originally described by Norman Good and colleagues in 1966 and very commonly used in tissue culture media because of its pK_a of 7.5. Related compounds are molluscicides so may be unsuitable for some invertebrate cultures.

HepG2 cells A cell line derived from a hepatic carcinoma. They are epithelial in morphology, produce a variety of proteins such as prothrombin, alpha-fetoprotein, C3 activator, and fibrinogen, and express a wide variety of liver-specific metabolic functions, including those related to cholesterol and triglyceride metabolism, making them a useful *in vitro* model system.

hephaestin A transmembrane ferroxidase (1158 aa) that acts in conjunction with *ferroportin and that has been implicated in duodenal iron export and possibly copper homeostasis. Mutations in the murine hephaestin gene produce microcytic, hypochromic anaemia that is refractory to oral iron therapy. Hephaestin has ~50% sequence identity with *ceruloplasmin and is predicted to bind six copper atoms per molecule. (Hephaestus was the Greek god of metalworking.)

heptad repeats A repeat sequence of seven amino acids that occurs many times in a protein sequence. Heptad motifs are the basis for most coiled-coil proteins, particularly *leucine zippers.

HER A family of receptors (HER-2, erbB2, 1255 aa; HER-3, erbB3, 1342 aa; HER-4, erbB4, 1308 aa) of the *epidermal growth factor receptor family of receptor *tyrosine kinases. The ligands are *neuregulins. The *HER2* gene is amplified and overexpressed in 25–30% of breast cancers and correlates with poor prognosis, although these are the tumours that respond to *herceptin. HER-3 is a receptor for heregulin.

heraclenin A furanocoumarin compound, extracted from sweet myrrh *Opopanax chironium*, that is an inhibitor of receptor-mediated proliferation in human primary T cells and induces DNA fragmentation at the G_2/M phases of the cell cycle. Despite a close structural similarity to *imperatorin, it induces apoptosis in a mechanistically different way.

herbimycin A An inhibitor of tyrosine kinases isolated from *Streptomyces hygroscopicus*. It is structurally related to *geldanamycin although less potent.

herceptin A humanized monoclonal antibody that blocks EGF-mediated cell proliferation by binding to the *epidermal growth factor receptor (*HER2; c-erbB2). Only the subset of breast carcinomas (25–30%) that overexpress HER2 can be treated with this antibody, one of the first examples of the importance of pharmacogenomics. *TN* Trastuzumab.

herculin The product (MYF6, muscle regulatory factor-4, MRF4, 242 aa) of the muscle regulatory gene *MYF6*, one of the **myoD* family of muscle regulatory genes, that probably plays a role in myotube differentiation. A mutation in *MYF6* is responsible for a mild myopathy in a single reported case.

hereditary angioedema An autosomal dominant disorder (hereditary angioneurotic oedema) in which there is episodic local subcutaneous oedema and oedema involving the upper respiratory and gastrointestinal tracts (which can have fatal consequences by restricting breathing). Hereditary angioedema **types I and II** are caused by mutation in the complement C1 inhibitor gene, the product of which is a serine peptidase inhibitor (serpin). As a result there is uncontrolled production of *C2-kinin. **Type III** occurs only in women and is precipitated or worsened by high oestrogen levels and is caused by mutation in the gene encoding coagulation factor XII (*Hageman factor).

hereditary ataxia A group of inherited central nervous system diseases in which there is ataxia, loss of motor control. *See* ATAXIA TELANGIECTASIA; EPISODIC ATAXIA; FRIEDREICH'S ATAXIA; SPINOCEREBELLAR ATAXIA.

hereditary haemorrhagic telangiectasia A set of vascular dysplasias (Osler–Rendu–Weber syndrome) characterized by thinning of blood vessel walls and thus a predisposition to bleeding (nosebleeds are a common symptom). Malformations (telangiectasia) in the arteriovenous system occur in skin, mucosa, and viscera. **Type 1** (HHT1) is an autosomal dominant disorder caused by mutation in the gene encoding *endoglin on chromosome 9, **HHT2** is caused by mutations in the activin receptor-like kinase-1 gene on chromosome 12, **HHT3** maps to 5q31 and **HHT4** to 7p14.

hereditary lymphoedema A complex of disorders in which there is lymphoedema. **Type 1A** (Milroy's disease) is an autosomal dominant disorder caused by mutation in the *FLT4* gene, the product of which is the vascular endothelial growth factor receptor-3. **Type IB** has a similar phenotype but the mutation is not in *FLT4* (a candidate locus has been identified). **Type II** (Meige lymphoedema) in which oedema, particularly severe below the waist, develops about the time of puberty, is caused by mutation in the forkhead family transcription factor gene *MFH1* (*FOXC2*). **Lymphoedema–distichiasis syndrome** is an autosomal dominant disorder in which there is lymphoedema of the limbs and double rows of eyelashes (distichiasis). It seems to be an allelic variant of type II lymphoedema caused by mutation in the same gene, as are lymphoedema and ptosis and lymphoedema and yellow nail syndrome.

hereditary nonpolyposis colon cancer *See* GTBP.

hereditary sensory and autonomic neuropathies A heterogeneous group of disorders associated with sensory dysfunction. Hereditary sensory neuropathy type I (**HSAN1**) is caused by mutation in the gene that encodes serine palmitoyltransferase (EC 2.3.1.50), the key enzyme in sphingolipid biosynthesis. **HSAN2** is caused by mutation in the *HSN2* gene that encodes a 434-aa protein of uncertain function. **HSAN3** (*see* DYSAUTONOMIA) is caused by mutation in the *IκB kinase-associated protein. **HSAN4** is caused by mutation in a *neurotrophin receptor. **HSAN5** is caused by mutation in the *nerve growth factor β subunit. **Adult-onset HSAN** with anosmia is believed to be a distinct form.

hereditary spastic paraplegias *See* SPASTIC PARAPLEGIAS.

hereditary stomatocytosis *See* STOMATOCYTOSIS.

heregulins *See* NEUREGULINS.

hERG A gene (human ether-a-go-go related gene, *ERG1, LQT2, SQT1, Kv11.1*, now officially *KCNH2*) encoding the pore-forming subunit of cardiac IKr, an inwardly rectifying potassium channel that may have a specific role in the normal suppression of arrhythmias. Mutations are associated with *long QT2 syndrome and *short QT syndrome.

Hermansky–Pudlak syndrome A genetically heterogeneous disorder characterized by oculocutaneous albinism, prolonged bleeding, and pulmonary fibrosis due to abnormal vesicle trafficking to lysosomes and related organelles, such as melanosomes and platelet dense granules. The syndrome can be caused by mutation in several genes, *HPS1–6*: The products of the *HPS1* and *HPS4* genes form a lysosomal complex termed *BLOC3 (biogenesis of lysosome-related organelles complex-3). Hermansky–Pudlak syndrome type 2 (**HPS2**), which is further complicated by immunodeficiency, is caused by mutation in the gene encoding the β-3A subunit of the *adaptor protein 3 complex. **HPS type 7** is caused by mutation in the *dystrobrevin binding protein gene and **type 8** by mutation in the *BLOC1* gene.

hernia The protrusion of an organ, or part of an organ, from the cavity in which it is normally enclosed, usually through an opening or weak spot in the surrounding tissue.

Herpesviridae A group of large double-stranded DNA viruses with 100–200 genes. **Herpes simplex** (human herpesvirus-1, HHV-1, HHV-2; HSV) causes cold sores and genital herpes; **varicella zoster** (HHV-3) causes chickenpox and shingles; ***Epstein–Barr virus** (EBV; HHV-4) causes glandular fever; **cytomegalovirus** (HHV-5) causes congenital abnormalities and is an opportunistic pathogen; **roseolovirus** (HHV-6) causes *sixth disease and **Kaposi's sarcoma-associated herpesvirus** is HHV-8. Herpes simplex type 2 is associated with cervical carcinoma and EBV with *Burkitt's lymphoma and nasopharyngeal carcinoma. Varicella becomes latent in sensory neurons of human dorsal root ganglia and can become reactivated as shingles if there is immunosuppression.

Herring bodies Granular regions towards the terminal part of axons in the posterior lobe of the pituitary gland. Contain, in separate bodies, the neurosecretory hormones *vasopressin and *oxytocin.

Hers' disease An autosomal recessive glycogen storage disease (glycogen storage disease VI) in which there is a deficiency of liver phosphorylase (EC 2.4.1.1). Affected individuals show mild to moderate hypoglycaemia, mild ketosis, growth retardation, and prominent hepatomegaly.

herstatin A secreted protein (419 aa) that is the product of an alternative ErbB2 transcript that retains intron 8 (*see* EPIDERMAL GROWTH FACTOR). Herstatin appears to be an inhibitor of p185-ErbB2, because it disrupts dimers, reduces tyrosine phosphorylation of p185, and inhibits the anchorage-independent growth of transformed cells that overexpress ErbB2.

Hes/Hey proteins *HES* (hairy and enhancer-of-split) and *HEY* (Hes-related repressor Herp, Hesr, Hrt, CHF, gridlock) genes are the mammalian counterparts of *Drosophila* genes and are the primary targets of the Delta-*notch signalling pathway. They are transcriptional repressors with basic *helix–loop–helix (bHLH) and Orange domains, and are implicated in preventing tissue-specific determination of stem cells.

HETE Derivatives of arachidonic acid (hydroxyeicosatetraenoic acids) produced by the action of lipoxygenase that have potent pharmacological activity; e.g. 20-HETE activates protein kinase C and thereby regulates renal sodium-potassium ATPase. *See also* HPETE.

hetero- Prefix denoting different or varied.

heteroagglutination The adhesion of erythrocytes to one another when blood of different groups is mixed. *Compare*

ISOAGGLUTINATION. The term can also be used more generally for agglutination of mixed cell types in experimental systems.

heterochromatin The region of the chromosome that is condensed, and thus stains differently, during interphase and during division; *compare* EUCHROMATIN. Heterochromatin is transcriptionally inactive and can be constitutive, present in all cells, or facultative, present in only some cell types, e.g. the inactive X chromosome of females.

heterochrony Lack of synchronization. In developmental biology, a change in the timing or duration of events when comparing the developmental trajectories of closely related species.

heteroclitic antibody An antibody with lower affinity for the immunogen than for some other antigen.

heterodimer A dimer in which the two subunits are different. There are many examples, but one of the best known is tubulin in which the *protomer, the subunit from which the microtubule is assembled, is a dimer of α- and β-tubulin.

heteroduplex A double-stranded nucleic acid molecule in which the two strands are different, either derived from different parents, formed *in vitro* by annealing similar strands with complementary sequences, or formed of mRNA and the corresponding DNA strand.

heterogenous nuclear RNA (hnRNA) A type of RNA, found in the nucleus but not the nucleolus, which is rapidly labelled in experimental systems and is very diverse in size. It comprises unprocessed *messenger RNAs from which introns are removed by splicing.

heterokaryon A cell that contains two or more genetically different nuclei. Heterokaryons can be produced experimentally by cell fusion techniques and are important in *hybridoma production and in testing for complementation. *See* SYNKARYON.

heterologous Derived from the tissues or DNA of a different individual or species. *Compare* AUTOLOGOUS; HOMOLOGOUS.

heterolysosome An obsolete name for a *secondary lysosome.

heterophile antibody An antibody raised against an antigen from one species that also reacts against antigens from other species. They usually arise because the antigen is poorly defined or impure and the antibody is of low affinity, and are a common cause of problems with immunoassays. The term can also be used of systems such as the *Forssman antigen where antibody against antigens from a variety of species is present without immunization.

heteroplasia The occurrence of a tissue in the wrong place in an organism, either because of inappropriate cellular differentiation or infiltration by cells of another tissue. *See* PROGRESSIVE OSSEOUS HETEROPLASIA.

heteroplasty The grafting of tissue from one individual to another.

heterosis The phenomenon of hybrid vigour, the superiority of a heterozygotic organism over the homozygote because a wider range of alleles gives an advantage in exploiting environmental variation. The opposite of inbreeding depression.

heterospecific antibody An antibody (heteroclitic antibody) produced against an antigen from one species that reacts, sometimes more strongly, with a different antigen in another species. Usually a consequence of rather weak affinity.

heterotopia Displacement of an organ or part of an organ from its normal anatomical position. **X-linked periventricular heterotopia**, a neurological disorder in which neuronal cell bodies are mislocated, is caused by mutation in the gene encoding *filamin-A. **Autosomal recessive periventricular heterotopia** with microcephaly (ARPHM) is caused by mutation in the gene encoding ADP ribosylation factor (ARF) guanine nucleotide exchange protein-2 (*ARFGEF2*). Both forms are often associated with seizures and epilepsy but can be asymptomatic. Other forms of periventricular heterotopia are known. *Lissencephaly type I is also a heterotopia.

heterotroph An organism that cannot synthesis its own food but must obtain carbon compounds from other plant or animal sources. The opposite of an *autotroph.

heterotypic Describing an interaction between dissimilar things; e.g. heterotypic adhesion is adhesion between different cell types.

heterozygosity index A measure of the number of gene loci that are heterozygous; a completely inbred individual, homozygous at all loci, would have an index value of zero. It can be estimated for individuals or for populations. For example, the ApaLI polymorphism in intron 5 of the *Wilms' tumour gene has a heterozygosity index of 24% in the USA but 59% in the Japanese population, with potential implications for the prevalence of the disorder.

heterozygote A nucleus, cell, or organism with different *alleles of one or more specific genes. Heterozygous organisms will produce unlike gametes and thus will not breed true, but may benefit from *heterosis.

heuristic Describing the process of arriving at an explanation or solution by trial and error or rule-of-thumb methods rather than using a formula or algorithm.

HEV *See* HEVEIN; HIGH ENDOTHELIAL VENULE.

hevein A protein, found in the rubber tree *Hevea brasiliensis*, that has a lectin-like domain that will bind chitin and GlcNAc-containing oligosaccharides and confers antifungal activity. Hevein (Hev b 6.02) is a major allergen in natural rubber latex, as used in gloves.

hexacyclinic acid An antitumour antibiotic that has weak cytotoxicity against different tumour cell lines. A polyketide isolated from *Streptomyces cellulosae* (but synthetic routes have been devised).

hexitol Any sugar alcohol with six carbon atoms; natural examples are sorbitol and mannitol.

hexokinases Enzymes (EC 2.7.1.1) responsible for the first step of the glycolytic pathway (*glycolysis), the transfer of phosphate from ATP to glucose, generating glucose 6-phosphate. Hexokinase (HK) is normally cytoplasmic or associated through a conserved domain with *porin in the outer mitochondrial membrane. **HK1** is a red-cell isoform, **HK2** is the major hexokinase expressed in skeletal muscle, **HK3** is an isoform in white blood cells and **HK4** is glucokinase, expressed only in mammalian liver and pancreatic islet β cells. Additionally, there is a spermatogenic cell-specific hexokinase without the porin-binding domain. Mutations in HK1 lead to hexokinase-deficiency haemolytic anaemia.

hexosaminidase An enzyme (*N*-acetyl-β-hexosaminidase, EC 3.2.1.52) involved in the metabolism of *gangliosides that hydrolyses terminal *N*-acetyl-D-hexosamine residues on *N*-acetylglucosides and *N*-acetylgalactosides. The α subunit (529 aa) is deficient in *Tay–Sachs disease and the β subunit (556 aa) in the clinically indistinguishable Sandhoff's disease.

hexose Any monosaccharide with six carbon atoms, e.g. *galactose, *glucose, *mannose.

hexose monophosphate shunt An important metabolic pathway (pentose phosphate pathway, phosphogluconate oxidative pathway) that generates a substantial fraction of the cytoplasmic NADPH required for biosynthetic reactions, and ribose 5-phosphate for nucleotide synthesis. It is the main metabolic pathway in activated neutrophils, so they are relatively insensitive to inhibitors of oxidative phosphorylation. Congenital deficiency of the first enzyme in the shunt, *glucose-6-phosphate dehydrogenase (EC 1.1.1.49), produces a condition similar to *chronic granulomatous disease and a variety of other disorders.

SEE WEB LINKS

- More details in the Reactome—a curated knowledgebase of biological pathways.

Heymann nephritis A rat model of human membranous nephropathy; an autoimmune disease induced by immunizing rats with kidney extracts. The antigen is megalin (lipoprotein receptor-related protein-2; glycoprotein 330).

HFE *See* HAEMOCHROMATOSIS.

Hfr A designation for a bacterial strain that has a conjugative plasmid (often the F plasmid) integrated into its genomic DNA. Hfr strains show high frequency recombination and will tend to transfer their entire DNA through the mating bridge at *conjugation.

HGF *See* HEPATOCYTE GROWTH FACTOR.

HGPRT The enzyme (HPRT1, hypoxanthine-guanine phosphoribosyl transferase, EC 2.4.2.8, 218 aa) that catalyses the conversion of hypoxanthine to inosine monophosphate and guanine to guanosine monophosphate as the first step in the purine salvage pathway. Normal cells can synthesize purines *de novo*, and are not totally dependent on the salvage pathway, but HGPRT-mutant cells are resistant to toxic purine analogues, which can be used as a basis for selection. *See* HAT MEDIUM. Mutation in the gene for HGPRT is responsible for *Lesch–Nyhan syndrome.

HHH syndrome A very rare autosomal recessive metabolic disorder (hyperornithinaemia–hyperammonaemia–homocitrullinuria syndrome) caused by mutation in the mitochondrial ornithine transporter (*SLC25A15*) gene. The clinical symptoms are related to hyperammonaemia and resemble those of other urea cycle disorders.

Hib *See* HAEMOPHILUS INFLUENZAE.

hidden Markov model A statistical model in which the system being modelled is assumed to be a Markov process with unknown parameters. Such models were originally developed for the study of machine learning and speech recognition, but have subsequently been applied to the analysis of gene sequences and to phylogenetics.

HIF-1 *See* HYPOXIA-INDUCIBLE FACTOR-1.

high density lipoproteins (HDL) *See* LIPOPROTEINS; *see also* VIGILIN.

high endothelial venule (HEV) A type of venule in which the endothelium, instead of being thin and squamous in morphology, is cuboidal. They are characteristic of lymph nodes where lymphocyte extravasation occurs in the normal trafficking. Morphologically similar endothelium is associated with some chronic inflammatory lesions. The adhesion molecules, GlyCAM-1 (MadCAM-1 in mucosal HEVs), *ICAM-1, and CD34 expressed on the apical surfaces account for preferential lymphocyte adhesion through *selectins and *LFA-1.

high-energy bond A somewhat misleading term applied to bonds in compounds that release more than 25 kJ/mol on hydrolysis: important examples of such compounds are ATP (30.5 kJ/mol on hydrolysis to ADP), phosphoenolpyruvate (53.2 kJ/mol) and creatine phosphate (42.7 kJ/mol), but the bonds themselves are no different: it is the compound that is important.

high-mannose oligosaccharides Oligosaccharides such as $(Glc)_3(Man)_9(GlcNAc)_2$ that are transferred from a lipid carrier to asparagine residues of nascent polypeptides destined for secretion or membrane insertion and that are subsequently trimmed by glucosidases that remove the glucose residues and mannosidases that remove all but three mannose residues. After trimming, terminal sugars such as *N*-acetylglucosamine, galactose, sialic acid, and fucose may be added to generate complex glycans. The trimming reactions are inhibited by compounds such as *deoxynojirimycin.

high-mobility group proteins A set of nonhistone chromosomal proteins involved in a wide variety of cellular processes including regulation of inducible gene transcription, integration of retroviruses into chromosomes, and the induction of neoplastic transformation and metastatic progression. Contain conserved DNA-binding peptide motifs (AT hooks) that preferentially bind to the minor groove of many AT-rich promoter and enhancer DNA regulatory elements. Lack of **HMGA1** (215 aa) is reported to cause insulin resistance and diabetes; ***HMGIY*** has been suggested as a potential oncogene. **HMG2** (208 aa) is a component of the *SET complex.

high-voltage electron microscopy (HVEM) *See* ELECTRON MICROSCOPY.

Hill coefficient A coefficient devised by A.V. Hill (1886–1977) to describe the binding of oxygen to haemoglobin (Hill coefficient of 2.8) and used more generally as a measure of cooperativity in a binding process. A Hill coefficient of 1 indicates independent binding, a value greater than 1 indicates positive cooperativity in which binding of one ligand facilitates binding of subsequent ligands at other sites; a value less than 1 indicates negative cooperativity. It can be estimated graphically using a Hill plot of initial reaction velocity (or fractional binding saturation) against substrate concentration.

Hind Type II *restriction endonucleases isolated from *Haemophilus influenzae* Rd. **HindII** cleaves 'GTPyPuAc' between the unspecified pyrimidine (Py) and purine (Pu) generating *blunt ends. **HindIII** cleaves the sequence 'AAGCTT' between the two A residues, generating *sticky ends.

hinge region A flexible region of a polypeptide chain that allows conformational change by steric rearrangement of adjacent domains.

hippocalcin A calcium-binding protein (193 aa) found in the hippocampus and related to *recoverin. **Hippocalcin-like-1** (visinin-like protein 3, 193 aa) is found in retina and brain.

hippocampus Paired regions of the forebrain, in the medial temporal lobes, that are important in short-term memory and spatial navigation. Once memories are consolidated, the hippocampus is no longer the locus for storage; this accounts for the pattern of memory loss in Alzheimer's disease, where loss of neurons in the hippocampus is one of the early changes. The hippocampus is also the site of long-term synaptic plasticity (*see* LONG-TERM POTENTIATION).

Hirano bodies Intracellular paracrystalline, eosinophilic aggregates of microfilament-associated proteins including actin, α-actinin, and vinculin, characteristic of neurodegenerative disorders such as Alzheimer's disease and Creutzfeldt–Jakob disease, although they occur in various neurons as a function of age, without apparent effect.

SEE WEB LINKS

- Additional information and picture of stained section.

Hirschsprung's disease (HSCR) A disorder, aganglionic megacolon, in which there is an absence of ganglion cells in the myenteric and submucosal neural plexuses of gut. The phenotype can result from mutation in any of several different genes. In some cases defect is due to mutation in RET receptor tyrosine kinase (**HSCR1**), in others to mutations in genes for endothelin-β receptor (**HSCR2**), *GDGF (**HSCR3**) and *endothelin-3 (**HSCR4**). Various other subtypes (**HSCR5–9**) are also known, and HSCR also occurs as a feature of several syndromes including Waardenburg–Shah syndrome (in which the mutation is sometimes in Sox10), *Mowat–Wilson syndrome, and *Goldberg–Shpritzen megacolon syndrome.

h

hirudin An anticoagulant peptide (65 aa), present in the saliva of the *medicinal leech *Hirudo medicinalis*, that potently inhibits the action of thrombin on fibrinogen. Hirudin-based anticoagulants such as lepirudin (*TN* Refludan) and desirudin (*TN*s Iprivask, Revasc) have been produced by recombinant methods, also a synthetic peptide (*bivalirudin).

His bundle *See* BUNDLE OF HIS.

his tag An *epitope tag (histidine tag) based on a short stretch (~6) of histidine residues added to either the N- or C-terminus of a protein, sometimes with an added region susceptible to endopeptidase cleavage to allow stripping of the tag from the recombinant protein. Addition of the tag allows detection with antipolyhistidine antibodies and protein purification on nickel-based affinity columns.

histamine A biogenic amine that is a potent natural mediator of various physiological reactions, acting through *histamine receptors in smooth muscle and in secretory systems. Histamine is stored in mast cell and basophil granules and is released in response to antigen (*see* HYPERSENSITIVITY (type 1). The early symptoms of *anaphylaxis are due to histamine. It is also a component of some venoms.

histamine receptors Receptors for histamine fall into two main classes, H_1 and H_2; the former are mostly in skin, nose, and airways and are targeted by antihistamine drugs (e.g. fexofenadine, mepyramine) that are used for allergic responses. **H_2-blockers** are antagonists of the H_2 receptors that reduce gastric acid secretion and are used in treatment of gastric and duodenal ulcers. Examples are cimetidine (Tagamet), famotidine (Pepcid), and ranitidine (Zantac).

histatins Family of small histidine-rich cationic proteins (7–38 aa) secreted into saliva by the parotid and submandibular glands that have potent antifungal activity, particularly against *Candida albicans*. They bind to a receptor on the fungal cell membrane and enter the cytoplasm where they target mitochondrial ATP synthesis. They also function to precipitate dietary tannins.

histidine (His, H) An amino acid (155 Da) with an imidazole side chain with a pK_a of 6–7 making it responsive to small physiological changes in pH. Can act as a proton donor or acceptor and has high potential reactivity and diversity of chemical function. Forms part of the catalytic site of many enzymes and is often involved in the coordination site for metals.

histidine ammonia lyase An enzyme (histidase, EC 4.3.1.3, 657 aa) that converts histidine into ammonia and urocanic acid. Mutations in the gene cause histidinaemia, an autosomal recessive disorder characterized by increased levels of histidine and decreased levels of urocanic acid.

histidine-rich calcium-binding protein A highly acidic protein (HCP, 699 aa) found in the sarcoplasmic reticulum that binds to *triadin and may affect calcium release through the ryanodine receptor. Binds low density *lipoprotein with high affinity. **Histidine rich protein II** is a pH-sensitive actin-binding protein, similar to hisactophilin, that is found in *Plasmodium falciparum* and that also binds phosphatidylinositol 4,5-bisphosphate.

histidine-rich glycoprotein A protein (525 aa) present in platelets and abundantly in plasma. It binds to heparin and to plasminogen, and inhibits fibrinolysis.

histidine triad A motif (HIT) related to the sequence H-ϕ-H-ϕ-H-ϕ-ϕ (where ϕ is a hydrophobic amino acid). Proteins containing HIT domains form a superfamily with nucleotide-binding and diadenosine polyphosphate hydrolase activities. Examples include *aprataxin, and the GalT subfamily of nucleoside monophosphate transferases (e.g. galactose-1-phosphate uridylyltransferase) that bind zinc and iron. The **fragile histidine triad** (*FHIT*) gene encodes a protein (*fragile histidine triad protein,147 aa) that has AP3A hydrolase (dinucleoside-triphosphatase, EC 3.6.1.29) activity and may be a tumour suppressor. **Histidine triad nucleotide-binding protein 5** (337 aa) is an mRNA decapping enzyme.

histiocytes Long-lived tissue *macrophages.

histiocytic lymphoma An obsolete name for non-Hodgkin's *lymphomas.

histiocytosis X *See* LANGERHANS CELL HISTIOCYTOSIS.

histioma (histoma) An obsolete term for a tumour derived from fully differentiated tissue.

histochemistry The use of specific staining reactions to study the composition of tissues.

histocompatibility antigens A set of cell surface glycoproteins encoded by the *MHC (the HLA system in humans and the H2 system in mice) that are crucial for T-cell recognition of antigens. **Class I** histocompatibility antigens are composed of two glycosylated subunits, a heavy chain of 44 kDa and β_2-microglobulin (12 kDa). The heavy chain may be coded by the A, B, or C genes of the human HLA complex (K, D, or L genes of mouse H2 system). Class I antigens are important in T-cell killing and are recognized in conjunction with the foreign cell surface antigens (MHC restriction). **Class II** antigens such as human HLA-DR antigens are heterodimeric (α, 254 aa; β, 266 aa) and are found mostly on B cells, macrophages, and accessory cells. T helper cells respond to foreign antigen presented in conjunction with the appropriate class II antigens. **Class III** MHC genes encode other immune components, such as complement components and some cytokines. Histocompatible tissues will not be rejected if transplanted from one individual to another.

histogenesis The developmental processes that give rise to tissues. The processes include differentiation, controlled proliferation, morphogenetic movements, and often infiltration by blood vessels.

histoma *See* HISTIOMA.

histone acetylation The post-translational modification of histones by histone acetyl transferases (HATs; EC 2.3.1.-) which add acetyl groups to N-terminal lysine residues. Acetylation reduces the binding affinity between histones and DNA, facilitating transcription by making it easier for RNA polymerase and transcription factors to access the promoter region. Acetylation of histone H4 is considered to be important in replication-dependent nucleosome assembly. Deacetylation (by *histone deacetylases (HDACs)) has the opposite effect.

histone deacetylases A family of eleven enzymes in humans (HDACs, EC 3.5.1.-) that remove acetyl groups from an ε-*N*-acetyl lysine amino acid on a histone, reversing the effect of histone acetyltransferases. Acetylation is thought to be increased in active genes, so deacetylation should switch off genes. Inhibitors of HDAC are being tested as antitumour drugs. The HDAC complex consists of HDAC1, HDAC2, and *retinoblastoma-binding proteins 4 and 7 (RBBP4 and RBBP7) which promotes histone deacetylation and transcriptional repression.

histone methyltransferase Euchromatic histone methyltransferase-1 (EHMT1, 1267 aa) is a component of the E2F6 transcriptional repressor and has specificity for core histone H3. Some cases of chromosome 9q subtelomeric deletion syndrome, a mental retardation syndrome, are caused by mutation in the gene for EHMT1. *See* JUMONJI.

histones A family of proteins that are complexed to DNA in eukaryotic *chromatin and *chromosomes. They are highly conserved basic proteins with a high arginine/lysine content. There are five major classes: two copies of H2A (130 aa), H2B (130 aa), H3 (136 aa), and H4 (103 aa) (constituting an octamer) bind to ~200 base pairs of DNA to form the repeating structure of chromatin, the *nucleosome, with histone H1 (194 aa) binding to the linker sequence. *See* HISTONE ACETYLATION.

histoplasmin An antigen derived from *Histoplasma capsulatum* culture filtrate used in a skin test for histoplasmosis.

histoplasmosis A disease (Darling's disease), primarily of the lungs, caused by infection with the fungus *Histoplasma capsulatum*, found very commonly in soil and animal litter.

histotope A region of *histocompatibility antigen that is bound by the T-cell receptor.

histrelin A synthetic nonapeptide analogue of gonadotropin-releasing hormone used to prevent the premature onset of puberty and in treatment of hormone-sensitive tumours. It reduces serum levels of testosterone or oestrogen by desensitizing pituitary gonadotropin-producing cells to the releasing hormone signals. *TN* Supprelin.

HIV (human immunodeficiency virus) A retrovirus (of the genus Lentiviridae) that causes acquired immunodeficiency syndrome (*AIDS) in humans by killing $CD4^+$ lymphocytes (Th cells). Early names were human lymphotrophic virus type III (HTLV-III) and lymphadenopathy-associated virus (LAV). There are two major strains: **HIV-1**, which is more virulent and responsible for most infections, and **HIV-2**, which is less transmissible and is largely confined to West Africa. Simian immunodeficiency virus (SIV) is similar and is used as an experimental model. *See also* FELINE IMMUNODEFICIENCY VIRUS.

HL60 cells A line of human promyelocytic cells established in 1977 from a patient with acute myeloid leukaemia. They are pluripotent and can be induced to differentiate into neutrophil or eosinophil-like cells by various treatments. The genome has an amplified c-*myc* proto-oncogene.

HLA Human leucocyte antigen. *See* HISTOCOMPATIBILITY ANTIGENS.

HMEC Human microvascular endothelial cells or (occasionally), human mammary epithelial cells.

HMG box A motif characteristic of *high-mobility group proteins, a domain (79 aa) involved in binding linear DNA in a sequence-specific manner, causing it to bend through a large angle, thereby influencing the binding of other transcription factors. *See* SOX; SRY.

h

HMG-CoA reductase An integral membrane protein (3-hydroxy-3-methylglutaryl-CoA reductase, EC 1.1.1.34, 888 aa) of endoplasmic reticulum and peroxisomes that catalyses the reaction between hydroxymethylglutaryl-CoA and two molecules of NADPH to produce mevalonate, the feedstock for the synthesis of cholesterol, other sterols and (e.g.) geranylgeraniol groups for post-translational modification of proteins. When the end products of mevalonate metabolism accumulate in cells the enzyme is rapidly degraded. As the rate-limiting enzyme in cholesterol synthesis, it is the target for cholesterol-lowering drugs (e.g. *fibrates).

HMM *See* MEROMYOSIN.

hMSH2 Human Mut S homologue 2. *See* MISMATCH REPAIR.

HNF A family of transcription factors (hepatocyte nuclear factors) enriched in liver. **HNF1** is predominantly expressed in liver and kidney and selectively interacts with the control regions of several liver-specific genes. Heterozygous mutations in hepatocyte nuclear factor (HNF)-1α and HNF-1β result in maturity-onset *diabetes of the young.

hnRNA *See* HETEROGENEOUS NUCLEAR RNA.

HNT 1. Human *neurotropin. **2.** A line of human neuroteratoma cells (hNT). **3.** *See* IgLONs.

Hodgkin's lymphoma (Hodgkin's disease) A human *lymphoma, a solid neoplasm that originates in lymphocytes, characterized by the spread of disease from one lymph node to another. A diagnostic feature is the presence of Reed–Sternberg cells, giant cells with mirror-image nuclei. Immunological depletion occurs and often leads to death through opportunistic infection. Four subtypes of the disease are recognized, depending on the morphology of Reed–Sternberg cells and the types of cells infiltrating the lymph nodes. It occurs most frequently in young adulthood (age 15–35) and in those over 55 years old.

hodology The study of pathways; an approach used in trying to understand brain function by mapping projections or pathways in the brain, rather than looking at the cellular architecture. The term is also used in philosophy, psychology, and geography.

Hoechst 33258 dye A fluorescent dye that emits blue light (~460 nm) when stimulated by UV at ~350 nm. It binds to the minor groove of DNA, and is often used to visualize chromosomes, and to check cell cultures for mycoplasmas. A related bis-benzimide is Hoechst 33342.

Hogness box *See* TATA BOX.

holandry Inheritance of characters expressed only in the male because the genes are on the Y chromosome.

Holliday junction The crossover region at meiosis that results in homologous genetic recombination. Robin Holliday first proposed the junction in 1964 as a structural intermediate in a mechanistic model and suggested that it was a four-way junction with separate DNA helices, or with stacked helices in either a parallel or an antiparallel orientation. Subsequently crystallographic studies have confirmed the antiparallel stacked-X conformation. Products of the *MSH4* and *MSH5* genes, homologues of the bacterial *MutS* mismatch repair genes, associate specifically with Holliday junctions but do not engage in repair.

holoblastic Describes those eggs that exhibit total cleavage, as in mammals.

holocentric Describes a chromosome with a diffuse rather than a discrete centromere so that microtubules associate during mitosis along the whole length, rather than at a single point.

holocrine A type of secretory process in which the whole cell is shed from the gland, usually after becoming packed with the main secretory product. Sebaceous glands are one of the few mammalian examples.

holoenzyme The complete enzyme complex comprising the protein portion (*apoenzyme) and a cofactor or coenzyme.

holoprosencephaly A craniofacial developmental disorder in which there is a failure of the prosencephalon (embryonic forebrain) to develop; severe forms may cause cyclopia. The genetic basis is diverse: holoprosencephaly-1 (**HPE1**) maps to 21q22.3. **HPE2** is caused by mutation in the *SIX3* homeobox gene, **HPE3** is caused by mutation in the *sonic hedgehog gene, **HPE4** is caused by mutation in the *TGIF* gene that encodes TGFβ-induced factor. **HPE5** is caused by mutation in the *ZIC2* gene that encodes zinc finger protein of the cerebellum-2. **HPE6** maps to 2q37.1. **HPE7** is caused by mutation in *PTCH1* (*see* PATCHED). **HPE8** maps to 14q13. **HPE9** is caused by mutation in the **gli2* oncogene.

SEE WEB LINKS

- Website created by parents of children with this condition.

homeobox (homoeobox) A conserved DNA sequence characteristic of *homeotic genes that were first identified as causing segmentation mutants in *Drosophila*. The homeobox is about 180 nucleotides long and encodes the **homeodomain** that binds DNA. Homeoboxes also occur in vertebrate genomes and have about 75% amino acid homology to the *Drosophila* homeodomain. Three subfamilies of homeobox-containing proteins can be identified, based on the archetypal *Drosophila* genes *engrailed, antennapedia*, and *paired*.

homeopathy (homoeopathy) A type of alternative medicine based on the idea that diseases can best be treated by drugs that produce physiological effects similar to those caused by the disease. The claimed curative effect is exhibited at extreme dilution (often dilutions at which no drug molecules are likely to remain, although the solvent is claimed to remember the former presence of the compound) so at least adverse side effects are avoided. Few double-blind trials have been carried out and most indicate no more than a *placebo effect.

homeostasis (homoeostasis) The maintenance of constancy. Homeostatic mechanisms keep the properties of the internal environment of organisms within fairly well-defined limits and generally require a sensor, a control centre, and positive or negative *feedback regulation.

homeotic genes An important and highly conserved set of genes (*hox* genes) that regulate the development of organ systems, appendages, body segments, etc. They are characterized by a ~180-nucleotide *homeobox region and the products are transcription factors which regulate the genes that are responsible for producing the particular organ system. Remarkably, the homeoboxes for insect systems (e.g. eye) can be replaced by the vertebrate equivalent even though the actual structures are totally different (compound eye vs single-lens vertebrate eye). Homeotic genes are regulated by positional cues and given their master switching role; mutations can cause major developmental abnormalities (**homeotic mutants**). Homeobox genes are clustered and activated in a 3′ to 5′ direction with genes that are involved in morphologically anterior systems, which are activated earlier in development, at the 3′ end. This colinearity of genes and morphological position is intriguing. Vertebrates have four duplicate sets (paralogues) of the ten ancient homeotic genes, known as *Hoxa, Hoxb, Hoxc*, and *Hoxd*. *Synpolydactyly is one of relatively few *developmental defects caused by mutations in *hox* genes.

homer The protein products of neuronal immediate-early genes (IEG) that are enriched at excitatory synapses and bind metabotropic glutamate receptors (mGluR). There are multiple splice variants of homer-1 and at least two isoforms, homer-2 and -3, with similar properties. They are involved in the clustering of subsets of mGluRs and may also regulate activity of the receptors. *See also* GRIP; POSTSYNAPTIC PROTEIN (postsynaptic density (PSD) proteins).

homocysteine The oxidized form of cysteine, a by-product of protein metabolism and an important intermediate in methionine metabolism. Elevated levels of plasma homocysteine are associated with a higher risk of coronary heart disease, stroke, and peripheral vascular disease although this does not imply a causative link. Deficiencies of folic acid, pyridoxine, or cyanocobalamin (*see* VITAMIN B) can lead to high homocysteine levels, as can *homocystinuria.

homocystinuria A recessive metabolic disorder due to deficiency of cystathionine β-synthase (EC 4.2.1.22). This is the enzyme that converts homocysteine and serine into cystathione, a precursor of cysteine, and deficiency has major effects on the eyes and the central nervous, skeletal, and vascular systems. Homocysteine levels in blood and urine are elevated.

homogentisic acid An intermediate in the metabolism of tyrosine and phenylalanine which accumulates in *alkaptonuria, an autosomal recessive condition in which there is homogentisic acid oxidase deficiency.

homograft Outmoded term for a graft from one individual to another of the same species, as opposed to a *xenograft. Imprecise because it includes *allogeneic grafts (allografts) between genetically dissimilar individuals, and syngeneic grafts between identical individuals (e.g. twins).

homokaryon A multinucleated cell (syncytium) in which all nuclei are identical. *See* HETEROKARYON.

homologous In general, the state of having the same relative position, proportion, value, or structure. *Compare* ANALOGOUS; AUTOLOGOUS; HETEROLOGOUS. More specifically, used to describe something derived from the tissues or DNA of a member of the same species. **Homologous chromosomes** are identical with respect to genetic loci, and the maternal and paternal homologues form pairs (*synapse) during mitosis. **Homologous genes** are similar in sequence and there are sophisticated algorithms to identify sequence homologues. **Homologous *recombination** involves exchange of homologous loci and is important for producing null alleles (*knockouts) in *transgenic mice.

homozygote A nucleus, cell, or organism in which the *alleles of one or more specific genes are identical. *Compare* HETEROZYGOTE.

Hoodia gordonii A succulent plant (not a cactus) found in the Kalahari desert. Used in traditional South African herbal medicine as an appetite suppressant and widely marketed as such in the West, although many preparations contain no active components and there is limited evidence for efficacy. The active constituent appears to be an oxypregnane steroidal glycoside, P57AS3 (P57).

hook proteins **1.** Cytosolic coiled-coil proteins with conserved N-terminal domains, which attach to microtubules, and more divergent C-terminal domains, which mediate binding to organelles. Human **hook1** (HK1; 719 aa) has 33% homology with the *Drosophila* protein, **hook2** binds to different organelles, and **hook3** is enriched in the *cis*-Golgi. Hook proteins are a subset of a larger family, **hook-related proteins** (HkRP) which all have the conserved domains found in hook proteins. **2.** *See* AT HOOK.

hookworms Parasitic strongyloid nematodes that attach to the small intestine of the host with hook-like organs on the mouth. The commonest hookworm infections of humans (ankylostomiasis) are by *Ankylostoma duodenale* or *Necator americanum*, which penetrate the bare feet, migrate to the lungs, and are then swallowed. In individuals with a poor diet the infection can cause anaemia.

hordeolum (stye) An infection of a sebaceous gland at the base of an eyelash.

horizontal transmission Disease transmission between individuals of the same generation rather than from mother to offspring. *Compare* VERTICAL TRANSMISSION.

hormesis An atypical dose–response relationship in which low doses stimulate, but high doses cause inhibition. The *dose–response curve may be J-shaped or an inverted U-shape. An example is the effect of chemotactic peptides on neutrophil adhesion; at low doses adhesion is reduced and movement stimulated, at high doses adhesion increases and the cells become trapped at the focus of the lesion.

hormone A substance that acts remotely on other cells and alters their behaviour. Hormones are produced by specialized secretory cells which may release the product into the blood (endocrine hormones) or into ducts (exocrine hormones). Implicit in the definition is the possession by the target cell of receptors, which may be on the cell surface if the hormone is hydrophilic, or in the cytoplasm when the hormone is lipophilic and can cross the plasma membrane (e.g. steroids). Neurotransmitters are not generally classed as hormones, although the mechanism of signalling at the synapse is effectively the same, and growth factors and cytokines are also excluded.

hormone disorders

Disorders in which there is under- or overproduction of a hormone which then affects systems regulated by the hormone in question, including, in some cases, the production of other hormones. In some cases the deficiency can arise because of a failure to process a prohormone; in other cases the defect may be in the receptor (as in *androgen insensitivity syndrome, *Reifenstein's syndrome). Other examples include *acromegaly; *Addison's disease; some types of *chondrodysplasia; *Cushing's syndrome; *diabetes; *dystrophia adiposogenitalis; *giantism; *hyperadrenalism; *hyperparathyroidism; *hyperthyroidism; *hypogonadism; *hypoparathyroidism; *hypopituitarism; *hypothyroidism; *Laron's syndrome; *lipoid congenital adrenal hyperplasia; *nesidioblast (nesidioblastosis); *triple A syndrome. Albright hereditary osteodystrophy, McCune–Albright syndrome, and *pseudohypoparathyroidism are caused by mutations in genes for *granins. Some disorders are described in the entries for the hormone involved: *see* ANGIOTENSIN; ANTIMÜLLERIAN HORMONE; ATRIAL NATRIURETIC PEPTIDE; ERYTHROPOIETIN; GLUCOCORTICOIDS; GONADOTROPIN-RELEASING HORMONE; GROWTH HORMONE; KISSPEPTINS; MELANIN-CONCENTRATING HORMONE; MELANOCORTIN; MELANOCYTE-STIMULATING HORMONE; THYMOPOIETIN; VASOPRESSIN.

hormone replacement therapy Treatment used to relieve symptoms that occur once the ovaries cease secreting hormones after the menopause or are surgically removed. Various pharmacological analogues of female hormones (oestrogens, progestogen) are used, although there are both advantages and disadvantages.

hormone-sensitive lipase An enzyme (EC 3.1.1.79, 1076 aa) that converts stored triglyceride into free fatty acids, the rate-limiting step in mobilizing energy from the adipocyte store. The enzyme is controlled hormonally (insulin stimulates dephosphorylation and inhibition) and neuronally (catecholamines stimulate).

hornerin A protein (2496 aa) of the fused-type S100 family (*see* FUSED GENE FAMILY); originally identified in mouse but human hornerin has been identified. Hornerin shares structural features, expression profiles, and intracellular localization with *profilaggrin, but has a distinct function from that of filaggrin in cornification (keratinization) of epidermal cells. *See also* REPETIN; TRICHOHYALIN.

Horner's syndrome A condition caused by unilateral damage to the cervical sympathetic nervous system. There is a combination of small pupil (miosis), sunken eye (enophthalmus), and drooping of upper eyelid (ptosis). In the congenital form the two eyes may be of different colour.

horripilation The condition in which the hairs on the skin stand on end, usually a result of cold (colloquially, gooseflesh).

horseradish peroxidase A stable enzyme (EC 1.11.1.7, 353 aa), isolated from horeseradish *Armoracia rusticana*, that is often coupled to antibody and used for an enzyme-linked amplification step in *ELISAs, *histochemistry, or western *blotting. The chromogenic substrate of choice is usually diaminobenzidine, which is oxidized to produce a brown deposit.

host-range mutant A mutant virus that is unable to grow in cells from some host species that are infected by the wild type, but not others.

host-versus-graft reaction The process that leads to rejection of allogeneic or xenogeneic cells in a graft. It is mediated by $CD4^+$ T cells that respond to differences in MHC class I antigens. The opposite of *graft-versus-host disease.

hotspot A colloquial term for a chromosomal region particularly prone to mutation or transposition.

housekeeping genes Genes that are constitutively active in all cells and code for proteins (housekeeping proteins) or RNAs that are essential for normal cell function. Differentiation leads to production of more specialized products, sometimes termed luxury proteins.

house mites *See* DERMATOPHAGOIDES PTERONYSSINUS.

HP1 A methyl-lysine binding protein (heterochromatin protein 1, chromobox homologue-5, 191 aa) localized at heterochromatin sites, where it mediates gene silencing. It binds to methylated histone H3 (methylated at lysine 9) through a single *chromodomain and will also homodimerize. HP1-α, -β, and -γ are released from chromatin during the M phase of the cell cycle. More recently it has been shown to participate in euchromatin gene silencing in association with DNA methyltransferases.

HPA axis *See* HYPOTHALAMO–PITUITARY–ADRENAL AXIS.

HPETE An intermediate in leukotriene synthesis (5-HPETE, 5-hydroperoxyeicosatetraenoic acid) that is produced from arachidonic acid by 5-lipoxygenase and is the starting point for the formation of 5-HETE or of leukotriene A_4.

hpf **1.** High power field (of a microscope). **2.** Hours post-fertilization. **3.** Human plasma fibrinogen. **4.** HPF-1 cells, a line of human fibroblasts. **5.** High-pressure freezing, a technique used in cryofixation. **6.** High pass filter.

HPLC A method (high performance (pressure) liquid *chromatography) used to separate, identify and quantify compounds using a tightly-packed column of finely divided particles that present a very large surface area to which the compounds bind. Solvent pumped through the column elutes compounds at different rates according to their properties. The rate of elution is characteristic of the compounds, which are detected as they leave the column, usually spectrophotometrically. The use of finely divided column packing necessitates high pressure to drive the solvent through the column and obtaining high performance requires high pressure. The original name was 'high pressure liquid chromatography' but manufacturers encouraged the name-change to 'high performance liquid chromatography'. As with any chromatographic method a range of different stationary phases, solvents, elution protocols, etc. have been developed and systems can be scaled up for high yield of pure compounds or scaled down to obtain high resolution with small samples. Pressures vary from 400 to 1000 atmospheres (4–10 MPa) and ultra-high pressure methods are coming into use.

h

HPRT *See* HGPRT.

HPV Human *papillomavirus, responsible for *warts.

HRG1 *See* ARIA; HEREGULINS.

hrk The product (harakiri, death protein 5, 91 aa) of the apoptosis-regulating gene *harakiri*, that interacts with the death-repressor proteins bcl-2 and bcl-X_L, but not with death-promoting homologues, bax or bak. Has a short *bcl-2 homology domain 3 (BH3 domain) but no BH1 or BH2 domains. Expression of hrk in mammalian cells induces rapid cell death, and hrk up-regulation is involved in gentamicin-induced ototoxicity.

HS1 **1.** An actin-binding protein (haematopoietic-specific protein-1, haematopoietic cell-specific LYN substrate 1, 486 aa) homologous to *cortactin and restricted to leucocytes. Becomes phosphorylated in response to ligation of the T-cell receptor and together with WASP, WAVE, and *cofilin is involved in triggering the formation of an F-actin scaffold at the *immunological synapse and may be involved in signalling through T-cell *lck or B-cell *lyn tyrosine kinases. Not required in platelets. **HS1-binding protein-3** (HS1-BP3, 392 aa) associates with *SH3 domains of HS1 and may be part of the signalling complex. Overexpression of mutant HS1BP3 protein in T-cell lines decreases IL-2 production. **2.** Tyrosine 3-monooxygenase/tryptophan 5-monooxygenase activation protein theta isoform (YWHAQ, 14-3-3 protein θ, HS1, 245 aa), one of the *14-3-3 family of signalling proteins, strongly expressed in brain and to a lesser extent in lung, kidney, liver, and spleen. Has significant sequence similarity with *stratifin. It is reported to be consistently up-regulated in the spinal cord of patients with *amyotrophic lateral sclerosis.

HSA **1.** Human serum *albumin. **2.** A surface glycoprotein adhesin (Hsa, 2178 aa) of *Streptococcus gordonii* that binds sialic acid moieties on platelet membrane glycoprotein Iba and on glycophorin A of the erythrocyte. *S. gordonii* is a pioneer colonizer of dental plaque and a significant causative bacterium for infective endocarditis. **3.** Prefix indicating a microRNA (*miRNA) is derived from *Homo sapiens*, e.g. hsa-miR-101.

SEE WEB LINKS

- Naming conventions of miRNAs, and further miRNA information.

HSC *See* HEDGEHOG SIGNALLING COMPLEX.

HSCR *See* HIRSCHSPRUNG'S DISEASE.

HSET A motor protein (kinesin-like 2, 1388 aa) of the *kinesin-14 family involved in spindle organization. Like the *Drosophila* protein Ncd, it moves towards the minus end of the microtubule (like cytoplasmic dynein).

HSP32 *See* HAEM OXYGENASES.

hsp60, hsp70, hsp90 *See* HEAT SHOCK PROTEINS.

HSR **1.** *Hypersensitivity reaction. **2.** Heat shock response (*see* HEAT SHOCK PROTEINS). **3.** Homogeneously staining region (of a chromosome). **4.** Health services research.

hst A human *oncogene (heparin-binding secretory transforming gene) that encodes fibroblast growth factor-4.

HT1080 cells A line of human fibrosarcoma cells derived in 1972 from a 35-year-old man.

hTERT Human *telomerase reverse transcriptase.

HTG *See* HYPERTRIGLYCERIDAEMIA.

Htk-L A membrane-anchored ligand (ephrin B2, 333 aa) for *hepatoma transmembrane kinase (HTK) that is tyrosine phosphorylated once bound to the receptor.

HTLV A family of retroviruses (human T-cell leukaemia/lymphoma viruses). **HTLV-1** (adult T-cell lymphoma virus type 1) causes leukaemia and sometimes a mild immunodeficiency. It activates production of Th1 cells which suppress Th2-based immune responsiveness. It is also associated with a progressive demyelinating upper motor neuron disease, tropical spastic paraparesis. **HTLV-2** has about 70% genomic homology with HTLV-1. Its pathological potential is uncertain. **HTLV-III** was the name originally given to *HIV, although this usage is obsolete, and a new virus, similar to simian T-lymphotropic virus-3, has been given this name.

HTRF An assay method (homogeneous time-resolved fluorescence) in which an absorbing fluorochrome is coupled to one component, an emitting fluorochrome with slow-release characteristics coupled to the other: binding will allow *fluorescence energy transfer between the fluorochromes and the light emission, analysed in a time-resolved fluorescence system, is a measure of binding. Because no separation is required the system lends itself to high-throughput screening for inhibitors of binding interactions.

HuH7 cells A line of cells derived from a well-differentiated human hepatocellular carcinoma that will grow in serum-free medium and have an epithelioid morphology.

human Generally ignored at the beginning of a noun phrase in this dictionary, thus 'human chorionic gonadotrophin' has its entry under 'chorionic gonadotophin'. Unless specified, proteins, etc. are the *H. sapiens* type; homologues in other species may differ (to a greater or lesser degree).

human double minute 2 *See* HDM2.

human endogenous retroviruses Long-established viruses (HERVs) integrated into the genome and no longer able to produce infective virus. HERVs comprise ~8% of the genome and proteins encoded by viral genes may be important in some autoimmune diseases, in particular multiple sclerosis. *HERV* genes also produce *syncytin and there may be an association between HERVs and the HELLP syndrome and pre-eclampsia.

Human Genome Diversity Project An international project, distinct from the Human Genome Project, that has the objective of analysing the DNA that varies between different ethnic groups, in order to understand the origins and migrations of human populations and the genetic basis of differing susceptibility to disease. Opposition from many groups, particularly those representing ethnic minorities (which were not consulted in the development of the project) has generated considerable controversy. *See* GENOME.

SEE WEB LINKS

- The Human Genome Diversity Project homepage.

human immunodeficiency virus *See* HIV.

humanized antibody In general, any nonhuman antibody that has been engineered to have human immunoglobulin domains substituted for all regions except the antigen-binding site. In practice, usually a modified mouse monoclonal antibody directed against a therapeutically important target (e.g. *infliximab directed against TNFα). The objective is to make the antibody minimally antigenic so that repeated treatment is possible.

humoral immune responses Immune responses mediated by antibody. *Compare* cell-mediated responses *at* CELL-MEDIATED IMMUNITY.

Humulin Trade name for recombinant human insulin.

Hunter's syndrome A recessive X-linked mucopolysaccharidosis type II, a *storage disease that arises from iduronate sulphatase deficiency. Both iduronate-2-sulphatase (EC 3.1.6.13) and α-iduronidase (EC 3.2.1.76) are required to degrade dermatan and heparan sulphates; fibroblasts from Hunter's syndrome will complement those from *Hurler's syndrome in culture, each producing the enzyme that the other lacks and recapturing the deficient lysosomal enzyme from the culture medium.

huntingtin The protein (3144 aa) encoded by the *IT15* gene that has variable numbers of polyglutamine repeats in *Huntington's disease. The *IT15* gene is widely expressed and required for normal development. The mutant Htt protein forms insoluble nuclear aggregates which may be responsible for neuronal cell death in the disease although a range of other effects have been suggested. The polyglutamine repeats (44 in the commonest form of the disease) increase the interaction of huntingtin with a membrane-bound **huntingtin-associated protein-1** (HAP-1, 671 aa) which is expressed more abundantly in the hypothalamus than in other brain regions. HAP1 may function as an adaptor protein mediating interaction with cytoskeletal and motor proteins and the HAP1-huntingtin complex may play a role in vesicle trafficking. **Huntingtin-interacting protein** (HIP-1, 1037 aa) is membrane-associated and has similarity to cytoskeleton proteins; it interacts with clathrin heavy chain and α-adaptins A and C, suggesting it may be involved in endocytotic vesicle activity. HIP-1 binds to the HIP1 protein interactor (HIPPI) to form a pro-apoptotic heterodimer. **Huntingtin-interacting protein-12** (HIP-1 related protein, 1068 aa) has some regions of similarity with HIP-1 with which it interacts. Other huntingtin-interacting proteins include the glycolytic enzyme GAPD, *pacsin1, and the ubiquitin-conjugating enzyme E2-25K (HIP2).

Huntington's disease A maturity-onset autosomal dominant neurodegenerative disorder (formerly known as Huntington's chorea), characterized by progressive loss of striatal neurons, leading to choreic movements and dementia. Caused by unstable amplification of a trinucleotide $(CAG)_n$ repeat within the coding region of the gene encoding *huntingtin.

HuP genes **1.** Genes coding for elements of the hydrogen uptake system in bacteria such as *Rhizobium leguminosarum* which produce hydrogen as a by-product of the activity of nitrogenase. **2.** (*less commonly*) The human homologue of the paired-box homeobox gene Pax-3 (HuP2) that is mutated in Waardenburg syndrome type 1.

h

HuR An RNA-binding protein that regulates the stability and/or the translation of mRNAs encoding cell stress response proteins.

Hurler–Scheie syndrome *See* HURLER'S SYNDROME.

Hurler's syndrome An autosomal recessive *storage disease in which the deficiency is in α-L-iduronidase (EC 3.2.1.76), leading to accumulation of heparan and dermatan sulphates. Extensive deposits of mucopolysaccharide are found in *gargoyle cells, and in neurons. Deficiency in this enzyme results in three different clinical entities, Hurler's (mucopolysaccharidosis IH, MPS IH), Scheie's (MPS IS), and Hurler–Scheie (MPS IH/S) syndromes. Hurler's and Scheie's syndromes represent phenotypes at the severe and mild ends of the clinical spectrum, respectively, and the Hurler–Scheie syndrome is intermediate. *See* HUNTER'S SYNDROME.

Hurthle cells Abnormal thyroid epithelial cells (Ashkenazy cells, oxyphil cells, oncocytes) found in *Hashimoto's thyroiditis and follicular thyroid cancer. The cells are enlarged, have abundant eosinophilic granular cytoplasm and often have bizarre nuclear morphology. Hurthle cell tumours are associated with chromosomal abnormalities or mutations in the *RAS* gene, the *PAX8/PPARG* fusion gene, or the mitochondrial NADH-ubiquinone oxidoreductase 1a subcomplex-13 gene (*NDUFA13*; gene associated with retinoid- and interferon-induced mortality-19, *GRIM19*).

HUT-78 cells Cells derived from blood of a 50-year-old man with *Sezary's syndrome. Cells exhibit the features of a mature T-cell line with inducer/helper phenotype and release IL-2.

Hutchinson–Gilford progeria syndrome *See* PROGERIA.

HUVEC Primary human endothelial cells (human umbilical vein endothelial cells) removed by mild enzymatic treatment of the large vein of the umbilical cord. They grow relatively easily in culture and retain their differentiated characteristics for several passages.

HVEM *See* HIGH-VOLTAGE ELECTRON MICROSCOPY.

hyaline Describing something that is clear, transparent, and granule-free. Examples are hyaline cartilage and the cytoplasm of the leading edge of amoeboid cells.

hyaline membrane disease An acute lung disease of the newborn (respiratory distress syndrome of prematurity) caused by deficiency in surfactant and associated with polymorphisms in several genes, including surfactant protein A1, B, and C. More common in premature babies.

hyaluronic acid A polymer with repeating dimeric units of glucuronic acid and *N*-acetylglucosamine that forms the core of complex proteoglycan aggregates found ubiquitously in extracellular matrix. May be of extremely high molecular weight (on average ~3–4 MDa) and is highly hydrated so that it effectively occupies a large volume (1000 times greater than its volume when dry) and solutions have a high viscosity. The vitreous humour of the eye is rich in hyaluronic acid and high molecular weight hyaluronate solutions (*TN* Healon) are used in ophthalmic surgery to replace lost vitreous fluid.

hyaluronidase A family of lysosomal endoglucosaminidases (EC 3.2.1.35) that degrade *hyaluronic acid. **Hyaluronidase 1** (435 aa) is deficient in mucopolysaccharidosis IX. **Hyaluronidase 2** (473 aa) is a glycosylphosphatidylinositol (GPI)-anchored protein on the cell surface that only hydrolyses high molecular weight hyaluronic acid into smaller fragments. Other hyaluronidases have been found, one with homolgy to PH-20, a sperm-specific protein.

hybrid antibody An antibody generated experimentally by fusing *hybridomas that produce two different antibodies; the hybrid cells produce three different antibodies, only one of which is a hybrid in the sense of having two different antigenic binding sites. Similar bispecific antibodies can be produced by the chemical reassociation of monovalent fragments derived from monoclonal mouse immunoglobulin G1. The term is sometimes also applied to antibodies covalently coupled to other molecules such as cytotoxic drugs. The ambiguity associated with the term means it should probably be avoided.

hybrid cells Any cell, other than a zygote, that contains elements from two or more genomes, usually as a result of cell fusion to produce a *heterokaryon. The production of hybrid cells is an essential step in generating monoclonal antibodies. Cybrid (cytoplasmic hybrid) cells are formed by the transfer of a nucleus from one species into the enucleated cytoplasm of an egg from another species, which of course retains its mitochondria.

hybridization In general, the production of a hybrid but specifically, in molecular biology, the process of allowing single-stranded nucleic acids to form a duplex. The kinetics of formation of a hybrid duplex depends upon the degree of similarity between the partner sequences and thus the extent of proper base pairing. This provides a method to assess sequence identity. Hybridization can be done in solution or with one component immobilized, either in a gel or bound to nitrocellulose paper or some other binding surface, as in arrays (*DNA chips). If one of the partners is labelled then formation of hybrids can be detected or, alternatively, nonhybridized strands can be selectively hydrolysed. All possible combinations are possible (DNA–DNA, DNA–RNA, or RNA-RNA). Single-stranded DNA can be produced by heat denaturation and the kinetics of reannealing were important in detecting the extent of repetitive sequences (*see* C_0T CURVE). It is also possible to use labelled nucleic acid as a probe for selectively staining tissue sections (**in situ* hybridization) or chromosomes (*FISH analysis).

hybridoma A hybrid cell with a tumour cell as one parent: usually refers to a hybrid between an antigen-producing B-cell and a myeloma cell which will secrete monoclonal antibody.

hydatid disease A disease (echinococcosis) caused by infection with the larval stage of the dog tapeworm *Echinococcus granulosus*, acquired by ingestion of tapeworm eggs. In the dog (and other carnivores) the eggs develop into intestinal worms but in sheep and occasionally humans, which are intermediate hosts, the larvae penetrate the intestinal wall and give rise to **hydatid cysts** in liver and other organs. The first-stage embryonated larva proliferates within the cyst and produces thousands of potentially infective larvae (scolices) which can themselves give rise to further cysts if released from the cyst. Not to be confused with *hydatidiform mole.

hydatidiform mole An abnormal form of pregnancy, characterized by an anomalous growth containing a nonviable embryo which implants and proliferates within the uterus. In most cases the tissue is diploid XX with both X chromosomes being of paternal origin, probably as a result of fertilization of an anucleate egg and a round of endoreduplication to make the sperm nucleus diploid. Occasionally the tissue is triploid, possibly as a result of polyspermy, and occasionally tetraploid. May occasionally become invasive, though the metastases regress following removal of the mole. Mutations in the *NALP7* gene (NLR family, pyrin-domain containing 7) which encodes a *NALP protein cause recurrent hydatidiform moles. Not to be confused with hydatid cyst.

hydragogue Describing a drug that has a purgative effect and produces watery faeces.

hydralazine An antihypertensive drug that causes relaxation of vascular smooth muscle and thus peripheral vasodilation. The mechanism of action is unclear. *TN* Apresoline.

hydrargyrism Mercury poisoning. *See* MINAMATA DISEASE.

hydroa A general term for any skin disease that is accompanied by formation of clear blisters. **Hydroa herpetiforme** is a chronic skin disease that may be associated with malignancy in the elderly. **Hydroa vacciniforme** (HV) is a rare, chronic photodermatosis of unknown origin occurring in childhood. Recurrent vesicles form on skin exposed to the sun and heal to form scars resembling those left by vaccination.

hydrocele A fluid-filled swelling in the scrotum.

hydrocephalus Accumulation of *cerebrospinal fluid (CSF) in the ventricles of the brain leading to increased intracranial pressure. Usually due to blockage of CSF outflow in the ventricles or in the subarachnoid space over the brain, although in some cases there is overproduction of CSF.

hydrochlorothiazide A thiazide diuretic used to treat hypertension.

hydrocodone A semisynthetic opioid (dihydrocodeinone) with analgesic and antitussive properties. Often combined with other drugs in proprietary medicines.

hydrocortisone The synthetic form (17-hydroxycorticosterone) of the natural steroid cortisol, the *corticosteroid produced by the adrenal cortex. Hydrocortisone has both mineralocorticoid and glucocorticoid activity, and is a potent anti-inflammatory drug.

hydrogel A *gel in which the bulk phase is water.

hydrogen peroxide (H_2O_2) An important component of the antibacterial system of phagocytes, the *myeloperoxidase-halide system. Hydrogen peroxide and superoxide anions are produced in the *metabolic burst.

hydrolases Enzymes (hydrolytic enzymes, EC class 3) that catalyse the hydrolysis of a variety of bonds, such as esters, glycosides, peptides.

hydrolytic enzymes *See* HYDROLASES.

h

hydropathy plot A representation of the hydrophobic character of a protein sequence which may be useful in predicting membrane-spanning domains, potential antigenic sites, and regions that are likely to be exposed on the surface. Various algorithms are used, based on a moving average spanning a 'window' of $2n + 1$ amino acids: Kyte-Doolittle is a widely applied scale in which the window is 7; a much larger window of 19–25 is generally thought appropriate for detecting transmembrane helices. The Hopp-Woods scale was designed for predicting potentially antigenic regions of polypeptides. Various websites enable an enquirer to input sequence and obtain a graphical plot. More sophisticated approaches involve estimating the proportion of the surface of a globular protein that will be occupied by a particular residue or the orientation of the residues in the protein to calculate a hydrophobic moment analysis.

SEE WEB LINKS

- Useful lead into web resources for hydropathy plots.

hydroperitoneum *See* ASCITES.

hydrophilic group A polar group or one that will form hydrogen bonds (e.g. -OH, -COOH, $-NH_2$) and that makes a compound readily soluble in water or, in larger molecules, determines which parts of the molecule are likely to be in contact with the aqueous phase.

hydrophobia Fear of water but, since this is a symptom of *rabies, is commonly used as a synonym for that disease.

hydrophobic bonding An interaction that is important in determining the conformation of proteins and lipid structures but is not a binding interaction, rather the maximizing of polar interactions and minimization of the exposure of nonpolar (hydrophobic) residues to the aqueous environment.

hydropic Describing one of the early signs of cellular degeneration in response to injury, swelling due to water uptake.

hydrops folliculi Accumulation of clear fluid in a graafian follicle forming an ovarian cyst.

hydroquinone A type of phenol that is sometimes used topically to whiten skin by inhibiting the oxidation of tyrosine to 3,4-dihydroxyphenylalanine (dopa) in the production of melanin.

hydroxyapatite A mineral (hydroxylapatite) composed of calcium and phosphate ($Ca_{10}(PO_4)_6(OH)_2$), that is found in rocks and as a component of bone and dentine. It is used as a packing material (matrix) for chromatography columns that will separate mixtures of double-stranded and single-stranded DNA and as a component of various materials designed for implantation in the body.

hydroxychloroquine An antimalarial drug that is also used to treat lupus erythematosus and rheumatoid arthritis. *TN* Plaquenil.

hydroxylysine A lysine residue that has been post-translationally hydroxylated by lysyl hydroxylase (procollagen-lysine 5-dioxygenase; EC 1.14.11.4). Found in *collagen where it may be glycosylated with galactose and then glucose residues added sequentially by glycosyl transferases.

hydroxymethylbilane synthase *See* UROPORPHYRINOGEN I SYNTHETASE.

hydroxyproline An unusual amino acid found most commonly as a component of *collagen. Proline residues on the amino side of a glycine residue in collagen are hydroxylated by *prolyl hydroxylase before the polypeptide becomes helical. The presence of hydroxyproline is essential for producing stable triple-helical tropocollagen and absence of ascorbate (vitamin C), which is an essential cofactor for the hydroxylating enzyme, leads to scurvy. It is also present in the major glycoprotein of primary plant cell walls.

hydroxytetraecosanoic acid *See* HETE.

hydroxytryptamine (5-hydroxytryptamine, 5HT) *See* SEROTONIN.

hydroxyurea An inhibitor (hydroxycarbamide) of DNA synthesis used as an antimetabolite in treating haematological neoplasia and in the treatment of sickle cell disease. Has some antiviral properties.

hydroxyzine An antihistamine drug with sedative (anxiolytic) properties.

hygromycin B An aminoglycoside antibiotic from *Streptomyces hygroscopicus* that induces misreading of aminoacyl-tRNA by distorting the ribosomal A site and the ribosomal translocation process. Toxic for both pro- and eukaryotic cells and used as an animal food additive (Hygromix) as an anthelmintic. It is a standard selection agent in eukaryotic and prokaryotic transfection; the vector-carried hygromycin-B-phosphotransferase gene, *hph*, conferring resistance.

Hymenolepis A genus of cestode parasites of the gut of mammals. *Hymenolepis nana* (dwarf tapeworm) is the commonest tapeworm infecting humans. Immunological responses to *H. diminuta* infection have been studied extensively as a model for gut immunity.

hyoscine *See* SCOPOLAMINE.

hyoscyamine An antimuscarinic drug that blocks the action of acetylcholine at parasympathetic sites in smooth muscle, secretory glands, and the central nervous system. It is found in many members of the Solanaceae (particularly *Datura stramonium*) and is the active component of *Atropa belladona* (belladonna) extract. Hyoscyamine sulphate is used for relief of gut spasm. It is the L-isomer of *atropine.

hypaesthesia Reduced sensitivity to touch.

hypalgesia Reduced pain sensitivity.

Hypaque™ A compound (sodium diatrizoate) used to prepare contrast media for radiology and as a component of media for blood cell separation by centrifugation (*see* FICOLL™).

hyper- Prefix denoting more than, greater than, excessive.

hyperacidity Excessive acidity, particularly of gastric juices. **Hyperchlorhydria** is increased secretion of hydrochloric acid by the acid-secreting cells of the stomach.

hyperactivity A behavioural state characterized by motor restlessness, poor attention, and excitability. *See* ATTENTION DEFICIT HYPERACTIVITY DISORDER (ADHD).

hyperadrenalism Overactivity of the adrenal glands. The symptoms depend on which hormone is involved and the degree of involvement. *See* CUSHING'S SYNDROME.

hyperaemia An excess of blood in a region of the body. **Active hyperaemia** is the increase flow of blood in response to increased metabolic activity of an organ or tissue. **Reactive hyperaemia** occurs in response to the release of occlusion of blood flow to an organ.

hyperaesthesia Enhanced sensitivity to tactile stimuli. **Hyperalgesia** is increased sensitivity to pain.

hypercalcaemia A condition in which calcium levels in the blood exceed normal limits (9–10.5 mg/dL). Most (90%) cases are due to *hyperparathyroidism or malignancy.

hypercapnia Excess of carbon dioxide in the lungs or the blood (>45 mmHg), generally a result of hypoventilation.

hypercatabolic hypoproteinaemia *See* MICROGLOBULIN.

hyperchlorhydria *See* HYPERACIDITY.

hypercholanaemia An inherited disorder characterized by elevated serum bile acid concentrations, itching, and fat malabsorption probably because bile acids leak through tight junctions into plasma. Associated with a mutation in tight junction protein 2 (*see* ZO-1).

hypercholesterolaemia The condition in which there are high serum levels of cholesterol (the average fasting blood cholesterol level in people living in the UK is 5.8 mmol/L). Often simply a result of excessive dietary lipid, but in some cases because of a defect in lipoprotein metabolism as in **familial hypercholesterolaemia**, where the defect is in the *LDL receptor. The classic autosomal dominant form of familial hypercholesterolaemia results from defects in the cell surface receptor that removes LDL particles from plasma. Autosomal dominant **type B** is caused by mutation in the LDL receptor-binding domain of *apolipoprotein B-100 and **type 3** is due to mutation in the gene encoding proprotein convertase subtilisin/kexin type 9 (PCSK9). **Autosomal recessive hypercholesterolaemia** is caused by mutations in the *ARH* gene that codes for *low-density lipoprotein receptor adaptor protein (LDLRAP1). In all cases the result is inadequate uptake of LDL into *coated vesicles for recycling. High cholesterol levels are one of the risk factors for *atherosclerosis.

hyperchromasia A histoathological term for cells in which the nuclei appear to be dark, smudged, or opaque when stained. Usually an indication of precancerous behaviour.

hyperCKmia A disease in which serum creatine kinase (CK) levels are persistently elevated but there is no muscle weakness. Caused by defects in *caveolin-3.

hyperemia *See* HYPERAEMIA.

hyperesthesia *See* HYPERAESTHESIA.

hyperexcitability Physiological state in which the threshold for neuronal firing is reduced and there are spontaneous action potentials. This may cause hyperalgesia and there are mutations in ion channels which lead to this condition. The term is also used of behaviour.

hyperferritinaemia–cataract syndrome An autosomal dominant condition in which there is congenital nuclear *cataract and elevated serum *ferritin unrelated to iron overload. It is caused by mutation in the iron-responsive element (IRE) in the 5′-noncoding region of the ferritin light chain gene.

hyperforin (HF) A compound isolated from St John's wort *Hypericum perforatum*. Often used as an antidepressant, apparently working by blocking neurotransmitter re-uptake (*see* SSRIs). Found to be inhibitory to multidrug-resistant *Staphylococcus aureus* and reported to promote apoptosis of B-cell chronic lymphocytic leukaemia cells.

hypergammaglobulinaemia A condition in which immunoglobulin concentrations in the blood exceed normal limits (IgA, 76–390 mg/dL; IgG: 650–1500 mg/dL; IgM, 40–345 mg/dL; IgE, 0–380 IU/ml). May result from continuous antigenic stimulation in chronic infections, from autoimmune diseases, or from abnormal proliferation of B cells as in Waldenstrom macroglobulinaemia or in myelomatosis. *See* HYPER-IGD SYNDROME.

hyperglycaemia A condition in which plasma glucose levels exceed the normal range (fasting: 70–110 mg/dL; postprandial, ~120 mg/dL), usually indicative of a deficiency in insulin production. *See* DIABETES.

hyperhidrosis (hyperidrosis) Excessive sweating.

hyperidrosis *See* HYPERHIDROSIS.

hyper-IgD syndrome An autosomal recessive disorder characterized by recurrent episodes of fever associated with lymphadenopathy, arthralgia, gastrointestinal disturbance, and skin rash, caused by mutations in the gene encoding *mevalonate kinase.

hyperimmune serum Serum with a high titre of specific polyclonal antibodies, usually from animals that have recently been exposed to the antigen.

hyperinosis Excess of fibrin in the blood, as opposed to hypinosis. Probably obsolete.

hyperinsulinism Hypoglycaemia caused by oversecretion of insulin, usually associated with pancreatic tumours. Familial hyperinsulinism is the most common cause of recurrent hypoglycaemia in neonates and can cause irreversible brain damage. Familial hyperinsulinaemic hypoglycemia-1 (**HHF1**) is caused by mutation in the gene encoding the SUR1 subunit of the inwardly rectifying potassium channel of pancreatic beta cells and **HHF2** by mutation in the gene encoding the Kir6.2 subunit. **Type 3** is caused by mutation in the glucokinase (EC 2.7.1.1) gene, **type 4** by mutation in the 3-hydroxyacyl-CoA dehydrogenase (HADH; EC 1.1.1.35) gene, **type 5** by mutation in the gene for the insulin receptor and **type 6** by mutation in the glutamate dehydrogenase (GDH; EC 1.4.1.3) gene. Further genetic heterogeneity of HHF probably exists and there are forms associated with enteropathy and deafness (caused by a 122-kb deletion on chromosome 11p15–p14).

hyperkalaemia A condition in which there are abnormally high levels of potassium in the blood (normal levels are 3.5–5.0 mmol/L). Most cases are caused by disorders that reduce potassium excretion by the kidney, either because of dysfunction or a deficiency of aldosterone (as in *Addison's disease). Very high levels are lethal. *See* PSEUDOHYPOALDOSTERONISM.

hyperkalaemic periodic paralysis *See* PERIODIC PARALYSIS; VOLTAGE-GATED SODIUM CHANNELS.

hyperkeratosis Overgrowth of the horny layer of the skin. Hyperkeratosis is the most troublesome feature of bullous ichthyosiform erythroderma (epidermolytic hyperkeratosis) which is associated with point mutations in *keratin genes. Keratinocytes move from the basal layer of the epidermis to the stratum corneum in as little as 4 days, a process that normally takes 2 weeks. Hyperkeratosis is also a feature of *palmoplantar hyperkeratoma and various *ichthyoses.

hyperkinesia (hyperkinesis) Excessive motility either of a person (hyperactivity) or of particular muscles (tics).

hyperleucine–isoleucinaemia A disorder in which there are elevated plasma levels of leucine, isoleucine, and proline although valine levels are normal (*see* HYPERVALINAEMIA) suggesting the defect is in a different branched chain *aminotransferase. The only reported patients had seizures and mental retardation and failed to thrive.

hyperlipidaemia A condition (hyperlipoproteinaemia) in which plasma lipid levels are elevated. Hyperlipidaemias can be classified according to the plasma lipids that are at high concentration and in most forms, except type IIa, triglyceride levels are high. Thus in **type I**, a rare condition due to a deficiency of lipoprotein lipase (LPL) or altered apolipoprotein C2, there are high levels of chylomicrons and extremely elevated triglycerides. In **type IIa** (familial *hypercholesterolaemia) low density lipoprotein (LDL) levels are raised, in some cases because of defects in the LDL receptor. **Type IIb** is the classic mixed hyperlipidaemia (high cholesterol and triglycerides) caused by elevations in both LDL and very low density lipoprotein (VLDL). **Type III** (dysbetalipoproteinaemia) is associated with high cholesterol and triglyceride levels and a defect in ApoE synthesis. **Type IV** (carbohydrate-inducible hyperlipoproteinaemia) is characterized by abnormal elevations of VLDL and triglyceride but serum cholesterol levels are normal. **Type V** is characterized by elevations of both chylomicrons and VLDL. **Familial combined hyperlipidaemia type 1** is associated with single-nucleotide polymorphisms in the gene encoding the transcription factor USF1 (upstream stimulatory factor-1) that controls expression of several genes involved in lipid and glucose homeostasis or, in **type 2**, mutation at a locus on chromosome 11.

hyperlipoproteinaemia *See* HYPERLIPIDAEMIA.

hypermetropia (hyperopia) A defect of vision (longsightedness) in which the image forms behind the plane of the retina, either because the eyeball is too small or the lens of the eye cannot become sufficiently spherical to focus the image onto the retina, and near objects appear blurred. Can usually be corrected with lenses, although mild forms are usually left untreated.

hypernephroma Renal cell carcinoma. *See* RENAL DISORDERS.

hyperopia *See* HYPERMETROPIA.

hyperosmotic Describing a liquid that has a higher osmotic pressure; usually the comparison is with the physiological level (275–295 mOsmol/kg).

hyperostosis corticalis generalisata *See* VAN BUCHEM'S DISEASE.

hyperoxaluria A disorder in which there is high urinary oxalate excretion, caused by mutation in the gene encoding alanine-glyoxylate aminotransferase (EC 2.6.1.44) an enzyme of intermediary metabolism found in liver peroxisomes.

hyperparathyroidism A condition in which there is overproduction of parathyroid hormone (*parathormone), usually because of hyperplasia of the gland, although in some cases there are parathyroid adenomas. The consequence is high blood calcium levels (*hypercalcaemia) and low blood phosphate levels. **Hyperparathyroidism-1** (HRPT1; familial isolated hyperparathyroidism, FIHP) can be caused by mutation in the *parafibromin (*HRPT2*) gene which also causes hyperparathyroidism–jaw tumour syndrome (**hyperparathyroidism 2**) or in the *MEN1* gene (mutations cause *multiple endocrine neoplasia type I). **Type 3** (HRPT3) has been mapped to chromosome 2p14–p13.3.

hyperphosphataemia A condition in which there are high levels of phosphate in the blood (normal levels are ~1.0–1.5 mmol/L), often associated with *hypocalcaemia. *See* HYPOPARATHYROIDISM.

hyperpituitarism Overactivity of the pituitary gland which can lead to *acromegaly, *Cushing's syndrome, or other conditions depending upon the cells involved and the hormone that is being overproduced.

hyperplasia A condition in which the numbers of cells in a tissue increase, and the tissue increases in size, as a result of enhanced division. The division rate returns to normal once the stimulus is removed. *Compare* NEOPLASIA where the increase in proliferation is independent of stimulus.

hyperpolarization A negative shift in a cell's *resting potential (e.g. from −70 mV to −90 mV) making the cell less likely to respond to depolarizing stimuli. The opposite of *depolarization.

hyperpyrexia Arbitrarily, a body temperature above ~41 °C. *See* HEAT EXHAUSTION; MALIGNANT HYPERTHERMIA.

hypersensitivity An excessive and potentially damaging immune response as a result of previous exposure to the antigen. Immediate hypersensitivity (types I–III) occurs rapidly: in **type I** immediate hypersensitivity the response involves antigen reacting with cell-bound IgE (usually mast cells) triggering histamine release; anaphylactic responses and urticaria. In a **type II** response circulating antibody reacts with cell surface or cell-bound antigen, and complement fixation can cause cytolysis. In **type III** reactions immune complexes are formed in solution and lead to damage (serum sickness, glomerulonephritis, *Arthus reaction). **Delayed-type hypersensitivity**, the **type IV** response, involves primed lymphocytes reacting with antigen and leads to the formation of a lymphocyte–macrophage granuloma without involvement of circulating antibody.

hypertelorism An unusually large separation between organs, generally referring to an increased distance between the eyes (orbital hypertelorism), an anatomical feature of several developmental abnormalities, e.g. *cri du chat syndrome.

hypertension An increase in blood pressure above the normal (usually >140/90), a risk factor for heart or renal failure, stroke, and myocardial infarction. A minority of cases have endocrine or renal causes. The pharmaceutical industry has produced a range of drugs for the treatment of hypertension (e.g. beta-blockers such as atenolol, angiotensin converting enzyme (ACE) inhibitors, etc.).

hyperthyroidism An autoimmune disorder (Graves' disease; Graves–Basedow disease) characterized by goitre, exophthalmia, and toxicity due to an excess of thyroid hormone (thyrotoxicisis). Antibodies to the *thyroid stimulating hormone (thyrotropin) receptor are agonistic and increase the production of thyroid hormone. In white people it is associated with HLA-B8 and DR3 and susceptibility may depend on polymorphisms in genes for CTLA4, vitamin D receptor, the vitamin D binding protein and IFIH1 (an interferon-inducible putative RNA helicase, mda-5). Other loci have also been implicated in susceptibility, including several associated with other forms of autoimmune thyroiditis, including *Hashimoto's thyroiditis.

hypertonic Describes a fluid that is sufficiently concentrated to cause osmotic shrinkage of cells. Since cells can regulate their volume by active transport they can tolerate some degree of hyperosmolarity in their environment without shrinkage. *See* HYPOTONIC; ISOTONIC.

hypertonus Increased muscle tension or muscle spasm.

hypertrichiasis (hypertrichosis) Excessive hairiness.

hypertriglyceridaemia A condition in which there are abnormally high levels of plasma triglycerides, which can cause acute pancreatitis and is a major risk factor for coronary heart disease. Most *hyperlidaemias are associated with raised triglyceride levels. Susceptibility to hypertriglyceridaemia has been associated with mutation in the apolipoprotein A5 gene and the lipase I gene, and with polymorphism in the RP1 retinitis pigmentosa gene. Normal levels are below 150 mg/dL.

hypertrophic cardiomyopathy A relatively common condition in which there is abnormal thickening of the muscle of the left ventricle. This can lead to sudden heart failure in young people (particularly athletes). Various types of hypertrophic cardiomyopathy have been described on the basis of the area of the myocardium that is hypertrophied. The condition can be caused by mutation in various genes associated with muscle proteins. **Cardiomyopathy-1** (CMH1) is caused by mutation in myosin heavy chain-7, **CMH2** by mutation in cardiac troponin T2, **CMH3** by mutation in tropomyosin 1, **CMH4** by mutation in cardiac *myosin-binding protein C. Other forms involve mutations in cardiac troponin I, myosin light chain 2 (MLC2), MLC3, *titin, cardiac muscle α-actin, cysteine-rich protein-3, myosin light chain kinase-2, *caveolin-3, and myosin heavy chain-6, with mutations in genes for mitochondrial tRNAs for glycine and isoleucine.

hypertrophy Increase in the size of a tissue or organ as a result of an increase in cell size rather than increased numbers of cells (*hyperplasia), though often both processes occur.

hyperuricaemia An abnormally high level of uric acid in the blood. The upper end of the normal range is 360–400 mol/L. *See* UROMODULIN.

hypervalinaemia An autosomal recessive disorder in which high levels of valine are present in plasma and urine. Due to a defect in branched-chain *aminotransferase.

hypervariable region Any region of a coding sequence that exhibits high variability. This occurs e.g. in viral genes encoding coat proteins. The best-known example (and often what is meant) is the variable regions of immunoglobulin *heavy chains or *light chains which determine the antigenic binding properties of a particular antibody.

hypervitaminosis Vitamin poisoning: the condition arising from an excess of a vitamin. The classic example is toxicity due to excess vitamin A from eating the liver of polar bears (and that of seals and huskies), but excess of other vitamins can cause toxicity.

hypinosis *See* HYPERINOSIS.

hypnotic Describing a drug that induces sleep.

hypo- Prefix indicating something that is low or reduced in magnitude; the opposite of hyper-.

hypoadrenalism Underactivity of the adrenal glands. *See* TRIPLE A SYNDROME.

hypoblast The inner of the two germinal layers at the blastula stage of the embryo, adjacent to the blastocoele, that gives rise to the extraembryonic endoderm (including yolk sac). The outer layer (epiblast) will give rise to all the tissues of the adult).

hypocalcaemia The condition in which blood calcium levels are below normal limits.

hypochlorhydria Diminished secretion of hydrochloric acid by gastric acid-secreting cells. Can be a feature of *Ménétrier's disease.

hypochondrogenesis A disorder (achondrogenesis–hypochondrogenesis type II) in which defects in *collagen type II lead to abnormal bone growth. The condition is characterized by a short body and limbs and abnormal bone formation in the spine and pelvis. Achondrogenesis **type IA** is a severe form, usually lethal in early neonatal life, **type IB** is caused by mutation in the *DTDST* gene that encodes a sulphate transporter (solute carrier family 26 member 2, SLC26A2, 739 aa) leading to undersulphation of cartilage proteoglycans and *diastrophic dysplasia.

hypocretin *See* OREXIN.

hypogammaglobulinaemia A condition in which the immunoglobulin levels in blood are below the normal range. This may be a result of deficient synthesis (primary hypogammaglobulinaemia) or a secondary consequence of increased breakdown. Congenital, chronic, and transient types are known and IgA deficiency is the most common form. Transient hypogammaglobulinaemia of infancy (THI) is due to slow maturation of the immune system, although it eventually functions entirely normally. *See* BRUTON'S AGAMMAGLOBULINAEMIA; HYPERGAMMAGLOBULINAEMIA.

hypogastrium The lower part of the abdomen, below the umbilicus.

hypoglycaemia Low blood glucose levels, usually below about 2.5 mmol/L.

hypogonadism A condition in which the gonads do not secrete normal levels of reproductive hormones, usually applied to permanent rather than transient or reversible defects. Hypogonadotropic hypogonadism can be caused by mutations in several genes, including those for the gonadotropin-releasing hormone receptor, the G-protein-coupled receptor-54, the nasal embryonic LHRH factor and the fibroblast growth factor receptor-1. *See also* KALLMANN'S SYNDROME. Hypogonadism is associated with various other disorders such as *testicular feminization.

hypohidrosis Abnormal decrease in production of sweat, sometimes restricted to parts of the body. Can arise through reduction in sweat glands or be due to a variety of autonomic or peripheral nervous system disorders. Focal areas of hypohidrosis may occur in patients with leprosy. **Hypohidrotic ectodermal dysplasia**, of which there are X-linked, autosomal recessive, and autosomal dominant forms, results in abnormal morphogenesis of teeth, hair, and eccrine sweat glands. The **X-linked form** (Christ–Siemens–Touraine syndrome) is due to mutation in the *ectodysplasin-A gene. The **autosomal recessive form** is a result of mutation in the ectodysplasin anhidrotic receptor (*EDAR*) gene or in the EDAR-associated death domain gene (*EDARADD*); mutations in these can also give rise to an autosomal dominant form.

hypomagnesaemia with hypocalcuria and nephrocalcinosis A familial disorder (FHHNC) of renal function associated with mutation in tight junction proteins (*claudins). The standard form is associated with mutation in *paracellin-1 (claudin 16), and a related form with severe ocular involvement is caused by mutations in the tight-junction gene claudin-19.

hyponatraemia A decreased level of sodium in the blood, below 135 mmol/L, unusual and potentially dangerous. Most cases are due to increased vasopressin levels (syndrome of inappropriate antidiuretic hormone secretion, **SIADH**), but can arise as a result of adrenal insufficiency, congenital adrenal hyperplasia, or hypothyroidism. Nephrogenic syndrome of inappropriate antidiuresis (**NSIAD**) is caused by gain-of-function mutations in the gene encoding the vasopressin V2 receptor and levels of circulating vasopressin are very low.

hypoparathyroidism A clinical disorder characterized by *hypocalcaemia and *hyperphosphataemia caused by deficiency of *parathormone (parathyroid hormone, PTH) or inability of PTH to function optimally in target tissues. **Familial isolated hypoparathyroidism** can be caused by mutation in the *calcium-sensing receptor gene, in the parathyroid hormone gene itself, or in the *GCM2* gene (a transcription factor important in parathyroid differentiation and a homologue of the *Drosophila* 'glial cells missing' (*gcm*) gene). There is also an **X-linked form**, possibly a defect in SOX3. **Hypoparathyroidism-retardation-dysmorphism syndrome** (Sanjad–Sakati syndrome) is caused by mutation in the gene encoding tubulin-specific chaperone E; mutations in this gene can also cause Kenny–Caffey syndrome. **Hypoparathyroidism, sensorineural deafness, and renal disease** (HDR) syndrome (Barakat's syndrome) is caused by haploinsufficiency of the gene for the transcriptional enhancer GATA3. Hypoparathyroidism can also be a feature of autoimmune polyendocrinopathy syndrome.

h

hypophosphatasia A metabolic disorder characterized clinically by defective bone mineralization, caused by mutation in the tissue-nonspecific isoenzyme of *alkaline phosphatase. Allelic variants have different ages of onset.

hypophysectomy Surgical removal of the pituitary gland.

hypopituitarism A general term for any condition in which the activity of the pituitary gland is diminished. This usually causes obesity and imperfect sexual development and a range of other disorders. Mutations causing combined pituitary hormone deficiency (**panhypopituitarism**) have been described in the genes for the pituitary-specific transcription factor (*PIT1*), for the paired-like homeodomain transcription factor, prophet of PIT1 (*PROP1*), the Rathke pouch homeobox (*HESX1*), and the LIM homeobox-3 (*LHX3*). Some cases of **X-linked panhypopituitarism** are associated with duplications in the *SOX3* gene. In **Simmonds' disease** there is a failure of the anterior lobe of the *pituitary to produce any one or more of its six hormones (ACTH, TSH, FSH, LH, GH, and prolactin), often due to a pituitary adenoma compressing the normal tissue in the gland.

hypotension Low blood pressure; a condition rather than a disease.

hypothalamo–pituitary–adrenal axis A set of direct influences and feedback interactions between the hypothalamus, the *pituitary gland and the *adrenal glands. The HPA axis is a neuroendocrine system that is responsible for controlling stress reactions and the levels of cortisol and other stress-related hormones. The system is involved in mood disorders such as anxiety disorder, bipolar disorder, insomnia, post-traumatic stress disorder, ADHD, major depressive disorder, burnout, chronic fatigue syndrome, fibromyalgia, irritable bowel syndrome, and alcoholism. Antidepressants are often used to regulate the responses.

hypothalamus The region of the brain that controls the autonomic nervous system and links the nervous and endocrine systems of the body. *See* HYPOTHALAMO–PITUITARY–ADRENAL AXIS.

hypothermia Low core body temperature.

hypothyroidism The condition in which levels of thyroid hormone are reduced. Iodine deficiency is the most common cause of hypothyroidism (*goitre) but the condition can also be caused by *Hashimoto's thyroiditis, agenesis or reduction in the thyroid gland or a deficiency of hypothalamic or pituitary hormones. Pendred's syndrome (goitre with deafness) is caused by mutation in the pendrin (anion transporter) gene. **Nongoitrous congenital hypothyroidism** can be caused by mutation in the gene for the thyroid-stimulating hormone receptor (**type 1**) or in the *PAX8* gene (**type 2**). **Athyroidal hypothyroidism** (Bamforth–Lazarus syndrome) is caused by mutations in the gene encoding thyroid transcription factor-2. Other forms can be caused by mutations in genes for iodotyrosine deiodinase, thyroid oxidase-2, thyroid peroxidase, the sodium–iodide symporter, and defects in thyroglobulin synthesis.

hypotonia A disorder in which muscle tone, the resistance to movement, is poor and there may be muscle weakness. Can arise from a variety of causes.

hypotonia–cystinuria syndrome An autosomal recessive syndrome with cystinuria and a range of neurological and developmental disorders, caused by a deletion on chromosome 2 that disrupts the *SLC3A1* gene (encoding an amino acid carrier) and the *PREPL* gene that encodes a prolyl oligopeptidase.

hypotonic Describing: **1.** A fluid that will cause osmotic shrinkage of cells immersed in it. Not necessarily hypo-osmotic. **2.** A muscle that has poor tone or diminished contraction.

hypotonus (*adj.* **hypotonic**) Decreased tension of a muscle, the opposite of *hypertonus.

hypotrichosis with juvenile macular dystrophy A rare autosomal recessive disorder characterized by early hair loss and severe degenerative changes of the retinal macula culminating in blindness, caused by defects in *cadherin-3.

hypovitaminosis Deficiency of a vitamin, because of poor diet.

hypovolaemia A state in which blood (plasma) volume is reduced. Can lead to hypovolaemic shock, in which blood pressure falls.

hypoxanthine A naturally occurring purine derivative (6-hydroxypurine), a spontaneous deamination product of adenine that can also be formed by oxidation of xanthine by xanthine oxidase. Hypoxanthine-guanine phosphoribosyltransferase (*HGPRT) converts hypoxanthine into inosine monophosphate.

hypoxanthine guanine phosphoribosyl transferase *See* HGPRT.

hypoxia State in which oxygen levels in the blood and tissues are abnormally low.

hypoxia-inducible factor-1 (HIF-1) A heterodimeric basic helix-loop-helix-PAS domain transcription factor (α, 826 aa; β, 798 aa) involved in responses to hypoxia and other environmental stressors by regulating the expression of genes that are involved in energy metabolism, angiogenesis, and apoptosis. The α subunits are normally rapidly degraded by the proteosome but are stabilized by hypoxia. An inhibitor, **factor-inhibiting HIF-1** (FIH-1, 349 aa), regulates HIF activity and plays a role in modulation of the hypoxic response by notch signalling. FIH-1 is widely expressed in invasive breast carcinoma and cytoplasmic rather than nuclear localization is an adverse prognostic indicator.

H zone The central portion of the A band of the *sarcomere in which there are no thin (actin) filaments when the muscle is only partially contracted. The *M line is in the centre of the H zone.

I-309 A *chemokine (CCL1, TCA3, 96 aa) secreted by activated T cells that is chemotactic for monocytes. The receptor is CCR8.

Ia antigen The name originally given to antigens encoded by the I-region of the murine MHC complex, class II histocompatibility antigens, each heterodimers with 34-kDa α and 29-kDa β chains. Similar molecules (Ia-like) were then found on B cells of other species, including humans, and the term Ia has become more general for class II MHC molecules.

***IAP gene* (inhibitor of apoptosis gene)** One of a family of evolutionarily conserved genes that inhibit apoptosis, originally identified as baculovirus genes that allow survival of the host cell whilst the virus replicates. Several human homologues have subsequently been identified. *See* XIAP.

IAPP *See* ISLET AMYLOID PEPTIDE.

iatrogenic disease A disease caused by medical treatment, usually a side effect of a drug.

I band The band of the *sarcomere which is isotropic in polarized light and where only thin filaments are found. The I band can vary in width depending upon the state of contraction of the muscle when fixed whereas the anisotropic (A) band is of fixed length.

IBD **1.** Identity by descent. Genes inherited from a common ancestor have IBD, an important factor in linkage studies for disease susceptibility. **2.** *See* INFLAMMATORY BOWEL DISEASE.

iberiotoxin A peptide toxin (37 aa), from the eastern Indian red scorpion *Buthus tamulus*, that is selective for the calcium-activated large conductance potassium channel (BK(Ca)). Similar (and highly homologous in sequence) to *charybdotoxin but more selective.

IBMPFD *See* INCLUSION-BODY MYOPATHY WITH PAGET'S DISEASE AND FRONTOTEMPORAL DEMENTIA; PAGET'S DISEASE OF BONE.

IBMX An inhibitor (isobutylmethylxanthine) of cAMP phosphodiesterase used experimentally to raise intracellular cAMP levels.

ibotenate (ibotenic acid) An *excitotoxin from *Amanita* sp., that acts on the NMDA class of *glutamate receptors. Causes hallucinations and was used in various shamanic rituals.

IBS *See* IRRITABLE BOWEL SYNDROME.

ibuprofen A *NSAID that inhibits both *cycloxygenase-1 (COX-1) and cyclooxygenase-2 (COX-2). Related compounds are ketoprofen, flurbiprofen, and naproxen.

iC3b Inactivated C3b (C3bi). *See* COMPLEMENT.

IC_{50} The concentration of an inhibitor that causes 50% inhibition of the response, strictly speaking only for *in vitro* test systems. Not the same as the concentration that has an effect in 50% of tests (ED_{50}).

ICaBP Calbindin 3. *See* CALBINDINS.

ICAD An inhibitor of caspase-activated DNase, the α subunit (DFF45, 45kDa, 331 aa) of *DNA fragmentation factor.

ICAMs Intercellular adhesion molecules, type I transmembrane glycoproteins, with immunoglobulin-like C2-type domains that are the ligands for leucocyte function antigen-1 (LFA-1; CD11a/CD18: a β_2-integrin) and to a lesser extent Mac-1 (CD11b/CD18). **ICAM-1** (CD54, 532 aa) is expressed on the luminal surface of endothelial cells and various haematopoietic cells and is up-regulated on activated T and B cells. It is an important ligand for leucocyte adhesion, the site to which rhinovirus binds and to which *Plasmodium falciparum*-infected erythrocytes adhere. **ICAM-2** (CD102, 275 aa) is constitutively expressed on endothelium and on resting lymphocytes and monocytes and is not up-regulated by inflammatory cytokines. **ICAM-3** (CD50, 547 aa) plays a role in the early stages of the immune response and crosslinking of ICAM-3 induces an rise in intracellular calcium and activates tyrosine kinases. **ICAM-4** (CD242, formerly LW blood group antigen, 271 aa) is an erythrocyte-specific adhesion ligand that also binds to the monocyte/macrophage-specific CD11c/CD18. **ICAM-5** (telencephalin, 924 aa) is only expressed in the forebrain and seems to play a role in neuronal targeting.

ICAP1 *See* INTEGRIN CYTOPLASMIC DOMAIN-ASSOCIATED PROTEIN-1.

ICE *See* INTERLEUKIN 1 CONVERTING ENZYME.

Icelandic-type cerebroarterial amyloidosis An autosomal dominant form of *cerebral amyloid angiopathy (amyloidosis VI). Symptoms often begin at age 30–40 with multiple brain haemorrhages (mostly in the basal ganglia), dementia, paralysis (weakness), and death in 10–20 years. Caused by mutation in the gene encoding *cystatin C.

I-cell disease *See* MUCOLIPIDOSES (mucolipidosis II) .

ICF **1.** International Classification of Functioning, Disability, and Health, a conceptual scheme defined by the World Health Organization. **2.** Intracortical facilitation. **3.** *See* ICF SYNDROME.

ICF syndrome A rare autosomal recessive disease, characterized by immunodeficiency, centromeric instability, and facial anomalies, caused by mutations in the DNA methyltransferase gene, *DNMT3B*.

ichthyin *See* NIPA1 GENE.

ichthyosiform erythroderma Disorders of the skin in which there is abnormal proliferation and hyperkeratinization, often accompanied by fragility and a tendency for blistering and infection. **Bullous ichthyosiform erythroderma** (epidermolytic hyperkeratosis) is caused by point mutations in keratin genes (*KRT1* and *KRT10*). **Nonbullous autosomal recessive ichthyoses** are divided into two major clinical entities, nonbullous congenital ichthyosiform erythroderma and lamellar *ichthyosis, although there seems to be some overlap in the clinical manifestations. Nonbullous congenital ichthyosiform erythroderma (NCIE) can be caused by mutation in the transglutaminase-1 gene (which also causes lamellar ichthyosis type 1), the 12R-lipoxygenase gene, and the lipoxygenase-3 gene. A rare form of NCIE, **Chanarin–Dorfman syndrome**, is caused by mutation in the *ABHD5* gene that encodes a 349-aa protein that has strong homology with a large protein family characterized by an α/β hydrolase fold.

ichthyosis A skin disorder in which there is abnormal scaling. **Ichthyosis vulgaris**, probably the commonest single-gene disorder, is caused by mutation in the gene for *filaggrin and is an autosomal dominant. **Lamellar ichthyosis type 1**, an autosomal recessive disorder, is caused by mutation in the gene for keratinocyte transglutaminase. Lamellar ichthyosis **type 2** and **congenital ichthyosis with harlequin fetus** (harlequin ichthyosis) are caused by mutations in the gene for an ABC transporter, *ABCA12*. Lamellar ichthyosis **type 3** can be caused by mutation in the *CYP4F22* gene, lamellar ichthyosis **type 5** by a mutation in a locus on chromosome 17. *See also* ICHTHYOSIFORM ERYTHRODERMA. **X-linked ichthyosis** (XLI) is caused by mutation in *arylsulphatase C or *multiple sulphatase deficiency. *See* NIPA1 GENE.

ichthyosis–sclerosing cholangitis neonatal syndrome A rare autosomal recessive complex ichthyosis syndrome (ichthyosis with leucocyte vacuoles, alopecia, and sclerosing cholangitis) characterized by scalp hypotrichosis, scarring alopecia, vulgar type ichthyosis, and sclerosing cholangitis. It is caused by mutation in the gene for *claudin-1.

icilin A synthetic agonist (331 Da) for the transient receptor potential channel-8 (*TRP-8) present on nociceptive sensory peripheral neurons, that gives the sensation of intense cold. It is nearly 200-fold more potent than menthol.

ICM **1.** Iodinated contrast medium. **2.** *Inner cell mass of the mammalian blastocyst. **3.** *See* ISCHAEMIC CARDIOMYOPATHY.

ICP Intracranial pressure.

IcsA An actin nucleating protein present asymmetrically in the outer membrane of virulent strains of *Shigella flexneri*. An F-actin bundle ('comet tail') assembles at the pole with IcsA as a result of N-WASP activation and involvement of the *Arp2/3 complex, and propels the bacterium through the cytoplasm.

ICSH *See* INTERSTITIAL CELL STIMULATING HORMONE.

ICSI Intracytoplasmic sperm injection, an *in vitro* fertilization method.

ictal Referring to the physiological state during a sudden attack, such as seizure (ictus) or stroke.

icterus (*adj.* **icteric**) Jaundice.

ictus A sudden event, such as a stroke, seizure, collapse, or faint.

Id proteins Proteins (inhibitor of DNA-binding proteins) that regulate tissue-specific transcription by binding to basic helix–loop–helix (bHLH) transcription factors to form a nonfunctional heterodimer. They themselves contain a helix–loop–helix motif through which they interact with the bHLH domain of the transcription factor. They play a part in cell growth, senescence, differentiation, and angiogenesis. For example, ID1 and ID2 are expressed at high levels in pro-B-cells and are down-regulated as cells differentiate into pre-B and mature B cells; ID4 inhibits the transactivation of the muscle creatine kinase E-box enhancer by MyoD and may be a tumour suppressor.

ID$_{50}$ The dose of an infectious organism that will produce infection in 50% of subjects.

idazoxan An α_2-adrenoreceptor antagonist that also binds to *imidazoline receptors; used experimentally.

IDDM *See* DIABETES.

idiogram (karyogram) A representation, often a photomontage, of the chromosome complement of a cell arranged to show the general morphology including relative sizes, positions of centromeres, etc.

idiopathic Describing a disease that arises spontaneously or is of unknown origin.

idiotope The antigenic determinants (*epitopes) that are unique to a particular immunoglobulin, one that is directed against a particular antigen and is produced by a clone of cells, or unique to a T-cell receptor. Each immunoglobulin or receptor may have multiple idiotopes which collectively constitute the *idiotype. The idiotope is often the antigen-binding site and located in the variable region of the immunoglobulin product of that clone.

idiotype The set of epitopes (*idiotopes) that characterize a particular immunoglobulin. Anti-idiotype antibodies combine with the specific sequences, often in the antigen-binding site, and may block immunological reactions. They may resemble the epitope to which the first antibody reacts. *See* NETWORK THEORY.

IDL *See* INTERMEDIATE DENSITY LIPOPROTEINS.

I-domain A domain (insertion or interaction domain, ~200 aa) found in the α chain of some closely related integrins. The I-domain has divalent cation-dependent ligand-binding activity.

idoxuridine A thymidine analogue that inhibits the replication of DNA viruses. Used as an antiviral drug for the treatment of *Herpes simplex* and varicella zoster. *TN* Herpid.

iduronic acid The major uronic acid component (α-L-iduronic acid) of dermatan sulphate and heparin. *See* IDURONIDASE.

iduronidase A lysosomal enzyme (α_1-iduronidase, EC 3.2.1.76, 653 aa) that hydrolyses the terminal α-L-iduronic acid residues of dermatan sulphate and of heparan sulphate. *See* HURLER'S SYNDROME.

IEF *See* ISOELECTRIC FOCUSING.

Iejimalides A group of macrolides (Iejimalides A–D), isolated from *Eudistoma* (a tunicate), that cause growth inhibition in a variety of cancer cell lines at nanomolar concentrations and depolymerize the microfilament network.

IFN *See* INTERFERONS.

ifosfamide A nitrogen mustard alkylating agent used chemotherapeutically for sarcomas, testicular tumours, and lymphomas. It requires metabolic activation by liver enzymes to produce biologically active metabolites which interact with DNA. *TN* Mitoxana.

IFs *See* EUKARYOTIC INITIATION FACTORS (eIFs); INITIATION FACTORS.

IgA The immunoglobulin class characteristic of external secretions where it occurs as a dimer (400 kDa) joined by a short J chain and linked to a *secretory piece or transport piece. It is also present in serum as a monomer. IgAs provide local immunity against infections in the gut or respiratory tract and probably act to reduce binding of IgA-coated microorganisms to the mucosal surface. It is present in human colostrum but not transferred across the placenta. The *heavy chains are α type.

IgD A major immunoglobulin (184 kDa) on the surface of *B cells where it may play a role in antigen recognition. Structurally similar to IgG but with δ heavy chains.

IgE The immunoglobulin (188 kDa) associated with immediate-type *hypersensitivity reactions and with helminth infections. Mostly bound to mast cells and basophils that have an IgE-specific Fc-receptor (FcεR) and only at very low amounts in serum. Heavy chain of ε-type.

IGF *See* INSULIN-LIKE GROWTH FACTOR. **IGF2R**: *see* MANNOSE-6-PHOSPHATE RECEPTOR.

IgG The major immunoglobulin class, present at 8–16 mg/mL in serum, also known as 7S IgG (150 kDa). Each IgG molecule comprises two identical *light chains and two identical heavy chains, usually schematically shown as Y-shaped with the two arms, composed of the variable regions of both the light and heavy chains, having the antigen-binding sites and the tail, comprising the constant regions of the γ-type heavy chains, forming the *Fc region that has cell-binding and complement-binding sites. By proteolytic cleavage the molecule can be split into two *Fab fragments and an Fc fragment. IgGs are functionally bivalent and will neutralize toxins, agglutinate, or *opsonize pathogens; binding of immunoglobulin can trigger complement activation. IgG can be transferred from mother to fetus, conferring immune resistance in the immediate neonatal period. Four main subclasses are known; IgG2 is not transferred across the placenta and IgG4 does not fix complement.

IGIF Interferon-inducing factor. An obsolete name for *interleukin-18.

IgLONs A family of GPI-anchored neural cell adhesion molecules that influence guidance of neurons to their appropriate targets. They belong to the immunoglobulin superfamily and founder members were *LAMP (limbic system-associated membrane protein), *OBCAM (opioid-binding cell adhesion molecule), and *neurotrimin, hence the name. NEGR1 is another member of the family. In rat a related molecule, kilon (46 kDa; kindred of IgLON) has been reported; neurotractin is the chick homologue.

IgM (macroglobulin) The polymeric immunoglobulin that is usually produced first in an immune response, before IgG production begins. IgM differs in having a larger heavy chain (μ chain) and the IgM molecule itself (970 kDa) is built up from five IgG-type monomers joined together, with the assistance of J chains, to form a cyclic pentamer. IgM binds complement and a single IgM molecule bound to a cell surface can lyse that cell. The human erythrocyte isoantibodies are IgM antibodies.

IGS *See* INTERGENIC SPACER.

IgSF *See* IMMUNOGLOBULIN SUPERFAMILY.

IHC Immunohistochemistry. *See* IMMUNOCYTOCHEMISTRY.

Ihh *See* INDIAN HEDGEHOG.

IK *See* IMMUNOCONGLUTININ.

IκB (IkappaB) A family of endogenous inhibitors of *NFκB that bind to the p65 subunit and prevents relocation to the nucleus. Multiple forms have been identified including IκBα (317 aa) and IκBβ (356 aa). Phosphorylation of serine residues on the IκB proteins by **IκB-kinases** (IKK) targets them for destruction, thereby releasing the inhibition. X-linked anhidrotic ectodermal dysplasia with immunodeficiency is caused by mutations in the gene encoding IKK-γ, the regulatory subunit of the kinase.

IKK *See* IκB.

IL-1, IL-2, etc. *See* INTERLEUKINS and individual *interleukin-*n* entries.

ile- Prefix indicating association with the ileum. **Ileitis** is inflammation of the ileum. **Ileostomy** is the surgical formation of an opening in the ileum. **Ileus** is disruption of normal propulsive gastrointestinal motor activity.

ILEI A cytokine (interleukin-like EMT inducer, ~25 kDa) involved in epithelial–mesenchymal transition (EMT). It is overexpressed or altered in intracellular localization in multiple human tumours, an event strongly correlated to invasion, metastasis, and survival in human colon and breast cancer.

ILK *See* INTEGRIN-LINKED KINASE.

imatinib mesylate A small-molecule protein tyrosine kinase inhibitor that inhibits bcr-abl kinase (and various growth-factor receptor tyrosine kinases) and is used to treat the Philadelphia chromosome positive form of chronic myeloid leukaemia. *TNs* Gleevec, Glivec.

Imd pathway One of two evolutionarily conserved defence mechanisms against infectious microorganisms, best defined in *Drosophila*, but not confined to insects. (The other pathway involves *toll-like receptors.) The Imd pathway is usually activated by Gram-negative bacteria, and in turn activates the *JNK signalling cascade and a *NFκB transcription factor, Relish, which induces the expression of a set of genes that encode antimicrobial peptides.

imidazoles (glyoxalines) A group of chemically related drugs with antifungal, antibacterial, and anthelmintic properties. Examples are *ketoconazole, tiabdazole. *Imidazolines are derived from imidazoles.

imidazoline Heterocyclic compound derived from imidazole. **Imidazoline receptors** are thought to play a role in hypertension, pancreatic function, cell proliferation, regulation of body fat, neuroprotection, inflammation, and some psychiatric disorders, such as depression. There appear to be several types of imidazoline receptors: the I_1 type has a high affinity for *clonidine and *agmatine and the I_2 type has a high affinity for *idazoxan. The receptors have proved difficult to identify and there remains some controversy.

imino acid An organic acid that has an imino (>C=NH) and a carboxyl group. Proline is often referred to as an imino acid although it is strictly an azacycloalkane carboxylic acid.

iminoglycinuria A condition in which there is abnormal excretion of glycine, proline, and hydroxyproline in the urine, due to a defect in the transport mechanism shared by these compounds. Intestinal absorption may also be inadequate.

imipramine A tricyclic antidepressant drug with very powerful action but severe side effects.

immediate early genes A class of genes (IEGs) rapidly expressed when quiescent cells are stimulated by growth factors, e.g. the proto-oncogenes c-*fos* and c-*myc*. Many IEGs encode *transcription factors which regulate the expression of other genes.

immortalization Describes the escape from the limitation of a finite number of division cycles (the Hayflick limit) which is characteristic of normal cells. Immortalization of cells in culture may be spontaneous, induced by mutagens or by *transfection of certain oncogenes. *See* CELL DEATH.

immotile cilia syndrome *See* PRIMARY CILIARY DYSKINESIA.

immune complex An antibody–antigen complex that may be soluble or insoluble depending on its size and whether or not complement is present. Immune complexes filtered from plasma can accumulate in the kidney where they can cause *glomerulonephritis. **Immune complex diseases**, e.g. *Arthus type hypersensitivity and serum sickness, are characterized by immune complexes in body fluids.

immune response A state of altered reactivity in the immune system as a result of exposure to antigen, usually conferring immunity. May involve antibody production, induction of cell-mediated immunity, complement activation, or development of immunological tolerance.

immune response gene *See* I-REGION.

immunization Induction of immunity by deliberate exposure to antigens associated with disease, either live attenuated bacteria or virus or purified products (such as bacterial toxins). There is optimism about DNA vaccines that would induce the production of disease-specific antigens for some diseases in which immunization has proved difficult (e.g. malaria, AIDS).

immunoadsorbent Any matrix with either an antigen or an antibody bound to it that can be used, e.g. in a column, to bind and remove the antibody or antigen from a solution.

immunoblotting A *blotting technique in which protein is detected by binding of labelled antibody.

immunoconglutinin (IK) Autoantibodies against complement components, usually against C3b or C4. Elevated serum IK levels can occur after infection and immunization, as well as in diseases such as *systemic lupus erythematosus which have an autoimmune basis.

immunocytochemistry A histological technique in which labelled antibodies are used to 'stain' the antigen of interest. Generally a double-antibody method (indirect immunofluorescence) is used in which the antigen-specific antibody (e.g. a rabbit IgG) is unlabelled and a second labelled antibody (anti-rabbit IgG) is used to locate the first. The second antibody, which can be used with a range of specific antibodies and serves to amplify the response, can be conjugated with fluorochromes (e.g. rhodamine), enzymes (e.g. peroxidase) for colorimetric reactions, gold beads (for electron microscopy), or with the biotin–avidin system.

immunodeficiency *See* AGAMMAGLOBULINAEMIA; AIDS; FELINE IMMUNODEFICIENCY VIRUS; HIV; IMMUNODEFICIENCY, CENTROMERIC INSTABILITY, AND FACIAL ANOMALIES SYNDROME; OMENN'S SYNDROME; SEVERE COMBINED IMMUNODEFICIENCY DISEASE.

immunodeficiency, centromeric instability, and facial anomalies syndrome A rare autosomal recessive disease (ICF syndrome) characterized by facial dysmorphism, immunoglobulin deficiency, and centromeric instability of chromosomes 1, 9, and 16. It is caused by mutations in the gene encoding DNA *methyltransferase-3B.

immunoelectrophoresis Technique in which molecules separated by *electrophoresis are located using a specific antibody.

immunofluorescence An immunocytochemical method in which the labelled antibody is coupled to a *fluorochrome. A very wide range of fluorochromes with different emission spectra are now commercially available.

SEE WEB LINKS

- Useful table of fluorochromes.
- Table of Fluorochrome Absorption Emission Wavelengths.

immunogenicity The property of being able to evoke an immune response within an organism. The immunogenicity of a substance depends partly upon size and partly upon the extent of difference from host molecules. Very highly conserved proteins such as collagen tend to have rather low immunogenicity.

immunoglobulins *See* IgA; IgD; IgE; IgG; IgM.

immunoglobulin superfamily (IgSF) A large group of proteins that all have immunoglobulin-like domains, many of which are involved with cell surface recognition events. Members of the superfamily include immunoglobulins, MHC molecules, some *cell adhesion molecules, and cytokine receptors.

immunological disorders

The immune system consists of two major elements, an innate nonspecific mechanism which includes the phagocytosis of foreign objects, including pathogens, and a superimposed highly specific system with *immunological memory that involves recognition of *epitopes expressed by pathogens (the humoral immune system involving *antibody) or on the surface of infected or altered cells (*cell-mediated immunity). There are also recognition mechanisms based on the *toll-like receptor system, the *Imd pathway and on carbohydrate recognition (*collectins, *C-reactive protein, *ficolins, *mannan-binding lectin) and the antiviral *interferon system (*see* OLIGOADENYLATE SYNTHETASES; RIG-1). Deficiences in any of the components can cause problems (*see* BRUTON'S AGAMMAGLOBULINAEMIA; COMPLEMENT DEFICIENCIES; HYPOGAMMAGLOBULINAEMIA) and the regulation of immune responses is complex and susceptible to error (*see* IMMUNOREGULATION; IMMUNOSUPPRESSION; INTERLEUKIN-10; OSTEOCLAST DIFFERENTIATION FACTOR; SLAM FAMILY; SLAP; T-HELPER CELLS; T-REGULATORY CELLS). Defects in the DNA-splicing system that generates diversity in the immunoglobulins and T-cell receptors can cause immunodeficiencies (*see* CERNUNNOS). The basic phagocytic killing mechanism is defective in *chronic granulomatous disease and *Chediak–Higashi syndrome, predisposing to infection and indicating the importance of the ancestral immune system. Defects in the more specific recognition systems are immunodeficiency disorders such as *severe combined immuno-deficiency, and can also be caused by virus infection (*see* AIDS; FELINE IMMUNODEFICIENCY VIRUS; HUMAN IMMUNODEFICIENCY VIRUS; HTLV). Other immunodeficiencies are *immunodeficiency, centromeric instability, and facial anomalies syndrome and *Omenn's syndrome. Overactivity of the immune system can lead to *autoimmune disorders although this may be a failure to develop *immunological tolerance. Many disorders have an immunological component: *see* ALLERGY; ANAPHYLAXIS; ANERGY; ECZEMA; GLOMERULONEPHRITIS; GOODPASTURE'S SYNDROME; GRISCELLI SYNDROME-2; HYPERSENSITIVITY; NEZELOF'S SYNDROME; SARCOIDOSIS; TIMOTHY'S SYNDROME.

immunological memory The system that enhances the response to a second or subsequent exposure to an antigen (although *immunological tolerance can occur) and thus is the basis for vaccination. The memory presumably resides in long-lived members of the clones of cells that were expanded during the first exposure (memory B cells), although there is a shorter-term T-cell memory (*see* T-MEMORY CELLS).

immunological network *See* NETWORK THEORY.

immunological surveillance The process, involving lymphocyte traffic through peripheral tissues and their recirculation, that ensures that all or nearly all parts of the body are surveyed by visiting lymphocytes in order to detect any altered-self material, foreign antigens, etc.

immunological synapse The contact site between a T cell and an antigen-presenting cell (APC) where T-cell receptors, coreceptors, signalling molecules, and adhesion receptors are concentrated on antigen recognition. Formation of the immunologic synapse is thought to be important for receptor signal transduction and full T-cell activation. Various cytoskeletal molecules such as ezrin, F-actin, and CD43 are also involved at the periphery of the synapse and presumably stabilize the structure.

immunological tolerance Condition in which specific antigens fail to elicit a response. Self-tolerance arises very early in life through suppression of self-reactive lymphocyte clones. In adults, tolerance to foreign antigens can be induced by exposure to antigens under conditions in which specific clones are suppressed. Immunological unresponsiveness differs from tolerance in being nonspecific.

immunomodulation Modification of the immune response, usually immunosuppression.

immunophilins Intracellular proteins that bind immunosuppressive drugs such as *ciclosporin, *tacrolimus, and *rapamycin. Both *cyclophilin and the receptor for tacrolimus (FK binding proteins; FKBPs; 100–1219 aa) are peptidyl prolyl *cis-trans* isomerases (rotamases) although their enzymic activity seems irrelevant to the binding activity. Immunophilins are thought to interact with *calcineurin.

immunoprecipitation The precipitation of a multivalent soluble antigen by a bivalent antibody, resulting in the formation of a large complex which precipitates when the ratios of antibody and antigen are near equivalence.

immunoreceptor tyrosine-based activation motif *See* ITAM; *see also* ITIM.

immunoregulation The regulation of the immune system, particularly the cellular and molecular mechanisms of activation, proliferation, and differentiation of human T and B cells. A complex set of feedback controls involving cell–cell interactions and through cytokines regulates the nature of the response that is elicited by antigenic challenge. There is also regulation as antigen is eliminated and, according to the immunological *network theory, through the formation of anti-idiotype antibodies.

immunostimulatory complexes *See* ISCOMS.

immunosuppression Process that reduces the immune response, such that T- and/or B-clones of lymphocytes are depleted in size or suppressed in their reactivity, expansion, or differentiation. It may be specific, with T-suppressor cells inhibiting T- or B-cell clones, or nonspecific. Drugs that affect lymphocyte proliferation have nonspecific effects; *ciclosporin and *tacrolimus act on T cells, as does antilymphocyte serum; alkylating agents such as cyclophosphamide are less specific in their action and damage DNA replication, while base analogues interfering with guanine metabolism act in a similar way. Immunosuppression is essential for the survival of transplanted tissues. *See* IMMUNOPHILIN.

immunotoxins A toxin conjugated to an immunoglobulin or a Fab fragment, thereby targeting the toxin to a site determined by the specificity of the antibody.

IMP **1.** Insulin-like growth factor (IGF2)-mRNA-binding proteins. A family of proteins that bind differentially to the various mRNAs with different 5′ untranslated regions that are generated from the *IGF2* gene, and repress translation. IMP1 (577 aa) has six RNA-binding modules, IMP2 (599 aa) and IMP3 (579 aa) have similar binding domains but the expression of the IMPs seems to be temporally and spatially regulated during development. **2.** Intramembrane protease-3 (IMP3; signal peptide peptidase-like-2A), a *presenilin-like membrane protease. **3.** U3 small nucleolar ribonucleoprotein (U3 snoRNPs; IMP3, 184 aa; IMP4, 291 aa), required for the early cleavages during pre-18S ribosomal RNA processing. **4.** Inositol monophosphate. **5.** Inosine-5′-monophosphate.

imperatorin A small-molecule inhibitor of receptor-mediated proliferation of primary T cells. Inhibits *NFAT binding to DNA and induces DNA fragmentation at the G_1/S phase of the cell cycle. *See* HERACLENIN.

impetigo A contagious skin disease usually caused by *Staphylococcus aureus*, with frequent secondary infection by *Streptococcus pyogenes*.

implant **1.** An organ or tissue grafted into an abnormal position. **2.** An engineered prosthetic device constructed of biocompatible materials, e.g. an artificial hip joint.

importins Proteins of the karyopherin family that regulate the movement of proteins from the cytoplasm into the nucleus. A heterodimer of importins α and β binds the *nuclear localization signals (NLS) on proteins destined for the nucleus, and importin β also binds ran-GTP and several nucleoporins that are involved in the transport process. The importins, once dissociated from their cargo, recycle to the cytoplasm. Multiple importins are known with differing predilections for their cargo. *Compare* EXPORTINS.

imprinting *See* GENOMIC IMPRINTING.

imprinting control region A chromosomal region that acts, in a methylation-sensitive way, to determine whether imprinted genes are expressed or not according to the parent from which the gene derived. The region is a regulated *transcriptional insulator that binds CTCF.

inactivation A neurophysiological term for the process by which *voltage-gated sodium channels that have been activated or opened by depolarization close during the depolarization. Inactivation is slower than activation.

inbred strain Organisms in which there is homozygosity at many loci as a result of a breeding strategy, e.g. brother–sister mating and back-crossing, that eliminates heterozygosity. *See* CONGENIC.

incidence The frequency of occurrence of new cases of a disease in a particular population during a specified period. *Compare* PREVALENCE.

inclusion bodies Structures with characteristic staining properties, usually found in the nucleus or cytoplasm at sites of virus multiplication. They are semicrystalline arrays of *virions, *capsids, or other viral components.

inclusion body myopathy with Paget's disease and frontotemporal dementia A disorder in which there is lower-body motor neuron degeneration and polyostotic skeletal disorganization resembling *Paget's disease of bone. It is caused by mutation in *valosin-containing protein.

incomplete dominance (codominance) A condition in which the F1 generation are phenotypically different from both parents, usually intermediate in character.

incretins Hormones released from the gastrointestinal tract after feeding, that stimulate insulin secretion. The two main hormones are probably *glucagon-like peptide-1 and *glucose-dependent insulinotropic polypeptide (GIP). Both these hormones are inactivated by dipeptidyl peptidase 4 and inhibitors (e.g. *sitagliptin) have potential as antidiabetic drugs. An injectable incretin mimetic (*exenatide) has been approved for the treatment of type 2 *diabetes.

incubation time The period between infection by a pathogen and the first appearance of symptoms.

indels Regions of DNA at which insertions or deletions occur, frequently polymorphic.

index case In epidemiology, the first or original case of a disease. Synonymous in genetics with the *proband or propositus.

Indian hedgehog (Ihh) One of the *hedgehog family of morphogens, involved in endochondral bone formation. Mutations in the Indian hedgehog gene cause *brachydactyly type A1 (BDA1).

indinavir A *protease inhibitor used to treat HIV/AIDS. *TN* Crixivan.

indirect immunofluorescence *See* IMMUNOCYTOCHEMISTRY.

indirubin-3′-monoxime A selective inhibitor of *cyclin-dependent kinases (CDKs) that arrests cells in the G_2/M phase of the cell cycle. Also inhibits *glycogen synthase kinase and *tau phosphorylation *in vitro* and *in vivo* at Alzheimer's disease-specific sites. It is the active ingredient of Danggui Longhui Wan, a herbal mixture used in traditional Chinese medicine.

indolicidin *See* CATHELICIDIN.

indometacin (indomethacin) A *NSAID that inhibits *cyclo-oxygenases and blocks the production of *arachidonic acid metabolites. Indometacin is the British Approved Name (BAN).

indomethacin *See* INDOMETACIN.

indoramin An α_1 selective adrenoceptor antagonist used for treatment of benign prostatic hyperplasia. *TN* Baratol.

inducer cells Cells that induce nearby cells to differentiate in specified pathways which may be predetermined, in which case the differentiation is evoked, or novel, a genuine induction.

induction *See* EMBRYONIC INDUCTION; ENZYME INDUCTION.

induration Hardness, a term often used to describe tumours.

infantile neuroaxonal dystrophy A neurodegenerative disease, characterized by axonal swelling and spheroid bodies in the central nervous system, caused, in many cases, by mutation in the gene for *phospholipase A2-G6. **Pantothenate kinase-associated neurodegeneration** (Hallervorden–Spatz disease) is a similar disorder with overlapping clinical and pathologic features caused by mutation in the pantothenate kinase 2 (*PANK2*) gene.

infantile paralysis Old synonym for *poliomyelitis.

infarct An area of tissue that has suffered anoxia, usually because of thrombotic occlusion of the vasculature, and become necrotic.

infectious diseases

Diseases caused by pathogenic organisms, viruses, bacteria, fungi, protozoa, or multicellular parasites. With the exception of viruses, the ease of treating the diseases depends on the extent of the difference between the metabolism of the pathogen and that of the host, so that infections with cestodes, nematodes, etc. can be more difficult to eliminate than those caused by bacteria, although the emergence of multidrug-resistant bacteria (*see* METICILLIN-RESISTANT STAPHYLOCOCCUS AUREUS) in response to selection pressure by the use of *antibiotics is a growing problem, particularly with *nosocomial infections. How readily the disease spreads depends on infectivity, transmissibility (sometimes upon the numbers of *vectors or *carriers in the environment), and virulence (*see* EPIDEMIC; PANDEMIC; ZOONOSIS). Parasites such as *schistosomes, *trypanosomes, and **Plasmodium* can evade immune responses by *antigenic variation; other organisms may be resistant because they occupy privileged sites within cells (e.g. various *mycobacteria) or in the central nervous system (*see* HERPESVIRIDAE). Many of the common infectious diseases or agents have separate entries (e.g. *encephalitis, *hepatitis viruses). There are a range of defence mechanisms against infectious diseases, ranging from nonspecific mechanisms (*see* DELETED IN MALIGNANT BRAIN TUMOURS-1; IMD PATHWAY; MANNAN-BINDING LECTIN; PROTEIN KINASE R; PYRIN DOMAIN PROTEINS; TOLL-LIKE RECEPTORS), phagocytic cells such as *neutrophils and *macrophages that use *oxygen-dependent killing mechanisms and the more sophisticated *immunoglobulin and *cell mediated immune systems. Resistance to viral diseases depends upon nonspecific mechanisms such as *interferons (*see also* CRYOPYRIN; MX PROTEINS; OLIGOADENYLATE SYNTHETASES; RIG1; VIPERIN) and on specific immunity, which can confer long-lasting resistance once a *primary immune response has been mounted. Increasingly, *antiviral drugs are being developed.

infectious endocarditis Infection of the endocardium of the heart valves, a more general term than subacute bacterial endocarditis which is mostly caused by *Streptococcus viridans*.

infectious hepatitis *See* HEPATITIS VIRUSES.

infectious jaundice *See* LEPTOSPIRA.

infectious mononucleosis *See* GLANDULAR FEVER.

infectious parotitis *See* MUMPS.

infiltration **1.** The abnormal accumulation of substances in cells. **2.** The gradual spread of infection through an organ (as in tuberculous infiltration of the lung).

inflammasome A protein complex (>700 kDa) that is responsible for the activation of *caspases 1 and 5, leading to the processing and secretion of the pro-inflammatory cytokines IL-1β and IL-18. Two types of inflammasome have been identified, the ***NALP1 inflammasome**, composed of NALP1, *ASC, caspase 1 and 5, and the **NALP2/3 inflammasomes** that contain, in addition to NALP2 or NALP3, the caspase recruitment domain (CARD)-containing protein, Cardinal, ASC, and caspase 1.

inflammation Response to injury. The **acute inflammatory response** involves local vascular dilation that leads to redness (rubor), swelling (tumor), heat (calor), and pain (dolor), the 'cardinal signs'. There is leucocyte adhesion in postcapillary venules and extravasation of phagocytes between the endothelial cells (neutrophils at first, monocytes later) towards the site of damage or infection. The endothelium becomes more permeable leading to plasma exudation, hence the swelling. In **chronic inflammation**, where the stimulus persists, the characteristic cells are *macrophages and *lymphocytes.

inflammatory bowel disease (IBD) A group of chronic relapsing intestinal inflammatory disorders generally subdivided into Crohn's disease and ulcerative colitis, both of which are considered to be autoimmune in origin with complex heritability. **Crohn's disease** usually affects the terminal ileum and colon, and the inflammation is discontinuous. Mutations in the *CARD15* gene (caspase recruitment domain-containing protein 15) are associated with susceptibility and a polymorphism in the promoter for interleukin-6 is associated with susceptibility to Crohn's disease-associated growth failure. There is an association (IBD10) between susceptibility to Crohn's disease and variation in the *ATG16L1* gene that encodes autophagy-16-like-1, a component of a large protein complex essential for *autophagy. **Ulcerative colitis**, which is restricted to rectal and colonic mucosa and has a different pattern of inflammation, has been associated with various genes, and there is often inflammation of other tissues (e.g. *uveitis). There is an association between inflammatory bowel disease (IBD17) and variation in the IL-23 receptor gene.

infliximab A humanized monoclonal antibody against TNFα. *See* ADALIMUMAB. *TN* Remicade.

influenza virus One of the *Orthomyxoviridae, responsible for the respiratory infection influenza (flu) in humans. Type A influenza virus causes the worldwide epidemics (*pandemics) and can infect other mammals and birds; type B only affects humans; type C causes only a mild infection. Types A and B virus evolve continuously, with changes in the antigenicity of their spike proteins preventing immune protection. The spike proteins, external haemagglutinin (HA) and *neuraminidase have been studied as models of membrane glycoproteins. *See* H5N1.

informosome Cytoplasmic inclusions that are masked messenger ribonucleoprotein particles (masked mRNPs) which are protected from degradation. The classic example is in the egg. Probably an obsolete term.

ING1 A tumour suppressor gene product (inhibitor of growth-1, p33ING1, 422 aa) functionally similar to *p53. Mutations have only been associated with squamous cell carcinomas of the head and neck.

inhibin Polypeptide hormone of the TGFβ family, produced by the granulosa cells of the ovary or Sertoli cells of testis, that suppress pituitary FSH (*follicle-stimulating hormone) secretion. There are two forms, $\alpha\beta_A$ and $\alpha\beta_B$, the β subunit being shared with *activin which counters the effects of inhibin. Inhibin is reportedly a tumour suppressor for testicular stromal cell tumours, the key gene being that for inhibin-α.

inhibition constant (K_i) The equilibrium dissociation constant for enzyme and inhibitor. The constant, $K_i = [E]*[I]/[EI]$ where $[E]$, $[I]$, and $[EI]$ are concentrations of enzyme, inhibitor and the complex respectively.

inhibitor of apoptosis gene *See* IAP GENE.

inhibitory postsynaptic potential *See* POSTSYNAPTIC POTENTIAL.

inhibitory synapse A synapse in which the release of an inhibitory neurotransmitter (commonly *GABA) from the presynaptic cell reduces the probability of an action potential in the postsynaptic cell. The inhibitory neurotransmitter opens channels in the postsynaptic cell that stabilize its *resting potential. *See* EXCITATORY SYNAPSE.

initiation codon (start codon) The *codon at which polypeptide synthesis is started (5′-AUG). It is recognized by formylmethionyl-tRNA in bacteria and by methionyl-tRNA in eukaryotes.

initiation complex Complex formed between *initiation factors, mRNA, the 30S ribosomal subunit, and formyl-methionyl-tRNA (in prokaryotes) or methionyl-tRNA (in eukaryotes).

initiation factors (IFs) Proteins that are essential for formation of an *initiation complex and for protein synthesis to begin. In bacteria three distinct proteins have been identified: IF-1 (72 aa), IF-2 (890 aa), and IF-3 (180 aa). IF-1 and IF-2 enhance the binding of initiator tRNA to the *initiation complex. *Eukaryotic initiation factors (eIFs) are more diverse.

INK4 A family of inhibitors of *cyclin-dependent kinase 4 which act as tumour suppressors. The family includes p14ARF (ARF here is 'alternative reading frame'), p15 (INK4b), p16 (INK4a), p18 (INK4c), and p19 (p19ARF, INK4d) which is induced by vitamin D and by retinoids and blocks the ubiquitin-ligase activity of *mdm2 from targeting and inactivating *p53. Mutations in p16 are more common in tumours than mutations in *p53. *See also* CIP/KIP.

inner cell mass Cells of the mammalian blastocyst that are potentially capable of forming all tissues, embryonic and extraembryonic, except the trophoblast.

innexins Highly conserved proteins that are the structural components of invertebrate *gap junctions. *Compare* CONNEXINS in vertebrates. Vertebrate *pannexins are distantly related but considered to be part of the same superfamily.

inoculum **1.** Cells added to start a culture. **2.** Viruses added to infect a culture. **3.** Antigenic material injected to induce an immune response.

inosine A nucleoside formed from hypoxanthine attached to a ribose ring. Commonly found in tRNAs; essential for proper translation of the genetic code because it fails to form specific pair bonds with the other bases. This allows matching of a single tRNA to several codons (*see* WOBBLE HYPOTHESIS). Primers for *PCR that contain inosine will tolerate some mismatch, which is useful when trying to clone homologous protein by using degenerate primers. **Inosine pranobex** (*TN* Imunovir, Isoprinosine, Inosiplex) is an antiviral drug used to treat herpes simplex.

inositol A cyclic hexahydric alcohol with six possible isomers. The biologically active form is myoinositol. *See* PHOSPHATIDYLINOSITOL and PI-3-KINASES.

inositol phosphoglycans A family of compounds (IPGs) released by heterotrimeric

G-protein-regulated hydrolysis of membrane-bound glycosylphosphatidylinositols. They are thought to be second messengers of insulin. There are two subfamilies, IPG-A and IPG-P; **IPG-A** inhibits PKA from bovine heart, decreases phosphoenolypyruvate carboxykinase mRNA levels in rat hepatoma cells, stimulates lipogenesis, and inhibits leptin release in rat adipocytes. **IPG-P** stimulates bovine heart pyruvate dehydrogenase phosphatase, and abnormal secretion may occur in pre-elampsia.

inotropic Describing something that alters the rate of heartbeat, e.g. *adrenaline which has a positive inotropic effect.

in ovo In the egg.

INR *See* INTERNATIONAL NORMALIZED RATIO.

i

insertional mutagenesis Mutation that arises through the insertion of one or more bases. This can be on a large scale, as with the integration of a retrovirus adjacent to a cellular proto-oncogene or involve *transposons.

insertion sequence (IS elements) *Transposon-like nucleotide sequences, usually less than 1500 bp, found in bacteria. They can shape and reshuffle the bacterial genome and may be important in the emergence of virulence.

inside-out patch *See* PATCH CLAMPING.

inside-out vesicle A vesicle produced from the plasma membrane by mechanical disruption that has the cytoplasmic face of the membrane on the outside.

in silico A term derived by analogy with *in vitro*, originally used rather facetiously, for 'experiments' that involve searching a computer database held on a silicon chip.

in situ Literally, in place. *See* IN SITU HYBRIDIZATION.

***in situ* hybridization** **1.** A technique used to show where particular genes are being expressed in a tissue. A labelled DNA or RNA probe is hybridized to the tissue section and binds to complementary mRNA sequences, indicating that the gene is being transcribed, though it does not, of course, prove that message is being translated. **2.** A variant of the technique is used to identify the position of a gene on a chromosome by hybridizing a probe to spread chromosomes. Probes are often fluorescently labelled for *fluorescence *in situ* hybridization (FISH) analysis.

inspissation Thickening by dehydration.

instructive theory A theory of antibody production in which antigen acted as template for the production of specific antibody. Now considered untenable and replaced by the *clonal selection theory.

insufflation Blowing gas, air, vapour, or powder into a body cavity, usually the lungs.

insulin Highly conserved peptide hormone (51 aa, 5808 Da) that is secreted by the *beta cells of the pancreas in response to high blood sugar levels. Insulin has two dissimilar polypeptide chains, A and B, linked by two disulphide bonds, which are derived from proinsulin by the enzymatic removal of the connecting (C) peptide that links the N-terminus of the A chain to the C-terminus of the B chain. C-peptide has no known function once released but has a plasma half-life longer than that of insulin and is a useful measure of the level of pancreatic insulin secretion. Insulin has sequence homologies with other *growth factors, and is a mitogen often added to cell culture media. *See* DIABETES; IRS (insulin receptor substrate).

insulin-dependent diabetes (IDDM) *See* DIABETES.

insulin-induced genes Genes that encode transmembrane proteins of the endoplasmic reticulum (insig-1, 277 aa; insig-2, 225 aa) that bind the sterol-sensing domain of the *SREBP cleavage-activating protein (*SCAP) and retain the SREPB–SCAP complex in the ER. This prevents proteolytic cleavage of SREBP to generate the active nuclear form. They are thought to play a central role in cholesterol homeostasis.

insulin-like growth factor A family of peptides (IGFs), with sequences similar to proinsulin, that are important growth factors. **IGF1** (somatomedin C, 153 aa) mediates many of the growth-promoting effects of growth hormone. **IGF2** (somatomedin A, 180 aa) is a mitogen for many cell types and an important modulator of muscle growth and differentiation. Increased levels of IGF2 are associated with the *Beckwith–Wiedemann syndrome. The activity of the gene is regulated by genomic *imprinting, the paternal gene normally being active. The product of the Wilms' tumour-1 gene is a potent repressor of IGF2 transcription *in vivo*. Expression is also regulated by **IGF2mRNA-binding proteins** (*IMPs). **Somatomedin B** is *vitronectin. The **IGF1 receptor** (IGFR1) is similar to the insulin receptor and is highly overexpressed in most malignant tissues. The **receptor for IGF2** (IGFR2) is also the receptor for mannose 6-phosphate, which is implicated in targeting of lysosomal enzymes, but the binding sites for the two ligands are distinct. IGFR2, unlike IGFR1, is imprinted and only the maternal allele is normally expressed. Mutations in IGFR2 are associated with hepatocellular carcinoma.

insulinoma A tumour of the beta cells of the pancreas, usually benign.

insulin sensitizer Drugs that enhance the effect of insulin but only work if some insulin is present. Two thiazolidinediones (TZDs), rosiglitazone and pioglitazone, are approved but another, troglitazone, was withdrawn after being shown to have adverse side effects on liver function in a few cases.

Integra™ A wound dressing made from crosslinked bovine collagen and shark-cartilage glycosaminoglycan (both very nonallergenic) with a semipermeable silicone outer layer.

integral membrane protein A protein that is firmly anchored in a membrane, either because it has a hydrophobic domain that preferentially associates with the inner region of the membrane, or because of post-translational modification with hydrophobic residues (e.g. a *GPI-anchor). Integral membrane proteins that cross the entire membrane are referred to as transmembrane proteins.

integrase protein A viral protein that is essential for integration of viral DNA into the host. Examples are found in bacteriophage lambda and retroviruses such as HIV-1.

integrin cytoplasmic domain-associated protein-1 A protein (alternatively spliced isoforms ICAP1α, 200 aa; ICAP1β, 150 aa) that interacts with the conserved NPXY (Asn-Pro-X-Tyr) sequence motif found in the C-terminal region of β_1-*integrin. It interacts with *KRIT1 in regulating vascular morphogenesis. The activity of ICAP1 is regulated by calcium/calmodulin-dependent protein kinase II (CAMK2).

integrin-linked kinase A serine-threonine protein kinase (EC 2.7.11.1, 452 aa) that associates with the cytoplasmic domain of β-integrins and regulates integrin-mediated signal transduction.

integrins A superfamily of cell surface proteins involved in adhesion to extracellular matrix or other cells. They are heterodimeric with a β subunit (~790 aa) that is conserved through the superfamily, and a more variable α subunit (~1100 aa). Different α chains will form heterodimers with a single β chain, but the converse is less common. Not only do they mediate adhesion but they are important in bidirectional signalling; binding to ligand can activate *integrin-linked kinase and induce intracellular events (outside-in signalling) and the activity of the ligand-binding site can be modified by the cell thereby altering interactions with the environment (inside-out signalling). Examples include the *fibronectin and *vitronectin receptors of fibroblasts, which bind an RGD (Arg-Gly-Asp) sequence in the ligand in a divalent cation-dependent manner, platelet IIb/IIIa surface glycoprotein (fibronectin and fibrinogen receptor), the *LFA-1 class of leucocyte surface protein and the *VLA proteins.

integron A gene capture system that involves an *integrase gene, a recombination/attachment site, and a promoter that will drive the inserted genes. Antibiotic resistance genes are often transferred by this means.

inteins Intervening sequences in proteins, by analogy with introns in DNA, but the intein also has the ability to proteolytically excise itself and splice the resultant exeins to produce the mature protein. Some intein polypeptides are site-specific endonucleases as well as protein splicing catalysts. The endonuclease activity allows insertion of the intein nucleic acid in a target site in DNA and the production of an intein-containing product. Inteins have been found in both prokaryotes and eukaryotes.

intelligent design The belief that biological systems are so complicated that they must have been designed by a rational creator (aka God). As complicated biological systems are gradually understood and their evolution clarified, so the proponents of this neocreationist doctrine change their ground. It is not a scientific hypothesis.

interactome The complete set of molecular interactions going on within cells, by analogy with the genome. The interactome, and to a lesser extent the proteome, differ from the genome in having a temporal aspect: not all possible interactions, or the proteins involved in them, are necessarily present at a particular moment, whereas the genome is unchanging.

SEE WEB LINKS

- The interactome portal.

inter-alpha-inhibitor A plasma serine peptidase inhibitor (inter-α-(trypsin) inhibitor, ~180 kDa) composed of *bikunin in association with one of a set of similar heavy chains (H1, 2, etc., 900–950 aa) encoded by separate genes (*ITH1–3*). The product of a fourth gene (*ITIH4*) has sequence similarities and a polymorphism in this gene is associated with susceptibility to hypercholesterolaemia.

intercalated disc The junctional complex between cardiac muscle cells that joins them end to end. The electron-dense discs contain *gap junctions, *desmosomes, and convoluted *adherens junctions into which the actin filaments of the terminal sarcomeres insert. The gap junctions couple the cells electrically.

intercalation Insertion into a pre-existing structure as opposed to addition of subunits to the end of a growing polymer.

intercellular Describing something that is between cells, either a connection (junction) or extracellular matrix material.

intercellular adhesion molecule *See* ICAMs.

interdigitating cells Cells with dendritic morphology found in thymus-dependent regions of lymph nodes. They are a subclass of *dendritic cells and are involved in *antigen presentation.

interferon regulatory factor-1 A transcription factor (IRF-1, ISGF3, STAT1, 750 aa) that activates genes that have an upstream *interferon-stimulated regulatory element. The cytoplasmic precursor is activated when the interferon receptor is occupied and the factor can then move into the nucleus.

interferons A family of glycoproteins (interferons) that elicit a virus-unspecific antiviral activity in cells. **Type 1 interferons** are interferon-α (leucocyte interferon), interferon-β (fibroblast interferon), -δ, -, and -τ. **Interferon-γ** is a type 2 interferon and has a range of cytokine-like activities. **Type 3 interferon** (interferon-λ) is a collective term referring to IL-28A, IL-28B, and IL-29. Multiple variants of interferon-α are known (at least 23) and range between 19 and 26 kDa; interferon-β is 187 aa. Interferon-γ is a dimer (166-aa subunits) produced by activated T cells, and has no homology with the type 1 interferons. Interferon-α and -β probably bind to the same cell-surface receptor, whereas the receptor of interferon-γ is distinct. *See* INTERFERON REGULATORY FACTOR-1.

interferon stimulated response element A *response element (ISRE) that is upstream of some interferon-α/β-responsive genes. It is the binding site for ISGF3, an interferon-dependent, positive-acting transcription factor that is cytoplasmically activated, possibly through direct interaction with the interferon receptor.

intergenic spacer (IGS) The region between ribosomal genes in eukaryotes and prokaryotes, commonly used in genotyping studies. In eukaryotes the IGS region varies in length from about 2 kb in yeast to 21 kb in mammals, and is highly variable, even among individuals of the same species. Variation is due to different numbers of internal subrepeats, some of which are duplications of the core promoter which enhances rDNA transcription.

intergenic suppression Restoration of the wild-type phenotype as a result of a mutation in a second gene that reverses the effect of the first mutation. *Compare* INTRAGENIC SUPPRESSION.

interleukin (IL) The interleukins are substances generally, although not exclusively, produced by leucocytes that function during inflammatory responses. They are a subset of *lymphokines and now more frequently considered as *cytokines. The human interleukins are often, but not always, similar to the mouse equivalent. *See* INTERLEUKIN-1, INTERLEUKIN-2, etc. Interleukins figure extensively in the valuable encylopaedia of cytokines and cells (22 600 entries) created, developed, and maintained by Prof. Dr H Ibelgaufts.

SEE WEB LINKS

- The cytokines encyclopaedia homepage.

interleukin-1 (IL-1) A cytokine (IL-1α, LAF, MCF; IL-1β, interferon-β-inducing factor, OAF, catabolin) involved in the activation of both T and B cells in response to antigens or mitogens, as well as affecting a wide range of other cell types. **IL-1α** (159 aa) is produced by keratinocytes and **IL-1β** (153 aa), the predominant form in humans, is mostly produced by monocytes. Although the two forms, encoded by different genes, have only 27% homology, they bind to the same receptor. The precursors lack a signal sequence. An endogenous **IL-1 receptor antagonist**, IL-1ra (152 aa, with a standard signal sequence), binds to the receptor without eliciting a response; soluble and intracellular forms are known. The receptor (p80, CD121a, 569 aa) is a member of the immunoglobulin superfamily and is expressed predominantly on T cells and cells of mesenchymal origin. *See also* CATABOLIN; ENDOGENOUS PYROGEN.

interleukin-1 converting enzyme (ICE) A cytoplasmic cysteine endopeptidase (*caspase-1, EC 3.4.22.36, 404 aa) that cleaves pro-IL-1β (269 aa) into mature IL-1β (153 aa) prior to release. The ICE gene has some homology with the *ced-9* gene of *Caenorhabditis elegans* that encodes a *caspase involved in apoptosis.

interleukin-2 (IL-2) An immunoregulatory cytokine (T-cell growth factor, TCGF; thymocyte stimulating factor, TSF), produced by activated T cells, that causes activation of other T cells independently of the antigen. The **receptor** (IL2R) has unique α (CD25, 272 aa; sometimes IL2RA, where 'RA' is not to be confused with 'ra' in *IL1ra) and β (CD122, 551 aa) subunits and a common γ chain (369 aa) shared by all receptors for the **IL-2 family** of cytokines (IL-4, IL-7, IL-9, IL-15, and IL-21). A recombinant form of IL-2 (Proleukin™) has been approved for the treatment of some tumours.

interleukin-2–inducible T-cell kinase (Itk) A protein tyrosine kinase of the Tec-family that is involved in T-cell activation; interacts with SLP-76 *transmembrane adaptor protein.

interleukin-3 (IL-3) A cytokine (HCGF, MCGF, multi-SCF, 140 aa) produced by mitogen-activated T cells that is a haematopoietic *colony-stimulating factor (HCGF) and has neurotrophic activity. It is species specific—the human form is ineffective in mouse systems and vice versa. The **receptor** is a heterodimer with a ligand-specific α chain (378 aa) and a β chain (897 aa) in common with the receptors for IL-5 and GM-CSF.

interleukin-4 (IL-4) A cytokine (B-cell stimulating factor, BSF-1, 129 aa), produced by a subset of activated T cells (Th2), that promotes the proliferation and differentiation of activated B cells, the expression of MHC class 2 antigens, and low-affinity IgE receptors in resting B cells. The receptor has a common subunit with the IL-2 receptor. Species specific—the human form is ineffective in mouse systems and vice versa.

interleukin-5 (IL-5) A selective eosinophil-activating growth hormone (eosinophil differentiation factor, EDF, T-cell replacing factor, TRF, B-cell differentiation factor I, homodimeric, 134 aa subunits). The **receptor** is heterodimeric, the α chain (420 aa) binds IL-5; the β chain (897 aa) is identical to the β chain of the human granulocyte-macrophage colony-stimulating factor (GM-CSF) high affinity receptor, increases the binding affinity, and is required for signal transduction.

interleukin-6 (IL-6) An immunoregulatory cytokine (BCSF, interferon-β2, 185 aa) that is an important mediator of fever and of the acute phase response. The **receptor** has a ligand-binding α chain (CD126, 468 aa), and the signal-transducing component (gp130; CD130, 918 aa) which is shared with receptors for other members of the *interleukin-6 cytokine family.

interleukin-6 cytokine family Cytokines that all act through receptors with a common gp130 signalling subunit. Members include *interleukin-6, *interleukin-11, *ciliary neurotrophic factor, *LIF, *oncostatin M, and *cardiotrophin-1. Degradation of gp 130 is regulated through a phosphorylation–dephosphorylation mechanism in which protein phosphatase-2A is crucially involved; gp 130 is a potential therapeutic target in cancers.

interleukin-7 (IL-7) Single-chain cytokine (lymphopoietin 1, 177 aa) secreted by monocytes, T cells, and NK cells that stimulates the proliferation of pre-B and pro-B cells without affecting their differentiation and has effects on a range of other cells, including T-cells. The receptor is an integral membrane protein (CD127, 459 aa) expressed on activated T cells, although a soluble form is also found.

interleukin-8 (IL-8) A cytokine of the CXC family (neutrophil activating protein, NAP-1, CXCL8, 72 aa), first isolated as a *chemokine, a potent chemotactic and chemokinetic factor for neutrophils and basophils. The 99 aa precursor is processed by cathepsin L to produce the active form. The **receptor** (CXCR1) is a dimeric G-protein-coupled glycoprotein with 350 aa (CD181) and 360 aa (CD182) subunits.

interleukin-9 (IL-9) A cytokine (mast cell growth factor, MCGF,144 aa) produced by mitogen-activated T cells that stimulates the proliferation of erythroid precursor cells (*BFU-E). May act synergistically with *erythropoietin and together with IL-3 promotes mast cell growth. The **receptor** (522 aa) belongs to the haematopoietic receptor superfamily.

interleukin-10 (IL-10) A cytokine (cytokine synthesis inhibiting factor, CSIF, TGIF, a homodimer of 160 aa subunits) produced by Th2 cells, some B cells, and LPS-activated monocytes. IL-10 regulates cytokine production by a range of other cells and is an inhibitor of immune responses. Individuals show marked variation in IL-10 secretion in response to LPS and the variation has a strong genetic component. The **IL-10 family** (IL-10, IL-19, IL-20, IL-22, IL-24, IL-26, IL-28, and IL-29) is highly pleiotropic and although there are shared receptors and conserved signalling cascades the members mediate a range of different activities, including immune suppression, enhanced antibacterial and antiviral immunity, antitumour activity, and promotion of self-tolerance in autoimmune diseases.

interleukin-11 (IL-11) A pleiotropic cytokine (megakaryocyte CSF, 199 aa) originally characterized as a haematopoietic cytokine, but subsequently shown to have activity in many other tissues, including brain, spinal cord neurons, gut, and testis. The **receptor** shares the gp130 subunit with other members of the *IL-6 cytokine family.

interleukin-12 (IL-12) A heterodimeric cytokine (NK stimulatory factor; cytotoxic lymphocyte maturation factor, IL12/p70, 219 aa and 328 aa) that promotes Th1 responses and regulates immunoglobulin isotype selection. The heterodimeric **receptor** (662 and 862 aa) activates the JAK–STAT pathway. High levels have been implicated in autoimmune disorders.

interleukin-13 (IL-13) An anti-inflammatory cytokine (NC30, p600) produced by activated T cells, a nonglycosylated protein (132 aa) that down-regulates macrophage activity; inhibits production of IL-6 and other pro-inflammatory cytokines such as TNFα, IL-1, and IL-8; and stimulates the production of IL1ra. The gene is located in cluster of genes on human chromosome 5q (*see* 5Q MINUS SYNDROME). A chimeric protein composed of IL-13 and a truncated *Pseudomonas* exotoxin (IL13 PE38QQR) is specifically cytotoxic to cells of Kaposi's sarcoma. There are two forms of the **receptor** (427 aa and 380 aa) that interact with *TRAF3IP1.

interleukin-14 (IL-14) A protein (high molecular weight B-cell growth factor, HMW-BCGF, 546 aa) originally thought to be a cytokine that enhanced proliferation of activated B cells, now redesignated as α-*taxilin.

interleukin-15 (IL-15) A cytokine (IL-T) of the IL-2 family that stimulates T-cell and NK-cell proliferation and activation as well as enhancing B-cell expansion and antibody production. The **receptor** shares β and γ subunits with the IL-2 receptor but has has a unique α subunit (267 aa).

interleukin-16 (IL-16) A cytokine (lymphocyte chemoattractant factor, LCF) produced mainly, but not exclusively, by $CD8^+$ T cells that is a chemoattractant for cells with cell-surface CD4. The IL-16 protein (121 aa) is derived from a much larger biologically inactive precursor (1332 aa) by caspase-3 cleavage. May bind to CC-CKR-5 and contribute to the blocking of *HIV internalization.

interleukin-17 (IL-17) A pro-inflammatory cytokine (cytotoxic T-lymphocyte-associated serine esterase 8, CTLA-8, 155 aa) produced by T cells that activates *NFκB, induces expression of IL-6, IL-8, and ICAM-1 in fibroblasts, and enhances T-cell proliferation. Levels of IL-17 in synovial fluids are significantly higher in rheumatoid arthritis than in osteoarthritis. The **receptor** (CD217, 866 aa) is a type I transmembrane protein, though a soluble form is also found, and has no homology with other known sequences.

interleukin-18 (IL-18) A pro-inflammatory cytokine (interferon-γ inducing factor, IGIF, IIF) of the IL-1 family, that induces interferon-γ production from T cells, augments T-cell proliferation and NK activity in cultures of human peripheral blood mononuclear cells, and has a role in angiogenesis. The precursor is processed by caspase-1 to produce the mature protein (157 aa). The receptor (541 aa) activates NFκB.

interleukin-19 (IL-19) A cytokine of the IL-10 family (melanoma differentiation associated protein-like protein, 177 aa), produced by activated monocytes and B cells; induces IL-6 and TNFα production by monocytes and the production of IL-4, IL-5, IL-10, and IL-13 by activated T-cells; shares its **receptor** with IL-20 and IL-24.

interleukin-20 (IL-20) One of the IL-10 family of cytokines, produced by monocytes and keratinocytes, and an autocrine factor for keratinocytes. The receptor is up-regulated dramatically in psoriatic skin.

interleukin-21 (IL-21) An IL-2-like cytokine (131 aa), produced by activated T cells, that induces the proliferation of T cells and B cells and differentiation of natural killer cells. The receptor shares a common γ chain with the IL-2 receptor. In some early reports *IL-22 was mistakenly named IL-21.

interleukin-22 (IL-22) A cytokine of the IL-10 family (IL-10-related T-cell-derived inducible factor, IL-TIF, 179 aa), produced by activated T cells. Binding to the **receptor**, which shares a common subunit with the IL-10 receptor, activates the STAT pathway. It has been suggested that IL-22, together with IL-17, mediates crosstalk between the immune system and epithelial cells and is involved in host defence against extracellular pathogens at mucosal sites.

interleukin-23 (IL-23) A heterodimeric cytokine composed of one IL-12 subunit (p40) and a different subunit, p19 (189 aa). It is pro-inflammatory, involved in the recruitment of inflammatory cells in Th1-mediated diseases, and may play a role in immune-mediated demyelination of the peripheral nerves in Landry–Guillain–Barré syndrome. The **receptor** (629 aa) is distinct from that for IL-12 and probably activates the JAK–STAT signalling pathway.

interleukin-24 (IL-24) A cytokine of the IL-10 family (ST16, melanoma differentiation-associated gene-7 product, FISP, 206 aa). Induces IL-6 and TNFα production by monocytes and is reported to have tumour-suppressive activity.

interleukin-25 (IL-25) A member of the IL-17 family of immunoregulatory cytokines (IL-17E; SF20, a homodimer of 177 aa subunits), produced by Th2 cells. It has a role in allergic inflammation by up-regulating IgG and IgE production, eosinophil levels, and inflammatory responses, through induction of IL-4, IL-5, and IL-13.

interleukin-26 (IL-26) A cytokine (AK155, a homodimer of 171-aa subunits) of the IL-10 family expressed in Th1 and Th17 cells. Induces secretion of IL-8 and IL-10. The **receptor** is highly specific for IL-26.

interleukin-27 (IL-27) A heterodimeric member of the IL-6/IL-12 family cytokines, with one subunit related to the p40 subunit of IL-12 and IL-23 and the other (p28; IL-30) related to the p35 chain of IL-12. Has both anti- and proinflammatory properties and is expressed by monocytes, endothelial cells, and dendritic cells.

interleukin-28 (IL-28) One of the IL-10 family of cytokines, also called interferon-λ2 (200 aa). Like interferon-I it has antiviral activity and up-regulates MHC class I antigen expression. IL-28A, IL-28B, and IL-29 all signal through a heterodimeric receptor complex composed of the IL-10 receptor β (IL-10 Rb) and a novel IL-28 receptor β (interferon-λR, 520 aa). IL-28B (interferon-λ3) and IL-29 (interferon-λ1) are very similar to IL28A.

interleukin-29 (IL-29, interferon-λ1) *See* INTERLEUKIN-28.

interleukin-30 (IL-30) The new name for the p28 subunit of *IL27.

interleukin-31 (IL-31) A cytokine (164 aa) produced by Th2 cells, closely related to the IL-6 family of cytokines. Signals through a **receptor** composed of IL-31 receptor A (732 aa) and oncostatin M receptor B (979 aa) that is constitutively expressed by keratinocytes. IL-31 is involved in allergic skin reactions and possibly *pruritus.

interleukin-32 (IL-32) An inflammatory cytokine (natural killer cell transcript 4, NK4, 234 aa) that induces production of TNFα, IL-8, and MIP-2. Apparently unrelated to any other cytokines but activates typical cytokine signal pathways. At least four splice variants (α, β, γ, δ) are reported.

interleukin-33 (IL-33) A cytokine (nuclear factor from high endothelial venules, NFHEV, 270 aa) constitutively expressed by smooth muscle cells and bronchial epithelial cells. Expression in primary lung or dermal fibroblasts and keratinocytes is induced by treatment with TNFα and IL-1β. The **receptor** is IL-1 receptor-like-1 (ST2, 556 aa), activates NFκB and MAP kinases, and leads to expression of IL-4, IL-5, and IL-13.

interleukin-34 (IL-34) A cytokine (242 aa) that binds to the colony-stimulating factor 1 (CSF-1) receptor and reportedly stimulates monocyte viability.

interleukin-35 (IL-35) An inhibitory cytokine that may be specifically produced by regulatory T cells. It is a heterodimer of IL-27β (encoded by Epstein–Barr-virus-induced gene 3 (Ebi3)) and IL-12α (p35).

intermediate density lipoproteins A class of plasma lipoproteins formed by triglyceride depletion of very low density *lipoproteins (VLDL), that contain the same apolipoproteins as VLDL, 24–30% triglycerides, cholesterol, and cholesteryl esters. They are ~25–35 nm diameter and are rapidly cleared by receptor-mediated endocytosis. Further triglyceride depletion in the liver produces low density lipoproteins (LDL).

intermediate filaments Cytoplasmic filaments of animal cells with a diameter intermediate between that of the thick and thin filaments of striated muscle. The protein subunits are diverse and tissue-specific. *See* CYTOKERATINS; DESMIN; GLIAL FIBRILLARY ACIDIC PROTEIN; NESTIN; NEUROFILAMENT proteins; VIMENTIN.

intermedin (adrenomedullin-2) *See* ADRENOMEDULLIN.

intermittent claudication A symptom of peripheral artery disease, pain when walking as a result of inadequate blood supply to the leg muscles.

internal bias A term used in analysis of the motile behaviour of crawling cells which show short-term *persistence and do not behave as true random walkers.

internalin Surface proteins (InlA, InlB) of **Listeria monocytogenes* that bind to E-cadherin or LCAM and mediate bacterial entry into epithelial cells. *See* CELL ADHESION MOLECULE.

internal membranes General and noncommital term for intracellular membrane systems such as *endoplasmic reticulum.

internal transcribed spacers Nontranscribed spacer sequences (ITS1 and ITS2) found in the tandemly repeated ribosomal RNA coding units of the nucleolar organizer. Each unit consists of 5′ external transcribed sequence (5′ ETS), 18S rRNA gene, internal transcribed spacer 1 (ITS1), 5.8S rRNA gene, ITS2, 28S rRNA gene, and 3′ ETS. The internal and external spacers are removed and degraded during rRNA maturation. The ITS region is highly conserved intraspecifically, but not interspecifically, and is therefore useful in molecular taxonomy.

international normalized ratio (INR) A system established by the World Health Organization to correct for variability in blood coagulation (clotting) tests. Variability in *prothrombin time (PT) arise because of differences in the *thromboplastin reagents used by laboratories. *See* INTERNATIONAL SENSITIVITY INDEX.

international sensitivity index (ISI) The value assigned to a batch of thromboplastin reagent by reference to a 'working reference' reagent which has been calibrated against internationally accepted standard reference preparations with an ISI value of 1. The more sensitive the thromboplastin reagent, the longer the resulting *prothrombin time (PT); its ISI will be less than 1. The ISI value allows comparison of coagulation times estimated in different laboratories. *See* INTERNATIONAL NORMALIZED RATIO.

interneurons Neurons that connect only with other neurons, and not with sensory cells or muscles. They are involved in the intermediate processing of signals.

internexin The *intermediate filament protein (α-internexin, 499 aa) from which type IV filaments of central nervous system neurons are assembled.

interphase The phase of the *cell cycle between mitotic divisions.

intersectins Multidomain adaptor (scaffold) proteins involved in clathrin-mediated endocytosis, regulation of actin polymerization, and Ras/MAPK signalling. Intersectin-1 (ITSN-1) is required for fission of caveolae and their internalization in endothelial cells and has also been reported to be an antiapoptotic protein. Through its dynamin binding it is involved in synapse formation and endocytosis in the nervous system.

interstitial cells **1.** A general term for relatively undifferentiated cells found in a tissue and distinct from the major tissue cells. **2.** Cells in the testis (*Leydig cells) that are responsible for the secretion of testosterone.

interstitial cell stimulating hormone The name given to *luteinizing hormone in the male, where it stimulates interstitial cells (Leydig cells) of the testis to produce testosterone.

intervening sequence *See* INTRON.

intestinal calcium-binding protein (ICaBP) Calbindin 3. *See* CALBINDINS.

intestinal L cells Specialized endocrine cells, mostly located in the distal gut, predominantly the ileum and colon, that express the *glucagon gene and also have *toll-like receptors important in immune responsiveness. Subpopulations may produce *glucagon-like peptides and other gut hormones. Distinct from the murine L cell line.

intima The inner layer of the blood vessel wall, comprising the endothelial lining that is in contact with blood and an elastic extracellular matrix with a few smooth muscle cells below the endothelial monolayer. Surrounding the intima is the *media, with the outermost layer being the *adventitia.

intimin A protein (*adhesin) of the outer bacterial membrane that binds to the translocated intimin receptor, a bacterially-produced protein that is transferred to the mammalian host cell and expressed on the plasma membrane. Intimins are important in pathology of pathogens such as enterohaemorrhagic *E. coli* O157:H7. Intimin has four Ig-like domains and a C-terminal lectin-like domain similar to that of *invasin.

int* oncogenes** Oncogenes first identified as genes activated by integration of mouse mammary tumour virus and causing the development of carcinomas. The nomenclature is therefore misleading: they are simply genes adjacent to insertion sites 1, 2, 3, etc. **Int1** (Wnt-1) is a member of the *Wnt family; the ***int2 product is fibroblast growth factor-3; the ***int3*** product is *notch4. The ***int4*** gene encodes Wnt-3, ***int-6*** the E-subunit of eukaryotic initiation factor-3 (EIF3).

intragenic suppression Restoration of the wild-type phenotype as a result of a second mutation in the affected gene.

intramembranous particles Particles (IMPs) seen in *freeze fractured membranes that are assumed to represent *integral membrane proteins. The complementary fracture face may have pits rather than particles.

intrathecal Describing the region surrounding the brain and spinal cord and within the meninges. The subarachnoid or subdural space.

intrinsic factor A glycoprotein (gastric intrinsic factor; transcobalamin III, 417 aa) produced by the parietal cells of the stomach that binds vitamin B_{12}. The intrinsic factor–B_{12} complex is selectively absorbed by the distal ileum, though only the vitamin is taken into the cell. Mutation in the gene causes congenital pernicious anaemia.

intrinsic pathway *See* CONTACT ACTIVATION PATHWAY.

intron (intervening sequence) Poorly conserved noncoding DNA sequences of variable length that are a feature of eukaryotic genes, interspersed among the coding *exons; as many as 80 may occur within a single gene. The whole DNA sequence is transcribed into *heterogeneous nuclear RNA (hnRNA) and the introns then removed by RNA splicing, leaving a mature mRNA composed entirely of exons that is then translated in the cytoplasm. The sequences at the 5′ and 3′ ends of the intron, the donor and acceptor splice sites, are self-complementary, which allows a hairpin structure, the cue for removal, to form naturally. Nucleotides in introns are numbered from +1 at the 5′ start to approximately the middle and from −1 at the 3′ end back to the approximate middle. Conventional notation for a mutation in an intron has IVS as part of the name: e.g. IVS4+1G>T indicates substitution of T for G at position 1 of the 5′ donor splice junction of intron 4; IVS4−2A>C indicates substitution of 1 nucleotide (C for A) at position −2 of the 3′ end (acceptor splice junction) of intron 4.

intubation The introduction of a tube, e.g. through the larynx into the trachea to assist breathing or into the oesophagus to allow feeding.

intumescence A swelling, or the process of swelling.

intussusception **1.** The insertion of new material into the thickness of an existing structure rather than at the ends. **2.** In medicine, used of the condition in which a part of the small intestine has invaginated ('telescoped') into another section, causing blockage.

inulin A variable-length plant-derived polysaccharide composed of fructofuranose subunits (~5 kDa). Used to help measure kidney function by determining the glomerular filtration rate (inulin is filtered out but not reabsorbed) and as a dietary additive because of its relatively low energy content.

invadopodia Actin-rich protrusions, similar to *podosomes, that are used by tumour cells to degrade *extracellular matrix and invade tissues. *Supervillin reportedly potentiates the function of invadopodia. A neologism that deserves to fade into obscurity.

invariant chain A small protein (Ia antigen-associated invariant chain, CD74, MHC class II γ chain, 296 aa) of invariant sequence, containing a single transmembrane domain, that is essential for the assembly of MHC class II molecules. The invariant chain assembles into a trimer which then associates with three class II $\alpha\beta$ MHC heterodimers; the nine-membered complex is then moved through the *endoplasmic reticulum system to compartments where peptide loading of class II takes place, the invariant chain is released and degraded, and the antigen-binding groove of the MHC dimer is exposed.

invasins Proteins that promote penetration into mammalian cells. Unlike simple adhesins, which are often lectin-like, interaction of the invasins with host cell surface molecules promotes active internalization involving cytoskeletal rearrangement. For example, the AfaD invasin of pathogenic *E. coli* interacts with α_5–β_1 integrin and thus probably activates integrin-linked kinase. The *internalins of *Listeria monocytogenes* are invasins and functionally similar molecules (*see* THROMBOSPONDIN-RELATED ANONYMOUS PROTEIN) are found on apicomplexan parasites.

invasion A term that in cell biology usually implies the movement of one cell type into a territory normally occupied by a different cell type. *See* INVASION INDEX.

invasion index An index used by M. Abercrombie and J.E.M. Heaysman to quantify the invasiveness of cells *in vitro*. The index is based on measurements on confronted explants of the putatively invasive cells and 'normal' embryonic chick heart fibroblasts growing in tissue culture: it is the ratio of the estimated and actual movement in the collision zone: a value of less than 1 indicates hindrance, which is interpreted as being due to *contact inhibition of locomotion.

• A discussion of the significance of this work.

inverse agonist (reverse antagonist) A ligand that binds to receptors and reduces the proportion in the active form. May actually reduce the background level of activity. Not the same as a *partial agonist.

inversin A protein (nephrocystin-2, 1062 aa) that interacts with the *anaphase-promoting complex subunit-2, has a role in primary cilia function, and is involved in the *cell cycle. Mutations lead to reversal of left-right polarity (*situs inversus) and cyst formation in the kidneys (nephronophthisis) but is distinct from *Kartagener's syndrome.

inversion heterozygote Condition in which one chromosome contains an inversion whereas the homologous chromosome does not.

invertase (sucrase) The enzyme (EC 3.2.1.26) that catalyses the hydrolysis of sucrose to glucose and fructose, a reaction that changes the solution from dextrorotatory to laevorotatory. DNA invertases are a class of *recombinases (resolvases).

in vitro Literally, in glass, but used more generally for cell and tissue culture experiments or tests, even though these are now almost always done in plastic containers. *Compare* IN VIVO.

***in vitro* fertilization (IVF)** Fertilization of an ovum outside the body, in culture, either by allowing sperm–egg fusion or by intracytoplasmic sperm injection using a micropipette. General practice is to produce several zygotes which are allowed to undergo a few division cycles before being implanted into the uterus. Preimplantation genetic diagnosis offers the opportunity to identify embryos with abnormal genetic characteristics, or to select female embryos to avoid X-linked disorders.

in vivo Literally, in life, but used generally for procedures or tests done on intact organisms rather than on isolated cells in culture (**in vitro*).

involucrin A keratinocyte protein (585 aa) that appears first in the cytoplasm but ultimately becomes crosslinked to membrane proteins by transglutaminase and, together with *trichohyalin and *loricrin, forms the scaffold for the cell envelope. First appears in the upper spinous layer of the *epidermis.

involuntary muscle Muscle that is not under conscious control; smooth muscle and cardiac muscle.

involution A term that has a specific meaning in mathematics but in biology generally means a reversal or restoration of an organ to an earlier state or, in embryology, the infolding (rolling) of the edges of a sheet of cells, as in gastrulation.

ion channel A hydrophilic channel that allows ions to move across a lipid bilayer down their electrochemical gradients. There is usually some ion specificity and flow rates are rapid, up to a million ions per second. Some channels are permanently open but others are *voltage-gated (e.g. *voltage-gated sodium channels) or *ligand-gated (e.g. the acetylcholine receptor).

ion exchange Interchange of ions of similar charge between a solution and a solid phase (an ion exchange resin). **Ion exchange resins** are used clinically to reduce potassium concentrations in patients with renal failure.

ion exchange chromatography A separation method, often used for protein purification, in which charged molecules are absorbed and selectively desorbed from a polymer matrix that has charged residues such as carboxymethyl (CM) or diethylaminoethyl (DEAE). The supporting matrix may be polystyrene, cellulose, acrylamide, or agarose.

ionic coupling *See* ELECTRICAL COUPLING.

ionizing radiation Radiation that can ionize, directly or indirectly, the substances through which it passes. α and β radiation are more effective at producing ionization than γ radiation or neutrons and are more likely to cause tissue or cell damage.

ionomycin An antibiotic that is a selective calcium ionophore and is more potent than A23187.

ionophore A small molecule that will facilitate passage of ions across lipid bilayers. Carriers like *valinomycin form cage-like structures around specific ions, and diffuse through the hydrophobic regions of the bilayer, whereas channel-forming ionophores (e.g. *gramicidin) assemble to form continuous aqueous pores through which ions can diffuse.

ionotropic receptors Transmembrane ion channels that open or close in response to a ligand (ligand-gated ion channels). They are often relatively ion-selective, e.g. the *nicotinic acetylcholine receptor allows sodium ions to move across the synaptic membrane causing depolarization. Other examples include some *glutamate receptors and *GABA receptors. *Compare* METABOTROPIC.

ion-selective electrode An electrode half-cell, with a membrane that is permeable only to a single ion species. The electrical potential measured between this and a reference half-cell such as a calomel electrode is the *Nernst potential for the ion concerned and the activity (rather than concentration) of the ion in the unknown solution can be measured if the ionic content of the solution used to fill the electrode is known. Ion-selective electrodes often have a hydrophobic membrane incorporating an *ionophore such as *valinomycin or *monensin. Proton-permeable glass is used in pH electrodes.

iontophoresis Movement of ions caused by the application of an electric field. A useful application is the controlled delivery of charged molecules from the end of a micropipette without hydraulic flow.

IP-10 A cytokine (interferon-inducible protein-10; CXCL10, 10kDa, 98 aa) produced by various cells in response to interferon-γ. It is a T-cell chemokine selectively chemotactic for Th1 cells and monocytes, and inhibits cytokine-stimulated haematopoietic progenitor cell proliferation and angiogenesis. The receptor is CXCR3.

IPG *See* INOSITOL PHOSPHOGLYCANS.

iporin A cytoplasmic protein (interacting protein of rab1, RUN, and SH3 domain-containing protein 2, 1516 aa) that is widely expressed, with highest levels in brain and testis. Has a punctate distribution and interacts with the *GM130 protein.

ipratropium bromide An anticholinergic drug, an atropine derivative, that blocks muscarinic receptors in the lung, inhibiting bronchoconstriction. Used by inhalation to treat obstructive lung disease. *TNs* Combivent, Atrovent.

IPS-1 *See* MAVS.

IPSP *See* INHIBITORY POSTSYNAPTIC POTENTIAL.

IPTG A compound (isopropyl β-D-thiogalactoside) used experimentally to induce expression of genes engineered to be under the control of the *gal promoter.

IQGAPs Multidomain scaffold proteins (IQ motif-containing GTPase activating proteins, p195, IQGAP1, 1657 aa; IQGAP2, 1575 aa) that bind to a wide variety of targets, including actin, catenin, rho family GTPases, and calmodulin (through the IQ motif) and that modulate cell–cell adhesion, transcription, cytoskeletal architecture, and signalling pathways.

IQ motif A protein motif involved in *calmodulin binding in the absence of calcium. Usually present in tandem copies and found e.g. in *neuromodulin and *myosin. Calmodulin binding to incomplete IQ motifs that lack the second basic residue is calcium dependent. *See* IQGAPs.

IRAKs A family of kinases (IL-1 receptor associated kinases) that mediate *toll-like receptor (TLR) signalling. IRAKs associate with the IL-1 receptor following IL-1 binding and are involved in the activation of *NFκB. Multiple isoforms have been isolated and all have homology with *Drosophila* pelle. **IRAK1** and **IRAK2** are essential for the initial responses to TLR stimulation and IRAK2 is critical for late-phase responses. Mutations in **IRAK3** (IRAK-M) are associated with susceptibility to early-onset asthma. Deficiency of **IRAK4** causes pyogenic bacterial and fungal infections in childhood.

IRE *See* IRON-RESPONSIVE ELEMENT.

I region The region (immune region) of the murine genome that encodes the I-A and I-E immune response (Ia) antigens. These polymorphic cell-surface MHC class II proteins are involved in the regulation of the T-cell-dependent immune response and are expressed mainly on B cells and antigen presenting cells.

IRFs Transcription factors (interferon regulatory factors) that control *interferon gene expression and are themselves interferon-inducible. **IRF1** (325 aa) is a transcriptional activator for the type I interferon genes, **IRF2** (349 aa) represses the action of IRF1. IRF1 maps to 5q31 (*see* 5Q MINUS SYNDROME) and is a tumour suppressor; loss of heterozygosity at the IRF1 locus occurs frequently in human gastric cancer. The ***IRF3*** gene encodes a protein (427 aa) that binds specifically to the interferon-stimulated response element (ISRE) but not to the IRF-1 binding site. **IRF4** (lymphocyte-specific IRF; multiple myeloma oncogene-1, 451 aa) negatively regulates *toll-like receptor signalling by selectively competing with **IRF5** (495 aa). Mutation in the ***IRF6*** gene can cause *van der Woude's syndrome and *popliteal pterygium syndrome. **IRF7** (503 aa) is critical for induction of type I interferon. **IRF-8** (interferon consensus sequence-binding protein, ICSBP, 426 aa) is only expressed in cells of the immune system. **IRF-9** (interferon-stimulated transcription factor 3-γ, ISGF3G, 393 aa) is activated in the cytoplasm following interferon-α binding.

Ir genes Immune response genes that map within the region of the MHC that controls T-cell help and suppression (*I region).

iridokeratitis Inflammation of the iris and cornea.

irinotecan A semisynthetic analogue of *camptothecin that inhibits *topoisomerase I and used as a chemotherapeutic agent. It is a prodrug and is enzymically converted to the active metabolite SN-38, which is in turn inactivated by the enzyme UDP glucoronosyltransferase 1A1 (UGT1A1) by glucuronidation. People with variants of UGT1A1 called TA7 (*see* GILBERT'S SYNDROME) have less enzyme in the liver and thus clear the drug more slowly; pharmacogenomic testing is therefore required to prevent excessive dosage. *TN* Camptosar.

iron-deficiency anaemia The commonest form of anaemia, usually due to dietary deficiency of iron or chronic blood-loss.

iron-responsive element (IRE) A translational-control sequence in the 5′ *untranslated region (UTR) of *ferritin mRNA and the 3′ UTR of *transferrin receptor mRNA recognized by iron-responsive element binding proteins (IRE-BP) such as *aconitase. Important in the regulation of iron metabolism.

irritable bowel syndrome (IBS) A common condition in which there are abnormal muscular contractions in the intestine, without any obvious physical cause. *See* INFLAMMATORY BOWEL DISEASE.

IRS (insulin receptor substrate) Multisite docking proteins (IRS-1, 1242 aa; IRS-2, 1338; IRS-4, 1257 aa) that act as an interface between signalling proteins with src homology-2 domains (SH2 proteins) such as p85 of *PI-3-kinase, *Grb-2, and *PTP-2, and the receptors for insulin, IGF2, growth hormone, several interleukins (IL-4, IL-9, IL-13), and other cytokines. They are tyrosine-phosphorylated following ligand binding to the receptor. Association with *14-3-3 proteins may, however, interrupt the association between the insulin receptor and IRS and regulate insulin sensitivity. Mutations in IRS-1 and IRS-2 are associated with increased susceptibility to type 2 *diabetes.

ischaemia Hypoxia as a result of inadequate blood flow.

ischaemic cardiomyopathy Weakness in the muscle of the heart (cardiomyopathy) because of inadequate oxygen delivery. The most common cause is coronary artery disease.

ISCOMS (immunostimulatory complexes) An adjuvant-antigen complex consisting of open cage-like complexes (~40 nm diameter) of cholesterol, lipid, immunogen, and saponins from the bark of the Amazonian oak tree *Quillaia saponaria*.

Is element *See* INSERTION SEQUENCE.

ISI *See* INTERNATIONAL SENSITIVITY INDEX.

islet amyloid peptide (IAPP) A peptide (amylin, 37 aa), structurally related to *calcitonin gene related peptide, that selectively inhibits insulin-stimulated glucose uptake in muscle. Originally isolated from pancreatic amyloid that is commonly found in the islets of patients with type 2 *diabetes.

islet cell autoantigen An autoantigen (ICA69, 69kDa, 483 aa) associated with type 1 *diabetes. It binds to the small GTPase rab2 and has an extended BAR domain that interacts with membrane phospholipids. ICA69 is enriched in the Golgi complex and probably has a role in regulating secretory granule transport.

SEE WEB LINKS

- Additional information about BAR domains.

islet cells *See* ISLETS OF LANGERHANS.

islets of Langerhans Groups of cells (islet cells) found within the pancreas: A cells and B (beta) cells secrete *glucagon and *insulin respectively. *See also* D CELLS.

isoagglutination **1.** The agglutination of spermatozoa by substances produced by ova of the same species. **2.** The agglutination of erythrocytes of the same blood group. *Compare* HETEROAGGLUTINATION.

isoantibody Antibody produced to antigen (an isoantigen) from another individual of the same species.

isobavachalcone A *chalcone isolated from *Angelica keiskei* and *Psoralea corylifolia* that is reported to induce apoptotic death in neuroblastoma cells and has a range of other inhibitory effects on platelet aggregation, Epstein–Barr virus early antigen induction, and MMP-2 activity. Displays DNA strand-scission (cleaving) activity.

isochores Very long stretches (≫300 kb) of DNA that are homogeneous in base composition; the complete human genome can be subdivided into around 3200 such isochores. Isochores can be classified into a small number of families that have different GC levels, ranging from <40% (gene-poor) to >52% (gene-rich). Replicons in a given isochore usually have similar replication timing; early-replicating isochores are short and GC-rich, late-replicating isochores are long and GC-poor.

isodesmosine A crosslinked amino acid formed from four lysine molecules linked into a pyridinium ring. Isodesmosine and *desmosine are involved in the intramolecular crosslinks between elastin chains.

isoelectric focusing (IEF) A method for separating large charge-carrying molecules, particularly proteins, by electrophoresis in a stabilized pH gradient. Migration ceases when the molecule is at the pH corresponding to its *isoelectric point.

isoelectric point The pH at which a molecule carries no net charge, at which point it will cease to migrate in an electrical field. Most proteins have an excess of weak acid residues and are negatively charged at neutral pH. Proteins are generally least soluble at the isoelectric point. *See* ISOELECTRIC FOCUSING.

isoenzymes Enzymes that catalyse the same reaction, but have slightly different amino acid sequences. Tissues often differ in the isoenzyme present and this may reflect subtle differences in enzyme kinetics that are particularly appropriate to the tissue in question.

isoflavones A class of plant-derived compounds related to flavonoids and that are *phytoestrogens. Soybeans are a major dietary source (the major isoflavones in soybean are *genistein and *daidzein) but their nutritional benefit is unclear.

isoform The product of a different gene that has the same function and similar (or identical sequence), sometimes a result of gene duplication which will eventually allow divergence of the properties in a tissue-specific fashion. Many protein isoforms arise from single nucleotide polymorphisms (*SNPs), small genetic differences between alleles of the same gene, and others from *alternative splicing.

isohaemagglutinins Antibodies that react against blood-group antigens of other individuals.

isoleucine (Ileu, I) A hydrophobic amino acid (131 Da) that cannot be synthesized by animals and must be obtained from dietary sources.

isologue (isolog) Describing a protein that has similarity rather than identity in genetic sequence.

isomers Different conformations of molecules containing the same atoms. Requires asymmetry so that different steric conformations are possible.

isometheptene A sympatheticomimetic drug sometimes used to treat migraine.

isometric tension Tension generated without a change in length. In molecular terms, cross-bridges are being re-formed with the same site on the thin filament, and in striated muscle the tension is proportional to the overlap between thick and thin filaments.

isoniazid An antibiotic used in combination therapy for tuberculosis, and for other mycobacterial infections. It inhibits the synthesis of *mycolic acids. *TN* Rimifon.

isopeptide bond A peptide bond that does not involve the α-amino group of an amino acid (the standard eupeptide bond in proteins) but a side group, e.g. the ε-amino group of lysine.

isoprenaline (isoproterenol) A synthetic β-adrenergic agonist (isopropyl-noradrenaline) that causes peripheral vasodilation, bronchodilation, and increased cardiac output. Superseded as a drug for asthma by more specific β_2-agonists such as *salbuterol.

isoprenoids A large family of molecules that includes *carotenoids, phytoids, prenols, steroids, terpenoids, and tocopherols. **Isoprenylation** is a post-translational modification of some proteins, e.g. the addition of geranylgeranyl- or farnesyl-moieties.

isoprostanes *Prostaglandin-like compounds that are produced by nonenzymic peroxidation of lipids, particularly arachidonic acid. They are potent inflammatory mediators and markers of oxidative stress.

isoproterenol *See* ISOPRENALINE.

isopycnic Having equal density. In equilibrium density gradient centrifugation molecules accumulate at the level where their density is the same as that of the solution because they cease to move.

isosbestic Describing the wavelength at which two (or more) chemical species have the same molar absorptivity (absorption coefficients of equimolar solutions).

isosmotic Having the same osmotic pressure.

isosorbide mononitrate *See* NITRATES.

isotonic Describing a fluid that will not cause osmotic volume changes of cells immersed in it. Isotonic solutions are not necessarily *isosmotic. *See* HYPOTONIC, HYPERTONIC.

isotonic contraction Describing contraction in which the tension remains constant. Since the overlap of thick and thin filaments in striated muscle changes as the muscle contracts, this means that the numbers of cross-bridges being formed remains the same, even though potentially more could be made. *Compare* ISOMETRIC TENSION.

isotretinoin A retinoid (vitamin A derivative) used to treat severe acne, but hazardous because it is a teratogen. *TN* Roaccutane.

isotropic Describing an environment in which there are no vectorial or axial cues and the properties are the same at all points.

isotype In immunology, describing an immunoglobulin on the basis of the constant regions of the heavy chains of which it is composed. There are nine isotypes, IgA-1, IgA-2, IgD, IgGs 1–4, IgE, and IgM.

isotype switching The change of immunoglobulin *isotype from IgM to IgG that occurs as the immune response progresses. The *light chain and variable region of the *heavy chain remain the same (and thus antigen specificity is unchanged) but the constant region of the heavy chain is switched from μ to γ. Similarly, IgM and IgD with the same variable region of the heavy chain, but with different heavy chain constant regions (μ and δ), coexist on the surface of some lymphocytes.

isozyme *See* ISOENZYMES.

ISRE *See* INTERFERON STIMULATED RESPONSE ELEMENT.

ISSR Inter-simple sequence repeat, the region of the genome that lies between *microsatellite loci.

I-TAC A CXC chemokine (CXCL11, interferon-inducible T-cell α chemoattractant, small inducible cytokine B11, IP9, 94 aa) that is potently chemotactic for IL-2 activated T cells and may be important in central nervous system diseases that involve T-cell recruitment (e.g. experimental autoimmune encephalomyelitis). Production from macrophages is induced by interferon-γ and down-regulated by anti-inflammatory steroids. The receptor is CXCR3.

ITAM A motif (immunoreceptor tyrosine-based activation motif) in the cytoplasmic domain of immunoglobulin (Ig) superfamily molecules that becomes phosphorylated when the receptor binds ligand. When phosphorylated it can bind tyrosine kinases, such as *zap-70 and *syk, through tandem SH2 domains. Examples are found in T- and B-cell receptors, Fc receptors, NFAM1 (NFAT activating protein with ITAM motif-1) that regulates B-cell development and *dectin-1 that is involved in dendritic cell function. Contrast with *ITIMs.

ITIM A motif (immunoreceptor tyrosine-based inhibitory motif) found in the cytoplasmic tail of various inhibitory receptors on T and B cells (e.g. killer inhibitory receptors, *KIRs and CD22, a B-cell transmembrane protein that inhibits B-cell signalling). The motif, when tyrosine phosphorylated, recruits inhibitory phosphatases (e.g. *SHP-1) that have SH2 domains that preferentially bind the ITIM.

Itk *See* INTERLEUKIN-2–INDUCIBLE T-CELL KINASE.

Ito cells Liver pericytes (hepatic stellate cells) that are activated in liver fibrosis. In chronic injury they are the major source of the collagens involved in fibrosis and cirrhosis, as well as of the tissue inhibitors of metalloproteinases (TIMPs) which inhibit collagen degradation.

itraconazole An antifungal drug that inhibits cytochrome P450-mediated synthesis of ergosterol. Effective against a broad range of fungi. *TN* Sporanox.

ITS *See* INTERNAL TRANSCRIBED SPACERS.

IUBMB The International Union of Biochemistry and Molecular Biology.

IUPAC The International Union of Pure and Applied Chemistry.

ivermectin A semisynthetic anthelminthic derived from *avermectins, used to treat infection with parasitic nematodes such as *Strongylus* and **Onchocerca*. Extensively used in veterinary practice.

IVET *In vivo* expression technology. A promoter-trapping technique that selects microbial promoters that are active in a specified niche, e.g. during the interaction of a microorganism with its host. Bacterial DNA is cloned randomly into a plasmid carrying a promoter trap which consists of a promoterless essential growth factor or antibiotic resistance gene and a linked reporter gene. This plasmid is then used to infect a host strain that is deficient in the growth factor or resistance gene and survival indicates that the promoter is specifically active under the conditions being experienced by the host. Constitutive promoters can be excluded by looking at expression of the reporter gene. The original technique has been modified in various ways and is a powerful approach for functional genomics.

IVS **1.** Abbreviation for the interventricular septum that separates the heart ventricles.
2. *See* INTRON.

J 1. The joule, the SI unit of energy. **2.** The single-letter code for the amino acid trimethyl lysine.

J774.2 cells A mouse (Balb/c) monocyte/macrophage cell line derived by subcloning from the original ascites and solid tumour J774.1. Produces IL-1 and has IgG and complement receptors.

JAB 1. Jun activation domain-binding protein-1 (JAB1), the fifth subunit (37 kDa) of the [COP9 signalosome], that selectively potentiates transactivation by Jun or JunD but not JunB. It interacts with the cytoplasmic domain of the β_2 subunit of the $\alpha_L\beta_2$ integrin, LFA-1, thus linking adhesion proteins with transcriptional regulation. **2.** JAK-binding protein. *See* SOCS.

jacalin A lectin, from the jackfruit *Artocarpus integrifolia*, that is used to bind *O*-glycoproteins such as mucins and IgA_1.

Jaeken's syndrome An autosomal recessive disorder of glycosylation type Ia (**carbohydrate deficient-glycoprotein syndrome Ia**, CDGIa) caused by mutation in the gene encoding phosphomannomutase-2 (EC 5.4.2.8, 246 aa). This leads to deficiencies in the synthesis and processing of *N*-linked glycans or oligosaccharides on glycoproteins, seriously affecting a range of cellular activities. **CDGIb** is caused by mutation in the gene encoding mannosephosphate isomerase (EC 5.3.1.8, 423 aa), **CDGIc** by mutation in the gene encoding asparagine-linked glycosylation protein 6 (ALG6, EC 2.4.1.-, 507 aa).

jagged-1 A ligand (JAG-1) of the notch receptor, originally identified in rat although the human homologue is known and is associated with the *tetralogy of Fallot and with *Alagille's syndrome-1.

jail fever *See* TYPHUS.

JAKs A family of cytoplasmic tyrosine kinases (Janus kinases, 120–140 kDa) that associate with cytokine receptors and are involved in the JAK/STAT signalling pathway that links cell surface receptors and cytokine responses. They have a second phosphotransferase-related domain immediately N-terminal to the protein tyrosine kinase (PTK) domain and are 'two headed'. This may be the origin of the name since Janus, the gatekeeper in Roman mythology, was two-headed, although some cynics believe the name to have originated as 'just another kinase'. JAKs have neither SH2 nor SH3 domains. Mutations in **JAK2** are associated with *polycythaemia vera, acute myelogenous leukaemia and other myeloproliferative disorders. Mutations in **JAK3** are involved in some forms of *severe combined immunodeficiency disease (SCID). *See* JH DOMAINS.

JAMs A family of transmembrane molecules with extracellular immunoglobulin domains, involved in epithelial junctions (junctional adhesion molecules). **JAM-1** (JAM-A, 299 aa) plays a role in regulating epithelial *tight junctions, is a ligand of the integrin, LFA-1, and is a reovirus receptor. It is also found on blood leucocytes and platelets. **JAM-2** (JAM-B, vascular endothelial JAM, VEJAM) is important in endothelial integrity and interacts with T cells but not other leucocytes. **JAM-3** (JAM-C) is a counter-receptor for the leucocyte integrin Mac-1 and promotes neutrophil transendothelial migration *in vitro* and *in vivo*.

janiemycin *See* RAMOPLANIN.

Janus kinase *See* JAKs.

Japanese encephalitis *See* ENCEPHALITIS.

jasplakinolide A cyclic peptide (~710 Da) isolated from the marine sponge *Jaspis johnstoni*. Stabilizes F-actin *in vitro* and binds to the same or similar site as *phalloidin, but permeates into cells more readily. *In vivo*, may disrupt the cytoskeletal microfilament meshwork by inducing formation of amorphous actin polymers.

jaundice Yellowing of the skin (and whites of eyes) by *bilirubin.

JAZF1 A zinc finger transcription factor (TAK1-interacting protein 27, 243 aa). The ***JAZF1–SUZ12* oncogene** is the product of a chromosomal translocation that is reported to occur in endometrial stromal sarcomas. The oncogene consists of the N-terminal portion of JAZF1 (juxtaposed with another zinc finger gene-1) and the C-terminal part of *SUZ12 (suppressor of zeste-12).

J chain A short (J piece, joining chain, 137 aa) polypeptide, found in each dimer of IgA and each pentamer of IgM, linked by disulphide bonds. May have a role in their polymerization and transport across epithelial cells. Not coded by the **J* gene, but by a single gene unlinked to other immunoglobulin structural genes, nor is it identical with the *J region.

JC virus A human *retrovirus similar to *polyoma virus, one of the *Papovaviridae named after a patient (John Cunningham) with progressive multifocal leukoencephalopathy (PML). Only causes PML in immunocompromised patients.

j

J domain A sequence of ~73 aa that is characteristic of DnaJ-like proteins. The J domain is involved in regulating the ATPase activity of heat shock protein 70 (Hsp70). (DnaJ is a bacterial heat shock protein.)

JE *See* MONOCYTE CHEMOTACTIC AND ACTIVATING FACTOR.

jejun- Prefix indicating association with the jejunum, the portion of the small intestine that connects the duodenum to the ileum. **Jejunitis** is inflammation of the jejunum.

Jensen's syndrome *See* MOHR–TRANEBJAERG SYNDROME.

Jervell–Lange-Nielsen syndrome An autosomal recessive disorder characterized by congenital deafness, prolongation of the QT interval (*long QT), ventricular arrhythmias, and a high risk of sudden death. The cause is mutation in the genes for KCNQ1 or KCNE1, which are components of the delayed rectifier potassium channel.

***J* genes** Genes that encode the joining segment of polypeptide which links the V (variable regions) to the C (constant) regions of both light and heavy chains of immunoglobulins. Recombination between the *V* gene, one of the 27 *D* genes and one of the 6 *J* genes produces a rearranged *VDJ* gene (only *VJ* in the case of the light chains) to which is added the relevant constant region to produce a composite from which the whole heavy or light chain is produced.

JH domains Domains (JAK homology domains) found in *JAKs. **JH1** is the the catalytic protein tyrosine kinase domain, the **JH2** domain is a catalytically inactive pseudokinase that inhibits basal Jak activity. JH1 and JH3 domains of Jak2 bind directly to protein tyrosine phosphatase-2A.

Jijoye cells A line of human lymphoblastic cells established from the tumour tissue of a 7-year-old boy with *Burkitt's lymphoma in Africa in 1963. They are $CD23^{+}$ and used as a model for B cells.

jimpy *See* PELIZAEUS–MERZBACHER DISEASE.

jinggangmycin (validamycin A) An aminocyclitol antifungal antibiotic, a trehalase inhibitor, from *Streptomyces hygroscopicus*.

jingzhaotoxin-III A peptide toxin (JZTX-III, 34 aa), isolated from the venom of the Chinese spider *Chilobrachys jingzhao*, that inhibits voltage-gated sodium channels of rat cardiac myocytes by modifying voltage-dependent gating and also binds to Kv2.1 channel, acting in a similar fashion to *hanatoxin1 and *SGTx1.

JNKs A family of serine/threonine protein kinases (jun kinases, c-jun N-terminal kinases, stress-activated protein kinases, SAPKs) involved in intracellular signalling cascades. JNKs, of which there are ten isoforms, are distantly related to *ERKs and activated by dual phosphorylation on tyrosine and threonine residues. In addition to phosphorylating c-jun they also phosphorylate *p53. **JNK1** (mitogen-activated protein kinase-8, EC 2.7.11.24, 427 aa) and **JNK2** (MAPK-9, 424 aa) are expressed in a range of different cell types; **JNK3** (464 aa) is mainly found in neurons.

Jnm1p *See* DYNAMITIN.

Job's syndrome An autosomal dominant syndrome (hyperimmunoglobulin E (hyper-IgE) recurrent infection syndrome, HEIS), caused by mutation in the transcription factor *STAT3 (acute-phase response factor), characterized by chronic eczema, recurrent staphylococcal infections, increased serum IgE, and eosinophilia. Patients have a distinctive facial appearance, abnormal dentition, hyperextensibility of the joints, and bone fractures. An autosomal recessive form is probably a distinct entity but shares most features except skeletal involvement. Originally ascribed to defective neutrophil chemotaxis; at one time, but no longer, all patients described were female, with red hair and elevated plasma IgE levels.

joining chain *See* J CHAIN.

Jones–Mote hypersensitivity A form of delayed-type hypersensitivity, mediated by $CD4^+$ T cells, that is induced by intradermal injection of protein in adjuvant or protein conjugated to haptens. The response is identical in time course to the tuberculin response but is characterized by infiltration of the skin by basophils.

jouberin A protein with multiple SH3 binding sites, an SH3 domain, and multiple WD40 repeat domains, the product of the Abelson helper integration site gene (*AHI1*). The down-regulation of jouberin expression is important in early differentiation of haematopoietic cells. Mutation is associated with *Joubert's syndrome type 3.

Joubert's syndrome A genetically heterogeneous autosomal recessive disorder presenting with psychomotor delay, decreased muscle tone, ataxia, oculomotor apraxia, and malformation of the midbrain–hindbrain junction. **Joubert's syndrome type 3** (JBTS3) is caused by mutation in *jouberin, **JBTS4** by mutation in *nephrocystin 1. The phenotype in **JBTS5** overlaps that of *Senior–Loken syndrome and both syndromes are caused by mutations in proteins of the centrosome or primary cilium (nephrocystin 6, 290-kDa centrosomal protein). **JBTS6** is caused by mutation in the *meckelin gene, **JBTS7** by mutation in the *RPGRIP1L* gene that encodes a protein (1315 aa) having strong similarity to RPGRIP1, a protein present at the photoreceptor connecting cilium and mutated in *Leber congenital amaurosis type VI.

J piece Joining chain. *See* J CHAIN.

J region The polypeptide encoded by **J* genes.

jumonji The jumonji (*jmj*) gene encodes a transcriptional repressor critical for normal cardiovascular development in the mouse and was subsequently shown to be important in development of various tissues. The **jumonji protein** (1246 aa) has a DNA-binding domain, *ARID, and two conserved jmj domains (JMJD1 and 2). The **jumonji 2** (JMJD2) proteins, of which several are known, are histone demethylases and are expressed at high level in a subset of human tumours; knockdown of **JMJD2C** (gene amplified in squamous cell carcinoma 1,GASC1) retards tumour cell growth *in vitro*. The H3K9-specific demethylase **Jhdm2a** (also known as Jmjd1a and Kdm3a) has an important role in nuclear hormone receptor-mediated gene activation and male germ-cell development, and regulates the expression of metabolic genes. Disruption of the *Jhdm2a* gene in mice results in obesity and hyperlipidemia.

jumping gene *See* TRANSPOSON.

jun An *oncogene from avian *sarcoma virus 17 the protein product of which, jun (the name comes from the Japanese 'ju-nana', meaning 17), dimerizes with *fos via a *zipper motif to form the *transcription factor AP1. The transcriptional activator activity of c-jun (the cellular homologue of viral v-jun) is attenuated and sometimes antagonized by JUNB which is an NFκB-regulated jun-like protein that can also dimerize with fos. *See* JNKs (Jun kinase).

junctin One of three proteins encoded by the *ASPH* gene (the others being ASPH and junctate (humbug)). **Aspartate β-hydroxylase** (ASPH, 744 aa) catalyses the post-translational hydroxylation of aspartic acid or asparagine residues within epidermal growth factor-like domains of numerous proteins. **Junctin** (225 aa) is a transmembrane *calsequestrin-binding protein found in junctional sarcoplasmic reticulum from cardiac and striated muscle tissue colocalized with the *ryanodine receptor and *triadin. **Junctate** (299 aa) shares the first 93 aa of the long isoform of junctin but has a much wider tissue distribution; it binds calcium with high capacity and moderate affinity.

junctional adhesion molecule *See* JAMs.

junctional basal lamina Specialized *extracellular matrix at the *neuromuscular junction that interacts with acetylcholinesterase and localizes it to the junction.

junctions *See* ADHERENS JUNCTION; DESMOSOME; GAP JUNCTION; ZONULA OCCLUDENS.

junk DNA Pejorative, and probably inaccurate, term for genomic DNA that serves no known function.

Jurkat cells A human T-cell line derived from a 14-year-old boy with acute T-cell leukaemia. A convenient model for studies of IL-2 production but differing from real T-cells, particularly in activation behaviour. Various subclones are available.

just noticeable difference *See* DIFFERENCE THRESHOLD.

juvenile polyposis An autosomal dominant condition that predisposes to various types of tumours and can be caused by mutations in the **SMAD4* gene or in the gene encoding bone morphogenetic protein receptor-1A (*ALK3). There are hamartomatous gastrointestinal polyps that turn into malignant lesions in ~20% of cases.

juxta- Prefix indicating contact or proximity. **Juxtaposition** is the state of being in close proximity.

juxtacrine activation Activation that requires contact (juxtaposition), e.g. activation of target cells by membrane-anchored growth factors or of leucocytes by *PAF bound to endothelial cell surface. *Compare* PARACRINE.

K562 cells A line of multipotential blast cells, established from a white woman with chronic myelogenous leukaemia, that spontaneously differentiate into progenitors of the erythrocyte, granulocyte, and monocytic series.

***K*$_a$** **1.** Acid *dissociation constant; pK_a is $-\log_{10}K_a$. **2.** Association constant (K_{ass}). The equilibrium constant for association, the reciprocal of K_d, with dimensions of litres/mole.

kainic acid A natural compound, originally isolated from the red alga *Digenea simplex* (kainin-sou), an agonist for the K-type *excitatory amino acid receptors. It is used as an anthelmintic drug and can act as an *excitotoxin. The kainate receptor is an amino acid-gated ion channel. *See* GLUTAMATE (glutamate receptors).

kaiso A zinc finger transcriptional regulator (zinc finger- and BTB domain-containing protein 33, ZBTB33, 672 aa) that interacts with *NPRAP.

KAL The gene (*KAL1*) responsible for the X-linked form of *Kallmann's syndrome, that encodes *anosmin.

kala-azar A disease (visceral leishmaniasis) in which there is enlargement of the liver and spleen, anaemia, wasting, and fever. Caused by infection with *Leishmania donovani* and spread by the bite of infected sand flies.

kalicludines Peptides (57–60 aa), isolated from the sea anemone *Anemonia sulcata*, that are structurally homologous to *dendrotoxins and to Kunitz-type peptidase inhibitors. They will block voltage-sensitive potassium channels and have peptidase inhibitory activity.

SEE WEB LINKS

- Article on kalicludines in the *Journal of Biological Chemistry* (1995).

kalinin The major glycoprotein of epidermal basement membrane (laminin 5β3, epiligrin, nicein, BM600,1172 aa, with smaller fragments when reduced), particularly prominent in the lamina lucida of the skin. It is the major ligand for $\alpha_3\beta_1$ integrin and forms an asymmetric 170-nm rod with two globules at one end, one at the other. It is absent in patients with junctional *epidermolysis bullosa.

kaliotoxin (KTX) An inhibitor of the high conductance Ca^{2+}-activated K^+ channels (KCa), isolated from the venom of the scorpion *Androctonus mauretanicus mauretanicus.* Has homology with *agitoxins, *charybdotoxin, *iberiotoxin, and *noxiustoxin.

kalirin A multidomain guanine nucleotide exchange factor (P-CIP10, huntingtin-associated protein-interacting protein, DUO), one of the *Dbl family, that interacts with small GTP-binding proteins of the rho family. Multiple kalirin isoforms with different functional domains are expressed in the brain and the gene has 60 coding exons. Kalirin-7 localizes to the postsynaptic density and is involved in formation of dendritic spines. Other isoforms associate with secretory granules and through the interaction with G proteins are involved in regulating actin organization, endocytosis, exocytosis, and free-radical production. Single-nucleotide polymorphisms in the first intron of an alternative transcript of the gene are associated with susceptibility to early-onset coronary artery disease.

kaliseptine A peptide (36 aa) that binds to the same receptor site as *dendrotoxins and *kalicludines and is equally effective as a K^+ channel inhibitor. There is no sequence homology with kalicludines or with dendrotoxins.

kallidin A decapeptide *kinin (lysyl-bradykinin) produced in kidney that causes dilation of renal blood vessels and increased excretion of water.

kallikreins Serine endopeptidases that act on *kininogens to produce *kinins and plasminogen to produce plasmin. They are normally present in plasma as prekallikreins which are activated by *Hageman factor. At least thirteen kallikrein genes (and five pseudogenes) are clustered on chromosome 19. **Kallikrein 1** (tissue kallikrein, EC 3.4.21.35, 262 aa) generates Lys-bradykinin by specific proteolysis of kininogen-1, **kallikrein 2** (261 aa) is glandular or prostatic kallikrein, and

kallikrein 3 is prostate-specific antigen (PSA; EC 3.4.21.77, 261 aa). Mutations in **kallikrein 4** (enamel matrix serine peptidase-1, 254 aa) lead to autosomal recessive pigmented hypomaturation *amelogenesis imperfecta. **Kallikrein 8** (260 aa) is the human homologue of mouse neuropsin. **Plasma kallikrein** (plasma prekallikrein. kininogenin, EC 3.4.21.34, 638 aa, formerly kallikrein 3) is the main form in plasma and when activated by cleavage converts kininogen to bradykinin. The γ subunit of nerve growth factor (NGF) is a kallikrein involved in the processing of NGF.

Kallmann's syndrome A syndrome in which there is hypogonadism and anosmia and, in some cases, cleft lip and cleft palate. The X-linked form (**KAL1**) is caused by a defect in the gene that encodes *anosmin. Autosomal Kallmann's syndrome (**KAL2**) is caused by mutation in the gene encoding fibroblast growth factor receptor-1. **KAL3** is caused by mutation in the G-protein-coupled *prokineticin receptor-2 gene, **KAL4** is caused by mutation in the prokineticin-2 gene, **KAL5** by mutation in **CHD7*.

Kanagawa haemolysin A thermostable monomeric haemolysin (189 aa) produced by strains of *Vibrio parahaemolyticus* that can cause food- and water-borne gastroenteritis, wound infections, and septicaemia. The toxin induces cation permeability.

kanamycin An *aminoglycoside antibiotic isolated from the soil bacterium *Streptomyces kanamyceticus*. Used in the treatment of Gram-negative bacterial infections but can have adverse side effects.

K antigen **1.** Capsular polysaccharide antigens of Gram-negative bacteria, identified by *serotyping. K serotyping has been a preferred method, e.g. for *Klebsiella*, because many types (77) are known and there is a correlation between K antigen type and pathogenicity. K antigens are subdivided into type I and II on the basis of chemical and genetic criteria. **2.** *See* KELL ANTIGENS.

Kaposi's sarcoma A tumour caused by human herpesvirus 8 and associated particularly with immunosuppression caused by HIV. Probably not a true sarcoma; consists of spindle-shaped cells derived from lymphatic endothelium with extensive vascularization.

kappa chain (κ-light chain) *See* L CHAIN.

kappa toxin (κ-toxin) **1.** An exotoxin produced by **Clostridium perfringens*; a collagenase responsible for the tissue damage in gas gangrene. **2.** κ-Hefutoxin-1 and -2, isolated from the venom of the scorpion *Heterometrus fulvipes*, block the voltage-gated potassium channels Kv1.3 and Kv1.2 and slow the activation kinetics of Kv1.3 currents; they have no homology to any other toxins.

kaptin An F-actin-associated protein from human blood platelets also found in lamellipodia and the tips of the stereocilia of the inner ear. Mutation in the gene may be responsible for hearing loss.

Karak's syndrome A disorder caused by mutation in the gene for *phospholipase A2-G6 in which there is early-onset progressive cerebellar ataxia, dystonia, spasticity, and intellectual decline. *See also* INFANTILE NEUROAXONAL DYSTROPHY.

KARP-1 A *leucine zipper protein (Ku86 autoantigen related protein-1, 156 aa) related to the *Ku autoantigen that appears to play a role in DNA double-strand break repair.

Kartagener's syndrome *See* SITUS INVERSUS.

karyogram *See* IDIOGRAM.

karyokinesis (mitosis) Division of the nucleus. *Compare* CYTOKINESIS.

karyopherins A superfamily of proteins (*importins and *exportins) involved in regulating movement of proteins across the nuclear envelope. They bind to transport signals on the cargo protein and interact with *nuclear pore complex (NPC) proteins (nucleoporins).

karyoplast A nucleus surrounded by a very thin layer of cytoplasm and a plasma membrane. The remainder of the cell is a **cytoplast**.

karyorrhexis Degeneration of the nucleus of a cell. The chromatin contracts into small pieces and becomes distributed through the cytoplasm.

karyosome An organelle typical of oocyte nuclei in insects, consisting of tightly packed oocyte chromosomes, arrested at the diplotene of meiotic prophase.

karyotype The complete set of chromosomes of a cell or organism. In cytogenetics the chromosomes are often arranged as photomontage (a **karyogram**) in which homologous pairs of mitotic chromosomes are ordered by size and position of centromere.

kassinin A *neuromedin dodecapeptide from the *Kassina* frog that binds preferentially to the mammalian *tachykinin 2 (NK-2) receptor.

katanin An ATP-dependent microtubule-severing protein composed of p60 and p80 subunits, originally found in the sea urchin. The p60 subunit has enzymatic activity and the p80 subunit targets the enzyme to the centrosome. Similar to *spastin.

katanosins (lysobactins) Katanosins A and B are naturally occurring cyclic depsipeptide antibiotics from strains of *Cytophaga*. They are effective against methicillin-resistant *Staphylococcus aureus* and VanA-type vancomycin-resistant enterococci. The mode of action is through inhibition of transglycosylation in bacterial cell wall peptidoglycan synthesis by blocking the formation of lipid intermediates.

Katayama disease (Katayama fever) The condition characteristic of the early stages of acute infection with *Schistosoma japonicum*. There is an inflammatory response in the liver resulting from immune complex formation involving antigens from the schistosome eggs.

Kawasaki's disease An acute, self-limited vasculitis (mucocutaneous lymph node syndrome) of infants and children which causes a skin rash, glandular enlargement and in some cases, severe dilatation of the coronary arteries. A functional polymorphism of the gene for inositol-1,4,5-trisphosphate 3-kinase C (*ITPKC*) is significantly associated with susceptibility.

Kayser–Fleischer ring A dark ring that appears to encircle the iris of the eye due to copper deposition. Symptomatic of *Wilson's disease.

kazal proteins A family of *serine peptidase inhibitors (MEROPS inhibitor family I1, clan IA) that inhibit peptidases of the S1 family. Examples include pancreatic secretory trypsin inhibitor (PSTI), bdellin B-3 from leech, avian ovomucoid, seminal *acrosin inhibitor, and elastase inhibitor.

K_{cat} A constant (catalytic constant, turnover number) that indicates the number of reactions catalysed per unit time by each active site of an enzyme.

K cells **1.** *See* KILLER CELLS. **2.** A subpopulation of enteroendocrine gut cells that secrete glucose-dependent insulinotropic polypeptide.

K_d The equilibrium dissociation constant for a reaction. For the reaction: A + B = C, at equilibrium, $K_d = ([A].[B])/[C]$, with dimensions of moles/litre. K_d is the reciprocal of the association constant, K_a. In ligand binding (where A is the receptor, B the ligand, and C the receptor–ligand complex), K_d is the ligand concentration that produces half-maximal response.

KDEL The consensus motif of four amino acids found at the C-terminus of proteins destined for the endoplasmic reticulum. Variants in some plants and other phyla include HDEL, DDEL, ADEL, and SDEL. The KDEL receptors (~212 aa) are Golgi/intermediate compartment-located integral membrane proteins with seven hydrophobic domains.

KDR *See* VASCULAR ENDOTHELIAL GROWTH FACTOR.

Kearns–Sayre syndrome A complex disorder with ophthalmoplegia, pigmentary degeneration of the retina, and cardiomyopathy as leading features, caused by various mitochondrial deletions.

Kelch proteins A family of proteins with diverse functions, characterized by a 50-aa repeat (Kelch domain) named after the *Drosophila* mutant in which it was first identified. Kelch repeats are found in *Drosophila* egg-chamber regulatory protein, α- and β-scruin, and galactose oxidase from the fungus *Dactylium dendroides*.

Kell antigens Blood group antigens (K antigens) that are relatively uncommon (9%) but second only to the *rhesus antigens as the cause of haemolytic disease of the newborn.

SEE WEB LINKS

- The Blood Group Antigen Gene Mutation database.

keloid A bulging scar, the result of excess collagen production. The tendency to produce keloids seems to be heritable (particularly in black Africans) and involves the expression of the *TGFB1* gene by the neovascular endothelial cells, activating adjacent fibroblasts to overexpress TGFβ, as well as type I and VI collagen genes.

Kenacid blue *See* COOMASSIE BRILLIANT BLUE.

kendrin *See* PERICENTRIN.

Kenny–Caffey syndrome *See* HYPOPARATHYROIDISM.

K_{eq} The equilibrium constant for a reversible reaction, e.g. $A + B \rightleftarrows C$. $K_{eq} = [C]/[A] \times [B]$.

keratan sulphate A *glycosaminoglycan in which the repeating disaccharide is galactose β1–4 linked to *N*-acetylglucosamine that is sulphated at C-6 of either or both of the Gal or GlcNAc monosaccharides. The polysaccharide chain is linked to various core proteins including *lumican, *keratocan, *mimecan, *fibromodulin, *proline arginine-rich end leucine-rich repeat protein (PRELP), *osteoadherin, and *aggrecan. **Keratan sulphate I** (KSI) is *N*-linked to asparagine residues via *N*-acetylglucosamine and **KSII** is *O*-linked to specific serine or threonine residues via *N*-acetylgalactosamine. Both are found in a range of tissues although originally KSI was thought to be specific to cornea.

keratectomy Excision of part of the cornea to adjust the curvature and improve vision.

keratinizing epithelium A stratified epithelium, such as skin, in which the outermost cells die and are shed after becoming filled with cytokeratin.

keratinocyte The typical cells of the keratinized layer of epidermis. The *intermediate filaments are composed of *cytokeratin.

keratinocyte growth factor A single polypeptide chain (KGF, fibroblast growth factor 7, 194 aa) that is a potent mitogen for epithelial cells (including keratinocytes) but nonmitogenic for fibroblasts or endothelial cells. The receptor is FGF-receptor 2.

keratins Insoluble fibrous proteins of high α-helical content that are constituents of the outer layer of vertebrate skin and of skin-related structures such as hair, wool, hoof and horn, claws, beaks, and feathers. Extracellular keratins are derived from the cytoplasmic intermediate filament *cytokeratins.

keratitis Inflammation of the cornea.

keratoacanthoma A benign epithelial tumour that usually regresses spontaneously. The cause is unknown but is probably associated with ultraviolet irradiation. Some authors consider it to be 'pseudo-benign' and to be a malignant tumour that fails to progress.

keratocan Core protein (352 aa) to which keratan sulphate glycosaminoglycan chains are attached, especially in the cornea. Mutations in the gene are responsible for autosomal recessive cornea plana (CNA2).

keratoconus An abnormality in which the cornea thins and is conical rather than hemispherical, leading to a loss of visual acuity. One form of keratoconus (**keratoconus 1**, KTCN1) can be caused by mutation in the visual system homeobox-1 (*VSX1*) gene. Other loci for keratoconus have been mapped to chromosomes 16 (*KTCN2*), 3 (*KTCN3*), and 2 (*KTCN4*). Keratoconus is frequent in cases of *Leber's congential amaurosis.

keratohyalin granules Granules found in the cells of the stratum granulosum of the epidermis and which contribute to the *keratin content of the cornified cells. Morphologically variable, depending on the tissue. Autosomal dominant *ichthyosis vulgaris is characterized by defective keratohyalin synthesis and lack of filaggrin.

keratoma A callus, not a tumour. **Keratoma malignum** is a synonym for congenital *ichthyosiform erythroderma.

keratomalacia An eye disorder in which the cornea becomes dry; associated with vitamin A deficiency.

keratoscleritis Inflammation of the cornea and sclera of the eye.

keratoses Benign but precancerous lesions (actinic keratoses, solar keratosis) of skin associated with ultraviolet irradiation, particularly in fair-skinned people. Can progress to squamous cell carcinomata, so are usually excised.

kernicterus Brain damage in neonates caused by excessive jaundice and neuronal damage due to bilirubin that crosses the blood–brain barrier.

ketamine An *NMDA receptor antagonist used as an anaesthetic and analgesic, particularly in veterinary medicine. Can have hallucinogenic side effects and these have made it a recreational drug. *TNs* Ketanest, Ketaset, Ketalar.

ketoacidosis A form of *acidosis in which there is excess production of *ketone bodies. A common complication of untreated diabetes.

ketoconazole An antifungal drug used in treating resistant candidiasis, gastrointestinal infections, and infections of skin and nails.

ketogenesis (ketosis) The production of *ketone bodies, mostly occurring in liver mitochondria.

ketone body Despite the name, actually compounds (acetoacetate, β-hydroxybutyrate, or acetone) that accumulate following starvation, in diabetes, and in some disorders of carbohydrate metabolism.

ketonuria (acetonuria) The presence of *ketone bodies in the urine.

ketoprofen A *nonsteroidal anti-inflammatory drug.

ketosis *See* KETOGENESIS.

ketothiolase deficiency A disorder of isoleucine catabolism (α-methylacetoacetic aciduria, 3-oxothiolase deficiency) caused by mutation in the mitochondrial acetyl-CoA acetyltransferase-1 gene (**ACAT1*). There are intermittent episodes of ketoacidosis and unconsciousness.

Keutel's syndrome A disorder in which there is abnormal (ectopic) calcification of cartilage as a result of mutation in the gene for matrix Gla protein (*MGP).

kexins A family of subtilisin-like peptidases, of which *furin is another example. All members are from eukaryotes although there are various kexin-like peptidases from prokaryotes. Kexin itself is a yeast serine peptidase of the S8 family (EC 3.4.21.61) that processes the precursors of α-factors and killer toxin; the mammalian **kexin-like peptidases** are involved in processing of various proproteins. **Proprotein convertase-1** (PC1; proprotein convertase subtilisin/kexin-type 1, PCSK1; EC 3.4.21.93, 753 aa) and **PC2** (EC 3.4.21.94, 638 aa) are neuroendocrine convertases that differentially cleave *pro-opiomelanocortin and process proinsulin and proglucagon in pancreatic islets. **PC4** (proprotein convertase subtilisin/kexin-type 4, PCSK4, 755 aa) is involved in processing pro-IGF2 (pro-*insulin-like growth factor II) to produce the intermediate IGF2 (1–102) that is further cleaved to the active form by another peptidase. Mutation in *PCSK4* is associated with intrauterine growth restriction. **PC5** (PCSK5, 913 aa) mediates post-translational endoproteolytic processing of several integrin α subunits. The gene encoding the proprotein convertase subtilisin/kexin type 9 (PCSK9, 692 aa) is linked to familial hypercholesterolemia and the product is involved in hepatocyte-specific low-density lipoprotein receptor degradation.

keyhole limpet haemocyanin (KLH) A large oxygen-carrying blood pigment (~3400 aa), from the keyhole limpet *Megathura crenulata*. Widely used as a carrier protein for conjugation to haptens and other antigens to make them more immunogenic.

KGF *See* KERATINOCYTE GROWTH FACTOR.

KH domain An evolutionarily conserved sequence (K homology domain, ~70 aa) present in a variety of nucleic acid-binding proteins. The KH domain binds RNA in a sequence-specific fashion and may be present in multiple copies.

KIAA genes Genes that have been identified in the cDNA sequencing project being carried out at the Kazusa DNA Research Institute, Japan.

SEE WEB LINKS

- Homepage for the Human Unidentified Gene-Encoded (HUGE) protein database.

Ki antigen A nuclear antigen (pKi-67, PA28γ), recognized by commercially available monoclonal antibody Ki-67, that is expressed in proliferating but not in quiescent cells and often used as a marker for tumour cells. It is a component of the *proteosome activator protein complex PA28. Patients with *systemic lupus erythematosus (SLE) produce autoantibodies against a variety of nuclear antigens including Ki antigen.

KIF **1.** *Keratin intermediate filaments. **2.** *See* KINESIN SUPERFAMILY PROTEINS (KIFs). **3.** *See* KI ANTIGEN (KiF). **4.** A form of *Cannabis sativa* smoked in the Rif mountains of northern Morocco (kif).

killer cells (K cells) **1.** Cells, probably of the mononuclear phagocyte lineage, that have Fc receptors and can lyse antibody-coated target cells. In the early literature they were sometimes confused with *cytotoxic T cells (CTL). **2.** Natural killer (NK) cells are large granular lymphocytes ($CD3^-$) that do not require expression of class I or II MHC antigens for cytolysis of target cells. **3.** Lymphokine-activated killer cells (LAK cells) are NK cells activated by *interleukin-2.

kilobase (kb, kbp) 1000 base pairs of DNA. Strictly should be kbp (kilobase pairs), but kb is the common usage.

kinase Strictly speaking, an enzyme that speeds up a process, but has come to be commonly used as shorthand for a 'phosphokinase', an enzyme catalysing transfer of phosphate from donor molecules such as ATP to a second substrate, e.g. creatine phosphokinase (*creatine kinase). Phosphotransferases have kinase activity but are membrane-bound and are also involved in transferring the substrate molecule from one side to the other. A further contraction of the usage is to mean *protein kinase, because of their importance in intracellular control and signalling, and then a further evolution to 'serine/threonine kinases', which phosphorylate proteins on serine or threonine residues, and 'tyrosine kinases' where the phosphate is added to a tyrosine residue.

Kindler's syndrome An autosomal dominant disorder (hereditary acrokeratotic poikiloderma) characterized by skin blistering, poikiloderma, photosensitivity, and predisposition to carcinogenesis. Caused by defects in *kindlin-1.

kindlin A family of adaptor proteins associated with focal adhesions and involved in activation of integrins by *talin (inside-out signalling) and for the outside-in signalling through integrins that enables adhesion and spreading. **Kindlin-1**

(kindlerin, 677 aa) is localized in basal epidermal keratinocytes and defects are associated with *Kindler's syndrome. **Kindlin-2** (mitogen induced gene 1, MIG1, 680 aa, *see* MIGFILIN) is an essential component of the intercalated discs of cardiac muscle and cell-matrix adhesion sites. **Kindlin-3** (667 aa) is associated with activation of platelet integrins and deficiency is associated with severe bleeding.

kinectin An integral membrane protein of the endoplasmic reticulum (1364 aa) that binds to *kinesin and is thought to be the membrane anchor for kinesin-driven vesicle movement. A kinectin autoantibody is a potential biomarker in *Behçet's disease.

kinesin A cytoplasmic motor protein (110 kDa) that transports vesicles and particles along microtubules towards the distal (plus) end, the opposite direction to *dynein. It has two heavy chains and two light chains (in kinesin-1, 963 aa and 573 aa). A large number of related gene products (**kinesin superfamily proteins**, KIFs) are involved in mitosis and in axonal transport. **Kinesin-14** proteins (formerly C-terminal motor proteins) differ from others of the superfamily and transport cargo towards the minus end of microtubules. There are at least four members of the group (Ncd, KAR3, CgCHO2, AtKCBP).

kinesis A change in the movement of a cell, without any directional (vectorial) bias. An increase or decrease in speed is an *orthokinesis and a change in turning behaviour is a *klinokinesis. *See* CHEMOKINESIS.

kinetics The study of the rates of chemical reactions or biological processes.

kinetochore A complex multilayered structure involving many different proteins (*see* CENPs (centromere associated proteins)) that forms on each of the paired mitotic chromosomes, adjacent to the *centromere. The inner layer is associated with centromeric DNA and the outer layer with microtubules, motor proteins, and various regulatory molecules such as *septin 7.

kinetoplast A mass of mitochondrial DNA, usually located near the flagellar *basal body, in kinetoplastid protozoa such as *Trypanosoma brucei.*

kinetosome *See* BASAL BODY.

kingdom Any of the major groupings used to classify living organisms. Most modern schemes have six kingdoms: Monera, Archaea, Protista, Plantae, Animalia, Fungi, but in some schemata Archaea are included within Monera, making only five kingdoms.

kininogen The inactive plasma precursor of *kinin. **High molecular weight kininogen** (Williams–Fitzgerald–Flaujeac factor, 626 aa) is enzymically inactive but acts as a cofactor for the activation of kallikrein and Hageman factor in the intrinsic pathway of coagulation. **Low molecular weight kininogen** is an alternatively spliced product of the same gene and, like the high molecular weight form, is an inhibitor of *cysteine peptidases.

kinins Small peptides produced from *kininogen by *kallikrein that act as inflammatory mediators by increasing arachidonic acid release and thus prostaglandin (PGE_2) production. They are broken down by kininases. *See* BRADYKININ; C2-KININ; KALLIDIN.

KIR **1.** Killer cell inhibitory receptor or killer cell immunoglobulin-like receptor. Immunoglobulin-like receptors that recognize HLA class I molecules and inhibit natural killer (NK) cell cytotoxicity through their *ITIM-mediated interaction with protein tyrosine phosphatases. KIR3D binds to HLA-B, KIR2D to HLA-C. They confer self-tolerance by an active signalling in the killer cell. **2.** Kir channels are a superfamily of inwardly rectifying potassium channels that regulate cell excitability.

Kirby–Bauer test A test assay for antibiotics that has been largely superseded except for bacteria that are difficult to handle by automated methods. Bacteria are grown in agar and discs impregnated with various antibiotics are placed on the plate. Inhibition is scored by measuring the clear zone around each disc.

Kirsten sarcoma virus A replication-defective murine retrovirus, generated by passaging a murine erythoblastosis virus in newborn rats. Source of the Ki-*ras* oncogene that encodes a GDP/GTP-binding protein (*see* RAS). *KRAS1* is a pseudogene but *KRAS2* is mutated in 17–25% of all human tumours.

KIS A serine/threonine kinase (kinase interacting with stathmin, 419 aa) expressed in all adult tissues, that phosphorylates cyclin-dependent kinase inhibitor *p27(Kip1) and *stathmin.

KISS1 A suppressor gene that inhibits metastasis of human melanomas and breast carcinomas although it does not inhibit proliferation. The human *KISS1* gene encodes *kisspeptin.

kisspeptins Peptide ligands for the G-protein-coupled receptor GPR54 that promote the central release of gonadotropin-releasing hormone

(GnRH) that, in turn, leads to reproductive maturation. The active form (54 aa) is derived from the **KISS* gene, originally identified as a metastasis suppressor gene.

kistrin A *disintegrin (68 aa) isolated from the venom of the Malayan pit viper *Agkistrodon rhodostoma*. It has an RGD motif that selectively binds to $\alpha_v\beta_3$ integrin and competes for the platelet IIb/IIIa glycoprotein, thereby inhibiting platelet aggregation.

kit A viral *oncogene originally identified in Hardy–Zuckerman 4 feline sarcoma virus. The cellular homologue encodes the receptor tyrosine kinase (mast cell growth factor) for which *stem cell factor is the ligand. Mutations are associated with piebaldism, mast cell disease, and *gastrointestinal stromal tumours.

Kjellin's syndrome *See* SPASTIZIN.

Klebsiella pneumoniae A Gram-negative, nonmotile, encapsulated, rod-shaped bacterium, facultatively anaerobic, associated with respiratory, intestinal, and urinogenital tract infections. Can cause pneumonia in humans.

Klebs–Löffler bacillus *See* CORYNEBACTERIA.

Klein–Waardenburg syndrome *See* WAARDENBURG'S SYNDROME.

Klenow fragment The larger part of *E. coli* DNA polymerase I (928 aa) that remains after *subtilisin treatment. It has DNA polymerase and $3' \rightarrow 5'$ exonuclease activities, but $5' \rightarrow 3'$ exonuclease activity is in the smaller fragment removed by the peptidase. Widely used as a reagent in molecular biology.

KLF 1. *See* KRÜPPEL-LIKE FACTORS. **2.** klf6, *see* ZF9.

KLH *See* KEYHOLE LIMPET HAEMOCYANIN.

Klinefelter's syndrome An aneuploidy in which there are three sex chromosomes (XXY). Affected individuals phenotypically appear male.

klinokinesis *See* KINESIS. Bacterial chemotaxis can be considered as an adaptive klinokinesis; the probability of turning is a function of the change in concentration of the substance eliciting the response.

Klippel–Feil syndrome A congenital shortening of the neck. Autosomal dominant and autosomal recessive inheritance have been reported.

K_m (Michaelis constant) A parameter used to characterize the properties of an an enzyme, the concentration of substrate that permits half-maximal rate of reaction. An analogous constant *K_a is used to describe binding reactions, in which case it is the concentration at which half the receptors are occupied.

Kniest's syndrome A moderately severe *chondrodysplasia caused by mutations in the collagen type II gene (*COL2A1*). The trunk and extremities are short and there is midface hypoplasia, cleft palate, myopia, retinal detachment, and hearing loss.

Knobloch's syndrome An autosomal recessive disorder defined by the occurrence of high myopia, vitreoretinal degeneration with retinal detachment, macular abnormalities, and occipital encephalocele. It is caused by a mutation in *collagen XVIII (*COL18A*).

knock-in A transgenic animal to which a new gene has been added rather than eliminated (knocked out).

knockout (gene knockout) An informal term for an organism in which the function of a particular gene has been completely eliminated ('knocked out'). For example, a knockout mouse is a transgenic animal from which a particular gene has been deleted. These animals often show disappointingly little phenotypic change—usually an indication of an alternative mechanisms or a second isoform of the protein product of the gene that has been targeted. Alternatively, it may be that the wrong challenge is being made: some genes are probably unnecessary for the survival of a well-fed laboratory mouse in well-regulated surroundings. *See also* HOMOLOGOUS RECOMBINATION; TRANSPOSON.

knottins Small proteins with at least three disulphide linkages that have a 'disulphide through disulphide' knot in which one disulphide bridge crosses the macrocycle formed by the two other disulphides and the interconnecting backbone. Examples include *agouti-related peptide, *conotoxin, and various antimicrobial peptides.

SEE WEB LINKS

- The knottin database.

Koch's postulates The criteria, proposed in 1890 by Robert Koch (1843–1910), that unambiguously identify the causative agent of a disease. The agent must: (1) be present in all cases, (2) be capable of being isolated in pure culture, (3) cause the disease if the pure isolated organism is inoculated into a disease-free host and, (4) be observable in the experimentally infected host. Although these postulates are generally accepted, there are some diseases, such as transmissible *spongiform encephalopathies, in which not all of them are satisfied.

Kohler's disease A paediatric osteochondrosis in which there is aseptic necrosis of the navicular bone, one of the small bones of the mid foot.

koilocytes Large squamous cells seen in stained cervical smears and indicative of precancerous lesions. They have enlarged pyknotic nuclei often with a clear halo surrounding the nucleus. **Koilocytosis** is induced by human papillomavirus infection of the superficial epithelial cells of the uterine cervix.

kosinostatin A quinocycline antibiotic, isolated from the culture broth of a marine actinomycete *Micromonospora* sp., that is effective against Gram-positive bacteria and to a lesser extent Gram-negative bacteria and yeasts. It is cytotoxic for various cancer cell lines and inhibits topoisomerase IIa.

k

Kostmann's disease Autosomal recessive *neutropenia, caused by mutation in the **HAX1* gene.

Kozak consensus A consensus nucleotide sequence in messenger RNA at which translation begins. In most cases the sequence is gccRccAUGG where R is a purine (either A or G) and AUG, the *start codon, codes for methionine.

KRAB (Krüppel associated box) A subset of transcription factors of the *zinc finger type. The KRAB domain of 75 aa is a transcriptional repression domain.

Krabbe's disease A disorder involving the white matter of the central and peripheral nervous system caused by mutation in the galactosylceramidase (galactocerebrosidase) gene. An atypical form is caused by deficiency in *saposin A.

k-*ras* *See* KIRSTEN SARCOMA VIRUS.

Krebs cycle The *tricarboxylic acid cycle (TCA cycle) or citric acid cycle.

Krev1 *See* RAP 1A.

kringle domain A triple-looped, disulphide-linked protein domain, resembling the eponymous Scandinavian pastry, that is found in some serine peptidases and other plasma proteins, including plasminogen (5 copies), tissue plasminogen activator (2 copies), thrombin (2 copies), hepatocyte growth factor (4 copies) and apolipoprotein A (38 copies).

KRIT A protein (KREV interaction trapped 1, CCM1 gene product, 529 aa) that interacts with integrin cytoplasmic domain-associated protein-1 and is involved in determining endothelial cell shape and function in response to cell–cell and cell–matrix interactions by an effect on cytoskeletal structure. Mutations lead to abnormal endothelial tube formation and the formation of *cerebral cavernous malformations (CCM1) lesions. KRIT1 also interacts with *malcaverin, a protein that is defective in CCM2.

Krukenberg's tumour A malignant tumour of the ovary, usually bilateral, most frequently secondary to malignancy of the gastrointestinal tract.

Krüppel associated box *See* KRAB.

Krüppel-like factors (KLF) A subfamily of transcription factors that have three *Krüppel-like zinc fingers. Tissue-specific forms include *erythroid Krüppel-like factor (**KLF1**); gut-enriched Krüppel-like factor (GKLF; **KLF4**). **KLF2** regulates T-cell quiescence and survival and is essential for T-cell trafficking; **KLF5** is important in regulating myosin production in vascular smooth muscle; **KLF6** is mutated in a subset of human prostate and gastric cancers; **KLF7** is ubiquitious; **KLF8** mediates cell cycle progression downstream of *focal adhesion kinase (FAK) by upregulating *cyclin D1. Mutations in **KLF11** are associated with early-onset type 2 diabetes. **KLF13** regulates **RANTES* genes; **KLF15** is an inhibitor of cardiac hypertrophy; **KLF16** regulates the expression of dopamine receptor genes.

KTX *See* KALIOTOXIN.

Ku autoantigen A heterodimeric protein (732 aa and 609 aa) that has ATP-dependent DNA unwinding activity and binds DNA double-strand breaks facilitating repair by the nonhomologous end joining (*NHEJ) pathway. *See* KARP-1. The subunits are autoantigens in systemic lupus erythematosus and thyroid lupus.

Kulchitsky cells *See* ENTEROCHROMAFFIN CELLS.

Kunitz domain Protein domain characteristic of inhibitors of the S1 serine peptidase family. Examples are *aprotinin, trypstatin, a rat mast cell inhibitor of trypsin, and tissue factor pathway inhibitor (TFPI). Kunitz-type serine protease inhibitors SPINT1 (478 aa) and SPINT2 (252 aa) are specific inhibitors of the peptidase that activates *hepatocyte growth factor.

Kupffer cells Specialized macrophages found in liver sinusoids that are responsible for the removal of particulate matter, particularly old erythrocytes, from the circulation.

kuru A *spongiform encephalopathy formerly endemic in the Fore tribe of New Guinea. Transmission was attributed to ritual funerary cannibalism.

Kveim test A skin test for detecting *sarcoidosis; now no longer performed because of the risk of transmissible encephalopathies.

kwashiorkor A tropical disease of children due to protein and calorie deficiency (protein–energy malnutrition).

Kyasanur forest disease A tick-borne haemorrhagic fever, endemic to South Asia, caused by a flavivirus.

kynurenine pathway The major route of L-tryptophan catabolism that generates *nicotinamide adenine dinucleotide and various neuroactive intermediates, in particular *quinolinic acid. The pathway is activated by IFN-γ and IFN-α.

kyotorphin An endogenous analgesic (L-tyrosyl-L-arginine) that does not bind to opiate receptors but acts by releasing Met-*enkephalin and stabilizing it from degradation.

kyphoscoliosis peptidase A peptidase that may specifically degrade *filamin C and is found only in skeletal muscle and heart. Defects are thought to be involved in several limb-girdle muscular dystrophies.

kyphosis Abnormal rearward curvature of the spine.

Kyte–Doolitle An algorithm used to produce a *hydropathy plot for a protein.

L1 (NgCAM) A neural adhesion molecule of the *immunoglobulin superfamily. Culture dishes coated with L1 provide a good substratum for neurite outgrowth. *See* NCAM.

L2 An abbreviation used variously for: **1.** The second lumbar vertebra. **2.** Inner lipoyl domain (L2) of dihydrolipoyl acetyltransferase (EC 2.3.1.12). **3.** A line of rat lung epithelial cells (L2 cells). **4.** Reovirus lambda-2 core spike (*L2*) gene. **5.** *Toxocara canis* L2 excretory/secretory antigen.

L32 protein A component of the 50S ribosomal subunit in both prokaryotes and eukaryotes that specifically represses splicing by binding to a purine-rich asymmetric loop adjacent to the 5′ splice site of its own transcript.

labetalol A drug consisting of a racemic mixture of four isomers; two are inactive, one is a powerful α_1-adrenergic blocker, and the fourth has nonselective β-blocking activity and is also a selective β_2-agonist. It is used to treat hypertension, particularly that associated with pre-eclampsia. *TNs* Normodyne, Trandate.

labyrinthitis Inflammation of the labyrinth of the ear.

lac operon *See* LACTOSE OPERON.

lacrimation The shedding of tears.

lactacystin A compound isolated from *Streptomyces* sp., widely used experimentally as an inhibitor of the 20S *proteosome.

lactadherin Mucin-associated glycoprotein (MFG-E8, milk-fat globule E8, BA46, 387 aa) found in milk; a phosphatidylserine-binding glycoprotein secreted by macrophages. Promotes adhesion through integrins and has been used as a tumour marker because it is expressed in most breast cancer cells. Binds to rotavirus and thus has a protective role for infants. Aortic medial amyloid (medin) consists of a 50-aa fragment of lactadherin.

lactalbumin A major milk protein (123 aa) that is structurally similar to lysozyme but without enzyme activity. Lactalbumin is the regulatory B chain of *lactose synthetase.

lactate (lactic acid) The end product (2-hydroxypropionic acid) of anaerobic glycolysis. In tissues working anaerobically, as with muscles during sprinting, lactate accumulates and is responsible for the so-called oxygen debt.

lactate dehydrogenase The enzyme (LDH, EC 1.1.2.3, 332 aa) responsible for the formation and removal of lactate. Often used as a marker for cell death since cytoplasmic enzymes in the culture medium indicate a breakdown of the vital barrier function of the plasma membrane.

lacticin A two-peptide *lantibiotic produced by *Lactococcus lactis* in which both peptides, LtnA1 and LtnA2, must interact to produce antibiotic activity. LtnA1 interacts specifically with lipid II in the outer leaflet of the bacterial cytoplasmic membrane and then recruits LtnA2 to generate a three-component complex which inhibits cell-wall biosynthesis and forms a potassium-permeable pore.

Lactobacillus A genus of Gram-positive anaerobic or facultatively aerobic bacilli that produce lactate from glucose. They are important in cheese- and yoghurt-making and the production of sauerkraut and silage.

lactoferrin (lactotransferrin) An iron-binding protein (690 aa) of the transferrin family that has a very high affinity for iron (K_d 10^{-19} at pH 6.4, 26-fold greater than that of *transferrin). It is present in milk, hence the name, but also in the specific granules of neutrophil leucocytes where it may serve to produce a bacteriostatic effect in inflammatory lesions by reducing the availability of iron for bacterial growth. It is also said to down-regulate the production of proinflammatory cytokines and reactive oxygen species.

lactogenic hormone *See* PROLACTIN.

lactoglobulin A major milk protein (178 aa; 3g/L in bovine milk) except in humans. Apart from its nutritional value, it has no clear function.

lactoperoxidase A natural antibacterial protein (EC 1.11.1.7, 721 aa) in bovine milk, and that of many other species, but in human present only in colostrum. Has considerable homology to myeloperoxidase and eosinophil peroxidase.

lactose The major sugar in human and bovine milk. Most adult humans have reduced levels of **lactase** (EC 3.2.1.23, 1927 aa), the enzyme required for digestion of lactose, and may be lactose intolerant as a result (the undigested lactose is fermented by gut bacteria, with uncomfortable consequences). Adult whites, however, maintain lactase and this persistence is a heritable autosomal dominant trait, apparently due to a noncoding variation in the *MCM6* (minichromosome maintenance-6) gene upstream of the lactase gene, which enhances activation of the lactase promoter.

lactose operon (lac operon) A group of adjacent and coordinately controlled genes for lactose metabolism in *E. coli*, chiefly famous because it was the first example of this type of gene regulation. The *operator region of the operon has a **lactose repressor**, a tetrameric protein (subunits 360 aa), bound until lactose becomes available in the medium. Allolactose, formed by transglycosylation within the cell, binds to the repressor protein and inhibits its interaction with the operator region, lifting the block to transcription of the suite of genes required for lactose metabolism and uptake. *LacZ codes for β-galactosidase, lacY for the lactose carrier protein (permease), lacA for the transacetylase.

lactose synthetase A heterodimeric enzyme responsible for the synthesis of lactose from glucose and UDP-galactose. The catalytic A chain (EC 2.4.1.22) is derived from β-**1,4-galactosyltransferase 1** (398 aa) by processing; the regulatory B subunit is α-*lactalbumin. The Golgi complex form catalyses the production of lactose in the lactating mammary gland. The cell surface form functions as a recognition molecule.

lactosuria The presence of lactose in the urine, quite common in pregnancy and during lactation.

lactotransferrin *See* LACTOFERRIN.

lacuna (*pl.* lacunae) Any small cavity or depression, e.g. the cavity in bone where an *osteoblast is found.

LacZ *See* BETA-GALACTOSIDASE.

laddering The regular pattern of oligonucleotides on a gel that is characteristic of apoptotic cells, a result of the cleavage of the DNA strand between *nucleosome beads by *endonucleases.

LAD syndrome *See* LFA-1.

LAG-1 *See* *LASS1 (longevity-assurance protein-1); LYMPHOCYTE ACTIVATION GENE-1 PROTEIN.

laidlomycin An ionophore that preferentially allows the passage of monovalent ions.

LALF peptide A cyclic peptide (*Limulus* anti-LPS factor peptide) that binds *lipid A and neutralizes LPS *in vivo* and *in vitro*. Derived from amino acids 31–52 of *Limulus* anti-LPS factor.

LAL test *See* LIMULUS POLYPHEMUS.

LAMB2 *See* LAMININS.

lambda bacteriophage (λ phage) A DNA virus first isolated from *E. coli*, structurally similar to *T even phages. It shows a *lytic cycle and a *lysogenic cycle, and is extensively used as a cloning vector, capable of carrying DNA fragments up to 15 kb long. The *cosmid vector, which will transfer larger sequences, was constructed from its ends.

lambda chain (λ chain) *See* L CHAIN.

lamellar phase *See* PHOSPHOLIPID BILAYER.

lamellipodium A flattened projection from the surface of a cell, usually at the front of a moving cell such as a fibroblast.

lamina A flat sheet. The **lamina propria** is a fibrous layer of connective tissue underlying the *basal lamina (basement membrane) of an epithelium. Unlike the basal lamina, may have cellular elements such as smooth muscle cells and lymphoid tissue in addition to fibroblasts and extracellular matrix. *See* NUCLEAR LAMINA.

lamina-associated proteins Proteins of the *nuclear lamina. *See* EMERIN; LAMINS.

laminins Link proteins of the *basal lamina consisting of an A chain (3075 aa) and two B chains (1786 aa). Each subunit contains at least 12 repeats of the *EGF-like domain. Laminin was first isolated from mouse *EHS cells and subsequently tissue-specific forms have been identified. In placental laminin the A chain is replaced with merosin, in the neuromuscular junction the B1 chain is replaced by laminin B2 (LAMB2), formerly s-laminin (synapse laminin), which is also found in kidney and smooth muscle basal lamina. Mutations in *LAMB2* are responsible for *Pierson's syndrome. Laminin induces adhesion and spreading of many cell types and promotes the outgrowth of neurites in culture. The three chains A, B1, and B2 have been renamed α, β, and γ respectively. Laminin 5 is *kalinin.

lamins Proteins of the *nuclear lamina. A-type lamins (lamin A/C, 664 aa) have C-terminal sequences homologous to the head and tail domains of *keratins and are expressed in a developmentally controlled manner. Mutations affecting A-type lamins have been associated with a variety of human diseases, including Emery–Dreifuss *muscular dystrophy; *cardiomyopathy; *lipodystrophy, and *progeria (*see also* THYMOPOIETIN). B-type lamins (lamins B1, 586 aa; B2, 600 aa) are expressed ubiquitously and have a role in spindle assembly. A duplication of the lamin B1 gene has been reported in autosomal dominant adult-onset *leukodystrophy and antibodies to lamin B are found in some autoimmune diseases. *See* EMERIN.

lamivudine A nucleoside analogue (2′,3′-dideoxy-3′-thiacytidine) that inhibits reverse transcriptase and is used for the treatment of chronic hepatitis B. *TNs* Epivir, Heptovir, Zeffix.

lamotrigine An anticonvulsant phenyltriazine drug, used to treat partial epileptic seizures. Thought to stabilize neuronal membranes by acting on voltage-sensitive sodium channels and thereby inhibiting the release of excitatory amino acid neurotransmitters (e.g. glutamate, aspartate).

LAMP **1.** An adhesion molecule (limbic system-associated membrane protein, LSAMP, 338 aa) of the *IgLON family; involved in specifying regional identity during development; promotes the outgrowth of limbic axons. There is some evidence for the association between polymorphisms in the *LAMP* gene and behaviour. **2.** *See* LAMPs (lysosome-associated membrane glycoproteins).

lampbrush chromosomes Very large chromosomes, up to 1 μm long, seen during prophase of the extended meiosis in the oocytes of some amphibians. They are actually meiotic *bivalents, and paired loops of DNA along the sides of the sister chromosomes, sites of very active RNA synthesis, give them a brush-like appearance.

LAMPs Heavily glycosylated proteins of lysosomal and plasma membranes (lysosomal-associated membrane glycoproteins), between 90 and 140 kDa depending on the extent of glycosylation. They may protect membrane from attack by lysosomal enzymes and regulate fusion of lysosomes with primary autophagocytic vacuoles. **LAMP-1** is also known as LEP100 (lysosomes, endosomes and plasma membrane, 100 kDa) and LGP120. Defects in **LAMP-2** cause *Danon's disease.

Landry–Guillain–Barré syndrome *See* GUILLAIN–BARRÉ SYNDROME.

Langendorff perfused heart A rodent heart maintained *in vitro* by perfusion of the aorta with oxygenated fluid so that the fluid passes into the coronary arteries. Used to study the pharmacology and metabolism of cardiac muscle.

Langerhans *See* DENDRITIC CELLS (5); ISLETS OF LANGERHANS.

Langerhans cell histiocytosis A rare disorder (histiocytosis X) in which there is dysregulated proliferation of Langerhans cells and granulomatous deposits occur at multiple sites within the body. Patients have high levels of soluble *RANKL and *IL-17A. Some authors consider it to be a set of clinical syndromes including, in decreasing severity: *Letterer-Siwe disease, *Hand–Schuller–Christian disease, and eosinophilic granuloma of bone.

Langerhans cells *See* DENDRITIC CELLS (5).

Langhans giant cells *See* FOREIGN BODY GIANT CELL.

lansoprazole A proton pump inhibitor used to treat gastric ulcers. Sold under a range of brand names.

lanthionine A nonprotein amino acid composed of two alanine residues thioether-linked at the β-carbon atoms. Found in human hair, lactalbumin, and feathers, also in bacterial cell walls. *See* LANTIBIOTICS.

lanthiopeptin *See* ANCOVENIN.

lantibiotics Lanthionine-containing peptide antibiotics. Type A lantibiotics (e.g. *nisin, *subtilin, epidermin) are flexible, elongated, amphipathic molecules which act mainly by forming pores in the bacterial cytoplasmic membrane. Type B lantibiotics (e.g. *ancovenin, mersacidin, actagardine, mutacin, and *duramycin) have a rigid globular shape and inhibit particular enzymes by forming a complex with their membrane-bound substrates.

LAP A transcription factor (liver activator protein, IL-6-dependent DNA-binding protein, CCAAT/enhancer-binding protein β, CEBPB, 345 aa) that stimulates transcription of genes for *acute phase proteins and is important in regulation of genes affected by cyclin D1. Unusually, the LAP gene contains no introns.

lapatinib A dual inhibitor of *epidermal growth factor receptor (EGFR) and human epidermal growth factor receptor 2 (HER2) tyrosine kinases, used in combination with capecitabine for the treatment of advanced breast cancer or metastatic breast cancer overexpressing HER2 (ErbB2). *TN* Tyverb.

LAPF An adaptor protein (lysosome-associated apoptosis-inducing protein containing PH and FYVE domains, 279 aa) that recruits phosphorylated p53 to lysosomes. Overexpression of LAPF in L929 cells induces apoptosis and also increases cell sensitivity to TNFα-induced apoptosis. A member of the *phafin family.

SEE WEB LINKS

- Article on LAPF in the *Journal of Biological Chemistry* (2005).

LA-PF4 *See* CONNECTIVE TISSUE-ACTIVATING PEPTIDE III.

La protein A protein (La autoantigen, La ribonucleoprotein domain family 3, LARP3, 408 aa) that is involved in diverse aspects of RNA metabolism, including binding and protecting RNA that has been newly synthesized by polymerase III, stabilizing nascent pre-tRNAs from nuclease degradation, acting as an RNA chaperone, and binding viral RNAs associated with hepatitis C virus. It is the nuclear autoantigen in patients suffering from *Sjögren's syndrome, *systemic lupus erythematosus, and neonatal lupus. La-related protein (LARP1) has an RNA-binding domain similar to that of La protein and a second conserved domain, the LARP1 domain, which also has RNA-binding ability. Other LARPs have been found; e.g. LARP7 is a negative transcriptional regulator of polymerase II genes.

LAR A membrane protein (leucocyte antigen-related protein, LAR-PTP) with a cytoplasmic domain having homology to protein-tyrosine phosphatase 1B and an extracellular domain homologous to neural cellular adhesion molecule (NCAM). The 200-kDa precursor is proteolytically processed into noncovalently associated subunits: a 150-kDa extracellular subunit with the cell adhesion domains, and an 85-kDa phosphatase subunit with extracellular, transmembrane and cytoplasmic domains. LAR negatively regulates the insulin signalling pathway. Other LAR-family phosphatases (PTPδ, PTPσ) are involved in axon guidance and mammary gland development. *See* LIPRINS.

LARGE A glycosyltransferase (756 aa) that is required for normal glycosylation of *dystroglycan. Defects in the *LARGE* gene, or that for dystroglycan, cause abnormal neuronal migration. Mutations in *LARGE* can cause congenital muscular dystrophy type 1D (MDC1D) and *Walker–Warburg syndrome.

large cell lymphoma Under the Revised European–American Lymphoma (REAL) classification schema, the non-Hodgkin's lymphomas are subdivided. **Diffuse large cell lymphomas** (DLCL) are mostly of B-cell origin and are the commonest of the non-Hodgkin's lymphomas; mutations in p53 are common. Less common is **anaplastic large cell lymphoma** (cutaneous $CD30^+$ ALCL) which is clinically and pathologically heterogeneous but generally arises from T cells. In many cases there is up-regulation of a tyrosine kinase (anaplastic lymphoma kinase, ALK) due to a chromosomal translocation involving the nucleophosmin gene on chromosome 5.

large T antigen *See* T ANTIGENS.

Lariam® *See* MEFLOQUINE.

Laron's syndrome A growth defect due to mutation in the gene for *growth hormone receptor. A Laron syndrome-like phenotype, associated with immunodeficiency, is caused by a postreceptor defect, a mutation in the *STAT5B* (signal transducer and activator of transcription 5B) gene.

Larsen's syndrome An autosomal dominant skeletal disorder caused by mutation in the gene for *filamin B. An autosomal recessive form is also known but the genetic basis is unclear.

LASPs *See* LIM AND SH3 PROTEINS.

LASS1 The human homologue of the yeast longevity assurance gene. In *S. cerevisiae* the *LAG* genes encode transmembrane proteins differentially expressed during the replicative life span that are necessary for *N*-stearoyl-sphinganine (C_{18}-(dihydro)ceramide) synthesis and play a role in determining yeast longevity, although the molecular mechanism is unclear. All Lag1 homologues contain a highly conserved stretch of 52 amino acids known as the Lag1p motif.

Lassa fever An acute haemorrhagic fever caused by a virulent and highly transmissible member of the *Arenaviridae, Lassa virus, first recorded from the town of Lassa in Nigeria. The normal host is a rodent (*Mastomys natalensis*).

LAT adaptor An adaptor protein (linker for activation of T-cells, 262- and 243-aa isoforms) involved in T-cell receptor (TCR)-mediated signalling. The intracytoplasmic domain is tyrosine-phosphorylated by ZAP-70 when the TCR aggregates and recruits SH2 domain-containing cytosolic enzymes and adaptors. Also found in platelets, where it becomes phosphorylated in response to collagen, collagen-related peptide (CRP), and FcgRIIA crosslinking. A similar adaptor, growth factor receptor-bound protein 2 (Grb2)-binding adaptor protein, transmembrane (GAPT), is expressed in B cells and myeloid cells.

latanoprost A prostaglandin F2α analogue that works as a receptor agonist, used to reduce intraocular pressure in glaucoma (ocular hypertension) by increasing the efflux of aqueous fluid. *TN* Xalatan.

late genes Genes that are expressed relatively late after infection of a host cell by a virus, mostly encoding structural proteins of the viral coat.

latency Decribing the period between initial infection and production of overt symptoms or, in electrophysiology, the time between onset of a stimulus and peak of the ensuing *action potential.

latent membrane proteins (LMPs) Oncogenic viral proteins expressed in *Epstein–Barr virus (EBV)-associated Hodgkin's lymphoma and nasopharyngeal carcinoma. They interact with various intracellular signalling systems to induce proliferation and metastasis. LMP-1 (BNFL1, 386 aa) is a member of the TNF-receptor family and interacts with tumour necrosis factor (TNF)-receptor 1-associated death domain protein (TRADD) but does not induce apoptosis, rather it up-regulates anti-apoptotic genes, including bcl-2. LMP2A signals through the src-kinase *lyn and up-regulates the *survivin gene through the NFκB pathway, making the cells resistant to apoptotic stimuli.

latent virus A virus that is integrated within a host genome but is inactive. An example is herpesvirus, which can remain latent for long periods unless activated by stress.

lateral diffusion Diffusion restricted to two dimensions, as in the case of membrane lipids or integral membrane proteins. The rate of diffusion can be measured by the technique of *fluorescence recovery after photobleaching (FRAP).

lateral inhibition The phenomenon found e.g. in photoreceptors which are adjacent to an excited receptor. This accentuates boundaries and edges and this peripheral processing reduces the information that needs to be processed centrally.

lathyrism A neurological disease of humans and domestic animals, caused by eating seeds of *Lathyrus sativus* (chickling pea, grass pea) that contain high levels of the neurotoxin β-oxalyl-L-α,β-diaminopropionic acid (ODAP, also known as β-*N*-oxalylamino-L-alanine, BOAA), a glutamate analogue. There is paralysis, particularly of the lower limbs. Lathyrism caused by eating seeds of the sweet pea *Lathyrus odoratus*; odoratism) is a disorder of collagen crosslinking as a result of copper sequestration by β-aminopropionitrile and inhibition of the metalloenzyme, *lysyl oxidase.

latrotoxins (α-latrotoxin) Toxins found in the venom of widow spiders *Latrodectus* spp. There are five insecticidal toxins, termed α, β, γ, δ, and ε-latroinsectotoxins, and a vertebrate-specific neurotoxin, α-latrotoxin (1401 aa). α-Latrotoxin forms multimers and creates pores that allow calcium influx and release of neurotransmitters from synapses.

latrunculins Inhibitors of actin polymerization (latrunculins A and B), isolated from the Red Sea sponge *Negombata magnifica*, that binds G actin rendering it assembly incompetent; more potent than *cytochalasin. Latrunculin inhibits binding of *thymosin β4 but not *profilin.

laulimalide A microtubule-stabilizing cytotoxic agent, isolated from the marine sponge *Cacospongia mycofijiensis*, that acts in a similar, but not identical, manner to *paclitaxel. A related compound, isolaulimalide, is also found.

SEE WEB LINKS

- Article on laulimalide in *Cancer Research* (1999).

Laurence–Moon syndrome A disorder in which there is mental retardation, pigmentary retinopathy, hypogenitalism, and spastic paraplegia. Probably not distinct from *Bardet–Biedl syndrome.

lavage Washing out of a cavity such as the peritoneal cavity, in order to remove loosely adherent cells.

lazy leucocyte syndrome A rare human complaint in which leucocyte phagocytosis and bactericidal activity are normal but both random motility and chemotaxis are defective. There is neutropenia and recurrent infection. Probably due to a defect in the cytoplasmic actomyosin system.

Lbc *See* DBL.

LBP *See* LIPOPOLYSACCHARIDE-BINDING PROTEIN.

LC-1 *See* FORMIMINOTRANSFERASE CYCLODEAMINASE.

LC_{50} The concentration of a chemical which kills 50% of a sample population. Usually used in reference to inhalational toxicity *Compare* LD_{50} which is used for toxins that are ingested, injected, or absorbed through the skin.

LCAT A soluble enzyme (lecithin:cholesterol acyltransferase, EC 2.3.1.43, 416 aa) that converts cholesterol and phosphatidylcholines to cholesteryl esters and lysophosphatidylcholines on the surface of high density lipoproteins. Complete loss of LCAT causes autosomal recessive LCAT deficiency; partial loss causes the autosomal recessive disorders *fish eye disease and *Norum's disease.

LCCL domain A conserved protein module (~100 aa) found in *Limulus* factor C, *cochlin) and late-gestation lung protein Lgl1. It is thought to be an autonomously folding domain that has been used for the construction of various modular proteins.

L cells A cell line established in 1940 by W.R. Earle from normal subcutaneous areolar and adipose tissue of a 100-day-old male mouse. L929 cells are a clone of this original line.

L chain (light chain) General term for the smaller chain (*light chain) of a multimeric protein, but usually refers to the 22-kDa light chains of immunoglobulins which can be of two types, kappa (κ) or lambda (λ). A single immunoglobulin has identical light chains (2 κ or 2 λ) each with one variable and one constant region. There are *isotype variants of both κ and λ.

lck A nonreceptor protein-tyrosine kinase (lymphocyte-specific protein tyrosine kinase, p56 (Lck), 509 aa) of the *src family that is involved in transduction of *T-cell receptor-mediated activation by tyrosine phosphorylation of *STAT5. Interacts with *zap70 and *fyn.

LCR *See* LOCUS CONTROL REGION.

LD_{50} The concentration of a substance that causes death in 50% of the test organisms. *See* LC_{50}. The regulatory requirement for LD_{50} tests is now being replaced by more sensitive methods involving fewer animals.

LD78 Chemokines (CCL3, ~92 aa) of the small chemokine (SCY) family that probably play an inhibitory role in haematopoiesis. LD78-α promotes migration of human peripheral T lymphocytes and enhances the adherence of monocytes to endothelial cells. LD78-β strongly down-regulates the expression of CCR5 in monocytes and macrophages and is a potent HIV-1 inhibitor (acting through CCR5), neutrophil activator (through CCR1), and eosinophil activator (through CCR1 and CCR3).

LDH *See* LACTATE DEHYDROGENASE.

LDL *See* LIPOPROTEINS.

LDL receptor A cell surface receptor (low density lipoprotein receptor) that mediates the endocytosis of LDL by cells. Mutations lead to abnormal serum levels of LDL and *hypercholesterolaemia. **LDL receptor adaptor protein 1** (LDLRAP1) is the product of the *ARH* gene, defective in autosomal recessive familial hypercholesterolaemia, which encodes a *transmembrane adaptor protein that interacts with the cytoplasmic tail of the LDL receptor, phospholipids, and components of the endocytic machinery (*clathrin or AP2). The adaptor is required for internalization of the LDL–LDLR complex, but also for efficient binding of LDL to the receptor. **LDL receptor related proteins** (LRP) are a family of receptors for multiple ligands that play a wide variety of roles in normal cell function and development. **LRP1** (CD91; apolipoprotein E receptor; α_2-macroglobulin receptor, 4544 aa) is involved in the endocytic clearance of triglyceride-rich lipoproteins from plasma and the uptake of apolipoprotein E-containing lipoprotein particles by neurons. It also mediates the endocytosis and degradation of secreted amyloid precursor protein and, via α_2-macroglobulin, the clearance and degradation of APP-generated β-amyloid. **LRP1B** has ~60% amino acid identity with LRP1 and binds receptor-associated protein, urokinase *plasminogen activator, tissue-type plasminogen activator, and plasminogen activator inhibitor-1. **LRP2** (megalin; glycoprotein-330) binds lipoprotein lipase, apolipoprotein E-enriched β-VLDL and apolipoprotein J/clusterin. It is the target antigen of *Heymann nephritis and is mutated in *Donnai–Barrow syndrome. **LRP3** (770 aa) does not bind lipoproteins and its function is unclear. **LRP4** has two EGF-like domains as well as similarity with the other LRPs. Mutations in the murine *Lrp4* gene are associated with syndactyly. **LRP5** (LRP7) is expressed by osteoblasts and can transduce Wnt signalling *in vitro*. Mutations are associated with osteoporosis–pseudoglioma syndrome, exudative vitreoretinopathy, and various disorders of bone mineralization. **LRP6** is a coreceptor for Wnt with LRP5 and mutations are associated with early coronary artery disease and metabolic syndrome. **LRP8** (apolipoprotein E receptor 2) is involved in transmitting the extracellular *reelin signal to intracellular signalling processes initiated by disabled (*DAB1). **LDL receptor-related protein associated protein 1** (LRPAP1, α_2-macroglobulin receptor-associated protein (MRAP), 357 aa) may function as a chaperone in the synthesis of LRPs.

L-DOPA *See* LEVODOPA.

leader peptide *See* LEADER SEQUENCE.

leader sequence The 5′ untranslated region (5′ UTR) of mRNA, upstream of the translational start site and containing the ribosome binding motif (Shine–Dalgarno sequence) in bacteria. It may have control elements such as the *iron-responsive element or other binding regions that are sensitive to the levels of particular metabolites through a binding protein which changes its properties when metabolite is available and bound. In some cases the leader sequence is actually transcribed and the protein has a leader peptide (e.g. *signal sequence) which may determine its fate or destination.

l

leading lamella The flattened front of a crawling cell, such as a fibroblast, that is pushed forward and supported by a meshwork of microfilaments that cause most cytoplasmic granules to be excluded.

leaky mutation A mutation that does not completely destroy the function, e.g. one in which the encoded enzyme has reduced activity or decreased stability.

Leber's congenital amaurosis A group of autosomal recessive retinal dystrophies that are the most common causes of congenital visual impairment. **Leber's congenital amaurosis type 1** (LCA1) is caused by mutation in the gene encoding retinal guanylate cyclase; **LCA2** by mutation in the *RPE65* gene (encoding a retinal pigmented epithelium-specific protein of 65 kDa); **LCA3** is caused by mutation in a retinol dehydrogenase gene (*RDH12*); **LCA4** is caused by mutation in the *AIPL1* gene (arylhydrocarbon receptor interacting protein-like 1); **LCA5** is caused by mutation in the gene encoding *lebercilin; **LCA6** is caused by mutation in the *RPGRIP1* gene (retinitis pigmentosa GTPase regulator-interacting protein 1); **LCA7** is caused by mutation in the CRX (cone-rod homeobox-containing) gene that encodes a transcription factor required for photoreceptor differentiation; **LCA8** is caused by mutation in the *CRB1* gene (*Drosophila crumbs* homologue 1); **LCA9** is mapped to a distinct locus on chromosome 1, but the gene is unidentified; **LCA10** is caused by mutation in the 290-kDa centrosomal protein encoded by the *CEP290* gene. **LCA11** is caused by mutation in the *IMPDH1* gene that encodes inosine-5-monophosphate dehydrogenase (EC 1.1.1.205).

Leber's optic atrophy A disorder (Leber's hereditary optic neuritis (LHON), hereditary optic atrophy, Leber's disease) in which there is acute or subacute loss of central vision, usually in middle age, that leads to blindness. The disease is associated with many missense mutations in mitochondrial DNA and is maternally inherited. *See* LEBER'S CONGENITAL AMAUROSIS.

lebercilin A protein (697 aa) that is mutated in *Leber's congenital amaurosis type 5 (LCA5). It is highly expressed in adult retina, testis, kidney, and heart, and in ciliated cell lines localizes to the ciliary axoneme. Interacts with many proteins associated with centrosomal or ciliary functions, including cytoplasmic *dynein, *nucleophosmin, nucleolin, 14-3-3-ε, and HSP70.

LE body *See* LE CELLS.

LEC A CC chemokine (liver-expressed chemokine, CCL16, novel CC chemokine (NCC)-4, SCYA16, LCC-1, HCC4, LMC, Ck-β-12, monotactin-1, SCYL4, 120 aa), chemotactic for monocytes and constitutively expressed by hepatocytes. Receptors are CCR1, CCR2, and CCR5.

LECAM *See* SELECTINS.

LE cells Cells found in smears of blood mixed with serum samples from patients with *systemic lupus erythematosus (SLE). They are phagocytes, usually neutrophils, with one or more large inclusion bodies (**LE bodies**), composed of DNA and antinuclear antibody, which displace the nucleus to one side. The **LE cell reaction** is an *in vitro* test for SLE in which leucocytes are incubated with test serum.

lecithin A mixture of phospholipids that may include phosphatidylcholine, phosphatidylethanolamine, and phosphatidylinositol, although sometimes used to refer to pure phosphatidylcholine. Usually extracted from egg yolk.

lecithinase *See* PHOSPHOLIPASES.

lecticans A family of chondroitin sulphate proteoglycans that includes *aggrecan, *versican, neurocan, and brevican. They have a hyaluronan-binding and C-type lectin domains and act as crosslinkers in the extracellular matrix.

lectin A protein that will bind specific mono- or oligosaccharide groups. There are two main types, *S-type lectins and *C-type lectins. Originally isolated from the seeds of leguminous plants, as agglutinins for red cells of particular blood groups, but subsequently found in various plants and animals. Lectins such as *concanavalin A and *wheatgerm agglutinin are used as analytical and preparative tools in glycobiology. *See also* SELECTINS.

leflunomide A *disease-modifying antirheumatic drug (DMARD), a pyrimidine synthesis inhibitor, used to treat rheumatoid arthritis. It has immunosuppressive activity and has a direct inhibitory effect on osteoclast differentiation by inhibiting the induction of NFATc1. TN Arava.

left–right asymmetry Although basically bilaterally symmetrical, the body has various L–R asymmetries, particularly in the heart and viscera. The classic condition, *situs inversus, is apparently due to a *dynein defect but a number of other genes are known to be implicated in the specification of the L–R axis. These include *nodal; *lefty; *rotatin, the transcription factor *ptx2 and *ZIC3. An autosomal form of visceral heterotaxy (HTX2) can be caused by mutation in the *CFC1* gene that encodes the CRYPTIC protein involved in nodal signalling, or in the type IIB *activin A receptor. There is also an X-linked form of the disorder.

LEFTY Genes involved in the determination of *left–right asymmetry; *LEFTY1* and *-2* encode proteins (366 aa) of the *TGFβ family which may antagonize *nodal. In mice the equivalent genes are expressed only on the left-hand side of the embryo and mutations in *LEFTY2* are associated with human L–R axis malformations.

Legg–Calvé–Perthes disease *See* PERTHES' DISEASE.

Legionella A genus of *Gram-negative bacteria, most species of which are pathogenic in humans, especially *L. pneumophila.* Named after an outbreak of pneumonia (legionnaires' disease) among those attending an American Legion reunion. Transmission is usually by aerosols derived from cooling systems, domestic hot water systems, etc.

legionnaires' disease *See* LEGIONELLA.

Leidig cells *See* LEYDIG CELLS.

Leigh's syndrome An early-onset progressive neurodegenerative disorder with considerable genetic heterogeneity; mutations have been identified in both nuclear- and mitochondrial-encoded genes that are involved in energy metabolism.

leiomodin An actin-binding protein (572 aa) that is expressed in muscle, eye, and thryoid and is the autoantigen in some cases of *Graves' disease where there is the complication of thyroid-associated ophthalmopathy. There are sequence similarities that suggest leiomodin, and *tropmodulin may have evolved from a common ancestral gene. May nucleate sarcomere assembly in striated muscle.

leiomyoma A benign tumour in which parallel arrays of smooth muscle cells form bundles arranged in a whorled pattern. Leiomyoma of the uterus (*fibroid) is the commonest form.

Leishman's stain A Romanovsky-type stain, containing both basic and acid dyes, that will differentially stain the various leucocytes in blood smears.

SEE WEB LINKS

- Useful resource on histological stains.

leishmaniasis A disease caused by protozoan parasites of the genus *Leishmania* that live intracellularly in macrophages. Transmitted by biting flies (particularly sandflies). Various forms of the disease occur: in particular visceral leishmaniasis (*kala-azar), and mucocutaneous leishmaniasis (*espundia).

LEM domain A motif (lamina-associated polypeptide–emerin–MAN1 domain) shared by *lamin-interacting proteins of the inner nuclear membrane (INM) and present in the nucleoplasm. The LEM domain mediates binding to a DNA-crosslinking protein, *barrier-to-autointegration factor (BAF).

lengsin A lens-specific member of the glutamine synthetase superfamily but devoid of enzyme activity, probably having a structural role. Binds intermediate filament proteins such as *vimentin and *phakinin.

lentigo (*pl.* lentigines) A small, sharply circumscribed, pigmented lesion of the skin in which there are variable numbers of melanocytes. Generally benign but can indicate past overexposure to sun and predisposition to melanoma.

Lentivirinae A subfamily of nononcogenic single-stranded RNA retroviruses responsible for so-called slow diseases that have long incubation periods and chronic progressive phases. *HIV is in this family; another example is *visna-maedi virus.

lentoid **1.** Lens-shaped or lens-like. **2.** Lentoid bodies may form *in vitro* through aggregation of lens epithelial cells to form a spherical mass with a central core of lens-like cells containing crystallins. They may also form ectopically in the eye through defects in the β-catenin mediated *Wnt signalling pathway.

leontiasis ossea Disorder characterized by diffuse hypertrophy of the bones of the skull; probably what is now known as *van Buchem's disease.

LEOPARD syndrome A syndrome named for an acronym for the clinical effects: multiple lentigines, electrocardiographic conduction abnormalities, ocular hypertelorism, pulmonic stenosis, abnormal genitalia, retardation of growth, and sensorineural deafness. Can be caused by mutations in the *PTPN11* or *RAF1* genes (*see* NOONAN'S SYNDROME).

Lepore haemoglobin Variant haemoglobin found in a rare *thalassaemia: there is a normal α chain but a composite δ–β chain as a result of an unequal *crossing-over event. The composite chain is functional but synthesized at reduced rate. Anti-Lepore haemoglobins are mutants with the opposite composite chain—the N-terminus of β-globin and the C-terminus of δ-globin. Several variants of Lepore and anti-Lepore haemoglobin are known.

lepromin test A skin test used to determine the type of leprosy that involves a subcutaneous injection of heat-killed *Mycobacterium leprae.* Patients with lepromatous leprosy do not react (nor do uninfected patients), but a tuberculin-type response occurs in those with tuberculoid leprosy.

leprosy A granulomatous disease (Hansen's disease) of the peripheral nerves, mucosa of the upper respiratory tract, and skin. The causative organism is *Mycobacterium leprae*, which survives lysosomal enzyme attack by having a resistant waxy coat. The bacterium requires a temperature lower than 37°C, hence the peripheral location, and destroys peripheral innervation so that patients fail to perceive pain associated with tissue damage. Only humans, a few primates, and the nine-banded armadillo *Dasypus novemcinctus* are susceptible. Multidrug treatment with *dapsone, rifampicin, and clofazimine is required because dapsone-resistant strains emerge very quickly.

leptin A protein (167 aa) involved in regulating body weight by inhibiting food intake and stimulating energy expenditure. Mutations of the *ob* gene, which encodes leptin, lead to obesity but this seems only very rarely to be the case in humans. The receptor is a single-transmembrane-domain receptor of the cytokine receptor family highly expressed in the hypothalamus, the site of appetite regulation in the brain. Several alternatively spliced forms are found in various other tissues and leptin may have a range of activities. Mutations in the receptor are responsible for the obese phenotype of the fa/fa rat.

leptinotarsins (leptinotoxin) Toxins (45–47 kDa), isolated from the haemolymph of the potato beetle *Leptinotarsa haldemani*, that rapidly stimulate Ca^{2+} influx and neurotransmitter release by acting on calcium channels.

leptocyte An abnormal erythrocyte that is thin, large in diameter, and has a peripheral ring of haemoglobin. Found in some types of anaemia.

leptomycin B An antifungal antibiotic first identified in *Streptomyces* sp. Inhibits the export of proteins from the nucleus by binding to *exportin-1 and preventing the formation of the complex between exportin and the nuclear export signal of cargo proteins.

leptonema *See* LEPTOTENE.

Leptospira A genus of *spirochaetes that cause a mild chronic infection in rats and many domestic animals. Infected animals excrete virus in urine and contact with contaminated water can lead to infectious jaundice (leptospirosis or Weil's disease), an occupational hazard for agricultural and sewerage workers.

leptospirosis *See* LEPTOSPIRA.

leptotene (leptonema) The first stage of *prophase I of *meiosis, during which the chromosomes first become condensed and microscopically visible.

Lesch–Nyhan syndrome An X-linked recessive inherited disorder caused by mutation in the gene for the purine salvage enzyme *HGPRT. Results in severe mental retardation and distressing behavioural abnormalities, such as compulsive self-mutilation, apparently because of perturbations in the dopaminergic system.

let-7 (lethal-7) One of the first *microRNAs (21 nucleotides) discovered (the other is lin-4). In *Caenorhabditis elegans* let-7 regulates the timing of development; loss of function leads to early death, over-expression to premature expression of adult traits. Various let-7 homologues are reported to map to regions deleted in human cancers and lung tumours have lower let-7 expression and higher ras protein levels than normal lung tissue.

letrozole A nonsteroidal *aromatase inhibitor used in treatment of hormone-dependent breast carcinoma. *TN* Femara.

Letterer–Siwe disease A subtype of *Langerhans cell histiocytosis (although opinions differ).

leuc-, leuco- (leuk-, leuko-) Prefix meaning white, e.g. *leucocyte. The c and k forms are interchangeable, although in the UK **leucocyte** and **leukaemia** are probably the more common forms.

leucine (leu, L) The most abundant amino acid (2-amino-4-methylpentanoic acid, 131 Da) in proteins. Confers hydrophobicity and has a structural rather than a chemical role.

leucine aminopeptidase A metallopeptidase (EC3.4.11.1, 1025 aa) that removes neutral amino acid residues from the N-terminus of proteins (an *exopeptidase).

leucine-rich repeat A short motif (LRR motif) of ~24 aa with leucine generally at positions 2, 5, 7, 12, 21, 24. Forms an amphipathic region and is probably involved in protein–protein interactions. Structurally distinct from a *leucine zipper.

leucine zipper A motif commonly found in transcription factors such as *AP1 that involves a stretch of ~35 aa with leucine in every seventh position, forming a short α-helix that will readily dimerize with a similar motif in another protein.

leucistic Describing an individual lacking pigmentation in the skin but with eye pigment and thus not a full albino.

leuco *See* LEUC-, LEUCO-.

leucocidin (Panton–Valentine leucocidin) Distinct exotoxins from staphylococcal and streptococcal species of bacteria that cause killing or lysis of myeloid but not lymphoid cells. The two subunits, S and F, (slow and fast on chromatographic columns) are inactive alone but interact to form a pore. The PVL genes are carried by two phages, phiPVL and phiSLT, and there is sequence homology with γ-haemolysin. Panton–Valentine leucocidin (PVL)-producing *Staphylococcus aureus* are an emerging problem causing necrotizing lung infections.

• A historical account of leucocidin.

leucocytes Generic term for white blood cells. *See* BASOPHIL; EOSINOPHIL; LYMPHOCYTE; MONOCYTE; NEUTROPHIL. **Leucocytolysis** is lysis specifically of leucocytes; **leucocytopenia** (leucopenia) is a diminished number of circulating leucocytes, sometimes restricted to particular types (e.g. *neutropenia), whereas a **leucocytosis** is an excess. **Leucopoiesis** is the production of leucocytes in the bone marrow.

leucoderma (leucodermia) *See* VITILIGO.

leuco-erythroblastic anaemia Anaemia characterized by the presence of nucleated erythrocytes and early myeloid cells in the peripheral blood.

Leu enkephalin *See* ENKEPHALINS.

leukaemia

A malignant *neoplasia affecting *leucocytes. The nomenclature reflects the affected stem cell and whether the condition is acute or chronic. In **acute lymphoblastic leukaemia** (ALL) there are large numbers of primitive lymphocytes (with a high nuclear/cytoplasmic ratio characteristic of dividing cells and few specific surface antigens expressed); tends to be common in the young. In **acute myeloblastic leukaemia** (AML) the proliferating cells are of the *myeloid series and the cells appearing in the blood are primitive *granulocytes or *monocytes; it is more common in adults. In **chronic lymphocytic leukaemia** (CLL), a disease of middle or old age, there are excessive numbers of normal, mature lymphocytes in the circulation, usually B cells. In **chronic myelogenous leukaemia** (CML) there are excessive numbers of circulating leucocytes, usually neutrophils (or precursors), but occasionally eosinophils or basophils. It is commonest in middle-aged or older people. **Hairy cell leukaemia** is a rare chronic lymphoid leukaemia in which the proliferating B cells have hair-like cytoplasmic projections on their surfaces. Several **virus-induced leukaemias** are known, including those caused by: *Abelson leukaemia virus; *avian leukaemia virus; *Friend murine leukaemia virus; *HTLV; *Moloney murine leukaemia virus.

leukaemia inhibitory factor (LIF) A cytokine (202 aa) that regulates growth and differentiation of primordial germ cells and embryonic stem cells but also has effects on peripheral neurons, osteoblasts, adipocytes, and various cells of the myeloid lineage. Can induce weight loss, behavioural disorders, and bone abnormalities in adults. Many of the effects of LIF *in vitro* can be mimicked by *interleukin-6, *oncostatin M, and *ciliary neurotrophic factor, all of which interact indirectly with gp130, a shared tranducer subunit. Not to be confused with leucocyte inhibitory factor (leucocyte migration inhibitory factor), an operational definition of an activity rather than a distinct factor.

leukemia *See* LEUKAEMIA.

leuko- *See also* LEUCO-.

leukodystrophy A duplication of the *laminB1 gene (*LMNB1*) is associated with adult-onset **autosomal dominant leukodystrophy**, a slowly progressive neurological disorder in which there is symmetrical and widespread myelin loss in the central nervous system. **Adrenoleukodystrophy** (Siemerling–Creutzfedt disease, bronze Schilder disease) is an X-linked disorder in which there is a defect in peroxisomal β-oxidation leading to the accumulation of saturated very long chain fatty acids in tissues, with particularly serious consequences in the nervous system. It is caused by mutation in the ATPase binding cassette protein (*ABCD1*) gene. **Neonatal adrenoleukodystrophy** is a result of mutation in the peroxisome receptor gene (*PTS1R, PEX5*) or *peroxin genes. **Metachromatic leukodystrophies** are a group of disorders with allelic variants (late infantile, juvenile, and adult forms, partial cerebroside sulphate deficiency, and pseudoarylsulphatase A deficiency) caused by deficiencies in *aryl sulphatase A and nonallelic forms caused by *saposin B deficiency and by multiple sulphatase deficiency. **Adrenomyeloneuropathy** is a form of the disorder in which muscular and neural side effects are prominent.

leukoplakia (leucoplakia) A thick white patch occurring on mucous membranes, e.g. in the mouth. In some cases may be a premalignant squamous metaplasia.

leukosialin (CD43, sialophorin) A major sialoglycoprotein with extensive *O*-linked glycosylation (75–85 oligosaccharides on the 239-aa extracellular domain) found on the surface of human T cells, some B cells, and myeloid cells. It is a transmembrane protein and may be involved in T-cell activation; clearance of CD43 from the region of the immunological synapse may be necessary to trigger activation. Similar but not homologous to *episialin.

leukotrienes (LTA_4, LTB_4, LTC_4, LTD_4, LTE_4) Derivatives of hydroxyeicosatetraenoic (HETE) acid. LTA_4 and LTB_4 are modified lipids. In leukotriene C the lipid is conjugated to glutathione (LTC_4) and in LTD_4, LTE_4 to cysteine. A mixture of the peptidyl leukotrienes (LTC_4, LTD_4, LTE_4) is SRS-A, the slow reacting substance of anaphylaxis, which has potent bronchoconstrictive effects. LTB_4 is a potent neutrophil chemotactic factor and inflammatory mediator.

leupeptin A reversible competitive inhibitor of serine and thiol peptidases such as calpain, cathepsin and trypsin. The commonest of the family is *N*-acetyl-Leu-Leu-argininal.

Leu-phyllolitorin *See* PHYLLOLITORIN.

levamisole An anthelmintic drug use to treat *Ascaris* infections.

levodopa (L-DOPA) The precursor of dopamine (3,4-dihydroxy-L-phenylalanine) that is used as a prodrug to increase dopamine levels in the basal ganglia thereby improving mobility in *Parkinson's disease. Usually administered together with a dopa-decarboxylase inhibitor (*carbidopa) to prevent peripheral effects of dopamine.

levofloxacin A *fluoroquinolone antibiotic that has a broad spectrum of antibacterial action and can be administered orally or intravenously. *TN* Levaquin

levonorgestrel (l-norgestrel) A synthetic *progestogen used in oral contraceptives.

levothyroxine (l-thyroxine sodium) A preparation of thyroid hormone (thyroxine) used to treat hypothyroidism.

Lewis blood group Erythrocyte surface antigens acquired by adsorption of circulating glycosphingolipids. The determinants are terminal fucose residues added to the H antigen (of the A,B, H system) through the interaction of two different fucosyltransferases, FUT2 and FUT3. There are only three phenotypes: $Le^{(a-b-)}$, $Le^{(a+b-)}$, and $Le^{(a-b+)}$.

SEE WEB LINKS

- The Blood Group Antigen Gene Mutation database.

Lewy body Intracellular aggregates of *α-synuclein associated with other proteins such as ubiquitin, neurofilament protein, synphilin, and αB crystallin, found in neurons in the *substantia nigra and locus coeruleus of patients with *Parkinson's disease and Lewy body dementia. **Classical Lewy bodies** consist of a dense core surrounded by a halo of 10-nm wide radiating fibrils that stain strongly with eosin. **Cortical Lewy bodies** are less structured, but also consist of α-synuclein. Dementia with Lewy bodies is the third most common cause of dementia.

Leydig cells (interstitial cells of Leydig) Interstitial cells of the mammalian testis that synthesize testosterone, androstenedione, and dehydroepiandrosterone in response to luteinizing hormone.

LFA-1 One of the *integrin superfamily of adhesion molecules (lymphocyte function-related antigen-1, CD11a/CD18, α_L 1170 aa; β, 95kDa, 769 aa), found on lymphocytes and that binds *ICAM-1. Deficiency of the β subunit (CD18) causes leucocyte adhesion deficiency (LAD) syndrome with serious immunodeficiency and a poor prognosis. The LFA-1 class of adhesion molecules have a common β subunit and include Mac-1 (α_M, CD11b, 1152 aa) and p150,95 (α_X, CD11c, 150 kDa, 1163 aa); they are also defective in severe forms of LAD. Mac-1 (Mo-1 in earlier literature) is the complement C3bi receptor (CR3) and is present on mononuclear phagocytes and on neutrophils; p150,95 is less well characterized, but is particularly abundant on macrophages. Integrin α_D/β_2 (CD11d, 1162 aa) is a receptor for ICAM3 and VCAM1 and may be involved in clearing lipoproteins from atherosclerotic plaques and phagocytosis of blood-borne particles.

LFA-3 A membrane-anchored protein (lymphocyte function-related antigen-3, CD58, 222 aa), expressed on cytolytic T cells, that is the ligand or counter-receptor for CD2, an adhesion mechanism distinct from the *LFA-1/*ICAM-1 system. Both LFA-3 and CD2 are members of the immunoglobulin superfamily and are thought to have arisen by duplication of a common evolutionary precursor. Binding of erythrocyte LFA-3 to T-cell CD2 is the basis of *E-rosetting.

Lfc *See* DBL.

LGP2 A cytoplasmic RNA helicase (678 aa) related to *RIG-I and *mda-5 but that functions as a negative regulator of RIG-I by blocking the dimerization of RIG-I and preventing the induction of interferon production. LGP2 apparently competes with the kinase IKKi (IKKε) for a common interaction site on *MAVS.

LH *See* LUTEINIZING HORMONE.

LHRF Luteinizing hormone releasing factor. *See* LUTEINIZING HORMONE.

library *See* GENOMIC LIBRARY.

lichen planus A rare skin disorder in which there is T-cell-mediated inflammation of the epidermis and hyperkeratosis. A familial form has been reported and is associated with HLA-B7.

Liddle's disease An autosomal dominant form of salt-sensitive human hypertension caused by mutation in the β or γ subunit of the multisubunit epithelial *amiloride-sensitive sodium channel (ENaC).

lidocaine (formerly **lignocaine**) A local anaesthetic that works by blocking fast voltage-gated sodium channels in neurons. Also used to treat ventricular arrhythmias.

LIF *See* LEUKAEMIA INHIBITORY FACTOR. Also used for leucocyte inhibitory factor.

Li–Fraumeni syndrome A clinically and genetically diverse predisposition to tumours inherited as an autosomal dominant in which there is early onset of tumours, multiple tumours within an individual, and multiple affected family members. **Li–Fraumeni syndrome-1** (LFS1) is caused by mutation in the *p53 gene. **Li–Fraumeni syndrome-2** is caused by mutations in the CHEK2 (*checkpoint kinase-2) gene and a third form (**LFS3**) maps to a locus on 1q23.

ligand Any molecule that binds to another; generally the smaller of the two, especially if there is no signal transduction by the receptor. Ligands are not necessarily small molecules such as hormones or neurotransmitters; sugar residues attached to proteins or lipids and incorporated in the membrane are often referred to as ligands for *lectins.

ligand-gated ion channel An *ion channel that responds to binding of a specific *ligand, typically a neurotransmitter at a *chemical synapse, by changing its properties and allowing an increase in the flow of ions across the membrane. The change may be from almost zero permeability when closed, to allowing passage of 10^7 ions per second when open. Examples are the *acetylcholine and *GABA receptors.

ligand-induced endocytosis The process by which many soluble extracellular ligands are taken up by cells by the formation of coated pits and then *coated vesicles. Binding of the ligand to the receptor is usually, but not invariably, the trigger for *clathrin and associated proteins (coatomers) to bind the cytoplasmic portion of the receptor.

ligase amplification reaction A method analogous to the *polymerase chain reaction (PCR) used to detect small quantities of a target DNA. The method relies upon the requirement, by *DNA ligase, for sequence similarity to allow ligation of synthetic oligonucleotides (probe) bound to target DNA. Various ligase-based techniques have been developed for the high-throughput detection of DNA, homogeneous assays, single-molecular detection, and scanning of unknown mutations.

ligases (synthetases) The class of *enzymes that catalyse the linking together of two molecules (category 6 in the *E classification), e.g. DNA ligases that link two fragments of DNA by forming a *phosphodiester bond.

ligatin A peripheral membrane protein (584 aa) that acts as a receptor for phosphoglycoproteins, and triggers endocytosis, a trafficking receptor. Glutamate receptor activation and calcium entry into hippocampal neurons causes a long-lasting down-regulation of ligatin.

LIGHT A TNF superfamily member (TNFSF14, 240 aa), a type II transmembrane protein produced by activated T cells that forms a membrane-anchored homotrimeric complex. Herpes simplex virus (HSV) envelope glycoprotein D binds to this complex and the virus gets internalized. A sensible abbreviation for homologous to lymphotoxins, exhibits inducible expression, and competes with HSV glycoprotein D for HVEM, a receptor expressed by T lymphocytes'.

light chain A general term for the smaller subunits of a multimeric protein, e.g. *immunoglobulin, *myosin, *dynein, and *clathrin light chains. *See also* L CHAIN.

Lightcycler® A *real time PCR machine that uses fluorescent detection during the PCR reaction to quantify the DNA present at each cycle.

Light Green An anionic counterstain for cytoplasm, often used together with iron haematoxylin and a component of Masson's trichrome stain.

ligneous conjunctivitis A form of conjunctivitis (inflammation of the conjunctiva) that progresses to the formation of white, yellow-white, or red masses with a wood-like consistency that replace the normal mucosa. It is the most common manifestation of type I congenital *plasminogen deficiency.

lignocaine *See* LIDOCAINE.

like Sm proteins A subfamily of proteins (LSm proteins) homologous to *Sm proteins that have a bipartite sequence motif called the Sm domain. Lsm proteins form heptameric complexes, either Lsm1p-7p or Lsm2p-8p (localized in the cytoplasm and the nucleus respectively). Lsm1p-7p is involved in the RNA *decapping process, a prelude to RNA degradation; the Lsm2p-8p complex binds to U6 snRNA and functions in RNA splicing.

LIM and SH3 proteins Proteins (LASPs) with a *LIM domain and an *SH3 domain that are overexpressed in breast and ovarian cancer. LASP2 (LIM/nebulette) is a splice variant of *nebulin. Both proteins are ubiquitously expressed and involved in cytoskeletal architecture, especially in the organization of focal adhesions. LASP-1 is involved in neuronal differentiation and plays a role in the migration and proliferation of certain cancer cells; LASP-2 is more structural.

limatin Retinal homologue of *dematin.

limb bud A bulge of mesenchyme surrounded by a simple epithelium that is the precursor of a limb in the early embryo. The limb bud of the chick embryo has been an important model system for studying the positional information that specifies the proximal–distal pattern of bone development and the anterior–posterior specification of digits.

limbic system Those regions of the central nervous system responsible for emotion, behaviour, and long-term memory. Includes hippocampus, amygdala, anterior thalamic nuclei, and limbic cortex.

limbus A distinct edge or border, e.g. the boundary between cornea and sclera in the eye.

LIM domain Highly conserved double zinc finger motif that will bind to a second LIM domain to mediate protein–protein interactions. Found in many proteins involved in cytoskeleton organization, cell lineage specification, organ development, and oncogenesis. Named after the first three members: Lin-11 (required for asymmetric division of blast cells in *Caenorhabditis elegans*), IsI-1 (mammalian insulin-gene binding enhancer protein), and mec-3 (required for differentiation of a set of sensory neurons in *C. elegans*).

LIME Lck-interacting *transmembrane adaptor (295 aa) that becomes tyrosine-phosphorylated by *lyn when the B-cell receptor is crosslinked and triggers B-cell activation. LIME is associated with lyn, *Grb2, PLC-γ2, and *PI3K.

limen The threshold of a physiological or psychological response below which a stimulus is not perceived. The adjective 'liminal' describes something situated at a sensory threshold, hence barely perceptible, and is most commonly encountered in the compound 'subliminal', i.e. undetectable.

limitin An interferon-like cytokine (IFN-ζ, 182 aa) that has some sequence homology with IFN-α, -β, and -ω. Suppresses B-cell lymphopoiesis through ligation of the IFN-α/β receptor, activation of *tyk2, and the up-regulation and nuclear translocation of *DAXX. Also affects activity of other lymphoid cells, but not myeloid cells.

LIM kinases (LIMKs) A small subfamily with a unique combination of two LIM motifs and a C-terminal protein kinase domain. They are probably components of an intracellular signalling pathway and may be involved in brain development, regulate actin treadmilling by phosphorylating members of the ADF/cofilin family of proteins, and are themselves activated by *PAK. LIMK1 hemizygosity is implicated in the impaired visuospatial constructive cognition of *Williams' syndrome. **LIMS1** (LIM and senescent cell antigen-like domains-1; particularly interesting new Cys-His protein, PINCH1; 314 aa) has five LIM domains and is an obligate partner of *integrin-linked kinase both of which are necessary for proper control of cell shape change, motility, and survival.

LIM mineralization protein (LMP-1) An intracellular osteogenic factor that has been implicated in the *bone morphogenetic protein (BMP) pathway. Has three LIM domains. Three splice variants have been identified: LMP-1, LMP-2, and LMP-3.

Limulus polyphemus The king crab or horseshoe crab, now renamed *Xiphosura*, although the old name persists. It is more closely related to the arachnids than the crustacea. The compound eyes have been widely used in studies on visual systems, but best known to nonzoologists as the source of the very sensitive **limulus-amoebocyte lysate** (LAL) test used to detect *endotoxin which causes rapid clotting. *See* FACTOR C.

LIN2 *See* CASK.

LINC complex A multiprotein complex of *sun proteins and *nesprins that links the nucleoskeleton and cytoskeleton across the nuclear envelope.

SEE WEB LINKS

- More detail about the role of LINC complexes in coupling the nucleus and cytoplasm.

lincomycin A *macrolide antibiotic isolated from *Streptomyces lincolnensis*. Clindamycin, a semisynthetic derivative of lincomycin, is used as an antimalarial drug.

lindane An organochlorine pesticide (benzene hexachloride, γ-hexachlorocyclohexane) used for treating head lice and scabies and as a general pesticide. Neurotoxic and probably carcinogenic.

Lindau's tumour A tumour of brain or spinal cord that derives from blood vessels of the meninges and that is considered a *haemangioblastoma.

LINE *See* LONG INTERSPERSED NUCLEOTIDE ELEMENT.

Lineweaver–Burke plot A graphical method for deriving V_{max} (maximal reaction rate) and K_m (rate constant, Michaelis–Menten constant) for an enzyme-catalysed reaction. The reciprocal of the reaction rate ($1/V$) is plotted against the reciprocal of the substrate concentration ($1/S$) and the y-intercept is equivalent to the inverse of V_{max} and the x-intercept represents $-1/K_m$. Because the method gives undue weight to the least accurate points, other methods of analysis, such as the *Eadie–Hofstee plot, are preferable.

linezolid An antibiotic used against antibiotic-resistant Gram-positive streptococci. The first clinically available oxazolidinone, chemically unrelated to other antibiotics. Acts by blocking the binding of 30S and 50S ribosomal subunits. *TN* Zyvox.

lining epithelium An unspecialized epithelium of the type that lines ducts, cavities, or vessels.

linkage *See* GENETIC LINKAGE.

linkage disequilibrium The occurrence of some alleles together more or less often than would be expected. The deviation of the probability of the two occurring together from the probability that would be expected from random assortment (linkage equilibrium), will disappear in time due to recombination and the smaller the distance between the two loci, the lower the rate at which equilibrium is approached. If, however, there is a selective advantage to the linkage, it will tend to persist for much longer. Thus, in the HLA system of *histocompatibility antigens, HLA-A1 is commonly associated with B8 and DR3, and A2 with B7 and DR2, presumably because the combination confers some selective advantage.

linoleic acid An essential *fatty acid (9,12-octadecadienoic acid), one of the ω-6 fatty acids. Vegetable oils are a major dietary source and linoleic acid is important as a precursor for prostaglandins.

linolenic acid An 18-carbon fatty acid with three double bonds (9,12,15-octadecatrienoic acid) and α- and γ-isomers. An essential dietary component for mammals. α-Linolenic acid is an ω-3 fatty acid, γ-linolenic is an ω-6.

linopirdine A selective blocker of the KCNQ-type potassium channels (*see* M CHANNELS) and a putative cognition-enhancing drug. *TN* DuP 996.

lipaemia Abnormal levels of blood lipid. *See* HYPERLIPIDAEMIA.

lipases Enzymes that break down mono-, di- or triglycerides to release fatty acids and glycerol, a subclass of esterases. Calcium ions are usually required. Triglyceride lipases (EC 3.1.1.3) hydrolyze the ester bond of triglycerides; *phospholipases break down phospholipids.

lipid A The lipid portion of the *lipopolysaccharide (LPS) of the cell wall of Gram-negative bacteria.

lipid bilayer *See* PHOSPHOLIPID BILAYER.

lipid disorders

The major lipid disorders (lipidoses) are ones in which plasma lipid levels are abnormal, as in *hyperlipidaemia, *hypercholesterolaemia and *chylomicron retention disease. *See also* APOLIPOPROTEIN B; LDL RECEPTOR. Abnormal accumulations of lipids in tissues (lipidoses) can occur in lysosomal *storage diseases (e.g. *Gaucher's, *Krabbe's, *Tay–Sachs, *Niemann–Pick diseases) or where there are defects in enzymes involved in lipid metabolism (*cerebrotendinous xanthomatosis). High levels of *cholesterol are a risk factor for *atherosclerosis. *See also* ANDERSON'S DISEASE; SITOSTEROLAEMIA; HYPERCHOLANAEMIA. More subtle defects in lipids may affect the behaviour of receptors or ion channels that are embedded in a *phospholipid bilayer and are sensitive to its physical properties. Whether obesity is a lipid disorder is questionable (*but see* ADIPOKINE).

lipid II A membrane-anchored cell-wall precursor (undecaprenyl diphosphate-MurNAc-pentapeptide-GlcNAc) that is essential for bacterial cell-wall biosynthesis. The target molecule for at least four different classes of antibiotic, including *vancomycin and several *lantibiotics.

lipidoses *See* LIPID DISORDERS.

lipids A heterogeneous group of compounds that includes fats, oils, waxes, and terpenes. The shared characteristic is insolubility in polar solvents such as water.

SEE WEB LINKS

- The Lipid Library is a useful resource.

lipins A family of homologous proteins (lipins1, 2, and 3). Lipin-1 is a Mg^{2+}-dependent phosphatidic acid (PA) phosphohydrolase (EC 3.1.3.4, 890 aa) that catalyses the dephosphorylation of PA to yield diacylglycerol and inorganic phosphate. Lipin is required for normal adipose tissue development, and in the mouse mutations are associated with fatty liver dystrophy, although similar associations have not been found in humans. Mutations in the lipin-1 gene are associated with autosomal recessive recurrent myoglobinuria which is postulated to arise through accumulation of lysophosphatidate and other lysophospholipids in muscle tissue, resulting in rhabdomyolysis during stress. Mutations in lipin-2 are found in patients with *Majeed's syndrome.

lipoamide The functional form of *lipoic acid in which the carboxyl group is attached to protein by an amide linkage with the amino side group of lysine. **Lipoamide dehydrogenase** (EC.1.8.1.4, 509 aa) regenerates lipoamide from the reduced form, dihydrolipoamide.

lipocalins A superfamily of extracellular proteins (18–20 kDa) that bind lipophilic molecules, such as retinol, porphyrins, and odorants, by enclosure within their structures, minimizing solvent contact. Examples include retinol-binding protein, β-lactoglobulin, and *orosomucoid. A number of lipocalin isoforms are found in tears and lipocalins have been found in many tissues. **Neutrophil-gelatinase-associated lipocalin** (NGAL, lipocalin-2) has been used as a biomarker in the detection of acute renal failure in children after cardiac surgery.

lipocortins A group of calcium-binding proteins believed to act as an inhibitors of phospholipase A2 enzymes and thus as endogenous anti-inflammatory agents. Several different lipocortins have been described although their function has been contested. They are members of the *annexin family.

lipocytes Fat-storing cells of the liver. Hepatic stellate cells seem to have two phenotypes, myofibroblast-like and fat-storing, which can interconvert.

lipodystrophy A disorder in which the distribution of adipose tissue in the body is altered and in extreme cases there may be a near-absence of adipocytes. Congenital generalized **lipodystrophy type 1** is caused by mutation in the *AGPAT2* gene that encodes lysophosphatidic acid acyltransferase (LPAAT; 1-acyl-sn-glycerol-3-phosphate acetyltransferase, EC 2.3.1.51). Congenital **lipodystrophy type 2** (Berardinelli–Seip syndrome) is caused by mutation in the gene encoding *seipin. **Familial partial lipodystrophy** comprises several heterogeneous syndromes, including FPLD1 (Kobberling type), FPLD2 (Dunnigan type), caused by mutation in the gene encoding *lamin A/C, and FPLD3, caused by mutation in the peroxisome proliferator-activated receptor-γ *PPARG* gene. In some cases of partial lipodystrophy, which affects predominantly the face and trunk, often with excess accumulation of fat in the lower part of the body, there is an association with C3 nephritic factor, an IgG antibody against complement components. Adipocytes in the upper part of the body produce more factor D (*adipsin) and it seems that the C3 nephritic factor stabilizes factor D sufficiently to lead to the formation of attack complex and destruction of adipocytes. **Partial acquired lipodystrophy** (Barraquer–Simons syndrome; partial progressive lipodystrophy) can be caused by mutations in the gene for lamin B2. *See also* MANDIBULOACRAL DYSPLASIA.

lipofectamine (lipofect amine, lipofectin) A proprietary preparation of liposomes made from cationic lipids, used for *transfection of cultured cells.

lipofuscin Yellow-brown granular cytoplasmic pigment that accumulates in ageing cells. It is the product of peroxidation of unsaturated fatty acids, although other compounds are present in small amounts. Abnormal accumulation of lipofuscin is a feature of various diseases such as *Batten's disease although there is no evidence that lipofuscin is itself deleterious.

lipoic acid (thioctic acid) A coenzyme in the oxoglutarate dehydrogenase complex of the *tricarboxylic acid cycle and involved in oxidative decarboxylations of α-keto acids.

lipoid congenital adrenal hyperplasia The most severe form of congenital adrenal hyperplasia in which a defect in *steroidogenic acute regulatory protein blocks all synthesis of steroid hormones. All patients are phenotypic females with a severe salt-losing syndrome.

lipolysis The hydrolysis of lipids to produce fatty acids and glycerol.

lipoma A benign tumour composed of fatty tissue. There are rare lipomatoses with a familial component (e.g. familial benign cervical lipomatosis and multiple lipomatosis) but the precise defects in these conditions are unclear.

lipoma-preferred partner A protein (LIM domain-containing preferred translocation partner in lipoma, LPP, 612 aa) with considerable homology to *zyxin and containing three *LIM domains. It is involved in the translocation t(3;12) in benign lipomas in which the *LPP* gene from chromosome 3 becomes fused to the DNA-binding molecule HMGIC (high mobility group 1C). It can also become fused to the mixed lineage leukaemia gene on chromosome 11.

lipomodulin Obsolete name for *lipocortin isolated from neutrophils.

lipophilic Having an affinity for lipids and therefore hydrophobic or with hydrophobic regions.

lipophilin *See* MYELIN PROTEOLIPID PROTEIN.

lipopolysaccharide (LPS) The major component of Gram-negative cell wall, an endotoxin that is highly immunogenic and binds the CD14/TLR4/MD2 receptor complex triggering the the production of endogenous pyrogen *interleukin-1 and *tumour necrosis factor (TNF) following infection. The molecule has three distinct regions: *lipid A, core polysaccharide to which are attached variable polysaccharide side chains (O antigen). Full-length O antigen makes the LPS smooth and the surface of the bacteria more hydrophilic and less easily phagocytosed. Rough LPS is more hydrophobic.

lipopolysaccharide-binding protein An *acute phase protein (LBP, 452 aa) produced during infection with Gram-negative bacteria. The LBP–lipopolysaccharide complex is bound specifically by CD14 on myeloid cells and is presented to the toll-like receptor-4 (TLR4) leading to up-regulation of various inflammatory genes.

lipoprotein glomerulopathy A rare kidney disorder caused by mutation in the *APOE* gene, characterized by proteinuria, progressive kidney failure, and distinctive lipoprotein deposits in glomerular capillaries. It mainly affects people of Japanese and Chinese origin.

lipoprotein lipase The major enzyme (LPL, EC 3.1.1.34, 475 aa) for hydrolysis of circulating triglyceride-rich lipoproteins, one of a family of enzymes that includes *hepatic lipase, and pancreatic lipase. Familial lipoprotein lipase deficiency is characterized by increased plasma trigylceride levels because the clearance of chylomicrons from the plasma after digestion of dietary fat is delayed.

lipoproteins Serum proteins (*apolipoproteins) with a surface coat of phospholipid monolayer. They are classified according to density: the least dense being *chylomicrons, then very low density (*VLDL); low density (LDL), *intermediate density, and high density (*HDL and *VHDL) species. They are important for lipid transport, especially of cholesterol. **Lipoprotein(a)** is formed from apolipoprotein(a) (Apo(a), 4548 aa), disulphide linked to apolipoprotein B100 (ApoB) of a low-density lipoprotein particle. It has serine peptidase activity, inhibits tissue-type plasminogen activator and may be a ligand for *megalin. High levels of plasma Apo(a) and its naturally occurring proteolytic fragments are correlated with atherosclerosis. Apo(a) is proteolytically cleaved to form mini-Lp(a) and Apo(a) fragments that accumulate in atherosclerotic lesions, where they may promote thrombogenesis. *See also* LDL RECEPTOR.

l

liposomes Single- or multilayered spherical lipid bilayer structures produced from lipids dissolved in organic solvents and then dispersed in aqueous media. Experimentally and therapeutically used for delivering drugs, etc. to cells because liposomes will fuse with cell membranes so the contents are transferred into the cytoplasm (*see* LIPOFECTAMINE). Size varies from submicron diameters to (in a few record-breaking cases) centimetres.

lipoteichoic acid (LTA) A major component of the cell wall of Gram-positive bacteria. It is amphiphilic with *teichoic acid linked to glycolipid and, like *lipopolysaccharide, is a virulence factor. LTA binds CD14 and CD36 and activates the production of inflammatory cytokines through toll-like receptor-2. It may also function as an *adhesin.

lipotropin A polypeptide hormone (LPH, lipotropic hormone, adipokinetic hormone; one of the *adrenocorticotrophin group) derived from *proopimelanocortin (β form: 91 aa; the γ form has only residues 1–58). Not only does it stimulate melanocytes to produce melanin, but it is the precursor of *endorphins and promotes lipolysis and steroidogenesis. Receptors are of the seven-spanning membrane receptor type (melanocortin receptor1, MCR2, 3, 4, and 5).

lipoxins Short-lived anti-inflammatory *eicosanoids, generated by *lipoxygenases from arachidonic acid, that are produced as the inflammatory response resolves. Two lipoxins, LXA_4 and LXB_4, are known and, as with the *leukotrienes, cysteinyl-lipoxins LXC_4, LXD_4 and LXE_4 are also found. The cysteinyl-lipoxins are antagonists for the leukotriene receptors. LXA_4 acts through a *formyl-peptide-like receptor and inhibits neutrophil chemotaxis, transmigration, superoxide generation, and NFκB activation. *Epi-lipoxins, *resolvins, and *protectins are analogous.

lipoxygenase The enzyme (5-lipoxygenase, 5-LO, EC 1.13.11.34, 674 aa) that catalyses the first step in *leukotriene synthesis.

liprins A family of proteins that act as scaffolds for the recruitment and anchoring of *LAR family PTPases. The C-terminal regions of α-liprins bind to the membrane-distal phosphatase domains of LAR family members as well as to β-liprins; α- and β-liprins also homodimerize via their N-terminal regions. Some liprins are widely distributed, others are more tissue specific.

lipstatin *See* ORLISTAT.

Lis1 The product (**platelet-activating factor acetylhydrolase IB subunit** α, 410 aa) of a gene that is deleted or mutated in patients with *lissencephaly. It interacts with *nuclear distribution gene E homologue-like 1 and is essential for proper activation of Rho GTPases and actin polymerization at the leading edge of locomoting cerebellar neurons and postmigratory hippocampal neurons.

lisinopril The lysine analogue of *enalapril, an angiotensin-converting enzyme inhibitor (*ACE inhibitor) used in treatment of heart failure and hypertension. *TNs* Carace, Prinivil, Tensopril, Zestril.

lissencephaly A severe human neuronal migration defect characterized by a smooth cerebral surface, mental retardation, and seizures, caused by mutation in the gene for *Lis-1. *Miller–Dieker lissencephaly syndrome is of greater severity and is associated with additional mutations in genes adjacent to Lis-1 on chromosome 17.

Listeria monocytogenes A rod-shaped Gram-positive bacterium responsible for serious food poisoning (listeriosis). It is normally *saprophytic but is an opportunistic parasite, and will survive within cells (particularly leucocytes) and can be transmitted transplacentally.

listeriolysin O A *cholesterol binding toxin that forms pores and is haemolytic. A virulence factor for **Listeria monocytogenes*.

lithium (Li) The lightest of the alkali metals, although it has the largest hydrated cation. Lithium salts (carbonate, citrate) are used for treatment and prevention of mania, bipolar disorder, and recurrent depression, probably acting by inhibiting the regeneration of *inositol from IP3 and down-regulating the *phosphatidyl inositol signalling pathways. Lithium succinate has anti-inflammatory and antifungal activity.

lithotomy Surgical removal of stones (calculi) from the bladder or ureter. An alternative is the use of ultrasound (lithotripsy) to fragment the stones.

litorin A *bombesin-like peptide with mitogenic effects; has a C-terminal octapeptide in common with bombesin. *See* PHYLLOLITORIN.

Little's disease A form of *cerebral palsy.

livedo A vascular disorder that causes blueish mottling of the skin. Mostly affects women and may be aggravated by exposure to cold.

liver activator protein *See* LAP.

liver cells An imprecise term, usually implying *hepatocytes, even though other cell types are present (e.g. *Ito cells, *Kupffer cells, *lipocytes).

liver cytosol antigen (LC-1) *See* FORMIMINOTRANSFERASE CYCLODEAMINASE.

liver disorders (hepatic disorders)

The liver is the major site of many basic metabolic functions and the cells (*hepatocytes) appear relatively unspecialized, the archetypal animal cell of textbooks. Liver is one of the few human tissues to have some capacity to regenerate (*see* OVAL CELLS). Not only does the liver produce many of the plasma proteins but it is an important site for detoxification of drugs and other xenobiotics (*see* CYTOCHROME P450; GLUTATHIONE S TRANSFERASES) and a major site of glycogen storage. *Glycogen storage diseases have a serious impact on the function of liver and many other tissues (*see also* PHOSPHORYLASE KINASE). *Kuppfer cells are important for removal of particles and effete erythrocytes from the circulation. Inflammation of the liver (*hepatitis) can be caused by various viruses, and hepatitis B and C virus infection is a major risk factor for hepatocarcinoma. Chronic inflammation, often a result of excess alcohol intake, can lead to *cirrhosis and *fibrosis (*see* ITO CELLS). Markers of liver dysfunction include *alanine aminotransferase and *formiminotransferase-cyclodeaminase. Accumulation of various substances in the liver can also lead to cirrhosis (e.g. in *haemochromatosis and *Wilson's disease) and lipid accumulation in *lipocytes can be a feature of dietary overload (as in production of foie gras). There are various disorders of the *bile system, including *Alagille's syndrome, primary sclerosing *cholangitis, and *Gilbert's syndrome. Some toxins (e.g. *aflatoxins) particularly affect the liver. *See* TORRES BODY.

liver-expressed chemokine *See* LEC.

liver X receptor *See* LXR.

Ljungan virus A picornavirus that causes myocarditis and diabetes-like symptoms in wild rodents (isolated from bank voles *Clethrionomys glareolus*, in Sweden) and occasionally in people who have been in contact (the first cases were among elite orienteers). The virus is closely related to human parechoviruses but forms a distinct subgroup of the picornaviridae. It has been suggested that it may play a role in type 1 *diabetes, although this is still contentious.

LKB1 A kinase (serine/threonine kinase 11, STK11, 433 aa) mutated in *Peutz–Jeghers syndrome that interacts with *p53; deficiency is associated with defective apoptosis.

LKR Abbreviation for: **1.** Lysine-α-ketoglutarate reductase, EC 4.6.1.10. **2.** LKR-13 is a line of lung adenocarcinoma cells derived from K-ras (LA1) mice. **3.** Leukokinin receptor (Lkr) from *Drosophila*.

LL-37 (hCAP18) Human *cathelicidin, found in leucocytes, particularly neutrophils, and up-regulated in inflammation.

LMM *See* MEROMYOSIN.

LMP **1.** Low molecular weight proteins (LMPs), components of the *proteosome. **2.** Low malignant potential. **3.** Last menstrual period. **4.** *See* LATENT MEMBRANE PROTEINS. **5.** *See* LIM MINERALIZATION PROTEIN (LMP-1).

L-myc A gene identified in a line of cells derived from a small-cell lung cancer with homology to a small region of both *myc and *N-myc*. Frequently overexpressed in metastatic lung tumours.

LNCaP cells A line of human cells used in the study of prostatic cancer. They are androgen-sensitive and derived from a lymph node metastatic lesion of prostatic adenocarcinoma.

Loa loa A filarial nematode that infects humans and causes swellings below the skin (Calabar swellings) and inflammatory lesions in other tissues (loiasis). The vectors are blood-sucking insects of the genus *Chrysops*.

lobar pneumonia *See* PNEUMONIA.

lobopodia Protrusions from the front of a moving cell which are roughly hemispherical.

local circuit theory The generally accepted model for the propagation of an action potential, in which depolarization of a small region of the neuronal plasma membrane produces transmembrane currents in adjacent regions, tending to depolarize them and causing *voltage-gated sodium channels to open.

lock and key model Recognition mechanism that depends upon precise steric matching of receptor and ligand, or enzyme and substrate. In biological systems the degree of precision varies.

lockjaw *See* TETANUS TOXIN.

l

locomotion Movement from place to place as opposed to flattening, shape-changing, dividing (*cytokinesis), or moving a protrusion such as a cilium.

locus (*pl.* **loci**) **1.** Generally, a place or site, a location. **2.** In genetics, the site in a linkage map or on a chromosome where the gene for a particular trait is located. Different alleles may be present at this site.

locus coeruleus A dense cluster of neurons in the brain that is the major source of *noradrenaline. There are neuronal projections throughout most of the central nervous system and the locus coeruleus is thought to be a key brain centre for anxiety and fear. The blue colour that gives it the name derives from melanin granules.

locus control region (LCR) The nontranscribed region that contains the *promoters and *enhancers which regulate the expression of a particular gene. Often taken to be a single region 0–2 kb upstream of the transcriptional *start site.

Lod score A statistically derived parameter (logarithm of the odds score) that indicates the probability (logarithm to base 10) that there is linkage between traits or markers. For non-X-linked genetic disorders a Lod score of +3 (1000:1) is usually taken to indicate linkage and a Lod score less than −2.0 is evidence to exclude linkage.

Loeffler's medium A culture medium containing horse serum, beef extract, dextrose, and proteose peptones that is used in diagnostic bacteriology for the culture of **Corynebacterium diphtheriae*.

Loeffler's syndrome An acute but mild and self-limiting eosinophilic pneumonia, usually due to an allergic reaction.

lofepramine A third-generation tricyclic antidepressant with anxiolytic properties. *TNs* Gamanil, Lomont, Tymelyt.

lofexidine A centrally acting α_2-adrenergic agonist used to relieve the symptoms of withdrawal from opiates. *TN* BritLofex.

logarithm of the odds score *See* LOD SCORE.

log-normal distribution A distribution function $F(y)$, in which the logarithm of a quantity has a gaussian (normal) distribution. A log-normal distribution results if the variable is the product of a large number of independent, identically distributed variables rather than the sum of the variables (which gives a normal distribution).

log *P* (log K_{ow}) The octanol–water *partition coefficient, often used as a quantitative indicator of lipophilicity, which is an important consideration for drug delivery and for predicting distribution of a chemical in the environment.

LOH *See* LOSS OF HETEROZYGOSITY.

longevity assurance genes (*LAG-1*) *See* LASS1.

long interspersed nucleotide element (LINE) An autonomously replicating *transposon of 6–7 kb that constitutes a major part (~15%) of the mammalian genome. Unlike viral retrotransposons they lack *long terminal repeats. **LINE1** has two open reading frames, one encoding a reverse transcriptase that is also used by non-autonomous *short interspersed nuclear elements (SINEs). An older family of transposons, **LINE2**, is no longer being replicated. *See* MAMMALIAN-WIDE INTERSPERSED REPEAT.

SEE WEB LINKS

- A list of recent publications on LINEs can be downloaded from the BioPortfolio website.

longitudinal study *See* CROSS-SECTIONAL STUDY.

long QT syndrome A group of disorders due to defects in ion channels (*channelopathies) which increase the risk of sudden heart failure and are probably responsible for most cases of sudden adult death syndrome. The QT interval is defined from the characteristics of the electrocardiogram and represents the time taken for electrical activation and inactivation of the ventricles. **Long QT syndrome-1** (LQT1) can be caused by mutation in the KQT-like voltage-gated *potassium channel-1 (*KCNQ1*) gene, which is also mutated in *Jervell and Lange-Nielsen syndrome; **LQT2** is caused by mutation in the *KCNH2* gene (the human homologue of the *Drosophila* gene *ether-a-go-go*) that encodes an inwardly rectifying potassium channel; **LQT3** is caused by mutation in the *SCN5A* gene for a voltage-gated sodium channel subunit); **LQT4** is caused by mutation in the *ankyrin-2 (*ANK2*) gene; **LQT5** is caused by mutation in the *KCNE1* gene for a transmembrane protein known to associate with the product of the *KCNQ1* gene to form the delayed rectifier potassium channel; **LQT6** is caused by mutation in the *KCNE2* gene; **LQT7** (*Andersen's syndrome) is caused by mutation in the *KCNJ2* gene for the inwardly rectifying potassium channel, Kir2.1; **LQT8** (*Timothy's syndrome) is caused by mutation in the *CACNA1C* gene for a subunit of a voltage-dependent L-type calcium channel); **LQT9** by mutation in the *CAV3* (*caveolin-3) gene ; **LQT10** by mutation in the *SCN4B* gene for the β subunit of the type IV voltage-gated sodium channel; and **LQT11** by mutation in the *AKAP9* gene (encoding one of the *AKAP family of anchoring proteins).

long-sightedness *See* HYPERMETROPIA.

long-terminal repeats (LTRs) DNA sequences, several hundred nucleotides long, that are found at either end of *transposons and proviral DNA, formed by reverse transcription of *retroviral RNA. They contain inverted repeats (identical when read in opposite directions) that are important for integration into host DNA. In proviruses the upstream LTR acts as a promoter and enhancer and the downstream LTR as a polyadenylation site. They comprise about 8% of the human genome.

long-term potentiation (LTP) A sustained increase in the strength of transmission at a *synapse that has been repeatedly used. It is a form of *synaptic plasticity that could play a part in learning and memory storage. Selective inhibition of *NMDA receptor channels has been shown to block LTP, and to block spatial learning, but potentiation occurs through a variety of mechanisms throughout the nervous system.

loop diuretics *See* DIURETICS.

loop of Henle The hairpin-loop portion of the nephron that leads from the proximal convoluted tubule to the distal convoluted tubule in the renal medulla. It is the major site of reabsorption of ions and water from the ultrafiltrate formed in the glomerulus. Four distinct regions can be recognized; the descending limb is highly permeable to water but not ions; the thin ascending limb is permeable to ions but not water; the medullary portion of the thick ascending limb is the major location of active transport of ions and the cortical portion of the thick ascending limb drains urine into the distal convoluted tubule.

loperamide An opioid receptor agonist that acts selectively on the μ-opioid receptors of the myenteric plexus of the large intestine, reducing peristalsis. Used to treat diarrhoea. *TN* Imodium.

lopinavir A *protease inhibitor used as an antiviral drug. *TNs* Aluvia, Kaletra.

loprazolam A short-acting *benzodiazepine used to treat insomnia.

loratadine An antihistamine used to relieve symptoms of allergy. Non-sedating at low doses. *TN* Clarityn.

lorazepam A short-acting *benzodiazepine anxiolytic and sedative, also used as an antiemetic in chemotherapy. *TNs* Ativan, Temesta.

lorglumide A specific antagonist for the cholecystokinin 1 (CCK1) receptor. Has been reported to inhibit experimentally induced pancreatic carcinoma in rats.

loricrin A major *keratinocyte protein that becomes localized to the cell periphery (cell envelope) of fully differentiated stratum corneum cells. Mutations in the gene for loricrin are responsible for some forms of *Vohwinkel's syndrome and progressive symmetric erythrokeratoderma.

losartan An *angiotensin II antagonist used to treat hypertension. *TN* Cozaar.

loss-of-function mutation A mutation that disrupts normal function in a protein or system. Since there are more ways of damaging function than adding functionality or restoring normality, such mutations are more common than *gain-of-function mutations.

loss of heterozygosity (LOH) Describes the situation in which the normal allele is lost at a locus which is heterozygous for a normal and a mutant allele: the converse, which leaves only the normal allele, is unlikely to be deleterious. It is a frequent cause of tumour progression when the affected gene is a tumour suppressor.

Lou Gehrig's disease *See* AMYOTROPHIC LATERAL SCLEROSIS.

Louis Bar syndrome *See* ATAXIA TELANGIECTASIA.

lovastatin (mevinolin) A *statin (6α-methylcompactin) isolated from cultures of *Aspergillus terreus.*

low affinity platelet factor IV *See* CONNECTIVE TISSUE ACTIVATING PEPTIDE III.

low density lipoprotein (LDL) *See* LIPOPROTEINS.

low density lipoprotein receptor *See* LDL RECEPTOR.

lowest observed adverse effect level The lowest concentration or amount of a substance that causes an adverse effect on morphology, functional capacity, growth, development, or lifespan of a target organism (LOAEL). Changes in testing methods can potentially change the value.

lowest observed effect level The lowest concentration or dose of a substance, that has an observable effect under defined conditions (LOEL). The level depends upon the sensitivity of the observational technique.

Lowry assay At one time the most commonly used assay to quantify protein content, based on the interaction of *Folin–Ciocalteau reagent with tyrosine or phenylalanine. Those proteins that have less than the average content of these amino acids will be underestimated and for proteins like collagen the *Bradford assay gives a more accurate result.

lox-Cre system A site-specific recombination system derived from the *bacteriophage P1, used in transgenic animals to produce conditional mutants. If lox sites are engineered at the beginning and end of a transgene, the intervening DNA is spliced out if active *cre recombinase is expressed in the animal or in the tissue.

LPA *See* LYSOPHOSPHATIDIC ACID.

LPL *See* LIPOPROTEIN LIPASE.

LPS *See* LIPOPOLYSACCHARIDE.

LRP *See* LDL RECEPTOR (LDL receptor related proteins).

LRR *See* LEUCINE-RICH REPEAT.

Lsc *See* DBL.

LSD *See* LYSERGIC ACID DIETHYLAMIDE.

LSM *See* LIKE SM PROTEINS.

LST8 A *WD-repeat scaffold protein (326 aa) that is part of the *raptor-mTOR complex (*TORC) and resembles G protein β subunits.

LTA *See* LIPOTEICHOIC ACID.

LTA_4, LTB_4, LTC_4, etc. *See* LEUKOTRIENES.

LTP *See* LONG-TERM POTENTIATION.

L-type channels A class of *voltage-sensitive calcium channels, activated at membrane potentials more positive than −30 mV, found in neurons, neuroendocrine cells, smooth, cardiac, and striated muscle. Their slow inactivation and possible role in *long-term potentiation is the basis for the designation.They are inhibited by dihydropyridines, benzodiazepines, and phenylalkylamines but are insensitive to ω-*conotoxin. Mutation can cause *hypokalaemic periodic paralysis.

luciferase The enzyme (EC 1.13.12.7) that is responsible for light production in the firefly and used experimentally in a *chemiluminescence bioassay for ATP which reacts with *luciferin to produce light. It is also used as a reporter gene.

luciferins Small molecule substrates for *luciferase which catalyses the oxidation of luciferin to oxyluciferin with the concomitant emission of light (bioluminescence).

lucifer yellow A fluorescent marker molecule (similar to fluorescein) that will not cross cell membranes but is used by *microinjection in developmental biology and neuroscience to study the outline of cells, in *cell lineage studies, or as an indicator of *dye coupling between cells. In practice it is often compounded with carbohydrazide so that it will covalently attach to proteins.

lucigenin A bystander substrate (bis-*N*-methylacridinium nitrate) used in assays for the *metabolic burst of leucocytes. When oxidized by *superoxide it emits light at 483 nm (*chemiluminescence).

Ludwig's angina A serious, potentially life-threatening infection of the tissues of the floor of the mouth leading to swelling which may cause strangulation. Caused by *Actinomyces israelii* or related species. There is no connection with *angina pectoris.

lumefantrine A synthetic antimalarial drug (*TN* Benflumetol) frequently used in combination with artemether, a drug related to *artemisinin. The combined therapy is advantageous because artemether is rapid-acting but short-lived and lumefantrine is slow-acting but with a long half-life in patients. The combination is distributed as Riamet™.

lumen A cavity or space.

lumican The core protein (338 aa) of corneal *keratan sulphate, also found in the arterial wall and many other tissues.

luminol A bystander substrate (*o*-aminophthaloyl hydrazide) used, like *lucigenin, in chemiluminescence assays for the *metabolic burst of leucocytes. When oxidized by the myeloperoxidase/hydrogen peroxide system, it emits light at 425 nm. Used by forensic scientists to detect traces of blood.

luminometer A laboratory instrument used to measure light emission, e.g. the chemiluminescence generated by the oxidation of *luminol or *lucigenin by activated leucocytes.

lumirhodopsin The conformationally changed form of *rhodopsin produced when retinal isomerizes having absorbed light.

lupus erythematosus A dermatological condition (discoid lupus erythematosus, DLE) in which there are red scaly patches, especially over the nose and cheeks (localized DLE), although it may be more widespread. Can be a symptom of *systemic lupus erythematosus.

lupus vulgaris Cutaneous lesions with a nodular appearance, particularly of the face, caused by infection with *Mycobacterium tuberculosis*.

lusitropic Describing something that relaxes the heart. *Compare* INOTROPIC.

lutein An orange-red carotenoid pigment found in many plants and animals and an important antioxidant. It is found in the macula where it may have light-filtering and anti-oxidant functions and is claimed (not wholly convincingly) to be a useful dietary supplement for reducing the risk of macular degeneration.

luteinizing hormone (LH) One of the *gonadotropin hormones composed of an α-chain (92 aa) identical to that of other gonadotropins, and a hormone-specific β-chain (141 aa). It stimulates sex hormone release in conjunction with *follicle-stimulating hormone (FSH). **Luteinizing hormone releasing factor** (LHRF, gonadoliberin-1) is a decapeptide cleaved from progonadoliberin-1 (92 aa) that stimulates LH and FSH release.

luteinoma (luteoma) A benign ovarian tumour composed of cells resembling those of the corpus luteum, usually an asymptomatic tumour-like ovarian lesion of pregnancy that may secrete androgens and cause virilization in ~25% of cases.

luteotrophic hormone *See* PROLACTIN.

lutropin (lutropin alpha) A drug, used in fertility treatment, that works like human pituitary luteinizing hormone, (*prolactin). *TN* Luveris.

luxury protein A protein that is produced only in specialized cells and is not necessary for general cell maintenance, unlike the so-called *housekeeping proteins.

LXR (liver X receptor) Ligand-activated transcription factors (LXRα and LXRβ), belonging to nuclear receptor family-1, that induce genes involved in cholesterol homeostasis and in lipogenesis, through their interaction with specific, naturally occurring oxysterols. They are expressed in liver, intestine, and adrenal gland, and can complex with retinoid-X receptor (RXR). LXRA specifically interacts with RXRA to form a functional heterodimer in which RXRA is the ligand-binding subunit and transcription is activated by retinoids. In experimental animals elimination of LXR activity mimics many aspects of *Tangier disease.

lyases Enzymes of the EC class 4 (*see* E CLASSIFICATION) that catalyse the nonhydrolytic removal of a group from a substrate with the resulting formation of a double bond; or the reverse reaction, in which case the enzyme is acting as a synthetase. The lyases include decarboxylases, aldolases, and dehydratases.

Lyb antigen Obsolete name for murine B-cell surface antigens, superseded by *CD antigen nomenclature.

lycopene A *carotenoid (536 Da) that is the major red pigment in some fruit and also an antioxidant.

Lyme disease A disease transmitted by ticks and caused by **Borrelia burgdorferi*.

lymph The fluid derived from the transendothelial ultrafiltrate from blood into tissues that is carried through the lymphatic system and returned to the blood through the thoracic duct, along with recirculating lymphocytes (*see* LYMPH NODES). **Lymphadenitis**

is inflammation of *lymph nodes and **lymphadenopathy** is a disorder of lymph nodes. Lymphadenopathy-associated virus (LAV) was the name originally given to *HIV by the Pasteur Institute group.

lymphangioma A benign hamartomatous hyperplasia of lymphatic vessels, with most cases occurring in the head and neck region. The lesion consists of extensive lymphatic channels poorly connected to the lymphatic system, embedded in loose connective tissue. In lymphangiomatosis, the lymphatic counterpart to angiomatosis of blood vessels, there are multiple lesions which may be life-threatening.

lymph nodes (lymph glands) Small organs that act as drainage points for tissue fluids and comprise a loose meshwork of reticular tissue with embedded lymphocytes, macrophages, and accessory *antigen presenting cells. Recirculating lymphocytes leave the blood through the specialized *high endothelial venules of the lymph node and pass through the node, where they may have antigens derived from tissue fluids presented to them by the accessory cells, before being returned to the blood through the lymphatic system.

lymphoblast (blast cell) The differentiated product of a B or T cell that develops in response to an antigenic stimulus. The quiescent lymphocyte enlarges and re-enters the cell cycle. Note that this usage of '-blast' is atypical: blasts are usually precursors of more differentiated cells (e.g. erythroblasts).

lymphocyte activation The altered shape and behaviour of lymphocytes which have been exposed to a mitogen or an antigen to which they have been primed. The activation leads to the production of *lymphoblasts which divide and form a population of effector cells. Sometimes referred to as lymphocyte transformation, but the transformation is unlike that caused by oncogenic viruses (*see* TRANSFORMATION) and activation is a less confusing term.

lymphocyte activation gene-1 protein A chemokine of the CC family (LAG-1, CCL4L1, MIP-1β, 92 aa) that binds to both CCR5 and CCR8.

lymphocyte function-related antigen *See* LFA-1; LFA-3.

lymphocytes (lymphoid cells) White blood cells of the lymphoid series. There are two main classes, T and B lymphocytes (*T cells and *B cells); the former subdivided into subsets (*helper, *suppressor, *cytotoxic T cells) and variously involved in cell-mediated immunity and stimulating B lymphocytes. B cells, when activated, secrete antibody.

lymphocyte transformation *See* LYMPHOCYTE ACTIVATION.

lymphocytic leukaemia *See* LEUKAEMIA.

lymphocytopenia A condition in which the number of lymphocytes in the blood is abnormally low (normal levels are $1\text{–}5 \times 10^9$/L).

lymphocytosis An abnormally high blood lymphocyte count.

lymphogranuloma venereum A sexually transmitted disease caused by some serovars of **Chlamydia trachomatis*.

lymphoid cell *See* LYMPHOCYTES.

lymphoid tissue General term for tissue that is rich in lymphocytes and accessory cells such as macrophages and reticular cells. The primary lymphoid tissues are the *thymus and bone marrow, where lymphocytes differentiate; the secondary (misnamed peripheral) lymphoid tissue is where lymphocytes encounter antigens and includes the *lymph nodes, spleen, *Peyer's patches, pharyngeal tonsils, adenoids, various tissue-specific lymphoid tissues and (in birds) the *bursa of Fabricius. *See also* BALT; GALT; MALT; SALT (bronchus, gut, mucosal, and skin-associated lymphoid tissues).

lymphokine A term, probably now almost obsolete, for a subset of the cytokines, those produced by leucocytes. Examples are *interferons, *interleukins, *lymphotoxin, granulocyte-monocyte colony-stimulating factor (*GM-CSF).

lymphoma A malignant tumour of lymphoreticular tissue in which there is a distinct tumour mass, as opposed to *leukaemia where the cells are in the circulation. Includes tumours derived both from the lymphoid lineage and from mononuclear phagocytes; lymphomas usually arise in lymph nodes, spleen, or other areas rich in lymphoid tissue. Lymphomas are subclassified as *Hodgkin's disease, and non-Hodgkin's lymphomas (e.g. *Burkitt's lymphoma, *large-cell lymphoma, *mantle cell lymphoma).

lymphosarcoma Synonym for a *lymphoma.

lymphotoxin A cytotoxic product of T cells; normally refers to *tumour necrosis factor β.

lymphotropic Having an affinity for lymphocytes. *See* HIV.

lyn A nonreceptor tyrosine *kinase of the *src-family that is important in B-cell signalling.

lyngbyastatin *See* DOLASTATINS.

Lyon hypothesis The hypothesis, first advanced by Mary Lyon (b.1925), that there is random inactivation of one of the two X chromosomes in cells of females. As a result, women are chimaeric for the products of the X chromosomes, which has been used as a means of demonstrating the monoclonal origin of papillomas and atherosclerotic plaques, using heterozygotes for isozymes of glucose-6-phosphate dehydrogenase.

lyophilization Freeze drying, the removal of water by sublimation under vacuum.

lysergic acid diethylamide (lysergide, LSD) A semisynthetic psychedelic (hallucinogenic) drug related to *ergot. Affects a large range of G-protein-coupled receptors, including dopamine receptors, adrenoreceptors and in particular the serotonin 5-HT2A receptor.

lysine (Lys, K) An amino acid (146 Da); the only one in proteins that has a primary amino side group; important both structurally and chemically.

lysis The process in which the cell membrane becomes permeable and the cell is disrupted.

lysogenic bacteriophage *See* LYSOGENY.

lysogeny The situation in which a lysogenic bacteriophage integrates its DNA into the host *chromosome as a prophage. The phage remains almost dormant because a virus encoded *repressor protein suppresses the lytic activity of the phage. Environmental stressors, e.g. ultraviolet irradiation, may inhibit repressor production and the phage then develops and the bacterium eventually lyses (the lytic cycle). The best studied example is the *lambda bacteriophage.

lysophosphatides Monoacyl derivatives of diacyl phospholipids produced during cyclic deacylation and reacylation of membrane phospholipids. At high concentrations can be membranolytic, and at slightly sublytic levels may promote cell fusion. *See* LYSOPHOSPHATIDIC ACID.

lysophosphatidic acid (LPA, lysoPA) A phospholipid derivative with a single acyl chain, produced by a lysophospholipase D (autotaxin), that has pleiotropic signalling activity causing mitogenesis, cell survival, inhibition of adenylyl cyclase, and calcium mobilization. It will also induce platelet aggregation, smooth muscle contraction, inhibition of neuroblastoma cell differentiation, chemotaxis, and tumour cell invasion. The **lysophosphatidic acid receptor-1** (LPAR 1, endothelial differentiation gene 2, EDG2; 364 aa), is a G-protein-coupled receptor through which most of its activities are mediated; the receptor also binds sphingosine 1-phosphate. LPAR1 is also reported to be expressed in the nucleus and to traffic between plasma membrane and nucleus in response to LPA. There are related receptors, LPAR2 (EDG4) and LPAR3 (EDG7). LPAR4 is a member of the purinergic GPCR cluster.

lysophospholipase D An enzyme (EC 3.1.4.39, 863 aa) found in plasma and serum, that hydrolyses lysoPE and lysoPC with different fatty acyl groups to the corresponding lysophosphoric acids. *See* AUTOTAXIN and LYSOPHOSPHATIDIC ACID.

lysosomal diseases *See* STORAGE DISEASES.

lysosomal enzymes Generic term for a range of degradative enzymes found in the lysosome. Most have acid pH optima. Standard lysosomal marker enzymes are *acid phosphatase and *β-glucuronidase.

lysosomal trafficking regulator A protein (2186 aa) that is involved in sorting endosomal proteins. Mutations in the gene cause *Chediak–Higashi syndrome.

lysosome A membrane-bounded cytoplasmic organelle containing hydrolytic enzymes that can fuse with a phagosome to form a secondary lysosome in which the contents are digested. Release of lysosomal enzymes into the cytoplasm causes autolysis and the tenderizing of hung meat. Lysosomes are part of the GERL complex or *trans*-Golgi network.

lysosome associated membrane glycoproteins *See* LAMPs.

lysosome–phagosome fusion The process in which primary *lysosomes fuse with a phagosome to form a secondary lysosome or phagosome. By interfering with this process, some intracellular parasites (e.g. *Mycobacterium microti*) avoid damage.

lysosome related organelles A family of cell-type-specific organelles (LROs) that include melanosomes, platelet dense bodies, and cytotoxic T-cell granules. These are the organelles that are affected in *Hermansky-Pudlak syndrome.

lysosomotropic Substances that have an affinity for *lysosomes, e.g. NH_4Cl, chloroquine, and methylamine.

lysozyme (muramidase) An important defensive enzyme (EC 3.2.1.17, 148 aa) that hydrolyses the bond between *N*-acetylmuramic acid and *N*-acetylglucosamine, in Gram-positive bacterial cell walls. Found in tears and saliva, and in the *lysosomes of phagocytic cells.

lysyl oxidase An extracellular enzyme (EC 1.4.3.13, 417 aa) that deaminates lysine and hydroxylysine residues in collagen or elastin to form aldehydes, which interact with each other or with other lysyl side chains to form crosslinks. It is a copper-containing metalloenzyme and is inhibited by copper chelators in some forms of *lathyrism.

Lyt antigen An almost obsolete name, superseded by the *CD antigen nomenclature, for surface glycoproteins of mouse T cells.

lytic A substance or process that causes lysis. The lytic complex (2000 kDa) formed through *complement activation is a complex of C5b6789.

lytic infection The normal cycle of viral infection, in which mature virus or phage particles are produced and the infected cell lyses, releasing virus particles.

LZTR-1 A transcriptional regulator of the *BTB/*kelch family that is deleted in most patients with *DiGeorge's syndrome.

Mab Monoclonal antibody. *See* MONOCLONAL.

MAC Membrane attack complex (C9). *See* COMPLEMENT.

Mac-1 (CR3, CD11b/CD18) *See* LFA-1.

MacConkey's agar An agar-based medium for bacterial culture containing lactose and neutral red as an indicator. Lactose-fermenting bacteria will produce red-pink colonies.

macedocin An anticlostridial bacteriocin *lantibiotic (22 aa) produced by *Streptococcus macedonicus.*

Machado–Joseph disease *See* SPINOCEREBELLAR ATAXIA.

Machupo virus The causative agent of Bolivian haemorrhagic fever, one of the *Arenaviridae. The main reservoir is the rodent *Calomys callosus*, and transmission, though rare, is probably through aerosol droplets of urine.

- More detailed information on Machupo virus.

macrocytes Abnormally large erythrocytes, characteristic of *pernicious anaemia. Hypochromic macrocytes are large cells with a reduced haemoglobin content.

macroglia A general term for glial cells other than *microglia, namely *oligodendrocytes and *astrocytes.

macroglobulin Nonspecific term for any large plasma globulin. **α_2-Macroglobulin** (α_2M) is a large (725 kDa, 1474 aa) plasma antipeptidase that will inhibit a wide range of peptidases. The inhibitory mechanism involves trapping of the peptidase within a 'cage' that closes when the peptidase-sensitive 'bait' region is cleaved. The peptidase is still active against small-molecule substrates within the cage but the conformational change in the α_2M causes it to be rapidly cleared from circulation. *Plasminogen activator is one of the few peptidases not inhibited by α_2M. In **macroglobulinaemia** there are abnormally high levels of IgM in the blood, usually a result of B-cell overproliferation.

macrolide antibiotics Broad-spectrum bacteriostatic antibiotics that inhibit bacterial protein synthesis by blocking the 50S ribosomal subunit. They are active against most aerobic and anaerobic Gram-positive cocci and Gram-negative anaerobes. Examples are erythromycin, lincomycin, azithromycin, and clindamycin.

macrophage activation syndrome A severe, potentially lethal condition in which there is excessive activation of well-differentiated macrophages. It is associated with *Still's disease and considered by some to be autoimmune disease-associated reactive *haemophagocytic lymphohistiocytosis (ReHLH), a subset of haemophagocytic lymphohistiocytosis. *See* MAF (macrophage activating factor).

macrophage colony-stimulating factor *See* MCSF.

macrophage inflammatory proteins (MIPs) Murine cytokines with closely related human homologues. **MIP1** was originally identified as a murine cytokine with inflammatory and chemokinetic properties. Two peptides (69 aa) are responsible, MIP1-α (CCL3; SIS-α, L2G25B, SCYA3, TY5, endogenous pyrogen; identical to human 464.1) and MIP1-β (CCL4; SIS-γ; L2G25B; identical to human 744.1). Subsequently other related forms have been isolated: MIP1-γ (CCL9) and MIP1-δ (CCL15). The latter is chemotactic for T cells and monocytes, but not for neutrophils, eosinophils, or B cells. **MIP2** (CXCL1, 6 kDa) is *chemotactic for *neutrophils. **MIP3-α** (CCL20; LARC, Exodus) and **MIP3-β** (CCL19) are CC-chemokines.

macrophage inhibition factor A substance (MIF, macrophage migration inhibition factor) originally claimed to be a lymphocyte product that inhibited macrophage emigration from capillary tubes, although this may have been a result of adhesive trapping. More recently it has been suggested that MIF is a protein secreted by anterior *pituitary cells in response to LPS stimulation.

macrophage stimulating protein A pleiotropic growth factor (hepatocyte growth factor-like protein, HGFL, 711 aa) with ~50% sequence homology to hepatocyte growth factor and identical domain structure. It is synthesized and secreted from hepatocytes as an inactive precursor and is activated at the cell surface. The receptor is *ron receptor tyrosine kinase.

macrophages Long-lived phagocytes, derived from blood *monocytes, found in tissues. Their properties vary according to the tissue in which they are resident. The main types are peritoneal and alveolar macrophages, tissue macrophages (*histiocytes), *Kupffer cells of the liver, and *osteoclasts. Activation may occur in response to foreign antigens, and macrophages play an important role in killing of some bacteria, protozoa, and tumour cells. They release various *cytokines and are involved in *antigen presentation. Within chronic inflammatory lesions they may further differentiate to epithelioid cells or may fuse to form *foreign body giant cells or *Langhans giant cells.

macrosialin A transmembrane glycoprotein (CD68, 110 kDa) that is highly expressed by human monocytes and tissue macrophages, a *scavenger receptor that binds oxidized *LDL. Belongs to the mucin-like class that includes *LAMP-1 and -2, *leukosialin, and haematopoietic progenitor cell antigen (CD34).

MACS **1.** Membrane-anchored C peptides derived from the C-terminal heptad repeat domain of HIV-1 gp41. **2.** Medium-chain acyl-coenzyme A synthetases that catalyse the ligation of medium chain fatty acids with CoA. **3.** Magnetic activated cell separation or magnetic assisted cell sorting. **4.** Membrane-associated adenylyl cyclases (mACs). **5.** Mammalian artificial chromosomes (MACs).

macula (*pl.* maculae) An anatomically distinct area. A **macula adherens** is a spot *desmosome. The **macula densa** is an area of closely packed specialized cells in the epithelium of the distal convoluted tubule. The cells are responsive to ion concentration in the luminal fluid and, by releasing prostaglandins, regulate the secretion of *renin. The **macula lutea** is the central region of the retina surrounding the *fovea. A **macule** is a spot on the skin. **Macular degeneration** is an ophthalmological disorder affecting the macula lutea (*see* AGE-RELATED MACULAR DEGENERATION).

macular dystrophy A group of disorders in which there is patchy (macular) degeneration of retinal cells or other parts of the visual system. **Juvenile-onset vitelliform macular dystrophy** (Best's disease) is caused by mutation in the *bestrophin gene. The adult form of **vitelliform macular dystrophy** is considered distinct and is usually due to mutation in the *peripherin 2 gene, which is responsible for some forms of *retinitis pigmentosa, although some cases have mutations in bestrophin. Type I and type II **macular corneal dystrophy** (MCD) are due to mutations in the gene for a specific sulphotransferase gene involved in extracellular matrix biosynthesis. *See also* STARGARDT DISEASE.

Madin–Darby canine kidney cells A line of canine cells (MDCK cells) often used as a general model for epithelial cells because they grow readily in culture and form confluent monolayers with relatively low trans-monolayer permeability, although clones vary.

madindolines Noncytotoxic indole alkaloids originally isolated from a fermentation broth of *Streptomyces nitrosporeus* K93-0711. Madindoline A (MadA; MDL-A) binds competitively but noncovalently to the extracellular domain of the membrane glycoprotein gp130 and inhibits the homodimerization of the trimeric IL-6/IL-6R/gp130 or the IL-11/gp130 complex, thereby blocking activation of the JAK–STAT signal transduction pathway. Analogues of madindolines have been synthesized as potent IL-6 inhibitors.

MAD-related proteins A family of proteins (mothers against decapentaplegic-related proteins, Sma- and MAD-related proteins, Smads) coded by genes that are homologous to the *Drosophila* gene *mothers against decapentaplegic* and the *Caenorhabditis elegans* gene *Sma*. The MAD-related proteins (MADRs) play critical roles in signal transduction pathways involving the TGFβ superfamily. At least eight have been identified (MADR1–7 and MADR9). **Smad1** (MADR1, 465 aa) and **Smad5** are activated (serine/threonine phosphorylated) by *BMP receptors, **Smad2 and 3** by *activin and TGFβ receptors. Smad2 (MADR2, 465 aa) is important for the establishment of anterior-posterior patterning. **Smad4** is important in gastrulation. Smads activated by occupied receptors form complexes with Smad4/DPC4 and move into the nucleus where they regulate gene expression and interact with the *forkhead activated signal transducer. *See* JUVENILE POLYPOSIS.

Madurella A genus of fungi responsible for chronic inflammatory infection of the foot (Madura foot) although this tropical disease can also be caused by bacteria such as *Nocardia* and *Streptomyces*.

MAF **1.** An imprecise term for a *cytokine that activates macrophages (macrophage activating factor). Main example is *interferon-γ although many other cytokines have this activity. **2.** An oncogene identified in an avian musculoaponeurotic fibrosarcoma (v-maf) that encodes an activator subunit involved in binding to the nuclear factor erythroid-2 (NFE2) site. *See* C-MAF; P18.

MAG An adhesion molecule (myelin-associated glycoprotein) of the *immunoglobulin superfamily involved in regulating neurite outgrowth. *See* NOGO; OLIGODENDROCYTE-MYELIN GLYCOPROTEIN.

magainins A family of small peptide antibiotics (~23 aa) that form anion-permeable pores across membranes of many bacteria, fungi, and some tumour cells, although they are not haemolytic. Originally isolated from *Xenopus* skin. The sequence is similar to that of *melittin.

MAGE antigens A superfamily of proteins (melanoma associated antigens) associated with tumours and part of the larger 'cancer/testis antigen' family. The MAGE superfamily includes multiple subfamilies (MAGEs A–E, and *necdin). Six *MAGE* genes are expressed at a high level in a number of histologically diverse tumours. The MAGE domain (~200 aa) has been identified in 32 genes on the X chromosome and in 4 autosomal genes and may mediate interaction with *neurotrophin. They are potential targets for immunotherapy.

SEE WEB LINKS

- More information in the Panther database.

MAGI proteins A family of *MAGUK proteins (membrane-associated guanylate kinase inverted proteins) with an inverted arrangement of protein–protein interaction domains. **MAGI-1** is present at adherens and tight junctions, where it acts as a structural and signalling scaffold and associates with *megalin and *synaptopodin. **MAGI-2** (*atrophin-1-interacting protein-1, AIP1) coimmunoprecipitates with *stargazin and interacts with *PTEN. Mutations may be responsible for infantile spasms in some *Williams–Beuren syndrome patients. **MAGI-3** has been found in mouse and interacts with the TGFα cytoplasmic region.

magnesium (Mg) An essential divalent cation, the chelated ion in ATP, the central ion in the chlorophyll molecule, and a component of bone. The only biologically active form of ATP is the Mg^{2+} complex. The serum concentration of magnesium is 1.8–2.5 mEq/L, approximately one-third of which is protein-bound. Cytoplasmic levels are ~5 mM, although only ~0.4 mM is unbound.

magnetic resonance imaging *See* MRI.

magnetoencephalography Measurement of the weak magnetic signals generated by electrical activity in the brain.

magnetotaxis A directional response to magnetic field. Magnetotactic bacteria apparently use the Earth's magnetic field as a guide to 'up' and 'down' in deep sediment.

magnocellular neuron Large, electrically excitable neurosecretory cells found in the supraoptic nucleus and paraventricular nucleus of the hypothalamus. Some secrete *oxytocin, others *vasopressin, and a few secrete both.

MAGUKs A large family of proteins (membrane-associated guanylate kinases) involved in the assembly of multiprotein complexes on the inner surface of the plasma membrane at regions of cell-cell contact. All have one or three PDZ domains, a src homology 3 (SH3) domain, and a C-terminal guanylate kinase (GuK) domain and are subdivided into four subfamilies, *DLG-like, *ZO-1-like, *p55-like, and LIN2-like, based on their size and the presence of additional domains. Caspase recruitment domain-containing proteins are mostly members (*see* CARD DOMAIN; CASK). *See also* MAGI.

maitotoxin A toxin that activates L-type *voltage-sensitive calcium channels and mobilizes intracellular calcium stores. Isolated from the dinoflagellate *Gambierdiscus toxicus*, which is responsible for 'red tides'.

Majeed's syndrome A disorder in which there is chronic recurrent multifocal osteomyelitis and congenital dyserythropoietic anaemia caused by mutation in the *lipin 2 gene.

major facilitator superfamily domain-containing proteins A large family of transporters (MFSDs) with ten to twelve membrane-spanning domains found in bacteria, archaea, and eukaryotes that act as permeases for simple sugars, oligosaccharides, inositols, drugs, amino acids, nucleosides, organophosphate esters, tricarboxylic acid cycle metabolites, and various organic and inorganic anions and cations. Many families (at least eighteen) have been recognized. In humans they participate in various important physiological processes. MFSD2 is the receptor for *syncytin-2; mutation in MFSD8 causes neuronal ceroid lipofuscinosis type 7.

- A review of MFSDs.

major histocompatibility complex *See* MHC.

major intrinsic proteins (MIPs) A family of structurally related proteins with six transmembrane segments, associated with gap junctions or vacuoles. Two distinct types of channels are formed by MIPs, those specific for water transport (*aquaporins) and those that transport small neutral solutes such as glycerol. The major intrinsic protein of the ocular lens fibre membrane (MIP26; aquaporin 0; 26 kDa) appears during differentiation of the ocular lens; mutations cause inherited forms of cataract.

major urinary protein A murine pheromone-carrying protein (MUP-1) of the *lipocalin family.

majusculamide C A microfilament-depolymerizing agent from the cyanobacterium *Lyngbya majuscula*; an analogue of *dolastatin.

malabsorption syndrome A general term for disorders in which there is impaired absorption in the small intestine. Can arise for a variety of reasons including lymphoma, amyloid, and other infiltrations, *Crohn's disease, gluten-sensitive enteropathy; and the sprue syndrome in which the villi atrophy for unknown reasons.

m

malacia A general term for the pathological softening of an organ or tissue.

malaria *See* PLASMODIUM.

Malassez cells Epithelial cells that derive from the root sheath and form the 'epithelial rests of Malassez' of the periodontal ligament.

malcaverin A protein involved in the pathway of integrin signalling regulating vascular morphogenesis. It interacts with *KRIT1, the *CCM3* gene product (PCD10) and the cytoplasmic domain of integrins. Mutation in malcaverin is associated with cerebral cavernous malformation type 2 (CCM2).

MALDI-TOF Abbreviation for an important method for analysing biological samples. Molecular ions are generated by 'matrix assisted laser desorption ionization' (MALDI) and analysed using 'time of flight' (TOF) mass spectroscopy.

maleate The ion from maleic acid (*cis*-butenedioic acid), a weak acid used in buffers.

malignant In pathology, describing tumours that have the capacity to show *metastatic spread (metastasize) from the primary tumour. Not only have the cells lost *growth control but they no longer respect positional controls. The term can also be applied more generally to any disease that can become life-threatening if untreated.

malignant hyperthermia A *channelopathy in which calcium channels are defective, increasing susceptibility to severe and fatal hypothermia under anaesthesia. Malignant hyperthermia type 1 (MHS1) is caused by mutation in the *ryanodine receptor gene. Several other loci for MHS have been identified: MHS2 on chromosome 17q, MHS3 on 7q, MHS4 on 3q, and MHS6 on 5p. MHS5 is caused by mutation in the *CACNA1S* gene that encodes a subunit of the voltage-dependent *L-type calcium channel.

malonate The ion from malonic acid (propanedioic acid). Malonate is a competitive inhibitor for succinate dehydrogenase in the *tricarboxylic acid cycle and malonyl-SCoA is an important precursor for fatty acid synthesis.

MALT *See* MUCOSAL ASSOCIATED LYMPHOID TISSUE.

maltase The lysosomal enzyme (α-glucosidase, EC 3.2.1.20, 1857 aa), also secreted by the cells of intestinal villi, that hydrolyses maltose to glucose during the enzymic breakdown of starch. Mutations lead to glycogen storage disease type II (*Pompe's disease).

MAML proteins A family of three cotranscriptional regulators (mastermind-like proteins) that bind to the ankyrin repeat domain of *Notch and are essential for signalling. They have distinct tissue-specific distributions; **MAML1** (1016 aa) localizes to nuclear bodies and interacts with *MEF2C. **MAML2** (1173 aa) is widely expressed. A *MAML2/*MECT1* fusion gene arising from a chromosomal translocation characterizes mucoepidermoid carcinoma (a common malignant salivary gland tumour). **MAML3** (1133 aa) acts selectively on different notch promoters.

mammalian expression vector A *vector used to engineer the production of large amounts of eukaryotic protein (not necessarily mammalian protein).

mammalian-wide interspersed repeat The second most common interspersed repeat in primates (~300 000 copies, 1–2% of the total DNA). They are 260-bp tRNA-derived *short interspersed nucleotide elements (SINEs). *See* ALU; LONG INTERSPERSED NUCLEOTIDE ELEMENT.

mammary-derived growth inhibitor A *fatty acid binding protein (MDGI, FABP3, 133 aa) of the heart/muscle type that inhibits proliferation of mammary carcinoma cells and may be a tumour suppressor.

mammary gland The milk-producing gland of female mammals. An adapted sweat gland made up of milk-producing alveolar cells, surrounded by contractile myoepithelial cells and many fat cells. Secretion of milk fat is unusual in being *apocrine, the fat globule being budded from the apex of the cell and remaining enclosed in plasma membrane. Milk production is hormonally controlled.

mammary tumour virus (Bittner agent) An endogenous *retrovirus isolated from very inbred strains of mice that had a high incidence of transmissible mammary tumours. The provirus is present in the germ line and transcription is regulated by a viral promoter that responds to glucocorticoid hormones. May transform cells by activating the cellular **int-1* oncogene.

mandibuloacral dysplasia A progeria-like disorder associated with partial *lipodystrophy. Type A can be caused by mutation in the gene encoding lamin A/C, which is also involved in familial partial lipodystrophy of the Dunnigan type (FPLD2), and type B by mutation in the zinc metallopeptidase *ZMPSTE24* gene. *See* HUTCHINSON–GILFORD PROGERIA SYNDROME.

manganese (Mn) An essential trace element that is found in cells at concentrations of ~0.01 mM. Manganese is the metal in several enzymes, e.g. *pyruvate carboxylase, *superoxide dismutase-2, and can replace magnesium in many others, although this may alter the substrate specificity. Manganese substituted for magnesium in physiological saline makes cells very adhesive.

mannan-binding lectin One of the acute phase proteins (MBL, formerly mannan-binding protein, MBP, collectin-1, 248 aa) secreted by the liver, structurally related to *complement C1. MBL binds *mannans on the surface of various microorganisms including bacteria, yeasts, parasitic protozoa, and viruses, causing opsonization and activating the complement cascade through *MASP. Deficiency is associated with frequent infections in childhood, one polymorphism is associated with susceptibility to gestational diabetes mellitus, and the D variant is significantly more common in preterm infants

mannopeptimycins Cyclic glycopeptide antibiotics produced by *Streptomyces hygroscopicus* LL-AC98 that inhibit cell wall biosynthesis through lipid II binding. Five variants (α, β, γ, δ, and ε) are known, with differing properties: mannopeptimycin-ε is effective in treating infection due to meticillin-resistant staphylococci and vancomycin-resistant enterococci.

mannose (D-mannose) A hexose found in many polysaccharides and glycoproteins, identical to D-glucose except that the orientation of the -H and -OH on C-2 are interchanged. **Mannose 6-phosphate** is found in *N*-glycan chains of lysosomal enzymes and is considered to be a targeting signal. Two distinct and unrelated *mannose-6-phosphate receptors have been identified.

mannose-6-phosphate receptor There are two unrelated receptors, both required for proper sorting of lysosomal enzymes. The cation-dependent mannose-6-phosphate receptor is a transmembrane protein of ~46 kDa and functions as a dimer. The cation-independent receptor (215 kDa) is identical to the receptor for insulin-like growth factor II (IGF2R) and has a role in the intracellular trafficking of lysosomal enzymes, the activation of *transforming growth factor β, and the degradation of IGF2, which is a growth factor that is often overproduced in tumours. It is also apparently the site to which *granzyme B binds in T-cell killing.

mannosidases Enzymes that hydrolyse the glycosidic bond between mannose residues and a variety of hydroxyl-containing groups. α-Mannosidases (EC 3.2.1.113) remove four mannose residues during the synthesis of the complex-type *N*-linked glycan chains of glycoproteins as they pass from rough endoplasmic reticulum into the *cis*-Golgi. β-Mannosidase (EC 3.2.1.25) is a lysosomal enzyme that catalyses the final exoglycosidase step in the degradation pathway for *N*-linked oligosaccharide moieties of glycoproteins.

mannosidosis An autosomal recessive *storage disease. **α-Mannosidosis** is caused by mutation in the gene encoding α-mannosidase, class 2B1 (EC 3.2.1.24), and clinical effects vary from early fatality to later neuronal degeneration; **β-mannosidosis** is caused by a deficiency of lysosomal β-mannosidase (EC 3.2.1.25) activity and most patients show mental retardation.

mantle-cell lymphoma A rare subtype of non-Hodgkin's lymphoma derived from $CD5^+$ antigen-naive B-cells from within the mantle zone that surrounds normal germinal centre follicles.

Mantoux test A skin test for current or previous infection with *Mycobacterium tuberculosis* in which tuberculin PPD (purified protein derivative) is injected intracutaneously. A positive reaction, in an individual not inoculated with BCG, is a swollen and reddened area caused by T-cell reactivity.

MAO inhibitor *See* MONOAMINE OXIDASES.

MAP kinases Serine/threonine kinases (mitogen-activated protein kinases, externally regulated kinases, ERKs, EC 2.7.11.24) that are activated through a kinase cascade when quiescent cells are treated with mitogens, and that therefore potentially transmit the signal for entry into cell cycle. Mitogen activates MAP kinase kinase kinase (MAPKKK), which activates MAP kinase kinase (MAPKK), which phosphorylates and activates MAP kinase. Rather confusingly, MAP kinases also phosphorylate *microtubule associated proteins. Six groups of MAP kinases are recognized: 1. extracellular signal-regulated kinases, ERK1 (MAPK3) and ERK2 (MAPK1), or classical MAP kinases; 2. c-Jun N-terminal kinases (*JNKs; MAPK8, MAPK9, MAPK 10), also known as stress-activated protein kinases (SAPKs); 3. *p38 isoforms (MAPK11, MAPK12/ERK6), MAPK13, MAPK14); 4. ERK5 (MAPK7) which is critical for endothelial function and maintenance of blood vessel integrity; 5. ERK3/4 atypical MAPKs. ERK3 (MAPK6) and ERK4 (MAPK4); 6. ERK7/8. (MAPK15). The complexity of the kinase cascades and their role in multiple systems suggests that this classification is unlikely to remain definitive.

MAPKK MAP kinase kinase. *See* MAP KINASES; MEKs.

MAPKKK MAP kinase kinase kinase. *See* MAP KINASES.

MAPK phosphatase A family of phosphatases (MKPs) that regulate *MAP kinases (MAPKs) by removing phosphate on tyrosine and threonine in the -pTXpY- activation loop (dual specificity phosphatases). Different members of the family have specificities for various MAPKs, are located in different tissues and subcellular sites, and are induced by different extracellular stimuli.

mappine *See* BUFOTENINE.

maprotiline A tetracyclic antidepressant. *TNs* Deprilept, Ludiomil, Psymion.

MAPs *See* MICROTUBULE ASSOCIATED PROTEINS and mitogen activated kinases (*MEKs).

MAPTAM A calcium-sensitive fluorescent indicator which readily enters cells as the ester and is trapped in the de-esterified form (5-methyl *BAPTA).

marasmus Wasting because of an inadequate energy-giving diet, as opposed to protein deficiency (kwashiokor).

Marburg virus A *filovirus that causes Marburg haemorrhagic fever (green monkey disease). It is related to *Ebola virus and comparably dangerous. The reservoir may be Egyptian fruit bats *Rousettus aegyptiacus*.

MARCKS A membrane-associated calmodulin- and actin-binding protein (myristoylated alanine-rich protein kinase C-substrate, 332 aa) implicated in macrophage activation, neurosecretion and growth-factor dependent mitogenesis. When phosphorylated by protein kinase C (PKC) it is displaced from the membrane; dephosphorylation by protein phosphatase-2 allows it to reassociate through the myristoyl residue. It may link mucin granule membranes to the contractile cytoskeleton, mediating the movement of granules to the cell periphery and subsequent exocytosis.

marenostrin *See* PYRIN.

Marfan's syndrome An inherited dominant disorder of connective tissue in which limbs are excessively long and loose-jointed and there are ocular and cardiovascular defects. True Marfan's syndrome is due to mutation in *fibrillin-1, important for collagen fibril assembly, although the skeletal abnormalities are mimicked by trisomy 8 syndrome.

margaratoxin A peptide toxin (39 aa), from the New World scorpion *Centruroides margaritatus*, that blocks mammalian voltage-gated potassium channels (Kv1.3) in neural tissues and lymphocytes. Related to *charybdotoxin and similar to *noxiustoxin and *kaliotoxin.

marginal band A bundle of ~50 equatorially located microtubules that stabilizes the biconvex shape of platelets, avian erythrocytes, and immature nucleated mammalian erythrocytes. They do not derive from the centrosomal *microtubule organizing centre.

margination The phenomenon in which leucocytes adhere to the endothelial lining of blood vessels, particularly of postcapillary venules where the wall shear stress is least. Usually a prelude to leaving the circulation and entering the tissues.

Marie–Strumpell spondylitis *See* ANKYLOSING SPONDYLITIS.

Marinesco–Sjögren syndrome A disorder characterized by cerebellar ataxia, congenital cataracts, and retarded somatic and mental maturation; caused by disruption of the *SIL1* gene that encodes a protein of the endoplasmic reticulum (ER) that is a nucleotide exchange factor for the chaperone (BiP) involved in translocation of secreted proteins into the ER lumen.

marker gene A gene that encodes something that is readily detectable. Typical markers are enzymes that will generate a coloured product (e.g. *β-galactosidase) or proteins such as *green fluorescent protein. Alternatively the marker may confer antibiotic resistance, or encode a membrane protein readily detected with a standard antibody.

marker rescue The restoration of gene function by replacing a defective gene with a normal one by recombination. The method can be used to isolate a gene for a particular enzyme provided a mutant form exists and a selective medium is available (this works well for biosynthetic enzymes). A common application is as a mapping technique to determine if a viral mutation in a DNA virus occurs within a particular region of the viral genome: cells are cotransfected with the mutant virus and a wild-type genomic fragment; recombination repairs the replication defect and the recombinant derivative of the mutant phage can reproduce—it has been rescued.

Maroteaux–Lamy syndrome A lysosomal storage disease, *mucopolysaccharidosis type VI, in which there is a deficiency of the lysosomal enzyme arylsulphatase B; resembles *Hurler's disease in some respects.

MARPS **1.** A conserved family of 'muscle ankyrin repeat proteins' that includes cardiac ankyrin repeat protein (CARP) and its two close homologues ankrd2 (Arpp) and DARP. All three genes are induced by stress/strain injury signals and can associate with the elastic region of *titin/connectin. **2.** Microtubule repetitive proteins MARP-1 and MARP-2 isolated from membrane skeleton of **Trypanosoma brucei*. They are large (~320 kDa) with many tandem 38-residue repeats, but have not appeared in recent literature.

Marshall's syndrome An autosomal dominant disorder with ocular, orofacial, auditory, and skeletal manifestations with several features in common with *Stickler's syndrome. It is caused by defects in *collagen type XI (COL11A1).

***mas* (*mas1*)** An oncogene isolated from the DNA of a human epidermoid carcinoma cell line. The *mas1* gene product is a G-protein-coupled receptor for the peptide angiotensin 1–7, a peptide that has the opposite effects to *angiotensin II. Angiotensin 1–7 binding to mas1 elicits arachidonic acid release and stimulates the proliferation of multipotential and differentiated progenitor cells in cultured bone marrow and human cord blood.

maskin A member of the *TACC (transforming acidic coiled-coil) domain protein family, found in *Xenopus laevis* oocytes and embryos. It is apparently required for microtubule anchoring at the centrosome and has been reported to mediate translational repression of cyclin B1 mRNA, although the 'maskin hypothesis' for transcriptional control is being questioned by some authors (2008).

maspardin A cytosolic protein (33-kDa acidic cluster protein) that colocalizes with CD4 in intracellular endosomal/*trans*-Golgi transportation vesicles and is mutated in autosomal recessive *hereditary spastic paraplegia type 21 (Mast syndrome; SPG21). The name derives from 'Mast syndrome, spastic paraplegia, autosomal recessive, with dementia'.

maspin A serine peptidase inhibitor (*serpin-B5, 375 aa) that is down-regulated, but not mutated, in mammary tumours.

MASPs Peptidases (MBL-associated serine proteases, EC 3.4.21.104, MASP-1, 699 aa; MASP-2; 686 aa) that are probably responsible for the cleavage and activation of *complement C2 and C4 triggered by the binding of mannan to *mannan-binding lectin (MBL). MASPs are similar to the *C1q-associated proteases, C1r and C1s.

mast cells Cells found in connective and mucosal tissue that contain many *histamine and heparan sulphate-rich granules. Mast cells are similar to *basophils and may derive from the same stem cells. Mucosal mast cells are distinct and are T-cell dependent. Histamine released from mast cells mediates the vascular changes that cause reddening in the weal-and-flare response. *See* MASTOCYTOSIS.

mastermind-like proteins *See* MAML PROTEINS.

mastocytoma A benign nodular skin tumour that is infiltrated by *mast cells. Usually resolves spontaneously. *See* MASTOCYTOSIS.

mastocytosis (mast cell disease) A condition in which there is infiltration of tissue by mast cells. In **urticaria pigmentosa**, the most common form of childhood mastocytosis, there are small aggregates of mast cells within characteristic salmon-brown itchy patches of skin. In **telangiectasia perstans**, more common in adults, there are multiple hyperpigmented macules with telangiectases. **Diffuse mastocytosis** (diffuse cutaneous mastocytosis) affects the entire skin; **systemic mastocytosis** involves a wider range of tissues, including lymph nodes, liver, spleen, gastrointestinal tract, bones, and joints. *See* MASTOCYTOMA.

mastoparans Basic tetradecapeptides found in wasp venoms. Like *melittin they can act as *phospholipase A2 activators and some have antimicrobial properties.

maternal antibody Antibody transferred from mother to fetus across the placenta or in colostrum. *See* IgG.

maternal effect gene A gene, usually one required for early embryonic development, whose product is specified by a maternal gene and incorporated in the egg. The maternal genome therefore specifies the embryo's phenotype. *Compare* ZYGOTIC EFFECT GENE.

maternal inheritance Inheritance through the maternal cell line rather than chromosomally. Mitochondrial genes are maternally inherited and various other non-Mendelian forms of inheritance may also appear as maternal inheritance. Not the same as *genomic imprinting.

Matrigel™ A gel-forming extracellular matrix material derived from *EHS cells. The physical and chemical nature of the culture environment can dramatically alter the behaviour of cells.

matrilysin A matrix metallopeptidases (MMP-7, EC 3.4.24.23, 267 aa) involved in tissue remodelling and in the cellular invasion of tissues. Matrilysin-2 is MMP-26.

matrix A general term for a material or substance in which things are embedded or that fills a space (e.g. nuclear matrix). Often used as a shorthand for extracellular matrix, the loose meshwork within which cells are embedded.

matrix attachment region The chromatin sequence (MAR, scaffold/matrix attachment regions) in intergenic DNA that binds directly to proteins of the nuclear matrix. Often flanks the 5′ ends of genes or clusters of genes and may have a sequence motif (MAR/SAR recognition signature sequence).

matrix Gla protein A vitamin K-dependent, (γ-carboxyglutamic acid (Gla)-containing protein (84 aa) that inhibits calcification. It is produced and secreted by vascular smooth muscle cells and chondrocytes and up-regulated by vitamin D. Comparable in function to bone Gla protein (Bgp or *osteocalcin). Mutations in MGP are associated with *Keutel's syndrome.

matrix metallopeptidases A family of enzymes (MMPs, matrix metalloproteinases) that degrade collagen, elastin, and other components of the extracellular matrix. The Merops database lists 22 examples, including collagenase 1 (MMP1), gelatinase A (MMP2), stromelysin 1 (MMP3, EC 3.4.24.17), stromelysin-2 (MMP-10, EC 3.4.24.22), stromelysin-3 (MMP-11, EC 3.4.24.-), *matrilysin (MMP7), neutrophil collagenase (MMP8), and epilysin (MMP28). Inhibitors are potentially beneficial in arthritis and as blockers of metastasis, although this is unproven. *See* TISSUE INHIBITORS OF METALLOPEPTIDASES (TIMPs).

maturation promoting factor (MPF) *See* CYCLINS.

Mauthner neurons Large neurons found in the *mesencephalon of fishes and amphibians. Because they are individually identifiable, they have been extensively used by neurophysiologists.

MAVS A mitochondrial membrane protein (mitochondrial antiviral signalling protein, IPS-1, Cardif, VISA, 542 aa) that plays a part in the signalling cascade involved in virus recognition and interferon production.

max A transcription factor (myc associated factor X) that as a homodimer is a repressor, but as a heterodimer with *myc is an activator of myc protein production. Max dimerization factor (MAD) and max-interacting protein-1 (MXI1) compete with myc for binding to max and the heterodimers are repressors.

Maxam–Gilbert method A DNA sequencing method in which DNA is degraded, in a set of independent, nucleotide-specific reactions, to produce fragments with characteristic sizes that can be resolved on a sequencing gel. Maxam–Gilbert sequencing has advantages, e.g. for oligonucleotides or covalently modified DNA, although *dideoxy sequencing is more commonly used.

maximum tolerable concentration (MTC, LC_0) The highest environmental concentration of a substance that is nonlethal.

maximum tolerable dose (MTD, LD_0) The highest amount of a substance that, when internalized, is nonlethal.

maximum tolerable exposure level (MTEL) The maximum amount (dose) or concentration of a substance that does not cause adverse effects after prolonged exposure.

maximum tolerated dose (MTD) The highest dose of a substance that should cause only limited toxicity when administered chronically for the duration of the test period. Open to considerable interpretative flexibility.

maxiprep A colloquial term for the large-scale (100–500 mL) purification of *plasmid from a bacterial culture. *See also* MEGAPREP; MINIPREP.

May–Hegglin anomaly A disorder which, like the Fechtner and Sebastian syndromes, is caused by mutation in the gene encoding nonmuscle myosin heavy chain-9. There is thrombocytopenia, enlargement of platelets, and characteristic leucocyte inclusions (Dohle bodies).

maytansine A cytotoxic antibiotic originally isolated from the Ethiopian shrub *Maytenus serrata*. It binds to tubulin at the same site as *rhizoxin, blocking microtubule assembly.

mayven An actin-binding protein (kelch-like 2) with six *kelch repeats that is expressed mostly in the central nervous system. May be involved in the dynamic organization of the actin cytoskeleton in brain. Associates with the SH3 domain of the **fyn* oncogene product and mayven is abundantly and diffusely expressed in primary human epithelial breast tumour cells. *See* ACTINFILIN.

M band The central region of the A band of the *sarcomere.

MBC **1.** Metastatic breast cancer or male breast cancer. **2.** Minimum (minimal) bactericidal concentration.

MBL *See* MANNAN-BINDING LECTIN.

MCAF *See* MONOCYTE CHEMOTACTIC AND ACTIVATING FACTOR.

McArdle's disease A glycogen storage-disease (type V) in which the defective enzyme is muscle glycogen phosphorylase (EC 2.4.1.1). Usually begins in young adulthood with exercise intolerance and muscle cramps and is a relatively benign disorder.

mcb **1.** Metaplastic carcinoma of the breast, a rare form of cancer containing a mixture of epithelial and mesenchymal elements in variable combinations. **2.** Multicolour banding (of chromosomes).

M cells Cells of the follicle-associated epithelium of *Peyer's patches that are the principal sites for the sampling of gut luminal antigens. They lack microvilli on their apical surface, but have broader microfolds.

McCune–Albright syndrome *See* GRANINS.

MCF7 cells A line of oestrogen receptor-positive human breast adenocarcinoma cells.

MCH **1.** *See* MELANIN-CONCENTRATING HORMONE. **2.** mean (erythrocyte) haemoglobin content. **3.** Methacholine (MCh). **4.** microcell hybrid.

M channels A set of slowly activating and deactivating voltage-sensitive *potassium channels, responsible for the M current in neurons, encoded by *KV7.2–KV7.5/KCNQ2–KCNQ5* genes. They are inactivated by *acetylcholine and regulate the sensitivity of neurons to synaptic input. Mutations in *KCNQ2* and *KCNQ3*, channels that interact functionally with each other, are associated with benign neonatal epilepsy. The *KCNQ4* gene product is expressed in cochlear hair cells and mutations are associated with deafness. The *KCNQ5* gene product seems to be widely expressed and presumably contributes to variation in the properties of M channels in different neurons. Linopiridine is a selective M-current blocker.

MCHC A standard haematological measure (mean corpuscular haemoglobin concentration); the average haemoglobin content of erythrocytes calculated by dividing the haemoglobin measurement by the haematocrit. The normal range is 32–36 g/dL.

MCM **1.** *See* MINICHROMOSOME MAINTENANCE PROTEINS. **2.** Modified Chee's medium, a rich cell culture medium sometimes used for hepatocytes. **3.** Microglia-conditioned medium. **4.** *See* METHYLMALONYL-COENZYME A MUTASE.

MCP-1 Monocyte chemoattractant protein-1. *See* MONOCYTE CHEMOTACTIC AND ACTIVATING FACTOR.

MCS *See* POLYCLONING SITE.

MCSF A homodimeric glycoprotein (macrophage colony-stimulating factor, CSF-1) that plays an important role in stimulating the growth of colonies of macrophages and granulocytes and the growth, survival, and differentiation of monocytes. Isoforms M-CSF-α (256 aa), M-CSF-β (554 aa), M-CSF-γ (438 aa) and CSF4 arise by alternative splicing. MCSF is important in adipocyte hyperplasia, in osteoclast differentiation, and in monocyte-to-macrophage differentiation in the arterial intima in atherosclerosis. The receptor for MCSF (c-fms) has a cytoplasmic tyrosine kinase domain and is expressed on the pluripotent precursor and mature osteoclasts and macrophages. Mutation in MCSF leads to *osteopetrosis because of osteoclast deficiency.

M current *See* M CHANNELS.

MCV **1.** Mean cell volume or mean corpuscular volume, a standard parameter in haematological analyses. **2.** Measles-containing vaccine. **3.** *See* MOLLUSCUM CONTAGIOSUM VIRUS.

MDCK cells *See* MADIN–DARBY CANINE KIDNEY.

MDGI *See* MAMMARY-DERIVED GROWTH INHIBITOR.

Mdm A gene (*mouse double minute*) encoding Mdm2 (491 aa), a protein that binds to the transcriptional activator domain of the tumour suppressor *p53 and targets it for destruction by the proteosome. Expression of *Mdm* is, however, activated by p53 so there is a feedback loop. Has a similar effect on the *retinoblastoma gene product. A second protein, Mdm4 (490 aa), has similar properties.

MDMA *See* ECSTASY.

Mdr **1.** *See* MULTIDRUG TRANSPORTER. **2.** mdr-TB is multidrug-resistant tuberculosis.

MDSC *See* MYELOID DERIVED SUPPRESSOR CELLS.

Mdx mouse A mouse model for Duchenne *muscular dystrophy. The defect is a deficiency of *dystrophin.

ME *See* MYALGIC ENCEPHALOMYELITIS.

Meacham's syndrome *See* DENYS–DRASH SYNDROME.

mean residence time The average length of time a drug molecule remains in the body or an organ after rapid intravenous injection, an important parameter in pharmacokinetics.

measles virus A morbillivirus of the *Paramyxoviridae family, responsible for childhood measles. A complication, more common in adults, is subacute sclerosing panencephalitis.

mechanoreceptor A sense organ or cell that responds to mechanical stimulation.

meckelin The product (995 aa) of a gene that is mutated in *Meckel's syndrome type 3 and localizes both to the primary cilium and to the plasma membrane in ciliated cells. It interacts with the product of the gene mutated in type 1 Meckel's syndrome, and appears to mediate a fundamental developmental stage of ciliary formation and epithelial morphogenesis. The predicted structure of meckelin has some topological similarity with the Frizzled (FZD) family of receptor proteins.

Meckel's syndrome (Meckel–Gruber syndrome) An autosomal recessive developmental disorder characterized by a combination of renal cysts and developmental anomalies of the central nervous system (typically encephalocele), hepatic ductal dysplasia and cysts, and polydactyly. **Meckel's syndrome type 1** is caused by mutation in a gene encoding a conserved component of the flagellar apparatus basal body proteome (MKS1) that interacts with *meckelin. **Type 2** maps to a region of chromosome 11 where the homeobox gene *PHOX2A* is also located. **Type 3** is caused by mutation in meckelin, **Type 4** is caused by mutation in the *CEP290* gene, the product of which (nephrocystin-6) is localized in centrosomes and cilia and is also associated with Joubert's syndrome-5, Leber congenital amaurosis, and *Senior–Loken syndrome.

meclozine HCl An antihistamine drug commonly used to prevent motion sickness.

MECT1 A protein (mucoepidermoid carcinoma-translocated 1, transducer of regulated cAMP response element-binding protein 1, TORC1, 634 aa) that activates genes with a cre promoter, e.g. genes for IL-8 and BDNF. A fusion protein made up of exon 1 from MECT1 and exons 2–5 of the **MAML2* gene binds CREB and is characteristic of mucoepidermoid carcinoma.

media (tunica media) The avascular middle layer of the artery wall, lying between the intima, which is in contact with blood, and the adventitia (tunica adventitia) which has nerves and vasculature supplying the arterial wall. The media is the thickest of the three layers and is composed of elastic fibres and smooth muscle cells.

median effective concentration (EC_{50}) The concentration of a substance in an environmental medium calculated to produce a certain effect under a defined set of conditions.

median effective dose (ED_{50}) The dose of a chemical substance or a physical agent (e.g. radiation) that is expected to produce a certain effect or to produce a half-maximal effect in a biological system under defined conditions.

median lethal concentration (LC_{50}) The concentration of a substance that would be expected to kill 50% of organisms. Despite its deficiencies, the LC_{50} estimate was a regulatory requirement in drug testing until it was superseded by more sensitive measurements. The **LD_{50}** is the median lethal dose, effectively the same measurement. The median lethal time (TL_{50}) is the average time interval during which 50% of a given population will die following acute administration of a chemical or physical agent. In all cases it is essential to specify the conditions.

mediator complex A multiprotein complex that acts as a coactivator for nearly all RNA polymerase II-dependent genes, conveying information from gene-specific regulatory proteins to the basal RNA polymerase II transcription machinery. The complex is recruited to promoters by direct interactions with regulatory proteins and serves as a scaffold for the assembly of a functional preinitiation complex with RNA polymerase II and the general transcription factors.

medicinal leech The freshwater leech *Hirudo medicinalis*, formerly used for bloodletting. The simplicity of the nervous system and the relatively large size of the animal have made it a convenient choice for neurophysiological studies. *See* EGLIN C.

medin *See* LACTADHERIN.

medium (*pl.* **media)** Common colloquial shorthand for culture medium or growth medium, the nutrient-rich solution used to grow or maintain cells or organs *in vitro*.

medroxyprogesterone A synthetic *progestagen used to treat menstrual disorders and as a depot contraceptive when given as a deep intramuscular injection. Also used in treatment of some tumours, particularly of breast and uterus. *TNs* Depo-Provera, Farlutal, Provera.

medulla The inner region of any tissue or organ, surrounded by the cortex. The **medulla oblongata** is the lowest part of the brainstem which tapers into the spinal cord and is involved in fundamental autonomic functions such as the regulation of breathing and heartbeat.

medulloblastoma A highly malignant type of brain tumour that originates in the cerebellum or posterior fossa and is thought to arise from cerebellar granule cells. Medulloblastomas are more common in children than in adults and can be caused by mutation in the **SUFU* (suppressor of fused) or **BRCA2* genes. In sporadic forms somatic mutations in several genes, including patched (*see* GORLIN'S SYNDROME), β-catenin, and *APC, have been found.

MEF 1. Mouse embryonic fibroblasts. **2.** *See* MYOCYTE ENHANCER FACTOR 2.

mefenamic acid An *NSAID. *TN* Ponstan.

mefloquine An antimalarial drug related to quinine; appears to interfere with the transport of haemoglobin and other substances from the erythrocytes to the food vacuoles of the malaria parasite. *TN* Lariam.

megakaryocyte (myeloplax) The polyploid cell in bone marrow from which platelets are produced. Each megakaryocyte may give rise to 3000–4000 platelets. *See* MESOTHELIN (megakaryocyte potentiating factor).

megalin (gp330) *See* LDL RECEPTOR (LRP2).

megaloblast An abnormally large erythrocyte progenitor cell in the bone marrow, often associated with *megaloblastic anaemia.

megaloblastic anaemia (pernicious anaemia) A group of disorders with common morphological characteristics, usually arising as a result of deficiency in vitamin B_{12} or folic acid. Erythrocytes are larger than normal, neutrophils can be hypersegmented, and megakaryocytes are abnormal. *Megaloblasts are found in the bone marrow, and often macrocytes in the peripheral blood.

megaprep A colloquialism for a plasmid preparation from more than 500 mL of culture medium. *See also* MAXIPREP; MINIPREP.

megesterol A synthetic *progestogen. *TN* Megace.

meglitinides A group of drugs used to treat type 2 diabetes by stimulating insulin release. They bind to a potassium channel at a different site from *sulphonylureas. Examples are repaglinide (Prandin) and nateglinide (Starlix).

Meige lymphoedema *See* HEREDITARY LYMPHOEDEMA (type II).

meiosis The specialized form of nuclear division that produces haploid gametes from the tetraploid parent cell in which chromosomes have been replicated. It involves two successive nuclear divisions (meiosis I and II) with no chromosome replication between them. Each meiotic division can be divided into four phases, as in mitosis (prophase, metaphase, anaphase, and telophase). The four daughter cells receive only one of each homologous chromosome pair, with the maternal and paternal chromosomes being distributed randomly between the cells, thereby segregating alleles. During the prophase of meiosis I (subdivided into *leptotene, *zygotene, *pachytene, *diplotene, and *diakinesis), homologous chromosomes form *bivalents, and *crossing-over, the interchange of chromatid segments, can occur—the process of *recombination.

meiotic spindle The equivalent of the mitotic spindle in *meiosis.

MEKs Mitogen-activated dual threonine/tyrosine kinases (MAPK/ERK kinase, MAPKK, MAP2K) that phosphorylate and activate *MAP kinases (MAP kinase kinases). MEK1 (393 aa) shares 80% sequence identity with MEK2 (400 aa). MEK3 (MKK3; MAP2K3), MEK4 (MKK4; MAP2K4), and MEK6 (MKK6) activate JNK and p38 MAP kinases with variable selectivity, but not ERK. If MEKs are constitutively activated (e.g. by *raf) the result is cell *transformation. Mutation in MEK1 leads to *cardiofaciocutaneous syndrome.

Mel-14 A monoclonal antibody that reacts with L-selectin (CD62L) and will block lymphocyte binding to *HEV both *in vitro* and *in vivo*.

melagatran A small synthetic peptidomimetic with direct thrombin inhibitory actions and anticoagulant activity. It is produced *in vivo* from the prodrug *ximelagatran.

melanin A group of pigments found in feathers, cuttle ink, human skin, hair and eyes, some neurons, and a few other locations. Colours vary but include black/brown, yellow, red, and violet. All the melanins are high molecular weight polymers of indole quinone. **Eumelanin** is one of the two forms found in skin and hair and in the pigmented retinal epithelium of the eye. The oligomer is produced from 5,6-dihydroxyindole-2-carboxylic acid (DHICA) and is black or brown. **Phaeomelanin**, the other skin and hair form, is chemically distinct with its oligomer incorporating L-cysteine, as well as DHI (5,6-indolequinone) and DHICA units, and is responsible for red coloration of hair. **Neuromelanin** is the dark pigment present in neurons, e.g. in the substantia nigra, and may be derived from metabolism of monoamine neurotransmitters. *See* MELANOCYTES.

melanin-concentrating hormone (MCH) A hypothalamic cyclic neuropeptide, with key central and peripheral actions on the regulation of energy balance. The gene that encodes the precursor molecule, pro-MCH, is located in the region of the gene for *spinocerebellar ataxia type 2. The receptors are G-protein coupled (MCHR1 and -2) and pharmacological antagonism at MCH1R in rodents diminishes food intake and results in weight loss. MCH1R antagonists have been shown to have anxiolytic and antidepressant properties.

melanocortin A general term for the hormones derived from *pro-opiomelanocortin. Melanocortin receptors are G-protein coupled and have specificity for various of the pro-opiomelanocortin-derived hormones. **Melanocortin-1 receptor** (MC1R) binds *melanocyte stimulating hormone. **Melanocortin-2 receptor** (MC2R; corticotropin receptor) binds *ACTH. Mutations in the *MC2R* gene result in glucocorticoid deficiency-1. **MC3R** recognizes the core heptapeptide sequence of melanocortins and, like **MC4R**, is expressed primarily in the brain; mutations in both are associated with obesity. **MC5R** appears to be involved in regulating exocrine glands. **Melanocortin-2 receptor accessory protein** (MRAPα, 172 aa, MRAPβ, 102 aa; fat tissue-specific low molecular weight protein, FALP) interacts with MC2R and is essential for its function: mutations, like those in MC2R, can cause familial glucocorticoid deficiency.

melanocytes Cells derived from neural crest and found in the epidermis and the retina, which synthesize and store *melanin pigments. The pigments are stored in *melanosomes but these granules cannot be rapidly relocated within the cell, unlike those of *chromatophores. Skin coloration varies depending upon the level of melanin production rather than differences in melanocyte numbers.

melanocyte-stimulating hormone A releasing hormone (MSH, melanotropin) derived from *pro-opiomelanocortin by proteolytic cleavage in the mammalian hypophysis. There are three forms, α-MSH (13 aa, identical to the first thirteen residues of *ACTH), β-MSH (22 aa), and γ-MSH (12 aa). MSH, particularly α-MSH, causes darkening of the skin and mutations in the *melanocortin-1 receptor, for which MSH is a ligand, are associated with red hair and fair skin. In animal models α-MSH plays an important part in regulating food intake by activation of the brain melanocortin-4 receptor.

melanoma A neoplasia of *melanocytes; benign forms are moles, but melanomas are often highly malignant. Uveal melanoma is the most common primary intraocular malignancy and mutations in several genes, including **BRCA2*, may underlie susceptibility.

melanoma associated antigens *See* MELANOMA DIFFERENTIATION ASSOCIATED GENES; *see also* MAGE ANTIGENS.

melanoma differentiation associated genes A miscellaneous set of genes associated with *melanoma cells, their products being melanoma-associated antigens. **Mda-5** is a cytoplasmic RNA-helicase that, like *RIG-I serves as a receptor for dsRNA and plays a part in antiviral responses. The expression of **Mda-7** (IL-24; 206 aa) is up-regulated in terminally differentiated melanoma cells and will suppress cell proliferation. **Mda-9**/syntenin (syndecan-binding protein) interacts with c-*src and promotes metastasis. Other differentiation antigens include gp100/PMel17, MART-1/MelanA (an antigen that is recognized by a high proportion of cytolytic T-cells), tyrosinase, and a large set of melanoma associated antigens coded by genes of the **MAGE* family which are recognized by autologous cytolytic T cells.

melanoma growth stimulatory activity A CXC-type chemokine (MGSA, neutrophil-activating protein 3, NAP-3, gro, KC, N51, CINC, CXCL1, 107 aa) secreted by melanoma cells and structurally related to platelet-derived β-thromboglobulin. It is mitogenic and will activate, and is chemotactic for, neutrophils.

melanophilin A member of the *rab family that interacts with *myosin5A in melanosome transport within the melanocyte.

melanophore *See* CHROMATOPHORES.

melanopsin (opsin 4) A photopigment (534 aa) found in a subpopulation of retinal ganglion cells that are intrinsically photosensitive and are involved in photo-entrainment of behaviour. *See* OPSIN (opsin subfamilies).

melanosome A membrane-bounded organelle found in *melanocytes, *pigmented retinal epithelium, and some connective tissue cells. The internal structure is characteristic, with melanofilaments arranged in a parallel array and with a periodicity of ~9 nm. In mature melanosomes, which are transferred to *keratinocytes in the skin, the filaments are obscured by dense melanin deposits.

melanotropin *See* MELANOCYTE-STIMULATING HORMONE.

MELAS syndrome A disorder (mitochondrial myopathy, encephalopathy, lactic acidosis, and stroke-like episodes) caused by mutations in mitochondrial tRNA genes, as is *MERRF syndrome.

melatonin A hormone (*N*-acetyl 5-methoxytryptamine) secreted by the pineal gland. In humans may play a role in establishing of circadian rhythms, and is claimed to help overcoming the effects of jet lag. **Melatonin receptors** are inhibitory G-protein-coupled receptors: melatonin receptor 1A (MTR1A; 350 aa) is of high affinity and is predominantly found in the suprachiasmatic nucleus of the hypothalamus, where the circadian clock resides; MTR1B is found in retina and to a lesser extent in brain.

Meleda disease A disorder (mal de Meleda) in which there is symmetrical cornification of the palms and soles, with ichthyotic changes elsewhere. It occurs with relatively high frequency among inhabitants of the island of Meleda in the Adriatic and is caused by mutation in the gene for secreted Ly6/PLAUR domain-containing protein 1 (*SLURP1).

melioidosis A fatal infectious disease (Whitmore's disease) caused by the bacterium *Burkholderia* (formerly *Pseudomonas*) *pseudomallei* that is found extensively in tropical soils. Affects lymph nodes and viscera and symptoms may include pneumonia-like lung pain as well as fever. In equiids, responsible for glanders. Has been considered a potential terrorist bioweapon.

melittin The major component (26 aa) of the venom of the bee *Apis mellifera*, mostly responsible for the pain of the sting. Can be membranolytic and activates phospholipase A2 enzymes. Has a very high affinity for calcium *calmodulin.

meloxicam An *NSAID used to treat rheumatoid arthritis and for the short-term treatment of acute osteoarthritis and ankylosing spondylitis. *TNs* Melox, Movalis, Recoxa.

melphalan A nitrogen mustard alkylating agent used as a chemotherapeutic drug to treat multiple myeloma, ovarian cancer, and occasionally malignant melanoma. *TN* Alkeran.

melusin A muscle specific protein (347 aa) that binds to the cytoplasmic domain of integrin-1β and is responsible for inducing hypertrophy of the heart in response to mechanical overload.

memantine A noncompetitive *NMDA antagonist (1-amino-3,5-dimethyladamantane) used in the treatment of *Parkinson's disease and that has regulatory approval for the treatment of moderate to severe Alzheimer's disease. *TNs* Abixa, Akatinol, Axura, Ebixa, Memox, Namenda.

membrane Generally, a sheet or skin. In cell biology it almost invariably means a modified *phospholipid bilayer with integral and peripheral proteins, the standard compartmentalizing barrier in and around cells. This usage is so widespread that it is probably worth avoiding other uses, particularly in histology or ultrastructure (thus *basal lamina is less ambiguous than *basement membrane).

membrane attack complex *See* COMPLEMENT (C9).

membrane capacitance The measure of the quantity of charge, measured in farads (F), that must be moved across unit area of the membrane to produce unit change in membrane potential. Plasma membranes are good insulators and have a capacitance around 1 μF cm^{-2}.

membrane depolarization *See* DEPOLARIZATION.

membrane fluidity An important property of cell membranes which behave as viscous two-dimensional fluids within their physiological temperature range. Alteration in the lipid composition may alter fluidity and the fluidity is thought to be important for facilitating conformational changes in integral membrane proteins. If fluidity alters then ion channels may malfunction and this has been speculated to be important in anaesthesia.

m

membrane potential A colloquial term for what is, strictly, the transmembrane potential difference, the electrical potential difference across a plasma membrane. *See* ACTION POTENTIAL; RESTING POTENTIAL.

membrane protein Any protein that is attached to a membrane, either as a peripheral membrane protein, where attachment may be e.g. through a *GPI-anchor, or through a protein–protein interaction with an integral membrane protein. Integral membrane proteins have some (hydrophobic) portion of the peptide chain embedded in the *phospholipid bilayer and some, but not all, may be *transmembrane proteins with extracellular and cytoplasmic domains in the hydrophilic external and internal regions. *Transport proteins are integral membrane proteins.

membrane recycling The process whereby membrane is constantly being internalized, fused with an internal membranous compartment, and then reincorporated into the plasma membrane. In exocrine secretion the mechanism allows retrieval of the membrane that surrounded the secreted material prior to fusion; not only is this energy-efficient recycling, but it prevents the cell surface from continuing to expand in area. Phagocytic cells have the opposite need, replenishment of the surface membrane used to surround phagocytosed material.

membrane vesicles Small closed 'compartments' surrounded by a single layer of membrane (a single *phospholipid bilayer). Important in physiological transport processes. Membrane vesicles form spontaneously when membrane is disrupted because the free ends of a lipid bilayer are hydrophobic and therefore unstable in an aqueous environment.

memory cells *See* IMMUNOLOGICAL MEMORY.

MEN1 *See* MENIN; MULTIPLE ENDOCRINE NEOPLASIA (type 1).

mena A component (mammalian enabled, Enah, Ndpp1, 541 aa) of focal adhesions, one of the EVL family (Ena-VASP like), homologous to *Drosophila* enabled, that binds to the putative tumour suppressor *Tes. Overproduction may facilitate metastatic invasion.

menadione (vitamin K_3) A precursor to various types of *vitamin K.

menaquinone One of the naphthaquinones (methylnapthoquinone, *see* VITAMIN K) produced by intestinal bacteria and important in blood clotting.

Mendelian inheritance The classical form of inheritance of characters as originally described by Gregor Mendel (1822–1884). In sexually reproducing organisms, the term is applied to heredity that can be explained by chromosomal segregation, independent assortment and homologous exchange. *Compare* GENOMIC IMPRINTING; MATERNAL INHERITANCE.

Ménétrier's disease A hyperproliferative disorder of the stomach in which there is enlargement of the gastric mucosal folds and often *hypochlorhydria. The cause is overexpression of TGFα in the stomach, but not elsewhere. TGFα is a ligand for the EGF receptor.

SEE WEB LINKS

- Further details from the National Digestive Diseases Information Clearing house.

Ménière's disease An autosomal dominant disorder characterized by hearing loss associated with episodic vertigo. There are autoantibodies to *cochlin in some patients.

SEE WEB LINKS

- Factsheet on the disease.

menin A protein (610 aa) with nuclear localization signals, product of the putative tumour suppressor *MEN1* gene, that interacts with the JunD transcription factor and blocks its activity. Defects in MEN1 lead to the autosomal dominant *multiple endocrine neoplasia type 1.

meninges The three layers of tissue that surround the brain: *See* ARACHNOID LAYER; DURA MATER; PIA MATER.

meningitis Inflammation of the *meninges of the brain and spinal cord. The most serious form is due to infection by the meningococcus *Neisseria meningitidis*, and can be rapidly fatal if untreated. Less serious forms can be caused by viral infections, by lymphocytic infiltration, and by other bacteria.

meningococcus *See* NEISSERIA.

Menkes' disease (*incorrectly* Menke's disease) An X-linked defect in copper uptake from the intestine (kinky hair disease, steely hair disease, copper transport disease) caused by mutation in the gene encoding the α polypeptide of Cu^{2+}-transporting ATPase. Copper-containing enzymes are important in the synthesis of various connective tissue components and the disorder is usually lethal at an early age. Occipital horn syndrome (X-linked cutis laxa) in which there is abnormality of connective tissue, but also mental retardation, is caused by mutation in the same gene.

mepacrine (quinacrine) An antimalarial drug also used as a treatment for infection by **Giardia*.

meperidine *See* PETHIDINE.

meprins Zinc metalloendopeptidases of the *astacin family (endopeptidase 2, EC 3.4.24.18). that degrade a wide variety of signalling peptides such as bradykinin and TGFα, and extracellular matrix components. Meprin A has both α and β subunits and is secreted; meprin B contains only β subunits and is plasma membrane associated.

meprobamate A mild anxiolytic drug, generally superseded by benzodiazepines. *TNs* Equanil, Meprospan, Miltown.

mepyramine (pyrilamine) An antihistamine drug used to treat insect bites and nettle stings by topical application. Appears to be an *inverse agonist of the histamine H_1 receptor (mepyramine binds with high affinity to a $G_{q/11}$-protein-coupled form of the receptor and reduces G-protein availability for other receptors in the signalling pathway).

merbendazole An anthelmintic drug. *TN* Vermox.

merbromin An organometallic salt with covalently linked mercury. Formerly used extensively as an antiseptic but now considered potentially toxic because of the mercury content. *TN* Mercurochrome.

mercaptoethanol A water-soluble thiol with a characteristic and pungent smell. Used as a reagent to cleave disulphide bonds in proteins or to protect sulphydryl groups from oxidation.

mercaptopurine A drug that inhibits purine nucleotide synthesis and metabolism and is immunosuppressive; used to treat leukaemia. *TN* Purinethol.

Mercurochrome® *See* MERBROMIN.

MERF A general opioid receptor agonist, methionine enkephalin-Arg-Phe. *Compare* DTLET.

Merkel cell carcinoma A rare and highly malignant skin tumour (primary neuroendocrine carcinoma of the skin) that arises from Merkel cells, pluripotential basal epidermal cells, and exhibits neuroendocrine differentiation.

merlin The protein product (schwannomin, neurofibromin 2, 587 aa) of the *neurofibromatosis 2 (*NF2*) tumour suppressor gene, one of the ERM (ezrin–radixin–moesin) family of proteins that link the cytoskeleton to membrane proteins. Merlin is defective or absent in schwannomas and meningiomas and in most cases mutations in merlin result in loss of adhesion. The tumour suppressor function of merlin is blocked by phosphorylation and activated by protein phosphatase I (myosin phosphatase target subunit 1, MYPT1).

merocrine The type of secretory process in which a membrane-bounded vesicle fuses with the plasma membrane and releases its contents to the exterior. Probably the commonest secretory mechanism.

meromyosin Fragments of skeletal-muscle myosin II generated by trypsin cleavage. **Heavy meromyosin** (HMM) has the ATPase activity and the hinge region; **light meromyosin** (LMM) is the tail region, mostly α-helical, that interacts with other LMMs to bundle myosin into the thick filament. HMM will bind to F-actin in the absence of ATP to produce a characteristic *arrowhead pattern in electronmicrographs. Further cleavage of HMM with papain yields S1 and S2 subfragments, the former having the ATPase activity, the latter acting as a flexible hinge.

MEROPS A very comprehensive and actively curated database of peptidases classified into 'families' on the basis of their catalytic site or '*clans of peptidases' on the basis of evolutionary relationships. The database has sequence data as well as links to structural information. Inhibitors of peptidases are also listed.

SEE WEB LINKS

- The MEROPS homepage.

merosin *See* LAMININS.

merozoite The phase of the life cycle of the malaria parasite (**Plasmodium*) that is produced by asexual division of the schizont within an erythrocyte and invades other cells when released.

MERRF syndrome A phenotype (myoclonic epilepsy associated with ragged red fibres) that can be produced by mutation in several mitochondrial genes including those encoding various tRNAs (for lysine, leucine, histidine, serine, and phenylalanine) and for the NADH-ubiquitone oxidoreductase subunit of complex I.

mersacidin A type B *lantibiotic (20-aa peptide) produced by *Bacillus* sp. strain HIL Y-85,54728 that inhibits cell wall biosynthesis and is effective against meticillin-resistant *Staphylococcus aureus* strains (at present).

mesangial cells Specialized cells found in the mammalian kidney. Extraglomerular mesangial cells (lacis cells) are found outside the glomerulus, and form the juxtaglomerular apparatus in combination with cells of the macula densa of the distal convoluted tubule and juxtaglomerular cells of the afferent arteriole. The juxtaglomerular apparatus controls blood pressure through the renin–angiotensin–aldosterone system. Intraglomerular mesangial cells are specialized pericytes, involved in filtration, structural support, and phagocytosis. They may also act as antigen-presenting *accessory cells.

mescaline A hallucinogenic compound (3,4,5-trimethoxy-β-phenethylamine) derived from the Mexican peyote cactus *Lophophora williamsii.*

mesencephalon (midbrain) The brain region that derives from the middle of the three cerebral vesicles of the embryonic nervous system. Includes the superior and inferior colliculi and cerebral peduncles and is involved in control of body and eye movement and hearing.

mesenchyme (*adj.* mesenchymal) Embryonic tissue that derives from *mesoderm.

mesna A protective drug used in conjunction with the chemotherapeutic drugs ifosfamide or cyclophsphamide to bind their toxic metabolites in the bladder. *TN* Uromitexan.

mesoderm The middle of the three *germ layers lying between the outer *ectoderm and inner *endoderm; gives rise to the musculoskeletal, blood vascular, and urinogenital systems, to connective tissue (including that of the dermis) and contributes to some glands.

mesothelin A GPI-linked cell surface glycoprotein (megakaryocyte potentiating factor, 630 aa), first identified as the target for monoclonal antibody K1, strongly expressed in normal mesothelial cells, mesotheliomas, nonmucinous ovarian carcinomas, and some other malignancies. Mesothelin function is not essential for growth or reproduction in mice.

mesothelioma A malignant tumour of the *mesothelium, usually of the lung; frequently caused by exposure to asbestos fibres (*see* ASBESTOSIS).

mesothelium The simple squamous epithelium that lines the peritoneal, pericardial, and pleural cavities and the synovial space of joints. The cells are of mesodermal origin.

mesotocin The oxytocin homologue in noneutherian tetrapods and birds.

messenger RNA (mRNA) The single-stranded RNA molecule that is transcribed from DNA and processed before being released to the cytoplasm, where it is translated into a polypeptide. In eukaryotes the processing of the primary transcripts (*hnRNA) includes the addition of a 'cap' at the 5′ end, a *poly-A tail at the 3′ end, the removal of any *introns and the splicing together of *exons; only 10% of hnRNA leaves the nucleus. Eukaryote mRNAs can be quite long-lived, with a half-life ranging from 30 min to 24 h, although regulating the stability of mRNA is beginning to be seen as an important control system (*see* RNA INTERFERENCE).

mestranol A synthetic oestrogen used in oral contraceptives. *TN* Norinyl-1.

met An oncogene from a chemically transformed human osteosarcoma-derived cell line. The alpha subunit of the c-Met proto-oncogene product is extracellular, the beta subunit is the *hepatocyte growth factor receptor, a tyrosine kinase. It is overexpressed in a significant percentage of human cancers, particularly renal and hepatocellular carcinomas.

meta-analysis The analysis of multiple data sets, e.g. published clinical trials, in order to produce a single statistically valid result. The approach has aroused controversy but with an appropriate reduction in the degrees of freedom is generally thought to be acceptable. There is, however, a risk that the nonpublication of negative results will skew the analysis.

metabolic burst (respiratory burst) The enhanced metabolic activity and oxygen uptake exhibited by phagocytes that have taken up particles (particularly particles coated with *opsonin) or been exposed to agonists such as *formyl peptides and *phorbol esters. An NADH-dependent system produces so-called *reactive oxygen species which are bactericidal, but which can also cause tissue damage. Defects in the metabolic burst, as in *chronic granulomatous disease, allow frequent infections and severely reduce life expectancy.

metabolic cooperation (metabolic coupling) The phenomenon observed *in vitro* in which low molecular weight compounds (e.g. nucleotides, ions) can be transferred from one cell to another through *gap junctions. The phenomenon is also assumed to take place *in vivo* and the linkage of cells within a tissue will buffer the levels of metabolites and of ions and other signalling molecules.

metabolic disorders

A large class of disorders that arise generally from a deficiency in an essential enzyme in a metabolic pathway. They can be subdivided according to whether the deficient enzyme is in nucleic acid (e.g. *Lesch–Nyhan syndrome), protein (*phenylketonuria; *porphyria), carbohydrate (e.g. *favism), or lipid metabolism (e.g. *fish-eye disease) but they have little in common except that they tend to be mutations in a single gene, rather than multifactorial, and are usually recessive. A separate class of disorders, the *storage diseases, are due to a deficiency in a lysosomal enzyme so that nonmetabolized material accumulates and gradually causes problems, particularly in the nervous system. Other examples include *alkaptonuria where the defect is in organic acid metabolism, *lipoid congenital adrenal hyperplasia with defective steroid metabolism, various *mitochondrial diseases, peroxisomal disorders (e.g. *Zellweger's syndrome, one of the *peroxisome biogenesis disorders). Defects in lipid transport systems rather than synthesis or breakdown lead to disorders such as *familial hypercholesterolaemia. *See* entries for ADDISON'S DISEASE; AMINOACYLASE; BUTYRYLCHOLINESTERASE; ETHYLMALONIC ENCEPHALOPATHY; GLYCOGEN STORAGE DISEASES; HEXOSE MONOPHOSPHATE SHUNT; HHH SYNDROME; HOMOCYSTINURIA; LEIGH'S SYNDROME; METABOLIC SYNDROME; NORUM'S DISEASE; ORNITHINE TRANSCARBAMYLASE; PSEUDOXANTHOMA ELASTICUM; THYMIDINE PHOSPHORYLASE; UROCANIC ACID; WERNICKE'S ENCEPHALOPATHY; XANTHINE DEHYDROGENASE.

metabolic half-life ($t_{1/2}$) The time taken for 50% of a substance to be metabolized.

metabolic syndrome A condition, associated with high risk of coronary heart disease and of diabetes, in which an individual has a collection of metabolic risk factors. The risk factors include: abdominal obesity, high triglyceride levels, low HDL cholesterol and high LDL cholesterol levels, hypertension, insulin resistance, high fibrinogen or plasminogen activator inhibitor-1 levels in the blood and elevated C-reactive protein levels. Gene loci associated with the condition have been identified.

metabolism The complete set of chemical changes that maintain life. Can be subdivided into anabolism (synthesis) and catabolism (breakdown).

metabolomics The study of the small-molecule metabolite profile (metabolic profiling), a quantitative analysis of all the low molecular weight molecules present in cells in a particular physiological or developmental state. The **metabolome** is the collection of all metabolites in an organism, the end product of expression of the genome. Unlike the genome, however, it has short-term and long-term (developmental) aspects. **Metabonomics** is the study of dynamic changes in the metabolome.

SEE WEB LINKS

- Brief information on metabolomics.

metabotropic Describing a neurotransmitter receptor which affects cell activity but does not act on ion channel properties (*ionotropic). May be G-protein coupled or a receptor tyrosine kinase. Examples include *glutamate receptors, *muscarinic acetylcholine receptors, and most *serotonin receptors.

metacentric Describing a chromosome with its centromere (*kinetochore) at or near the midpoint of the chromosome. In submetacentric chromosomes the two arms (p and q) are unequal, in **telocentric** chromosomes the centromere is near one end and in **acrocentric** chromosomes the short (p) arm is hardly distinguishable.

metachromasia (metachromatic staining) Describing the colour change exhibited by a stain when it is bound to cells or tissue components.

metachromatic leukodystrophy *See* LEUKODYSTROPHY.

metacyclic The infective form of a trypanosome that is transferred to the blood of the mammalian host when the tsetse fly feeds.

metafemales Human females with 4 X chromosomes in addition to 44 autosomes.

metalloenzyme An enzyme with a bound metal ion as part of its structure. The metal ion may be required for enzymic activity, participating directly in the catalytic process, or may stabilize the active conformation of the protein.

m

metallopeptidases The most diverse clan of peptidases (metalloproteases, metalloproteinases, MMPs), with more than 30 families identified. One or more divalent cations, held in place by the charged side groups of several amino acids, often histidine and glutamate, activate a water molecule and facilitate hydrolysis. The commonest ion is zinc.

metalloprotein A protein that contains a bound metal ion as part of its structure. Major examples are *haemoglobin and *metallopeptidases, but many other metalloproteins are known.

metalloproteinase *See* METALLOPEPTIDASES.

metallothioneins A ubiquitous family of low molecular weight, heavy metal-binding proteins with a high cysteine content. They bind seven to twelve heavy metal atoms per molecule of protein. Their synthesis can be induced by heavy metals such as zinc, cadmium, copper, and mercury, and they probably serve a protective function. In humans, metallothioneins are encoded by at least ten to twelve genes separated into two groups designated MT-I and MT-II. Metallothionein-I (MT-I) and MT-II are expressed in most organs, MT-III (neurotrophic growth inhibitory factor) expression is restricted to the brain, and MT-IV is only expressed in certain stratified squamous epithelia and may be involved in keratin assembly.

metaphase The second phase of *mitosis or one of the divisions of *meiosis at which the nuclear envelope has dispersed and chromosomes are fully condensed and aligned along the equatorial plane of the spindle (the metaphase plate). A **metaphase spread** is a cytological preparation that flattens the cell and makes it possible to count and analyse the chromosomes.

metaplasia An alteration of the differentiated phenotype of a cell, e.g. the change of simple squamous endothelium of the postcapillary venule to a cuboidal high endothelial form in sites of chronic inflammation.

metaraminol A potent *sympathomimetic amine, an α_1-adrenergic receptor agonist, that causes an increase in systolic and diastolic blood pressure and is used to treat hypotension. *TN* Aramine.

metastasis The spread of malignant cells from the primary tumour to secondary site(s). Cell biologists view this as a loss of the positional control that keeps cells in their appropriate place.

metastatic spread The formation of secondary tumours involving local invasion (in most cases), passive transport, lodgement, and proliferation at a remote site.

metavinculin *See* VINCULIN.

metaxolone A skeletal muscle relaxant that does not act directly on contractile systems. *TN* Skelaxin.

met-enkephalin *See* ENKEPHALINS.

metformin HCl An oral biguanide drug used for treatment of type 2 diabetes. *TNs* Diabex, Diaformin, Dianben, Fortamet, Glucophage, Glumetza, Obimet, Riomet.

methacholine A nonselective *muscarinic receptor agonist. Used to diagnose bronchial hyperreactivity in asthma (methacholine challenge test).

methadone A synthetic opioid used in treating opioid addiction, although it can itself be habit-forming.

methaemoglobin A nonfunctional form of haemoglobin, containing ferric iron, produced by the action of oxidizing poisons.

methenamine (hexamine) An antibacterial drug used as the hippurate to treat infections of the urinary tract. *TN* Hiprex.

methicillin *See* METICILLIN.

methimazole *See* CARBIMAZOLE.

methionine (Met, M) An essential sulphur-containing amino acid (149 Da) that can act as a methyl donor (*see* S-ADENOSYL METHIONINE). *See* FORMYL PEPTIDES.

methionine aminopeptidase A mitochondrial metallopeptidase (MAP1, EC 3.4.11.18, 478 aa) that removes the N-terminal methionine from proteins, facilitating further modification. Methionine aminopeptidase 2 (eukaryotic initiation factor-2 (eIF-2)-associated protein; p67) is the target for the antiangiogenic drug *fumagillin.

methisazone An antiviral drug (*N*-methylisatin) that blocks the translation of late viral mRNA in poxvirus infection. Used in the past as a treatment for smallpox.

methotrexate (amethopterin) A dihydrofolate analogue that inhibits *dihydrofolate reductase and kills rapidly growing cells. Used in treatment of acute leukaemias and some autoimmune diseases. It is less toxic than *aminopterin.

methylcholanthrene A highly carcinogenic polycyclic hydrocarbon that can be formed during incomplete combustion of organic material. The commonest isomer, 3-methylcholanthrene, is used experimentally as a mutagen.

methyl-CpG-binding protein 2 A protein (MeCP2, 486 aa) that binds to methylated CpG-rich regions of DNA and that contributes to silencing (*see* DNA METHYLATION and GENOMIC IMPRINTING). MeCP2 regulates gene expression in neurons and is important in neurogenesis. The X-linked gene is mutated in *Rett's syndrome and some cases of *Angelman's syndrome.

methyldopa An α_2 receptor agonist used as an antihypertensive drug, especially during pregnancy.

methylene blue A water-soluble dye (Swiss blue, Basic Blue 9) that is colourless when reduced and can be oxidized by atmospheric oxygen. Used in bacteriology and histology as a stain.

methylmalonyl-coenzyme A mutase A mitochondrial enzyme (EC.5.4.99.2, 750 aa) that catalyses the rearrangement of methylmalonyl-CoA to succinyl-CoA with adenosylcobalamin as a coenzyme. Mutations in the gene cause methylmalonic aciduria that is unresponsive to vitamin B_{12} therapy, and the severity depends on the extent of the deficiency. Other forms of the disorder arise from mutations in the enzymes involved in adenosylcobalamin synthesis.

methylphenidate A *dexamfetamine-like drug that stimulates the central nervous system and is used in *attention deficit hyperactivity disorder (ADHD). It is thought to block the reuptake of norepinephrine and dopamine and increase the release of these monoamines. *TN* Ritalin.

methylprednisolone A synthetic corticosteroid (*glucocorticoid) used as an anti-inflammatory drug.

methyltransferases Enzymes that transfer a methyl group from *S*-adenosyl methionine to a substrate. DNA (cytosine-5)-methyltransferase (EC 2.1.1.37; DNMT1) has a strong preference for hemimethylated DNA, so the pattern of DNA methylation tends to persist. Other minor forms (DNMT2, DNMT3A, and DNMT3B) are known, the latter possibly being *de novo* methyl transferases.

methylxanthines Naturally occurring purine alkaloids, theobromine, theophylline, and caffeine (trimethylxanthine), that are the active compounds in tea and coffee. By inhibiting cAMP phosphodiesterase they cause an increase in intracellular cAMP levels.

methysergide A synthetic *ergot alkaloid, a serotonin-receptor antagonist, used for prophylaxis of migraine. A known, though rare, side effect is retroperitoneal fibrosis. *TNs* Sansert, Deseril.

meticillin (formerly **methicillin**) Synthetic penicillinase-resistant β-lactam antibiotic. No longer used therapeutically but the name persists in *meticillin-resistant *Staphylococcus aureus* (MRSA).

meticillin-resistant *Staphylococcus aureus* Bacteria (MRSA, multidrug-resistant *Staphylococcus aureus*) that are resistant to β-lactam antibiotics (penicillins and cephalosporins), not just *meticillin. An increasing problem, particularly in hospitals, and many are now also resistant to the newer antibiotics such as *vancomycin.

metoclopramide A dopamine D_2-receptor antagonist used mostly to treat nausea and vomiting.

metolazone A diuretic drug used to treat oedema and hypertension. *TN* Metenix 5.

metopirone *See* METYRAPONE.

metoprolol tartrate A cardioselective *β-adrenoceptor blocking drug.

metorphamide An endogenous *opioid octapeptide proteolytically cleaved from proenkephalin.

metrizoate A radio-opaque compound (3-acetimido-5-(*N*-methyl-acetamido)-triiodobenzoate, *TN* Isopaque) used in diagnostic radiography. Solutions of the sodium salt have a high density and are used for cell density gradient separations. *See* FICOLL.

metronidazole An antiprotozoal and antibacterial drug. It is metabolized by anaerobic organisms to an active form that inhibits nucleic acid synthesis and also inhibits the production of reactive oxygen species. *TN* Flagyl.

metyrapone (metopirone) A drug that inhibits the 11-β-hydroxylation reaction in corticosterone synthesis. Used in treating some adrenal tumours and as a diagnostic drug for testing hypothalamic–pituitary ACTH function.

metzincins A clan of metallopeptidases (zinc endopeptidases) of which *pappalysins are one family. Other members include *astacins and *matrix metallopeptidase.

m

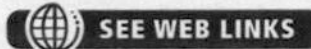

- The Cytokines & Cells Online Pathfinder Encyclopedia contains further details.

mevalonate kinase A key enzyme (396 aa) involved in synthesis of isoprenoids and sterols. Mutation in the gene can lead to *mevalonic aciduria and *hyper-IgD syndrome. In *Zellweger's syndrome and *neonatal adrenoleukodystrophy there is reduced peroxisomal mevalonate kinase activity.

mevalonic acid An important intermediate compound in polyprenyl biosynthesis and cholesterol synthesis. Derived from hydroxymethylglutaryl-CoA (HMG-CoA) by a reaction inhibited by *statins. *See* COMPACTIN.

mevalonic aciduria An autosomal recessive disorder characterized by psychomotor retardation, cerebellar ataxia, recurrent fever attacks, and death in early childhood. Caused by mutation in the gene for *mevalonate kinase.

mevastatin (compactin) A *statin isolated from *Penicillium citrinum* that is not used therapeutically but as a source for pravastatin.

mevinolin An analogue of *mevastatin isolated from cultures of *Aspergillus terreus*. *TN* Lovastatin.

mexiletine A drug that blocks sodium influx into neurons thereby slowing the development of an action potential; used to treat ventricular arrhythmia. *TN* Mexitil.

MFSDs *See* MAJOR FACILITATOR SUPERFAMILY DOMAIN-CONTAINING PROTEINS.

MG132 A potent and reversible inhibitor (carbobenzoxy-L-leucyl-L-leucyl-L-leucinal) of the *proteosome system that prevents *β-secretase cleavage.

MGP *See* MATRIX GLA PROTEIN.

MGSA *See* MELANOMA GROWTH STIMULATORY ACTIVITY.

MHC (major histocompatibility complex) The set of gene loci specifying major *histocompatibility antigens, e.g. HLA in humans, H2 in mice, RLA in rabbits, RT-1 in rats, DLA in dogs, SLA in pigs, etc.

MHC antigens The major histocompatibility antigens are involved in rapid graft rejection whereas the minor histocompatibility antigens are involved in much slower rejection phenomena. *See* HISTOCOMPATIBILITY ANTIGENS; MHC RESTRICTION.

MHC restriction A phenomenon that arises because foreign antigens are only recognized in association with *MHC antigens and cytotoxic T-cells are restricted to killing virally infected cells that have the same class I antigens as themselves.

mianserin A tetracyclic antidepressant with few antimuscarinic effects. Inhibits norepinephrine reuptake and strongly stimulates the release of norepinephrine. A relatively selective 5-HT_2 receptor antagonist. *TN* Tolvon.

***mib* genes** Human homologues of the *mindbomb* genes in *Drosophila*. The protein product of *MIB1* (DAPK-interacting protein-1, DIP1; 1006 aa) has E3 ubiquitin ligase activity and regulates the cellular levels of *death-associated protein kinase-1 (Dapk1). The *MIB2* product (1013 aa) is *skeletropin.

mibolerone A synthetic *androgen, an anabolic steroid.

micelle A spherical structure in which the hydrophobic portions of amphipathic molecules such as phospholipids are inwardly directed, with only hydrophilic portions in contact with the surrounding aqueous phase. The converse arrangement would exist if the major phase were nonpolar.

Michaelis constant *See* K_M; MICHAELIS–MENTEN EQUATION.

Michaelis–Menten equation An equation in enzyme kinetics that successfully accounts for the hyperbolic relationship between substrate concentration S and reaction rate V. $V = V_{max} \times S/(S + K_m)$, where K_m is the Michaelis constant and V_{max} is the maximum reaction rate at very high substrate concentrations.

miconazole An antifungal drug used to treat ringworm and candidiasis. *TN* Daktarin.

microaerophiles Organisms that thrive in low-oxygen concentrations but do not survive at normal oxygen concentrations. Examples are *Borrelia burgdorferi, Helicobacter pylori, Lactobacillus* spp., and *Treponema pallidum*.

microangiopathic haemolytic anaemia A type of anaemia in which circulating erythrocyte fragments (microcytes) are produced as blood is forced through a fibrin meshwork. A consequence of *disseminated intravascular coagulation (DIC).

microarray A solid support, often a glass slide, to which samples of biological material (DNA, protein, etc.) are attached in a regular pattern. The slide can then be used to look for binding activity by nucleic acid probes or antibodies.

microbicide A substance that kills microbes.

microbody (microperoxisome) *See* PEROXISOMES.

microcarrier Small beads, either solid or of immiscible liquid, on which cells are grown in suspension culture. They allow the production of large numbers of *anchorage-dependent cells in small volumes. An imprecise term, with other potential meanings.

microcells Small fragments of cells containing a subdiploid number of chromosomes, prepared by cytochalasin treatment and centrifugation. Hybrid cells (microcell hybrids), constructed by fusing microcells with full-size cells of interest, are useful in chromosome mapping.

microchimerism The presence in a single individual of two genetically distinct populations of cells, one population being very much in the minority. Can arise through transfer of cells between mother and fetus or between dizygotic twins *in utero*, or as a result of blood transfusion or transplantation.

Micrococcus A genus of small Gram-positive aerobic bacteria, ~1–2 μm in diameter. *M. lysodeikticus* (now *M. luteus*) was commonly used as the source of bacterial cell wall suspension for measuring lysozyme activity by a decrease in turbidity.

microcolliculus A broad swelling (~0.5 μm) on the upper surface of a cultured epithelial cell that is moving. The microcolliculi, like *ruffles on a fibroblast, move towards the rear as the cell moves forward.

microcystins Cyanobacterial toxins that are important environmental hazards. They are cyclic heptapeptides containing an unusual hydrophobic amino acid, 3-amino-9-methoxy-10-phenyl-2,6,8-trimethyl-deca-4,6-dienoic acid (ADDA), which is important for their activity. They are potent inhibitors of protein phosphatases, acting at the same site as *okadaic acid. Microcystin MC-LR is the most studied and is hepatotoxic and probably carcinogenic. Other cyanobacterial toxins include *anatoxins, *cylindrospermopsin, *domoic acid, and *nodularins.

microcytes *See* MICROANGIOPATHIC HAEMOLYTIC ANAEMIA.

microdialysis A technique used to sample the fluid in the extracellular space of tissues. A small semipermeable probe is inserted into the tissue and the contents later analysed.

microelectrode An electrode that is small enough (10-200μm diameter) to be inserted into a cell without compromising the impermeability of the plasma membrane. Microelectrodes can be used to record *resting and *action potentials, to measure intracellular ion and pH levels with *ion-selective electrodes, or to *microinject substances into the cytoplasm. Standard microelectrodes are pulled from glass capillaries, filled with conducting solutions of potassium chloride or potassium acetate and connected via a silver chloride-coated wire.

microfilament A cytoplasmic F-actin filament, 5–7 nm thick, that can be decorated with *HMM. A meshwork of microfilaments provides mechanical support to the cell and microfilaments may be laterally associated with other proteins (tropomyosin, α-actinin), and may be anchored to the membrane, notably at *adherens junctions.

microfilaria The motile embryos (150–300 μm long) of nematodes such as *Wuchereria*, the infective forms in the insect vector.

microglial cells (microglia) Small glial cells of the monocyte lineage that are found throughout grey and white matter of the brain. They constitute ~20% of the glial cell population and provide a primary immune response system behind the *blood-brain barrier. Microglia invade neural tissue in early development and can differentiate into macrophages, present antigen, secrete cytokines, and act as scavengers.

microglobulin Any small globular plasma protein. **α_1-Microglobulin** is a plasma protein (184 aa), one of the *lipocalin superfamily, the product of a gene that encodes a precursor (352 aa) which is proteolytically cleaved into α_1-microglobulin and *bikunin. Present in free form and in a complex with IgA; involved in regulation of the inflammatory process. **β_2-Microglobulin** (b2M) is an immunoglobulin-like serum protein (119 aa) that is also the β chain of MHC class I molecules (*histocompatibility antigens). Defects in b2M cause hypercatabolic hypoproteinaemia in which there is marked reduction in serum concentrations of immunoglobulin and albumin because of abnormal rates of degradation.

microinjection The introduction of substances into the cytoplasm of a cell through a *microelectrode. Hydrostatic pressure can be used or, in some cases, a small current (ionophoresis). The method can be used to introduce drugs, histochemical markers (e.g. *horseradish peroxidase or lucifer yellow), RNA, or DNA.

micromere A small cell in the upper (animal) hemisphere of a fertilized egg during *holoblastic cleavage.

micrometer eyepiece *See* EYEPIECE GRATICULE.

microperoxidase The haem-containing peptide portion of cytochrome *c*, generated by proteolytic cleavage, that retains *peroxidase activity.

microperoxisome *See* PEROXISOMES.

micropinocytosis The formation of small pinocytotic vesicles (~100 nm in diameter), a process that is insensitive to inhibitors of microfilaments and may involve *dynamin 2.

microplate reader An instrument for spectroscopic analysis of multiwell plates. Can be used to measure radioactivity, optical absorbance, fluorescence, or luminescence.

micropore filters Filters that are composed of a meshwork of cellulose acetate or nitrate ~150 μm thick. They can be sterilized by autoclaving, and the smaller pore sizes (0.22 μm, 0.45 μm) are used to filter out microorganisms from heat-labile solutions, such as culture medium. Filters with pore sizes of several micrometres that allow cells to move through them are used in *Boyden chambers. *Nucleopore filters are quite different. *TN* Millipore filter.

microprocessor complex A protein complex both necessary and sufficient for processing long primary miRNAs (pri-miRNAs) to the intermediate 60–70-nt precursor miRNAs (pre-miRNAs). It comprises *drosha and the double-stranded RNA-binding protein DGCR8 (product of the *di George's syndrome critical region 8 gene). The pre-miRNAs are cleaved by another RNase III (*dicer) to generate mature miRNAs.

microRNA (miRNA) Small noncoding RNA gene products (usually ~22 nt) generated by *dicer from precursors with a characteristic hairpin secondary structure. They associate with proteins such as *argonaute to form *RISCs and regulate gene expression by inducing mRNA degradation or interfering with translation. Aberrant miRNA expression has been described for several human malignancies. The first examples were lin-4 and let-7, which control developmental timing in *Caenorhabditis elegans* (formerly referred to as 'small temporal RNAs', stRNAs) but many miRNA sequences are evolutionarily conserved and it appears that, as a class, miRNAs have broad regulatory functions. MicroRNAs cannot be distinguished from *siRNAs on the basis of function but can be distinguished on the basis of their biogenesis: miRNAs come from endogenous transcripts that can form local hairpin structures, which ordinarily are processed such that a single miRNA molecule accumulates from one arm of a hairpin precursor molecule, whereas biological siRNAs come from long exogenous or endogenous dsRNA molecules (very long hairpins or bimolecular duplexes), processed such that numerous siRNAs accumulate from both strands of the dsRNA.

SEE WEB LINKS

- RNA website for micro RNA naming conventions.
- A resource for predicted microRNA targets and expression.

microsatellites *See* SATELLITE DNA.

microscopic polyangiitis A necrotizing *vasculitis that affects small blood vessels in many organs. There are few or no immune deposits and the disorder is considered distinct from *polyarteritis nodosa.

microscopy The use of various methods to visualize small objects. Two major forms are important, light microscopy and *electron microscopy. In the commonest form of light microscopy, **bright-field microscopy**, the image is formed mainly by absorption of light by the specimen, although there is a minor contribution from diffraction. Histological stains that bind selectively to different parts of the specimen enhance the absorption. In **phase contrast** or **interference** methods the image is formed as a result of constructive or destructive interference between light that has passed through the specimen, which has a different refractive index, with a reference beam. **Dark-field microscopy** is used to image small objects that diffract light and appear bright against a dark background, the direct light path being blocked. **Polarization** methods involve illumination with polarized light and the introduction of a polarizing filter in the eyepiece so that the background is dark and only birefringent objects are bright. **Fluorescence microscopy** involves illumination with light at a wavelength (usually UV) that will excite fluorescence in dyes used to stain the specimen which emit at a different wavelength (usually in the visible spectrum): a system of filters blocks the exciting light. The illumination is often through the objective (epi-illumination) rather than through a condenser. In **confocal microscopy** a fine laser beam of light is scanned over the object through the objective lens. The technique is particularly good at rejecting light from outside the plane of focus, and so produces higher effective resolution than is normally achieved.

microsequencing Imprecise colloquial term used for the sequencing of very small amounts of protein, usually to produce an oligonucleotide probe ('oligo') with which to screen a cDNA library.

microsomal fraction *See* MICROSOMES.

microsomal triglyceride transfer protein A heterodimeric protein (MTTP), found in liver and intestine, composed of *protein disulfide isomerase (EC 5.3.4.1, 504 aa) and a large subunit (894 aa) that is a member of the vitellogenin gene family. It catalyses the transport of triglyceride, cholesteryl ester, and phospholipid from phospholipid surfaces, and the large subunit is defective in *abetalipoproteinaemia.

microsomes Small vesicles (20–200 nm diameter) produced when cells are disrupted. Unless fractionated may contain membrane from various organelles, though mostly the *endoplasmic reticulum.

microspikes (filopodia) Projections from the leading edge of some cells, usually ~100 nm diameter, 5–10 μm long, with a loosely bundled microfilament core. They are often a feature of nerve *growth cones and are a linear version of the leading lamella of fibroblasts.

Microsporidia Obligate intracellular parasites, probably extremely reduced fung (1.5–2.0 μm in size). The infective spore grows within the host cell and then divides, either by binary fission (merogony) or multiple fission (schizogony), to produce spores (sporogony) which can then infect other cells. They cause a range of diseases in vertebrates and invertebrates and at least nine genera infect humans: *Encephalitozoon* spp., *Septata intestinalis*, and *Enterocytozoon bieneusi*; less common infections are with *Brachiola* spp., *Microsporidium* spp., *Nosema* spp., *Pleistophora* spp., *Trachipleistophora* spp., and *Vittaforma* spp.

microtome A device used for cutting thin sections of an embedded specimen, for either light or electron *microscopy.

microtubule (MT) A component of the cytoskeleton, a tubule of 25 nm outside diameter with a wall 5 nm thick, assembled from heterodimers of *tubulin packed in a regular fashion and associated with various other proteins (including *microtubule associated proteins (MAPs), *dynein, and *kinesin). Microtubules of the ciliary *axoneme are more permanent than cytoplasmic and spindle microtubules which are in a state of dynamic equilibrium, constantly having new subunits added at one end (the plus end) and removed at the minus end ('treadmilling').

microtubule associated proteins (MAPs) A general term for a heterogeneous set of proteins that share the property of being associated with microtubules. Some contribute to the electron-lucent zone that surrounds microtubules in EM pictures. **MAP1A** (2805 aa) and **1B** (2468 aa) and **MAP2A** and **2B** (1827 aa) form projections from brain microtubules. MAP2 is characteristic of axonal processes of vertebrate nerve cells and distinguishes them from dendritic processes. **MAP4** (1152 aa) copurifies with MAP1 and MAP2 and is found in many tissues. **MAP1C** is a microtubule associated motor protein with two heavy chains (cytoplasmic dynein 1 heavy chain, 4646 aa; CD2HC, 4307 aa) associated with six or seven light and intermediate chains of about 350–500 aa, the two-headed cytoplasmic equivalent of ciliary dynein responsible for retrograde transport (transport towards the centrosome). NB Not to be confused with mitogen activated protein kinases (*MAP kinases).

microtubule organizing centre (MTOC) A rather amorphous region of cytoplasm which determines the number and pattern of microtubules in the cytoplasm and is the location from which microtubules radiate. In animal cells the *pericentriolar region is the major organizing centre for cytoplasmic and spindle microtubules. Basal bodies are organizing centres for axonemal microtubules. The MTOC has γ-tubulin which serves as a marker.

microvillus (*pl.* **microvilli**) A long thin projection from the apical surface of an epithelial cell, supported by a central bundle of actin microfilaments associated with *villin and *fimbrin. In the intestinal *brush border the microvilli increase absorptive surface area, whereas cochlear stereovilli (*see* STEREOCILIUM) have a distinct mechanical role in sensory transduction.

Microviridae A family of single-stranded DNA bacteriophages.

midazolam A *benzodiazepine drug. *TN* Hypnovel.

midbody A structure formed at the cleavage furrow during *cytokinesis that disappears before cell division is completed.

m

midbrain *See* MESENCEPHALON.

midecamycin A macrolide antibiotic with a sixteen-member ring that has anti-inflammatory activity, probably through inhibition of cytokine production.

midkine A heparin-binding growth factor (neurite growth-promoting factor 2, NEGF2, 121 aa) of the TGFβ superfamily that is 46% homologous with NEGF1 (pleiotropin). Both NEGF1 and NEGF2 have neurite outgrowth-promoting activity and may play a role in nervous system development. *Nucleolin binds midkine.

MIF *See* MACROPHAGE INHIBITION FACTOR; MIGRATION INHIBITORY FACTOR.

mifepristone A synthetic steroid (RU-486) that inhibits the action of *progesterone, used as a pharmacological abortifacient up to the twentieth week of pregnancy. *TNs* Mifegyne, Mifeprex.

migfilin A *LIM-domain containing protein (filamin-binding LIM protein 1, FBLP1, 373 aa) that colocalizes with MIG2 (mitogen inducible gene-2; *see* KINDLIN) at cell–extracellular matrix adhesion sites where actin stress fibres are anchored, and interacts with *filamin.

migraine A neurological syndrome characterized by painful headaches, often unilateral (hemicrania), together with nausea and/or photophobia and phonophobia. Thought to be due to malfunction in the serotonin system and can be treated with the *triptans which are serotonin 5-$HT_{1B/1D}$ receptor agonists. **Visual migraine** (migraine with aura) is a brief condition often with flickering in the visual field; in some cases it may presage a classical migraine with headache. **Familial hemiplegic migraine** (FHM) is an autosomal dominant classical migraine subtype with hemiparesis during the aura phase. **FHM1** is caused by mutations in a gene coding for the P/Q-type calcium channel α subunit; **FHM2** by mutations in the Na^+,K^+-ATPase gene *ATP1A2* and **FHM3** is a rare subtype caused by mutations in a gene coding for the sodium channel α subunit (*SCNA1*).

migrastatin A compound isolated from *Streptomyces platensis,* closely related to the *dorrigocins. Migrastatin and its analogues may inhibit metastasis.

migration inhibitory factor *See* MACROPHAGE INHIBITION FACTOR.

miliary tuberculosis A relatively uncommon form of tuberculosis in which small ('millet-like') lesions are found in various organs of the body, especially the meninges and the lungs.

Miller–Dieker lissencephaly syndrome A disorder characterized by microcephaly and a thickened brain cortex caused by haploinsufficiency of genes on a deleted region of chromsome 17, including the *LIS1* gene (*see* LISSENCEPHALY).

Miller unit A unit used in assays for β-galactosidase, using *o*-nitrophenyl-β-D-galactoside (ONPG) as a substrate. Named for Jeffrey Miller who published the first protocol in 1972.

Millipore filter *See* MICROPORE FILTERS.

milrinone An inhibitor of phosphodiesterase III which leads to elevation of *cAMP levels. Has positive inotropic and vasodilatory effects and is sometimes used in treatment of heart failure if other drugs fail. *TN* Primacor.

Milroy's disease *See* HEREDITARY LYMPHOEDEMA (type IA).

MIM A G-actin binding protein (missing in metastasis, metastasis suppressor 1, 356 aa) with greater affinity for ATP-actin than for ADP-actin. Contains a *WH2 domain and interacts directly with the SH3 domain of *cortactin. MIM appears to regulate cell motility by modulation of *Arp2/3 activators.

mimecan (osteoglycin) A secretory protein expressed in the anterior pituitary gland that belongs to a family of small leucine-rich proteoglycans (SLRPs). The predicted 298-aa precursor is processed into a 103-aa mature protein. Appears to be a key regulator of left ventricular mass.

Mimivirus The largest known virus (*Acanthamoeba polyphaga* mimivirus, APMV) with a capsid diameter of 400 nm and a linear double-stranded DNA genome (1.2 Mbp) encoding more than 900 genes (comparable to some rickettsias). Probably one of the nucleocytoplasmic large DNA virus group. It may cause some forms of pneumonia.

mimosine A nonprotein amino acid similar to tyrosine, first isolated from *Mimosa pudica* and found in tropical legumes of the genus *Leucaena*. It is toxic to animals and potently inhibits DNA replication in mammalian cells by blocking the formation of replication forks. It is an iron/zinc chelator and may inhibit iron-dependent ribonucleotide reductase and the zinc-sensitive transcription of the cytoplasmic serine hydroxymethyltransferase gene (SHMT).

mimotope A term usually applied to peptides from *phage display libraries that mimic the structure of a conformational *epitope and elicit an identical antibody response. They may have potential as vaccines.

Minamata disease A disease characterized by sensory and motor disturbances and fetal damage in pregnant women, caused by methylmercury poisoning. First described in 1956 in humans consuming fish and shellfish contaminated by waste water from a chemical plant in Minamata, Japan.

mindin (spondin-2) One of a small subfamily of thrombospondin type I proteins, the other member being spondin-1 (*F-spondin). Like F-spondin, mindin is involved in axonal guidance in the developing brain. Mindin-deficient mice exhibit defective inflammatory and immune responses probably because mindin is a ligand for neutrophil and macrophage integrins and binding via mindin regulates the expression of rho GTPases in dendritic cells.

mineralocorticoid A corticosteroid that acts on water and electrolyte balance by promoting retention of sodium ions and excretion of potassium ions in the kidney. Aldosterone is the most potent natural example. Many steroids also have *glucocorticoid activity.

miniature end plate potential (MEPP) *See* POSTSYNAPTIC POTENTIAL.

minichromosome **1.** A chromatin structure, resembling a small chromosome, found in some virus-infected cells. **2.** A plasmid with a chromosomal *origin of replication. **3.** *See* MINICHROMOSOME MAINTENANCE PROTEINS.

minichromosome maintenance proteins A conserved family of proteins (MCM proteins) involved in the initiation and elongation of DNA replication forks in archaea and eukaryotes. The eukaryotic MCM2–7 complex is a heterohexameric *helicase, with six related subunits, that assembles at the DNA replication fork. Various MCMs are phosphorylated during the cell cycle, at least in part by *cyclin-dependent kinases. MCM-2, like Ki-67, is often used as a marker for grading tumours and MCM-7 is up-regulated in a variety of tumours including neuroblastoma, prostate, cervical, and hypopharyngeal carcinomas.

minimal medium The simplest tissue culture medium (minimal essential medium, MEM) that will support the proliferation of normal cells *in vitro*.

miniprep Colloquialism describing a small-scale purification of *plasmid from a bacterial culture of ~1–10 mL. *See also* MAXIPREP; MEGAPREP.

minisatellite Variable number tandem repeat (VNTR). *See* SATELLITE DNA.

minK **1.** Minimal potassium channel. A widely expressed protein (129 aa) that interacts with other membrane proteins to form *potassium channels and may regulate the structure and activity of the channels; e.g. minK and KVLQT1 proteins co-assemble to form the cardiac I(Ks) channel. **2.** MINK1 (misshapen/NIK-related kinase 1, 1300 aa), one of the *germinal centre kinase (GCK) family.

minocycline A tetracycline antibiotic. *TNs* Minocin, Sebomin.

minoxidil A vasodilatory drug used to treat severe hypertension and topically to stimulate the regrowth of hair in cases of male pattern baldness.

MIP **1.** *See* MAJOR INTRINSIC PROTEINS. **2.** *See* MACROPHAGE INFLAMMATORY PROTEINS. **3.** MIP90 (microtubule interacting protein 90) is a protein that associates with microtubules and colocalizes with actin in the leading lamella of fibroblasts.

MIR *See* MAMMALIAN-WIDE INTERSPERSED REPEAT.

miracidium The free-living ciliated first-stage larva of a trematode. Miracidia infect the intermediate host; in the case of *Schistosoma* sp., aquatic snails.

miRNA *See* MICRORNA.

mirtazapine A noradrenergic and specific serotonergic antidepressant drug used to treat moderate to severe depression. *TNs* Remeron, Zispin.

MIS12 A component (protein MIS12 homologue, 215 aa) of the MIS12 complex that is required for normal chromosome alignment and segregation and for kinetochore formation during mitosis. The complex consists of MIS12 together with DSN1 (kinetochore-associated protein DSN1 homologue, 356 aa), NSL1 (kinetochore-associated protein NSL1 homologue, 281 aa) and PMF1 (polyamine-modulated factor 1, 205 aa). (The human proteins are homologues of proteins originally described in yeast.)

mismatch repair A system that corrects mispairing of bases in newly replicated DNA or that arise due to DNA damage. Long patch repair involves excision of tracts of a few kilobases, whereas short patch repair is used for mismatches due to damage and only around ten nucleotides are involved. Not only must the enzyme systems involved recognize the mismatch but they must also recognize the parental (template) strand and correct the copy, not the original. The excision is filled by DNA polymerase. The enzymes involved are highly conserved from prokaryotes to eukaryotes. MLH genes are homologous to the *E. coli MutL* gene and mutations result in some types of hereditary nonpolyposis colorectal cancer, as do mutations in the *MSH* genes (homologues of *MutS* of *E. coli*). Mutation in MSH3 is associated with endometrial cancer. *PMS* genes are homologues of the yeast mismatch repair enzymes (postmeiotic segregation increased genes) and mutations in these or the *MSH* and *MLH* genes lead to the mismatch repair cancer syndrome. *See also* MUIR–TORRE SYNDROME; TURCOT'S SYNDROME.

misoprostol A synthetic *prostaglandin (PGE1) analogue used in the prevention of NSAID-induced gastric ulcers. It inhibits the secretion of acid by gastric parietal cells, via G-protein-coupled receptor-mediated inhibition of adenylate cyclase. *TN* Cytotec.

missense mutation A point mutation that changes a *codon so that a different amino acid is specified.

mitochondrial diseases

A wide range of disorders, frequently neurological, caused by defects in mitochondrial function. If the defect is in the mitochondrial genome then the pattern of inheritance is non-Mendelian and passes down the maternal line. Examples include various myopathies (lethal infantile mitochondrial myopathy (LIMM) is caused by mutation in tRNA for threonine), *Bjornstad syndrome, *Kearns–Sayre syndrome, *Leber's optic atrophy, *Leigh syndrome, *MELAS syndrome, *MERRF syndrome, *Navajo neurohepatopathy, and *sideroblastic anaemia with spinocerebellar ataxia. Ageing is considered to be due, in part at least, to progressive damage to mitochondrial DNA. *See also* MITOCHONDRIAL DNA DELETION SYNDROME.

mitochondrial DNA deletion syndrome The hepatocerebral form of mitochondrial DNA (mtDNA) depletion syndrome can be caused by mutation in the nuclear-encoded mitochondrial deoxyguanosine kinase gene, the *MPV17* gene or the *C10ORF2* gene that encodes *twinkle. The myopathic form of mtDNA depletion syndrome is caused by mutation in the nuclear-encoded mitochondrial thymidine kinase gene and the encephalomyopathic form is caused by mutation in the succinyl-CoA synthase gene.

Mitochondrial Eve The populist name for the woman thought to be the most recent common matrilineal ancestor of all living humans, who was one of a population living ~150 000 years ago that gave rise to multiple lineages, only one of which remains. The supposition depends upon all mitochondria deriving from the egg, and there is some evidence that this may not be correct.

mitochondrial neurogastrointestinal encephalopathy A multisystem disorder that can be caused by mutations in the genes encoding either *gliostatin or DNA polymerase γ. There are multiple deletions of mitochondrial DNA.

mitochondrial trifunctional protein (TFP) A mitochondrial multienzyme complex that catalyses the last three steps in the long-chain fatty acid β-oxidation pathway. There are four hydroxyacyl-CoA dehydrogenase-α (HADHA, 79 kDa) and four hydroxyacyl-CoA dehydrogenase-β (HADHB, 48 kDa) subunits. TFP deficiency leads to a wide clinical spectrum of disorders.

mitochondrion (*pl.* mitochondria) The organelle that generates ATP by oxidative phosphorylation (*see* CHEMIOSMOTIC HYPOTHESIS). Morphologically very variable and may be short and rod-like or long and branched. Contains mitochondrial DNA and *mitoribosomes and is surrounded by a double membrane, the inner one, which contains the components of the *electron transport chain, often infolded to form cristae. The inner fluid phase has most of the enzymes of the *tricarboxylic acid cycle and some of the urea cycle. Mitochondria are inherited almost entirely from the mother, and disorders due to mutations in the mitochondrial genome are inherited in a non-Mendelian fashion. *See* MITOCHONDRIAL DISEASES.

mitofusin Transmembrane GTPases, (mitofusin-1, 741 aa; mitofusin-2, 757 aa) which mediate mitochondrial fusion independently of the cytoskeleton. Overexpression induces the formation of mitochondrial networks. Mitofusin-2 is important in regulating proliferation of vascular smooth muscle cells and defects in mitofusin-2 cause *Charcot–Marie–Tooth disease type 2A2.

mitogen activated kinases *See* MEKs.

mitogenesis The process of stimulating cell proliferation; progression through the whole cell cycle, not just mitosis. Resting cells are usually considered to be in the G_0 phase and so mitogens act to stimulate cells into G_1. A range of *growth factors are mitogenic and in the immune system antigens are mitogenic.

mitogen inducible gene-2 *See* KINDLIN.

mitogillin One of a group of related aspergillus toxins (*aspergillins) that are ribonucleases which cleave the 23–28S RNA of the large ribosomal subunit.

mitomycin C An antibiotic isolated from *Streptomyces caespitosus* and *S. lavendulae*, used as a chemotherapeutic drug in bladder, oesophageal, and breast cancer. Crosslinks DNA and inhibits DNA synthesis. Mitomycin-treated cells are sometimes used as feeder layers.

SEE WEB LINKS

- Entry in National Cancer Institute Drug Dictionary.

mitoribosomes Mitochondrial ribosomes, more like prokaryotic ribosomes than cytoplasmic ribosomes but with a higher protein/RNA ratio. All the ribosomal proteins are encoded in the nuclear genome. The size (55S) is distinct from bacterial (70S) and cytoplasmic (80S) ribosomes.

mitosis The process of nuclear division in the somatic cells of *eukaryotes in which the genomic information, copied during S phase of the cell cycle, is distributed equally between two daughter cells so that each contains a *diploid set of chromosomes identical to that of the parent cell. In **prophase** the *nuclear envelope breaks down, the chromosomes condense, and the two chromatids become visible. The centriole divides and the two daughter centrioles separate to define the two poles of the mitotic spindle. The spindle consists of interdigitated *microtubules (spindle fibres) nucleated from the two pericentriolar organizing centres together with microtubules attached to kinetochores of chromosomes. In the next stage, **metaphase**, the chromosomes align on the equatorial plane of the spindle, the metaphase plate. Then in **anaphase** the paired chromatids separate, one toward each pole, a process that is completed during **telophase**. The chromatids, now considered to be chromosomes, lengthen and become diffuse, new nuclear envelopes form round the two sets of chromosomes, and the spindle disassembles. Mitosis is usually followed by cell division or cytokinesis in which the cytoplasm is also divided to give two daughter cells.

mitotic apparatus *See* MITOSIS.

mitotic death The death of cells that have been damaged by ionizing radiation, which occurs only when they enter the next mitosis.

mitotic index The proportion of cells in a population that are undergoing mitosis. If all the cells are actively moving through the cell cycle then it is a measure of the relative length of the mitotic phase of the *cell cycle.

mitotic recombination A *crossing-over (somatic crossing-over) that occurs between *homologous chromosomes during mitosis, a very rare occurrence because, unlike meiosis, the chromosomes do not normally pair. It leads to loss of heterozygosity at important gene loci.

mitotic segregation The phenomenon that can arise after *mitotic recombination; two alleles of a heterozygous individual are rendered homozygous in the diploid progeny of the mitosis and give rise to two distinct populations of cells.

mitotic shake-off method A method for selecting cells undergoing mitosis to allow karyotyping. Many cells round up during mitosis and are less firmly attached to the culture substratum, so that gentle shaking of the culture vessel will detach the mitotic cells. The proportion in mitosis is usually increased

m

by adding a drug that blocks cells at metaphase (e.g. *colchicine). The method can also be used to obtain a reasonably well synchronized population.

mitotic spindle *See* MITOSIS.

mitoxantrone An anthracenedione drug, a type II *topoisomerase inhibitor; used to treat some cancers and to slow the progression of secondary progressive multiple sclerosis. *TNs* Novantrone, Onkotrone.

mitsugumin29 A protein (trimeric intracellular cation channel type A, 33 kDa) with three putative transmembrane segments that forms homotrimers. It is preferentially expressed in excitable tissues, including striated muscle and brain, where it may act as a counter-ion channel synchronizing functions including the release of calcium from intracellular stores.

mix A human gene (*MIXL1*) encoding a paired class *homeobox transcription factor (identified on the basis of homology to the *Xenopus* homeobox genes *mix1* and *mix2*). The transcription factor responds to *transforming growth factor β superfamily signals. There is high expression of *MIXL1* in many aggressive B-cell and T-cell non-Hodgkin's lymphomas and in Hodgkin's lymphoma.

mixed lymphocyte reaction (MLR) A mitogenic response in T cells when cocultured with allogeneic lymphocytes, mismatched in histocompatibility loci. Although originally used to test for possible graft compatibility, a negative reaction is a poor predictor of graft acceptance.

MKK MAPK kinase. *See* MEKs.

MKPs *See* MAPK PHOSPHATASE.

MLCK *See* MYOSIN LIGHT CHAIN KINASE.

MLEE A method (multilocus enzyme electrophoresis) used to distinguish strains of organisms based on two-dimensional *electrophoresis of proteins. *See* MLST.

M line The midline of the *M band which lies at the middle of the A band of the sarcomere. M line protein (*myomesin), *creatine kinase, and *glycogen phosphorylase b are located at the M line, which is involved in controlling thick filament spacing.

MLKs A family of serine/threonine kinases (mixed lineage kinases, MAP3K9, 10, 11) that regulate signalling by the *JNK and *p38 MAPK pathways. They are considered to be mitogen-activated protein kinase kinase kinases (MAPKKKs, MAP3Ks). There are two subgroups, one containing the highly related MLK1 (MAP3K9), MLK2/MST (MAP3K10), and MLK3/SPRK/PTK (MAP3K11) kinases. Overexpression of MLKs effectively induces apoptotic death of cultured neuronal PC12 cells

MLL A gene (myeloid/lymphoid leukaemia gene, mixed lineage leukaemia gene) homologous to the *Drosophila* gene *trithorax*, encoding a DNA-binding protein (3910 aa) that methylates histone H3 and positively regulates the expression of target genes, including many homeobox (*HOX*) genes. It is frequently translocated in leukaemias. MLL is cleaved by *taspase-1 to produce fragments that heterodimerize to stabilize the complex and define its intranuclear destination.

MLR *See* MIXED LYMPHOCYTE REACTION.

MLST A method (multilocus sequence typing) for characterizing isolates of bacteria or other organisms using nucleotide sequences of fragments derived from several (usually seven) housekeeping genes. Alleles at each locus are defined by nucleotide sequencing, rather than indirectly from the electrophoretic mobility of their gene products.

MLV Usually **1.** *Murine leukaemia virus. Less commonly: **2.** Modified live virus. **3.** Multilamellar liposomes or multilamellar vesicles. **4.** Mechanical lung ventilation.

MLZE A cytoplasmic protein (melanoma-derived leucine zipper-containing extranuclear factor, gasdermin C, 508 aa) related to *pejvakin and of unknown function, although up-regulated in melanoma.

MMR **1.** Measles, mumps, and rubella vaccine, a combined vaccine given in infancy. **2.** *See* MISMATCH REPAIR.

MMTV Mouse (murine) *mammary tumour virus.

MN blood group antigens A pair of blood group antigens independent of the ABO locus. The alleles are codominant and both can be expressed in an individual, so there are three phenotypes: MM, NN, and MN. *Glycophorin has M or N activity depending upon carbohydrate epitopes attached to the N-terminal region. The MN phenotype has been associated with essential hypertension.

SEE WEB LINKS

- The Blood Group Antigen Gene Mutation database.

mnemiopsin A calcium-activated photoprotein, a protein–luciferin complex, from the ctenophore *Mnemiopsis*. *See* AEQUORIN.

MNNG A potent mutagen and carcinogen (*N*-methyl-*N*′-nitro-*N*-nitrosoguanidine) used experimentally to induce mutations.

Mnt A nuclear protein (582 aa) that will heterodimerize with *Max and functions as a transcriptional repressor. Binds to same site on Max as does *Sin3.

Mo-1 *See* LFA-1.

mobile genetic elements *See* TRANSPOSON.

mobile ion carrier *See* IONOPHORE.

moclobemide A *monoamine oxidase inhibitor used as an antidepressant. *TN* Manerix.

modafinil A stimulant drug that enhances wakefulness and does not appear to act in the same way as amphetamines. Used to treat narcolepsy and other sleep disorders. *TN* Provigil.

modification enzymes Enzymes involved in the bacterial restriction/modification system (*see* RESTRICTION ENDONUCLEASES (restriction enzymes)) that introduces minor bases into DNA or RNA, or alters existing bases, thereby removing binding sites for restriction enzymes.

modulation *See* NEUROMODULATION.

moesin One of the *FERM domain protein family (membrane-organizing extension spike protein, pasin2, 577 aa) widely expressed in different tissues and localized to filopodia and other membranous protrusions that are important for cell–cell recognition and signalling and for cell movement, where it acts as a cytoskeletal adaptor. *Pasin-2 is identical to moesin. Moesin is excluded from the region of T-cell/APC contact (the immunological synapse) and colocalizes with *leukosialin (CD43). Platelet moesin levels are secondarily reduced in *Wiskott–Aldrich syndrome.

Mohr–Tranebjaerg syndrome A progressive form of deafness (dystonia–deafness syndrome) caused by mutation in the *TIMM8A* (DDP) gene, the product of which is involved in the import and insertion of hydrophobic membrane proteins into the mitochondrial inner membrane (*see* GENERAL IMPORT PORE). Mutation in the same gene causes Jensen's syndrome, in which the deafness is accompanied eventually by dementia.

molecular clock **1.** The rate of fixation of mutations in DNA, a measure of the time since species diverged. **2.** A biological system capable of maintaining a timing rhythm or pulse. *See* CIRCADIAN RHYTHM.

molluscum contagiosum virus (MCV) A double-stranded DNA poxvirus that causes a benign disease of skin and mucous membranes. Four types are recognized, MCV-1 to -4, with MCV-1 being the most prevalent The virus is large, 240–320 nm diameter. **Molluscum bodies** are intracellular inclusions of virus found in epidermal cells.

Moloney murine leukaemia virus A replication-competent retrovirus that causes leukaemia in mice, first isolated by John B. Moloney from cell-free extracts of a transplantable mouse sarcoma. *See* MOLONEY MURINE SARCOMA VIRUS.

Moloney murine sarcoma virus A replication-defective retrovirus that will induce fibrosarcomas *in vivo*, and transform cells in culture. It arose by recombination between the *Moloney murine leukaemia virus and a sequence (*v-mos* oncogene) derived from mouse cells and required for induction and maintenance of viral transformation; the human homologue was subsequently identified. *See* MOS.

Moloney test Obsolete skin test for immunity to diphtheria.

MOLT-4 cells A human T-cell leukaemia cell line that grows in suspension and was derived from a male with acute lymphoblastic leukaemia.

molybdenum cofactor deficiency A molybdenum-containing cofactor is essential to the function of sulphite oxidase, xanthine dehydrogenase, and aldehyde oxidase; deficiency leads to severe neurological damage and death in early childhood. Three distinct complementation groups exist: type A with defective products of the *MOCS1* locus, type B with defective molybdopterin synthase (MOCS2) and a third complementation group (C) with defects in *gephyrin. *See* XANTHINURIA.

MOM **1.** Mitochondrial outer membrane. **2.** Prefix designating proteins of the outer mitochondrial membrane, e.g. MOM19 and MOM72, proteins of 19 and 72 kDa respectively. **3.** Multiples of the median, a measure of how far an individual test result deviates from the median. MoM is commonly used to report the results of medical screening tests, although the statistical validity of threshold values is doubtful.

mometasone furoate A moderately potent corticosteroid used topically for inflammatory skin disorders and by inhalation for allergic rhinitis and asthma. *TNs* Elocon, Nasonex.

Mona **1.** Monocytic adaptor (Mona; Gads) A Grb2-like adaptor protein that links the adaptors *Slp-76 and *LAT when the T-cell receptor is activated. Also involved in monocytes and platelets when they are activated. **2.** Multicentric osteolysis, nodulosis and arthropathy (MONA), an autosomal recessive bone disorder caused by mutation in the gene encoding matrix metallopeptidase 2 (neutrophil gelatinase). Winchester syndrome is caused by mutation in the same gene.

monastrol A mitotic inhibitor that acts on *kinesin (Eg5) and leads to the formation of monopolar spindles.

monellin A nonglycosylated heterodimeric protein (44 aa and 50 aa protomers) that is around 1000-fold sweeter than sucrose. Originally identified in the fruit of the West African shrub *Dioscoreophyllum cumminsii* (serendipity berry). *See* THAUMATIN.

monensin A sodium *ionophore (671 Da) from *Streptomyces cinnamonensis* that forms pseudomacrocyclic complexes with cations and acts as a transporter. It is used as a feed additive in chickens because of its antibiotic properties and also in *ion-selective electrodes.

Monera The kingdom that contained all prokaryotes, now generally split into two kingdoms, *Archaea and *Bacteria.

mongolism *See* DOWN'S SYNDROME.

moniliasis *See* CANDIDIASIS.

monoamine oxidases The enzymes (MAOs, EC:1.4.3.4, 527 aa, 520 aa) that break down *biogenic amines, such as serotonin, adrenaline, noradrenaline and dopamine. Monoamine oxidase inhibitors were used to treat very severe depressive illness but are now rarely used except in patients unresponsive to *tricyclic antidepressants or *SSRIs because of their dangerous interactions with foods containing tyramine and with *sympathomimetic drugs.

monocentric chromosome A chromosome with a single *centromere, as is the case for most chromosomes.

monocistronic RNA A *messenger RNA (mRNA) that encodes a single polypeptide chain when translated. All eukaryotic mRNAs are monocistronic, but *alternative splicing means that a single gene may produce multiple variant mRNAs. Some bacterial mRNAs are polycistronic.

monoclonal Describing a lineage (clone) of cells derived from a single progenitor. Monoclonal antibodies (Mabs) are the product of a single clone of *hybridoma cells, and are therefore a single species of antibody molecule.

monocyte A circulating mononuclear phagocyte. Monocytes differentiate into *macrophages having emigrated into tissues.

monocyte chemoattractant protein-1 *See* MONOCYTE CHEMOTACTIC AND ACTIVATING FACTOR.

monocyte chemotactic and activating factor A C-C *chemokine (MCAF, JE, monocyte chemoattractant protein-1, MCP-1, CCL2, 76 aa), coinduced with *interleukin-8 on stimulation of endothelial cells, fibroblasts or monocytes, in response to IL-1, TNF, lipopolysaccharide, and various other inflammatory mediators. It is a *chemokine and potently chemotactic for monocytes.

SEE WEB LINKS

- Entry in the Cytokines & Cells Online Pathfinder Encyclopedia.

monoglyceride lipase The enzyme (monoacyl-glycerol lipase, EC 3.1.1.23, 303 aa) that hydrolyses stored intracellular triglyceride to fatty acids and glycerol. It hydrolyses 2-arachidonoyl glycerol, an endogenous agonist of the central cannabinoid (CB_1) receptor, to arachidonic acid and glycerol.

monokines An obsolete term for *cytokines produced by macrophages that act on other cells.

monolayer A single layer of any molecule, but most commonly applied to polar lipids. Can be formed at an air/water interface in experimental systems. The term should be avoided in describing one layer of a lipid bilayer, which is generally termed a 'leaflet'. *See also* MONOLAYERING OF CELLS.

monolayering of cells The tendency of cells grown in culture to form a single layer of cells covering the surface (substratum). Most normal (nontransformed) cells do not form multilayers in culture, partly because contact inhibition of locomotion causes cells to change their direction of movement when they collide with another cell, partly because proliferation is limited by the supply of growth factors that must reach the cells by diffusion through the medium above them (the diffusion boundary layer). Proliferation continues after a confluent monolayer is formed and the limit is a density effect, rather than a contact phenomenon (*see* DENSITY DEPENDENT INHIBITION OF GROWTH).

mononuclear cells Blood monocytes and lymphocytes as opposed to erythrocytes and *polymorphonulear leucocytes. Mononuclear phagocytes are monocytes and their differentiated products, macrophages.

monosaccharide A simple sugar that cannot be hydrolysed to smaller sugar units. They range in size from trioses ($(CH_2O)_n$, where $n = 3$) to heptoses ($n = 7$).

monosome **1.** A single ribosome attached to a mRNA molecule. **2.** A ribosome that has dissociated from a polysome. **3.** A chromosome in an aneuploid set that does not have a homologue.

monosomy The situation in which only one of a *homologous pair of chromosomes is present in a diploid cell or organism. In *Turner's syndrome only one X chromosome is present. Partial monosomy can arise if there is a deletion of part of one chromosome, as in *cri-du-chat syndrome.

montelukast A *leukotriene antagonist that blocks the action of leukotriene D4 on the cysteinyl leukotriene receptor in the airways. Used to treat asthma, though not acute attacks. *TN* Singulair.

MOPC Cell lines (eg MOPC-21, MOPC-315) derived from a murine (BALB/CJ) myeloma (plasmacytoma) that secrete immunoglobulin light and heavy chains. Extensively used as an experimental model system.

moracizine A drug used as a last resort in treating ventricular arrhythmia. It inhibits the rapid inward sodium current across myocardial cell membranes. *TN* Ethmozine.

Moraxella catarrhalis A Gram-negative, aerobic, oxidase-positive diplococcus, a commensal in the upper respiratory tract but that can cause otitis media, sinusitis, and occasionally laryngitis. Formerly *Micrococcus catarrhalis, Neisseria catarrhalis, Branhamella catarrhalis.*

Morbillivirus A genus of single-stranded RNA viruses in the *Paramyxoviridae family. The type species is *measles virus; other species include canine distemper virus (CDV), the related seal virus (phocine distemper virus (PDV)), and rinderpest virus.

morphiceptin A tetrapeptide fragment (Tyr-Pro-Phe-Pro-NH_2), derived from α-casein, that is a μ-selective *opioid receptor ligand. May occur naturally. *Endomorphin-2 (YPFF-NH_2) is a related compound.

morphine An *opioid alkaloid, isolated from opium, that has powerful analgesic properties but is highly addictive. Binds to receptors for endogenous neurotransmitters (*endorphins and *enkephalins) but is not broken down by the peptidases that inactivate the endogenous compounds.

morphogen A diffusible substance involved in morphogenetic processes, specifiying positional information and thus the developmental fate of cells that perceive the signal.

morphogenesis The process of 'shape formation' in development. Complex morphogenetic movements transform the simple ball of cells formed by early cleavage of the egg into an embryo.

morphometry The measurement of shape. **Morphometrics** is the study of variation and change in the form (size and shape) of organisms or objects. There are various methods for extracting data from shapes and for using data on the distribution of objects in a two-dimensional section to predict the shapes and the distribution of these objects in three dimensions.

Morquio–Brailsford disease (Morquio's syndrome) A rare autosomal recessive *storage disease characterized by dwarfism, kyphosis, and skeletal defects in the hip joint. The defect is due to failure to degrade keratan sulphate. There are two forms: in type A (mucopolysaccharidosis type IVA) the mutation is in galactosamine-6-sulphate sulphatase; in type B (mucopolysaccharidosis type IVB) it is in β-galactosidase, as in GM1-gangliosidosis.

mortalin One of the *HSP70 family of proteins (75-kD glucose regulated protein, GRP75). Mortal fibroblasts (i.e. those with a finite division potential) have MOT-1 (679 aa) uniformly distributed in the cytoplasm, immortalized fibroblasts have MOT-2 in a juxtanuclear concentration. Interacts with *p53 and promotes its dissociation from centrosomes. The gene is assigned to 5q31.1, the locus of a putative tumour suppressor gene involved in myeloid malignancies.

morula An early developmental stage once individual blastomeres can no longer be distinguished easily (~96 h after fertilization). Name derives from a supposed likeness to a mulberry.

mos An *oncogene, (Moloney murine sarcoma viral oncogene homologue), originally identified in *Moloney murine sarcoma, that encodes a serine/threonine *protein kinase. Normal *c-mos* is expressed only in the germ cells of both testis and ovary. Overexpression of the c-*mos* proto-oncogene product stimulates activity of *jun through a *MAP kinase.

mosaicism A condition in which an individual is composed of a mixture of cells that differ in karyotype or genotype.

motif A small structural element or domain that is recognizable in several proteins.

motilin A peptide (22 aa) synthesized in the small intestine that stimulates intestinal motility. The receptor is G-protein coupled.

motogen A generic name for substances that stimulate cell motility, by analogy with *mitogens. *Scatter factor (hepatocyte growth factor) is a good example because a factor may be a motogen, a mitogen, or both, according to the cell type and the situation.

motoneuron *See* MOTOR NEURON.

motor end plate *See* NEUROMUSCULAR JUNCTION.

motor neuron (motoneuron) A *neuron that synapses on to a *muscle fibre.

motor neuron disease *See* AMYOTROPHIC LATERAL SCLEROSIS.

m

motor protein A protein that will undergo a conformational change when bound ATP is hydrolysed. The shape change is generally used to generate gross movement by arranging that the conformational change puts the motor protein in a new position where it can attach to the next molecule in a linear array, so that the motor protein moves progressively along the linear polymer, each step costing one ATP. Thus myosin 'walks' along F-actin, dynein moves along microtubules. Motor proteins may be attached to 'cargo' that is moved around the cell or be incorporated into more complex structures such as the thick filaments of striated muscle.

Mott cells Abnormal *plasma cells containing **Mott bodies** or Russell bodies, rough endoplasmic reticulum-derived vesicles containing immunoglobulins of the IgM class. Found in lymphoid tissues in autoimmune diseases, in multiple myeloma, and in the late stages of African trypanosomiasis.

Mounier–Kuhn syndrome A rare congenital abnormality (tracheobronchomegaly) in which there is atrophy or absence of elastic fibres and thinning of the smooth muscle layer in the trachea and main bronchia which become dilated.

mouse double minute *See* MDM.

Mowat–Wilson syndrome An autosomal dominant developmental disorder caused by mutation in the zinc finger homeobox 1B gene (Smad-interacting protein 1). There is mental retardation and a range of other abnormalities, sometimes including *Hirschsprung's disease.

moxonidine *See* CLONIDINE.

MPF *See* CYCLINS.

M phase The mitotic phase of the eukaryotic cell cycle. *See* MITOSIS.

M phase promoting factor (MPF) The original name for the *cyclinB2–Cdc2 complex.

MPO *See* MYELOPEROXIDASE.

M protein **1.** A galactoside carrier in *E. coli*. **2.** A cell surface antigen of *Brucella*. **3.** *See* MYOMESIN. **4.** *See* STREPTOCOCCAL M PROTEIN. **5.** *See* PARAPROTEIN.

MPTP A compound (1-methyl-4-phenyl-1,2,3,6-tetrahydropyridine) that is selectively toxic to dopaminergic neurons and will induce a condition in mice that resembles Parkinson's disease ('MPTP mouse model'). Inbred mouse strains differ in their susceptibility to MPTP, indicating a genetic element.

MPV **1.** Mean platelet volume. The occurence of large platelets (macrothrombocytes, >10 fl) can be a symptom of autoimmune *thrombocytopenic purpura, *Bernard–Soulier syndrome, or the *May–Hegglin anomaly, whereas small platelets (microthrombocytes, <6 fl) can indicate *aplastic anaemia; *Wiskott-Aldrich syndrome, thrombocytopenia-absent radii (*TAR syndrome), or *storage pool disease. **2.** Mouse parvovirus-1.

MRC-5 cells A cell line derived from normal human male fetal lung tissue. The cells become senescent after 50–60 doublings, having reached the Hayflick limit (*see* CELL DEATH).

MRF-4 *See* HERCULIN.

MRI (magnetic resonance imaging) An imaging technique that uses nuclear magnetic resonance and that can provide information about soft tissues. Formerly known as nuclear magnetic resonance imaging (NMRI). MRI scanning has become increasingly common and functional studies are possible: *see* FMRI.

mRNA *See* MESSENGER RNA.

MRN complex A multiprotein complex (MRE11–RAD50–NBN complex) involved in recruiting *ATM to double-strand breaks in DNA.

The complex has single-strand endonuclease activity and double-strand-specific 3′–5′-exonuclease activity, provided by MRE11A (708 aa). RAD50 (DNA repair protein RAD50, 1312 aa) may be required to bind DNA ends and hold them in close proximity. The complex consists of two heterodimers of RAD50 and MRE11A associated with a single *nibrin. The MRN complex is also part of the *BASC complex.

MRP *See* CALGRANULINS.

MRSA *See* METICILLIN-RESISTANT STAPHYLOCOCCUS AUREUS.

Mrt1 *See* PATIP.

MSC **1.** Mesenchymal stem cells. **2.** Bone marrow stromal cells. The latter is a confusing usage since there are bone marrow mesenchymal stem cells.

MSDS **1.** Musculoskeletal disorders (MSDs). **2.** Material safety data sheet. **3.** Mean square displacements, of atoms, particles, etc. **4.** Membrane-spanning domains.

MSH *See* MELANOCYTE-STIMULATING HORMONE or MSH GENES.

***MSH* genes** A multigene family that encodes proteins involved in *mismatch repair. They are the human homologues of the *MutS* genes in *S. cerevisiae*, which are themselves homologues of the *E. coli* MutHLS mismatch repair system that involves the *MutH, MutL, MutS*, and *MutU* genes. The protein products of the human genes are MSH2, 3, 4, 5, and *GTBP (MSH6). Defects in mismatch repair may arise by mutation in these genes or through imbalance in the expression of some relative to others. Mutations in the *MSH2* gene result in *hereditary nonpolyposis colorectal cancer-1 and an *MSH3* frameshift mutation has been observed in an endometrial carcinoma. MSH4–MSH5 heterodimers bind uniquely to *Holliday junctions. *MLH1*, PMS1 (*MLH2*), and PMS2 (*MLH4*) are homologues of the *E. coli MutL* gene and the yeast *PMS* (postmeiotic segregation increased-1 and -2) genes respectively.

MSI **1.** Microsatellite instability (*see* SATELLITE DNA). **2.** Magnetic source imaging, the reconstruction of current sources in the heart or brain from the measurements of external magnetic fields. **3.** Amphipathic antimicrobial peptides, MSI-78 and MSI-594, derived from *magainin-2 and *melittin, respectively.

MSM **1.** Minimal salt medium. **2.** Men who have sex with men, a common shorthand encountered in reports on sexually transmitted diseases.

MSO *See* MSX.

MSS4 **1.** Mammalian suppressor of SEC4 (Rab interacting factor; ras-specific guanine-releasing factor, RASGRF3; 123 aa) A small GTP-binding protein involved in the regulation of intracellular vesicular transport that binds integrin α subunits. *See* TRANSLATIONALLY CONTROLLED TUMOUR PROTEIN. **2.** The human homologue of the yeast *MSS4* gene, encoding a phosphatidylinositol-4-phosphate-5-kinase that synthesizes phosphatidylinositol (4,5)-bisphosphate.

Msx **1.** A family of vertebrate homeobox (*HOX*) genes (*msx* genes) originally isolated by homology to the *Drosophila* muscle segment homeobox (*msh*) gene. Mutations are associated with tooth agenesis and *Witkop's syndrome. **2.** L-methionine-DL-sulfoximine, an inhibitor of glutamine synthetase and a potent convulsant which metabolically and morphologically affects astroglia (*sometimes* MSO).

MT *See* MICROTUBULE.

MTC *See* MAXIMUM TOLERABLE CONCENTRATION.

MTD *See* MAXIMUM TOLERABLE DOSE; MAXIMUM TOLERATED DOSE.

MTEL *See* MAXIMUM TOLERABLE EXPOSURE LEVEL.

MTOC *See* MICROTUBULE ORGANIZING CENTRE.

mTOR A serine/threonine kinase (mammalian/molecular target of rapamycin, 2549 aa), one of the phosphoinositide 3-kinase related kinase (PIKK) family, an important modulator of the size of proliferating cells. The target of rapamycin proteins (TOR1, TOR2) in *Saccharomyces cerevisiae* regulate growth in a rapamycin-sensitive manner and TOR2 regulates polarization of the actin cytoskeleton. The activity of mTOR is regulated by *rheb, a Ras-like small GTPase, in response to growth factor stimulation and nutrient availability. The TOR complex 1 (TORC1) contains mTOR1 or mTOR2, *raptor and *LST8 and binds *FKBP-rapamycin; whereas TORC2, which contains mTOR2, *rictor, and *LST8 does not. TORC2 disruption causes an actin defect.

MTS **1.** *See* MICROTUBULE (MT). **2.** *See* METALLOTHIONEINS. **3.** Mitochondrial targeting sequence. **4.** *See* MTS ASSAY. **5.** *See* MUIR–TORRE SYNDROME.

MTS assay A cell viability assay, often used to assay toxicity of compounds, that depends upon the production of formazan, a reduction product of MTS, by live cells, The intensity of the colour (at 492 nm) due to formazan is proportional to the number of live cells. An alternative to the *MTT assay in which formazan is produced from MTT by a mitochondrial dehydrogenase, which only happens in live cells.

MTTP *See* MICROSOMAL TRIGLYCERIDE TRANSFER PROTEIN.

MUC-1 *See* EPISIALIN; EPITECTIN.

Muckle–Wells syndrome A periodic fever syndrome (urticaria-deafness-amyloidosis syndrome) in which there is chronic recurrent urticaria, arthralgia, progressive sensorineural deafness, and reactive renal amyloidosis. It is caused by mutation in *cryopyrin.

mucolipidoses Autosomal recessive *storage diseases which differ from *mucopolysaccharidoses in that there is not excessive secretion of sugars in the urine. Mucolipidosis-I (sialidosis types I and II, the former milder than the latter) is caused by mutation in the gene encoding neuraminidase (EC 3.2.1.18). Mucolipidosis IIα/β (I-cell disease) is caused by complete absence of the $\alpha\beta$ subunits of *GlcNAc-phosphotransferase, whereas mucolipidosis IIIα/β (classic pseudo-Hurler polydystrophy) is caused by severely reduced levels of the $\alpha\beta$ subunits. The γ subunits are normal in both forms but are mutated in mucolipidosis III. Mucolipidosis IV is caused by mutation in the gene encoding *mucolipin-1 and, unusually for a storage disease, lysosomal hydrolases are normal. *See also* GOLDBERG'S SYNDROME.

mucolipin A 580-aa protein, deficient in *mucolipidosis IV, that has homology to *polycystin-2 and and to the family of transient receptor potential calcium channels. It is thought that mucolipin may be responsible for the calcium release involved in endocytic vesicle fusion and thus in lysosomal function.

mucopeptide *See* PEPTIDOGLYCAN.

mucopolysaccharide *See* GLYCOSAMINOGLYCANS.

mucopolysaccharidoses Autosomal recessive *storage diseases in which the deficiency is in lysosomal enzymes needed to hydrolyse glycosaminoglycans. Mucopolysaccharidosis I is *Hurler's syndrome, mucopolysaccharidosis II is *Hunter's syndrome, mucopolysaccharidosis III is *Sanfillipo's syndrome. Mucopolysaccharidosis IVA (Morquio's syndrome A) is due to a deficiency in galactosamine-6-sulphatase (EC 3.1.6.4) and the milder Morquio's IVB is due to β-galactosidase deficiency: both enzymes are needed for the breakdown of keratan sulphate. Mucopolysaccharidosis VI is *Maroteaux–Lamy syndrome, mucopolysaccharidosis VII is *Sly's syndrome, mucopolysaccharidosis IX is caused by mutation in the hyaluronidase I gene. Types V and VIII are no longer recognized. *See* MUCOLIPIDOSES.

mucosal associated lymphoid tissue (MALT) Aggregates of lymphoid tissue, antigen-presenting cells, and lymphocytes associated with mucosal surfaces such as the gastrointestinal tract (*gut associated lymphoid tissue, GALT), thyroid, breast, lung (*bronchus associated lymphoid tissue, BALT), salivary glands, eye, and skin (*skin associated lymphoid tissue).

mucous gland A *merocrine gland that produces a thick (mucopolysaccharide-rich) secretion. *Compare* SEROUS GLAND.

mucous membrane (mucosa) An endodermally derived epithelium that lines body cavities that are exposed to the external environment and internal organs. Many, but not all, have specialized cells that produce a thick (mucous) secretion.

mucus A viscous, glycoprotein-rich solution.

Muir–Torre syndrome An autosomal dominant disorder, part of the Lynch cancer family syndrome II, characterized by sebaceous neoplasms and visceral malignancies; the cause is mutation in *mismatch repair genes (*hMSH2* or *MLH1*).

Müller cell (Müller glia) The main glial cell of the neural retina. The cell body and nucleus lie in the middle of the inner nuclear region and the bases of the cells form the internal and external limiting membranes. May be capable of dedifferentiation to produce multipotent progenitor cells.

multidrug transporter (Mdr, P-glycoprotein) A family of *ABC proteins in both prokaryotes and eukaryotes that transport xenobiotics, including drugs, from the cytoplasm and thereby confer resistance.

multienzyme complex A general term for a set of distinct enzymes, usually catalysing consecutive reactions in a metabolic pathway, that are physically associated and can be purified as a complex. They differ from multifunctional enzymes in which the various enzymic activities are associated with different domains of a single polypeptide.

multigene family *See* GENE FAMILY.

multilocus enzyme electrophoresis *See* MLEE.

multiple cloning site *See* POLYCLONING SITE.

multiple endocrine neoplasia An autosomal dominant disorder characterized by a high frequency of peptic ulcer disease and primary endocrine abnormalities involving the pituitary, parathyroid, and pancreas. **Multiple endocrine neoplasia type I** (MEN1) is caused by mutation in the gene encoding *menin and mutation in MEN1 is the basis for inherited forms of *Zollinger–Ellison syndrome. **MEN2A** and **MEN2B** are caused by mutation in the **ret* protooncogene, and **MEN4** is caused by mutation in the cyclin-dependent kinase inhibitor 1B (CDKN1B; p27/KIP1) gene. *See also* GIANTISM.

multiple epiphyseal dysplasia A generalized skeletal dysplasia associated with significant morbidity. There is deformity and pain in the joints and a short stature. Multiple epiphyseal dysplasia with myopia and conductive deafness (EDMMD) is an autosomal dominant disorder caused by mutation in the *collagen *COL2A1* gene.

multiple myeloma *See* MYELOMA CELL.

multiple myeloma oncogene-1 An oncogene encoding a protein (interferon regulatory factor 4, IRF4, 405 aa) that interacts with *myd88 and is involved in a regulatory pathway for B-cell proliferation and differentiation. This regulatory system is aberrant in most *myelomas, although the IRF4 is not altered itself. In some cases IRF4 is overproduced as a result of a chromosomal translocation that puts it next to the immunoglobulin heavy chain locus. *See* IRFs.

multiple paragangliomata Tumours (glomus tumours) derived from paraganglia located throughout the body. Those in which the cells are chromaffin types have endocrine activity and are usually referred to as *phaeochromocytomas. **Familial paragangliomas-1** (PGL1) is caused by mutation in the *SDHD* gene, which encodes the small subunit of cytochrome B in succinate-ubiquinone oxidoreductase. **PGL2** has been mapped to chromosome 11q13. **PGL3** and **PGL4** are a result of mutations in genes for other subunits of the oxidoreductase.

multiple sclerosis (disseminated sclerosis) A neurodegenerative disease characterized by demyelination, particularly in the periventricular areas of the brain. Peripheral nerves are not affected. Onset is usually in the third or fourth decade with intermittent progression over an extended period. The cause is uncertain but a number of specific genes, including those for protein tyrosine phosphatase (CD45), the IL-7 receptor, and CD24, are involved in susceptibility. There is also an association with a particular HLA haplotype in high-risk northern European populations.

SEE WEB LINKS

- The MS Society homepage.

multiple sulphatase deficiency A *lysosomal storage disease caused by mutation in the *sulphatase-modifying factor-1 gene that has features of metachromatic *leukodystrophy and of a *mucopolysaccharidosis.

multipotent cell A progenitor or precursor cell, or *stem cell, that can give rise to a range of different cell types.

multipotential colony stimulating factor *See* INTERLEUKIN-3.

multivesicular body A secondary *lysosome (MVB) that contains many vesicles of ~50 nm diameter. MVB formation is important in receptor down-regulation, viral budding, antigen presentation, and the generation of lysosome-related organelles.

mulundocandin A lipopeptide of the *echinocandin type, isolated from the culture broth of a strain of *Aspergillus sydowi*, that has a wide spectrum of antifungal activity.

mumps Infection of the parotid salivary glands by a paramyxovirus. The virus is highly neurotropic, and before widespread vaccination was a major cause of viral meningitis.

munc-13 One of the *CAZ proteins (homologue of *Caenorhabditis elegans* UNC-13 protein, 1591 aa) that promotes the priming of synaptic vesicles by acting through *syntaxin.

MUP **1.** 4-methylumbelliferylphosphate, a fluorogenic substrate used in phosphatase assays. **2.** Motor unit potential. An action potential in a muscle unit. **3.** *See* MAJOR URINARY PROTEIN. **4.** *See* MUP GENES.

***mup* genes** The genes that encode type I polyketide synthase and other enzymes that convert polyketides into the active antibiotic *mupirocin.

mupirocin A polyketide-derived antibiotic from *Pseudomonas fluorescens*, that inhibits bacterial cell wall synthesis by blocking isoleucyl-tRNA synthase. *TNs* Bactroban, Centany.

m

muramic acid A subunit of *peptidoglycan in bacterial cell walls.

muramidase *See* LYSOZYME.

muramyl dipeptide A *peptidoglycan fragment (*N*-acetylmuramic acid-L-Ala-D-isoGln) from mycobacterial cell walls, used as an *adjuvant.

murein *See* PEPTIDOGLYCAN.

muscarine A toxic alkaloid, from fly agaric *Amanita muscaria*, that binds to *muscarinic acetylcholine receptors.

muscarinic acetylcholine receptors *G-protein-coupled receptors for acetylcholine that, unlike the *nicotinic acetylcholine receptors, are not ion channels. Five subtypes of muscarinic receptors have been identified (M1–M5). The M1 receptor is abundant in the hippocampus and cerebral cortex where it may modulate NMDA receptor-meditated excitatory synaptic transmission and have a role in memory and attentiveness. The M3 receptor may be important in regulating food intake.

muscle Specialized contractile tissue. *See* CARDIAC (cardiac muscle); SMOOTH MUSCLE; STRIATED MUSCLE; TWITCH MUSCLE. Striated (skeletal) muscle is a *syncytium formed by the fusion of embryonic *myoblasts; cardiac muscle has distinct cells but they are electrically linked by specialized junctional complexes (*intercalated discs); smooth muscle is composed of separate cells that have large amounts of actin and myosin and are capable of contracting to a small fraction of their resting length.

muscle–eye–brain disease A disorder in which there is severe early-onset muscle weakness, mental retardation and pathological change in the eye. Caused by mutations in a *N*-acetylglucosaminyl transferase (EC 2.4.1.-) which is involved in *O*-mannosyl glycan synthesis; hypoglycosylation of α-*dystroglycan alters the binding of laminin, neurexin, and agrin. A similar glycosylation defect apparently underlies Fukuyama's *muscular dystrophy and the *Walker–Warburg syndrome.

muscle fibre A subunit of a skeletal muscle comprising a single syncytial cell with peripheral nuclei and cytoplasm packed with *myofibrils.

muscle spindle A specialized bundle of muscle fibres innervated by sensory neurons that act as a stretch receptor within the muscle.

muscular dystrophy (MD) A group of diseases characterized by progressive degeneration and/or loss of muscle fibres leading to muscle weakness. In some cases there are also abnormalities of the nervous system. Nearly all of the dystrophies are hereditable, but the genetic defect and the prognosis for the disease vary from type to type. **Duchenne MD** (pseudohypertrophic MD) is the most common form with an incidence of about 1 in 4000 male births, of which one-third are estimated to be new mutational events. It is a sex-linked recessive disorder caused by mutation in the gene for *dystrophin; there is degeneration and necrosis of skeletal muscle fibres with extensive but insufficient muscle fibre reformation from *satellite cells. **Becker MD** is a benign X-linked muscular dystrophy with later onset and lower severity than Duchenne. The mutation is also in the *dystrophin gene. **Emery–Dreifuss MD** is usually an X-linked late-onset dystrophy caused by a mutation in the gene for *emerin although autosomal variants exist where the mutation is in *lamin and affects the emerin–lamin interaction. **Facioscapulohumeral MD**, the third most common hereditary disease of muscle, initially involves the face and the scapulae and maps to the distal end of chromosome 4q. **Fukuyama type MD** is caused by mutation in *fukutin. **Limb girdle MD** generally causes weakness in the shoulder and pelvic girdles, particularly the large muscles nearest the girdles, with leg weakness often preceding that in the arm. There are two classes, autosomal dominant (class 1) and autosomal recessive (class 2), both classes being genetically heterogeneous. Autosomal dominant forms arise from mutations in the genes for *myotilin (*LGMD1A*), *lamin A/C (*1B*), *caveolin-3 (*1C*), and others that are mapped but not identified (*1D–1G*). Myotiliopathy has similar clinical features. The recessive forms arise from mutations in *calpain-3 (*LGMD2A*), *dysferlin (*2B*: also Miyoshi myopathy) γ-*sarcoglycan (*2C*), α-sarcoglycan (*2D*), β-sarcoglycan (*2E*), δ-sarcoglycan (*2F*), *telethonin (*2G*), tripartite motif-containing protein-32 (*2H*), fukutin-related protein (*2I*), *titin (*2J*), and protein *O*-mannosyltransferases-1 (*2K*). **Merosin-deficient congenital MD** type 1A is caused by mutation in the *laminin α_2 gene. **Rigid spine MD-1**, severe classic multiminicore myopathy, and desmin-related myopathy with Mallory bodies, which are believed to be part of the same disease spectrum, are caused by mutation in the *selenoprotein N gene. **Tibial MD** is caused by heterozygous mutation in the gene for titin (homozygous mutation causes limb girdle MD2J). **Ullrich congenital MD** can be caused by recessive mutations in any of the three genes that encode the subunits of *collagen type VI and *Bethlem myopathy is a benign form of this disorder. Other degenerative muscle disorders include **myotonic dystrophy-1** (DM1), caused by mutation in the *dystrophia myotonica protein kinase gene; DM2 is caused by mutation in the zinc finger protein 9 gene. **Myofibrillar myopathy** (MFM) is a noncommittal term that refers to a group of morphologically homogeneous, but genetically heterogeneous chronic neuromuscular disorders that can arise from mutations in the genes for αB-*crystallin, LIM domain-binding protein 3, *filamin C, *myotilin, and *desmin.

SEE WEB LINKS

- The Muscular Dystrophy Campaign homepage.

MuSK A receptor-like tyrosine kinase apparently restricted to skeletal muscle in its expression and localized at the neuromuscular junction. In experimental systems, MuSK-*dishevelled interaction regulates the *agrin-stimulated acetylcholine receptor clustering that is important for formation of the neuromuscular junction. A missense mutation that altered expression of MuSK, but not its catalytic activity, led to decreased agrin-dependent acetylcholine receptor aggregation in a patient with congenital myasthenic syndrome.

mut **1.** A family of genes encoding proteins involved in DNA repair (*see* MISMATCH REPAIR). Originally described in bacteria, but homologues are found in eukaryotes; e.g. the *mut7* gene in *Caenorhabditis elegans*, which is homologous to the human gene mutated in *Werner's syndrome, encodes a protein implicated in RNA interference. **2.** Methylmalonyl-CoA mutase (MUT) (EC 5.4.99.2) is a mitochondrial enzyme that catalyses the isomerization of methylmalonyl-CoA to succinyl-CoA.

mutagen An agent that causes an increase in the rate of *mutation, usually by causing damage to DNA. Ionizing radiation, UV irradiation (260 nm), and various chemicals are mutagenic. The *Ames test is a test for mutagenicity.

mutation An alteration in the DNA sequence of an organism, which can arise for many different reasons. *See* FRAME-SHIFT MUTATION, nonsense mutation (at *nonsense codons), and *missense mutation. The mutation rate can either be the frequency with which a particular mutation appears in a population or the frequency with which any mutation appears in the whole genome of a population. The context should make it clear which meaning is intended. *See* FLUCTUATION ANALYSIS.

muted The human homologue of the mouse gene (the mice are of a muted brown colour), shown to be mutated in the *Hermansky–Pudlak syndrome gene. The human protein (187 aa, 76% homologous to the mouse protein but with a vacuolar targeting motif) together with *pallidin, is a component of the *BLOC-1 complex (biogenesis of lysosome-related organelles complex-1).

MVB *See* MULTIVESICULAR BODY.

Mx proteins Proteins belonging to the large GTPase family (70–100 kDa) that were first identified in mice resistant to infection with myxovirus (influenza A and B) and subsequently found in all eukaryotes that have been investigated. Mx1 (MxA; p78; interferon induced 78-kDa protein, 662 aa) expression is up-regulated by interferon and it blocks transcription of the viral RNA genome. Other Mx proteins are cytoplasmic and are related to dynamin.

myalgia Muscle pain.

myalgic encephalomyelitis A long-term postviral syndrome (ME, chronic fatigue syndrome) in which there is chronic fatigue and muscle pain (myalgia) on exercise. Diagnostic markers are poor but it is now accepted as a real disorder after many years of being dismissed as being psychosomatic ('yuppie flu'). It is unclear whether it is a single disorder.

myasthenia gravis A disorder in which there is progressive muscular weakness on exercise. Autoantibodies to the acetylcholine receptor (usually the α subunit) at the neuromuscular junction are present and are considered to be responsible for the symptoms; there is often an association with other autoimmune disorders. Susceptibility is associated with the HLA-D2 antigen and with a variant in the gene for the acetylcholine receptor α subunit which leads to a failure of thymic-induced tolerance of the antigen. **Congenital myasthenic syndrome with episodic apnea** (CMS-EA; formerly familial infantile myasthenia gravis) is caused by mutation in the choline acetyltransferase gene (EC 2.3.1.6), which is responsible for the biosynthesis of acetylcholine.

myasthenic syndrome Muscular weakness that can arise through inherited defects in components of the neuromuscular junction (congenital myasthenic syndrome) or through autoimmune disorders that act on the acethylcholine receptors (*myasthenia gravis) or the calcium channels at the neuromuscular junction (Lambert–Eaton myasthenic syndrome, LEMS). Congenital forms can arise through mutations in presynaptic release of acetylcholine, postsynaptic defects in acetylcholine receptors or, in the synaptic form, a deficiency of *acetylcholine esterase. *See* RAPSYN.

myb An *oncogene, originally identified in avian myeloblastosis, encoding proteins that are critical for haematopoietic cell proliferation and development. Some patients with T-cell acute lymphoblastic leukaemia have duplications of *myb*.

myc A *proto-oncogene, originally identified in several avian tumours, encoding transcription factors (65 kDa) with a C-terminal basic helix–loop–helix–zipper domain. Myc-*Max heterodimers activate transcription but Myc in association with Miz1 ('Myc-interacting zinc finger protein-1) can repress transcription. N-myc is found in neuroblastomas. *See* BURKITT'S LYMPHOMA, *see also* MAX (myc associated factor X).

Mycobacteria A genus of bacteria of the order Actinomycetales that have unusual waxy cell walls rich in lipid, especially esterified *mycolic acids, and as a result stain differently from Gram-positive and -negative bacteria. *Muramyl dipeptide (MDP) in the wall makes it strongly immunogenic. The cell walls resist digestion within lysosomes, making them serious infectious agents. *Mycobacterium leprae* causes *leprosy and *M. tuberculosis* causes *tuberculosis.

m

mycolactone A family of toxins produced by *Mycobacterium ulcerans*; strains from different geographic areas produce distinct patterns of mycolactone congeners. Mycolactone has significant immunosuppressive effects and inhibits production of macrophage inflammatory protein (MIP-1α), MIP-1β, RANTES, interferon-γ-inducible protein 10, and monocyte chemoattractant protein 1, but not IL-12, TNFα, or IL-6. *See* BURULI ULCER.

mycolic acids Saturated fatty acids, with chain lengths of 60–90 carbons, found in the cell walls of *mycobacteria, **Nocardia*, and **Corynebacteria*.

mycophenolic acid An antibacterial and antitumour compound from *Penicillium brevicompactum* that inhibits *de novo* nucleotide synthesis. Mycophenolate mofetil (*TN* CellCept) is derived from the fungus *Penicillium stoloniferum* and is used in combination with *ciclosporin and *corticosteroids to prevent organ transplant rejection. Mycophenolate sodium (*TN* Myfortic) has similar properties.

mycoplasma Mollicute bacteria that lack a cell wall and are resistant to many antibiotics. Formerly known as pleuropneumonia-like organisms (PPLO). *Mycoplasma pneumoniae* is a causative agent of pneumonia. Mycoplasma can contaminate cell cultures and alter their properties in a misleading fashion, and it is important to monitor for their presence.

mycosides Complex glycopeptidolipids that are major immunogens in the cell walls of many *mycobacteria. Mycoside C is a monoglycosylated fatty-acylated peptide to which small variable oligosaccharides are attached. Variability in the oligosaccharides accounts for serological differences between strains.

SEE WEB LINKS

- Information on glycopeptidolipids on the CyberLipid website.

mycosis A disease caused by a fungus. Superficial mycoses are limited to the outermost layers of the skin and hair (e.g. tinea); cutaneous mycoses extend deeper into the epidermis (e.g. ringworm); subcutaneous mycoses involve the dermis and deeper tissues; systemic mycoses are usually opportunistic infections and can be serious in immunocompromised individuals (e.g. *candidiasis). **Mycosis fungoides** is a T-cell tumour in which T-cells accumulate in the dermis and epidermis and there are often secondary fungal infections.

myc tag *See* EPITOPE TAG.

myd **1.** Myotonic dystrophy (MyD). *See* MUSCULAR DYSTROPHY. **2.** A gene (*myd*) involved in the *determination of muscle cells, but *compare* MYOD. **3.** A strain of mice, myodystrophy (myd) mouse (Large mice), in which there is a mutation in a glycosyltransferase which leads to incomplete glycosylation of *dystroglycan, resulting in altered properties of the sarcolemma of fast-twitch skeletal muscle fibres. **4.** Myeloid differentiation factor (MyD)-88 is an *adaptor protein that is important in innate immunity.

mydriatic A drug that causes the pupil of the eye to dilate.

mydricaine hydrochloride A solution containing procaine hydrochloride (anaesthetic), atropine sulphate (to cause dilation of the pupil), and adrenaline, used particularly in ophthalmology.

myelin The major component of the *myelin sheath of nerve axons. Myelin basic protein (MBP, 304 aa) exists as three isoforms (21, 18, and 17 kDa) produced by alternative splicing of the primary transcript; the 18-kDa form is myelin membrane encephalitogenic protein, which is used as an antigen to induce *experimental allergic encephalomyelitis. The 'shiverer' (shi) mouse has a mutation in the mouse *MBP* gene. *See also* MYELIN PROTEOLIPID PROTEIN and MAG (myelin-associated glycoprotein).

myelin proteolipid protein (lipophilin) The primary constituent of myelin in the central nervous system. There are two isofoms, PLP and DM20, produced by alternative splicing; both are very hydrophobic integral membrane proteins, tetraspanins, with two extracellar loops and one intracellular loop. Mutation in the *PLP* gene leads to X-linked *hereditary spastic paraplegia type 2 (SPG2) and to the X-linked dysmyelinating disorder of the central nervous system, *Pelizaeus–Merzbacher disease.

myelin sheath The insulating layer that surrounds peripheral *neurons and increases the speed of conduction. The sheath is actually of specialized *Schwann cells that wrap around the neuron up to 50 times. Areas of the neuron that are not encased by the sheath are *nodes of Ranvier.

myeloblasts Cells of the bone marrow that produce *myelocytes and thus the myeloid series of cells.

myelocytes The cells that give rise to *myeloid cells. They are derived from *myeloblasts and are found in bone marrow.

myelodysplasia (myelodysplastic syndrome) A group of haematological disorders in which there is reduced production of blood cells. Several different forms exist: refractory anaemia with or without ring sideroblasts (RA or RARS); refractory cytopenia with multilineage dysplasia (RCMD); *5q minus syndrome; refractory anaemia with excess blasts (RAEB); unclassified myelodysplasia that does not fit into one of the other categories.

myeloid cells One of the two main classes of cells produced by the bone marrow, the others being *lymphoid cells. The myeloid series includes *megakaryocytes, erythrocyte precursors, *mononuclear phagocytes, and all the *polymorphonuclear leucocytes. Myeloid cells all derive from a single stem cell and the *Philadelphia chromosome is found in the whole series. Myeloid is often used in a narrower sense to mean mononuclear phagocytes and granulocytes, excluding the erythroid lineage.

myeloid derived suppressor cells Cells of the myeloid lineage that infiltrate tumours, induce tumour-specific T-cell tolerance, and facilitate tumour growth and metastasis. They show increased arginase activity and production of reactive oxygen species.

myeloid nuclear differentiation antigen A protein (406 aa), found only in nuclei of cells of the granulocyte–monocyte lineage, that has regions of sequence homology with several interferon-induced proteins and probably has a role in the response to interferon. Levels are reduced in familial and sporadic cases of *myelodysplasia. MNDA has a *pyrin domain.

myeloma cell A neoplastic *plasma cell. In multiple myeloma the proliferating plasma cells dominate the marrow, leading to immune deficiency and often destruction of surrounding bone. Various molecular events can cause the disorder, including dysregulation of oncogenes by translocations to the IgH locus, mutations in genes encoding cyclin D1 and fibroblast growth factor receptor-3. (*See also* MULTIPLE MYELOMA ONCOGENE-1.) Myeloma cells are monoclonal and therefore secrete a single type of immunoglobulin, a characteristic exploited in the production of monoclonal antibodies by *hybridoma cells; see BENCE JONES PROTEIN.

myeloperoxidase A lysosomal haemoprotein (MPO, EC 1.11.1.7), a heterodimer of 60-kDa heavy and 12-kDa light chains, found in azurophil granules of neutrophil leucocytes and in macrophages. In the presence of halide ions, MPO hydrolyses hydrogen peroxide generated in the *metabolic burst and is important for bactericidal activity, but individuals deficient in MPO are asymptomatic and chickens manage without the enzyme. A polymorphism in the *MPO* gene is reportedly a risk factor for Alzheimer's disease.

myeloplax *See* MEGAKARYOCYTE.

myf A group of transcription factors (myogenic factors) involved in differentiation of muscle. The product of the *MYF5* gene is structurally related to *MyoD1. Mutation in *MYF6* (herculin) causes a mild centronuclear myopathy.

myoblast A cell that will fuse with other myoblasts to form a *myotube which differentiates into a skeletal muscle fibre. The term can be applied to any cell recognizable as a precursor of skeletal muscle fibres, although some authors reserve the term for postmitotic cells capable of fusion and refer to their precursors as presumptive myoblasts.

myobrevin *See* SYNAPTOBREVINS.

myocardial infarction A condition in which the blood supply to part of the heart is interrupted, often as a result of a thrombus blocking a coronary artery. If the ischaemia is prolonged then there will be necrotic damage to the heart (infarction).

myocardium The central muscular layer of the wall of the heart. **Myocarditis** is inflammation of the cardiac muscle.

myocilin A cytoskeletal protein (trabecular meshwork glucocorticoid-inducible response protein, TIGR, 504 aa) that is found in ocular tissues, including the *trabecular meshwork, but also in most tissues other than brain, placenta, liver, kidney, spleen, and leucocytes. Levels of myocilin in the trabecular network can be up-regulated by steroids which leads to an increase in intraocular pressure. *Optimedin has 40% amino acid similarity to myocilin and a comparable pattern of tissue expression. Mutations in the *MYOC* gene encoding myocilin are responsible for primary open-angle *glaucoma (POAG). Interacts with *flotillin-1.

myoclonic epilepsy of Unverricht and Lundborg A convulsive disorder (EPM1) with childhood onset caused by mutation in the *stefin B (cystatin B) gene. EPM1B is caused by mutation in the **PRICKLE1* gene that encodes REST-interacting LIM domain protein (831 aa), probably a nuclear receptor.

myoclonus **1.** An abrupt, spasmodic contraction of a muscle. *Adj.* myoclonic. **2.** Paramyoclonus multiplex is a disorder in which sudden shock-like contractions of muscles occur and is often associated with epilepsy.

myocyte enhancer factor 2 (MEF2) A family of transcription factors in the *MADS superfamily that bind to the MEF2 locus present in most muscle-specific genes. MEF-2A is calcium-regulated and promotes cell survival during development and induction of myogenesis, whereas MEF2C may be more involved with maintenance of the differentiated state. Mutations in MEF2A are responsible for autosomal dominant coronary artery disease with acute myocardial infarction. MEF2B is important in the regulation of T-cell apoptosis, normally inactive because bound to *cabin1 but released when T-cell receptor activation increases intracellular calcium. *MEF2C* is expressed in skeletal muscle, spleen, brain, and various myeloid cells. In monocytic cells *MEF2C* has its transactivation activity enhanced by *LPS acting through the MAP kinase *p38 and is itself regulated by myeloid-specific microRNA-223 (miR223). MEF2D is present in undifferentiated myoblasts and in the cerebellum and cerebrum of developing mouse brains, suggesting that it may be involved in the early stages of differentiation of both nerves and muscles.

MYOD An important regulatory gene (myogenic determination gene) involved in *determination of muscle cells and normally only expressed in myoblasts and skeletal muscle cells. The gene product (myoD) triggers the transcription of many muscle specific genes. *MYOD* is one of a family of genes that encode nuclear proteins that bind to a consensus sequence (-CANNTG-) and form homodimers, or heterodimers, with other members of the basic *helix-loop-helix transcription factor superfamily which includes *myf.

myoepithelial cell A type of cell (basket cell, basal cell) that resembles smooth muscle cells and is found in exocrine glands (e.g. salivary, sweat, mammary, and mucous glands).

myoferlin *See* DYSFERLIN; FERLIN.

myofibril The highly organized array of thick and thin filaments, arranged as *sarcomeres, that is the contractile element in striated muscle.

myofibroblasts A class of fibroblast-like cells resembling smooth muscle cells, that have arrays of actin microfilaments, myosin, and other muscle proteins arranged in such a way as to indicate they are capable of contraction. They are commonly found in granulation tissue formed during wound healing and are also found in the intima of some thickened arteries.

myogenesis The process of striated muscle development in which *myoblasts fuse to form *myotubes that increase in size by further myoblast recruitment, form *myofibrils within their cytoplasm, and establish functional *neuromuscular junctions with *motoneurons. The basic process is similar in cardiac muscle although there is not complete fusion, just the formation of specialized junctional complexes (*intercalated discs).

myogenin A member of the *MyoD family of transcription factors (myf4, 224 aa) involved in regulating muscle-specific genes and some microRNAs.

myoglobin A haem protein (153 aa) found in red striated muscle. Myoglobin has a higher affinity for oxygen and so oxygen is transferred from blood haemoglobin into the mucle where it forms a storage pool of oxygen. Muscles that work constantly tend to have higher concentrations of myoglobin.

myokymia A form of involuntary muscular movement, that can be visualized on the skin as worm-like twitching under the skin or continuous rippling of the skin surface.

myomesin (skelemin) A linker protein, found in the *M line of the *sarcomere, that interacts with *titin. Two isoforms occur, myomesin 1 (skelemin, 1451 aa) and myomesin 2 (1465 aa).

myometrium The middle layer of the uterine wall consisting mainly of smooth muscle cells with associated matrix and vasculature. It lies between the inner *endometrium and the outer perimetrium. The myometrium increases in thickness by growth and division of smooth muscle cells during pregnancy.

myopalladin A protein (1320 aa) present in the Z line of the sarcomere that links skeletal muscle *nebulin and cardiac muscle *nebulette to α-actinin. It also colocalized with cardiac ankyrin repeat protein (Carp) in the I band where it may be involved in structural integrity.

myopathic carnitine deficiency *See* SYSTEMIC CARNITINE DEFICIENCY.

myopathy A disorder in which muscle fibres do not function properly and there is muscle weakness. *See* B CRYSTALLINOPATHY; BETHLEM MYOPATHY; CENTRONUCLEAR MYOPATHY; COMPTON–NORTH CONGENITAL MYOPATHY; DANON'S DISEASE; DERMATOMYOSITIS; DILATED CARDIOMYOPATHY; HYPERTROPHIC CARDIOMYOPATHY; INCLUSION BODY MYOPATHY WITH PAGET'S DISEASE AND FRONTOTEMPORAL DEMENTIA; ISCHAEMIC CARDIOMYOPATHY; MELAS SYNDROME; MYOSCLEROSIS, and NEMALINE MYOPATHY. *See also* NEUROMUSCULAR DISORDERS.

myopodin An actin bundling protein (three isoforms, 1093, 1109, and 1261 aa respectively) with a nuclear localization signal that can shuttle between nucleus and cytoplasm. Myopodin can become hypermethylated, especially in some bladder cancers, and the hypermethylation apparently prevents it from activating gene expression. Myopodin is a tumour suppressor that inhibits prostate cancer growth and metastasis and has significant homology with *synaptopodin.

myosclerosis Autosomal recessive myosclerosis (myosclerotic myopathy, congenital myosclerosis of Lowenthal) consists of a chronic inflammation of skeletal muscle with hyperplasia of the interstitial connective tissue caused by mutations in *collagen type VI (*COL6A2*).

myosin A large family of motor ATPases that interact with filamentous actin, either in *thin filaments or cytoplasmic *microfilaments. There are at least fifteen classes of myosin and the 'classical' myosin of striated muscle (myosin II) is actually unusual in forming an ordered multimolecular array in the thick filament. Myosin II is hexameric with two *myosin heavy chains and two pairs of *myosin light chains (*see also* MEROMYOSIN). The unconventional myosins do not self-assemble but all have similar N-terminal motor domains, with actin-binding and ATPase activity, a linking region that binds *myosin light chains (and has an *IQ motif) and a tail region that determines the functional properties and tissue location, often interacting with specific types of vesicles. Most are probably capable of processive movement along a microfilament by virtue of being dimeric and thus able to 'walk' along the filament. Myosin I has a tail that interacts with membrane phospholipids and is therefore probably involved in transporting 'cargo' vesicles. Myosin VA (myosin 5A, myoxin) is associated with the centrosomal region and mutations in myosin VA are implicated in some types of *Griscelli's syndrome. Myosin 7A is expressed in pigmented retinal epithelium and photoreceptors and mutations may be the cause of *Usher's syndrome (type IB). Myosins 6 and 15 are also essential for hearing; myosin 15 interacts with *whirlin at the tips of stereovilli. Myosin 10 is associated with *filopodia.

myosin-binding proteins A family of muscle proteins that bind myosin (MyBPs). The most abundant type is **MyBPC,** which is transversely arrayed in the cross-bridge-bearing (C region) of the A band of the sarcomere in a fixed molar ratio with myosin heavy chains. **MyBPC2** (1142 aa) is found in fast skeletal muscle, **MyBPC1** in slow muscle. **MyBPC3**, the cardiac isoform, is expressed exclusively in heart muscle. Protein kinase A-mediated (PKA) phosphorylation of MyBPC3 accelerates the kinetics of cross-bridge cycling and regulates myocyte power output. Mutations are associated with ~15% of cases of familial *hypertrophic cardiomyopathy. *See* OBSCURIN. Myosin-binding protein H (**MyBPH**, 52 kDa) is found in skeletal muscle in a specific pattern.

myosin heavy chain The large subunit of *myosin (~2000 aa) that has the ATPase activity. The head region undergoes conformational change during the actin attachment/detachment cycle, each cycle involving hydrolysis of one ATP molecule, and it is this conformational change that converts energy into movement. In conventional filament-forming myosins (type II myosin) the head is attached via a hinge region to a tail domain that interacts laterally with other myosin tails to form the backbone of the thick filament In the nonmuscle myosins, the tail may attach to 'cargo' vesicles. In skeletal muscle there are seven heavy chain isoforms, two developmental, three in adult skeletal muscle, one also expressed in cardiac muscle, one expressed primarily in extrinsic eye muscles.

Mutations in the gene for myosin heavy chain 9 lead to the *May–Heggelin anomaly. *Heavy meromyosin is a subfragment of the heavy chain of myosin II. *See also* MYOSIN LIGHT CHAINS.

myosin light chain kinase (MLCK) A *calmodulin-regulated serine/threonine kinase that phosphorylates *myosin II light chains and activates the myosin motor. At least three different isoforms exist in heart, skeletal, and smooth muscle, and the smooth muscle form is found in nonmuscle cells. Mutations in *MLCK2* are associated with midventricular hypertrophic cardiomyopathy.

myosin light chains Two pairs of smaller subunits (17–22 kDa) of hexameric *myosin II, each pair being different. The two regulatory light chains (LC-2, DNTB light chains, *MYL2* gene products) may directly confer calcium sensitivity on the ATPase activity, or indirectly when they are phosphorylated by calcium-regulated *myosin light chain kinase. Essential light chains (LC-1, LC-3; alkali light chains, *MYL3* gene products) may have a partly structural role. Multiple tissue-specific variants exist. Mutation in the gene for the essential light chains causes some types of *hypertrophic cardiomyopathy.

myositis Muscle inflammation that can arise from infection by *Clostridium welchii* (gas gangrene), viral infection (epidemic myalgia; usually due to Coxsackie B virus) or infection with the nematode *Trichinella.*

myostatin A member of the TGFβ superfamily (growth/differentiation factor-8a, GDF8, 375 aa) that regulates muscle mass. It is present in skeletal muscle as a 26-kDa mature glycoprotein (myostatin-immunoreactive protein) and is secreted into the plasma. Mutation of the myostatin gene can lead to hypertrophy of muscle and is common in some of the 'double-muscled' strains of cattle and sheep.

myotilin A cytoskeletal protein (titin immunoglobulin domain protein, TTID, 498 aa) that colocalizes with α-actinin in the I band of the sarcomere and has two immunoglobulin domains similar to those in *titin. Mutations in the gene lead to myotilinopathy, a form of myofibrillar myopathy (*see* MUSCULAR DYSTROPHY) in which there is myofibrillar degradation and progressive muscle weakness.

myotonia A skeletal muscle disorder characterized by muscle stiffness and an inability of the muscle to relax after voluntary contraction. There is no atrophy or hypertrophy of the muscle. Both autosomal dominant myotonia congenita (Thomsen's disease) and the more severe autosomal recessive myotonia congenita (Becker's disease) are due to mutations in the muscle *chloride channel, CLCN1, which regulates the electric excitability of the skeletal muscle membrane. **Potassium-aggravated myotonia** is an autosomal dominant disorder provoked by fasting and oral potassium administration. It is caused by a mutation in the *SCN4A* gene that encodes the α subunit of a *voltage-gated sodium channel and that is also mutated in **paramyotonia congenita**, a nonprogressive disorder in which there is myotonia exacerbated by cold, intermittent muscle weakness and lability of serum potassium. Mutation in *SCN4A* also causes hyperkalaemic *periodic paralysis.

myotonic dystrophy *See* MUSCULAR DYSTROPHY.

myotoxins Small basic peptide toxins (42–45 aa), found in rattlesnake venom, that induce rapid paralysis and muscle necrosis.

myotube An elongated multinucleate cell formed by the fusion of several *myoblasts. There may be some *myofibrils and they are an immature form of muscle fibres.

myotubularin One of a family of putative *tyrosine phosphatases required for muscle cell differentiation. Myotubularin 1 (621 aa) is only one of eight myotublarin gene products. Mutations in myotubularin 1 lead to X-linked myotubular myopathy-1; a close homologue, myotubularin 2 is mutated in a recessive form of *Charcot–Marie–Tooth neuropathy.

myozenin (calsarcin) A set of striated muscle-specific *calcineurin-interacting proteins that colocalize with α-actinin, tether calcineurin to the sarcomere and may be important for the structural integrity of the Z-disc. Myozenin 1 (calsarcin, 2299 aa) is expressed only in adult fast skeletal muscle, myozenin 2 (calsarcin-1) is expressed specifically in adult cardiac and slow-twitch skeletal muscle. Myozenin 3 (calsarcin 3) is restricted to skeletal muscle.

myristic acid A C-14 fatty acid that is a relatively uncommon acyl residue in phospholipids but is often post-translationally added to the amino group of N-terminal glycine in membrane-associated proteins and many viral proteins by myristoyl CoA:protein *N*-myristoyl transferase. Myristoylated proteins (e.g. small G proteins) bind strongly to membranes although the targeting mechanism is not clear. *See* MARCKS.

SEE WEB LINKS

- Further information on the CyberLipid website.

myxoedema A skin and tissue disorder, usually caused by prolonged hypothyroidism as in *Hashimoto's thryroiditis.

myxomatous Describing a form of tissue degradation in which there is accumulation of mucus.

Myxoviridae An obsolete name for a group of single-stranded RNA viruses that is now subdivided into *Orthomyxoviridae and *Paramyxoviridae.

NAAG A peptide neurotransmitter (*N*-acetyl-L-aspartyl-L-glutamate) in the central nervous system that suppresses glutamate transmission through selective activation of presynaptic group II metabotropic glutamate receptor subtype 3 (mGluR3). *See* GLUTAMATE CARBOXYPEPTIDASE II; ZJ-43.

nabilone A synthetic cannabinoid used as an antiemetic and to treat neuropathic pain. *TN* Cesamet.

nabothian cyst A benign mucus-filled cyst on the surface of the cervix.

NAC *See* NASCENT POLYPEPTIDE ASSOCIATED COMPLEX.

***N*-acetylglucosamine** A sugar moiety (2-acetamidoglucose) that is found in many glycoproteins and in polysaccharides such as *chitin, bacterial *peptidoglycan, and *hyaluronic acid.

***N*-acetylmuramic acid** The subunit of bacterial peptidoglycan, which has repeat units of *N*-acetylmuramic acid linked to *N*-acetylglucosamine via a β(1–4)-glycosidic bond. *Lysozyme cleaves the linkage between the two sugars.

***N*-acetylneuraminic acid** A C-9 amino sugar, the commonest of the *sialic acids. Confers negative charge to *glycolipids, especially *gangliosides, and *glycoproteins, and is responsible for most of the net negative charge on cell surfaces.

nAChR *See* NICOTINIC ACETYLCHOLINE RECEPTOR.

NAD *See* NICOTINAMIDE ADENINE DINUCLEOTIDE.

nadolol A nonselective *β-blocker used to treat hypertension. *TN* Corgard.

NADP *See* NICOTINAMIDE ADENINE DINUCLEOTIDE.

NADPH oxidases (NOX1, NOX2) Enzymes that catalyse the transfer of electrons from NADPH to molecular oxygen to generate reactive oxygen species (ROS). The NADPH oxidase activator, NOXA1 (p51-NOX, 476 aa), can stimulate both NOX1 and NOX2 (p91phox, 570 aa). The **NADPH oxidase organizer** (NOXO1, p41-NOX, 371 aa) targets NOX activators to NOX and to different subcellular compartments. *See* PHOX.

nadrin (RICH-1) A neuron-specific rho GTPase-activating protein involved in regulated exocytosis.

Naegleria A genus of soil-dwelling amoebae that can cause opportunistic infections.

naevus A benign *hamartoma of skin. A **vascular naevus** has an extensive capillary network and as a result is red ('strawberry birthmark'; sometimes the much more extensive 'port wine stain'). A **mole** is a nonmalignant pigmented naevus, a cluster of highly pigmented *melanocytes. A **blue naevus** is an unusual form of mole in which the melanocytes are deep in the *dermis.

Nagler's reaction A standard microbiological method for identifying *Clostridium perfringens* in which bacteria are cultured on agar containing egg yolk; an opalescent halo is formed around colonies that produce lecithinase (clostridial α-toxin).

Na^+/H^+ exchanger regulatory factor *See* EBP50.

nail–patella syndrome A disorder (Turner–Kieser syndrome, Fong's disease) in which there is dysplasia of the nails and the patellas are absent or hypoplastic. It is caused by mutation in the limb homeobox transcription factor gene (*LMX 1B*).

Nairovirus A genus of the *Bunyaviridae. One of the group is responsible for *Crimean–Congo haemorrhagic fever.

nalidixic acid An antibiotic that is effective against both Gram-positive and -negative bacteria, often used in treating urinary tract infections and experimentally in selective media. It inhibits prokaryotic *DNA gyrase. *TN* NegGram.

naloxone A drug used to treat opiate overdoses. It has a very high affinity for μ-opioid receptors in the central nervous system and acts as a competitive antagonist. *TNs* Nalone, Narcan, Narcanti.

NALP proteins A subfamily of the caterpillar protein family, cytoplasmic proteins that are implicated in the activation of proinflammatory caspases and are involved in *inflammasomes. NALP 7 may play a role in cell proliferation and is a negative regulator of IL-1β; mutations are associated with recurrent *hydatidiform mole.

naltrexone An *opioid receptor antagonist. *TNs* Depade, Nalorex, Revia.

Namalwa cells A line of human B cells, derived from a patient with *Burkitt's lymphoma, that grow in suspension and are used to produce interferon.

nandrolone An anabolic steroid that may occur naturally and is used to treat aplastic anaemia and, formerly, osteoporosis. *TNs* Deca-Durabolin, Durabolin.

nanobacteria A group of very small (~200 nm) bacteria, although this is controversial and many authors have disputed that they are actually living organisms. Proponents claim that they are responsible for producing aggregates of apatite and are associated with diseases in which there is tissue calcification.

NAP-1 *See* INTERLEUKIN-8.

NAP-3 *See* MELANOMA GROWTH STIMULATORY ACTIVITY.

naphthylamine (β-naphthylamine) A compound used in production of aniline dyes and associated with bladder cancer. The carcinogenic action is due to a metabolite produced by hydroxylation (1-hydroxy-2-aminonaphthalene) which is detoxified in the liver but reactivated by a glucuronidase in the bladder.

naproxen A *NSAID that inhibits cyclooxygenases 1 and 2. *TN* Naprosyn.

naratriptan A *triptan drug used to treat migraine, an agonist for the $5HT_1$ subclass of serotonin receptors. *TN* Amerge.

NASBA A method (nucleic acid sequence-based amplification) used to amplify nucleic acids in a single mixture at one temperature, most commonly for amplification of single-stranded RNA.

SEE WEB LINKS

- The PCR Encyclopedia contains further details.

nascent polypeptide associated complex (NAC) A heterodimeric component (α 215 aa, β ~150 aa) of cytoplasmic ribosomes, that interacts with nascent polypeptides and prevents those without a *signal sequence from being directed to the endoplasmic reticulum. The β subunit is a transcription factor (basic transcription factor 3b). NAC is highly conserved, and mutations cause embryonic lethality in mice.

nasopharyngeal carcinoma A carcinoma, highly prevalent in southern China, for which both genetic and nongenetic factors have been identified. These factors include *Epstein–Barr virus infection, the HLA-Bw46 locus, mutations in the *TP53* gene, and a major susceptibility locus on chromosome 4. Environmental factors probably include inhaled cocarcinogens.

Nasu–Hakola disease A recessive disorder (polycystic lipomembranous osteodysplasia with sclerosing leukoencephalopathy, presenile dementia with bone cysts) characterized by a combination of psychotic symptoms, rapidly progressing to presenile dementia, and bone cysts restricted to wrists and ankles.

natriuretic peptides A family of peptides, with sequence and structural homology, that stimulate secretion of sodium in the urine (natriuresis). The family includes ***atrial natriuretic peptide** (ANP), **brain-type natriuretic peptide** (BNP) which is produced in the ventricles of the heart, **C-type natriuretic peptide** (CNP), and possibly osteocrin/musclin. C-type natriuretic peptide (CNP-22) is a vasodilator and is an important regulator of blood pressure, renal function, volume homeostasis, and long-bone growth, possibly counteracting the effects of ANP. It is produced by endothelial and renal cells, regulates endothelial function and is a neuropeptide. **Natriuretic peptide receptor-A** (NPR1,EC 4.6.1.1, 1061 aa) is a membrane-bound guanylate cyclase that binds both ANP and BNP. **NPR-B** (NPR2, EC 4.6.1.2, 1047 aa) is the receptor for CNP; mutations cause an autosomal recessive skeletal dysplasia (Maroteaux-type *acromesomelic dysplasia). The **NPR-C** (NPR3, 541 aa) is the atrial natriuretic peptide clearance receptor. *See also* DENDROASPIS NATRIURETIC PEPTIDE.

natural killer cells *See* KILLER CELLS.

n

natural selection The process that confers reproductive advantage on individuals with particular phenotypes so that there is an increase in the proportion of the population that has the alleles responsible for the phenotype. Darwin used the term in contrast to the 'artificial selection' practised by animal and plant breeders.

SEE WEB LINKS

- Evolution 101's pages on natural selection.

Na_V channels *See* VOLTAGE-GATED SODIUM CHANNELS.

Navajo neurohepatopathy An autosomal recessive multisystem disorder prevalent in the Navajo population of the southwestern USA. Abnormal regulation of mitochondrial DNA copy number may be the primary defect.

Naxos disease An autosomal recessive disorder (palmoplantar keratoderma with arrhythmogenic right ventricular cardiomyopathy and woolly hair) first reported in families on the Greek island of Naxos and due to mutation in *plakoglobin. Related disorders are associated with mutations in *desmoplakin but there appears to be great heterogeneity and other genes are probably involved.

Nbk *See* BIK.

***NBLRR* genes** *See* NUCLEOTIDE BINDING SITE/ LEUCINE-RICH REPEAT GENES.

NBQX A compound (6-nitro-7-sulphamoylbenzo(*f*)-quinoxaline-2,3-dione) that blocks *AMPA receptors.

NBT *See* NITROBLUE TETRAZOLIUM.

N-cadherin *See* CADHERINS.

NCAM *See* CELL ADHESION MOLECULE.

NCAP **1.** *N*-Acetyl-4-cystaminylphenol; used topically for the treatment of hyperpigmentation. **2.** The N-terminal residue of a protein. **3.** The phase of sleep where there is not a cyclic alternating pattern (CAP) in the EEG.

Nck A member of a family (Nck family) of small adaptor proteins with SH2 and SH3 domains that link receptor-mediated tyrosine phosphorylation with downstream regulators of actin dynamics. There are two isoforms, Nck1 (Nck-α, 377 aa) and Nck2 (Nck-β, Grb4, 380 aa), both widely distributed but with slightly different activities. Similar to *Crk and *Grb2.

N-CoR (nuclear receptor corepressor) A family of transcriptional *repressor proteins regulated by thyroid hormone and retinoic acid receptors. The corepressor, once bound, initiates the assembly of a larger complex that may include *Sin3 and *histone deacetylases (HDAC). Members of the family include receptor interacting protein 140 (RIP140), which is important in regulating energy expenditure and fat accumulation in adipocytes, and SMRT (nuclear corepressor 2) a silencing mediator for retinoid and thyroid-hormone receptors.

NCP1 One of a family of cell adhesion molecules present at the vertebrate axoglial synaptic junction. Other members include *neurexin IV, contactin associated protein (*Caspr), and *paranodin.

NCS family A family of neuronal calcium sensors. *See* FREQUENIN; VISININ-LIKE PROTEINS.

NDE1 A protein (nuclear distribution protein, nudE homologue 1, 346 aa) that is required for centrosome duplication and for the formation and function of the mitotic spindle. NDE1 is essential for the development of the cerebral cortex where it controls the orientation of the mitotic spindle during division of cortical neuronal progenitors. *See* CENPs.

nearest neighbour analysis A statistical method for analysing spatial distributions to detect clustering or regularity. Originally applied to the distribution of cell or organisms, but also used to show (e.g.) a deficiency of the nucleotide pair CG in eukaryotic DNA.

nebulette A protein (1014 aa) found in cardiac muscle. It has sequence and structural homology with *nebulin and possibly an analogous role. *See* MYOPALLIDIN.

nebulin A giant protein (6669 aa) found in the *N line of the *sarcomere of striated muscle, accounting for 3–4% of the total myofibrillar protein. It has multiple (>200) actin-binding repeats and may determine the length of thin filaments. Multiple tissue- and developmental-stage specific isoforms are found. Mutations are associated with *nemaline myopathy 2. **Nebulin-related anchoring protein** (NRAP, 1730 aa) is found in the myotendinous junction in skeletal muscle and the intercalated disc in cardiac muscle where it may anchor terminal thin filaments. *See* NEBULETTE.

necdin A growth suppressor protein (321 aa) that interacts with the nerve growth factor receptor and induces terminal differentiation of postmitotic neurons in the brain. Necdin-like proteins are a family within the *MAGE superfamily. Expression of the necdin gene is restricted to the paternal allele (*see* GENOMIC

IMPRINTING), and necdin is one of several genes disrupted in *Prader–Willi syndrome.

necrosis The death of tissue cells as a result of injury, infection, or ischaemia. Necrotic death elicits an inflammatory response, unlike *apoptosis.

necrotizing fasciitis A rare but aggressive infection of the deeper layers of skin and subcutaneous tissues leading to extensive tissue destruction. Type 1 involves infection with several species of bacteria; type 2 involves a single species, often *meticillin-resistant *Staphylococcus aureus*.

nectin **1.** A protein component of the stalk of mitochondrial ATPase. **2.** An immunoglobulin-like cell adhesion molecule that is calcium-independent and causes adhesion by interaction with other nectin molecules. The cytoplasmic domain interacts with *afadin and, in conjunction with the *cadherin–*catenin system, is important in organizing *adherens junctions. Nectins are receptors for various viruses, including poliovirus. Mutation in the gene for nectin 1 (poliovirus receptor-like 1) is responsible for *cleft lip/palate–ectodermal dysplasia syndrome. **Nectin-like protein 1** is *synCAM 3.

nedd2 (caspase-2, Ich-10) *See* CASPASES.

Nedd8 A developmentally regulated protein (neural-precursor-cell-expressed and developmentally down-regulated 8, 81 aa) found in brain that shares 60% sequence identity with ubiquitin and is found primarily in the nucleus. Coupling of Nedd8 (neddylation) to *cullins generates a recognition site for a ubiquitin-conjugating enzyme.

nedocromil A drug related to *sodium cromoglicate and with comparable properties, blocking degranulation of mast cells. *TN* Tilade.

nef A viral protein that is important for enhancing infectivity of *HIV. Nef regulates the sorting of at least two cellular transmembrane proteins, CD4 and MHC class I, and interacts with various signalling kinases.

negative feedback A mechanism of regulation in which the products of a process or reaction act to inhibit their own formation. Negative feedback tends to stabilize systems whereas *positive feedback amplifies.

negative stranded RNA virus The class V *viruses with an RNA genome that is complementary to the mRNA produced by their own virus-specific *RNA polymerase following infection. Includes *Rhabdoviridae, *Paramyxoviridae.

NEGR1 A neuronal member of the *IgLON family of cell adhesion molecules.

Negri body An acidophilic cytoplasmic inclusion in neurons, characteristic of rabies virus infection. Consists of a mass of *nucleocapsids.

Neisseria A genus of Gram-negative nonmotile pyogenic cocci. Two species are serious pathogens, *N. meningitidis* (meningococcus; *see* MENINGITIS) and *N. gonorrhoeae* (gonococcus). The latter associates specifically with urinogenital epithelium through surface *pili. Both species seem to evade the normal consequences of attack by phagocytes.

NEL *See* NO EFFECT LEVEL.

nemaline myopathy (NEM) A set of clinically and genetically heterogeneous disorders characterized by abnormal thread- or rod-like (nemaline) structures (nemaline bodies) in muscle and progressive muscle weakening. All are due to defects in genes encoding muscle proteins. The mutation in **NEM1** is in tropomyosin-3, in **NEM2** (the commonest form) in *nebulin, in **NEM3** in α-actin-1, in **NEM4** in β-tropomyosin, in **NEM5** (Amish nemaline myopathy) in troponin T1, in **NEM7** in *cofilin-2. Loci for other forms have been identified.

nematosome An electron-dense aggregate, 0.3–0.5 μm in diameter and of variable length, found in cytotrophoblast cells of the placenta. Morphologically similar inclusions are found in some neurons.

N-end rule A rule, conserved in pro- and eukaryotes, that holds that the identity of the N-terminal residues of proteins determines the *in vivo* half-life. **N-recognins**, a class of E3 ligases (*see* UBIQUITIN CONJUGATING ENZYMES), bind to a destabilizing N-terminal residue of a substrate protein and are involved in ubiquitination and subsequent proteasomal degradation.

neobavaisoflavone An antifungal compound, isolated from the Indian herb *Psoralea corylifolia*, that inhibits platelet aggregation and DNA polymerase.

neoendorphin An opioid peptide (*endorphin) derived from prodynorphin.

neogenin A multifunctional transmembrane receptor of the immunoglobulin superfamily that is found in two isoforms (1461 and 1408 aa) in many tissues. It has 50% homology with the receptor for repulsive neuronal guidance molecules, *deleted in colorectal cancer (DCC), but does not seem to be associated with tumours. Neogenin is a receptor for *netrin and other repulsive guidance molecules and also binds *hemojuvelin. *See also* PROTOGENIN.

neointima The new intimal lining of a blood vessel that has been dilated by *angioplasty. Proliferation of the neointima is a probable cause of restenosis.

neomycins Aminoglycoside antibiotics with broad activity produced by *Streptomyces fradiae* and used topically. Neomycin resistance, produced by expression of an aminoglycoside phosphotransferase (neo) gene, is often used in DNA plasmids as a selective marker when establishing stable mammalian cell lines.

neonatal adrenoleukodystrophy *See* LEUKODYSTROPHY.

neoplasia

Literally, a term meaning 'new growth' but referring to abnormal new growth that persists in the absence of the original stimulus, unlike *hyperplasia. The term covers both tumours, where there is an actual swelling, and other proliferative disorders, such as *leukaemias, all colloquially referred to as 'cancer', although this term strictly refers to *carcinoma (*see* GLIOMAS; LYMPHOMA; MEDULLOBLASTOMA; MULTIPLE ENDOCRINE NEOPLASIA; SARCOMA; TERATOCARCINOMA). The cells in *benign tumours do not spread and such tumours are not life-threatening unless the growth pressure causes damage. Malignant neoplasia involves the loss of both growth control and positional control and the malignant cells invade territory normally occupied by other cells (*see* CONTACT INHIBITION OF LOCOMOTION/MOVEMENT) and, distributed through the vasculature or the lymphatics, establish secondary tumours (*metastases) at sites that may be tumour-type specific or in the largest area of downstream vascularization (lung or liver). The growth of solid tumours is restricted unless there is vascularization of the tumour (*see* TUMOUR ANGIOGENESIS FACTOR). The original tumour is probably clonal but the progression to increased malignancy may involve further mutation (*loss of heterozygosity, *tumour progression). The change in cell behaviour in tumour cells is thought to be modelled in *transformation and can be brought about by various agents that damage DNA (radiation, mutagens) and deficiencies in *DNA repair mechanisms increase the probability of tumours developing. There may be genetic susceptibility to tumours (*see* BRCA) although this is usually through loss of an endogenous control or surveillance system. Neoplasia can arise through activation of *growth factor genes, e.g. through chromosomal translocation to a region under the control of an active promoter site (e.g. *Philadelphia chromosome). Some *tumour viruses carry growth factor genes (*see* ONCOGENE), and mutations in growth factor genes can make the product unresponsive to normal controls (e.g. *trk). Another route to tumorigenicity is through defects in *tumour suppressors, increasing numbers of which are being identified, and possibly alterations in *telomerase. Because tumour cells differ little from normal cells, treatment is difficult and most *cytotoxic drugs used in chemotherapy target rapidly proliferating cells, which unfortunately also affects normal proliferating cells in the skin, intestine, and bone marrow. Tumour antigens are often just features of cells that are proliferating so fast that (e.g.) glycosylation of surface proteins is incomplete, and attempts to activate cell-mediated immune responses to destroy tumours have not been particularly successful to date.

neopterin A pteridine derivative of GTP that is produced by human monocytes/macrophages in response to interferon-γ. Increased neopterin concentrations in human serum and urine are a marker of cell-mediated (Th1-type) immune responses.

neostigmine An anticholinesterase drug used to improve muscle tone in patients with *myasthenia gravis. *TN* Prostigmin.

neoteny (paedomorphosis) The persistence of juvenile characteristics in the reproductively-mature adult.

NEP A transmembrane zinc endopeptidase (neutral endopeptidase, neprilysin, enkephalinase, common acute lymphocytic leukaemia antigen, CALLA, CD10, EC 3.4.24.11, 750 aa) that hydrolyses regulatory peptides such as *ANP. Widely distributed but particularly highly expressed in kidney. It is an important cell surface marker in the diagnosis of human acute lymphocytic leukaemia.

nephelometry A light-scattering method for estimating the concentration of particles in a dilute solution. A nephelometer can be used for following aggregation of stirred cell suspensions.

nephr-, nephro- Prefix denoting kidney-related.

nephrin A protein (1241 aa) that is an important component of the glomerular filtration barrier formed by the filtration slits between interdigitated pedicels of *podocytes and the associated extracellular matrix. Mutations in the nephrin gene are associated with congenital nephrotic syndrome of the Finnish type (NPHS1).

nephroblastoma *See* WILMS' TUMOUR.

nephrocystins Proteins originally identified as being associated with *nephronophthisis. **Nephrocystin-1** (732 aa) is associated with nephronophthisis-1 and mutation in the *NPH1* gene are responsible for ~85% of the purely renal forms of juvenile nephronophthisis. It interacts with *p130Cas (BCAR1), proline-rich tyrosine kinase-2 (PTK2B), and *tensin in embryonic kidney and testis. A related protein, **nephrocystin-4** (nephroretinin,1426 aa), is mutated in juvenile nephrophthisis-4 and is involved in similar signalling complexes; the proteins interact with one another and colocalize with α-tubulin especially in primary cilia and the microtubule organizing centre. **Nephrocystin-2** (*inversin, 1065) is mutated in juvenile nephrophthisis-2, **nephrocystin-3** (1330 aa) in adolescent nephrophthisis (nephrophthisis-3). **Nephrocystin-5** (IQ motif-containing protein B1, IQCB1; p53- and DNA damage-regulated IQ motif protein, PIQ, 598 aa) interacts with calmodulin and the retinitis pigmentosa GTPase regulator and mutations are associated with *Senior–Loken syndrome. **Nephrocystin-6** (2479 aa) is associated with *Meckel's syndrome type 4.

nephron The basic subunit of the kidney consisting of the glomerulus, Bowman's capsule, and the convoluted tubule.

nephronophthisis A hereditary cystic kidney disorder that leads to kidney failure but is associated with various nonrenal dysfunctions, all linked through dysfunctions in *primary cilia. Various forms are known: *see* NEPHROCYSTINS.

nephropathia epidemica A disease (vole fever) caused by Puumala virus, a hantavirus which is carried by the bank vole *Clethrionomys glareolus*. Infection is usually from aerosolized droppings and is not passed between humans.

neprilysin *See* NEP.

Nernst equation An equation from which it is possible to calculate the potential difference, at equilibrium, across a semipermeable membrane if the concentrations of the permeating ions on either side are known. The potential (in volts) on side 2 relative to side 1, $E = (RT/lzF).\ln([C_1]/[C_2])$, where R is the gas constant (8.314 J K^{-1} mol^{-1}), T is the absolute temperature, z is the charge on the permeant ion, F is the Faraday constant (96 500°C mol^{-1}) and $[C_1]$ and $[C_2]$ are the concentrations (more correctly activities) of the ions on sides 1 and 2 of the membrane. *See* RESTING POTENTIAL.

nerve cell *See* NEURON.

nerve ending *See* SYNAPSE.

nerve growth cone *See* GROWTH CONE.

nerve growth factor (NGF) A polypeptide involved in growth, differentiation, and survival of nerve cells. It has three types of subunits, α, β, and γ, which interact to form a 7S, 130-kDa complex. The β subunits (118 aa), two per complex, have the nerve growth stimulating activity. NGF has chemotropic effects so that neurites grow towards sources and form synapses which require continuing supplies of NGF to be sustained. The receptor is one of the TNF receptor superfamily (TNFRSF16) and binds other *neurotropins such as brain-derived neurotropic factor.

nerve impulse *See* ACTION POTENTIAL.

nesfatin A polypeptide (82 aa) produced by cleavage of *nucleobindin-2 that is a satiety molecule associated with melanocortin signalling in the hypothalamus of rats. Whether human nesfatin-1 has the same role is unproven.

nesidioblast A cell that gives rise to pancreatic *beta cells. **Nesidioblastosis** (familial hyperinsulinaemic hypoglyaemia) is a rare childhood condition in which these cells fail to mature properly and promiscuously secrete various hormones. **Type 1** is caused by mutation in the gene that encodes the regulatory *SUR1 subunit of Kir6.2, the inwardly rectifying potassium channel of the beta cell. Type 2 is caused by mutation in the channel subunit.

nesprins A family of α-actinin type actin-binding proteins (nuclear envelope spectrin repeat proteins), distantly related to *spectrin, that are transmembrane scaffolding proteins of the nuclear envelope. Nesprin-1 (synaptic nuclear envelope protein-1, syne-1, enaptin, myne, MSP-300, Ank-1, 8797 aa) and nesprin-2 (6885 aa) connect to the actin cytoskeleton through an N-terminal actin-binding domain and a C-terminal region that binds *Sun1 of the inner membrane. Nesprin-3 (975 aa), an outer nuclear envelope membrane protein, lacks an actin-binding domain but associates with the linker protein *plectin. A large isoform of nesprin-2 (Nesprin-2 Giant; 800- kDa) is also found on the cytoplasmic membrane of the nuclear envelope. Mutations in nesprin-1 cause *spinocerebellar ataxia type 8.

nestin An intermediate filament protein (neural stem cell protein, 1621 aa) expressed in stem cells of the central nervous system.

Netherton's syndrome A disorder of hair ('bamboo hair', trichorrhexis nodosa) and skin (ichthyosiform erythroderma) caused by mutation in the gene for the serine protease inhibitor, *LEKTI* (lymphoepithelial Kazal-type-related inhibitor), expressed in epithelial and mucosal surfaces and in the thymus.

netrins A family of large soluble proteins (450–600 aa) involved in neuronal guidance in the developing nervous system. Receptors include members of the *deleted in colorectal cancer (DCC) protein family including *neurogenin.

netropsin An antibiotic with antitumour and antiviral activity from *Streptomyces netropsis*. It binds to the minor groove of DNA in spots rich in AT. Synthetic analogues are the lexitropsins. *TNs* Congocidin, Sinanomycin.

network theory **1.** A general term for theories about the properties of networks, of which signalling pathways and the internet are classic examples. The important feature of a network is that it is robust and alternative pathways can compensate for damage or deletion. **2.** In immunology, the 'network theory' was proposed in 1974 by N. K. Jerne, who suggested that the immune system was a network in which there were antigen binding sites (*paratopes) on antibodies which would bind an *epitope on an external antigen and an idiotope, a structurally similar region, on another immunoglobulin molecule. The theory has generally fallen out of favour.

***neu* (*erb*-B2)** An *oncogene that encodes a receptor *tyrosine kinase of the EGF-receptor family that binds *neuregulin. It was first identified in a *neuroblastoma.

neublastin *See* ARTEMIN.

NeuN (neuronal nuclei) An antigenic marker commonly used for mature neurons.

neur-, neuro- Prefix for terms relating to nerves or the nervous system.

neurabin A neuron-specific actin-binding protein (neural tissue-specific F-actin binding protein, spinophilin) that tethers protein phosphatase-1 to regions of actin-rich postsynaptic density and modulates its activity. Neurabin-1 (1095 aa) and spinophilin (neurabin-2, 817 aa) appear to have distinct roles, the latter being highly enriched in dendritic spines.

neural crest An important class of embryonic cells, derived from the *neural plate during neurulation, that migrate to give rise to spinal and autonomic ganglia, the *glial cells of the peripheral nervous system, and non-neuronal cells, such as *chromaffin cells, *melanocytes, and some haematopoietic cells.

neural induction The triggering of neural differentiation in the *ectoderm of the early embryo by signals derived from the underlying *mesoderm of the archenteron roof. The exact mechanism remains unclear.

neural plate The region of embryonic ectoderm (neuroectoderm) lying dorsal to the *notochord. During *neurulation, the cells change shape and the plate infolds to produce the neural fold, which eventually closes to form the neural tube.

neural retina The inner layer of the retina that lies in front of the photoreceptors and adjacent to the vitreous, that contains the innervating neurons. Embryologically it is part of the central nervous system.

neuraminidase An enzyme (sialidase, EC3.2.1.18), usually an exoenzyme, that removes **N*-acetylneuraminic acid residues from the oligosaccharide chains of glycoproteins and glycolipids. The influenza virus transmembrane neuraminidase is the N antigen (*see* H5N1) and a target for antiviral drugs.

neuraxis The neural axis of the body consisting of the brain and spinal cord.

neuregulins A family of structurally related glycoproteins (NRGs, neu differentiation factors, NDFs, neuregulins1–4) that are growth and differentiation factors related to *epidermal growth factor (EGF) and are ligands for the ErbB family of tyrosine kinase transmembrane receptors. They induce growth and differentiation of epithelial, glial and muscle cells in culture. Gene disruption is lethal during embryogenesis, with heart malformation and defects in Schwann cells and neural ganglia. There are at least fourteen alternatively spliced forms of **neuregulin-1**, although all contain an EGF-like domain. Type I NRG1, is also known as heregulin-α, type II NRG1 is glial growth factor-2, type III is sensory and motor neuron-derived factor (SMDF), and type 4 may have a role in brain differentiation. Neuregulin-2 (neural- and thymus-derived activator for the ErbB kinase, NTAK) has some differences in domain structure but considerable homology with neuregulin-1. Neuregulin-2 stimulates AChR transcription in cultured myotubes expressing ErbB4 (*see* ARIA). Neuregulin-3 binds to ErbB4, but not ErbB2 or ErbB3. Neuregulin-4 (heregulin-4) down-regulation is strongly correlated with stage, grade, and type of bladder carcinoma.

neurexin A family (NCP family) of highly variable cell-surface molecules involved in synaptic transmission and/or synapse formation. There are three neurexin genes, each producing two variants (α and β) that have identical cytoplasmic domains but differ in their extracellular domain. Neurexin Iα binds *α-latrotoxin and α-neurexins regulate presynaptic N- and P/Q-type calcium channels and are required for normal neurotransmitter release. β-Neurexin binds to presynaptic *neuroligin and is involved in adhesion.

neurite A general term for an outgrowth from a neuron, especially in culture where distinguishing *dendrites from *axons is difficult.

neuro- *See* NEUR-, NEURO-.

neuroblast The cell, derived from *neurectoderm, that differentiates into a neuron. **Neuroblastomas** are malignant tumours derived from primitive ganglion cells and occur mostly in childhood, most commonly in the adrenal medulla and in retroperitoneal tissue. Neuroblastoma cells may partially differentiate into cells rather like immature neurons.

neurocalcin A member of the *EF-hand calcium-binding protein superfamily (dimeric, 193 aa) found in neurons, especially in the central nervous system. Other members of the family include *recoverin, visin, VILIP, and *hippocalcin. Neurocalcin is myristoylated, and also binds clathrin and tubulin.

neuroectoderm The *ectoderm of the dorsal surface of the early embryo (neuroepithelium, neural plate) that gives rise to the cells (neurons and glia) of the nervous system.

neuroendocrine cell *See* NEUROHORMONE.

neurofascin An immunoglobulin superfamily *cell adhesion molecule of the L1 subgroup (1347 aa) involved in axon subcellular targeting and synapse formation during neural development. Neurofascin-155 (NF155) is found in oligodendrocytes and NF-186 in neurons.

neurofibrillary tangle An intracellular tangle of *neurofibrils found within large neurons of the cerebral cortex in patients with *Alzheimer's disease. Whether they are a cause or a secondary consequence of neurodegeneration remains controversial.

neurofibrils A noncommittal term, used especially by early light microscopists, for intracellular fibrillar elements within neurons, not necessarily *neurofilaments.

neurofibromatosis An autosomal dominant disorder characterized by cafe-au-lait spots and fibromatous tumours of the skin. **Type 1** (von Recklinghausen's disease) is caused by mutation in *neurofibromin, a *GTPase activating protein which interacts with the *ras family proteins. Mutations in neurofibromin 1 also underly familial spinal neurofibromatosis. **Neurofibromatosis type II**, characterized by tumours of the eighth (acoustic) cranial nerve, meningiomas of the brain, and schwannomas of the dorsal roots of the spinal cord, is caused by mutation in the gene encoding neurofibromin-2 (*merlin). Other variants of the disease are recognized. **Neurofibromatosis 1-like syndrome** is caused by mutations in the **SPRED1* gene.

neurofibromin The product (NF1, 2839 aa) of a tumour-suppressor gene that is a *ras-activating GTPase. Mutations lead to *neurofibromatosis type 1. Mutations in neurofibromin 2 (*merlin, 595 aa) cause neurofibromatosis type II).

neurofilament A type of *intermediate filament found in neurons. There are three distinct protein subunits (NF-L, 543 aa; NF-M, 916 aa; NF-H, 1026 aa) which will incorporate into the vimentin filament system if expressed in fibroblasts. Mutations in NF-L are associated with *Charcot–Marie–Tooth disease-2E; mutations in NF-H seem to increase susceptibility to *amyotrophic lateral sclerosis.

neurogenesis The development of the nervous system from the *neuroectoderm of the early embryo.

neurogenins A family of basic helix–loop–helix transcription factors involved in neuronal differentiation. **Neurogenin-1** (237 aa) is required for specification of dopaminergic progenitor cells and inhibits the differentiation of neural stem cells into astrocytes. **Neurogenins 2 and 3** (272 aa and 214 aa) control distinct phases of neurogenesis involved in differentiation of different classes of sensory neurons. Mutation in the neurogenin 3 gene is associated with congenital malabsorptive diarrhoea.

n

neuroglia *See* GLIAL CELLS.

neurogranin A brain-specific protein (RC3, 78 aa) expressed in telencephalic neurons. It is a substrate for protein kinase C and expression is related to the development of dendritic spines and in postsynaptic plasticity. It is also a direct target for thyroid hormone. Neurogranin binds calmodulin and apparently enhances *long-term potentiation and learning by promoting calcium-mediated signalling; knockout mice have learning difficulties.

neurohormone Any hormone secreted by specialized neuroendocrine cells, e.g. gonadotropin releasing hormone.

neurokinin *See* TACHYKININS.

neuroleptic drugs *See* ANTIPSYCHOTIC DRUGS.

neuroligins A family of neuronal cell surface proteins that are enriched in postsynaptic plasma membranes and are ligands for *neurexins. **Neuroligin 1** (823 aa) and **neuroligin 2** (550 aa) promote postsynaptic differentiation and interact with the multidomain scaffolding protein SAP90/PSD95 which anchors proteins at postsynaptic sites. Mutations in **neuroligin 3 and 4** seem to be associated with susceptibility to X-linked autism.

neurolin A cell surface glycoprotein (BEN, ALCAM, SC1, DM-GRASP, CD166, 500 aa) originally identified in goldfish and zebrafish as being involved in axonal path finding. A member of the immunoglobulin superfamily of cell adhesion proteins, with significant homology to *activated leucocyte cell adhesion molecule.

neurological disorders

The nervous system is subject to a wide range of different disorders, caused by *developmental defects (e.g. *spina bifida), infectious agents (e.g. *encephalitis), neoplasia (e.g. *glioma), *cerebrovascular accidents (stroke), and degenerative processes (e.g. *Alzheimer's disease). Others are caused by inappropriate neuronal activity (e.g. *epilepsy), or are problems with higher-order cognitive activities (e.g. *aphasia); although these may have an underlying physiological basis; some may have a psychological component (e.g. *attention deficit hyperactivity disorder), but are important nonetheless. Since the function of neurons depends on *ion channels and *neurotransmitters, defects in any of these systems affect neurological function. Many disorders have separate entries that lead in turn to other entries. *See* particularly: *neuromuscular disorders; *spastic paraplegias; *spinocerebellar ataxia; *spongiform encephalopathies; *startle disease with epilepsy.

SEE WEB LINKS

- The US National Institute for Neurological Disorders and Stroke provides an extensive list of neurological disorders.

neuromedins A family of neuropeptides (100–200 aa). Four classes are recognized: *kassinin-like, *bombesin-like, *neurotensin-like, and neuromedin U. **Neuromedin U** (NMU) acts on smooth muscle through G-protein-coupled receptors (NMU1R, 426 aa, in peripheral tissues, NMU2R in the brain) and can increase blood pressure, alter ion transport in the gut, control local blood flow, and regulate adrenocortical function. Levels of NMU in the ventromedial hypothalamus are reduced following fasting.

neuromelanin *See* MELANIN.

neuromeres Morphologically identifiable segments of the developing neural tube. Neuromeres in the hindbrain region are called *rhombomeres and each gives rise to a specific region of the hindbrain.

neuromodulation The process by which *voltage-gated ion channels or *ligand-gated ion channels are regulated and their current flow altered, probably through *second messenger systems.

neuromodulin A protein (growth associated protein 43, GAP-43, pp46, B-50, F1, P-57, 238 aa) expressed in neuronal tissue, linked to the synaptosomal membrane, that is important in nerve regeneration and is highly expressed in nerve growth cones. May be important in neuronal plasticity. Binds *calmodulin, and phosphorylation of neuromodulin by *protein kinase C correlates with *long-term potentiation.

neuromuscular disorders

Disorders that affect the function of muscle itself (*myopathies) are caused by defects in the various proteins involved in contractility (*see* SARCOMERE) or its control, often restricted to a subset of muscle (cardiac myopathy), or in the neuromuscular system that controls muscle activity. The muscle diseases include *Charcot–Marie–Tooth disease, *dyssynergia, *hypertrophic cardiomyopathy, *Joubert's syndrome, *muscle–eye–brain disease, *muscular dystrophies, *myotonia congenita, *periodic paralysis, *rippling muscle disease, *spinal muscular atrophy, *stiff person syndrome, *torsins. Some disorders are caused by defects in the neuromuscular junction, the *synapse between nerve and muscle that activates *excitation–contraction coupling. These include *myasthenia gravis and *myasthenic syndrome. Many toxins target the neuromuscular junction including: *ammodytoxins, *botulinum toxin, *bungarotoxins, *crotoxin, *curare, *notexins, *taipoxin, *textilinin, and *textilotoxin. Various neoplasias affect muscle including *leiomyoma and *rhabdomyosarcoma (*see* FORKHEAD). The development of muscle (*myogenesis) has been very extensively studied and various important regulatory factors identified (*see* MYD; MYF; MYOCYTE ENHANCER FACTOR 2; MYOD; MYOGENIN; MYOSTATIN; MYOTUBULARIN). *See also* FIBRILLATION; MYOKYMIA.

neuromuscular junction (motor end plate) The *chemical synapse formed between a motor neuron and a muscle fibre.

neuron (neurone, nerve cell) An *excitable cell that transmits electrical signals over long distances, up to several metres in large animals. **Sensory neurons** receive input from sensory cells or other neurons, and transmit a signal to muscles or other neurons. **Motor neurons** (motoneurons) innervate muscle; **interneurons** connect only with other neurons. Connections are usually through chemical *synapses and signals are usually sent as *action potentials, although some neurons are *nonspiking.

neuronal calcium sensor-1 *See* FREQUENIN.

neuronal ceroid lipofuscinoses A range of progressive neurological disorders primarily affecting children, caused by lysosomal storage defects. **Neuronal ceroid lipofuscinosis-1** (CLN1) is caused by mutation in the gene encoding palmitoyl-protein thioesterase-1. **CLN2** is caused by mutation in a lysosomal tripeptidyl peptidase (EC 3.4.14.9), **CLN3** by mutation in battenin (*see* BATTEN'S DISEASE), **CLN5** by mutation in the *CLN5* gene encoding a soluble lysosomal glycoprotein, **CLN6** by mutation in the *CLN6* gene which encodes a voltage-gated potassium channel in the brain, **CLN7** by mutation in *major facilitator superfamily domain-containing protein-8, **CLN8** by mutations in the *CLN8* gene which encodes an ER-resident protein, **CLN10** by mutations in *cathepsin D. **CLN4 and CLN9** have unknown mutations. Despite many causative genes being known, the basis for the pathology is unclear.

neuronal guidance *See* AXONAL GUIDANCE.

neuronal plasticity The change in properties of nerve cells as a result of new input. May involve production of new processes, establishing new synapses or altering the strength of existing synapses. *See* LONG-TERM POTENTIATION; SYNAPTIC PLASTICITY.

neuropeptides Peptides that have direct effects at synapses (peptide neurotransmitters) or indirect modulatory effects (peptide neuromodulators). **Neuropeptide AF** (NPAF, 18 aa) is an antiopiate with analgesic effects acting through a G-protein-coupled receptor. **Neuropeptides FF and SF** are similar and also members of the RFamide class. **Neuropeptide Y** (NPY, melanostatin, 36 aa) is a neurotransmitter found in adrenals, heart, and brain. It stimulates feeding and regulates the secretion of gonadotropin-releasing hormone through G-protein-coupled receptors. *Leptin inhibits the expression and release of NPY.

neurophysins Carrier proteins that bind hormones within secretory vesicles for transport along axons. **Neurophysin I** (90–97 aa) binds *oxytocin, **neurophysin II** binds *vasopressin. Both the hormone and the carrier are part of the same precursor polypeptide.

neuropil (neuropile) A tangled network of nonmyelinated axons and dendrites within the central nervous system.

neuropilin Neuropilin-1 (NRP1) is a membrane-bound coreceptor for *vascular endothelial growth factor and *semaphorin. **Neuropilin 2** has similar properties but a subtly different tissue distribution. **Neuropilin and tolloid like-1** (NETO1) is a retina- and brain-specific transmembrane protein (three isoforms) with homology to both neuropilin and *Drosophila* tolloid (a dorsal–ventral patterning protein).

neuropore The anterior and posterior openings of the neural tube of the early embryo. Failure of the posterior neuropore to close causes *spina bifida.

neuropsin (opsin 5) A secreted serine peptidase of the *chymotrypsin family (kallikrein-8, KLK8, EC 3.4.21.118, 260 aa) involved in neuronal plasticity and regeneration following injury. The involvement with plasticity is thought to bring a role in learning and memory. Neuropsin has 25–30% amino acid identity with other opsins and is expressed in the eye, brain, testis, and spinal cord. *See* NEUROSIN; OPSIN.

neurosecretory cells Cells that behave like neurons, carrying electrical impulses, but secrete hormones into the bloodstream rather than neurotransmitters into the synaptic cleft.

neurosin A serine endopeptidase of the kallikrein family (protease M, kallikrein 6, KLK6, EC 3.4.21.-, 244 aa), expressed in brain, pancreas, and several primary tumours (but not metastases). May be important in degrading α-*synuclein and thus preventing its accumulation in pathological aggregates.

neurosteroids Steroids that are synthesized in the brain and affect neuronal excitability. Examples include *epalons and *dehydroepiandrosterone.

neurotactin **1.** *See* FRACTALKINE. **2.** *Drosophila* neurotactin is a transmembrane glycoprotein (135 kDa) expressed in neuronal and epithelial tissues during development and is involved in cell adhesion and axon fasciculation. The extracellular domain has a catalytically inactive cholinesterase-like domain.

neurotensin A tridecapeptide hormone that is widely distributed throughout the central nervous system where it is localized to catecholamine-containing neurons. Lithium salts potentiate *NTS* gene expression in a catecholamine-producing cell line.

n

neurotoxin Any toxin that acts on neuronal function. Many are found in venoms; examples include *tetrodotoxin, *curare, and *bungarotoxin.

neurotransmitter A substance released from the presynaptic terminal of a *chemical synapse in response to depolarization. The transmitter diffuses across the synaptic cleft and binds a ligand-gated ion channel on the postsynaptic cell, triggering a response if sufficient channels are affected. Examples include *acetylcholine, *GABA, *noradrenaline, *serotonin, *dopamine, and *excitatory amino acids.

neurotrimin One of the *IgLON family of GPI-anchored cell adhesion molecules (345 aa), expressed at high levels during brain development.

neurotrophic Describing a substance that is involved in the nutrition or maintenance of neural tissue. *See* NERVE GROWTH FACTOR; NEUROTROPHINS; *compare* NEUROTROPIC.

neurotrophins A family of basic proteins (200–300 aa) important for the survival of different classes of embryonic neurons. Neurotrophin-3 (NT-3; hippocampal-derived neurotrophic factor, NGF-2. 257 aa) is similar to *nerve growth factor (NGF) and *brain-derived neurotrophic factor (BDNF) but has a different pattern of neuronal specificity and regional expression. The receptors are a related family of tyrosine kinases. *See also* GDGF and CILIARY NEUROTROPHIC FACTOR.

neurotropic Having an affinity for, or growing towards, neural tissue. Some neurotropins, such as *nerve growth factor also have neurotrophic properties and nerves will grow towards sources of NGF. Rabies virus, which localizes in neurons, is referred to as neurotropic. *Compare* NEUROTROPHIC.

neurotubules Neuronal *microtubules.

neurturin A potent neurotrophic factor (197 aa), closely related to glial cell line-derived neurotrophic factor (*GDNF), the two forming a distinct TGFβ subfamily (TRNs, TGFβ-related neurotrophins). Like GDNF, neurturin signals through the tyrosine kinase receptor RET and a GPI-linked coreceptor.

neurula The stage in embryogenesis when the *neural plate closes, the process referred to as neurulation.

neutral mutation A mutation that causes no selective advantage or disadvantage. The idea has been controversial and many authors doubt whether such mutations can exist.

neutral protease A peptidase that has a pH optimum around pH 7 (neutral). May be a member of any family or clan, and current peptidase classifications (*see* MEROPS) do not use this as a classificatory characteristic. An example is *calpain.

neutropenia A condition in which the number of circulating *neutrophils in the blood is below the normal level of 4.3–10.8 × 10^9 cells/L. Autosomal dominant or sporadic congenital neutropenia is associated with mutation in the neutrophil elastase gene, but this is not the case with the autosomal recessive form (*Kostmann's syndrome). Other recessively inherited neutropenic syndromes include congenital neutropenia with eosinophilia, *Chediak–Higashi syndrome, and Fanconi pancytopenic syndrome.

neutrophil (neutrophil granulocyte, polymorphonuclear leucocyte, PMN, PMNL) A short-lived phagocytic cell of the blood, of the *myeloid series, present in relatively large numbers (2500–7500/mm^3), and responsible for general tissue 'housekeeping' by removal of damaged macromolecules. It is also the cell chiefly involved in the early stages of the acute inflammatory response where it adheres to the endothelium of postcapillary venules (*margination), and then migrates into tissue, possibly up a gradient of chemotactic factors released from the inflammatory focus. The granules that characterize the neutrophil are of two types, specific and *azurophil.

neutrophil-activating protein *See* INTERLEUKIN-8; MELANOMA GROWTH STIMULATORY ACTIVITY.

nevus *See* NAEVUS.

nexilin An actin filament-binding protein that was identified in rat cell-matrix adherens junctions, colocalizing with *vinculin, *talin, and *paxillin. Two splice variants were isolated from brain (b-nexilin, 656 aa) and fibroblasts (s-nexilin). A nexilin-like protein (NELIN, 675 aa) has been found in human heart, skeletal muscle, artery, and vein.

nexin **1.** A protein (165 kDa) that links adjacent microtubule doublets in the *axoneme. **2.** *See* PROTEASE NEXIN-1. **3.** *See* SORTING NEXINS.

nexus A bond or link.

Nezelof's syndrome An immune deficiency in which there is thymic hypoplasia and T-cell deficiency but normal immunoglobulin levels, thus distinguishing it from *Bruton's agammaglobulinaemia and from severe combined immunodeficiency (*SCID).

NF-1 **1.** Nuclear factor-1, a family of dimeric transcription factors (~500 aa) the products of four genes (*NF1A*, *-B*, *-C*, and *-X*) with alternatively spliced tissue-specific isoforms. *See* CTF. **2.** Not to be confused with *neurofibromatosis type 1 (NF1). *See* NEUROFIBROMIN.

NFAR Evolutionarily conserved proteins (nuclear factors associated with dsRNA, IL enhancer-binding factor 3, ILF3) associated with the *spliceosome and that may regulate gene expression in response to dsRNA-signalling events in the cell. The NFAR gene encodes a 671-aa polypeptide (NF90) and an alternatively spliced 404-aa polypeptide (NF45). NF90 and NF45 together form an *NFAT DNA-binding activity that is enhanced by T-cell stimulation and inhibited by *ciclosporin and *tacrolimus.

NFAT A family of transcription factors originally identified as a DNA-binding complex activated when T cells are stimulated by antigen (nuclear factor of activated T cells) but now known to be involved in regulating gene expression in non-immune tissues including skeletal muscle. The complex consists of several components, including homodimers or heterodimers of *fos and *jun family proteins, and a pre-existing cytosolic component (NFATP, NFAT1, or NFATC2) that translocates to the nucleus when dephosphorylated by *calcineurin, a step that is inhibited by *ciclosporin and *tacrolimus.

SEE WEB LINKS

- Abstract of a paper on NFAT in the *Journal of Cell Biology* (2002).

NF-B *See* NFκB.

NF-E1 *See* ERYTHROID TRANSCRIPTION FACTOR.

NFκB (NFkappaB, NF-B) A heterodimeric transcription factor (p50 and p65 subunits) that was first identified as regulating genes for the κ class of immunoglobulins in B cells, but now known to activate the transcription of genes in a variety of cells and tissues. Inactive NFκB is bound to the protein *IκB in the cytoplasm but various stimuli, including TNF, phorbol esters, and bacterial lipopolysaccharide trigger its release, allowing it to enter the nucleus and bind to DNA.

NgCAM Neural–glial *cell adhesion molecule.

NGF *See* NERVE GROWTH FACTOR.

***N*-glycanase** An enzyme (EC 3.5.1.52) that will cleave *N-glycans from glycoproteins.

***N*-glycans** Oligosaccharides that are post-translationally linked to protein through an amide bond with asparagine. They have a common core pentasaccharide $Man_3GlcNAc_2$, which is transferred from the dolichol precursor by oligosaccharyltransferase (OST) in the cisternal space of the endoplasmic reticulum, where further modification may also take place. High mannose *N*-glycans are also modified in the Golgi. The *N*-glycosylation site, where sugars are added, has a consensus -Asn-X-Ser/Thr-X, where X can be any amino acid except proline. *Compare* O-GLYCANS.

NHEJ *See* NONHOMOLOGOUS END-JOINING.

NHE-RF A phosphoprotein (Na^+/H^+ exchanger regulatory factor, ezrin–radixin–moesin binding protein 50, solute carrier family 9 isoform A3 regulatory factor 1, SLC9A3R1, 358 aa) that binds to *ERM proteins and is involved in *protein kinase A (PKA)-mediated regulation of ion transport and in recruiting *PTEN to the cytoplasmic region of the PDGF receptor, thereby inhibiting PI3kinase signalling. Expression is up-regulated in response to oestrogen in oestrogen receptor-positive breast carcinoma cell lines. Mutations in the gene are found in some patients with calcium-containing renal stones and/or bone demineralization, apparently because of renal phosphate loss.

SEE WEB LINKS

- Article on NHE-RF in the *American Journal of Pathology* (2001).

niacin (nicotinic acid) *See* VITAMIN B.

nibrin A protein component (p95 protein of the MRE11/RAD50 complex) of the *MRN double-strand break repair complex. Nibrin is thought to be defective in *Nijmegen breakage syndrome but has two modules that are also found in cell cycle checkpoint proteins.

nicalin A membrane protein (*nicastrin-like protein, 563 aa) that is an antagonist of *nodal signalling and interacts with nodal modulator (NOMO). Nomo and nicastrin mutually stabilize one another.

nicardipine A *calcium channel blocker. *TN* Cardene.

nicastrin A transmembrane glycoprotein (709 aa) that forms a high molecular mass *gamma-secretase complex with *presenilin-1 or presenilin-2 that is probably involved in processing of βAPP to amyloid-β-peptide and in the processing of *notch.

nick A single-strand break in a DNA molecule.

nick translation A method for radioactively labelling DNA by using *E. coli* DNA polymerase I to add labelled nucleotide to the free 3′-OH group at a *nick, at the same time as its exonuclease activity removes the 5′-terminus. The enzyme then processes along the strand from the nick, translating nucleotides into their labelled equivalents. Enzyme (*deoxyribonuclease) at trace levels may be used to generate nicks and a high level of labelling can be achieved.

niclosamide An anthelmintic drug effective against cestodes (tapeworms). *TN* Niclocide.

nicotinamide adenine dinucleotide (NAD, formerly **DPN)** A coenzyme that is cyclically reduced to $NADH^+$ and oxidized back to NAD by *dehydrogenases. $NADH^+$ is an important source of reducing equivalents for the electron transport chain and NAD is the source of ADP-ribose (*see* ADP-RIBOSYLATION). **Nicotinamide adenine dinucleotide phosphate** (NADP) is an analogue of NAD mostly generated by the *hexose monophosphate shunt, important in many biosynthetic pathways and the *metabolic burst in neutrophils. **Nicotinic acid adenine dinucleotide phosphate** (NAADP) is a calcium-releasing second messenger.

nicotine An alkaloid from tobacco that blocks the *nicotinic acetylcholine receptor.

nicotinic acetylcholine receptor (nAChR) An integral membrane *ion channel in the postsynaptic membrane that opens when it binds *acetylcholine. The nicotinic acetylcholine receptor, which is blocked by *nicotine, initiates muscle contraction at the *neuromuscular junction. The channel is a heteropentamer of related subunits with both the acetylcholine binding site and the ionic channel. The nAChR mediates rapid transduction events (1 ms), unlike *G-protein-coupled receptors which operate on millisecond to second time scales. Mutations in various of the subunits cause congenital *myasthenic syndrome with either fast or slow channels being affected according to which subunit is affected. *See also* MYASTHENIA GRAVIS.

nicotinic acid A product (pyridine 3-carboxylic acid) of the oxidation of nicotine and a precursor of *nicotinamide adenine dinucleotide (NAD).

NIDDM *See* DIABETES.

nidogen (entactin) A sulphated glycoprotein (150 kDa) found in *basement membranes, with three globular regions, G1 and G2 connected by a thread-like structure and G3 connected to G2 by a rod-like region. The G2 region binds to collagen IV and *perlecan. There are two isoforms, nidogen-1 and nidogen-2 (osteonidogen), with broadly similar properties.

Niemann–Pick disease A group of lysosomal *storage diseases characterized by progressive neurodegeneration. **Types A and B** are caused by deficiency in *sphingomyelinase and excess sphingomyelin accumulates in 'foam' cells (macrophages) in spleen, bone marrow, and lymphoid tissue. **Types C1 and D** are due to a mutation in the *NPC1* gene that encodes a protein (1278 aa) with a critical role in regulating intracellular cholesterol trafficking. NPC1 has sequence similarity to the morphogen receptor *patched and with sterol-sensing regions of *SREBP and *HMGCoA reductase. *Compare* PICK'S DISEASE.

nifedipine A *calcium channel blocker (BAYa1041). *TNs* Nifedin; Procardia.

niflumic acid A rather nonspecific inhibitor of calcium-activated chloride channels.

nigericin An antibiotic derived from *Streptomyces hygroscopicus*, an ionophore which will act as a carrier for K^+ or Rb^+ or as an exchange carrier for H^+ with K^+.

NIH 3T3 cells A mouse fibroblast cell line, derived from the parent line at the US National Institute of Health.

Nijmegen breakage syndrome An autosomal recessive chromosomal instability syndrome characterized by microencephaly, growth retardation, immunodeficiency, and predisposition to tumours; caused by a defect in the DNA double-strand repair system making cells hypersensitive to ionizing radiation. *See* NIBRIN.

NIK A *MAP kinase kinase kinase (NFκB-inducing kinase, MAP3K14, 947 aa) that activates the CD28 responsive element (CD28RE) of the IL-2 promoter and is downstream of the TNFα, IL-1, and CD95 signalling cascades that activate *NFκB. A NIK and IκB-binding protein (NIBP) plays a role in the neuronal NFκB signalling pathway.

ninein A protein (glycogen synthase kinase 3 β-interacting protein, 2090 aa) found in the centrosome and required for the positioning and anchorage of the minus end of microtubules in epithelial cells. Autoantibodies against ninein are found in autoimmune disease such as *CREST syndrome.

niosomes Artificial multilamellate *liposomes produced from nonionic lipids and used for drug delivery.

***NIPA1* gene** A gene (nonimprinted gene in Prader–Willi syndrome/Angelman's syndrome chromosome region 1) that encodes a putative membrane transporter or receptor (deduced 329 aa) and is defective in *hereditary spastic paraplegia (SPG6). It is highly expressed in neuronal tissues. The function is unknown but the *Drosophila* orthologue spichthyin seems to inhibit *bone morphogenetic protein (BMP) signalling to the microtubule cytoskeleton and promotes the internalization of BMP receptors from the plasma membrane. Spichthyin is also an orthologue of ichthyin which is mutated in autosomal recessive congenital *ichthyosis.

Nipah virus A virus of the Paramyxoviridae family that causes encephalitis, frequently fatal, in rural Bangladesh and adjacent areas. It is one of two viruses in the genus *Henipavirus* (the other being *Hendravirus*). The natural reservoir is pteropid fruit bats (flying foxes).

nisin A *lantibiotic (34 aa) used as a food preservative. Gallidermin and epidermin, though smaller (22 aa), possess the same putative lipid II binding motif as nisin.

Nissl granules Large basophilic granules in the perikaryon of some neurons, particularly motor neurons. They are RNA-rich regions of rough endoplasmic reticulum and following neuronal damage become dispersed through the cytoplasm.

nitrates A class of vasodilatory drugs used to treat angina and congestive heart failure. Examples are *glyceryl trinitrate (nitroglycerine) and isosorbide mononitrate.

nitrazepam A nonbarbiturate hypnotic used to treat insomnia, although it can be addictive. *TN* Mogadon.

nitric oxide (NO) An intra- and intercellular messenger (endothelium-derived relaxation factor, EDRF), particularly in the vascular and nervous systems, that is produced from L-arginine by the enzyme *nitric oxide synthase. It activates soluble (cytoplasmic) *guanylate cyclase and is produced in large amounts as a cytotoxic attack mechanism.

nitric oxide synthase The enzyme (NO synthase, EC 1.14.13.39) that produces the vasorelaxant *nitric oxide from L-arginine. The constitutive isoform (NOS1, 1434 aa) is calmodulin dependent but the inducible (iNOS, 1153 aa) isoform is not. Other isoforms are found.

nitroblue tetrazolium A yellow dye that can be taken up by phagocytosing neutrophils and reduced to deep blue insoluble formazan if the *metabolic burst is normal.

nitrocellulose paper A type of paper that will bind various biological macromolecules (proteins or nucleic acids) which can then be stained or probed. Commonly used in blot-transfer methods where chromatographically or electrophoretically separated molecules are transferred and immobilized.

nitrofurantoin An antibiotic used in treating urinary tract infections. *TNs* Furadantin, Macrobid, Macrodantin.

nitrogen mustards Tertiary amine compounds with vesicant (blistering) properties similar to mustard gas. They will alkylate compounds such as DNA, and derivatives have been used in cancer chemotherapy.

nitroglycerine *See* GLYCERYL TRINITRATE.

nitrosamines A class of molecules containing the N-N=O group (*N*-nitrosamines), many of which are carcinogens.

nizatidine An *H_2-receptor antagonist used to treat gastric and duodenal ulcers. *TN* Axid.

NKCC1 A cation-coupled chloride cotransporter (Na^+,K^+,Cl^- cotransporter-1) found in kidney and other transporting epithelia. It is regulated by *SPAK and *OSR1 and inhibited by *thiazide diuretics.

NK cells *See* KILLER CELLS.

N line A thin band within the I band of the sarcomere in striated muscle where proteins such as *nebulin are located. The N1 line is near the Z disc and the N2 line at the end of the A band.

NM23H1 A DNase (nonmetastatic cells 1) that is part of the *SET complex where it is inhibited by SET until released by granzyme A. When released and activated it relocates to the nucleus where it generates single-strand breaks.

NMDA A potent agonist (*N*-methyl-D-aspartic acid) for a subclass of *glutamate receptors (NMDA receptor) that seem to be potentiated by intracellular *arachidonic acid. *See* EXCITATORY AMINO ACID.

***N*-methyl-D-aspartate** *See* NMDA.

NMRI *See* MAGNETIC RESONANCE IMAGING.

N-myc *See* MYC.

NO *See* NITRIC OXIDE.

Nocardia A genus of Gram-positive bacteria that form a mycelium but can fragment into rod- or coccoid-shaped cells. They are common soil saprophytes but can be opportunistic pathogens, causing **nocardiosis** which is characterized by abscesses, particularly of the jaw.

nociceptin A neuropeptide (orphanin FQ, 17 aa) that is the endogenous agonist of the G-protein-coupled receptor, opioid receptor-like-1, and mediates pain signals. It is derived from prepronociceptin which also gives rise to *nocistatin. A fragment (amino acids 1–13) has the same potency as the full-sized protein.

nociception The detection of pain. Nociceptive sensory nerve endings send signals about painful stimuli. *See* CAPSAICIN.

nocistatin A neuropeptide (EQKQLQ) involved in pain transmission. The precursor molecule, prepronociceptin (ppNCP), also gives rise to *nociceptin, although nocistatin antagonizes several effects of nociceptin by acting on a different receptor.

nocodazole A compound that binds to the *tubulin heterodimer, rendering it assembly-incompetent and thus disrupting microtubules.

nodal A signalling molecule involved in induction of mesoderm and endoderm and a member of the TGFβ family. In mice nodal is expressed during mouse gastrulation in a left-sided pattern and appears to be important in development of *left-right asymmetry. A mutation in nodal has been observed in a female with situs ambiguus. *See* CERBERUS; ZIC3. **Nodal modulators** (NOMO1, 1222 aa; NOMO2, 1267 aa; and NOMO3, 1222 aa) apparently down-regulate nodal signalling and are transmembrane proteins that interact with *nicalin.

node of Ranvier A nonmyelinated region of neuronal plasma membrane in a myelinated axon. They are the sites where action potentials are propagated by saltatory conduction and have high concentrations of *voltage-gated ion channels.

NOD mice A strain of mice (nonobese diabetic mice) with unique histocompatibility antigens in which an autoimmune response destroys pancreatic beta cells, as in type 1 *diabetes.

NOD proteins A family of cytoplasmic proteins (nucleotide-binding oligomerization domain proteins, NODs, caspase recruitment domain proteins, CARDs) that are pathogen recognition receptors, bind breakdown products of bacterial peptidoglycan, and through receptor-interacting protein 2 (RIP2) stimulate *NFκB activation. They have a caspase recruitment domain (*CARD domain), a nucleotide-binding domain, and a C-terminal regulatory domain. **NOD1** (CARD4, 953 aa) detects a *N*-acetylglucosamine-*N*-acetylmuramic acid tripeptide motif found in Gram-negative bacterial peptidoglycan, and has structural similarity to a class of disease-resistance plant proteins that induce localized cell death at the site of pathogen invasion. **NOD2** (CARD15) recognizes *muramyl dipeptide.

nodularins Hepatotoxic monocylic pentapeptides, produced by the cyanobacterium *Nodularia spumigena*, that are potent inhibitors of *protein phosphatase types 1 (PP1) and 2A (PP2A) and bind to the same site as *okadaic acid.

no effect level (NEL) The maximum dose of a substance that produces no detectable changes under defined conditions of exposure. Improved sensitivity of the detection system can change the value and the terms ‘no observed effect level’ and ‘no observed adverse effect level’ are more honest.

noelins A family of extracellular proteins within the larger *olfactomedin family, with proposed roles in neural and neural crest development. Noelin 1(135 aa) is olfactomedin 1; noelin 3 (olfactomedin 3) is also known as optimedin.

noggin An embryonic inducing factor (232 aa) highly homologous to the dorsalizing factor produced by the Spemann's organizer region of the amphibian embryo. Noggin modulates *bone morphogenetic protein (BMP) signalling by binding and inactivating BMP and is essential for normal skeletal development. Mutations are associated with bone or joint fusion abnormalities (synostoses). Noggin expression overlaps with that of *chordin.

nogo A myelin-associated neurite growth inhibitory protein (reticulon 4A), expressed by oligodendrocytes but not Schwann cells, involved in diverse processes that include axonal fasciculation and apoptosis. There are three splice variants, nogo-A (1192 aa), nogo-B, and nogo-C, the latter two lacking the axon-inhibiting extracellular domain of 66 aa. Nogo is thought to be one of the molecules responsible for blocking axonal regeneration in the central nervous system and, like *MAG and *OMgp, can cause *growth cone collapse *in vitro*. Although Nogo, MAG, and OMgp lack sequence homologies, they all bind to the **Nogo receptor** (NgR, 473 aa), a GPI-linked cell surface molecule widely expressed in brain which, in turn, binds p75 (neurotrophin receptor) and activates *rhoA.

nojirimycin *See* DEOXYNOJIRIMYCIN.

noncoding DNA DNA that does not encode a polypeptide chain or RNA. Noncoding DNA includes *introns and *pseudogenes and one of the two strands of DNA, the nonsense or noncoding strand.

noncoding RNA Any *RNA that does not encode a protein. The category includes *microRNA, *ribosomal RNA, *transfer RNA, *catalytic RNA (ribozyme), *piwi-interacting RNA (piRNA), small nuclear RNAs and small nucleolar RNAs, small interfering RNA (*siRNA), and long noncoding RNAs.

noncompetitive inhibitor An inhibitor that binds at a site other than the substrate-binding site (an *allosteric site) and does not compete with the normal substrate.

nondepolarizing muscle relaxants A class of drugs which compete with acetylcholine for binding to the acetylcholine receptor and block neuromuscular transmission. They are used to cause muscle relaxation during anaesthesia. A common example is tubocurarine.

nondisjunction The failure of separation of homologous chromosomes at *meiosis or sister *chromatids at mitosis, causing *aneuploidy.

nonequivalence In embryology, describing *determined cells that will give rise to the same sorts of differentiated tissues but in different locations, e.g. cells of forelimb and hindlimb buds.

nonhistone chromosomal proteins A heterogeneous mixture of proteins other than histones that are found in *chromatin. They include DNA polymerases, transcription factors, etc. They are sometimes termed acidic proteins, to distinguish them from the basic histones.

non-Hodgkin's lymphoma *See* LYMPHOMA.

nonhomologous end-joining (NHEJ) The primary mammalian repair mechanism for reconnecting double-stranded breaks in DNA. A number of different proteins are involved, including DNA-dependent protein kinase catalytic subunit (DNA-PKcs), the *Ku autoantigen, XRCC4, *cernunnos, and DNA ligase IV. *Synapsis results in the autophosphorylation of DNA-PKcs, making the DNA ends available for ligation. An equivalent mechanism is found in prokaryotes (*see* DNA LIGASES (DNA ligase D)).

Nonidet™ Nonionic detergents, often used for isolating membrane proteins.

non-Mendelian inheritance A pattern of transmission of genetic characters that cannot be explained by Mendelian mechanisms of segregation, independent assortment and linkage. May be a result of *cytoplasmic inheritance or *gene conversion.

nonpermissive cell Describing a cell from a tissue or a species that does not permit a particular virus to replicate, although early stages of the viral cycle may occur, leading to cell *transformation. The term is often used less specifically for cells that do not respond to various treatments or agents.

nonreciprocal contact inhibition Describing the situation in which two different cells collide and one shows contact inhibition of locomotion, and the other does not. This will allow the noninhibited cells to invade territory normally occupied by the others, possibly an important factor in *metastasis.

nonsense codons (nonsense triplet) The three *codons, ochre (UAA), amber (UAG), and opal (UGA), that act as termination signals for protein synthesis, rather than specifying an amino acid. A nonsense mutation generates a stop signal and thus prevents synthesis of the protein.

nonsense-mediated mRNA decay A mechanism for regulating gene expression in which mRNA is degraded because one exon contains a premature termination codon (PTC), in mammals, a termination codon more than ~50 nucleotides upstream of the final intron. It has been estimated that ~30% of alternatively spliced variants result in mRNA with such codons; if the exon containing the PTC is skipped, then the message is preserved.

SEE WEB LINKS

- Abstract of an article in *Genome Biology* (2004).

nonsense strand *See* NONCODING DNA.

nonspiking neuron A neuron that does not generate action potentials but transmits information over short distances, e.g. *interneurons in the central nervous system.

nonsteroidal anti-inflammatory drugs *See* NSAIDs.

nontranscribed spacer *See* INTERNAL TRANSCRIBED SPACERS.

n

Noonan's syndrome Noonan's syndrome **type 1** (NS1; male Turner syndrome; female pseudo-Turner syndrome) is an autosomal dominant dysmorphic syndrome, due to mutations in the gene (*PTPN11*) that encodes the nonreceptor protein tyrosine phosphatase *Shp2. Mutations in this gene also cause *LEOPARD syndrome, although another form of LEOPARD syndrome is caused by mutation in the *raf1 oncogene, which also causes Noonan's syndrome **type 5**. **NS2** is autosomal recessive. **NS3** is caused by mutations in the Kirsten-*ras gene. **NS4** is caused by mutation in the **SOS1* gene. **Neurofibromatosis–Noonan's syndrome** is caused by mutation in the *neurofibromin gene.

nootropic drugs A class of drugs, so-called 'smart drugs', that act as cognitive enhancers.

noradrenaline (norepinephrine, arterenol) The main neurotransmitter for most of the adrenergic neurons of the sympathetic nervous system, and also a catecholamine neurohormone. Noradrenaline binds more strongly to α-adrenergic receptors than to β-adrenergic receptors. It is stored and released from *chromaffin cells of the adrenal medulla.

norepinephrine *See* NORADRENALINE.

norethisterone A synthetic *progestogen. *TNs* Micronor, Primolut N, Utovlan.

norfloxacin A quinolone antibiotic. *TN* Utinor.

norleucine (Ahx, Nle) A nonprotein amino acid (2-aminohexanoic acid). Formyl-norleucyl-leucyl-phenylalanine has been used in studies on neutrophil chemotaxis because it is less readily oxidized than *fMLP.

normoblast A normal *erythroblast.

normocyte An erythrocyte of normal size and shape.

Norovirus A genus within the family Caliciviridae that cause intestinal disorders. Noroviruses have a positive strand RNA genome of ~7.5 kb. **Norwalk virus**, responsible for acute infectious gastroenteritis, is a norovirus.

northern blot *See* BLOTTING.

northwestern blot *See* BLOTTING.

nortriptyline A tricyclic antidepressant. *TN* Allegron.

Norum's disease An autosomal recessive disorder of lipoprotein metabolism caused by mutation in the **LCAT* gene.

Norwalk virus *See* NOROVIRUS.

nosocomial infections Infections acquired in hospital, mostly involving *Staphylococcus aureus, Pseudomonas aeruginosa, E. coli, Klebsiella pneumoniae, Serratia marcescens*, or *Proteus mirabilis*. Antibiotic resistance (*see* METICILLIN-RESISTANT STAPHYLOCOCCUS AUREUS) is a major problem.

NO synthase *See* NITRIC OXIDE SYNTHASE.

Notch A family of four transmembrane receptor proteins (~350 kDa, Notch1–Notch4) important in cell fate determination. They have an extracellular domain with multiple epidermal growth factor-like repeats and an intracellular region containing the *RAM domain, *ankyrin repeats, and a C-terminal *PEST sequence. Ligands include *jagged-1 and -2 and *delta-like-1, -3, and -4. Mutations in Notch have been associated with leukaemia, breast cancer, stroke, and dementia (*see* CADASIL). Notch maturation involves proteolytic processing by *furin-like convertase and activation, following ligand binding, involves proteolysis by *TACE to release the extracellular portion and remove the inhibition of a further proteolytic step that releases an intracellular portion which is a transcriptional activator. The Notch signalling pathway is further modulated by proteins such as *dishevelled and in *Drosophila*, where the pathway has received much attention, its complexity is remarkable.

SEE WEB LINKS

- Review in *Human Molecular Genetics* (2003), 'Notch signalling and inherited disease syndromes'.

notexins Phospholipase A2 isoforms (notexins Np and Ns, 119 aa) that are found in the venom of the tiger snake *Notexis scutatus scutatus*. They act (slowly) by blocking acetylcholine release at the neuromuscular junction.

SEE WEB LINKS

- Review of toxins, including notexin.

notochord A long, rod-shaped mass of vacuolated cells lying immediately below the nerve cord in the early embryonic stages of chordates and protochordates. The notochord is mesodermal in origin and has a supportive skeletal role.

nov An oncogene (nephroblastoma overexpressed) originally identified in an avian nephroblastoma (a model of *Wilms' tumour) and sequences homologous to *nov* were subsequently found to be expressed in normal haematopoietic cells and in a human nephroblastoma. The product (CCN3) is an immediate early gene product and a member of the *CCN family of growth factors; aberrant expression is associated with vascular injury and reduced expression with malignant adrenocortical tumours.

novobiocin (albamycin, cathomycin) An aminocoumarin antibiotic isolated from *Streptomyces niveus* and other *Streptomyces* species, used clinically against staphylococci and Gram-positive bacteria. Acts as a competitive inhibitor of the ATP-binding site of the GyrB subunit of bacterial DNA gyrase.

NOX *See* NADPH OXIDASES.

Noxa *See* BCL (bcl-2 homology domain 3).

noxiustoxin A toxin (39 aa) from the scorpion *Centruroides noxius*. It is related to *charybdotoxin and blocks mammalian voltage-gated potassium channels and high-conductance calcium-activated potassium channels.

NP-40 Nonidet P40, a nonionic detergent used for the isolation and purification of functional membrane proteins.

NPAF *See* NEUROPEPTIDES.

NPC **1.** *See* NIEMANN–PICK DISEASE (type C1). **2.** *See* NUCLEAR PORE (nuclear pore complex). **3.** Nasopharyngeal carcinoma (*see* EPSTEIN–BARR VIRUS). **4.** Neural precursor cells.

NPFF *See* NEUROPEPTIDES.

NPRAP *See* CATENINS (δ-*catenin-2).

N protein **1.** The protein product of one of the early genes expressed by bacteriophage. **2.** Obsolete name for *GTP-binding proteins (G proteins).

NPSF *See* NEUROPEPTIDES.

Nramps Integral membrane proteins (natural resistance associated macrophage proteins), of the solute carrier family, expressed only in cells of the myeloid series and recruited to the phagosome membrane following phagocytosis. Mutations in **Nramp1** (SLC11A1, 550 aa) impair macrophage killing of intracellular parasites such as *Mycobacteria, Salmonella,* and *Leishmania* and are also associated with the onset of rheumatoid arthritis. **Nramp2** (SLC11A2, 568 aa) is very similar to Nramp1 but more widely expressed and is known to be involved in cellular iron absorption at the luminal surface of the duodenum.

N-recognins *See* N-END RULE.

NRG *See* NEUREGULIN.

NS1 **1.** A nonstructural protein, a nonessential virulence factor of influenza virus, that is found as a dimer in nuclei of infected cells. It inhibits immune responses, interferes with host mRNA processing, and inhibits production of inflammatory cytokines. The cellular **NS1 binding protein** (NS1BP, 619 aa) is found in nuclei as a dimer that colocalizes with spliceosomes. In infected cells NS1BP becomes more diffusely distributed. **2.** *See* NOONAN'S SYNDROME.

NSAIDs A general category of anti-inflammatory drugs (nonsteroidal anti-inflammatory drugs) that includes *aspirin, *ibuprofen, and many derivatives. Most act on cyclooxygenases (*COX-1, etc.).

NSF A homotetrameric protein (*N*-ethylmaleimide sensitive factor, 744 aa), a member of the *AAA family, involved, together with three *SNAPs, in mediating vesicle traffic between medial and *trans*-Golgi compartments.

NT2 cells A line of human teratocarcinoma cells with properties similar to progenitor cells of the central nervous system, able to differentiate into any of the three major lineages—neurons, astrocytes, and oligodendrocytes.

NT3 *See* NEUROTROPHINS.

NTAL A *transmembrane adaptor protein (TRAP), non-T-cell activation linker, found in mature B cells, that is phosphorylated following immunoreceptor engagement and then recruits *Grb2.

NTG **1.** Normal tension *glaucoma. **2.** *See* NITROGLYCERIN. **3.** Nontransgenic.

Ntk **1.** General abbreviation for nonreceptor tyrosine kinases such as *src, *yes, and *fyn. **2.** An obsolete name for 'nervous tissue and T-lymphocyte kinase' very similar to *Csk.

NTRK1 One of a family of receptor tyrosine kinases (neurotrophic tyrosine kinase receptor) for which the ligands are *neurotrophins. NTRK1 (790 aa) binds *nerve growth factor with high affinity. Mutations are associated with hereditary sensory neuropathy type IV (congenital insensitivity to pain and anhidrosis). Somatic rearrangements of NTRK1, producing chimeric oncogenes with constitutive tyrosine kinase activity (**trk* oncogene), occur in some papillary thyroid tumours.

N-type channels A class of *voltage-sensitive calcium channels found only in neurons and neuroendocrine cells where they regulate neurotransmitter or neurohormone release. A substantial depolarization is required to activate the channels and they become inactivated with time. Potently inhibited by ω-conotoxin and the analgesic drug ziconotide.

nuclear distribution gene E homologue-like 1 A protein (nudE) which may function as both a cysteine protease and a centrosomal structural protein. Binds *disrupted-in-schizophrenia 1. Homologous to *Nde1 which similarly interacts with dynein and the dynein-regulator Lis1. *See* NDE.

nuclear envelope The double membrane system separating the nuclear compartment from the cytoplasm during interphase. The inner and outer membranes are separated by perinuclear space and perforated by *nuclear pores. A less confusing term than 'nuclear membrane'.

nuclear export signal (NES) A short leucine-rich amino acid motif in proteins destined for export from the nucleus by *exportins. The consensus sequence is poorly defined.

SEE WEB LINKS

- A database of experimentally validated leucine-rich NES curated from literature.

nuclear factor 1 (NF1) A family (CTF/NF1 family) of transcription factors (~500 aa) that bind to a consensus CCAAT sequence (*see* CTF). There are several types (A, B, C, X) involved in tissue-specific transcriptional regulation.

nuclear lamina (fibrous lamina) A fibrous protein network that lines the inner surface of the nuclear envelope. The extent to which this system also provides a scaffold within the nucleus is controversial. *See* LAMINS.

nuclear localization signal (NLS) A peptide motif that is found in proteins that are imported into the nuclear compartment by *importins. Has the opposite effect to a *nuclear export signal.

nuclear magnetic resonance imaging *See* MAGNETIC RESONANCE IMAGING.

nuclear matrix The meshwork of proteins that anchor *DNA replication and *transcription complexes within the nucleus.

nuclear membrane *See* NUCLEAR ENVELOPE.

nuclear mitotic apparatus protein A microtubule-binding protein (Numa1, formerly centrophilin, 2115 aa) that associates with the minus ends of spindle microtubules and appears to be important in spindle organization. It dissociates from condensing chromosomes during early prophase and becomes concentrated at the spindle pole; as mitosis progresses, it reassociates with telophase chromosomes during nuclear reformation.

nuclear polyadenylation hexanucleotide The six-nucleotide sequence (AAUAAA) in mRNA that is required for polyadenylation in the nucleus. *See* CYTOPLASMIC POLYADENYLATION ELEMENT.

nuclear pore An opening, ~10 nm diameter, in the membranes of the *nuclear envelope, through which macromolecules pass from nucleus to cytoplasm and vice versa. Pores involve a multiprotein assembly, the nuclear pore complex. *See* NUCLEOPORINS.

nuclear receptor A general term for a diffusible signal molecule that enters the nucleus once ligand has bound, e.g. receptors for *steroid hormones. *See* N-CoR (nuclear receptor corepressor).

nuclear RNA RNA that is not released from the nucleus, or not released without processing. Includes heterogenous nuclear RNA (*hnRNA) and small nuclear ribonucleoproteins *snRNPs.

nuclear run-on (nuclear run-off) Describing an assay used to identify the genes being transcribed at a particular instant. Rapidly isolated nuclei are incubated with labelled nucleotides which generates labelled RNAs from those that were stalled because of lack of nucleotides. The labelled RNA can then be used as a probe to identify corresponding cDNAs. Probably more accurately described as a run-off assay, but the terms are used interchangeably.

nuclear transplantation An important experimental approach used in studies of the interactions between nucleus and cytoplasm; a nucleus is transferred from one cell to the cytoplasm, often anucleate, of a second.

nuclear transport (nucleocytoplasmic transport) The import or export of molecules into or from the nucleus via *nuclear pores. *See* IMPORTINS and EXPORTINS.

nuclear transport factor A protein (NTF 2, placental protein 15, PP15,127 aa) required for nuclear protein import after the association of the *nuclear localization signal with nucleoporins of the nuclear envelope.

nuclease An enzyme (EC 3.1.33.1) that will cleave the phosphodiester bonds between the nucleotide subunits of nucleic acids.

nucleation A general term used in polymerization or assembly reactions in which, as is often the case, the first steps are energetically less favoured than the further addition of subunits. A preformed 'seed' may be used, or some sort of assembly template.

nucleic acids Linear *nucleotide polymers with 3′,5′-phosphodiester linkages. *DNA encodes the genome whereas *RNA mainly acts as an information intermediate (as *messenger RNA) although it has a range of other functions (*see* RIBOSOMAL RNA (rRNA); RIBOZYME; RNA INTERFERENCE; SMALL INTERFERING RNA (siRNA); TRANSFER RNA (tRNA)).

nucleobindin A DNA- and calcium-binding leucine zipper protein with two isoforms, nucleobindin-1 (nuc-1, 461 aa) and nucleobindin-2 (nuc-2, 420 aa). Nuc-1 is found in the nucleus and associated with endoplasmic reticulum and Golgi apparatus. The secreted form is a minor constituent of bone extracellular matrix and may be involved in mineralization. A secreted cleavage product of nuc-2 is *nesfatin.

nucleocapsid The *capsid coat of a virus together with the enclosed nucleic acid genome.

nucleocytoplasmic transport *See* NUCLEAR TRANSPORT.

nucleolar organizer A loop of DNA with multiple copies of the ribosomal RNA (*rRNA) genes. *See* NUCLEOLUS.

nucleolin A highly expressed acidic phosphoprotein (707 aa) found mainly in dense fibrillar regions of the nucleolus. It is involved in the control of transcription of ribosomal RNA (*rRNA) genes by RNA polymerase I, in ribosome maturation and assembly, and in transport of ribosomal components into the cytoplasm. Nucleolin binds *midkine, *endostatin, and heparin binding growth associated molecule (HB-GAM).

nucleolus A small dense body (suborganelle) within the nucleus, where ribosomal RNA (*rRNA) is synthesized from the *nucleolar organizer region. The protein and nucleic acid-rich nucleolus can be seen in live cells using phase contrast or interference microscopy throughout interphase.

nucleophosmin An abundant, predominantly nucleolar phosphoprotein (NPM, B23, numatrin, NO38, 294 aa) involved in ribosome assembly and transport, trafficking from cytoplasm to nucleus, regulation of DNA polymerase α activity, centrosome duplication, regulation of ARF/p53 tumour suppressors and cellular responses to stress. Mutations in exon 12 of the nucleophosmin (*NPM1*) gene occur in about ~60% of cases of adult acute myeloid leukaemia (AML) with normal karyotype. Nucleophosmins 2 and 3 have similar properties. All three are related to *Xenopus* nucleoplasmin.

nucleoplasm The nuclear contents other than the nucleolus at interphase, by analogy with cytoplasm.

nucleopore filter A type of filter made by etching a thin polycarbonate film that has been bombarded by neutrons, the pore size being determined by the extent of etching. The pores are circular holes rather than gaps in a meshwork. *Compare* MICROPORE FILTERS.

nucleoporins A large family (50–100 members) of glycoprotein components (NUP*xx*, where *xx* is the size in kDa, ranging from 37 to 214) of the *nuclear pore complex characterized by cytoplasmically oriented *O*-linked *N*-acetylglucosamine residues and numerous repeats of the pentapeptide sequence XFXFG. The *NUP98* gene is fused to one of a considerable number of other genes by chromosomal translocation in various haematological malignancies. Pericentrin is NUP75.

nucleoproteins A protein structurally associated with nucleic acid, e.g. *histones, protamines, *ribonucleoproteins.

nucleoside A purine base (adenine, guanine) or pyrimidine base (cytosine, thymine, uridine) linked glycosidically to ribose or deoxyribose, but lacking the phosphate residues that would make it a nucleotide. The major ribonucleosides are adenosine, guanosine, cytidine, and uridine. Deoxyribosides are deoxyadenosine, deoxyguanosine, deoxycytidine, and deoxythymidine (commonly thymidine).

nucleoskeletal DNA Noncoding DNA that is thought to serve a structural or skeletal role.

nucleosome A repeating unit in chromatin fibres, consisting of ~200 bp of DNA associated with two molecules each of the *histones H2A, H2B, H3, and H4. About 140 of the 200 bp are wrapped around a roughly spherical histone complex, the remainder serving to link adjacent nucleosomes to form a structure resembling a string of beads. The nucleosome remodelling and histone deacetylase complex (NuRD complex) promotes transcriptional repression by histone deacetylation and nucleosome remodelling and is composed of the core *histone deacetylase (HDAC) complex together with MTA2, MBD3, CHD3, and CHD4.

n

nucleosome remodelling factor complex A multisubunit complex (NURF complex) consisting of *SMARCA1, BPTF (bromodomain and PHD finger-containing transcription factor, 3046 aa) and the *retinoblastoma-binding proteins, RBBP4 and RBBP7, that uses the energy of ATP hydrolysis to catalyse nucleosome sliding, thereby altering chromatin structure and regulating transcription.

nucleotidase An enzyme (5′-nucleotidase, EC 3.1.3.5) that will cleave the 5′ monoester linkage of a *nucleotide converting it to the corresponding *nucleoside.

nucleotide A phosphate ester of a *nucleoside. The monoester with phosphate on C-5 of the pentose (known as 5′ to indicate it is on the sugar, not the base) is the basic unit of nucleic acid, but there are other nucleotides, such as adenosine 3′,5′-cyclic monophosphate (cyclic AMP, cAMP), and others with two or three phosphates.

nucleotide binding fold A protein motif that forms a fold or pocket that will bind nucleotides.

nucleotide binding site/leucine-rich repeat genes A large group of plant disease resistance genes (*nbs-lrr* genes) which also mediate pathogen recognition in animals. Two major subclasses are the *tir-nbs-lrr* genes that encode proteins with similarity to *toll and *interleukin-1 receptors, while the non-TIR class typically contains a coiled-coil (CC) domain in the N-terminal region.

nucleotide excision repair factor (NEF) *See* EXCISION REPAIR.

nucleus **1.** The major organelle of eukaryotic cells, bounded by the *nuclear envelope, in which the chromosomes are located. **2.** In neuroanatomy a distinct region of neurons, e.g. the *caudate nucleus.

nucleus accumbens A region of the forebrain thought to be involved in motivation, reward, pleasure, addiction, and fear.

nudE *See* NUCLEAR DISTRIBUTION GENE E HOMOLOGUE-LIKE 1.

nude mice A strain of mice with the recessive allele *nu/nu* (a mutation in the *forkhead homologue 11 gene) which are largely hairless (nude) and generally athymic. They lack all or most T cells and do not reject allografts or xenografts. Some strains do, however, have nearly normal numbers of T cells. Knockout mice with a deletion of the *FOXN1* gene also show the 'nude' phenotype.

null cell A lymphocyte that has neither T- or B-cell markers but is capable of cell-mediated lysis of various tumour cells or virus-infected cells without obvious antigenic stimulation. Can effect *antibody-dependent cell lysis, and have the CD16 marker.

null mutant A mutant in which the gene product is absent rather than depleted or defective.

NuMA *See* CENTROPHILIN.

numatrin *See* NUCLEOPHOSMIN.

Numb The human homologue of *Drosophila* numb (mammalian numb, mNumb, 603 aa) is a membrane-associated protein that predominantly segregates to only one of the daughter cells of a dividing neural precursor and has an important role in neural development, by a signalling pathway that antagonizes the *Notch pathway. There are multiple splicing isoforms. Numb also blocks the ubiquitination and degradation of *p53, and Numb-defective breast tumours have a poor prognosis. The cytoplasmic ligands of Numb are the PDZ domain-containing ring finger proteins 1 and 2 (632 aa), *E3 ligases involved in ubiquitination.

NURF complex *See* NUCLEOSOME REMODELLING FACTOR COMPLEX.

nurse cells Cells that support developing egg and/or sperm progenitors, synthesizing important substances which are transferred to the developing gamete.

nutlins A group of *cis*- imidazoline derivatives that selectively inhibit binding of Mdm2-ligase (*see* HDM2) to *p53 and that drive tumour cells into apoptosis.

nutraceutical A foodstuff engineered to have therapeutic properties. An example is golden rice, engineered to have higher vitamin A content.

NXT (p15) A cofactor for *Tap protein-mediated mRNA export, related to nuclear transport factor 2 (NTF2). Stimulates the binding of a Tap-RNA complex to *nucleoporin.

Nycodenz® A nonionic substance that is the basis of dense, heat-stable solutions for density gradients used in centrifugal separations. A density of 2.1 g/mL can be achieved.

nyctalopin A *small leucine-rich proteoglycan (SLRP) that is defective in an X-linked form of *stationary night blindness.

nystatin An antifungal polyene antibiotic produced by *Streptomyces noursei.* The name is derived from 'New York State Health Department' where it was discovered. It is an ionophore selective for Na^+ and increases the activity of the Na^+-K^+ pump. *TN* Nystan.

NZB, NZW mice (New Zealand Black, New Zealand White) Highly inbred strains of mice which spontaneously develop autoimmune diseases and B-cell lymphomas. There is thought to be an underlying retroviral aetiology and a putative NZB virus has been isolated.

O-antigens Cell wall antigens of Gram-negative bacteria consisting of tetra- and pentasaccharide repeat units of *lipopolysaccharide.

SEE WEB LINKS

- A review of endotoxins in K.Todar *Online Textbook of Bacteriology*.

oat cell carcinoma A type of lung carcinoma (small-cell carcinoma) in which the cells are small, spindle-shaped, and dark-staining. They may derive from argyrophilic *APUD cells and secrete one or more of a range of hormones such as *ACTH or *bombesin.

OBCAM (OPCML) An opioid-binding *cell adhesion molecule (345 aa) of the *IgLON family, probably GPI-linked and possibly an accessory molecule in opioid receptor function.

obestatin A peptide (23 aa) with appetite-suppressing activity, an endogenous ligand for the orphan G-protein-coupled receptor GPR39, produced by post-translational modification of a precursor that also gives rise to *ghrelin, an appetite stimulator.

obscurin A multidomain muscle protein (6620 aa) with significant similarity to *myosin-binding protein C and with *titin, with which it coassembles. Interacts with the Z disc and probably has a role in the spatial organization of sarcomeres within the myofibrils.

OCA2 An integral membrane protein (833 aa) of the melanosome that has a role in regulating melanosome pH. The mouse equivalent is mutated in the 'pink-eyed dilution' (p) mouse. Mutations in the gene lead to oculocutaneous albinism type 2.

occipital horn syndrome *See* MENKES' DISEASE.

occipital lobe The posterior lobe of each cerebral hemisphere, particularly involved in visual processing.

occludens junction *See* ZONULA OCCLUDENS.

occludin A four-pass integral plasma membrane protein (522 aa) which is a component of the *zonula occludens. The C-terminal cytoplasmic domain interacts with *ZO-1.

occlusion bodies Large, protein-rich granules (0.3×0.5 μm) formed late in the infection of cells by *baculovirus.

ochnaflavone A naturally occurring biflavonoid, the active ingredient of the herbal product isolated from *Lonicera japonica*; inhibits the production of a range of inflammatory mediators and the serum-induced proliferation of vascular smooth muscle cells, and has antifungal activity.

ochre *See* TERMINATION CODONS.

ochronosis *See* ALKAPTONURIA.

OCIF *See* OSTEOPROTEGERIN.

oct **1.** A family of *leucine zipper transcription factor genes (*oct* genes), the products of which act as RNA polymerase II promoters and bind to *octamer sequences. Oct4 maintains the pluripotency of embryonic stem cells and in the expanding blastocyst is confined to the inner cell mass; as development proceeds it gradually disappears as the three germ layers become progressively differentiated. **2.** OctN1 (solute carrier family 22, member 4, SLC22A4, 551 aa) is a proton/organic cation transporter involved in active secretion of cationic compounds, including xenobiotics, across the renal epithelial brush-border membrane. Mutations in OctN2 (SLC22A5), the sodium ion-dependent carnitine transporter, are responsible for *systemic carnitine deficiency.

octamer **1.** The complex of eight histone proteins, two each of H2A, H2B, H3, and H4, that forms the core of the *nucleosome. **2.** The eight-base motif common in eukaryotic *promoters, for which the consensus is ATTTGCAT. *See* OCT.

octopamine A *biogenic amine closely related to noradrenaline with noradrenergic and dopaminergic effects. First identified in the salivary gland of *Octopus*. Its role in humans is unclear.

octyl glucoside A biological detergent used to solubilize membrane proteins.

ocular coloboma A congenital abnormality in which there is defective closure of the optic cup in the fifth to seventh week of development. Can be caused by mutation in the *PAX6* homoeobox gene and is associated with various complex malformation syndromes.

ocular micrometer *See* EYEPIECE GRATICULE.

oculodentodigital dysplasia A developmental disorder (oculodento-osseous dysplasia) caused by mutation in the gap junction α_1 protein (*connexin-43). There are craniofacial and limb dysmorphisms, spastic paraplegia, and neurodegeneration. There may also be ocular defects, syndactyly type III, and conductive deafness.

ODC **1.** *See* ORNITHINE DECARBOXYLASE. **2.** Oxygen (oxyhaemoglobin) dissociation curve.

ODF **1.** The orientation distribution function, a mathematical formalism used in describing the orientation of fibres, crystals, etc. **2.** *See* OSTEOCLAST DIFFERENTIATION FACTOR. **3.** *See* OUTER DENSE FIBRES.

odontoblasts Columnar cells that are derived from the dental papilla after *ameloblasts have differentiated and that produce the *dentin matrix underlying the enamel of a tooth.

odoratism *See* LATHYRISM.

oedema (edema) Tissue swelling as a result of either increased vascular permeability or increased blood pressure.

oestradiol (follicular hormone) A female sex hormone (272 Da) produced mainly by the granulosa cells of the ovaries, but also in the placenta, testis, and possibly adrenal cortex. A potent *oestrogen. The synthetic form is used in hormone replacement therapy. Oestradiol is also produced in the Sertoli cells of the testes and it may be important in inhibiting apoptosis of developing male germ cells. There are two main **oestrogen receptors**, ERα (595 aa) and ERβ (530 aa) that are ligand-activated transcription factors which move from cytoplasm to nucleus once oestrogen or oestradiol is bound. Additional isoforms, generated by alternative mRNA splicing, have been identified in several tissues.

oestrogen A type of hormone that induces oestrus ('heat') in female animals. It controls changes in the uterus that precede ovulation, and is responsible for development of secondary sexual characteristics in pubescent girls. Although present in females in greater amounts and generally thought of as female hormones, oestrogens are also produced in males. In humans the three main oestrogens are *oestradiol, oestriol, and oestrone, all of which are produced from androgens. Approximately 80% of mammary carcinomas are oestrogen-dependent and treatment is therefore by suppression of oestrogen production.

oestrogen-responsive B box protein A protein (EBBP, 564 aa) structurally related to *pyrin, a downstream target of oestrogen receptor α, and a member of RING finger-B box-Coiled Coil family. Its function is unclear although it binds IL1β and may have a role in its secretion. It has also been suggested as a regulator of the retinoid anticancer signal.

oestrone *See* OESTROGEN.

off-label A colloquial term, mainly in the USA, for a drug used to treat a condition other than that for which it has been approved by the Food and Drug Administration (FDA).

ofloxacin A fluoroquinolone antibiotic, the active isomer being *levofloxacin. *TNs* Floxin, Tarivid.

***O*-glycans** Carbohydrate moieties linked post-translationally to protein through the hydroxyl group of serine or threonine, usually with an initial GalNAc residue. Unlike the **N*-glycans there is no consensus sequence for addition and no oligosaccharide precursor is required for protein transfer. *See* N-GLYCANS.

Oguchi's disease A recessively inherited form of *stationary night blindness in which there is abnormally slow dark adaptation due to mutation in either the *arrestin or the *rhodopsin kinase gene, the products of which are involved in reactivating rhodopsin that has absorbed a photon.

okadaic acid A toxin produced by various species of dinoflagellates, although first isolated from the sponge *Halichondria okadai*. It is an inhibitor of protein *phosphatases 1 and 2A and can act as a tumour promoter. *Microcystins and *nodularins bind to the same site on the phosphatase.

Okazaki fragments Short fragments, a few hundred nucleotides long, discontinously synthesized in the 5′ to 3′ direction on the lagging strand of DNA that is being replicated by *DNA polymerase. Synthesis on the leading strand occurs continuously at the replication fork (which is progressing in the 5′ to 3′ direction) as the duplex unzips. The Okazaki fragments are covalently linked by *ligases to produce a continuous strand.

olanzapine An atypical antipsychotic drug of the thienobenzodiazepine class used for treatment of schizophrenia. It is a selective monoaminergic antagonist and may act through a combined antagonism of dopamine and serotonin type 2 ($5HT_2$) receptors. *TN* Zyprexa.

oleic acid A monounsaturated C-18 ω-9 fatty acid.

olfactomedin family A family of extracellular proteins with diverse roles but sharing a conserved protein motif through which protein–protein interactions occur. The family includes amassin, *myocilin, *noelins, *pancortins, *photomedins, and tiarin.

olfactory epithelium The *epithelium lining the nose that has the diverse G-protein-coupled-receptors that are involved in the sense of smell.

olfactory neuron A *sensory neuron that obtains input from the smell receptors of the *olfactory epithelium. Unusually, they continue to proliferate in adults.

oligoadenylate synthetases A conserved family of interferon-induced proteins (EC 2.7.7.-) that are critical components of the innate immune response to viruses. They are encoded by *OAS1, OAS2, OAS3* genes and the 2′–5′ oligoadenylate synthetase-like gene (*OASL*). When activated by double-stranded RNA they polymerize ATP which then activates a latent endoribonuclease L that will degrade viral and cellular RNAs. Mutations increase susceptibility to viral infection.

oligodendrocyte A glial cell of the central nervous system responsible for myelinating axons.

oligodendrocyte–myelin glycoprotein An integral membrane protein (OMgp, myelin–oligodendrocyte glycoprotein, MOG, 440 aa) expressed in oligodendrocytes and outer myelin lamellae. It is a ligand for the *nogo receptor (as is *MAG and nogo itself) and will inhibit neurite outgrowth.

oligodontia–colorectal cancer syndrome *See* AXINS.

oligomycin A bacterial toxin that inhibits mitochondrial oxidative phosphorylation. The oligomycin sensitivity-conferral protein (OSCP) to which oligomycin binds is the δ subunit of the *F-type ATP synthase which is thought to link the F1 catalytic segment to the Fo proton-conduction segment.

oligonucleotide A linear sequence of *nucleotides joined by phosphodiester bonds, conventionally up to 20 nucleotides, longer polymers being referred to as *polynucleotides.

oligopeptide A peptide of 3–40 amino acids; larger peptides are usually called *polypeptides.

oligophrenin *See* P21-ACTIVATED KINASES.

oligosaccharide A saccharide with an undefined but small number of component sugars, either *O*-linked or *N*-linked.

olomoucine A purine derivative that inhibits *cyclin-dependent kinase 5 (cdk5), but is inactive against cdk4 and cdk6. A similar compound, roscovitine, is more potent.

olsalazine An anti-inflammatory drug, a salicylic acid derivative, used to treat inflammatory bowel disease. *TN* Dipentum.

omalizumab A humanized monoclonal antibody that selectively binds immunoglobulin E (IgE), used to treat allergic asthma. *TN* Xolair.

omega fatty acid An unsaturated fatty acid with a double bond in the n-3 or n-6 position, typically found in unsaturated fat and generally regarded as being helpful in reducing the risk of heart disease.

omega-oxidation An alternative metabolic pathway for medium chain-length fatty acids that occurs in the endoplasmic reticulum. The dicarboxylic acid product can enter the mitochondrial *beta-oxidation pathway.

Omenn's syndrome A *severe combined immunodeficiency disease (severe combined immunodeficiency with hypereosinophilia) caused by mutation in the **RAG* genes or in *artemis.

omentin A visceral fat depot-specific *adipokine that enhances insulin-mediated glucose-uptake and activates *Akt. Omentin 1 (intelectin-1, endothelial lectin HL-1, intestinal lactoferrin receptor, galactofuranose-binding lectin, 313 aa) and omentin 2 (intelectin-2) expression levels are decreased with obesity.

omentum The thin infolding of the peritoneal lining that connects two or more folds of the alimentary canal. May be enlarged by abundant *adipocytes.

omeprazole A proton pump inhibitor used in the treatment of dyspepsia, peptic ulcer, and *Zollinger–Ellison syndrome. *TNs* Losec, Prilosec.

OMgp *See* OLIGODENDROCYTE–MYELIN GLYCOPROTEIN.

OMIM An important database, Online Mendelian Inheritance in Man, that catalogues more than 12 000 human genes and genetic disorders. It is authored and edited by Dr V.A. McKusick and his colleagues at Johns Hopkins University, and made freely available through NCBI, the National Center for Biotechnology Information. Many of the entries in this dictionary are covered in much greater detail in OMIM.

SEE WEB LINKS

- The OMIM homepage.

Onchocerca A genus of filarial nematodes that cause river blindness (onchocerciasis).

oncocytes *See* HURTHLE CELLS.

oncogene A normal cellular gene (a *proto-oncogene) that is mutated or is overexpressed so that normal restraints on proliferation and sometimes of positional control are lost. The mutation may (e.g.) make the gene product constitutively active, or insensitive to normal regulation. Viral versions of cellular genes, often prefixed v- (e.g. v-*src*) to distinguish them from the normal cellular version (e.g. c-*src*), are responsible for transformation by retroviruses. *See* ONCOVIRINAE and individual oncogenes: AKT; ARG; AXL; BCL; BRX; CBL; C-MAF; CRK; DBL; E1; E5; ECTODERM-NEURAL CORTEX-1; ERB; ETS; FELINE SARCOMA ONCOGENE HOMOLOGUE; FES; FMS; FOS; FYN; GASC1; GIP2; GLI ONCOGENES; HST; INT ONCOGENES; JAZF1; JUN; KIT; MAF; MAS; MET; MOS; MULTIPLE MYELOMA ONCOGENE-1; MYB; NEU; NOV; PIM; RAF; RAL; RAS; REL; RET ONCOGENE; ROS; SEA; SIS; SKI; SRC gene and the SRC-FAMILY of kinases; TRE; TRK and YES.

oncomodulin *See* PARVALBUMINS.

OncoMouse™ The proprietary name for a line of transgenic mice which have been genetically engineered to have the activated v-Ha-*ras* oncogene driven by a mouse ζ-globin promoter. They are predisposed to papillomas and are used in screening for carcinogens, etc. The patenting of these animals generated considerable controversy.

oncoprotein 18 *See* STATHMIN.

oncostatin M A member (252 aa) of the *interleukin-6 cytokine family that inhibits the growth of a variety of cancer cells and is also an inflammatory mediator. OSM binds to the gp130 receptor subunit common to many of the IL-6 family, in conjunction with either the *leukaemia inhibitory factor receptor subunit or a specific subunit (**OSM receptor** β, OSMRβ, 979 aa) with characteristic motifs of the haematopoietin receptor family. Mutations in *OSMRB* are responsible for *familial primary localized cutaneous amyloidosis.

Oncovirinae A taxonomically doubtful group of unrelated genera of retroviruses (*Retroviridae) that can cause tumours.

ondansetron An antiemetic drug often used to alleviate the side effects of chemotherapy. It selectively blocks serotonin 5-HT_3 receptors. *TN* Zofran.

onychogryphosis Hypertrophy of the nails.

oocyte The developing female gamete before it matures (the process of oogenesis) and is released.

oocyte expression A method used to produce functional protein from mRNA or an expression vector by introducing it into an amphibian oocyte (typically *Xenopus*) which then translates the message. In conjunction with *patch clamping, has been used to characterize various ion channels.

OPA1 *See* PARAPLEGIN.

opal *See* TERMINATION CODONS.

OPC **1.** Oropharyngeal candidiasis, a common opportunistic infection in immunosuppressed patients. **2.** Oral and pharyngeal cancer. **3.** Oligodendrocyte precursor cell, a lineage-restricted neural precursor cell that expresses most neurotransmitter receptors. **4.** Organophosphorus compounds. **5.** OPC-14523 (OPC), a compound with high affinity for sigma and 5-HT_{1A} receptors and potential antidepressant properties.

OPCML *See* OBCAM.

open reading frame A DNA *reading frame (ORF) which could be translated into protein and does not have *termination codons. This does not necessarily mean that the protein is produced.

operator The site on DNA, upstream of a particular gene or *operon, to which a specific *repressor protein binds, thereby blocking the initiation of transcription at the adjacent *promoter.

operon A set of bacterial genes controlled as a unit with a common *promotor and a single *polycistronic mRNA. The operon usually encodes proteins involved in linked metabolic activities. The first described example was the *lactose operon and although famously its discoverer, Jaques Monod (1910–1976), claimed that what was true for *E. coli* was true for the elephant, eukaryotic gene regulation has proved a little more complex.

OPG *See* OSTEOPROTEGERIN.

opiates *See* OPIOIDS.

opioids Naturally occuring alkaloids, from the opium poppy *Papaver somniferum*, that are very effective painkillers but tend to be addictive and are drugs of abuse. Modified opium derivatives are opiates, including morphine, but opioid is a more general term covering natural, semisynthetic, and synthetic compounds. Weak opioids include codeine and dihydrocodeine; strong opioids include morphine, diamorphine, fentanyl, and phenazocine. **Opioid receptors** fall into four distinct classes: δ, μ, and κ together with an orphan receptor, the ligand for which is *orphanin FQ. All are G-protein coupled and the endogenous ligands are *enkephalins and *endorphins.

opportunistic infection An infection caused by an agent which is usually innocuous unless the immune system is compromised, as in *AIDS, or when immunosuppression is required following tissue transplantion. A wide range of organisms, some of them common commensals, can be opportunistic pathogens.

opsin A general term for the apoproteins of the *rhodopsin family, G-protein-coupled receptors of which many variants are known. They can be divided into subfamilies: 1. the vertebrate visual (transducin-coupled) and nonvisual opsins, 2. the *encephalopsin/tmt-opsin subfamily, 3. the G_q-coupled opsin/*melanopsin subfamily, 4. the G_o-coupled opsins, 5. the *neuropsins, 6. the *peropsins, and 7. the retinal photoisomerase subfamily. *See* COLOUR BLINDNESS.

opsonin A substance that binds to a particle and enhances its uptake by a phagocyte. Important opsonins in mammals derive from *complement (C3b or C3bi) or immunoglobulins (which are bound through the *Fc receptor).

optic atrophy Type 1 optic atrophy (OPA1) is an autosomal dominant disorder in which there is progressive visual impairment in childhood and is caused by mutation in the gene encoding the human homologue of the yeast *dynamin-related protein Msp1. Other loci are known to be responsible for other types of OPA. *See* GLAUCOMA.

optic nerve The bundle of neurons that receive input from retinal photoreceptors and interface with the *optic tectum (the 'retinotectal projection'). Embryologically, the optic nerve is a tract of the central rather than the peripheral nervous system. Because of the 'inverted' nature of the eye, with the photoreceptors behind the neural retina, there is a blind spot (optic disc) where the elements of the optic nerve converge and leave the eye.

optic tectum A region of the midbrain in which input from the *optic nerve is processed. The way in which retinally derived neurons are specifically 'mapped' onto the optic tectum has been a long-standing problem in cell biology and although there is some evidence for adhesion gradients and specificity in adhesion, the problem is unresolved.

optimedin A secreted protein (olfactomedin 3, ~460 aa) of the *olfactomedin family that appears to be a downstream target of the **PAX6* gene product in eye development. *See* MYOCILIN.

Orbivirus A genus of the *Reoviridae that mostly cause diseases in animals, including bluetongue disease. They have a double-stranded DNA genome and characteristic doughnut-shaped capsomeres.

ORC *See* ORIGIN OF REPLICATION.

orcein A purple dye originally extracted from lichens, used as a histological stain for elastic fibres and for *ground-glass cells.

orcinol A compound used in Bial's orcinol test to distinguish pentoses from hexoses. Pentoses produce a green to deep blue colour but hexoses give a muddy brown/grey colour.

orexin (hypocretin) Potent orexigenic (appetite-stimulating) peptides, orexin-A (hypocretin-1, 33 aa) and orexin B (hypocretin-2, 28 aa), derived from the same precursor peptide. They are specifically localized in neurons in the lateral hypothalamic area, which is implicated in feeding behaviour, but the physiological role is likely to be more complex and mutations in orexin are associated with narcolepsy. The receptor is G-protein coupled.

orf **1.** A skin disease of sheep and goats, caused by a parapox virus, and characterized by pustular or 'scabby' lesions around the mouth and nostrils of lambs. Can cause lesions on hands and arms of humans who handle sheep. **2.** *See* OPEN READING FRAME (ORF).

organ culture The culture of pieces of tissue *in vitro*, rather than single cells, so that some normal spatial relationships are retained. *Compare* TISSUE CULTURE.

organelle Any structurally discrete component of a cell.

organizing centre *See* MICROTUBULE ORGANIZING CENTRE.

organogenesis The developmental process involving cell differentiation and morphogenesis to produce specific organs.

organophosphate-induced delayed neurotoxicity *See* PNPLA6.

oriental sore A localized skin form of *leishmaniasis caused by *Leishmania tropica*.

origin of replication The DNA sequence that marks the site where the pre-replication complex (pre-RC) will assemble and from which replication commences. The **origin recognition complex** (ORC) is a six-subunit complex which recruits *cdc6 and *cdt1 during interphase; phosphorylation by *cyclin-dependent kinase releases these and allows *DNA helicase to begin the process of replication. The proteins involved (ORC1–6) are highly conserved.

orlistat A saturated derivative of lipstatin (tetrahydrolipstatin), a natural inhibitor of pancreatic lipases isolated from *Streptomyces toxytricini*. It is used to treat obesity by inhibiting hydrolysis of dietary triglycerides thereby reducing the absorption of fat. *TN* alli, Xenical.

ornithine decarboxylase The first enzyme (EC 4.1.1.17, 461 aa) in polyamine synthesis; it converts ornithine to putrescine, the rate-limiting step in production of spermidine and *spermine, which regulate DNA synthesis.

ornithine transcarbamylase A nuclear-encoded mitochondrial matrix enzyme (OTC, EC 2.1.3.3, 354 aa) that catalyses the second step of the urea cycle in mammals. **OTC deficiency** (an X-linked metabolic defect) leads to hyperammonaemia. Other urea cycle disorders have been described in which the deficient enzymes are variously: carbamyl phosphate synthetase, argininosuccinate synthetase (causing citrullinaemia), argininosuccinate lyase and arginase.

ornithosis *See* PSITTACOSIS.

orosomucoid A highly glycosylated plasma protein (α_1-seromucoid, α_1-acid glycoprotein, 210 aa, 0.6–1.2 mg/mL), one of the *acute phase proteins, of which there are three phenotypes, SS, FF, and FS, (electrophoretically slow and fast) determined by two codominant alleles and that vary between populations. A second isoform, orosomucoid 2, is present at about one-third of the concentration of orosomucoid 1. Orosomucoid probably acts as a carrier for basic compounds in plasma.

orphan drug A drug given special status and financial support by govenmental or charitable agencies to enable it to be produced for an uneconomically small patient population.

orphanin FQ *See* NOCICEPTIN.

orphan receptor A receptor with an unknown ligand.

orphenadrine An antimuscarinic drug used to treat muscle spasm. *TNs* Banflex, Biorphen, Brocasipal, Disipal, Flexon, Mephenamin, Norflex.

orthodromic Describing impulses that pass down a nerve fibre from the cell body towards the distal presynaptic region. The opposite of antidromic.

orthogonal Describing elements (fibres, cells) that are arranged at approximately right angles to one another, e.g. collagen fibres in the cornea.

orthograde transport Transport of material from the cell body of a neuron towards the synaptic terminal involving *kinesin interacting with microtubules. Retrograde transport, in the opposite direction, uses a different motor protein.

orthokinesis A *kinesis response in which the speed or frequency of movement is increased (positive orthokinesis) or decreased, but the direction of movement is not affected.

orthologous genes Describing genes that have a common phylogenetic descent. *Compare* PARALOGOUS GENES.

Orthomyxoviridae The class V viruses, mostly influenza viruses. Their genome is a negative strand of RNA present as several separate segments, each a template for a single mRNA synthesized by a virus-specific RNA polymerase. The *nucleocapsid is helical and they are budded from cells and enveloped in host plasma membrane in which virus-coded spike proteins are enriched. The spike proteins have *haemagglutinin and *neuraminidase activity (the basis for classification, e.g. *H5N1), both being important in the invasion of cells by the virus.

Orthopoxviridae A genus of viruses with a double-stranded DNA genome that preferentially infect epithelial cells. Includes variola (smallpox) and *vaccinia.

oscillin A soluble protein (289 aa) originally identified as a sperm-derived factor responsible for oocyte calcium oscillations. It is, however, highly conserved and has a ubiquitous tissue distribution and a high level of homology with bacterial glucosamine-6-phosphate isomerase, suggesting a more general housekeeping role.

OSCP (oligomycin sensitivity-conferral protein) *See* OLIGOMYCIN.

oseltamivir An inhibitor of viral neuraminidase. *TN* Tamiflu.

Osler–Rendu–Weber disease *See* HEREDITARY HAEMORRHAGIC TELANGIECTASIA (type 1).

osmiophilic Having an affinity for osmium tetroxide, which is used as a postfixative/stain in electron microscopy.

osmole The amount of a solute that, when dissolved in water, produces a solution with the same osmotic pressure as that produced by 1 mol/L of an ideal nonionized solute. Isotonic saline is ~290 milliosmolar, seawater ~1000 milliosmolar.

osmoreceptors Cells that are specialized to react to osmotic changes, e.g. those involved in regulating the secretion of antidiuretic hormone by the neurohypophysis.

osmoregulation The processes by which a cell or an organism regulates its internal osmotic pressure. These may include water transport, ion accumulation, or loss or the synthesis of osmotically active substances.

osmosis The movement of solvent through a membrane impermeable to solute, driven by an imbalance in chemical potential (mainly due to concentration difference) across the membrane. Widely misused in general writing.

osmotic pressure The pressure required to prevent *osmosis (solvent flow) across a semipermeable membrane separating two solutions of different solute concentration.

osmotic shock Cell lysis as a result of water influx driven by *osmosis when cells are placed in a hypotonic solution.

OSR1 *See* SPAK; WNKs.

osteo- A prefix indicating bone-related. **Osteoblasts** are mesodermal cells that produce bone. **Osteoclasts** are cells of the myeloid series that are responsible for bone breakdown. **Osteocytes** are quiescent osteoblasts resident within bone. **Osteoid** is the nonmineralized extracellular matrix produced by osteoblasts which forms the basis of bone. The main fibrillar component is collagen, but also contains *osteonectin. An **osteoma** is a benign tumour of bone. **Osteomalacia** is softening of bone as a result of vitamin D deficiency, the adult equivalent of rickets. **Osteomyelitis** is inflammation of the bone marrow. **Osteopetrosis** is the formation of abnormally dense bone (*see* BONE DISORDERS) as opposed to **osteoporosis**, the demineralization that can occur particularly in postmenopausal women. An **osteosarcoma** is a malignant tumour of bone, probably derived from osteocytes.

osteoadherin (osteomodulin) One of the small leucine-rich repeat protein family of proteoglycans found in mineralized extracellular matrix. It has *keratan sulphate side chains and binds osteoblasts via the integrin $\alpha_v\beta_3$. May be important in mineralization since it binds hydroxyapatite crystals.

osteoarthritis The degradation of joints as an eventual consequence of mechanical trauma. There is cartilage degradation by *matrix metallopeptidases and progressive loss of joint function. Unlike *rheumatoid arthritis there is not an autoimmune element and the inflammation is generally considered to be secondary rather than causative. Osteoarthritis with mild chondrodysplasia is associated with mutations in the *collagen (*COL2A1*) gene.

osteocalcin A small, highly conserved protein (bone γ-carboxyglutamic acid protein, bone gla protein, BGP, 50 aa) associated with the mineralized matrix of bone. It has three γ-carboxyglutamic acid residues, formed post-translationally by a vitamin K-dependent process, which are essential for its binding interaction with *hydroxyapatite. Genetic variation at the osteocalcin locus may predispose some women to osteoporosis.

osteochondrosis (*pl.* osteochondroses) An orthopaedic disorder in which abnormal growth of cartilage or bone leads to joint damage. *See* ECCHONDROMA.

osteoclast differentiation factor One of the tumour necrosis factor (TNF) superfamily, a cytokine (ODF, receptor activator of NFκB ligand, RANKL, 317 aa) that regulates differentiation of the immune system as well as bone metabolism. Mice deficient in RANKL have defects in early thymocyte development. RANKL-RANK signalling is essential for *osteoclast development and bone destruction. Inhibition of RANKL either by the decoy receptor *osteoprotegerin or by anti-RANKL antibody, may be therapeutically beneficial in *osteoporosis and rheumatoid arthritis.

osteodystrophia fibrosa (osteofibrosis) A disorder in which fibrous tissue replaces bone which has become demineralized as a result of *hyperparathyroidism, nutritional problems or excretory dysfunction.

osteogenesis The production of bone. **Osteogenesis imperfecta** is a heterogenous group of disorders that affect connective tissue in bone, cartilage, and tendon. Characteristically bones are very brittle and fracture easily. Type I is a dominantly inherited, generalized connective tissue disorder caused by mutation in either of the collagen type 1A genes (*COL1A1* and *COL1A2*); other types also involve mutations in one or the other of these genes but have phenotypic differences.

osteogenin *See* BONE MORPHOGENETIC PROTEINS.

osteoglycin *See* MIMECAN.

osteomodulin *See* OSTEOADHERIN.

osteonectin A calcium-binding protein (basement membrane protein BM-40, secreted protein acidic and rich in cysteine, SPARC, 303 aa) of bone that binds to both collagen and hydroxyapatite.

osteopetrosis A disorder (Albers–Schoenberg disease) in which there is excessive mineralization and brittleness of bone. **Autosomal dominant osteopetrosis type 1** (OPTA1) is caused by mutation in the *LRP5* gene, encoding the LDL receptor related protein 5 that mediates *wnt signalling in association with *axin; **type 2** (OPTA2) is caused by mutation in the *CLCN7* gene encoding chloride channel 7. **Autosomal recessive osteopetrosis type 1** (OPTB1, infantile malignant form) is caused by mutation in the TCIRG1 subunit of the vacuolar proton pump. Other forms of autosomal recessive infantile malignant osteopetrosis include **OPTB4** caused by mutation in the *CLCN7* gene, and **OPTB5** caused by mutation in the *OSTM1* gene that encodes **osteopetrosis-associated transmembrane protein 1** (334 aa). The CLC7 and OSTM1 gene products colocalize in late endosomes and lysosomes of various tissues, as well as in the ruffled border of bone-resorbing *osteoclasts. A milder, osteoclast-poor form of autosomal recessive osteopetrosis (**OPTB2**) is caused by mutation in the *TNFSF11* gene encoding the *osteoprotegerin ligand, an intermediate form (**OPTB6**) is caused by mutation in the *PLEKHM1* gene encoding the pleckstrin homology domain-containing protein M1 (926 aa), and a severe osteoclast-poor form associated with hypogammaglobulinaemia caused by mutation in the *TNFRSF11A* gene (*see* PAGET'S DISEASE OF BONE). A form of autosomal recessive osteopetrosis (**OPTB3**) is associated with renal tubular acidosis caused by mutation in the gene encoding *carbonic anhydrase II. Mutation in *MCSF leads to osteopetrosis because of osteoclast deficiency.

osteopontin The principal phosphorylated glycoprotein of bone, a sialoprotein (bone sialoprotein 1, secreted phosphoprotein 1, SPP-1, nephropontin, uropontin, 40 kDa, 314 aa) that links cells and the hydroxyapatite of mineralized matrix. It is produced by osteoblasts, to which it binds via $\alpha_v\beta_3$ integrin. Synthesis is stimulated by *calcitriol (vitamin D_3).

osteoporosis A condition in which there is reduced bone mineral density, disruption of bone microarchitecture and changes in the amount and variety of noncollagenous proteins in bone. Osteoporotic bones are more at risk of fracture. **Involutional osteoporosis** (senile osteoporosis or postmenopausal osteoporosis) is associated with variations in type 1 *collagen.

osteoporosis–pseudoglioma syndrome *See* LDL RECEPTOR (LDL receptor-related protein-5, LRP5).

osteoprotegerin A soluble decoy receptor of the tumour necrosis factor (TNF) receptor superfamily (OPG, osteoclastogenesis inhibitory factor, OCIF, TNFRSF11B, 401 aa) that binds *osteoclast differentiation factor/RANKL and inhibits the differentiation and function of osteoclasts and therefore bone destruction. Deficiency of OPG results in juvenile *Paget's disease.

OTC **1.** OTC drugs are drugs sold 'over the counter' for which a doctor's prescription is not required. **2.** *See* ORNITHINE TRANCARBAMYLASE.

otocadherin A large single-pass transmembrane protein (cadherin 23, 3354 aa) that is involved, together with *protocadherin 15 (PCDH15) in forming the tip links at the tips of stereocilia of the hair cells of the inner ear. These tip links are thought to gate the mechanoelectrical channel that transduces movement into electrical impulse. The extracellular domain contains 27 repeats with significant homology to the cadherin ectodomain, and otocadherin also interacts with *harmonin. Mutations in otocadherin are involved in *Usher's syndrome (type 1D).

otoferlin The product of a gene found in the critical region of chromosome 2p23.1 for a form of nonsyndromic deafness. Like *ferlin, a transmembrane protein with multiple C2 domains probably involved in calcium-dependent membrane fusion events. Otoferlin expression in mouse hair cells correlates with afferent synaptogenesis, and otoferlin interacts with *syntaxin-1.

otosclerosis A disorder in which there is abnormal growth of bone of the middle ear, which can result in deafness. It is a common age-onset condition and multiple loci for susceptibility have been mapped, although the gene has not yet been identified. Mutations in *reelin have been implicated.

otospondylomegaepiphyseal dysplasia A skeletal dysplasia accompanied by severe hearing loss caused by defects in *collagen type XI (COL11A2).

ouabain *See* STROPHANTHIN.

Ouchterlony assay An immunological assay for antigen–antibody reactions in which radial diffusion of soluble antigen and antibody from wells cut in a gel leads to the formation of an antigen–antibody complex, visible usually as a whitish band, where the antibody/antigen ratio is appropriate. The method is tolerant of wide variation in concentrations of antibody and antigen.

outer dense fibres Fibres (ODFs) located on the outside of the axoneme in the midpiece and principal piece of the mammalian sperm tail, probably contributing to the mechanical properties. The proteins composing the fibres include **Odf1** (ODF27, 241 aa) which interacts with sperm associated antigen 4 (SPAG4) through a leucine zipper, **Odf2** which shares partial homology with centriolar *cenexins, **Odf3** (254 aa), and **Odf4** (257 aa). **Odf5** is *rhophilin 1, a rho GTPase binding factor.

outron A nucleotide sequence at the 5′ end of pre-mRNA containg an intron-like sequence, followed by a splice acceptor, that is the signal for *trans*-*splicing.

outside-out patch *See* PATCH CLAMPING.

ovalbumin *See* ALBUMEN.

oval cells Stem cells of the adult liver that can produce hepatocytes or bile duct epithelial cells. They are normally quiescent, but proliferate following severe liver trauma and may be implicated in hepatocellular carcinoma.

ovalocytosis Hereditary disorder of erythrocytes relatively common in areas where malaria is endemic. Not only are the erythrocytes more rigid, but there is a mutation in *band III, the anion transporter. Other forms, such as **hereditary haemolytic ovalocytosis**, may be caused by mutation in other genes involved in maintaining erythrocyte shape and known to be associated with *elliptocytosis: band III, *spectrin, *protein 4.1, or *protein 4.2.

ovarian follicle *See* FOLLICLE.

ovarian granulosa cells The cells of the stratum granulosum or ovarian granulosa that line the Graafian *follicle which contains the developing ovum.

overlap index A measure of the extent of multilayering of cells in culture. The actual extent of overlapping, usually of nuclei, is measured on fixed and stained preparations and compared to the predicted level based on cell density, the projected area of the nucleus and the assumption of a Poisson distribution. A value of 1 implies a random distribution but a constraint on overlapping, as with normal fibroblasts that exhibit *contact inhibition of locomotion, may produce a value as low as 0.05. The reduction of overlapping may also arise though the relative nonadhesiveness of the dorsal surfaces of cells.

overlapping Describing the situation in which the *leading lamella of one cell moves actively over the dorsal surface of another cell. *Compare* UNDERLAPPING.

ovotransferrin *See* CONALBUMIN.

ovum (*pl.* ova) A female gamete, an egg.

owl-eye cells Cells that are infected with *cytomegalovirus and have large inclusion bodies surrounded by a halo, fancifully said to resemble owl's eyes.

Oxa1 A translocase (435 aa) required for cytochrome *c* oxidase assembly in the inner mitochondrial membrane. Similar to Albino3 in plants and YidC.

oxacladiellanes A class of microtubule-stabilizing compounds that includes *sarcodictyin A and *eleutherobin. The valvidones and *eleuthosides are anti-inflammatory.

oxaliplatin A chemotherapeutic drug, an analogue of *cisplatin. *TN* Eloxatin.

oxaloacetate An intermediate in the *tricarboxylic acid cycle that couples with acetyl CoA to form citrate.

oxazepam A benzodiazepine drug used to treat anxiety.

oxidation The process in which an electron is transferred to an oxidizing agent or, alternatively, when a compound is combined with oxygen or has hydrogen removed.

oxidation–reduction potential *See* REDOX POTENTIAL.

oxidative metabolism The process of controlled oxidation of compounds to generate energy for metabolic use, respiration in the biochemical sense. **Oxidative phosphorylation** is the process of generating ATP from ADP by phosphorylation coupled to the *respiratory chain.

oxidative stress The overproduction of *reactive oxygen species that may occur e.g. as a result of inflammation.

oxidoreductase *See* DEHYDROGENASES.

oxybutinin (oxybutynin) An *antimuscarinic drug that is used to treat urinary incontinence by acting on musculature of the bladder wall.

oxycodone An *opioid analgesic derived from *thebaine. *TNs* Oxycontin, Endocet.

oxygenases The subclass (EC 1.13) of dehydrogenase/oxidoreductase enzymes that catalyse the incorporation of oxygen from molecular oxygen into organic substrates. Dioxygenases (EC 1.13.11.-), incorporate two oxygen atoms, mono-oxygenases (EC 1.13.12.-) only one.

oxygen-dependent killing The major bactericidal mechanism employed by phagocytes as the primary defence against infection. Various toxic oxygen species (hydrogen peroxide, superoxide, singlet oxygen, hydroxyl radicals) are generated through the *metabolic burst. Defects in the system are responsible for *chronic granulomatous disease. *See* CHEMILUMINESCENCE; MYELOPEROXIDASE.

oxygen electrode An electrode that will detect changes in oxygen concentration and can be used to monitor oxygen consumption. It uses an oxygen-permeable PTFE (Teflon) membrane to separate the test chamber from a compartment containing a saturated KCl solution and two electrodes, a platinum cathode and a silver anode. A fixed polarizing voltage is applied between the two electrodes and the resulting current (~1 μA) is proportional to oxygen concentration.

oxygen radical Any oxygen species, other than free oxygen, that has an unpaired electron. Oxygen radicals are highly reactive and can cause damage to many biological macromolecules. They mediate the damaging effects of ionizing radiation. *See* OXYGEN-DEPENDENT KILLING.

oxyntic cell (parietal cell) A gastric epithelial cell that secretes 0.1 mol/L HCl, by means of H^+ *antiport ATPases on the luminal cell surface.

oxyntomodulin A peptide (37 aa) that contains the 29-aa sequence of glucagon followed by an 8-aa C-terminal extension. It inhibits meal-stimulated gastric acid secretion, has weak affinity for the glucagon receptor, decreases appetite, and mimics the effects of glucagon-like peptides 1 and 2 (GLP-1, GLP-2) on gastric acid secretion and gut motility. No specific receptor is known.

oxyphil cells *See* HURTHLE CELLS.

oxyprenolol A *β-adrenergic receptor antagonist. *TN* Trasicor.

oxysome The multimolecular complex that acts as a unit in oxidative phosphorylation.

oxytetracycline A broad-spectrum *tetracycline-class antibiotic from *Streptomyces rimosus*.

oxytocin A nonapeptide hormone (1007 Da) synthesized in the hypothalamus together with the carrier protein, *neurophysin and the related hormone, *vasopressin. Induces smooth muscle contraction in uterus and mammary glands. The receptor is G-protein coupled.

p A prefix, usually of a number, indicating: **1.** A protein, sometimes a phosphoprotein, e.g. *p53. **2.** A region of a chromosome (*see* P REGION). **3.** Postnatal age in days (P1, P2, etc.).

p0071 A protein (plakophilin-4, 1211 aa) of the *catenin/*plakophilin family that has ten *armadillo repeats. It is mostly localized in epithelial cells at regions of cell–cell contact and may colocalize with *desmoplakin in desmosomes. The C-terminal fragment of *presenilin 1 binds to p0071. *See* ERBIN; PAPIN.

p14 *See* INK4.

p15 **1.** Activated RNA polymerase II transcription cofactor 4 (p15, 127 aa) that mediates transcriptional activation of class II genes (those transcribed by RNA polymerase II). **2.** PCNA-associated factor p15 (PAF) (111 aa) which binds *proliferating cell nuclear antigen (PCNA) and is overexpressed in several types of tumour. **3.** The binding partner of the nuclear export receptor *TAP, the NTF2-like export factor 2 (NXT2). **4.** *See* INK4.

p16 INK4a *See* INK4.

p17 A structural viral matrix protein of *HIV-1 that acts as a cytokine, promoting cell proliferation, proinflammatory cytokine release, and HIV-1 replication of activated T cells.

p18 **1.** p18maf: One of the *AP1 superfamily encoded by v-*maf* (an avian musculoaponeurotic fibrosarcoma oncogene) which forms heterodimers with *p45 that block transcription at NFE2 sites. **2.** *See* INK4. **3.** An N-terminal truncated version of *Bax (p18 Bax.). **4.** An eighteen-residue antimicrobial peptide, P18, based on a *cecropin A–*magainin 2 hybrid.

p19 **1.** A line of murine embryonal carcinoma stem cells (P19 cells). **2.** One subunit of the cytokine *interleukin-23, the other subunit being the p40 subunit of IL-12. **3.** *See* INK4.

p20 **1.** p20-ARC One of the subunits of *ARP2/3. **2.** p20-CGGBP (CGG-binding protein1) A protein that binds to the unmethylated form of the trinucleotide repeat in the 5′ UTR that is a feature of *fragile X syndrome. **3.** p20-CCAAT enhancer-binding protein β (*CEBPβ), a truncated C/EBPβ isoform.

p21 An inhibitor (Waf1, Cip1, CDKN1A, 164 aa) of many *cyclin dependent kinases that is transcriptionally regulated by *p53. *See* P21-ACTIVATED KINASES.

p21-activated kinases (p21-associated kinases) A family of serine/threonine protein kinases of the *STE20 group. **PAK1** (545 aa) homodimers show *transinhibition but dissociate and are activated by GTP-cdc42 or GTP-rac1 and are then able to act on various targets. PAK1 binds to the *SH3 domains of phospholipase Cγ and of *Nck, and promotes the disassembly of stress fibres and focal adhesions. **PAK2** (507 aa) can be activated by rac or cdc42 and stimulates cell survival, but a *caspase-generated constitutively active catalytic fragment induces an apoptotic response. **PAK3** (oligophrenin 3, 559 aa) has three tissue-restricted isoforms, and mutations in *PAK3* are associated with X-linked mental retardation-30.

p22 phox *See* PHOX.

p23 **1.** One of the p24 family (Tmp21, p24δ, 219 aa) of transmembrane proteins that are involved in vesicular trafficking between the Golgi complex and the endoplasmic reticulum; the family also includes p25. **2.** A cochaperone (p23, prostaglandin E synthase 3, PTGES3, 160 aa) that forms a complex with the chaperone *Hsp90. **3.** *See* PANCREATITIS ASSOCIATED PROTEIN I.

p24 **1.** A family of transmembrane proteins involved in the secretory pathway. **2.** Tubulin polymerization-promoting protein (TPPP, 214 aa).

p25 **1.** A proteolytic fragment of *p35 (cdk5). **2.** *See* P23. **3.** P25 and P28 are major surface proteins of *Plasmodium vivax* ookinetes, essential for infection of mosquitoes, against which transmission-blocking vaccines have been produced.

p27kip1 A member of the *Cip1/Kip1 family of cyclin-dependent kinase inhibitors.

p28 **1.** One of the subunits of *interleukin-27. **2.** The poxviral *RING protein p28 is a virulence factor of unknown function. **3.** A subunit (p28) of *eukaryotic initiation factor 3 (eIF3k). **4.** One of the replicase proteins (p28 and p65) of mouse hepatitis virus. **5.** *See* P25 (3).

p29 **1.** A subline (P29) of the mouse Lewis lung cancer line. **2.** Type IV collagen-binding protein (p29) of ML-SN2 (murine *fyn* cDNA-transfected clone). **3.** An *adhesin from *Mycoplasma fermentans.* **4.** An active *calpain cleavage product of the cdk5 neuronal activator p39 (*compare* P25). **5.** An antigen (P29) found in the dense granules of *Toxoplasma gondii.* **6.** Oestrogen-receptor-related protein, p29. **7.** GCIP-interacting protein (243 aa), a nuclear protein of uncertain function. **8** An autoantigen, p29, in *Wegener's granulomatosis. **9.** *See* BAG FAMILY (BAG-1). **10.** *See* SYNAPTOGYRINS.

p30 **1.** A protein produced by HTLV-I that interferes with *toll-like receptor (TLR)-4 signalling. **2.** A major surface protein, P30, from *Toxoplasma gondii.* **3.** Terminal organelle protein (P30) of *Mycoplasma pneumoniae* that is involved in gliding motility. **4.** A serine-arginine-rich protein that directs the alternative splicing of glucocorticoid receptor pre-mRNA to glucocorticoid receptor β in neutrophils. **5.** An integral membrane protein, NKp30, on natural *killer cells; involved in activating the cytotoxicity.

P32 **1.** The β-emitting isotope of phosphorus (^{32}P), extensively used for labelling biomolecules. **2.** A cofactor of *splicing factor ASF/SF-2. **3.** A mitochondrial matrix protein that binds the capsid of rubella virus. **4.** The α polypeptide of the CD8 antigen.

p33 **1.** Complement and kininogen binding protein gC1qR/p33 (gC1qR), distinct from *calreticulin. **2.** *See* ING1 (p33ING1).

p34 **1.** A cyclin-dependent kinase, cdc2. *See* CYCLINS. **2.** Zinc finger CCCH domain-containing protein 12D (ZC3H12D, p34, 312 aa) that may be associated with susceptibility to lung cancer. **3.** *See* ARP 2/3 COMPLEX.

p35 **1.** A neuronal activator of *cyclin-dependent kinase 5 (Cdk5). **2.** One of the two subunits of *interleukin-12. **3.** A *Baculovirus* caspase inhibitor, p35 protein. **4.** A surface antigen (P35) of *Toxoplasma gondii.* **5.** *uroplakin3B.

p36 **1.** A subunit of *eukaryotic initiation factor 3. **2.** One of a family of secreted mycobacterial proteins; P36 (exported repetitive protein, Erp) of *Mycobacterium tuberculosis,* associated with virulence. **3.** A sperm antigen identified by antibodies eluted from the spermatozoa of infertile men. **4.** *See* ENDONEXINS. **5.** *See* BAG FAMILY (BAG-1).

p37 **1.** A membrane lipoprotein of *Mycoplasma hyorhinis.* **2.** One of the isoforms of *AUF1. **3.** One of two RNA-binding proteins (p34 and p37) from *Trypanosoma brucei.* **4.** A platelet-agglutinating protein, p37, probably prethrombin 2. **5.** A major envelope protein of vaccinia virus. **6.** A protein (P37) of *Borrelia burgdorferi* that elicits an early IgM response in *Lyme disease patients. **7.** A *nucleoporin, NUC37.

p38 **1.** A subset of the *MAP kinases (p38MAPK). **2.** tRNA synthetase cofactor (312 aa), one of three auxiliary proteins (p18, p38, and p43) associated with the multi-tRNA synthetase complex.

p39 **1.** A neuronal activator of cdk5, like *p35. *See* P29. **2.** A subunit of *eukaryotic initiation factor 3.

p40 **1.** A subunit shared by both *interleukin-12 and *interleukin-23. **2.** *Long interspersed nucleotide element-1 (LINE-1) p40 protein that is expressed in childhood malignant germ-cell tumours. **3.** A cytoplasmic protein (372 aa) that preferentially interacts with the GTP-bound form of rab9 and acts as an effector, stimulating endosome-to-transGolgi transport of the mannose-6-phosphate receptor. **4.** A subunit of the *Arp2/3 complex.

p41 **1.** One of the isoforms (p31 and p41) of the MHC II-associated chaperone molecule of *invariant chain (inhibitory p41 Ii). **2.** A subunit of the *Arp2/3 complex, p41-Arc. **3.** Human herpesvirus 6 (HHV-6) early protein, p41. **4.** Polypeptide p41 of a Norwalk-like virus, a nucleic acid-independent nucleoside triphosphatase. **5.** *NADPH oxidase organizer 1, p41-NOX.

p42 One of the *MAP kinases, p42/MAP-kinase is erk2, p44MAP-kinase is erk1.

p43 **1.** One of the tRNA synthetase cofactors (*see* P38) but apparently identical to a fragment of *endothelial monocyte-activating polypeptide II. **2.** The mitochondrial triiodothyronine receptor. **3.** A subunit (p43) of placental isoferritin.

p44 (p44/42) **1.** A *MAP kinase, also called erk1. **2.** The 44-kDa major outer membrane proteins (P44s) of **Anaplasma phagocytophilum.* **3.** A splice variant of *arrestin, p44. **4.** Interferon-induced hepatitis C-associated microtubule aggregate protein, p44 (444 aa). **5.** A subunit of *eukaryotic initiation factor-3. **6.** WD repeat-containing protein 77 (WDR77, androgen receptor-associated protein, 637 aa) a component of the 20S *PRMT5-complex. **7.** Subunit 9 of the 26S *proteosome, p44.5. **8.** A subunit of the transcription factor IIH (TFIIH).

p45 **1.** Nuclear factor E2 p45-related factor 2 (Nrf2). **2.** S-phase kinase *F-Box protein p45 (SKP2) the substrate-specific receptor of the ubiquitin-protein ligase involved in the degradation of p27(Kip1).

p46 **1.** An isoform of *Shc, p46Shc. **2.** An isoform of cell division kinase, cdk11. **3.** A 46-kDa glucose-6-phosphate translocase (P46), part of the glucose-6-phosphatase enzyme complex of the ER. **4.** An isoform of *JNK/SAPK, p46JNK. **5.** Natural cytotoxicity triggering receptor 1 (NKp46) of natural killer cells. *See* P30.

p47 **1.** A family of GTPases (p47 GTPases, immunity-related GTPases (IRG) family) that are important interferon-inducible resistance factors in mice although possibly less important in humans. **2.** The constitutively expressed *heat shock protein, p47. **3.** *See* PHOX.

p48 **1.** A splice variant of visual *arrestin (p48: the other is p44). **2.** A nucleotide *excision repair protein, p48. **3.** The DNA binding subunit of the transcription factor PTF1 (pancreas specific transcription factor 1a). **4.** A surface lipoprotein of *Mycoplasma bovis*. **5.** One subunit (p48) of *chromatin assembly factor-1. **6.** Norwalk virus nonstructural protein p48 that may interfere with intracellular protein trafficking. **7.** The e-subunit (p48) of mammalian initiation factor 3 (eIF3e). **8.** *See* P49.

p49 **1.** A 49-kDa protein (p49/STRAP, serum response factor binding protein 1, SRFBP1) that specifically interacts with a motif in the N-terminus of *GLUT4. **2.** A member of the *p35 family that inhibits *caspases and will block apoptosis triggered by treatment with *Fas ligand (FasL), *TRAIL, or UV radiation, but not apoptosis induced by *cisplatin. **3.** An inhibitory receptor present on natural *killer cells from placenta but not on those in peripheral blood. **4.** A subunit of the DNA polymerase-(alpha/primase complex, PRIM1 (sometimes p48), the others being p180 (POLA), p68 (POLA2), and the primase p58 (PRIM2A).

p50 **1.** Subunit 1 of *NFκB. **2.** *See* DYNAMITIN.

p51 **1.** A homologue of *p53, p51/p63, p73L, p40/KET. **2.** A subunit, p51, of *HIV-1 reverse transcriptase. The other subunit is p66. **3.** A major antigen in *Neorickettsia risticii*, P51.

p52 **1.** A product of *NFκB2 (p100) when *IKKα activates a noncanonical pathway. p52 can enter the nucleus and induces genes important to adaptive immunity. **2.** An isoform of *Shc. **3.** A transcriptional coactivator (*see* P75). **4.** A repressor of the inhibitor of protein kinase, a protein (492 aa) that inhibits p58(IPK) (*protein kinase R). **5.** Polypeptide 4 of general transcription factor IIH (TFIIH).

p53 An important tumour suppressor (393 aa) that responds to cell stress by regulating genes that induce cell cycle arrest, apoptosis, senescence, DNA repair, or changes in metabolism. In unstressed cells p53 levels are kept low by the action of hmdm2 ubiquitin ligase. Mutations in p53 are found in many tumours and if only one p53 gene is functional (*see* LI–FRAUMENI SYNDROME) there is a predisposition to tumours. p53 protein binds DNA and stimulates *p21 production: p21 interacts with cdk2 and the complex inhibits progression through the cell cycle. The p53 family consists of p53, *p63, and *p73, with multiple isoforms and splice variants of each.

p55 **1.** A membrane-associated guanylate kinase (*MAGUK) protein, erythrocyte protein p55, that is expressed in the *stereocilia of outer hair cells in the inner ear. **2.** *TNF receptor α, p55. **3.** A human homologue of *fascin. **4.** A subunit of the 26S *proteosome. **5.** A regulatory subunit of *phosphatidyl-inositol 3-kinase, p55γ. **6.** A multifunctional protein, the (beta subunit of procollagen-proline 2-oxoglutarate-4-dioxygenase (prolyl 4-hydroxylase, EC 1.14.11.2) which is also the cellular thyroid hormone-binding protein p55 and glutathione-insulin transhydrogenase.

p56 **1.** The *src-family kinase, p56(lck). **2.** Cyclin-dependent kinase-like 2. **3.** A surface glycoprotein (p56) of *Borna disease virus, implicated in viral entry. **4.** A multiple docking protein, p56(*dok-2), that acts downstream of receptor or non-receptor tyrosine kinases.

p57 **1.** A glycoside, P57AS3 (P57), from **Hoodia gordonii*, possibly the only active ingredient. **2.** A *coronin-like actin-binding protein. **3.** p57/kip2 is a *cyclin-dependent kinase inhibitor of the p21CIP1, p27KIP1 family. Mice deficient in this inhibitor have developmental defects similar to those seen in *Beckwith–Wiedemann syndrome. It is encoded by a maternally imprinted gene in both human and mouse. **4.** p57lck (lck) is the isoform of the *src-family kinases that is expressed predominantly in thymocytes and peripheral T cells. It associates with the cytoplasmic domains of CD4 and CD8, and with the β-chain of the IL-2 receptor. **5.** *See* NEUROMODULIN.

p58 **1.** Cyclin-dependent protein kinase 1 (*see* PITSLRE KINASES). **2.** A calcium-dependent animal lectin, p58/ERGIC-53, that acts as a cargo receptor involved in endoplasmic reticulum to Golgi transport. **3.** An inhibitor, P58(IPK), of *protein kinase R (PKR). **4.** A substrate for the insulin receptor (p53/p58, IRSp53) that regulates the cytoskeleton. **5.** A subunit of *DNA polymerase α-primase complex). **6.** One of the *KIR family of natural killer cell immunoglobulin-like receptors, KIR2DL3.

p59 **1.** The *src-family kinase, p59(fyn). **2.** The product of the *oligoadenylate synthetase-like gene. **3.** A protein expressed in testis and developmentally regulated during spermatogenesis, p59(scr). **4.** The human homologue of *GRASP55. **5.** An *immunophilin, FK506 binding protein-59. **6.** *See* INTEGRIN-LINKED KINASE (ILK).

p60 **1.** An autolysin secreted by *Listeria monocytogenes* that digests peptidoglycan and promotes infection of macrophages. **2.** One of the two subunits of *katanin. **3.** The type 1 TNF receptor, p60. The type 2 receptor is p80. **4.** One of the subunits of *chromatin assembly factor 1, CAF-1p60. **5.** Tyrosine kinase p60(c-*src). **6.** *Caveolin isoform, cav-p60. **7.** The early T-cell activation antigen p60, CD69.

p61 **1.** A *STE20-like serine/threonine kinase (STK4, kinase responsive to stress 2), very similar to p63 (STK3). **2.** A catalase, P61, that is an immunodominant antigen from **Nocardia brasiliensis*.

p62 **1.** A subunit of *dynactin. **2.** A signalling adaptor (p62/*DOK1) downstream of tyrosine kinases. **3.** Polyubiquitin-binding protein component of the *sequestosome, p62/SQSTM1. **4.** A subunit of the protein kinase C ζ (PKCζ)–p62–Kvβ complex, a potassium channel-modulating complex.

p63 **1.** One of the *p53 family of transcription factors. p63 mutations are involved in five distinct malformation syndromes: ectrodactyly, ectodermal dysplasia, and facial clefts syndrome (EEC3), split-hand/foot malformation-4 (SHFM4), limb–mammary syndrome, Hay–Wells syndrome, and acro–dermato–ungual–lacrimal–tooth (ADULT) syndrome. **2.** A *surfactant protein A binding protein on type II pneumocytes, (CKAP4/ERGIC-63/CLIMP-63). **3.** *See* P61.

p64 **1.** A chloride channel. *See* PARCHORIN. **2.** *Interleukin 2-receptor γ chain, p64.

p65 **1.** *See* PLASTINS. **2.** RelA/p65, a subunit of *NFκB. **3.** *See* SYNAPTOTAGMIN.

p66 **1.** An accessory subunit (POLD3) in the *DNA polymerase δ complex. **2.** A subunit of *eukaryotic initiation factor-3. **3.** A 66-kDa protein encoded by the *shc locus.

p67 **1.** Methionine aminopeptidase 2 (478 aa, EC 3.4.11.18), an enzyme involved in N-terminal protein processing and that also binds eukaryotic initiation factor-2 (eIF-2) and prevents phosphorylation and inactivation of eIF-2α by eIF-2 kinases and thus promotes protein synthesis. It is also a target for inhibition by the antiangiogenic compound, *fumagillin. **2.** CD33, a sialic acid-binding immunoglobulin-like lectin, SIGLEC3. **3.** *See* PHOX.

p68 **1.** One of the *DEAD-box helicases. **2.** A subunit of the *DNA polymerase-(alpha/primase complex. **3.** *See* ANNEXINS (annexin A6).

p69 An *oligoadenylate synthetase, OAS2.

p70 **1.** The heterodimeric form of *interleukin-12, p70. **2.** Ubiquitin-associated and SH3 domain-containing protein B, UBASH3B (suppressor of T-cell receptor signalling 1, STS1, 650 aa), one of the *phosphoglycerate mutase family of proteins that interacts with *cbl. **3.** A subunit of the p70/p80 thyroid autoantigen (lupus autoantigen, Ku antigen, *Ki antigen) that forms a 10S DNA-binding complex. **4.** p70 *S6 kinase. **5.** *See* ANNEXINS (annexin VI).

p72 One of the *DEAD-box RNA helicases, DDX17.

p73 **1.** One of the *p53 family of transcription factors. Overexpression of p73 protein and aberrant expression of its particular isoforms is associated with malignant myeloproliferations, including acute myeloblastic leukaemia. Hypermethylation of the p73 gene and underexpression is common in malignant lymphoproliferative disorders, particularly acute lymphoblastic leukaemia (ALL) and non-Hodgkin's lymphomas. **2.** The rho GTPase-activating protein 24 (p73 rhoGAP, filamin A-associated rhoGAP, FILGAP, 748 aa) involved in regulating actin remodelling, cell polarity, and cell migration.

p75 **1.** Adhesion inhibitory receptor molecule-1, one of the sialic acid-binding immunoglobulin-like lectins, SIGLEC7 (467 aa) found mostly in natural killer cells. **2.** A coactivator of transcription and pre-mRNA splicing (lens epithelium-derived growth factor, LEDGF), an alternatively-spliced isoform of *p52. **3.** *Nerve growth factor receptor-associated protein 1

(NGFRAP1, p75(NTR)-associated cell death executor, NADE, 111 aa).

p76 A member of the transmembrane 9 superfamily of proteins, TM9SF2, that may function as an endosomal ion channel.

p78 **1.** The product of prostate tumour overexpressed gene 2 (mediator of RNA polymerase II transcription subunit 25, MED25, activator-recruited cofactor, ARC92, 645 aa) a component of the *RNA polymerase mediator complex. **2.** Microspherule protein 1, MCRS1 (534 aa) that colocalizes with *fibrillarin in nucleolar microspherules and is probably involved in activation of rRNA transcription. **3.** Interferon-α-inducible proteins (myxovirus resistance 1 and and 2, MX1, MX2, MxA, MxB) that are members of the large GTPase family and have antiviral activity.

p80 **1.** The type I *interleukin-1 receptor (IL-1RI). **2.** The homodimer of *interleukin-12 p40 subunits that will antagonize IL-12 and IL-23. **3.** A nuclear autoantigen, p80 *coilin that accumulates in *Cajal bodies. **4.** The type 2 TNF receptor, p80. **5.** One of the subunits of *katanin that is concentrated at centrosomes during interphase. **6.** WD repeat containing protein 48 (WDR48, 607 aa) that may inhibit T-cell signal transduction. **7.** CD44, a membrane glycoprotein with a role in matrix adhesion, lymphocyte activation, and lymph node homing.

p84 **1.** A regulatory subunit for *phosphatidyl inositol-3-kinase γ. **2.** *STAT1β. **3.** A tyrosine-phosphorylated protein (SIRP-α-1, SHP substrate 1, SHPS1, MYD1, macrophage fusion receptor, 503 aa) that is a substrate for the phosphatase *shp. **4.** A nuclear matrix protein that is part of the *TREX complex.

p85 **1.** The regulatory subunit (724 aa) of type I *PI-3-kinase. There are two isoforms, p85α and p85β. **2.** A *pluronic block copolymer (P85).

p87 (p87 PIKAP) A regulatory subunit of *PI-3-kinase γ (PI3Kγ adapter protein) that binds to both p110γ and the heterotrimeric G protein βγ subunit and mediates activation of p110γ downstream of G-protein-coupled receptors. Highly expressed in heart.

p91 phox *See* PHOX.

p100 (NFκB2) *See* NFκB.

p101 *See* PI-3-KINASES.

p107 A protein (936 aa) with extensive homology to the *retinoblastoma gene product and that is a tumour suppressor in the context of activated H-**ras*. Binds to *E2F and is found in the *cyclin/E2F complex together with p33cdk2.

p110 *See* PI-3-KINASES.

p120 *See* CATENINS.

p125 **1.** A phospholipase A(1)-like protein that interacts with Sec23p and is involved in endoplasmic reticulum exit sites. **2.** One of the four subunits of DNA polymerase δ. **3.** p125(FAK, focal adhesion kinases): *see* FOCAL ADHESIONS.

p130cas An adaptor protein that can be phosphorylated by *src-family kinases, allowing it to act as docking protein for proteins with *SH2 domains. It is a key mediator of focal adhesion turnover and cell migration. The founder member of the *Cas-family proteins.

p150 **1.** A replicase protein of rubella virus, P150. **2.** A subunit (p150) of *dynactin (p150 (Glued)). **3.** Part of the PI-3-kinase complex involved in endosomal protein sorting (hVPS34/p150). **4.** One subunit of leucocyte β_2 integrin, CD11c/CD18 (p150,95). **5.** p150(Sal2), a vertebrate homologue of the *Drosophila* homeotic transcription factor Spalt.

p185 (p185/neu) The **Erb-B2* gene product, the HER-2 (185 kDa) *epidermal growth factor receptor.

p300 A *histone acetyltransferase that regulates transcription via chromatin remodelling. It is targeted by various viral oncoproteins and mutations are associated with various tumours and with *Rubinstein–Taybi syndrome.

p400 protein **1.** The *inositol 1,4,5-trisphosphate receptor (InsP3-R), mutations in which are associated with *spinocerebellar ataxia-15. **2.** Adenovirus E1A-associated p400, one of the *SWI/SNF family of chromatin remodelling proteins.

PA *See* PHOSPHATIDIC ACID; PLASMINOGEN ACTIVATOR.

PA28 *See* PROTEOSOME.

PA700 subcomplex *See* PROTEOSOME.

PAAD domain A conserved domain (pyrin, AIM, ASC, and death domain) of the *death domain superfamily found in many of the proteins involved in apoptosis and inflammatory signalling pathways. *See* AIM; ASC; DEATH (death domain); PYRIN.

PABA An intermediate (*p*-aminobenzoic acid) in the synthesis of *folic acid by yeast. Sometimes called vitamin B_X, although it is not really a vitamin. A common UV-blocking ingredient in sunscreen.

PABPs *See* POLY-A BINDING PROTEINS.

PAC **1.** A protein phosphatase (PAC-1, EC 3.1.3.48, 314 aa) that is a positive regulator of inflammatory cell signalling and effector functions, mediated through *Jnk and Erk *MAP-kinase crosstalk. **2.** Polyaluminium chloride, a gel matrix for column chromatography. **3.** Protein antigen c (PAc) of *Streptococcus mutans.* **4.** A monoclonal antibody, PAC-1, that mimics the ligand for integrin α_{IIb}/β_3.

PACAP One of the *secretin superfamily of neuropeptides (pituitary adenylate cyclase-activating peptide) with neurotrophic and neurodevelopmental effects. It is expressed in both central and peripheral nervous system. The receptor is G-protein coupled.

pachy- A prefix indicating thickening.

pachynema *See* PACHYTENE.

pachytene (pachynema) The third stage of meiotic prophase, during which the homologous chromosomes are closely paired and *crossing-over takes place.

pacinian corpuscles Pressure sensors in the skin in which the nerve ending is surrounded by concentric layers of connective tissue that resist deformation.

paclitaxel A *taxane, originally isolated from the Pacific yew *Taxus brevifolia* and called taxol. Taxol became a registered trademark, so the compound should now be called paclitaxel. It blocks microtubule disassembly and is antimitotic.

pacsins A family of cytoplasmic adaptor phosphoproteins (protein kinase C and casein kinase substrate in neurons) that interact with dynamin, synaptojanin, and N-WASP, and are involved in vesicle trafficking. **Pacsin 1** (434 aa), which is more strongly expressed in brain, interacts with *huntingtin in a repeat length-sensitive manner, interacting more strongly with mutant huntingtin. **Pacsin 2** (486 aa) and **pacsin 3** (424 aa) are more ubiquitously expressed and do not interact with huntingtin.

SEE WEB LINKS

- Structural details of BAR domains.

pactamycin An antitumour antibiotic, isolated from *Streptomyces pactum*, that is a translation-initiation inhibitor in both pro- and eukaryotes. The binding site on the 30S ribosomal subunit is distinct from the tetracycline and hygromycin B sites.

PADGEM *See* SELECTINS.

paedomorphosis *See* NEOTENY.

PAF **1.** *See* PLATELET ACTIVATING FACTOR. **2.** *See* PAF PROTEIN COMPLEX.

PAF protein complex A multiprotein transcriptional regulatory complex that associates with the RNA polymerase II subunit POLR2A and with a *histone methyltransferase complex. It is normally associated with RNA polymerase II (Pol II) throughout the transcription cycle. The components include *parafibromin, Leo1 RNA polymerase II associated factor, SH2 domain-binding protein 1 (p150) and PAF1 (pancreatic differentiation protein 2).

PAGE *See* POLYACRYLAMIDE GEL ELECTROPHORESIS.

Paget's disease An eczema-like change in the skin of the nipple, usually with an underlying breast carcinoma. Extramammary forms do occur. *See* PAGET'S DISEASE OF BONE.

Paget's disease of bone A chronic disease (osteitis deformans) characterized by focal areas of increased bone turnover, due to activated *osteoclasts, affecting one or several bones throughout the skeleton. Several different mutations can give rise to the condition. **Types 1 and 4** have been mapped to specific loci but the genes are unidentified. **PDB2** is caused by mutation in *RANK, one of the TNF receptor superfamily (TNFRSF11A). **PDB3** is caused by mutation in the *sequestosome 1 (*SQSTM1*) gene, the product of which is associated with the RANK pathway. **Juvenile Paget's disease** can result from *osteoprotegerin deficiency caused by mutation in the *TNFRSF11B* gene. **Inclusion-body myopathy with Paget's disease and frontotemporal dementia** (IBMPFD) is a disease of muscle, bone, and brain that results from mutations in the gene encoding *valosin-containing protein (VCP). *See* PAGET'S DISEASE.

PAH *See* POLYCYCLIC AROMATIC HYDROCARBONS.

PAI *See* PLASMINOGEN ACTIVATOR INHIBITOR.

paired box domain A conserved domain (124 aa) that is a feature of the *PAX homeobox gene family. The function is unknown. The name derives from the original paired box genes expressed in alternate segments of developing *Drosophila*.

PAKs *See* P21-ACTIVATED KINASES.

PAL **1.** Pyothorax-associated lymphoma, a rare form of non-Hodgkin's lymphoma.
2. Peptidoglycan-associated lipoprotein, a structural protein in the outer membrane of Gram-negative bacteria. **3.** A motif found in the C-terminal region of the aspartyl peptidases (*gamma-secretase and *signal peptide peptidase) that contributes to the active site conformation.

palatoschisis *See* CLEFT PALATE.

palindromic sequence A nucleic acid sequence that is identical to its complementary strand when each is read in the correct direction (e.g. TGGCCA), often a recognition site for *restriction enzymes.

palladin A microfilament-associated proline-rich protein (77 aa) that has sequence similarity with *myotilin and plays a role in regulating cell shape and motility. Binds to vasodilator-stimulated phosphoprotein (*VASP), α-actinin, *ezrin, and *profilin. *See* pallidin.

pallidin The 172-aa protein encoded by the homologue of the mouse *pallid* gene. Interacts with *syntaxin 13 and defects are responsible for a storage pool disorder of platelets (probably a defect in organelle biogenesis). *See* BLOCs. Do not confuse with *palladin.

Pallister–Hall syndrome A pleiotropic autosomal dominant disorder comprising hypothalamic hamartoma, pituitary dysfunction, central polydactyly, and visceral malformations caused by mutation in the **gli3* oncogene.

Pallister–Killian syndrome A rare disorder in which there are multiple congenital abnormalities, seizures, and mental retardation; caused by mosaicism for tetrasomy of chromosome 12p in fibroblasts.

palmdelphin A protein (551 aa) with similarity to *paralemmin and the paralemmin-like N-terminal domain of *AKAP2. It is expressed in many tissues, particularly cardiac and skeletal muscle, where it may regulate intracellular signalling and membrane–cytoskeletal interactions.

palmitic acid A common acyl residue (n-hexadecanoic acid) in membrane phospholipids and also attached to a cysteine residue as a post-translational modification (palmitoylation) of some membrane proteins. *Compare* MYRISTIC ACID.

palmoplantar keratoderma A condition in which there is abnormal keratinization of the skin of the palm of the hand and sole of the foot. **Epidermolytic palmoplantar keratoderma** can be caused by mutation in the keratin 9 gene which is responsible for producing intermediate filaments that insert into *desmosomes in palmar and plantar epidermis. **Striate keratosis palmoplantaris type 1** is due to mutations in *desmoglein, **type 2** to mutation in *desmoplakin or mutation in the epidermal cytokeratin 1 gene.

palonosetron A second generation 5-HT_3 antagonist, used to treat chemotherapy-induced nausea and vomiting. *TN* Aloxi.

palytoxin An extremely complex linear peptide (2670 Da) with 64 chiral centres, isolated from corals of *Palythoa* spp. though probably a product of dinoflagellates of the genus *Ostreopsis.* It binds to Na^+,K^+-ATPase and converts it into a channel. Possibly the most potent animal-derived toxin known.

pamidronate A bisphosphonate drug used to treat hypercalcaemia and *Paget's disease of bone. *TN* Aredia.

pancortins Proteins (478 aa) of the *olfactomedin family that are found in the extracellular matrix of the central nervous system and highly expressed during brain development. There are four alternatively spliced variants.

pancreastatin A peptide hormone (49 aa) that has a general inhibitory effect on secretion in many exocrine and endocrine systems, including inhibition of insulin release from the pancreas. It is derived from *chromogranin A by proteolytic processing.

pancreatic acinar cells The pancreatic cells that secrete digestive enzymes (as opposed to the cells of the *islets of Langerhans that secrete insulin). Much of the early work on the secretory pathway was done on these cells.

pancreatic peptide A peptide hormone (PP, PPY, pancreatic hormone, 36 aa), secreted by cells of the *islets of Langherhans, that is involved in the regulation of exocrine pancreatic secretion and biliary tract motility. It may also be important in regulation of food intake. PPY2 inhibits secretion of enzymes and bicarbonate from the exocrine pancreas and has homology with *peptide YY, one of the *neuropeptide Y (NPY) family.

pancreatic triglyceride lipase The enzyme (PNLIP, EC 3.1.1.3, 465 aa) that hydrolyses dietary triglycerides to fatty acids in the intestine. The cofactor is *colipase. Similar hepatic and gastric/lingual isozymes exist.

pancreatitis associated protein I A protein (PAP I, HIP, p23, Reg2, islet neogenesis-associated protein, INGAP, 175 aa) normally present at very low levels but overexpressed in acute pancreatitis. It is also associated with various inflammatory diseases, such as *Crohn's disease. The product of the hepatocarcinoma-intestine-pancreas (*HIP*) gene, which is activated in most hepatocellular carcinomas and is identical to *PAP1*.

pancreozymin *See* CHOLECYSTOKININ.

pancytopenia A decrease in the numbers of all types of blood cells which can occur as a result of aplastic anaemia, hypersplenism, or tumours of the bone marrow.

pandemic Describing an epidemic of an infectious disease that is spreading through human populations across a large region. The World Health Organization (WHO) defines a pandemic as occurring when three conditions are satisfied: the emergence of a disease new to a population, the infectious agent causes serious illness, and the infection is spread easily and sustainably among humans.

Paneth cells Cells located in the basal regions of crypts in the small intestine, particularly the distal region, that secrete α-*defensin microbicidal peptides.

panmictic Describing a population in which there is random mating.

pannexins A family of a widely expressed proteins (~45 kDa) with structural, but not amino acid, homology with gap junction proteins, the *connexins. Pannexin 1 does not form gap junctions but it forms mechanosensitive ATP-permeable channels in erythrocytes. There are three isoforms (pannexins 1–3) which have different distributions.

panniculitis A group of diseases in which there is inflammation of subcutaneous adipose tissue.

pannus **1.** Vascularized *granulation tissue, derived from synovial tissue, that grows over the bearing surface of the joint in *rheumatoid arthritis and leads to breakdown of the articular surface. **2.** Granulation tissue that invades the cornea from the conjunctiva.

pantetheine An intermediate (*N*-pantothenylcysteamine) in CoA biosynthesis. Pantetheinases (EC 3.5.1.-) hydrolyse pantotheine to pantothenic acid (vitamin B_5) and cysteamine, a potent antioxidant. *See* PANTOTHENATE KINASE; VANINS.

PANTHER A database (protein analysis through evolutionary relationships) that classifies proteins on the basis of function. Version 6.1, released in August 2008, contained 5547 protein families divided into 24582 functionally distinct protein subfamilies.

SEE WEB LINKS

- The Panther homepage.

P antigen The human P blood group system consists of three antigens: the globoside antigens P and P(k), and the oligosaccharide, or paragloboside, antigen P1. They are synthesized by the sequential addition of monosaccharides to ceramide by different glycosyltransferases. The Donath–Landsteiner antibody reacts with P antigen (*see* PAROXYSMAL COLD HAEMOGLOBINURIA) and P antigen is the cellular receptor for parvovirus B19, which causes *erythema infectiosum.

Panton–Valentine leucocidin *See* LEUCOCIDIN.

pantophysin A *synaptophysin homologue (259 aa) found in cells of non-neuroendocrine origin.

pantothenate kinase A set of regulatory enzymes (PANK1–4) in CoA biosynthesis that catalyses the phosphorylation of pantothenate (vitamin B_5). Mutation in the *PANK2* gene causes pantothenate kinase-associated neurodegeneration (Hallervorden–Spatz disease: *see* INFANTILE NEUROAXONAL DYSTROPHY).

papain A cysteine peptidase (EC 3.4.22.2, 345 aa), from pawpaw *Carica papaya*, that is thermostable and will act in the presence of denaturing agents. The preferential cleavage site is at one residue on the C-terminal side of a phenylalanine.

Papanicolaou's stain A complex histochemical stain (Pap stain) used for cervical smears (*Pap test).

SEE WEB LINKS

- The recipe for PAP stain.

papaverine An opium alkaloid distinct from other opiates, that acts as a smooth muscle relaxant, probably by inhibiting cAMP phosphodiesterase.

papilla A small nipple-like projection. A range of functionally and morphologically distinct papillae are found e.g. on the tongue.

papilloma A benign epithelial tumour. The commonest example is a wart, caused by a papillomavirus and a clone of the original virus-infected cell.

Papillomaviridae A family of double-stranded DNA viruses that infect vertebrates and cause benign epithelial tumours (e.g. warts). They are not enveloped and are isometric (the virions have icosahedral symmetry).

SEE WEB LINKS

- Further details of papilloma viruses.

Papillon–Lefevre syndrome An autosomal recessive disorder (keratosis palmoplantaris with periodontopathia) characterized by *palmoplantar keratosis and severe periodontitis affecting both deciduous and permanent dentitions and causing premature tooth loss. There is a defect in *cathepsin C.

PAPIN A protein (plakophilin-related armadillo-repeat protein-interacting protein, activated in prostate cancer, AIPC, 2642 aa) with six *PDZ domains that is highly expressed in prostate cancer. It interacts with *p0071, a *catenin-related protein. *See* PLAKOPHILINS.

Papovaviridae A family of small nonenveloped DNA viruses now usually split into two families, *Papillomaviridae and *Polyomaviridae.

pappalysin-1 A metalloendopeptidase (insulin-like growth factor binding protein-4 protease, pregnancy-associated plasma protein A, PAPP-A, EC 3.4.24.79, 1627 aa) of the *metzincin family that cleaves *insulin-like growth factor (IGF) binding proteins 2 and 4 (IGFBP-2, -4), causing a dramatic reduction in its affinity for IGF-I and -II and thus increasing the levels of growth factor activity. In pregnancy, PAPP-A circulates as a complex with the proform of eosinophil major basic protein (proMBP), which inhibits its proteolytic activity; decreased levels are indicative of *Down's syndrome. **Pappalysin-2** may be a IGFBP-5 proteinase in many tissues.

Pap technique **1.** Casual term for the use of *Papanicolaou's stain. **2.** The peroxidase-antiperoxidase method used to detect sparse antigen by immunohistochemical staining. The primary antibody (against the antigen of interest, e.g. rat-antiX) has a second antibody (e.g. rabbit anti-rat IgG) bound in excess so that there are unoccupied binding sites to which a third antibody (rat-antiperoxidase) will bind; peroxidase is added and binds to the third antibody and generates a colour reaction from the standard peroxidase method.

Pap test (Pap smear) A procedure in which cervical cells are stained with *Papanicolaou's stain to screen for precancerous changes, cancer, infection, or inflammation.

papule A small raised spot on the skin.

PAR *See* PROTEASE ACTIVATED RECEPTORS.

paracellin One of the *claudin protein family (claudin-16, 305 aa) expressed at tight junctions of renal epithelial cells of the thick ascending limb of the *loop of Henle, with an important role in the reabsorption of divalent cations. Mutations in the paracellin-1 gene cause familial *hypomagnesaemia with hypercalciuria and nephrocalcinosis (FHHNC).

paracentric Describing a region of a chromosome that does not include the *centromere. A **paracentric inversion** is a chromosomal rearrangement in which a paracentric portion has been rotated through 180° and reinserted in the same location.

paracetamol (acetaminophen) An analgesic and antipyretic drug that may work by inhibiting *COX-3.

paracortex The midcortical region of a lymph node, the thymus-dependent area.

paracrine Decribing a type of signalling in which the target cell is close to the cell releasing the signal. Neurotransmitters and neurohormones are examples of paracrine signals as opposed to *endocrine or *juxtacrine signals.

paradominance A cryptic condition in which heterozygous individuals are phenotypically normal and the trait only becomes obvious if there is loss of heterozygosity in some lineages during development, so that some cells become homozygous or hemizygous for the mutation and the individual is a mosaic of normal and abnormal cells. A few rare disorders, particularly of skin, are thought to arise through paradominant inheritance. Cutis marmorata telangiectatica congenita (CMTC) is a rare congenital vascular anomaly also thought to arise through this route.

parafibromin A tumour suppressor protein, an evolutionarily conserved component (cdc73, 531 aa) of the *PAF protein complex. It is encoded by the gene (*HRPT2*) that is mutated in *hyperparathyroidism-2.

paraganglia (chromaffin bodies) Small groups of chromaphil cells derived from neural crest, that are associated with the ganglia of the sympathetic trunk and the ganglia of the coeliac, renal, suprarenal, aortic, and hypogastric nerve plexuses. Can give rise to *multiple paragangliomata in which the cells are nonchromaffin (chemosensory) type and *phaeochromocytomas in which the cells secrete hormone.

paragloboside *See* P ANTIGEN.

parainfluenza viruses A group of viruses within the *Paramyxoviridae. Type 1 is also known as *Sendai virus, types 2–4 cause mild respiratory infections in humans.

parakeratosis A condition in which nuclei are retained in cells of the *stratum corneum. This is normal on mucous membranes and also occurs in *psoriasis.

paralemmin A phosphoprotein (387 aa) associated with brain synaptic plasma membranes. It is abundant in several endocrine tissues and is concentrated at sites of plasma membrane activity, such as filopodia and microspikes, and induces cell expansion and process formation. It is post-translationally modified by prenylation and palmitoylation and the lipid anchor is necessary for function. *See* PALMDELPHIN.

paralogous genes Genes that have arisen by gene duplication followed by divergence of function. *Compare* ORTHOLOGOUS GENES.

paramorphine *See* THEBAINE.

paramutation An epigenetic phenomenon involving an interaction between two alleles at a single locus, resulting in a heritable change of one allele induced by the other. The change may persist in future generations despite the absence of the causative allele, possibly because of some sort of silencing by cytoplasmically inherited interfering RNA.

paramyotonia congenita *See* MYOTONIA.

Paramyxoviridae The class V viruses of vertebrates. The genome is a single negative strand of RNA that is transcribed by a virus-coded, virus-specific RNA polymerase (transcriptase). Major examples are *measles virus and the *parainfluenza viruses.

paranode The region that flanks the *node of Ranvier in myelinated fibres where glial cells are closely apposed to the axon. *See* CASPRs.

paranodin *See* CASPRs.

paraoxonase A family of serum proteins (EC 3.1.1.2, 40–45-kDa glycoproteins) that associate with high density *lipoprotein (HDL) and are important in detoxification of organophosphates, aromatic esters, nerve agents such as sarin, and various other organic compounds. The isoforms have different substrate specificities. There are multiple polymorphisms in the *PON1* and *PON2* genes, some associated with altered risk of coronary artery disease and sensitivity to organophosphates.

paraplegin A nuclear-encoded mitochondrial metalloprotease protein (795 aa) mutated in *hereditary spastic paraplegia type 7 (SPG7), causing a mitochondrial dysfunction that appears to disrupt axonal transport. Paraplegin is involved in processing of OPA1, a dynamin-like GTPase of the mitochondrial inner membrane.

paraprotein (M protein) An abnormal protein found in blood or urine, usually as a result of monoclonal *gammopathy or multiple *myeloma. *Bence Jones proteins are paraproteins.

SEE WEB LINKS

- Further details of paraproteins and assays.

parasitaemia The infection of a host by a parasite or the level of infection by the parasite.

parasympathetic nervous system One of the two major and complementary divisions of the *autonomic nervous sytem. The parasympathetic system is responsible for rest and repose as opposed to the *sympathetic nervous system, which is generally characterized as being responsible for fight or flight responses. Thus the parasympathetic system innervates salivary glands, thoracic and abdominal viscera, bladder, and genitalia. Most neurons are cholinergic and responses are mediated by *muscarinic acetylcholine receptors.

parathormone *See* PARATHYROID HORMONE.

parathyroid hormone A peptide hormone (parathormone, parathyrin, 84 aa) secreted by the parathyroid glands. Parathyroid hormone acts on osteoclasts and causes an increase in blood calcium ion concentrations; *compare* CALCITONIN. Mutations lead to *hypoparathyroidism. **Parathyroid hormone-related protein** (PTHRP, 141 aa), regulates endochondral bone development and epithelial-mesenchymal interactions during the formation of the mammary glands and teeth. PTHRP is responsible for most cases of humoral hypercalcaemia of malignancy. **Parathyroid hormone receptor 1** (PTHR1, 585 aa) binds both parathyroid hormone and PTHRP.

paratope 1. The region of an antibody that is the antigen-binding site (i.e. binds a particular *epitope) and is an antigen in its own right; in *network theory, an idiotope. 2. The site on a ligand to which a cell surface receptor binds.

paratyphoid An enteric fever, similar to typhoid fever but milder, usually caused by *Salmonella enterica* serovar *Paratyphi* (rather than by *S. typhi*). Transmission is by ingestion of contaminated food or water.

parchorin A protein (637 aa) that is one of a group of related *chloride intracellular channel (CLIC) proteins. Parchorin is highly enriched in rabbit tissues that secrete water, such as parietal cells, choroid plexus, salivary duct, lacrimal gland, kidney, airway epithelia, and chorioretinal epithelia.

- Details in an article in the *Ion Channels Newsletter*.

pardaxin A channel-forming neurotoxic peptide (33 aa) from *Pardachirus marmoratus* (a Red Sea flatfish) that induces neurotransmitter release and has high antibacterial activity. It is used as a shark repellent.

parenchyma The functional tissue of an organ, as opposed to the connective tissue (*stroma).

parenteral Referring to the administration of a substance by any route other than the alimentary canal.

paresis A mild or incomplete paralysis.

parietal cell *See* OXYNTIC CELL.

parkin An E3 ubiquitin-protein ligase (465 aa). The *PARK2* gene is mutated in autosomal recessive juvenile *Parkinson's disease. The **parkin coregulated gene** (*PACRG*) encodes a protein (257 aa), probably also linked to the ubiquitin/proteosome system. **p53-associated parkin-like cytoplasmic protein** (PARC, 2517 aa) has a C-terminal region with high homology to parkin and a domain with homology to *cullin and anchors *p53 in the cytoplasm of lung carcinoma cells. *See* SYNPHILIN.

Parkinson's disease A disease (paralysis agitans) that is characterized by tremor and associated with the underproduction of L-DOPA (dihydroxyphenylalanine) in the substantia nigra of the brain and the progressive loss of dopaminergic neurons. Administration of L-DOPA is an effective treatment. Parkinson's disease (PD, PARK) can have more than one genetic and/or environmental cause. **Autosomal dominant Parkinson's disease** can be caused by mutation in the (alpha-*synuclein gene (type 1, **PARK1**) or triplication of the gene in **PARK4. PARK5** is due to mutation in the *ubiquitin carboxylC-terminal esterase L1 (UCHL1), **PARK8** is due to mutation in the leucine-rich repeat kinase 2 (LRRK2, *dardarin) gene, **PARK13** is due to mutation in a gene encoding a serine peptidase (HTRA2) that may inhibit the inhibitor of apoptosis (*IAP) gene product. Mutation in the gene for *synphilin-1 has been observed in cases of sporadic Parkinson's disease. Several loci for autosomal recessive early-onset Parkinson's disease have been identified. **PARK2** is caused by mutation in the gene encoding *parkin; **PARK7** by mutation in the DJ1 oncogene. There is also evidence that mitochondrial mutations may cause or contribute to Parkinson's disease. Other forms of both dominant and recessive Parkinson's disease have been mapped to particular loci but the specific mutations are not yet known. Polymorphisms in various genes have been associated with susceptibility to Parkinson's disease. *See* PARKINSONISM.

parkinsonism A neurological syndrome in which there is tremor, hypokinesia, rigidity, and postural instability. The most common cause is neurodegeneration (*Parkinson's disease) but the condition can arise for other reasons.

paroxetine A potent antidepressant drug, a *selective serotonin reuptake inhibitor, extensively used to treat major depressive disorder, social anxiety disorder, obsessive compulsive disorder, panic disorder, generalized anxiety disorder, and post-traumatic stress disorder.

paroxysmal cold haemoglobinuria An antibody-induced anaemia (Donath–Landsteiner syndrome) caused by the Donath–Landsteiner autoantibody, a cold-reacting polyclonal IgG which reacts with *P antigen. Antibody binding occurs at temperatures below normal body temperature and complement-dependent lysis follows after warming. Autoantibody production may be a result of infection with a parvovirus. *Compare* PAROXYSMAL NOCTURNAL HAEMOGLOBINURIA.

paroxysmal extreme pain disorder A disorder characterized by paroxysms of rectal, ocular, or submandibular pain caused by mutation in the gene for the α subunit of the *voltage-gated sodium channel-9 (SCN9A).

paroxysmal nocturnal haemoglobinuria An acquired haematological disorder in which the abnormal progeny of a multipotential haematopoietic stem cell produce a condition of haematological mosaicism. The mutated cells (an X-linked somatic mutation) have deficient surface expression of multiple GPI-anchored proteins, such as *decay-acceleration factor and CD59, both of which help to protect red cells from complement-mediated lysis, which leads to the characteristic feature of the disorder: black urine on arising from sleep.The proportion of abnormal erythrocytes in the blood determines the severity of the disease. It seems that patients with PNH are predisposed to somatic mutation in haematopoietic cells. *Compare* PAROXYSMAL COLD HAEMOGLOBINURIA.

PARP *See* POLY(ADP-RIBOSE) POLYMERASE.

parrot disease *See* PSITTACOSIS.

parthenogenesis A mode of development in which the ovum develops without being fertilized.

parthenolide A sesquiterpene lactone which is the bioactive constituent of extracts from feverfew *Tanacetum parthenium.* It inhibits *NFκB and has anti-inflammatory activity.

partial agonist An agonist that will not produce a full response even if all receptors are occupied.

partition coefficient The equilibrium constant for partitioning of a molecule between hydrophobic and hydrophilic phases; a guide to the ease with which the molecule will cross cell membranes.

parvalbumins High affinity calcium-ion binding proteins (110 aa), part of a superfamily that includes *calmodulin and *troponin C. Parvalbumin is expressed at high levels in fast skeletal muscle and at low level in brain and some endocrine tissues. **Parvalbumi** n-β (oncomodulin) is found in early embryonic cells, in the placenta, in outer cochlear hair cells in the organ of Corti, and in tumours. Oncomodulin can act as a growth factor for neurons of the mature central and peripheral nervous systems.

SEE WEB LINKS

- Further details in the Washington University Inner Ear Protein Database.

parvins Actin-binding proteins of the α-actinin superfamily that are associated with *focal adhesions and form a complex with *integrin-linked kinase (ILK). **α-Parvin** (actopaxin, 372 aa) and **β-parvin** (affixin, 365 aa) are colocalized with actin filaments at membrane ruffles, focal contacts, and *tensin-rich fibres in the centre of fibroblasts. **γ-Parvin** (331 aa) lacks the nuclear-localization signals found in parvins α and β, and is expressed in leucocytes.

Parvoviridae Class II DNA viruses with a single-stranded DNA genome and an icosahedral nucleocapsid. Parvovirus B19 causes *erythema infectiosum, the only human infection caused by a parvovirus. Parvovirinae are a subfamily of the Parvoviridae.

SEE WEB LINKS

- More about parvoviruses can be found in the MicrobeWiki.

parvulins A subfamily of the *peptidyl prolyl cis/trans isomerases (PPIases), highly conserved in all metazoans. Parvulin 14 (PIN4) is preferentially located in the mitochondrial matrix. The human parvulin PIN1 (EC 5.2.1.8) is a nuclear protein implicated in the regulation of mitosis through interaction with CDC25 and *polo-like kinase-1.

PAS **1.** *See* PERIODIC ACID–SCHIFF REACTION. **2.** *p*-aminosalicylic acid. **3.** The preautophagosomal structure, from which the autophagosome is thought to originate. **4.** *See* PAS DOMAIN; PAS GENES.

PAS domain A ubiquitous protein module (Per–Arnt–Sim domain) in which the structure but not sequence is conserved. PAS domains regulate the function of many intracellular signalling pathways through protein-protein interactions or by binding cofactors. *See* PERIOD and *ARNT (aryl hydrocarbon receptor nuclear translocator). **PAS kinase** (PASK, 1323 aa) is an evolutionarily conserved serine/threonine kinase that may phosphorylate and inactivate muscle glycogen synthase.

SEE WEB LINKS

- Structure and details.

***PAS* genes** Genes that are essential for the biogenesis and proliferation of *peroxisomes in yeast. The human gene *PXR1* (peroxisome biogenesis factor 5, PEX5) is the homologue of *Pichia pastoris* PAS8 and encodes the receptor for the *peroxisomal targeting signal type 1. Mutations in PEX5 are associated with neonatal *adrenoleukodystrophy and *Zellweger's syndrome.

pasins Cytoskeletal adaptor molecules that interact with the Na^+,K^+-ATPase. Pasin-2 is identical to *moesin, pasin-1 is not mentioned in recent literature.

passage A term originally describing serial infection of animals with a parasite in order to maintain the parasite strain but more recently applied to serial subculturing of cells *in vitro.*

Each passage involves multiple rounds of cell division, so passage number is not the same as the number of divisions the original cells have undergone.

passive immunity Immunity that is acquired by the transfer of antibody or sensitized lymphocytes from another animal. Maternal transfer of antibody to the fetus is important for neonatal immunity.

passive transport Movement of a substance by a mechanism that does not require metabolic energy, just a difference in electrochemical potential. Passive transport across the plasma membrane by diffusion is countered by various *active transport mechanisms or enhanced by *ion channels (*facilitated diffusion).

Pasteurella pestis Now **Yersinia pestis*.

Pat1p (Mrt1) Partner of the Like Sm protein (Lsm) complex in the *decapping activator complex.

Patau's syndrome (trisomy 13) A range of developmental defects arising as a result of trisomy of chromosome 13 or, in some cases, translocation of a portion of chromosome 13 resulting in partial trisomy. Usually fatal in the neonatal period.

patch clamping A variant of the *voltage clamp method, in which current flowing through individual *ion channels in a small area of plasma membrane can be measured. Several further variants depend on the geometry, but the first step in all is to press the patch electrode (~5 μm tip diameter) against the plasma membrane of a cell to form an electrically tight, 'gigohm' seal. This is a **cell-attached patch**. Detaching the cell but leaving a small disc of plasma membrane covering the tip of the electrode produces an **inside-out patch**. If suction is applied to burst the plasma membrane under the electrode, this produces a **whole cell patch** (effectively the same as inserting an electrode into the cytoplasm). By withdrawing the electrode from a whole cell patch, the membrane fragments may reform a seal across the electrode, an **outside-out patch**.

patched The human homologue of the *Drosophila* gene *patched*. *PTCH1* encodes a transmembrane protein with homology to bacterial proton-driven transmembrane molecular transporters and represses genes encoding members of the *transforming growth factor β and *Wnt families of signalling proteins. *Smoothened and patched mediate the cellular response to the *hedgehog secreted protein signal. A second gene, *PTCH2*, has high homology. *See* GORLIN'S SYNDROME.

patching A process of diffusion and trapping of integral membrane components in small clusters crosslinked by an internal or external polyvalent ligand such as antibody or lectin. *See* CAPPING.

PAT family A family of proteins (*perilipins; *adipophilin, and *TIP47) localized in the outer membrane monolayer enveloping lipid droplets and involved in regulating lipid deposition and mobilization.

pathogenic Describing something capable of causing disease.

pathognomonic A sign or symptom that is characteristic of a disease.

pattern formation One of the classic problems in developmental biology, the production of complex patterns from an apparently uniform field of cells. Various hypotheses have been energetically propounded and in some cases there is evidence for gradients of diffusible substances (morphogens) specifying the differentiative pathway according to the concentration of the *morphogen around the cell.

patulin A mycotoxin, probably carcinogenic, produced by certain species of *Penicillium, Aspergillus*, and *Byssochlamys* growing on fruit, particularly apples, pears, and grapes. Inhibits aminoacyl-tRNA synthetase and causes DNA strand breakage.

PAUP* A software package (Phylogenetic Analysis Using Parsimony) used in analysing phylogenetic relationships among a group of organisms using DNA sequence data.

SEE WEB LINKS

- The PAUP* homepage.

pavementing Describing the *margination of large numbers of leucocytes on vascular endothelium near a site of damage.

PAX genes A highly conserved family of genes (paired domain genes, paired box genes) encoding transcription factors that are important in *pattern formation during development. **Pax1** (Hup48) is involved in development of the vertebral column. **Pax2** is important in nephrogenesis and is mutated in *renal-coloboma syndrome, **Pax3** (Hup2) is involved in neurogenesis and mutations lead to *Waardenburg's syndrome. The differentiation of endoderm-derived endocrine pancreas is mediated through **Pax4** and Pax6; mutations in Pax4 are associated with susceptibility to diabetes. **Pax5** has a role in B-cell differentiation and in neural development and spermatogenesis. **Pax6** has a role in oculogenesis, and mutations cause aniridia and various other eye-releated disorders. **Pax7** is required for neural crest formation, **Pax8** is essential for the formation of thyroxine-producing follicular cells and mutation can cause hypothyroidism. Splice variants of some of the *PAX* gene products are known and it is likely that each factor has multiple tissue-specific roles.

paxillin A cytoskeletal protein (557 aa) that localizes to regions where actin microfilaments interact with the plasma membrane. Alternative splicing produces β and γ isoforms with different affinities for *vinculin and *focal adhesion kinase.

PAZ domain PIWI-Argonaute-Zwille domain. *See* PIWI GENES; ARGONAUTE PROTEINS.

PBMC Peripheral blood mononuclear cells. The mononuclear cells of the blood: *monocytes and *lymphocytes.

P bodies *See* GW BODIES.

PBP **1.** Nuclear receptor coactivator PBP (*peroxisome proliferator-activated receptor (PPAR)-binding protein) that is a coactivator for PPARs and other nuclear receptors. **2.** *See* PLATELET BASIC PROTEIN. **3.** *See* PENICILLIN-BINDING PROTEINS.

pBR322 A plasmid commonly used as a *cloning vector.

PC *See* PHOSPHATIDYLCHOLINE.

PC12 A rat *phaeochromocytoma cell line often used in studies of *stimulus-secretion coupling. When treated with *nerve growth factor they differentiate to resemble sympathetic neurons.

PCA **1.** Principal components analysis, an analytical technique that can be applied to a variety of complex systems. **2.** Patient-controlled analgesia. **3.** *p*-Chloroamphetamine. *See* AMPHETAMINE. **4.** Primary cutaneous *aspergillosis. **5.** Prostate cancer antigen-1 (PCA-1), a potentially useful marker but of unknown function.

PCAF A histone acetyltransferase (p300/CREB-binding protein-associated factor, HAT, 832 aa, *see* HISTONE ACETYLATION) which plays a role in the remodelling of chromatin and is thought to have a role in transcriptional regulation.

PCD **1.** Programmed cell death (*see* APOPTOSIS). **2.** Premature centromere division. **3.** A strain of mice (pcd mice) which exhibit adult-onset degeneration of cerebellar Purkinje neurons and other neuronal tissues caused by a mutation in a putative nuclear protein containing a zinc carboxypeptidase domain. **4.** *See* PRIMARY CILIARY DYSKINESIA.

PCDHα An adhesion molecule (protocadherin-α) involved in synapses. *See* CONTACTINS; NEUREXIN; NEUROLIGIN.

PCGF A transcriptional repressor (polycomb group RING finger protein 2, 344 aa) involved in *polycomb repressive complex 1 (PRC1). It has tumour suppressor activity and may play a role in control of cell proliferation and/or neural cell development. Regulates proliferation of early T progenitor cells by maintaining expression of HES1 and plays a role in anteroposterior specification of the axial skeleton and negative regulation of the self-renewal activity of haematopoietic stem cells.

pCMBS An inhibitor (*p*-chloromercuriphenylsulphonic acid) of water movement through *aquaporin-1.

PCNA *See* PROLIFERATING CELL NUCLEAR ANTIGEN.

PCP *See* PHENCYCLIDINE.

PCP pathway *See* PLANAR CELL POLARITY PATHWAY.

PCR *See* POLYMERASE CHAIN REACTION.

PCR *in situ* hybridization A technique for detection of very rare mRNA or viral transcripts in a tissue in which a *polymerase chain reaction is used to amplify the transcript.

PCTA-1 A secreted *galectin (prostate carcinoma tumour antigen-1, galectin 8, 316 aa) that is highly expressed in prostate cancer. *See* STEAP; PCA (PCA-1).

PD-1 *See* PROGRAMMED CELL DEATH.

PD 184 352 A potent and selective inhibitor of *MAP kinase (MAPK) that suppresses activation of MAPK but does not block its activity. Strongly suppresses the growth of human implanted colon tumours in mice.

PDE *See* PHOSPHODIESTERASE.

PDGF (platelet-derived growth factor) The major growth factor for mesenchymal cells in serum derived from whole blood (and therefore containing platelet derived factors). It is a heterodimer of disulphide-linked polypeptides, PDGF-A and PDGF-B, the latter being responsible for the mitogenic properties and identical to the *sis* oncogene product, the transforming protein of simian sarcoma virus. PDGF-A (211 aa) and -B

(241 aa) are homologous and exist in various isoforms. The PDGF receptor is a *tyrosine kinase (~1100 aa) that signals through the PI-3-kinase cascade. A fusion between intron 43 of the type I collagen gene and intron 1 of the platelet-derived growth factor β gene results in *dermatofibrosarcoma protuberans through ectopic overproduction of PDGF.

PDK 1. *See* PYRUVATE (pyruvate dehydrogenase kinase). 2. *See* PHOSPHOINOSITIDE-DEPENDENT KINASE.

P domain *See* TREFOIL MOTIF.

PDX-1 A transcription factor (pancreas-duodenum homeobox-1, insulin promoter factor 1, somatostatin transcription factor-1, 283 aa) that regulates the development of both the exocrine and endocrine pancreas and transcription of the insulin gene. Some mutations can lead to pancreatic agenesis, others to increased susceptibility to type 2 diabetes.

PDZ domains Protein domains that are a characteristic feature of many intracellular membrane-associated signalling proteins and that mediate the formation of multimolecular complexes by interacting with short linear C-terminal sequences in their binding partners. The name derives from the first proteins in which they were observed, the postsynaptic density, disc-large, ZO-1 proteins.

PE *See* PHOSPHATIDYLETHANOLAMINE.

p

PEA-15 An acidic serine-phosphorylated protein (phosphoprotein enriched in astrocytes, 15 kDa, 130 aa) containing a death effector domain that colocalizes with microtubules. It is highly expressed in the central nervous system, where it apparently protects against cytokine-induced apoptosis. It is phosphorylated in astrocytes by CaMKII (or a related kinase) and by protein kinase C in response to endothelin; the phosphorylation state may determine whether PEA-15 influences proliferation or apoptosis.

peanut agglutinin A *lectin from *Arachis hypogaea* (a homotetramer of 273 aa subunits). It binds to galactosyl (β-1,3)-*N*-acetylgalactosamine (T-antigen) a moiety present in M and N blood groups, gangliosides, and many other soluble and membrane-associated glycoproteins and glycolipids, although binding is usually blocked by terminal sialic acid. Used histologically in distinguishing between normal and tumour tissues and in assessing malignancy in transitional mucosa.

PEDF A noninhibitory *serpin (pigment epithelium-derived factor) secreted by pigmented retinal epithelial cells that is a potent inhibitor of angiogenesis and has neuronal differentiating activity.

pedicels *See* PODOCYTES.

PEG A hydrophilic polymer (polyethylene glycol) that interacts with cell membranes and promotes fusion of cells to produce viable hybrids, as in the production of *hybridomas. **Pegylation** is the covalent coupling of PEG to a protein thereby improving stability, biological half-life, water solubility, and immunological characteristics following injection. The addition of 40–50 kDa of PEG increases the size of small molecules so they are less readily excreted and less susceptible to enzymic degradation.

pejvakin A protein (352 aa) that is a member of a family that includes DFNA5, the *gasdermins, and *MLZE. It has a nuclear localization signal and a zinc-binding motif. Mutations in the gene cause autosomal recessive neuronal deafness.

Pelizaeus–Merzbacher disease An X-linked recessive dysmyelinating disorder of the central nervous system in which myelin is improperly formed because of mutation in the proteolipid protein (*PLP1*) gene. Mouse model is jimpy. Autosomal recessive **Pelizaeus–Merzbacher-like disorders** (PMLD) have also been described: PMLD1, caused by mutation in the *GJA12* gene that encodes *connexin 46, and PMLD2 in which the mutation is at a different (unknown) locus.

pellagra A chronic disease caused by a deficiency of vitamin B_3 (niacin) or of tryptophan. Can result from a diet predominently of maize in which nicotinic acid is in bound form and there is little tryptophan, the precursor of nicotinic acid.

PEM Polymorphic epithelial mucin. *See* EPISIALIN.

pemetrexed A chemotherapeutic drug (a folate antimetabolite) used to treat pleural mesothelioma and non-small-cell lung cancer. *TN* Alimta.

pemphigus A group of autoimmune blistering diseases affecting the skin and mucous membranes. In **bullous pemphigoid** the blistering is subepidermal, whereas in pemphigus the blistering is intraepidermal. **Pemphigus vulgaris**, the commonest form, is caused by antibodies against *desmoglein 3. **Pemphigus foliaceus** is less serious because only the outer layers of skin are affected. The autoantigen is desmoglein 1. **Paraneoplastic pemphigus** is rare and a complication of lymphoma and *Castleman's disease. **Benign chronic**

pemphigus, *Hailey–Hailey disease, is not an autoimmune disorder.

PEN-2 One of the components (presenilin enhancer) of the *γ-secretase complex which also includes *presenilin and *nicastrin. PEN is required for the endoproteolysis of presenilin that confers γ-secretase activity.

penaeidins A family of small (5–6 kDa) antimicrobial peptides, originally identified in the haemolymph of the Pacific white shrimp *Litopenaeus vannamei*. Activity is predominantly directed against Gram-positive bacteria. A curated database (Penbase) of all penaeidins has been established.

SEE WEB LINKS

- The Penbase homepage.

Pendred's syndrome *See* HYPOTHYROIDISM.

pendrin A protein (780 aa) of the solute carrier family 26 (SLC26A4) with eleven transmembrane domains. Its function is unknown and although it has high sequence homology to several sulphate transporters it does not seem to act as one. *See* HYPOTHYROIDISM (Pendred's syndrome).

penetrance The proportion of individuals that have a particular allele and express the associated phenotype.

penicillamine A metabolite of *penicillin that acts as a chelator of heavy metals. The D-form is used to treat heavy metal poisoning. It is also used as an immunosuppressant to treat rheumatoid arthritis although the mode of action is unclear. *TNs* Cuprimine, Depen.

penicillin The archetypal *β-lactam antibiotic, isolated from the mould *Penicillium notatum*. *See* PENICILLIN-BINDING PROTEINS.

penicillin-binding proteins Enzymes (glycosyltransferases and transpeptidases) involved respectively in forming glycan polymers and crosslinking them in the biosynthesis of the bacterial cell wall peptidoglycans. PBPs are the targets for *beta-lactam antibiotics and antibiotic resistance is often achieved by modification in these enzymes.

pentagastrin A synthetic polypeptide that binds to the *cholecystokinin-B receptor and has *gastrin-like effects when given parenterally.

pentamidine An antimicrobial drug used to treat **Pneumocystis* pneumonia.

pentazocine A powerful synthetic opiate analgesic drug, often combined with *naloxone to inhibit recreational use. *TNs* Fortal, Talwin.

pentobarbital A short-lived barbiturate with anticonvulsant and anaesthetic properties, usually used as the sodium or calcium salt. *TN* Nembutal.

pentose phosphate pathway *See* HEXOSE MONOPHOSPHATE SHUNT.

pentoses Monosaccharides that have five carbon atoms. Examples include *ribose and *deoxyribose, the aldoses *arabinose and *xylose, and the ketoses ribulose and *xylulose.

pentostatin A cytotoxic drug, a purine analogue that inhibits adenosine deaminase, used to treat hairy-cell leukaemia. *TN* Nipent.

pentoxifylline A xanthine derivative used to reduce blood viscosity, thereby improving blood flow, in patients with intermittent claudication. The mode of action is unclear. *TN* Trental.

pentraxins A family of proteins that can form pentameric (or decameric) complexes, have the ability to bind numerous ligands and may be involved in nonspecific uptake of bacteria and cell debris. **Pentraxin 1** is expressed mostly in brain and binds snake venom toxin *taipoxin. **Pentraxin 2** is more widely distributed. **Pentraxin 3** (pentaxin 3) binds various pathogens, including *Aspergillus fumigatus*, *Pseudomonas aeruginosa*, and *Salmonella typhimurium*. Other members of the family are *C-reactive protein and *serum amyloid P.

PEP *See* PHOSPHOENOLPYRUVATE.

pepducins Lipid-conjugated cell-penetrating peptides that can act as agonists or antagonists of their cognate receptor. Pepducins based on the cleaved portion of the thrombin receptor PAR1 (*see* PROTEASE ACTIVATED RECEPTORS) can act as antagonists at that receptor.

peplomers Glycoprotein components of the outer viral envelope that are particularly large and conspicuous in *Coronaviridae.

pepsin An *aspartic peptidase (EC 3.4.23.1, formerly EC 3.4.4.1) from the stomach that preferentially cleaves peptide chains between two hydrophobic amino acids. The active enzyme is a single-chain phosphoprotein (327 aa) autocatalytically cleaved from the zymogen, **pepsinogen** (373 aa), at acid pH in the presence of HCl. Another peptide generated in this cleavage is a pepsin inhibitor which must be further hydrolysed to allow pepsin to work effectively. Pepsin is the type example of peptidase family A1.

pepstatin A potent inhibitor of *aspartic peptidases, a hexapeptide containing the unusual amino acid statine. Isolated from *Streptomyces* spp.

peptidase Alternative and preferred name for a *protease or proteinase. Peptidases can be grouped into clans on the basis of common ancestry and into families on the basis of the type of catalysis. The families are the *aspartic peptidases, *cysteine peptidases, *glutamic peptidases, *metallopeptidases, *serine peptidases, *threonine peptidases, and unknown-type peptidases. A rationalized classification is provided in the *MEROPS database.

peptide bond The covalent bond between the carboxyl group of one amino acid and the amino group of another. The linkage can be either *cis* or *trans*, with *trans* the more frequent in natural peptides.

peptide histidine methionine One of the *secretin family of neuropeptides, a 27-aa peptide with N-terminal histidine and C-terminal methionine generated from pre-pro-*vasoactive intestinal peptide. PHM has vasodilatory activity and probably effects on cell proliferation and differentiation.

peptide map The characteristic set of fragments produced by proteolytic cleavage of a protein, used to test for identity (or otherwise). In practice a gel band is excised, digested, and then run on a second gel.

p

peptide neurotransmitter *See* NEUROPEPTIDES.

peptide nucleic acid (PNA) A synthetic analogue of *nucleic acid, in which the sugar-phosphate backbone is replaced by a peptide-like polyamide. PNAs will bind to complementary DNA or RNA and are resistant to both *nucleases and *peptidases. Extensively used as experimental tools, but early optimism about their potential therapeutic uses has diminished.

peptide receptor A receptor for a *neuropeptide neurotransmitter.

peptide YY A peptide hormone (PYY) with anorectic properties secreted from endocrine cells in the lower small intestine, colon, and pancreas. It inhibits gastric acid secretion, gastric emptying, pancreatic enzyme secretion and gut motility. Has marked homology with *pancreatic peptide PPY2. **PYY1** is a 98-aa peptide but **PYY2** has only 33 aa because of its alternative start site.

peptidoglycan (murein) A crosslinked polysaccharide–peptide complex (mucopeptide) of variable size found in the inner cell walls of all bacteria although there is more (50%) in Gram-negatives than in Gram-positives (10%). The carbohydrate portion is a polymer of ~20 residues of β(1–4)-linked *N*-acetylglucosamine and *N*-acetylmuramic acid crosslinked by small peptides (4–10 aa).

peptidomimetics Compounds that mimic the properties of (small) peptides, usually synthetic peptides with non-natural amino acids incorporated to inhibit proteolysis.

peptidyl arginine deiminase The enzyme (EC 3.5.3.15, 663 aa) responsible for the post-translational modification of arginine to *citrulline in *trichohyalin, altering its mechanical properties.

peptidyl prolyl *cis/trans*-isomerases Enzymes (PPIases, EC 5.2.1.8) that catalyse the *cis–trans* isomerization of prolyl bonds in oligopeptides, an important factor in correct protein folding. The prolyl isomerase activity is apparently unimportant for the immunosuppressive role of the *immunophilins. *Parvulins also have PPIase activity.

peptidyl transferase An enzyme activity (EC 2.3.2.12) integral to the large subunit of the ribosome that catalyses *peptide bond formation between the C-terminus of the growing peptide and the amino group of an incoming tRNA-associated amino acid.

peptoid A subclass of *peptidomimetics, oligomers of N-substituted glycines, in which sidechains are attached to nitrogen atoms along the molecule's backbone.

Percoll™ Colloidal silica coated with polyvinylpyrrolidone that will quickly form a density gradient when centrifuged. Used to produce gradients for separating cells, viruses, and subcellular organelles.

perforins Proteins (PRFs, ~555 aa) found in the cytolytic granules of cytotoxic T cells and natural *killer cells that form tubular transmembrane complexes at the sites of target cell lysis. There are structural similarities with complement *C9, part of the membrane attack complex. Perforin (*PRF1*) gene alterations are responsible for familial *haemophagocytic lymphohistiocytosis type 2 (FHLH2).

perfringolysin O A *cholesterol binding toxin (theta toxin, θ-toxin) from *Clostridium perfringens*, one of the *thiol-activated haemolysins.

perfusion Passage of a fluid through a compartment, chamber, or tissue.

pergolide A dopamine receptor agonist used for the treatment of *Parkinson's disease but withdrawn in the USA. *TN* Celance.

periaxins Protein products of a single gene, found in the plasma membrane (L-periaxin, 1461 aa) and cytoplasm (S-periaxin, 147 aa) of *Schwann cells and necessary for maintenance of peripheral nerve myelin. Both isoforms have *PDZ domains. Mutations in the periaxin gene are responsible for some types of *Dejerine–Sottas neuropathy. *See also* CHARCOT–MARIE–TOOTH DISEASE.

pericanicular dense bodies Old name for small, electron-dense lysosomes in hepatocytes.

pericentric inversion A chromosomal inversion that includes the kinetochore.

pericentrin A conserved coiled-coil protein (NUP75, 3336 aa) of the pericentriolar region involved in organization of microtubules during meiosis and mitosis. It is also a member of the *nucleoporin family; kendrin is pericentrin-B. *AKAP450 (AKAP350, CG-NAP, hyperion) has a 90-aa domain near the C-terminus, similar to that in pericentrin.

pericentriolar region An amorphous region of electron-dense material surrounding the centriole that is the major *microtubule organizing centre of the cell.

perichondrium The fibrous connective tissue surrounding the cartilage of developing bone. The outer layer is fibroblast-rich, the inner layer has relatively undifferentiated chondroblasts and chondrocytes. It becomes vascularized as the periostium of mature bone.

pericyte A cell type associated with the walls of small blood vessels: neither a smooth muscle cell nor an endothelial cell.

perikaryon The cell body of a neuron, excluding axonal and dendritic processes.

perilipin A hormonally regulated phosphoprotein that coats lipid storage droplets in adipocytes. There are two alternatively spliced forms, perilipin-A (517 aa) and -B (422 aa). Perilipin is the major protein kinase A (PKA) substrate in adipocytes and phosphorylation is apparently essential for the translocation of *hormone-sensitive lipase from the cytosol to the lipid droplet in stimulated lipolysis.

perimysium The connective tissue sheath binding muscle fibres into bundles.

perinuclear space The gap (10–40 nm wide) that separates the two membranes of the *nuclear envelope.

period A *Drosophila* gene involved in regulating circadian rhythm. The mammalian homologues *PER1* (*rigui*) and *PER2* encode proteins with basic helix-loop-helix (bHLH) and Per-ARNT-Sim (PAS) domains that in the case of the mouse genes are respectively 44% and 53% homologous to the *Drosophila* gene. Expression of the proteins in the suprachiasmatic nucleus of the brain shows marked circadian periodicity. Mouse and human genes are highly homologous, especially in the conserved domains. *See* DEC.

periodic acid–Schiff reaction (PAS) A histochemical method for staining carbohydrates in which hydroxyl groups are oxidized by periodic acid (HIO_4) to aldehydes which react with Schiff's reagent (basic fuchsin decolorized by sulphurous acid) to give a purple colour.

periodic fever An autosomal dominant disorder (familial Hibernian fever) characterized by recurrent attacks of fever, abdominal pain, localized tender skin lesions, and myalgia. It is caused by mutation in the TNFα receptor type 1.

periodic paralysis A set of disorders in which there are episodes of partial paralysis of muscles. **Hyperkalaemic periodic paralysis** (HYPP) is caused by mutation in the *voltage-gated sodium channel gene *SCN4A* (*see* PARAMYOTONIA CONGENITA). **Hypokalaemic periodic paralysis** (HOKPP) can be caused by mutations in the gene encoding a subunit of the *L-type calcium channel (*CACNL1A3*) or in the *SCN4A* gene. **Thyrotoxic periodic paralysis** can be caused by a mutation in the voltage-regulated potassium channel, KCNE3. *See also* ANDERSEN'S SYNDROME.

periodontal Describing the region around the teeth, gums, and gingival crevice. Periodontal disease often arises if there is inadequate phagocyte function.

peripheral lymphoid tissue *See* LYMPHOID TISSUE. It is secondary in function rather than peripherally located.

peripheral membrane proteins Proteins that are bound to a membrane but not embedded within the hydrophobic region. Many are covalently linked to molecules that are part of the membrane bilayer (*see* GLYPIATION), and others are linked indirectly but nevertheless firmly, forming the membrane cytoskeleton (e.g. *protein

4.1 and *spectrin) or linking with the extracellular matrix (e.g. *fibronectin).

peripherin 1. A type III intermediate filament protein (470 aa) coexpressed with *neurofilament triplet proteins in the peripheral nervous system. Mutations seem to increase susceptibility to *amyotrophic lateral sclerosis. **2.** A photoreceptor-specific glycoprotein (346 aa) found in the rim region of rod outer segment disk membranes. Mutations in the *RDS* gene (retinal degeneration, slow) can cause *retinitis pigmentosa or *macular dystrophy.

periphilin A protein (373 aa) that is incorporated into the cornified cell envelope during the terminal differentiation of keratinocytes.

periplasmic space A structureless region between the plasma membrane and the cell wall of Gram-negative bacteria. **Periplasmic-binding proteins** are transport proteins located within the periplasmic space and some act as receptors for bacterial chemotaxis.

peritoneal exudate *See* EXUDATE.

periventricular heterotopia An X-linked neurological disorder in which neurons fail to migrate into the cerebral cortex and nodules are formed in the ventricular and subventricular zones. The disorder is caused by mutation in the gene for *filamin A but since some neurons do migrate it is likely that filamin B (which normally forms heterodimers with filamin A) partially compensates by forming homodimers. An autosomal recessive form is caused by mutation in the *ARFGEF2* gene (ADP ribosylation factor guanine nucleotide exchange protein-2) and other forms are associated with chromosome 5p anomalies or have *Ehlers–Danlos features.

perivitelline space The space between the ovum and the zona pellucida.

perlecan A ubiquitous heparan sulphate proteoglycan of basement membranes where it interacts with *nidogen, helps maintain structural integrity, and contributes to the ultrafiltration properties in kidney basement membrane. It also has angiogenic and growth-promoting attributes through acting as a coreceptor for basic fibroblast growth factor, bFGF. There are domains homologous to *LDL receptor, *laminin, neuronal cell adhesion molecule (*N-CAM), and *epidermal growth factor (EGF). Mutations of perlecan cause lethal chondrodysplasia and dyssegmental dysplasia, Silverman–Handmaker type; partially functional mutations of perlecan also cause *Schwartz–Jampel syndrome.

permease A general term for a membrane protein that selectively increases the permeability of the plasma membrane to a particular molecule. *See* FACILITATED DIFFUSION.

permissive cells Cells in which a particular virus can replicate successfully.

permissive temperature The temperature at which a mutated gene product behaves normally so the temperature-sensitive mutant operates as though wild type. At the **restrictive temperature** the gene product expresses its mutant phenotype.

pernicious anaemia *See* MEGALOBLASTIC ANAEMIA.

peropsin A rhodopsin homologue (337 aa) synthesized in pigment retinal epithelium and expressed exclusively on the apical face. *See* OPSIN.

peroxidase A family of enzymes (EC 1.11.1.n) many of which catalyse the oxidation of a substrate by hydrogen peroxide. Many are haem enzymes but others, such as glutathione peroxidase, have a selenocysteine residue (glutathione is the electron donor). Haloperoxidases generate reactive halogen species and are important in bacterial killing in the *metabolic burst. Peroxisomes are often, but not invariably, the intracellular site for peroxidase. Experimentally, peroxidases are often used to produce coloured reaction products that allow detection of the enzyme with high sensitivity, either microscopically (e.g. using peroxidase-coupled antibody) or in biochemical assays (e.g. *ELISA). *Lactoperoxidase is used in the catalytic surface labelling of cells by radioactive iodine.

peroxins General name for products of the *PEX* genes that are essential for *peroxisome biogenesis. At least 29 are known and molecular defects have been defined in 10 *PEX* genes, 8 related to protein import and 2 related to membrane synthesis. For example, peroxin 1 encodes one of the *AAA family, peroxin 5 encodes the receptor that recognizes the *peroxisomal targeting signal PTS1. *See* PEROXISOME BIOGENESIS DISORDERS.

peroxisomal targeting sequence 1 A C-terminal consensus Ser-Lys-Leu (SKL) motif on proteins that will be imported into peroxisomes.

peroxisome biogenesis disorders Fatal autosomal recessive diseases (PBDs) in which peroxisomes are defective or deficient which affects brain development and myelin production. *Zellweger's syndrome is the most severe form, infantile Refsum's disease is the mildest, and neonatal adrenoleukodystrophy and rhizomelic chondrodysplasia have similar but less severe symptoms. *See* PEROXINS.

peroxisomes Ubiquitous membrane-bounded organelles (0.5–1.5 μm) that contain peroxidase and catalase, sometimes as a large crystal. Microperoxisomes (microbodies) are 150–250 nm in diameter. Peroxisomes are important in lipid metabolism: *see* PEROXISOME BIOGENESIS DISORDERS; PEROXINS; PPARs (peroxisome proliferator-activated receptors).

perphenazine A phenothiazine antipsychotic drug. *TN* Fentazin.

persephin A neurotrophic factor (156 aa) for multiple populations of neurons that is the ligand for growth factor receptor α- 4 (GFRα-4; *see* GDNF).

persistence **1.** The tendency of a cell or organism to continue moving in one direction rather than turning. Persistence introduces an internal bias to the pure random walk that would be predicted in a uniform environment. **2.** A phenomenon exhibited e.g. by viruses that persist in cells, animals, plants, or populations for long periods, often without replication or transmission. Persistence may arise through integration into the host DNA.

Perthes' disease A type of avascular necrosis (Legg–Calvé–Perthes disease) of the femoral head that can be caused by mutation in the gene encoding type II collagen.

pertussis toxin An essential virulence factor from *Bordetella pertussis*, the causative agent of whooping cough, an *AB toxin. The active (A) subunit (269 aa) has ADP-ribosylating activity against inhibitory *GTP-binding protein (Gi). The binding (B) subunit is a complex heteropentamer.

PESKY The nonsecreted form of *CTACK (CCL27) in which the signal peptide has been replaced with a nuclear targeting sequence so that it acts in an intracrine fashion.

PEST sequence A tetrapeptide motif (Pro-Glu-Ser-Thr) that is found in many rapidly degraded proteins and may be the signal for proteolysis.

petechia (*pl.* petechiae) A small, flat, red-purple spot less than 3 mm in diameter, caused by intradermal haemorrhage.

pethidine (meperidine) A fast-acting synthetic opioid analgesic. *TN* Demerol.

Peutz–Jeghers syndrome An autosomal dominant disorder caused by mutation in the serine/threonine kinase 11 gene (**LKB1*), in which benign polyps develop in the gastrointestinal tract and there is a fifteen-fold increased risk of developing malignant tumours in other tissues.

PEV **1.** Position-effect variegation, a gene-silencing phenomenon caused by the abnormal juxtaposition of a gene to heterochromatin. **2.** Porcine enterovirus 1 (PEV-1).

PEVK domains A tetrapeptide domain (Pro-Glu-Val-Lys) first recognized in *titin and associated with the elasticity of that protein. Also interacts with actin.

Peyer's patches Specialized regions of *lymphoid tissue, found in the submucosa of the gut, involved in gut-associated immunity. They have large numbers of IgA-secreting precursor cells, B- and T-dependent regions, and germinal centres.

PFA A standard laboratory test (PFA-100) that estimates platelet function by measuring the time (in seconds) taken for whole blood to occlude a membrane impregnated with platelet activators (adrenaline or ADP).

Pfam A large database of protein families, each represented with multiple sequence alignments and hidden Markov models. Can be used e.g. to view the domain organization of proteins. Pfam release 23.0 (August 2008) included 10 340 families (vs 8183 at the end of 2005).

SEE WEB LINKS

- The Pfam homepage.

PFK *See* PHOSPHOFRUCTOKINASE.

PGC Peroxisome proliferator-activated receptor γ coactivator-1. *See* PPARs.

PGD **1.** *See* PREIMPLANTATION GENETIC DIAGNOSIS. **2.** Prostaglandin D (*see* PROSTAGLANDINS).

PGK *See* PHOSPHOGLYCERATE (phosphoglycerate kinase).

P glycoprotein *See* MULTIDRUG TRANSPORTER.

PGP P glycoprotein. *See* MULTIDRUG TRANSPORTER *but compare* PGP 9.5.

PGP 9.5 A commonly used immunohistochemical marker (protein gene product 9.5) for nerves, a *ubiquitin hydrolase confined to neural and neuroendocrine cells.

PGs (PGE, PGD, etc.) *See* PROSTAGLANDINS.

pH A logarithmic scale ($-\log_{10}[H^+]$) of hydrogen ion concentration (which determines the acidity or alkalinity of an aqueous solution). Neutrality corresponds to pH 7, strong acids approach pH 0, strong alkalis pH 14.

PH-30 One subunit (β-fertilin, ADAM2) of a heterodimeric sperm surface protein involved in sperm–egg fusion. The α subunit (ADAM1) has some similarities to viral fusion proteins, and the β subunit has an ADAM domain (a disintegrin and metalloprotease domain) similar to that in soluble integrin ligands (*disintegrins). The first 382 aa (16 aa of signal sequence and the metalloprotease domain) are cleaved from the 735-aa precursor.

PHA 1. *See* PHYTOHAEMAGGLUTININ. **2.** Polyhydroxyalkanoates, the basis for biodegradable plastics.

phaeo- (US **pheo-**) Prefix meaning dark-coloured. The UK English spelling is used throughout this dictionary.

phaeochromocytoma A normally benign *neuroblastoma derived from the *chromaffin tissue of the adrenal medulla. The cells secrete *catecholamines, and will form neuron-like cells in culture when treated with cAMP or nerve growth factor. Overproduction of adrenaline and noradrenaline *in vivo* causes secondary hypertension, sometimes paroxysmal.

phaeomelanin *See* MELANIN.

phafins A family of proteins that have *PH (pleckstrin homology) and *FYVE domains. *See* LAPF. The FYVE, rhoGEF, and PH domain-containing protein 1 (FGD1) is associated with *Aarskog–Scott syndrome (faciogenital dysplasia).

phage *See* BACTERIOPHAGES.

phage display library *See* EPITOPE LIBRARY.

phage integrase family *See* RECOMBINASE; SITE-SPECIFIC RECOMBINATION.

phagocyte Any cell that is capable of *phagocytosis; in humans, *neutrophils and *macrophages.

phagocytic vesicle A membrane-bounded vesicle (phagosome) surrounding an internalized (phagocytosed) particle. *Lysosomes fuse with the primary phagosome to form a secondary phagosome in which digestion will occur.

phagocytosis The uptake of particulate material by a cell into a *phagocytic vesicle. *See* OPSONIN.

phakinin A cytoskeletal protein (phakosin, beaded filament structural protein 2, 415 aa), found in eye lens, that coassembles with *filensin in a 3:1 ratio to form beaded-chain *intermediate filaments. Phakinin has very strong sequence homology with *cytokeratins but only forms filaments in association with filensin. Mutations are associated with autosomal dominant juvenile-onset *cataract.

phalangeal cells Epithelial cells of the organ of Corti (in the inner ear). There is one row of inner phalangeal cells and three rows of outer phalangeal cells (Deiter's cells). They produce neurotrophins and nerve growth factor which are essential for the survival of *hair cells.

SEE WEB LINKS

- Entry in the Cytokines & Cells Online Pathfinder Encyclopedia.

phalloidin A cyclic peptide (789 Da), from the death cap *Amanita phalloides*, that binds to, and stabilizes, F-actin. Fluorescent derivatives can be used to stain actin in fixed and permeabilized cells.

phallotoxins A range of toxins (bicyclic peptides), of which *phalloidin is one, that are produced by the death cap *Amanita phalloides*. They bind to F-actin and block depolymerization; hepatotoxicity is the primary cause of problems.

pharmacodynamics The study of the effect of drugs on the body: *compare* PHARMACOKINETICS.

pharmacogenetics The study of the genetic causes of variations in the response to drugs, a term often used interchangeably with *pharmacogenomics.

pharmacogenomics The study of the effect of the whole genome on the efficacy and toxicity of drugs, as opposed to the study of the effect of polymorphism in single genes (*pharmacogenetics).

pharmacokinetics The study of what happens to a drug in the body; *compare* PHARMACODYNAMICS. The main aspects relate to the rate and extent of uptake (absorption), the kinetics of distribution of the drug and its metabolites between compartments, metabolism of the drug, and the mode of excretion. Commonly abbreviated as *ADME (absorption, distribution, metabolism, and excretion).

phase variation The change that can occur in the expression of surface antigens by bacteria.

PHB **1.** The *prohibitin homology (PHB) domain, characteristic of a family of proteins. **2.** poly(β_3-hydroxybutyrate), a bacterially synthesized biodegradable polymer. **3.** A class of *ubiquitin conjugating enzymes (E3 ligases).

PHC Components (polyhomeotic-like protein 1 and 2, early development regulatory protein 1 and 2, 1004 aa and 858 aa) of the *polycomb repressive complex-1.

PH domain A domain (pleckstrin homology domain, ~120 aa) found in various proteins involved in intracellular signalling (e.g. *pleckstrin, *tec family kinases) and the cytoskeleton. PH domains interact with membrane phospholipids, particularly phosphatidyl inositol phosphate (PIP3).

phenanthridinone A potent inhibitor of *poly(ADP-ribose) polymerase (PARP) that has immunosuppressant activity.

phenanthroline (*o*-phenanthroline) An aromatic hydrocarbon used as an analytical reagent for photometric determination of iron(II), with which it forms a red complex.

phenazone *See* ANTIPYRINE.

phenazopyridine An analgesic used for pain relief in infections of the lower urinary tract. It is excreted unchanged into the urine and acts as a topical analgesic through an unknown mechanism.

phencyclidine (PCP, angel dust) An anaesthetic that interacts with the *NMDA receptor and can produce marked behavioural effects, hence its recreational abuse.

phenelzine A monoamine oxidase inhibitor used as an antidepressant drug. *TN* Nardil.

phenindione An anticoagulant drug, a vitamin K antagonist.

phenobarbital (formerly **phenobarbitone)** A barbiturate used as an anticonvulsant in the treatment of epilepsy.

phenobarbitone *Now* *phenobarbital.

phenocopy A phenotype that arises through environmental effects but resembles that produced by a particular genotype.

phenolphthalein An indicator dye that is colourless at neutral pH and red-pink in slightly alkaline solutions. Used as a laxative.

phenol red An indicator dye that changes from yellow to red in the pH range 6.8–8.4 and is commonly added to tissue culture medium which turns acid and obviously yellow as it becomes exhausted. It can interfere with some luminescence assays.

phenome The set of all phenotypes expressed by cells, tissues, organs, individuals, or species, including those due to both genetic and environmental influences. The phenotypic equivalent of the genome.

phenothiazines One of the main classes of neuroleptic antipsychotic drugs, thought to act by blocking dopaminergic transmission in the brain. Examples are *chlorpromazine and *trifluoperazine.

phenotype The observable characters, including morphology and behaviour of an organism, regardless of the actual genotype of the organism. Identical genotypes do not necessarily produce identical phenotypes.

phenoxybenzamine A nonspecific, irreversible *alpha-blocker used to treat hypertension caused by *phaeochromocytoma. *TN* Dibenzyline.

phentermine An appetite-suppressing drug with amphetamine-like activity, used in treating obesity. It was combined with *fenfluramine or dexfenfluramine as Fen-Phen (now withdrawn).

phentolamine A reversible nonselective α-adrenergic antagonist used as an emergency treatment for hypertension. *TN* Regitine.

phenylalanine (Phe, F) An essential aromatic amino acid (165 Da).

phenylbutazone A *NSAID used in the treatment of rheumatoid arthritis.

phenylephrine An α_1-adrenergic agonist, used as a decongestant, to dilate the pupil of the eye, and to increase blood pressure.

phenylethylamine A naturally occurring compound biosynthesized from phenylalanine. May function as a neuromodulator or neurotransmitter in the brain. Also found in chocolate.

phenylketonuria (PKU) A disorder in which phenylalanine hydroxylase (EC 1.14.16.1, 452 aa), the enzyme that converts phenylalanine into tyrosine, is congenitally deficient. The accumulation of phenylalanine seriously impairs early neuronal development but dietary control mitigates the problem.

phenylmethylsulphonyl fluoride (PMSF) A broad-spectrum serine peptidase inhibitor used experimentally.

phenylthiourea An inhibitor (PTU, phenylthiocarbamide, PTC) of *tyrosinase and melanin synthesis. The ability to taste PTC is a common polymorphism.

phenytoin An antiepileptic drug, also used to control abnormal heart rhythms. It stabilizes against neuronal hyperexcitability by acting as a sodium channel blocker. *TNs* Phenytek, Dilantin.

pheo- *See* PHAEO-.

pheresis A method of separating blood to remove either cells or plasma, the unwanted portion being returned to the circulation. In **leukapheresis**, white cells are removed; in **plasmapheresis**, plasma is removed and all cellular components returned.

pheromone A volatile hormone or behaviour-modifying substance. The classic example is the sex attractant bombesin for the moth *Bombyx.*

PHF1 A *polycomb group protein (PHD finger protein 1, polycomb-like protein 1, 567 aa) that is a transcriptional repressor.

Philadelphia chromosome The chromosome formed by a reciprocal translocation between chromosomes 9 and 22 causing chronic myelogenous *leukaemia. Part of the *bcr* (breakpoint cluster region) gene of chromosome 22 (region q11) is fused with part of the *abl* gene from chromosome 9 (region q34) and encodes a protein on chromosome 22 (p185, bcr/abl) that has unregulated tyrosine kinase activity.

phlebolith A venous thrombus (venous calculus) that has become calcified, most commonly found in pelvic veins.

phlebotomy Taking blood by venepuncture.

phlyctenular Describing a condition in which there are small pustules, or whitish elevations. In **phlyctenular conjunctivitis** (phlyctenular keratoconjunctivitis) there are small vesicles surrounded by a reddened zone on the inflamed conjunctiva of the eye, possibly as an allergic response to bacterial proteins, as in eczematous conjunctivitis.

phomopsin A An antimitotic cyclic peptide from the fungus *Phomopsis leptostromiformis.* It binds to tubulin at the *rhizoxin/maytansine site and blocks assembly into microtubules. Used in chemotherapy.

phorbol esters A range of polycyclic compounds isolated from croton oil (from the seeds of *Croton tigliumin*), the commonest of which is phorbol myristoyl acetate (PMA). They are potent *cocarcinogens or tumour promotors, acting as diacyl glycerol analogues to irreversibly activate *protein kinase C.

phosducin (PDC) A phosphoprotein (246 aa) found in retina, pineal gland, and many other tissues, that inhibits Gs-GTPase activity by binding the $G\beta\gamma$ subunit of heterotrimeric G proteins, making them unavailable for signalling. Phosphorylation by protein kinase A inhibits its activity. The phosducin gene is a potential candidate gene for *retinitis pigmentosa and *Usher's syndrome (type II). Three phosducin-like proteins with extensive amino acid sequence homology to phosducin have been identified.

phosphatases A wide range of enzymes that remove a phosphate group from phosphomonoesters. **Acid phosphatases** specifically remove single-charged phosphate groups whereas **alkaline phosphatases** remove double-charged groups. *Protein phosphatases dephosphorylate specific phosphoproteins, countering the activity of *protein kinases, and have specificity for phosphate groups on serine/threonine or tyrosine residues. They may be targeted to specific locations through interactions with other proteins. There are also **lipid phosphatases** which act on signalling molecules such as phosphatidylinositol phosphates.

phosphatides A family of phospholipids based on 1,2-diacyl-3-phosphoglyceric acid. *See* PHOSPHOLIPID.

phosphatidic acid (PA) The simplest of the *phospholipids (diacylglycerol 3-phosphate) consisting of a glycerol backbone usually with a saturated fatty acid covalently linked to C-1, an unsaturated fatty acid on C-2, and a phosphate group on C-3. It is present in low concentrations in membranes, where it may affect membrane curvature, and is an intermediate in the synthesis of diacylglycerol and most phospholipids other than phosphatidyl inositol. *See* LYSOPHOSPHATIDIC ACID.

phosphatidylcholine (PC) The major phospholipid of most mammalian cell membranes with *choline attached to *phosphatidic acid by a phosphodiester linkage. It will form monolayers at an air/water interface, and forms bilayer structures (*liposomes) when dispersed in aqueous medium. The 1-acyl residue is normally saturated and the 2-acyl residue unsaturated.

phosphatidylethanolamine (PE) A major structural phospholipid, with ethanolamine attached to *phosphatidic acid by a phosphodiester linkage. It is more abundant than phosphatidylcholine in internal cell membranes.

phosphatidylinositol (PI) A minor but important phospholipid with myo-inositol linked through the 1-hydroxyl group to phosphatidic acid. The 4-phosphate (PIP) and 4,5-bisphosphate derivatives (PIP2) are formed and broken down in membranes by the action of specific kinases and phosphatases (futile cycles). A ligand-activated PIP2-specific phosphodiesterase (phospholipase Cγ) breaks down PIP2 to form diacylglycerol, which stimulates protein kinase C, and inositol 1,4,5-trisphosphate (InsP3) which releases calcium from the endoplasmic store. *See* PI-3-KINASES.

phosphatidylinositol-3-kinase *See* PI-3-KINASES.

phosphatidylserine (PS) An important minor species of phospholipid with serine attached to *phosphatidic acid by a phosphodiester linkage. It is asymmetrically distributed and found only on the cytoplasmic leaflet of the cell membrane. It is negatively charged at physiological pH, interacts with divalent cations, and is involved in calcium-dependent interactions of proteins with membranes (e.g. protein kinase C).

phosphocreatine (creatine phosphate) A compound found at high concentration (~20 mmol/L) in striated muscle where it acts as an energy reserve, buffering ATP levels through the action of *creatine kinase.

phosphodiesterase (PDE) Any enzyme that catalyses the hydrolysis of one of the two ester linkages in a phosphodiester. PDE-I (EC 3.1.4.1) catalyses the removal of 5-nucleotides from the 3′-end of an oligonucleotide. PDE-II (EC 3.1.16.1) catalyses removal of 3-nucleotides from the 5′-end of a nucleic acid. The term is often used as a shorthand for cAMP-phosphodiesterase. *See* CYCLIC NUCLEOTIDE PHOSPHODIESTERASES.

phosphodiesterase inhibitors Any inhibitor of a phosphodiesterase although usually refers to an inhibitor of cAMP-phosphodiesterase (EC 3.1.4.17). Such inhibitors (e.g. methylxanthines such as theophylline) cause accumulation of intracellular cAMP which will potentiate the action of the sympathetic nervous system.

phosphodiester bond A linkage between two parts of a molecule through a phosphate group. Examples are found in RNA, DNA, phospholipids, cyclic nucleotides, nucleotide diphosphates, and triphosphates. *Phosphodiesterases hydrolyse the bond.

phosphoenolpyruvate (PEP) An important metabolic intermediate that can be broken down to pyruvate by pyruvate kinase, generating an ATP molecule.

phosphofructokinase A key regulatory enzyme (6-phosphofructo-1-kinase, EC 2.7.1.11, 780 aa) in glycolysis that catalyses the conversion of fructose 6-phosphate to fructose 1,6-bisphosphate. It is allosterically regulated by the ATP/ADP ratio. There are three isoenzymes in humans, encoded by different genes: muscle *M, liver *L, and platelet *P forms. Deficiency of the muscle form leads to glycogen *storage disease type VII with associated myopathy.

phosphoglycerate Intermediates in glycolysis (2-phosphoglycerate and 3-phosphoglycerate) and, in the case of 3-phosphoglycerate, the precursor for synthesis of *phosphatidic acid and *diacylglycerol. Phosphoglycerate kinase (PGK, EC 2.7.2.3, 417 aa) is an X-linked enzyme that catalyses the phosphorylation of 3-phospho-D-glycerate to 3-phosphohydroxypyruvate in glycolysis. The phosphoglycerate mutase (EC 2.7.5.3, 254 aa) family is widely distributed with tissue-specific isoforms; deficiency of the muscle form leads to myopathy similar to that caused by deficiency in *phosphofructokinase.

phosphoinositide-dependent kinase A protein kinase (PDK-1, EC 2.7.1.37, 556 aa) that activates protein kinase B in a PIP2- or PIP3-dependent manner (*see* PHOSPHATIDYLINOSITOL) and many other protein kinases in the *AGC kinase superfamily.

phosphokinase *See* KINASE.

phospholamban An integral membrane protein (a homopentamer with subunits of 52 aa) that is the endogenous regulator of the calcium ATPase of sarcoplasmic reticulum (*SERCA). Phosphorylation by protein kinase A blocks the inhibition which can be reversed by protein phosphatase 1 dephosphorylation. Mutations affect control of cardiac muscle and cause *dilated cardiomyopathy. *See* SARCOLIPIN.

phospholemmans Small transmembrane proteins (FXYD domain-containing ion transport regulator, FXYD1, 92 aa) that are phosphorylated in response to insulin and adrenergic stimulation. Like *phospholamban they interact with P-type ATPases and regulate ion transport in cardiac cells and other tissues. **Phospholemman-like protein** (FXYD3, 67 aa, mammary tumour 8-kDa protein) lacks the consensus phosphorylation site in the cytoplasmic domain and is overexpressed in many mammary tumours.

phospholipase A Aliphatic esterases, PLA1 and PLA2, that release fatty acids from phospholipids. Phosphatidic acid-selective phospholipase A1 (lipase H, 451 aa) generates *lysophosphatidic acid and is mutated in autosomal recessive hypotrichosis. PLA1-α (456 aa) acts specifically on phosphatidylserine. Phospholipases A2 (EC 3.1.1.4) are diverse, with secreted, cytosolic, and lipoprotein-associated classes. They are divided into groups: Group I includes pancreatic, PLA2G1B, and is also found in the venom of cobras and kraits; group II are extracellular and require calcium, examples are found in synovial fluid (PLA2G2A) and in the venom of rattlesnakes and vipers; group III is found in bee and lizard venom; group IV are cytosolic and will generate arachidoic acid, the precursor for *eicosanoids, and are therefore important in inflammation. A number of other groups are recognized and have differing substrate specificities or tissue distributions. Lipoprotein-associated PLA2s (group VII, EC 3.1.1.47, lp-PLA2) hydrolyse *platelet activating factor and deficiency increases susceptibility to asthma and atopy. PLA2G6 (806 aa) is a calcium-independent PLA2 mutated in *infantile neuroaxonal dystrophy and *Karak's syndrome. A PLA2 receptor has been described (1465 aa) and may be involved in internalizing secreted forms of the enzyme for degradation. Phospholipase A2-activating protein (PLAP, 738 aa) from monocytes may modulate the availability of substrate for eicosanoid synthesis. Various patatin-like phospholipase domain-containing proteins are also known and are intracellular, calcium-independent enzymes with PLA2-like activity implicated in regulation of adipocyte differentiation and are responsive to metabolic demand (*see* PNPLA6).

phospholipase D A family of *phospholipases that are found in a wide variety of organisms. Phosphatidylcholine (PC)-specific phospholipases D (EC 3.1.4.4: PLD1, 1072 aa; PLD2, 931 aa) hydrolyse PC to phosphatidic acid and choline and their activity, stimulated through G-protein-coupled receptors and receptor tyrosine kinases, has been implicated in signal transduction, membrane trafficking, and the regulation of mitosis. Glycosylphosphatidylinositol-specific phospholipase D (GPI-PLD, 840 aa) is abundant in serum and will release *GPI-anchored proteins from membranes. A specific phospholipase D type enzyme with no homology to PLD1 is important in the biosynthesis of *N*-acylethanolamines such as *anandamide.

phospholipases Enzymes that hydrolyse ester bonds in phospholipids. Aliphatic esterases (phospholipase A1, A2, and B) release fatty acids, whereas phosphodiesterases (types C and D) release diacylglycerol or phosphatidic acid respectively. Types A1, A2, and C are present in all mammalian tissues. Type C, specific for *phosphatidylinositol, is also found as a bacterial toxin (e.g. *Clostridium perfringens* α-toxin). Type B attacks monoacyl phospholipids and is poorly characterized. *See* PHOSPHOLIPASE A; PHOSPHOLIPASE D.

phospholipid The major structural lipid of most cellular membranes, forming a *phospholipid bilayer in which membrane proteins are embedded. *See* PHOSPHATIDIC ACID, PHOSPHATIDYLCHOLINE, PHOSPHATIDYLETHANOLAMINE, PHOSPHATIDYLINOSITOL, PHOSPHATIDYLSERINE, PLASMALOGENS, and SPHINGOMYELINS.

phospholipid bilayer A lamellar organization (lamellar phase) of phospholipids packed as a bilayer ~7 nm thick with the hydrophobic acyl tails inwardly directed and polar head groups on the outside surfaces. It is this fluid bilayer that forms the basis of membranes in cells, though in most cellular membranes a very substantial proportion of the area may be occupied by *integral membrane proteins that are free to move in the plane of the membrane. Some proteins are anchored, some may be surrounded by atypical lipid domains, lipid domains (rafts) may also exist, and the inner and outler leaflets may have differences in phospholipid composition. Other proteins (*peripheral membrane proteins) are anchored to the membrane but the protein portion is not embedded within the hydrophobic region. The ubiquitous triple-layered appearance of membranes (the so-called unit membrane) seen in electron microscopy is thought to arise because osmium tetroxide binds to the polar regions leaving a central, unstained, hydrophobic region.

phospholipid transfer proteins A family of plasma proteins (lipid transfer proteins, LTPII, 476 aa) that bind phospholipids and facilitate their transfer between cellular membranes. They have homology with *cholesteryl ester transfer protein (lipid transfer protein I). **Phosphatidylinositol transfer protein** (PIPTP, 271 aa) is specific for PI. **Glycolipid transfer protein** is a soluble protein (209 aa) that selectively accelerates the transfer of glycosphingolipids and glycoglycerolipids between donor and acceptor membranes.

phosphomannose *See* MANNOSE-6-PHOSPHATE RECEPTOR.

phosphoprotein Any protein that has a phosphate group bound through an ester linkage to serine, threonine, or tyrosine. The addition or removal of phosphate often affects function.

phosphoramidite Casual term used by molecular biologists for nucleoside phosphoramidites in which a phosphite with an amino group is substituted for one of the hydroxyls, instead of the normal phosphate. Nucleoside phosphoramidites can be used to synthesize short nucleic acid chains.

phosphorescence **1.** The emission of light following absorption of radiation. The emission wavelength is longer than that of the excitatory radiation. Phosphorescence is longer-lasting than fluorescence and occurs after a longer delay. **2.** Inaccurate term often applied to biological *luminescence.

phosphorylase kinase The enzyme (EC 2.7.1.38) that integrates the hormonal and calcium signals in muscle and regulates the activity of *glycogen phosphorylase and glycogen synthetase by addition of phosphate groups. A large and complex enzyme, itself regulated by phosphorylation, consisting of four copies of an α-β-γ-δ tetramer with several isoforms of the α, β, and γ subunits. The α (1235 aa) and β (1093 aa) subunits have regulatory functions, the γ subunit (406 aa) has the catalytic activity, the δ subunit is *calmodulin. Mutations lead to X-linked liver glycogenosis (*glycogen storage disease type IXa, GSDIXa). X-linked muscle phosphorylase kinase deficiency (GSDIXd) is caused by mutation in the gene encoding the α subunit. Autosomal recessive liver glycogenosis (GSDIXc) is caused by mutation in the γ_2 isoform.

phosphotransferases The class of enzymes (EC 2.7.-.-) which catalyse phosphorylation reactions. Includes *kinases.

phosphotyrosine In general usage, the phosphate ester of a tyrosine residue in a protein, though strictly it should mean tyrosine phosphate. Protein *tyrosine kinases, which catalyse this phosphorylation, are important in signalling and regulatory systems.

photoadduct A compound formed by a light-induced reaction. Commonest examples are covalent modifications of DNA as a result of UV irradiation. This can be deliberate, as in *photodynamic therapy where adducts with *psoralens are generated. *See* PHOTOAFFINITY LABELLING.

photoaffinity labelling The covalent coupling of a label or marker molecule onto another molecule such as a protein triggered by illumination (usually with UV light). The label is activated by light and binds to molecules in close proximity.

photobleaching The loss of the ability of a *chromophore to absorb light that occurs after a period of illumination. This fading of the signal is a common problem in fluorescence microscopy.

photodynamic therapy A treatment in which a light-sensitive prodrug is administered and the target area (usually a tumour) is illuminated to activate the drug locally.

photolyases A ubiquitous family of enzymes involved in the light-dependent repair of UV-induced DNA damage in many organisms. The human homologues, *cryptochromes, do not seem to have this repair capacity.

photolysis The light-induced cleavage of a chemical bond.

photomedins Retinal proteins of the *olfactomedin-family; photomedin-1 is selectively expressed in the outer segment of photoreceptor cells and photomedin-2 in all retinal neurons. They interact with proteoglycans, preferentially binding to chondroitin sulphate and heparin.

photoperiodism Cycles of activity or behaviour linked to the duration of illumination or the pattern of light/dark cycles. *See* CIRCADIAN RHYTHM.

photophosphorylation The synthesis of ATP that occurs in photosynthesis.

photoreceptor A system that is light sensitive; in multicellular organisms, a cell that responds to light. In the eye *retinal rods and *retinal cones are the primary receptors but other photosensitive systems exist in the pineal gland. *See also* CRYPTOCHROME.

phototransduction The process of transforming light energy into a change in electrical transmembrane potential as happens in *retinal rods and *retinal cones.

phox The *NADPH oxidase system in phagocytes, that produces reactive oxygen species through the *metabolic burst and is important for bacterial killing. The membrane-associated portion is a heterodimer of p22phox and p91phox, the A and B subunits of cytochrome *b*245. Two other components, p47phox and p67phox, are cytoplasmic and only become membrane-associated following activation. Both p47 and p67 contain *SH3 domains. Deletion or mutation in any of the components can cause *chronic granulomatous disease although the commonest form involves mutation in p91.

phyllolitorin A *bombesin-like peptide, originally isolated from the skin of the South American frog *Phyllomedusa sauvagei.* The precursor peptide (90 aa) has a signal sequence, and the 9-aa phyllolitorin peptide flanked by N- and C-terminal extension peptides. Leu8-phyllolitorin increases branching of developing airways. Central injection of Leu8-phyllolitorin has been shown to produce hypothermia in animals exposed to a cold environment.

phylogenetic profiling An approach to identifying functional interactions between proteins based upon their invariable joint presence or absence in different genomes. If they do not interact then it is argued that there is no reason for them not to appear alone. *See* *PAUP; ROSETTA STONE METHOD.

physaliphorous cells Cells from a chordoma, a rare malignant tumour derived from embryonic notochordal remnants, that contain large intracytoplasmic droplets of mucoid material resembling vacuoles.

physical mapping The process of sorting genomic DNA clones in order to select those that give a complete coverage of a genetic locus. Candidate clones are screened for marker sequences and clones sharing particular markers are assumed to overlap. A computational approach can then be used to select the smallest set of clones that will cover the whole region.

physins A subgroup of the *tetraspan vesicle membrane proteins (tetraspanins). Examples include *synaptophysin, *synaptoporin, *pantophysin, and *mitsugumin29.

physostigmine *See* ESERINE.

phytanic acid A branched-chain fatty acid, derived from dietary chlorophyll, that accumulates in *Refsum's syndrome and *Zellweger's syndrome. *See* PEROXISOMAL BIOGENESIS DISORDERS.

phytoestrogen Any plant-derived compound that has weak *oestrogen-like properties.

phytohaemagglutinin (PHA) A *lectin, isolated from the red kidney bean *Phaseolus vulgaris,* that binds to *N*-acetylgalactosyl residues in oligosaccharides. Although it binds to B and T cells it is only mitogenic for the latter.

phytonutrient Any plant-derived substance believed to have health-giving properties.

phytotoxin Any plant-derived toxin, usually present to deter herbivores or pests rather than to poison people. A very diverse set of substances including aristolochic acids; pyrrolizidine alkaloids; (beta-*carotene, *coumarin, the alkenylbenzenes safrole, methyleugenol, and estragole, *ephedrine alkaloids and synephrine, kavalactones, anisatin, St. John's wort ingredients (*hyperforin), cyanogenic glycosides, *picrotoxin, *solanine, and *chaconine, thujone, and glycyrrhizinic acid.

PI *See* PHOSPHATIDYLINOSITOL.

PI-3-kinases A family of kinases (phosphatidylinositol-3-kinases, PI kinases) that phosphorylate phosphatidylinositol phosphate on the 3 position. They are involved in signalling cascades downstream of many *growth factor receptors, particularly *receptor tyrosine kinases (RTKs). The classical form (class I) has a p110 enzymatic subunit which is regulated by p85 (p85α, p85β, p55γ, and splicing variants). p110 is recruited to the membrane and activated via the interaction of SH2 domains on the regulatory subunit which bind phosphotyrosine motifs on the stimulated RTK. An increasing family is being identified, including the PI3Kγ isoform which is activated by receptor-stimulated *G proteins and recruited to the membrane by p101. Others are regulated by calcium. Most, but not all, are inhibited by *wortmannin. *See* AKT.

PI-103 A cell-permeable ATP competitor that selectively inhibits *DNA-activated protein kinase, *PI-3-kinase, the rapamycin-sensitive (mTORC1), and rapamycin-insensitive (mTORC2) complexes of the protein kinase *mTOR, but has little activity against a range of other kinases. Blocks cell proliferation in glioma cell lines both *in vitro* and *in vivo.*

pia mater The well-vascularized inner meningeal membrane that is immediately adjacent to the brain. In the spinal cord the pia mater is connected to the *dura mater by denticular ligaments that pass through the *arachnoid membrane.

piccolo A *CAZ protein of the presynaptic cytoskeletal matrix (PCM) that regulates synaptic vesicle traffic. Interacts with other proteins in the presynaptic region including RIM (*retinoblastoma-interacting myosin-like) and *bassoon.

PICK1 A neuronal protein (protein interacting with C kinase-1) that is phosphorylated by *protein kinase C and interacts with the GTP-bound forms of ADP-ribosylation factor-1. It binds AMPA receptors and acid-sensing ion channels and may regulate Golgi-to-endoplasmic reticulum vesicle transport.

SEE WEB LINKS

- Further details in the BAR Superfamily website.

Pick's disease A *frontotemporal dementia caused by mutation in the gene for the microtubule-associated protein *tau or, in some cases, the gene for *presenilin-1. *See* NIEMANN–PICK DISEASE.

Picornaviridae A family of small RNA viruses (class IV) that have a genome composed of a single positive strand of RNA and an icosahedral capsid. There are two main classes, the *enteroviruses and the *rhinoviruses.

picrotoxin (cocculin) A plant alkaloid isolated from the fruit (fish-berries) of *Anamirta cocculus*, an East Indian woody vine. Picrotoxin is a noncompetitive antagonist of inhibitory $GABA_A$ receptors and has a stimulative effect. Has been used as an antidote to poisoning by central nervous system depressants, especially the barbiturates.

Pierson's syndrome An autosomal recessive disorder (microcoria–congenital nephrosis syndrome) characterized by severe nephrotic syndrome and extreme nonreactive narrowing of the pupils (microcoria). Caused by mutation in the gene encoding *laminin β_2.

pigeon fancier's disease An allergic alveolitis common among pigeon keepers.

pigmentation The colour of skin, hair, and eyes are affected by a number of genes. Phenotypes influenced by variation in the **OCA2* gene are termed **SHEP1** (skin, hair, eye pigmentation 1). The **SHEP2** association is determined by variation in the *melanocortin 1 receptor (MC1R) locus and produces a phenotype characterized by red hair and fair skin. **SHEP3** encompasses pigment variation influenced by the *tyrosinase gene. **SHEP4** is influenced by the *SLC24A5* gene that encodes a potassium-dependent sodium/calcium exchanger. Variation in the *SLC45A2* (melanoma antigen *AIM-1) and *SLC24A4* (similar to SLC24A5) genes produce **SHEP5** and **SHEP6** phenotypes. Variation in *kit ligand expression results in the **SHEP7** phenotypic association; **SHEP8** is associated with *single-nucleotide polymorphisms (SNPs) at chromosome 6p25.3. Polymorphism in the agouti signalling protein gene (*ASIP) influences **SHEP9**, **SHEP10** is due to variation in the *two-pore channel 2 gene, and **SHEP11** is associated with polymorphism near the tyrosinase-related protein gene.

pigment cells *See* MELANOCYTES; CHROMATOPHORES.

pigmented retinal epithelium The layer of phagocytic epithelial cells (PRE; retinal pigmented epithelium, RPE) containing many pigment granules that are apposed to the photoreceptors of the eye and internalize discs shed from the rod outer segment. Dysfunction is involved in *age-related macular degeneration and *retinitis pigmentosa.

pigtail *See* GPI-ANCHOR.

pilin **1.** In general, protein subunits composing a *pilus. **2.** More specifically, the subunit (121 aa) of *sex pili encoded by the F-plasmid.

pilocarpine A muscarinic alkaloid, isolated from the Jaborandi tree *Pilocarpus jaborandi*, used in the treatment of glaucoma.

pilomatrixoma A firm, circumscribed tumour (epithelioma calcificans of Malherbe), usually in the head and neck area, which feels like a button. Caused in some cases by mutation in the β-*catenin gene.

pilus (*pl.* pili, fimbrium, *pl.* fimbria) A hairlike projection composed of *pilin subunits, found on the surface of some bacteria. Pili are involved in adhesion and may be important determinants of virulence; specialized *sex pili are involved in bacterial *conjugation.

pim A family of oncogenes that encode serine/threonine *protein kinases. **Pim-1** (404 aa) is up-regulated in prostate cancer; **pim-2** (334 aa) is highly expressed in human leukaemic and lymphoma cell lines and a colorectal adenocarcinoma cell line; **pim-3** (326 aa) is aberrantly expressed in human pancreatic cancer and phosphorylates *bad.

pimozide An antipsychotic drug with side effects that make it a drug of last resort. *TN* Orap.

PIN-1 *See* PARVULINS.

pindolol A nonselective beta-blocker (*β-adrenoceptor blocking drug) with partial β-agonist activity, used to treat hypertension and angina pectoris.

pinitol The active principle of the traditional antidiabetic plant *Bougainvillea spectabilis*, claimed to exert insulin-like effects. Double-blind clinical trials reveal neither toxic nor therapeutic effects.

pinocytosis The endocytotic uptake of fluid-filled vesicles (pinocytotic vesicles, pinosomes), usually less than 150 nm diameter.

p

Micropinocytosis (vesicles of ~70 nm) is distinct from **macropinocytosis**, being energy independent and involving receptor–ligand clustering on the cell surface and interaction with cytoplasmic *clathrin.

pinosome *See* PINOCYTOSIS.

PINX1 One of the proteins (TRF1-interacting protein 1, 328 aa) involved in the *shelterin complex that binds telomeric repeat binding factor 1 (TRF1); it has *telomerase-inhibiting activity and is potentially a tumour suppressor. Overexpression of *PINX1* causes shortening of telomeres; depletion has the opposite effect and increases tumorigenicity in animal model systems.

pioglitazone A thiazolidinedione drug that stimulates nuclear receptor peroxisome proliferator-activated receptor γ (*PPARγ) and is used in treatment of type 2 diabetes. *TN* Actos.

PIP2 Phosphatidylinositol-4,5-bisphosphate. *See* PHOSPHATIDYLINOSITOL.

piperazine An anthelmintic drug. Piperazines are a class of compounds with a core piperazine group; many have useful pharmacological properties.

piroplasm A class of apicomplexan protzoa, of which **Babesia* is a member.

piroxicam A *NSAID, a nonselective cyclo-oxygenase (*COX) inhibitor. *TNs* Brexidol, Feldene.

PIST *See* RHOTEKIN.

PITSLRE kinases A family of cell division control-related protein kinases (CDK11 family kinases, CDKs), the regulatory regions of which are characterized by the amino acid motif PITSLRE.

pituicytes Cells of the neural lobe of the hypophysis that resemble *neuroglia and secrete *antidiuretic hormone.

pituitary (hypophysis) An endocrine gland that protrudes from the bottom of the hypothalamus at the base of the brain. It secretes a wide range of hormones involved in homeostasis, including *somatotropin, *follicle stimulating hormone, *gonadotropins, *thyroid stimulating hormone, and *lipotropin. There are two morphologically and functionally distinct lobes, the anterior adenohypophysis and the posterior neurohypophysis.

PITX* genes** A family of genes (*PTX1–3*) that encode paired-like class homeodomain transcription factors expressed in normal pituitary and aberrently expressed in various pituitary adenomas. ***PITX2 (solurshin) mutations are associated with *Rieger's syndrome, ***PITX3*** mutations with various eye abnormalities such as autosomal dominant posterior polar cataract-4.

pivmecillinam An antibiotic used to treat urinary-tract infections.

***piwi* genes** A family of genes, first identified in *Drosophila,* that are important in stem cell self-renewal and gametogenesis in diverse multicellular organisms ranging from *Arabidopsis* to mouse (*miwi* genes) and human (*HIWI* genes). The gene products contain a highly conserved motif (piwi domain) which is important for the endonuclease cleavage activity of *RISCs. **Piwi interacting RNAs** (piRNAs) are germ-cell-specific RNAs essential for male germ-cell development that are bound to a class of *argonaute proteins which may repress or activate gene expression.

pizotifen An antihistamine drug used to treat migraine. *TN* Sanomigran.

pK_a *See* ASSOCIATION CONSTANT.

PKA, PKB, etc. *See* PROTEIN KINASE A; PROTEIN KINASE B, etc.

PKI **1.** The antigen (pKi-67, *Ki antigen) recognized by the Ki67 monoclonal antibody. **2.** An inhibitor of *Her-2/Her-1 (PKI-166). **3.** The negative logarithmic dissociation constant of a competitive inhibitor (pK_i). **4.** *See* PROTEIN KINASE INHIBITOR PROTEINS.

pKi-67 *See* KI ANTIGEN.

PKN *See* PROTEIN KINASE N.

PKU *See* PHENYLKETONURIA.

PLA2 *See* PHOSPHOLIPASE A.

placebo An inactive compound, used in place of a drug that is being tested in a *clinical trial so that patients are unaware of whether they are in the treatment or the control group. The term is also used for inactive drugs given with the assurance that they will help a condition. The so-called 'placebo effect' can account for 30% or more of the positive outcomes in a trial—patients who respond positively even though they have not received a real drug.

placental calcium-binding protein *See* S100 CALCIUM-BINDING PROTEINS.

plakins A family of giant multidomain proteins that link cytoskeletal elements together and connect them to junctional complexes. Plakins were first identified in epithelial cells connecting intermediate filaments to desmosomes and hemidesmosomes, and splicing isoforms have subsequently been shown to be important for the integrity of muscle cells and crosslinking of microtubules (MTs) and actin filaments in neurons. One member is **bullous pemphigoid antigen 1** (Bpag1)/dystonin, which has neuronal and muscle isoforms that have actin-binding and microtubule-binding domains at either end separated by a plakin domain and several spectrin repeats. *See* EPIPLAKIN.

plakoglobin (γ-catenin) *See* CATENINS.

plakophilins A family of proteins with *armadillo repeats found in desmosomal plaques and the cell nucleus. There are at least two isoforms of **plakophilin 1**, PKP1a (726 aa) and PKP1b (747 aa), the latter located exclusively in nuclei, whereas PKP1a is in nuclei and desmosomal plaques of stratified and complex epithelia. Mutations in PKP1 are the underlying cause of *ectodermal dysplasia–skin fragility syndrome. Desmosomal plakophilins, like **plakophilin 2** (837 aa and 881 aa isoforms), link the cytoplasmic tail of *cadherins and the intermediate filament cytoskeleton. Mutations in the plakophilin-2 gene (*PKP2*) have been found in patients with *arrhythmogenic right ventricular dysplasia/cardiomyopathy (ARVC9). **Plakophilin-3** (PKP3, 797 aa) interacts with *catenin γ (plakoglobin), *desmoplakin, and the epithelial keratin 18 and has been shown to bind all three *desmogleins, *desmocollin-3a and -3b, and possibly also desmocollin-1a and -2a. **Plakophilin 4** (1149 aa) is *p0071.

planar cell polarity pathway (PCP pathway) A highly conserved signalling cascade that coordinates both epithelial and axonal morphogenic movements during development. Inhibition of PCP signalling disrupts endothelial cell growth, polarity, and migration, but this can be reversed through downstream activation of the pathway by expression of either Daam-1, Diversin, or Inversin. In *Drosophila* the important genes are *frizzled* (fz), *van gogh* (vang), *strabismus* (stbm), *prickle* (Pk), *dishevelled* (dsh), *flamingo* (fmi), and *diego*.

plantaricin C A *lantibiotic produced by *Lactobacillus plantarum* LL441 that has type A and type B lantibiotic properties, forming pores and blocking cell wall synthesis.

plaque assay **1.** An assay in which a dilute solution of a virus is applied to a culture of the host cells and the number of plaques (areas of dead or transformed cells) that develop indicates the number of infective viruses (plaque-forming units, pfu) in the solution. Secondary infection by convective spread is inhibited by making the medium viscous. **2.** An assay for the number of cells producing antibody against erythrocytes or against antigen bound to the erythrocytes. A clear plaque of haemolysis surrounds cells that secrete antibody (plaque-forming cells).

plasma The protein-rich fluid in which blood cells are suspended in the circulation and from which *serum is derived.

plasma cell A terminally differentiated B lymphocyte that produces and secretes antibody.

plasmacytoma A malignant tumour derived from *plasma cells, very similar to a *myeloma. Can easily be induced in rodents and immunologists by the injection of complete *Freund's adjuvant. Plasmacytoma cells fused with primed lymphocytes produce monoclonal antibodies.

plasma kallikrein A *serine peptidase composed of a heavy chain (371 aa) and a disulphide-linked light chain (248 aa) produced by proteolytic cleavage of inactive plasma prekallikrein by factor XIIa. The four kallikrein *apple domains mediate high affinity binding to its major substrate, high molecular weight *kininogen.

plasmalemma (*adj.* plasmalemmal) *See* PLASMA MEMBRANE.

plasmalogens A class of widely distributed glycerol-based phospholipids in which the aliphatic side chains are not attached by ester linkages. Plasmalogen synthesis can be defective in *Zellweger's syndrome.

plasmal reaction The reaction between aldehydes, derived from *plasmalogens by treatment with mercuric chloride, and *Schiff's reagent that produces coloured compounds. Used, rarely, as a histochemical procedure.

plasma membrane (plasmalemma) The outer membrane of a cell, a *phospholipid bilayer with associated membrane proteins.

plasmid A small, autonomously replicating, piece of DNA that can be transferred from one organism to another. May be linear or circular and can sometimes become integrated into the host's genomic DNA. Low copy number plasmids are referred to as stringent plasmids; relaxed plasmids occur at higher copy number (10–30 copies). Plasmids can transfer genes between individuals and between species and are the basis for *cloning vectors. Antibiotic resistance genes can spread rapidly as a result of being plasmid encoded. *See* F-FACTOR

plasmin A *serine endopeptidase (fibrinolysin, fibrinase, thrombolysin, EC 3.4.21.7) of the S1 peptidase family that digests *fibrin in blood clots, activated *Hageman factor, and complement. Inactive *plasminogen is cleaved into the active form by *plasminogen activator.

plasminogen The inactive precursor (810 aa, a zymogen) of *plasmin present at ~200 mg/L in plasma. Mutations in the gene can lead to *ligneous conjunctivitis and occlusive *hydrocephalus.

plasminogen activator (PA) A *serine peptidase that converts *plasminogen to active *plasmin. There are two forms, urinary PA (urokinase, uPA, EC 3.4.21.73, 431 aa) and **tissue PA** (tPA, EC 3.4.21.68, 562 aa). The production of plasminogen activator may contribute to degradation of extracellular matrix and facilitate cell invasion. Polymorphisms in uPA have been associated with susceptibility to late-onset Alzheimer's disease. The **uPA receptor**, CD87, controls matrix degradation in tissue remodelling and is important for cell migration and the metastasis of tumour cells. *See* SLURP; STREPTOKINASE.

plasminogen activator inhibitor A member of the *serpin family (PAI-1, mesosecrin, 402 aa) that is found in tissues where it inhibits *plasminogen activator. Plasminogen activator inhibitor-2 (PAI-2) is only secreted by the placenta, presumably to regulate trophoblast invasiveness. *See* PROTEASE NEXIN-1.

Plasmodium The parasitic protozoan responsible for malaria. *Plasmodium vivax* causes the tertian type, *P. malariae* the quartan type, and *P. falciparum* the quotidian or irregular type of disease. In humans the predominant form is the *merozoite which undergoes a form of multiple cell division (*schizogony) within erythrocytes, causing them to burst and release parasites which then infect other erythrocytes. Eventually some cells develop into gametes that, when taken up by a female *Anopheles* mosquito, will fuse in her gut to form a zygote (ookinete). Multiple cell divisions within the resultant oocyte produce infective sporozoites which migrate to the salivary glands and are transferred to a new vertebrate host when the mosquito takes its next blood meal. Insecticide-resistant mosquitoes and drug-resistant parasites are making malaria an increasing problem.

plastins A family of actin-bundling and crosslinking proteins differentially expressed in normal and malignant cell. **I-plastin** (intestinal-plastin; plastin-1; fimbrin; 68 kDa) is also found in intestinal microvilli. Fimbrin was originally identified in the microvilli of avian epithelial brush border and the plastins are the human homologues. **L-plastin** (lymphocyte cytosolic protein-1; plastin-2; p65) is expressed in leucocytes, transformed fibroblasts, and various tumour cell lines. **T-plastin** (plastin 3), which is similar to L-plastin, is expressed in normal cells of solid tissues and in transformed fibroblasts.

plate count The number of bacterial colonies that grow on a nutrient agar plate under defined conditions. A standard method to estimate bacterial contamination.

platelet An anucleate discoid blood cell (3 μm diameter, 150–400 × 10^6/mL) important for haemostasis. Platelet α-granules contain lysosomal enzymes; dense granules contain ADP (a potent platelet aggregating factor) and *serotonin. They also release *platelet derived growth factor.

platelet activating factor (PAF, PAFacether) A phospholipid (1-*O*-hexadecyl-2-acetyl-sn-glycero-3-phosphorylcholine) that mediates diverse pathologic processes including allergy, asthma, septic shock, arterial thrombosis, and inflammation, not simply a platelet activator. The PAF receptor is a rhodopsin-type GTP-binding protein that produces inositol 1,4,5-trisphosphate in response to ligand binding.

platelet basic protein A protein (PBP, 94 aa), found in the α granules of platelets, that is proteolytically derived from a pro-PBP (128 aa) and further processed to yield various cytokines, including connective tissue activating peptide-III (85 aa; CTAP-III), *β-thromboglobulin (81 aa), and neutrophil activating peptide-2 (NAP-2; 70 aa).

platelet-derived endothelial cell growth factor (PD-ECGF) *See* THYMIDINE PHOSPHORYLASE.

platelet-derived growth factor *See* PDGF.

platelet factor 3 Particles of 70–170 nm diameter that are derived from platelets, although probably not platelet membranes, being deficient in carbohydrates. They are reportedly 40% protein, 42% phospholipids, 13% cholesterol, and 5% triacylglycerols and may function as a surface for attachment of other coagulation factors leading to activation of *prothrombin, rather than having a catalytic role.

platelet factor 4 A *cytokine (CXCL4), released from the α-granules of activated platelets during platelet aggregation, that promotes blood coagulation by moderating the effects of heparin-like molecules.

platyspondylic lethal skeletal dysplasias A heterogeneous group of *chondrodysplasias characterized by severe *platyspondyly and limb shortening. The Torrance type is caused by mutation in the gene for *collagen type II (*COL2A1*) and is usually lethal in the neonatal period.

platyspondyly Describing a condition in which the vertebrae have a flattened shape with reduced distance beween the endplates. It may be restricted to one vertebra (vertebra plana) or be generalized. The condition occurs in *Morquio-Brailsford disease, *spondyloepiphyseal dysplasia tarda and *Kniest's syndrome.

PLD **1.** *See* PHOSPHOLIPASE D. **2.** Pegylated liposomal *doxorubicin. **3.** PLD-118 is an antifungal drug, a synthetic derivative of the β-amino acid cispentacin. **4.** The Protein Ligand Database (PLD) is a web-based resource devoted to protein–ligand interactions (13138 protein–ligand structures in 2008).

SEE WEB LINKS

- The protein–ligand structure database homepage.

pleckstrin A protein (platelet, and leucocyte C kinase substrate, pleckstrin-1, 350 aa) that is phosphorylated by protein kinase C in platelets stimulated by agonists that cause phosphoinositide turnover. **Pleckstrin-2** (353 aa) is similar but is not phosphorylated to the same extent and is bound to the cell membrane where its PH domains appear to contribute to the formation of lamellipodia. Pleckstrin homology domains (*PH domains) are found in a number of proteins.

plectin A very large (4684 aa) intermediate filament-binding protein that may provide mechanical strength to cells and tissues by acting as a crosslinking element between the intermediate and microfilament cytoskeleton. The N-terminal region has homology to the actin-binding domain of *dystrophins and the C-terminal region has homology to *desmoplakin. Mutations are responsible for *epidermolysis bullosa simplex with various additional complications depending upon the particular mutation.

pleiotrophin *See* MIDKINE.

pleiotropic Describing a substance or a gene that has multiple effects, which may differ according to the location. A standard example is cAMP which inhibits secretion in some cells and enhances it in others. Many genes have pleiotropic effects when mutated.

plesiomorphic *See* APOMORPHIC.

pleurodynia A disease (Bornholm disease, epidemic pleurodynia, epidemic myalgia) characterized by attacks of severe pain in the lower chest, often on one side. Caused by infection with Coxsackie B viruses.

pleuropneumonia-like organism (PPLO) *See* MYCOPLASMA.

plexin domain-containing proteins Proteins found in endothelial cells in tumours but not normal tissue and that may be involved in tumour angiogenesis. PLXDC1 (tumour endothelial marker 3 and 7, 500 aa), PXDC2 (529 aa).

plexins Transmembrane receptors for the *semaphorins with homology to the MET-hepatocyte growth factor receptor tyrosine kinase family. At least nine members of the family are known although the specific roles of each are poorly defined. Nomenclature is idiosyncratic (plexin 4 is SEX, plexin 5 is SEP, plexin 2 is OCT, and plexin 1 is NOV) but will probably be rationalized. Plexin domain-containing protein 2 (tumour endothelial marker 7-related protein, 529 aa) may be involved in tumour angiogenesis.

PLGF A growth factor (placental growth factor, 149 aa) similar in activity to *VEGF and involved in the regulation of angiogenesis in various tissues.

P light chain (DNTB light chain) *See* MYOSIN LIGHT CHAINS.

P loop *See* ATP BINDING SITE.

pluripotent stem cell *See* STEM CELL.

pluronic block copolymers Amphiphilic polymers of poly(ethylene oxide)-poly(propylene oxide)-poly(ethylene oxide) that can form micelles and are useful as drug carriers. They are often identified as P84, P85, etc., referring to their size and composition.

plusbacins Depsipeptide antibiotics, similar to *katanosins, isolated from *Pseudomonas* spp. and effective against vancomycin-resistant enterococci. Probably inhibit cell wall synthesis.

PM1 **1.** Airborne particulate matter (PM) with a diameter less than 1 μm. PM5, PM10, etc. are particles with correspondingly larger diameters. **2.** A transformed $CD4^+$ T-cell clone derived from the Hut78 T-cell line. **3.** A monoclonal antibody (precursor marker 1: PM1) that labels most neuroepithelial cells in day 4 embryonic chick retinal sections.

PMA Phorbol myristate acetate, a *phorbol ester, confusingly called tumour promotor activity (TPA) in the early literature.

PMAT Plasma membrane monoamine transporter. *See* EQUILIBRATIVE NUCLEOSIDE TRANSPORTERS.

PMF *See* PROTON MOTIVE FORCE.

PML body A nuclear structure (promyelocytic leukaemia body, Kremer body) containing multimers of promyelocytic leukaemia (PML) protein and a range of other nucleoproteins including the *Nijmegen breakage disease syndrome protein (p95/Nbs1, *nibrin) which assists in repair of double-strand breaks in DNA. **PML protein** is a *RING finger motif protein that acts as a tumour suppressor and is implicated in the pathogenesis of a variety of tumours.

PMN (PMNL) Polymorphonuclear leucocyte. In casual usage usually a neutrophil *granulocyte.

PMSF *See* PHENYLMETHYLSULPHONYL FLUORIDE.

PNA *See* PEPTIDE NUCLEIC ACID.

pneumococci *See* STREPTOCOCCUS PNEUMONIAE.

pneumoconiosis A group of lung diseases caused by occupational exposure to dust. Includes variants such as silicosis, asbestosis, and anthrocosis.

Pneumocystis jiroveci **(formerly *Pneumocystis carinii*)** A yeast-like fungus, often responsible for opportunistic pneumonia in immunocompromised patients.

pneumocyte The cell type that forms the lining of lung alveoli. Type I pneumocytes are squamous whereas type II are smaller, roughly cuboidal cells, usually found at the alveolar septal junctions. Type II pneumocytes secrete *surfactant and can divide and differentiate to replace damaged type I pneumocytes.

pneumolysin A *cholesterol binding toxin from *Streptococcus pneumoniae*.

pneumonia An inflammation of the lungs that can be a result of infection with bacteria (often **Streptococcus pneumoniae* or **Klebsiella pneumoniae*), although other bacteria can be responsible. Viral infections can also cause pneumonia, notably a coronavirus in *SARS. In **lobar pneumonia** only some lobes of the lung are affected; in **bronchial pneumonia** the inflammation affects the bronchi or bronchioles.

PNMT The enzyme (phenylethanolamine *N*-methyl transferase, EC 2.1.1.28, 282 aa) in the catecholamine biosynthetic pathway that converts noradrenaline to adrenaline in the adrenal medulla.

PNPLA6 An esterase, one of a family of patatin domain-containing proteins, a glycoprotein (patatin-like phospholipase domain-containing protein-6, neuropathy target esterase, 1366 aa) that hydrolyses lysophosphatidylcholine and is involved in neuronal development. It is the target for neurodegeneration induced by some organophosphorus pesticides (organophosphate-induced delayed neurotoxicity (OPIDN)). Mutations lead to autosomal recessive *hereditary spastic paraplegia type 39 (SPG39).

podocalyxin A heavily glycosylated single-pass transmembrane protein (528 aa) mainly found on the apical membrane of renal glomerular epithelial cells (*podocytes) but also on vascular endothelium and some tumour cells. It is linked to *ezrin and the actin cytoskeleton through the scaffold protein, NHERF2 (sodium/hydrogen exchanger regulatory factor).

podocytes Visceral epithelial cells that form part of the cells of the glomerular filtration barrier in the kidney. The cell body is separated from the *basal lamina by club-shaped protrusions (pedicels) that form a meshwork with the pedicels of adjacent cells, the mesh being surrounded by extracellular matrix that filters out large macromolecules. The matrix is associated with *nephrin, *podocalyxin, and *P-cadherin.

podophyllotoxin A toxic glucoside (414 Da), isolated from the roots of the American mayapple *Podophyllum peltatum*. It blocks microtubule assembly by binding to tubulin. It also binds to the E2 protein of human papillomavirus (HPV) type 1a, which causes plantar warts, and this may be the basis of its effectiveness in treating warts. *Etoposide is a derivative of podophyllotoxin; other analogues retain the capacity to block tubulin polymerization, still others are topoisomerase II inhibitors.

podoplanin A type I membrane glycoprotein (T1A, aggrus, OTS8, GP36, 162 aa) that is a marker for normal lymphatic endothelial cells, although expression is increased in many intestinal tumours. Can be detected immunochemically by the monoclonal antibody D2-40 and is a potential tumour marker. One domain has platelet-aggregating properties and is involved in tumour cell-induced platelet aggregation.

podosomes Specialized adhesion complexes of osteoclasts that form a broad ring of contacts with the underlying bone; the encircled area below the cell is then absorbed. Contain *vinculin, *talin, *fimbrin, and *F-actin.

poikilocytosis A condition in which erythrocytes are of irregular shape.

point mutation A *mutation in which a single base pair is altered.

pokeweed mitogen Any of the *lectins from *Phytolacca americana* (pokeweed), all of which are mitogenic for T cells. They bind β-D-acetylglucosamine.

polar body The small cell derived from the meiotic division that produces the (large) oocyte.

polarity Describing anything that has distinct poles, e.g. a planet, magnet, or battery, but in the context of cells, having defined apical and basolateral regions, as in epithelia that are attached to a basal lamina, or having a distinct front and rear as in the case (temporarily at least) of a moving cell. The underlying molecular basis is not always clear.

pole fibres Microtubules that originate from the *microtubule organizing centre around the *centrioles that lie at the poles of the mitotic spindle.

***pol* genes** The genes encoding *DNA polymerases. In *E. coli* there are three, *polA, polB,* and *polC,* coding respectively for polymerases I, II, and III. Retroviral *pol* genes encode *reverse transcriptase.

poliomyelitis An acute inflammation of the central nervous system, caused by infection with poliovirus, that can lead to paralysis (infantile paralysis). A global vaccination programme has almost eradicated the disease.

poliovirus *See* PICORNAVIRIDAE and POLIOMYELITIS.

polo-like kinases A conserved family of serine/threonine kinases (Plks) that have a C-terminal Polo-box domain (PBD) and an N-terminal kinase domain. There are four human Plks, (Plk1 (603 aa); Plk2 (685 aa), serum-inducible kinase, Snk; Plk3 (646 aa), cytokine-inducible kinase, proliferation-related kinase, Prk; Plk4 (970 aa), homologue of mouse Sak, although some species have fewer, as is the case for polo in *Drosophila,* mutations in which cause aberrant spindle pole behaviour. Plks are important in centriole duplication. *See* PROTEIN KINASE N.

poly-A binding proteins Proteins (PABPs) that are complexed to the 3′ poly-A tail of mRNA and are required for poly-A shortening and initiation of translation through an interaction with translation initiation factor EIF4G1. PABP1 is involved in nuclear events associated with the formation and transport of mRNP to the cytoplasm. There are **PABP-interacting proteins** (PAIP1, 480 aa) that are also involved in initiating translation. PAIP2 (127 aa) competes with PAIP1 for binding to PABP.

polyacrylamide gel electrophoresis (PAGE) A common form of *electrophoresis in which the gel is of polyacrylamide, sometimes with a concentration gradient to assist the resolution of larger molecules. Usually combined with sodium dodecylsulphate (SDS) solubilization of proteins (SDS-PAGE), a process that separates noncovalently linked subunits and adds negative charge approximately in proportion to polypeptide chain length.

poly(ADP-ribose) polymerase A chromatin-associated enzyme (PARP, EC 2.4.2.30, 1013 aa) that catalyses the covalent transfer of 60–80 ADP-ribose units to a variety of nuclear proteins. It is activated by DNA nicks and is important in DNA repair. Other members of the PARP family may regulate gene transcription by altering chromatin organization as a result of histone polyadenylation.

poly-A polymerase The enzyme (PAP, PAPOL, EC 2.7.7.19) that adds the poly-A tail to mRNA. There are two isoforms, PAPOLA (745 aa) and PAPOLB (636 aa).

polyarteritis nodosa A vasculitis of medium-sized arteries, without glomerulonephritis or vasculitis elsewhere. It is an autoimmune disease, although the autoantigen is unknown; in some cases antineutrophil cytoplasmic antibodies (*ANCA) are present. *Compare* MICROSCOPIC POLYANGIITIS.

poly-A tail A polyadenylic acid sequence of variable length (60–200 residues) added post-transcriptionally to the 3′ end of most eukaryotic mRNAs to produce *hnRNA, although there are some exceptions (e.g. histone mRNAs). The poly-A tail on mRNA is gradually reduced in length in the cytoplasm. Affinity chromatography with oligo(U) or oligo(dT) is a standard technique for purifying mRNA.

polycistronic mRNA A single *mRNA molecule transcribed from several tandemly arranged genes, typically those in an *operon. Almost all eukaryotic mRNA is monocistronic.

polyclonal antibody An antibody produced by several clones of B lymphocytes and potentially directed at multiple epitopes. Antibody produced by immunization of an animal will be polyclonal. *Compare* MONOCLONAL.

polycloning site An engineered region (multiple cloning site, MCS) in a phage or plasmid vector with a series of unique *restriction sites making it easy to insert or excise DNA fragments for subcloning.

polycomb repressive complexes Multiprotein complexes involving products of the polycomb group (*PcG*) genes which were initially identified as regulators of homeotic genes in *Drosophila* and are implicated in regulation of stem cell self-renewal and in cancer development. The PcG complexes are mostly associated with heterochromatin, where they inhibit gene expression by histone modification. There are numerous human polycomb group homologues and at least two distinct sets of complexes, PRC1 and PRC2. *See* AEBP2; EED; EZH; PCGF; PHC; RETINOBLASTOMA-BINDING PROTEIN-4; SUZ12.

polycyclic aromatic hydrocarbons (PAH) Compounds that consist of fused aromatic rings and do not contain heteroatoms or carry substituents. Many are carcinogenic, mutagenic, and teratogenic.

polycystic kidney disease Adult polycystic kidney disease is an autosomal dominant disorder in which there are renal and hepatic cysts and intracranial aneurysm, although there is considerable variability. Mutations in several genes can give rise to the disorder including those for *polycystins 1 and 2. An infantile form of the disease is an autosomal recessive disorder caused by mutation in the gene encoding *fibrocystin.

polycystic ovary syndrome A metabolic syndrome (Stein–Leventhal syndrome) that is one of the most common causes of female infertility. It is characterized by enlargement of the ovaries with multiple ovarian cysts, elevated levels of androstenedione and/or testosterone and of luteinizing hormone, with reduced levels of oestradiol and follicle-stimulating hormone. There is clearly a genetic component but this is not well defined. Complications include insulin resistance, hyperinsulinaemia, and often hyperlipidaemia and obesity.

polycystins Large transmembrane proteins (polycystin 1, PC1, 4303 aa; polycystin-2, 968 aa) with multiple splice variants that regulate calcium channels, are involved in cell-matrix interactions and may modulate G-protein-coupled signal transduction pathways. Polycystins are involved in mechanosensing, receiving signals from the primary cilia, neighbouring cells, and extracellular matrix and transducing them into cellular responses that regulate proliferation, adhesion, differentiation of renal tubules, and kidney morphogenesis. Mutations in polycystin 1 are responsible for most cases of autosomal dominant *polycystic kidney disease (ADPKD).

polycythaemia (erythrocytosis) An increase in the haemoglobin content of the blood, which may arise through a reduction of plasma volume or an increase in erythrocyte numbers. Erythrocytosis is often used as a synonym for polycythaemia but implies more strongly that the cells are fully differentiated. **Polycythaemia vera** (Vaquez–Osler disease), in which there is increased production of red cells, is usually caused by a somatic mutation in the **JAK2* gene in a single haematopoietic stem cell. **Familial erythrocytosis-1** is caused by mutations in the *erythropoietin receptor. **Familial erythrocytosis-2** is caused by mutation in the *VHL tumour suppressor gene.

polyductin *See* FIBROCYSTIN.

polyendocrine syndrome An autoimmune disorder (autoimmune polyendocrinopathy syndrome) in which there is *Addison's disease and/or *hypoparathyroidism and/or chronic mucocutaneous *candidiasis. **Type I** is caused by mutation in the autoimmune regulator gene that encodes a protein (545 aa) that is probably a transcription factor. In **type 2** (Schmidt's syndrome), patients have Addison's disease with autoimmune thyroid disease and/or type 1 diabetes, but do not have hypoparathyroidism or candidiasis. In **type 3**, patients have autoimmune thyroid disease and other autoimmune disorders but do not have Addison's disease. In **IPEX** (X-linked immunodysregulation, polyendocrinopathy, and enteropathy) there is a mutation in the **foxP3* gene.

polyene antibiotics A group of structurally related antibiotics that interact with sterols in eukaryotic membranes. Examples include *amphotericin B and *nystatin.

polyethylene glycol *See* PEG.

polygenic Describing something controlled or caused by many genes rather than by mutation in a single gene. This may be the case for many of the major noninfectious diseases, although mutations in individual genes may contribute to susceptibility.

polyglutamine disorders *See* POLY-Q DISEASES.

polyimmunoglobulin receptor A member of the immunoglobulin superfamily (764 aa) that is expressed on the basolateral surface of various glandular epithelia where it binds polymeric IgA and IgM; the complex is transported across the cell and secreted at the apical surface. A portion of the molecule is proteolytically cleaved and released as the *secretory piece of IgA.

polyisoprenylation *See* GERANYL; FARNESYLATION.

polylysine A polymer of 20–30 lysine residues that carries multiple positive charges (is polycationic) and can be used to enhance the (nonphysiological) adhesion of cells (which have negative charge on their surfaces) to acellular substrata.

polymer A macromolecule formed from repeating units (protomers) by the process of polymerization. In some cases nucleation may be required to start the process and if the polymerization is reversible a critical protomer concentration is necessary. Many biological polymers (e.g. microfilaments, microtubules) exhibit different kinetics for binding and detachment at the two ends, so that assembly may occur primarily at one end, disassembly at the other. In a dynamic system, making the protomer assembly-incompetent will cause gradual disassembly of the polymer (as when colchicine is added to cells, binds tubulin, and blocks assembly).

polymerase chain reaction (PCR) A method for *in vitro* amplification of DNA in which oligonucleotide primers complementary to two regions of the target DNA (one for each strand) to be amplified, are added to the sample containing the target DNA, together with excess deoxynucleotides and heat-stable *Taq polymerase. The mixture is heated to ~90 °C to denature the target DNA, primers are allowed to anneal (at ~50–60 °C), and a daughter strand is extended from the primers at ~72 °C). The daughter strands then act as templates for subsequent temperature cycles, giving an exponential amplification only of the target DNA, so that ~30 cycles will produce sufficient DNA to be analysed. The original DNA need not be pure or abundant, and the PCR reaction is a powerful tool in research, clinical diagnostics, and forensic science. Many modifications of the original method have been made, extending the capacity.

polymorphic epithelial mucin *See* EPISIALIN.

polymorphism The occurrence of two or more alleles of a gene in a population with the frequency of the rarer alleles being larger than could plausibly be explained by recurrent mutation (generally >1%). Balanced polymorphisms are often maintained because the heterozygote has a selective advantage. The *MHC has a very high level of polymorphism, generating a desirable diversity.

polymorphonuclear leucocyte (PMNL, PMN) A blood leucocyte of the myeloid series (granulocytes) with a nucleus of irregular form, not a simple circle or oval in plan view. *See* BASOPHIL; EOSINOPHIL; NEUTROPHIL. *Compare* MONONUCLEAR CELLS.

polymyxins Cyclic peptide antibiotics (1–2 kDa) produced by *Bacillus* spp. Effective against many Gram-negative bacteria, working apparently by increasing membrane permeability. Polymyxin E is *colistin.

polynucleotide A linear polymer composed of *nucleotides, in which the 5′-linked phosphate on one sugar group is linked to the 3′ position on the adjacent sugars (the sugar is *deoxyribose in DNA and *ribose in RNA). They may be double-stranded or single-stranded and there may be intrastrand hybridization of complementary sequences, producing loops and hairpins.

Polyomaviridae A family of the *Papovaviridae, DNA tumour viruses with a small circular genome. Polyomavirus, the only genus within the family, was isolated from mice; infection of adults is asymptomatic but experimental injection of high titres of the virus into neonatal mice causes histologically diverse tumours (hence poly-oma). Mouse cells are permissive for virus replication *in vitro*, and are killed, but hamster cells undergo *abortive infection, and may become *transformed.

polyp **1.** A growth, usually benign, that projects from a mucous membrane and resembles (vaguely) (2). **2.** The sessile stage of the life cycle of some coelenterates (e.g. *Hydra*), a cylindrical body attached to the substratum with an apical mouth surrounded by tentacles bearing *nematocysts.

polypeptide Polymers composed of α-*amino acids linked by peptide bonds. Polypeptides conventionally have more than ten residues, below which they are *oligopeptides.

Decapeptides are often described, unadecapeptides very rarely.

polypeptide antibiotics A class of antibiotics that are polypeptides and that have activity against Gram-negative bacilli, disrupting the membrane. Examples include *bacitracin, *colistin, and *polymyxins.

polyphemusin II An antagonist of the CXCR4 chemokine receptor, isolated from the American horseshoe crab *Limulus polyphemus*.

polyploid Describing a cell or organism that has more than two *haploid sets of *chromosomes. Triploid cells have three sets, tetraploid cells have four, etc.

polyposis coli An autosomal dominant disorder (adenomatous polyposis coli, familial adenomatous polyposis, FAP) in which there is a predisposition to cancer. Affected individuals develop hundreds to thousands of adenomatous *polyps of the colon and rectum which may progress to malignancy. The mutated gene (*APC* gene) encodes a multidomain tumour suppressor that antagonizes the *wnt signalling pathway and is implicated in other carcinomas. Gardner's syndrome is a variant of FAP, caused by mutation in the same gene, in which a variety of other tumours develop. *Compare* HEREDITARY NONPOLYPOSIS COLON CANCER (*see* GTBP).

polyprotein A protein that is post-translationally cleaved to produce functionally distinct polypeptides. Some polypeptide hormones are generated from a single precursor polyprotein (e.g. *pro-opiomelanocortin).

polypyrimidine tract binding protein 1 A nuclear protein (PTBP1, heterogeneous nuclear ribonucleoprotein type I, hnRNP I, 531 aa) that binds pre-mRNAs and is considered to have a role in pre-mRNA splicing. PTBP2 is neuron-specific and the level of PTBP2 is controlled by a *miRNA (microRNA124A1) that targets PTBP1 and leads to its degradation; this switches the cell to a neuron-specific pattern of alternative splicing. *IMP1 competes with PTBP1 binding to IGF2 leader 3 mRNA in a Mg^{2+}-dependent manner.

poly-Q diseases A miscellaneous set of disorders (polyglutamine disorders), including *Huntington's disease, caused by expansion of polyglutamine (poly-Q)-encoding repeats in unrelated genes. The mutant protein accumulates as insoluble aggregates and causes cell death.

polysaccharide A polymer composed of more than about ten monosaccharide residues linked by glycosidic bonds. Polysaccharides may be branched or unbranched.

polysialic acid A polysaccharide in which there are multiple *N*-acetylneuraminic acid (sialic acid) subunits. Such polymers are polyanionic and the effect of the negative charge may be to inhibit adhesion through electrostatic repulsion.

polysome (polyribosome) Several *ribosomes attached to a mRNA molecule.

polysomy Describing the condition in which all chromosomes are present but some at more than the diploid number, e.g. trisomy 21 (*Down's syndrome).

polyspermy Condition that is not uncommon in large, yolky eggs, in which several spermatozoa penetrate the ovum although only one male pronucleus fuses with the egg nucleus. Various mechanisms usually block polyspermy in mammals.

polytene chromosomes Giant *chromosomes consisting of many (up to 1000) identical chromosomes (strictly, chromatids) running parallel and in strict register (synapsed). They are sufficiently large to remain obvious during interphase, and distinct bands, each roughly equivalent to one genetic locus, can be identified. The banding pattern in the polytene chromosomes of *Drosophila* salivary gland were important in relating chromosomal morphology to genetic linkage maps. Bands where extensive transcription is occurring appear as *puffs.

polytropic virus *See* XENOTROPIC VIRUS.

polyunsaturated fatty acid *See* PUFA.

polyvinylpyrrolidone (PVP) An inert, water-soluble polymer occasionally used to produce viscous solutions for gradient centrifugation.

POMC *See* PRO-OPIOMELANOCORTIN.

POMC/CART neurons *See* arcuate nucleus at ARCUATE.

Pompe's disease A *glycogen storage disease (type II) caused by deficiency in lysosomal α(1–4)-glucosidase; glycogen accumulates in lysosomes even though the nonlysosomal glycogenolytic system is normal.

Pontiac fever A mild influenza-like illness caused by *Legionella*.

popliteal pterygium syndrome A disorder caused by mutation in the **IRF6* gene that exhibits a similar orofacial phenotype to *van der Woude's syndrome but includes skin and genital anomalies.

population diffusion coefficient A coefficient that describes the diffusion of motile elements (cells or organisms) through the environment. It is based on the assumption that the movement is random (an unbiased random walk).

porencephaly Describing any cavitation or cerebrospinal fluid-filled cyst in the brain. **Type 1** (encephaloclastic porencephaly) is usually unilateral and results from focal destructive lesions, **type 2** (schizencephalic porencephaly) is usually symmetric and represents a primary developmental defect. *See* BRAIN SMALL VESSEL DISEASE WITH HAEMORRHAGE.

porins Transmembrane proteins (outer membrane channel proteins, 200–500 aa) of the outer membranes of Gram-positive bacteria that interact as trimers to form channels (1 nm pore size) through which small (<600 Da) hydrophilic molecules can pass. Similar porins occur in outer membranes of mitochondria and form the voltage-dependent anion-selective channel (VDAC, 283 aa), the major pathway for movement of adenine nucleotides through the outer membrane. Two isoforms of VDAC have been identified (VDAC1 and VDAC2).

porphobilinogen deaminase *See* UROPORPHYRINOGEN I SYNTHETASE.

porphyria A range of disorders characterized by excessive excretion of porphyrins or porphyrin precursors, usually because of deficiencies in enzymes involved in haem biosynthesis. The deposition of porphyrins in the skin can cause photosensitive dermatitis. **Acute intermittent porphyria** is caused by mutation in the gene for *uroporphyrinogen I synthetase. **Coproporphyria** is caused by mutation in coproporphyrinogen oxidase (EC 1.3.3.3). **Erythropoietic porphyria** is caused by mutation in the gene encoding ferrochelatase (EC 4.99.1.1). **Gunther's disease** (congenital erythropoietic porphyria) is caused by mutation in the uroporphyrinogen III synthase gene (EC 4.2.1.75). **Porphyria cutanea tarda** is due to deficiency of uroporphyrinogen III cosynthase (EC 4.1.1.37). **Porphyria variagata** is caused by mutation in the gene for protoporphyrinogen oxidase (EC 1.3.3.4).

porphyrins Heterocyclic pigments that form chelates with metals (iron, magnesium, cobalt, zinc, copper, nickel) and are found in haemoglobin, chlorophyll, and cytochromes.

***Porphyromonas gingivalis* collagenase** An unusual peptidase (proteinase C, prtC gene product), one of the U-family, that is produced by *Porphyromonas* (formerly *Bacteroides*) *gingivalis*, a Gram-negative anaerobe associated with periodontal lesions. It is inhibited by EDTA and thiol blocking agents.

POSH A RING finger protein (plenty of SH3s, 888 aa) with four SH3 domains that acts as a scaffold for the Jun N-terminal kinases (*JNK). It is an important component of the *Imd pathway of immunity in *Drosophila* and pathways leading to apoptosis in mammals.

positional cloning A standard approach to identifying a gene in which a *linkage analysis provides an indication of the nearest-neighbour genes and then a *chromosome walk is undertaken from this known sequence.

positional information In development, the information that allows cells to differentiate appropriately for their position within the body, e.g. ensuring that limb bones are formed in the anatomically correct position.

position effect An effect on gene expression that depends on its chromosomal location. Genes translocated to transcriptionally active regions can have their expression increased, as happens in the *Philadelphia chromosome abnormality.

positive control A transcriptional control mechanism in which a regulatory protein interacts with the promoter region of the gene to cause transcription, as opposed to a mechanism in which a repressor is removed.

positive feedback *See* FEEDBACK REGULATION.

positive strand RNA viruses Viruses of classes IV and VI that have a single-stranded RNA *genome that can act as mRNA and is itself infectious. Examples include *Picornaviridae, *Togaviridae, and *Retroviridae.

postcapillary venule The vasculature immediately downstream of the capillary network which has the lowest wall shear stress, and is therefore the main site for adhesion (margination) of leucocytes and their transendothelial migration (diapedesis).

postsynaptic cell The cell that receives a signal from the presynaptic cell and responds with depolarization. In a chemical *synapse the presynaptic cell releases neurotransmitter molecules that bind to receptors on the postsynaptic cell. In an electrical synapse it is the cell that responds and, unless it is a *rectifying synapse (as most are), either cell could be postsynaptic according to the direction of information flow.

postsynaptic potential The change in the *resting potential of the membrane of a postsynaptic cell following stimulation by the presynaptic cell. In a chemical synapse the binding of neurotransmitter causes a channel to open allowing ions to flow, causing depolarization. The release of a single synaptic vesicle (quantal release) elicits a small depolarization (~0.5 mV), a **miniature end plate potential** (mepp). Multiple mepps additively generate an **excitatory postsynaptic potential** (epsp). An **inhibitory postsynaptic potential** (ipsp) is a hyperpolarization elicited by inhibitory neurotransmitter receptors such as *GABA receptors and *glycine receptors, which reduces the probability of generating an action potential and reduces the firing rate of the neuron.

postsynaptic protein A highly conserved peripheral membrane protein (rapsyn, 412 aa) that anchors *nicotinic acetylcholine receptors in the postsynaptic membrane. Mutations in rapsyn can be one cause of congenital myasthenic syndrome. **Postsynaptic density proteins** (PSD proteins) are a family of proteins involved in the clustering of receptors on the postsynaptic cell. The PSD-95 protein (synapse-associated protein 90, 723 aa) is a membrane-associated guanylate kinase (*MAGUK) responsible for the clustering of NMDA receptors and potassium channels; a similar protein is responsible for the formation of synaptic complexes. *See* AKAP FAMILY; NEURABIN; NEUROLIGINS.

post-transcriptional gene silencing A mechanism of gene inactivation that operates after transcription has already taken place by destruction of the mRNA. *See* RNA INTERFERENCE.

post-translational modification Alterations made to proteins after the primary sequence of peptide-bonded amino acids has been assembled. In some cases portions of the polypeptide are removed (e.g. signal sequences) or additional side groups are added permanently or temporarily (*acetylation, *glycosylation, *glypiation, phosphorylation).

POT1 A protein of the *shelterin complex (protection of telomeres 1, 634 aa but multiple isoforms) that binds to the G-rich strand of human telomeric DNA.

potassium-aggravated myotonia *See* MYOTONIA.

potassium channels Ion channels that are selective for potassium ions. They are tetrameric, with hybrid combinations of subunits possible in some cases. More than 80 genes encode components of these channels and even more are involved in regulating their activity. There are four main classes: 1. Calcium-activated channels (*BK and *SK subtypes). 2. Inwardly-rectifying channels (*ROMK (Kir1.1), G-protein-coupled receptor-regulated (*GIRKs, Kir3.x), and ATP-sensitive (Kir6.x, associated with *sulphonyl urea receptors). 3. Tandem-pore domain channels (TWI, TRAAK, *TREK and TASK). 4. Voltage-gated channels (*hERG, (Kv11.1); KvLQT1 (Kv7.1)). *See* A-TYPE CHANNELS; DELAYED RECTIFIER CHANNELS; DLGS; MINK; SHAKER; WEAVER. Disorders of potassium channels are associated with *Andersen's syndrome, *Bartter's syndrome, *episodic ataxia-1, *hyperinsulinism, *Jervell and Lange-Nielsen syndrome, *long QT syndrome, *nesidioblastosis, *neuronal ceroid lipofuscinosis-6, *short QT syndrome, and *spinocerebellar ataxia-13. Toxins that act on potassium channels include *agitoxin, *apamin, *charybdotoxin, *iberiotoxin, *margaratoxin, *noxiustoxin, *scyllatoxin, *SGTx1, and *ShK toxin.

potassium-sparing diuretics *See* DIURETICS.

potency The relationship between the therapeutic effect of a drug and the dose required to achieve that effect or, in toxicology, the relative toxicity of an agent as compared to some reference standard or compound.

potentiation **1.** The phenomenon whereby exposure to a substance or agent, at a concentration or dose that does not elicit an effect, enhances the response to subsequent exposure to the same or different substance or agent. Sometimes referred to as priming. **2.** The increase in neurotransmitter release at a synapse after repetitive stimulation. Release is quantal so the effect is due to a greater number of vesicle fusion events. Potentiation may last for minutes or hours; *compare* FACILITATION which operates for only a few hundred milliseconds.

potocytosis The transmembrane movement of small molecules in *caveolae rather than *coated vesicles.

Potter's syndrome A condition (renal adysplasia) in which kidney development is abnormal or, in severe cases, absent. May be caused by mutation in the **ret* protooncogene or in the uroplakin IIIA gene.

POU domain A conserved DNA-binding domain (Pit-Oct-Unc domain, ~150 aa) with a 20-aa *homeobox region and a larger POU-specific domain found in various transcription factors. Named after three factors with the domain: Pit-1 which is pituitary-specific; Oct-1 and 2, which bind an octamer sequence; and Unc-86, a neural transcription factor from *Caenorhabditis elegans*.

Poxviridae Class I viruses with a double-stranded DNA genome that encodes more than 30 polypeptides. They are the largest viruses and have a complex multilayered coat composed of lipid and enzymes, including a DNA-dependent RNA polymerase. They multiply in the cytoplasm of the cell. Four genera cause human infections: orthopox, parapox, yatapox, and molluscipox. Important examples are *vaccinia, *variola (smallpox), and *myxoma.

POZ domain A protein-binding domain (poxvirus zinc finger domain) characteristic of a family of transcription factors with an N-terminal POZ domain and C-terminal zinc fingers. They are involved in growth arrest and differentiation in several mesenchymal cells. *See* BTB/POZ DOMAIN.

pp- Prefix, usually to a number, identifying a phosphoprotein. The number is generally the molecular mass in kDa.

PP1 *See* PROTEIN PHOSPHATASES (type 1).

pp32/PHAP-I A potent and specific inhibitor (putative HLA class II-associated protein 1) of protein phosphatase 2A and a member of the *SET complex. Up-regulated levels increase apoptosis of tumour cells.

pp46 *See* NEUROMODULIN.

pp60src The **src* oncogene product, a phosphoprotein (60 kDa, 526 aa) that is a protein tyrosine kinase. *See* SRC FAMILY.

PPARs A family of nuclear hormone receptors (peroxisome proliferator-activated receptors, PPARα, γ1, γ2, and δ) that are implicated in metabolic disorders predisposing to atherosclerosis and inflammation. PPARα stimulates β-oxidative degradation of fatty acids and is activated by gemfibrozil and other fibrate drugs. PPARγ promotes lipid storage by regulating adipocyte differentiation. **PPAR-binding protein** (PPARBP, 1581 aa) is a transcriptional coactivator for PPAR and for other nuclear receptors. **PPARγ coactivator-1** (PGC, 910 aa) is a transcriptional coactivator involved in various aspects of energy metabolism including regulation of the expression of *uncoupling proteins. Various other coactivators have been identified. The **PPARα-interacting cofactor complex** (PRIC285, 2649 aa) is a nuclear transcriptional coactivator for PPARα and -γ, as well as various other nuclear receptors.

PPD A protein (purified protein derivative), purified from the culture supernatant of *Mycobacterium tuberculosis*, that is used as a test antigen in *Heaf and *Mantoux tests.

PPFIA1 One of the *liprin family of proteins (protein tyrosine phosphatase receptor type f, polypeptide-interacting protein α1, LAR-interacting protein 1, 1202 aa) involved in anchoring the *LAR protein-tyrosine phosphatase to *focal adhesions and regulating their disassembly. Several alternatively spliced variants have been described and the expression of one, PPFIA2, is downregulated by androgens in a prostate cancer cell line.

PPI *See* PROTON PUMP INHIBITORS.

PPIases Enzymes (peptidyl-prolyl *cis-trans* isomerases) that accelerate protein folding by catalysing *cis-trans* isomerization of proline. *Immunophilins are PPIases but PPIase activity is not necessary for immunosuppression.

PPLO Pleuropneumonia-like organisms. *See* MYCOPLASMA.

Prader–Willi syndrome A syndrome characterized by short stature, obesity, and mild mental retardation; caused by deletion of the paternal copies of the imprinted **SNRPN* gene, the **NECDIN* gene, and possibly other genes within the chromosome region 15q11–q13. In *Angelman's syndrome the equivalent maternal region is deleted. *See* GENOMIC IMPRINTING.

Prausnitz–Kustner reaction An obsolete skin test for IgE levels.

pravastatin A cholesterol-lowering *statin. *TNs* Pravachol, Selektine.

praziquantel An anthelminthic drug used to treat schistosomiasis and infection with liver flukes such as *Clonorchis*.

prazosin An antagonist selective for the α_1-*adrenergic receptors on vascular smooth muscle. Used to treat hypertension. *TNs* Minipress,Vasoflex, Hypovase.

pRb The **retinoblastoma* gene product (928 aa).

PRC **1.** A progesterone receptor isoform, PRc. **2.** Plasma renin concentration. **3.** *See* POLYCOMB REPRESSIVE COMPLEXES.

PRC2/EED-EZH complexes Multiprotein complexes which promote repression of homeotic genes during development. The *polycomb repressive complex-2 (PRC2)/EED-EZH1 complex is a complex with *EED, *EZH, *SUZ12, *retinoblastoma-binding protein-4 (RBBP4), and *AEBP2 and methylates lysine residues on histone H3, leading to transcriptional repression of the affected target gene. The PRC2/EED-EZH2 complex may also associate with *histone deacetylase, HDAC1.

PRD1-BF1 A transcriptional repressor (positive regulatory domain I binding factor 1, B lymphocyte-induced maturation protein 1, BLIMP1, 789 aa) that plays an important role in determining whether immature B cells differentiate into plasma cells rather than memory cells.

preBCR The receptor (preB-cell antigen receptor) on immature B cells that incorporates the immunoglobulin μ heavy chain (Ig μ) and signals through tyrosine kinases including *Blk.

precipitin Obsolete name for an antibody that will form a precipitating complex (a precipitin line) with a multivalent antigen.

prednisolone A *glucocorticoid, very similar to *prednisone, with anti-inflammatory properties.

prednisone A synthetic steroid (*glucocorticoid) that has potent anti-inflammatory and antiallergic effects.

p

pre-eclampsia A condition that can arise during pregnancy in which there is oedema, high blood pressure, and albuminuria. Without treatment can cause *eclampsia.

p region The smaller of the two regions of a chromosome that are separated by the centromere. The larger region is the q region. For example, the gene for insulin is located at 11p15.5 (chromosome 11, p region, subregion 15.5).

pregnanediol A steroid formed by metabolism of *progesterone.

pregnenolone A steroid prohormone, synthesized from cholesterol, in the synthetic pathway leading to progesterone, mineralocorticoids, glucocorticoids, androgens, and oestrogens.

preimplantation genetic diagnosis (PGD) A technique for screening *in vitro* fertilized embryos for congenital defects before implantation in the uterus.

prenylation The post-translational modification of protein by the addition of prenyl (farnesyl, geranyl, or geranylgeranyl) groups, which promotes association with membranes.

pre-pro-protein The full-length product of mRNA that must be processed to generate the mature protein. The pre-protein has a *signal sequence, a pro-protein is inactive (a *zymogen) until an inhibitory sequence is removed by proteolysis. A pre-pro-protein has both sequences still present.

presecretory granules Vesicles (Golgi condensing vacuoles) that are near the maturation face of the Golgi but are not fully mature secretory granules.

presenilins A family of multi-pass transmembrane proteins (467 aa) that are localized to the nuclear envelope and are the catalytic aspartyl peptidase core of γ-*secretase, but also interact with *catenins. In *Drosophila*, presenilins have a role in *notch signalling. Mutations in the gene encoding PS1 are associated with 25% of cases of early-onset *Alzheimer's disease (type 3). **Presenilin enhancer** 2 (PEN2, 101 aa) is another component of the γ-secretase complex. **Presenilin-associated rhomboid-like protein** (PARL, 379 aa) seems to be involved in suppression of apoptosis in lymphocytes and neurons.

prestin A motor protein (solute carrier family 26 member 5, 744 aa) that converts auditory stimuli to length changes in the outer *cochlear hair cells and mediates sound amplification in the ear. Prestin is a bidirectional voltage-to-force converter; changes in the distribution of anions in response to changes in transmembrane voltage trigger conformational changes in the protein, with rapid (microsecond) mechanical consequences.

presynaptic cell *See* POSTSYNAPTIC CELL.

prevalence The proportion of a population that exhibit a disease (are classified as cases) at a point in time, approximately the product of *incidence and the average duration of the disease.

PRH *See* PROLINE-RICH HOMEODOMAIN PROTEIN.

Pribnow box *See* PROMOTER.

prickle Human homologues of the *Drosophila* polarity-determining protein prickle, which regulates cell movement through its association with the *dishevelled (Dsh) protein. The ***Prickle 1*** gene product is a LIM-domain protein (RE-1 silencing transcription factor/NRSF-interacting protein, 831 aa) and is mutated in *myoclonic epilepsy of Unverricht and Lundborg 1B. **Prickle 2** (844 aa) is coexpressed with prickle 1 in brain, eye, and testis, and alone in various other tissues. (NRSF is neuron-restrictive silencer factor.)

prickle cell Large flattened polygonal cells immediately above the basal stem cells of the epidermis. They appear to have fine spines (prickles) projecting from their surfaces but these are, in fact, processes linked to adjacent cells through desmosomes and filled with *cytokeratin tonofilaments.

prilocaine A local anaesthetic, often used in dentistry. *TN* Citanest.

primaquine An 8-aminoquinoline drug used in conjunction with quinine or chloroquine to treat **Plasmodium vivax* or *P. ovale* malaria.

primary cell culture Tissue-derived cells that will be subcultured once they have proliferated.

primary cilia Cilia or ciliary rudiments, that, together with the basal body, act as mechanosensory organelles in a range of tissues: if a propulsive cilium is a motor then, by analogy, this is a generator. Many genetic diseases have been linked to the dysfunction of primary cilia, including *Kartagener's syndrome, *polycystic kidney disease, *nephronophthisis, *Bardet–Biedl syndrome, and *Meckel's syndrome.

primary ciliary dyskinesia A disorder (immotile cilia syndrome) in which ciliary function is abnormal as a result of mutations in various different genes including several for axonemal dynein components and the gene for thioredoxin domain-containing protein-3, which has a role in the axoneme. The consequence can be Kartagener's syndrome and *situs inversus but there is also association with *retinitis pigmentosa.

primary immune response The response that follows the first exposure to a particular antigen, usually slower and less extensive than subsequent secondary responses to the same antigen.

primary lateral sclerosis Juvenile primary lateral sclerosis, a progressive paralytic neurodegenerative disorder, can be caused by mutations in the gene encoding *alsin and is an allelic disorder with juvenile *amyotrophic lateral sclerosis-2 and infantile-onset ascending spastic paralysis. An adult form is not associated with mutations in alsin.

primary lymphoid tissue *See* LYMPHOID TISSUE.

primary lysosome A *lysosome that has not (yet) fused with a vesicle or vacuole.

primary oocyte The diploid ovum before it divides asymmetrically to produce the haploid secondary oocyte and first polar body. The secondary oocyte halts in metaphase of the second meiotic division before fertilization triggers division to produce the ovum and second polar body.

primary pigmented nodular adrenocortical disease A rare bilateral adrenal defect causing ACTH-independent *Cushing's syndrome caused by mutations in PDE11A (dual 3′,5′-cAMP and -cGMP phosphodiesterase 11A). The adrenal glands are of normal size but have multiple small yellow-to-dark brown nodules.

primary spermatocyte The differentiated diploid product of the division of a spermatogonium that will give rise to haploid secondary spermatocytes which will complete the meiotic division process to produce spermatids.

primary structure The lowest level of structural organization in a macromolecule. In proteins, the primary structure is the amino acid sequence, the secondary structure is the folding of the peptide chain determined by interactions between amino acids of the chain (into α-helical coils or β-pleated sheets), the tertiary structure is the way in which the helices or sheets are folded or arranged to give the three-dimensional structure of the protein, and quaternary structure refers to the arrangement of protomers in a multimeric protein. Comparable hierarchical levels of organization are seen in nucleic acids, e.g. the secondary level of cloverleaf structure in tRNA.

primary transcript The unprocessed mRNA transcript. *See* MESSENGER RNA; RNA PROCESSING.

primary tumour The clonal mass of cells at the site where the first proliferating cell arose. *Metastasis leads to secondary tumours at remote sites.

primase *See* DNA PRIMASES.

primer extension A method used to identify the site where transcription of a gene actually begins. A labelled antisense oligonucleotide primer that is complementary to the mRNA near the 5′ end is used to prime a *reverse transcription reaction and, by sequencing the products, the putative start site can be deduced.

primidone An anticonvulsant drug that acts on voltage-gated sodium channels; now rarely used to treat epilepsy. *TN* Mysoline.

priming *See* POTENTIATION.

primitive erythroblast The cells in the early embryo that give rise, by enucleation, to erythrocytes containing fetal haemoglobin. They are derived from the yolk sac region.

primitive streak The thickened elongated region that marks the location of the embryonic axis in avian and mammalian embryos before gastrulation begins. The inward morphogenetic movements of gastrulation cause the anterior end of the streak, Hensen's node, to regress caudally.

SEE WEB LINKS

- Modelling the Formation of the Primitive Streak, including a simulation of the process.

primordial germ cells The embryonic precursors of germ cells that migrate, dividing as they do so, from their extraembryonic site of origin to the gonadal primordia, a process often thought to involve chemotaxis, although it may actually be random movement with trapping.

primosome A multiprotein complex involving *DNA primase, *DNA helicase, and other proteins that moves as a unit with the *replication fork.

prions Proteinaceous infective particles that are generally considered to be the causative agents of various transmissible *spongiform encephalopathies such as *scrapie (in sheep), kuru, *Creutzfeldt–Jakob disease and *Gerstmann–Straussler–Scheinker syndrome in humans. Prions apparently contain no nucleic acid, only prion protein (PrP, 253 aa) which is anchored to the outer surface of neurons and, to a lesser extent, the surfaces of other cells, including lymphocytes. Normal PrP (PrPc) is conformationally altered in the encephalopathies. The exact role of PrP is unclear but the normal form may be neuroprotective and this property may be lost with the conformational change. *See* DOPPEL; SHADOO.

pristane A saturated terpenoid alkane that is extracted from shark liver. It is used experimentally to induce a lupus-like syndrome in nonautoimmune mice.

PRK *See* POLO-LIKE KINASES; PROTEIN KINASE N.

PRL 1. Protein tyrosine phosphatases found in regenerating liver Prl-1, Prl-2, and Prl-3 (PTP4A1, PTP4A2, and PTP4A3, respectively). The gene encoding Prl-3 is induced by DNA-damaging stimuli in a p53-dependent manner and Prl-3 promotes tumour metastasis. Overproduction of Prl-3 in human fibrosarcoma cells initially activates *Akt kinase, but as Prl-3 abundance increases with time, it inhibits Akt activity through a negative-feedback mechanism, leading to cell-cycle arrest. **2.** *See* PROLACTIN.

PRMT5 A subunit (protein arginine *N*-methyl transferase 5) of the 20S methyltransferase complex (the methylosome) that modifies specific arginines to dimethylarginines in several spliceosomal Sm proteins, targeting them to the survival of motor neurons (SMN) complex.

proacrosin The inactive proenzyme of *acrosin.

proadrenomedullin The precursor polypeptide from which two biologically active peptides, *adrenomedullin (AM) and proadrenomedullin N-terminal 20 peptide (PAMP) are produced by proteolytic cleavage. PAMP is hypotensive, angiogenic, and has some antimicrobial capability. The receptor is G-protein coupled.

proanthocyanidins Flavonoid precursors (oligomeric proanthocyanidins, OPCs) of blue and red plant pigments, found in grapes. Their antioxidant properties are not sufficient to justify the red wine consumption that would be required to obtain a benefit.

proband (index case) The first individual in whom a condition, usually hereditable, is diagnosed and from whom the inheritance can be traced.

probe In molecular biology, a shorthand term for a labelled complementary sequence of DNA or RNA that will hybridize with the relevant sequence in a test sample. The label chosen depends on the detection system to be employed (e.g. radioactive or fluorescent labels). *See* BLOTTING.

probenecid A drug that inhibits the reabsorption of urate (and many drugs) by the organic anion transporter in the distal tubule of the kidney. Used to treat gout. *TN* Benuryl.

procainamide An antiarrhythmic agent that blocks open sodium channels and prolongs the cardiac action potential. *TNs* Pronestyl, Procan.

procaine An organic base (234 Da) that is a local anaesthetic. Less potent than *lidocaine. *TN* Novocaine.

procarbazine A chemotherapeutic drug (an alkylating agent) used to treat Hodgkin's lymphoma and glioblastoma. *TNs* Matulane, Natulan, Indicarb.

procaryote *See* PROKARYOTES.

procentriole The precursor of a daughter centriole (centrosome) that forms initially as a ring of nine singlet microtubules adjacent to the existing centriole at the start of the S phase of the *cell cycle. The way this is specified is unclear but a centrosome-associated kinase, ZYG-1, plays an important role in triggering centriole duplication.

prochlorperazine One of the phenothiazine class of antipsychotic drugs. Used for the treatment of nausea and vertigo and rarely for treating psychosis. *TNs* Compazine, Buccastem, Stemetil, Phenotil.

procolipase The polypeptide (112 aa) that is cleaved to produce *colipase, the cofactor for pancreatic lipase, and an anorectic N-terminal pentapeptide, *enterostatin.

procollagen A trimer of collagen molecules, arranged as a triple helix, in which the terminal extension peptides are disulphide-linked. The terminal peptides are removed by *procollagen peptidases to produce the mature *tropocollagen molecule.

procollagen peptidases The enzymes (EC 3.4.24.19, of the *astacin M12 peptidase family) that are responsible for removing the terminal extension peptides from procollagen to produce tropocollagen. Deficiency of the procollagen I N-terminal peptidase causes *Ehlers–Danlos syndrome type VIIC.

procyclidine An antimuscarinic drug used to reduce muscle tremor in Parkinson's disease.

prodigiosin *See* SERRATIA MARCESCENS.

prodromal Describing an early indication of a disease, often preceding the classical symptoms.

prodrug A compound that is pharmacologically inactive (or relatively inactive) until it is converted to the active form in the body.

proenzyme *See* ZYMOGEN.

profilaggrin A large polyprotein precursor (400 kDa) of *filaggrin with ten to twelve tandemly repeated units (324 aa), that is dephosphorylated and proteolytically cleaved during terminal differentiation of the epidermal cells. One of the *fused gene family.

profilin An actin-monomer binding protein (140 aa) that forms an assembly-incompetent complex. The profilin–actin complex is dissociated by phosphatidylinositol bisphosphate. There are two isoforms, profilin-1 and -2, with very similar properties.

proflavine A synthetic acridine dye, deep orange in colour, formerly used as an antiseptic wound dressing. Proflavine intercalates into DNA and is a mutagen.

progenitor cell *See* STEM CELL.

progeria An accelerated ageing syndrome in which there is premature senescence and early death. The Hutchinson–Gilford progeria syndrome is caused by mutation in the *lamin A gene.

progestagens Hormones (progestogens, gestagens) that produce effects similar to those of the natural hormone *progesterone. All other progestogens are synthetic and sometimes referred to as progestins.

progesterone A natural progestogen hormone (luteohormone, 314 Da) that is produced in the *corpus luteum and acts as an *oestrogen antagonist. It promotes the proliferation of the uterine mucosa in readiness for blastocyst implantation and blocks follicular development. The receptor is one of the nuclear hormone receptors (NR3C3).

programmed cell death A developmental mechanism in which certain cells die, usually by apoptosis, as a programmed morphogenetic shaping mechanism, e.g. separating digits of the developing hand and foot. **Programmed death-1** (PD-1, 288 aa) is an inhibitory receptor of the *B7 family on T cells that causes immune dysfunction when it is up-regulated, as happens in HIV infection. A polymorphism in a regulatory region is associated with susceptibility to *systemic lupus erythematosus. The ligands are themselves membrane proteins, homologues of B7, (B7H1, 290 aa; PDL2, 182 aa) and members of the immunoglobulin superfamily, but do not bind to *CTLA4 or CD28. **Programmed cell death 10** (PCD10, 212 aa) is defective in *cerebral cavernous malformations type 3 and seems to be involved in vascular morphogenesis.

progranulin A glycoprotein growth factor (epithelin precursor, proepithelin, acrogranin, 593 aa) that has seven similar cysteine-rich domains; these are cleaved from progranulin and some were originally isolated as granulins (granulins A–D). Progranulin is expressed in many epithelial cells, macrophages, and monocyte-derived dendritic cells and is involved in multiple physiological and pathological processes. The granulins are also active in various processes. Mutations in the progranulin gene have been shown to cause familial frontotemporal lobar degeneration with ubiquitin-positive inclusions (FTDL-U). Progranulin mediates proteolytic cleavage of *TDP-43 to generate ~35-kDa and ~25-kDa species.

progressive external ophthalmoplegia A set of disorders characterized by multiple mitochondrial DNA (mtDNA) deletions in skeletal muscle, in which the clinical features are adult onset of weakness of the external eye muscles and exercise intolerance. Additional symptoms are variable. Autosomal dominant progressive external ophthalmoplegia (adPEO) with mtDNA deletions-1 (**PEOA1**) is caused by mutation in the nuclear-encoded DNA polymerase-γ gene, as is an autosomal recessive form (**PEOB1**). **PEOA2** is caused by mutation in the adenine nucleotide translocator (ANT) gene; **PEOA3** by mutation in the *twinkle gene; **PEOA4** by mutation in the nuclear-encoded DNA polymerase-γ_2 gene (POLG2).

progressive osseous heteroplasia A rare autosomal dominant disorder in which bone forms in the dermis in infancy and subsequently within deep muscle and fascia. It is caused by paternally inherited inactivating mutations of the *GNAS1* gene that encodes the α subunit of the Gs stimulatory heterotrimeric G protein. Mutations in this gene can also give rise to *Albright hereditary osteodystrophy, pseudohypoparathyroidism Ia, and pseudopseudohypoparathyroidism.

progressive supranuclear palsy A Parkinson's disease-like dementia (Steele–Richardson–Olszewski syndrome) that can be caused by mutation in *tau (atypical Steele–Richardson–Olszewski syndrome) although other loci are implicated in the typical form of the disorder.

proguanil An antimalarial drug used to treat falciparum malaria and as a prophylactic. Usually given in combination with atovaquone (as Malarone). *TN* Paludrine.

prohibitins Highly conserved membrane proteins. **Prohibitin-1** (272 aa) has antiproliferative and tumour suppressor activities, and mutations have been linked to sporadic breast cancer. **Prohibitin 2** (repressor of oestrogen receptor activity, REA, 299 aa) is involved in transcriptional repression by nuclear hormone receptors. The **prohibitin homology domain** (PHB domain) is evolutionarily conserved and proteins with this domain have affinity for lipid raft domains.

prohormone An inactive form of a protein hormone that must be proteolytically cleaved to become active.

proinsulin *See* INSULIN.

prokaryotes The unicellular organisms, *bacteria and *archaea, that form two of the three main domains of living organisms (the third being *eukaryotes).

prokineticins Small cysteine-rich secreted proteins that regulate diverse biological processes. **Prokineticin 1** (endocrine gland vascular endothelial growth factor, 105 aa) induces proliferation, migration, and fenestration in capillary endothelial cells derived from endocrine glands. **Prokineticin 2** (PK2, Bv8, 81 aa) is rhythmically expressed in the *suprachiasmatic nucleus, and may have effects on gastrointestinal activity. Mutations in the *PK2* gene cause some forms of *Kallmann's syndrome, other forms being caused by mutations in one of the two closely related G-protein-coupled receptors (PKR1 and PKR2).

prolactin The pituitary hormone (lactogenic hormone, lactotropin, mammotropin, luteotropic hormone, LTH, luteotropin, 199 aa) that stimulates production of milk and is also associated with postcoital sexual gratification. Prolactin is synthesized as pre-prolactin and the signal peptide is cleaved to generate the mature form. Prolactin in serum apparently exists in three molecular forms: big-big prolactin (macroprolactin, ~100 kDa), big prolactin (~48 kDa), and little prolactin (~22 kDa), the last having had some additional residues removed. Macroprolactin is a complex of prolactin with antiprolactin autoantibody and is biologically inactive. The receptor (598 aa) is a cytokine-type receptor that signals through the JAK–STAT pathway.

prolactin releasing peptide A peptide releasing hormone (prolactin releasing hormone, 87 aa) that stimulates *prolactin release from cells in the anterior lobe of the pituitary gland and has a range of other effects in the central nervous system, inhibiting food intake, stimulating sympathetic tone, and activating stress hormone secretion. The receptor is a G-protein-coupled receptor (GPR10).

proliferating cell nuclear antigen (PCNA) A nuclear protein (261 aa) first identified as a marker for proliferating cells, later shown to be an auxiliary factor for DNA polymerases δ and ε that are involved in repair and replication. Transcription of PCNA is modulated by *p53. **PCNA-associated factor** (PAF, p15PA, overexpressed in anaplastic thyroid carcinoma-1, OEATC-1, 111 aa) constitutively associates with the tumour suppressor p33ING1B. PCNA is well conserved between plants and animals and there are homologues in prokaryotes.

proliferative unit A column of epidermal cells derived from a single basal proliferating cell.

proliferins A family of proteins (mitogen-regulated proteins, ~224 aa) encoded by genes of the *prolactin family that are involved in angiogenesis of the uterus and placenta in mice.

proline (Pro, P) One of the 20 amino acids directly coded for in proteins. Strictly speaking, proline (115 Da) is an imino acid because its side chain is bonded to the nitrogen of the α-amino group, as well as the α-carbon. Proline strongly influences the secondary structure of proteins. Collagen is particularly rich in proline and its derivative, hydroxyproline.

proline arginine-rich end leucine-rich repeat protein A connective tissue glycoprotein (PRELP, 382 aa) of the leucine-rich repeat (LRR) family with some similarities to *fibromodulin and *lumican. Abundantly expressed in juvenile and adult, but not neonatal, cartilage.

proline-rich homeodomain protein A transcription factor (PRH, haematopoietically expressed homeobox, Hex, PRH/Hex, 270 aa) that regulates cell differentiation and cell proliferation, and is required for the formation of the vertebrate body axis, the haematopoietic and vascular systems, and the formation of many vital organs. It is thought to form octamers and controls gene expression at both transcriptional and translational levels.

proline-rich polypeptide (PRP) A peptide isolated from ovine colostrum that has immunoregulatory effects and is reported to have a stabilizing effect on cognitive function in Alzheimer's disease patients. A nonapeptide fragment seems to have full activity. *TN* Colostrinin.

proline-rich proteins (PRPs) A miscellaneous category of proteins that includes several with unusual structure (collagens, complement 1q) and those with proline-rich domains that are associated with protein–protein interactions in signalling cascades. There are also a group of salivary PRPs (100–150 aa) which form 70% of the protein content in saliva and are thought to protect against dietary tannins. *See* PROLINE-RICH HOMEODOMAIN PROTEIN.

prolyl hydroxylase The enzyme (EC 1.14.11.2, 534 aa) that hydroxylates proline to *hydroxyproline in procollagen. It is a heterotetramer of two α_2 chains and two β chains; the active site has a ferrous ion, and a reducing agent such as ascorbate is necessary as a cofactor.

promastigote A stage in the life cycle of certain trypanosomatid protozoa (e.g. *Leishmania*). The cell is elongated or pear-shaped with a central nucleus, an anterior *kinetoplast, and a basal body with a single long flagellum.

promazine A phenothiazine antipsychotic drug.

promethazine A histamine H_1-receptor antagonist used for treating hay fever, urticaria, as an antiemetic, and because of its sedative properties, to relieve insomnia. Often combined with other drugs in proprietary remedies. *TN* Phenergan.

prominin A conserved transmembrane glycoprotein (CD133, 865 aa) found on apical protrusions of neuroepithelial cells and on haematopoietic stem cells. Mutations are associated with *retinitis pigmentosa type 41 and *cone–rod dystrophy-12.

promoter The segment of DNA, upstream of a gene, where RNA polymerase binds to initiate *transcription. The first nucleotide to be transcribed is denoted +1; upstream (untranscribed) nucleotides have negative numbers. In bacteria there are two *consensus sequences required for polymerase binding, the Pribnow box at about −10 and another centred at about −35. Less is known about eukaryote promoters: RNA polymerase I recognizes a single promoter for the precursor of rRNA, but polymerase II recognizes many thousands of promoters, most of which have the Goldberg–Hogness or *TATA box at around position −25. Several promoters have a CAAT box around −90 and promoters for *housekeeping proteins have multiple copies of a GC-rich element. Polymerase II transcription is further modulated by enhancers. Polymerase III, which synthesizes 5S ribosomal RNA, all tRNAs, and a number of small RNAs, has a third type of promoter. Transcription factors bind at or near the promoter and affect initiation.

promoter insertion A form of *insertional mutagenesis in which a host gene is activated by the *long terminal repeat sequence of a virus that has integrated nearby.

promyelocytes The cells that give rise to *myelocytes.

pronucleus The haploid nucleus resulting from a meiotic division. The female pronucleus is the nucleus of the ovum before fusion with the male pronucleus of the sperm.

pro-opiomelanocortin (POMC) A polyprotein (241 aa) that is converted by serine peptidases (prohormone convertases) into several different hormones including *adrenocorticotropin (ACTH), β-*lipotropin (β-LPH), corticotropin-like intermediate lobe peptide (CLIP), γ-LPH, α-melanotropin (α-MSH, α-*melanocyte-stimulating hormone), β-endorphin, γ-MSH, and β-MSH. Mutations in the gene can cause a variety of disorders, since point mutations may affect the cleavage sites for some derived hormones but not others.

propantheline bromide An antimuscarinic drug that blocks parasympathetic innervation and is used for the treatment of excessive sweating (hyperhidrosis) and cramps or spasms of the stomach, intestines, or bladder.

properdin (factor P) A plasma protein involved in the alternative complement pathway where it binds to many microbial surfaces and stabilizes C3b,Bb convertase that cleaves C3. Mutation in the properdin gene, an X-linked recessive trait, is associated with increased susceptibility to infection by *Neisseria* species.

prophage The genome of a *lysogenic bacteriophage that is integrated into, and replicates with, the chromosome of the host bacterium.

p

prophase The first phase of mitosis or of the first meiotic division in which the chromosomes condense and become visible.

prophylaxis An action or behaviour that prevents infection.

propidium iodide A fluorescent molecule (668.4 Da) that can be used to stain DNA, into which it intercalates. When excited by light at 488 nm it emits fluorescence at 562–588 nm. It is excluded from live cells and so can be used to detect dead cells, which are stained.

propolis A resinous substance used by bees as a gap sealant. It has some antibiotic properties and is claimed to have beneficial effects on health.

propranolol A potent antagonist of β_1- and β_2-adrenergic receptors, a nonselective *beta-blocker.

proprietary name The trade name (TN) of a drug, which may vary from country to country. For example, Pepcid is the proprietary name for famotidine in the USA and Europe, whereas in Japan it is marketed as Gaster. Entries in this dictionary refer mostly to the generic names of drugs.

proprioception Sensory awareness of body position, essential for motor control.

prorenin The inactive precursor of *renin.

prosaposin The glycoprotein (sphingolipid activator protein, sulphated glycoprotein-1, SGP-1, 524 aa) produced in testis, spleen, and brain from which *saposins are produced by proteolysis. There are two alternative transcripts, one targeted to the lysosome in a mannose-6-phosphate-independent manner apparently involving *sortilin, whereas the other is secreted. Inhibitors of sphingolipid biosynthesis, such as *fumonisin B1 and *tricyclodecan-9-yl xanthate potassium salt, interfere with the trafficking of prosaposin to lysosomes.

Prosite A searchable database of protein domains, families, and functional sites provided as a service to the scientific community by the Swiss-Prot and Proteome Informatics groups of the Swiss Institute of Bioinformatics. It can provide useful clues to the likely function of a protein.

SEE WEB LINKS

- The Prosite home page.

prosome Alternative and probably obsolete name for a *proteosome (proteasome).

prostacyclin A short-lived *prostaglandin (PGI_2) released by mast cells and endothelium, a potent inhibitor of platelet aggregation; also causes vasodilation and increased vascular permeability. Release of PGI_2 is increased by *bradykinin. Used in treatment of pulmonary hypertension. *TN* Flolan.

prostaglandins (PGs) Compounds originally purified from prostate but ubiquitous in tissues that are important inflammatory mediators and hyperalgesics. They are derived from arachidonic acid by the action of *cyclo-oxygenases that produces cyclic endoperoxides (PGG_2 and PGH_2) which in turn can give rise to *prostacyclin or *thromboxanes as well as prostaglandins.. Prostaglandins have a short half-life in circulation and are rapidly degraded in the lungs. Prostaglandin E_2 (PGE_2) acts on *adenylate cyclase to enhance the production of *cyclic AMP. PGI is more commonly known as *prostacyclin.

prostanoids A collective term for *prostaglandins, *prostacyclins, and *thromboxanes. *Compare* EICOSANOIDS.

prostate specific antigen *See* PSA; PSCA; PSM; PCTA-1; STEAP.

prosthetic group A nonprotein component of an enzyme or other protein that is essential for its activity. The prosthetic group can be organic (e.g. a vitamin) or inorganic (e.g. a metal). Unlike coenzymes, they are permanently bound to the apoprotein.

protamine The major DNA-binding proteins, very basic and arginine-rich, in the nucleus of sperm that allow the DNA to be packed into a volume less than 5% of a somatic cell nucleus. Most mammals have only protamine P1 (50 aa), but humans, mice, and horses also have P2 protamines (57 aa). In humans there are three members of the P2 family, HP2, HP3, and HP4, that differ only at their N-termini. *See also* TRANSITION PROTEINS.

protanopia The form of colour blindness (red colour blindness) due to deficiency in the red cone pigment (opsin 1, long-wavelength sensitive cone pigment, 164 aa). Unlike the green cone pigment (*see* DEUTERANOPIA) there is only one X-linked gene. If the mutation causes a change in the absorption spectrum, rather than deficiency of the pigment, then the disorder is referred to as protanomaly.

protease A term used for endopeptidases with broad specificity that would cleave most proteins into small fragments. Examples include digestive enzymes such as *trypsin and *pepsin, enzymes of plant origin such as ficin and *papain, and bacterial enzymes such as pronase. Proteases are used for *peptide mapping. *See* PEPTIDASE.

protease activated receptors A family of unusual G-protein-coupled receptors (PARs) that are activated by cleavage of part of the extracellular domain: the cleaved fragment then acts as a ligand. **PAR1** and **PAR4** are *thrombin receptors, **PAR2** responds to trypsin but not thrombin and is up-regulated in chronic inflammation. **PAR3** does not signal but acts as a cofactor for the cleavage and activation of PAR4. NB The abbreviation PAR4 has also been used for prostate apoptosis response protein-4.

protease inhibitors Any inhibitor of a peptidase (protease) that breaks down proteins, but now commonly a shorthand name for drugs that inhibit the protease involved in virus maturation. An example is indinavir, used in combination therapy for AIDS.

protease M *See* NEUROSIN.

protease nexin-1 A serine protease inhibitor (PN-1, plasminogen activator inhibitor type 1 member 2, glial derived neurite promoting factor, glia-derived nexin, GDN, *serpin E2, 397 aa), that is an important physiological regulator of α-thrombin in tissues and also inhibits plasmin and *plasminogen activators. It is neuroprotective in a number of assay systems and is highly expressed and developmentally regulated in the nervous system.

proteasome *See* PROTEOSOME.

protectin A membrane-bound (GPI-anchored) glycoprotein (CD59), which protects cells from bystander attack by autologous complement by preventing the formation of the membrane attack complex. To avoid confusion with anti-inflammatory *protectins, the name has been dropped in recent literature.

protectins A family of bioactive products generated from omega-3 fatty acids (eicosapentaenoic acid and docosahexaenoic acid) and analogous to *lipoxins and *resolvins. They act as antagonists of the inflammatory mediators produced from omega-6 fatty acids. Protectin D1 (PD1, neuroprotectin D1) has marked anti-inflammatory, neuroprotective and immunomodulatory effects. *See* OMEGA FATTY ACID; *compare* PROTECTIN.

protegrins A family of *cathelin-associated antimicrobial peptides found in mammalian leucocytes. Will kill a range of bacteria including Gram-positives by forming multimeric anion channels and are active against multidrug resistant strains.

protein A linear polymer of amino acids linked by peptide bonds in a sequence specified by mRNA.

protein 4.1 (band 4.1) An abundant phosphoprotein (864 aa) of the human erythrocyte membrane cytoskeleton which interacts with *spectrin and actin. It is a member of the ERM family of proteins (which have a characteristic *FERM domain) and there are multiple tissue-specific isoforms: erythrocyte 4.1 (4.1R, EPB41), a neuronal form (4.1N, EPB41L1), a general form (4.1G, EPB41L2) encoded by another gene and widely distributed; a brain form (4.1B, EPBL3, DAL1, *differentially expressed in adenocarcinoma of the lung) is another variant. Brain protein 4.1 (*synapsin I) is the best characterized of the nonerythroid forms of protein 4.1. Mutations in 4.1R can cause hereditary elliptocytosis.

protein 4.2 A major erythrocyte membrane skeletal protein (EPB42, 691 aa) that links protein 3 with *ankyrin. It has homology with transglutaminase and the α subunit of

p

coagulation factor XIII, but has no enzymatic activity. Protein 4.2 is mutated in the Japanese type of recessive spherocytic *elliptocytosis and in some forms of recessive haemolytic anaemia. At one time was mistakenly thought to be *pallidin.

protein A A protein (508 aa) from the cell wall of **Staphylococcus aureus* that binds the Fc region of most classes of immunoglobulin and can be used to affinity-purify immunoglobulins from serum (or culture medium in the case of monoclonal antibodies) and, especially when coupled to label, to detect antibody. It is a very effective B-cell mitogen.

proteinase inhibitors A variety of endogenous inhibitors that regulate the activity of peptidases. An antileukoprotease that inhibits neutrophil lysosomal elastase and cathepsin G is found in seminal fluid (human seminal plasma inhibitor-I, HUSI-I, 138 aa) and in various mucous secretions (secretory leucocyte protease inhibitor). *Protease inhibitors have become the casual name for antiviral drugs. *See* ALPHA-1-ANTITRYPSIN; ALPHA-2-MACROGLOBULIN; SERPINS.

protein B **1.** A name that has been applied to the β antigen of the C-protein complex (Cβ protein, Bac) in the cell wall of group B streptococci, by analogy with *protein A. It binds the Fc region of IgA1 and IgA2. **2.** Centromere protein B (CENPB, 599 aa) is a centromere autoantigen and interacts with centromeric heterochromatin and binds to the CENP-B box in centromeric α-satellite DNA (alphoid DNA). **3.** A transferrin-binding protein (703 aa) in *Neisseria meningitidis* serogroup B.

protein C A vitamin K-dependent plasma glycoprotein (262 aa), the zymogen of a serine endopeptidase (EC 3.4.21.69) that is a key component of the anticoagulation system. The *thrombin–*thrombomodulin complex will remove a 42-aa pre-pro sequence and activate protein C that will, in conjunction with *protein S, hydrolyse factors Va and VIIIa, inhibiting clotting. Heterozygous mutation in the gene leads to autosomal dominant hereditary thrombophilia; homozygous mutation causes the more severe autosomal recessive form. The latter is characterized by recurrent venous thrombosis whereas the former may be asymptomatic.

protein disulphide isomerase A multifunctional protein (cellular thyroid hormone-binding protein, EC 5.3.4.1, 504 aa) that catalyses the formation, breakage, and rearrangement of disulphide bonds. At high concentrations it functions as a chaperone by inhibiting the aggregation of misfolded proteins but at low concentrations it may have the opposite effect. It is a structural subunit of various multisubunit enzymes such as prolyl 4-hydroxylase and *microsomal triacylglycerol transfer protein.

protein engineering The production of proteins with specific modifications in the primary sequence by means of recombinant DNA technology. *See* SITE-SPECIFIC MUTAGENESIS.

protein G An immunoglobulin-binding protein (60 aa) expressed in group C and G streptococci. Similar to *protein A but has a higher affinity for IgG and will bind IgGs from a wider range of species.

protein kinase An enzyme that catalyses the phosphorylation of proteins, usually causing a change in function. The serine/threonine kinases transfer phosphate to the hydroxyl side group of serine (commonly) or threonine (less frequently). The tyrosine kinases phosphorylate tyrosine residues, accounting for very little of the overall level of protein phosphorylation but with major effects in signalling systems. *See* PROTEIN KINASE A, PROTEIN KINASE C, PROTEIN KINASE G; RECEPTOR TYROSINE KINASES.

protein kinase A (PKA) A family of serine/threonine kinases (PKA, cAMP-dependent protein kinase, EC 2.7.11.1) regulated by the level of *cyclic AMP in the cell. They have two catalytic subunits and two regulatory subunits that undergo a conformational change when they bind cAMP and activate the catalytic subunits. A diverse range of substrates gives pleiotropic effects.

protein kinase B (PKB) *See* AKT.

protein kinase C (PKC) A family of serine/threonine kinases (EC 2.7.11.1) that are activated by phospholipids and have an important signalling function. The classical PKCs (α, β_1, β_2, γ) are calcium dependent and can be activated by diacylglycerol produced by the action of *phospholipase C or, nonphysiologically, by *phorbol esters. An increasing number of nonclassical calcium-independent isoforms are being found; these have a highly conserved catalytic domain but a variety of regulatory domains. PKC tends to phosphorylate serine or threonine residues near a C-terminal basic residue.

protein kinase G (PKG) A member of the *AGC kinase family, a cGMP-regulated serine-threonine kinase (671 aa) involved in relaxation of vascular smooth muscle and inhibition of platelet aggregation. Type I is soluble, type II is membrane-bound.

protein kinase inhibitor proteins (PKIs) Small, specific and potent heat-stable inhibitors of the catalytic subunit of cAMP-dependent protein kinases. There are several isoforms (PKIα,75 aa; PKIβ, PKIγ) all expressed in the brain. Some alternatively spliced forms of PKIβ are also inhibitory for cGMP-dependent kinase.

protein kinase IV A multifunctional calcium/calmodulin-dependent serine/threonine protein kinase (brain CaM kinase IV, 503 aa) implicated in transcriptional regulation in lymphocytes, neurons, and male germ cells by phosphorylating and activating *CREB and CREM-τ.

protein kinase N (PKN) A family of protein kinase C (PKC)-related serine/threonine kinases, probably regulated by *rho-dependent phosphorylation. **PKN1** (PRK-1, 942 aa) is ubiquitously expressed and is activated by phospholipids and arachidonic acid. **PKN2** (PRK-2, PAK-2, 984 aa) is activated by cardiolipin and acidic phospholipids, and interacts with *Nck and PLCγ. **PKN3** is reportedly required for invasive prostate cell growth and is regulated by PI3K. *See* POLO-LIKE KINASES.

protein kinase R (PKR) A serine/threonine protein kinase (eukaryotic translation initiation factor 2α kinase 2, 551 aa) that is activated by viral double-stranded RNA which causes it to autophosphorylate and to inhibit protein synthesis in the cell. It is induced by interferon and is part of the antiviral defence mechanism.

protein L **1.** A protein (36 kDa) from *Peptostreptococcus magnus* that has an affinity for immunoglobulin κ light chains from various species. It can be used to purify monoclonal or polyclonal IgG, IgA, and IgM as well as Fab, F(ab) 2, and recombinant scFv fragments that contain κ light chains. (NB Bovine, goat, sheep, and horse Igs contain almost exclusively λ chains and do not bind.) **2.** One of the proteins involved in the centromere complex that is required for proper kinetochore function (centromeric protein L).

protein phosphatases Enzymes that remove phosphate groups from proteins and reverse the effects of *protein kinases. Protein phosphatase types 1 (PP1) and 2A (PP2A) remove phosphate from serine or threonine; *protein tyrosine phosphatases act on phosphorylated tyrosine residues.

protein phosphorylation The addition of phosphate groups to a protein by a protein kinase to hydroxyl groups either on serine or threonine (by serine/threonine kinases) or on tyrosine (by tyrosine kinases). A common method for regulating activity, the effect being reversed by *protein phosphatases.

protein S A vitamin K-dependent plasma protein (635 aa) that inhibits blood clotting by acting as a nonenzymatic cofactor for activated *protein C. About 40% of protein S in plasma is free and functionally active; the remainder is complexed with C4b-binding protein and inactive. Mutation leads to a form of thrombophilia.

protein sequencing The determination of the sequence of linked amino acids in a protein, the *primary structure; once a major undertaking but now relatively routine. The standard method was Edman degradation in which peptide fragments were tethered and the N-terminal residues sequentially reacted with *Edman reagent, removed, and identified. More recently mass spectroscopy has become more common; the protein is proteolytically cleaved and fragments are identified by their characteristic mass using electrospray injection into the spectrometer. Extensive databases of proteins with known sequences make the approach increasingly powerful. In both cases it is necessary to combine the information about fragments to generate the complete sequence.

protein tyrosine phosphatase (PTP) A phosphatase that removes the phosphate from a tyrosine residue in a protein. Examples include CD45, *shp, *density-enhanced phosphatase-1, *PTEN, and *LAR. *See also* TYROSINE KINASE.

protein zero A transmembrane protein (myelin protein zero, 248 aa), synthesized by *Schwann cells, that constitutes ~50% of the total protein in the myelin sheath of peripheral nerves. Mutations in the *MPZ* gene that encodes it are associated with the autosomal dominant form of *Charcot–Marie–Tooth disease type 1 (CMT1), *Dejerine–Sottas syndrome, congenital hypomyelinating neuropathy, and several types of axonal CMT2.

proteoglycan Large complexes composed of a protein core with many *glycosaminoglycan side chains that are important interstitial components of connective tissue and that confer viscoelastic properties on materials such as synovial fluid. They are also found on cell surfaces. The properties vary depending upon the types, sizes, and numbers of polysaccharides.

- Further details about proteoglycans.

proteolipid protein (lipophilin) Highly conserved tetraspanin membrane proteins that

p

account for about half of the protein content of adult central nervous system myelin (*compare* PROTEIN ZERO). Proteolipid protein 1 (PLP1, 276 aa) has an alternatively spliced isoform, DM20, and another form, PLP2, has also been described. The cellular function is unclear, although it may be purely structural, but mutations are lethal. *See* PELIZAEUS–MERZBACHER DISEASE.

proteolysis Hydrolysis of proteins by peptidases (proteases). Very limited proteolysis may be a way of activating enzymes that have inactive *zymogens, of generating active fragments (as with some prohormones, e.g. *pro-opiomelanocortin), or of signalling (e.g. *protease-activated receptors and several cascades (e.g. *complement activation, blood clotting) involve proteolytic activation of proteases, giving amplification at each stage.

proteolytic enzyme *See* PEPTIDASE; PROTEASE.

proteome **1.** The entire set of proteins encoded by the genome. Although the primary structure (sequence) of the proteins is coded, the proteins present in a particular cell depend upon differential gene expression, processing of the mRNA, and a range of post-translational modifications which may include both additions (e.g. glycosylation) and deletions (proteolytic cleavage). By analogy with genomics, proteomics is the study of the proteome. **2.** The set of proteins present in a cell at a particular time and under specific environmental conditions.

p

SEE WEB LINKS

- Further information on the proteome.

proteosome (proteasome) The **20S proteosome** has 28 protein subunits arranged as an $(\alpha_1\text{–}\alpha_7, \beta_1\text{–}\beta_7)_2$ complex in four stacked rings. The interior of the complex has the active sites. The β-type subunits are synthesized as pro-proteins and are proteolytically cleaved before assembly. The **26S proteosome** is a 2000-kDa protein complex composed of the catalytic 20S proteosome and the regulatory PA700 subcomplex that consists of approximately 20 heterogeneous proteins. The **proteosome activator complex** (PA28 complex, 11S regulator) is an alternative proteosome activator that does not involve *ubiquitin. The complex is composed of two homologous subunits called PA28α (Reg1a, 249 aa) and -β (Reg1b, 239 aa), which form a hexameric ring, and is expressed constitutively in antigen-presenting cells. A third subunit, PA28γ (254 aa), is the *Ki antigen. Expression of the PA28 complex is up-regulated by interferon-γ and enhances the generation of MHC class I binding peptides by altering the cleavage pattern of the proteosome.

Proteus A genus of highly motile Gram-negative bacteria. They occur normally in soil and in the intestinal tract, but can be opportunistic pathogens. *P. mirabilis* is a common cause of urinary tract infections.

Proteus syndrome A variable disorder in which there is asymmetric and disproportionate overgrowth of body parts, together with connective tissue and epidermal naevi, dysregulated adipose tissue, and vascular malformations. It may be a form of somatic mosaicism but the underlying cause remains obscure. The suggestion that it is associated with *PTEN mutations seems to have been excluded.

prothrombin A vitamin K-dependent plasma protein (622 aa, 100 μg/mL), the inactive precursor of *thrombin, synthesized in the liver. The prothrombin time is the time taken, in seconds, for a fibrin clot to form after a specific volume of thromboplastin reagent is added to the sample of citrated blood plasma. Used in calculating the *international normalized ratio (INR).

protirelin *See* THYROTROPIN RELEASING HORMONE.

protocadherin 15 A cadherin that is expressed in stereocilia of the inner ear and in retinal photoreceptors. Together with cadherin 23 (*otocadherin) forms tip links, extracellular filaments that connect the stereocilia and are thought to gate the mechanoelectrical transduction channel. Mutations cause *Usher's syndrome (type1F).

protofilaments The linear arrays of tubulin subunits in a microtubule, which can be visualized as being composed of (normally) thirteen such parallel strands.

protogenin A cell surface protein (1150 aa) of the immunoglobulin superfamily, closely related to the *deleted in colorectal cancer/*neogenin family. It is expressed early in development and may have a role in elongation of the embryonic axis.

protomers A general term for the subunits that are assembled to form a larger structure. Cell biological examples are G-actin, which assembles to filamentous F-actin, and the tubulin heterodimer which is the subunit for microtubules. The latter example illustrates the utility of a term that does not assume protomers to be single molecules.

proton ATPase (H^+-ATPase) An active transport mechanism that uses ATP to move hydrogen ions across a lipid bilayer. The *F-type ATPases can also run in reverse to synthesize ATP ('ATP synthase'). The *V-type ATPases are found in intracellular vesicles with an acidic lumen, and on certain epithelial cells (e.g. kidney intercalated cells). Another example is the H^+,K^+-ATPase involved in secretion of gastric acid, the target for *proton pump inhibitors.

proton motive force (PMF) The proton (pH) gradient across a prokaryotic or mitochondrial membrane that couples oxidation and ATP synthesis or, in bacteria, the rotation of the flagellar motor. *See* CHEMIOSMOSIS.

protonophore An *ionophore that allows protons to cross a membrane. Many *uncoupling agents are protonophores that dissipate the proton gradient across the mitochondrial membrane.

proton pump inhibitors (PPIs) A class of drugs, e.g. *omeprazole, that inhibit the proton ATPase of the gastric epithelium and reduce acid secretion Used as a short-term treatment for gastric and duodenal ulcers.

proto-oncogene A gene that is involved in regulation of cell proliferation which, if overexpressed, has the capacity to cause oncogenesis. The overexpression can occur if the gene has been incorporated in a virus (as a viral *oncogene) and then reintegrated into the genome under the control of a viral promoter.

protoplast A bacterial cell without its cell wall. Protoplasts can be produced experimentally by growing bacteria in isotonic medium (to prevent lysis) in the presence of antibiotics that block the synthesis of the wall. The term is also applied to plant cells from which the cell wall has been enzymatically stripped.

protoporphyrin A porphyrin ring structure without metal ions. Protoporphyrin IX is the immediate precursor of *haem.

protrudin A protein (zinc finger FYVE domain-containing protein-27, ZFYVE27, 404 aa) that promotes neurite formation by regulating directed membrane trafficking. Phosphorylation of protrudin in response to nerve growth factor promotes its association with *Rab11-GDP. Mutation is associated with autosomal dominant *hereditary spastic paraplegia type 33 (SPG33). Protrudin interacts with *spastin and the chaperone FKBP38.

provirus The genome of a virus integrated into the host DNA where it may be expressed or remain latent.

Prozac Proprietary name for a *selective serotonin reuptake inhibitor (SSRI), fluoxetine hydrochloride, widely used as an antidepressant.

prozymogen granule (condensing vacuole) An immature *secretory vesicle (zymogen granule).

PRP **1.** Platelet-rich plasma. **2.** *See* PROLINE-RICH POLYPEPTIDE. **3.** *See* PROLINE-RICH PROTEINS. **4.** *Prion proteins. **5.** *See* PRP PROTEINS.

Prp proteins Proteins (precursor RNA processing proteins) involved in the multiprotein Prp19-associated complex (nineteen complex, NTC) that is associated with the *spliceosome during spliceosome activation. There are at least eight proteins involved in the Prp-19 complex: Prp19 itself is one of the U-box family of E3 ubiquitin ligases, Prp3 is a U4/U6-associated splicing factor, Prp8 (2335 aa) is one of the larger proteins involved and is mutated in *retinitis pigmentosa-13. Not to be confused with PrP (*prion protein) or proline-rich proteins *PRPs.

pruritus Severe and persistent itching.

PS *See* PHOSPHATIDYLSERINE.

PS2 **1.** *See* PRESENILINS (presenilin-2). **2.** *See* TREFOIL FACTORS (trefoil factor 1) (pS2/TFF1).

PSA A *kallikrein-like protease (prostate specific antigen, semenogelase, kallikrein 3, EC 3.4.21.77, 261 aa) present in seminal plasma; when present in blood it seems to be a relatively reliable marker for prostatic hyperplasia or carcinoma.

PSCA A prostate-specific cell-surface antigen (prostate stem cell antigen, 123 aa) that is one of the Thy-1/Ly-6 family of GPI-anchored surface proteins. It is primarily expressed on normal basal cells in the prostate and is up-regulated in most cases of prostatic carcinoma. *See* STEAP.

PSD proteins *See* POSTSYNAPTIC PROTEIN.

P-selectin *See* SELECTINS.

pseudoachondroplasia A growth disorder caused by mutation in *cartilage oligomeric matrix protein in which there is disproportionately short stature, deformity of the lower limbs, various joint disorders, and progressive osteoarthritis. An allelic disorder, epiphyseal dysplasia, has milder effects.

p

pseudogene A DNA sequence very similar to that of a known gene but nonfunctional. May arise through gene duplication followed by the loss of promoters or occurrence of mutations that introduce stop codons or prevent correct mRNA processing. Some pseudogenes have a *poly-A tail which may indicate they originated as mRNA that has been reverse transcribed and integrated.

pseudohypoaldosteronism A disorder marked by salt wasting in infancy despite high levels of *aldosterone. **Autosomal recessive pseudohypoaldosteronism type I** can be caused by mutation in the α, β, or γ subunits of the epithelial sodium channel (ENaC). The **autosomal dominant form of type I pseudohypoaldosteronism** is caused by mutations in the mineralocorticoid receptor gene. **Type II pseudohypoaldosteronism** (Gordon hyperkalaemia–hypertension) is a *hyperkalaemia that arises despite normal renal glomerular filtration and is caused by a defect in *WNK kinases.

pseudohypoparathyroidism *See* ALBRIGHT HEREDITARY OSTEODYSTROPHY.

Pseudomonas A genus of rod-shaped motile Gram-negative bacteria. *P. aeruginosa*, normally found in soil, is an opportunistic pathogen of immunocompromised patients.

pseudopod A blunt projection from a cell, characteristically seen in amoeboid movement.

pseudopseudohypoparathyroidism *See* ALBRIGHT HEREDITARY OSTEODYSTROPHY.

pseudopterosins Anti-inflammatory and analgesic compounds isolated from the soft coral *Pseudopterogorgonia elisabethae*. The mechanism of action is unclear, but may be by blocking eicosanoid release rather than biosynthesis.

pseudoxanthoma elasticum A disorder (Gronblad–Strandberg syndrome) of elastic tissue that affects the skin, eyes, and vasculature. Most cases are due to mutation in the gene for the ATP-binding cassette C6 protein (ABCC6, 1503 aa) which may be causing a systemic metabolic disorder that secondarily affects elastic tissue. Heterozygotes exhibit a less serious form of the disorder. Polymorphisms in the xylosyltransferases decrease the severity of the phenotype.

PSGL-1 The high affinity counter-receptor (P-selectin glycoprotein ligand, CD162, 412 aa) for P-selectin expressed on myeloid cells and stimulated T-cells. *See* PADGEM.

psicofuranine An antibiotic produced by *Streptomyces hygroscopicus* var. *decoyicus* that inhibits bacterial nucleic acid synthesis by blocking GMP biosynthesis.

P-site The site on the *ribosome to which the growing chain is attached, the peptidyl-tRNA binding site. The next *aminoacyl-tRNA attaches to the A-site.

psittacosis A disease of wild and domestic birds caused by *Chlamydia psittaci* that can also cause respiratory and systemic infections in humans.

PSM A cell-surface antigen (prostate-specific membrane antigen, 750 aa) expressed in the prostate, a type II transmembrane protein with folate hydrolase and *N*-acetylated-α-linked-acidic dipeptidase activity, also expressed in the small intestine and the brain. May have a role in neuropeptide catabolism.

psoralens Drugs that will form adducts with nucleic acids if UV-irradiated. Mutagenic, but used in combination with UV to treat some skin diseases.

psoralidin A cytotoxic drug isolated from the medicinal herb *Psoralea corylifolia* (bakuchi). It inhibits tyrosine phosphatase 1B and has antidepressant properties.

psoriasis A chronic inflammatory skin disease in which there is epidermal hyperplasia in areas of variable extent; it may also be associated with severe arthritis. There is T-cell involvement and a fairly strong genetic predisposition associated with the HLA-Cw6 allele.

psychotomimetic Describing a class of drugs that cause bizarre psychotic effects including hallucinations and delusions.

PTB domain A domain (phosphotyrosine-binding domain) that is present in many proteins that interact with receptors that become tyrosine-phosphorylated when they bind ligand (e.g. *shc, *IRS-1, and *dok proteins).

PTEN A *protein tyrosine phosphatase (phosphatase and tensin homologue deleted on chromosome 10, mutated in multiple advanced cancers 1/phosphatase and tensin homologue, MMAC1/PTEN, TEP-1, 403 aa) that is a tumour suppressor gene on chromosome 10q23. Has homology to *tensin. Somatic mutations in PTEN are found in many tumours, particularly glioblastomas. Germ-line mutations in PTEN are responsible for *Cowden's disease.

PTGS Post-transcriptional gene silencing. *See* RNA INTERFERENCE.

PtK2 cells A cell line established from the kidney of potoroo or kangaroo rat *Potorous tridactylis* that is often used in studies on mitosis because there is only a small number of chromosomes and the cells do not round up during the M phase.

ptosis Drooping of the upper eyelid that may be a result of damage to the third cranial nerve, to *myasthenia gravis, to *Horner's syndrome, or merely idiosyncratic.

PTP *See* PROTEIN TYROSINE PHOSPHATASE.

ptRNA (precursor tRNA) *See* RNASE P.

PTX **1.** *See* PACLITAXEL. **2.** *See* PERTUSSIS TOXIN. **3.** *See* PENTOXIFYLLINE. **4.** Pectenotoxins, dinoflagellate toxins from *Dinophysis* sp. that accumulate in shellfish. **5.** *Palytoxin, another dinoflagellate toxin from *Ostreopsis* sp. **6.** *See* PICROTOXIN. **7.** A family of paired homeobox transcription factors, Ptx-1–3 (*see* LEFT–RIGHT ASYMMETRY).

ptyalin *See* AMYLASE.

P-type ATPase *See* E1–E2-TYPE ATPASE. *Compare* F-TYPE ATPASE; V-TYPE ATPASE.

P-type calcium channels A class of *voltage-sensitive calcium channels ($CaV_{2.1}$, CACNA1A) that are found in various neurons but especially in Purkinje cells of the cerebellum. They require substantial depolarization to activate and are slow to inactivate. The α_{1A} subunit forms the pore and the beta, gamma, and delta subunits have regulatory functions. Mutations are associated with *episodic ataxia-2, familial hemiplegic *migraine, *spinocerebellar ataxia-6, and idiopathic generalized *epilepsy. They are inhibited by *agatoxin and a polyamine FTX, through a G-protein-coupled mechanism.

PubMed A service provided by the National Center for Biotechnology Information (NCBI) at the National Library of Medicine, located at the US National Institutes of Health. PubMed provides access to the peer-reviewed biomedical literature and can be searched. In many cases the full text of an article can be accessed via PubMed: for almost all articles the abstract is freely available. There are also links to other NCBI databases such as *OMIM.

SEE WEB LINKS

- The PubMed homepage.

PU box A purine-rich motif recognized by the Sp-1 transcription factor.

PUFA A *fatty acid (polyunsaturated fatty acid) with more than one double bond, generally with a lower melting point. Cell behaviour can be affected by modifying the ratio of unsaturated to saturated fatty acids; exactly how the ratio affects health is less clear.

puffs Regions of a *polytene chromosome where the chromatin is less condensed and there is active RNA transcription. A puff usually involves unwinding at a single band, although multiple bands are associated with Balbiani rings.

pulchellin A ribosome-inactivating toxic protein isolated from seeds of the liana *Abrus pulchellus tenuiflorus*.

pull-down assay A purification method that uses a known protein, often immobilized, to capture proteins with which it interacts. An alternative approach to the *yeast two-hybrid screening method.

pulse-chase An experimental protocol used to follow the progress of a process: a radiolabelled molecule is given for a brief period (the pulse), then substituted with excess unlabelled molecule (the cold chase). By examining the system at different times the sequential fate of the labelled molecule can be determined.

pulse-field electrophoresis A high-resolution electrophoretic separation method used to analyse large (megabase) fragments of DNA. The applied electric field is alternated so that molecules do not become aligned.

Puma *See* BCL (bcl-2 homology domain 3).

punctin *ADAM-related glycoproteins (punctin 1, ADAMTSL1, 497 aa; punctin 2, ADAMTSL3, 1690 aa) with thrombospondin repeats, deposited in the extracellular matrix in a punctate fashion and excluded from focal contacts.

puratrophin-1 A protein (Purkinje cell atrophy-associated protein 1, 1192 aa) that is probably involved in intracellular signalling and actin dynamics in the Golgi apparatus. It is the product of a gene mutated in a form of cerebellar ataxia mapping to chromosome 16q.

purine A heterocyclic compound with a fused pyrimidine/imidazole ring. The purines *adenine and *guanine are parent compounds of nucleotides. Other important purines are *hypoxanthine and *xanthine.

purinergic receptors Receptors for purine nucleotides such *adenosine or *ATP.

Purkinje cell A class of cerebellar neurons that convey signals away from the cerebellum. *Compare* PURKINJE FIBRES.

Purkinje fibres Specialized cardiac muscle cells involved in regulating heartbeat.

puromycin An antibiotic that acts as an *aminoacyl tRNA analogue, binds to the A-site on the *ribosome, and then causes premature termination of the polypeptide chain.

purpura Condition in which small spontaneous haemorrhages occur, forming purple patches on the skin and mucous membranes.

purpurin **1.** A photosensitizer used in photodynamic therapy, purpurin-18. **2.** An anthraquinone from the root of madder *Rubia tinctorum*, reportedly antimutagenic.

pus A viscous fluid, usually yellow-white, formed mostly from dead neutrophils around a persistent infected object or as a result of local infection with bacteria that are resistant to phagocytic killing. There is sometimes a greenish colour because of the presence of myeloperoxidase, and blue pus is found in certain *Pseudomonas aeruginosa* infections that produce the blue pigment pyocyanin.

putamen *See* BASAL GANGLIA.

putrescine An amine associated with putrifying tissue that has been suggested as a growth factor for mammalian cells in culture. It binds strongly to DNA and is the metabolic precursor of the polyamines *spermine and *spermidine.

p

Puumala virus *See* NEPHROPATHIA EPIDEMICA.

PVDF *See* POLYVINYLIDENE FLUORIDE.

PVP *See* POLYVINYLPYRROLIDONE.

pyaemia Condition in which there are pyogenic organisms in the bloodstream (blood poisoning).

pycnodysostosis An autosomal recessive osteochondrodysplasia characterized by osteosclerosis and short stature, caused by a defect in *cathepsin K.

PYK2 A calcium-dependent proline-rich *tyrosine kinase (focal adhesion kinase 2, FAK2, 1009 aa), with 42% identity to *focal adhesion kinase 1, that is activated by tyrosine phosphorylation triggered by binding of neuropeptides or elevation of cellular calcium. It is activated in neurons following NMDA receptor stimulation and is thought to be involved in hippocampal *long-term potentiation.

pyknosis A condition in which chromatin condenses to form a deep-staining irregular mass, usually followed by *karyorrhexis and indicative of apoptosis.

pyocyanin A blue-green antibiotic pigment produced by *Pseudomonas aeruginosa*.

pyogenic Describing bacterial infections, usually by streptococci or staphylococci, that cause the formation of pus. Bacterial toxins kill the neutrophils which are the main component of pus.

PYPAFs Proteins (PYRIN-containing apoptotic protease-activating factor 1-like proteins, NALPs) involved in inflammatory signalling by regulating NFκB activation and cytokine processing, and implicated in autoimmune and inflammatory disorders.

pyramidal cells The major neuronal cell type in the cerebral cortex.

pyramidal system A collection of neurons in the central nervous system that link the motor cortex to the spinal cord and are involved in initiating movement.

pyrazinamide An antibacterial drug, the pyrazine analogue of nicotinamide, used in combination with other drugs to treat tuberculosis. Its mode of action is unclear.

pyridoxal phosphate The coenzyme derivative of vitamin B_6, important in amino acid metabolism.

pyrilamine *See* MEPYRAMINE.

pyrimethamine An antimalarial drug used together with sulphonamide to treat falciparum malaria and with sulfadiazine to treat toxoplasmosis.

pyrimidine A heterocyclic organic compound with two nitrogen atoms in the six-membered ring. *Cytosine, *thymine, and *uracil are pyrimidine derivatives

pyrin A protein (marenostrin, predicted 781 aa) that is found in mature granulocytes and is a member of a family of nuclear factors homologous to the Ro52 antigen and the RET finger protein. Both pyrin and *cryopyrin contain an N-terminal domain, the pyrin domain (PyD) that is responsible for the binding interaction with a common adaptor protein, *ASC. Pyrin affects IL1-β production through control of caspase-1 activation and is defective in familial Mediterranean fever.

pyrin domain proteins Signalling molecules involved in the development of innate immunity against intracellular pathogens through activation of inflammatory mediator pathways. The adaptor protein *ASC links pathogen recognition by pyrin domain (PYD)-containing proteins (PYD-NLR, PAN, PYPAF, NALP, Nod, and Caterpillar proteins), to the activation of downstream effectors. Activation of these effectors occurs when specific protein complexes, known as *inflammasomes, are formed. The pyrin domain is involved in protein–protein interactions and is a subfamily of the death domain (DD) superfamily.

pyrogen (*adj.* pyrogenic) A substance or agent that causes fever. Endogenous pyrogen is *interleukin-1.

pyroninophilic cells Cells that have their cytoplasm stained bright red by methyl green pyronin, indicating the presence of a lot of RNA and active protein synthesis. *Plasma cells are a good example.

pyrosequencing A method for sequencing DNA based on detecting pyrophosphate that is released when the complementary base is incorporated on the strand being synthesized using the unknown DNA as a template. Pyrophosphate release is detected with *luciferase.

pyruvate A three-carbon intermediate (2-oxopropanoate) important in glucose metabolism and amino acid synthesis. Pyruvate produced in glycolysis (the *Embden–Meyerhof pathway) feeds into the (*tricarboxylic acid cycle). **Pyruvate carboxylase** (EC 6.4.1.1) catalyses the formation of four-carbon oxaloacetate from pyruvate, CO_2, and ATP in *gluconeogenesis whereas **pyruvate dehydrogenase** (PDH, EC 1.2.1.51) uses pyruvate to produce acetyl-CoA. Pyruvate carboxylase is a key regulatory enzyme in gluconeogenesis, lipogenesis, and neurotransmitter synthesis. Defects in the pyruvate dehydrogenase complex are a common cause of primary lactic acidosis in children, which can cause mental retardation and death, and can arise through mutations in either the gene for pyruvate decarboxylase or the gene encoding the E1α polypeptide of the pyruvate dehydrogenase (EC 1.2.4.1) complex. Phosphorylation of PDH by a specific **pyruvate dehydrogenase kinase** (PDK, EC 2.7.11.2) results in inactivation and is an important control in energy metabolism. There are multiple tissue-specific PDK isozymes. **Pyruvate dehydrogenase phosphatase** deficiency can also lead to lactic acidosis.

Q_{10} The ratio of the velocity of reaction at one temperature to that at a temperature 10 °C lower. For many biological reactions the rate approximately doubles when the temperature rises by 10° C, but the value depends on the temperature range chosen.

Q banding *See* BANDING PATTERNS; QUINACRINE.

Q fever A rare typhus-like illness caused by *Coxiella burneti.* Mainly a zoonosis and tends to be caught by people in contact with animals (vets, abattoir workers, etc.).

QH2-cytochrome c reductase The membrane-bound complex (ubiquinol–cytochrome-*c* reductase, EC 1.10.2.2) in the inner membrane of the mitochondrion that is responsible for electron transfer from reduced coenzyme Q to cytochrome *c*.

q region *See* P REGION.

QSAR *See* STRUCTURE–ACTIVITY RELATIONSHIP.

QT interval The time between the start of the Q wave and the end of the T wave in one cycle of electrical activity in the heart. The shorter the interval the faster the heart rate, and a rate-corrected value is usually calculated. *See* LONG QT SYNDROME.

Q-type channels A class of *voltage-sensitive calcium channels (CaV2.1 (P/Q-type)) very similar to *P-type channels with the pore-forming subunit being of the α-1A type (CACNA1A, 2506 aa). They are mostly expressed in neuronal tissue and regulate the release of neurotransmitters. They are inhibited by neurotransmitters that act through G-protein-coupled receptors and by high concentrations of ω-*conotoxin and ω-*agatoxin. Mutations in *CACNA1A* are associated with familial hemiplegic *migraine, *episodic ataxia type 2, and *spinocerebellar ataxia type 6.

quaking One of a family of RNA-binding proteins with a *KH domain embedded in a GSG domain. QK1 is the human homologue of the mouse quaking mutation, which causes body tremor and severe dysmyelination. Other members of this family include *SAM68. It regulates the mRNA levels of several genes involved in oligodendrocyte differentiation and maturation in the human brain and may be a candidate gene associated with schizophrenia.

quantal mitosis The concept that certain mitotic divisions are crucial ones in which reprogramming of the determination of the daughter cells takes place. A growing understanding of the complex control of gene expression makes it overly simplistic.

quantitative character A character such as height which shows continuous variation and is likely to be controlled by several genes, not just one.

quantitative structure–activity relationship (QSAR) *See* STRUCTURE–ACTIVITY RELATIONSHIP.

quantum dot A small particle of semiconductor material, typically a few nanometers in diameter. Quantum dots of the same material, but of different sizes, can emit light of different colours, which allows them to be used for intravital staining.

quasi-equivalence A term that describes the subunit packing in a quasi-crystalline array, e.g. a virus coat where there is some strain but the overall structure is very stable.

quaternary structure *See* PRIMARY STRUCTURE.

Quellung reaction The swelling of a bacterial capsule caused by interaction with antibody. The effect is to make the capsule more conspicuous.

quercetin A flavonol pigment found in many plants, said to be the basis for a variety of health benefits. Inhibits *F-type ATPases.

quiescence Describing a state of relative inaction: in cells, the nondividing state; in neurons, not firing.

quiescent stem cell A 'resting' stem cell that has the potential to divide if stimulated. The satellite cells in skeletal muscles are quiescent myoblasts that will proliferate after wounding.

quin2 A fluorescent dye that changes its absorption and emission characteristics when calcium is bound. The membrane-permeable acetoxymethyl ester (quin2AM) can be loaded into cells, being de-esterified in the cytoplasm and trapped, and used as an intracellular indicator, although it does sequester calcium and can perturb normal function.

quinacrine A fluorescent dye that will intercalate into DNA and gives a characteristic banding pattern to stained chromosomes (Q banding). *See* BANDING PATTERNS.

quinapril An *angiotensin-converting enzyme inhibitor used to treat hypertension. *TN* Accupril.

quinine An antipyretic, antimalarial alkaloid isolated from the bark of cinchona trees. Originally effective against *Plasmodium falciparum*, although resistance has become common. The mechanism seems to be toxic accumulation of haem, produced by the breakdown of haemoglobin in the infected erythrocyte, because the formation of haemozoin from haem is inhibited.

quinolinic acid An endogenous NMDA receptor agonist and neurotoxin produced in the *kynurenine pathway. May be responsible for the pathogenesis of several major neuroinflammatory diseases, particularly AIDS dementia.

SEE WEB LINKS

- Article in the *Journal of Neuroinflammation* (2005).

quinolone antibiotics A group of compounds (quinolones and fluoroquinolones), derived from nalidixic acid, that inhibit the activity of DNA gyrase. The fluoroquinolones (e.g. ciprofloxacin) have a broad spectrum of antibacterial activity.

quinupristin An antibiotic, produced by *Streptomyces pristinaepiralis*, of the *streptogramin family. Quinupristin/dalfopristin is active against most Gram-positive bacteria, many Gram-negatives, and mycoplasmas.

quisqualate A glutamate analogue that is the agonist for Q-type *glutamate receptors.

rab A family of small *GTP-binding proteins of the *ras superfamily that control intracellular vesicle budding, cargo sorting, transport, tethering, and fusion. Different members of the family are involved in particular transport systems, e.g. **rab7** is a marker of vesicles directed to endosomal/lysosomal compartments and mutations are associated with *Charcot–Marie–Tooth disease type 2B. **Rab11** plays a central role in plasma membrane recycling and there is a large family of rab11 interacting proteins. The **rab escort protein-1** (REP1, 653 aa) is component A of rab geranylgeranyl transferase (EC 2.5.1.60), a heterodimeric enzyme (A and B subunits) that attaches a geranylgeranyl moiety to rab1A and rab3A: mutations cause *choroideaemia. **Rim1α** (1692 aa), a rab3a-interacting molecule, is involved in long-term potentiation in the hippocampus.

rabaptin A protein (862 aa) recruited to endosomes by *rab5-GTP and essential for rab-mediated endosomal fusion.

rabies An encephalomyelitis, frequently fatal, caused by a neurotopic lyssavirus (of the *Rhabdoviridae).

rabphilin A receptor (704 aa) for *rab3A that is implicated in regulated secretion, particularly of neurotransmitters. Rabphilin may inhibit the GTPase activity of rab3A, keeping it in the active state.

rac One of the three subtypes of rho GTPases in the ras superfamily of small G proteins (the others are *rho and *cdc42). They are substrates for botulinum toxin C3 ADP-ribosylation. The active GTP-bound form of rac seems to induce membrane ruffling in motile cells, whereas *rho acts on stress fibres. *Chimerins are rac-specific GTPase activating proteins (rac-*GAPs).

RACE *See* RAPID AMPLIFICATION OF DNA ENDS.

racemic mixture (racemate) A mixture containing equal amounts of the two optical isomers (D- and L-forms) of a chiral molecule.

rachitic *See* RICKETS.

RACKs Proteins (receptors for activated C kinase) that selectively bind activated *protein kinase C and thus control their attachment to membranes or other elements at various locations in the cell. RACK1 (guanine nucleotide-binding protein β_2-like 1, 317 aa) is a conserved protein implicated in numerous signalling pathways and is also a component of the 40S ribosomal subunit.

rad **1.** Abbreviation for radian. **2.** A unit of radiation, 1 rad = 0.01 Gy. **3.** The *rad1* gene in *Schizosaccharomyces pombe* is a *checkpoint control gene; various *rad* genes are of comparable function in other organisms (*Hrad1* from humans, *Mrad1* from mouse, *RAD17* from *Saccharomyces cerevisiae*). **4.** *See* RAD PROTEINS.

RAD21 One of the *rad protein family (double-strand-break repair protein, rad21 homologue, 631 aa) that is a component of the *cohesin complex which is proteolytically cleaved at the metaphase–anaphase transition by *separin, allowing sister chromatids to separate.

radial glial cells Glial cells arranged in parallel tracts between the inner and outer surfaces of the developing cortex, possibly providing a *neuronal guidance cue during development. *See* CONTACT GUIDANCE.

radial spoke The structure in the ciliary axoneme that links an outer microtubule doublet with the sheath around the central pair of microtubules. Spokes are thought to restrict the sliding of doublets relative to one another. There are human homologues of many spoke proteins identified in *Chlamydomonas*; several have domains associated with signal transduction.

radiation inactivation An old method used to explore the size, structure, and function of soluble and membrane-bound enzymes by examining the dependence of enzyme activity on radiation dose.

radicicol An antifungal compound that inhibits *src kinase and competes with *geldanamycin for a binding site on *Hsp90 and inhibits its ATPase activity. It has antiangiogenic activity.

radioimmunoassay (RIA) A sensitive method for measuring the concentration of a substance in a complex mixture. A small (known) amount of radiolabelled analyte is added and the extent of labelled immune complex formation is measured: analyte in the sample competes with the labelled material and its concentration can be calculated. The analyte may be antigen or antibody. A **radioimmunoprecipitation assay** (RIPA) is a qualitative assay used to confirm the presence of antiviral antibodies. Radiolabelled antigen fragments are added to the serum sample, antigen–antibody complexes are precipitated using *protein A and run on a gel; autoradiography shows the antigens to which antibodies exist.

radixin A cytoskeletal protein (627 aa) that binds the barbed ends of microfilaments and may link them to the plasma membrane. Found in *adherens junctions and in the cleavage furrow. Has homology with *ezrin and *moesin (*see* FERM DOMAIN). Mutations lead to autosomal recessive deafness-24.

rad proteins A range of proteins involved in DNA repair. The rad proteins in yeast were originally identified as products of genes that were very sensitive to X-irradiation and were involved in mitotic and meiotic recombination and mating-type switching. Subsequently human homologues of many of these proteins have been identified. The homologues of yeast RAD23, HHR23A and -B, interact with the product of the *xeroderma pigmentosum gene to form a nucleotide *excision repair complex. Rad51 (339 aa) is involved (together with the product of a related gene, *dmc1*) in the repair of double-strand breaks. Mutations to *RAD51* modify susceptibility to breast carcinoma. Various others are known, and defects in DNA repair systems are usually associated with tumour susceptibility. *See* RAD21.

raf An oncogene (v-*raf*), originally identified in a murine sarcoma virus, that encodes a serine/threonine kinase (EC 2.7.11.1, 648 aa). In humans Raf-1 is ubiquitously expressed, whereas A-Raf is predominantly found in urogenital tissue and B-Raf shows the highest expression in neural tissue and testis. All three forms can be activated by GTP-*ras and phosphorylate *MEK, though their potency differs. Mutations in *BRAF* are associated with various tumours, including melanomas, carcinomas, lymphomas, and also *cardiofaciocutaneous syndrome. The *mil* oncogene is the avian equivalent of the murine *raf* oncogene. *See* NOONAN'S SYNDROME.

rag **1.** Rag proteins are a family of four small GTP-binding proteins (RagA–D) that interact with the mTORC1 protein kinase complex and are necessary for the activation of the mTORC1 pathway by amino acids. *Raptor is the key mediator of the rag–mTORC1 interaction.
2. RAG1 and RAG2 (recombination-activating genes 1 and 2) encode proteins (1043 and 1040 aa) that together form a transposase capable of excising a piece of DNA containing recombination signals from a donor site and inserting it into a target DNA molecule. The RAG complex activates the V(D)J recombination process that generates diversity in immunoglobulins and T-cell receptors. Mutations in *RAG1* and *RAG2* can lead to forms of *severe combined immunodeficiency disease and *Omenn's syndrome.

RAGE **1.** A member of the immunoglobulin superfamily of cell surface molecules (receptor of advanced glycation endproducts, 237 aa) that will bind a broad range of ligands, including advanced glycation endproducts (AGEs), amyloid fibrils, *S100/calgranulins, and *amphoterin. It is present on many cells of the immune system and signalling triggered by receptor occupancy may be involved in development of inflammatory responses, diabetes, and Alzheimer's disease.
2. Renal tumour antigen (419 aa), an antigen recognized by cytolytic T cells and encoded by *MAGE-type genes.

RAIDD An adaptor protein (receptor-interacting protein *RIP-associated ICH-1/CED-3-homologous protein with a death domain; caspase and RIP adaptor with death domain, CRADD,199 aa) that mediates the recruitment of *caspase-2 to *tumour necrosis factor receptor-1 (TNF-R1) signalling complex through *RIP kinase.

Raji cells A line of *EBV-transformed lymphocytes derived from a patient with *Burkitt's lymphoma. They will grow in suspension culture and have Fc receptors.

rak A *src-family tyrosine kinase (fyn-related kinase, FRK, 505 aa) found in the nucleus that binds to retinoblastoma (*Rb) protein and suppresses cell growth. Previously named GTK/Bsk/IYK (GTK is gut tyrosine kinase, Bsk is β-cell src-homology kinase, and IYK is intestinal tyrosine kinase).

ral An *oncogene related to **ras* that encodes a multifunctional small GTPase involved in tumorigenesis and in controlling intracellular membrane trafficking.

raloxifine A selective oestrogen receptor modulator, mimicking oestrogen in some tissues and having antioestrogen activity in others, used in the prevention of postmenopausal

r

osteoporosis and to reduce the risk of hormone-positive breast cancer. *TN* Evista.

RAM domain An N-terminal region (RBP-Jκ-associated module) in *notch essential for interaction with other proteins in the signalling pathway. NB There is a prokaryotic RAM domain which is involved in regulating amino acid metabolism.

ramoplanin A peptide antibiotic that sequesters lipid intermediates involved in peptidoglycan biosynthesis. Ramoplanin is structurally related to two lipodepsipeptide antibiotics that also interfere with wall synthesis, janiemycin and enduracidin, and is functionally related to *lantibiotics and glycopeptide antibiotics (e.g. *vancomycin and *teicoplanin).

RAMPs Transmembrane proteins (receptor activity modifying proteins) that interact with G-protein-coupled receptors and alter their specificity. If **RAMP-1** (148 aa) transports the calcitonin receptor-like receptor to the membrane it behaves as a *calcitonin gene related peptide-receptor (CGRPR), but if **RAMP2** (175 aa) is involved it acts as an *adrenomedullin-receptor. **RAMP3** is similar to RAMP1 but does not potentiate responses to CGRP.

ran A small *GTP-binding protein involved, together with *importins and pp15, in protein transport into the nucleus. The nucleotide exchange factor for ran is nuclear (RCC1) and the GTPase-activating factor is cytoplasmic, thus ensuring vectorial transport.

ranatensin A *bombesin-like peptide (11 aa) from the skin of the frog *Rana pipiens.*

r

random amplification of polymorphic DNA (RAPD) A variant of the *polymerase chain reaction using short, nonspecific primers to compare DNA samples. Has been applied to phylogenetic studies. *Compare* DIFFERENTIAL DISPLAY PCR; SUBTRACTIVE HYBRIDIZATION.

SEE WEB LINKS

- A description of the RAPD method.

ranitidine An H_2-receptor antagonist used to treat gastric and duodenal ulcers. *TN* Zantac.

RANK One of the tumour necrosis factor receptor superfamily (TNFRSF11A, 616 aa) that activates NFκB when it binds RANKL (*osteoclast differentiation factor). Mutations cause familial expansile osteolysis, which is similar to *Paget's disease of bone, and *osteopetrosis.

RANKL *See* OSTEOCLAST DIFFERENTIATION FACTOR.

RANTES One of the natural ligands for the CCR5 chemokine receptor (regulated upon activation normal T-expressed and secreted, SCYA5, CCL5), produced by T cells, and chemotactic for monocytes, memory T cells, and eosinophils. It will suppress *in vitro* replication of the R5 strains of HIV-1, which uses CCR5 as a coreceptor.

RAP 1. *Nebulin-related anchoring protein, N-RAP (1175 aa) that occurs at the myotendinous junction in skeletal muscle and the intercalated disc in cardiac muscle. **2.** *See* RECEPTOR ASSOCIATED PROTEIN (RAP). **3.** Rhoptry-associated protein 1 (rap1) of *Plasmodium falciparum,* a potential antigen in an antimalarial vaccine. **4.** RAP74 is the large subunit of the transcription initiation factor (*TFIIF).

rap1A A small 'ras-related protein' (KREV1, 184 aa); a GTP-binding protein that apparently antagonizes *ras, has antimitogenic activity, and negatively regulates IL-2 transcription. Mutations in rap1A disrupt cell migration and cause abnormalities in cell shape. Very similar proteins, **rap1B** (184 aa) and **rap2** (183 aa), interact with the cytoskeleton in platelets and neurons and may be involved in NGF-induced neuronal differentiation and T-cell activation. The rap family are activated by *CRK adaptor proteins and rap-specific *GEFs. *See* TUBEROUS SCLEROSIS.

rapamycin An antibiotic with immunosuppressive activity and structural similarity to *tacrolimus. It inhibits lymphocyte proliferation at a later stage than tacrolimus, although they share an *immunophilin. *See* MTOR.

RAPD *See* RANDOM AMPLIFICATION OF POLYMORPHIC DNA.

rapid amplification of DNA ends A method (RACE) for amplifying either the 5′ end (5′ RACE) or 3′ end (3′ RACE) of a cDNA molecule, if some of the sequence in the middle is already known and knowing that the 5′ end has a poly-A tail; an artificial tail is added to the 3′ end. Variants of the technique have been developed.

SEE WEB LINKS

- A diagram of the RACE method.

rapsyn *See* POSTSYNAPTIC PROTEIN.

raptor A conserved protein (regulatory associated protein of mTOR, 1335 aa) that forms a complex with the mammalian target of *rapamycin (*mTOR) and regulates cell size in response to nutrient levels. *See* RICTOR.

ras A family of *oncogenes, first identified as transforming genes of Harvey and *Kirsten murine *sarcoma viruses. ('ras' from rat sarcoma because the murine Harvey virus obtained its transforming gene during passage in a rat). The gene product is a GTP-binding protein (p21ras) with GTPase activity, that resembles regulatory G-proteins. *See* RAS-LIKE GTPASES. The *KRAS, HRAS,* and *NRAS* genes encode the human cellular homologues of the Kirsten, Harvey, and neuroblastoma oncogenes which are immunologically distinct proteins, but similar in function. The ras superfamily is usually divided into three families, ras, *rab, and *rho.

ras-like GTPases An extensive family of small *GTP-binding proteins (rab, rac, rad, rag, ral, ran, rheb, rho, gem, kir, ric, rin, rit, Ypt). The ***rab** subfamily is involved in regulation of membrane trafficking; *rac and *rho regulate the cytoskeleton. The **rad** subfamily (not related to *rad proteins) includes rad (ras associated with diabetes), gem (immediate early gene expressed in mitogen-stimulated T cells), and kir (tyrosine kinase-inducible ras-like) and are regulated themselves by serine kinases. The ***rag GTPases** are linked to the *mTOR system, ***ral** has been associated with oncogenic transformation, and ***ran** is involved in traffic into the nucleus. ***Rheb** is enriched in brain. **Rin, ric,** and **rit** lack prenylation sequences and are well conserved between *Drosophila* and humans; **rin** is confined to neuronal cells. **Ypts** are the yeast homologues, of which eleven are known.

Rauber's cells Cells found in Rauber's layer, a thin layer of trophoblast that overlies the inner cell mass in the early mammalian embryo.

Rauwolfia serpentina A plant (Indian snake-root) that is the source of various alkaloids, e.g. *reserpine, serpentine, sarpagine, and ajmalicine, that are used in traditional Ayurvedic medicine.

Raynaud's disease (Raynaud's phenomenon) A disorder of the peripheral circulation in which the blood supply, particularly to the fingers, becomes restricted. There appears to be some hereditable component. Similar effects are seen in 'white finger disease', a response to cold and vibration and an occupational hazard e.g. for forestry workers using chain saws.

Rb *See* RETINOBLASTOMA.

RBC, rbc *See* ERYTHROCYTE.

RBD **1.** The *ras-binding domain. **2.** REM sleep behaviour disorder. **3.** The receptor-binding domain found e.g. on spike protein of severe acute respiratory syndrome-associated coronavirus.

RBL-1 cells A line of rat basophilic leukaemia cells that are used as a model for basophils.

RC3 *See* NEUROGRANIN.

rDNA The DNA coding for ribosomal RNA.

RDP **1.** The Ribosomal Database Project, a database of aligned and annotated rRNA gene sequences. **2.** Rapid-onset dystonia-*parkinsonism.

SEE WEB LINKS

- The Ribosomal Database Project homepage.

RDW **1.** The red cell distribution width, a standard parameter in haematology. **2.** RDW rats are a strain of dwarf rats with a missense mutation in the *thyroglobulin (Tg) gene.

RE1-silencing transcription factor A transcriptional repressor (REST, neuron-restrictive silencer factor, 1097 aa) that represses neuronal genes in non-neuronal tissues by binding to the neuron-restrictive silencer element. Also blocks the expression of an miRNA (miR-21) that prevents embryonic stem cells from reproducing themselves and causes them to differentiate into specific cell types.

reactive oxygen species (ROS) General term for the various oxygen-containing molecules produced in the *respiratory burst and responsible for bacterial killing and collateral damage to tissue. Include oxygen radicals, reactive ions such as superoxide, singlet oxygen, and hydroxyl radicals.

reactome A curated resource of core pathways and reactions in human biology developed through collaboration between Cold Spring Harbor Laboratory, the European Bioinformatics Institute, and the Gene Ontology Consortium. The free website provides much more detail than is possible here.

SEE WEB LINKS

- The Reactome homepage.

reading frame One of the three possible ways of reading a nucleotide sequence depending upon which nucleotide is chosen as the first, which defines the subsequent set of triplet codons. Thus AUGGAGACC might be read AUG (start), GAG (Glu), ACC (Thr) or UGG (Trp), AGA (stop) or GGA (Val), GAC (Asp). Usually only one reading frame avoids premature stop codons.

reagin Obsolete name for *IgE.

r

real time PCR A *polymerase chain reaction method (RT-PCR) that can be used to give semiquantitative information about the amount of DNA originally present. It involves measuring the rate of accumulation of PCR product using a fluorescent marker. Should not be confused with reverse transcriptase PCR (also abbreviated as RT-PCR).

reannealing The process of reformation of a double-stranded DNA molecule from single strands produced by thermal dissociation. The rate of reannealing depends upon the probability of getting good base pairing. Highly repetitive sequences will reanneal quickly because there are more ways of getting alignment (they have low values on the *Cot curve).

rebeccamycin An antitumour antibiotic produced by *Saccharothrix aerocolonigenes*. It is a weak topoisomerase I inhibitor similar to staurosporine.

reboxetine An antidepressant drug that acts as a norepinephrine reuptake inhibitor. *TNs* Davedax, Edronax, Norebox, Prolift, Solvex, Vestra.

receptor At a cellular level, an immobilized molecule, usually a membrane-bound or membrane-enclosed protein, that binds to, or responds to, a smaller and more mobile ligand. Specificity of the interaction is implicit. There are rare exceptions in which both receptor and ligand are large and membrane-associated (e.g. *plexins and their ligands, *semaphorins) but more standard examples are *acetylcholine receptors, *photoreceptors, and *nuclear receptors.

receptor activator of NFκB *See* RANK.

receptor associated protein (RAP) A range of different proteins associated with various receptors have been given this name (e.g. *TRAFs). The **LDL receptor associated protein** is a chaperone/escort protein that binds to *lipoprotein lipase (LPL) and is a general antagonist for binding of ligands to LDL receptor family members. The LDL receptor associated protein is also involved in secretion of thyroglobulin.

receptor disorders

A significant number of disorders arise because of mutation in receptors, in some cases blocking the normal signalling pathway simply because the receptor is absent or does not bind ligand, in other cases because the receptor is unregulated and constitutively active (e.g. *stem cell myeloproliferative disorder). There are many types of receptors, ranging from the *G-protein-coupled cell surface receptors, *integrins, nuclear receptors, peroxisome proliferator-activated receptors *PPAR, and receptors for neurotransmitters, hormones, signal sequences, and so on; mutations in any of these can cause problems. Receptors for light and sound, and defects in these systems, are covered under *visual disorders and *auditory disorders. Disorders in which a receptor dysfunction is implicated include *acromegaly, Maroteaux-type *acromesomelic dysplasia, *Alagille's syndrome-2, *Angelman's syndrome, *Apert's syndrome, *arrhythmogenic right ventricular dysplasia/cardiomyopathy-2, *Bernard–Soulier syndrome, *brachydactyly-A2, metaphyseal *chondrodysplasia, *Elejalde's syndrome, *epidermolysis bullosa, *familial hypercholesterolaemia, *hereditary haemorrhagic telangiectasia, *familial primary localized cutaneous amyloidosis, *familial startle disease, *fibrodysplasia ossificans progressiva, *Glanzmann's thrombasthenia, *haemochromatosis-3 and -4, *hereditary lymphedema-1A, *hereditary sensory and autonomic neuropathy-4, *Hirschsprung's disease, various *hormone disorders, *hyperthyroidism and *myasthenia gravis (in which there are antibodies to the receptor), autosomal recessive *hypohidrosis, neonatal *leukodystrophy, *periodic fever, autosomal dominant *pseudo-hypoaldosteronism, *renal tubular dysgenesis, *Schwartz–Jampel syndrome 2, some forms of *stationary night blindness, congenital *stiff person syndrome, and *thrombocythaemia.

receptor down-regulation A response in which the number of receptors on the cell surface is reduced after exposure to ligand because receptor–ligand complexes are internalized rapidly and only replaced slowly. The term is often misapplied.

receptor expression-enhancing protein-1 A nuclear-encoded mitochondrial protein (REEP1, 201 aa) that promotes functional expression of G-protein-coupled odorant receptors on the cell surface. Mutations are associated with autosomal dominant *hereditary spastic paraplegia type 31 (SPG31).

receptor interacting protein 1 (RIP1) This name is given to a variety of proteins and there is scope for considerable ambiguity. **1.** A component of the NFκB signalling network that activates NFκB, leading to an increase in expression of *mdm2 which inhibits *p53. Has been suggested as a prognostic marker for glioblastoma, high levels being associated with poor outlook. **2.** Receptor-interacting protein-140 (RIP140, nuclear receptor interacting protein 1, NRIP1, 1158 aa) associates with a transcriptional activation domain of the oestrogen receptor in the presence of oestrogen. **3.** Eukaryotic initiation factor 3 (eIF3) is TGFβ receptor interacting protein 1 (TRIP1, 325 aa). **4.** Nuclear receptor coactivator 2 is *glucocorticoid receptor-interacting protein 1 (GRIP1). **5.** GRIP1 is also used for *glutamate receptor-interacting protein 1 (1128 aa). **6.** There is a receptor-interacting protein 1 (620 aa) associated with the leucine-rich repeat Ig domain-containing *Nogo receptor. **7.** Another RIP1 interacts specifically with the C-terminal tail of the angiotensin II type 2 receptor (angiotensin II type 2 receptor-interacting protein 1, ATIP1). **8.** There is a CB_1 cannabinoid receptor-interacting protein 1 (CRIP-1, 164 aa). **9.** Thyroid hormone receptor associated protein 1 (TRIP1, 406 aa) is the 26S protease regulatory subunit 8 (proteasome 26S subunit ATPase 5). **10.** *See* RECEPTOR INTERACTING PROTEIN KINASE 1.

receptor interacting protein kinase 1 A serine/threonine protein kinase (RIP kinase) involved in the signal transduction machinery that mediates programmed cell death. RIP interacts with the death domain of *TRADD (TNFR1-associated death domain protein) and may then recruit *RAIDD, leading to apoptosis/inflammation) or NFκB, leading to survival/proliferation. It also interacts with *TRAF2 via either the kinase or the intermediate domain. Multiple isoforms (RIP1 (671 aa), 2, 3, and 4) have been described. A RIP-like kinase (540 aa, *RICK, RIP2) may be involved in mediating Fas signalling.

receptor mediated endocytosis The process by which molecules are taken up by cells using a specific receptor protein that becomes associated with a *coated pit when it binds its ligand. The pit is internalized as a coated vesicle which fuses with endosomes and may release its contents to the cytoplasm or to a lysosome. Many bacterial toxins and viruses enter cells by this route.

receptor potential The *resting potential in a sensory cell, which will change gradually in response to the appropriate stimulus but without generating an action potental.

receptors for activated C kinase *See* RACKs.

receptor tyrosine kinase A large class of transmembrane receptors with cytoplasmic domains that act as tyrosine kinases when extracellular ligand is bound. Many of the growth factor receptors are receptor tyrosine kinases.

receptosome *See* ENDOSOME.

recessive Describing an *allele or *mutation that only affects the phenotype when homozygous. If only a single copy is present then the effect is hidden by the dominant allele.

recombinant DNA DNA produced by linking sequences from different sources, usually before incorporating them in a *vector.

recombinant protein A protein produced using the methods of recombinant DNA technology (genetic engineering) usually by cloning the gene and putting it into an appropriate expression system.

recombinase An enzyme that mediates *site-specific recombination in prokaryotes. *See* LOX-CRE SYSTEM.

recombination The rearrangement of a pair of nonidentical chromosomes so that information (alleles) on one are exchanged for equivalent loci on the other so that the information from maternal and paternal sources is shuffled ('crossing-over'). Recombination actually involves cutting and splicing the DNA duplex and this probably occurs at recombination nodules, transient structures (~90 nm diameter) associated with the *synaptonemal complex during pachytene. The number of nodules corresponds to the number of genetic exchanges. *See* SITE-SPECIFIC RECOMBINATION.

recon A unit of genetic *recombination, the smallest portion of a chromosome in which recombination can occur.

recoverin A calcium-binding protein (cancer-associated retinopathy, CAR, p26, RCVRN, RCV1, 200 aa), one of a family of neuronal calcium sensors that mediates the recovery of the dark current after photoactivation in the retina. Calcium causes a myristoyl group to move away from a binding pocket and allows new protein–protein interactions (calcium-myristoyl switch). NB Recoverin is no longer thought to be an activator of photoreceptor guanylate cyclase. It may be the antigen involved with cancer-associated retinopathy. Other members of the recoverin family are *visinin, *hippocalcin, *neurocalcin, S-modulin, visinin-like protein, and *frequenin.

SEE WEB LINKS

- Further information about recoverin and its role in vision.

rec proteins A set of proteins encoded by *rec* (recombination) genes and involved in genomic repair and recombination in all organisms. RecA (339 aa) aligns a single strand of DNA with a duplex DNA and mediates a DNA strand switch; recB, C, and D are subunits of the helicase that unwinds and fragments double-stranded DNA allowing recA to bind single-stranded fragments.

RECQ-like helicases Human homologues of the *E. coli* RECQ helicase that is involved in *mismatch repair. Five are known: the **RECQL2** gene is defective in Werner's syndrome, **RECQL3** in *Bloom's syndrome, and **RECQL4** in *Rothmund–Thomson syndrome. Mutations in *RECQL1* and *RECQL5* are not associated with known disorders.

rectifying synapse An *electrical synapse at which current flow is unidirectional.

red blood cell *See* ERYTHROCYTE.

redox potential (oxidation/reduction potential) A measure of the tendency of a solution to either gain or lose electrons when it is subject to change by introduction of a new species. In practice, the potential difference between an inert indicator electrode in contact with the solution and a stable reference electrode is then related to the standard hydrogen electrode, which is defined as having a potential of zero.

Reed–Sternberg cells Large lymphoid cells with abundant pale cytoplasm and two or more oval lobulated nuclei containing large nucleoli. They are a common, even diagnostic, feature of *Hodgkin's lymphoma.

reeler A mouse autosomal recessive mutant, deficient in *reelin and with disruption in large areas of the brain. Reelin, a large extracellular protein, is secreted by Cajal–Retzius cells in the forebrain and by granule neurons in the cerebellum. The reeler mouse has been proposed as a neurodevelopmental model for certain neurological and psychiatric conditions; the structural features of the brain most closely resemble those of patients with Norman–Roberts lissencephaly (lissencephaly 2).

reelin The protein product (3460 aa) of the human homologue of the murine *reeler gene, an extracellular matrix component produced by pioneer neurons and important in cortical neuronal migration. Reelin protein is composed of an N-terminal *F-spondin-like domain, eight reelin repeats, and a short and highly basic C-terminal region.

refractile Describing something, that scatters (refracts) light, often applied to intracellular granules. Not to be confused with refractory.

refractory period A term commonly applied to the interval (~1 ms) after an *action potential during which an axon is unresponsive to a second stimulus because the sodium channels are temporarily inactivated. This sets the maximum firing rate at around a few hundred hertz. The term can be applied to any system where a similar insensitive period follows stimulation.

Refsum's disease A relatively mild *peroxisomal biogenesis disorder, characterized by *retinitis pigmentosa and chronic polyneuropathy, that can be caused by mutation in the gene encoding phytanoyl-CoA hydroxylase or the gene encoding *peroxin-7. The infantile form of Refsum's disease can be caused by mutation in the genes encoding *peroxins 1–3.

Reg 1. A family of small secreted proteins of the C-lectin type that are implicated in regeneration, proliferation, and differentiation of the pancreas, liver, and gastrointestinal mucosa. **Reg1** (lithostathine, pancreatic stone protein, 166 aa) is the product of the *REG1* gene (regenerating gene 1) and is a mitogen for pancreatic islet cells. **Reg3β** (HIP/PAP1) is a Schwann cell mitogen (175 aa) produced by motor and sensory neurons during development. **2.** *See* PROTEOSOME.

regulatory sequence (control element) A DNA sequence upstream of a coding region to which molecules such as transcription factors bind and regulate expression.

regulatory T cells (Treg) *See* T-REGULATORY CELLS.

Reifenstein's syndrome A form of male pseudohermaphroditism (partial androgen insensitivity) caused by mutation in the androgen receptor gene on the X chromosome.

rejection In the context of tissue transplantation, the processes that lead to the destruction or detachment of the graft.

relapsing fever A zoonotic disease caused by several species of **Borrelia*.

relative excess risk (RER) A measure used in evaluating risks, e.g. the probability of adverse reactions to a diagnostic dose of radiation, taking account of the background exposure experienced by everybody. The risk due solely to the exposure being evaluated.

relative risk **1.** The ratio of the risk of disease or death among those exposed to the risk among those unexposed. *Synonym*: risk ratio. **2.** The ratio of the cumulative incidence rate in those exposed compared to those who are not exposed. *Synonym*: rate ratio.

relaxation time The time period required for a system to return to normal (or equilibrium) after having been perturbed. In receptor systems that have a *refractory period, the relaxation time includes the time taken for sensitivity to return to normal. More precisely defined formulations are applied in systems such as *magnetic resonance imaging where the spin–lattice and spin–spin relaxation times are important time constants.

relaxin A peptide hormone belonging to the insulin-like hormone superfamily, which includes insulin, relaxin, and insulin-like growth factors. **Relaxin I** is produced by the corpora lutea of ovaries during pregnancy and consists of two peptide chains, A (24 aa) and B (29 aa), covalently linked by disulphide bonds and synthesized as a single-chain precursor. It is released just before parturition and acts to relax muscles and facilitate birth, but also acts in various other organ systems, regulating matrix metalloproteinases (MMPs) and thus the properties of extracellular matrix. A second relaxin gene is expressed specifically in the ovary and **relaxin 3** produces a dose-dependent increase in cAMP production in a human monocytic cell line. The receptors are G-protein coupled. **Relaxin-like factor** (insulin-like peptide 3) is produced by *Leydig cells and seems to have a role in regulating the onset of puberty in males. Mutations in relaxin-like factor lead to cryptorchidism.

relay cell An interneuron in the central nervous system that links the afferent and efferent neurons of a reflex arc.

release factor **1.** A component of the *vitamin B_{12} (cobalamin) transport system in the intestine that dissociates the vitamin B_{12}–*intrinsic factor complex. **2.** A protein that recognizes a *termination codon and causes release of the polypeptide strand from the ribosome–mRNA complex. There are three release factors in *E. coli*, but there is only a single GTP-requiring release factor (eRF, eukaryotic translation termination factor 1, ETF1, 404 aa) in eukaryotes (although there may be a different release factor for mitochondrial protein synthesis).

rel homology domain A conserved DNA-binding domain (~300 aa), first identified in the product of the *rel* oncogene, a transcription factor, and subsequently found to be characteristic of eukaryotic transcription factors such as *NFκB and *NFAT. Phosphorylation of the domain may regulate activity of the factors.

***rel* oncogene** A retroviral oncogene, first isolated from an avian reticuloendotheliosis virus, that encodes a transcription factor of the Rel/NFκB family. Rearrangement of the c-*rel* gene has been detected in various tumours, including breast carcinoma.

renal Describing something associated with the kidney.

renal disorders

The kidneys are primarily involved in the filtration of blood and the pumping of ions and molecules, and many disorders arise through defects in specific transporters, especially *ion channels or *aquaporins. The normal action of the kidneys is important for regulating blood volume and hypertension is often treated with *diuretics that stimulate excretion of water. The *natriuretic peptides and *kallidin, *angiotensin, *renin, *aldosterone, *vitamin D, and *parathyroid hormone regulate renal function. Renal tubular acidosis can arise from mutation in *carbonic anhydrase or in *V-type ATPase. The filtration function depends on specialized basal lamina in the glomerulus (*see* PODOCYTES) and this can be 'clogged' if there are circulating immune complexes (*see* GLOMERULONEPHRITIS, GOODPASTURE'S SYNDROME). Many disorders have renal dysfunction as one of a range of symptoms, e.g. *arthrogryposis, renal dysfunction and cholestasis (ARC) syndrome, *Barttin's syndrome 4, *cystinosis, *Fanconi's syndrome, *focal segmental glomerulosclerosis, *GATA-3, *Henoch–Schönlein purpura, *hypomagneaemia with hypocalcuria and nephrocalcinosis, *nephronophthisis, *polycystic kidney disease, *pseudohypoaldosteronism, *tuberous sclerosis, *hyperuricaemia, *uromodulin, *Wilms' tumour. Development of the kidneys is defective in *Potter's syndrome and *renal tubular dysgenesis. Renal cell carcinomas arise from cells of the proximal tubule cells and are the commonest of the tumours of the kidney in adults. Obsolete names include hypernephroma and Grawitz's tumour. *Pax 2 is mutated in renal–coloboma syndrome (papillorenal syndrome) in which there is a gap in the structure (coloboma) of the optic nerve and renal disease.

renal tubular dysgenesis An autosomal recessive disorder of renal tubular development characterized by persistent fetal anuria and perinatal death. Can be caused by mutations in genes encoding components of the renin–angiotensin system: *renin, *angiotensinogen, *angiotensin-converting enzyme, or angiotensin II receptor type 1.

renaturation The restoration of normal conformation in a denatured protein or DNA. Renaturation of proteins is relatively unusual, but DNA denatured by heating will undergo *reannealling quite readily.

renin (angiotensinogenase) An *aspartic peptidase (EC 3.4.23.15, 406 aa) released from afferent arterioles in the kidney in response to reduced blood flow, a drop in plasma sodium levels, or a diminution in plasma volume. Renin cleaves *angiotensin I from *angiotensinogen and renin inhibitors (e.g. *aliskiren) have potential for the treatment of hypertension.

Reoviridae Class III viruses that have a segmented double-stranded RNA genome with eight to ten segments each coding for a different polypeptide. Only one of the RNA strands (the minus strand) acts as template for *mRNA (plus strand). The capsid is icosahedral and the *virion has all the enzymes needed to synthesize mRNA. Reoviruses are found in the gut and respiratory tract where they seem to be nonpathogenic ('respiratory, enteric, orphan viruses'), although some pathogenic viruses including *orbivirus and *rotavirus are now classed as reoviruses.

repair nucleases Enzymes involved in DNA repair; *endonucleases that recognize a site of damage or an incorrect base pairing and excise them, and exonucleases that remove the neighbouring nucleotides on one strand. *See* CERNUNNOS; DAMAGED DNA-BINDING PROTEINS; DNA LIGASE; DNA POLYMERASE; EXCISION REPAIR; MISMATCH REPAIR; NONHOMOLOGOUS END-JOINING; SOS SYSTEM, also diseases in which repair is deficient such as *ataxia telangiectasia and *xeroderma pigmentosum.

repellent guidance molecule General term for molecules that inhibit *growth cone movement and regulate the developing nervous system. *See* COLLAPSIN; REPULSIVE GUIDANCE MOLECULE.

repetin One of the *'fused' gene family of proteins (784 aa) that is associated with cytokeratin intermediate filaments and partly crosslinked to the cell envelope of keratinocytes. Has calcium-binding *EF-hands of the *S100 type. Expressed throughout the epidermis but at higher levels in the *acrosyringium, the inner hair root sheath, and the filiform papilli of the tongue.

repetitive DNA DNA sequences that are present in multiple copies and will therefore reanneal more rapidly (*see* C_0T CURVE). Highly repetitive DNA is found as short sequences (5–100 nucleotides) that may be repeated thousands of times in a single long stretch. It is mostly *satellite DNA, makes up 10–15% of mammalian DNA, and includes tandem repeats. Moderately repetitive DNA comprises 25–40% of the DNA, consists mostly of sequences ~150–300 nucleotides in length dispersed evenly throughout the genome, and includes *Alu sequences and *transposons.

replica plating A technique for producing multiple copies of a set of bacterial colonies plated on a petri dish. Multiple tests (e.g. for nutritional requirements or antibiotic sensitivity) can then be carried out. The replicas are made using a velvet pad to inoculate a series of plates.

replicase Generic name for any enzyme that duplicates a polynucleotide sequence (either RNA or DNA). The term is mostly used for the enzyme involved in the replication of single-stranded viral RNA.

replication A term usually applied to the production of daughter strands of nucleic acid from the parental template.

replication factors Proteins involved in DNA replication. **Replication factor A** (replication protein A) is a heterotrimeric single-stranded-DNA-binding protein (subunits p70, 616 aa; p32, 270 aa; p14, 121 aa) that associates, together with the kinase *ATM, at sites where homologous regions of DNA interact during meiotic prophase and at breaks associated with meiotic recombination. It is essential for replication of SV40 virus. **Replication factor C** (activator 1) is a five-subunit accessory protein (clamp loader) that loads, in an ATP-dependent manner, various components of the DNA replicating machinery at the replication fork. and interacts with proliferating cell nuclear antigen (PCNA).

replication fork The region where the DNA duplex is separated into single strands and complementary daughter strands are synthesized. Various proteins (helicases, polymerases, ligases, etc.) constituting the replisome, are associated with the fork.

replicative intermediate A copy of the original RNA strand of an RNA virus formed during the process of replication.

replicons The segment of DNA associated with an *origin of replication. Replicons, each ~30 µm long, are arranged in tandem along the chromosome.

replisome *See* REPLICATION FORK.

reporter gene A gene encoding a protein that can be visualized easily, either directly (e.g. *green fluorescent protein) or indirectly (e.g. *CAT), that is engineered to have the promoter of the gene of interest. The reporter construct, once transfected, shows where the gene would normally be expressed.

repressor protein A DNA-binding protein that binds to the *operator region and blocks *transcription. Originally described as a control mechanism for regulating bacterial gene expression (*see* LACTOSE OPERON) but analogous mechanisms exist for eukaryotes, although they are more usually thought of as transcription factors (e.g. *SMRT, silencing mediator of retinoic acid and thyroid hormone receptors).

reproductive cloning The cloning of an organism by transferring a nucleus into an enucleated oocyte *in vitro* and allowing the embryo to implant and develop to full term in a surrogate mother. Most countries ban the application of the technique to humans, but it has been used successfully in sheep and a few other species. The embryo is chimeric since the mitochondria (and crucial morphogenetic information) are derived from the enucleated egg.

repulsive guidance molecule A GPI-linked protein, originally identified in the chick, that is involved in neuronal differentiation and survival and selectively affects the outgrowth of particular classes of axons through an interaction with *neogenin. The cDNA for a human form has been reported. The more general term is *repellent guidance molecule.

RER *See* RELATIVE EXCESS RISK; ROUGH ENDOPLASMIC RETICULUM.

resealed ghosts The remnants of erythrocytes that have been lysed, released their contents, and then allowed to seal by altering the composition of the medium. They are relatively impermeable and have been considered as vehicles for drug delivery.

residual body **1.** A *secondary lysosome containing indigestible material. **2.** The surplus cytoplasm shed during the differentiation of spermatozoa from spermatids. **3.** Cytoplasm containing pigment that remains after the production of merozoites during schizogony of *malaria parasites.

resiniferatoxin An analogue of *capsaicin from *Euphorbia resinifera* (a flowering succulent) that is an ultrapotent agonist for *vanilloid receptor-1.

resistin A cysteine-rich secreted protein, an *adipokine (found in inflammatory zone 3, FIZZ3, 108 aa) produced by adipocytes and by immunocompetent cells that may modulate insulin resistance. Polymorphisms are associated with susceptibility to type 2 diabetes. **FIZZ2** (resistin-like protein β) has ~50% sequence homology and probably has similar functions.

resolvase An enzyme complex that sorts out the DNA entanglements that are generated by recombination events (*see* SITE-SPECIFIC RECOMBINATION). The resolvases are a family of *recombinases sharing various structural features including a conserved serine residue involved in the transient covalent attachment to DNA. The Holliday junction resolvase complex converts a *Holliday junction into two separate duplex DNA molecules.

resolving power **1.** Describing the properties of an optical system, the distance between two objects that allows them to be seen as distinct entities. The resolving power of a microscope depends on the numerical aperture (NA) of the objective, the contrast between the objects and the background, and the shape of the objects. **2.** In genetics, the smallest *chromosome map distance that can be experimentally determined with a defined number of recombinant progeny.

resolvins A family of atypical eicosanoids analogous to *lipoxins and associated with resolution of the inflammatory response; derived from omega-3 fatty acids, eicosapentaenoic acid (resolvin E series) and docosahexaenoic acid (resolvin D series) by the *cyclo-oxygenase pathway, especially in the presence of aspirin.

resonance energy transfer *See* FLUORESCENCE energy transfer.

resorcinol (resorcin) A compound used as a topical treatment for dermatological conditions such as acne, eczema, and psoriasis, also used in the synthesis of resins and dyestuffs.

resorufin A pink fluorescent dye. A nonfluorescent form (caged resorufin) coupled to G-actin has been used to investigate microfilament dynamics; the fluorescent form is 'released' by UV irradiation.

respiration **1.** In physiology, the process of breathing. **2.** In biochemistry, the intracellular oxidation of substrates coupled to ATP production. Anaerobic respiration occurs in *glycolysis; aerobic respiration involves the *tricarboxylic acid cycle and the *electron transport chain.

respiratory burst *See* METABOLIC BURST.

respiratory chain *See* ELECTRON TRANSPORT CHAIN.

respiratory disorders

The gaseous exchange surface of the lung is exposed to many environmental insults and several respiratory disease are a result of inflammation induced by allergens (generically, *pneumoconiosis, with distinct forms such as *anthracosis, *aspergillosis, *asbestosis, *bagassosis, *beryllicosis, *farmer's lung, *siderosis, and *silicosis.) Asthma may be triggered by allergens, *Goodpasture's syndrome can affect the lung. The lung is protected by the *bronchus associated lymphoid tissue and *surfactant proteins produced by *Clara cells are essential to prevent the lung from filling with fluid (*see* HYALINE MEMBRANE DISEASE). Persistent inflammation in the lung can cause *chronic obstructive pulmonary disease or emphysema (*see* ALPHA-1-ANTITRYPSIN) and when triggered by asbestos can lead to *mesothelioma. Lung infections include *histoplasmosis, *pneumonia, *SARS, *tuberculosis, and *influenza. Systemic inflammation can cause *acute respiratory distress syndrome. The lungs are one of the major systems affected in *cystic fibrosis.

respiratory enzyme complex The series of mitochondrial enzyme complexes involved in the *electron transport chain, the final one being *cytochrome oxidase.

respiratory quotient (RQ) The dimensionless molar ratio of carbon dioxide produced to oxygen consumed in respiration. Values range from 1.0 (pure carbohydrate oxidation) to ~0.7 (pure fat oxidation).

respiratory syncytial virus An enveloped RNA virus of the *parainfluenza family, responsible for coughs and colds in the winter period.

response elements The DNA binding sites for transcription factors such as *CREB and *interferon stimulated response element. Usually within 1 kb of the transcriptional start site.

resting potential The electrical potential of the inside of a cell, relative to its surroundings, usually in the range −20 to −100 mV, with −70 mV being a typical value. The resting potential is mainly a result of the leakage of potassium ions down the concentration gradient produced by the *sodium-potassium ATPase pump and values are close to the *Nernst potential for potassium. *See* ACTION POTENTIAL.

restriction endonucleases Enzymes (restriction enzymes, restriction nucleases) that cut single or double-stranded DNA at specific sites (restriction sites). They are important in destroying foreign DNA in bacteria and are also a key experimental tool for genetic engineering. **Type I restriction endonucleases** cut at a site remote from the recognition site and their activity is restricted (blocked) by methylation. **Type II** restriction endonucleases have very specific recognition and cutting sites. The recognition sites are of four to eight nucleotides, usually *palindromic sequences, so that both strands of a duplex can be cut. If the site of cleavage is offset from the recognition site the resulting fragments have short single-stranded tails which can hybridize to the tails of other fragments ('sticky ends' as opposed to 'blunt ends'). **Type III** restriction enzymes recognize two separate nonpalindromic sequences. The nomenclature is based on the bacterium from which they were isolated (first letter of genus name and the first two letters of the specific name), the strain of bacterium, and, if necessary, a number. Thus the two enzymes from the R strain of *E. coli* are designated Eco RI and Eco RII.

restriction fragment length polymorphism (RFLP) A method of identifying individuals based on polymorphisms in genomic DNA ('DNA fingerprinting'). The DNA is amplified using the *polymerase chain reaction and then cut by a *restriction endonuclease (different enzymes will, of course, generate different fragments). Differences in the DNA sequence determine where cleavages occur, and the pattern of fragments (restriction fragments) on a gel is characteristic of the the individual and similar in close relatives. The technique has important uses in genealogy and in forensic analysis of biological samples. A polymorphism close to the locus of a genetic defect can be a useful marker for the defective allele.

restriction map A diagram of a DNA sequence showing the position of recognition and cleavage sites for various *restriction endonucleases.

restriction point The point in the cell cycle that marks irreversible commitment to completion of the cycle. Usually near the end of *G_1. *See* CHECKPOINT.

restrictive temperature *See* PERMISSIVE TEMPERATURE.

restrictocin *See* ASPERGILLIN.

retention signal Usually used to refer to the amino acid sequence (C-terminal KDEL) that labels a protein for retention in the secretory processing system, although it can be applied to a situation in which macromolecules are selectively retained in a subcellular compartment.

reticular fibres A histological term for fibres formed from type III collagen (reticulin) and a feature of the extracellular matrix of lymph nodes, spleen, liver, kidneys, and muscles.

reticular lamina The layer of fibrillar extracellular matrix immediately below the *basal lamina of epithelial cells. The reticular lamina contains collagen and elastin and is secreted by connective tissue fibroblasts. The reticular lamina and the basal lamina together constitute the so-called 'basement membrane' (a term better avoided).

reticulin *See* RETICULAR FIBRES.

reticulocalbins Members of the *CREC family of low-affinity calcium-binding proteins found in the lumen of the endoplasmic reticulum. Include: reticulocalbin-1 (331 aa), reticulocalbin-2 (calcium-binding protein ERC-55, E6-binding protein, 317 aa), reticulocalbin-3 (EF-hand calcium-binding protein RLP49, 328 aa).

reticulocytes Immature erythrocytes in the bone marrow and, rarely, in the circulation. Reticulocyte lysate, a cell homogenate from reticulocytes, is used as an *in vitro* translation sytem.

reticuloendothelial system An old term coming back into use to refer to the phagocytic system of the body, including the fixed macrophages of tissues, liver, and spleen.

reticulol A cytotoxic antibiotic isolated from a strain of *Streptoverticillium* sp. that inhibits topoisomerase I and cAMP-phosphodiesterase.

reticulum cells A class of peripheral dendritic cells with interdigitating cytoplasmic surface projections, found in T-cell areas of lymph nodes, spleen, thymus, and other lymphoid organs.

retina The light-sensing layer of the eye. Looking through the cornea into the eye the first layer is the neural retina, containing neurons (ganglion cells, *amacrine cells, *bipolar cells) and retinal blood vessels; light must traverse this layer to reach the single layer of rods and cones, the photoreceptors themselves. Underlying the photoreceptor layer is the *pigmented retinal epithelium (PRE or RPE) and below this the choroid, composed of connective tissue, fibroblasts, and a well-vascularized layer, the chorio capillaris, immediately below the basal lamina of the PRE. The outermost layer of the eye is the sclera, a thick organ capsule. *See* RETINAL CONE; RETINAL ROD; RHODOPSIN.

retinal The aldehyde of *retinoic acid (vitamin A) which is complexed with opsin to form the photosensitive pigment *rhodopsin. Absorption of light causes retinal to shift from the 11-*cis* to the all-*trans* configuration, and through a complex amplification cascade excites neurons that are synapsed with the *retinal rods and cones.

retinal cone The photoreceptor cells essential for colour vision and high acuity vision, present in large numbers in the fovea. In the human eye there are three types of cones, differentially sensitive to particular wavelengths of light depending on the opsin variant expressed. The first is most sensitive at long wavelengths (564–580 nm, yellow), the second to medium wavelengths (534–545 nm, green), and the third to shorter wavelengths (420–440 nm, violet). Neural processing allows a full spectrum of colours to be perceived. *See* COLOUR BLINDNESS; *compare* RETINAL ROD.

retinal ganglion cell *See* GANGLION (ganglion cell).

retinal pigmented epithelial cell *See* PIGMENTED RETINAL EPITHELIUM; RETINA.

retinal rod The major photoreceptor cell of the *retina (~125 million in a human eye). They are columnar cells (~40 μm high, 1 μm diameter) with three distinct regions: nearest the vitreous and adjacent to the neural retina, a region that forms synapses with neurons of the neural retina and contains the nucleus and other cytoplasmic organelles; below this is the mitochondrion-rich inner segment that is connected through a thin 'neck' (in which is located a *ciliary body) to the outer segment. The outer segment consists of a stack of discs (actually membrane infoldings that are incompletely separated in cones) which are produced next to the inner segment and progressively move towards the distal end where they are shed and phagocytosed by the pigmented epithelium. The membranes of the discs contain the *rhodopsin. Rods are more light-sensitive than cones and respond slightly less quickly; they are important for night vision.

retinitis pigmentosa A disorder in which abnormalities of rods, cones, or pigmented retinal epithelial cells lead progressively to loss of vision. There is considerable genetic heterogeneity with at least 35 different genes being involved; the following list is not exhaustive, but illustrates the diversity. The mutation in **retinitis pigmentosa 1** (RP1) is in an oxygen-regulated photoreceptor protein called ORP1; **X-linked retinitis pigmentosa 3** (RP3) is caused by mutations in the retinitis pigmentosa GTPase regulator; in **RP7** the mutation is in the gene encoding the photoreceptor type of *peripherin. **RP9** is due to mutation in the gene encoding *PIM1-associated protein (*PAP1*). **RP11** is caused by mutations in the *PRPF31* gene that encodes a spliceosome component; in **RP12** the mutation is in the gene for the homologue of the *Drosophila* gene *crumbs* that is involved in epithelial development; **RP14** is caused by mutations in the *tubby-like protein, TULP1. **RP37** is caused by mutation in a photoreceptor cell-specific nuclear receptor. Other forms are known, e.g. a retinitis pigmentosa–deafness syndrome due to mutation in the *MTTS2* gene (encoding the tRNA for serine).

retinoblastoma A malignant tumour of the retina, usually arising in the inner nuclear layer of the neural retina. In a few cases it can be caused by an autosomal dominant mutation, in which case it may be bilateral. The **retinoblastoma gene product** is a tumour suppressor (Rb), a nuclear protein (928 aa) that is a negative regulator of cellular proliferation. The Rb protein forms inactive complexes with various transcription factors such as *myc and *E2F. Mutation in the *RB* gene markedly increases the probability of developing cancer, classically retinoblastoma, but also other *sarcomas and *carcinomas.

retinoblastoma-binding proteins A family of proteins that bind to the retinoblastoma gene product. RBBP4 (histone-binding protein RBBP4, chromatin assembly factor 1 subunit C, CAF-I p48, nucleosome-remodelling factor subunit RBAP48, 425 aa) is a component of several complexes that regulate chromatin metabolism, including the chromatin assembly factor 1 (*CAF-1) complex, the core *histone deacetylase (HDAC) complex, the nucleosome remodelling and histone deacetylase complex (the NuRD complex), the PRC2/EED-EZH2 complex, and the NURF (nucleosome remodelling factor) complex. **Retinoblastoma interacting myosin-like protein** (retinoblastoma-binding protein 8, RBBP8, RIM, 897 aa) interacts with Rb and may be a tumour suppressor; mutations are found in some carcinomas.

retinoic acid The oxidized form of retinol (vitamin A) which has a variety of roles. It acts by binding to heterodimers of the retinoic acid receptor (RAR) and the retinoid X receptor (RXR), which regulate genes, including some homeobox genes, that have a **retinoic acid response element** (RARE). Retinoic acid is thought to be a morphogen in chick limb-bud development.

retinoic acid receptor related orphan receptor A family of nuclear receptors involved in many pathophysiological processes such as cerebellar ataxia, inflammation, atherosclerosis, and angiogenesis. **RORα** (RORA, 468 aa) may regulate production of a bone matrix component. **RORβ** (RORB, 459 aa) shows a marked circadian rhythm of expression in the brain and may be involved with clock-related functions. **RORγ** (RORG, 560 aa) is expressed at highest level in skeletal muscle and also in T cells where it seems that the decision of antigen-stimulated cells to differentiate into either *Th17 or *Treg cells depends on the cytokine-regulated balance of RORγ and *FOXP3.

retinoic acid regulated nuclear matrix associated protein A protein (RAMP, denticleless protein homologue, 730 aa) that is required for *cdt1 proteolysis in response to DNA damage and that helps ensure cell cycle regulation of DNA replication. Increased levels have been observed in gastric carcinoma.

***ret* oncogene** An *oncogene (rearranged during transfection proto-oncogene) that encodes a receptor *tyrosine kinase which forms part of the complex (GFRα1, RET, and *NCAM) through which glial-cell-line-derived neurotrophic factor (*GDNF) acts. Autosomal dominant gain-of-function mutations cause hereditary medullary thyroid carcinoma; loss-of-function mutations are associated with *Hirschsprung's disease and some *ret* mutations are associated with *multiple endocrine neoplasia.

retrograde axonal transport The transport system that moves vesicles from the synapse towards the cell body of a neuron. The mechanism depends on cytoplasmic dynein interacting with microtubules.

retrolental fibroplasia (retinopathy of prematurity) *See* FIBROPLASIA.

retromer A multiprotein complex involved in recycling transmembrane receptors, such as the cation-independent mannose-6-phosphate receptor, from endosomes to the *trans*-Golgi network. *See* SORTING NEXINS.

retrotransposon *See* TRANSPOSON.

retroviral vector Vectors based on retroviruses (*see* RETROVIRIDAE) that have the ability to integrate into the genome of the infected cell. Retroviral vectors can be used to introduce *reporter genes into cells and, because they are integrated, they are copied together with the cell's genome. This allows labelling of cell lineages.

Retroviridae A class of single-stranded RNA viruses (class VI) which, following infection of a host cell, generate a DNA replica of their genome using virally coded *reverse transcriptase. The mature virus is enclosed in host-derived plasma membrane with virally coded membrane proteins, and is released without cell lysis. A heterogeneous set of retroviruses will cause tumours and are sometimes grouped as the Oncovirinae, although the viruses involved are not closely related. Other subclasses are the *Lentivirinae and Spumavirinae. Endogenous retroviruses are integrated into the host genome. *See* RETROVIRAL VECTOR.

Rett's syndrome A severe progressive neurological disorder, X-linked and lethal in males, that causes arrested development, loss of some acquired skills, and often some stereotypical movements. The cause is a mutation in the gene encoding *methyl-CpG-binding protein 2 (MeCP2). An atypical form of Rett's syndrome can be caused by mutation in the cyclin-dependent kinase-like 5 (*CDKL5*) gene and a variant of Rett's syndrome may sometimes be caused by mutation in the *FOXG1* gene that encodes a forkhead-box transcription factor.

reverse antagonist *See* INVERSE AGONIST.

reverse genetics A casual term to describe the approach to determining a gene's function by first sequencing the gene and then trying to deduce function from phenotypic changes when the gene is mutated or silenced.

reverse passive haemagglutination A *haemagglutination response that occurs when antibody-coated erythrocytes are exposed to soluble multivalent antigen. The method can be used e.g. to detect hepatitis B virus in serum.

reverse transcriptase The enzyme that synthesizes double-stranded DNA using an RNA template. It was first discovered in retroviruses but subsequently shown to be involved in transposon movement, in the replication of some other viruses, and possibly in the production of pseudogenes from mRNA. Experimentally it is an important enzyme, e.g. for the construction of *cDNA libraries.

reversion Describing the restoration of normal function that can occur when a second mutation reverses the effects of a first mutation or compensates in some way.

Reynolds number A dimensionless constant that relates the resistance to movement through a fluid, derived from inertial and viscous drag. At cellular dimensions viscous drag is the major source of drag and the Reynolds number is very small.

RF *See* RELEASE FACTOR.

RFC **1.** Reduced-folate carrier (solute carrier family 19 member 1, SLC19A1, 591 aa) a transport protein for folate. **2.** *See* REPLICATION FACTORS (replication factor C).

RFLP *See* RESTRICTION FRAGMENT LENGTH POLYMORPHISM.

RGD motif A tripeptide motif that is the binding site for *integrins. It is found in *fibronectin and in *disintegrins. In many cases the consensus is RGDS (Arg-Gly-Asp-Ser).

RGS proteins A large and diverse family of proteins that regulate G-protein signalling. Originally identified as GTPase activating proteins (GAPs) for the Gα-subunits of heterotrimeric G-proteins but later shown to have a broader range of activities. A characteristic feature is a **RGS domain**, a conserved stretch of 120 aa that binds to activated Gα subunits.

SEE WEB LINKS

- RGS domain structure.

rhabdomyosarcoma A rare malignant tumour (sarcoma), usually occurring in childhood, that is derived from striated muscle precursor cells (satellite cells of muscle).

Rhabdoviridae Class V viruses that have a single negative strand RNA genome and a virus-specific RNA polymerase. The capsid is bullet shaped and covered by an envelope, derived from the host cell when the virus buds off, with host lipids but only viral glycoproteins. The glycoproteins (usually 1–3 species) appear ultrastructurally as regularly arranged spikes ~10 nm long (spike glycoproteins). The class includes *rabies virus, *vesicular stomatitis virus, and a number of plant viruses.

rheb A ras-related GTP-binding protein (ras homologue enriched in brain, 184 aa) that regulates *mTOR through FK506 binding protein-8 (FKBP38). *Rheb* is a candidate gene for sacral agenesis (*see* CURRARINO'S SYNDROME), which maps to this region.

rhegmatogenous retinal detachment A form of retinal detachment that usually results from a break or tear in the retina, allowing fluid from the vitreous humour to infiltrate and displace the retina. An autosomal dominant form is caused by defects in *collagen COL2A1.

rheotaxis A directed movement (*taxis) in response to fluid flow.

rhesus blood group A human blood group system with five main antigens, C, c, D, E, and e with the D antigen (rhesus factor, Rh) being the most important. Rhesus-positive (Rh^+) cells have the D antigen, rhesus-negative cells (Rh^-) do not. Rh^+ fetal cells can cross the placenta and sensitize a Rh^- mother; maternal anti-D antibody can then cause haemolytic disease of the newborn (*erythroblastosis fetalis) in a subsequent pregnancy if the fetus is Rh^+.

SEE WEB LINKS

- The Blood Group Antigen Gene Mutation database.

rheumatoid arthritis A chronic, systemic autoimmune disorder that causes inflammation and tissue damage in joints (arthritis) and tendon sheaths, although there may be more extensive tissue involvement and systemic effects. The autoantigen in unknown but there is an association with HLA-DR4 and related allotypes of MHC class II. Many (85%) patients have circulating *rheumatoid factor.

rheumatoid factor An antibody directed against the Fc portion of IgG found in ~85% of patients with *rheumatoid arthritis and in sera from patients with other autoimmune diseases. Can be of any class but is often IgM.

Rh factor *See* RHESUS BLOOD GROUP.

rhinovirus A class of viruses of the *Picornaviridae that infect the upper respiratory tract. Main examples are the common cold virus and foot-and-mouth disease virus.

rhizomelic chondrodysplasia punctata *See* CHONDRODYSPLASIA.

rhizopodin An actin-binding macrolide isolated from *Myxococcus stipitatus* that is weakly cytostatic and causes cells to produce long, narrow, branched extensions (resembling rhizopods) in the same way as *latrunculin.

rhizoxin An antimitotic drug with potential antitumour activity that binds to β-tubulin and blocks microtubule assembly. It is isolated from a pathogenic plant fungus, *Rhizopus microsporus*, although is actually synthesized by *Burkholderia rhizoxina*, an endosymbiotic bacterium of the fungus.

rhizoxin/maytansine site The binding site on *tubulin for various microtubule-disrupting agents including *rhizoxin, maytansine, *phomopsin A, *cryptophycin-1 and -52, *dolastatin 10, and *hemiasterlin. It is distinct from the binding site for *vinca alkaloids.

rhodamines A group of red fluorescent dyes used in labelling proteins and membrane probes.

SEE WEB LINKS

- Absorption and emission maxima for various fluorochromes.

rhodopsin (visual purple) The light-sensitive pigment involved in vision (maximal absorption at 495 nm), formed from *retinal and *opsin. It is an integral membrane protein (seven transmembrane spanning, G-protein coupled) and makes up 40% of the membrane in the discs of retinal rods. The visual pigments of cones are about 40% homologous (*see* COLOUR BLINDNESS). *Compare* BACTERIORHODOPSIN.

rhodopsin kinase A *G-protein dependent receptor kinase (GRK1, 564 aa) that mediates rapid desensitization of rod photoreceptors to light by phosphorylating *rhodopsin which is then bound by cytosolic *arrestin, and uncoupled from *transducin. *See* RECOVERIN. Some forms of Oguchi's disease, a recessively inherited form of stationary night blindness, are due to defects in GRK1, others to mutations in the arrestin gene.

rho factor (ρ factor) A hexameric ATP-dependent helicase involved in release of RNA transcripts from the elongation complex in prokaryotes. Has no connection with eukaryotic rho.

***rho* genes** Genes coding for small *GTP-binding proteins that regulate the interaction of the actin cytoskeleton with focal adhesions. *See also* RAS and RAB.

rho kinases Serine/threonine kinases (ROCKs) that are activated when bound to GTP-rho and are involved in effecting rho-mediated control of stress fibres and focal adhesions by phosphorylating and and activating *LIM kinase. Abnormal activation of the rhoA/ROCK pathway may be important in cardiovascular disorders. A selective inhibitor, fasudil, has been reported. ROCK2 (1388 aa) is an isozyme of ROCK1 (1354 aa).

rhombencephalon The hindbrain, subdivided into metencephalon, myelencephalon, and the reticular formation.

Involved in regulating very basic activities: sleep, autonomic functions, motor coordination, reflex movements, attention, and simple learning.

rhombomeres Regions of the hindbrain that derive from morphologically distinct developmental segments (*neuromeres). Cells of adjacent rhombomeres do not mix with each other and different regulatory genes are expressed.

rhophilin The GTPase (643 aa) that acts on *rho. It has a C-terminal *PDZ domain suggesting that it might act as an adaptor molecule. **Rhophilin-associated tail protein 1** (Ropn1, ropporin, 212 aa) is localized in the principal piece and the end piece of sperm flagella, where it may form a complex with rhophilin. Ropn1-like protein (AKAP-associated sperm protein) has homology with ropporin and with two other sperm-specific proteins, SPA17 and fibrousheathin II.

rhotekin A protein (551 aa) that has a rho-binding domain and is a target (effector) for the rho-GTPases. Rhotekin is highly expressed in brain, found at low level in normal cells and is overexpressed in many cancer-derived cell lines. A rhotekin-interacting protein PIST (PDZ domain protein interacting specifically with TC10 (a rho-family small GTPase)) may be involved in determining cell polarity.

RIA *See* RADIOIMMUNOASSAY.

ribavirin A prodrug that is metabolically activated to generate a guanoside analogue that inhibits inosine monophosphate dehydrogenase and blocks the replication of various RNA and DNA viruses. Used to treat hepatitis C infections. *TNs* Copegus, Rebetol, Ribasphere, Vilona, Virazole.

ribbon synapse A structurally distinct synapse found in a variety of sensory receptor cells where transmitter is continuously released in response to small graded changes in potential, rather than by an action potential. Synaptic vesicles are attached to the presynaptic membrane in dense bars or ribbons and there may be multivesicular release. The ribbon-specific proteins *ribeye and *bassoon are responsible for the physical integrity of the photoreceptor ribbon complex.

ribeye An alternate product (985 aa) of the C-terminal binding protein 2 (*CTBP2*) locus, a specific component of *synaptic ribbons in the retina. The CTBPs are ubiquitously distributed cellular proteins that bind to the C-terminal portion of adenovirus E1A proteins and are important for the oncogenic properties of adenoviruses

riboflavin *See* VITAMIN B (vitamin B_2).

ribonuclease (RNAse, RNAase) Any enzyme that will hydrolyse RNA. They can be endonucleases or exonucleases and apparently recognize their substrates on the basis of tertiary structure rather than sequence. They are involved in degradation of RNA in both turnover and digestion, as well as in normal processing of RNA and in defence against viruses. A nonspecific RNAse is involved in angiogenesis (*see* ANGIOGENIN). **RNAse A** (EC 3.1.27.5) is a pancreatic enzyme, extremely heat-stable and often used experimentally. **RNAse E** is involved in the formation of 5S ribosomal RNA from pre-rRNA. **RNAses F and L** are induced by interferons and involved in immunity to viruses. **RNAse H** (EC 3.1.26.4) specifically cleaves RNA base-paired to a complementary DNA strand, leaving single-stranded DNA. **RNAse P** (EC 3.1.26.5) is a *ribozyme and acts as an endonuclease in cleaving t-RNAs from their precursors. **RNAse T** is an endonuclease and removes the terminal AMP from the 3′ end of a nonaminoacylated tRNA. **RNAase T1** cleaves RNA specifically at guanosine residues. **RNAase III** cleaves double-stranded regions of RNA molecules.

ribonucleic acid *See* RNA.

ribonucleoprotein (RNP) A complex of RNA and protein. The classic example is the *ribosome but increasing numbers are becoming known, especially those involved in processing mRNA (*see* SPLICEOSOMES) and in terminating transcription. Other examples include the *signal recognition particle that targets proteins to the endoplasmic reticulum.

ribophorin Ribophorins I and II are abundant, highly conserved transmembrane glycoproteins located on the rough endoplasmic reticulum that interact with ribosomes while cotranslational insertion of membrane or secreted proteins is taking place. They form part of an *N*-oligosaccharyl transferase complex involved in post-translational glycosylation and also form part of the regulatory subunit of the 26S *proteosome.

riboprobe An RNA segment, labelled appropriately, used to probe for a complementary nucleotide sequence in mRNA or DNA.

ribose A five-carbon monosaccharide (D-ribose) found as a component of various molecules, notably RNA.

r

ribosomal proteins The proteins that are associated with rRNA in *ribosomes. The eukaryotic ribosome has 50 proteins in the large (60S) subunit and 33 in the small (40S) subunit. Mitochondrial ribosomes have 48 in the large (39S) subunit and 29 in the small (28S) subunit.

SEE WEB LINKS

- The ribosomal protein gene database.

ribosomal RNA (rRNA) The structural RNA of the *ribosome and 80% of the total RNA in a cell. Eukaryotic ribosomes have 5S, 5.8S, and 28S species in the large (60S) subunit and an 18S species in the small (40S) subunit. Prokaryotes have 5S and 23S species in the large subunit and a 16S species in the small subunit. Mitochondrial ribosomes have 16S rRNA in the large (39S) subunit and 12S in the small (28S) subunit.

ribosome A multi-subunit complex of RNA and *ribosomal proteins that is responsible for the translation of mRNA into protein. The nascent protein is held at the P-site (peptidyl-tRNA complex), while aminoacyl-tRNAs bearing new amino acids are bound at the A-site. Eukaryotic ribosomes (80S) have two subunits of 60S and 40S; prokaryotic ribosomes (70S) have subunits of 50S and 30S; mitochondrial ribosomes (55S) have 39S and 28S subunits.

SEE WEB LINKS

- The Ribosomal protein gene database.

riboswitches Part of an mRNA that can bind a small molecule (e.g. a metabolite) and as a result regulate whether translation of the message occurs. Originally identified in bacteria, but eukaryotic examples that regulate *alternative splicing have been found.

r

ribotype A strain-specific feature of an organism based on variation in ribosomal RNA. Ribotyping is used to identify bacterial isolates, e.g. using polymorphism in the intergenic spacer region of 16S–23S rRNA.

ribozyme An RNA molecule that has catalytic (enzymic) properties. A naturally occurring example is *ribonuclease P. Ribozymes have been experimentally synthesized, mostly with the ability to cleave specific RNA sequences.

RICH-1 *See* NADRIN.

ricin *See* RICINUS COMMUNIS AGGLUTININS.

***Ricinus communis* agglutinins** Glycoprotein lectins isolated from seeds of the castor bean *Ricinus communis*. **Agglutinin I** (RCA-I, 564 aa) has binding specificity similar to ricin, but is much less toxic. It binds preferentially to oligosaccharides ending in galactose, but may also interact with *N*-acetylgalactosamine. **Agglutinin-II** (ricin) is an *AB toxin; the A subunit (267 aa) enzymically inactivates ribosomes (at a rate of a few thousand per minute, faster than they can be replaced), and is disulphide-linked to the carbohydrate-binding B subunit (262 aa) that binds β-galactosyl residues.

RICK A *RIP-like interacting CLARP (caspase-like apoptosis regulatory protein) kinase, a serine/threonine kinase (RIP2, 540 aa), similar to RIP, that may mediate apoptotic signalling via Fas. It has a C-terminal caspase activation and recruitment domain (CARD domain) and is important in the innate and adaptive immune responses.

Ricker's syndrome *See* DYSTROPHIA MYOTONICA.

rickets (rachitis, *adj.* rachitic) A vitamin D deficiency disorder that leads to defective mineralization of bone.

Rickettsia A genus of Gram-negative bacteria that are obligate intracellular parasites and the cause of several insect-borne diseases of humans, including scrub typhus and *Rocky Mountain spotted fever.

rictor A binding partner of *mTOR (rapamycin-insensitive companion of mTOR, 1708 aa) that forms a complex which integrates nutrient- and growth factor-derived signals to regulate cell growth. It modulates the phosphorylation of PKCα and the actin cytoskeleton. There is an inverse correlation between the relative amounts of *raptor and rictor in several human cell lines.

Riedel's disease (Riedel's thyroiditis) A rare chronic inflammatory disorder of the thyroid, of unknown aetiology.

Rieger's syndrome An autosomal dominant disorder of morphogenesis in which there is abnormal development of the anterior segment of the eye, causing glaucoma in many cases. Other developmental defects may also occur. Rieger's syndrome type 1 is caused by mutation in the paired homeobox transcription factor *ptx2, but other types have been distinguished.

SEE WEB LINKS

- Entry for Rieger's syndrome on Who Named It? website.

rifamycin An antibiotic produced by *Streptomyces mediterranei* that inhibits prokaryotic RNA synthesis by blocking initiation of transcription. It is effective against mycobacteria and various derivatives have been synthesized (rifampicin, rifabutin, and rifapentine).

RIG-1 **1.** Retinoic acid-inducible gene I (RIG1, 925 aa) is an RNA helicase of the DEAD/H box family that is the receptor for intracellular viral double-stranded RNA. It activates NFκB and interferon regulatory factors (*IRFs) through the adaptor protein *MAVS. RIG-I must dimerize to signal and this self-association is blocked by *LGP2. **2.** Retinoid-inducible gene I (RIG1, 164 aa) is a growth regulator protein that suppresses cell growth and induces cellular differentiation and apoptosis.

rigor Muscle stiffening that occurs when intracellular calcium levels rise and ATP levels are depleted so that the actin–myosin links remain unbroken.

rilmenidine *See* CLONIDINE.

riluzole A drug used to treat *amyotrophic lateral sclerosis. It preferentially blocks *tetrodotoxin-sensitive sodium channels, which are associated with damaged neurons. *TN* Rilutek.

RIM **1.** Regulating synaptic membrane exocytosis 1 (Rim, Rim1, Rims1) a *CAZ protein (1053 aa) that interacts with others to regulate release of synaptic vesicles at synapses in the central nervous system during short-term plasticity. Mutations in Rims1 are associated with cone–rod dystrophy 7. Closely related proteins Rim2, Rim3 have been identified. **2.** Retinoblastoma-interacting myosin-like protein, RIM. *See* RETINOBLASTOMA.

RING *See* E3 LIGASE; RING FINGER MOTIF.

Ringer's solution An isotonic salt solution used for mammalian tissues. Ringer's original solution was for frog tissues and has been extensively modified. The term is sometimes used generically for any balanced salt solution.

RING finger motif A specialized type of *zinc finger that is probably involved in protein–protein interactions. The motif is of 40–60 aa and binds two zinc atoms. The RING (really interesting new gene) finger proteins are important in differentiation, oncogenesis, and signal transduction.

SEE WEB LINKS

- Details of RING finger motif.

ringworm (tinea) A contagious dermatological infection by various fungi in which ring-shaped patches of inflammation may form.

RIPA *See* RADIOIMMUNOASSAY.

rippling muscle disease A rare disorder in which mechanical stimulation of skeletal muscle causes electrically silent muscle contractions that spread to neighbouring fibres and cause visible ripples to move over the muscle. Caused by a defect in *caveolin-3.

RISCs Multiprotein complexes (RNA-induced silencing complexes) associated with *RNA interference. The protein components, which include *argonaute family members, *dicer, nucleases, and other factors, regulate sequence-specific ribonuclease activity according to the *small interfering RNA (siRNA) or *microRNA incorporated into the complex.

risk In general, the probability of an unwanted outcome, something that is rarely perceived rationally. In toxicology it is the probability that an agent will cause adverse effects in an organism, a population, or an ecological system, or the expected frequency of such an event. **Risk assessment** is the process of trying to identify and quantify risk in order to mitigate the potential for harm, although many people consider excessive risk-aversion to be a sure way to hinder progress.

risperidone An atypical antipsychotic, a dopamine antagonist with high affinity for D_2 dopaminergic receptors, used to treat schizophrenia. *TN* Risperdal.

ristocetin An antibiotic, a mixture of ristocetins A and B, from *Amycolatopsis* (*Nocardia*) *lurida*, formerly used to treat staphylococcal infections but now considered too toxic. It induces platelet agglutination and blood coagulation and is used experimentally (and in some diagnostic tests) for this purpose.

rituximab A monclonal antibody, directed against CD20, used to treat non-Hodgkin's lymphoma. *TN* Rituxan.

rivastigmine A reversible *acetylcholine esterase inhibitor used to treat Alzheimer's disease. May increase the concentration of acetylcholine in the cholinergic neuronal pathways involved in memory, attention, learning, and other cognitive processes.

R-loop The structure formed by RNA that has displaced one strand of DNA and hybridized to the other. R-loops may be primers for mitochondrial leading-strand DNA replication but have the potential to disrupt the genome and a variety of mechanisms exist to inhibit their formation.

RLTPR A protein (RGD, leucine-rich repeat, tropomodulin domain and proline-rich domain containing protein, 1435 aa) that is down-regulated in psoriatic skin but secreted in most tissues.

RNA (ribonucleic acid) A linear polymer of ribose units linked through 3′ and 5′ positions by a phosphodiester bond and with a *purine or *pyrimidine base attached in the 1′ position. The purines are adenine and guanine, as in DNA, but the pyrimidines are cytosine and uracil (rather than thymine). The newly transcribed form of RNA has only these bases but mature RNAs contain various modified bases including pseudouridine, ribothymidine, and hypoxanthine, a deaminated adenine base whose nucleoside is called inosine. The latter is important for the *wobble hypothesis. RNA has informational, structural and enzymic roles and is, in this respect, more versatile than DNA. *See* CATALYTIC RNA; DOUBLE-STRANDED RNA; GUIDE RNA; HETEROGENOUS NUCLEAR RNA; MESSENGER RNA; MICRORNA; NONCODING RNA; NUCLEAR RNA; ONCOVIRINAE; RIBOSOMAL RNA; RNA EDITING; RNA INTERFERENCE; SHORT HAIRPIN RNA; SMALL INTERFERING RNA; SMALL NUCLEAR RNA; TAR; TRANSFER RNA.

RNA editing The processing of newly transcribed RNA that alters the information content. Does not include the *splicing out of introns to produce *messenger RNA, nor various modifications to *transfer RNA. RNA editing in mRNAs effectively alters the amino acid sequence of the encoded protein. The commonest form of editing is the substitution of uridine (U) for cytidine (C); other examples are A-to-I editing by *ADARs (adenosine deaminases acting on RNA). The process often involves *guide RNA. Editing occurs in the nucleus and in organelles but has not been reported in prokaryotes.

SEE WEB LINKS

- The RNA editing website.

RNA helicases A large family of highly conserved enzymes that utilize the energy derived from nucleotide triphosphate hydrolysis to modulate the structure of RNA. RNA helicases are involved in all aspects of RNA metabolism including transcription, splicing, and translation. A major subfamily is the *DEAD-box helicases. *See* RIG-1.

RNA interference An important mechanism of translational regulation that has only been recognized relatively recently. mRNA is selectively degraded, following the binding of *small interfering RNA (siRNA) and formation of *RISCs, thereby silencing the gene. It is proving a valuable experimental tool for selective deletion of particular proteins, and there may be therapeutic opportunities.

RNA polymerase mediator complex A multiprotein complex of ~20 polypeptides, physically associated with RNA polymerase II and essential for transcriptional activation.

RNA polymerases Enzymes that catalyse transcription, the formation of an RNA molecule complementary to a DNA sequence. In eukaryotes there are three main types: **type I** synthesizes all *ribosomal RNA except the 5S component; **type II** (EC 2.7.7.6), a complex multisubunit enzyme, synthesizes *messenger RNA and *heterogenous nuclear RNA; **type III** synthesizes *transfer RNA and the 5S component of rRNA. Many of the subunits in the three types are highly conserved.

RNA primase *See* DNA PRIMASES.

RNA processing Modifications of primary RNA transcripts including splicing, cleavage, base modification, capping, and the addition of *poly-A tails. Unlike *RNA editing, these processes do not alter the information content.

RNA recognition motif (RRM) An abundant protein domain, typically consisting of four anti-parallel β-strands and two α-helices arranged in a βαββαβ fold. Proteins with RRMs have a variety of RNA binding preferences and functions, and include heterogeneous nuclear ribonucleoproteins (hnRNPs), proteins implicated in regulation of alternative splicing, protein components of small nuclear ribonucleoproteins (snRNPs), and proteins that regulate RNA stability and translation.

RNase P The *ribonuclease (a ribozyme) responsible for processing the 5′ end of immature transfer RNAs. It consists of an RNA species (H1 RNA) that has the enzymic activity, the POP1 protein (processing of precursor 1), and at least seven proteins called RPPs. Patients with *scleroderma have serum reactive with RNase P, the Th antigen, RPP30, and RPP38.

RNAse protection assay A method for quantitatively determining the level of expression of a particular gene. Labelled antisense cRNA is hybridized with an mRNA sample, and RNAse-treated: only double-stranded RNA survives hydrolysis, and the amount of labelled polynucleotide is proportional to the amount of specific mRNA in the sample.

RNA splicing *See* SPLICING.

RNA tumour virus *See* RETROVIRIDAE.

RNP *See* RIBONUCLEOPROTEIN.

RO-3306 An inhibitor of cyclin-dependent kinase CDK1 that blocks cells at the G_2/M border in the *cell cycle.

Robertsonian translocation A nonreciprocal chromosomal translocation in which the long arms of two nonhomologous acrocentric chromosomes are attached to a single centromere and the short arms to another. The two short arms often disappear at a subsequent division. Robertsonian translocations in humans are confined to the acrocentric chromosomes 13, 14, 15, 21, and 22, the short arms of which have no essential genetic material.

Robinow's syndrome A genetically heterogeneous developmental disorder characterized by mesomelic limb shortening, characteristic facial features, orodental abnormalities, and hypoplastic genitalia. The autosomal recessive form is allelic with *brachydactyly type B. There is also an autosomal dominant form.

robo *See* SLIT.

ROCKs *See* RHO KINASES.

Rocky Mountain spotted fever A tick-borne rickettsial disease (black fever) caused by *Rickettsia rickettsii.*

rod cell *See* RETINAL ROD.

rodent ulcer *See* BASAL CELL CARCINOMA.

rolipram An anti-inflammatory and possibly antidepressant drug that inhibits cAMP-specific phosphodiesterase IV. This has the effect of enhancing noradrenergic transmission in the central nervous system and suppressing the production of proinflammatory cytokines and other inflammatory mediators.

Romanovsky-type stain A histological stain that has several components, including methylene blue, Azure A or B, and eosin. Romanovsky stains are often used on blood films because they differentiate the various leucocyte classes. Examples are Giemsa, Wright's, and Leishman's stains.

SEE WEB LINKS

- Useful resource on histological stains.

ROMK An inwardly-rectifying ATP-dependent potassium channel (renal outer medullary potassium channel, Kir1.1, KCNJ1) that transports potassium out of cells and is expressed in the thick ascending limb of Henle and throughout the distal nephron of the kidney. Mutations lead to antenatal *Bartter's syndrome type 2 and decrease the strength of channel-PIP2 interactions which are important for normal function.

ron A receptor tyrosine kinase (recepteur d'origine nantais, CD136), an αβ heterodimer derived from a single 1400 aa precursor, for which the ligand is *macrophage stimulating protein (hepatocyte growth factor-like protein). It is one of the family that includes the proto-oncogene **met* and the avian oncogene **sea*.

ropinirole An agonist for D_2 and D_3 dopamine receptors, used to treat Parkinson's disease and restless legs syndrome. *TNs* Requip, Ropark.

ros **1.** An oncogene, *ros*, highly expressed in a variety of tumour cell lines, that encodes one of the sevenless subfamily of *receptor tyrosine kinases. **2.** Reactive oxygen species (ROS); *see* METABOLIC BURST.

rosacea A chronic condition (acne rosacea) in which there is facial *erythema (redness) and sometimes pimples. One factor that has been suggested is excessive production of *cathelicidin.

roscovitine *See* OLOMOUCINE.

Rosetta stone method In bioinformatics, a method for inferring functional linkage between two proteins. If proteins A and B are separate in some species but found as a fused protein in others, then it is deemed probable that the proteins are functionally associated.

Rotarix A vaccine that produces immunity to *rotavirus and prevents viral gastroenteritis.

rotatin A gene required for axial rotation and left–right specification in mouse embryos that encodes a transmembrane protein. The human homologue (2226 aa) has now been identified.

Rotavirus A genus of the *Reoviridae. They have a wheel-like appearance in the electron microscope, and cause acute diarrhoeal disease in their mammalian and avian hosts.

rotenone An inhibitor of the *electron transport chain, found naturally in the roots of the South American barbasco shrub *Lonchocarpus nicou* (formerly *Robinia nicou*). A very potent poison for fish, and causes Parkinson's disease-like symptoms in rats.

Rothmund–Thomson syndrome An autosomal recessive dermatosis (poilioderma atrophicans and cataract) associated with multiple developmental defects including short stature, bone defects, disturbances of hair growth, and hypogonadism. Some cases are caused by mutation in the *RECQ-like helicase, RECQL4.

rottlerin A widely selective protein kinase C δ (PKCδ) inhibitor isolated from *Mallotus philippinensis* (the monkey-face tree). It is effective against several human tumour cell lines and in potentiating chemotherapy-induced cytotoxicity.

rough endoplasmic reticulum (RER) The ramifying intracellular membrane system (endoplasmic reticulum) that is associated with *ribosomes which bind to the *signal recognition particle of RER when translating mRNA for secreted and transmembrane proteins. It is similar to the nuclear envelope outer membrane and is also the site for membrane lipid synthesis. Within the lumen are proteins with the *KDEL motif.

rough microsome Small vesicles with attached ribosomes, derived from the rough endoplasmic reticulum by sonication. Can be used to study protein synthesis.

rough strain Bacterial strains that have a dull appearance as colonies grown on agar. They have much less hydrophilic cell wall surface carbohydrate than smooth strains and are more readily phagocytosed. In streptococci smooth strains are virulent and rough strains are not.

rouleau (*pl.* rouleaux) A cylindrical mass of aggregated erythrocytes. Rouleaux formation can be induced experimentally by reducing the repulsion forces between erythrocytes, e.g. by adding dextran.

Rous sarcoma virus (RSV) A virus that causes a transmissible chicken *sarcoma, as shown in 1911 by F. Peyton Rous (1879–1970). It is a *C-type oncorna virus and carries the **src* oncogene.

RPMI 1640 A cell culture medium developed at Roswell Park Memorial Institute that utilizies a bicarbonate buffering system. Widely used for the culture of human leucocyte cultures.

R point of cell cycle *See* RESTRICTION POINT.

RQ *See* RESPIRATORY QUOTIENT.

RRM A common feature (RNA recognition motif, RNA-binding domain, RBD, ribonucleoprotein domain, RNP) in many eukaryotic RNA-binding proteins.

rRNA *See* RIBOSOMAL RNA.

RSC complex A chromatin-remodelling complex, related to the *SWI/SNF complex, that facilitates opening up of the chromatin to allow transcription and other metabolic reactions to occur. Different chromatin-remodelling complexes may perform quite specific and noninterchangeable functions.

RS domain An arginine-serine-rich domain required for RNA binding by splicing factors such as *U2AF. *See* SR-PROTEINS.

R-spondin *See* SPONDIN.

RSV *See* ROUS SARCOMA VIRUS.

RT-PCR A polymerase chain reaction (reverse transcriptase polymerase chain reaction, reverse transcription PCR) in which the starting point is RNA which is reverse-transcribed to produce the DNA template. An important approach to investigating the mRNA of a cell. *Compare* REAL TIME PCR.

RTX toxins A family of related Gram-negative bacterial cytolysins and cytotoxins that contain a characteristic repeat domain with multiple glycine and aspartate-rich motifs. They are calcium-dependent pore-forming toxins and are produced in inactive pro-form that must be post-translationally modified to generate an active toxin. Examples include *E. coli* haemolysin, *Proteus vulgaris* haemolysin, *Pasteurella haemolytica* leukotoxin, and adenylate cyclase-haemolysin from *Bordetella pertussis*. *Vibrio cholerae* RTX toxin causes the depolymerization of actin stress fibers.

R-type channels A subclass of *voltage-sensitive calcium channels in which the α subunit is α-1E (2251 aa). R-type channels are unaffected by dihydropyridines, phenylalkylamines, and *conotoxins and have a high activation threshold. SNX-482 is a selective R channel blocker. They are expressed in neuronal tissues, kidney, spleen, and pancreatic islet cells and appear to have a role in insulin release.

rubella A so-called 'childhood disease' (German measles) caused by *Rubella* virus, a *togavirus. Although the disease itself is mild, infection during pregnancy, particularly during the first trimester, can cause severe fetal abnormalities.

Rubinstein–Taybi syndrome A syndrome characterized by mental retardation, broad thumbs and toes, and facial abnormalities that can be caused by mutation in the gene encoding CREB-binding protein or in the *EP300* gene (*see* P300). A severe form of the syndrome is a result of chromosome 16p13.3 deletion.

Ruffinini's corpuscles Subcutaneous sensory nerve endings surrounded by an ovoid connective tissue capsule, probably mechanosensors.

ruffles Thin lamellar protrusions, supported by a microfilament meshwork and devoid of organelles, at the front of a crawling cell. The leading edge appears to 'ruffle' in time-lapse film and ruffles can move centripetally over the dorsal surface of a cell in culture.

rugae Wrinkles, usually in a series like furrows. *Adj.* rugose, or, in the case of small wrinkles, rugulose.

Runx2 A transcription factor (runt-related transcription factor 2) involved in early osteoblast differentiation. It is inhibited by *twist.

Russell–Silver syndrome A disorder (Silver–Russell dwarfism) involving growth retardation and characteristic 'triangular' facial features. Associated with loci on chromosomes 7 and 11.

ruthenium red A polycationic stain that binds tightly to tubulin dimers and the *ryanodine receptor. Used in electron microscopy to stain acid mucopolysaccharides on the outer surfaces of cells.

ryanodine An alkaloid from the South American shrub *Ryania speciosa* that blocks calcium release from the sarcoplasmic reticulum of skeletal muscle. The **ryanodine receptor** is a large transmembrane protein with skeletal muscle (RyR1, 5032 aa), cardiac muscle (RyR2, 4965 aa) and brain (RyR3, 4872 aa) isoforms. The cardiac channel is a tetramer comprised of four RYR2 polypeptides and four *calstabin (FK506-binding protein 1B) subunits and is opened by PKA-mediated phosphorylation; the other channels are similar. The ryanodine-receptor channels exhibit a positive feedback in which increased intracellular calcium triggers opening and gives a very rapid response. RyR1 mutations are associated with *malignant hyperthermia and *central core disease, RyR2 mutations with stress-induced ventricular *tachycardia. RYR3 is thought to have a role in hippocampal synaptic plasticity, particularly the modification of acquired memory in response to external stimuli.

ryk An atypical receptor protein tyrosine kinase that functions as a coreceptor along with Frizzled for *Wnt ligands and binds to *dishevelled, through which it activates the canonical Wnt pathway.

S1 **1.** A soluble fragment of *meromyosin. **2.** A ribosomal protein, S1, that is involved in initiation of translation in *E. coli.* **3.** S1 nuclease (EC 3.1.30.1, ~300 aa) is a fungal nuclease that degrades single-stranded nucleic acids and is preferentially active against DNA. Used experimentally to analyse the structure of DNA: RNA hybrids (S1 nuclease mapping, *Berk–Sharp technique), and to remove single-stranded extensions from DNA to produce blunt ends (*see* RESTRICTION ENDONUCLEASES).

S2 *See* MEROMYOSIN.

S6 kinases (p70) A family of serine/threonine kinases (EC 2.7.11.1) that phosphorylate ribosomal S6 protein and stimulate protein production in mitogen-activated cells. Various forms are known (α1–6, β1–2, δ1): p70 is S6 kinase β1 (525 aa) or β2 (482 aa).

S9 **1.** A supernatant fraction of liver homogenate, rich in drug-metabolizing enzymes (P450s) and sometimes used in the *Ames test. **2.** A ribosomal protein. **3.** One of the *dermaseptins.

S100 calcium-binding proteins A large family of calcium-binding proteins that contain an *EF-hand. **S100A1** (94 aa, a dimer with three forms, αα, αβ, ββ) is produced by a wide variety of normal and neoplastic cells of mesodermal, neuroectodermal, and epithelial origin. It improves cardiac contractile performance by regulating sarcoplasmic reticulum calcium ion handling and myofibrillar calcium ion responsiveness. **S100A4** (placental calcium-binding protein, Mts1, metastasin, p9Ka, pEL98, CAPL, calvasculin, Fsp-1) is found in placenta, uterus, and the vasculature and its expression is upregulated in various pathological conditions. It apparently contributes to tumour cell motility and metastasis. **S100A6** is *calcyclin. Calprotectin is a heterodimer composed of **S100A8** (*calgranulin A) and **S100A9** (calgranulin B) and is a major calcium- and zinc-binding protein in the cytosol of neutrophils, monocytes, and keratinocytes. **S100A10** is the light chain of *calpactin, **S100A12** is calgranulin C. **S100β** has neurotrophic and mitogenic activity. Other members of the family include *calmodulin and *troponin.

S180 (sarcoma 180) A line of highly malignant mouse sarcoma cells, often passaged in *ascites form, that were used in classical studies on *contact inhibition of locomotion.

Sab A protein (SH3 binding protein 5, SH3BP5, 425 aa) that selectively binds the SH3 domain of *btk and inhibits its activity.

saccade An abrupt eye movement, in which the focus of attention jumps from one limited portion of the visual field to another. Our impression of seeing a complete picture is a neurally generated illusion built of limited snapshots taken at each saccade.

saccharomicins Oligosaccharide antibiotics (saccharomicins A and B) isolated from the actinomycete *Saccharothrix espanaensis.* They are active against bacteria and yeast, probably being membranolytic.

saccharopine An intermediate in the synthesis or degradation of lysine, formed from lysine and α-ketoglutarate.

sacsin A protein (4579 aa) that is highly expressed in the central nervous system and may have a chaperone-type role. Mutations in the gene cause *autosomal recessive spastic ataxia of Charlevoix-Saguenay (ARSACS). ARSACS is an early onset neurodegenerative disease with high prevalence in the Charlevoix-Saguenay–Lac-Saint-Jean region of Quebec, Canada. It is characterized by absent sensory nerve conduction, reduced motor nerve velocity, and hypermyelination of retinal nerve fibres.

***S*-adenosyl methionine** A derivative of *methionine, a coenzyme involved in methylation reactions in various metabolic pathways. It is produced by methionine adenosyltransferase (EC 2.5.1.6, 395 aa) from ATP and methionine.

Saethre–Chotzen syndrome An autosomal dominant disorder in which bones in the skull that are normally separate are fused (craniosynostosis). It is caused by a loss-of-function mutation in *twist-1.

SAGA A multiprotein complex that activates transcription by remodelling chromatin and mediating histone acetylation and deubiquitination and is required for transcription of a subset of RNA polymerase II-dependent genes. The SAGA complex is recruited to specific gene promoters by activators such as myc and is required for nuclear receptor-mediated transactivation. Components include *ataxin-7-like protein 3.

sagittal section A cross-section through the median vertical longitudinal plane.

salbutamol A short-acting β_2-adrenergic receptor agonist used to treat bronchospasm in asthma and chronic obstructive pulmonary disease. *TNs* Asthalin, ProAir, Proventil, Ventolin.

salicylic acid An antiseptic and anti-inflammatory compound (2-hydroxybenzoic acid) found naturally at high levels in willow (*Salix*) bark. The acetylated form is *aspirin (acetylsalicylic acid). Oil of wintergreen is ethyl salicylate.

Salla disease *See* SIALIN.

salmeterol A long-acting β_2-adrenergic receptor agonist used to treat asthma and chronic obstructive pulmonary disease.

Salmonella A genus of the Enterobacteriaceae, motile Gram-negative bacteria that can cause enteric fevers (e.g. typhoid, caused by *S. typhi*), food poisoning (usually *S. typhimurium* or *S. enteridis*) and occasionally septicaemia in nonintestinal tissues.

salping-, salpingo- Prefix for anything associated with the fallopian tubes or the auditory meatus.

SALT *See* SKIN ASSOCIATED LYMPHOID TISSUE.

saltatory Describing a movement that occurs as a series of jumps (like a kangaroo). Some intracellular particles exhibit saltatory movements, although the mechanism is unclear. **Saltatory conduction** is the rapid method by which nerve impulses jump along an axon with excitation occurring only at *nodes of Ranvier. *See* MYELIN; SCHWANN CELLS. **Saltatory replication** is a term applied to the sudden amplification (in generational terms) of a DNA sequence to generate many tandem copies, possibly the way in which *satellite DNA arose.

salusins *See* TORSINS.

salvage pathways The metabolic pathways that recycle important intermediates from materials that would otherwise be waste products. An important salvage pathway is that from hypoxanthine to nucleotides. *See* HGPRT.

Sam68 A protein (src-associated in mitosis, 68 kDa, 443 aa) that is tyrosine-phosphorylated by *src-family kinases in mitotic cells. It interacts with RNA, src-family kinases, grb2, and PLCγ. *Radicicol will inhibit src-mediated phosphorylation of Sam68 and block exit from mitosis.

SAM domain A protein interaction module (sterile α motif domain, ~70 aa) that is present in a variety of signalling molecules including *EPH-related receptor tyrosine kinases, serine/threonine protein kinases, cytoplasmic scaffolding, and adaptor proteins, regulators of lipid metabolism, and GTPases as well as members of the ETS family of transcription factors. SAM domains form homo- and hetero-oligomers.

SEE WEB LINKS

- A very detailed website on the SAM domain.

Sandhoff's disease A *storage disease in which there is accumulation of GM2 gangliosides, particularly in neurons, leading to progressive neurodegeneration. Clinically indistinguishable from *Tay–Sachs disease but caused by mutation in the β subunit of hexosaminidase (EC 3.2.1.52).

Sanfillipo's syndrome A *storage disease in which the defect is in either keratan sulphate sulphatase or *N*-acetyl-α-D-glucosaminidase. Cocultured fibroblasts from apparently clinically identical patients can cross-complement if a different enzyme is missing in each.

Sanger–Coulson method *See* DIDEOXYNUCLEOTIDE.

Sanjad–Sakati syndrome *See* HYPOPARATHYROIDISM.

SANS A scaffold protein (scaffold protein containing ankyrin repeats and SAM domain) localized to the apical region of cochlear and vestibular hair cell bodies but not found in stereocilia. It interacts with *harmonin and myosin 7A and it has been proposed that it controls hair bundle cohesion and development through its interactions with these stereocilia-associated proteins. Mutations cause *Usher's syndrome (type 1G). The murine *sans* gene is in the critical region of the Jackson shaker mouse mutation on chromosome 10 which has syntenic homology with a region of human chromosome 17 to which the Usher's syndrome type IG maps.

S-antigen **1.** *Arrestin (409 aa), the protein found in the retina and pineal gland that elicits experimental autoimmune uveitis. **2.** Soluble heat-stable surface antigens of *Plasmodium falciparum* that are responsible for antigenic heterogeneity.

SAP **1.** *See* SERUM AMYLOID (P protein). **2.** *See* SLAM-ASSOCIATED PROTEINS.

sapintoxin D (SAPD) A fluorescent phorbol ester that is a potent and selective activator of protein kinase C (PKC)-α.

saponins A class of glycosidic surfactants produced by plant cells that are often used experimentally to permeabilize cells in a fairly gentle manner.

saporin A protein toxin (253 aa), from seeds of the soapwort *Saponaria officinalis*, that enzymically inactivates ribosomes.

saposins A set of small Trp-free, multifunctional glycoproteins (co-β-glucosidase, A1 activator, glucosylceramidase activator, SAP-A–D, 81 aa) derived from a single precursor, prosaposin, that act as sphingolipid activator proteins (SAPs) and assist in the lysosomal hydrolysis of sphingolipids.Other functions include neuritogenic/neuroprotection effects and induction of membrane fusion. Deficiency of **saposin A** causes atypical *Krabbe's disease, deficiency of **saposin B** causes metachromatic *leukodystrophy, deficiency of **saposin C** causes an atypical form of *Gaucher's disease. **Saposin-like proteins** (SAPLIPs) include surfactant protein B (SP-B), *Entamoeba histolytica* pore-forming peptide, *granulysin, NK-lysin, acid sphingomyelinase, and acyloxyacyl hydrolase.

sapoviruses A genus of the Calicivirus family that cause relatively mild gastroenteritis in young children.

SAR **1.** Secretion-associated and ras-related protein (sar1p). In yeast, a small GTPase (190 aa) that controls the assembly of the *coat protein complex II (COPII) that surrounds vesicles that are involved in export from the endoplasmic reticulum. The human homologue is SAR1 (SAR1A and SAR1B, both 198 aa); mutations in SAR1B are associated with *Anderson's disease and chylomicron retention disease. **2.** *See* STRUCTURE–ACTIVITY RELATIONSHIP.

sarafotoxins Small peptides (21 aa), isolated from the venom of the burrowing asp *Atractaspis engaddensis*, that are structurally and functionally related to *endothelins and cause constriction of cardiac and smooth muscle.

sarcin/ricin loop A highly conserved sequence found in the 28S RNA of the large subunit of the ribosome. It is cleaved by both *α-sarcin and *ricin which prevents elongation factor from binding and thus blocks translation.

sarcodictyins A group of compounds, isolated from the Japanese soft coral *Bellonella albiflora*, that, like *paclitaxel, will stabilize microtubule bundles.

sarcoglycans A family of transmembrane proteins involved in the *dystrophin-associated protein complex and that, together with dystrophin, *dystroglycans, and *syntrophins, link the contractile system of muscle to extracellular matrix. At least six sarcoglycans are known including α-sarcoglycan (50DAG, A2, adhalin, 387 aa), β-sarcoglycan (43DAG; A3b), and γ-sarcoglycan (35DAG; A4). Mutations in sarcoglycan genes are responsible for some forms of *muscular dystrophy.

sarcoidosis A granulomatous disorder (Besnier–Boeck disease) associated with an accumulation of $CD4^+$ T cells and a Th1 immune response. Various loci are associated with susceptibility, in particular allelic variation at the HLA-DRB1 locus.

sarcolemma The plasma membrane of skeletal and cardiac muscle fibres.

sarcolipin A small protein (31 aa) with homology to *phospholamban that acts as an inhibitor of the calcium ATPase of sarcoplasmic reticulum (*SERCA) by direct binding and as a binary complex with phospholamban.

sarcoma A malignant tumour derived from connective tissue cells, e.g. osteosarcoma (from bone). Some sarcomas are of viral origin: *see* ROUS SARCOMA VIRUS; S180; SRC GENE.

sarcomere The repeating subunit of the *myofibrils in striated muscle. There are two main bands (A bands and *I bands), the I band being subdivided by the *Z disc, and the A band being split by the *M line and the H zone.

sarcoplasm The cytoplasm of a striated muscle fibre.

sarcoplasmic–endoplasmic reticulum calcium-ATPase The calcium-ATPase pump (SERCA, EC 3.6.3.8, 1043 aa) in the sarcoplasmic reticulum (SR) that lowers sarcoplasmic levels of calcium and allows muscle to relax. SERCA moves calcium ions into the SR and a different ATPase, the plasma-membrane calcium ATPase (PMCA, 1258 aa), pumps calcium to the extracellular space. Various tissue-specific forms and alternatively-spliced variants exist in the sarcoplasmic reticulum of different types of muscle and in the endoplasmic reticulum of nonmuscle cells.

sarcoplasmic reticulum The endoplasmic reticulum of striated muscle that is specialized for storing calcium ions which are released to trigger contraction of the muscle when *T tubules relay a signal from the neuromuscular junction.

sarcosine A natural amino acid (*N*-methylglycine) found in muscle, most tissues, and blood at around 1.6 mmol/L. It inhibits the type 1 glycine transporter.

sarcosyl (sarkosyl) A mild anionic detergent (*N*-lauroylsarcosine), often used in preparing solubilized fractions of biological materials.

sarin A nerve gas that inhibits *acetylcholine esterase.

SARS A viral pneumonia (severe acquired (acute) respiratory syndrome) caused by one of the *Coronaviridae.

satellite cells **1.** Mononucleated cells in vertebrate skeletal muscle that seem to be quiescent stem cells and are important in regeneration. **2.** A synonym for glial cells.

satellite chromosome A small segment of a chromosome that is connected, usually to the short arm of an acrocentric chromosome, by a narrow neck. *Compare* SATELLITE DNA.

satellite DNA Tandem repeats of DNA sequences widely distributed through the genome that may constitute around 10% of the total DNA. The repetitive sequences may have a base composition (and thus density) sufficiently different from that of normal DNA that they sediment as a distinct band in caesium chloride density gradients. Satellites can be subgrouped on the basis of the size of the repeat region: minisatellites (variable number of tandem repeats, VNTRs) have core repeats with 9–80 bp, microsatellites (short tandem repeats, STRs; simple sequence repeats, SSRs) have repeats of only 2–6 bp. Tandem repeats are easily detectable with probes and are frequently polymorphic, so make good markers for genetic mapping, linkage analysis, and human identity testing. *See also* HUNTINGTON'S DISEASE; FRAGILE X SYNDROME; ISSR.

Sativex® A *cannabinoid preparation containing tetrahydrocannabinol and cannabidiol, used to relieve spasticity in multiple sclerosis and for the relief of neuropathic pain.

saturated fatty acids *Fatty acids that do not contain double bonds. In eukaryotes the main examples are stearic, palmitic, and myristic acids.

saturation density The cell density at which further proliferation ceases under particular *in vitro* culture conditions. An important factor for some cells may be the availability of free substratum on which to spread (*see* DENSITY DEPENDENT INHIBITION OF GROWTH) in order to be able to absorb sufficient growth factor (or limiting nutrient) from the medium. Transformed cells generally achieve higher saturation densities than normal cells but even so far lower than those found *in vivo*.

saxitoxin A *neurotoxin produced by dinoflagellates (e.g. *Alexandrium catenella*) responsible for the 'red tide' phenomenon. Filter-feeding shellfish, such as the clam *Saxidomus giganteus*, can accumulate the toxin and cause paralytic shellfish poisoning if eaten. The toxin binds to *sodium channels and blocks development of action potentials.

scabies A contagious infection of skin by the mite *Sarcoptes scabiei* which burrows in the horny layer of the skin and elicits an inflammatory response.

scaffolding proteins A general name for proteins that contribute to the formation of large multimolecular complexes, e.g. clusters of receptors at synapses (*caveolins; *flotillins, etc.), viral capsids, chromosomes. *See also* AKAP FAMILY (Akap 79); *involucrin; *titin.

scaffold/radial loop model A common model for the organization of a metaphase chromosome that postulates a nonhistone core (scaffold) composed of structural maintenance of chromosomes 2 (SMC2) protein and topoisomerase II to which radial loops of DNA are attached at regular intervals. The model remains somewhat controversial and there are some inconsistencies, notably that nucleases but not proteases rapidly reduce the tension resistance of the chromosome to zero.

SCAMP *See* SECRETORY CARRIER ASSOCIATED MEMBRANE PROTEINS.

SCAP A membrane protein (SREBP cleavage-activating protein, 1277 aa) that escorts *SREBPs from the endoplasmic reticulum to the Golgi, where they are cleaved by site-1 protease to release their transcription factor domain which activates genes for lipid synthesis. If sterols are present this transport is blocked, providing a feedback control for lipid synthesis.

SCAPER A regulatory protein (S-phase cyclin A-associated protein in the endoplasmic reticulum, 1399 aa) that transiently maintains the *cyclinA2–*CDK2 complex in the cytoplasm.

Scar An endogenous activator of actin polymerization. *See* WAVE/SCAR PROTEINS.

Scatchard plot A graphical method for analysing reversible ligand/receptor binding interactions. The plot of (bound ligand/free ligand) against (bound ligand) gives a line with a slope that is the negative reciprocal of the binding affinity and an intercept on the *x*-axis that is the number of receptors (at infinite ligand concentration bound/free becomes zero). The Scatchard plot is better than the *Eadie–Hoffstee plot for binding data because it depends more on the values at high ligand concentration, the most reliable values. Nonlinearity can arise through heterogeneity of receptors, although there are other possible causes.

scatter factor A motility factor (*motogen) that stimulates dispersal (scattering) of colonies of epithelial and endothelial cells as single cells. It is identical to *hepatocyte growth factor, but is not mitogenic for all cell types.

scavenger receptors A structurally diverse family of macrophage receptors involved in the uptake of a variety of things including modified *LDL. Six classes are recognized, with different binding preferences: Class A bind a wide range of ligands, including bacteria and possibly apoptotic cells. They have also been implicated in the development of atherosclerotic lesions. *Collectins are considered to be scavenger receptors.

S cells *See* SECRETIN.

S

SCF **1.** SCF complexes (a multiprotein aggregate of Skp1p-cdc53p-F-box) are a class of E3 ubiquitin protein ligases that play a role in regulation of cell division. *See* SKP and *cullins. **2.** *See* STEM CELL FACTOR.

scFv A recombinant protein (single-chain variable fragment) that is a heterodimer of the variable regions of an immunoglobulin heavy and immunoglobulin light chain linked together with a short peptide. The construct has comparable binding properties to the monoclonal antibody from which the variable regions were derived.

SCG10 A gene (superior cervical ganglion-10 gene) that encodes a growth-associated protein (179 aa) that is expressed early in the development of neural crest derived neurons and is similar to *stathmin. **SCG10-like protein** (SCLIP, stathmin-like 3, 180 aa) forms a complex with two tubulins for every stathmin-like protein and interferes with microtubule dynamics.

Scheie's syndrome *See* HURLER'S SYNDROME.

Schick test A test for immunity to diphtheria by challenging intradermally with a small amount of diphtheria toxin; immune individuals do not show an inflammatory response.

Schiff's reagent *See* PERIODIC ACID–SCHIFF REACTION.

Schilder's disease A rare progressive demyelinating disorder (sudanophilic cerebral sclerosis, myelinoclastic diffuse sclerosis, diffuse sclerosis, encephalitis periaxialis), a variant of *multiple sclerosis, which usually begins in childhood. Probably autosomal recessive.

schistocytes Irregularly shaped erythrocyte fragments in the circulation, often a result of mechanical damage (artificial heart valves) or haemolytic anaemia.

schistosomiasis (bilharzia) An infectious disease, endemic in many tropical countries, caused by trematodes (flukes) of the genus *Schistosoma*. The larvae develop in freshwater snails and cercariae are released into the water and burrow into the skin where they transform to the schistosomulum stage, and migrate to the urinary tract (*S. haematobium*), liver, or intestine (*S. japonicum, S.mansoni*) where the adult worms develop. The adults release eggs into the urinary tract or the intestine and these hatch to form miracidia which then infect snails, completing the life cycle. Adult worms cause substantial tissue damage and resist immune clearance by poorly understood means.

schizogony An asexual reproductive process in many protozoa.

Schizosaccharomyces pombe A species of fission yeast often used for studies on control of the cell cycle because it has a distinct G_2 phase in its cycle. Distantly related to the budding yeast **Saccharomyces cerevisiae*.

Schmidt's syndrome *See* POLYENDOCRINE SYNDROME (type 2).

Schulman–Upshaw syndrome *See* THROMBOTIC THROMBOCYTOPENIC PURPURA.

Schultz–Charlton test A skin test for scarlet fever that uses antitoxin to the *erythrogenic toxin of *Streptococcus pyogenes* subcutaneously: a positive reaction is blanching of the rash in the area around the injection site.

Schwann cells Specialized glial cells that produce the *myelin sheath around peripheral axons.

schwannoma derived growth factor One of the *EGF-family of growth factors, mitogenic for astrocytes, Schwann cells, and fibroblasts and similar to *amphiregulin.

schwannomin *See* MERLIN.

Schwartz–Jampel syndrome A progressive disorder characterized by short stature, myotonic myopathy, dystrophy of epiphyseal cartilages, joint contractures, blepharophimosis, unusual pinnae, myopia, and pigeon breast. **Type 1** (SJS1) is caused by mutation in the gene encoding *perlecan. The Silverman–Handmaker type of dyssegmental dysplasia is caused by a different mutation in the same gene and has a more severe phenotype. Neonatal Schwartz–Jampel syndrome **type 2** (Stuve–Wiedemann syndrome) is a genetically distinct disorder with a more severe phenotype caused by mutation in the leukaemia inhibitory factor receptor (*LIFR*) gene.

Schwartzmann reaction Misspelling of *Shwartzman reaction.

SCID *See* SEVERE COMBINED IMMUNODEFICIENCY DISEASE.

scinderin *See* ADSEVERIN.

SCIP (Oct-6, Tst-1) A *POU domain transcription factor that regulates the differentiation of *Schwann cells.

scirrhous carcinoma A *carcinoma in which there is extensive production of dense connective tissue making the tumour mass hard.

scleritis Inflammation of the sclera of the eye and adjacent tissues which can threaten vision. It can be associated with systemic diseases such as *Wegener's granulomatosis or *rheumatoid arthritis. Three types of scleritis are distinguished: diffuse, nodular, and necrotizing.

scleroderma A chronic autoimmune connective tissue disorder (systematic sclerosis) characterized by immune activation, vascular damage, and fibrosis of the skin and major internal organs, not just hardening of skin. The causes are unclear, although there are some genetic predisposing factors; antibodies to *fibrillarin are found in some cases. *See* CREST SYNDROME; *RNase P.

sclerosis Abnormal hardening of tissue. *But see* AMYOTROPHIC LATERAL SCLEROSIS; BALO'S CONCENTRIC SCLEROSIS; MULTIPLE SCLEROSIS; SCHILDER'S DISEASE; SCLERODERMA; TUBEROUS SCLEROSIS.

sclerosteosis An uncommon autosomal recessive bone dysplasia characterized by generalized osteosclerosis and hyperostosis of the skeleton, affecting mainly the skull and mandible and leading to facial paralysis and hearing loss. Caused by mutation in the coding region of the gene for *sclerostin. Unlike *van Buchem's disease, in which there is downregulation of sclerostin production, there are gigantism and hand abnormalities.

sclerostin A *bone morphogenic protein (BMP) antagonist (213 aa), the gene for which is mutated in *sclerosteosis and down-regulated in *van Buchem's disease. Sclerostin apparently represses BMP-induced osteoblast differentiation and function. Sclerostin has a cystine knot motif (residues 80–167) similar to that in the *cer/dan family of secreted glycoproteins which are antagonists of members of the TGFβ superfamily. Sclerostin is inactivated by *noggin. **Sclerostin-domain containing protein1** (ectodin, 206 aa) inhibits BMP2, BMP4, BMP6, and BMP7.

SCN **1.** The suprachiasmatic nucleus of the hypothalamus. **2.** The thiocyanate anion (SCN^-). **3.** *See* VOLTAGE-GATED SODIUM CHANNELS.

SCO The subcommissural organ, a highly conserved brain gland that secretes glycoproteins into the cerebrospinal fluid, where they aggregate to form Reissner's fibres. A major secreted protein is a form of *spondin, **SCO-spondin**, a large multidomain protein probably involved in axonal growth and/or guidance.

scoliosis A structurally fixed lateral curvature of the spine (>10°). Loci for susceptibility to idiopathic scoliosis have been mapped to chromosomes 19, 17, 8, and 9 and it may also occur as a secondary feature in other disorders including *Marfan's syndrome, *dysautonomia, *neurofibromatosis, *Friedreich's ataxia, and *muscular dystrophies.

scombrotoxin The causative agent of scombroid poisoning (histamine poisoning), probably a misnomer because the poisoning is caused by eating foods that have high levels of histamine and possibly other vasoactive amines.

scopolamine (hyoscine) An alkaloid structurally and functionally similar to *atropine that acts as a *muscarinic acetylcholine receptor antagonist. Occurs naturally in the thorn apple *Datura stramonium*.

scopoletin A naturally occurring fluorescent plant growth inhibitor (7-hydroxy-6-

methoxycoumarin) that is said to be valuable in both hypertension and hypotension, to be bacteriostatic and anti-inflammatory. It inhibits *acetylcholine esterase.

scorpion toxins A family of polypeptide toxins (~7 kDa) that inhibit voltage-sensitive *sodium channels of nerve and muscle. The α-toxins are found in the venom of Old World scorpions, β-toxins in those of the New World.

scotophobin A peptide (15 aa) isolated from the brains of rats that were trained to fear the dark. The claim that this dark-aversion was transferable to naive animals has been considered rather improbable, although there does seem to be some positive experimental data for pharmacological effects of the peptide.

SEE WEB LINKS

- A discussion of the controversy over scotophobin.

scramblases A family of cytoplasmic membrane-associated proteins that mediate a calcium-dependent transbilayer flip/flop of membrane lipids, changing the normal asymmetric distribution in the plasma membrane. They have some similarities to *tubby.

scrapie A chronic progressive and fatal neurological disease of sheep and goats, similar to other *spongiform encephalopathies. There is a polymorphism in the prion protein that confers resistance and this is being selected for in the British sheep flock, despite fears that the scrapie-susceptible genotype may have other benefits, possibly the ability to thrive in poor conditions. Atypical forms of the disease seem to be emerging in sheep of the genotypes that are resistant to the classical form.

S

scRNP Small cytoplasmic ribonucleoprotein. *See* SMALL INTERFERING RNA.

scrofula An infection of the skin, usually of the neck, by *Mycobacterium tuberculosis* in adults. The infection may also involve cervical lymph nodes. In England until the 18th century the sovereign's touch was thought to cure the disease (hence known as the King's evil).

scrub typhus *See* SHIMAMUSHI FEVER.

scurfy A murine X-linked lymphoproliferative autoimmune disease similar to *Wiskott–Aldrich syndrome. The cause is a loss of function mutation of the **FOXP3* gene, which is essential for development and maintenance of *regulatory T cells ($CD4^+$).

scurvy The disorder caused by deficiency of vitamin C. The major effects are connective tissue weakness because of the failure of proline hydroxylation and the consequent inability to produce mature collagen. *See* HYDROXYPROLINE.

SCY cytokine superfamily A superfamily of small cytokines. The **SCYA** family of CC-chemokines are designated CCL1–28, the **SCYB** family (CXC chemokines) are designated CXCL1–16, the **SCYC** family are the C-chemokines XCL1 and XCL2. **SCYD1** is *neurotactin (fractalkine; CX3CL1).

scyllatoxin A toxin (31 aa), from the venom of the scorpion *Leiurus quinquestriatus hebraeus*, that specifically blocks low conductance calcium-dependent potassium channels (*SK channels).

SDF-1 A chemokine (stromal cell derived factor-1, CXCL12, 93 aa) that is a chemoattractant for lymphocytes and plays a part in regulating haematopoietic stem cell function. It was originally known as pre-B-cell growth-stimulating factor and is identical to human intercrine reduced in hepatomas (hIRH).The receptor is *CXCR4 which is a coreceptor with CD4 for T-tropic HIV-1; a polymorphism in SCF-1α confers some resistance to infection by T-trophic HIV-1.

SDGF *See* SCHWANNOMA-DERIVED GROWTH FACTOR.

SDH *See* SERINE DEHYDRATASE; SUCCINATE DEHYDROGENASE.

SD sequence *See* SHINE–DALGARNO REGION.

sea-blue histiocyte disease A disorder caused by a defect in *apolipoprotein E, characterized by splenomegaly, mild thrombocytopenia, and numerous *histiocytes in the bone marrow that contain cytoplasmic granules which stain bright blue with haematologic stains.

***sea* oncogene** An *oncogene (S13 avian erythroblastosis oncogene homologue) that encodes a *receptor tyrosine kinase of the Met/ *hepatocyte growth factor/scatter factor family.

sebaceous cyst **1.** A cyst just below the surface of the skin derived, usually, from a sebaceous gland. **2.** An epidermoid cyst, a dome-shape swelling filled with keratin.

Sebastian syndrome *See* MAY–HEGGLIN ANOMALY.

seborrhoea (*adj.* seborrheic) Excessive production of oily sebum by the sebaceous glands of the skin.

SEC **1.** The *serpin–enzyme complex, formed by a *serpin and a *serine peptidase. The SEC receptor is expressed on hepatocytes and binds to a conserved sequence in α_1-antitrypsin and other serpins. It mediates catabolism of the complex and stimulates serpin production. **2.** Selenocysteine (Sec). **3.** Size-exclusion chromatography (SEC). **4.** The general secretory (sec) pathway in bacteria. **5.** sec proteins (often with a numerical suffix, e.g. Sec 7, sec65) are involved in the secretory pathway for proteins. **6.** sec61 is a channel (translocon) that allows movement of proteins across cellular membranes and integrates transmembrane proteins into lipid bilayers. (It is homologous to the SecY channel in bacteria and archaea.) The channel is a heterotrimeric complex (α, β, and γ) that associates with the subcomplex Sec62/Sec63 and interacts with the *signal peptidase complex.

secondary granules *See* SPECIFIC GRANULES.

secondary immune response The response of the immune system when an antigen that has been encountered before (and elicited a *primary immune response) is met on a subsequent occasion. The response is generally more rapid.

secondary lymphoid tissue *See* LYMPHOID TISSUE.

secondary lysosome An intracellular vacuole formed by the fusion of lysosomes with organelles (*autosomes) or with primary phagosomes (to produce digestive vacuoles). Indigestible material may persist in *residual bodies.

secondary structure *See* PRIMARY STRUCTURE.

second messenger An intracellular mediator of signals that involve the binding of a hormone to the extracellular portion of a receptor, receptor occupancy then stimulating production of the second messenger. The same intracellular second messenger can be used by multiple receptors, thus allowing signal integration and a general cell response (e.g. secretion) mediated by a second messenger can be made tissue specific by restricting receptor expression. Examples include *cyclic AMP, *cyclic GMP, *diacylglycerol, and *inositol trisphosphate.

secretagogue A substance that stimulates cells to release material from secretory vesicles. The term was originally applied to peptides inducing gastric and pancreatic secretion. *See* STIMULUS–SECRETION COUPLING.

secretases A family of peptidases involved in processing of membrane-bound precursor molecules. **α-Secretase** (*ADAM10, EC 3.4.24.81, 748 aa) cleaves *amyloid precursor protein (APP) to produce the soluble nonamyloidogenic product sAPPα; this prevents the formation of amyloid (beta peptide. It also has TNF-convertase activity. (Beta-**Secretase** (BACE-1, beta-site APP cleaving enzyme 1, EC 3.4.23.46, 501 aa) is an integral membrane aspartic peptidase that generates the N-terminus of the β-amyloid protein from APP. Further cleavage is then carried out by **γ-secretase**, a multiprotein complex responsible for the intramembranous cleavage of the amyloid precursor protein (APP) and other type I transmembrane proteins such as *notch and *E-cadherin. There are four components, *presenilin, *nicastrin, *APH-1, and *PEN-2, all of which are required for activity; a fifth component, CD147, seems to have a regulatory role. Mutations in presenilin are a risk factor for Alzheimer's disease.

secretin A peptide hormone (27 aa) produced by endocrine cells, S cells, located in the mucosa of the proximal small intestine. Through its G-protein-coupled receptor (449 aa) it stimulates pancreatic secretion and inhibits gastric acid secretion. The secretin family consists of *vasoactive intestinal peptide (VIP), pituitary adenylate cyclase-activating polypeptide (*PACAP), secretin, *glucagon, *glucagon like peptide-1 (GLP(1), GLP(2)), *gastric inhibitory peptide (GIP), *growth hormone releasing hormone (GHRH or GRF), *peptide histidine methionine (PHM), and *helodermin. Most of the family members occur both in the central nervous system and in various peripheral tissues.

secretion The release of synthesized product from cells, as opposed to the excretion of waste products. Material may be released in membrane-bounded vesicles (merocrine secretion) or the vesicle content may be released following fusion of the vesicle with the plasma membrane (apocrine secretion). Holocrine secretion involves the release of whole cells.

secretogranins *See* GRANINS.

secretoneurin A peptide (33 aa) generated by proteolytic cleavage of *secretogranin II, widely distributed throughout the central and peripheral nervous systems; stimulates dopamine release from striatal neurons, and gonadotropin II secretion from the pituitary; inhibits serotonin and melatonin release from pinealocytes. Activates monocyte migration, and probably has a role in neurogenic inflammation. The receptor is G-protein coupled.

S

secretor A person whose mucous secretions (e.g. saliva) contain ABO blood group substances. Around 80% of people are secretors.

secretory carrier associated membrane proteins Integral membrane proteins (SCAMPs), *tetraspan vesicle membrane proteins) that act as carriers, recycling proteins to the cell surface. At least three members of the family have been identified in humans: SCAMP1 (338 aa), SCAMP2 (329 aa), and SCAMP3 (347 aa).

secretory cells Cells that are specialized for secretion, usually epithelial. If the secreted material is protein the *rough endoplasmic reticulum is usually conspicuous, whereas lipid- or lipid product-secreting cells have predominantly *smooth endoplasmic reticulum.

secretory piece A polypeptide (70 kDa), produced by epithelial cells from the *polyimmunoglobulin receptor, associated with oligomeric *IgA secreted from mucosal surfaces.

secretory proteins A diverse group of proteins, nearly all of which are synthesized on the *rough endoplasmic reticulum, cotranslationally transported into the lumen of the endoplasmic reticulum (*see* SIGNAL RECOGNITION PARTICLE) and then moved, via the Golgi, to secretory vesicles. The majority are glycosylated. There are few exceptions; some are not glycosylated, others (e.g. collagen) do not go through the same processing pathway. In prokaryotes, secreted proteins may be synthesized on ribosomes associated with the plasma membrane or exported from the cytoplasm after translation has been completed.

secretory vesicle A membrane-bounded vesicle derived from the *Golgi apparatus in which material to be secreted is stored prior to release. The contents are often densely packed and sometimes are in an inactive form (*zymogen).

securin The product of the pituitary tumour transforming gene (*PTTG*), an inhibitor (202 aa) of the protease *separin (separase/Esp1p) that breaks down the *cohesin-mediated adhesion between sister chromatids and allows them to separate at anaphase. The *anaphase-promoting complex ubiquitinates securin and degradation of securin relieves the inhibition of separase. Cyclin-dependent kinase-1 (CDK1) phosphorylates securin and blocks ubiquitination by the APC. Defects in securin are thought to account for the chromosomal instability often seen in tumours.

sedimentation coefficient *See* SVEDBERG UNITS.

sedimentation test *See* ERYTHROCYTE (erythrocyte sedimentation rate).

sedlin A widely expressed and conserved protein (140 aa) probably involved in transport from the endoplasmic reticulum to Golgi. Mutations in the encoding gene cause spondyloepiphyseal dysplasia tarda, a progressive skeletal disorder which mainly involves vertebral bodies and hips. The intracellular chloride channel protein CLIC1 has been shown to associate with sedlin by yeast two-hybrid screening.

Segawa's syndrome A disorder (infantile Parkinsonism) in which there is severe motor retardation. The **autosomal recessive form** is caused by mutation in the *tyrosine hydroxylase gene and an **autosomal dominant form** by mutation in the *GCH1* gene that encodes GTP cyclohydrolase I (EC 3.5.4.16), responsible for the rate-limiting step in tetrahydrobiopterin synthesis. (Tetrahydrobiopterin is essential for catecholamine synthesis.)

segmentation The organization of the body into repeating units that can arise in a variety of ways. Segmentation occurs during embryonic development of vertebrates, most noticeably the subdivision of mesoderm into *somites, but is also a feature of the development of the central nervous system. *See* NEUROMERES; RHOMBOMERES.

segment long-spacing collagen *See* SLS COLLAGEN.

seipin An endoplasmic reticulum-resident membrane protein (398 aa), that is *N*-glycosylated and proteolytically cleaved into N- and C-terminal fragments. Highly expressed in the nervous system. Seipin mutations are involved in various disorders including congenital *lipodystrophy type 2, Silver's syndrome/spastic paraplegia 17, and distal hereditary motor neuropathy type V.

Seip's syndrome A rare autosomal recessive disease (Berardinelli–Seip congenital lipodystrophy type 2) in which there is an almost complete absence of adipose tissue from birth or early infancy and severe insulin resistance. It is caused by mutation in the gene encoding *seipin. *See* LIPODYSTROPHY.

selectins (addressins) A family of cell adhesion molecules that bind carbohydrates via a *lectin-like domain. They are integral membrane glycoproteins with an N-terminal, *C-type lectin domain. The prefix in the nomenclature reflects the tissue in which the selectin was first identified, and is not necessarily indicative of a functional role. **E-selectin** (endothelial selectin, CD62E, endothelial leucocyte adhesion molecule-1, ELAM-1, 610 aa) is up-regulated on endothelial cells at sites of inflammation where it contributes to trapping of neutrophils. **L-selectin** (leucocyte–endothelial cell adhesion molecule, LECAM, CD62L, LAM-1, MEL-14 antigen, leu-8, 372 aa) is important for the initial adhesion of circulating leucocytes to endothelium, the process of margination, and for lymphocyte homing to *high endothelial venules in peripheral lymph nodes, Peyer's patches, and areas of inflammation. It is present on the surface of most haematopoietic cells and on mature monocytes, eosinophils, and neutrophils. Once the cell has left the circulation L-selectin is removed from the cell surface by a membrane-associated metallopeptidase, a sheddase. It binds to carbohydrates on CD34, CD162, GlyCam, and MAdCAM. Polymorphism in L-selectin is associated with diabetic nephropathy in type 2 diabetes. L-selectin on the trophoblast binds to ligands on the uterine wall and is important in establishing implantation. **P-selectin** (platelet selectin CD62P, PADGEM, GMP-140, LECAM-3, 830 aa) is found on megakaryocytes and is rapidly up-regulated in platelets and endothelial cells when activated, although only transiently expressed. It mediates rolling of neutrophils, platelets, and some T-cell subsets along the luminal surface of blood vessels.

selective serotonin reuptake inhibitors (SSRIs) A class of antidepressant drugs used to treat depression, anxiety disorders, and some personality disorders. They inhibit the uptake of serotonin into presynaptic cells, thus increasing the duration of the signal but also reducing serotonin production through a feedback mechanism. Examples include *fluoxetine (*TN* Prozac), *fluvoxamine, *paroxetine, and *sertraline (*TN* Zoloft).

selegiline A selective irreversible monoamine oxidase B inhibitor used to treat early-stage Parkinson's disease. *TNs* Eldepryl, Zelapar.

selenium (Se) An essential trace element needed in serum-free culture media for most animal cells. *See* SELENOCYSTEINE.

selenocysteine (Sec, U) An unusual amino acid of proteins, the analogue of *cysteine in which selenium replaces sulphur. Involved in the catalytic mechanism of selenoenzymes such as *glutathione peroxidase. May be cotranslationally coded by a special *opal suppressor tRNA that recognizes certain UGA nonsense codons. **Selenoproteins** contain selenocysteine. **Selenoprotein N** (590 aa) is an endoplasmic reticulum glycoprotein with a single selenocysteine residue. It is highly expressed in myoblasts and down-regulated in differentiating myotubes. Mutations in the gene lead to rigid-spine muscular dystrophy. **Selenoprotein P** (381 aa) is a plasma heparin-binding protein that appears to be associated with endothelial cells and has been implicated as an oxidant defence. It has multiple selenocysteine residues. **Mitochondrial capsule selenoprotein** (116 aa) is important for the maintenance and stabilization of the crescent structure of the sperm mitochondria. **Selenoprotein Z** (524 aa) is thioredoxin reductase 2 (EC 1.6.4.5). Various other selenoproteins are known but are of uncertain function.

self-antigen (autoantigen) Any antigen within the body to which an immune response could be mounted. Most of these do not elicit a response because self-reactive clones of T cells are eliminated as the immune system matures, but an autoimmune response (against self-antigens) can occur and is responsible for a number of diseases. *See* AUTOIMMUNE DISORDERS.

self-assembly The process by which multimeric structures assemble from subunits (protomers) in a manner which is defined by the properties of the protomers, not by a template. In many cases (as with microtubules) the basic self-assembly process is regulated by initiation (nucleation) sites which determine where the multimers will form.

self-splicing The self catalysed removal of certain classes of *introns in the maturation of RNA, without the involvement of the *spliceosome.

semaphorins A conserved family of proteins that inhibit the movement of the *nerve growth cone and influence the architecture of the nervous system. Some are transmembrane proteins, others are secreted. Most are relatively large (~750 aa) with a conserved extracellular 'sema' domain (~500 aa) which is the ligand for the receptors, *plexins. *See also* COLLAPSIN (collapsin response-mediator proteins).

semiautonomous Describing systems or processes that are only partly independent of other systems or processes.

semiconservative replication The replication method that produces daughter copies of DNA; each of the daughter molecules has one newly synthesized nucleotide strand and one old (parental) strand.

semipermeable membrane A membrane that is only permeable to some solutes, usually determined by the size or charge of the solute molecule although the incorporation of selective carrier molecules can alter the properties of the membrane.

Semliki forest virus An alphavirus of the *Togaviridae family. It has a positive-stranded RNA genome and an icosahedral capsid enveloped by a host-cell derived lipid bilayer and has been extensively used as a model system for studying the synthesis and incorporation of the transmembrane spike glycoproteins. It is spread by mosquitos and was first isolated from mosquitos in the Semliki forest of Uganda. It causes a lethal encephalitis in rodents but only one human infection, of an immunocompromised individual, has been reported.

senataxin A protein (EC 3.6.1.-, 2677 aa) that, on the basis of homology, may have both RNA and DNA helicase activities and be involved in the DNA repair pathway. Mutations in the gene for sentaxin are associated with autosomal recessive *spinocerebellar ataxia-1 (SCAR1) and juvenile *amyotrophic lateral sclerosis (ALS4).

Sendai virus A type 1 parainfluenza virus (haemagglutinating virus of Japan, HVJ), one of the *Paramyxoviridae. It can cause respiratory disease in humans and causes a fatal pneumonia in mice. Inactivated virus can be used to cause cell–cell fusion and has been used to produce *heterokaryons and *hybrid cell lines.

senescent cell antigen A glycosylated polypeptide antigen that appears on the surface of senescent and damaged red cells as result of *calpain proteolysis of the cytoplasmic region of *band III. Circulating IgG binds to the antigen and triggers the removal of aged erythrocytes from the circulation by Kuppfer cells of the liver.

senile plaque Extracellular deposits of amyloid fibrils in the grey matter of the brain, a feature of *Alzheimer's disease although they occur in the brain with increasing frequency with ageing. The amyloid fibrils are composed of B/A4, a peptide (~4 kDa) derived from *amyloid precursor protein (APP).

Senior–Loken syndrome An autosomal recessive disease characterized by *nephronophthisis and *Leber congenital amaurosis. **Senior–Loken syndrome-1** (SLSN1) is caused by mutation in the gene encoding *nephrocystin 1, **SLSN4** is associated with mutation in nephrocystin 4, **SLSN5** with mutation in nephrocystin 5. **SLSN6** is caused by mutation in nephrocystin 6, a centrosomal protein that is also mutated in *Joubert's syndrome type 5.

sensitization A state of heightened responsiveness, used by immunologists to refer to the effect of exposure to antigen, eliciting a *primary immune response but also in the context of *hypersensitivity.

sensory ataxic neuropathy A mitochondrial disorder that causes clinically variable systemic effects. It is phenotypically similar to autosomal recessive progressive external ophthalmoplegia and the mitochondrial neurogastrointestinal encephalopathy syndrome. Sensory ataxic neuropathy, dysarthria, and ophthalmoparesis (SANDO) is caused by mutation in the nuclear gene that encodes DNA polymerase-γ or by mutation in the gene for *twinkle. Spinocerebellar ataxia with epilepsy (SCAE) is a similar disorder.

sensory neuron **1.** A *neuron that receives input from sensory cells. **2.** Sense cells such as the mechanoreceptors in skin and muscle.

separin A cysteine endopeptidase (separase, Esp1, EC 3.4.22.49, 2120 aa), related to caspases, that cleaves the *RAD21 component of the *cohesin complex that holds sister chromatids together, allowing the chromatids to separate in anaphase. It is inhibited by *securin and this inhibition is relieved when the *anaphase-promoting complex ubiquitinates securin leading to its degradation.

Sephadex A trademarked name for a crosslinked dextran gel used in bead form for *gel filtration columns.

Sepharose *See* AGAROSE.

septicaemia A serious condition, blood poisoning, in which there are many bacteria in the blood. *Compare* BACTERAEMIA.

septic shock An acute systemic inflammatory response triggered by endotoxin in the circulation which activates the complement system and leads to massive overproduction of various cytokines. Can arise from massive bacterial infection and has a very high mortality rate (~50%). *See* ACUTE RESPIRATORY DISTRESS SYNDROME.

septins A family of cytoplasmic cytoskeletal filament-forming proteins (~350 aa) with a conserved GTP-binding domain. They were first identified in the ring of 10-nm filaments that forms where daughter cells are budded off in *Saccharomyces cerevisiae*, but homologues are found associated with the cleavage furrow in insects and vertebrates. Various septins are involved in membrane dynamics, vesicle trafficking, apoptosis, and cytoskeletal remodelling and there are at least twelve human septins. Septin-7 interacts with *CENP-E at the kinetochore. Mutations in septin-9 are associated with hereditary neuralgic amyotrophy.

septum Literally, a separating wall. In prokaryotes and fungi, the wall that forms between daughter cells after cell division.

sequence homology A term that is often used rather casually to mean an observable sequence similarity in nucleic acids or proteins, although strictly homology implies a (hypothetical) common evolutionary origin.

sequestosome A perinuclear inclusion body (aggresome) containing aggregated, misfolded, ubiquitinated protein, destined for degradation. May arise if the capacity of the *proteosome is exceeded, which may also be the case for *Lewy bodies. Cells that over-express wild-type *parkin form fewer aggresomes and disruption of microtubules blocks the formation of aggresomes. The sequestosome-1 gene encodes ubiquitin-binding protein p62 and *Paget's disease of bone type 3 is due to mutation in this gene.

sequon A term most commonly used of the tripeptide motif, Asn-Xaa-Ser/Thr, that is the site for *N*-linked glycosylation.

SER *See* SMOOTH ENDOPLASMIC RETICULUM.

SERCA *See* SARCOPLASMIC–ENDOPLASMIC RETICULUM CALCIUM ATPASE.

serglycin The core protein (proteoglycan 1, 153 aa) of the proteoglycan that is found in the storage granules of platelets, leucocytes and mast cells. The protein has 24 serine-glycine repeats to which various (~15) *glycosaminoglycan chains, usually heparin or highly sulphated chondroitin sulphate, are attached. The negative charge on the proteoglycan is probably important for packing positively charged proteases, histamine, etc. within the granules.

serine (Ser, S) An amino acid (105 Da) found in proteins. Serine can be phosphorylated by the serine/threonine *protein kinases.

serine dehydratase An enzyme (EC 4.2.1.13, 328 aa) that catalyses the deamination of L-serine and L-threonine to yield pyruvate or 2-oxobutyrate and thus is involved in generating glucose from amino acids.

serine hydroxymethyltransferase An enzyme (SHMT, EC 2.1.2.1) that catalyses the reversible conversion of serine and tetrahydrofolate to glycine and 5,10-methylene tetrahydrofolate. The gene for the cytosolic enzyme (SHMT1, 444 and 483 aa isoforms) maps to the critical interval for *Smith–Magenis syndrome. A mitochondrial form of the enzyme (SHMT2, 494 aa) also exists.

serine peptidases (serine proteases) A diverse clan of peptidases with a common reaction mechanism involving an active serine residue. Most are inhibited by serine peptidase inhibitors (*serpins) and irreversibly inactivated by organophosphorus esters, such as di-isopropylfluorophosphate (DFP). Members of the clan include *trypsin, *chymotrypsin, and bacterial *subtilisin.

seroconversion Immunological jargon for having responded to immunization by producing antibody, becoming seropositive.

serogroup Describing bacteria or other microorganisms that have an antigen (serotype) in common.

serosa **1.** A serous epithelium with *serous glands or cells. *Compare* MUCOUS MEMBRANE. **2.** The thin infolding of the lining of the peritoneal cavity that forms the *omentum.

serotonin (5-hydroxytryptamine, 5-HT) An important *neurotransmitter in the central nervous system where it is involved in modulating anger, aggression, body temperature, mood, sleep, appetite, and metabolism; drugs that affect the levels of serotonin (e.g. *selective serotonin reutake inhibitors) are used extensively to treat depression and various psychiatric conditions. Serotonin also has an important role as a peripheral hormone and is produced by enterochromaffin cells in the gut as well as being stored in platelets. Most receptors are G-protein coupled, except for the 5-HT_3 receptor which is a ligand-gated ion channel permeable to sodium, potassium, and calcium ions.

serotype *See* SEROGROUP.

serous gland An exocrine gland that secretes a watery, protein-rich solution rather than a carbohydrate-rich mucous secretion.

serpentine receptors *See* G-PROTEIN-COUPLED RECEPTORS.

serpins A superfamily of structurally similar proteins, most of which are *serine peptidase inhibitors. Examples include *alpha-1-antitrypsin, *plasminogen activator inhibitor 1, *maspin,

*PEDF, *protease nexin-1, and *vaspin. Deficiency in complement C1 inhibitor (serpin G1) causes *hereditary angioedema. The serpin–enzyme complex (SEC) is the complex formed by a serpin and a serine peptidase for which there is a receptor (SEC receptor) that is expressed on hepatocytes. The receptor binds to a conserved sequence in α_1-antitrypsin and other serpins. It mediates catabolism of the complex and stimulates serpin production.

Serratia marcescens A Gram-negative bacterium that is common in soil and water but is an opportunistic pathogen, increasingly a problem with the emergence of antibiotic resistance. Most strains produce a characteristic red pigment, prodigiosin. *Serratia marcescens* haemolysin is a pore-forming toxin quite distinct from other pore-forming toxins (e.g. *RTX toxins). It is a heterodimer of ShlB and ShlA that is activated by a conformational change that requires phosphatidylethanolamine as a cofactor.

Sertoli cells Tall columnar cells in the testis that provide the microenvironment for developing spermatocytes and spermatids. They also have the capacity to phagocytose degenerate sperm.

sertraline An antidepressant drug of the *selective serotonin reuptake inhibitor type. *TNs* Lustral, Zoloft.

serum The fluid that remains when blood is allowed to clot. The cells become entangled in a *fibrin meshwork which contracts (clot retraction) by the activity of platelets. Unlike *plasma it has lost proteins involved in the formation of the clot (*fibrinogen, *prothrombin, *blood clotting factors) but has gained various platelet-released factors, particularly *platelet derived growth factor, which makes serum much better than defibrinated plasma (plasma derived serum) for supporting the growth of cells in culture.

serum amyloid A family of apolipoproteins (SAA, SAP) associated with high-density lipoprotein that are *acute phase proteins. Partial proteolysis of serum amyloid A produces an N-terminal (1–76) fragment, **amyloid A protein** (AA protein) which forms the amyloid fibrils that are deposited in tissues in secondary *amyloidosis. **Amyloid P protein** is derived from serum amyloid P, a *pentraxin similar to *C-reactive protein, and is a minor component of fibrils in both primary and secondary amyloidosis.

SEE WEB LINKS

- Additional information on serum amyloid.

serum- and glucocorticoid-inducible kinases Serine/threonine protein kinases (SGKs) that are transcriptionally regulated by corticoids, serum, and cell volume. Alternatively spliced variants, under separate promoters, have been identified. Sgk1 (431 aa) regulates epithelial ion transport by phosphorylating and inactivating Nedd4-2, an E3 ubiquitin-protein ligase that targets the epithelial Na^+ channel (ENaC) and the excitatory amino acid transporter (EAAT)2 for degradation. The latter may account for the role of SGK in facilitating spatial learning in rats. SGKs are related to *Akt (PKB).

serum hepatitis *See* HEPATITIS VIRUSES.

serum response element (SRE) A promoter site found in many growth-related genes, a region of 20 bp upstream of the c-fos transcription initiation site, where it was first described as a dyad symmetry element (DSE). **Serum response factor** (SRF; p67SRF) is a transcription factor which interacts with Elk-1 (p62TCF) to bind the SRE.

SET A protein (suppressor of variegation, enhancer of zeste, and trithorax, 290 aa) that is a specific inhibitor of *protein phosphatase 2A and that modifies phosphorylation of histone H3 with effects on nucleosome structure. It has been shown to be involved in the regulation of renal cell proliferation and tumorigenesis. SET mRNA is expressed at much higher levels in developing kidney and in transformed cell lines and expression is markedly reduced in quiescent cells. It is highly expressed in *Wilms' tumour. The SET domain is found in a number of proteins involved in embryonic development.

SET complex An *endoplasmic reticulum associated complex which contains three *granzyme A substrates, the nucleosome assembly protein *SET, the DNA bending protein *high-mobility group 2 (HMG-2), and the base excision repair endonuclease *Ape1 as well as the tumour suppressor protein *pp32/PHAP-I and the granzyme A-activated DNase *NM23-H1, which is inhibited by SET. The complex is a target of the granzyme A cell death pathway.

seven-membrane spanning receptors *See* G-PROTEIN-COUPLED RECEPTORS.

severe combined immunodeficiency disease (severe congenital immunodeficiency disease, SCID) A range of disorders in which the immune system is seriously impaired. T cells are deficient in all types (T^-) but B cells and natural killer (NK) cells may or may not be affected ($T^-B^-NK^+$ SCID, etc.). The commonest form is caused by mutation in the IL-2 receptor (*IL2RG*) gene (X-linked $T^- B^+ NK^-$ SCID). Autosomal recessive SCID includes $T^- B^+ NK^-$ SCID caused by mutation in the *JAK3* gene; $T^- B^+ NK^+$ SCID is caused by mutation in the *IL7R* gene, the *CD45* gene, or the *CD3D* gene; $T^- B^- NK^-$ SCID is caused by mutation in the *ADA* (*adenosine deaminase) gene; $T^- B^- NK^+$ SCID with sensitivity to ionizing radiation is caused by mutation in the *artemis gene; $T^- B^- NK^+$ SCID is caused by mutation in the *RAG1 and -2 genes. Gene therapy has been used to treat some patients with ADA deficiency.

sex chromosomes The chromosomes that are involved in sex determination. In humans the two sex chromosomes (X and Y) are dissimilar, the female has two X chromosomes (one of which is condensed and inactive, the Barr body; *see* LYON HYPOTHESIS), and the male is heterogametic (XY). In birds the opposite is the case, the male being XX and the female XY; in many organisms there is only one sex chromosome, one sex being XX, the other X0. A portion of the X and Y chromosomes is similar and is referred to as the pseudoautosomal region.

sex hormone A hormone secreted by the gonads, or that influences development of the gonads. Examples are *oestrogen, *testosterone, *gonadotropins.

sex-linked disorder A genetic defect, usually a mutated gene on the X chromosome which, because only a single copy is present in the male, has its full effect. Most sex-linked disorders are seen in males; only homozygous females will be affected.

sex pili The filamentous projections (*pili) from the surface that are important for conjugation in bacteria. In some cases they are encoded on plasmids that make conjugation possible and facilitate the spread of the plasmid. The F-plasmid encodes F-pili 8–9 nm in diameter and several micrometres long, composed of *pilin.

Sezary's syndrome A type of cutaneous lymphoma in which the affected cells are peripheral $CD4^+$ T-cells that have pathological quantities of mucopolysaccharides. Sezary's disease is sometimes considered a late stage of mycosis fungoides, but the cause is unknown. *HuT-78 cells were derived from a patient with this syndrome.

SGK *See* SERUM- AND GLUCOCORTICOID-INDUCIBLE KINASES.

SGOT The enzyme (serum glutamic-oxaloacetic transaminase, aspartate transaminase, EC 2.6.1.1, 413 aa) that catalyses the transfer of an amine group from glutamic acid to oxaloacetic acid to produce α-ketoglutaric acid and aspartic acid. There are cytoplasmic and mitochondrial isozymes. Elevated serum levels can indicate liver or heart damage.

SGPT *See* ALANINE AMINOTRANSFERASE.

SGTx1 A peptide toxin (34 aa), isolated from the venom of the tarantula *Scodra griseipes*, that has been shown to inhibit Kv2.1 potassium channels in rat cerebellar granule neurons. Like *hanatoxin, with which it is homologous, it is a gating modifier.

SH2, SH3 *See* SH DOMAINS.

SH3BP4 A protein (SH3-domain binding protein 4, TTP, 963 aa) that is involved with regulating the uptake of *transferrin receptors into *clathrin-coated vesicles. It has various protein recognition domains, including a SH3 domain, and a nuclear targeting signal and interacts with endocytic proteins, including *clathrin and *dynamin.

shadoo A GPI-linked protein (~140 aa) expressed in the brain, that has some similarities to *prion proteins. The gene (*SPRN*, shadow of prion protein) is highly conserved from fish to mammals.

SEE WEB LINKS

- Article in *Molecular Biology and Evolution* (2004).

shaker A *Drosophila* gene that encodes a potassium channel (Kv1.1). Related genes *shab*, *shal*, and *shaw* are known in flies and humans and encode homologous potassium channels. The mutation is so called because the fly's legs shake under ether anaesthesia and in humans mutations lead to *episodic ataxia/*myokymia syndrome (episodic ataxia type 1).

SHANKs A family of adaptor proteins (SH3 and multiple ankyrin repeat domains proteins). **SHANK1** (somatostatin receptor-interacting protein, 2161 aa), **SHANK2** (cortactin-binding protein 1, 1253 aa) and **SHANK3** (proline-rich synapse-associated protein 2, 1741 aa) are found in the postsynaptic density of excitatory synapses and may play a role in the structural and functional organization of the dendritic spine and synaptic junction. A chromosomal aberration that disrupts SHANK3/PSAP2 is responsible for the clinical features of chromosome 22q13.3 deletion syndrome, and defects in SHANK3 are a cause of autism spectrum disorders. **Shank-interacting protein-like 1** (sharpin, SHANK-associated RH domain-interacting protein, 387 aa) may have a role in normal immune development and control of inflammation.

sharpin *See* SHANKs.

shc A family of proteins (~580 aa) with SH2 and SH3 domains and a tyrosine motif which, when phosphorylated, binds *grb2, linking receptor tyrosine kinases with the ras signalling pathway. The *shc* gene encodes overlapping proteins of 46 and 52 kDa and p66Shc that is involved in signal transduction pathways that regulate the cellular response to oxidative stress and life span. Several shc-related proteins act as adaptors for other growth factor receptors and have different tissue distributions.

SH domains Protein domains (src homology domains) homologous to those in *src. **SH1** is a kinase domain, **SH2** domains are involved in binding to phosphorylated tyrosine residues on other proteins, **SH3** domains bind to proline-rich motifs of other proteins, and the **SH4** domain has myristoylation and membrane localization sites. Interactions between the domains are affected by phosphorylation.

shear stress response element (SSRE) A response element that activates gene expression and cell division in response to the mechanical stress derived from fluid shear, as experienced by vascular cells.

Sheldon–Hall syndrome *See* ARTHROGRYPOSIS.

shelterin A protein complex that protects *telomeres from the attention of DNA repair systems. Components include *telomeric repeat binding factors (TRF1 and TRF2), TRF1-interacting factors 1 and 2 (*TIN2 and *PINX1), *TRF2-interacting protein, protection of telomeres 1 (*POT1), *TIN2-interacting protein and accessory proteins such as *apollo.

SEE WEB LINKS

- Article in *Genes and Development* (2005).

S

shiga toxin A toxin (verotoxin) from *Shigella dysenteriae*. Shiga toxin and the shiga-like toxins (SLTs) are structurally related *AB toxins with a pentameric B subunit that binds globotriaosylceramide. The A subunit is internalized and cleaved into two parts and the A1 component then binds to the ribosome, disrupting protein synthesis.

Shigella A genus of nonmotile Gram-negative enterobacteria of the Escherichiae group that causes dysentery. *See* SHIGA TOXIN.

shimamushi fever An acute fever (scrub typhus, flood fever, Japanese river fever, tsutsugamushi fever) caused by *Rickettsia tsutsugamushi*, transmitted by the bite of an infected larval mite (chigger) *Leptotrombidium akamushi*.

Shine–Dalgarno region A polypurine sequence (SD sequence) in the ribosome binding site of many prokaryotic mRNAs, seven nucleotides upstream of the *initiation codon. A complementary sequence at the 3′ end of 16S rRNA is probably involved in binding.

shingles The disease caused in adults by varicella zoster virus (*Herpesviridae), the virus responsible for chickenpox in childhood. In most cases latent virus in spinal or cranial sensory nerve ganglia is reactivated, usually because of reduced cell-mediated immune surveillance.

SHIP A phosphatase (SH2-containing inositol phosphatase, 1188 aa) that acts on 5-inositol phosphates and is important in modulating the inositol signalling pathway.

ShK toxin A peptide toxin (35 aa), from the sea anemone *Stichodactyla helianthus*, that blocks the voltage-gated potassium channel in T cells, Kv1.3.

shock The clinical condition associated with circulatory collapse that can be caused by blood loss, *bacteraemia, anaphylaxis, or emotional stress. *See* ACUTE RESPIRATORY DISTRESS SYNDROME (septic shock).

Shope fibroma virus A DNA virus of the *Poxviridae family that causes skin tumours in rabbits.

Shope papilloma virus A member of the *Papillomaviridae that causes keratinous carcinomas near the head in cottontail rabbits. It has been used as an experimental model for human papillomaviruses.

short hairpin RNA (shRNA) A short RNA sequence that has a tight hairpin loop and that is cleaved to produce *siRNA. Plasmids encoding shRNAs are used experimentally to suppress gene expression.

short interfering RNA *See* SIRNA.

short interspersed nucleotide elements (SINEs) Retrotransposons without *long terminal repeats that use the reverse transcriptase of the *long interspersed nucleotide element LINE1 to replicate. They are ~300 bp long and present in very large numbers. Most SINEs other than *Alu are derived from tRNAs, including the MIRs (*mammalian-wide interspersed repeats).

short QT syndrome A disorder of potassium channels in which the QT interval in the cardiac cycle is short and there can be paroxysmal atrial fibrillation and sudden cardiac failure. **Short QT**

syndrome-1 (SQT1) is caused by mutation in the *KCNH2* gene (*hERG that encodes a voltage-sensitive potassium channel); **SQT2** by mutation in the *KCNQ1* gene for the *shaker-related subfamily of potassium channels which is also associated with *long QT syndrome, and **SQT3** by mutation in the *KCNJ2* gene (Kir2.1).

short tandem repeat *See* SATELLITE DNA.

shotgun approach A colloquial term for any approach based on the analysis of large numbers of randomly acquired data points and their integration. Often used in the context of sequence analysis (**shotgun sequencing**) in which DNA (or even the whole genome) is cut with restriction enzymes into small sections of random length which are all sequenced, sometimes more than once, and then the full sequence is assembled using software that recognizes overlap.

Shp **1.** Protein tyrosine phosphatases (Shp-1, Shp-2) with SH2 domains. Are recruited to *ITIM motifs of receptor tyrosine kinases and play an important role in the control of cytokine signalling. **Shp-1** (haematopoietic cell phosphatase, PTP1C, PTPN6, 595 aa) is important in regulating antigen responses in T-cells and the mouse mutant (motheaten) is immunosuppressed. **Shp-2** (Syp, PTP2C, PTPN11, 593 aa) is more ubiquitously expressed and functions downstream of a variety of growth factor receptors and has a role in cell spreading and migration; the homozygous mouse knockout is embryonic lethal. An important substrate for both phosphatases is **shps1** (shp-substrate 1, signal regulatory protein α1, SIRP-α1, 503 aa) when it has been phosphorylated by receptor tyrosine kinases. **2.** Small heterodimer partner (shp, 257 aa), an orphan nuclear receptor belonging to the nuclear receptor superfamily of transcription factors. Interacts with various other receptors such as the retinoid receptor and seems to be a negative regulator of signalling pathways. Mutations in the gene are associated with early-onset mild obesity.

shRNA *See* SHORT HAIRPIN RNA.

shrooms A family of actin-binding proteins with PDZ domains that are involved in control of various developmental processes through the regulation of cell shape. The original mouse mutation was designated 'shroom' (shrm), because the neural folds 'mushroom' outward and do not converge at the dorsal midline. The common feature is the *Apx/Shrm domain 2 (ASD2) and shroom-related proteins are found in a wide range of species. **Shroom 1** (852 aa) may affect microtubule architecture, **shroom 2** (1616 aa) binds cortical actin and is expressed in cells that exhibit polarized growth, including vascular endothelial cells and epithelial cells. **Shroom 3** (1986 aa) interacts with microfilament bundles and also γ-tubulin. **Shroom 4** (1498 aa) may regulate the spatial distribution of myosin II.

SEE WEB LINKS

- Article on standardization of nomenclature in *BMC Cell Biology* (2006).

shuttle vector A *cloning vector that is able to replicate in more than one organism, e.g. allowing a eukaryotic gene to be grown and amplified in *E. coli* and then reintroduced to a eukaryotic cell for *complementation testing.

Shwachman–Diamond syndrome A disorder (Shwachman–Bodian–Diamond syndrome) characterized by exocrine pancreatic insufficiency, haematological abnormalities, including increased risk of malignant transformation, and skeletal abnormalities. It is caused by compound heterozygous or homozygous mutations in the *SBDS* gene which may perhaps function in RNA metabolism.

Shwartzman reaction A response to repeated exposure to *endotoxin that is mediated by platelets and neutrophils. In the local Shwartzman reaction the first injection is given intradermally, the second intravenously 24 h later; a haemorrhagic reaction develops at the dermal site. If both injections are intravenous the result is a generalized Shwartzman reaction, often with *disseminated intravascular coagulation.

sialic acids A generic term for *N*- or *O*-substituted derivatives of neuraminic acid, the commonest being **N*-acetylneuraminic acid.

sialidase *See* NEURAMINIDASE.

sialidosis *See* MUCOLIPIDOSES.

sialin A transmembrane (tetraspanin) lysosomal protein (sodium phosphate cotransporter, solute carrier family 17 member 5, **SLC17A5,** 495 aa) that is responsible for sialic acid export. Mutations cause Salla's disease or infantile free sialic acid storage disease, both neurodegenerative conditions.

sialoglycoprotein A glycoprotein with terminal *neuraminic acid on *N*- or *O*-glycan chains.

sialophorin *See* LEUKOSIALIN.

sialyl Lewis^x An important blood group antigen (sLe^x, CD15s), a tetrasaccharide that is usually attached to *O*-glycans on the cell surface (the sialylated form of CD15). It is the ligand to which *selectins bind and is expressed on neutrophils, basophils, and monocytes; some lymphocytes; and *high endothelial venules (HEV). Leucocyte adhesion deficiency type II is caused by the absence of sialyl Lewis^x.

sibutramine A serotonin-noradrenaline reuptake inhibitor that acts as an appetite suppressant. *TNs* Meridia, Reductil.

sickle cell anaemia A disorder caused by a point mutation in beta-globin that leads to substitution of valine for glutamic acid at position 6, producing haemoglobin S (HbS). The mutation is relatively common in people from areas in which *Plasmodium falciparum* malaria is endemic. The altered haemoglobin crystallizes at low oxygen tension, causing erythrocytes to change to a sickled shape which causes them to become trapped or mechanically damaged, leading to anaemia. In heterozygotes, the disadvantages are apparently outweighed by increased resistance to malaria, probably because parasitized cells tend to sickle and be removed.

sideramines Naturally occurring iron-binding compounds produced by fungi that bind ferric ions. Examples include ferricrocin and fusigen. *Compare* ENTEROCHELIN.

sideroblasts Nucleated erythrocytes containing small, basophilic, ferric iron-rich granules (Pappenheimer bodies) because of a defect in haem synthesis. **X-linked sideroblastic anaemia** (XLSA) is caused by mutation in the gene encoding δ-aminolevulinate synthase-2, the enzyme that catalyses the first committed step of haem biosynthesis. **Sideroblastic anaemia with spinocerebellar ataxia** is an X-linked mitochondrial disease caused by mutation in the nuclear gene for ABC7, an ATP-binding cassette (ABC) transporter that localizes to the mitochondrial inner membrane and is involved in iron homeostasis.

siderochromes *See* FERRICHROMES.

sideromycins Antibiotics that are linked to *siderophores and are actively transported into Gram-positive and Gram-negative bacteria. An example is albomycin, a derivative of ferrichrome with a bound thioribosyl-pyrimidine moiety, that inhibits seryl-tRNA synthetase.

siderophilin *See* TRANSFERRIN.

siderophores Natural iron-binding compounds that chelate ferric ions (which form insoluble complexes and are unavailable) and are internalized together with the metal ion. *See* SIDERAMINES.

siderosis **1.** A lung disease (pneumonoconiosis) caused by the inhalation of metallic particles. An occupational hazard for workers in tin, copper, lead, and iron mines, and steel grinders. **2.** Excessive deposition of iron in the body tissues. African iron load (formerly Bantu siderosis) results from a genetic predisposition to iron loading exacerbated by excessive intake of dietary iron from traditional beer brewed in nongalvanized steel drums.

sigma receptors Originally described as one of the opiate receptor subtypes but subsequently shown to be nonopioid receptors. **Sigma-1** receptors have high to moderate affinities for (+) benzomorphans, many psychotrophic drugs, and neurosteroids such as *dehydroepiandrosterone sulphate. **Sigma-2** receptors, unlike sigma-1, are not coupled to Gi/o proteins and sigma-2 receptor agonists have been shown to induce cell death via both caspase-dependent and -independent pathways.

signal peptidase complex A complex located in the endoplasmic reticulum membrane that cleaves the *signal sequence from proteins that are destined for export as soon as the cleavage site is exposed in the endoplasmic reticulum lumen. The complex is composed of five polypeptides: SPC12 and SPC25 have substantial cytoplasmic domains and span the membrane twice; SPC18, SPC21, SPC22/23 are single-spanning membrane proteins mostly exposed to the lumen of the endoplasmic reticulum. *See also* SEC (6).

signal peptide *See* SIGNAL SEQUENCE.

signal peptide peptidase An aspartyl peptidase (SPP, EC 3.4.23.-, 377 aa) that catalyses the intramembrane proteolysis of some signal peptides after they have been cleaved from a preprotein. It has sequence motifs characteristic of the *presenilin-type aspartic proteases. More than fifteen proteins homologous to human SPP have been identified and can be subdivided into at least five subfamilies. Their function is unclear.

signal recognition particle (SRP) A complex of 7S RNA and six proteins that binds to the nascent polypeptide chain of eukaryotic proteins that have a *signal sequence and halts further translation until the ribosome has become bound to rough endoplasmic reticulum. One of the SRP proteins (srp54) binds GTP and in association with 7SRNA and srp19 has GTPase activity. *See* ALU repeats.

SEE WEB LINKS

- The Signal Recognition Particle Database.

signal recognition particle receptor (docking protein) The heterodimeric receptor for the *signal recognition particle (SRP) in the membrane of the endoplasmic reticulum (ER). Both protomers have GTP-binding capacity. Once the SRP, mRNA, ribosome, and nascent polypeptide bind the SRP receptor the block to translation is lifted and the SRP is released. The growing polypeptide is cotranslationally transported to the lumen of the ER.

signal sequence **1.** A sequence of amino acids at the N-terminus of a protein that indicates that further synthesis should occur on membrane-bound ribosomes (*rough endoplasmic reticulum) and that the proteins are to be integral membrane proteins or are to be processed through the Golgi and secreted. The sequence (usually hydrophobic but with some positively charged residues and ~20 aa long) is synthesized in the cytoplasm and is recognized by the *signal recognition particle which blocks further translation until the ribosome has bound to the *signal recognition particle receptor. The signal sequence is later removed by the *signal peptidase complex in the lumen of the ER. **2.** Any sequence that determines post-translational uptake by organelles (e.g. *KDEL motif).

signal-transducing adaptor proteins Adaptor molecules (signal-transducing adaptor molecules, STAMs, 540 aa, 525 aa) that link cytokine receptors and *STATs.

signal transduction (signal–response coupling) The sequence of events by which an extracellular signal such as a hormone or light brings about a change in cellular activity. It may involve binding of a molecule to a cell surface receptor that changes the properties of that receptor triggering, in turn, the production of a *second messenger which then alters the activity of some effector system. In many cases the intermediate steps provide some amplification so that a single binding event triggers the flow of many ions or the production of many second messenger molecules.

signet-ring cell A cell with a large central vacuole that displaces the nucleus to one side and produces an appearance (vaguely) reminiscent of a signet ring. May be adipocytes with lipid vesicles, or epithelial cells with mucin vesicles as in **signet-ring cell adenocarcinomas** (where the number of signet-ring cells correlates with the degree of malignancy).

silanization The chemical modification of a silica or glass surface in which OH^- groups are replaced by $-O-SiR_3$ groups. By reducing surface charge, nonspecific binding is reduced. Surfaces for the binding of DNA fragments to produce a microarray are often silanized.

sildenafil An inhibitor of cGMP-specific phosphodiesterase type 5, used to treat erectile dysfunction. *TN* Viagra.

silent gene A gene that is not expressed. This can be a result of a mutation that leads to premature termination of transcription, a defect in the promoter, the blocking of translation by RNA interference, or a mutation that makes the product inactive. A **silent mutation** is one that does not (appreciably) alter the *phenotype, e.g. a point mutation that changes a codon but not the amino acid that is encoded (because of multiple codon assignments to some amino acids).

silicosis A chronic inflammation of the lung caused by inhalation of silica particles. It results in fibrosis but is less damaging than *asbestosis.

Silverman–Handmaker type dyssegmental dysplasia *See* SCHWARTZ–JAMPEL SYNDROME.

Silver–Russell syndrome *See* H19 GENE.

simian virus 40 *See* SV40.

Simmonds' disease *See* HYPOPITUITARISM.

simple epithelium Any epithelium in which all cells are in contact with the basal lamina. The cells can be cuboidal, columnar, squamous, or pseudostratified. *Compare* STRATIFIED EPITHELIUM.

simple sequence length polymorphism (SSLP) Tandem repeat sequences that show polymorphism in the number of repeats in different individuals. Since the flanking sequences are the same in all individuals, PCR primers that will selectively amplify the repeat sequences can be used to detect differences in repeat number (length of the amplified sequence). *See* SATELLITE DNA.

Simpson–Golabi–Behmel syndrome An X-linked condition characterized by pre- and postnatal overgrowth, coarse facies, congenital heart defects, and other congenital abnormalities; phenotypically similar to *Beckwith–Wiedemann syndrome. **Type I** is caused by mutation in the *glycipan-3 gene, probably reducing the extent to which insulin-like growth factor 2 is bound and unavailable for growth stimulation. **Type 2** has been associated with a mutation in the *CXORF5* gene, an open reading frame on the X chromosome.

simvastatin *See* STATINS.

Sin3 Transcriptional corepressors (Sin3A, 1219 aa; Sin3B, 1130 aa), homologues of yeast Sin3, that have multiple protein–protein interaction domains and serve as a scaffold on which corepressor complexes assemble. They do not bind to DNA but may be tethered by forming a complex with *mad and *max. Sin3A interacts with histone deacetylase.

Sindbis virus An enveloped virus of the alphavirus group of *Togaviridae. It is not thought to be infectious in humans and has been extensively used as a model system in which to study the synthesis and export of the spike proteins, via the endoplasmic reticulum and Golgi complex.

SINEs *See* SHORT INTERSPERSED NUCLEOTIDE ELEMENTS.

single channel recording An electrophysiological record of the ion flow through a single channel, usually involving *patch clamping.

single nucleotide polymorphism *See* SNP.

single-stranded conformational polymorphism (SSCP) A method for detecting point mutations (in amplified genomic DNA) by the electrophoretic separation of single-stranded nucleic acids. For conformationally unstable single-stranded DNA even single nucleotide changes are sufficient to alter secondary structure and have a detectable effect on mobility.

single-stranded DNA (ssDNA) A DNA molecule in which there is only one chain of nucleotides, not the standard complementary *base-paired strands of the double-helical form. Heat denaturation of double-stranded DNA causes the strands to separate and the kinetics of *reannealing (*see* C_0T CURVE) are informative about the degree of repetition. *Parvoviridae have a single-stranded DNA genome.

singlet oxygen ($^1O_2$) One of the reactive oxygen species (ROS) produced in the *metabolic burst of leucocytes. It is an energized but uncharged form of oxygen (the diamagnetic form of molecular oxygen) that is cytotoxic.

SipA An actin-binding protein (*Salmonella* invasion protein A) from *Salmonella enterica*, that stabilizes F-actin bundles by increasing polymerization and decreasing depolymerization. The effect is to facilitate internalization of the bacterium, an important factor in virulence.

siRNA *See* SMALL INTERFERING RNA.

sirtuins A family of NAD^+-dependent protein *histone deacetylases (silent information regulator 2 (Sir2) enzymes) that are important in epigenetic gene silencing, DNA recombination, cellular differentiation and metabolism, and the regulation of ageing. The **sirtuin 1** (SIRT1) sequence has greatest homology with *S. cerevisiae* Sir2 protein; **SIRT4** and **SIRT5** more closely resemble prokaryotic sirtuin sequences. SIRTs 1–5 are widely expressed in fetal and adult tissues. **SIRT6** is a class IV sirtuin, a chromatin-associated protein involved in DNA repair. **SIRT7** (also class IV) is associated with active rRNA genes (rDNA) and histones; depletion of SIRT7 stops cell proliferation and triggers apoptosis.

sis **1.** An *oncogene, *sis*, found in a monkey *sarcoma, and later shown to encode the *PDGF B-chain. Human c-sis is overexpressed in many human tumour cells. **2.** *See* MACROPHAGE INFLAMMATORY PROTEINS. **3.** Small intestinal submucosa: cell-free porcine SIS has been used for surgical repair purposes.

sister chromatid One of the two semiconservative copies of the original *chromatid that form a *bivalent and are separated at anaphase.

SIT A transmembrane adaptor protein (SHP2-interacting transmembrane adaptor protein, 196 aa) found in lymphocytes that is a homodimer of glycoprotein subunits. It has an immunoreceptor tyrosine-based inhibition motif (*ITIM), that interacts with *SHP2 phosphatase and *grb2.

sitagliptin An inhibitor of dipeptidyl peptidase-4, that hydrolyses *incretins and so has the potential for use as an appetite suppressant. *TN* Januvia.

site-1 protease *See* SKI.

site-specific mutagenesis (site-directed mutagenesis) An experimental approach to determining which residues in a protein are important for function. The gene is cloned, specific nucleotide pairs changed to alter a particular codon, and the modified gene is reintroduced into the cell in an expression vector. An **alanine scan** is a systematic exchange, using this methodology, of every amino acid in turn for alanine, a procedure that is unlikely to make major changes to secondary structure but removes side groups.

site-specific recombination Genetic *recombination in which DNA strand exchange occurs between segments that have limited sequence homology. It involves site-specific recombinases that recognize short sites where they bind, cleave, excise, exchange, and rejoin the strands. Examples include the *lox-Cre system, *resolvases, Flp-Frt recombination, lambda integrase.

sitosterolaemia A disorder in which affected individuals absorb cholesterol and all other sterols from the intestine. The major plant sterol, sitosterol, is found at very high levels in the plasma and leads to the development of tendon and tuberous *xanthomas, accelerated atherosclerosis, and premature coronary artery disease. Dietary reduction in sterol-rich foods reduces the symptoms. It is caused by a defect in ABC transporters (ABCG8 or ABCG5) that normally cooperate to limit intestinal absorption and to promote biliary excretion of sterols.

situs inversus A condition in which the normal asymmetry of the circulatory system and intestinal coiling is reversed. It is a feature of Kartagener's syndrome and of Ivemark's syndrome (asplenia with cardiovascular anomalies). It occurs in approximately 50% of patients with *primary ciliary dyskinesia (immotile cilia syndrome), indicating that the normal asymmetry depends on ciliary function during development.

sixth disease A benign disease (roseola, roseola infantum) of children under 2 years old characterized by a transient rash ('exanthem') that occurs after a brief fever of about 3 day's duration. Caused by human herpesvirus-6 and -7 (roseolovirus).

Sjögren's syndrome An autoimmune disease that mainly affects the exocrine glands, characterized by inflammation of the conjunctiva and cornea. There is an association with HLA alleles DRB1*03 and DQB1*02.

SK channels A subfamily of calcium-activated *potassium channels (small conductance calcium-activated potassium channels) of which there are four members (SK1 (KCa2.1), SK2 (KCa2.2), SK3 (KCa2.3), and SK4 (KCa3.1), the products of genes *KCNN1-4*). The calcium sensitivity comes from association with *calmodulin. Because they are activated when intracellular calcium levels rise they limit the firing frequency of neurons and other electrically excitable cells. SK channels are thought to be involved in *synaptic plasticity. They are selectively inhibited by *apamin and *scyllatoxin.

skelemin *See* MYOMESIN.

skeletal disorders *See* BONE DISORDERS.

skeletal muscle Striated muscle that is under voluntary control. The muscle fibres are a syncytium formed by *myoblast fusion during development and contain the *myofibrils which have tandem arrays of *sarcomeres.

skeletrophin A RING finger-dependent ubiquitin ligase (mindbomb homologue 2, MIB2, 1013 aa), which targets the intracellular region of *notch ligands. It has a *dystrophin-like domain, five *ankyrin repeats, interacts with α-actin, and is induced by the overexpression of truncated human SWI1 (SMARCF1), a subunit of the chromatin remodelling complex.

ski **1.** An oncogene, *ski* (Sloan-Kettering Institute proto-oncogene), found in an avian carcinoma. The cellular form, c-ski, is a corepressor of *Smad3 and regulates fibroblast proliferation. **2.** *Subtilisin/kexin isozyme-1 (SKI-1, site-1 protease, S1P), a Golgi proteinase mediating the proteolytic activation of the precursor to sterol-regulated element-binding proteins (*SREBPs 1 and 2). **3.** The homologue of *Drosophila* skinny hedgehog (SKI1, hedgehog acyltransferase, HHAT), an enzyme that catalyses N-terminal palmitoylation of *hedgehog.

skin associated lymphoid tissue (SALT) The subset of the lymphoid system associated with the dermis and epidermis. Includes *Langerhans cells and resident phagocytes. Some authors consider it to be conceptual rather than a discrete entity and others would include cutaneous nerve termini containing *calcitonin gene-related peptide (CGRP) that stimulate cytokine release from mast cells. *Compare* BRONCHUS ASSOCIATED LYMPHOID TISSUE; GUT ASSOCIATED LYMPHOID TISSUE; MUCOSAL ASSOCIATED LYMPHOID TISSUE.

skl *See* PEROXISOMAL TARGETING SEQUENCE 1.

skp Proteins associated with the CDK2/cyclin A kinase complex (S-phase kinase-associated proteins) involved in regulating cell cycle progression; the complex includes a 19-kDa protein (p19, Skp1, 163 aa) and a 45-kDa protein (p45, Skp2, 424 aa). **Skp1** is also a component of the SCF (Skp1-*cullin- F-box) ubiquitin ligase complex, in which Skp2 is also involved. The *SCF complex regulates the function of *NFκB. **Skp2** can deregulate growth *in vivo* and may be involved in the pathogenesis of lymphomas.

SL1 **1.** A transcription factor required for the activity of RNA polymerase I. It consists of *TATA box-binding protein (TBP) and three TBP-associated factors (TAFs). **2.** Stem–loop structure 1 (SL1), a characteristic feature of some viral RNAs.

SLAM family A family of immunoglobulin superfamily receptors (signalling lymphocyte activation molecules) expressed by a wide range of immune cells. They interact with various signalling proteins through their cytoplasmic domains to coactivate immune responses in T and B cells. **SLAM-associated proteins** (SAPs) are adaptors with SH2 domains. SAP (128 aa) is T-cell-specific and blocks recruitment of *Shp2, it is mutated in X-linked lymphoproliferative disease (*XLP). A set of SLAM family receptors, including SLAM1 (CD150, 335 aa), SLAM4 (CD244, natural killer cell activation inducing ligand), and SLAM5 (CD84) among others, play important roles in innate and adaptive immunity.

SLAP **1.** Src-like adaptor proteins. Dimeric adaptor proteins, SLA (281 aa) and SLA2 (261 aa), that are negative regulators of T-cell and B-cell immune responses. Through their *SH3 and *SH2 domains they interact with the T- or B-cell receptor complexes and with the ubiquitin ligase, *cbl. **2.** Sarcolemmal-associated protein, an integral membrane protein (828 aa with smaller isoforms) that may interact with myosin and is found in transverse tubules, near the junctional sarcoplasmic reticulum and along the Z- and M-lines in cardiomyocytes. NB The term 'sarcolemmal associated protein' is also used loosely of *dystrophin

SLC **1.** A chemokine (secondary lymphoid cytokine, slc, CCL21) that promotes coclustering of T cells and dendritic cells in lymph nodes and spleen. **2.** A G-protein-coupled receptor (SLC1, 402 aa) for *melanin-concentrating hormone (MCH). **3.** *See* SODIUM LITHIUM COUNTERTRANSPORT.

sleeping sickness *See* TRYPANOSOMA.

slicer activity The cleavage of targeted RNA by *argonaute proteins in *RNA interference.

sliding filament model The accepted model for contraction in striated muscle; the *thick filaments (myosin) slide relative to the *thin filaments in the *sarcomere. None of the filaments changes in length but the amount of overlap increases and the I band becomes narrower while the A band remains the same.

SLIP1 A protein (SLBP (stem-loop binding protein)-interacting protein 1, 222 aa) belonging to the MIF4GD family that is involved in replication-dependent translation of histone mRNAs that end with a stem-loop, not a poly-A tail.

slipped strand mispairing Mispairing of complementary DNA strands that can occur during replication. If the slippage is forwards then there will be deletion of bases, if backwards then there will be insertion. The phenomenon is most likely when there are *short tandem repeats.

slit A small family of proteins homologous to the *Drosophila* protein (slit) that plays an important part in morphogenesis of the central nervous system. They are diffusible chemorepellants and both **slit1** (1534 aa) and **slit2** (1529 aa) are inhibitors of retinal axon growth and of spinal axons. Slit genes are expressed in the ventral midline and prevent axons from crossing between hemispheres. **Slit3** (1532 aa) is expressed more strongly in peripheral tissues, particularly thyroid, than in the central nervous system. The receptor is robo (roundabout) and is expressed on the *nerve growth cone.

SLK A microtubule-associated *STE20-like kinase (1235 aa, of the germinal centre subfamily of STE20-like kinases) that is required for *focal adhesion turnover and cell migration downstream of the *FAK/*c-src complex.

SLM *STAR proteins, (Sam68-like mammalian proteins, 346 aa and 349 aa), related to *SAM68, that regulate splice site selection. They are targeted for degradation by the E3 *ubiquitin ligases, SIAH1, and SIAH2. Expression of SLM2 is reported to be up-regulated in patients with various kidney diseases.

slow muscle Myoglobin-rich striated muscle (slow twitch, type I muscle) that is used for long-term activity such as maintenance of posture and that relies on oxidative metabolism. Fast twitch muscles (type IIa) are also myoglobin rich and have a faster contraction speed than type I. A second type of fast twitch muscle (type IIb) is used for rapid but relatively infrequent contraction, has less myoglobin, and relies more on glycolysis.

slow reacting substance of anaphylaxis (SRS-A) *See* LEUKOTRIENES.

slow virus **1.** One of the *Lentivirinae. **2.** Any virus causing a disease with a very slow onset. Such viruses were postulated to be responsible for *spongiform encephalopathies.

SLP-76 A *transmembrane adaptor protein required for T-cell receptor (TCR) signalling.

SLRPs *See* SMALL LEUCINE-RICH REPEAT PROTEOGLYCANS.

SLS collagen An abnormal packing pattern of collagen molecules (segment long spacing collagen) in which the molecules are all in register rather than quarter-staggered as in normal collagen fibres.

slug A zinc finger transcription factor (SNAI-2, 268 aa), related to *snail, that regulates the development of neural crest-derived cells. Mutations in slug cause piebaldism and *Waardenburg's syndrome type 2.

SLURP A family of leucocyte antigen-6 (secreted Ly6)/urokinase-type *plasminogen activator (uPA or PLAUR domain-containing protein) proteins. Members of the superfamily have been implicated in transmembrane signal transduction, cell activation, and cell adhesion. **SLURP1** does not have a GPI-anchor but may interact with neuronal acetylcholine receptors. **SLURP2** (LY6/neurotoxin 1, LYNX1) will enhance nicotinic acetylcholine receptor function in the presence of acetylcholine but is not a ligand. *See* MELEDA DISEASE.

Sly syndrome An autosomal recessive lysosomal *storage disease (mucopolysaccharidosis type VII) caused by mutation in the gene encoding β-*glucuronidase and an inability to degrade glucuronic acid-containing glycosaminoglycans. The phenotype is highly variable.

smac (second mitochondria-derived activator of caspases) *See* DIABLO/SMAC.

Smad proteins *See* MAD-RELATED PROTEINS.

small cell carcinoma *See* OAT CELL CARCINOMA.

small interfering RNA (short interfering RNA, siRNA) Small double-stranded RNA molecules (20–25 bp) that interfere with the expression of particular mRNAs (*RNA interference) and can also activate innate immune systems that operate to block viral infection (*see* RIG-1). They may also induce gene expression and be termed small activating RNAs (saRNAs). *Compare* MICRORNA.

small leucine-rich repeat proteoglycans A family of proteoglycans (SLRPs) that includes *asporin, *decorin, *biglycan, osteomodulin, *fibromodulin, and *osteoadherin/osteoglycin.

small nuclear RNA (snRNA) An abundant class of RNA molecules (100–300 nucleotides) found in the nucleus and mostly complexed with proteins (**small nuclear ribonucleproteins**, snRNPs). Conventionally includes RNAs with sedimentation coefficients of 7S or less but excluding 5S rRNA and tRNA. Some are involved in processing *heterogenous nuclear RNA. *See* SNIP.

small nucleolar RNA (snoRNA) A class of small RNA molecules that guide chemical modifications (methylation or pseudouridylation) of ribosomal and transfer RNAs and other small nuclear RNAs (snRNAs). Sometimes (misleadingly) called guide RNA. *See* FIBRILLARIN.

smallpox *See* VARIOLA VIRUS (smallpox virus).

small temporal RNAs (stRNAs) Early name for *microRNAs.

SMARCs *SWI/SNF complex-related, matrix-associated, actin-dependent regulators of chromatin (SMARCs), also called BRG1-associated factors (*BAFs). Mutation in the *SMARCB1* gene can result in schwannomatosis, rhabdoid tumours, atypical teratoid tumours, and familial posterior fossa tumours of infancy.

SMART A free web-based resource, a Simple Modular Architecture Research Tool, that allows the identification and annotation of protein domains. SMART-6 (2008) has models for 784 protein domains.

SEE WEB LINKS

- The SMART homepage.

SMC proteins A family of proteins (structural maintenance of chromosomes proteins, including SMC1A, 1233; SMC3, 1217 aa; SMC6, 1091 aa) that are components of the *cohesin complex; they have a flexible hinge domain, which separates the large intramolecular coiled coil regions, and form heterodimers that are then linked into a ring structure, which traps chromatids, by RAD21 protein. Defects in SMCs are the cause of *Cornelia de Lange's syndrome.

SMCT A high-affinity sodium-coupled lactate transporter (sodium-coupled monocarboxylate transporter, SLC5A8, 610 aa) involved in reabsorption of lactate in the kidney and maintenance of blood lactate levels. It also transports iodide passively. It is the product of a tumour suppressor gene silenced by methylation in human colon cancer.

Smith–Magenis syndrome A set of complex developmental abnormalities affecting a wide range of neural, skeletal, and organ systems. Caused in most cases by a deletion in chromosome 17p11.2, but can also be caused by mutations in the *RAI1* gene (retinoic acid inducible-1) and a milder phenotype is associated with duplication of the same chromosomal region.

Smith–McCort dysplasia A rare autosomal recessive osteochondrodysplasia characterized by short limbs and trunk with barrel-shaped chest, with many features similar to *Dyggve–Melchior-Clausen syndrome which is also caused by defects in *dymeclin.

SMN complex A large macromolecular complex (survival of motor neurons complex) that plays an essential role in the production of spliceosomal small nuclear ribonucleoproteins (snRNPs). It is present in both cytoplasm and nucleus, and is enriched in Gems (gemini of the coiled bodies), nuclear structures that resemble Cajal bodies in number and size. At least seven *gemins are components, as well as the product of the survival of motor neurons (*SMN*) gene (defective in *spinal muscular atrophy). SMN complex substrates have a domain rich in arginine and glycine residues (RG) and include Sm and Sm-like (LSm) proteins, RNA helicase A, *fibrillarin, the RNP proteins, and p80-*coilin.

SMO **1.** Sulfamethoxazole, an antibacterial sulfonamide, used in combination with *trimethoprim. **2.** *See* SMOOTHENED. **3.** *See* SPERMINE (spermine oxidase).

smoothelins Smooth muscle specific actin-binding proteins that are abundant in visceral (smoothelin-A) and vascular (smoothelin-B) smooth muscle. Smoothelins colocalize with α-smooth muscle actin in stress fibres and are involved in the contractile process. **Smoothelin-like 1** (SMTNL1, CHASM) has a *calponin homology domain and shares sequence similarity with smoothelin but does not associate with F-actin *in vitro*, and its role is unclear.

smooth endoplasmic reticulum (SER) The intracellular membrane complex that encloses a separate compartment which is more tubular than the sheet-like *rough endoplasmic reticulum (RER), although the membranes are similar. SER is conspicuous in cells concerned with lipid metabolism and in hepatocytes following administration of lipophilic drugs.

smoothened (Smo) A G-protein-coupled receptor, the homologue of the *Drosophila* protein, that is translocated to the *primary cilium when acted upon by *patched that has bound the *sonic hedgehog ligand. Smoothened (Smo) is then thought to convert the Gli family of transcription factors (*see* GLI ONCOGENES) from transcriptional repressors to transcriptional activators. The interaction of Smo and the kinesin motor protein KIF3A is regulated by β-*arrestin. Mutations in Smo are associated with some basal cell carcinomas. *See* CYCLOPAMINE; HEDGEHOG SIGNALLING COMPLEX.

smooth microsomes A subcellular fraction of membrane vesicles, derived mostly from the *smooth endoplasmic reticulum, produced by ultracentrifugation of a cellular homogenate.

smooth muscle Muscle tissue formed of individual cells (20–500 μm long) that lack conspicuous sarcomeres and have a contractile system similar to that in nonmuscle cells. Smooth muscle is generally involuntary, and can contract to a much smaller fraction of its resting length than striated muscle. Smooth muscle is found in blood vessel walls, surrounding the intestine, and in the uterus. *See* DENSE BODIES.

smooth strain *See* ROUGH STRAIN.

SM proteins **1.** Proteins (Sec1/Munc18-like proteins) that interact with *SNARE complexes and regulate membrane fusion events. **2.** *See* SMN COMPLEX.

SMRT Silencing mediator of retinoic acid and thyroid hormone receptors. *See* N-CoR.

snail A zinc finger transcription factor (264 aa) that down-regulates E-*cadherin and is involved in mesoderm formation and the epithelial-mesenchymal transition. *See* SLUG.

SNAP **1.** Soluble *NSF attachment (accessory) proteins (SNAPs) that are involved in the control of vesicle transport. **SNAPα** (295 aa) and **SNAPγ** are found in a wide range of tissues. **SNAPβ** is a brain-specific isoform of SNAPα. SNAPs and NSF bind to *SNAREs. SNAP-25 (synaptosomal-associated 25-kDa protein, 206 aa) is involved in the regulation of neurotransmitter release. *See* SNIP (2). **2.** *S*-nitroso-*N*-acetyl*penicillamine.

snapin A ubiquitous *SNARE associated protein (136 aa), a component of the multiprotein SNARE complex required for synaptic vesicle docking and fusion and also a component of the *BLOC1 complex.

SNAREs The receptors for *SNAPs that are found on vesicles and the membrane targets with which they are destined to fuse. The neuronal receptor for vesicle-SNAPs, v-SNARE, is *synaptobrevin, also known as VAMP-2. The target (t-SNARE) associated with the plasma membrane of the axonal terminal is *syntaxin. Neurotoxins such as *tetanus toxin and *botulinum toxin selectively cleave SNAREs or SNAPs. *See also* CELLUBREVIN.

SNIP A hydrophilic protein (SNAP-25-interacting protein, 1055 aa), that codistributes with *SNAP-25 in most brain regions.

snoRNA *See* SMALL NUCLEOLAR RNA.

SNP (single nucleotide polymorphism) A change in a single nucleotide in a DNA sequence, something that occurs about once in every 1200 bases, on average. The difference in sequence may have little effect (because of the degeneracy of codon assignment) but it contributes to differences betwen individuals. Conventionally, a variation is only considered to be a SNP if it occurs in 1% or more of the population. SNP mapping is a powerful approach to finding genetic associations of disease because they act as convenient markers of alleles and there are ~10^6 known SNPs. The SNP consortium started as a collaborative project that included pharmaceutical companies and medical charities (Wellcome Trust) and morphed into the International HapMap Project which makes the data freely available.

SEE WEB LINKS

- The International HapMap Project homepage.

snRNA *See* SMALL NUCLEAR RNA.

snRNP Small nuclear ribonucleoprotein. *See* SMALL NUCLEAR RNA.

SNRPN An imprinted gene that encodes two polypeptides (i.e. is bicistronic), the **SmN splicing factor**, (small nuclear ribonucleoprotein polypeptide N) which is involved in *RNA processing, and the ***SNRPN*** **upstream reading frame** (SNURF) polypeptide. Deletion of the paternal copy leads to *Prader–Willi syndrome, that of the maternal copy to *Angelman's syndrome.

SOCS A family of at least eight proteins (suppressor of cytokine signalling, SOCS1 to SOCS7 and CIS (cytokine-inducible SH2-containing protein); all ~210 aa; rapidly induced in response to IL-6 and other cytokines, and apparently switch off *JAK signalling. All contain a central *SH2 domain. **JAB** (JAK binding protein; STAT-induced STAT inhibitor-1, SSI-1) is SOCS1. Aberrant methylation of SOCS1 correlates with its transcriptional silencing in hepatocellular carcinoma cell lines.

sodium channels *See* AMILORIDE-SENSITIVE SODIUM CHANNELS; VOLTAGE-GATED SODIUM CHANNELS.

sodium cromoglicate (formerly **sodium cromoglycate)** A drug used prophylactically by inhalation for allergic asthma. It acts on mast cells to inhibit release of bronchoconstrictors although the exact mechanism is still unknown. *TN* Intal.

sodium cromoglycate *See* SODIUM CROMOGLICATE.

sodium lithium countertransport The sodium-stimulated lithium efflux (sodium lithium countertransport, SLC, Na/LiCT) from lithium-loaded erythrocytes, is elevated in patients with essential hypertension and diabetes mellitus. A candidate molecule is solute-carrier SLC34A2. (NB 'SLC' usually indicates molecules of the *solute carrier family and not sodium lithium countertransport).

sodium-potassium ATPase (Na^+,K^+-ATPase, sodium pump) The ion transporter (EC 3.6.1.3, not a *sodium channel) that maintains the electrochemical gradients of Na^+ and K^+ ions across the plasma membrane. It has two subunits, a catalytic α subunit (multiple isoforms; ATP1A4 is 1029 aa) that has the catalytic activity and a smaller β subunit of uncertain function. For each ATP hydrolysed it moves three sodium ions out of the cell, and two potassium ions in. The sodium gradient is used to drive various cotransport systems (*see* FACILITATED DIFFUSION) and is essential for the *action potential in excitable cells. The potassium gradient dissipates through the potassium leak channel and is the main basis for the normal *resting potential. The α2 isoform in neural and muscle tissues is mutated in familial hemiplegic *migraine-2.

sodium pump *See* SODIUM-POTASSIUM ATPASE.

sodoku An infectious fever (rat bite fever) caused by *Streptobacillus moniliformis* or *Spirillum minus*, usually caused by a rat bite or contact with rat saliva. It can be recurrent.

SOK1 A *Ste20 protein kinase of the germinal centre kinase (GCK) family that is activated by oxidant stress and chemical anoxia, a cell culture model of ischaemia. It is localized in the Golgi apparatus, where it functions in a signalling pathway required for cell migration and polarization.

solanine A highly toxic glycoalkaloid from plants of the Solanaceae family (including potatoes). It has both fungicidal and pesticidal properties. *See* CHACONINES.

solifenacin An anticholinergic drug used to treat urinary frequency, urgency, and incontinence. *TN* Vesicare.

solurshin *See* PITX GENES.

solute carrier family proteins (SLCs) A large superfamily of proteins that transport small molecules across membranes. At least 300 are known and 47 families are recognized and there is a formalized nomenclature: e.g. solute carrier family 1 (SLC1) is of high-affinity glutamate and neutral amino acid transporters, SLC28 family is of sodium-coupled nucleoside transporters, SLC47 is involved in multidrug and toxin extrusion. The SLC superfamily does not include primary active transporters (e.g. ATP binding cassette transporters), ion channels or *aquaporins.

SEE WEB LINKS

- The Transporter Classification database.

somatic cell A cell in a multicellular organism that does not produce gametes, i.e. the cells of the body with the exception of the *germ cells. *Compare* SOMATIC MESODERM.

somatic cell genetics A method used to identify which chromosome carries a particular gene, without sexual crossing. The general principle is to produce unstable *heterokaryons (often human–mouse heterokaryons) between the cell with the gene of interest and another cell and then isolate a series of clones in which different chromosomes have been lost. It is then possible to deduce which chromosome carries the gene.

somatic cell nuclear transfer A *cloning method in which the nucleus of a *somatic cell is transferred into an enucleated egg. Has been used, with mixed success, to produce embryonic stem cells and in a few cases a cloned animal (famously, Dolly the sheep).

somatic hybrid A *heterokaryon of two somatic cells, usually from different species. *See* SOMATIC CELL GENETICS.

somatic mesoderm Embryonic mesoderm that is associated with the body wall and separated from splanchnic (visceral) mesoderm by the coelomic cavity. *Compare* SOMATIC CELL.

somatic mutation A mutation in a somatic cell, which is therefore not heritable. Depending upon the time in development at which the mutation occurs, a variable amount of tissue is affected (and the extent to which the organism is *chimeric). A somatic mutation that confers the ability to escape normal controls on proliferation will give rise to a tumour. Somatic mutation is probably important in generating diversity in the variable regions of immunoglobulins.

somatic recombination Recombination via homologous crossing-over during mitosis. It is a rare event in most tissues, but important in generating diversity in variable regions of immunoglobulins.

somatocrinin (growth hormone releasing factor, somatotropin releasing factor, somatoliberin) *See* GROWTH HORMONE RELEASING HORMONE.

somatomedins *See* INSULIN-LIKE GROWTH FACTORS.

somatostatin A cyclic peptide (14 aa) that is an important regulator of endocrine and nervous system function. It inhibits gastric secretion and motility and the release of *somatotropin. There are five G-protein-coupled receptor subtypes (SSTR1–5), with tissue-specific distribution and multiple subtypes present on many cells. *See* CORTISTATIN.

somatotrope A type of cell in the anterior pituitary which secretes *growth hormone in response to releasing hormone (GHRH, somatocrinin); release is inhibited by *somatostatin.

somatotropin (somatotrophin) *See* GROWTH HORMONE.

somites Blocks of mesoderm arrayed on either side of the notochord and neural tube during development of the vertebrate embryo. They are formed sequentially, starting at the head, and each somite gives rise to muscle (from the myotome region), spinal column (from the sclerotome), and dermis (from the dermatome).

sonic hedgehog A secreted morphogen (465 aa) that is the inductive signal for patterning of the ventral neural tube, the anterior–posterior limb axis, and the ventral *somites. Defects in shh or in its signalling pathway are a cause of *holoprosencephaly 3 and *VACTERL syndrome. *See* DESERT HEDGEHOG; HEDGEHOG; INDIAN HEDGEHOG.

son-of-sevenless *See* SOS.

sortilin A lysosomal sorting receptor, important for the sorting and trafficking of sphingolipid activator proteins (*saposins).

sorting nexins A diverse group of proteins involved in trafficking and sorting of transmembrane proteins. They all have a phospholipid-binding (phox homology, PX) domain and tend to form protein–protein complexes, in some cases with receptor tyrosine kinases, in other cases with TGFβ family receptors. Sorting nexin-1 is a component of the mammalian *retromer complex.

sorting out The reorganization of a mixed cell aggregate formed in cuture into distinct homotypic domains, usually with one internal to the other. It has been argued that quantitative differences in homotypic and heterotypic adhesion are sufficient to account for the sorting (the *differential adhesion hypothesis) without the need to invoke cell-type specific adhesion molecules.

sos **1.** A guanine nucleotide exchange factor (*GEF, 1336 aa) that is a positive regulator of *ras and the human homologue of *Drosophila* son-of-sevenless. Binds to the *SH3 domain of *grb2. A family of related proteins includes *vav, *C3G, Ost, NET1, Ect2, RCC1, *tiam, RalGDS, and *Dbl. **2.** *See* SOS SYSTEM.

SOS system A bacterial DNA repair system (error-prone repair) that allows replication to bypass lesions or errors in the DNA. The SOS system involves the RecA protein which has a human homologue in Rad51 (*see* RAD PROTEINS).

sotalol A *β-adrenoceptor blocking drug that also inhibits potassium channels in the heart and lengthens the *QT interval. *TNs* Betapace, Sotacor, Sotalex.

Sotos' syndrome A growth disorder (cerebral gigantism) caused by mutation in the *NSD1* gene (nuclear receptor-binding suppressor of variation, enhancer of zeste, trithorax domain protein 1) that encodes a coregulator of the androgen receptor. *See* SET.

sox The Sox family of genes: (Sry-related HMG-box genes) are involved in many developmental processes and their products, like the *Sry proteins, have a DNA-binding domain, the *HMG box. **Sox-2** regulates transcription of the *FGF4* gene; **sox-3** is involved in neural tube closure and lens specification; **sox-4** is a transcriptional activator in lymphocytes; **sox-9** plays an essential role in sex determination and may act immediately downstream of *sry, and mutations cause *campomelic dysplasia; **sox-10** is important in neural crest development. Many other *sox* genes are known but their functions are not yet determined.

soybean trypsin inhibitor A *Kunitz-type trypsin inhibitor (SBTI, 181 aa), from the soybean *Glycine max*, that forms an inactive complex with trypsin. It also inhibits chymotrypsin, plasmin, and trypsin-like peptidases.

SP1 A zinc finger *transcription factor (specificity protein 1, 785 aa) that regulates gene expression in early development.

SPA, SPD *See* SURFACTANT PROTEINS A AND D.

spacer DNA A general term for DNA sequences between genes. The term is used more specifically for the DNA between the tandemly repeated copies of the ribosomal RNA genes.

SPAK A serine/threonine kinase (Ste20/SPS-1-related proline, alanine-rich kinase, 547 aa) that will phosphorylate and activate *NKCC1 (Na^+-K^+-$2Cl^-$ cotransporter-1). SPAK and OSR1 (oxidative stress-responsive kinase-1), are activated by *WNKs.

SPARC *See* OSTEONECTIN.

sparsomycin An antibiotic isolated from the fermentation broth of *Streptomyces sparsogenes* that inhibits *peptidyl transferase in both prokaryotes and eukaryotes.

spartin A protein (666 aa) that is is both cytosolic and membrane-associated and apparently binds *Eps15. It has sequence homology with *spastin and other proteins involved in endosomal trafficking. Named because a frameshift mutation causes spastic paraplegia autosomal recessive Troyer's syndrome (SPG20). *See* SPASTIC PARAPLEGIAS.

spastic paraplegias A large group of neurodegenerative disorders characterized by lower-extremity spasticity and weakness; inheritance is most commonly autosomal dominant, but X-linked and autosomal recessive forms also occur. The affected protein is shown in parentheses: **Autosomal dominant forms** include **SPG3A** (Strumpell's disease) (*atlastin), **SPG4** (*spastin), **SPG6** (*NIPA1), **SPG8** (*strumpellin), **SPG10** (kinesin-5A), **SPG13** (HSP60), **SPG31** (*receptor expression-enhancing protein-1), and **SPG33** (*protrudin). **X-linked forms** include **SPG2** (*myelin proteolipid protein) allelic to *Pelizaeus–Merzbacher disease. **Autosomal recessive forms** of SPG include **SPG7** (*paraplegin), **SPG11** (*spatacsin), **SPG15** (*spastizin), **SPG20** (*spartin), **SPG21** (*maspardin), and **SPG39** (*PNPLA6). Other autosomal dominant and recessive forms have been mapped.

spastin One of the *AAA family of ATPases that is mutated in autosomal dominant hereditary *spastic paraplegia type 4 (SPG4). It interacts with microtubules and probably has a role in microtubule dynamics. It is highly homologous to the microtubule-severing protein *katanin and there is a domain of ~80 aa found also in *spartin. Binds to *atlastin.

spastizin A widely distributed protein (zinc finger FYVE domain-containing protein-26, ZFYVE26, 2539 aa) that probably has a role in endosomal trafficking. Mutations in spastizin cause autosomal recessive hereditary *spastic paraplegia type 15 (SPG15; Kjellin's syndrome).

spatacsin A protein (predicted 2443 aa) that is defective in hereditary *spastic paraplegia type 11 (SPG11). The name is from 'spasticity with thin or atrophied corpus callosum syndrome'. There are four putative transmembrane domains, suggesting it may be a receptor or transporter. The *SPG11* gene is expressed ubiquitously in the nervous system but most prominently in the cerebellum, cerebral cortex, hippocampus, and pineal gland.

specific activity The number of activity units per unit of mass, volume, or molarity. In enzymology, the moles converted per unit time per unit mass of enzyme (enzyme activity/total mass of protein). Perhaps most often encountered in the context of radiochemicals, the number of decays per second per amount of substance.

specific granules (secondary granules) One of the two types of granules found in *neutrophils that contain *lactoferrin, *lysozyme, vitamin B_{12} binding protein, and *elastase. Although they are referred to as secondary granules they are actually released more readily than the primary (*azurophil) granules.

spectral karyotyping A method of facilitating the examination of chromosome spreads in which each pair of chromosomes is given a unique fluorescent stain and *translocations become very conspicuous because the affected chromosomes are multicoloured.

spectraplakins A superfamily of giant cytoskeletal linker proteins that bind actin, tubulin, and intermediate filaments. The common features are two *calponin homology domains, a *plakin domain, a series of *plectin repeats, numerous *spectrin repeats, and finally a GAS2 domain. The human spectraplakin (macrophin 1, trabeculin-α, actin crosslinking factor 7, 5373 aa) is found in all tissues.

spectrin The major protein of the erythrocyte membrane cytoskeleton that determines its mechanical properties. It is heterodimeric (α, 2419 aa; β, 2137 aa) and forms multimers as well as interacting with *protein 4.1, *ankyrin, *actin, *band III, and other proteins to form a meshwork that can restrict the lateral mobility of integral proteins. Isoforms have been described from other tissues (*fodrin, *TW-240/260-kDa protein), where they probably have a comparable function. Defects in proteins of the erythrocyte membrane skeleton can cause *spherocytosis.

spermatids The haploid products of the second meiotic division in spermatogenesis that differentiate into mature spermatozoa.

spermatocytes The cells that will undergo two meiotic divisions to produce haploid *spermatids.

spermatogenesis The sequence of cell divisions that eventually produces mature spermatozoa from primordial *germ cells.

spermatozoon (*pl.* spermatozoa) A mature male gamete (sperm cell).

spermine A basic polyamine involved in packaging nucleic acids in sperm. Also found in ribosomes and some viruses. Synthesis is regulated by *ornithine decarboxylase. **Spermine oxidase** (SMO, EC 1.5.3.-, 555 aa) will oxidize spermine but not spermidine and may also be important in the cellular response to antitumour polyamine analogues.

S phase The phase of the cell cycle during which DNA replication (synthesis) takes place.

spherocytosis A condition in which erythrocytes are spherical, not biconcave. It occurs as the red cells age but in hereditary spherocytosis there is a defect in β-*spectrin or, in some cases, *ankyrin. Autosomal recessive spherocytosis is due to mutation in the α-spectrin gene. Spherocytosis can lead to severe haemolytic anaemia. *See* ELLIPTOCYTOSIS.

sphingolipid A structural lipid in which an acyl chain is attached to *sphingosine to form a ceramide. Further modification (glycosylation) gives rise to cerebrosides and gangliosides, major glycolipids in the nervous system. *See* SAPOSINS.

sphingomyelin A *sphingolipid in which phosphorylcholine or phosphoroethanolamine is ether-linked to ceramide. The concentrations of sphingomyelin and phosphatidyl choline in the plasma membrane often seem to be reciprocally related. *See* LIPIDOSES.

sphingomyelinase A stress-activated enzyme (sphingomyelin phosphodiesterase-1, acid sphingomyelinase, EC 3.1.4.12, 629 aa) that will generate ceramide, which acts as a second messenger in initiating apoptosis. Deficiency in the enzyme leads to *Niemann–Pick disease types A and B.

sphingosine An 18-carbon amino-alcohol that is acylated to form *sphingolipids.

sphingosine kinase The enzymes (SPHK, sphinganine kinase, EC 2.7.1.91, SPHK1, 384 aa; SPHK2, 654 aa) that phosphorylate sphingosine to produce sphingosine 1-phosphate (SPP), a lipid second messenger acting intracellularly, and extracellularly through the G-protein-coupled receptor, EDG1.

Spi-1 **1.** A proto-oncogene, *spi-1*, encoding an ETS-domain transcription factor (PU1) that is essential for the development of myeloid and B-lymphoid cells. **2.** *Salmonella typhimurium* pathogenicity island 1 (SPI-1), the region of the bacterial genome that contains genes involved in mediating invasion into intestinal epithelial cells. **3.** Vaccinia virus host range/antiapoptosis genes, *SPI1* and *SPI2*.

spina bifida A neural tube defect in which the bones of the spinal canal fail to meet and fuse during development. In the more serious form (open spina bifida, **spina bifida cystica** (SBC) or myelomeningocele) there is complete failure of fusion, usually in posterior segments (lumbar and sacral). **Spina bifida occulta** (SBO) is a bony defect of the spine covered by normal skin; it is much less serious and may be asymptomatic. Dietary insufficiency of folic acid during pregnancy may be a risk factor, and various neural tube defects have been associated with variations in genes involved in folate and homocysteine metabolism.

spinal cord The portion of the central nervous system that lies in the vertebral canal, and from which spinal nerves emerge.

spinal ganglion (dorsal root ganglion) An enlargement of the dorsal root of the spinal cord that contains the cell bodies of afferent spinal neurons.

spinal muscular atrophy A group of autosomal recessive neuromuscular disorders in which there is degeneration of the anterior horn cells of the spinal cord, leading to symmetrical muscle weakness and atrophy. They are all caused by mutation in the telomeric copy of the survival of motor neuron gene (*SMN1*; *see* SMN COMPLEX) but four types are recognized: **type I** is severe infantile acute spinal muscular atrophy (SMA, Werdnig–Hoffman disease); **type II** is infantile chronic SMA; **type III** is juvenile SMA (Wohlfart–Kugelberg–Welander disease); and **type IV** is adult-onset SMA. Expression of the centromeric *SMN2* gene affects the phenotype.

spindle *See* MITOSIS.

spindle fibres The microtubules of the mitotic spindle that derive from the *microtubule organizing centre of the spindle poles and interdigitate at the equatorial plane. *Kinetochore fibres are microtubules that link the poles with the kinetochore.

spindle pole body (SPB) The yeast equivalent of the *centrosome, the microtubule organizing centre anchored to the nuclear envelope that organizes both the spindle and cytoplasmic microtubules.

spinocerebellar ataxia A group of genetic disorders, with variable involvement of the brainstem and spinal cord, in which gait becomes uncoordinated, and there is slow and progressive loss of control of hands, speech, and eye movements. There is often atrophy of the cerebellum. Multiple genetic defects causing the condition have been identified. In **autosomal recessive type 1** (SCAR1) *senataxin is affected; in **SCAR8** (recessive ataxia of Beauce), *nesprin 1; and in **SCAR9**, CABC1 chaperone-like protein. The **autosomal dominant forms** (ADCAs) are also heterogeneous, many caused by trinucleotide (CAG) repeats in the affected genes (shown in parentheses): spinocerebellar ataxia-1 (**SCA1**) (*ataxin-1), **SCA2** (ataxin-2), **SCA3,** Machado–Joseph disease (ataxin-3), **SCA7** (ataxin-7), **SCA8** (nontranslated *SCA8* gene), **SCA12** (a brain-specific regulatory subunit of the protein phosphatase PP2A), **SCA17** (TATA box-binding protein). **SCA10** is caused by an expanded 5-bp repeat in the ataxin-10 gene. In other forms there are defects in the proteins shown in parentheses: **SCA4** (*puratropin-1, in some cases), **SCA5** (β-III spectrin), **SCA6** (a voltage-dependent calcium channel), **SCA11** (tau tubulin kinase-2), **SCA13** (a voltage-gated potassium channel), **SCA14** (protein kinase Cγ), **SCA15** (type 1 inositol 1,4,5-trisphosphate receptor), **SCA27** (fibroblast growth factor-14). An autosomal recessive form of **spinocerebellar ataxia with axonal neuropathy** (SCAN1) can be caused by homozygosity for a mutation in the tyrosyl-DNA phosphodiesterase-1 gene. **Infantile-onset spinocerebellar ataxia** (IOSCA; formerly SCA8) is caused by defects in *twinkle and twinky proteins. Several **X-linked forms** have been described (SCAX1-5) but they are clinically and genetically heterogeneous. *See also* SENSORY ATAXIC NEUROPATHY (sensory ataxic neuropathy, dysarthria, and ophthalmoparesis); SIDEROBLASTIC ANAEMIA WITH SPINOCEREBELLAR ATAXIA.

spinophilin *See* NEURABIN.

spirochaete An elongated, corkscrew-shaped bacterium (spirillum). Important examples are *Treponema pallidum*, which causes syphilis and *Vibrio cholerae*, which causes cholera.

splanchnic A term for anything relating to the viscera. **Splanchnic mesoderm** is the embryonic *mesoderm associated with the inner (endodermal) part of the body. *Compare* SOMATIC MESODERM.

splenocytes A rather imprecise term for a phagocytic cell found in the spleen.

spliceosomes The macromolecular RNA-protein complexes involved in the processing (*splicing) of the primary transcript of a gene to produce mRNA. Introns are removed and exons ligated (or skipped, in some alternative splicing variants). Components include U2, U5, and U6 *snRNAs and the essential spliceosomal protein Prp8 (*see* PRP PROTEINS). There are subclasses of spliceosomes that process particular classes of introns, e.g. U12-dependent introns are spliced by the so-called minor spliceosome.

splice variants *See* SPLICING.

splicing The processing of heterogeneous nuclear RNA to produce mature mRNA by removal of introns and splicing together of the exons. Alternative splicing occurs in many proteins (current estimates are that ~30% of gene transcripts are alternatively spliced) and alternative exon usage generates a set of related proteins (splice variants) from a single gene, often in a tissue or developmental stage-specific manner. *See* SPLICEOSOMES. **Trans-splicing** is the process by which an identical short leader sequence, the spliced leader (SL), is spliced onto the 5' ends of multiple mRNAs.

SEE WEB LINKS

- The Alternative Splicing and Transcript Diversity database.

split gene *See* INTRON.

split ratio A measure of how many subcultures can be set up from a fully grown culture of cells. Many cell types require a substantial population as an inoculum, probably to condition the medium with growth factors that are absent in the medium itself.

S

spokein A name originally given to the protein thought to form the radial spokes of the *axoneme. It is now known that at least 23 *radial spoke proteins are required for normal axonemal motility.

spondin Proteins involved in patterning axonal growth by inhibiting or promoting adhesion of embryonic nerve cells. **Spondin 1** (F-spondin, 624 aa) is a secreted extracellular matrix protein that interacts with a central sequence of amyloid precursor protein (APP) and prevents proteolytic processing. **Spondin 2** (M-spondin, 331 aa) is *mindin. The **R-spondins** are a family of secreted proteins that regulate β-*catenin signalling. **R-spondin 1** (RSPO1, 263 aa) is expressed in enteroendocrine cells as well as in epithelial cells from various tissues. Mutations are associated with palmoplantar hyperkeratosis with squamous cell carcinoma of skin, and sex reversal. **R-spondin 2** (cristin 2, 243 aa) apparently functions in a positive feedback loop to stimulate the WNT/β-catenin cascade. **RSPO3** (273 aa) regulates kidney cell proliferation. **RSPO4** and a RSPO4 splice variant induce epithelial proliferation in the gastrointestinal tract; mutations are associated with *anonychia congenita.

spondyl-, spondylo- Prefix meaning pertaining to the spine or to vertebrae.

spondyloarthropathy *See* ANKYLOSING SPONDYLITIS.

spondylocarpotarsal syndrome *See* FILAMINS (filamin B).

spondyloepiphyseal dysplasia A range of disorders due to cartilage defects, usually causing short stature and a variable range of malformations. **Spondyloepiphyseal dysplasia congenita** is caused by mutation in the gene for collagen IIA2 (cartilage type). The **Strudwick type** is caused by a dominant mutation in collagen IIA1. The **Omani type** is caused by mutation in the chondroitin 6-sulphotransferase 3 gene. A mild autosomal dominant form, **Kimberley type**, is caused by a mutation in *aggrecan. **Spondyloepiphyseal dysplasia tarda** is caused by mutation in *sedlin. *See also* PSEUDOACHONDROPLASIA.

spondyloperipheral dysplasia A disorder in which there is short stature, midface hypoplasia, sensorineural hearing loss, spondyloepiphyseal dysplasia, platyspondyly, and brachydactyly. It is caused by mutations in the *collagen *COL2A1* gene.

spongiform encephalopathies A group of diseases in which there is a characteristic spongiform degeneration of grey matter of the cortex. They have a long incubation period and a slow but fatal progressive course. There are two main forms, *kuru and *Creutzfeldt–Jakob disease, with a new variant (vCJD or nvCJD) of the latter. There is clear similarity to bovine spongiform encephalopathy (BSE) and diseases such as *scrapie and mink encephalopathy. Cross-species transfer may occur. There is continuing controversy about the causative agent, although current (2009) opinion generally favours Prusiner's *prion hypothesis. *See also* FATAL FAMILIAL INSOMNIA; GERSTMANN–STRAUSSLER–SCHEINKER SYNDROME.

spongioblast The cell of the developing nervous system that gives rise to *astrocytes and *oligodendrocytes. **Spongioblastomas** are rare tumours of childhood and adolescence and are probably not a distinct clinical entity but may be ependymomas or neuroblastomas.

spongiocytes Cells found in the adrenal cortex that contain many lipid droplets and have unusual vesicular mitochondria.

spontaneous transformation The *transformation of a cell in culture without the deliberate addition of a transforming agent. Rodent cells are particularly prone to such spontaneous transformations whereas human cells rarely transform without intervention.

sporadic Describing a disorder that arises *de novo*, rather than as a result of an inherited mutation. Many diseases occur as both sporadic and inherited forms, presumably because of mutational events.

spore A resistant dehydrated reproductive cell produced to withstand environmental stress. Bacterial spores can survive quite extraordinary extremes of temperature, dehydration, or chemical insult.

sporotrichosis (rose thorn disease) A disease caused by infection with the fungus *Sporothrix schenckii*. An occupational skin disease of farmers and gardeners, usually contracted through puncture wounds, although it can occasionally infect lungs or cause a disseminated disease affecting many tissues.

sporozoite The infective stage in the life cycle of *Apicomplexa protozoans such as **Plasmodium* and **Cryptosporidia*.

spot desmosome *See* DESMOSOME.

SPREDs Members of the *sprouty family of proteins (sprouty-related EVH1 domain-containg proteins, SPREDs 1–3) that regulate growth factor-induced activation of the MAP kinase cascade. Mutations in SPRED1 cause *neurofibromatosis 1-like syndrome.

sprouting The outgrowth of new processes from neurons either during development or in response to damage. It can also be observed in neurons in culture.

sprouty A gene family encoding proteins that negatively regulate *fibroblast growth factor signalling in a variety of systems. The original *Drosophila* gene is involved in regulating branching in the tracheal system, but the mammalian homologues (*SPRY1–3*) have comparable growth-factor regulatory functions. SPRY1 is down-regulated in ~40% of prostate cancers. A fourth member of the family has also been identified that suppresses the insulin receptor and epidermal growth factor receptor signalling pathways. *See* SPREDs.

squalene A 30-carbon lipid that is a key intermediate in cholesterol biosynthesis. Abundant in shark liver oil and to a lesser extent in various other natural oils.

squames The flattened remnants of keratinized epithelial cells that are shed from the outermost surface of a squamous *stratified epithelium.

squamous epithelium An epithelium in which the cells have a flattened morphology (*compare* cuboidal epithelium and columnar epithelia at EPITHELIUM). They can be *simple epithelial monolayers (e.g. *endothelium) or *stratified (e.g. *epidermis). Squamous cell carcinomas derive from this epithelium and are considerably more serious than *basal cell carcinomas and can metastasize.

SRBC Sheep red blood cells, a common reagent in some immunological tests.

src family A family of protein tyrosine kinases involved in cellular control, src having been the first one described (*see* SRC GENE). All cells have one at least of these kinases which have characteristic domains (*see* SH DOMAINS). Phosphorylation of a src tyrosine residue by *csk allows the SH2 domain to bind and block activity. Members of the family include *blk, *chk, *fgr, *fyn, *hck, *lyn, *p57lck, and *yes.

***src* gene** The transforming (sarcoma-inducing) oncogene of Rous sarcoma virus that encodes an unregulated tyrosine kinase (v-src, pp60vsrc). The normal c-src kinase is subject to regulation and does not trigger neoplastic growth.

SRE **1.** *See* SERUM RESPONSE ELEMENT. **2.** Sterol regulatory element. *See* STEROL REGULATORY ELEMENT-BINDING PROTEINS.

SREBP *See* STEROL REGULATORY ELEMENT-BINDING PROTEINS.

SRF Serum response factor. *See* SERUM RESPONSE ELEMENT.

SRP *See* SIGNAL RECOGNITION PARTICLE.

SR proteins A family of highly conserved nuclear phosphoproteins of the *spliceosome that have a C-terminal SR domain enriched in serine-arginine (SR) dipeptides. They are necessary for pre-mRNA splicing and also influence *alternative splicing by affecting splice-site selection.

SRS-A *See* LEUKOTRIENES.

SRTX *See* SARAFOTOXINS.

sry A transcription factor that has an *HMG-box, the product of *sry*, the primary testis-determining gene on the Y chromosome (sex-related gene on Y, testis determining factor). *See* SOX.

SSB **1.** A single-strand break (SSB) in DNA. **2.** *Sjogren's syndrome antigen (La/SSB phosphoprotein), the target of autoantibodies in sera of patients with Sjögren's syndrome and *systemic lupus erythematosus. *See* LA PROTEIN.

SSCP *See* SINGLE-STRANDED CONFORMATIONAL POLYMORPHISM.

SSEA **1.** A virulence determinant in *Salmonella* (SseA), a type III secretion system chaperone for other products of the SPI-2 pathogenicity island such as SseB and SseD that are required for bacterial replication inside macrophages. **2.** *See* STAGE-SPECIFIC EMBRYONIC ANTIGEN.

SSH (suppression subtractive hybridization) *See* SUBTRACTIVE HYBRIDIZATION.

SSI-1 STAT-induced STAT inhibitor-1 (JAB). *See* SOCS.

SSLP *See* SIMPLE SEQUENCE LENGTH POLYMORPHISM.

SSRE *See* SHEAR STRESS RESPONSE ELEMENT.

SSRIs *See* SELECTIVE SEROTONIN REUPTAKE INHIBITORS.

stable transfection *See* TRANSFECTION.

STAGA A multiprotein transcription coactivator-histone acetyltransferase complex that interacts with pre-mRNA splicing factors and DNA damage-binding factors. Most of the protein components are homologues of proteins originally identified in yeast. *See* ATAXINS (ataxin 7).

stage-specific embryonic antigen A cell surface antigen (SSEA, CD15) used to define a stage in the developmental pathway. SSEA-1 is a commonly used marker for undifferentiated murine embryonic cells.

staggered cut *See* STICKY ENDS.

STAG proteins Components of the *cohesin complex, STAG1 (stromal antigen-1, 1258 aa), STAG2 (1231 aa), and STAG3 (1225 aa).

stanniocalcin (STC) Homologues of a calcium-regulated hormone in teleost fish, stanniocalcin 1 (STC1, 247 aa) and STC2 (302 aa), that are involved in calcium and phosphate homeostasis and are regulated by protein kinase C.

stanols Plant-derived compounds (24-α-ethylcholestanols) that in esterified form will reduce the level of low-density lipoprotein (LDL) cholesterol in blood by competing with cholesterol for absorption in the intestine. Added as a dietary supplement (*TN* Benecol).

***Staphylococcal* scalded skin syndrome** *See* EXFOLIATIN.

staphylococcal toxins *Staphylococcus aureus* produces five membranolytic toxins, four haemolysins (alpha-, beta-, gamma-, and delta) and *leucocidin. Alpha-toxin (293 aa) is a pore-forming toxin that preferentially attacks platelets and cultured monocytes. Beta-toxin is a Mg^{2+}magnesium-dependent *sphingomyelinase C. The gamma-toxin locus expresses three proteins, two class S components (HlgA and HlgC) and one class F component (HlgB) which form S/F heterodimers, have potent proinflammatory effects, and may be important in pathogenesis of toxic shock syndrome. Delta-toxin is a small peptide (26 aa) that is very amphipathic and surface active and has properties similar to *melittin.

staphylococcins Staphylococcal *bacteriocins.

Staphylococcus A genus of nonmotile Gram-positive bacteria that are opportunistic pathogens responsible for a range of infections. They produce a range of exotoxins and secrete various enzymes (*coagulase, hyaluronidase, lipase, and *staphylokinase). The cell wall has *protein A on the outer surface.

staphylokinase (streptokinase) A staphylococcal *plasminogen activator (EC 3.4.99.22, 163 aa).

Stargardt disease One of the most frequent causes of macular degeneration in childhood. **Stargardt disease-1** (fundus flavimaculatus) is an autosomal recessive disorder caused by mutation in the retina-specific ABC transporter-A4 or the cyclic-nucleotide gated ion channel β_4. Yellow or white pigment is deposited in the pigmented retinal epithelium, even though ABC-A4 is in rods. Visual acuity is seriously diminished but peripheral visual fields remain normal throughout life. Fundus albipunctatus is an allelic variant of Stargardt disease-1 with mutations in ABC-A4 and *peripherin-2 and onset later in life. **Stargardt-2** was thought to map to chromosome 13q but subsequently turned out to be Stargardt-3.

Stargardt disease-3 (autosomal macular degeneration with flecks type 3) is caused by mutation in the *ELOVL4* gene on chromosome 6 (elongation of very long-chain fatty acids-like 4). A locus for autosomal dominant **Stargardt disease-4** has been mapped to chromosome 4.

stargazin A brain-specific 36-kDa protein, mutated in the mouse ataxic and epileptic mutant stargazin, that has structural similarity to the γ subunit of skeletal muscle voltage-gated calcium channels. Synaptic trafficking of AMPA receptors is apparently controlled by stargazin. There are structural similarities with *clarins.

STAR proteins **1.** The signal transduction and activation of RNA (STAR) family of RNA-binding proteins, an evolutionarily conserved family that are involved in regulating developmental processes. Characteristically have a *KH domain within a *GSG domain. *See* QUAKING. **2.** StAR, *steroidogenic acute regulatory protein. **3.** STaR, an intestine-specific membrane-bound guanylate cyclase that binds *guanylin and bacterial heat-stable enterotoxins such as the *E. coli* ST toxin, STa.

STAR syndrome A developmental disorder in which there is toe syndactyly, telecanthus, and anogenital and renal malformations. It is caused by mutation in *cyclin M.

start codon *See* INITIATION CODON.

startle disease with epilepsy A genetically heterogeneous neurologic disorder (hyperekplexia with epilepsy) in which there is muscular rigidity of central nervous system origin, particularly in the neonatal period, and by an exaggerated startle response to unexpected acoustic or tactile stimuli. Can be caused by mutation in *collybistin.

start site An ambiguous term for either a *transcriptional start site or a *translational start site.

statherin A stable salivary protein (43 aa) that binds calcium and may prevent build-up of harmful deposits in the salivary glands and on the tooth surfaces. Binds *fimbrillin.

stathmin A cytosolic phosphoprotein (oncoprotein 18, Op18, leukaemia-associated phosphoprotein p18, LAP18, metablastin, 149 aa) that stoichiometrically binds two tubulin heterodimers and blocks assembly of microtubules unless inactivated by phosphorylation by *KIS (kinase interacting with stathmin). Several other kinases target stathmin, including PKA, MAPK, CDK2, p34cdc2, and CaM kinase II and IV. It is overexpressed in some tumours. **Stathmin-like 2** (superior cervical ganglion-10, 179 aa) is a neuronal growth-associated protein with significant sequence similarity to stathmin and **stathmin-like 3** (180 aa) and both have the microtubule-destabilizing functions.

statins A class of drugs that inhibit *HMG-CoA reductase, a key enzyme in cholesterol biosynthesis. Examples are atorvastatin, lovastatin, mevastatin, and simvastatin.

stationary night blindness A nonprogressive retinal disorder characterized by impaired night vision because of reduced or absent retinal rods. An X-linked form (**type 1**) is caused by mutation in *nyctalopin; the **type 2** X-linked form by mutation in the retina-specific calcium channel α1-subunit. Autosomal recessive forms can be caused by mutation in the metabotropic *glutamate receptor-6 or in calcium-binding protein-4. Autosomal dominant forms can be caused by mutations in *rhodopsin, *cyclic nucleotide phosphodiesterase-6B or in rod-specific *transducin. The Oguchi-type stationary night blindness, in which dark adaptation is abnormally slow, is caused by mutation in *arrestin or in rhodopsin kinase.

STATs A family of proteins (signal transducers and activators of transcription) that interact with receptors through a phosphotyrosine/SH2 domain interaction, are then phosphorylated by *JAKs, dimerize, and translocate to the nucleus where they act as transcription factors. Many STATs are known with varying receptor specificity and differences in the binding sites on DNA to which they bind, e.g. **STAT1** is activated by interferon-α, interferon-γ, EGF, PDGF, and IL-6 whereas **STAT2** is only activated by interferon-α. **STAT3, STAT5**, and **STAT6** are involved in signalling from the leptin receptor, as well as from various interleukin receptors. Mutations in STAT1 affect innate immunity to viruses and mycobacteria. Mutations in STAT3 are associated with hyper-IgE syndrome (*Job's syndrome). *See* SIGNAL-TRANSDUCING ADAPTOR PROTEINS.

staufen A *Drosophila* protein that binds to the 3′ untranslated region of specific mRNAs and is involved in localizing them in the oocyte. The human homologues (staufen1, 496- and 577-aa isoforms, and staufen 2, 479 aa) differ in having an additional microtubule-binding domain and are thought to be involved in targeting RNA to its site of translation.

staurosporine An inhibitor of a range of PKC-like protein kinases, derived from *Streptomyces* sp.

STE20 kinases A group of serine/threonine kinases involved in regulating a range of processes including the cell cycle, apoptosis, and stress responses. Yeast Ste20p (sterile 20 protein) is a mitogen-activated protein kinase kinase kinase kinase (MAP4K) involved in the mating pathway. The human homologues, of which there are 28, have a conserved kinase domain but a diversity of additional domains involved in interactions with signalling systems and cytoskeletal-regulatory proteins. The Ste20 group kinases are further divided into the *p21-activated kinase (PAK) and *germinal centre kinase (GCK) families.

STEAP A cell-surface antigen (six-transmembrane epithelial antigen of the prostate, 339 aa) expressed at cell–cell junctions in the secretory epithelium of prostate and strongly expressed in prostate cancer cells. STEAP may be a channel or transporter protein. *See also* PAPIN; PSCA; PSM.

steatoblasts Cells from which fat cells (*adipocytes) are derived.

Steele–Richardson–Olszewski syndrome *See* PROGRESSIVE SUPRANUCLEAR PALSY.

steel factor *See* STEM CELL FACTOR.

stefins A subfamily of the *cystatins which are *cysteine peptidase inhibitors. **Stefin A** (cystatin A, keratolinin, 98 aa) inhibits *cathepsins D, B, H, and L and is found in neutrophils, spleen, liver, and epidermis (as keratolinin). **Stefin B** (cystatin B, 98 aa) has antipeptidase activity but also interacts with several proteins and coimmunoprecipitates with *RACK1, β-spectrin, and neurofilament light chain (NFL) in rat cerebellum. Mutations in stefin B are associated with *myoclonic epilepsy of Unverricht and Lundborg.

S

stem-and-loop structure Describes the secondary structure of tRNA in which there are four base-paired stems and three loops (single-stranded), one of which contains the *anticodon.

stem cell **1.** Any cell that gives rise to a lineage of cells. **2.** More commonly, a cell that divides to produce dissimilar daughter cells, one that replaces the original stem cell, the other differentiating further. *Embryonic stem cells are totipotent and can produce any cell type, **pluripotent stem cells** can produce only a limited range of differentiated cells (as with haematopoietic stem cells). May exist as *quiescent stem cells with a repair function. **Induced pluripotent stem cells** (iPSCs) are derived from more differentiated cells that have been artificially stimulated (by transfection of genes) to revert to a pluripotent state.

stem cell derived tyrosine kinase A receptor tyrosine kinase of the hepatocyte growth factor receptor family, the murine homologue of the human *ron receptor tyrosine kinase. Expressed on macrophages and binds *macrophage stimulating protein, a serum protein activated by the coagulation cascade.

stem cell factor A pleiotropic growth factor (SCF, steel factor in mice, mast cell growth factor, c-kit ligand, 245 aa) that acts *in utero* in germ cell and neural cell development, and in the adult is primarily a haematopoietic growth factor, for both myeloid and lymphoid series. It is the ligand for the *kit receptor tyrosine kinase.

stem cell myeloproliferative disorder A myeloproliferative disorder in which there is myeloid hyperplasia, eosinophilia, T-cell or B-cell lymphoblastic lymphoma, and progression to acute myeloid *leukaemia. A fusion protein of *centriolin with the fibroblast growth factor receptor-1 has constitutive kinase activity and may be responsible.

stereocilium *See* STEREOVILLUS.

stereotaxis A system to identify the three-dimensional coordinates of a locus. An important technique in studies of the brain that may involve imaging to locate the target and then using the *x, y, z* coordinates to focus an electromagnetic beam for radiotherapy or to guide the insertion of a microelectrode.

stereovillus (*pl.* stereovilli) One of the set of long, stiff projections, several micrometres long, from the apical surface of *hair cells in the inner ear. They resemble *microvilli and are supported by bundled microfilaments: their role is probably to restrict the movement of the sensory cilium, which has an *axoneme and is sensitive to bending. Morphologically similar projections are also found on the apical surface of the pseudostratified epithelium in the epididymal duct. Often misleadingly called a stereocilium even though there is no axoneme.

Sternberg–Reed cells *See* HODGKIN'S LYMPHOMA.

steroid finger motif A characteristic structural motif of two zinc fingers found in the DNA-binding region of steroid receptors.

steroid hormones A group of structurally related hormones synthesized from cholesterol that control various physiological functions.

There are five major subgroups, *glucocorticoids, *mineralocorticoids, androgens, oestrogens, and *progestagens. Vitamin D is a sterol, but closely related. They are lipophilic and can readily cross the plasma membrane. Some **steroid hormone receptors** (type II) are part of the nuclear receptor family (which also includes receptors for nonsteroid ligands such as thyroid hormones and vitamin A), others are cytoplasmic but translocate to the nucleus and act as transcription factors once they bind hormone and release a chaperone heat shock protein (type I receptors).

steroidogenic acute regulatory protein A mitochondrial protein (StAR protein, 285 aa) that enhances the conversion of cholesterol into *pregnenolone and promotes steroid hormone production. Mutations in the gene cause *lipoid congenital adrenal hyperplasia. The herbicide Roundup inhibits steroidogenesis by disrupting expression of *STAR*.

sterol regulatory element (SRE) *See* STEROL REGULATORY ELEMENT-BINDING PROTEINS.

sterol regulatory element-binding proteins (SREBPs) A family of transcription factors that bind to the sterol regulatory element (SRE) in the promoter regions of genes that are involved in cholesterol and fatty acid metabolism. SREBPs are regulated by proteolytic cleavage by the ubiquitin-proteasome pathway and *sumoylation (*see* INSULIN-INDUCED GENES; SCAP).

sterols Steroid alcohols synthesized from acetyl-coenzyme A via the *HMG-CoA reductase pathway.The commonest example is *cholesterol.

STI Sexually transmitted infection.

Stickler's syndrome A connective tissue disorder in which there is progressive myopia beginning in the first decade of life, resulting in retinal detachment and blindness. There are also premature degenerative changes in various joints with abnormal epiphyseal development. An atypical autosomal dominant form of Stickler's syndrome (type 1), in which the effects are mainly confined to the eye, can be caused by mutation in the *collagen *COL2A1* gene. Other forms of the syndrome are caused by mutations in other types of collagen (COL11A1, COL11A2, COL9A1).

SEE WEB LINKS

- NIH information website.

sticky ends Short stretches of single-stranded DNA at the ends of a DNA duplex that has been cut with a *restriction enzyme that makes offset cuts. The two single-stranded ends are complementary and will hybridize to stick the two pieces (or pieces from elsewhere cut by the same enzyme) together.

stiff person syndrome An adult-onset sporadic acquired disorder in which there is progressive muscle stiffness with painful muscle spasms. In ~60% of cases there are antibodies to glutamic acid decarboxylase, the rate-limiting enzyme in the synthesis of γ-aminobutyric acid (*GABA); a smaller proportion have antibodies to *amphiphysin. The congenital form is caused by mutations in the α_1 subunit of the *glycine receptor.

stilboestrol A synthetic oestrogen that was used to prevent threatened miscarriage but was subsequently shown to be a risk factor for vaginal clear-cell carcinoma in women exposed *in utero*. Occasionally used to treat prostate carcinoma.

Still's disease A systemic inflammatory disorder of unknown aetiology with high fever, an evanescent skin rash, arthritis, and hyperleukocytosis. Adult-onset Still's disease is now generally regarded as systemic onset juvenile idiopathic arthritis and is associated with *macrophage activation syndrome.

stimulus–secretion coupling The cascade of events that link a stimulus such as membrane depolarization at the presynaptic terminal with the release of the contents of membrane-bounded vesicles (e.g. neurotransmitter release). The term draws an analogy with excitation–contraction coupling in muscle contraction since both involve elevations of intracellular calcium.

STK *See* CYCLIN-DEPENDENT KINASE (7); STEM CELL DERIVED TYROSINE KINASE; STREPTOKINASE.

Stokes' radius The apparent radius of a molecule sedimenting under centrifugal force, calculated from Stokes' law (which defines the frictional coefficient for a particle moving through a fluid). The Stokes' radius depends on the tertiary structure of the molecule.

stomatin An integral membrane protein (288 aa) of the erythrocyte membrane (band 7.2b) that is defective in overhydrated hereditary *stomatocytosis. Several stomatin-like proteins have been identified; **stomatin-like 1** (STOML1, SLP1, 394 aa), **STOML2** (356 aa), which is overexpressed in a range of epithelial tumours and **STOML3** (287 aa), which may be involved in in modulating odorant signals in olfactory epithelium.

stomatocytosis A condition in which the passive leakage of ions across the erythrocyte plasma membrane is increased, leading to shortened cell survival, increased osmotic fragility, and haemolytic anaemia. The cells take on an abnormal shape, resembling a mouth or 'stoma'. In **hereditary stomatocytosis I** (overhydrated stomatocytosis) there is variation in the extent of sodium influx and osmotic swelling and three subtypes with marked, moderate, and normal sodium influx have been suggested. There appears to be a defect in *stomatin, although the aetiology is still obscure. A second form, **stomatocytosis II,** in which there is no increase in fragility, has been reported. A different disorder, **dehydrated hereditary stomatocytosis** (xerocytosis; desiccytosis), is caused by excessive leakage of potassium from the erythrocytes which lose water and become more fragile.

STOP (stable tubulin-only protein) A calmodulin-binding and calmodulin-regulated protein (microtubule associated protein 6, 813 aa) that is involved in stabilizing microtubules to disassembly induced by drugs or cold temperatures, particularly in neurites. At least two isoforms have been identified.

stop codon *See* TERMINATION CODON.

stop transfer sequence (membrane anchor sequence) A sequence of hydrophobic amino acids that causes cotranslational transport of a protein across the RER to stop, leaving the protein embedded in the membrane.

storage diseases (lysosomal diseases)

Diseases in which there is a deficiency of a lysosomal enzyme and undigested substrate for that enzyme accumulates within cells. The storage diseases are not immediately fatal, but within a few years some can cause serious neurological and skeletal disorders leading eventually to death. See the following diseases or syndromes: *Batten's; *cystinosis; *Danon's; *Fabry's; *Farber's; *Gaucher's; *Hers's; *Hunter's; *Hurler's; liposes (*see* LIPID DISORDERS); *McArdle's; *mannosidosis; *Morquio-Brailsford; *mucopolysaccharidoses; *Niemann–Pick; *Pompe's; *Sanfillipos's; *Salla; *Sandhoff's; *Scheie's; *Sly; *Tay–Sachs. *See also* GLYCOGEN STORAGE DISEASES.

storage granules Membrane-bounded vesicles (zymogen granules, condensing vacuoles) with condensed secretory materials, often in an inactive (zymogen) form.

storage pool disease A range of platelet disorders encompassing variable degrees of reduction in the numbers and contents of dense granules (δ-granules), α-granules, or both. In **α-SPD** (grey platelet syndrome) there are severe reductions in the α-granules and their contents, in **δ-SPD** the defect is only in the dense granules, and in a third form (**α/δ-SPD**) both types of granule are affected.

STR *See* SHORT TANDEM REPEAT.

stratified epithelium An epithelium in which there are multiple layers of cells and only the basal layer of cells are in contact with the *basal lamina. The basal *stem cells divide to produce another basal cell and a cell that joins the upper layers and will eventually be shed. The classic example is skin, where the cells become progressively more heavily keratinized as they move up and eventually die and are shed as squames. Stratified epithelia are usually found where abrasion is likely.

stratifin A *14-3-3 protein (14-3-3-σ, 248 aa) found diffusely distributed in the cytoplasm of epithelial cells, particularly stratified keratinizing epithelium, that mediates cell cycle arrest. *See* HS1.

stratum corneum The outermost layer of the skin, consisting of dead, scale-like cells (*squames) with little left of their contents other than *keratin.

stratum granulosum The layer of granular cells below the *stratum corneum, sometimes separated from it by the *stratum lucidum. The cells progressively accumulate keratin and become compressed as they move outwards and away from the basal layer.

stratum lucidum A thin layer of dead cells separating the *stratum granulosum and *stratum corneum in areas where the skin is thick, such as the palms of the hands and the soles of the feet.

stratum Malpighii A layer of cells (malpighian layer, prickle cell layer) lying between the proliferating cells of the basal layer (*stratum germinatum) and the stratum granulosum where keratin deposition occurs.

streptavidin A tetrameric protein (subunits of ~160 aa) isolated from *Streptomycetes avidinii* that has a very high 1:1 binding affinity for *biotin. Much used as an experimental tool to detect biotin. *See* AVIDIN.

Streptobacillus A Gram-negative bacterium (*Streptobacillus moniliformis*) that causes rat-bite fever (*sodoku).

streptococcal M-protein A cell wall protein of streptococci that is used as a marker for serotyping of group A streptococci (>100 serotypes are known). It is present as hair-like *fimbriae and by making the cell resistant to opsonization is an important virulence factor.

streptococcal pyrogenic exotoxins (SPE, formerly **erythrogenic toxin)** A group of toxins produced by *Streptococcus pyogenes* and responsible for many of the pathological sequelae of infection. Several (SPEA, SPEC, SPEX) are *superantigens. SPEB is a cysteine peptidase (streptopain: peptidase 3.4.22.10, 398 aa) which possesses mitogenic activity but, when pure, is not a superantigen. SPEA (streptococcal pyrogenic exotoxin A) is a phage-encoded exotoxin. SPEA and SPEC show strong sequence homology to the staphylococcal enterotoxins.

streptococcal toxins A range of virulence factors released by streptococci that include *haemolysins α–δ similar to the *staphylococcal toxins, *leucocidin, *streptolysins O and S, and *streptococcal pyrogenic exotoxins. Various enzymes (*streptokinase, hyaluronidase, *streptodornase) also contribute to virulence although they are not toxins as such.

Streptococcus A genus of Gram-positive cocci that grow in chains. Many are responsible for diseases of varying severity: *Streptococcus pyogenes* is responsible for pharyngitis, scarlet fever, and rheumatic fever, *Strep. pneumoniae* is the main cause of lobar and bronchopneumonia. The virulence depends upon a range of *staphylococcal toxins and associated virulence factors such as *streptococcal M-protein, *streptococcal pyrogenic toxins, *streptolysins O and S, enzymes (*streptokinase, *streptodornase, hyaluronidase, and proteinase) and often a capsule that inhibits phagocytosis.

Streptococcus pneumoniae **(**formerly **Pneumococcus pneumoniae, Diplococcus pneumoniae, Fraenkel's bacillus)** The Gram-positive pneumococcus responsible for bacterial pneumonia and otitis media (middle ear infections) and an important contributor to bacterial meningitis.

streptodornase A mixture of four DNAases (EC 3.1.21.1, ~270 aa) released by streptococci. It reduces the viscosity of pus by degrading DNA released from dead leucocytes, allows the bacterium greater motility, and makes phagocytosis more difficult.

streptogramins A group of antibiotics that act on the bacterial ribosome to inhibit translation. There are two structurally distinct compounds involved which are separately bacteriostatic, but bactericidal in appropriate ratios. Examples include *dalfopristin, *quinupristin.

streptokinase A *plasminogen activator released by β-haemolytic streptococci that is used as a fibrinolytic. *TN* Streptase.

streptolydigin An antibiotic produced by *Streptomyces lydicus* that inhibits bacterial RNA polymerase.

streptolysins Streptococcal haemolytic exotoxins with sequence homologies to *bacteriocins. **Streptolysin O** is oxygen-labile and only acts on membranes that contain cholesterol, with which it interacts to form pores that are large enough to allow protein leakage. **Streptolysin S** is the main haemolysin of β-haemolytic strains, is oxygen stable, has sequence homologies with *bacteriocins and is nonimmunogenic.

Streptomyces A genus of Gram-positive spore-forming bacteria that grow as a branching filamentous mycelium similar to that of fungi. They are an important source of many antibiotics, e.g. *actinomycin; *neomycin; *staurosporine; *streptomycin; *tetracycline.

streptomycin An *aminoglycoside antibiotic derived from *Streptomyces griseus* that blocks initiation of protein synthesis on the bacterial ribosome. It is used extensively in veterinary practice and in laboratory culture work.

streptopain *See* STREPTOCOCCAL PYROGENIC EXOTOXINS.

streptovaricins A group of antibiotics of the *ansamycin type that block initiation of transcription in prokaryotes; isolated from various actinomycetes.

streptozotocin An antibiotic that is effective against growing Gram-positive and Gram-negative organisms. Streptozotocin will cause pancreatic *beta cell death if used in high doses and is used to induce a form of diabetes in mice that is a model for type 1 *diabetes.

stresscopin *See* UROCORTIN III.

stress fibres Bundles of actin microfilaments and associated proteins found in fibroblasts, particularly slow-moving fibroblasts grown on nondeformable substrata. They are contractile and have some periodicity though are much less organized than sarcomeres. They are attached to the membrane and indirectly to the environment, through *focal adhesions.

stress-induced proteins A better name for *heat shock proteins, a group of proteins that are produced as a response to heat and other stresses.

striated muscle Muscle in which the contractile *myofibrils are arranged with the *sarcomeres in register so that the whole mucle fibre has transverse or oblique striations visible in the light microscope. Both *skeletal muscle and *cardiac muscle are striated.

stringency A term used in describing nucleic acid *hybridization reactions. The removal of unbound probe can be carried out with low stringency (low temperature, high ionic strength) which will permit some mismatching of probe and target. High stringency conditions allow only closely matching sequences to remain base-paired, making it less likely that related but nonidentical sequences will be detected.

stroma Loose connective tissue that has few resident cells (stromal cells). *Compare* PARENCHYMA.

stromelysins *See* MATRIX METALLO-PEPTIDASES.

strong promoter An imprecise term, often used, that refers to a promoter site that will strongly activate expression of the gene.

strongyloidiasis (strongyloidosis) Intestinal infestation with the nematode, *Strongyloides stercoralis.*

strophanthin A mixture of glycosides, from a tropical liana *Strophanthus kombe*, that has properties similar to digoxin. Strophanthin G is ouabain, which inhibits the sodium-potassium ATPase. *See* DIGITALIS.

STRP Short tandem repeat polymorphism. *See* SATELLITE DNA.

S

structural gene A gene that codes for a product (e.g. an enzyme, structural protein, tRNA) that does not have a regulatory function.

structure–activity relationship (SAR) The correlation between the structure of a molecule and its biological action derived from systematically altering the structure and assaying for activity. The **quantitative structure–activity relationship** (QSAR) takes this further by using computed parameters of shape and charge distribution to describe the structure. A common hope is that the use of QSAR will facilitate rational drug design.

strumpellin A protein (1159 aa) with putative transmembrane domains and highest expression in skeletal muscle. Function currently unknown. *See* SPASTIC PARAPLEGIAS.

Strumpell's disease *See* SPASTIC PARAPLEGIAS.

struvite A magnesium ammonium phosphate mineral found in some kidney stones, particularly those associated with infections by urease-producing bacteria such as *Ureaplasma urealyticum* and *Proteus.*

strychnine An *alkaloid toxin, from the Indian tree *Strychnos nux-vomica*, that blocks the activity of *glycine as a neurotransmitter. The characteristic convulsions are probably the result of blocking inhibitory synapses on spinal cord motoneurons.

Stuve–Wiedemann syndrome *See* SCHWARTZ–JAMPEL SYNDROME.

STX **1.** Sialyltransferase, a family of enzymes that transfer sialic acid to nascent oligosaccharides. **2.** *See* SAXITOXIN. **3.** *See* SHIGA TOXIN.

S-type lectins *See* GALECTINS.

subacute Describing a disease that is more rapid than a chronic disease but does not become acute.

subacute sclerosing panencephalitis A chronic progressive illness in children, some years after an infection with measles virus, that involves demyelination of the cerebral cortex. A *slow virus disease in which the virus, possibly mutated, persists in the brain.

substance K Old name for neurokinin A (neuromedin L). *See* TACHYKININS.

substance P A *vasoactive intestinal peptide (11 aa) found in the brain, spinal ganglia, and intestine, derived from a precursor containing both substance P and neurokinin A (*see* TACHYKININS). It induces vasodilation, salivation, and increased capillary permeability and is involved in inflammation and pain. Binds to the neurokinin 1 (tachykinin 1) receptor, a G-protein-coupled receptor (407 aa), that is highly expressed in areas of the brain that are implicated in depression, anxiety, and stress, as well as areas associated with motivation in response to natural rewards and drugs of abuse.

substantia nigra An area of darkly pigmented dopaminergic neurons in the mesencephalon (midbrain) that plays an important role in reward, addiction, and movement. It is particularly affected in *Parkinson's disease.

substrate **1.** A substance that is acted upon by an enzyme. **2.** A culture medium that can support growth of a particular species of bacterium. **3.** A surface (*but see* SUBSTRATUM).

substratum The solid surface over which a cell moves, or upon which a cell grows. The term avoids the ambiguity that can arise if *substrate is used in this sense.

subtilin A pore-forming *lantibiotic produced by *Bacillus subtilis*. Acts preferentially on Gram-positive microorganisms, but bacteria that produce subtilin are resistant because they have an *ABC transporter-2 subfamily member that confers multidrug resistance.

subtilisin An extracellular serine endopeptidase (EC 3.4.21.62, ~380 aa) released by *Bacillus* spp.

subtilysin *See* SUBTILISIN. The lipopeptide surfactant produced by *Bacillus subtilis*, *surfactin, has erroneously been ascribed this name in some dictionaries.

subtractive hybridization (subtraction cloning) A method for identifying differentially expressed genes. An excess of *mRNA from one sample is hybridized to cDNA from the other, and double-stranded hybrids removed. cDNA that has not hybridized is presumably only expressed in the second sample. *See* DIFFERENTIAL SCREENING.

succinate dehydrogenase **1.** Succinate-coenzyme Q reductase (SDH, EC 1.3.5.1), complex II in electron transport, succinate-ubiquinone oxidoreductase. An enzyme complex of four subunits located in the inner mitochondrial membrane, involved in the tricarboxylic acid cycle and the electron transport chain. **2.** Fumarate reductase (fumarate dehydrogenase, EC 1.3.99.1), a bacterial enzyme or a degraded form of the mitochondrial enzyme that oxidizes succinate to fumarate.

succinylcholine *See* SUXAMETHONIUM.

sucrose A nonreducing disaccharide, table sugar, α-D-glucopyranosyl-β-D-fructofuranose.

Sudan stains Histochemical stains for lipids.

SEE WEB LINKS

- Useful resource on histological stains.

sudden infant death syndrome The sudden and inexplicable death of an infant (cot death). Although many possible explanations have been offered, none is entirely satisfactory.

sudorific Describing something involved in the secretion of sweat or, in the case of a drug, one that stimulates sweating.

SUFU The major negative regulator of the *sonic hedgehog/*patched signalling pathway in vertebrates (suppressor of fused). The human gene has 40% homology with the *Drosophila* gene and the product (484 aa) inhibits the transcriptional activity of **gli-1* (glioma-associated oncogene-1) and osteogenic differentiation in response to sonic hedgehog signalling.

SUGT-1 The homologue of yeast SGT1, the product of which (suppressor of G2 allele of SKP1, 365 aa) is involved in kinetochore function and the G_1/S and G_2/M transitions: the human protein will work in yeast. A cochaperone that binds to Hsp90, Hsp70, or Hsc70, and is elevated in HEp-2 cells as a result of stress conditions such as heat shock. *See* SKP.

sulcus (*pl.* sulci) A groove or furrow, e.g. on the surface of the cerebrum.

sulfa drugs (sulpha drugs) General name for *sulfonamides.

sulfasalazine (sulphasalazine) A mixture of aminosalicylic acid and the *sulphonamide sulfapyridene. Used to treat inflammatory bowel disease and some cases of rheumatoid arthritis.

sulfinpyrazone (formerly **sulphinpyrazone)** A compound related to phenylbutazone, but with no anti-inflammatory activity. Does not affect platelet aggregation *in vitro*, but inhibits the platelet adhesion and release reactions. It also inhibits uric acid resorption in the kidney.

sulfur, sulfo- The UK English spelling **sulphur, sulpho-** is used throughout this dictionary, except where the British Approved Name (BAN) for a drug has the sulf- spelling.

sulphatase-modifying factor 1 The enzyme (SUMF1, 374 aa) in the endoplasmic reticulum that catalyses the post-translational formation of C-α-formylglycine (FGly), the catalytic residue in the active site of *sulphatases, from a cysteine. Mutations lead to *multiple sulphatase deficiency, a lysosomal storage disease.

sulphatases *Esterases that catalyse the hydrolysis of sulphate esters; e.g. *aryl sulphatase A (cerebroside-sulphatase, EC 3.1.6.8, 507 aa) is a lysosomal enzyme that breaks down cerebroside 3-sulphate, and *N*-acetylgalactosamine-6-sulphatase (EC 3.1.6.4, 522 aa) hydrolyses sulphate groups of *chondroitin sulphate and *keratan sulphate. *See* MULTIPLE SULPHATASE DEFICIENCY.

sulpholipids Any lipids that contains sulphur, although the term is usually used for sulphate esters of glycolipids. **Sulpholipid-1** (SL-1) in the cell wall of *Mycobacterium tuberculosis* is a glycolipid antigenic marker used in diagnosis. **Sulpholipid immobilizing protein 1** (SLIP1) is a conserved plasma membrane protein (68 kDa), involved in sperm–zona pellucida binding (*but see* SLIP1).

SEE WEB LINKS

- Article on sulpholipids in *Molecular Human Reproduction* (2001).

sulphonamides (sulfonamides) A group of drugs that contain the sulphonamide group. The antibacterial sulphonamides (sulpha drugs, e.g. prontosil) were some of the first effective antibacterial compounds, but cause adverse effects in ~3% of people. They act as competitive inhibitors of dihydropteroate synthetase, an enzyme involved in folate biosynthesis. Many derivatives have been produced, some without antibacterial activity (e.g. sultiame), and the *sulphonylureas and thiazide diuretics are based on the antibacterial sulphonamides.

sulphonylurea receptor (SUR) An ABC transporter family member (ABCC8, SUR1, 1581 aa) that regulates glucose-induced insulin secretion by an effect on inwardly rectifying potassium–ATP channels (Kir6.2) in pancreatic beta cells. Four Kir6.2 subunits form the pore and interact with four regulatory subunits (SUR1) each with two nucleotide-binding folds (NBFs) that sense changes in the ATP:ADP ratio in the cell. Tritiated *glibenclamide is often used to detect their tissue distribution.

sulphonylureas (sulfonylureas) A group of drugs that increase insulin secretion and are widely used in the treatment of type 2 diabetes. Examples include *tolbutamide, *glipizide, and glibenclamide, the latter often used in experimental studies. *See* SULPHONYLUREA RECEPTOR.

sulphydryl reagents Compounds which interact with disulphide bonds and sulphydryl groups in proteins to promote or inhibit the formation of disulphide crosslinks; usually reducing agents which break disulphide bonds Examples are *p*-chlormercuribenzoate, *N*-ethylmaleimide, iodoacetamide.

sumatriptan A $5HT_1$ agonist (triptan) used to treat migraine. *TN* Imigran.

SUMO (small ubiquitin-related modifier protein) *See* SUMOYLATION.

sumoylation The post-transcriptional modification of a protein by the addition of small ubiquitin-related modifier (sumo) proteins. The effect may be to stabilize or to alter subcellular localization. **SUMO-1** (101 aa) conjugates as a monomer, **SUMO-2** (95 aa) and **SUMO-3** (103 aa), which are closely related, are conjugated as higher molecular weight polymers. Targets include p53, ran-Gap. Several transcription factors (e.g. C/EBP and c-myb) are regulated by sumoylation.

sun proteins A conserved family of proteins (Sun1, 824 aa) of the inner nuclear membrane that have a **sun domain** (Sad1/UNC-84 homology domain) and interact with *lamins on the inner face of the inner nuclear membrane and with *nesprins in the outer nuclear membrane (the LINC complex that links nucleoskeleton with cytoskeleton). The *Caenorhabditis elegans* protein UNC-84 is involved in nuclear migration/positioning. **Nsun2** (misu, 767 aa) is another nuclear protein with a Sun domain and is up-regulated following activation of Myc by 4-hydroxy-tamoxifen.

superantigen An antigen that will activate all T cells with a T-cell receptor that has a particular Vβ sequence (of which there are ~50), a far greater number than activated by a normal antigen. Superantigens are presented on MHC class II but are not processed and do not bind to the normal peptide binding site. The best-known examples are staphylococcal and streptococcal enterotoxins, which excite such a broad response that the immune system is overwhelmed.

supercoiling A phenomenon that occurs when the closure of a loop of DNA occurs after a rotation of the two ends relative to one another. If the rotation is in the opposite direction to that of the duplex (anticlockwise) then either the double helix has to unwind or the loop has to twist—negative supercoiling. If rotation is clockwise, the helix becomes more tightly coiled or there is positive supercoiling of the loop. DNA that shows no supercoiling is relaxed. Bacterial and mitochondrial circular DNA is usually negatively supercoiled and nuclear DNA mostly exists as supercoils associated with protein in the *nucleosome. *Topoisomerases will alter the degree of supercoiling.

superhelix The result of *supercoiling a molecule that is already coiled.

superoxide (superoxide radical) A term used for both the superoxide anion. O_2^-, or the weak acid $HO_2^{\cdot}$, an important product of the *metabolic burst of neutrophil leucocytes. It is highly reactive and chronic inflammation may lead to superoxide-mediated inactivation of plasma antiproteases, e.g. contributing to the fibrosis seen in emphysema.

superoxide dismutase (SOD) A metalloenzyme (EC 1.15.1.1, 154 aa) that catalyses the formation of hydrogen peroxide and oxygen from superoxide, and prevents oxidative damage. Most eukaryotic forms have copper or zinc as the metal cation. *See* AMYOTROPHIC LATERAL SCLEROSIS (1).

supershift A phenomenon seen in *bandshift assays. The shift, the reduction in mobility on a gel that occurs when the nucleic acid interacts with a protein, is further enhanced (supershifted) if an antibody (or any other protein) binds to the protein.

supervillin A highly conserved actin-binding protein (1788 aa) with homology to the *gelsolin/villin family. It can modify the extent of crosslinking of actin in the cytoplasm and binds to myosin-II, but also has a nuclear localization signal and is a coregulator of the androgen receptor. **Archvillin** (2214 aa) is a splice variant found in smooth muscle.

suppressor factor **1.** A generic name for uncharacterized factors that nonspecifically suppress immune responses. **2.** *See* SUPPRESSOR MUTATION; TERMINATION CODONS.

suppressor mutation A mutation that counteracts the effect of a mutation at another locus. This can arise in a variety of ways, one of the more surprising being the modification of a tRNA anticodon to misread a *termination signal.

suppressor T cells *See* T-REGULATORY CELLS.

suprachiasmatic nucleus A region of the hypothalamus, immediately above the optic chiasm, that generates the endogenous *circadian rhythm. It receives inputs from specialized *melanopsin-containing retinal ganglion cells and is the master regulator for other oscillator systems.

SUR *See* SULPHONYLUREA RECEPTOR.

suramin An antiprotozoal and anthelmintic drug that will block the binding of various growth factors, uncouple G proteins from receptors, inhibit phospholipase D and act as an antagonist at purinergic receptors. Has been tested as a treatment for hormone-refractory prostate carcinoma. *TN* Germanin.

surface-active compound In biological systems, a molecule that is *amphipathic and thus has detergent-like properties.

surfactant A *surface-active compound with detergent-like properties. *See* SURFACTANT PROTEINS A AND D; SURFACTIN.

surfactant proteins A and D (SP-A, SP-D) Proteins of the *collectin family that have an important role in making the alveolar surface hydrophobic and thereby preventing the lungs from filling with water by capillary action. The production of this surfactant just before parturition is important for the transition to independent breathing, and defects have been speculatively linked to cot death. They also have a role in defence against microorganisms. SP-A is oligomeric, consisting of eighteeen protomers with collagen and lectin-like domains that recognizes glycoconjugates, lipids, and protein determinants on both host cells and invading microorganisms. SP-D also occurs on the luminal surface of the gastric mucosa.

surfactin A powerful lipopeptide surfactant produced by *Bacillus subtilis* that has potent antimicrobial properties and may have anti-inflammatory and antitumour activity. A range of similar surfactants are produced by other bacteria and are mostly heptapeptides coupled to an acyl (C13–C15) residue. They may have potential for bioremediation of contaminated environments.

- Article in the *Brazilian Journal of Microbiology* (2007).

surrogate marker A symptom or parameter associated with a disease state that can be monitored easily and can substitute for a more fundamental but less easily quantified clinical measure. The validity is often questioned and needs to be well justified.

survivin An inhibitor of apoptosis (baculoviral *IAP repeat-containing protein, BIRC-5, 142 aa) that is selectively overexpressed in common human cancers. It is expressed in the G_2/M phase of the cell cycle and associates with spindle microtubules; disruption of the survivin-microtubule interaction results in increased activity of *caspase-3 and apoptotic death. Several splice variants have been described. Interacts with *XIAP.

sushi domains Protein modules (complement control protein (CCP) modules, short consensus repeats SCR, ~60 aa) that structurally are said to resemble Japanese sushi. They are involved in protein–protein interactions and are common in many proteins involved in the regulation of the complement system, blood coagulation, cell surface proteins (IL2 and IL15 receptors), and some selectins. *See* SUSHI PEPTIDES.

sushi peptides Peptides (S1 and S3, 34 aa] derived from the *lipopolysaccharide-binding domain of *factor C from the horseshoe crab *Carcinoscorpius rotundicauda*. They bind LPS with high affinity and neutralize its toxicity. *See* SUSHI DOMAINS.

SEE WEB LINKS

- Article in *Biochemical Society Transactions* (2006).

sustentacular Describing something that supports or maintains. Sustentacular cells (Sertoli cells) of the testis support and may provide nutrition for developing sperm.

suxamethonium (succinylcholine) A muscle relaxant that binds to nicotinic acetylcholine receptors but is only slowly broken down by plasma butyrylcholinesterase, and persists so that the muscle depolarizes and relaxes. Chemically it consists of two acetylcholine molecules linked by their acetyl groups, and is usually used as the chloride. *TNs* Anectine, Scoline.

SUZ12 A *polycomb group protein (suppressor of zeste 12 protein homologue, 739 aa) that is a component of PRC2/EED-EZH complexes. A chromosomal translocation involving *SUZ12* and **JAZF1* may be a cause of endometrial stromal tumours.

SV2 **1.** Integral membrane proteins of synaptic vesicles and secretory vesicles with a regulatory role. **SV2A** (742 aa) and **SV2C** are associated with insulin-containing granules and synaptic-like microvesicles, whereas **SV2B** is present only on synaptic-like microvesicles in neural tissue. SV2 is the protein receptor for botulinum neurotoxin A. **SV2 related protein** (SVOP, 548 aa) is probably an anion transporter (SLC22 family) and is expressed in most regions of the brain, but not elsewhere. **2.** A line of nontumorigenic human lung fibroblasts, MRC-5 SV2 cells.

SV3T3 *Swiss 3T3 cells transformed with *SV40 virus.

SV40 A small DNA *tumour virus (simian virus 40, simian vacuolating virus 40), one of the Polyomaviridae. Originally identified in monkey cells used for preparing polio vaccine. It will induce tumours in newborn hamsters and will transform many cells in culture. *See also* T ANTIGENS.

Svedberg units The units used for the sedimentation coefficient of particles in ultracentrifugation, which depends on the size, shape, and density of the particle. Svedberg units are nonadditive; e.g.eukaryotic ribosomes have a sedimentation coefficient of 80S but the subunits are of 60S and 40S.

swainsonine An indolizidine alkaloid that is the principal toxin in locoweeds and plants of the genus *Swainsona*. It is a reversible inhibitor of mannosidases and inhibits the processing of asparagine-linked glycoproteins.

SWI/SNF complex A multiprotein complex involved in the ATP-dependent remodelling of *nucleosomes which may involve chaperone-like activities. The complex in *S. cerevisiae* and *Drosophila* is thought to facilitate transcriptional activation of specific genes by antagonizing chromatin-mediated transcriptional repression. *See* BAF COMPLEX; RSC COMPLEX.

Swiss 3T3 cells A line of fibroblast-like cells derived from whole trypsinized embryos of the outbred Swiss strain of mice.

switch regions The nucleotide sequences in the introns at the 5′ end of the constant region of immunoglobulin heavy chain genes that are important in the recombination event that allows (e.g.) the switch from IgM to IgG to occur as the immune response matures. *See* ISOTYPE SWITCHING.

Sydenham's chorea A disease (chorea minor, Saint Vitus' dance) characterized by rapid, uncoordinated jerking movements (chorea) affecting primarily the face, feet, and hands. It is caused by childhood infection with group A β-haemolytic streptococci.

syk A nonreceptor tyrosine kinase (spleen tyrosine kinase, 630 aa) more widely expressed in myeloid and lymphoid cells than *zap70 and involved in coupling activated immunoreceptors to downstream signalling events controlling proliferation, differentiation, and phagocytosis. It has two tandem *SH2 domains that bind phosphorylated *ITAM motifs.

symbiosis The mutually beneficial but obligatory cohabitation exhibited by some organisms (and probably by mitochondria in the ancestral eukaryotic cell).

sympathetic nervous system The subdivision of the *autonomic nervous system that innervates the heart and blood vessels, sweat glands, viscera, and the adrenal medulla. Most neurons are noradrenergic. *Compare* PARASYMPATHETIC NERVOUS SYSTEM.

sympathomimetic Describing something (usually a drug) that mimics the effect of stimulation through the *sympathetic nervous system. Sympathomimetic drugs are adrenergic receptor agonists.

symplesiomorphic *See* APOMORPHIC.

symport A transport mechanism in which two different molecules move in the same direction, generally with one moving down an electrochemical gradient and driving the other, as in the case of the sodium/glucose cotransport system. *Compare* ANTIPORT; UNIPORT.

synaphin *See* COMPLEXINS.

synapse A connection between *excitable cells through which an impulse can be transmitted. In **chemical synapses** transmission occurs by the release of neurotransmitter from the presynaptic cell when an action potential arrives. The neurotransmitter substance diffuses across the synaptic cleft and binds to *ligand-gated ion channels on the postsynaptic cell, triggering an ion influx and a depolarization that leads to a second action potential or a response such as muscle contraction. The signalling is one-way, with only the presynaptic cell having the capacity to release neurotransmitter. In **electrical synapses** there is a direct electrical connection through *gap junctions. In **rectifying synapses** action potentials can only pass in one direction (all chemical and some electrical synapses). **Excitatory synapses** act to increase the probability of response in the postsynaptic cell whereas **inhibitory synapses** do the opposite.

synapsins A family of phosphoproteins associated with synaptic vesicles and that interact with small G proteins such as *rab to regulate neurotransmitter release and are phosphorylated by various kinases. They are also implicated in axonogenesis and synaptogenesis. **Synapsin Ia** (Brain protein 4.1, 705 aa) and **Ib** are alternatively spliced variants, as are **synapsins IIa** and **IIb** (582 aa). A third synapsin (**synapsin III**, 580 aa) has a more restricted distribution. Mutations in synapsin I may be associated with X-linked disorders with primary neuronal degeneration such as Rett's syndrome.

synapsis **1.** The pairing of the chromatids of homologous chromosomes during *prophase I of meiosis that allows *recombination to take place. **2.** The process involved in bringing together the ends of double-strand breaks in DNA, prior to nonhomologous end-joining (*NHEJ).

synaptic cleft *See* SYNAPSE.

synaptic facilitation *See* FACILITATION.

synaptic plasticity Changes in the properties of synapses, thought to be important in learning and memory. Small changes in the efficiency of a synapse, brought about by modifying the release of neurotransmitter or the number of receptors on the postsynaptic cell, can alter the properties of a neuronal circuit. *See also* NEURONAL PLASTICITY.

synaptic vesicle A vesicle containing a neurotransmitter, located in the terminal region of the presynaptic neurons of a chemical *synapse.

synaptobrevins (vSNAREs, VAMPs) Integral membrane proteins of *synaptic vesicles that bind *SNAPs and interact with target-SNARE (*syntaxin) in the process of exocytosis. Two isoforms, **VAMP-1** (118 aa) and **VAMP-2** (116 aa) are known. **VAMP-3** is *cellubrevin. **VAMP-5** (myobrevin, 102 aa) is found in muscle. **VAMP-8** (endobrevin, 100 aa) is associated with secretory vesicles in other tissues. A **synaptobrevin-like protein** (220 aa) is encoded by a gene on the X chromosome and may be subject to complex silencing by histone methylation and acetylation.

synaptogenesis The formation of a *synapse.

synaptogyrins (p29) Integral membrane proteins with four transmembrane domains, abundant in membranes of *synaptic vesicles. Synaptogyrin 1 exists as multiple isoforms (1a, 1b, and 1c; 234 aa, 191 aa, and 192 aa). Synaptogyrins 2, 3, and 4 are also known.

synaptojanin A major presynaptic protein, a polyphosphoinositide phosphatase, concentrated together with *endophilin at clathrin-coated endocytic pits in nerve terminals. **Synaptojanin-1** (1575 aa and 1311 aa) interacts with *dynamin and *amphiphysin in the process of vesicle recycling. **Synaptojanin-2** (1496 aa) has a much wider tissue distribution.

synaptomorphic *See* APOMORPHIC.

synaptonemal complex The meiosis-specific scaffolding structure that lies between chromosomes during *synapsis. There appear to be two lateral plates closely apposed to the chromosomes and connected to a central plate by filaments. Various synaptonemal proteins have been identified (SYCP1, 976 aa; SYCP2, 1530 aa; SYCP3, 236 aa and synaptosomal central element proteins, Syce 1, 329 aa and Syce2, 171 aa). The complex does not appear to be essential for *recombination.

synaptophysin An integral membrane protein (313 aa) of small synaptic vesicles in brain and endocrine cells. The function is unclear but it has structural similarities with *connexins. **Synaptophysin 2** (265 aa, synaptoporin) is found in different subsets of neurons.

synaptopodin An F-actin-associated proline-rich protein (685 aa) found in kidney podocytes and some dendritic spines (with tissue-specific isoforms). May play a role in *synaptic plasticity and is up-regulated during the late phase of *long-term potentiation.

synaptoporin *See* SYNAPTOPHYSIN.

synaptosome A subcellular fraction, consisting mainly of synaptic vesicles, prepared from tissues rich in chemical *synapses.

synaptotagmin (p65) A family of integral membrane proteins found in synaptic vesicles where they may serve as calcium sensors that regulate vesicular trafficking and exocytosis.

synCAMs A group of four immunoglobulin (Ig) superfamily members (synaptic cell adhesion molecules) involved in synaptic adhesion and in triggering presynaptic differentiation. They interact with *β-neurexin/*neuroligin.

synchronous cell population A cell culture in which all the cells are at the same stage in the cell cycle and will all divide at the same time. The synchronization tends to break down after only a few cycles.

syncoilin An intermediate filament-type III protein (482 aa), highly expressed in striated and cardiac muscle, that binds to α-dystrobrevin and desmin and may be involved in linking the intermediate filament cytoskeleton to the *dystrophin-associated protein complex.

S

syncollin A membrane-associated protein (134 aa) found within zymogen granules in the exocrine pancreas and in azurophil granules of neutrophils and required for exocytosis.

syncytin Membrane proteins (both 538 aa) encoded by envelope genes from the *human endogenous retroviruses (HERV-W in the case of syncytin-1, HERV-FRD for syncytin-2) that are specifically expressed in the placenta where they are involved in the formation of the *syncytiotrophoblast. The **syncytin 2 receptor** is the *major facilitator superfamily domain-containing 2 (MFSD2), a presumptive carbohydrate transporter with ten to twelve membrane-spanning domains.

syncytiotrophoblast The outermost layer of fetal tissue in the placenta, a syncytium that is the interface with maternal tissue. *See* SYNCYTIN.

syncytium A multinucleated tissue formed by the fusion of cells (or potentially, by failure of cytokinesis). Major examples are striated muscle (formed by the fusion of myoblasts) and the syncytiotrophoblast.

syndactyly An autosomal dominant trait, the most common congenital anomaly of the hand or foot, marked by persistence of the webbing between adjacent digits. Type 3 is caused by mutation in *connexin-43. *See* SYNPOLYDACTYLY.

syndecans Integral transmembrane proteoglycans (syndecan-1, 310 aa) that link the extracellular matrix with the actin cytoskelton and act as receptors or coreceptors in intracellular communication. The extracellular domain has multiple heparan sulphate and two chondroitin or dermatan sulphate chains plus an *N*-linked oligosaccharide, and binds to collagens, fibronectin, and *tenascin. Binding of heparin-binding growth-associated molecule to **N-syndecan** (syndecan-3, 442 aa) increases phosphorylation of c-*src and *cortactin, and N-syndecan may act as a neurite outgrowth receptor. Four members of the syndecan family show a remarkable physical relationship with four members of the *myc gene family. Syndecans bind gp120 of HIV and act as mediators for viral adsorption. **Syndecan-4** (amphiglycan, ryudocan, 198 aa) was isolated from human epithelial and fibroblastic cells. **Syndecan-binding protein** (syntenin, melanoma differentiation-associated protein 9, scaffold protein Pbp1, 298 aa) is an adaptor protein with tandem repeat of PDZ domains that binds the cytoplasmic domain of syndecans and of pro-TGFα (*see* TRANSFORMING GROWTH FACTOR).

***SYNE* gene family** A family of genes encoding synaptic nuclear envelope proteins. **Syne1** (8797 aa) is *nesprin1 and mutations are associated with autosomal recessive *spinocerebellar ataxia type 8 (recessive ataxia of Beauce). **Syne2** (6885 aa) is nesprin2.

synemin An intermediate filament protein found in mammalian muscle. Colocalizes with *desmin near myofibrillar *Z-discs. Three synemin isoforms (339 aa, 1251 aa, and 1563 aa) are regulated differently during development.

synergistic Describing two things (e.g. drugs) that interact to produce an effect that is more than additive. Often applied very loosely: demonstrating true synergy requires that there is an additional response when one of the two effectors is already producing its maximal effect.

synexin *See* ANNEXINS (annexin 7).

syngamy The fusion of two haploid nuclei to produce the diploid zygote nucleus.

syngeneic Describing organisms that do not differ in histocompatibility antigens and will accept grafts from each other. Monozygotic twins are syngeneic, as are some highly inbred strains of animals.

synkaryon A hybrid cell with a nucleus containing chromosomes from two different somatic cells. *Compare* HETEROKARYON.

synoviocytes Fibroblastic cells that form the synovial membrane that lines joints. They produce both the synovial fluid and the extracellular matrix of the bearing surface.

synovium The connective tissue that forms the bearing surface of a joint. Arthritis is the erosion of the synovium.

synphilin A neural protein (synphilin-1, 919 aa with alternatively spliced isoforms) that interacts with α-*synuclein, and is ubiquitinylated by *parkin to produce cytoplasmic inclusions similar to *Lewy bodies. Mutation has been associated with *Parkinson's disease.

synpolydactyly (syndactyly) A condition in which adjacent digits are joined, either by soft tissue webbing or by fusion of some bones. **Synpolydactyly-1** (SPD1) is caused by mutation in the *HOXD13* gene, **synpolydactyly-2** by mutation in the *fibulin-1 gene, and **SPD3** by a different mutation that has been mapped to chromosome 14q11.2-q12.

syntaxins (t-SNAREs) An extensive family of integral membrane proteins that are the receptors (*SNAREs) for intracellular transport vesicles. They bind *synaptotagmin in a calcium-dependent fashion and interact with voltage-dependent calcium and potassium channels. **Syntaxin 1** (288 aa) functions specifically in neurotransmitter release in the brain, whereas **syntaxins 2, 3**, and **4** have a wider tissue distribution. Syntaxin-1a and -4 preferentially interact with β- and β-*taxilin respectively. **Syntaxin 5** is a Golgi-localized *SNARE protein that has been shown to be required for endoplasmic reticulum–Golgi traffic. Syntaxins are the target for *botulinum neurotoxin type C.

syntenic Describing genes on the same chromosome. **Shared synteny** is the colocalization of genes on chromosomes of related species, which may be because the preservation of *linkage is advantageous.

syntenin *See* SYNDECANS.

synthetase An enzyme of class 6 in the *E classification that couples the breakdown of nucleotide triphosphate to the synthesis of other molecules. *See* LIGASES.

syntrophins A family of peripheral membrane proteins involved in the *dystrophin-associated protein complex together with *dystrophin, *dystrobrevin, and diacylglycerol kinase (DGK-ζ) in muscle and nerve. There are three different but highly conserved syntrophin isoforms (~500 aa), differentially expressed. The DGK-ζ, syntrophin, and Rac1 complex controls polarized outgrowth in neuronal cells.

synuclein A family of structurally related proteins (α, β, and γ-synuclein) that are abundant in neurons and form aggregates in various neurodegenerative diseases. α-Synuclein (140 aa) is a component of plaque amyloid in *Alzheimer's disease and accumulates in *Lewy bodies, and an α-synuclein allele is linked to various familial cases of *Parkinson's disease, interacts with *synphilin, and is degraded by *neurosin.

Syp *See* SHP (2).

syphilis A sexually transmitted infection with the spirochaete *Treponema pallidum*.

systematic sclerosis *See* SCLERODERMA.

systemic Describing something that occurs throughout the body, not just locally.

systemic carnitine deficiency A disorder in which there are low carnitine concentrations in tissues other than muscle, caused by mutations in the *SLC22A5* gene which encodes *OctN2, leading to increased urinary losses of carnitine. **Myopathic carnitine deficiency** is caused by mutation in a different locus and the carnitine deficiency is restricted to muscle.

systemic lupus erythematosus A chronic remitting, relapsing, inflammatory, and often febrile multisystemic disorder of connective tissue with an autoimmune basis. The classic symptom is a facial butterfly rash, although this can occur with other forms of lupus (*see* LUPUS ERYTHEMATOSUS). The antigens are often nuclear (often anti-RNA) and multiple suceptibility loci have been identified. Resistance to systemic lupus erythematosus is associated with a polymorphism in the *toll-like receptor-5 gene.

systems biology An integrated approach to biology that looks at effects on the whole organism and avoids the reductionism of which molecular bioscience is sometimes accused. Practitioners need to be able to handle the jargon of a range of biological subspecialities, and this dictionary may be helpful to them.

T_3, T_4 *See* THYROID HORMONES.

Tacaribe complex A group of *Arenaviridae that cause severe haemorrhagic disease in humans and have been isolated from bats in South America. There are three lineages: **A** (Flexal, Parana, Pichinde, and Tamiami viruses), **B** (Amapari, Guanarito, Junin, Machupo, Sabia, and Tacaribe viruses) and **C** (Latino and Oliveros viruses). Most of the B lineage are highly pathogenic.

TACCs A family of proteins (transforming, acidic, coiled-coil-containing proteins) that are rich in serine, proline, and acidic residues, have nuclear localization signals but no DNA- or RNA-binding domains. Expression of **TACC1** (805 aa) in stable cell lines induces a transformed phenotype and anchorage-independent growth. TACC1 interacts with products of CHTOG (colonic and hepatic tumour overexpressed gene, cytoskeleton-associated protein 5, CKAP5) and GAS41 (glioma-amplified sequence 41). **TACC2** (652 aa) is very similar to TACC1; **TACC3** (838 aa) lacks nuclear localization signals and is strongly expressed in most tumour cell lines tested. Phosphorylation of TACC3 by *aurora A is essential for proper localization of TACC3 to centrosomes and proximal mitotic spindles.

TACE A zinc metallopeptidase (TNFα converting enzyme, ADAM17, EC 3.4.24.-, 827 aa) of the *ADAM family that releases soluble TNFα from the inactive membrane-bound precursor. It is also involved in cleavage and release of *fractalkine (CX3CL1) and in *notch signalling.

tachycardia An abnormal increase in heart rate. Usually refers to the ventricular rate, but can also refer to an elevated atrial rate while the ventricular rate remains unaffected. **Ventricular tachycardia** (VT) is a potentially life-threatening arrhythmia that can progress to ventricular fibrillation and sudden death. The abnormal heart beats may be uniform (monomorphic VT) or show variation from beat to beat (polymorphic VT or torsade de pointes).

tachykinins A group of neuropeptide hormones encoded by two genes, one encoding a precursor containing both *substance P and neurokinin A (formerly substance K), the other encoding a precursor containing only neurokinin B. All have ten or eleven residues with a common -FXGLM-NH_2 ending. They have effects similar to *bradykinin and *serotonin and act on a range of tissues.

tachyphylaxis A diminished response to an agonist following repeated exposure.

tachyzoite An asexual proliferative stage in the life cycle of coccidia such as *Toxoplasma gondii* that occurs within a parasitophorous vacuole in an infected cell.

tacrolimus *See* FK506.

TAF **1.** *See* TUMOUR ANGIOGENESIS FACTOR. **2.** *TATA-binding protein-associated factor.

taicatoxin An oligomeric protein toxin, from the taipan snake *Oxyuranus scutelatus scutelatus*, that blocks voltage-dependent L-type calcium channels and small-conductance calcium-activated potassium channels. The oligomer has three components, a neurotoxin-like peptide (8 kDa), a phospholipase (EC 3.1.1.4, probably *taipoxin), and a serine-peptidase inhibitor (88 aa).

taipoxin Heterotrimeric phospholipase toxin (EC 3.1.1.4), from the taipan snake *Oxyuranus scutelatus scutelatus*. All three subunits (α, β, γ) have homology with pancreatic phospholipase A2. It binds to neuronal *pentraxins and blocks release of acetylcholine at the neuromuscular junction. Said to be the most potent vertebrate toxin.

talin A protein (2540 aa) that links *vinculin to integrins and thus the cytoskeleton to the extracellular matrix. Has a *FERM domain. A second talin gene has been identified.

Tamiami virus *See* TACARIBE COMPLEX.

Tamiflu *See* OSELTAMIVIR.

tamoxifen A synthetic antioestrogen used to treat oestrogen-receptor-positive breast carcinoma. *TNs* Istubal, Nolvadex, Valodex.

tamsulosin A selective α_1 alpha-blocker used to treat benign prostatic hyperplasia. TN Flomax.

tandem pore domain channels A family of potassium channels in which each α subunit has two pore-forming domains and the standard tetrameric structure of potassium channels requires only dimerization. They form 'leak channels'. **TWIK1** (product of the *KCNK1* gene; 336 aa) is weakly inwardly rectifying. **TREK1** (*KCNK2* product; TWIK-related potassium channel) is expressed throughout the central nervous system, is opened by polyunsaturated fatty acids, lysophospholipids, and volatile anesthetics, and is inhibited by neurotransmitters that increase intracellular cAMP or activate the Gq signalling pathway. **TASK** (*KCNK3* product; TWIK-related acid-sensitive potassium channel) has a role in cellular responses to changes in extracellular pH. **TRAAK** (*KCNK4* product) is a polyunsaturated fatty acid activated and mechanosensitive channel, highly expressed in brain and placenta.

tandem repeats Copies of DNA sequences (sometimes whole genes) or protein motifs that are adjacent to one another. *See* SATELLITE DNA.

Tangier disease A disorder (high density lipoprotein (HDL) deficiency type 1) in which there are low levels of HDL in plasma; hypercholesterolaemia; enlarged liver, spleen, and lymph nodes; and peripheral neuropathy. The cause is a mutation in the ATP-binding cassette-1 gene (*ABC1*) (familial HDL deficiency is also caused by mutation in this gene) which leads indirectly to hypercatabolism of HDL. *See also* LXR.

tankyrase-1 A *poly (ADP-ribose) polymerase (TRF1-interacting ankyrin-related ADP-ribose polymerase, 1327 aa) that regulates the binding of *telomeric repeat binding protein 1 (TRF1) to telomeric DNA.

tannic acid A soluble tannin used as a contrast-enhancing agent in electron microscopy.

T antigens Virally encoded proteins orginally detected as tumour (T) antigens using antisera from animals with tumours. They are associated with cell transformation and viral replication. In SV-40 there are two, small-t and large-T, the former having a range of effects, possibly through stimulating NFκB-responsive genes, the latter essential for normal viral replication. They are expressed in nonpermissive cells transformed by these viruses. Polyoma virus has three T antigens, large, middle, and small.

tapasin A transmembrane protein (428 aa) of the immunoglobulin superfamily that is required for the interaction of MHC class I with the transporters associated with antigen processing (*TAPs).

tapetum A layer of reflective tissue just behind the pigmented retinal epithelium of many vertebrate eyes, especially in nocturnal animals. Absent in humans.

TAPs 1. Transporters associated with antigen processing: *ABC proteins that transport protein fragments across endoplasmic reticulum membranes during antigen processing. TAP is a heterodimer of TAP1 and TAP2, both having transmembrane and nucleotide-binding domains. TAP-like proteins have also been found. *See* TAPASIN. **2.** Tap protein is one of the evolutionarily conserved nuclear RNA export factor (NXF) family that mediates the sequence-nonspecific nuclear export of cellular mRNAs as well as the sequence-specific export of retroviral mRNAs bearing the *constitutive transport element (CTE). Binds to *nucleoporins and *NXT1.

Taq polymerase A heat-stable *DNA polymerase with a temperature optimum for activity around 75–80 °C, isolated from *Thermus aquaticus*. Used in the *polymerase chain reaction although it is being replaced with other heat-stable enzymes with greater replication fidelity.

TAR A region (*trans*-activating-responsive region) of nontranslated mRNA at the 5′ end of the virion RNA of *HIV that binds the transactivator protein (*tat).

tardive dyskinesia A condition in which there are involuntary repetitive movements, often a side effect of antipsychotic drugs.

target cell (codocyte) An abnormal erythrocyte that resembles a target with a central bullseye. They have an increased surface area to volume ratio and can be a symptom of liver disease, iron deficiency, or *thalassaemia.

targeting signal A peptide motif that determines where a protein will be located, e.g. *nuclear localization signals. *See* KDEL.

TAR syndrome A rare disorder (thrombocytopenia–absent radii) in which there is thrombocytopenia and bilateral aplasia of the radius of the arm. Generally considered a contiguous gene deletion syndrome.

t

Tarui disease *See* GLYCOGEN STORAGE DISEASES.

tasidotin *See* DOLASTATINS.

taspase-1 An aspartyl endopeptidase (threonine aspartase 1, clan PB, EC 3.4.25.-, 420 aa) responsible for cleaving the histone methyltransferase *MLL.

tastin *See* TROPHININ.

TATA box (Goldberg–Hogness box) A consensus sequence ~25 nucleotides before the site of initiation of transcription in most genes that are transcribed by *RNA polymerase II. The **TATA-binding protein** (TBP, 339 aa) is a component of the RNA polymerase II transcription factor D (*TFIID) complex and binds to the consensus 5′-TATAAAA-3′ sequence, thereby positioning the polymerase. *See also* SL1. *Spinocerebellar ataxia type 17 is caused by an expanded polymorphic polyglutamine-encoding trinucleotide repeat in the gene for TBP.

tat protein A *transactivator protein that recognizes the *TAR sequence on viral RNA. Peptides derived from tat are potent neurotoxins.

tauopathy Any neuropathy in which there are brain lesions with an excess of *tau protein. In Alzheimer's disease the neuronal cytoskeleton is progressively disrupted and replaced by neurofibrillary tangles of paired helical filaments (PHFs) composed mainly of hyperphosphorylated forms of tau. Mutations in tau are associated with familial frontotemporal dementia with parkinsonism, and with *Pick's disease.

tau protein (tau factor) A protein (352 aa with multiple isoforms) that copurifies with *tubulin and coassembles into microtubules. It has tandem repeats of a tubulin binding domain. Tau isoforms are found in all cells and are highly expressed in neurons where they associate with axonal microtubules. *See* MAPs; TAUOPATHY. **Tau tubulin kinase** (1244 aa) is a casein kinase type enzyme that will phosphorylate both tau and tubulin. Mutations are associated with *spinocerebellar ataxia (SCA11).

taurine An organic acid derived from cysteine, found in bile and in the cytoplasm of some cells (notably neutrophils) at high concentrations. Taurocholate is a bile salt formed by conjugation of taurine with cholate.

tautomycin An antibiotic, isolated from *Streptomyces spiroverticillatus*, that inhibits *protein phosphatases type 1 and type 2a.

tax-1 *See* CONTACTINS.

taxanes Cytotoxic drugs, originally derived from *Taxus* sp. (yew trees), that inhibit mitosis by stabilizing GDP-tubulin and blocking microtubule disassembly. Best known is *paclitaxel.

taxilins Proteins that interact selectively with *syntaxins and are involved in the targeting of intracellular vesicles. **α-Taxilin** (formerly IL-14, 546 aa) interacts with syntaxin-3 but also with the *nascent polypeptide-associated complex. **β-Taxilin** (684 aa) interacts with syntaxin-1a and -4, but not other syntaxins. **γ-Taxilin** (528 aa) preferentially interacts with syntaxin-4A, but is also reported to be a transcriptional regulator of osteoblast function.

taxis A response to vectorial environmental cues that affects, positively or negatively, the direction of movement. *Compare* KINESIS.

tax oncoprotein A protein (*trans*-activating transcriptional regulatory protein of HTLV-1, 353 aa) encoded by human T-cell leukaemia virus type 1 (HTLV-1) and involved in transcriptional regulation, cell cycle control, and transformation. Interacts with *TIP1 (Tax interacting protein 1).

Tay–Sachs disease A *storage disease in which lysosomal hexosaminidase A (EC 3.2.1.52), which degrades *ganglioside GM2, is absent. Mostly affects brain, where ganglion cells become swollen and die, and is usually fatal within the first few years of life.

tazarotene A synthetic *retinoid that is selective for retinoic acid receptor β/γ. *See* CHEMERIN.

TB *See* TUBERCULOSIS.

TBC domains The GTPase-activating (catalytic) domains (Tre-2/Bub2/Cdc16-domains) of G proteins of the *rab family.

T-box genes A family of highly conserved genes encoding transcription factors that all have a DNA-binding T-domain. Mutations cause developmental defects. *See* BRACHYURY; THYMIC HYPOPLASIA; VELOCARDIOFACIAL SYNDROME.

TBP **1.** *TATA-binding protein. **2.** *Thioredoxin binding protein-2 (TBP-2).

TC10 A *rho-family small GTPase (205–213 aa) involved in insulin-stimulated glucose uptake and GLUT4 translocation.

TCA cycle *See* TRICARBOXYLIC ACID CYCLE.

T cell (T lymphocyte) One of the two major lymphocyte classes, those that are of thymic origin. T cells are involved in cell-mediated immune responses and regulate B-cell development. They have antigen receptors (*see* T-CELL RECEPTOR) but not Fc or C3b receptors. Various subsets are now recognized: *T helper cells, *T regulatory cells, *cytotoxic T-cells, *regulatory T cells, *gamma-delta cells, *Th17 cells, *T memory cells.

T-cell factor (TCF) A family of transcription factors (lymphoid enhancer-binding factor (LEF), LEF/TCFs) that are important in the *wnt/β-*catenin signalling cascade. The nomenclature is very confusing: TCF1 (T-cell specific transcription factor) is officially TCF7 (transcription factor 7) and is actually found in many tissues, not just T cells. The 'official' TCF1 (transcription factor 1) is hepatocyte nuclear factor 1 (*HNF1, a homeobox transcription factor involved in regulating liver-specific genes and mutated in maturity onset diabetes of the young. TCF1α (LEF1) is expressed in pre-B and -T cells and mutations are associated with sebaceous tumours. TCF4 (T-cell transcription factor 4, TCF7-like 2) is implicated in blood glucose homeostasis. Activation of various genes by the β-catenin/TCF4 complex is a key step in tumorigenesis and variations in the *TCF4* gene are associated with susceptibility to type 2 diabetes.

T-cell growth factor *See* INTERLEUKIN-2.

T-cell leukaemia/lymphoma viruses *See* HTLV.

T-cell receptor (TCR) The antigen-recognizing receptor of *T cells that binds antigen in association with the *MHC, leading to the activation of the cell. The major class of TCR has α and β subunits, both of the immunoglobulin superfamily and with variable and constant regions produced by recombination to generate diversity in antigen recognition. A minority have γ and δ subunits and are not *MHC-restricted. The γδ T-cell receptors (TCRs) are formed on very early T cells in the thymus. The **T-cell receptor complex** consists of the TCR and accessory proteins (the gamma, delta, epsilon, and zeta components of the *CD3 complex, also of the immunoglobulin superfamily but invariant, and the coreceptor, CD4).

$TCID_{50}$ (50% tissue culture infective dose) A figure derived from an assay, usually for viruses, in which serial dilutions of the test solution are added to wells containing susceptible cells; the titre that causes infection in half the wells.

TCOF1 *See* TREACLE.

TCP-1 A chaperonin complex (T-complex polypeptide 1, chaperonin containing T-complex, CCT) consisting of eight related ~60-kDa proteins arranged into two stacked multimeric rings, with a central cavity on each side of the molecule, that is important for the correct folding of a range of molecules, including tubulin and actin. CCT is localized to the leading edge of fibroblasts and neurons. Mutations in elements of the complex are suspected to be responsible for some developmental abnormalities, especially neural tube defects.

TCR *See* T-CELL RECEPTOR.

TDG *See* THYMINE-DNA GLYCOSYLASE.

t-DNA DNA coding for *transfer RNA (tRNA).

TDP-43 A highly conserved heterogeneous nuclear ribonucleoprotein (TAR DNA-binding protein-43, 43 kDa, 414 aa) that binds to the transcription-activating response region (*see* TAR). It is a major component of ubiquitin-positive, tau-negative inclusions in AMYOTROPHIC LATERAL SCLEROSIS (ALS), and frontotemporal lobar degeneration (FTLD-U).

TDT 1. Transmission disequilibrium test. A test for linkage between a genetic marker and a disease susceptibility locus. **2.** *See* TERMINAL DEOXYNUCLEOTIDYL TRANSFERASE (TdT).

Tec family A family of protein *tyrosine kinases involved in signalling, mostly in haematopoietic cells. Members include Tec, found in T cells, B cells, and liver cells; Bruton's tyrosine kinase (Btk); and *interleukin-2-inducible T-cell kinase (Itk/Emt/Tsk).

teichoic acids Acidic polymers crosslinked to peptidoglycan in the cell walls of Gram-positive bacteria. *See* LIPOTEICHOIC ACID.

teicoplanins A group of glycopeptide antibiotics used to treat *meticillin-resistant *Staphlococcus aureus*. They interfere with cell wall formation in Gram-positive bacteria by inhibiting the formation of crosslinkages.

tektins A family of highly conserved filamentous proteins (tektin1, 418 aa; tektin2, 430 aa) associated with axonemal microtubules in spermatozoa. They are structurally similar to some *intermediate filament proteins.

telangiectasia A condition in which capillary vessels are chronically dilated, causing elevated dark red blotches on the skin. *See* ATAXIA TELANGIECTASIA. **Telangiectasia perstans** is a form of *mastocytosis.

telecanthus A condition in which the distance between medial corners (canthi) of the eyes is increased although the interpupillary distance is normal. *Compare* HYPERTELORISM.

telencephalin *See* ICAMs.

telethonin (T-cap) A sarcomeric protein (167 aa) localized in the *Z disc in striated and cardiac muscle that links two titin molecules into an antiparallel sandwich complex and is important in organizing sarcomere structure. It is mutated in some forms of *muscular dystrophy.

telocentric *See* METACENTRIC.

telodendria The branched (dendritic) ends of axons, often slightly enlarged to form synaptic bulbs. A potentially confusing term, best avoided.

telokin The C-terminal 154 aa of *myosin light chain kinase (1914 aa) expressed as an independent protein in smooth muscle. When phosphorylated it inhibits myosin phosphatase, binds to unphosphorylated myosin, and helps maintain smooth muscle relaxation.

telomerase A ribonucleoprotein complex (telomere terminal transferase) that adds multiple telomeric repeats (TTAGGG) to the 3′ end of DNA by using an RNA template, the **telomerase RNA component** (TERC: encoded by a 451-nucleotide gene). It can be considered a reverse transcriptase, and **telomerase reverse transcriptase** (TERT; EC 2.7.7.49) is the catalytic subunit. There are also **telomerase associated proteins** (*see* TELOMERIC REPEAT BINDING FACTORS; SHELTERIN). Telomerase will only elongate oligonucleotides from the telomere and not other sequences. Ectopic expression of telomerase in normal human cells extends their replicative lifespan, although does not necessarily transform them, but some oncogenes, such as *myc*, stimulate expression of *TERT*. Mutations in the *TERC* or *TERT* genes cause short telomeres in congenital aplastic anaemia of *dyskeratosis congenita and some cases of acquired aplastic anaemia, and increased susceptibility to adult-onset idiopathic pulmonary fibrosis. Heterozygous deletion of *TERT* is found in patients with *cri-du-chat syndrome.

telomere A region of repetitive DNA at the end of a chromosome, which protects it from destruction during DNA replication. Each division cycle depletes the telomere repeats, which are partially replenished by the action of *telomerase, but as cells age their telomeres progressively shorten. A complex regulatory system controls telomere length. *See* SHELTERIN.

telomeric repeat amplification protocol assay An assay (TRAP assay) for *telomerase activity in cells using a synthetic biotinylated oligonucleotide as substrate and then amplifying the isolated products by PCR. The products differ by 6-bp increments, each increment being a telomeric repeat.

telomeric repeat binding factors Proteins (TRFs) that bind to the TTAGGG tandem repeats of *telomeres as homodimers, recruit other proteins into the *shelterin complex, and control telomere length by inhibiting *telomerase. **TRF1** (439 aa) is related to the proto-oncogene **myb*. **TRF2** is similar and coexpressed: it is presumed to have a slightly different role. TRF2 dysfunction results in the exposure of the telomere ends and activation of ATM (*ataxia telangiectasia mutated)-mediated DNA damage response. The binding of TRF1 to DNA is inhibited by the *poly(ADP-ribose) polymerase (PARP) activity of *tankyrase-1.

telomestatin A *telomerase inhibitor, isolated from *Streptomyces anulatus* 3533-SV4. It causes telomeric repeat factor 2 (TRF2) to dissociate from the telomeres in cancer cells.

telopeptides Terminal regions of proteins that are removed during maturation; generally refers to the N- and C-terminal telopeptides of procollagen that contribute to generating the quaternary structure and are subsequently removed by *procollagen peptidases.

telophase The final stage of mitosis or meiosis at which chromosome separation is complete.

temazepam A benzodiazepine used to treat insomnia and as a premedication before minor surgery.

temozolomide A prodrug that produces MTIC (3-methyl-(triazen-1-yl)imidazole-4-carboxamide), a potent alkylating agent that will cross the blood–brain barrier. Used in the treatment of recurrent malignant glioma. *TN* Temodar.

temperate phage A bacteriophage that integrates its DNA into the host genome (*lysogeny) and does not proliferate and cause lysis (the *lytic cycle).

temperature-sensitive mutation A conditional mutation (ts mutation), one that is only expressed at certain temperatures, usually because the nonpermissive temperature causes conformational instability in an essential protein. By shifting cells in culture to the permissive temperature the gene is 'switched' on or off, depending on the temperature-sensitive gene involved.

template A structure that determines the patterning of a second structure by some sort of physical interaction. Although this is certainly true of the semiconservative replication of DNA, it can also be true for more complex structures for which there is continuity between generations.

temporal sensing A mechanism for sensing a gradient of some environmental factor that involves sampling the current value with the value before having moved. An increase indicates that the movement was been up-gradient. Until the first pair of readings have been compared there is no directional information and movement should be random. Bacterial chemotaxis (so called) is based on this mechanism.

tenascins Extracellular matrix proteins (~240 kDa) that form six-armed hexamers (hexabrachions). The tenascins have different numbers of fibronectin and EGF repeats and are important determinants of matrix properties, regulated temporally and spatially during development. Four tenascin genes encode tenascin-C, tenascin-R, tenascin-X, and tenascin-W. **Tenascin C** (cytotactin, myotendinous antigen) was the first to be described and is found in tendons and embryonic extracellular matrix. **Tenascin R** is found primarily in the central nervous system. **Tenascin X** was first identified as 'gene X' in the MHC class III gene region, and mutations in the gene cause autosomal recessive *Ehlers–Danlos-like syndrome. **Tenascin W** is expressed in kidney and at sites of bone and smooth muscle development. **Tenascins M1–M4** have homology with *Drosophila* ODZ and mouse Doc, proteins involved in developmental signalling. Tenascin M1 (teneurin-1), for example, is found only in fetal brain.

tenofovir An anti-AIDS drug, a nucleotide analogue that inhibits viral reverse transcriptase.

tenoxicam A *NSAID. *TN* Mobiflex.

tensin An actin-binding protein (tensin1, 1735 aa) found in some *focal adhesions possibly regulating cell movement. It has an *SH2 domain and interacts with PI3-kinase and JNK signalling pathways. Other tensins (tensins 2–4) have variable tissue distribution. *See* PTEN.

TEP1 TGFα-regulated and epithelial cell-enriched phosphatase-1. *See* PTEN.

teratocarcinoma A malignant tumour (teratoma) that contains tissue normally derived from ectoderm, mesoderm, and endoderm (e.g. muscle, cartilage, nerve, tooth buds, and various glands) and thought therefore to be neoplastic primordial germ cells or misplaced blastomeres. There are also undifferentiated, pluripotent epithelial cells, *embryonal carcinoma cells.

teratogen An agent that causes developmental abnormalities; teratogens are thought to be responsible for ~10% of birth defects. The notorious example is *thalidomide, but many mutagens have this capacity. The effect of a teratogen is affected by fetal genotype, the stage of development at which exposure occurs, the tissue specificity of the teratogen, and the dose.

teratoma *See* TERATOCARCINOMA.

terbinafine An antifungal drug that inhibits ergosterol synthesis by inhibiting squalene epoxidase, an enzyme involved in synthesis of the fungal cell membrane.

terbutaline A β_2-adrenergic receptor agonist, used as a fast-acting bronchodilator for asthma and to delay premature labour. *TNs* Brethine, Brethaire, Bricanyl.

terlipressin A vasopressin analogue used in management of hypotension. *TN* Glypressin.

terminal buttons Small swellings at the axonal terminus that are rich in synaptic vesicles which can be released to signal to the postsynaptic cell.

terminal cisternae Portions of the *sarcoplasmic reticulum closely apposed to *T tubules in which calcium is stored and from which it is released upon receipt of a stimulus.

terminal deoxynucleotidyl transferase A DNA polymerase (EC 2.7.7.31, 509 aa) that catalyses the addition of deoxyribonucleotides to the 3′-OH end of DNA primers without a template being involved. It is involved in the addition of nucleotides at the junction (N region) of rearranged Ig heavy chain and T-cell receptor gene segments during the maturation of B and T cells. It is found only in immature lymphocytes and acute lymphoblastic leukaemia cells.

terminal web The cytoplasmic region that lies immediately below the microvilli in intestinal epithelial cells, rich in microfilaments derived both from the core of the microvillus and from *adherens junctions that link the epithelial cells. Also present are myosin and other actomyosin motor-related proteins.

termination codons The three codons, **ochre** (UAA), **amber** (UAG), and **opal** (UGA), that do not code for an amino acid but act as termination signals for protein synthesis. They are not recognized by a tRNA and termination is catalysed by *release factors. An **ochre mutation** is one that changes a codon into ochre (analogously amber, opal); an **ochre suppressor** is a gene that encodes a tRNA that recognizes the ochre codon and allows continuation of protein synthesis. Ochre suppressors will also suppress amber codons.

TERT *See* TELOMERASE.

tertiary structure *See* PRIMARY STRUCTURE.

tes The protein product of a putative tumour suppressor gene, similar to *zyxin, and a component of focal adhesions where it interacts with actin, *mena, and *vasodilator stimulated phosphoprotein (VASP). Expression negatively regulates proliferation of T47D breast carcinoma cells.

testicans A family of calcium-binding proteoglycans (SPARC/osteonectin, CWCV and kazal-like domain proteoglycans, SPOCKs) found in the extracellular matrix of brain. They are related to proteins involved in adhesion, migration, and cell proliferation. *See* CWCV DOMAIN; KAZAL PROTEINS; OSTEONECTIN.

testicular feminization *See* ANDROGEN.

testosterone An *androgen secreted by the interstitial cells of the testis and to a lesser extent by the ovary. It is an anabolic steroid and triggers the development of sperm and of many secondary sexual characteristics.

tetanolysin A cholesterol-binding pore-forming toxin (haemolysin, 50 kDa) produced by *Clostridium tetani*, the infectious agent of tetanus (lockjaw).

tetanospasmin *See* TETANUS TOXIN.

tetanus toxin (tetanospasmin) An *AB toxin produced by *Clostridium tetani* and responsible for the pathology in tetanus. The toxin (EC 3.4.24.68, 1315 aa) is activated by cleavage to form disulphide-linked heavy (100 kDa) and light (50 kDa) chains. The heavy B chain binds to disialogangliosides and forms a pore through which the light A chain, a zinc endopeptidase, enters cells where it specifically attacks *synaptobrevin and blocks neurotransmission. *See also* BOTULINUM TOXIN. **Tetanus antitoxin** is an antibody to tetanus toxin, usually from horses that have been hyperimmunized. Immune response to horse serum causes serum sickness.

tetrabenazine A drug used to treat hyperkinetic disorders (e.g. *Huntington's chorea, Tourette's syndrome, and tardive dyskinesia). It inhibits the *vesicular monoamine transporter (VMAT) and affects dopamine levels. *TNs* Nitoman, Xenazine.

tetracaine (amethocaine) A potent local anaesthetic. Used experimentally to inhibit the *ryanodine receptor.

tetracyclines A group of antibiotics produced by *Strepomyces* spp., effective against a wide range of bacteria.

tetrad Any group of four objects, often encountered referring to the four homologous chromatids paired during the first meiotic prophase.

tetraethylammonium ion A monovalent cation often used in experimental neurophysiology as a specific blocker of potassium channels.

tetrahydrobiopterin *See* BIOPTERIN.

tetrahydrocannabinol (THC) One of the more psychoactive *cannabinoids.

tetrahydrogestrinone (THG) A synthetic steroid used illicitly by some athletes as a performance-enhancing drug.

tetralogy of Fallot A relatively common type of congenital heart disease, occurring in 1 in 3000 live births, where there is an incomplete septum between left and right ventricles and the baby is cyanotic (blue baby). Can be caused by mutations in the human homologue of rat Jagged-1 (a ligand of the notch receptor), or in the gene encoding the cardiac-specific homeobox NKX2.5 that is encoded by the *CSX* gene.

tetranectin A tetrameric protein in human plasma (~10 mg/L) with four identical polypeptide chains (181 aa each), noncovalently bound. May regulate proteolytic processes by binding to plasminogen kringle 4. Plasma levels are reduced in many malignancies.

tetraploid A nucleus, cell, or organism with four copies of the normal *haploid complement of chromosomes.

tetraspan vesicle membrane proteins A widespread and abundant family of integral membrane proteins (tetraspanins) that have four transmembrane regions and both ends located in the cytoplasm. They interact with one another and with other membrane proteins to produce specific membrane microdomains in vesicles that shuttle between various membranous

t

compartments. They can be subdivided into three distinct groups: *physins, gyrins such as *synaptogyrin, and *secretory carrier-associated membrane proteins (SCAMPs). The **tetraspanin web** is the network of TVPs (tetraspanins) and their partner proteins (*ADAMs and *integrins) that facilitates cellular interactions and cell fusion.

tetrocarcin An antibiotic (tetrocarcin-A) isolated from actinomycetes (*Micromonospora* sp.) that inhibits the antiapoptotic function of *Bcl-2 and has potential as an antitumour drug.

tetrodotoxin A potent *neurotoxin (TTX, 319 Da), with no known antidote, that binds to the voltage-gated sodium channel and blocks action potentials. Originally isolated from the Japanese puffer fish *Fugu rubripes*, and found in various other marine animals (where it was called tarichatoxin and maculotoxin), but is actually a bacterial product from various species of *Vibrio* and other bacteria. *Compare* SAXITOXIN.

Teunissen–Cremers syndrome *See* ANKYLOSIS.

T even phages Bacteriophages (T2, T4,T6) of enterobacteria that are serologically related and have large dsDNA genomes. The T odd phages (T1, T3, T5, T7) are more diverse.

Texas red Rhodamine-type fluorophore often conjugated to antibodies for use in *flow cytometry.

textilinin A Kunitz-type serine peptidase inhibitor (59 aa), from the venom of *Pseudonaja textilis*, that is a potent inhibitor of *plasmin and slows fibrinolysis. *See* TEXTILOTOXIN.

textilotoxin A multimeric phospholipase A2 neurotoxin (subunits of ~120 aa), from the venom of the Australian eastern brown snake *Pseudonaja textilis*, that blocks neuromuscular transmission by interfering with the release of synaptic vesicles. Now thought to have six subunits. *See* TEXTILININ.

t

TFG **1.** *Trigonella foenum-graecum* (fenugreek; TFG), leaf extracts of which have analgesic, anti-inflammatory, and antipyretic effects in experimental models. **2.** The two larger subunits of the *TFIIF complex, TFG1 and TFG2. **3.** *See* TRK (TRK-fused gene). **4.** A surprisingly common typographical error for TGF.

TFII General transcription factors (TFIIA, TFIIB, TFIID, TFIIE, TFIIF, TFIIG/TFIIJ, and TFIIH) involved in the activity of *RNA polymerase II and formation of the RNA polymerase II preinitiation complex. **TFIID** is a multicomponent transcription factor that recognizes and binds to the TATA-box promoter with the *TATA-binding protein (TBP) subunit and has several TBP-associated factors (or TAFs). TFIID nucleates the transcription complex, recruiting the rest of the factors through a direct interaction with TFIIB. **TFIIA** interacts with the TBP subunit of TFIID and aids in the binding of TBP. **TFIIB** forms a complex with TFIID and TFIIA and bridges between IID and the polymerase. Mutations in TFIIB cause a shift in the transcription start site. **TFIIE** regulates initiation of transcription, **TFIIF** helps recruit polymerase II to the initiation complex in collaboration with TFIIB. **TFIIH** consists of ten subunits: some have helicase and ATPase activities, some phosphorylate serine residues in the polymerase, others are associated with cell cycle control. TFIIH is also involved in nucleotide excision repair. Other components (**TFIIG, TFIIJ**) also associate with the complex. **TFIIX** is a general name for various TBP accessory proteins.

TFIII Transcription factors for genes that are transcribed by *RNA polymerase III. The initiation complexes are multisubunit and are assembled in a stepwise fashion. **TFIIIC**, itself a multisubunit complex, binds to promoters of tRNA genes and mediates binding of **TFIIIB**, which consists of three polypeptides including the *TATA-binding protein, and TFIIIB recruits polymerase III. **TFIIIA** is a single-subunit protein that binds the promoters of the 5S RNA genes and, having done so, forms a complex with TFIIIC, which can then interact with TFIIIB and polymerase as with the tRNA initiation complex.

TFP **1.** *See* TRIFLUOPERAZINE. **2.** *See* MITOCHONDRIAL TRIFUNCTIONAL PROTEIN (TFP). **3.** Type IV pili (Tfp) of various bacteria.

TGF *See* TRANSFORMING GROWTH FACTOR.

TGN *See* TRANS-GOLGI NETWORK.

Th17 cells A lineage of $CD4^+$ effector *T-helper (Th) cells that selectively produce *interleukin-17 and are involved in autoimmune disease and mucosal immunity.

thalassaemia A haemoglobinopathy widespread in Mediterranean countries in which mutations in globin genes lead to reduced haemoglobin production and anaemia. In α-thalassemiaemias the mutation is in the globin α chain, in β-thalassemiaemias in the β chain. The relatively high incidence is probably because heterozygosity provides some protection against malaria.

thale cress *See* ARABIDOPSIS THALLIANA.

thalidomide A sedative drug that was used as an antiemetic in pregnancy until it was recognized as a potent *teratogen that could cause phocomelia (reduced limbs) or amelia (absence of limbs) if taken between the third and fifth weeks of pregnancy. Thalidomide inhibits TNFα production and has anti-inflammatory properties. It is used (under careful control) to treat leprosy and multiple myeloma. Although only the *S*-stereoisomer has the teratogenic effect, the drug racemizes in the body so the nonteratogenic (but sedative) *R*-isomer cannot safely be used.

thapsigargin An inhibitor of the sarcoplasmic calcium ATPase (*SERCA), isolated from the Drias plant *Thapsia garganica.* It is experimentally useful, being cell permeable, and acts independently of InsP3.

thaumatin An intensely sweet protein from the West African katemfe fruit *Thaumatococcus daniellii.*

THC *See* TETRAHYDROCANNABINOL.

TH domain A domain (Tec homology domain) found in *Tec family protein kinases, probably the region to which the *SH3 domains of other proteins bind.

thebaine (paramorphine) A natural alkaloid of the opium poppy *Papaver somniferum,* chemically similar to both morphine and codeine, but with stimulatory effects. A partial agonist of the μ-*opioid receptor.

T-helper cells (Th cells) Classically two subclasses of $CD4^+$ T-helper cells, Th1 and Th2, have been recognized. **Th1** cells are responsible for clearing intracellular pathogens and are involved in cell-mediated immunity. They produce IL-2, interferon-γ, and TNFα but not IL-4, IL-5, and IL-10. Selective activation of Th1 cells is promoted by interferon-γ and IL-12 and inhibited by IL-4 and IL-10, the products of Th2 cells. **Th2** cells are involved with the humoral immune response, produce IL-4, IL-5, and IL-10, and promote antibody production; IL-4 is essential for growth and differentiation of Th2 cells. There is cross-inhibition between the two classes; if one subclass is activated it will inhibit the activity of the other so that the response is polarized. More recently a third class of Th cells that produce IL-17 in response to autoimmune tissue damage has been identified (*see* TH17 CELLS).

theobromine The major alkaloid (3,7-dimethylxanthine) of the cacao bean, with properties similar to those of THEOPHYLLINE and CAFFEINE. An inhibitor of cAMP phosphodiesterase.

theophylline A nonspecific adenosine antagonist (1,3-dimethylxanthine) and an inhibitor of cAMP phosphodiesterase, used occasionally to treat asthma. Present in tea, though below the therapeutic dosage.

therapeutic cloning The production of a cloned embryo by *somatic cell nuclear transfer in order to use *stem cells from the very early embryo for therapeutic purposes, e.g. in the treatment of Parkinson's disease. *Compare* REPRODUCTIVE CLONING.

therapeutic index The relationship between the levels of a drug that cause toxicity and the dose required to give a therapeutic effect. A common value is the ratio of the concentration that causes toxicity in 50% of the population (TD_{50}) divided by the minimum effective dose for 50% of the population (ED_{50}). It is desirable for the value to be greater than 1.

thermal cycler (thermocycler) A laboratory instrument for conducting automated *polymerase chain reactions (PCR); it subjects the reaction tubes to a sequence of temperature cycles. *See* LIGHTCYCLER.

thermogenin (uncoupling protein 1, UCP) A protein (305 aa) of the inner membrane of mitochondria in *brown adipose tissue. It is a proton transporter and dissipates the proton gradient generated by oxidative phosphorylation and diverts energy from ATP synthesis to thermogenesis (uncouples the system). It is unclear whether polymorphisms in thermogenin have any relationship with 'metabolic efficiency'. Two other uncoupling proteins (UNC2 and UNC3) have been identified; UCP2 is widely expressed, whereas UCP3 expression seems to be restricted to skeletal muscle.

thermolysin A heat-stable zinc metallopeptidase (EC 3.4.24.27, 548 aa) of the M4 clan, produced by *Bacillus thermoproteolyticus.*

Thermus aquaticus An aerobic Gram-negative bacillus, originally found in hot springs in Yellowstone National Park, USA, the source of *Taq polymerase.

theta antigen *See* THY1.

THG *See* TETRAHYDROGESTRINONE.

thiamine *See* VITAMIN B (B_1).

thiazolidine diones A group of oral hypoglycaemic drugs used to treat type 2 diabetes. Examples include pioglitazone, rosiglitazone.

thick filaments Generally, the bipolar *myosin-II filaments (12–14 nm diameter, 1.6 μm long) in the sarcomeres of striated muscle, although may refer to myosin filaments elsewhere. The axis of the filament comprises antiparallel light meromyosin (*LMM) with the myosin heads projecting in a regular fashion at the ends and a central bare zone.

thigmotaxis (*adj.* thigmotactic) A directed response (*taxis) to mechanical contact (touch).

thin filaments The F-actin filaments (7–9 nm diameter) of the *sarcomere. They are attached to the *Z discs of striated muscle and have opposite polarity in each half-sarcomere. The actin is associated with *tropomyosin and *troponin which regulate the interaction with myosin of the *thick filaments.

thiocoraline A bioactive octathiodepsipeptide isolated from a marine actinomycete of the genus *Micromonospora* that has antitumour activity and is effective against Gram-positive bacteria. It binds to supercoiled DNA and inhibits transcription.

thioctic acid *See* LIPOIC ACID.

thiol-activated haemolysins Bacterial exotoxins (oxygen-labile haemolysins) that bind to cholesterol in cell membranes and cause cytolysis by forming ring-like pores. The SH-groups of these toxins must be in the reduced state for activity and oxidation renders them inactive. Examples include *tetanolysin, streptolysin O, θ-toxin (*perfringolysin), *cereolysin.

thiol endopeptidases *See* CYSTEINE (cysteine endopeptidases).

thiopentone (thiopental) A rapid-onset, short-acting barbiturate used as a general anaesthetic. *TN* Sodium Pentathol.

thioredoxin A ubiquitous protein (104 aa) that participates in various redox reactions through the reversible oxidation of its active centre dithiol to a disulphide and catalyses dithiol–disulphide exchange reactions. The active site (Trp-Cys-Gly-Pro-Cys) is common to bacterial and eukaryotic thioredoxins. Secreted by various cells, despite the lack of a *signal sequence. Adult T-cell leukaemia-derived factor is an isoform of thioredoxin and is an autocrine growth factor produced by HTLV-1 or *EBV-transformed T-cells that will upregulate interleukin 2 receptor-a (IL-2Ra).

Thomsen's disease *See* MYOTONIA.

thoracic duct The lymph duct into which lymph from most peripheral lymph nodes drains and through which recirculating lymphocytes return to the blood.

THP-1 cells A monocyte-like cell line derived from the peripheral blood of a 1-year-old boy with acute monocytic leukaemia. They have Fc and C3b receptors and are capable of differentiating in culture into macrophage-like cells.

Th-POK A zinc finger transcription factor (T-helper-inducing POZ/Krüppel-like factor, cKrox, 539 aa) that regulates the commitment of immature T-cell precursors to particular lineages. In mice the absence of Th-POK leads to complete conversion of $CD4^+$ cells to $CD8^+$ cells. It is also involved in development.

threonine (Thr, T) A polar amino acid (119 Da). The hydroxyl side chain can undergo *O*-linked glycosylation and can be phosphorylated by serine/threonine kinases.

threonine peptidases A family of peptidases that includes the *proteosome peptidases, *polycystin-1, and glycosylasparaginase (EC 3.5.1.26). The latter is deficient in *aspartylglycosaminuria.

thrombasthenia A condition in which platelet aggregation is deficient, although platelet adhesion is normal. *See* GLANZMANN'S THROMBASTHENIA.

thrombin (coagulation factor II) A serine endopeptidase of the chymotrypsin family (EC 3.4.21.5) that acts on *fibrinogen to produce *fibrin and activates platelets, leucocytes, and endothelium and mesenchymal cells at sites of vascular injury. Thrombin is produced from *prothrombin by the extrinsic system (*tissue factor + phospholipid) or the intrinsic system (contact of blood with a foreign surface or connective tissue), both of which activate plasma factor X to form factor Xa which cleaves prothrombin. Mutations in prothrombin can prevent normal thrombin production (hypoprothrombinaemia) or generate prothrombin that cannot be cleaved and is inactive (dysprothrombinaemia) and cause bleeding disorders. Elevated plasma prothrombin levels (hyperprothombinaemia) increase the risk of venous thrombosis. There are multiple G-protein-coupled, thrombin-activated receptors which are unusual in that cleavage of a portion of the receptor generates the autologous ligand (protease-activated receptors).

thrombocyte Old name for a platelet.

thrombocythaemia A chronic myeloproliferative disorder in which abnormally high platelet numbers occur in the circulation. Essential thrombocythaemia can be caused by mutation in the *thrombopoietin (*THPO*) gene or the THPO receptor gene. A somatic mutation in the [*JAK2*] gene is common.

thrombocytopenia An abnormally low number of blood platelets, below ~100 000/mm^3 (normal platelet counts are 150 000–450 000/mm^3). **Thrombocytopenic purpura** arises in severe thrombocytopenia and is caused by bleeding into the skin producing small petechial haemorrhages. In **primary thrombocytopenic purpura** an autoimmune mechanism seems to cause platelet destruction; **secondary thrombocytopenic purpura** may be a result of drug-induced type II *hypersensitivity in which platelets are destroyed in a complement-mediated reaction.

thromboglobulin (β-thromboglobulin) A protein (81 aa) released when platelets aggregate, derived from *platelet basic protein by proteolysis.

thrombomodulin A glycoprotein (575 aa) of the luminal surface of endothelial cells that is a high-affinity receptor for *thrombin. The thrombomodulin–thrombin complex activates *protein C which degrades clotting factors V and VIII and acts as an anticoagulant. Mutations cause *thrombophilia and a predisposition to myocardial infarction.

thrombophilia A complex multifactorial trait, susceptibility to venous thrombosis, that can be inherited as a result of mutation in any one of a considerable number of clotting, anticoagulant, or thrombolytic factors including *antithrombin 3, *protein C, *protein S, *factor V, *histidine-rich glycoprotein, *plasminogen, *plasminogen activator inhibitor, *fibrinogen, and *thrombomodulin. Environmental factors are also important.

thromboplastin The traditional name for the reagent used in assaying *prothrombin times, actually several different things including *tissue factor and phospholipid. The reagent is now usually a recombinant preparation of human tissue factor in combination with phospholipids. *See* INTERNATIONAL SENSITIVITY INDEX and INTERNATIONAL NORMALIZED RATIO.

thrombopoietin A growth factor (TPO, thrombopoiesis stimulating factor, TSF, 353 aa) that regulates the proliferation of *megakaryocytes and platelet production (thrombopoiesis). Mutation causes essential *thrombocythaemia. The receptor is c-mpl, a cytokine receptor that activates the JAK-STAT signalling pathway.

thrombosis The formation of a solid aggregate of platelets and fibrin (a **thrombus**), in the lumen of a blood vessel or the heart. A **cerebral thrombosis** is one within a vessel supplying the brain, causing a stroke. A **deep vein thrombosis** occurs in a deep vein, most commonly in the calf of the leg, immobility being one of several risk factors.

thrombospondin-related anonymous protein (TRAP) A protein *invasin of *Plasmodium falciparum* that is essential for sporozoite motility and for liver cell invasion. TRAP is a type 1 membrane protein that possesses multiple adhesive domains in its extracellular region and interacts, through aldolase, with parasite microfilament-based motor system. Related proteins are found in other apicomplexan parasites such as *Toxoplasma gondii*.

thrombospondins A family of five distinct gene products, thrombospondins (TSP) 1–4 and *cartilage oligomeric matrix protein, that are extracellular calcium-binding proteins involved in cell proliferation, adhesion, and migration. **Thrombospondin I** (TSP1, 1170 aa) is a homotrimeric glycoprotein (450 kDa) found in the α granules of platelets, and synthesized and secreted for incorporation into the extracellular matrix by a variety of cells including endothelial cells, fibroblasts, smooth muscle cells, and type II pneumocytes. It binds heparin, and various extracellular matrix components. **TSP 2** (1172 aa) is similar in properties to TSP1. TSP 1 and 2 promote synaptogenesis in the central nervous system. **TSP 3** (956 aa) is a developmentally regulated heparin binding protein. **TSP 4** (961 aa) also has heparin-binding properties but differs in being a potential integrin ligand.

thrombotic thrombocytopenic purpura A congenital condition (TTP, Schulman–Upshaw syndrome) characterized by haemolytic anaemia, thrombocytopenia, diffuse and nonfocal neurologic findings, decreased renal function, and fever; caused by mutation in the *ADAMTS13* gene which encodes the von Willebrand factor-cleaving protease (*see* ADAM FAMILY). **Acquired TTP**, which is usually sporadic, occurs in adults and is caused by an IgG inhibitor of the protease.

thromboxanes Compounds produced from prostaglandin cyclic endoperoxides (which are produced by cyclo-oxygenase acting on arachidonic acid) by thromboxane synthase. Thromboxane A2 (TxA2) is vasoconstrictive, stimulates the aggregation of platelets, the release of granule contents, and the production of more TxA2. Another product of endoperoxide metabolism, *prostacyclin, has the opposite effects to the thromboxanes. Thromboxane receptors are G-protein-coupled.

thrombus *See* THROMBOSIS.

THUMP domain An evolutionarily conserved domain (thiouridine synthases, RNA methyltransferases, and pseudouridine synthases domain, ~110 aa) found in eukaryotes and archaea, but not in bacteria. The domain is probably an RNA-binding domain involved in delivering various RNA modification enzymes to their targets.

thuringolysin O A *cholesterol binding toxin produced by *Bacillus thuringiensis*.

thy1 (theta antigen) An *N*-glycosylated, glycophosphatidylinositol (GPI) anchored cell surface protein (CD90, 161 aa) with a single V-like immunoglobulin domain found on T cells, axonal processes, endothelial cells, and fibroblasts.

thymic hypoplasia (di George's syndrome) A developmental failure of the thymus and parathyroid leading to an absence of *T cells and defective cell-mediated immunity. It is due to a hemizygous deletion of 1.5–3.0 Mbp from chromosome 22q11.2 leading to haploinsufficiency of the TBX1 (*T-box 1) transcription factor gene, or point mutations in the *TBX1* gene. Hypoplasia of the parathyroid causes hypocalcaemia and there are other developmental abnormalities associated with the syndrome and caused by defects in cervical neural crest migration.

thymidine Standard name for thymine deoxyriboside; not the riboside, despite the normal naming conventions.

thymidine block A method for synchronizing cells in culture by depriving them of thymidine, thereby preventing DNA synthesis. Release of the block, the addition of thymidine to the medium, allows synchronous entry into the S phase of the *cell cycle.

thymidine kinase An enzyme (TK, EC 2.7.1.21, 234 aa) in the pyrimidine salvage pathway that phosphorylates thymidine (thymine deoxyriboside) to the nucleotide thymidylate. Cells that do not have TK are resistant to bromodeoxyuridine, because they do not incorporate it into DNA. They are also unable to grow in *HAT medium, which is useful for selecting somatic hybrids.

thymidine phosphorylase A dimeric enzyme (EC 2.4.2.4, subunits 482 aa) that functions in the salvage pathway to catalyse the conversion of thymidine (and to a lesser extent, 2′-deoxyuridine) to produce thymine (or uracil) and 2′-deoxyribose-1-phosphate. It has also been isolated as platelet-derived endothelial cell growth factor (PD-ECGF), an angiogenic factor, and as gliostatin, an inhibitor of astrocyte and astrocytoma proliferation. Mutations are associated with mitochondrial neurogastrointestinal encephalomyopathy in which there are multiple deletions of mitochondrial DNA, possibly because of aberrant thymidine metabolism.

thymine One of the two pyrimidine bases found in DNA (replaced by uracil in RNA). Thymine dimers can be formed in DNA by the UV-induced covalent linkage of adjacent (*cis*) thymidine residues. This is potentially mutagenic, although normally corrected by repair enzymes. *See* XERODERMA PIGMENTOSUM.

thymine-DNA glycosylase A mismatch-specific DNA-binding glycosylase (EC 3.2.2.-, 410 aa) that corrects G/T mismatches to G/C base pairs.

thymocyte An immature lymphocyte within the *thymus.

thymopentin A synthetic pentapeptide corresponding to residues 32–36 of *thymopoietin that enhances the production of thymic T cells and is immunostimulatory.

thymopoietin (lamina-associated polypeptide-2) A thymic hormone that induces differentiation of thymocytes, although the gene encodes a nuclear protein (454 aa) present in three alternatively spliced isoforms that associate with *lamin A/C and are mutated in some forms of *dilated cardiomyopathy. *See* THYMOPENTIN.

thymosin The **α-thymosins** are polypeptide hormones, derived from prothymosin-α and secreted by the thymus, that control the maturation of T cells. Thymosin-α_1 (28 aa) primes dendritic cells for antifungal T-helper type 1 resistance through toll-like receptor (TLR)-9 signalling. **β-Thymosins** are quite different, and thymosin-β_4 (Fx peptide, 43 aa) is an abundant cytoplasmic protein (~0.2 mM in neutrophils) that binds G-actin and inhibits formation of microfilaments.

thymus The lymphoid organ where T-cell maturation takes place. It is composed of stroma (thymic epithelium) and lymphocytes and is anatomically just anterior to the heart within the rib cage. Thymus-derived lymphocytes are *T cells. The thymus regresses in adulthood, although thymectomy, the removal of the thymus by surgical or other means, has profound effects on the immune system. A **thymoma** is a tumour derived from the thymus.

thyroglobulin The glycoprotein precursor of the thyroid hormones, a homodimer of 2767 aa subunits. If the gene is absent or no thyroglobulin is synthesized, the effect is likely to be recessive, but an abnormal subunit will compromise the majority of the homodimers and the effect is dominant. Mutations lead to goitre or in some cases a predisposition to autoimmune thyroiditis. *See* THYROXINE BINDING GLOBULIN.

thyroid hormones Hormones secreted by the thyroid gland, **thyroxine** (T_4; tetra-iodothyronine) and the more potent **tri-iodothyronine** (T_3) which is derived from T_4. They are important regulators of growth and metabolism and are carried in the plasma by *throxine-binding globulin, thyretin, and albumin. *Calcitonin is also secreted by the thyroid.

thyroiditis Disease of the thyroid. *See* GRAVES' DISEASE; HASHIMOTO'S THYROIDITIS; HYPERTHYROIDISM; HYPOTHYROIDISM; RIEDEL'S DISEASE.

thyroid stimulating antibody An autoantibody that causes hyperplasia of the thyroid, found in sera of most patients with *Graves' disease.

thyroid stimulating hormone (thyrotropin stimulating hormone, TSH) A heterodimeric glycoprotein produced in the pituitary that has a specificity-conferring β subunit (112 aa) and an α subunit (116 aa) shared with other pituitary hormones (chorionic gonadotropin, follicle-stimulating hormone, and luteinizing hormone). Mutations cause nongoitrous hypothyroidism. The TSH receptor (764 aa) has seven transmembrane domains and activates cAMP production in thyroid cells which stimulates the production and release of *thyroid hormones. Mutations in the receptor cause various dysfunctions of the thyroid (*see* HYPERTHYROIDISM).

thyroliberin *See* THYROTROPIN RELEASING HORMONE.

thyrotoxicosis *See* HYPERTHYROIDISM.

thyrotropin *See* THYROID STIMULATING HORMONE.

thyrotropin releasing hormone (TRH, TRF, protirelin, thyroliberin) A tripeptide (pyroGlu-His-Pro-NH_2) that stimulates the release of thyrotropin (*thyroid stimulating hormone) and prolactin from the anterior pituitary. The receptor is G-protein linked. It may have neurotransmitter and paracrine functions.

thyroxine *See* THYROID HORMONES.

thyroxine binding globulin One of the proteins (TBG, 415 aa) that carries *thyroid hormones in the circulation. *Transthyretin and albumin also carry the hormones but TBG has the highest affinity and carries the majority of T_4. It is a *serpin, although it has no inhibitory function.

tiabendazole An anthelmintic drug used to treat strongyloidiasis, cutaneous larva migrans (creeping eruption), dracunculiasis (guinea worm), and visceral larva migrans. *TNs* Mintezol, Triasox.

Tiam-1 An exchange factor (a *GEF, T-lymphoma invasion and metastasis protein 1, 1591 aa) for the small GTPase *rac that is implicated in tumour invasion and metastasis.

tiaprofenic A *NSAID. *TNs* Surgam, Surgamyl, Tiaprofen.

tibolone A synthetic steroid with oestrogenic, progestational, and androgenic properties used to ameliorate postmenopausal vasomotor symptoms.

TICAM-1 An adaptor protein (toll-IL-1R homology domain-containing adaptor molecule-1, TIR domain-containing adaptor inducing interferon-β, TRIF, 712 aa), like *Myd88 and TIRAP (*TIR domain-containing adaptor protein), that induces interferon-β, specifically interacts with toll-like receptor 3, and activates *NFκB.

ticarcillin A carboxypenicillin antibiotic usually used in combination with clavulanate as Timentin, to treat Gram-negative infections, especially by *Pseudomonas aeruginosa*.

Tie Endothelium-specific receptor tyrosine kinases involved in vascular development and in tumour angiogenesis. They associate with p85 of *PI-3-kinase. The ligand for Tie2 (Tek) is *angiopoietin but that for Tie1 is unknown although Tie1 and Tie2 interact.

TIGAR A p53-inducible protein (Tp53-induced glycolysis and apoptosis regulator, 270 aa) that functions to regulate glycolysis and to protect against oxidative stress. Probably fructose-2,6-bisphosphatase (EC 3.2.3.46).

tigecycline *See* GLYCYLCYCLINES.

tight junction *See* ZONULA OCCLUDENS.

tiling A term used to describe the process (tiling, tessellation) of covering a surface completely and has come to be applied to the design of DNA arrays (*DNA chips) that give comprehensive unbiased coverage of genomic DNA.

Tillman's reagent *See* DCIP.

timeless A gene that encodes a transcription factor (1208 aa) which interacts with the *period gene product, moves into the nucleus, and specifically inhibits clock-BMal1 induced transactivation of the period promoter. *See* CIRCADIAN RHYTHM.

time-weighted average A parameter used in the context of exposure to an environmental contaminant that takes account of the concentration at several sampling periods, the sampling times and the total time of exposure.

timolol A *beta-blocker used to treat hypertension and glaucoma. *TN* Blocadren.

Timothy's syndrome A disorder characterized by multiorgan dysfunction, including lethal arrhythmias (*long QT-8), webbing of fingers and toes, congenital heart disease, immune deficiency, intermittent hypoglyaemia, cognitive abnormalities, and autism. Caused by a mutation in the gene for the α-1C subunit of the voltage-dependent L-type calcium channel from cardiac muscle (CACNL1A1).

TIMP *See* TISSUE INHIBITORS OF METALLOPEPTIDASES.

TIN2 A protein (TRF1-interacting nuclear factor-2, 354 aa) recruited to the *shelterin complex through interaction with *telomeric repeat binding factor 1 (TRF1) and thought to be a negative regulator of telomerase length. Mutations can be one cause of *dyskeratosis congenita. **TIN2-interacting protein** (POT1- and TIN2-organizing protein, PTOP, 544 aa) blocks the interaction of *POT1 with telomeres and allows telomere extension.

tioguanine An antimetabolite (thioguanine) used in treatment of leukaemia.

TIP **1.** An inhibitory regulator (Tip41-like protein, 272 aa) of protein phosphatases 2A, 4, and 6, homologous to yeast Tip41, with a role within the ATM/ATR signalling pathway that controls DNA replication and repair. **2.** A T-cell immunomodulatory protein (612 aa) that induces the secretion of interferon-γ, TNF, and IL-10. **3.** Tension induced/inhibited protein, induced by stretch in mesenchymal cells. **4.** Tax-interacting protein-1 (TIP-1) is an unusual signalling protein, containing a single PDZ domain. TIP-1 is able to bind β-*catenin with high affinity and inhibit its transcriptional activity; also functions as a negative regulator of PDZ-based scaffolding. **5.** TIP30 (Tat-interacting protein 30, 242 aa) is a proapoptotic factor that interacts with HIV-1 tat protein. **6.** TIP47 (tail-interacting protein of 47 kDa) is one of the *PAT family of proteins now known to be identical to mannose-6-phosphate receptor-binding protein 1 (434 aa) that binds the cytoplasmic domain of the receptor and is required for its transport from endosomes to the trans-Golgi network.

TIRAP *See* TIR DOMAIN.

TIR domain A domain (~200 aa) found in *toll-like receptors and the IL-1 receptor. There are also four mammalian TIR-domain-containing adaptor proteins, MyD88, TIR domain-containing adaptor inducing interferon-β (TRIF), TRIF-related adaptor molecule (TRAM), and TIR-domain containing adaptor protein (TIRAP). TIR domains are also found in some plant cytoplasmic proteins implicated in host defence.

SEE WEB LINKS

- The TIR Domain Signal Transduction Project homepage.

tissue **1.** A group of cells, often of different kinds, that are held together by extracellular matrix, often peculiar to the tissue, and perform a particular function. Tissues represent a level of organization intermediate between cells and organs; organs may be made up of several different tissues. **2.** More generally applied (e.g. epithelial tissue) where the common factor is the pattern of organization, or (e.g. connective tissue) where the common feature is the function.

tissue array An array (or microarray) of tissue sections on a microscope slide (analogous to a DNA chip, but rather larger) that can be used for immunocytochemical staining or *in situ* hybridization.

tissue culture The maintenance and growth of pieces of tissue taken from an organism in artificial medium. The term has, however, become much more widely used to mean cell culture, where the cells may have been obtained from dissociated tissue (primary culture) or, more often, are the distant descendants of such cells. What was originally tissue culture is now referred to as 'organ culture'.

tissue culture plastic The special plastic used for growing anchorage-dependent cells in culture. Bacteriological grade plastic dishes are nonadhesive for animal cells and need to be modified to make them wettable and adhesive. The original culture vessels were of glass (hence '*in vitro* culture') which was wettable if clean.

tissue engineering The production of complex tissues by the *in vitro* culture of cells on artificial support matrices. A major problem is the absence of vascularization, but a few tissues for transplantation have been produced, e.g. skin and bladder wall.

tissue factor (factor III) An integral membrane glycoprotein (CD142, 263 aa) that is normally segregated from the blood but can be exposed after tissue injury. Tissue factor binds factor VII and initiates the extrinsic clotting mechanism by conversion of factor VII to enzymatically active factor VIIa which activates factors X and IX and prothrombin and leads to the formation of fibrin. **Tissue factor pathway inhibitor** (lipoprotein-associated coagulation inhibitor, 304 aa) is a protease inhibitor that inhibits factor Xa, and the factor VIIa–tissue factor catalytic complex.

tissue inhibitors of metallopeptidases (TIMPs) A family of proteins (~200 aa) that inhibit *matrix metallopeptidases by forming 1:1 complexes. TIMP1 (human collagenase inhibitor) is secreted by platelets and alveolar macrophages. TIMPs vary in their affinity for peptidases and in their tissue distribution.

tissue plasminogen activator (tPA) *See* PLASMINOGEN ACTIVATOR.

tissue typing The identification of the *MHC antigens expressed on a tissue that determine whether a graft will be accepted or rejected. Usually done using a panel of antibodies or carrying out *mixed lymphocyte reactions with lymphocytes of the recipient and of potential donors.

titin (connectin) A family of enormous proteins (2000–3500 kDa, ~27 000 aa), found in the sarcomere of striated muscle, which are important for muscle assembly, force transmission at the Z line, and maintenance of resting tension in the I band region. Each titin molecule spans from M line to Z disc, and there are muscle-type specific isoforms. Titin has calcium-regulated kinase activity and can become autophosphorylated. Mutations are responsible for various forms of *hypertrophic cardiomyopathy, *myopathy, and *muscular dystrophy. *See* MYOTILIN.

TL1 A TNF-like ligand (tumour necrosis factor ligand superfamily member 15, TNFSF15, vascular endothelial growth inhibitor; VEGI, 251 aa) that binds to *death receptor DR3. TL1 is highly expressed in endothelial cells and may act as an angiogenesis inhibitor. Expression is up-regulated in *Crohn's disease and polymorphisms in the *TNFSF15* gene may be associated with susceptibility.

TLCK An inhibitor (tosyl lysyl chloromethylketone) of many serine and cysteine peptidases, particularly trypsin-like peptidases.

T-loop of RNA *See* TRANSFER RNA.

TLR *See* TOLL-LIKE RECEPTORS.

T lymphocyte *See* T CELL.

TMB-8 A small molecule inhibitor of calcium release from intracellular stores and a potent, noncompetitive, antagonist of various *nicotinic acetylcholine receptor subtypes.

TMEFF *See* TRANSMEMBRANE PROTEINS WITH EGF-LIKE AND 2 FOLLISTATIN-LIKE DOMAINS.

T-memory cells (Tm) T cells involved in *immunological memory of specific antigens. It is not entirely clear whether Tm derive from effector T cells that become quiescent until re-exposed to antigen (historically, the standard view) or whether the generation of effector T cells in the primary immune response is accompanied by the production of central memory cells with a capacity for self-renewal and that act as memory stem cells. There appear to be two distinct classes of Tm cells.

SEE WEB LINKS

- Article in the *Journal of Clinical Investigation* (2001).

TMPD (Wurster's reagent) An easily oxidized compound used to detect peroxidases. Dark blue when oxidized.

TMS *See* TRANSCRANIAL MAGNETIC STIMULATION.

TNF *See* TUMOUR NECROSIS FACTOR.

TNF receptors A superfamily (TNFRSF) of receptors that includes the receptors for *tumour necrosis factor as well as *Fas, CD40, CD27, and *RANK. The **type 1 receptor** (CD120a, TNFRSF1A, p60, 455 aa) is present on most cell types and **type 2** (CD120b, TNFRSF1B, p80, 461 aa) is mainly restricted to haematopoietic cells. Both bind TNFα and lymphotoxin (TNFβ) and have substantial sequence homology in the extracellular domain, but different signalling capacities. TNFRSF1A has a *death domain and binding of TNF to the extracellular domain leads to homotrimerization and interaction with a number of cytoplasmic proteins (*TRADD, *TRAF,

*FADD). Mutations in the type 1 receptor are associated with *periodic fever. The extracellular portion of the type 2 receptor can be produced by proteolysis and is used as a treatment for rheumatoid arthritis (*TN* Enbrel).

TNP **1.** Trinitrophenol. **2.** TNP-470 is a semisynthetic derivative of *fumagillin. **3.** *See* TNP-AMP.

TNP-AMP An ATP analogue (trinitrophenol-adenosine monophosphate) useful in studying ATPases. The TNP-nucleotide is only fluorescent when bound.

TNT **1.** Trinitrotoluene, an explosive. **2.** *Troponin T (TnT).

tobramycin An *aminoglycoside antibiotic.

tocopherol *See* VITAMIN (vitamin E).

toeprinting An assay method used to examine the formation of complexes between RNA and RNA-binding proteins. Essentially, bound protein blocks reverse transcriptase and the truncated transcript can be used to deduce the binding site (e.g. the transcription initiation site).

Togaviridae Class IV viruses that have a single (positive) strand RNA genome. The capsid is bullet-shaped and enveloped by membrane phospholipid from the host cell with embedded viral ('spike') glycoproteins. There are two main groups: alphaviruses, including *Semliki Forest virus and *Sindbis virus, and *Flaviviridae, including yellow fever virus and rubella (German measles) virus. Many have insect vectors and were previously classified as *arboviruses.

tolbutamide A *sulphonylurea.

tolerable daily ingestion A specified value for the amount of a food additive or contaminant that can be ingested on a daily basis without adverse effects. Usually given as mg/person, assuming a body weight of 60 kg.

tolerance *See* IMMUNOLOGICAL TOLERANCE.

toll-like receptors (TLR) A family of receptors, the first example having been the *Drosophila* toll receptor, that are involved in innate immunity. They are transmembrane proteins with significant homology in their cytoplasmic domains to the IL-1 receptor type I (*see* TIR DOMAIN) that recognize and respond to microbial components such as lipopolysaccharide.

tolnaftate An antifungal drug, probably inhibiting ergosterol synthesis.

tolterodine An antimuscarinic drug used to treat urinary incontinence. *TN* Detrusitol.

Tom1 One of a small family of proteins (target of Myb, Tom1, 492 aa; Tom1-like1, TomL1/Srcasm; Tom1L2) that have sequence similarity with proteins associated with vesicular trafficking at the endosome. Tom1L1 has potential binding sequences for Tsg101, a regulator of the formation of multivesicular bodies. The gene for Tom1 is a target for the *myb transcription factor.

Tom complex A multiprotein complex of the mitochondrial outer membrane (translocase of outer membrane) made up of eight different proteins. Tom40 (40 kDa, 362 aa) spans the outer membrane and forms a hydrophilic pore that permits proteins to pass. The other proteins are also embedded in the membrane, variously exposed on the cytosolic and inner faces. *Compare* TOM1.

tomoregulin *See* TRANSMEMBRANE PROTEINS WITH EGF-LIKE and 2 FOLLISTATIN-LIKE DOMAINS.

tonic *See* ADAPTATION.

tonofilaments The *intermediate filaments that are attached to *desmosomes.

tophus A mass of urate crystals that elicit a chronic inflammatory reaction: characteristic of gout.

topographical control Describing aspects of cell behaviour affected by the shape of the local environment, e.g. *contact guidance.

topoinhibition *See* DENSITY DEPENDENT INHIBITION OF GROWTH.

topoisomerases Enzymes that change the extent of *supercoiling in DNA by breaking and rejoining one or both strands. **Type I topoisomerases** (EC 5.99.1.2) are ATP-independent, cut only one strand of DNA, and are important during transcription.Topoisomerase 1 (765 aa) forms a protein clamp around the DNA duplex, creates a transient nick, and releases supercoils by a swivel mechanism that involves friction between the rotating DNA and the enzyme cavity. After each rotation there is a chance that religation will occur. Mutations in type I topoisomerase will confer *camptothecin resistance on leukaemic cells. **Type II topoisomerases** (EC 5.99.1.3, Top2A, 1531 aa; Top2B, 1621 aa) cut both strands of DNA and require ATP. Inhibitors of topoisomerase II will block transcription on chromatin templates, though not on naked DNA. Type II topoisomerase of *E. coli* (*DNA gyrase) is inhibited by several antibiotics, including *nalidixic acid and ovobiocin. **Type III topoisomerases** (Top3A, 1001 aa; Top3B, 862 aa) in humans have homology with the prokaryotic topoisomerase 1, but not with the eukaryotic Top1 enzyme. In association with *helicase it is thought to be important in recombination and in the alternative lengthening of telomeres (ALT) pathway that allows telomere recombination in the absence of *telomerase. Various topoisomerase inhibitors (e.g. *irinotecan) are used as antitumour drugs.

TOR Target of rapamycin. *See* MTOR.

torasemide A loop *diuretic used to treat oedema associated with congestive heart failure. *TN* Demadex.

TORC *See* MECT1.

Torres body An intranuclear inclusion body of host protein seen in liver cells infected with yellow fever virus (*Togaviridae). Not considered to be a reliable diagnostic indicator.

torsins A subfamily of the *AAA family of ATPases (torsin1A, 332 aa; torsin 1B, 289 aa; torsin 2A, 321 aa; torsin 3A, 397 aa) that are associated with the endoplasmic reticulum and the nuclear envelope. An alternatively spliced form of torsin2A (242 aa) has its C-terminus processed to produce bioactive peptides, salusin-α (28 aa) and salusin-β (20 aa), which can produce hypotension and bradycardia. Mutations in torsin A are associated with early-onset torsion dystonia, a movement disorder that begins in childhood and is characterized by twisting muscle contractures which arise from altered neuronal communication in the basal ganglia.

torsion dystonia *See* TORSINS.

torulosis An obsolete name for infection by *Torula histolytica*, now known as *Cryptococcus neoformans*. *See* CRYPTOCOCCOSIS.

totipotent *See* STEM CELL.

toxicity The capacity to cause injury to a living organism. Toxicity is sometimes quantified as the reciprocal of the absolute value of median lethal dose ($1/LD_{50}$) or lethal concentration ($1/LC_{50}$), although the route of administration, etc. need to be specified.

toxicodynamics By analogy with *pharmacodynamics, the study of the way in which toxins interact with target sites, and the biochemical and physiological consequences that cause the adverse effects.

toxicogenetics The study of the genetic basis of individual variations in the response to toxins, usually focused on polymorphisms in single genes, rather than on the overall genomic environment in which particular genes are being expressed (toxicogenomics).

toxicokinetics By analogy with *pharmacokinetics, the analysis of the uptake, biotransformation, distribution, and elimination of toxins.

toxigenicity The ability of a pathogen to produce *toxins.

toxin A naturally produced poison that will damage or kill other cells. The pathogenicity of many infectious bacteria derives from their production of various *endotoxins and *exotoxins.

Toxocara A genus of nematodes (roundworms) that cause intestinal infections, usually acquired from dogs (*Toxocara canis*) or cats (*T. cati*). The larval stages may migrate into tissues, with potentially serious consequences.

toxoid A bacterial exotoxin that has been treated to render it nontoxic but retains its antigenic properties and can be used as a vaccine.

Toxoplasma A genus of parasitic protozoa. *T. gondii* causes toxoplasmosis in humans, which are intermediate hosts, the final host being cats or other felines. Infection is by resistant oocysts in faeces. It is an intracellular parasite and can cause serious, even fatal, lesions in tissues such as brain, heart, or eye.

TPA **1.** A *phorbol ester tumour promoter; more correctly, phorbol myristyl acetate (PMA), a substance, rather than an activity ('tumour promoting activity', TPA). **2.** *See* TISSUE PLASMINOGEN ACTIVATOR.

TPCK A nonspecific inhibitor (tosyl phenyl chloromethyl ketone) of chymotrypsin, subtilisin, and many other serine peptidases; interacts with histidine residues at the active site.

TPO **1.** A protein (TPO1, tetracyline transporter-like protein, TETRAN, 455 aa) that has twelve transmembrane domains. The rat homologue is up-regulated in cultured oligodendrocytes during the stage when myelination occurs. **2.** Thyroid peroxidase (TPO), an important enzyme in the production of *thyroid hormones.

TPR motif A consensus sequence (tetratricopeptide motif, ~34 aa) found as multiple copies in various functionally different proteins where it facilitates protein–protein interactions. TPR motifs are important to chaperones and in protein complexes involved in the cell cycle, transcription, and protein transport.

trabecula A 'small beam', a strand of material that crosses a cavity, e.g. the small interconnecting rods that make up *cancellous bone (trabecular bone). The trabecular meshwork is a specialized tissue involved in regulating intraocular pressure. *See* MYOCILIN.

trace amine associated receptors A family of G-protein-coupled receptors (TAARs), encoded by at least seven human genes, for trace amines such as *p*-tyramine (derived from tyrosine) and β-phenylethylamine (derived from phenylalanine), although many do not have known ligands. Trace amines and their receptors may be implicated in the pathogenesis of psychiatric disorders.

trachoma An infection of the conjunctiva with **Chlamydia trachomatis*. The lacrimal glands and ducts are often affected, and the upper eyelid may turn inwards allowing the lashes to abrade the cornea and cause corneal ulceration. It is highly infectious, transmitted by houseflies, and associated with poor hygiene; without treatment it usually leads to blindness.

TRADD A protein (TNF-receptor-1 associated death domain protein, 312 aa) that interacts with the cytoplasmic tail of TNF receptor 1 (TNFR1) and with other proteins involved in downstream signalling (e.g. *FADD and *receptor interacting kinase). The interaction is through *death domains in both TRADD and TNFR1.

TRAFs Family of adaptor proteins (TNF receptor-associated proteins) implicated in the activation of NFκB by the TNF receptor superfamily. There are multiple isoforms.

TRAIL A member of the TNF ligand family (TNF-related apoptosis-inducing ligand, Apo2L, TNFSF10) found either as a transmembrane protein (281 aa) or in soluble form. TRAIL induces apoptosis in various tumour cell lines and is expressed at significant levels in most normal tissues. There are four receptors, two of which mediate apoptosis (TRAILR1 and 2, death receptors DR4 and DR5) and two are potentially *decoy receptors.

tram **1.** TRAM1 (thyroid hormone receptor activator molecule 1) is a nuclear receptor coactivator (NCOA3, steroid receptor coactivator-3) that binds nuclear receptors in a hormone-dependent fashion. NCOA3 recruits two other nuclear factors, CBP (*CREB binding protein) and PCAF (p300/CBP-associated factor), into a multisubunit complex that stimulates transcription. **2.** *See* TRANSLOCATING CHAIN-ASSOCIATING MEMBRANE PROTEIN. **3.** TRIF-related adaptor molecule (TRAM), *see* TIR DOMAIN.

tramadol An opioid analgesic.

TRAMP **1.** Transgenic adenocarcinoma of the mouse prostate (TRAMP). A tumour that arises in mice engineered as a model for human prostatic carcinoma. **2.** Tyrosine-rich acidic matrix protein (TRAMP); *see* DERMATOPONTIN. **3.** At one time a synonym for a TNF receptor superfamily member, but this seems to have fallen out of use.

transactivation An increase in gene expression brought about by the action of a transactivator, a transcription factor that up-regulates multiple genes. The transactivator may be endogenous or experimentally introduced and the latter are often controlled by an inducible promoter and linked to a reporter gene. Many viruses encode transactivators (e.g. *tat of HIV). A viral transactivator that activates a proto-oncogene will induce tumour formation.

transacylase An enzyme of the EC 2.3 class that catalyses the transfer of an acyl group.

transcranial magnetic stimulation (TMS) A noninvasive technique that induces a current in the brain using an external magnetic field. Multiple stimuli per second (**repetitive TMS**) can be produced and have interesting research potential.

transcriptase *See* REVERSE TRANSCRIPTASE.

transcription The synthesis of RNA from a DNA template. *See* RNA POLYMERASES.

transcriptional control The control of gene expression at the level of transcription, the number of RNA transcripts of the coding region. Although this is an important regulatory mechanism, other mechanisms are also involved in controlling the amount of (e.g.) protein that is actually expressed.

transcriptional insulators The DNA elements that set boundaries on the actions of enhancer and silencer elements and so partition the eukaryotic genome into regulatory domains. The transcriptional repressor CTCF (CCCTC-binding factor) binds through multiple zinc fingers (of which it has eleven) to a range of unrelated DNA sequences and functions as a transcriptional insulator, repressor, or activator, depending on the context of the binding site.

transcriptional silencing A method of transcriptional control in which the DNA is complexed as *heterochromatin and made permanently inaccessible to transcription. This long-term suppression presumably occurs in *determination.

t

transcription factor A protein that binds to a specific DNA sequence upstream of a coding region and triggers the assembly of an RNA polymerase complex and the production of mRNA or other RNA species. A range of different transcription factors are known. *See e.g.* TFII; TFIII.

transcription squelching The suppression of transcription as a result of overexpression of a transcription factor that would normally increase transcription: may arise by sequestration of a cofactor by the excess transcription factor.

transcription unit A DNA sequence that is transcribed into a single RNA transcript which is then processed.

transcriptome The complete set of transcripts present in a cell at a specific time. Analogous to the *proteome, but not identical.

- Further information in the Human Genome website.

transcriptosome A preassembled multiprotein complex that carries out transcription, and possibly RNA processing, as opposed to a complex that is assembled *in situ* after the binding of a transcription factor. The model has some problems, and the term should not be confused with the *transcriptome.

transcytosis The transport of material across an epithelium by uptake into a coated vesicle, and release on the opposite face. The material remains enclosed within a vesicle. There have been suggestions that there is actually a transepithelial pore, but this is disputed.

transdifferentiation The change from one differentiated state to another. It is rare, although some examples are known; if differentiated cells could be induced to revert to a less differentiated and determined state, then regeneration of tissues would be possible.

transducin A trimeric *GTP-binding protein (α, 350 aa; β, 340 aa; γ, 42 aa) in the disc membrane of *retinal rods and *retinal cones. The α subunits in rods and cones are encoded by different genes. *Rhodopsin, having absorbed a photon, interacts with transducin and promotes GDP/GTP exchange. The α subunit with GTP bound dissociates from the trimeric complex and activates a cGMP-phosphodiesterase as part of the 'cascade amplifier' that produces a neuronal signal in response to light. The α subunit of transducin is ADP-ribosylated by cholera toxin and pertussis toxin.

transduction **1.** The conversion of a signal from one form to another, e.g. an action potential is transduced into a mechanical movement by muscle; sensory cells transduce a range of signals into nerve impulses. **2.** The transfer of a gene from one bacterium to another by a *bacteriophage.

transfection The introduction of DNA into a host eukaryotic cell and its integration into chromosomal DNA (**stable transfection**), analogous to bacterial transformation. Various methods are employed to introduce the DNA (e.g. *electroporation) and usually only a small percentage (~1%) of cells become transfected and express the product of the introduced DNA. Frequently used more loosely to mean the introduction of DNA into a target cell even if the DNA does not become integrated (**transient transfection**).

transferase A suffix in the name of an enzyme indicating that it transfers a specific grouping from one molecule to another; EC 2 in the *E classification.

transfer factor **1.** A parameter that relates to the effectiveness of gas exchange across the alveolar membrane in the lung. Usually given as mL min^{-1} $mmHg^{-1}$ of alveolar pressure of CO which has a capillary pressure of zero, being irreversibly bound. **2.** A dialysable factor from sensitized T cells that was claimed to be immunostimulatory but seems to be sinking into obscurity, possibly deserved.

transferrin (siderophilin) The iron transport protein (678 aa) of plasma, a β-globulin with a very high binding affinity for ferric ions (K_{ass} ~21). It is an important constituent of growth media because iron is essential for all cells. Transferrin with bound iron enters cells by receptor-mediated endocytosis. The receptor (CD71, 760 aa) is a disulphide-linked homodimer found as an integral membrane protein and in soluble form in plasma as a result of cleavage. Soluble levels of the receptor rise in iron deficiency.

transfer RNA (tRNA) The family of RNA molecules (4S RNA, 70–80 bp) each of which binds a specific amino acid to form *aminoacyl tRNA, and has the appropriate triplet anticodon to bind to the *codon on *messenger RNA specifying that amino acid. In some cases the anticodon has inosine, allowing some 'wobble'. The **T-loop** (thymine pseudo-uracil loop; Tψ loop) of transfer RNA is responsible for ribosome recognition.

transformation A heritable alteration in the properties of a cell. Usually refers to malignant transformation of cells in culture (*but see* BLAST TRANSFORMATION) and causes changes in growth characteristics, particularly in requirements for macromolecular growth factors, and often changes in morphology. Transformation *in vitro* correlates reasonably well with tumorigenicity *in vivo* and can be brought about by some viruses (e.g. *SV40).

transforming growth factor (TGF) A family of growth factors secreted by transformed cells that induce the phenotypic characteristics of cell transformation (e.g. the ability to grow in semisolid agar), but do not cause hereditable changes. TGFα (50 aa) binds to the EGF receptor and will stimulate the growth of microvascular endothelial cells. TGFβ is a homodimer (two 112 aa chains) secreted by many different cell types and a growth inhibitor for some cell types. Mutations in TGFβ cause *Camurati–Engelmann disease. The TGF family includes many of the *bone morphogenetic proteins (BMPs). *See* ACTIVIN; SYNDECANS.

transforming viruses Viruses that will cause *transformation of animal cells in culture. Some of the *Oncovirinae lack oncogenes and can induce leukaemias in animals, but cannot transform *in vitro*. If they acquire oncogenes they become (acute) transforming viruses.

transgelin A protein (SM22-α, 201 aa) found exclusively in smooth muscle cells and fibroblasts that has structural similarity to *calponin. Binds to F-actin (1:6 transgelin:G-actin) and causes gelation. Levels are down-regulated when cells are transformed *in vitro*. Transgelin 2 (199 aa) has been found in a wider range of tissues.

transgene A *gene or a DNA fragment derived from one organism and incorporated into the *genome of another.

transglutaminase An extracellular enzyme (EC 2.3.2.13, protein-glutamine γ-glutamyltransferase, factor XIIIa, fibrinoligase, fibrin stabilizing factor, 732 aa) that crosslinks proteins by catalysing the formation of a stable amide bond between the side groups of glutamine and lysine. There are also tissue forms of the enzyme.

***trans*-Golgi network (TGN)** The vesicles associated with the export side of the *Golgi apparatus, the side from which vesicles and their processed contents move towards their destination, e.g. to lysosomes or to the plasma membrane for secretion.

transient expression (transient transfection) The temporary expression of a nonintegrated transgene carried as an episome and lost after a few cell divisions.

transient receptor potential channels Nonselective cation channels (TRP channels) of which 28 mammalian types are known, divided into 6 main subfamilies: the **TRPC** (canonical), **TRPV** (vanilloid), **TRPM** (melastatin), **TRPP** (polycystin), **TRPML** (mucolipin), and **TRPA** (ankyrin) groups. Some have high temperature-sensitivity (*$Q_{10} > 10$) but differ in their thermal thresholds for activation and are found in temperature-sensing cells; six thermo-TRP channels have been cloned; TRPV1–4 are heat activated, whereas TRPM8 and TRPA1 are activated by cold. *Capsaicin and resiniferatoxin are agonists for TRPV1, menthol for TRPM8 (cold receptor), and *icilin for both TRPM8 and TRPA1. Mutation in TRPV4 causes *brachyolmia 3. TRPC6 and TRPC4 have been implicated in late-onset Alzheimer's disease. *See* VANILLOID RECEPTOR-1.

SEE WEB LINKS

- Review of TRP channels and disease in *Physiological Reviews* (2006).

transin A zinc metallopeptidase (EC 3.4.24.17, matrix metalloproteinase 3, MMP-3, procollagenase activator, proteoglycanase, stromelysin 1, 477 aa) involved in matrix degradation and in cellular invasion.

transinhibition A situation where the homodimer is catalytically inactive because the inhibitory domain of one blocks the kinase domain of the other and vice versa; found (e.g.) in the *p21-activated kinases. Once the subunits dissociate, they are catalytically active.

transitional elements Transport vesicles (transitional endoplasmic reticulum) at the interface betwen the rough endoplasmic reticulum and the *cis*-face of the *Golgi apparatus.

transitional epithelium An epithelium adapted to being stretched, usually a *stratified epithelium, e.g. that of the bladder.

transition proteins A group of low molecular weight basic proteins that displace *histones from nuclear DNA during *spermatogenesis. They are themselves replaced by *protamines.

translation The incorporation of amino acids, carried by *transfer RNA, into a growing polypeptide chain using information encoded in *messenger RNA (mRNA). The growing chain, as it emerges from the *ribosome, interacts with other proteins which determine whether the ribosome becomes associated with *endoplasmic reticulum or remains free in the cytoplasm.

Translational control of protein synthesis can involve the selective usage of preformed mRNA or instability of the mRNA.

translationally controlled tumour protein An abundant protein (TCTP, IgE-dependent histamine-releasing factor, HRF, fortilin, 172 aa) of the *MSS4/DSS4 superfamily. TCTP is involved in both cell growth and human late allergy reactions, as well as having a calcium binding property. Similar proteins, with very divergent sequences, are found in a wide range of organisms.

translational research Research that is directed at translating a scientific discovery into a clinical treatment, particularly the process of moving a compound from discovery research, through development, clinical trials, and the regulatory process, so that it can be used on patients.

translin A DNA-binding protein (228 aa) that specifically binds as a multimer to consensus sequences at breakpoint junctions of chromosomal translocations in many cases of lymphoid malignancies. **Translin-associated factor X** (290 aa) interactes with translin and is the only protein with any similarity.

translocase **1.** A general term for a protein involved in assisting molecules to move from one location to another or assemble in the right place (e.g. *Oxa1). **2.** *See* ELONGATION FACTOR ELL.

translocating chain-associating membrane protein **1.** An eight-pass transmembrane glycoprotein (TRAM, 374 aa) of the endoplasmic reticulum that influences glycosylation and is stimulatory or required for the translocation of secretory proteins. A component of the *translocon and there are approximately as many TRAM molecules as there are associated ribosomes. **2.** *Compare* TRAM.

translocation *See* CHROMOSOME TRANSLOCATION.

translocon The multiprotein complex involved in the process of moving a nascent polypeptide into the cisternal space of the endoplasmic reticulum. The translocon regulates the ribosome/endoplasmic reticulum interaction, translocation, and the integration of membrane proteins in the right orientation. Tram (*translocating chain-associating membrane protein), *signal peptidase, and the *signal recognition particle are associated with the translocon.

transmembrane adaptor proteins A family of proteins (TRAPS) that share several common structural features, in particular multiple sites for tyrosine phosphorylation in their cytoplasmic tails. Examples include linker for activation of T cells (*LAT), *low-density lipoprotein receptor adaptor protein 1, the phosphoprotein associated with glycosphingolipid-enriched microdomains (PAG)/C-terminal src kinase (Csk) binding protein (Cbp), SHP2-interacting transmembrane adaptor protein (*SIT), T-cell receptor interacting molecule (*TRIM), non-T-cell activation linker (*NTAL), and pp30. *See also* ADAPTOR PROTEINS.

transmembrane protein An integral membrane protein with the polypeptide chain exposed on both sides of the membrane. **Types I and II** are single-pass molecules with the N-terminus to the exterior or in the cytoplasm respectively. **Types II and IV** are multipass proteins.

transmembrane proteins with EGF-like and 2 follistatin-like domains Proteins (TMEFFs) that may be tumour suppressors or regulators of signalling in embryogenesis. **Tomoregulin-1** (TMEFF1) is enriched in neuroendocrine tissue, **TMEFF2** is mainly expressed in the prostate and brain and three splice variants, TRa, TRb, and TRc, have been described. TRa (368 aa) promotes the survival of hippocampal and mesencephalic neurons in primary culture but not survival of cortical neurons. The *TMEFF2* gene is frequently hypermethylated in human tumour cells, and expression of *TMEFF2* has been reported to inhibit prostate cancer cell growth.

transmembrane transducer Any system that transmits (transduces) a chemical or electrical signal across a membrane. The transduction may involve a conformational change in a single transmembrane protein or a change in the interaction of two (or more) subunits of a receptor complex.

transmigration The migration of cells from one side of a cell monolayer to the other; often referring to the extravasation of leucocytes from the postcapillary venule into tissue.

transpeptidase An enzyme that catalyses the formation of an amide linkage between a free amino group and a carbonyl group. The crosslinking of peptidoglycan in bacterial cell walls, carried out by a transpeptidase (EC 3.4.16.4), is blocked by penicillin-type antibiotics. **Peptidyl transferase** is an aminoacyltransferase (EC 2.3.2.12), a function that is also carried out by ribosomes. **D-Glutamyl transpeptidase** (D-glutamyltransferase, EC 2.3.2.1) forms a link between D-glutamine and D-glutamyl-peptide.

transplantation antigen An antigen that elicits an immune response in graft rejection. The main antigens are those of the *MHC and the H-Y antigens. The minor histocompatibility antigens have a lesser role.

transplantation reaction The cellular responses to an allogeneic (mismatched) graft, that eventually destroy the grafted tissue. The main cells involved in the initial reaction are cytotoxic lymphocytes.

transporter *See* TRANSPORT PROTEIN.

transportins Proteins involved in the import of ribonucleoproteins into the nucleus. Transportin-1 (TNPO1, karyopherin β_2, 898 aa) and **TNPO2** (karyopherin β_{2B}, importin β_2, 897 aa) have different binding specificities for pre-mRNA/mRNA-binding proteins (heterogeneous nuclear ribonucleoproteins, hnRNP) that have an M9 domain. **Transportin-SR** (transportin-3, 923 aa), binds the phosphorylated RS domains of various proteins and is responsible for targeting *SR-proteins to the nucleus. Activity is regulated by Ran-GTP. *See* IMPORTINS.

transport protein A transmembrane protein that facilitates movement of molecules across a membrane. For most hydrophilic molecules diffusion is extremely slow and transport proteins are essential. A subset of transport proteins (transport ATPases) use energy for *active transport and can move molecules up a gradient of electrochemical potential. *See* ANTIPORT; FACILITATED DIFFUSION; SYMPORT.

transport vesicle A vesicle involved in moving material from the rough endoplasmic reticulum to the receiving (*cis*-) face of the Golgi.

transposable element *See* TRANSPOSON.

transposition Movement from one location to another, particularly of a DNA sequence (*transposon) within the genome.

transposon (transposable element, 'jumping gene') A small, mobile DNA sequence that can be replicated and inserted at other sites in the genome. Transposons characteristically have nearly identical sequences at each end, oppositely oriented (inverted) repeats, and code for transposase, the enzyme that catalyses their insertion. **Bacterial transposons** can be simple, with only the genes needed to jump, or complex, carrying additional genes not involved in transposition. Some **eukaryotic transposons** are like bacterial transposons; others (**retrotransposons**) are transcribed into RNA and reverse-transcribed back into DNA at another site.

***trans*-splicing** *See* SPLICING.

transthyretin A plasma protein (4.5 mmol/L in plasma, thyroxine-binding prealbumin, a homotetramer of 127-aa subunits) that is a transport protein for both *thyroxine and retinol (vitamin A). Transthyretin forms a complex with retinol-binding protein (2 mmol/L in plasma) and is found in neuritic plaques, neurofibrillary tangles, and microangiopathic lesions of senile cerebral amyloid. Point mutations lead to various distinct forms of *amyloidosis and amyloid polyneuropathy.

transudate The fluid within tissue that is derived from plasma. *Oedema arises from an increase in venous and capillary pressure, so that more transudate accumulates. Altered vascular permeability leads to the formation of cellular exudate.

transverse tubule *See* T TUBULE.

transversion A *point mutation in which a pyrimidine is replaced by a purine or vice versa, a more serious and rarer change than a transition involving substitution of one purine or one pyrimidine for another.

tranylcypromine A monoamine oxidase inhibitor used as an antidepressant. *TN* Parnate.

TRAP *See* TELOMERIC REPEAT AMPLIFICATION PROTOCOL ASSAY; THROMBOSPONDIN-RELATED ANONYMOUS PROTEIN; TRANSMEMBRANE ADAPTOR PROTEINS.

trastuzumab A therapeutic monoclonal antibody directed against the constitutively-active *HER-2/neu EGF receptor, used to treat HER-2-positive metastatic breast carcinoma. *TN* Herceptin.

trazodone A drug related to the tricyclic antidepressants with sedative, anxiolytic, and antidepressant properties. *TNs* Deprax, Desyrel, Molipaxin, Thombran, Trazorel, Trialodine, Trittico.

TRE 1. An *oncogene (*tre*), first isolated from NIH3T3 cells transfected with human Ewing's sarcoma DNA, that encodes ubiquitin-specific protease 6 (786 aa), only found in hominoids. **2.** Thyroid hormone response element (TRE), a DNA sequence recognized by the thyroid hormone receptor. **3.** *TPA responsive element.

Treacher Collins' syndrome A disorder of craniofacial development caused by mutation in the **treacle* gene.

***treacle* (*TCOF1*)** A gene mutated in *Treacher Collins' syndrome that encodes a protein (1411 aa) with motifs shared with nucleolar trafficking proteins and has putative nuclear and nucleolar localization signals. Haploinsufficiency may lead to inadequate ribosomal RNA production in the prefusion neural folds during the early stages of embryogenesis.

treadmilling The term applied to the continual assembly and disassembly that occurs in polymers such as the microtubule. Because assembly occurs at one end and disassembly at the other, there is a gradual progression of subunits along the structure, even though the length remains constant. A similar phenomenon probably occurs with microfilaments.

trefoil factors (trefoil peptides) A family of stable secretory polypeptides with at least one *trefoil motif that are thought to be proinvasive and angiogenic agents acting through *cyclo-oxygenase-2 (COX-2)- and thromboxane A2 receptor (TXA2-R)-dependent signalling pathways. Includes three members: **TFF1** (pS2, 84 aa) is up-regulated by oestrogen in many breast carcinomas, **TFF2** (spasmolytic peptide, 129 aa) and **TFF3** (intestinal trefoil factor, 80 aa) are associated with mucin-secreting epithelial cells and play a crucial role in mucosal defence and healing. TFF3 may be a *motogen.

trefoil motif A domain (P domain, 40 aa) formed by conserved cysteine residues that are disulphide bonded in such a way as to generate a trefoil structure (bonded 1–5, 2–4, 3–6) and characteristic of various secretory polypeptides.

T-regulatory cells (suppressor T cells, Treg) A class of T cells, now recognized to have subtypes ($CD8^+CD28^-$, $CD4^+CD25^+$) that suppress activation of the immune system in an antigen-specific manner and are essential for the induction and maintenance of immune tolerance. Their existence (as T-suppressor cells) was disputed for a long time, but they have become generally accepted.

TREK-1 A *tandem-pore domain channel (KCNK2, TWIK-related potassium channel) expressed throughout the central nervous system. It is opened by polyunsaturated fatty acids, lysophospholipids, and volatile anesthetics, and inhibited by neurotransmitters that increase intracellular cAMP or activate the Gq signalling pathway.

TREM-2 *See* TRIGGERING RECEPTOR EXPRESSED ON MYELOID CELLS 2.

Treponema A genus of *spirochaetes; *T. pallidum* causes syphilis.

treppe A gradual increase in the extent of muscular contraction when there is rapid repeated stimulation (staircase phenomenon). The phenomenon is seen in the heart, where an increase in heart rate progressively increases the force of ventricular contraction (Bowditch treppe).

TREX complex A multiprotein complex (transcription/export complex) that is recruited to a transcribing gene and travels with the polymerase during transcriptional elongation.

TRF **1.** Thyrotropin releasing factor. *See* THYROID STIMULATING HORMONE. **2.** *Telomeric repeat binding factors 1 and 2 (TRF-1, TRF-2).

TRF2-interacting protein A component (RAP1, 399 aa) of the *shelterin complex that interacts with *telomeric repeat binding protein 2 (TRF2), the homologue of the yeast telomeric RAP1 protein.

triacyl glycerols *See* TRIGLYCERIDES.

triad (triad junction) The region where a *T tubule is flanked by terminal cisternae of *sarcoplasmic reticulum on either side, forming a triplet of membrane profiles in cross-section (electron microscopy).

triadin A transmembrane protein (729 aa) found particularly where terminal cisternae of sarcoplasmic reticulum abut the T tubule. Thought to play an active role in calcium release. Colocalizes with the *ryanodine receptor and *junctin.

triamcinolone A synthetic corticosteroid used orally or topically. *TNs* Aristocort, Azmacort, Fougera, Kenalog, Nasacort, Tricortone, Triderm, Triesence, Trilone, Tri-Nasal, Tristoject, Volon A.

triamterene A potassium-sparing *diuretic used in combination with thiazide diuretics for the treatment of hypertension and oedema. *TN* Dyrenium.

tribbles A protein family (homologues of the *Drosophila* protein) that modulate protein kinase signalling pathways. **Tribbles-1** (372 aa) inhibits *AP-1 activation and interacts with several mitogen-activated protein kinases. **Tribbles-2** is a regulator of the inflammatory activation of monocytes. **Tribbles-3** (TRIB3) affects insulin signalling at the level of *Akt-2.

tricarboxylic acid cycle The central metabolic pathway of oxidative metabolism (TCA cycle, citric acid cycle, Krebs cycle) in which carbohydrates, fats, and proteins are converted into carbon dioxide and water in the mitochondrion, providing reducing equivalents (NADH or $FADH_2$) to power the electron transport chain. The cycle also provides intermediates for various biosynthetic processes.

trichiniasis (trichinosis) Infestation of the intestine with the nematode *Trichinella* (formerly *Trichina*) *spiralis*, the larvae of which migrate from the gut and become encysted in muscle. Can be picked up as a result of eating infected meat.

trichinosis *See* TRICHINIASIS.

Trichoderma viride A common ascomycete fungus found in soil and decaying vegetation that can cause infections in immunocompromised individuals.

trichohyalin A hair follicle protein (1943 aa) similar to *profilaggrin, *involucrin, and *loricrin. It forms intermediate filament-like structures in the mature cells of the hair follicle's inner root sheath and becomes extensively crosslinked by *transglutaminase.

trichomoniasis A sexually transmitted infection with *Trichomonas vaginalis*. Many other trichomonads infect domestic animals.

trichosetin An antibiotic produced when *Trichoderma harzianum* and *Catharanthus roseus* callus are cocultured. Has activity against Gram-positive bacteria. It is a homologue of *equisetin.

trichostatin A An antifungal antibiotic isolated from *Streptomyces platensis*. It is a potent inhibitor of *histone deacetylase (HDAC) and will activate the transcription of genes that have been silenced by methylation.

trichothecenes Mycotoxins (T-2 toxin, HT-2 toxin, diacetoxyscirpenol, deoxynivalenol) produced by various species of fungi that can occur as contaminants in foodstuffs. They are toxic for granulocytic and erythroblastic progenitor cells.

Trichuris trichiura An intestinal nematode parasite, particularly common in tropical Asia.

tricyclic antidepressants A group of drugs that were used to treat moderate to severe depression. They have adrenergic and serotonergic effects and probably work by inhibiting the reuptake of norepinephrine and serotonin. They also decrease the effects of histamine on H_1 receptors and have sedative effects. Have largely been replaced by *selective serotonin reuptake inhibitors (SSRIs).

tricyclodecan-9-yl xanthate An inhibitor (D609) of sphingolipid biosynthesis and phosphatidylcholine-specific phospholipase C. *See* PROSAPOSIN.

TRIF *See* TICAM-1.

triflavin *See* DISINTEGRINS.

trifluoperazine (trifluperazine) An antipsychotic drug that inhibits *calmodulin at near-cytotoxic concentrations. *TN* Stelazine.

trigeminal system The neurons associated with the fifth (trigeminal) nerve that provides sensory innervation to the face and mucous membrane of the oral cavity, and motor innervation to jaw muscles. There are three major peripheral branches: the ophthalmic, the maxillary, and the mandibular nerves.

triggering receptor expressed on myeloid cells 2 (TREM-2) One of a family of receptors that regulate myeloid cell function. TREM-2 is expressed on macrophages, microglia, and pre-osteoclasts and signals through the adaptor DAP12. Defects of TREM-2 and DAP12 result in a rare syndrome characterized by presenile dementia and bone cysts.

trigger protein *See* U PROTEIN (3).

triglycerides The storage fats in adipose tissue, mostly glycerol esters of saturated fatty acids. They are major components of very low density *lipoprotein (VLDL) and chylomicrons and are an important energy source. High levels in plasma have been linked to atherosclerosis (normal levels are <1.69 mmol/L).

trigramin *See* DISINTEGRINS.

trihexyphenidyl (benzhexol) A drug used to treat the symptoms of *Parkinson's disease that binds to the M_1 *muscarinic receptor. The exact mode of action is, however, unclear. *TNs* Aparkan, Artane.

triiodothyronine (T_3) *See* THYROID HORMONES.

TRIM **1.** T-cell receptor interacting molecule. A transmembrane adaptor (TRAP) in lymphocytes. **2.** TRIM5α is a member of the tripartite motif (TRIM) protein family, an antiviral protein that will bind retroviral core proteins and target them for premature disassembly or destruction. TRIM5 proteins have a tripartite motif comprising RING, B-Box and coiled-coil domains; the antiviral α splice variant additionally encodes a B30.2 domain which is recruited to incoming viral cores and determines antiviral specificity.

trimethaphan camsylate An anticholinergic drug used to acutely reduce blood pressure in emergency. *TN* Arfonad.

trimethoprim A drug that inhibits the reduction of dihydrofolate (DHF) to tetrahydrofolate and is selective for some bacterial DHF reductases. Often used in conjunction with *sulphonamides which inhibit an earlier step in bacterial folate metabolism.

trimipramine A tricyclic antidepressant. *TN* Surmontil.

trinucleotide repeat A repeated nucleotide motif that may occur in either exons or introns. The number of repetitions is often polymorphic and unstably amplified repeats appear to be the major cause of diseases such as *Huntington's disease, *fragile X syndrome, spinobulbar muscular atrophy, and *dystrophia myotonica.

trio A protein (2861 aa) with triple functional domains that may be important in several signalling pathways controlling cell proliferation. It binds the tyrosine phosphatase *LAR, has a serine/threonine protein kinase domain and separate rac- and rho-specific *GEF-domains.

triple A syndrome An autosomal recessive neuroendocrinological disease (alachrima–achalasia–adrenal insufficiency syndrome, addinsonianism) in which there is defective tear formation (alacrima), achalasia of the stomach cardia, and adrenal hormone deficiency (hypoadrenalism). It is caused by mutations in a gene that encodes a *nucleoporin, aladin (546 aa), a component of the nuclear pore complex.

triple response (weal and flare) The set of visible changes in the skin in response to mild mechanical injury. Increased blood flow leads to reddening (flare, rubor), warmth (calor) and swelling (weal, tumour). Redness, heat, and swelling are three of the 'cardinal signs' of *inflammation.

triple vaccine A vaccine used in infancy to induce immunity to diphtheria, pertussis (whooping cough), and tetanus (DPT), a sterile preparation of diphtheria and tetanus toxoids with acellular pertussis vaccine. The triple vaccine for measles, mumps, and rubella is usually referred to as the *MMR vaccine.

triploid Having three times the haploid chromosome complement.

triptans A family of drugs used for the acute treatment of migraine; serotonin 5-$HT_{1B}/_{1D}$ receptor agonists that induce vasoconstriction of extracerebral blood vessels and reduce neurogenic inflammation. Examples include sumatriptan (Imitrex®, Imigran®), zolmitriptan, and naratriptan.

triskelion A three-legged structure, sometimes seen in heraldic devices, the shape assumed by *clathrin isolated from *coated vesicles. It is a trimer of clathrin and probably the physiological subunit of the clathrin coat.

trisomy A condition in which there is an additional copy of a chromosome so that part of the genome is tetraploid. The classic example is trisomy 21 in *Down's syndrome. Trisomy 13 is *Patau's syndrome.

tristetraprolin A basic proline-rich protein containing three PPPP repeats, and an unusual zinc finger structure. It binds ARE motifs (*AU-rich elements) in mRNA and regulates the stability of ARE-containing mRNA.

tritanopia The rare form of *colour blindness that is due to deficiency of short-wave sensitive (420 nm) blue visual pigment. Unlike red–green forms of colour blindness the disorder is dominant, presumably because the mutated opsin interferes with cone function.

Triton X-100 A nonionic detergent often used when isolating membrane proteins: the detergent substitutes for the phospholipid environment in which the proteins are normally embedded and helps to preserve the conformation. There are other Triton-group detergents.

trk An oncogene that is formed through a somatic rearrangement of the neighbouring genes for neurotrophic tyrosine kinase receptor type 1 (*NTRK1) and tropomyosin-3. The ***trkA*** gene product (790 aa) is a receptor for *NGF, that of ***trkB*** for *neurotrophin 4 (NT-4) and *BDNF, and that of ***trkC*** for NT-3. The trk-fused gene product (400 aa) associates with NTRK1 tyrosine kinase in generating the thyroid *trk-T3* oncogene found in papillary thyroid carcinoma. There are several chimeric genes in tumours in which the tyrosine kinase domain of trk has become linked to 5′ sequences from unrelated loci, yielding products with ectopic, constitutive tyrosine kinase activity.

tRNA *See* TRANSFER RNA.

trophectoderm The extraembryonic portion of the ectoderm of the blastula that gives rise to the *trophoblast.

trophic Something associated with food or nutrition, often as combining form (e.g. *neurotrophic). Not to be confused with tropic (stimulatory).

trophinin An integral membrane protein (749 aa) that is involved in the adhesion of the blastocyst to the endometrial epithelium at the time of implantation. **Trophinin-assisting protein** (tastin, 778 aa) forms a complex with trophinin through *bystin. Trophinin and bystin (a trophinin- and tastin-binding protein) are found in the placenta from the sixth week of pregnancy but disappear after week 10.

trophoblast The extraembryonic layer that forms around the mammalian blastocyst, and attaches the embryo to the wall of the uterus. It forms the outer layer of the chorion, and contributes, together with maternal tissue, to the placenta.

trophozoite The feeding stage of a protozoan, as opposed to the reproductive or encysted stages.

tropicamide An antimuscarinic drug used in eye drops to dilate the pupil for ophthalmological purposes. *TN* Mydriacyl.

tropocollagen The trimeric subunit, derived from *procollagen by cleavage of the telopeptides, from which collagen fibrils self-assemble.

tropoelastin The soluble polypeptide precursor of *elastin. Secreted tropoelastin molecules are aligned on a scaffold of *fibrillin-rich microfibrils and crosslinked by the formation of *desmosine to form elastin fibres.

tropomodulin An F-actin-capping protein (359 aa) that interacts with *tropomyosin (TM) at the pointed end of actin filaments. Binds actin, tropomyosin, and *nebulin. In the erythrocyte membrane skeleton **tropomodulin 1** modulates the interaction of tropomyosin with the spectrin-actin complex and affects the viscoelastic properties of the cells. There are ubiquitous (**tropomodulin 3**, 352 aa) and neuron-specific (**tropomodulin-2**, 351 aa) forms. **Tropomodulin 4** (345 aa) is found in striated muscle. *See* LEIOMODIN.

t

tropomyosins A family of proteins encoded by four genes with diverse isoforms that are expressed in a tissue-specific manner and regulated by an alternative splicing mechanism. They are associated with the actin filaments of myofibrils and stress fibers. In thin filaments of striated muscle they are associated with *troponin. Each tropomyosin has six or seven similar domains and each domain interacts with a G-actin molecule. Mutations in **tropomyosin 1** are associated with a form of hypertrophic cardiomyopathy. **Tropomyosin 2** (β-tropomyosin) is mainly expressed in slow, type 1 muscle fibres and mutations are associated with nemaline *myopathy-4 and distal *arthrogryposis type 2B. **Tropomyosin 3** is found together with TPM2 in slow muscle fibres. TPM 3 is part of a chimeric oncogene (*see* TRK) and mutations in TPM3 cause nemaline myopathy-1. **TPM4** is found in skeletal muscle undergoing growth and repair and also in osteoclasts.

troponin A complex of three proteins, troponins C, I, and T, (TnC, TnI, TnT) that forms a 1:1 association with *tropomyosin. The binding of calcium by troponin C (161 aa) causes a conformational change in the troponin–tropomyosin complex which shifts position on the thin filament of the sarcomere, making the myosin-binding site on actin accessible and allowing a cross-bridge to form. **Troponin C** has a variable number of *EF-hand motifs and is very highly conserved, although there are different isoforms in slow and fast muscle. TnC binds TnI (cardiac, 210 aa; fast 182 aa, slow, 187 aa) and TnT, but not actin. **Troponin I** binds to actin and can inhibit the actin–myosin interaction on its own, but not in a calcium-sensitive manner. **Troponin T** (278 aa and other isoforms) binds strongly to tropomyosin. Mutation in troponins cause various *myopathies.

Troyer's syndrome *See* SPARTIN.

trp **1.** Tyrosinase-related proteins Trp1 (537 aa), Trp2 (517 aa), found in *melanosomes, involved in pigment formation in the hair and mutated in forms of albinism. **2.** *See* TRYPTOPHAN. **3.** *See* TRANSIENT RECEPTOR POTENTIAL CHANNELS.

TRT Telomerase reverse transcriptase. *See* TELOMERASE.

Truvada A fixed dose combination of two antiretroviral drugs, used for the treatment of HIV. The two drugs are tenofovir and emtricitabine, both inhibitors of reverse transcriptase.

Trypan blue An azo dye (diamine blue, Niagara blue 3B) often used to determine cell viability. Trypan blue is excluded from live cells but labels dead cells a strong blue colour.

Trypanosoma A genus of Protozoa responsible for serious infections (trypanosomiasis) of humans and domestic animals in tropical regions. African trypanosomes, of the brucei group, are carried by tsetse flies (*Glossina*) and go through a complex series of developmental stages in the blood of the mammalian host. The parasites evade immune response by *antigenic variation and there are recurrent bouts of parasitaemia. The South American trypanosomes (of which *T. cruzi* is the best known) are carried by reduviid bugs, and cause a chronic and incurable disease (Chagas' disease).

trypomastigote A trypanosome-like stage in the life cycle of certain protozoa, resembling the typical adult form of members of the genus **Trypanosoma*. The trypomastigote of *Trypanosoma* spp. is the infectious form carried by the insect vector.

trypsin A pancreatic serine peptidase (EC 3.4.21.4, 247 aa) that cleaves peptide bonds involving the amino groups of lysine or arginine. Three **trypsinogens** (the inactive zymogen) are found in human pancreatic secretions, **cationic trypsinogen** (PRSS1, ~65%), **anionic trypsinogen** (PRSS2, 30%) and **mesotrypsinogen** (5%) Mutations in the gene encoding cationic trypsinogen is one cause of hereditary pancreatitis, but a missense variant in the *PRSS2* gene confers protection against chronic pancreatitis.

tryptophan (Trp, W) One of the 20 amino acids (204 Da) found in proteins and an essential component of the diet for humans.

TSH *See* THYROID STIMULATING HORMONE.

TSH releasing factor *See* THYROTROPIN RELEASING HORMONE.

t-SNARE *See* SYNTAXINS.

Tst A gene in some strains of *Staphylococcus aureus* that codes for toxic shock syndrome toxin-1.

TTC **1.** Triphenyltetrazolium chloride, a compound that can be reduced by cells to produce formazan and used as a stain to evaluate the volume of infarction in tissue. **2.** Threshold of toxicological concern, a parameter used in risk assessment, the threshold exposure level below which there is a very low probability of an appreciable health risk. **3.** *Tetanus toxin C-fragment (TTC). **4.** A nucleotide triplet that is part of the GAATTC repeat sequence amplified in *Friedreich's ataxia. **5.** A stable prostacyclin analogue, TTC-909.

ttp *See* SH3BP4; THROMBOTIC THROMBOCYTOPENIC PURPURA; TRISTETRAPROLIN.

TTSS (T3SS) A bacterial secretory system, a multisubunit membrane-spanning macromolecular assembly, encoded in salmonella pathogenicity island 2 (SPI2), that enables many Gram-negative bacterial pathogens to translocate proteins into the eukaryotic host cells they infect.

T tubule (transverse tubule) An invagination of the sarcolemma of striated muscle that becomes depolarized when muscle is stimulated and triggers the release of calcium from the adjacent tubular portions of the sarcoplasmic reticulum (SR). The T-tubule is flanked by tubular portions of the SR and usually lies near the A-band/I-band junction.

TTX *See* TETRODOTOXIN.

T-type channels Low-voltage-activated calcium channels, so called because they have transient and tiny currents, are rapidly inactivated and have low conductance. They respond to relatively small depolarizations and are involved in cardiac pacemaking and regulation of blood flow. Three family members have been cloned (α1G, α1H, and α1I). Mutations in α1H are associated with idiopathic generalized epilepsy. The channels are modulated by various hormones and neurotransmitters. Efonidipine blocks both L- and T-type channels.

tuba A dynamin-binding protein (1577 aa) that activates *cdc42 and interacts with various actin-regulatory proteins such as *WASP. It is one of the *dbl family of guanine nucleotide exchange factors and is involved in regulation of tight junctions.

SEE WEB LINKS

- Additional information on the BAR domain superfamily.

tubby A small family of proteins associated with regulation of fat deposition. Mutation in the homologous mouse gene leads to progressive retinal degeneration, deafness, and maturity-onset obesity. **Tubby** (561 aa and 506 aa isoforms) is a transcription factor that translocates to the nucleus in response to phosphoinositide hydrolysis and shares some common features with *scramblases. **Tubby-like genes** (TULPs) have been found in plants, vertebrates, and invertebrates. **Tubby-like protein 1** (TULP1, 489 aa) is a photoreceptor-specific protein that, when mutated, may cause *retinitis pigmentosa. TULP2 (520 aa) and TULP3 (442 aa) have different tissue distributions.

tubercle A chronic granulomatous inflammatory focus caused by *Mycobacterium tuberculosis*.

tuberculin skin test *See* HEAF TEST; MANTOUX TEST.

tuberculosis (TB, formerly **consumption)** An infectious disease caused by *Mycobacterium tuberculosis* (or other *mycobacteria, particularly *Mycobacterium bovis* from cattle). The infection is usually pulmonary but can affect other tissues (notably *miliary tuberculosis). One feature of the disease that makes treatment more difficult is that the bacterium lives intracellularly and is therefore protected from the immune system. Although antibiotics reduced the incidence considerably, drug-resistant TB, particularly in immunocompromised individuals, is an increasing problem and TB is still one of the major causes of mortality and morbidity worldwide. *See* BACILLE CALMETTE–GUÉRIN; MYCOBACTERIA.

tuberin *See* TUBEROUS SCLEROSIS.

tuberous sclerosis An autosomal dominant disorder caused by mutation in tumour suppressor genes *TSC1* or *TSC2* on chromosomes 9 and 16. Other forms (*TSC3* and *TSC4*) are due to mutations in loci on chromosomes 11 and 12 respectively. The disease is characterized by a range of features including seizures, mental retardation, renal dysfunction, and dermatological abnormalities. *TSC1* encodes hamartin (1164 aa) which interacts with the product of *TSC2*, tuberin (1784 aa) to form a tumour suppressor complex. Tuberin has no homology with hamartin but acts in conjunction with it. Tuberin apparently stimulates the intrinsic GTPase activity of the ras-related protein rapP1A. Loss of the TSC genes leads to constitutive activation of *mTOR and downstream signalling elements, resulting in tumour development, neurological disorders, and severe insulin/IGF1 resistance.

tubocurarine *See* CURARE.

tubulin The cytoplasmic protein (~450 aa) from which *microtubules are assembled. The assembly subunit (protomer) is a heterodimer of α- and β-tubulin. There are multiple isoforms of tubulin expressed in tissue-restricted ways but functionally equivalent. **γ-Tubulin** is restricted to the microtubule organizing centre (*centrosome) and is involved in nucleation of microtubule assembly. Tubulin mutations can cause *lissencephaly. **γ-Tubulin complex protein 5** (978 aa) is a core component of the centrosome.

tularaemia A disease (deer-fly fever) caused by infection with the Gram-negative bacterium *Francisella tularensis* (formerly *Pasteurella tularensis*); an acute, febrile, granulomatous infection. A zoonosis spread readily from animals to humans, usually by ticks.

tumbu disease A disease in which the skin is invaded by larvae of the tumbu fly *Cordylobia anthropophaga*, producing a boil or a warble. Common in Central and West Africa.

tumor *See* TUMOUR.

tumorigenic Describing something capable of causing tumours, e.g. a carcinogen, an agent such as radiation, an oncogenic virus, or transformed cells that will produce tumours if introduced into animals.

tumour Strictly, any abnormal swelling, but usually a mass of neoplastic cells. *See* NEOPLASIA.

tumour angiogenesis factor (TAF) A substance, or more probably a variety of different substances, released from a tumour and that promotes vascularization of the tumour (angiogenesis). *See* ANGIOGENIN; VEGF.

tumour necrosis factor (TNF) A pro-inflammatory cytokine (TNFα, cachectin, 157 aa) originally found as a tumour-inhibiting factor in the blood of animals exposed to bacterial *lipopolysaccharide or *bacille Calmette–Guérin. (BCG). Although it kills tumour cells it also has a wide range of pro-inflammatory actions. Soluble TNFα is released from the cell surface by the action of *TACE (TNFα converting enzyme). **TNFβ** (*lymphotoxin) has 35% structural and sequence homology with TNFα and binds to the same *TNF receptors. Unlike TNFα, it is secreted in a conventional manner from activated T and B cells.

tumour progression The process that is thought to occur in the course of development of a tumour. Implicitly, the idea that more than one change must occur to cause full malignancy and that initiation must be followed by other changes. In many tumours heterogeneity develops as a result of further mutational events.

tumour promoters Compounds that increase the probability of tumour formation when applied after administration of a primary carcinogen, although not themselves considered carcinogenic. The concept arose in experimental studies on rodents and the best-known agent was croton oil in which the active ingredients were *phorbol esters (hence their curious misnomer as 'tumour promoting activity'). They differ from *cocarcinogens which are only active when administered concurrently with the primary carcinogen, although the distinction is rarely made. Most tumour promoters are probably carcinogens if tested more thoroughly, and may contribute to *tumour progression.

tumour specific antigen A tumour cell antigen (tumour specific transplantation antigen, TSTA) that elicits a cell-mediated immune response. In virus-transformed cells TSTA (unlike *T-antigen) differs between tumours, even when the same virus is involved. It seems very unlikely that any generic tumour-specific antigen exists.

tumour suppressor Generally a gene (antioncogene, cancer susceptibility gene) encoding a negative regulator of the cell cycle, e.g. an inhibitor of a growth factor signalling system, that must be mutated or otherwise inactivated for unregulated proliferation (*neoplasia). There are negative regulators of tumour suppressors (e.g. *hdm2, *mena) that, when overexpressed, increase suceptibility to tumours. The number of known and suspected tumour suppressors is growing: *See* BASC COMPLEX; BRIDGING INTEGRATORS; CTCF; DEATH-ASSOCIATED PROTEIN; DELETED IN COLORECTAL CANCER; DELETED IN MALIGNANT BRAIN TUMOURS-1; EXOSTOSINS; GCIP; H19 GENE; ING1; INK4; IRFs; MENIN; MERLIN; MYELOID DERIVED SUPPRESSOR CELLS; MYOPODIN; P53; P107; PARAFIBROMIN; PCGF; PINX1; PML BODY (PML PROTEIN); POLYPOSIS COLI; PROHIBITINS; PTEN; RETINOBLASTOMA (RB); SMCT; TES; TRANSMEMBRANE PROTEINS WITH EGF-LIKE AND 2 FOLLISTATIN-LIKE DOMAINS; TUBEROUS SCLEROSIS; VON HIPPEL–LINDAU TUMOUR SUPPRESSOR PROTEIN; WILMS' TUMOUR; WWOX; ZF9.

tumour suppressor complex (tuberous sclerosis complex) *See* TUBEROUS SCLEROSIS.

tumour virus *See* AVIAN LEUKAEMIA VIRUS; DNA TUMOUR VIRUS; EPSTEIN–BARR VIRUS; FRIEND MURINE LEUKAEMIA VIRUS; KIRSTEN SARCOMA VIRUS; MAMMARY TUMOUR VIRUS; SHOPE FIBROMA VIRUS, and YABA VIRUS. *Also* PAPILLOMAVIRIDAE; POLYOMAVIRIDAE, and RETROVIRIDAE.

tumstatin The N-terminal portion of an isoform of *collagen type IV (COL4A3) that is the autoantigen in *Goodpasture's syndrome.

TUNEL method A method (transferase-mediated dUTP nick-end labelling) used to visualize apoptotic cells by labelling the ends of their fragmented DNA.

tunica media *See* MEDIA.

tunicamycin A mixture of antibiotics produced by *Streptomyces lysosuperificus* that inhibit *N*-glycosylation by blocking the addition of *N*-acetylglucosamine to dolichol phosphate. A useful experimental tool.

Turcot's syndrome A combination of colorectal polyposis and primary tumours of the central nervous system caused by mutation in one of the components of the *mismatch repair system.

Turner–Kieser syndrome *See* NAIL–PATELLA SYNDROME.

Turner's syndrome A condition (Ullrich–Turner syndrome) in which all or part of the X chromosome is missing so that females are XO (monosomy X). A range of morphological, metabolic, and cognitive disorders arise and there may be X-linked diseases that are normally confined to males. *See also* NOONAN'S SYNDROME.

turnover number A parameter in enzyme kinetics (equivalent to V_{max}); the number of substrate molecules converted to product by one molecule of enzyme in unit time with excess substrate available.

TW-240/260 Obsolete name for a *spectrin-like protein (240/260 kDa) found in the *terminal web of intestinal epithelial cells.

TWEAK One of the TNF family (TNF-related weak inducer of apoptosis, 249 aa) expressed on interferon-γ-stimulated monocytes, that will induce cell death in some tumour cell lines. Probably the ligand for *TRAMP.

Tween A family of detergents often used for solubilizing membrane proteins or for lysing mammalian cells. Tween variants (Tween-20, -40, -60, -80) have fatty acid moieties of differing chain length.

twinfilin An actin-depolymerizing factor (350 aa) that has two *ADF-H domains and forms an assembly-incompetent 1:1 complex with G-actin. Found in the cortical actin cytoskeleton of fibroblasts.

twinkle A mitochondrial protein (684 aa) with structural similarity to various *helicases. The protein colocalizes with mtDNA and the unusual localization pattern is said to be reminiscent of twinkling stars; expression is relatively high in skeletal muscle and pancreas. A splice variant lacks residues 579–684 but localizes in a similar way and was christened **twinky**. Mutations are associated with infantile-onset *spinocerebellar ataxia and mutations in the same open reading frame are also associated with autosomal dominant *progressive external ophthalmoplegia (PEOA3), *sensory ataxia neuropathy, dysarthria, and ophthalmoparesis (SANDO) and some cases of autosomal recessive *mitochondrial DNA deletion syndrome of the hepatocellular type.

twinky *See* TWINKLE.

twist A basic helix–loop–helix transcription factor that is involved in promoting the *epithelial-mesenchymal transition. Twist-1 and -2 repress cytokine gene expression through interaction with RelA, and transiently inhibit *Runx2 function during development of the skeleton. Loss-of-function mutations of the *TWIST1* gene are responsible for *Saethre–Chotzen syndrome.

twitch muscle A striated muscle innervated by a single motoneuron that exhibits an all-or-none response. The sarcolemma is electrically excitable. Almost all skeletal muscles are twitch muscles and they are often subdivided into fast- and slow-twitch types, the former being associated with rapid movement.

two-hybrid screen A screening method (yeast two-hybrid system) used to find proteins that interact with one another. Both are expressed in yeast as *fusion proteins, one with the DNA-binding site of the *GAL4 transcription factor, and the other with the transcriptional activator domain of GAL4. If the two proteins interact then the two GAL4 domains are brought together and will trigger expression of a *reporter gene (usually *LacZ) downstream of the GAL4 promoter. Similar systems using other transcription factors have been developed. Interactions identified using the system need to be confirmed by other methods such as a *pull-down assay.

two-pore channels (two-pore segment channels) Ion channels similar to voltage-sensitive calcium and sodium channels but that have only two of the domains, with six transmembrane segments not four. They are related to *CATSPER and *transient receptor potential channels. Two-pore channel 1 (TPC1, 816 aa) and TPC2 (752 aa) are voltage-sensitive calcium channels, and variations in TPC2 affect skin *pigmentation.

TxA2 *See* THROMBOXANES.

TY-5 *See* MACROPHAGE INFLAMMATORY PROTEINS (1a).

tyk2 A non-receptor tyrosine kinase (EC 2.7.1.112, 1187 aa) upstream of STAT3, important in the initiation of type I interferon signalling and other cytokine signalling pathways and an important regulator of lymphoid tumour surveillance. Deficiency of tyk2 causes autosomal recessive hyper-IgE syndrome (HIES) with atypical mycobacteriosis, a primary immunodeficiency characterized by recurrent skin abscesses, pneumonia, and elevated serum IgE.

typhoid An enteric fever caused by infection with *Salmonella typhi*, usually through consumption of faecally contaminated food or water.

typhus A fever (spotted fever) caused by infection with *Rickettsia prowazekii*, transmitted by lice. A less serious infection, *scrub typhus, is caused by a related rickettsia.

tyrosinase A copper-containing monoxygenase (EC 1.14.18.1, 548 aa) that catalyses the oxidation of tyrosine and leads to *melanin production. Lack of tyrosinase activity is responsible for albinism.

tyrosine (Tyr, Y) An amino acid (181 Da) incorporated into protein. It is nonessential in humans because it can be synthesized from phenylalanine.

tyrosine hydroxylase The enzyme (EC 1.14.16.2, four splice variants of 497–528 aa) involved in the conversion of phenylalanine to dopamine and the rate-limiting enzyme in the synthesis of catecholamines (*noradrenaline and *dopamine). Mutations cause one form of *Segawa syndrome.

tyrosine kinase A protein kinase that phosphorylates tyrosine residues. There are two subfamilies: receptor tyrosine kinases with an extracellular ligand-binding domain and an intracellular tyrosine kinase domain; and nonreceptor tyrosine kinases, which are soluble, cytoplasmic kinases. Both forms play important roles in signalling systems.

tyrphostins Small-molecule inhibitors of *tyrosine kinases derived from the naturally occurring erbstatin. They competitively inhibit substrate binding without affecting the binding of ATP. Tyrphostin A25 inhibits the GTPase activity of transducin, blocks the induction of inducible nitric oxide synthase in glial cells and induces apoptosis in human leukaemic cell lines. Various other tyrphostins have different specificities.

SEE WEB LINKS

- Review in the *Annual Review of Biochemistry* (2006).

U1 snRNP One of the classes of *small nuclear RNAs. The U-type snRNPs have a high uridylic acid content; U1–U5 are synthesized by RNA polymerase II, U6 by RNA polymerase III. U2 snRNP auxiliary factor (U2AF) is a major determinant of 3′ splice-site selection, a heterodimeric protein composed of a large (475 aa) subunit (hU2AF65) and a small (240 aa) subunit (hU2AF35).

U2OS cells A cell line derived from an osteosarcoma.

U937 cells A human myelomonocytic cell line derived from a patient with histiocytic leukaemia. Often used as a model for myeloid cells, although the cells are rather undifferentiated.

UASG Upstream activation site G. *See* GAL PROMOTER.

UBCs **1.** UBC is ubiquitin C (polyubiquitin, 546 aa). The mRNA has nine ubiquitin coding units and is presumably proteolytically cleaved to produce ubiquitin monomers. **2.** *See* UBIQUITIN CONJUGATING ENZYMES

ubiquinone (coenzyme Q_{10}) A small-molecule electron carrier in the respiratory chain. It has a hydrocarbon chain (Q_{10} has ten isoprenoid side chains). Reduction produces ubiquisemiquinone (QH). **Primary coenzyme Q_{10} deficiency** is a rare, clinically heterogeneous autosomal recessive disorder, with five major phenotypes including a predominantly myopathic form with central nervous system involvement, an infantile encephalomyopathy with renal dysfunction, and an ataxic form with cerebellar atrophy. It can be caused by mutations in mitochondrial parahydroxybenzoid-polyprenyltransferase, *aprataxin, decaprenyl diphosphate synthase subunits 1 or 2, and a chaperone-like protein encoded by the *CABC1* gene.

ubiquitin A small protein (76 aa) found ubiquitously in eukaryotic cells. *Ubiquitin conjugating enzymes (Ubcs) catalyse the linkage of ubiquitin to the lysine side chains of proteins (ubiquitinoylation) in an ATP-requiring process making the protein/ubiquitin complex liable to rapid proteolytic degradation. **Ubiquitinylation** (ubiquitination) is important in a wide range of cellular activities. Ubiquitin C (UBC) is a multimeric ubiquitin which is proteolytically processed to produce nine monomers.

ubiquitin C-terminal hydrolase L1 A cysteine endopeptidase (UCH-L1, ubiquitin C-terminal esterase L1, EC 3.1.2.15, 212 aa) found in neurons (~2% of total protein) that is thought to cleave polyubiquitin (ubiquitin C) to monomers and hydrolyse bonds between ubiquitin and small adducts. Found in *Lewy bodies, and mutations in UCH-L1 have been associated with type 5 *Parkinson's disease; polymorphisms in the gene affect susceptibility to Parkinson's. A similar esterase **UCH-L3** (230 aa) is found in a wider range of tissues.

ubiquitin conjugating enzymes A family of enzymes (ubcs), more than 30 in humans, that are involved in the addition of ubiquitin to proteins thereby 'flagging' them for degradation. The E1 enzyme (ubiquitin-activating enzyme) adenylates ubiquitin and forms a thioester-linked complex that is transferred to the E2 enzyme which interacts with an E3 ligase that transfers the ubiquitin to the target protein. There are multiple E3 ligases generally subdivided into four classes: *HECT-type, RING-type, PHD-type, and U-box containing, each of which targets specific proteins. *See also* N-END RULE.

ubiquitin ligase (E3 ligase) *See* UBIQUITIN CONJUGATING ENZYMES.

ubiquitinylation (ubiquitinoylation) *See* UBIQUITIN.

ubisemiquinone *See* UBIQUINONE.

U-box *See* UBIQUITIN CONJUGATING ENZYMES.

UDG *See* URACIL-DNA GLYCOSYLASE.

UDP-galactose A sugar nucleotide (uridine diphosphate-galactose) used in galactosyl transfer reactions.

UDP-glucose A sugar nucleotide (uridine diphosphate-glucose) used in glucosyl transfer reactions.

UK1 **1.** The recombinant soluble kringle domain (UK1) of the urinary *plasminogen activator (uPA) that has antiangiogenic activity. **2.** A pathogenic strain of *Salmonella typhimurium* (UK1). **3.** A benzoxazole antibiotic isolated from *Streptomyces* sp. that inhibits *topoisomerase II. It has potent cytotoxic activity against B16, HeLa, and P388 cells.

ularitide A synthetic form of *urodilatin.

ulcer An area of inflammation where the epithelium and underlying tissue is eroded.

ulcerative colitis *See* INFLAMMATORY BOWEL DISEASE.

ultradian Describing cyclic processes or activities that occur with a frequency of less than 24 h. *Compare* CIRCADIAN RHYTHM.

ultrafiltration Pressure-driven filtration. An ultrafiltrate is formed in the kidney because blood is at higher pressure than the fluid in the glomerular lumen. Ultrafiltration is used experimentally to fractionate and concentrate solutions using selectively permeable artificial membranes.

ultraviolet (UV) The portion of the spectrum beyond the violet end of the visible spectrum (<400 nm) but above the X-ray wavelengths (>5 nm). Conventionally subdivided into **UV-A** (400–320 nm), **UV-B** (320–280 nm), and **UV-C** (280–100 nm). Nucleic acids absorb UV most strongly at ~260 nm, the wavelength most likely to cause the formation of thymine dimers and the UV component of sunlight that causes *keratoses. UV-A (and the shorter wavelengths) will cause damage to collagen in the dermis; UV-B is important for *vitamin D synthesis in the skin but causes sunburn in excess. UV light is absorbed by glass, but optical systems can be made of quartz.

umami One of the five basic tastes (salt, sweet, sour, bitter, and umami). Umami is generally described as the taste of monosodium glutamate.

UNC-5 family A family of receptors (~930 aa) for *netrins, involved in repulsion of growing axons but also in inhibition of angiogenesis. Down-regulation of UNC-5 proteins is a feature of several cancers. *See* DELETED IN COLORECTAL CANCER.

uncoupling agent Any agent capable of dissociating two linked processes, but most commonly applied to those that uncouple electron transport from oxidative phosphorylation. Ionophores do this by discharging the mitochondrial ion gradient required for *chemiosmosis. *See* THERMOGENIN.

uncoupling proteins *See* THERMOGENIN.

underlapping Describing the outcome of collision between two cells in culture where one cell crawls underneath the other, retaining contact with the substratum. Unlike the *overlapping situation, the underlapping cell retains tension-resisting anchorages.

unequal crossing-over Describing *crossing-over that results in nonequivalent exchange of material between imprecisely-paired homologous chromosomes. Tends to happen particularly in regions with extensive tandem repeat sequences.

uniport A transmembrane transport process in which only one species of substance moves across a membrane. *Compare* COTRANSPORT; COUNTER-TRANSPORT.

UniProt The Universal Protein Resource database, a comprehensive resource for protein sequence and annotation data. It is maintained through a collaboration between the European Bioinformatics Institute (EBI), the Swiss Institute of Bioinformatics (SIB) and the Protein Information Resource (PIR).

SEE WEB LINKS

- EBI homepage.
- Swiss Institute of Bioinformatics homepage.
- PIR homepage.

unit membrane *See* PHOSPHOLIPID BILAYER.

unsaturated fatty acid A *fatty acid containing one or more double (unsaturated) bonds.

untranslated region The portions of a *messenger RNA that are not translated into amino acid sequence. The 5′ leader sequence (5′ UTR) may be more than 100 nucleotides long and contains a ribosome binding site. The 3′ UTR, which may be several kilobases long, presumably has some function, although that function is unclear, and is terminated by the poly-A tail. The 5′ end also has a cap of a few bases.

uPA *See* PLASMINOGEN ACTIVATOR.

U protein **1.** Heterogeneous ribonuclear protein U (scaffold attachment factor A, 806 aa), the largest of the major hnRNP proteins: contains an RNA binding domain and a scaffold-associated region (SAR)-specific bipartite DNA-binding domain. It is thought to be involved in the packaging of hnRNA into large ribonucleoprotein complexes. **2.** Occasional abbreviation for 'urinary protein'. **3.** (*Obsolete*) Hypothetical protein (trigger protein) thought to regulate the transition of cells from G_0 to G_1 phase of the cell cycle. Superseded by *cyclins.

upstream Describing: **1.** Early events in any process that involves sequential reactions. **2.** DNA sequences preceding the transcription start site. **3.** mRNA sequences preceding the protein-coding sequence (5′ UTR).

uracil The pyrimidine base (2,6-dihydroxypyrimidine) to which ribose is added to produce *uridine.

uracil-DNA glycosylase An enzyme (UNG, UDG, EC 3.2.2.-) that removes uracil in DNA that has arisen from deamination of cytosine or incorporation of dUMP instead of dTMP, thereby suppressing GC-to-AT transition mutations. There are two alternatively spliced isoforms, UNG1M (313 aa) in mitochondria and UNG1N in the nucleus. Defects in **UNG1** are a cause of immunodeficiency with hyper-IgM type 5 syndrome because of impairment in the immunoglobulin (Ig) class-switching mechanism. A second gene encodes **UNG2** (327 aa) which has no homology with UNG1 but homology with the conserved cyclin box region of several cyclins.

uranyl acetate A uranium salt used as a stain in electron microscopy because of its electron density.

urea The final nitrogenous excretion product ($(NH_2)_2CO$) of many organisms (*compare* URIC ACID). Urea solutions can be used to denature proteins.

urea cycle (ornithine cycle) The cycle of biochemical reactions occurring in the liver that produces urea from ammonia. *See* ORNITHINE TRANSCARBAMYLASE.

uric acid The final nitrogenous excretion product in animals that need to conserve water. Uric acid has very low solubility in water, and crystals may be deposited in tissues as a *tophus.

uricase The liver enzyme (EC 1.7.3.3, urate oxidase) associated with *peroxisomes that catalyses the oxidation of uric acid to *allantoin, the final step of purine degradation in most mammals (although not primates; the human gene has nonsense codons and is nonfunctional). There are four subunits of 303 aa and one copper atom per molecule.

uridine The ribonucleoside formed from ribose and uracil. The uridyl group is formed by loss of a hydroxyl. *Compare* URIDYLYL.

uridylyl The uridine monophospho group derived from urydylic acid (*compare* URIDINE). Uridylylation is the post-translational addition of a uridylyl group to a protein, RNA, or sugar phosphate. Uridydyl transferase (EC 2.7.7.12) transfers UDP from glucose to galactose. Terminal uridylyl transferase 1 (TUT1, EC 2.7.7.52, 874 aa) specifically catalyses uridylylation of U6 snRNA and is essential for cell proliferation.

urinary plasminogen activator (uPA) *See* PLASMINOGEN ACTIVATOR.

urocanic acid An intermediate formed in the degradation of L-histidine found as the *trans* isomer (t-UA) at ~30 mg/cm^2 in the stratum corneum. Absorption of UV light will cause it to isomerize to the *cis*-isomer which may inhibit some immune responses, including contact and delayed hypersensitivity. **Urocanic aciduria** is an autosomal recessive metabolic disorder caused by a deficiency of the enzyme urocanase.

urocortin A small family of neuropeptides of the CRF (*corticotropin-releasing factor) family synthesized in the anterior pituitary. **Urocortin I** (40 aa; 123 aa precursor) stimulates secretion of ACTH through the type 1 *corticotropin-releasing hormone receptor (CRH receptor) that mediates the 'fight or flight' response. **Urocortin II** (urocortin-related peptide, stresscopin-related peptide, 43 aa) and **urocortin III** (stresscopin, 40 aa) are specific ligands for the type-2 CRH receptor which mediates stress-coping responses.

urodilatin A natriuretic peptide, related to atrial, brain, and C-type natriuretic peptides, that is synthesized and secreted from the distal tubules of the kidney. It is generated by a different post-translational processing of the atrial natriuretic peptide prohormone and because it is secreted into urine is not found in circulation.

urogastrone *See* EPIDERMAL GROWTH FACTOR.

uroguanylin *See* GUANYLIN.

urokinase (uPA) *See* PLASMINOGEN ACTIVATOR.

uromodulin A glycosylphosphatidylinositol-linked cell surface-associated form of of the Tamm–Horsfall glycoprotein (640 aa) which is found as a high molecular weight polymer in urine. It has potent immunosuppressive activity. Mutations cause juvenile hyperuricaemic nephropathy, an autosomal dominant renal disorder characterized by the juvenile onset of *hyperuricaemia, polyuria, progressive renal failure, and gout.

uroplakins A family of proteins found in the asymmetric unit membrane (AUM) of mammalian bladder transitional epithelium (urothelium) where they probably have a mechanical stabilizing function. The major proteins are uroplakins (UPK) 1A (260 aa), 1B, 2, and 3, with uroplakin 3B (p35) as a minor component. Mutations in UPK3A are associated with renal adysplasia (*see* POTTER'S SYNDROME).

uroporphyrinogen I synthetase The third enzyme (EC.4.3.1.8, porphobilinogen deaminase, hydroxymethylbilane synthase) in the biosynthetic pathway leading to haem. There are two isoforms (317 aa and 334 aa), one active in all tissues and the other restricted to erythrocytes. Mutations cause acute intermittent *porphyria. Uroporphyrinogen I (UP1) is converted to UP III by UP III synthetase (hydroxymethylbilane hydrolyase (cyclizing), EC 4.2.1.75, 265 aa) that is defective in autosomal recessive erythropoietic porphyria.

urotensin II (UII) A neuropeptide (11 aa) with potent cardiovascular effects, the endogenous ligand for the orphan G-protein-coupled receptor, GPR14, that is functionally coupled to calcium mobilization. Its sequence is strongly conserved among different species and has structural similarity to *somatostatin. Urotensin I in other species is homologous to *urocortin.

urticaria pigmentosa *See* MASTOCYTOSIS.

usherin A type I transmembrane protein (5202 aa) that is defective in *Usher's syndrome (type 2A) and *retinitis pigmentosa type 39. Usherin interacts with collagen IV and fibronectin in the basal lamina in many tissues and also with *harmonin, *whirlin, and *ninein-like protein (NINL). It is probably a component of the interstereocilia linkages in inner ear sensory cells; in the retina it is present in the basal lamina just beneath the retinal pigment epithelial cells.

Usher's syndrome A rare, autosomal recessive syndrome of congenital deafness and progressive blindness. **Type IB/1A** is caused by mutation in the myosin-7A gene that encodes an unconventional myosin that is expressed in hair cells, **type 1C** by mutations in *harmonin, **type 1D** by mutations in *otocadherin, **type 1F** by mutations in *protocadherin 15, **type 1G** by mutation in the **SANS* gene. Other type 1 variants have been mapped but specific genes have not been identified. **Usher's syndrome type 2** is similar but the hearing loss is less profound and vestibular function is normal. **Type 2A** is caused by mutation in the *usherin gene, **type 2B** is linked to chromosome 3p, **type 2C** involves mutation in the *GPR98* gene (a G-protein-coupled receptor), and **type 2D** by mutation in the *whirlin gene. **Usher's syndrome type 3**, in which there is progressive hearing loss, is caused by mutation in the *clarin 1 gene.

uteroglobin A homodimeric protein (blastokinin, Clara cell phospholipid-binding protein, CC16, secretoglobulin 1A1, 91-aa subunits) secreted by nonciliated secretory epithelial cells (*Clara cells) in the lung. It is a phospholipase A2 inhibitor and plays an important protective role against intrapulmonary inflammation. Uteroglobin was originally identified as a *progesterone-binding protein in lagomorphs and subsequently shown to be identical to the Clara cell secretory protein.

UTR *See* UNTRANSLATED REGION.

utrophin A protein (dystrophin associated protein, 3433 aa) with 80% sequence homology to *dystrophin, localized near the neuromuscular junction in adult muscle where it is associated with a complex of sarcolemmal glycoproteins that link to *laminin 2 in the extracellular matrix. In Duchenne *muscular dystrophy utrophin is also located on the cytoplasmic face of the sarcolemma.

uveitis An inflammation of the interior of the eye with different tissues affected in anterior uveitis, posterior uveitis, etc. In some cases it is a response to infection or injury but in other cases it is associated with autoimmune-type disorders.

uvomorulin *See* CADHERINS (cadherin 1).

V5 **1.** Part of the visual area of the brain, (V5/MT). **2.** An antiapoptotic pentapeptide V5, (VPMLK). **3.** Pentavalent vanadium, V^{5+}. **4.** One of the catalytic domains of *protein kinase C- (beta]. **5.** The envelope (Env) gene V3--V5 regions of feline immunodeficiency virus that encodes the neutralizing epitopes. **6.** One of the transient receptor potential channels, TRPV5, involved in active ca^{2+} calcium reabsorption. **7.** A splice variant of CD44.

V8 protease A serine peptidase (glutamyl endopeptidase, EC 3.4.21.19, 336 aa) from *Staphylococcus aureus* strain V8 that is used to cleave proteins prior to sequencing or peptide mapping. It cuts polypeptides only at the C-terminal side of aspartic and glutamic acid residues.

VAC Vascular anticoagulant: VACα is *annexin V; VACβ is annexin-8.

VacA A multisubunit toxin (vacuolating cytotoxin) produced by type I **Helicobacter pylori* that causes large vacuoles in epithelial cells. There are six or seven identical subunits (1287 aa) which are proteolytically cleaved to produce 37- and 58-kDa fragments that behave as an AB toxin. The toxin binds to receptor-like protein tyrosine phosphatases on the surface of target cells. Antibodies to VacA are common even in asymptomatic individuals.

vaccination The induction of protective immunity by prior exposure to antigens associated with a pathogenic organism. Various forms of vaccine can be used: an antigenically related nonpathogenic strain (attenuated strain) of the organism; a related nonpathogenic species; killed or chemically modified organisms of low pathogenicity; purified proteins that carry the antigenic determinants. *See also* DNA VACCINE.

vaccinia An Orthopoxvirus that was used in vaccination against smallpox and is related to cowpox. Vaccinia has been used experimentally as a vector for introducing relatively large sequences of DNA into animal cells.

VACTERL syndrome A disorder characterized by vertebral defects, anal atresia, cardiac anomalies, tracheoesophageal fistula with oesophageal atresia, radial and renal dysplasia, and limb abnormalities. *See* SONIC HEDGEHOG.

vacuolar ATPase *See* V-TYPE ATPASE.

vacuole Any membrane-bounded vesicle of eukaryotic cells; commonly used by botanists to describe the large vesicles found in plant cells.

vacuolins Small triazine-based molecules that affect membrane traffic in fibroblasts. Vacuolin-1 (577 Da) alters the morphology of lysosomes without affecting the ability of cells to reseal their plasma membrane after injury.

valaciclovir A prodrug of *aciclovir. *TNs* Valtrex, Zelitrex.

validamycin A *See* JINGGANGMYCIN.

valine (Val, V) An essential amino acid (117 Da).

valinomycin A macrocyclic ionophore consisting of twelve residues (D- and L-valine, D-hydroxyvaleric acid, and L-lactic acid) arranged so that their carboxyl groups line a central pore that is size-selective for potassium ions.

valosin-containing protein A structural protein (VCP, p97, 806 aa) of the *AAA family that is complexed with clathrin and, like *NSF, is involved in vesicle transport and fusion. Missense mutations in VCP are the cause of *inclusion body myopathy with Paget's disease of bone and frontotemporal dementia (IBMPFD), probably by compromising ubiquitin-mediated protein quality control.

VAMP *See* SYNAPTOBREVINS.

vanadate (VO_4^{3-}) An inorganic anion that inhibits many ATPases, probably by acting as an analogue of the transition state of the cleavage reaction that removes the terminal phosphate. *Tyrosine kinases are sensitive to vanadate, but serine/threonine kinases are insensitive.

van Buchem's disease A disorder (hyperostosis corticalis generalisata, hyperphosphataaemia tarda, leontiasis ossea) in which there is excessive bone growth, caused by a 52-kb deletion ~35 kb downstream of the *SOST* gene that encodes *sclerostin, a *bone morphogenic protein (BMP) antagonist. The deletion apparently removes a SOST-specific regulatory element and down-regulation of sclerostin production.

vancomycin A glycopeptide antibiotic used as a last resort for treatment of infections by Gram-positive bacteria, although resistance has emerged.

van der Woude's syndrome An autosomal dominant form of cleft lip and palate associated with mutation in *IRF-6.

vanilloid receptor-1 One of the *transient receptor potential channels found on sensory neurons that appears to be a thermoreceptor. Binding of *capsaicin activates the receptor, and induces death of the cell.

vanins A small family of closely related proteins (vascular noninflammatory molecules, vanins 1-3) that have similar functions. Vanin-1 (513 aa) is a glycosylphosphatidylinositol (GPI)-anchored molecule expressed by perivascular thymic stromal cells and thought to regulate late adhesion steps of normal (non-inflammatory) homing to the thymus. It is a pantetheinase that hydrolyses *pantetheine to pantothenic acid (vitamin B_5).

Vaquez–Osler disease *See* POLYCYTHAEMIA.

variable antigen *See* ANTIGENIC VARIATION.

variable gene *See* V REGION.

variable number tandem repeats *See* SATELLITE DNA.

variable region *See* V REGION.

varicella zoster *See* HERPESVIRIDAE.

variola virus (smallpox virus) A large DNA orthopox virus (brick-like, 250–390 nm × 200–260 nm) with complex outer and inner membranes that are not derived from host-cell plasma membrane. The World Health Organization certified the eradication of smallpox in December 1979.

varix (*pl.* varices) An enlarged and dilated vein, as in varicose veins.

vascular anticoagulant (VAC) *See* ANNEXIN.

vascular cell adhesion molecule *See* VCAM.

vascular endothelial growth factor A growth factor (VEGF, vascular permeability factor, VPF) specific for vascular endothelium, produced by vascular smooth muscle cells, that will stimulate angiogenesis and increases permeability of endothelial monolayers. The functional form is a dimer or heterodimer of splice variants or with *placental growth factor (PlGF). VEGF is distantly related to *platelet growth factor and has strong sequence homology with PlGF and VEGF-B. Tissue-specific splice variants (VEGF121, VEGF165, VEGF-C) are found, VEGF165 having heparin-binding activity which VEGF121 lacks. VEGF-related factor (VEGF-B) is produced in two alternatively-spliced forms (186 aa, 167 aa) and both isoforms show strong homology to VEGF at their N-termini. There are several **VEGF receptors**: VEGF-R1 (fms-related tyrosine kinase 1; flt-1) binds most forms but VEGF121 only binds to VEGF-R2 (kinase insert domain receptor, KDR; fetal liver kinase-1,flk-1); VEGF- R3 (flt-4), binds only VEGF-C and is mainly restricted to lymphatic endothelium during development. **VEGF coregulated chemokine-1** (VCC1; dendritic cell and monocyte chemokine-like protein, DMC) is up-regulated in breast and colon tumours and has similarities with CXCL8, CCL4 and CCL5; it apparently attracts nonactivated monocytes and dendritic cells.

vascularization The process of invasion of a tissue by blood vessels, stimulated by angiogenic factors which cause endothelial cell proliferation and migration. The vascularization of tumours allows rapid growth and is often a prelude to metastasis. Excessive vascularization of the retina occurs in diabetic retinopathy and retrolental *fibroplasia. *See* ANGIOGENIN; TUMOUR ANGIOGENESIS FACTOR.

vasoactive intestinal peptide (VIP) A peptide hormone (28 aa), with similarities to *glucagon, *secretin, and gastric inhibitory peptide, originally isolated from porcine intestine, but also a neuropeptide of the central nervous system where it is released by specific *interneurons. It modulates innate and adaptive immunity, and has an anti-inflammatory effect. There are two G-protein-coupled receptors (460 and 495 aa).

vasoconstrictor An agent that causes a reduction in the luminal diameter of blood vessels. Most are α-*adrenergic receptor agonists, e.g. *ephedrine.

vasodilator Any drug or substance that causes an increase in the diameter of blood vessels and increased blood flow. Examples include *nitrates, *adrenomedullin, *bradykinin, *atrial natriuretic peptide, *isoproterenol, *nifedipine, *prostacyclin, and *substance P. *See* VASODILATOR STIMULATED PHOSPHOPROTEIN.

vasodilator stimulated phosphoprotein (VASP) A platelet protein (380 aa) that is phosphorylated in response to both cAMP- and cGMP-elevating agents, and when phosphorylated inhibits platelet function. It is associated with microfilament bundles in many tissue cells.

vasopressin (antidiuretic hormone, ADH) A nonapeptide hormone synthesized as a much larger precursor which includes the vasopressin carrier, *neurophysin, and a glycoprotein. It is synthesized in the hypothalamus (as is *oxytocin), packaged into neurosecretory vesicles, and transported to the nerve endings in the neurohypophysis from where it is released. It has a wide range of effects mediated through three subtypes of G-protein-coupled **AVP receptors** (angiotensin/vasopressin receptors), V1a, V1b, and V2. The **V1a receptor** (418 aa) mediates cell contraction and proliferation, platelet aggregation, release of coagulation factor, and glycogenolysis and acts through phospholipase C causing an increase intracellular calcium. The **V1b receptor** (424 aa) mediates calcium-triggered release of *ACTH, β-*endorphin, and *prolactin from the anterior pituitary. The **V2 receptor** (371 aa) is found only in the kidney, where it is involved in water homeostasis; defects result in nephrogenic *diabetes insipidus (which can also be caused by mutation in vasopressin).

vasopressor Any compound or substance that causes contraction of blood vessels (vasoconstriction), reducing blood flow and increasing blood pressure.

vasostatins Peptides (vasostatin I and II) derived from chromogranin A (*see* GRANINS) that inhibit vasoconstriction, promote fibroblast adhesion, inhibit parathyroid hormone secretion, trigger microglial-cell-mediated neuronal apoptosis, and have bacteriolytic and antifungal effects.

vasp *See* VASODILATOR STIMULATED PHOSPHOPROTEIN.

vaspin A *adipokine (visceral adipose tissue-derived *serpin, 395 aa) that has ~40% homology to α_1-antitrypsin. It has insulin-sensitizing effects, but its role in obesity and as a risk factor in diabetes are unclear as yet.

vault A large cytoplasmic ribonucleoprotein particle that has the appearance of an octagonal dome. It is composed of three proteins: the major vault protein (MVP), the vault poly(ADP-ribose) polymerase (VPARP), and the *telomerase-associated protein 1, together with one or more small untranslated RNAs. MVP (lung resistance-related protein, 896 aa) is frequently overexpressed in multidrug resistant tumours and accumulates into lipid rafts during *Pseudomonas aeruginosa* infection.

vav A family of guanine nucleotide exchange factors (*GEFs) (vav1, vav2, vav3) for the rho/rac GTPases, members of the *dbl family. The *vav1* proto-oncogene product (p95vav, 845 aa) is found mainly in haematopoietic cells but vav2 (872 aa) has a wider tissue distribution. Activation of vav1 by phosphorylation in Bcr-Abl-expressing cells leads to the activation of Rac-1 and is associated with leukaemias. Vav3 (847 aa) is expressed at higher levels in glioblastoma vs low-grade glioma and is overexpressed in prostate cancer.

VCAM (vascular cell adhesion molecule) A cell surface glycoprotein (CD106, 739 aa) of the immunoglobulin superfamily, expressed by cytokine-activated endothelium, that mediates the adhesion of monocytes and lymphocytes through binding by the integrin VLA4.

VCA region The *Arp2/3-activating region (verprolin-like, cofilin-like, acidic region) of *WASP-family proteins that binds both the Arp2/3 complex and an actin monomer.

VDAC Voltage-dependent anion channel. *See* PORINS.

vector **1.** In mathematics, something that has both direction and magnitude. **2.** In molecular biology, a plasmid that can be used to transfer DNA sequences from one organism to another. *See* TRANSFECTION. Plasmids usually have selectable markers to identify successfully transfected cells. **3.** An animal that transmits an infectious disease; e.g. tsetse flies are vectors of African trypanosomes (*see* TRYPANOSOMA).

vectorial synthesis The process by which proteins destined for export from the cell are synthesized. As translation begins a *signal sequence triggers association of the ribosome with *rough endoplasmic reticulum (ER) and the nascent polypeptide moves through the ER membrane as it is synthesized. Proteins that will remain integral membrane proteins may have the process terminated before synthesis is complete so that they do not become free within the cisternal space of the ER.

vectorial transport The transport of an ion or molecule across an epithelium in only one direction (e.g. absorption of nutrients in the gut). Vectorial transport requires that *transport proteins are nonrandomly distributed between the apical and basolateral plasma membrane.

vecuronium bromide A competitive antagonist for neuromuscular cholinergic receptors, a muscle relaxant. *Compare* SUXAMETHONIUM. *TN* Norcuron.

vegetal pole The region of the egg opposite to the animal pole; cytoplasm in this region is destined for incorporation into endodermal tissues.

VEGF *See* VASCULAR ENDOTHELIAL GROWTH FACTOR.

veiled cell (dendritic cell) An *accessory cell of the afferent lymph that has a distinct morphology with large ruffled margins. Now normally referred to as a *dendritic cell. They migrate from the periphery (where they are called *Langerhans cells if in the skin) to the draining lymph node where they are sometines called interdigitating cells.

vein A blood vessel that returns blood from the microvasculature to the heart.

velocardiofacial syndrome A disorder in which there are frequently cleft palate, cardiac anomalies, typical facies, and learning disabilities, although other abnormalities may further exacerbate matters. It is caused by a 1.5–3.0-Mb hemizygous deletion of chromosome 22q11.2, and haploinsufficiency of the *T-box1 gene is responsible for most of the physical malformations (point mutations in the T-box1 gene may have similar consequences). *See* GOOSECOID; THYMIC HYPOPLASIA.

venom A toxic secretion that is actively delivered to the target organism. May paralyse or incapacitate the target animal, or may be defensive. Venoms contain a range of protein and peptide toxins.

V

verapamil HCl A drug that blocks *L-type channels, used in the treatment of hypertension, angina pectoris and cardiac arrhythmia. *TNs* Bosoptin, Calan, Covera-HS, Isoptin, Verelan.

veratridine An alkaloid neurotoxin that activates voltage-dependent sodium channels. Found in the seed of *Schoenocaulon officinale* (cevadilla) and in the rhizome of *Veratrum album* (white hellebore).

Vero cells An epithelial-like cell line derived from African green monkey kidney, much used in virology.

verotoxin *See* SHIGA TOXIN.

verprolins An highly conserved family of proteins that regulate the actin cytoskeleton either by binding to actin or by binding to actin-regulating proteins of the *WASP family. Verprolin (751 aa) was originally identified in budding yeast. *See* WAVE; WIP; WIRE/WICH.

versene Trivial name for *EDTA.

versican A large chondroitin sulphate proteoglycan (large fibroblast glycoprotein, chondroitin sulphate core protein, 264 kDa) that interacts with hyaluronan to form large supramolecular complexes in hyaline cartilage. The core protein of the longest isoform is 3396 aa. Versican is also involved in cell signalling and some versican isoforms have been implicated in migration and proliferation of cancer cells. *Decorin and *biglycan are other soft tissue proteoglycans. In *Wagner's syndrome, an autosomal dominant vitreoretinopathy, there is a mutation at the 3′ acceptor splice site of intron 7.

vertical transmission Transmission of an infectious agent from mother to offspring; *compare* HORIZONTAL TRANSMISSION. Can occur transplacentally or through minor skin lesions sustained during birth.

vesicle **1.** In cells, a closed compartment enclosed by a membrane either physiologically (as in intracellular vesicles involved in secretion, etc.) or generated by mechanical damage to cells (sonication). *See also* COATED VESICLE. **2.** A small fluid-filled blister on the skin.

vesicular monoamine transporters Members of the solute carrier 18 family (SLC18A1 and -2) that transport cytosolic monoamines (*dopamine, *serotonin) into synaptic vesicles. They are an important element of the monoaminergic neurotransmission systems, which are implicated in various neuropsychiatric disorders. They are inhibited by reserpine and *tetrabenazine.

vesicular stomatitis virus (VSV) A *rhabdovirus that is responsible for soremouth in cattle. Being nonpathogenic to humans it has been a favourite system for studies on the spike glycoprotein as a model for the synthesis, post-translational modification, and export of membrane proteins.

Vg1 (vitellogenin-1) A member of the TGFβ family encoded by the *VG1* gene and involved in induction of mesoderm. Not to be confused with *vitellogenin.

V gene *See* V REGION.

VH and VL genes/domains The genes that encode the variable regions of heavy (VH) and light (VL) regions of immunoglobulin molecules. *J* genes and, in the case of the heavy chain, a *D* (diversity) gene also contribute to these regions, which are rearranged during maturation of the antigen-specific B cell.

VHDL The fraction of plasma lipoprotein that has a density greater than 1.21 g/mL (very high density lipoprotein). VHDL is ~57% protein, 21% phospholipid, 17% cholesterol, and 5% triacylglycerols.

VHL tumour suppressor gene *See* VON HIPPEL–LINDAU TUMOUR SUPPRESSOR PROTEIN.

viability test A procedure used to determine the proportion of living individuals, cells, or organisms, in a sample. Simple viability tests often depend on the ability of live cells to exclude a dye, (an exclusion test), or to take it up (an inclusion test).

Vibrio cholerae A short, motile, curved rod-shaped Gram-negative bacterium that causes *cholera.

vidarabine An antiviral nucleoside analogue (adenine arabinoside, Ara-A) used to treat severe herpesvirus infections.

vigabatrin An anticonvulsant drug that blocks the catabolism of γ-aminobutyric acid (*GABA) by irreversibly inhibiting GABA-transaminase.

vigilin (high density lipoprotein-binding protein) An RNA-binding protein (1289 aa) with multiple (14) *KH domains found in the cytoplasm and the nucleus and implicated in heterochromatin formation, chromosome segregation and the localization of mRNAs to actively translating ribosomes. The conserved vigilin class of proteins have a high affinity for inosine-containing RNAs. **High density lipoprotein-binding protein**, also known as vigilin, seems to be the same protein but, oddly, is membrane associated and binds HDL molecules; it has been suggested that it may couple sterol metabolism to protein synthesis.

SEE WEB LINKS

- Both roles for HBP/vigilin are discussed in an article in *Arteriosclerosis, Thrombosis, and Vascular Biology* (1997).

villin A calcium-regulated, actin-binding protein (827 aa), a major structural component of the microfilament bundles of microvilli in the *brush border. It severs microfilaments at high calcium concentrations, caps them at lower levels. **Villin 2** (cytovillin, *ezrin) is expressed strongly in placental syncytiotrophoblasts and in certain human tumours.

vimentin The protein (466 aa) that forms *intermediate filaments in mesodermally-derived cells. *Desmin is the homologue found in muscle.

vinblastine *See* VINCA ALKALOIDS.

vinca alkaloids Alkaloids (~800Da) isolated from Madagascar periwinkle *Catharanthus roseus* (formerly *Vinca rosea*) that bind to the *tubulin heterodimer, induce the formation of paracrystals, and prevent assembly into microtubules which gradually disassemble and are not replaced. Used in tumour chemotherapy. The best-known examples are vinblastine and vincristin; others include vindesine and vinorelbine.

Vincent's angina A recurring periodontal disease (trench mouth, acute necrotizing ulcerative gingivitis) which may be due to infection with *Fusobacterium nucleatum* along with *Borrelia* or *Treponema*. *Compare* ANGINA PECTORIS.

vincristine *See* VINCA ALKALOIDS.

vinculin A cytosleletal protein (1134 aa) isolated from muscle, fibroblasts, and epithelial cells. Connects microfilaments to plasma membrane integral proteins through *talin at *focal adhesions. **Metavinculin** is a splice variant found in smooth and cardiac muscle where it connects microfilaments to the intercalated disc. Mutation (R975W), in the alternatively spliced exon 19, is associated with hypertrophic cardiomyopathy.

VIP *See* VASOACTIVE INTESTINAL PEPTIDE.

viperin A protein (virus inhibitory protein endoplasmic reticulum-associated interferon-inducible, 361 aa) that is induced in response to interferon, lipopolysaccharide, double-stranded RNA (poly(I-C)) or *Sendai virus. Infection with *cytomegalovirus causes redistribution of induced viperin from the endoplasmic reticulum to the Golgi and then to cytoplasmic vacuoles containing viral proteins.

viraemia The presence of virus in the blood.

viral antigens Antigens, often viral coat proteins, that are encoded by the viral genome and can be detected by a specific immunological response.

viral haemorrhagic fevers A group of illnesses caused by several distinct families of viruses: *Arenaviridae, *Filoviridae, *Bunyaviridae, and *Flaviviridae. In some cases the illness may be relatively mild but in some cases can be severe and life-threatening.

viral transformation *Transformation of a cell in culture, induced by a virus.

virgin lymphocyte Obsolete term for a lymphocyte that has not encountered the antigenic determinant for which it has receptors.

virion A single virus particle, including the coat.

virstatin A small molecule (4-(*N*-(1,8-naphthalimide))-n-butyric acid) that blocks virulence in *Vibrio cholerae* by inhibiting the virulence transcriptional activator ToxT, thereby preventing expression of *cholera toxin and the toxin coregulated pilus.

virus Obligate intracellular parasites (20–300 nm diameter) that are noncellular and consist only of genomic information (DNA or RNA) and a protein coat. Many are defined by the diseases they cause. In the Baltimore classification scheme, **class I viruses** have double-stranded DNA as their genome; **class II** have a single-stranded DNA genome; **class III** have a double-stranded RNA genome; **class IV** have a positive single-stranded RNA genome, the genome itself acting as mRNA; **class V** have a negative single-stranded RNA genome used as a template for mRNA synthesis; and **class VI** have a positive single-stranded RNA genome but with a DNA intermediate not only in replication but also in mRNA synthesis. *See* BACTERIOPHAGES; HUMAN ENDOGENOUS RETROVIRUSES.

viscoelastic Describing substances or structures that exhibit non-Newtonian viscous behaviour and have both elastic and viscous properties in response to mechanical stress.

V

visfatin An *adipokine found in omental (visceral) adipose tissue and, unlike *omentin, also in epicardial and subcutaneous fat.

visinin A cone-specific protein first characterized in chicken retina, but a human homologue (visinin-like 1,191 aa) is strongly expressed in granule cells of the cerebellum where it associates with membranes in a calcium-dependent manner and modulates intracellular signalling pathways of the central nervous system through adenylyl cyclase. *See* VISININ-LIKE PROTEINS.

visinin-like proteins A subfamily of neuronal intracellular *EF-hand calcium sensor proteins, the NCS (neuronal calcium sensor) family that modulate calcium-dependent signalling events. *Recoverin is one of the subfamily.

visna–maedi virus A *retrovirus of sheep and goats, one of the *Lentivirinae and related to *HIV. It was first recognized in Iceland as the cause of two diseases in sheep: maedi, a pulmonary infection, and visna, a neurological infection that causes a paralysis similar to *multiple sclerosis.

visual disorders

Defects in vision can arise from problems with the optics of the eye (*astigmatism, *hypermetropia, *keratoconus) or with the transparency of the *cornea, which may be temporary (as in *keratoscleritis and *keratomalacia) or lens (*see* CATARACT), with deficiencies of particular classes of photosensitive cells (*retinal rods and *retinal cones) causing *stationary night blindness or *colour blindness of various sorts (*see* DICHROMATISM, PROTANOPIA, TRITANOPIA). In some cases, there is progressive loss of retinal cells (*see* AGE-RELATED MACULAR DEGENERATION; CONE–ROD DYSTROPHY; DOYNE HONEYCOMB RETINAL DYSTROPHY; GYRATE ATROPHY; LEBER'S OPTIC ATROPHY; OGUCHI'S DISEASE; RETINITIS PIGMENTOSA), or the whole *retina can become detached, a problem with neovascularization of the retina in diabetes and in *Stickler's syndrome. Defects in the phototransduction system can occur at various levels in the signalling cascade (*see* ALAND ISLAND EYE DISEASE; ARRESTINS; OPSIN; RHODOPSIN; STARGARDT DISEASE; TRANSDUCIN) Other vision-related problems include *uveitis and *glaucoma; in some cases the defect is not obvious (*amaurosis). Infection with **Onchocerca* or **Trachoma* can cause blindness, and blindness can be a feature of more complex syndromes (*hypotrichosis with juvenile macular dystrophy, *Usher's syndrome). Problems can also occur with the eye muscles (*Duane retraction syndrome). More complex disorders such as cortical blindness can occur at a higher level of processing of the visual image, usually due to damage of the relevant parts of the brain, typically the visual cortex.

visual purple *See* RHODOPSIN.

vital dye (vital stain) A dye that can be used to stain cells without affecting their normal function and that tends to persist so that the distribution of the cells (e.g. during morphogenetic movements) can be traced.

vitamin B A diverse class of vitamins: **Vitamin B_1** (thiamine, aneurin) is a water-soluble vitamin, the phosphate derivatives of which are involved in many cellular processes. Thiamine pyrophosphate is a coenzyme for several important enzymes, including pyruvate dehydrogenase, and is synthesized by thiamine pyrophosphokinase (EC 2.7.6.2). Thiamine deficiency causes beri-beri. **Vitamin B_2** (riboflavin) is ribose attached to a flavin moiety that is the core component of *flavin nucleotides (FAD and FMN). Riboflavin deficiency (ariboflavinosis) causes cracked and red lips, inflammation of the lining of the mouth and tongue, dry and scaling skin, and iron-deficiency anaemia. **Vitamin B_3** (niacin, nicotinic acid) is deficient in pellagra. Niacin is converted to nicotinamide and then to NAD and NADP. **Vitamin B_5** (pantothenic acid) is needed to form *coenzyme A. **Vitamin B_6** (pyridoxine, pyridoxal, and pyridoxamine) is a cofactor in many reactions of amino acid metabolism, including transamination, deamination, and decarboxylation. The active form is pyridoxal phosphate. **Vitamin B_7** (biotin, vitamin H) is important in fatty acid biosynthesis and catabolism and an essential growth factor for many cells. Biotin is bound very strongly by *avidin or *streptavidin and this is often used experimentally by labelling avidin and using it to detect molecules conjugated to the acyl group derived from biotin (biotinylated). **Vitamin B_9** (folic acid, vitamin M) is a pteridine derivative that must be obtained in the diet or from intestinal microorganisms. The active form is tetrahydrofolate which acts as a carrier of one-carbon units in biosynthetic pathways such as those for *methionine, *thymine and *purines; deficiency leads to megaloblastic anaemia. *Aminopterin and *methotrexate are analogues of dihydrofolate and block the regeneration of tetrahydrofolate from dihydrofolate. There is good evidence that maternal deficiency of folate during pregnancy increases the risk of spinal abnormalities in the fetus. **Vitamin B_{12}** (various cobalamins) is a water-soluble vitamin, actually a class of related compounds, essential for various biological activities including the recycling of *folic acid. The physiologically active forms are methylcobalamin and adenosylcobalamin which can be produced from the common synthetic form of the vitamin, cyanocobalamin, which is used as a supplement in foods. A lack of *intrinsic factor, as seen in *megalobastic anaemia, causes a vitamin B_{12} deficiency.

vitamin C (ascorbic acid) An essential dietary vitamin for both humans and guinea-pigs that is involved in a range of metabolic processes but particularly in collagen synthesis. Defects in collagen are responsible for many of the pathological aspects of scurvy, vitamin C deficiency.

vitamin D A group of fat-soluble prohormones (collectively vitamin D_1 or calciferol), the two major forms of which are **vitamin D_2** (ergocalciferol) and **vitamin D_3** (cholecalciferol). Vitamin D_3 is produced in the skin in response to sunlight and deficiencies lead to rickets. Calcitriol is the 1α,25-dihydroxy form of vitamin D_3 and acts to increase gastrointestinal calcium absorption, stimulate osteoclastic calcium resorption from bone, facilitate the effect of parathyroid hormone on bone resorption, and increase renal absorption of calcium. Vitamin D_2 is very effective at absorbing UV radiation and may serve a protective function in the skin.

vitamin E The collective name for a set of eight related alpha-, beta-, gamma-, and delta-tocopherols and the corresponding four tocotrienols, which are fat-soluble vitamins with antioxidant properties that may prevent free-radical damage.

vitamin H *See* VITAMIN B (vitamin B_7).

vitamin K (menaquinone) A group of lipophilic vitamins required for the post-translational modification of proteins by the carboxylation of some glutamate residues to form γ-carboxyglutamate residues (Gla residues). Gla residues are usually involved in binding calcium and several blood clotting factors (*prothrombin, factors VII, IX, X, *protein C, *protein S, and protein Z) are Gla proteins. *See* COUMARIN; MGP; OSTEOCALCIN; WARFARIN.

vitamins Low-molecular-weight organic compounds that are essential minor components of food. For humans vitamin A, the B series, C, D_1 and D_2, E, and K are required in the diet and cannot be synthesized. Various vitamin deficiency diseases are known, more commonly in the developing world. There are separate entries for various vitamins.

vitellin The major protein in the yolk of many eggs.

vitelline membrane (zona pellucida) The fibrillar membrane that surrounds the plasmalemma of the ovum and is modified by the addition of proteins released from cortical granules to form the fertilization membrane that blocks polyspermy.

vitellogenin A family of proteins that provided nutritional reserves in the yolk of nonmammalian eggs. The three ancestral vitellogenin-encoding genes were progressively lost during mammalian evolution but proteins related to the vitellogenins are still found (e.g. the large subunit of *microsomal triglyceride transfer protein). *But see* VGI.

vitiligo A disorder associated with autoimmune disease in which there is patchy depigmentation of the skin, mucous membranes, and retina. The pattern of vitiligo on opposite sides of the body and in pairs of identical twins is generally similar. There are multiple loci associated with susceptibility to vitiligo, although vitiligo-multiple autoimmune disease susceptibility can be accounted for by variants in the *NALP1* gene which encodes one of the *NALP proteins that is a regulator of the innate immune system.

vitronectin A serum glycoprotein (serum spreading factor, complement S-protein, somatomedin B, 478 aa) that promotes adhesion and spreading of tissue cells in culture and has the RGD (Arg-Gly-Asp) motif to which integrins bind. It inhibits cytolysis by the complement C5b-9 complex.

VLA proteins (very late antigens) Members of the β-*integrin family of cell-surface adhesion molecules. VLAs 1–6 share a common β subunit. Some are receptors for collagen, laminin, or fibronectin, and are not restricted to leucocytes.

VLDL Plasma *lipoprotein with density of 0.94–1.006 g/L (very low density lipoprotein). The protein content (~10%) is lower than that of *VHDL but includes apoproteins B, C, and E. The VLDLs transport triacylglycerols to adipose tissue.

V_{max} The maximum initial velocity of an enzyme-catalysed reaction when substrate is present in excess.

Vohwinkel's syndrome A skin disorder (mutilating keratoderma) caused by mutation in the gene encoding *connexin-26. A variant form of Vohwinkel's syndrome, mutilating keratoderma with ichthyosis, is caused by a defect in the *loricrin gene.

voltage clamp An electrophysiological technique by which the transmembrane potential is held at a predetermined level by use of an intracellular *microelectrode. An important technique for the study of *ion channels. *See* PATCH CLAMPING.

voltage-gated ion channels Transmembrane *ion channels that alter their permeability according to the transmembrane potential difference. Voltage-gated channels are essential for neuronal signalling and for intracellular signal transduction. *See* VOLTAGE-GATED SODIUM CHANNELS; VOLTAGE-SENSITIVE CALCIUM CHANNELS.

voltage-gated sodium channels A family of multisubunit *ion channels with an aqueous pore ~0.4 nm diameter formed by the α subunits, with a negatively charged interior that inhibits the passage of anions. The various β subunits confer different responsiveness on channels, according to the tissue. The channel is responsible for electrical excitability of neurons, and a small depolarization of the cell (usually caused by an approaching action potential), triggers the channel to open. Within a millisecond ~1000 sodium ions pass through before the channel spontaneously closes. The channel is then *refractory until the membrane potential approaches the *resting potential. There are around 100 channels/μm^2 in unmyelinated axons, but in myelinated axons they are concentrated at the *nodes of Ranvier. Mutations in either the α or β subunits of the neuronal $NaV_{1.1}$ (SCN1A and B) cause various forms of *epilepsy. $NaV_{1.2}$ (SCN2A) is similar to $NaV_{1.1}$ but found in caudal regions of the brain; mutations cause febrile seizures. **$NaV_{1.3}$** is also found in the central nervous system. **$NaV_{1.4}$** (SCN4A and -B) is the skeletal muscle form and mutations in the gene are involved in various muscular disorders, including hyperkalemic *periodic paralysis, *myotonia, and hypokalemic periodic paralysis; mutations in SCN4B cause *long QT syndrome-10. **$NaV_{1.5}$** is expressed in the heart, brain, and gastrointestinal tract, and mutations in the gene *SCN5A* cause long QT syndrome-3 and *Brugada syndrome-1. **$NaV_{1.6}$** (alpha subunit encoded by SCN8A) may be defective in hereditary neurodegenerative diseases. SCN7A (previously SCN6A) is thought to be the sodium-level sensor in the brain. Mutations in SCN9A (alpha subunit of NaV1.7), the major sodium channel expressed in bronchial and arterial smooth muscle, cause primary *erythromelalgia, *congenital insensitivity to pain, and *paroxysmal extreme pain disorder. NaV1.8 (SCN10A) is specific to peripheral sensory neurons. NaV1.9 (SCN11A) mediates *brain-derived neurotrophic factor-evoked membrane depolarization through the receptor tyrosine kinase NTRK2 and is implicated in inflammatory but not acute pain. Many toxins (*see* ANTHOPLEURINS; BATRACHOTOXINS; BREVETOXINS; CALITOXINS; CIGUATOXINS; CONOTOXINS; JINGZHAOTOXIN-III; SAXITOXIN; SCORPION TOXINS; TETRODOTOXIN; VERATRIDINE) affect the channel with serious consequences. *Compare* AMILORIDE-SENSITIVE SODIUM CHANNELS.

voltage gradient The potential difference between two points, divided by the distance between them. Often casually used to mean the potential difference across a plasma membrane.

voltage-sensitive calcium channels (VSCC) Calcium-ion specific *voltage-gated ion channels that are subdivided into six classes on the basis of electrophysiological and pharmacological criteria. They allow calcium influx into the cell as a result of membrane depolarization. The majority (*L-type, *N-type, *P-type, and *Q-type channels) require substantial depolarization and are sometimes collectively known as high voltage-activated types. The *R-type channels activate after moderate depolarization and the *T-type channel opens at relatively negative potentials.

v-*onc* General abbreviation for the viral form of an *oncogene. In many cases the viral gene was discovered some time before the cellular *proto-oncogene (c-*onc*).

von Economo's disease *See* ENCEPHALITIS (encephalitis lethargica).

von Hippel–Lindau syndrome A dominantly inherited familial cancer syndrome caused by mutation in the gene encoding the *von Hippel–Lindau tumour suppressor protein. The syndrome involves a predisposition to a variety of malignant and benign neoplasms, most frequently retinal, cerebellar, and spinal haemangioblastoma, renal cell carcinoma (RCC), phaeochromocytoma, and pancreatic tumours. The highly vascular tumours associated with von Hippel–Lindau syndrome overproduce angiogenic peptides such as vascular endothelial growth factor (VEGF).

von Hippel–Lindau tumour suppressor protein The VHL gene encodes a 213-aa protein (VHL30) and a second active protein (VHL19) transcribed from an alternative start site. The proteins interact with *elongin and *cullin-2 and an important part of their role in tumour suppression may be by binding to hypoxia inducible factor-1 (HIF-1), targeting it for destruction when oxygen is present. There is also a role in up-regulating *p53. Expression during embryogenesis is ubiquitous but levels are particularly high in the urogenital system, brain, spinal cord, sensory ganglia, eyes, and bronchial epithelium.

von Recklinghausen's disease *See* NEUROFIBROMATOSIS (type 1).

von Willebrand factor (vWF) A large, multimeric plasma glycoprotein (2050 aa) produced in endothelium (found in the Weibel–Palade bodies), megakaryocytes (α-granules of platelets), and subendothelial connective tissue. vWF is important for platelet adhesion through an interaction with factor VIII. The receptor is the *Gp-Ib-IX-V complex and is defective in *Bernard–Soulier syndrome. The vWF domain is found in many extracellular proteins that form multiprotein complexes, e.g. in *complement factors B, C2, CR3 and CR4, integrins, *collagen types VI, VII, XII and XIV. *See* VON WILLEBRAND'S DISEASE.

von Willebrand's disease An autosomal dominant haemorrhagic disorder in which platelet adhesion to collagen is reduced but not platelet aggregation. Factor VIII levels are secondarily reduced. *See* VON WILLEBRAND FACTOR.

VP16 **1.** Virion protein 16, a protein found in the coat of the herpes simplex virion that is a transcriptional activator of the viral immediate-early genes. **2.** *See* ETOPOSIDE.

VP22 Virion protein 22, one of the most abundant proteins of the tegument (which lies between the nucleocapsid and the viral envelope) of herpes simplex virus types 1 and 2 (HSV-1 and HSV-2; *see* HERPESVIRIDAE). Vp22 has been shown to have a range of activities and mediates intercellular transport of proteins. It accumulates inside infected cells at late stages of infection and is required for optimal protein synthesis at this stage.

vpr A soluble protein (viral protein R, 96 aa) expressed at a late stage of the replication of HIV-1. It is cytotoxic and has proapoptotic activity, causing a rapid dissipation of the mitochondrial transmembrane potential, and the release of cytochrome *c* or apoptosis-inducing factor from mitochondria.

VRE *Vancomycin-resistant enterococci.

V region The highly variable regions of both heavy and light chains of immunoglobulin molecules that form the antigen binding site. The **variable gene** (V gene) products are recombined with the constant region (C region) for the immunoglobulin class. Immunoglobulin-type variable regions are also found in the *T-cell receptor.

VSCC *See* VOLTAGE-SENSITIVE CALCIUM CHANNELS.

VSG Variant surface glycoprotein, a trypanosome surface protein that is important in *antigenic variation of the parasite.

V-SNARE *See* SYNAPTOBREVINS.

VSV-G tag An *epitope tag, a peptide (11 aa), derived from the G protein of *vesicular stomatitis virus.

V-type ATPase The class of proton-pumping ATPases (vacuolar ATPases) found in intracellular acidic vacuoles and in some epithelia (e.g. intercalated cells of kidney). They have up to fourteen protein subunits arranged in a cytoplasmic V_1 complex, which mediates the hydrolysis of ATP, and a membrane-embedded V_0 complex, which translocates H^+ across the membrane. The pump is a molecular motor in which ATP hydrolysis turns a rotor consisting of one copy of subunits D and F of the V_1 complex and a ring of six or more copies of subunit C of the V_0 complex. The rotation of the ring is thought to deliver H^+ from the cytoplasmic to the endosomal or extracellular side of the membrane. V-type ATPases are inhibited by *bafilomycin. *Compare* F-TYPE ATPASE; P-TYPE ATPASE. Mutation in one of the V_0 subunits (ATP6V0A2) causes autosomal recessive *cutis laxa type II; mutation in ATP6V1B1 (a V_1 complex subunit) causes distal renal tubular acidosis with progressive deafness.

SEE WEB LINKS

- Review article in the *Journal of Experimental Biology* (2006).

vWF *See* VON WILLEBRAND FACTOR.

V

Waardenburg's syndrome A set of autosomal dominant disorders in which there is usually deafness and pigmentary disturbances of variable extent, as a result of *neural crest dysfunction. **Waardenburg's syndrome 1** (WS1) and **WS3** (Klein–Waardenburg syndrome) are caused by mutation in *Pax3, **WS2** can be caused by mutation in the gene for microphthalmia-associated transcription factor or the gene for *slug. Two other loci have also been indentified. **Waardenburg–Shah syndrome** (WS4), in which Waardenburg's syndrome is associated with *Hirschsprung's disease, is due to mutation in Sox10.

Waf1 An inhibitor (p21, cip1, cyclin-dependent kinase inhibitor 1A, 164 aa) of *cyclin-dependent kinase (cdk) activity that forms a complex with cyclins A, D_1, and E, and CDK2. Waf1 can inhibit all cdks, although to a variable extent, and its expression is regulated by the *p53 tumour suppressor.

Wagner's syndrome A disorder in which there is degeneration in the vitreous humour of the eye and often retinal detachment. **Type 1** Wagner's syndrome can be caused by mutation in the gene encoding chondroitin sulphate proteoglycan-2 (*versican), **type 2** by defects in *collagen COL2A1.

WAGR syndrome A syndrome in which there is susceptibility to *Wilm's tumour, aniridia, genitourinary abnormalities, and mental retardation that arises because of a constitutive deletion on chromosome 11p13 affecting several contiguous genes.

Walker–Warburg syndrome A congenital condition (HARD ± E syndrome) characterized by hydrocephalus, agyria, retinal dysplasia, with or without encephalocele. It is often associated with several distinct congenital muscular dystrophies. The disorder is caused by mutation in the genes encoding protein *O*-mannosyltransferase-1 and -2 (POMT1 and -2) but individual patients with WWS have been shown to have homozygous mutations in the **fukutin* and **LARGE* genes, respectively.

Wallerian degeneration The degeneration (anterograde degeneration) that occurs in distal parts of a nerve fibre bundle following injury; axons that are disconnected from the cell body degenerate.

WAP, follistatin, immunoglobulin, Kunitz, and NTR domain-containing proteins *See* WFIKKNs.

warfarin An antithrombotic compound, a synthetic *coumarin, that inhibits *vitamin K epoxide reductase, an enzyme that recycles oxidated vitamin K to its active reduced form. Has been superseded as a rat poison by more potent compounds.

warm antibodies Immunoglobulins, mostly IgG, that work optimally at body temperature, in contrast to *cold agglutinins. *See* WARM ANTIBODY HAEMOLYTIC ANAEMIA.

warm antibody haemolytic anaemia An autoimmune disorder in which there is erythrocyte destruction triggered by the binding of immunoglobulin (usually IgG). Can be triggered by drugs, which may act as haptens. *Compare* PAROXYSMAL COLD HAEMOGLOBINURIA.

wart A benign tumour that results from the infection of a single *basal cell of the skin by a papillomavirus, which causes excessive proliferation in the stratum corneum.

WASP family proteins A family of proteins (Wiskott–Aldrich syndrome proteins) involved in regulating actin polymerization that includes WASP (502 aa), N-WASP (WASP-like, WASL), WAVE1, 2, and 3 (WASP family, verprolin homology domain-containing proteins), WASP-interacting proteins (WIPF1 and 2). WASP has a GTP-binding site and links *cdc42 and the actin cytoskeleton. Mutations in the gene encoding WASP are responsible for *Wiskott–Aldrich syndrome. N-WASP is more broadly distributed than WASP itself which is mainly found in myeloid cells. *See* VERPROLINS.

WASP homology domain-2 An actin-binding motif (WH2 domain, ~35 aa) common in proteins involved in regulation of the actin cytoskeleton, including *thymosin-β_4, *ciboulot and the *WASP family proteins. *See also* BETA-THYMOSIN.

Wassermann reaction An obsolete complement-fixation test for syphilis, that used *cardiolipin as an antigen.

Watson's syndrome A disorder in which there is pulmonary stenosis, café-au-lait spots, short stature, and dull intelligence, caused by mutation in the *neurofibromin gene. Probably allelic to neurofibromatosis (NF1).

WAVE/SCAR proteins Members of the *WASP family (WASP family, verprolin homology domain-containing proteins, Wave1, 2, and 3) with homology to the scar protein of *Dictyostelium*. Wave1 and Wave3 are both strongly expressed in brain, Wave2 has a wide tissue distribution and is highly expressed in peripheral blood leucocytes.

WBC *See* WHITE BLOOD CELLS.

WD-repeat proteins Proteins that contain the WD40 motif, a conserved sequence (~40 aa) usually ending with tryptophan (W) and aspartic acid (D) (but with a nod to water dispersant 40, WD40, the well-known aerosol lubricant), that is implicated in protein–protein interactions. The seven WD-repeats in the GTP-binding protein β subunit form a propeller-like structure with seven blades.

weal and flare *See* TRIPLE RESPONSE.

weaver A mouse strain that carries a mutation in the G-protein-coupled inwardly rectifying potassium channel 2 (Girk2); dopaminergic neurons of the substantia nigra are lost after they differentiate. Since these are the neurons lost in Parkinson's disease, the weaver mouse is used as a model system for the human disorder.

Weber's law *See* FECHNER'S LAW.

Wegener's granulomatosis A granulomatous vasculitis in which there are granulomas of the upper and lower respiratory tract and necrotizing focal glomerulonephritis. Usually associated with autoantibodies to neutrophil azurophil granule protease 3 (c-*ANCA). Polymorphisms in the MHC predispose to the disease (HLA-DPB1*0401 allele).

Weibel–Palade body A cytoplasmic organelle in vascular endothelial cells that contains *von Willebrand factor and P-*selectin.

Weil–Felix reaction A serological test for typhus and certain other rickettsial diseases; it is relatively insensitive and is being superseded.

Weill–Marchesani syndrome A rare connective tissue disorder characterized by short stature, brachydactyly, joint stiffness, and lens abnormalities. The autosomal recessive form can be caused by mutations in the **ADAMTS10* gene and the similar but distinguishable autosomal dominant form by mutations in the *FBN1* gene (*fibrillin 1).

Weil's disease *See* LEPTOSPIRA.

Weismann's germ plasm theory The theory that genetic continuity is through the germ-line cells (germ plasm) that are separated from the somatic lineage at a very early stage in development. Generally accepted and seen as a way of minimizing the risk of somatic mutations entering gametes.

Weissenbacher–Zweymüller syndrome An autosomal dominant disorder, allelic with Stickler's syndrome type 3 and otospondylomegaepiphyseal dysplasia (OSMED), sometimes called heterozygous OSMED. It is caused by defects in *collagen type XI (COL11A2).

Werdnig–Hoffman disease *See* SPINAL MUSCULAR ATROPHY.

Werner's syndrome A syndrome of premature ageing due to a mutation in the gene that encodes the human homologue of *E. coli* RecQ DNA helicase, an enzyme involved in *mismatch repair. The defect in the repair system manifests in chromosomal instability and a high incidence of tumours. Werner's syndrome cells in culture usually achieve only ~20 population doublings compared to the normal Hayflick limit of ~60 (*see* CELL DEATH). A more severe phenotype is seen in 'atypical Werner's syndrome', where the defect is in the gene encoding nuclear *lamin A/C.

Wernicke's encephalopathy An inherited metabolic defect (Wernicke–Korsakoff syndrome) caused by an abnormal isoform of transketolase (EC 2.2.1.1), a thiamine-dependent enzyme, that is only clinically important when there is dietary insufficiency of thiamine. Deficiency, often a result of alcohol abuse, leads to brain damage.

western blot *See* BLOTTING.

West Nile virus One of the Flaviviridae family, for which the main host is birds, but that can (and increasingly does) infect humans, in some cases causing severe meningitis and encephalitis. The vector is a mosquito.

WFIKKNs Proteins (WAP, follistatin, immunoglobulin, Kunitz and NTR domain-containing proteins) that are thought to inhibit both serine peptidases and metallopeptidases; WFIKKN1 (548 aa) is highly expressed in pancreas, kidney, liver, placenta, and lung, WFIKKN2 (WFIKKN-related protein, 576 aa) is similar but has a different tissue distribution.

WGHA *See* WHOLE GENOME HOMOZYGOSITY ASSOCIATION.

Wharton's jelly A viscous hyaluronic acid-rich jelly found in the umbilical cord. It collapses postpartum and clamps the umbilical vessels.

wheatgerm The embryonic plant at the tip of the wheat grain. It is the starting material for a cell-free translation system. **Wheatgerm agglutinin** (WGA) is a *lectin from wheatgerm that binds to *N*-acetylglucosaminyl and sialic acid residues.

whirlin A protein (CASK-interacting protein, CIP98 907 aa) localized to the tips of stereocilia. In brain interacts with a calmodulin-dependent serine kinase (CASK, one of the membrane-associated guanylate kinase (MAGUK) family), colocalizes with CASK along the dendritic processes of neurons and may be involved in the formation of scaffolding protein complexes that facilitate synaptic transmission in the central nervous system. Mutations in the gene for whirlin cause *Usher's syndrome (type 2D) and autosomal recessive nonsyndromic deafness-31. The gene was originally identified in the mouse whirler mutant.

white An eye colour gene of *Drosophila* that encodes an ATP-binding cassette (ABC) transporter involved in the distribution of pigments found in the eyes. The human homologue (ABCG1, 638 aa) is expressed in brain, spleen, and lung and is thought to have a role in regulating cholesterol and phospholipid transport, especially in macrophages.

white adipose tissue Tissue that is mainly composed of *adipocytes but, unlike *brown adipose tissue, is not specialized for heat production, although it has insulating properties. It is the main storage site for triglycerides.

white blood cells (WBC) A generic term for both myeloid and lymphoid cells of the blood. *See* BASOPHIL; EOSINOPHIL; LYMPHOCYTE; MONOCYTE; NEUTROPHIL.

Whitmore's disease *See* MELIOIDOSIS.

whole cell patch *See* PATCH CLAMPING.

whole genome homozygosity association (WGHA) A mathematical technique developed to look at genetic information derived from a patient's mother and father and identify pieces of chromosomes that are identical.

Widal test A diagnostic test for *typhoid fever much used in developing countries. It demonstrates the presence of somatic (O) and flagellar (H) agglutinins to *Salmonella typhi* in the patient's serum, but is of limited value because of persistence of antibodies from previous infections and cross-reactivity.

Williams–Beuren syndrome A neurodevelopmental disorder (William's syndrome) that can involve supravalvular aortic stenosis (SVAS), multiple peripheral pulmonary arterial stenoses, elfin face, mental and statural deficiency, characteristic dental malformation, and infantile hypercalcaemia. It is caused by the hemizygous deletion of several genes (~1.5 Mbp) on chromosome 7q11.23, among which are those for *elastin, *LIM kinase-1, replication factor C, and *MAGI.

Wilms' tumour A kidney tumour (nephroblastoma) of childhood thought to derive from persistent renal stem cells. Both sporadic and inherited forms occur. The tumour arises from a mutation in the **Wilm's tumour 1** (*WT1*) gene that encodes a zinc finger DNA-binding protein (429 aa) that acts as a transcriptional activator or repressor depending on the cellular or chromosomal context and that is important in development of the genitourinary system and mesothelial tissues. Mutations in the **BRCA2* gene and the *glypican-3 gene have also been described in Wilms' tumour. *See* WAGR SYNDROME. **Wilms' tumour 2** (WT2) is associated with mutation of the imprinted *H19* gene that is also implicated in *Beckwith–Wiedemann syndrome in which Wilms' tumour is a common feature. Other forms of Wilms' tumour (WT3, 4, 5) are associated with other loci.

Wilson's disease A rare autosomal recessive disease (hepatolenticular degeneration) characterized by degenerative changes in the brain and cirrhosis of the liver. A defect in a copper-transporting ATPase coded by the *ATP7B* gene leads to excessive deposition of copper in liver, brain, and kidney.

Winchester syndrome *See* MONA (2).

winged helix transcription factors A large family of transcription factors (>80) characterized by a conserved DNA-binding domain (~110 aa), the *forkhead box.

wingless *See* WNT.

WIP A protein (WASP-interacting protein, WAS/WASL-interacting protein family member 1, WIPF, 1503 aa) that interacts with *WASP. The adaptor protein CRKL binds directly to WIP and following T-cell receptor ligation, a CRKL–WIP–WASP complex is recruited by *zap70 to lipid rafts and immunological synapses. A related protein (**WASP-interacting protein-related protein**, WIRE, WIPF2, 440 aa) interacts with WASP and *profilin downstream of the *platelet-derived growth factor receptor (PDGFR).

WIRE *See* WIP.

Wiskott–Aldrich syndrome An X-linked recessive immunodeficiency characterized by thrombocytopenia, eczema, and recurrent infections. Caused by mutation in the gene for the Wiskott–Aldrich syndrome protein (*WASP).

WISP Regulatory proteins (Wnt-induced secreted proteins) of the *CCN family of regulatory proteins expressed downstream of *wnt and involved in morphological transformation. WISP-1 (367 aa), WISP-2 (250 aa), and WISP-3 (354 aa) are expressed differentially in various tumours.

Witkop's syndrome A condition (tooth-and-nail syndrome) in which some teeth are missing and nails grow poorly; caused by mutation in the **MSX1* gene.

WNKs A family of serine/threonine kinases (with no K (lysine) protein kinases) of the STE20 family that have cysteine, not lysine, at a key position in the active site and are activated by hyperosmotic stress. WNKs in turn activate *SPAK and OSR1 and defects can cause *pseudohypoaldosteronism. Human **WNK-1** (1246 aa) is expressed in most tissues, but particularly in polarized epithelia involved in chloride ion transport. **WNK2** (2126 aa) is widely expressed, but its function is unclear. **WNK3** (1800 aa) regulates transport mediated by both the Na-K-2Cl cotransporter (SLC12A1) and the Na-Cl cotransporter (SLC12A3). **WNK4** (1243 aa) colocalizes with *ZO1 in specific regions of the kidney and regulates SLC12A3.

wnt A family of genes that encode secreted glycoproteins involved in various developmental processes, particularly specification of position and morphogenetic events, but also in carcinogenesis. The original example was the *Drosophila* gene *wingless*, but many vertebrate homologues have subsequently been discovered (there are at least ten) and a complex signalling system worked out. The *wnt-1* (formerly int-1, *see* INT ONCOGENES) product binds to cell-surface receptors of the *Frizzled family, causing the receptors to activate *Dishevelled family proteins and eventually causing a change in the amount of β-*catenin that reaches the nucleus.

wobble hypothesis The hypothesis advanced by Francis Crick (1916–2004) to explain why many *codons translate into a single amino acid and why there are fewer tRNAs (45) than there are codons (64). Essentially, the first (5′) base of the anticodon is less specific in its pairing requirements and G in this position will pair with C or U in the third codon position, U will pair with A or G, and inosine will pair with A, C, or U. There are, some exceptions, however, so that there are 45 tRNAs as opposed to the 31 needed if the hypothesis applied to all codons.

Wohlfart–Kugelberg–Welander disease *See* SPINAL MUSCULAR ATROPHY.

wolframin A transmembrane protein (890 aa) of the endoplasmic reticulum involved in the regulation of calcium levels. Mutations cause *Wolfram's syndrome and nonsyndromic sensorineural deafness autosomal dominant type 6 (DFNA6).

Wolfram's syndrome An autosomal recessive disorder (diabetes insipidus, diabetes mellitus, optic atrophy, and deafness, DIMOAD) in which there is growth retardation and a variety of developmental and neurological defects. One form is caused by a mutation in *wolframin, another by mutation in the *CISD2* gene that encodes *CDGSH iron–sulphur domain protein 2 that colocalizes with *calnexin in the endoplasmic reticulum.

Woodhouse–Sakati syndrome A disorder in which there is a combination of hypogonadism, partial alopecia, diabetes mellitus, mental retardation, and deafness. The cause is a mutation in the C2ORF37 gene (chromosome 2 open reading frame 37) although the function of the product, found in two alternatively spliced isoforms (240 aa and 520 aa), is uncertain.

woolsorter's disease *Anthrax that has been contracted by handling infected wool or hair.

wortmannin A fungal metabolite, isolated from *Penicillium wortmanni* and other fungi, that is a fairly specific, covalent inhibitor of *PI-3-kinases although it will inhibit some related kinases such as *polo-like kinases, *mTOR, and *myosin light-chain kinase. It has a relatively short half-life in physiological solutions.

Wurster's reagent *See* TMPD.

WW domain A small semiconserved protein domain (38 aa) that binds proline-rich sequences. Found in various proteins including *WWOX, *Yes-associated protein and *PIN-1.

WWOX A tumour suppressor protein (WW domain-containing oxidoreductase, 414 aa) that is a mediator of TNFα-induced apoptosis and is deleted or altered in several cancer types.

Xaf1 A protein (XIAP-associated factor 1, 301 aa) that antagonizes the effect of the X-linked inhibitor of apoptosis (*XIAP) by causing it to relocalize to the nucleus. Expression is reduced or absent in tumours.

xanthine A purine (2,6-dihydroxypurine) that is generated from guanine by guanine deaminase, or from hypoxanthine by xanthine oxidoreductase in the purine degradation pathway. Methylated derivatives (theophylline, theobromine, caffeine) are potent cAMP phosphodiesterase inhibitors.

xanthine dehydrogenase A molybdenum-containing hydroxylase (XDH, EC 1.1.1.204, 1333 aa) involved in the oxidative metabolism of purines. It can be converted to **xanthine oxidase** (XO; EC 1.2.3.2) by reversible sulphydryl oxidation or by irreversible proteolytic modification. XO-derived reactive oxygen species are thought to be mediators in the pathogenesis of cellular injury. Mutations in the *XDH* gene lead to *xanthinuria type 1.

xanthine oxidase *See* XANTHINE DEHYDROGENASE.

xanthinuria A disorder characterized by excretion of large amounts of xanthine in the urine. There are two distinct forms of the disorder: in **type 1** there is a deficiency of *xanthine dehydrogenase, and in **type 2** there is a dual deficiency of both xanthine dehydrogenase and aldehyde oxidase. Type 1 patients can metabolize allopurinol, whereas type 2 patients cannot. Xanthinuria can also arise through *molybdenum cofactor deficiency.

xanthoma A local accumulation of yellow cholesterol-rich material in various tissues (tendons, dermis). Xanthomas are symptomatic of various lipid-handling disorders. *See* CEREBROTENDINOUS XANTHOMATOSIS; SITOSTEROLAEMIA.

X chromosome The mammalian sex chromosome; females are XX, males are XY.

X chromosome inactivation centre (Xic) The region of the X chromosome that inactivates one of the two X chromosomes in females (*see* LYON HYPOTHESIS). Inactivation is by a nuclear RNA (17kb), encoded by Xist, which is expressed specifically from inactive, but not active, X chromosomes; the inactive X is coated with this transcript.

xenin A peptide (xenopsin-related peptide, 25 aa) cleaved from the α-*coatomer subunit. It stimulates secretion from the exocrine pancreas, inhibits *pentagastrin-stimulated secretion of acid and affects gut motility. Interacts with the *neurotensin receptor.

xenobiotic Any substance found in an organism but that is not produced by that organism and is not a normal constituent of its diet. More often used to describe substances foreign to an entire biological system, artificial substances that did not exist in nature before being synthesized by humans.

xenobiotic response element (XRE) An upstream regulatory sequence recognized by the transcription factors, particularly the *aryl hydrocarbon receptor, that regulate the production of detoxification enzymes.

xenogeneic Literally, having a foreign genome; usually describing tissue or cells from another species, as in xenogeneic transplantation (xenografting).

xenograft Tissue that has been transplanted between individuals of unlike species, genus, or family.

Xenopus laevis The African clawed toad, an animal extensively used in developmental biology because of its large embryo. The large eggs are often used as an expression system, particularly for ion channels which can be functionally expressed in the membrane and tested electrophysiologically. *Xenopus* was formerly used in pregnancy diagnosis because it ovulates under the influence of luteinizing hormone in urine; this attribute is useful for the laboratory breeding of the toads using pure human chorionic gonadotropin.

xenotropic virus An endogenous virus that replicates with the genome of the host and is benign but that will produce infective virus particles if it infects cells of a different species. Murine endogenous leukaemia viruses fall into three categories: *ecotropic, that will replicate only in mouse cells; xenotropic, that will replicate only in nonmouse cells; and amphotropic or polytropic, that will replicate in a wide range of mammalian cells including murine cells. The differences arise largely due to envelope proteins that determine binding and infectivity. The **xenotropic virus receptor** (xenotropic and polytropic retrovirus receptor, XPR1) has multiple membrane-spanning domains and may play a role in G-protein-coupled signal transduction.

xeroderma pigmentosum An autosomal recessive disorder in which there is increased sensitivity to sunlight with the development of carcinomas at an early age. Neurological symptoms may develop and a severe form is *de Sanctis–Cacchione syndrome. There are seven complementation groups, each arising from a different defect in the DNA repair system, particularly that required for excision of UV-induced *thymine dimers.

X-Gal A compound that is hydrolysed by β-galactosidase to form an intense blue precipitate. X-Gal in conjunction with the inducer *IPTG is used to detect recombinants (white) from nonrecombinants (blue) in blue/white colony screening and to detect β-galactosidase reporter gene activity in transfected cells.

XIAP An X-linked inhibitor of apoptosis (*IAP, baculovirus IAP repeat-containing protein 4, BIRC4, 487 aa) that directly inhibits *caspases-3, -7, and -9. XIAP interacts with *survivin and also regulates the levels of *COMMD. Mutations are associated with *X-linked lymphoproliferative syndrome.

Xic *See* X CHROMOSOME INACTIVATION CENTRE.

ximelagatran An oral prodrug (now withdrawn) of *melagatran, a synthetic small peptidomimetic with direct thrombin inhibitory actions and anticoagulant activity.

X-inactivation The random inactivation of one of the two *X chromosomes in a female somatic cell that occurs early in development and produces the Barr body. Either X chromosomes can be inactivated and females are therefore a mosaic for heterozygous genes on the X chromosome. *See* LYON HYPOTHESIS and X CHROMOSOME INACTIVATION CENTRE.

xipamide A sulfonamide diuretic drug that reduces sodium reabsorption in the distal convoluted tubule and increases the secretion of potassium in the distal tubule and collecting ducts. *TN* Aquaphor.

X-linked disease A disease caused by a mutation in a gene carried on an *X chromosome. Because males have only one X chromosome, recessive characteristics will be expressed; females may be affected but only because of haploinsufficiency and can carry the mutation, which may be expressed in some (potentially 50%) of their male progeny.

X-linked lymphoproliferative syndrome A disorder (XLP, Duncan's disease, Purtilo's syndrome) characterized by extreme sensitivity to infection with Epstein–Barr virus. Most cases (**XLP1**) are caused by mutation in the *SH2D1A* gene (*SLAM-associated protein). **XLP2** is caused by mutations in the gene encoding *XIAP.

XLP *See* X-LINKED LYMPHOPROLIFERATIVE SYNDROME.

XRE *See* XENOBIOTIC RESPONSE ELEMENT.

XRN1 An exonuclease (strand exchange protein 1, EC 3.1.11.-, 1694 aa) in *GW bodies that will degrade mRNA in the 5′ to 3′ direction once the cap has been removed.

xylulose A five-carbon sugar. The 5-phosphate is an intermediate in the *pentose phosphate pathway. L-Xylulose accumulates in the urine in patients with pentosuria, a disorder caused by a deficiency in L-xylulose reductase.

yaba virus A double-stranded DNA virus (yaba monkey tumour virus) that induces the formation of focal (benign) histiocytomas. It was isolated from African monkeys but can infect all primates. The genome encodes a TNFα binding protein (YMTV-2L, 341 aa) and inhibits TNF signalling. The **yatapoxvirus genus** has three members: tanapox virus (TPV), yaba-like disease virus (YLDV), and yaba monkey tumour virus.

YAC *See* YEAST ARTIFICIAL CHROMOSOME.

yaws (framboesia) A tropical infection of the skin, bones, and joints caused by the spirochaete *Treponema pallidum pertenue.*

Y chromosome The male chromosome that carries the primary determinant of male sexual development (*sry). The pseudoautosomal region of the Y chromosome is homologous to part of the *X chromosome with which it pairs.

yeast The common name for a diverse group of fungi of different families (ascomycetes, basidiomycetes, and imperfect fungi) that are unicellular for most of their life cycle. *Saccharomyces cerevisiae* and *Schizosaccharomyces pombe* are important experimental organisms, the former being commercially important in brewing and baking. *See* CANDIDIASIS (Candida albicans).

yeast artificial chromosome (YAC) An artificially constructed chromosome that contains the telomeric, centromeric, and replication origin sequences needed for replication and preservation in yeast cells, used as a vector for cloning long DNA sequences. Contiguous YACs covering the whole *Drosophila* genome and certain human chromosomes have been important in chromosome mapping.

yeast two-hybrid screening *See* TWO-HYBRID SCREEN.

yellow fever virus A positive-sense, single-stranded, encapsulated RNA virus (togavirus) of the Flaviviridae that causes yellow fever. It is transmitted by mosquitoes.

Yersinia A genus of Gram-negative rod-shaped bacteria of the *Enterobacteriaceae. They are facultative intracellular parasites and most are pathogenic. Plague (Black Death) is caused by infection with *Y. pestis* (formerly *Pasteurella pestis*). *Y. enterocolitica* causes yersiniosis.

yes An oncogene originally identified in Yamaguchi avian sarcoma virus. The product (p62yes) is a *src-family nonreceptor tyrosine kinase. There are multiple copies of *yes*-related genes in the human genome and YES1 may be translocated in follicular lymphoma. **Yes-associated protein** (YAP65) is a transcriptional coactivator that enhances p73-dependent apoptosis in response to DNA damage.

yolk sac A membranous sac connected to the embryo by the yolk stalk (vitelline duct), lined by extraembryonic endoderm and covered by extraembryonic mesenchyme. It appears early in gestation and provides early nourishment that is distributed through the vitelline circulatory system. In mouse it is the site of early haematopoiesis. The vitelline duct may persist in the adult as Meckel's diverticulum from the small intestine.

YPD The Yeast Proteome Database (YPDTM), a curated database dedicated to proteins of *Saccharomyces cerevisiae.* Access requires a subscription.

SEE WEB LINKS

- The Yeast Proteome Database.

YY1 A ubiquitous transcription factor (Yin Yang 1, 414 aa) important in embryogenesis, differentiation, replication, and cellular proliferation. It is involved in repressing and activating a diversity of promoters and may direct histone deacetylases and histone acetyltransferases to promoters.

zalcitabine A nucleoside analogue (dideoxycytidine) that inhibits viral reverse transcriptase and was formerly used to treat HIV. *TN* Hivid.

zanamivir A neuraminidase inhibitor used prophylactically and to treat influenza. *TN* Relenza.

Zantac *See* RANITIDINE.

zap70 A protein tyrosine kinase (ζ-chain-associated protein, 70 kDa, 619 aa) that associates with the ζ chain of the T-cell receptor and is tyrosine phosphorylated when the receptor binds ligand. Interacts with *lck and *fyn. Mutations in zap70 can cause a form of *SCID.

Z-DEVD A tetrapeptide (DEVD; L-aspartyl-L-glutamyl-L-valyl-L-aspartic acid amide) with a benzyloxycarbonyl (Z- or BOC-group) at the N-terminus that is a substrate for *caspase 3. Often conjugated with rhodamine or other fluorochromes for chromogenic assays or fluoromethyl ketone (FMK)-derivatized to act as an irreversible caspase inhibitor.

Z disc The densely-staining region lies between (*German* zwischen) *sarcomeres and to which the plus ends of *thin filaments from adjacent sarcomeres are attached. α-*actinin is a major component.

Z-DNA A left-handed helical form of DNA adopted by sequences of alternating *purines and *pyrimidines which winds in a zigzag pattern (hence Z) and there is a single deep groove. It may only exist as a transient conformation for relief of supercoiling or to allow access of transcriptional complexes. *See* A-DNA; B-DNA.

zearalenone A mycotoxin from *Fusarium* spp. that binds to oestrogen receptors. Mostly a problem in animal husbandry.

zeaxanthin The predominant carotenoid pigment in the central macula region of the retina (*see* LUTEIN). It is found naturally in dark green, leafy vegetables such as spinach, collard greens, and kale and is claimed to be a valuable nutritional supplement.

zeiosis A term occasionally used to describe extensive blebbing of the plasma membrane (cell boiling).

zeitgeber The environmental agent or event (*German* time-giver) that provides the cue for setting or resetting a biological clock mechanism.

Zellweger's syndrome An unusual malformation syndrome, one of a group of *peroxisome biogenesis disorders, part of a larger group of diseases known as the leukodystrophies. Zellweger's syndrome can be caused by mutations in any of several different genes involved in peroxisome biogenesis *See* MEVALONATE KINASE; PAS GENES; PEROXINS.

zf9 A *zinc finger transcription factor (core promoter element-binding protein, COPEB, Krüppel-like factor 6, KLF6) from rat stellate cells (*see* ITO CELLS) that transactivates a key promoter driving stellate cell fibrogenesis. The human homologue, KLF6 (283, 290 aa), can be alternatively spliced to produce at least four isoforms and is a tumour suppressor mutated in some forms of prostate and gastric carcinoma. Wildtype KLF6 upregulates p21 (WAF1/CIP1) in a p53-independent manner and significantly reduces cell proliferation.

ZIC3 A zinc finger transcription factor (zinc finger protein of cerebellum 3) that is mutated in some cases of heterotaxy (abnormal positioning of organs due to failure of determination of normal left–right asymmetry). *See also* NODAL.

zidovudine (AZT, retrovir) A nucleoside (thymidine) analogue (azidodeoxythymidine) that inhibits viral reverse transcriptase. Used to treat HIV infection.

zinc finger A small motif of 12 aa with two cysteines and two histidines (a cysteine–histidine zinc finger), or four cysteines (a cysteine–cysteine zinc finger), that directly coordinates a zinc atom and is a feature of many DNA-binding proteins. They are also involved in binding to RNA, protein, and/or lipid substrates, the binding properties depending on the amino acid sequence of the finger domains and of the linker. Zinc fingers are often found in clusters, where fingers can have different binding specificities.

zip kinase *See* DEATH ASSOCIATED PROTEIN (death-associated protein kinase-3).

zipper *See* LEUCINE ZIPPER.

zippering Process proposed to occur as the membrane of a phagocyte enfolds the particle being engulfed. A progressive adhesive interaction is postulated and supported by the observation that capped B-lymphocytes are only partially internalized, whereas those with a uniform opsonizing coat of the Fc moieties of anti-IgG are fully engulfed.

zithromax *See* AZITHROMYCIN

ZJ-43 An inhibitor of peptidases that break down the neurotransmitter N-acetylaspartylglutamate (*NAAG) which suppresses glutamate transmission through selective activation of the presynaptic group II metabotropic glutamate receptor subtype 3 ($mGluR_3$). It may potentially reduce neuronal and astrocyte damage associated with glutamate excitotoxicity after brain injury.

ZO-1 A protein (tight junction protein 1, TJP1, 1748 aa) of the *MAGUK family located on the cytoplasmic membrane surface of the *zonula occludens (tight junction) and interacting with *claudins. Two isoforms are known. *Cingulin, which is distinct, is found in the same region. Other tight junction proteins are known, **TJP2** (ZO-2, 1116 aa) is mutated in familial *hypercholanaemia. **TJP3** (ZO-3, 898 aa) colocalizes with ZO-1. **Tight junction associated protein** (TJAP1, ZO-4, TJP4, 547 aa) also colocalizes with ZO-1.

zoledronic acid (zolendronate) A bisphosphonate drug used to treat hypercalaemia of malignancy, Paget's disease of bone and osteoporosis. *TNs* Aclasta, Reclast, Zometa.

Zollinger–Ellison syndrome A disorder in which there is excessive production of *gastrin. The inherited form is due to mutation in the *MEN1* gene (*see* MULTIPLE ENDOCRINE NEOPLASIA).

zona pellucida A translucent, noncellular layer surrounding the ovum composed of three glycoproteins, ZP1 (638 aa), ZP2 (745 aa), and ZP3 (424 aa). **ZP3** is the primary sperm receptor. **ZP2**, the secondary receptor, is only involved after the acrosomal reaction has occurred. Changes in the zona pellucida after fertilization are responsible for the block to polyspermy. After fertilization, ZP2 undergoes limited proteolysis by an oocyte-derived protease probably released during exocytosis of cortical granules and ZP3 ceases to be able to induce an acrosomal reaction.

SEE WEB LINKS

- 'Perspectives' article in the *Journal of Clinical Investigation* (1992).

zone of polarizing activity (ZPA) A small group of mesenchymal cells at the posterior margin of the developing limb bud in avian embryos that is the source of a diffusible morphogen, possibly retinoic acid or *sonic hedgehog, that provides positional information. Whether this is a general model for vertebrate limb development is unresolved.

zonula adherens A specialized intercellular junction that resists mechanical disruption. The membranes of the two cells are separated by 15–25 nm (*compare* ZONULA OCCLUDENS), and microfilaments are attached at the cytoplasmic face by their plus ends. The junction resembles two apposed *focal adhesions, although the proteins involved may differ.

zonula occludens (tight junction) A specialized junction between epithelial cells that makes the epithelial sheet impermeable, even to small molecules, because the apical junctions form a complete seal around the periphery of the cell. The membranes of the two cells are separated by only 1–2 nm. Integral membrane proteins cannot diffuse laterally through the junctions and so the apical and basolateral membrane domains are isolated from one another.

zonula occludens toxin An enterotoxin (399 aa) released by *Vibrio cholerae* that binds to a receptor on intestinal epithelial cells and increases the permeability of tight junctions. The active domain of the toxin is a hexapeptide (AT-1002), that causes redistribution of *ZO-1 away from junctions.

zoonosis An infectious disease that normally exists in other animals, but also infects humans. A reverse zoonosis is a human disease that can cause disease in other animals.

zopliclone A drug (a cyclopyrrolone derivative) that acts on the same receptors as benzodiazepines and is used in the treatment of insomnia. *TNs* Imovane, Zimovane, Zopinox.

zovirax *See* ACICLOVIR.

ZPA *See* ZONE OF POLARIZING ACTIVITY.

zygins Proteins (fasciculation and elongation proteins zeta-1, and zeta-2, FEZ-1, FEZ-2, 392 aa and 353 aa) thought to be involved in axonal outgrowth, guidance, and fasciculation.

zygonema *See* ZYGOTENE.

zygote The *diploid cell formed by the fusion of haploid male and female gametes at fertilization.

zygotene The second stage of the *prophase of *meiosis I, during which homologous chromosomes begin pairing.

zygotic effect gene Genes of the zygote that are important in early development. Many early developmental cues are provided by maternally derived products (*see* MATERNAL EFFECT GENE) but some are translated products of mRNA transcribed from the zygotic DNA. *Genomic imprinting may, however, determine whether the maternal or paternal allele is expressed and the zinc finger protein Zfp57 (the product of a maternal zygotic effect gene) contributes to the embryonic maintenance of these imprints.

zymogen (proenzyme) An inactive precursor of an enzyme, often the storage form. Many peptidases are synthesized and secreted as zymogens and activated by limited proteolysis. **Zymogen granules** are secretory vesicles containing zymogens.

zymogenic cells Basophilic cells (gastric chief cells) with extensive rough endoplasmic reticulum found in the gastric glands of the stomach. They secrete pepsinogen, gastric lipase and rennin in response to cholinergic signals from the vagal nerve or hormonal secretagogues such as *gastrin and *secretin. *Compare* PARIETAL CELL.

zymosan A particulate preparation of mannan-rich yeast cell-wall polysaccharide that will activate *complement through the alternate pathway. The particles become coated with C3b/C3bi and are readily phagocytosed (opsonized zymosan) and will trigger production of complement C5a.

zyxin A protein (572 aa) of the *LIM protein family that is concentrated at focal adhesions and along actin filament bundles near the *zonula adherens. Interacts with α-*actinin. **Zyxin-related protein 1** (ZRP-1, 476 aa) interacts with the cytoplasmic domain of *endoglin. *See* LIPOMA-PREFERRED PARTNER.

Appendix 1
Prefixes for SI Units

Factor	*Prefix*	*Symbol*
10^{24}	yotta	Y
10^{21}	zetta	Z
10^{18}	exa	E
10^{15}	peta	P
10^{12}	tera	T
10^{9}	giga	G
10^{6}	mega	M
10^{3}	kilo	k
10^{2}	hecto	h
10^{1}	deca	da
10^{-1}	deci	d
10^{-2}	centi	c
10^{-3}	milli	m
10^{-6}	micro	μ
10^{-9}	nano	n
10^{-12}	pico	p
10^{-15}	femto	f
10^{-18}	atto	a
10^{-21}	zepto	z
10^{-24}	yocto	y

Appendix 2
Greek alphabet

A	α	alpha
B	β	beta
Γ	γ	gamma
Δ	δ	delta
E	ε	epsilon
Z	ζ	zeta
H	η	eta
Θ	θ	theta
I	ι	iota
K	κ	kappa
Λ	λ	lambda
M	μ	mu
N	ν	nu
Ξ	ξ	xi
O	*o*	omicron
Π	π	pi
P	ρ	rho
Σ	σ	sigma
T	τ	tau
Y	ν	upsilon
Φ	ϕ	phi
X	χ	chi
Ψ	ψ	psi
Ω	*ω*	omega

Appendix 3
Single-letter and three-letter codes for amino acids

A	Ala	Alanine
R	Arg	Arginine
N	Asn	Asparagine
D	Asp	Aspartic acid
B		Asparagine or aspartic acid
C	Cys	Cysteine
Q	Gln	Glutamine
E	Glu	Glutamic acid
Z		Glutamine or glutamic acid
G	Gly	Glycine
H	His	Histidine
I	Ileu	Isoleucine
L	Leu	Leucine
K	Lys	Lysine
M	Met	Methionine
J	Nle	Norleucine
F	Phe	Phenylalanine
P	Pro	Proline
S	Ser	Serine
T	Thr	Threonine
W	Trp	Tryptophan
Y	Tyr	Tyrosine
V	Val	Valine

Oxford Paperback Reference

Concise Medical Dictionary

Over 12,000 clear entries covering all the major medical and surgical specialities make this one of our best-selling dictionaries.

'"No home should be without one" certainly applies to this splendid medical dictionary'

Journal of the Institute of Health Education

'An extraordinary bargain'

New Scientist

'Excellent layout and jargon-free style'

Nursing Times

A Dictionary of Nursing

Comprehensive coverage of the ever-expanding vocabulary of the nursing professions. Features over 10,000 entries written by medical and nursing specialists.

An A–Z of Medicinal Drugs

Over 4,000 entries cover the full range of over-the-counter and prescription medicines available today. An ideal reference source for both the patient and the medical professional.

Oxford Paperback Reference

The Kings of Queens of Britain
John Cannon and Anne Hargreaves

A detailed, fully-illustrated history ranging from mythical and pre-conquest rulers to the present House of Windsor, featuring regional maps and genealogies.

A Dictionary of World History

Over 4,000 entries on everything from prehistory to recent changes in world affairs. An excellent overview of world history.

A Dictionary of British History
Edited by John Cannon

An invaluable source of information covering the history of Britain over the past two millennia. Over 3,000 entries written by more than 100 specialist contributors.

Review of the parent volume
'the range is impressive . . . truly (almost) all of human life is here'
Kenneth Morgan, *Observer*

OXFORD